International Conference on Applications of the Mossbauer Effect 2009

(ICAME 2009)

Journal of Physics: Conference Series Volume 217

Vienna, Austria
19-24 July 2009

ISBN: 978-1-61738-480-6
ISSN: 1742-6588

Printed from e-media with permission by:

Curran Associates, Inc.
57 Morehouse Lane
Red Hook, NY 12571

Some format issues inherent in the e-media version may also appear in this print version.

Printed by Curran Associates, Inc. (2010)

For permission requests, please contact the Institute of Physics at the address below.

Institute of Physics
Dirac House, Temple Back
Bristol BS1 6BE UK

Phone: 44 1 17 929 7481
Fax: 44 1 17 920 0979

techtracking@iop.org

Additional copies of this publication are available from:

Curran Associates, Inc.
57 Morehouse Lane
Red Hook, NY 12571 USA
Phone: 845-758-0400
Fax: 845-758-2634
Email: curran@proceedings.com
Web: www.proceedings.com

TABLE OF CONTENTS

BIOLOGICAL AND MEDICAL APPLICATIONS

CHEMICAL APPLICATIONS, CATALYSIS, STRUCTURE AND BONDING

EARTH SCIENCES, MINERALOGY, ARCHAEOLOGY

INDUSTRIAL APPLICATIONS

MATERIALS SCIENCE, METALLURGY

SURFACES, MULTILAYERS AND NANOSTRUCTURED MATERIALS

<u>SOLID STATE PHYSICS AND MAGNETISM</u>

DYNAMICS

PREFACE

The International Conference on the Applications of the Mössbauer Effect (ICAME2009) was held for the first time in Austria, at the Vienna University of Technology on 19–24 July 2009. This largest conference of the Mössbauer community attracted much interest, nearly 500 colleagues announced their intention to participate. Unfortunately, due to the world-wide finance crisis, in the end many could not manage to come. Nevertheless 380 abstracts were submitted. More than 300 participants from 38 countries were present at the conference. The programme contained 20 invited talks, selected by the international advisory board. 49 contributions were selected for oral presentation. Following the advice of the IBAME board, no parallel sessions were organized. Therefore most of the work was presented as posters, attached for the whole week, giving thus the opportunity for intense discussions outside of the poster sessions. A special tutorial session, dedicated to Mössbauer spectroscopy with synchrotron radiation, organized by R Rüffer was held on the first day. It was included in the conference programme giving all participants the opportunity to attend this meeting. Although lasting until late in the evening, the audience was always well occupied. New developments in the field of iron-containing superconductors were discussed in a micro symposium. The scientific level of the whole conference was high. After thorough peer-reviewing, 147 papers were accepted by the referees and are presented in this volume.

ICAME2009 was possible due to the efforts of many people, in the organizing committee, the programme committee and the editorial board and to the students who took care of the participants during the conference and all social activities.

ICAME2009 was in several ways a special conference. It was the 30th in the series of ICAME meetings, and the first after the celebration of 50 years of the Mössbauer Effect. In 2009 Rudolf Mössbauer also celebrated his 80th birthday. The positive atmosphere, the high attendance in the sessions and the lively discussions made the conference a great success and a memorable event. It was pointed out, that Mössbauer spectroscopy is still an interesting and powerful method with great opportunities in the future.

Herbert Müller (Secretary)
Michael Reissner (Chairman)

This book is dedicated to our colleagues

Nicol Malcom, who could not come, because he suddenly died a few weeks in advance to the conference

and

Hercilio Rechenberg, who died on his way home from Vienna.

ICAME 2009
International Conference on the Applications of the Mössbauer Effect
19-24 July 2009, Vienna, Austria

Conference Organisation

Local Organizing Committee

Reissner Michael (Chairman)
Amthauer Georg
Bauer Ernst
Müller Herbert (Conference Secretary)
Lottermoser Werner
Michor Herwig
Steiner Walter
Vogl Gero

Bühler-Paschen Silke
Grodzicki Michael
Grössinger Roland
Hilscher Gerfried
Müller Martin
Redhammer Günther
Sassik Herbert
Sepiol Bogdan
Waas Monika
Wiesinger Günter

International Programme Committee

Amthauer Georg
Baggio-Saitovich Elisa
Berry Frank
Felner Israel
Greneche Jean-Marc
Gütlich Philipp
Litterst Fred Jochen
Long Gary
Nagy Denes Lajos
Rüffer Rudolf
Steiner Walter
Trautwein Alfred Xaver
Vogl Gero
Yoshida Yutaka

International Advisory Board

Alp E Ercan
Baggio-Saitovitch Elisa
Balogh Judit
Bender Koch Christian
Berry Frank
Brown Dennis
Campbell Stewart
Carbucicchio Massimo
Croci Simonetta
Di Naili
Elzain Mohamed
Fabris José Domingos
Felner Israel
Fern George R
Forder Sue D
Gajbhiye Nandeo
Génin Jean-Marie
Greneche Jean-Marc
Grodzicki Michael
Gütlich Philipp
Häggström Lennart
Hanzel Darko
Hassaan Mohamed Yousri
Jumas Jean-Claude
Kadyrzhanov Kariat
Katila Toivo
Kim Chul Sung
Klingelhöfer Göstar
Langouche Guido
Lyubutin Igor S
Marco Jose F
Mašlaň Miroslav
Miglierini Marcel
Music Svetozar
Nagy Dénes Lajos
Nishida Tetsuaki
Pérez Alcázar German
Rüffer Rudolf
Ryan Dominic H
Sanchez Francisco
Schünemann Volker
Stanek Jan
Stevens John
Suzdalev Igor P
Szymanski Krzysztof
Waanders Frans
Yoshida Yutaka

All papers published in this volume of *Journal of Physics: Conference Series* have been peer reviewed through processes administered by the proceedings Editors. Reviews were conducted by expert referees to the professional and scientific standards expected of a proceedings journal published by IOP Publishing.

Journal of Physics: Conference Series **217** (2010) 012001 doi:10.1088/1742-6596/217/1/012001

Mössbauer Spectroscopy: Elegance and Versatility in Chemical Diagnostics

Philipp Gütlich[1] and Yann Garcia[2]

[1]Institut für Anorganische Chemie und Analytische Chemie, Johannes-Gutenberg-Universität, Staudingerweg 9, 55099 Mainz, Germany

[2]Unité de Chimie des Matériaux Inorganiques et Organiques, Département de Chimie, Université Catholique de Louvain, Place L. Pasteur 1, 1348 Louvain la Neuve, Belgium

guetlich@uni-mainz.de

Dedicated to Professor Rudolf Ludwig Mössbauer on the occasion of his 80th birthday.

Abstract. Soon after the discovery of the *recoilless nuclear resonance fluorescence* by Rudolf L. Mössbauer some fifty years ago a new spectroscopic technique developed quickly on the basis of this resonance phenomenon as an excellent tool for the investigation of materials through electric and magnetic hyperfine interactions between electrons and suitable Mössbauer nuclides. Many disciplines of solid state research have benefited from applications of the new tool for non-destructive phase analysis. Chemists in particular have recognized the information that can be derived from Mössbauer spectra regarding oxidation and spin state, molecular symmetry, bonding properties, magnetism, dynamic phenomena such as spin state switching, electronic transfer processes and other phase fluctuations to name a few. This *Keynote Lecture* presents highlights of chemical applications of Mössbauer spectroscopy selected from work during the past five decades which will demonstrate the elegance and versatility of the technique in the hands of (not only) chemists, but materials scientists in the broadest sense. The retrospect begins with studies in the early sixties on simple oxides and coordination compounds, where Mössbauer spectroscopy played a decisive role in solving unanswered questions, and concludes with fascinating current discoveries on the Planet Mars and most recent developments in the use of synchrotron radiation for nuclear resonance scattering (NFS, NIS).

1. Introduction: the beginning of Mössbauer spectroscopy

Shortly after the magnificent discovery of *nuclear resonance fluorescence of gamma radiation* by Rudolf Mössbauer [1], for which he has been awarded the Nobel prize in Physics in 1961, at age 32, many laboratories around the world started to repeat and confirm Mössbauer's experiments, and subsequently explored the possibility to use this effect as a basis for a new physical technique in materials science. One of the first publications, by Kistner and Sunyar [2] from Brookhaven National Laboratory, reported the magnetic hyperfine splitting of α-Fe_2O_3 (figure 1a). This was the first spectrum of a

[1] To whom any correspondence should be addressed.

material that reflected all three types of hyperfine interactions between nuclear moments and electrons penetrating the nucleus that can be observed in a Mössbauer spectrum: the electric monopole interaction resulting in the *isomer shift*, the electric quadrupole interaction causing the *quadrupole splitting* and the magnetic dipole interaction giving rise to *magnetic splitting of degenerate nuclear levels*. Indeed, these interactions are all taken into account in figure 1b (middle) in the case of Fe_2O_3, and may be compared on the left side with metallic iron which shows a magnetic splitting only and stainless steel, on the right side, that displays an isomer shift only. Hundreds of papers appeared soon after in a short period showing that Mössbauer spectroscopy could not only be used as a characterization technique of materials but could also be suitable to elucidate long lasting problems and incorrect statements in the literature. Two examples will be developed hereafter.

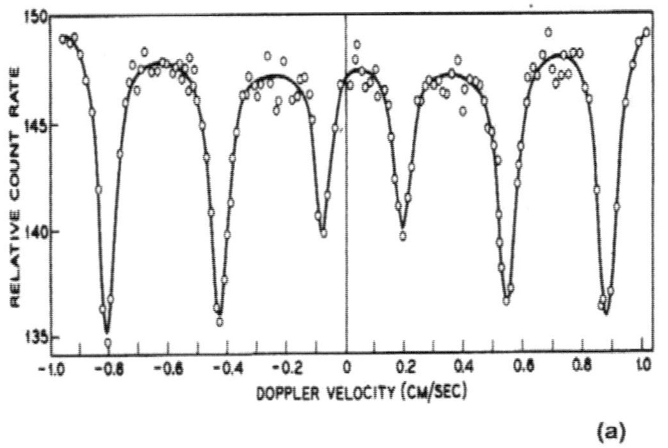

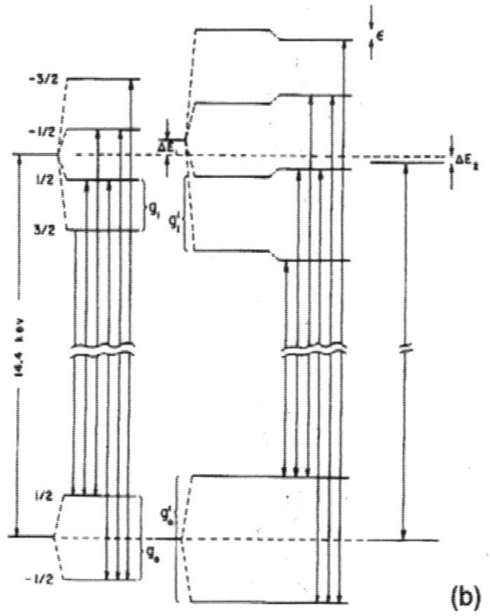

Figure 1. (a) Room temperature Mössbauer spectrum of Fe_2O_3. (b) Schematic representation of the ground and 14.4 keV excited states of ^{57}Fe bound in metallic iron (left), Fe_2O_3 (middle) and stainless steel (right) [2].

1.1. Ferrous sulfate

Ferrous sulfate, $FeSO_4{\cdot}7H_2O$, that can be better formulated according to the Werner classification as $[Fe^{II}(H_2O)_6]SO_4{\cdot}H_2O$, is a well known coordination compound. Its first crystal structure determination shown in figure 2a revealed a perfectly regular octahedron formed by 6 coordinated water molecules instead of 7 as initially thought [3], and bond lengths typical of an Fe(II) ion in the high-spin (HS) state. The Mössbauer spectrum recorded by De Benedettti et al. revealed, however, a large quadrupole splitting (figure 2b) indicating the presence of an electric field gradient (EFG) [4]. This wonderful textbook example clearly tells us that in this case the octahedron cannot have an O_h symmetry. Indeed, according to ligand field theory, we have a quintet $^5T_2$ ground state, with five orbitally degenerate orbitals, that split into e_g and t_{2g} orbitals (figure 3) with the six electrons of the d^6 configuration of Fe(II) accommodated according to Hund's rule of maximum spin multiplicity. As the four t_{2g} electrons are uniformly distributed over the three degenerate orbitals of the cubic subgroup, the EFG is zero. A nonzero EFG does, however, arise from further splitting due to a Jahn-Teller distortion. Indeed, a system that is orbitally degenerate is unstable and orbitals go lower in energy. As a result, the octahedron will distort and a preferred tetragonal compression along the z axis will be observed. Now the d_{xy} orbital lies lowest and is doubly occupied whereas all other d-orbitals are only singly occupied; this results in a nonzero EFG and explains the observed quadrupole splitting.

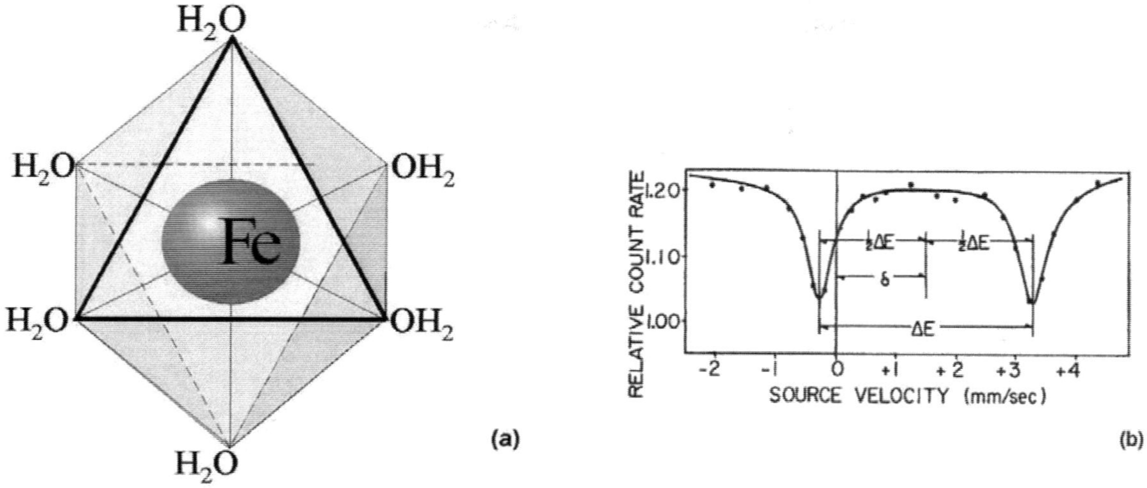

Figure 2. (a) $[Fe(H_2O)_6]^{2+}$ octahedron with x = y = z and all angles equal to 90° [3]. (b) ^{57}Fe Mössbauer spectrum of $[Fe^{II}(H_2O)_6]SO_4{\cdot}H_2O$ recorded at 77 K [4].

The structure determination was thus carefully repeated and revealed a distorted octahedron with x ≠ y ≠ z but all angles equal to 90° [5].

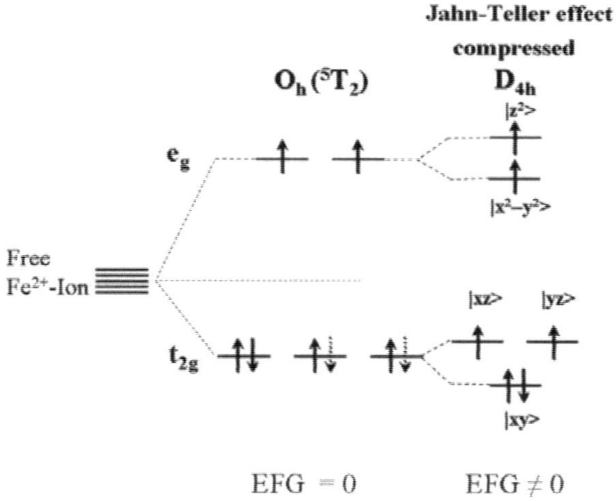

Figure 3. Ligand field splitting of $^5T_2$ ground state in O_h symmetry and Jahn-Teller effect leading to a compressed octahedron with D_{4h} symmetry for $[Fe(H_2O)_6]^{2+}$.

1.2. Prussian Blue vs. Turnbull's blue

It has been recognized for many years and stated in old textbooks of inorganic chemistry that Prussian Blue and Turnbull's blue are different substances depending on their synthetic pathways [6]. An original Mössbauer experiment was, however, performed by Fluck et al. [7] which demonstrated that the two end products prepared in different ways are chemically identical. In preparation A they prepared Prussian Blue, $Fe_4^{III}[Fe^{II}(CN)_6]_3$, by mixing equivalent amounts of ionic Fe^{3+} and $[Fe^{II}(CN)_6]^{4-}$. Their Mössbauer spectrum shown in figure 4a reveals a singlet (dark grey) attributed to low-spin (LS) Fe(II) ions and a quadrupole doublet (light grey) attributed to HS Fe^{3+} ions. In preparation B they mixed ionic Fe^{2+} with $[Fe^{III}(CN)_6]^{3-}$ and also obtained Fe_4^{III} $[Fe^{II}(CN)_6]_3$ as shown in the Mössbauer spectrum (figure 4a). Indeed, a fast electron transfer occurs from Fe^{2+} to Fe^{3+} during mixing the two components, the rate constant for electron transfer being $k_e >> 1/\tau_n$, where τ_n is the lifetime of the 14.4

keV state of ^{57}Fe). This experiment was elegantly confirmed a few years later by Maer et al (figure 4b) [8].

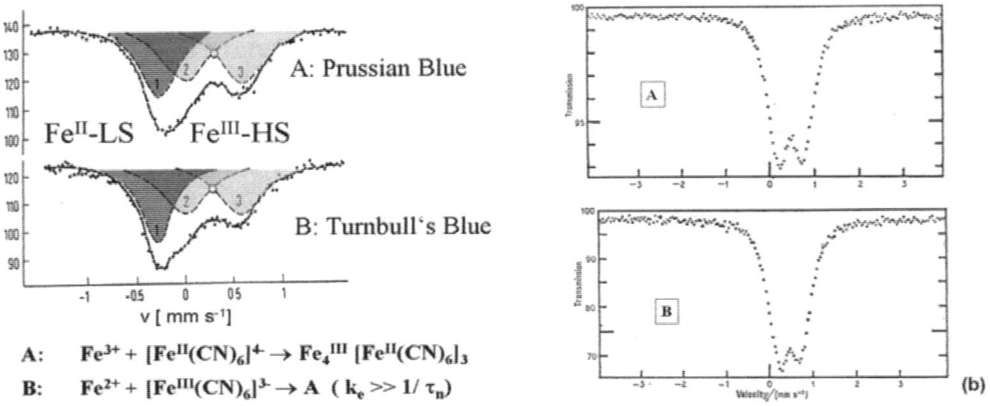

Figure 4. (a) Mössbauer spectra of Prussian blue and Turnbull's Blue [7]. (b) Mössbauer spectra (77 K) of "Prussian Blue" prepared from (i) $^{57}Fe_2^{III}(SO_4)_3$ (enriched) + $K_4[Fe^{II}(CN)_6]$ (unenriched) and (ii) $^{57}Fe^{II}Cl_2$ (enriched) + $K_3[Fe^{III}(CN)_6]$ (unenriched) [8].

2. Bioinorganic compounds

Bioinorganic chemistry has also benefited to great extent from the applications of the Mössbauer effect [9], also beginning from the sixties. As an early example, the oxidized and reduced form of ferredoxin, from a green alga, could be elegantly identified by Mössbauer spectroscopy (figure 5). The oxidized form shows a quadrupole doublet attributed to two ferric ions antiferromagnetically coupled and the reduced form shows a HS Fe(II) ion identified by a large quadrupole doublet and a HS Fe(III) ion with a much smaller quadrupole splitting [10].

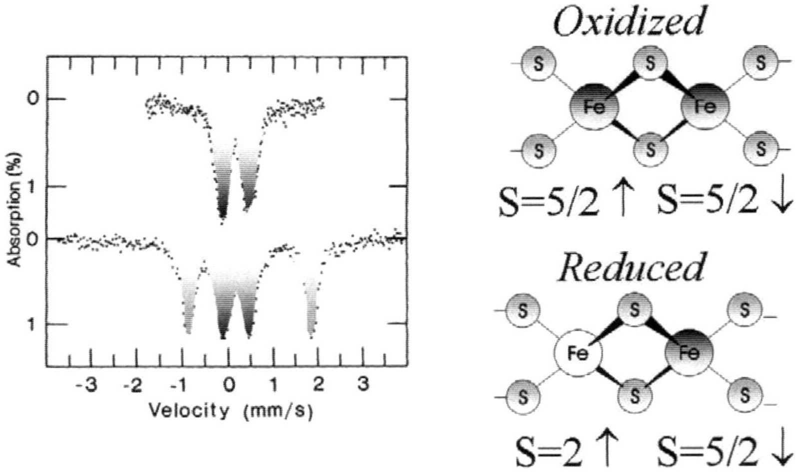

Figure 5. Mössbauer spectrum of ferredoxin: a two-iron centre protein in oxidized and reduced forms [10].

Aquaspirillum Magnetotacticum, a magnetotactic bacterium that behaves as a biomagnetic compass, was also investigated by Mössbauer spectroscopy. These microorganisms follow the weak geomagnetic field due to the presence of Fe_3O_4 magnetic nanoparticles (40 – 120 nm) of hexagonal and cubic shapes (figure 6a). Their Mössbauer spectrum reveals two magnetic sextets typical of Fe_3O_4, an inverse spinel (figure 6b). Other signals are noted in the middle of the spectrum representing a reservoir of Fe(III) out of which the bacteria are building the magnetic particles [11].

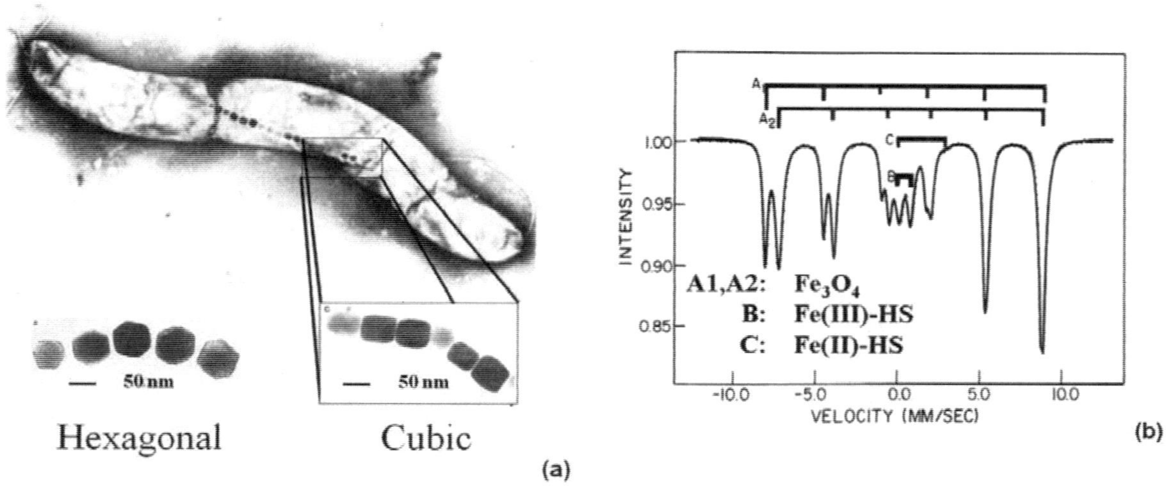

Figure 6. (a) Hexagonal and cubic shaped Fe_3O_4 nanoparticles identified in the magnetotactic bacteria (e.g. *Aquaspirillum Magnetotacticum*). (b) ^{57}Fe Mössbauer spectrum of magnetic particles extracted from the same bacteria [11].

3. Spin crossover in iron compounds

Mössbauer spectroscopy, that can provide an accurate determination of oxidation and spin states in iron compounds, is an outstanding technique to gain insight to the spin crossover (SCO) phenomenon.

3.1. Thermally induced spin state switching in Fe(II) compounds

Thermal spin crossover can be encountered for many Fe(II) and Fe(III) compounds, also numerous Co(II) compounds, and to a minor extent for compounds of some other ions of the 3d series [12]. Thermal spin state switching is, however, not observed in the 4d series and by no means in the 5d series, which can be nicely explained by ligand field theory [13]. We consider the case of $3d^6$ transition metal complexes (Fe(II) and Co(III)). At room temperature, the electrons are distributed in the t_{2g} and e_g orbitals with a HS ($^5T_2$) ground state, as already shown for $[Fe(H_2O)_6]^{2+}$ (figure 3). As the temperature is lowered, electrons rearrange in the t_{2g} orbitals and reveal a LS (^1A) ground state. The condition to fulfil in order to observe the phenomenon is when the energy difference between the lowest vibronic levels of the LS and HS potential wells is on the order of thermal energy, $\Delta E_{HL} \approx k_BT$, as depicted in figure 7 showing energy *vs.* the distance between the Fe centre and the nitrogen that belongs to the ligand molecules [14]. Suitable ligand molecules that form a N_6 chromophore can be composed for instance, from N-donor heterocycles such as phenanthroline, triazoles, tetrazoles, to name a few. A common feature of Fe(II) compounds is their colour change associated with SCO: from colourless in the HS state to dark pink in the LS state. Such thermochromism can be nicely followed by optical spectroscopy which is a macroscopic technique, and also, as we will read later, very effectively by Mössbauer spectroscopy, a microscopic technique [12]. We consider first a mononuclear Fe(II) complex, $[Fe(phen)_3]X_2$, whose crystal structure is depicted in figure 8b. The Mössbauer spectrum reveals only one signal, typical of the LS state (figure 8a). When replacing one phenanthroline by two isothiocyanato anions, a weaker ligand, as seen in figure 8b, the ligand field is tuned to reach the SCO region.

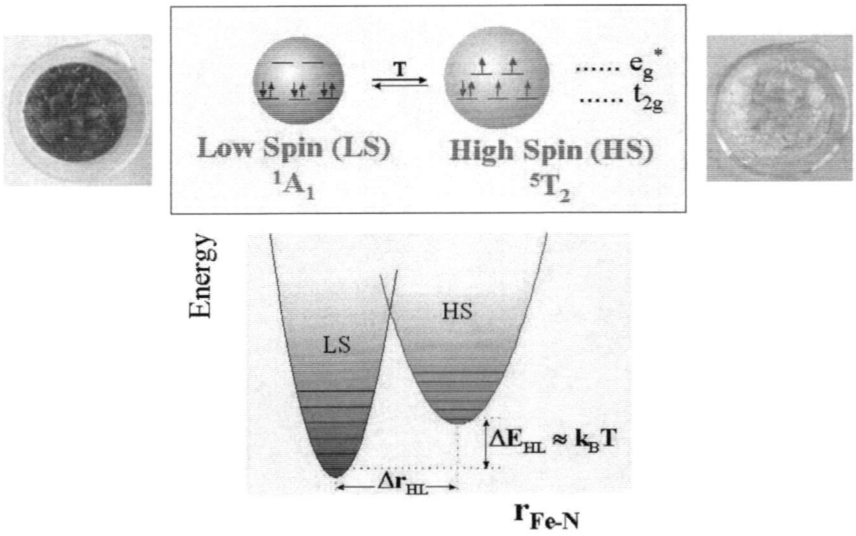

Figure 7. Thermal spin crossover in $3d^6$ transition metal complexes.

Indeed, at room temperature, one signal with a large quadrupole doublet, typical of HS Fe(II) ions is observed (figure 8c) [15]. Upon cooling, the spin sate converts to the LS one as observed at 77 K. We also learn that the lifetime of the spin states is longer than 10^{-7}s, the characteristic time window in ^{57}Fe Mössbauer spectroscopy, because well resolved resonance signals are observed [15, 16].

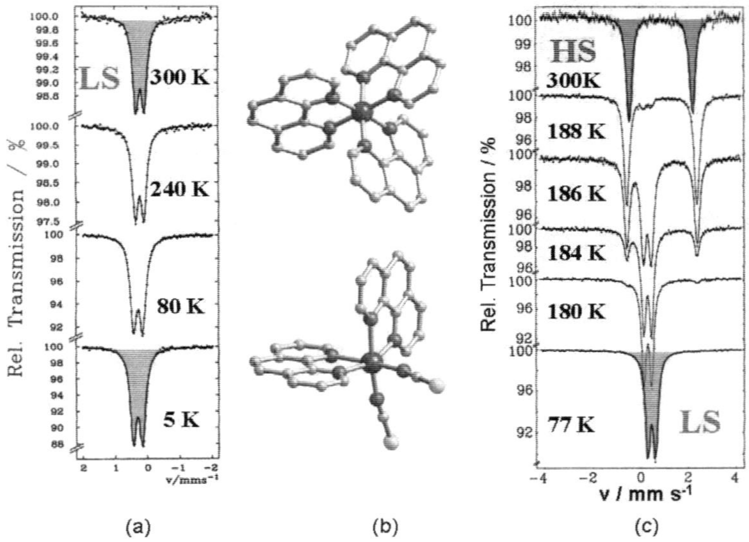

Figure 8. (a) Mössbauer spectra of $[Fe(phen)_3]X_2$ recorded over the temperature range 300-5 K. (b): View of the crystal structures of: (top) : $[Fe(phen)_3]^{2+}$ and (bottom): $[Fe(phen)_2(NCS)_2]$. (c) Mössbauer spectra of $[Fe(phen)_2(NCS)_2]$ recorded over the temperature range 300-77 K [16].

Another typical example concerns $[Fe(ptz)_6](BF_4)_2$ (ptz = 1-propyl-1H-tetrazole) where 6 N-donor ligand molecules surround the iron(II) ion. The compound remains in the HS state at room temperature and on cooling down to ~ 136 K where SCO to the LS state is observed [17]. At 98 K, the singlet attributed to LS Fe(II) clearly indicates a regular octahedron. This SCO material has been used to demonstrate the possibility to photo-address spin states, for the first time [18,19]. This topic will be presented in the next section [20].

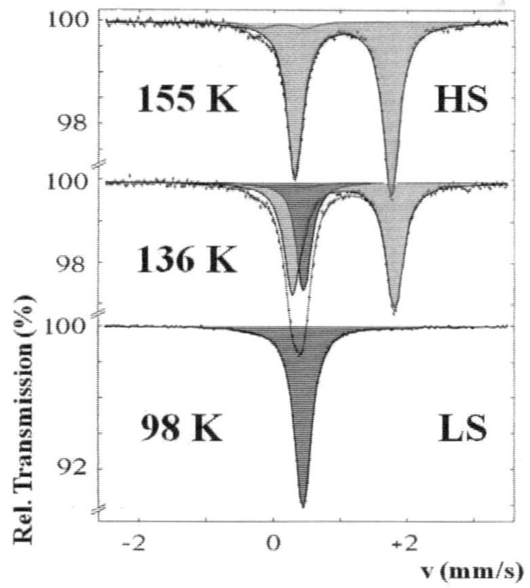

Figure 9. Selected Mössbauer spectra of $[Fe(ptz)_6](BF_4)_2$ recorded on cooling from room temperature down to 98 K [17].

3.2. Light induced spin state switching

A single crystal of $[Fe(ptz)_6](BF_4)_2$ was mounted on an optical cryostat. At 300 K, the crystal is colourless and does not absorb light. One weak absorption band is observed in the far IR, corresponding to the $^5T_2 \rightarrow {}^5E$ transition indicative of the HS state. Upon cooling to 80 K, the crystal changes its colour to dark violet and two more intense absorption bands corresponding to the singlet-singlet transitions to the $^1T_1$ and $^1T_2$ states, indicative of a LS state, are observed. Upon green light excitation at ~ 10 K, the centre of the single crystal becomes colourless as a result of the successful photo-switching of spin states by the laser beam. Thus, only one absorption band corresponding to the HS state is observed, as noted at 300 K (figure 10a). Of course, the absence of accidental temperature increase that could have caused a spin state transition from LS to HS was checked. This experiment thus translates a complete photo-switching process which was termed LIESST (Light Induced Excited Spin State Trapping) [18, 19, 20].

Light induced spin state switching was nicely confirmed by Mössbauer spectroscopy on a polycrystalline sample [18] (figure 10b). At 15 K, a LS signal is clearly observed. After green light irradiation, a HS quadrupole doublet with very long lifetime is observed indicating a complete LS → HS transition. Upon warming, relaxation to the LS state is observed at ca. 50 K after 15' and 30', respectively (figure 10b). Upon warming further to 100 K, a complete relaxation to the LS state is observed. At 150 K, the compound is fully in the HS state, as observed at 15 K after photoswitching, as a result of the LS → HS thermal switching. The LIESST phenomenon is well understood on the basis of LF theory [14, 19]. Upon cooling to low temperature, the LS state ($^1A_1$) prevails. The material converts from the LS to the HS state by irradiation with green light (λ = 514 nm) into the $^1A_1 \rightarrow {}^1T_{1,2}$ absorption bands, followed by fast relaxation with two successive intersystem crossing (ISC) pathways $^1T_{1,2} \rightarrow {}^3T_{1,2} \rightarrow {}^5T_2$ to populate the HS state (see. Fig. 11).

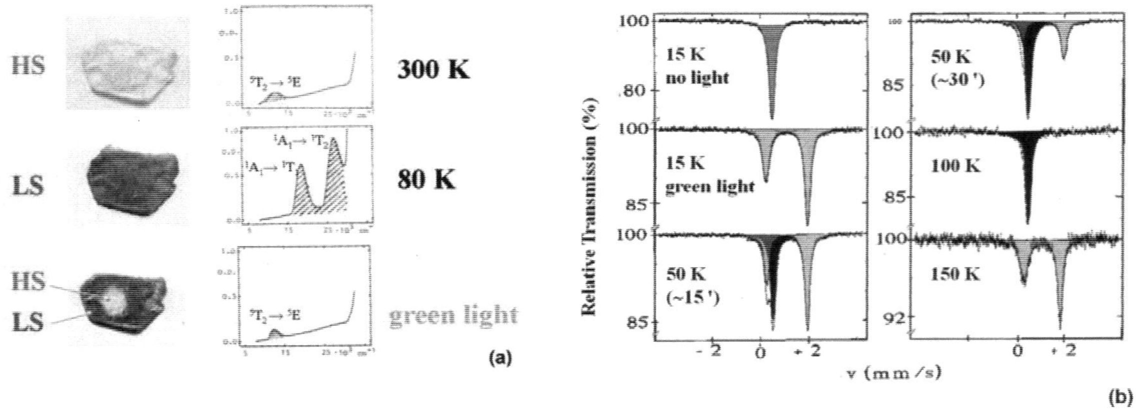

Figure 10. Thermally and light induced spin state switching in $[Fe(ptz)_6](BF_4)_2$ followed by (a) optical spectroscopy [21] and (b) Mössbauer spectroscopy [18].

Radiative decay $^5T_2 \rightarrow {}^1A_1$ is spin and parity forbidden, and due to the considerably larger M-L bond distance in the HS state as compared to the LS state, which builds up an energy barrier between the potential wells of the two states (see figure 7), the lifetime of the metastable HS state can be very long, e.g. weeks at 20 K in the case of $[Fe(ptz)_6](BF_4)_2$. When the material is irradiated with red light (λ = 820 nm), the HS $\rightarrow$ LS transition is fully reversible as first shown by Hauser [22].

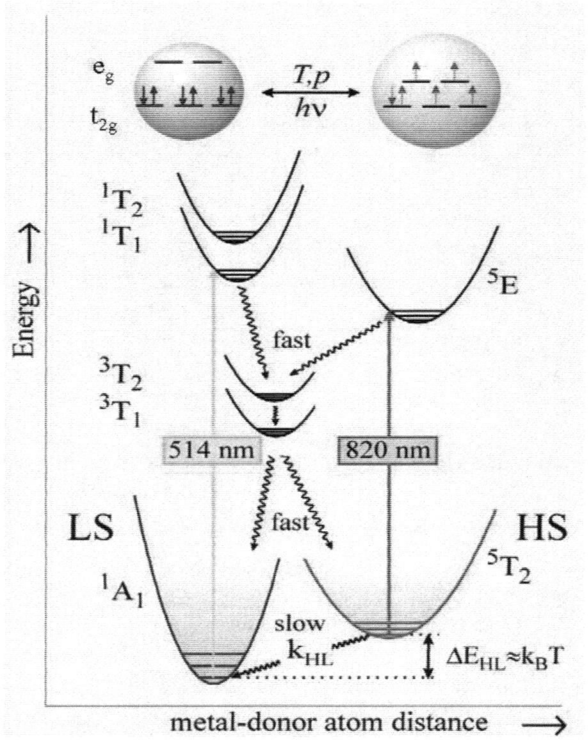

Figure 11. Jablonski diagram illustrating the LIESST and reverse-LIESST mechanism in Fe(II) complexes. Spin allowed d-d transitions are represented by arrows and the radiationless relaxation processes by wavy lines [14, 20].

Replacing propyl by methyl in the tetrazole molecule yields $[Fe(mtz)_6](BF_4)_2$ with mtz = 1-methyl-1H-tetrazole which undergoes thermal spin transition around 74 K. The crystal structure at room temperature shows two lattice sites called A and B that only differ slightly by their Fe-N bond lengths (Figure 12a) [23]. Mössbauer spectroscopy between room temperature and 160 K shows only one HS site. However, on further cooling, two sites in equal population become clearly distinguishable. Upon cooling to 60 K, the HS quadrupole doublet (green) attributed to B is not affected. However, a new signal (blue) appears at 85 K which is characteristics of the LS state indicating a SCO of the A site. At 60 K, the transition of the A sites is complete while the B sites fully remain in the HS state with a 50% population (figure 12b).

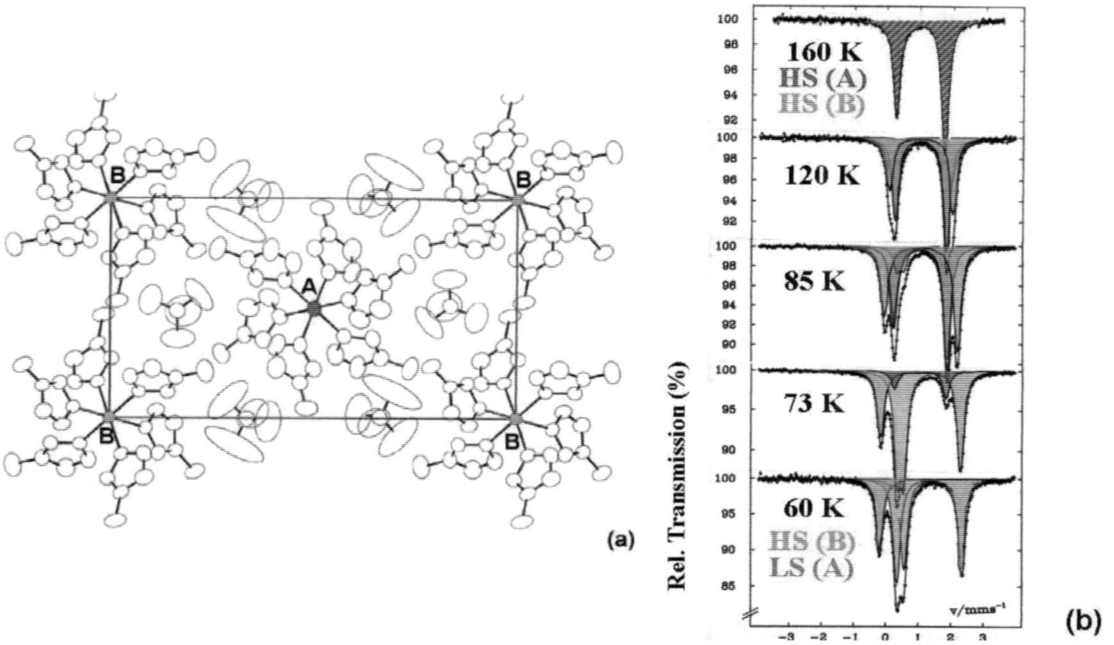

Figure 12. (a) View of the crystal structure of $[Fe(mtz)_6](BF_4)_2$ showing two lattice sites, A and B at 300 K. Fe(A)-N = 2.181(5), 2.181(5), 2.181(7) Å and Fe(B)-N = 2.161(5), 2.197(5), 2.207(4) Å [23]. (b) Temperature dependent Mössbauer spectra of $[Fe(mtz)_6](BF_4)_2$ over the temperature range 160 K-60 K [24].

When green light is shed on the sample at 20 K, a complete LS(A) $\rightarrow$ HS(A) photo-conversion is observed. Around 65 K, thermal relaxation back to the LS state is observed for the A sites. When the sample is bleached by red light at 20 K, the signal corresponding to the B site disappears almost totally, and a new resonance signal appears in the region typical for Fe(II) in the LS state. This experiment actually corresponds to the first example of light induced excitation of a stable HS state into a long lived metastable LS state. Again, upon warming to 65 K, complete thermal relaxation is observed [24].

Journal of Physics: Conference Series **217** (2010) 012001 doi:10.1088/1742-6596/217/1/012001

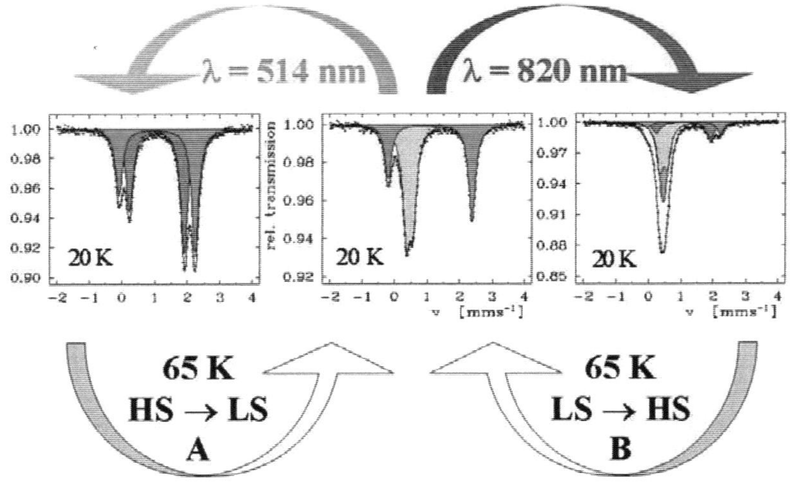

Figure 13. LIESST effect in sites A (LS to metastable HS with green light) and B sites (HS to metastable LS with red light) for $[Fe(mtz)_6](BF_4)_2$ and thermal back switching [24].

3.3. Mechanism of spin state switching in iron(II) dinuclear compounds

The crystal structure of $[Fe(bpym)(NCS)_2]_2$bpym is shown in figure 14a. This compound belongs to a family of materials that can combine both antiferromagnetic coupling and SCO phenomena. It consists of two Fe(II) centres, each of them coordinated to a bidentate bpym and two monodentate isothiocyonato anions, and bridged by a bipyrimidine ligand to form a dinuclear unit [25]. The temperature dependent magnetic properties of $[Fe(bpym)(NCSe)_2]_2$bpym reveals first weakly coupled HS Fe(II) ions. Upon cooling further to ca. 120 K, a dramatic decrease is observed to reach a plateau where a spin transition of half of the Fe(II) ions has occurred (figure 14b) [26]. An important question arises concerning the nature of the plateau region: Does the plateau originate from only [HS-LS] pairs or from a 1:1 mixture of [HS-HS] and [LS-LS] pairs?

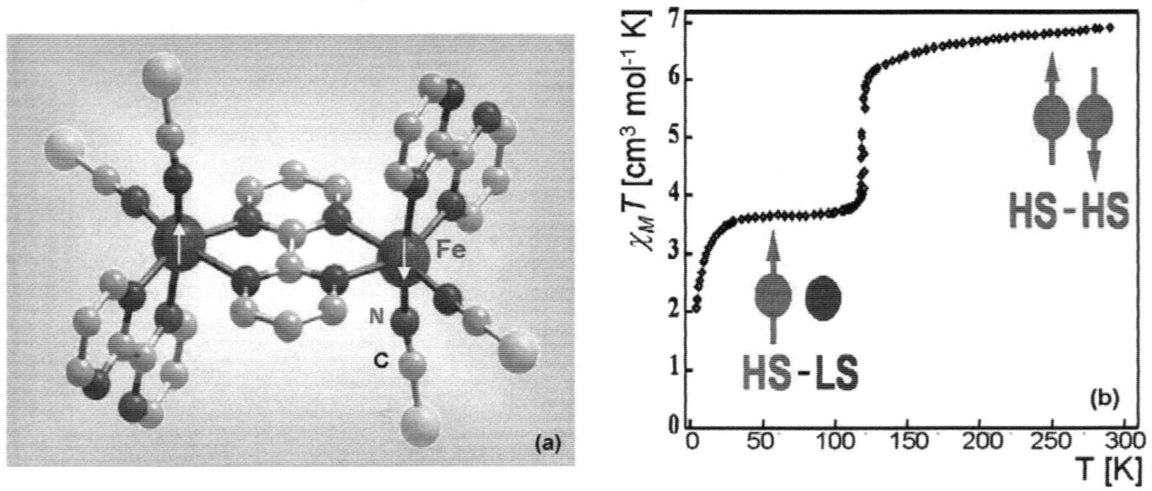

Figure 14. (a) View of the crystal structure of $[Fe(bpym)(NCS)_2]_2$bpym [25]. (b) Temperature dependent magnetic properties for $[Fe(bpym)(NCSe)_2]_2$bpym [26].

Magnetic susceptibility measurements cannot answer this question. Neither can it be answered by standard Mössbauer spectroscopy, because it would yield identical spectra in both cases (at least at higher temperatures); cf. figure 15a. However, the Mössbauer spectrum recorded at 4.2 K in applied magnetic field (figure 15b) gives a conclusive answer, that the plateau is predominantly due to the presence of [HS-LS] pairs [27].

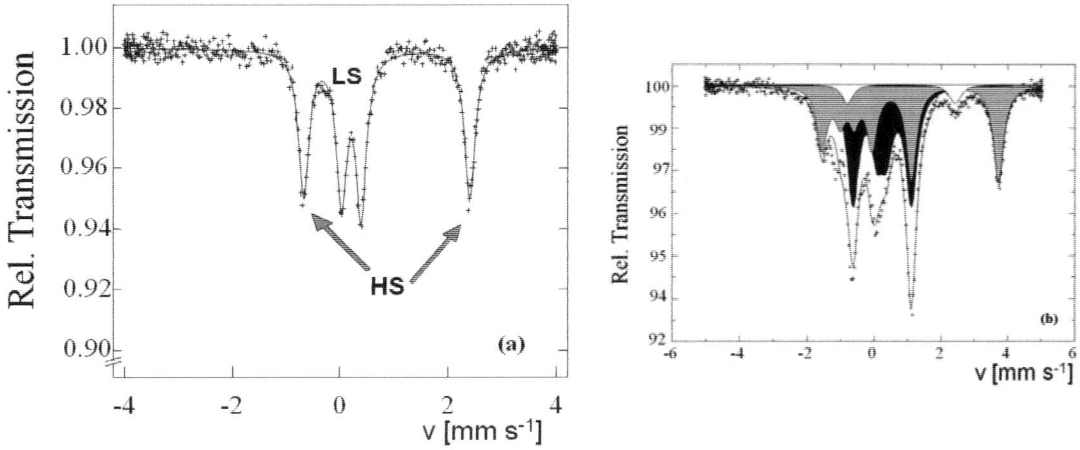

Figure 15. Mössbauer spectra of $[Fe(bpym)(NCSe)_2]_2bpym$ at: (a) 4.2 K and (b) 4.2 K and 5 T. LS in [HS–LS] and [LS–LS] pairs (black), HS in [HS–LS] pairs (grey), HS in [HS–HS] pairs (white) [27].

In another Mössbauer effect study the mechanism of the half-way spin transition in the dinuclear complex of formula $[Fe^{II}_2(PMAT)_2](BF_4)_4{\cdot}DMF$ with PMAT = 4-amino-3,5-bis{[(2-pyridyl-methyl)amino]methyl}-4H-1,2,4-triazole could be elucidated be recording the spectra solely as a function of temperature in zero field. The structure of this dinuclear compound is shown in figure 16a [28]. The Fe(II) ions are in the HS state (shown in red) at room temperature as expected for [HS-HS] pairs. On cooling, only half of the Fe(II) ions switch to the LS state as confirmed by the magnetic susceptibility measurements (figure 16b). This is possible in two ways: (i) Both Fe(II) complex ions have undergone spin transition but in only 50 % of the original [HS-HS] pairs; (ii) Only one Fe(II) centre in all [HS-HS] pairs has switched. Mössbauer spectroscopy has confirmed that the second switching manner holds, viz. a spin transition between [HS-HS] to [LS-HS] pairs. Such a spin transition from [HS-HS] to [LS-HS] pairs on cooling is in agreement with the temperature dependence of the magnetic properties (figure 16b). The HS to LS transition of half of the Fe(II) ions occurs around 225 K to reach a plateau of very large length. The drop in $\chi_M T$ at very low temperature is definitely not due to a further HS to LS transition, because the Mössbauer spectrum recorded at 4.2 K shows a 50 % HS population. The sharp decrease of $\chi_M T$ at low temperatures is due to zero-field splitting of HS Fe(II) ions [29]. The Mössbauer spectra recorded as a function of temperature and shown in figure 17 not only confirm a 50% spin state conversion, but also gives insight in structural distortion that accompanies the spin transition (figure 17). The quadrupole doublet measured at 298 K refers to the original [HS-HS] pairs. At 240 K new signals appear corresponding to LS Fe(II) ions (blue) in [HS-LS] pairs and HS Fe(II) ions (dark red) in [HS-LS] pairs, both with the same area fraction. Upon further cooling to 180 K, a complete transition of HS ions in [HS-HS] pairs to a 50 % mixture of LS and HS ions in [HS-LS] pairs is identified, for the first time in zero-field Mössbauer spectroscopy [30]. As seen from the 180 K spectrum the original quadrupole doublet (red) from [HS-HS] pairs has totally disappeared at the favour of the poorly resolved LS quadrupole doublet (blue) from [HS-LS] pairs and the new HS doublet (dark red) from [HS-LS] pairs.

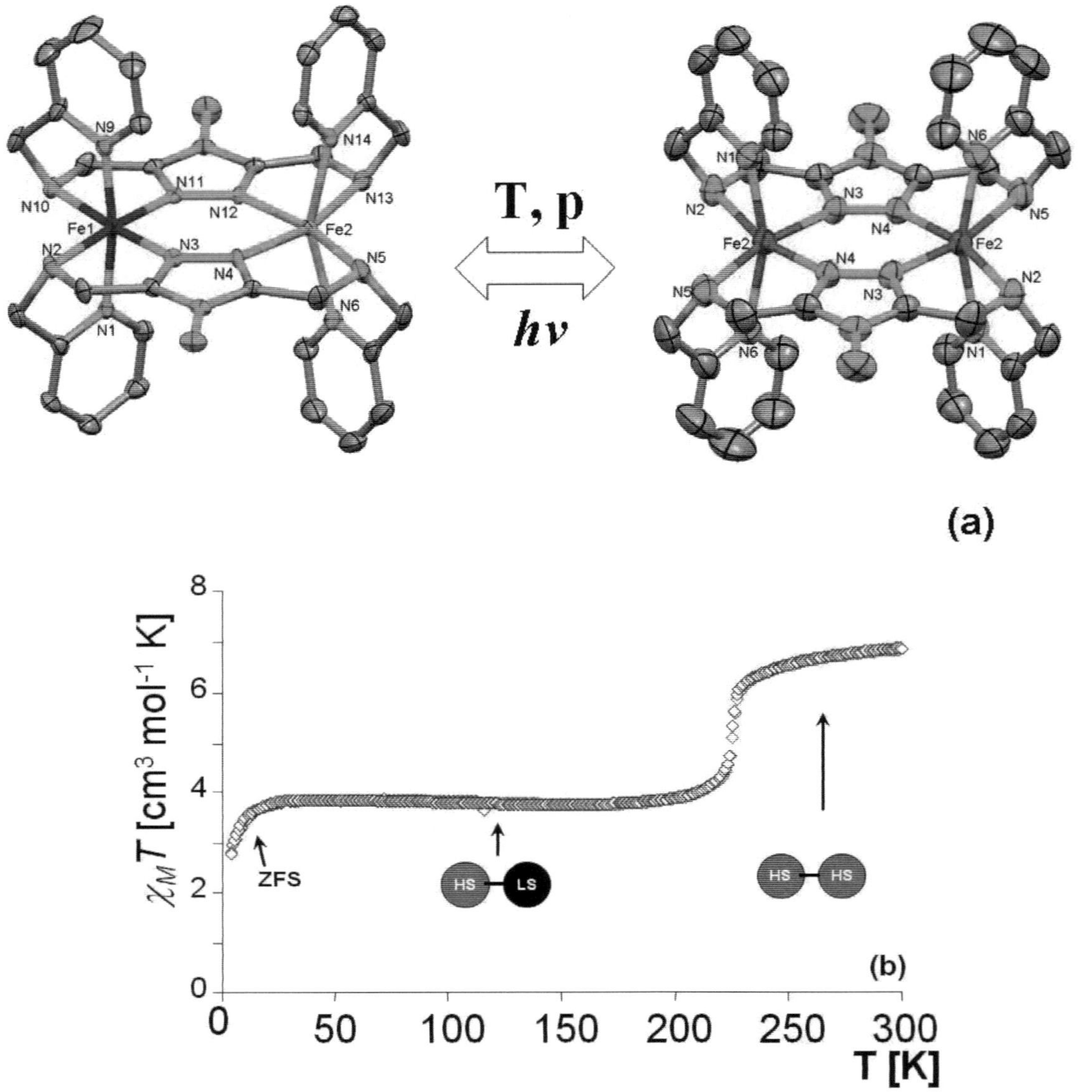

Figure 16. (a) View of the crystal structure of $[Fe^{II}_2(PMAT)_2](BF_4)_4{\cdot}DMF$ at 298 (right)and 123 K (left) [28]. (b) Magnetic susceptibility as a function of temperature for $[Fe^{II}_2(PMAT)_2](BF_4)_4{\cdot}DMF$ [30].

Its quadrupole splitting is significantly smaller than that in the original [HS-HS] pairs, and clearly points to the occurrence of a molecular distortion caused by the spin transition. In other words, the spin transition occurring in one Fe(II) centre is rapidly (in times characteristic of molecular vibrations) transmitted through the bridging ligand to the neighbouring Fe(II) centre where it causes some molecular distortion such that the ligand field strength is weakened to such an extent that no more thermal SCO takes place with this Fe(II) ion on further cooling. This is indicated in the structure (figure 16a) by the pink colour for the Fe(II) centre remaining permanently in HS state. Clearly, Mössbauer spectroscopy has revealed such extreme subtleties in this study which is hardly possible with any other physical technique.

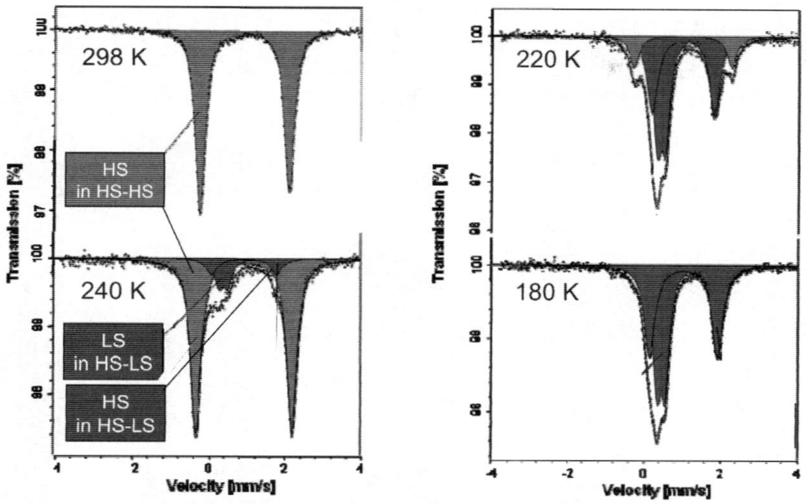

Figure 17. Proof of [HS-LS] pair formation by variable temperature ^{57}Fe Mössbauer spectroscopy over the temperature range 298-180 K [30].

3.4. Advanced functional materials: Spin crossover combined with liquid crystallinity

The last decade has seen an intense research activity concerning the combination of SCO phenomenon with another property in the same compound. Recent examples concern electrical conductivity, nonlinear optics, and liquid crystallinity to name a few [31]. We show here one example for the later property. It concerns a 1D polymeric metallomesogen of formula $[Fe(Cn\text{-}tba)_3](anion)_2$ with *Cn*-tba = 3,5-bis(alkoxy)-*N*-(4*H*-1,2,4-triazole-4-yl)benzamide. The bidentate bridging triazole ligand between Fe(II) ions carry long alkyl chains whose lengths are known to influence the liquid crystalline properties (figure 18a). For C_{10}, the material shows a thermally induced SCO around room temperature due to a glass transition occurring at $T_g = 295$ K. It also displays a thermochromic behaviour from white to pink (figure 18a). The structure can be seen as a succession of disks that are far apart in the HS state and become compressed in the LS state (figure 18b). Mössbauer spectra recorded for $[Fe(C_{12}\text{-}tba)_3](CF_3SO_3)_2$ nicely confirm the SCO upon warming (figure 18b) [32].

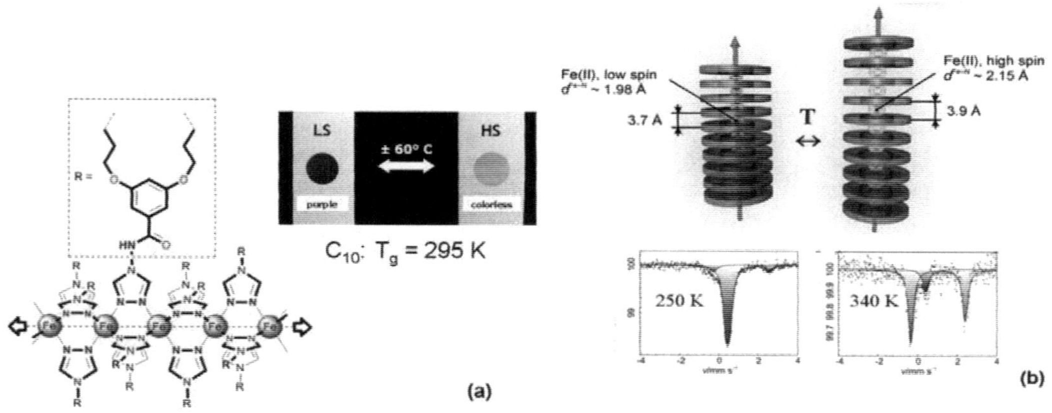

Figure 18. (a) Right: schematic view of 1D polymeric metallomesogens of Fe(II). Left: photograph of the thermochromic spin transition. (b) Top: schematic view of 1D chain with associated disks in the LS and HS states. Bottom: Mössbauer spectra of $[Fe(C_{12}\text{-}tba)_3](CF_3SO_3)_2$ at 250 K and 340 K [32].

The variation of the chain length on the triazole ligand in $[Fe(C_n\text{-tba})_3](BF_4)_2{\cdot}H_2O$ with C_n-tba = 5-bis(alkoxy)-*N*-(4*H*-1,2,4-triazol-4-yl)benzamide) can dramatically affect the spin state population as illustrated in Figure 19 where the blue and red signals refer to the LS and HS states, respectively [33].

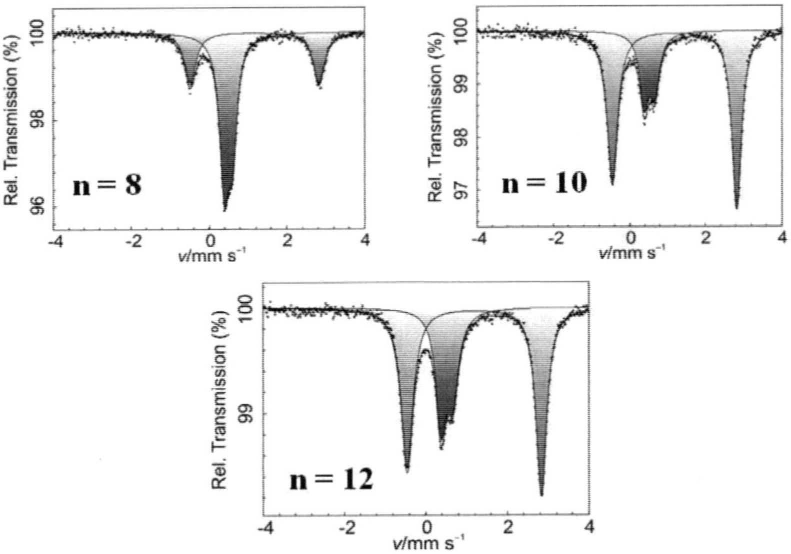

Figure 19. Effect of chain length on the investigated by Mössbauer spectroscopy on $[Fe(Cn\text{-}tba)_3](BF_4)_2{\cdot}H_2O$ (n = 8, 10, 12) at 80 K [33].

4. Mössbauer emission spectroscopy

Mössbauer emission spectroscopy research started in the early sixties and was used in order to learn something about the chemical after effects accompanying nuclear disintegration in solids. For instance, the substitution of iron in a SCO compound by ^{57}Co leads to ^{57}Fe which is in a highly excited state after electron capture decay. Several events such as bond rupture, ligand radiolysis, change of charge state or excited ligand field states may then occur. We shall show one example with the Mössbauer emission study of $[^{57}Co/Co(phen)_2(NCS)_2]$ recorded *vs.* the single line absorber $K_4[Fe(CN)_6]$ (figure 20). A HS state of the nucleogenic ^{57}Fe is mostly observed over the whole range of temperature [34] compared to the spectra of $[^{57}Fe/Fe(phen)_2(NCS)_2]$ which show a thermally induced SCO on cooling as seen in figure 8c. The surprising appearance of a HS state in the low temperature region shown in figure 21 for $[^{57}Co_x/Co_{1-x}(phen)_3](ClO_4)_2$ results from the relaxation following $^{57}Co(EC)^{57}Fe$ decay [35]. On the left of figure 21 are plotted the Mössbauer absorption spectra of $[Fe(phen)_3](ClO_4)_2$. The Mössbauer emission spectra of $[^{57}Co_x/Co_{1-x}(phen)_3](ClO_4)_2$ recorded versus $K_4[Fe(CN)_6]$ (298 K) as absorber are shown on the right of figure 21. These are very similar to the absorption spectra at higher temperatures because of fast relaxation (faster than the Mössbauer time scale of ^{57}Fe) of excited ligand field states. However, when the temperature is lowered to 80 K and below resonance signals characteristic of excited HS-Fe(II) ligand field states appear in the emission spectra with increasing intensity towards lower temperatures. Clearly, the lifetimes of such excited ligand field states induced by the nuclear decay of ^{57}Co are now comparable to the lifetime of the 14.4 keV state of ^{57}Fe. The mechanism of formation of the excited ligand field states is very similar to that of the LIESST phenomenon (described above), this phenomenon has therefore been termed NIESST for (Nuclear Decay-Induced Excited Spin State Trapping).

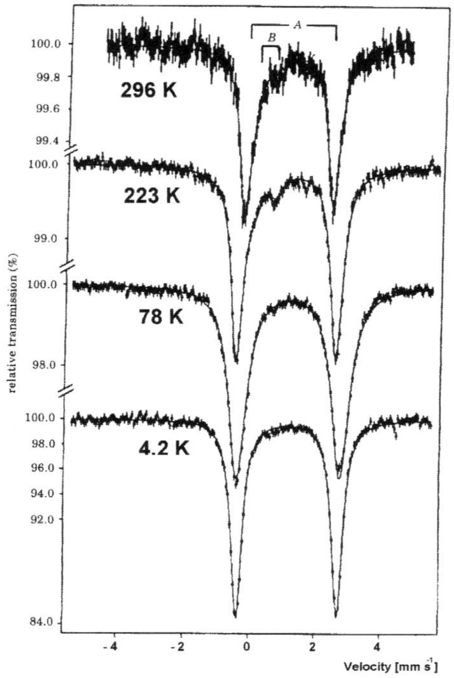

Figure 20. ^{57}Fe Mössbauer emission spectra of a [^{57}Co/Co(phen)$_2$(NCS)$_2$] source as a function of temperature vs. $K_4[Fe(CN)_6]$ (298 K) [34]. The large quadrupole doublet with increasing splitting at lower temperatures is characteristic of Fe(II) in HS state. The nucleogenic ^{57}Fe(II) ion arising from EC decay of ^{57}Co(II) in [^{57}Co/Co(phen)$_2$(NCS)$_2$] is observed in HS state at all temperatures under study, even in the region where the corresponding Fe(II) compound has undergone spin transition to the LS state.

Only the initial steps of electronic excitation are different, irradiation with an external light source in LIESST but nuclear decay as an intrinsic molecular light source in NIESST, but the mechanisms of the final electronic relaxation are the same (figure 11).

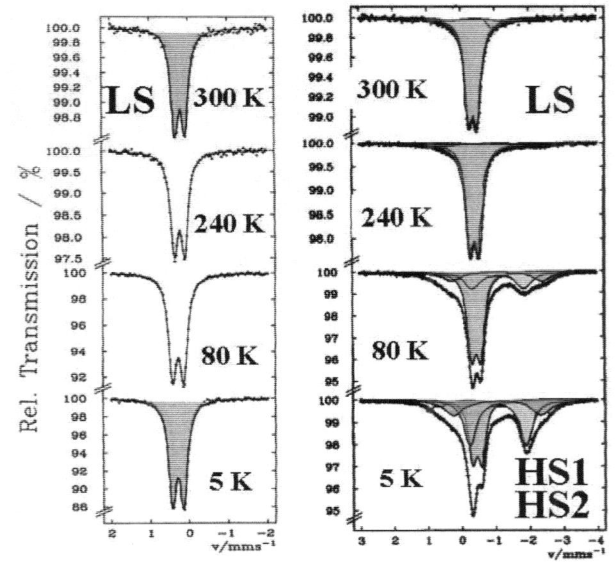

Figure 21. (a) Left: ^{57}Fe Mössbauer absorption spectra of [^{57}Fe$_x$/Co$_{1-x}$(phen)$_3$](ClO$_4$)$_2$ as a function of temperature vs. ^{57}Co/Rh (295 K) as source (x = 0.001). (b) Rigth: Time-integral ^{57}Fe Mössbauer emission spectra of a [^{57}Co$_x$/Co$_{1-x}$(phen)$_3$](ClO$_4$)$_2$ source as a function of temperature vs. $K_4[Fe(CN)_6]$ (295 K) (x = 0.001). In (a) the source was moved relative to the absorber and in (b) the absorber was moved relative to the fixed source mounted in the crystal [35].

5. Mobile Mössbauer spectroscopy

The Miniaturization of the Mössbauer spectrometer (MIMOS), initiated by E. Kankelet in the early nineties at the Technical University of Darmstadt and further developed and brought to completion at the Institute of Inorganic and Analytical Chemistry of Johannes Gutenberg-University Mainz by Dr G. Klingelhöfer, has provided a portable analytical tool for numerous in-field applications [36]. Most spectacular, for instance, is the application of MIMOS in the two current NASA missions to the planet Mars for investigations of the surface mineralogy. One MIMOS instrument each is mounted on the robotic arms of two Rover exploration vehicles from space agencies [37]. MIMOS was chosen by the ESA and NASA to participate in the Mars missions launched in 2003. The two NASA missions with the Mars exploration rovers (MER) "Spirit" and "Opportunity" on board landed successfully in different locations of the red planet Mars in January 2004. MIMOS has been functioning extremely well since then, although the quality of the radioactive sources mounted on the instruments have now weakened. It has recorded hundreds of Mossbauer spectra of high quality which have been sent to earth for data recording and evaluation. This work has benefited enormously from the huge data base accumulated by Mössbauer spectroscopists over the years [38].

Figure 22 presents views of MIMOS whose size is so small that the whole spectrometer can lie on a human hand. It is composed of several parts such as a drive, a drive control unit, a reference channel for constant velocity calibration mounted on the vibrator opposite to the front position with a very strong source (ca. 250 mCi at launch, 1 mCi ≙ 37 MBq) and four Si-PIN detectors for recording the scattered 14.4 keV gamma rays and the 6.4 keV X-rays. The two Rovers are indeed still operating on the surface of Mars in 2009 although the activity of the ^{57}Co sources has become very weak (a few mCi).

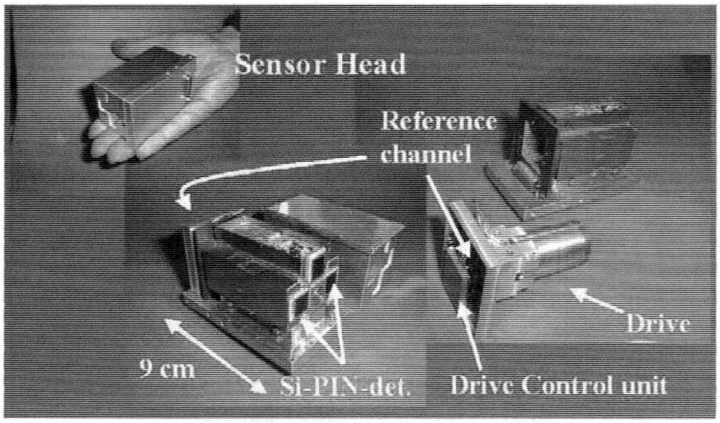

Figure 22. Views of the miniaturized Mössbauer spectrometer and of its electronic components after demounting [36].

The first Mössbauer spectrum from the surface of Mars was obtained in the Columbia Hills of Gusev Crater [39] close to the Rover landing site (figure 23). It clearly revealed Fe^{2+} containing species (olivine and pyroxene) as well as some Fe^{3+} species, not yet identified at that stage. The discovery of olivine, commonly found in lava on earth, was quite informative since this mineral is known to weather to clays and iron oxides in the presence of water [40,41].

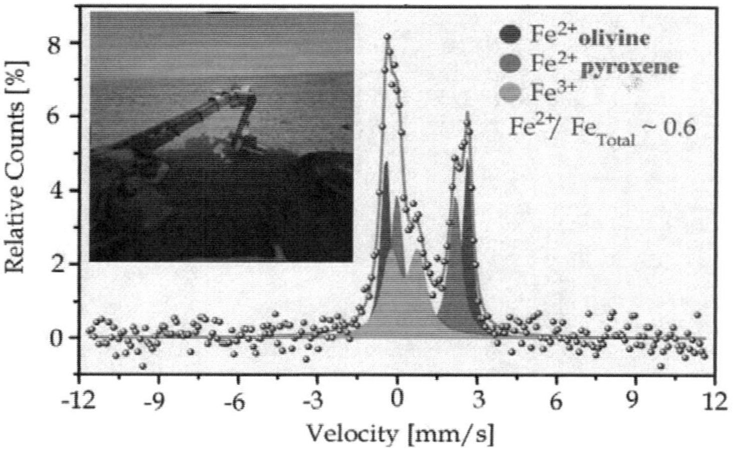

Figure 23. First Mössbauer spectrum recorded on the Martian surface at Gusev crater (January 17th 2004, measuring time 3h25min). The inset shows a view of the Rover and of the MIMOS spectrometer operating on Mars. A view of the ground of Mars is shown in the back [40,41].

Nests with bluish mineral spherules similar to blueberries [42] (figure 24a) were discovered in a region near the landing site of the Rover *Opportunity* in Meridiani Planum [43]. Mössbauer spectra were recorded in places with and without such 'Blueberry' spherules (figure 24b). Clearly, the 'blueberries' minerals are enriched in haematite (α-Fe_2O_3), thus demonstrating once again the usefulness of the technique.

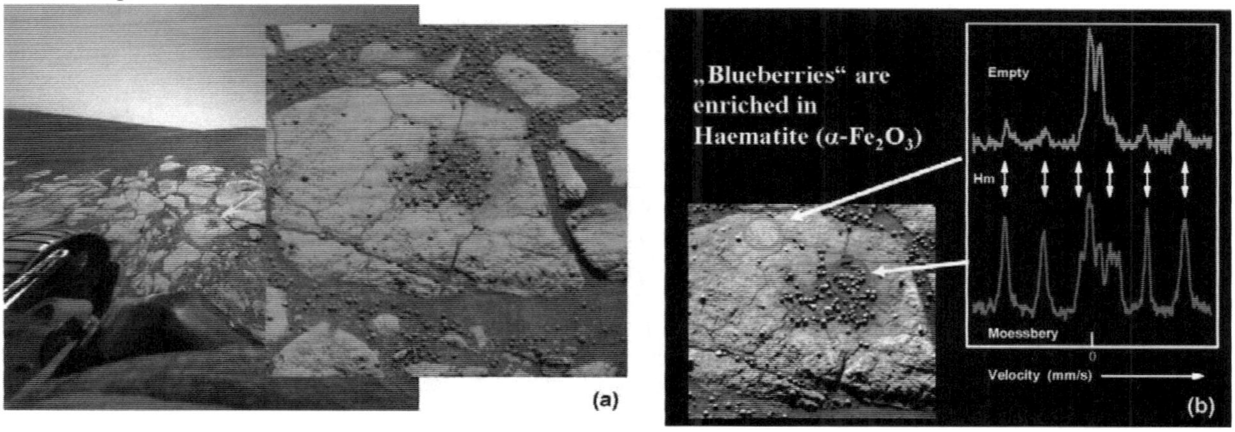

Figure 24. (a) View of a place called 'Berry bowl' on Mars planet, with bluish spherules similar to 'Blueberries' The front of the Rover is also seen. (b) Identification of haematite (α-Fe_2O_3) in the 'blueberries' by Mössbauer spectroscopy [42].

The breakthrough of the mission is the definite proof that there was, and still is, water on the surface of Mars. Indeed, in the Mössbauer spectrum of figure 25, beside several species including an Fe^{2+} silicate and a magnetic phase attributed to an iron-oxide mineral, the presence of the mineral *Jarosite* was confirmed. This Fe^{3+} mineral of formula $(K,Na,X^{+})Fe_3(SO_4)_2(OH)_6$ contains water in the form of hydroxyl groups. It is known to grow in wet sulfuric acidic surroundings like volcanic outcrops and constitutes a signature of the (past) presence of water on Mars [44].

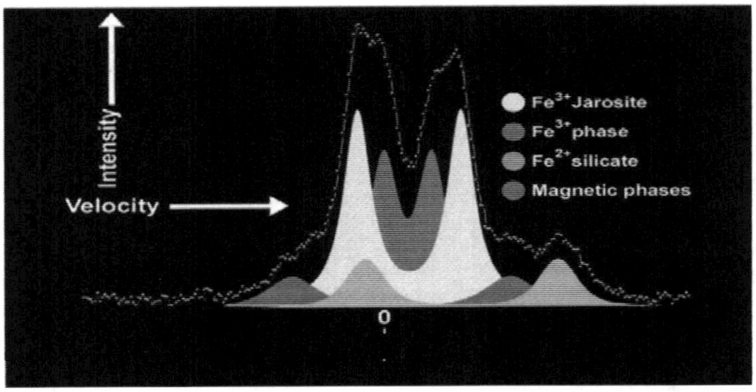

Figure 25. Mössbauer spectrum that identifies *Jarosite*, a mineral of formula $(K,Na,X^{+})Fe_3(SO_4)_2(OH)_6$ [44].

6. Conclusion and Outlook

Mössbauer spectroscopy has, from the very beginning in the early sixties up to now, very extensively employed in solid state research, mostly in conjunction with other physical techniques in order to reach deeper and more conclusive information, but also in cases where certain problems could not be solved with other techniques. More than fifty thousand publications dealing with Mössbauer effect studies have appeared [38] since the brilliant discovery of the *recoilless nuclear resonance fluorescence* by Rudolf Mössbauer [1]. Only a small selection of applications, particularly in chemistry, which in the authors' opinion may be considered highlights or textbook examples that demonstrate convincingly the elegance and versatility of this spectroscopic technique, could be included in this Keynote Lecture. In view of the limited space most of these examples are chosen from work performed in the authors' laboratories. The authors wish to extend their apologies to the many colleagues who have made excellent contributions to the Mössbauer spectroscopy literature which would have vastly enriched this presentation.

Looking ahead and asking the question "*quo vadis Mössbauer effect researchh?*", one cannot but focusing ones attention to the two outstanding developments who have opened new pathways in Mössbauer spectroscopy and will certainly play a remarkable role in future. These concern, without quality ranking, (i) the instrumental progress regarding MIMOS as briefly described above, and (ii) the use of synchrotron radiation for observing nuclear resonance fluorescence.

MIMOS has most spectacularly demonstrated its usefulness for extraterrestrial studies as discussed above, viz. the current NASA missions to the planet Mars. There are, of course, also hundreds of possibilities to use it on earth in mobile analytical studies outside the laboratory. Examples have been reported such as *in-situ* monitoring of airborne particles [45], archaeological artefacts [46], rockpainting in field [47], to name a few.

A real breakthrough in Mössbauer spectroscopy research was initiated by E. Gerdau et al. in 1985 who proposed an unconventional Mössbauer technique based on the possibility to use synchrotron radiation to observe nuclear resonance [48]. Two possibilities are pursued (figure 26): (i) looking at the elastic nuclear forward scattering (NFS) to study hyperfine interactions, as obtained with conventional Mössbauer spectroscopy, or nuclear inelastic scattering (NIS) to investigate local phonon spectra (partial density of states, PDOS) at the Mössbauer probe nucleus.

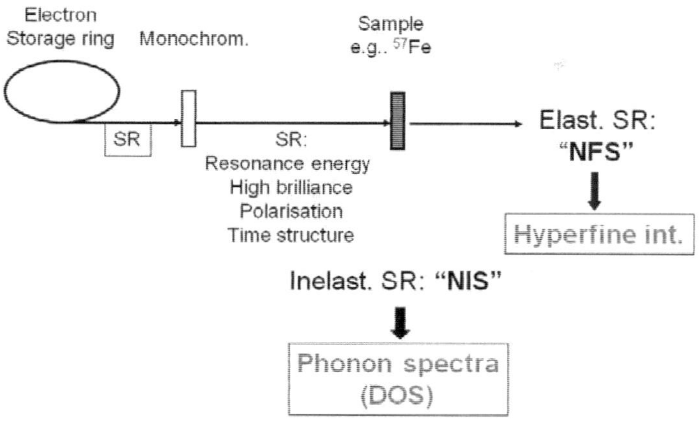

Figure 26. Mössbauer effect on ^{57}Fe with synchrotron radiation [48].

As an example, NIS spectra were recently recorded on $[Fe(ptz)_6](BF_4)_2$, a material that displays a thermally induced SCO as shown above in figure 9. Beautifully resolved spectra were obtained in the HS and LS states on cooling below room temperature and by green light irradiation (see figure 27) [49]. It clearly demonstrates the usefulness of NIS compared for instance to Raman spectroscopy that can by far not achieve such a resolution because of the perturbation by surrounding vibrations.
Both synchrotron radiation techniques, NFS and NIS, are certainly on their way to a great future.

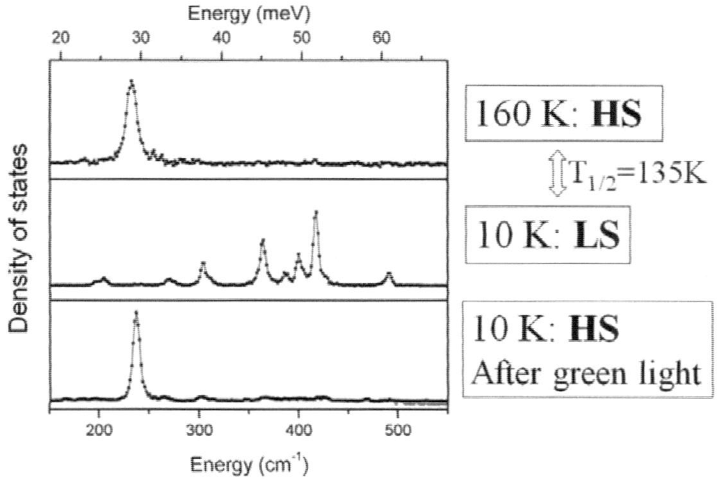

Figure 27. NIS spectra of $[Fe(ptz)_6](BF_4)_2$ recorded in the HS state at 160 K, in the LS state at 10 K, and in the metastable LIESST state after irradiation at 10 K with green light [49].

Acknowledgements
We acknowledge the financial support from the Deutsche Forschungsgemeinschaft, Fonds der Chemischen Industrie, Federal Government of Germany, University of Mainz, European Union: TMR, ESF, FNRS-FRFC (No 2.4508308) and IAP-VI (P6/17) INANOMAT.

References

[1] Mössbauer R L 1958 *Physik* **151** 124. Mössbauer R L 1958 *Naturwissenschaften* **22** 538. Mössbauer R L 1962 *Science* **137** 731. Mössbauer R L 2000 *Hyperfine Interact.* **126** 1
[2] Kistner O C and Sunyar A W 1960 *Phys. Rev. Lett.* **4** 412
[3] Leonhardt J and Ness I 1950 *Fortschritte Mineral.* **26** 83
[4] De Benedetti S, Lang G and Ingalls R 1961 *Phys. Rev. Lett.* **6** 60
[5] Baur W H 1964 *Acta. Cryst.* **17** 1167
[6] Müller E 1914 *Chemik. Zeit.* **38** 281
[7] Fluck E, Kerler W and Neuwirth W 1963 *Angew. Chem.* 461
[8] Maer K Jr., Beasley M L, Collins R L and Milligan W O 1968 *J. Am. Chem. Soc.* **90** 3201
[9] Münck E and Stubna A. 2004 *Compreh. Coord. Chem.* **2** 279.
[10] Rao K K, Cammack R, Hall D O and Johnson C E 1971 *Biochem. J.* **122** 257.
[11] Frankel R B, Papaefthymiou G C, Blakemore R P and O'Brien W 1983 *Biochim. Biophys. Acta* **763** 147
[12] Gütlich P, Garcia Y and Goodwin H A 2000 *Chem Soc Rev* **29** 419. Gütlich P, Garcia Y and Spiering H 2003 Magnetism: From Molecules to Materials IV, Eds: J. S. Miller and M. Drillon Wiley-VCH 271
[13] Garcia Y and Gütlich P 2004 *Top. Curr. Chem.* **234** 49.
[14] Gütlich P, Hauser A and Spiering H 1994 *Angew. Chem. Int., Ed. Engl.* ***33*** 2024. Gütlich P and Goodwin H A (Eds.) Spin Crossover in Transition Metal Compounds, *Top. Curr. Chem.* ***233-235***, Springer, Berlin-Heidelberg.
[15] Dezsi I, Molnar B, Tarnoczi T and Tompa K 1967 *J. Inorg. Nucl. Chem.* **29** 2486
[16] Gütlich P 1981 *Struct. Bonding* (Berlin) **44** 83
[17] Müller E W, Ensling J, Spiering H and Gütlich P 1983 *Inorg. Chem.* **22** 2074
[18] Decurtins S, Gütlich P, Köhler C P, Spiering H and Hauser A 1984 *Chem. Phys. Lett.* **105** 1
[19] Decurtins S, Gütlich P, Hasselbach K M, Spiering H and Hauser A 1985 *Inorg. Chem.* **24** 2174
[20] Gütlich P, Garcia Y and Woike Th 2001 *Coord. Chem. Rev.* **219-221** 839
[21] Hauser A 1991 *J. Chem. Phys.* **94** 2741
[22] Hauser A 1986 *Chem. Phys. Lett.* **124** 543.
[23] Wiehl L 1993 *Acta Cryst. B* **49** 289.
[24] Poganiuch P, Decurtins S and Gütlich P 1990 *J. Am. Chem. Soc.* **112** 3270
[25] De Munno G, Julve M, Real J A, Lloret F and Scopelliti R 1996 *Inorg. Chim. Acta* **250** 81
[26] Ksenofontov V, Gaspar A B, Real J A and Gütlich P 2001 *J. Phys. Chem. B* **105** 12266
[27] Ksenofontov V, Spiering H, Reiman S, Garcia Y, Gaspar A B, Moliner N, Real J A and Gütlich P 2001 *Chem. Phys. Lett.* **348**, 381
[28] Klingele M H, Moubaraki B, Cashion J D, Murray K S and Brooker S 2005 *Chem. Comm.* **8** 987
[29] Garcia Y, Grunert C M, Reiman S, van Campenhoudt O and Gütlich P 2006 *Eur. J. Inorg. Chem.* 3333.
[30] Grunert C M, Reiman S, Spiering H, Kitchen J A, Brooker S and Gütlich P 2008 *Angew. Chem. Inter. Ed.* **47** 2997.
[31] Gaspar A B, Ksenofontov V, Seredyuk M and Gütlich P 2005 *Coord. Chem. Rev.* ***249*** *2661*
[32] *Seredy*uk M, Gaspar A B, Ksenofontov M, Reiman S, Galyametdinov Y, Haas*e W, Rentschler* E and Gütlich P 2006 *Chem. Mater.* **18** 2513
[33] Seredyuk M, Gaspar A B, Ksenofontov V, Galyametdinov Y, Verdaguer M, Villain F and Gütlich P 2008 *Inorg. Chem.* **47** 10232
[34] Ensling J, Gütlich P, Hasselbach K M and Fitzsimmons B W 1976 *Chem. Phys. Lett.* **42** 232.
[35] Sano H and Gütlich P in: T. Matsuura (Ed.), *Hot Atom Chemistry*, Kodanski, Tokyo, 1984, p. 265.
[36] Klingelhöfer G, Bernhardt B, Foh J, Bonnes U, Rodionov D, de Souza P A Jr, Schröder Ch,

Gellert R, Kane S, Gütlich P and Kankeleit E 2003 *Hyperfine Interact.* **144/145** 371
[37] Klingelhöfer G 2004 *Hyperfine Interact.* **158** 117. Klingelhöfer G, Morris R V, Souza P A Jr., Rodionov D and Schröder C 2007 *Hyperfine Interact.* **170** 169. Schröder C, Klingelhöfer G, Morris R V, Rodionov D S, Fleischer I and Blumers M 2008 *Hyperfine Interact.* **182** 149
[38] Mössbauer Effect Data Center, University of North Carolina, Asheville
[39] Squyres SW et al 2004 *Science* **305** 794.
[40] Klingelhöfer G, De Grave E, Morris R V, Alboom A, Resende V G, Souza P A Jr, Rodionov D, Schroeder C, Ming D W and Yen A 2006 *Hyperfine Interact.* **166** 549.
[41] Morris RV, Klingelhofer G, Bernhardt B, Schroder C, Rodionov D S, De Souza P A Jr, Yen A, Gellert R, Evlanov E N, Kankeleit E, Gütlich P, Ming D W, Renz F, Wdowiak T, Squyres S W, Arvidson R E 2004 *Science* **305** 833
[42] Squyres SW et al 2004 *Science* **306** 1698. Soderblom LA et al 2004 *Science* **306** 1723
[43] Squyres SW et al 2004 *Science* **306** 1709. Squyres et al 2006 *Science* **313** 1403
[44] Klingelhöfer G, Morris R V, Bernhardt B, Schröder C, Rodionov D S, de Souza P A Jr, Yen A, Gellert R, Evlanov E N, Zubkov B, Foh J, Bonnes U, Kankeleit E, Gütlich P, Ming D W, Renz F, Wdowiak T, Squyres S W and Arvidson R E 2004 *Science* **306** 1740.
[45] De Souza Paulo A Jr., Klingelhöfer G and Morimoto T Proceedings of theInternational Conference on the Applications of the Mössbauer Effect (ICAME 2001), Oxford, UK, *Hyperfine Interact.* (C), Vol. 5, pp. 487-490 (2002); De Souza P A Jr 2004 PhD thesis, Johannes Gutenberg-University Mainz, Germany
[46] De Souza P A Jr., Bernhardt B, Klingelhöfer G and Gütlich P 2003 *Hyperfine Interact.* ***151*/152** 125. De Souza P A Jr 2004 PhD thesis, Johannes Gutenberg-University Mainz, Germany
[47] Klingelhöfer G, da Costa G M, Prous A and Bernhardt B Proceedings of the International Conference on the Applications of the Mössbauer Effect (ICAME 2001), Oxford, UK, *Hyperfine Interact. (C)*, Vol. 5, pp. 423-426
[48] Gerdau E, Rüffer R, Winkler H, Tolksdorf W, Klages C P and Hannon J P 1985 *Phys. Rev. Lett.* ***54*** 835.
[49] Böttger L H, Chumakov A I, Grunert C M, Gütlich P, Kusz J, Paulsen H, Ponkratz U, Rusanov V, Trautwein A, Wolny J A 2006 *Chem. Phys. Lett.* ***429*** 189

Mössbauer spectroscopy in the energy domain using synchrotron radiation

M Seto[1, 2, 4], R Masuda[1, 4], S Higashitaniguchi[1, 4], S Kitao[1, 4], Y Kobayashi[1, 4], C Inaba[1], T Mitsui[2, 4] and Y Yoda[3, 4]

[1]Research Reactor Institute, Kyoto University, Kumatori-cho, Sennan-gun, Osaka 590-04, Japan
[2]Japan Atomic Energy Agency, Mikazuki-cho, Sayo-gun, Hyogo 679-5148, Japan
[3]Japan Synchrotron Radiation Research Institute, Mikazuki-cho, Sayo-gun, Hyogo 679-5198, Japan
[4]CREST, Japan Science and Technology Agency, Saitama 332-0012, Japan

E-mail: seto@rri.kyoto-u.ac.jp

Abstract. We have developed a Mössbauer spectroscopic method that yields absorption-type spectra by using synchrotron radiation. Owing to the energy selectivity of synchrotron radiation, this method can be applied to almost all Mössbauer nuclides including those that are difficult to prepare their parent radioactive sources. This method offers the flexibility of measuring the sample in a transmission or in a scattering configuration depending on the sample condition. We have investigated the modulation and narrowing of the spectral shapes caused by the time-window effect, and we have confirmed that the previously developed theory reproduces the measured spectra well.

1. Introduction

The use of synchrotron radiation as an alternative source for Mössbauer spectroscopy was proposed in 1974 [1]. Energy-tunable, highly bright, well-collimated, pulsed, and polarized synchrotron radiation is distinct, and it affords unique opportunities to perform measurements of the hyperfine interactions in very small samples, samples under high pressures, and ultrathin films, which are difficult using conventional radioactive sources. The first conclusive result of nuclear resonant Bragg scattering of synchrotron radiation was published in 1985 [2] after some pioneering experiments. This result opened new possibilities for Mössbauer spectroscopy using synchrotron radiation. Time-differential nuclear resonant forward scattering of synchrotron radiation, a time analogue of Mössbauer spectroscopy that was developed in 1991 [3], is a standard method, and it has been employed in many scientific researches because it does not require single-crystal samples. The applicability of this method is as extensive as that of Mössbauer spectroscopy using radioactive sources. Furthermore, it was recently shown that the nuclear resonant forward scattering method is applicable to the high-energy nuclide ^{61}Ni (67.41 keV) using multielement (stacked) avalanche photodiodes (APDs) [4]. In addition, a method that uses the nuclear lighthouse effect was developed [5]. In this method, the time evolution of the decaying nuclei in a rotating sample is translated into an angular deviation. This method is effective for high-energy short-lifetime excited states, although the sample environment in this method is restricted. Furthermore, a heterodyne (stroboscopic) detection method for nuclear resonant forward scattering of synchrotron radiation [6][7] has also been developed. This method is effective for nuclides with very long lifetimes. Mössbauer spectroscopy can be carried out in a straightforward manner by using synchrotron Mössbauer sources obtained using pure nuclear Bragg scattering. In fact, single-line Mössbauer sources for ^{57}Fe were obtained [8][9], and the successful results of a high-pressure study were reported [10]. Because the spectrum obtained using synchrotron Mössbauer sources is basically identical to that obtained using conventional radioactive sources, this approach may be easily applicable and beneficial for a wide range of Mössbauer research; however, thus far, ^{57}Fe is the only available nuclide. Furthermore, the use of synchrotron radiation enables us to measure nuclear resonant inelastic scattering, which provides element-specific phonon energy spectra [11]. Recently, we developed a method that yields Mössbauer absorption spectra using synchrotron radiation and that is applicable to many Mössbauer nuclides [12]. This method enabled us to carry out unique Mössbauer spectroscopy with the excellent features of synchrotron radiation. We discuss the theoretical framework of this method and the spectra obtained using it.

2. Conceptualization and theoretical framework of developed method

The concept behind the developed method and its theoretical framework have been described previously [12]. We repeat only those details that are required here and use the same abbreviated notations as used previously.

We use two samples containing the same resonant Mössbauer nuclides; one is used as a transmitter and the other as a scatterer. The scatterer is placed behind the transmitter in the path of the synchrotron radiation. A schematic of the typical experimental setup is shown in figure 1. In order to measure the energy-domain Mössbauer spectra, we use a silicon APD detector placed above or below the scatterer to detect delayed 4π scattering containing internal conversion electrons and fluorescent X-rays following the de-excitation in addition to the emitted γ-rays, as a function of the Doppler velocity. It is easier to detect fluorescent X-rays than γ-rays using a silicon APD detector; this is because the energies of fluorescent X-rays are lower than those of γ-rays. Moreover, the detection of electrons is not difficult using APD detectors.

When the resonance energies of the scatterer and the transmitter are equal, there is a decrease in the nuclear resonance scattering from the scatterer; this is because of the strong nuclear resonance absorption of the resonance energy radiation in the transmitter. Therefore, by measuring the scattering from the scatterer as a function of the relative energy (velocity), we can observe a local minimum when the relative energy is zero. This velocity-dependent scattering intensity yields a spectrum in the energy domain. This description is simple and purely conceptual; however, an actual reproduction of the measured spectra will be more complex. In the scatterer, there exist two processes that depend on the relative resonance energy shift w_s in units of the natural width of the nuclear excited state between the transmitter and the scatterer—the nuclear resonant recoilless absorption of radiation (A-channel)

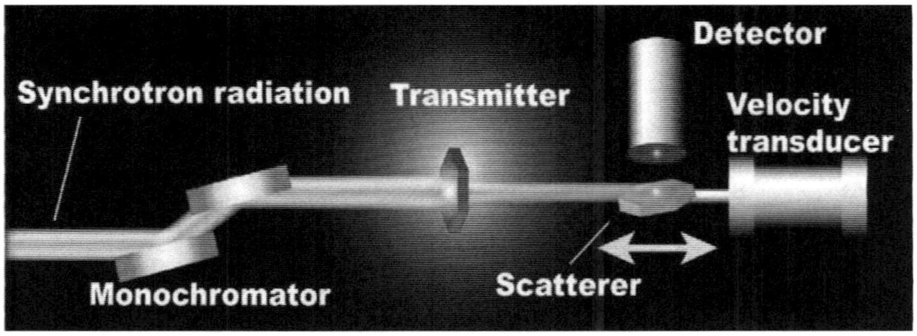

Figure 1. Schematic drawing of experimental setup.

and the scattering of the radiation owing to photoelectron absorption (C-channel). The dependence of other processes such as nuclear resonant absorption of radiation with recoil (B-channel) on the relative energy shift is small and can be ignored. Using the corresponding intensities (I_A, I_B, and I_C), the absorption spectrum intensity $I(w_s)$ can be expressed as follows:

$$I(w_s) = I_A(w_s) + I_C(w_s) + I_B, \tag{1}$$

$$I_A(w_s) = C_A \int_{\tau_1}^{\tau_2} d\tau \int_0^{z_s} dz \left| \int \frac{dw}{2\pi} \frac{\exp(-i w \tau)}{w - w_s + i/2} E_t(w) E_s(w, w_s, z) \right|^2, \tag{2}$$

$$I_C(w_s) = C_C \int_{\tau_1}^{\tau_2} d\tau \int_0^{z_s} dz \left| \int \frac{dw}{2\pi} \exp(-i w \tau)(E_t(w) E_s(w, w_s, z) - 1) \right|^2, \tag{3}$$

$$E_t(w) = E_{0t} \exp\left(-\frac{\mu_{et} z_t}{2}\right) \exp\left(-i \frac{\mu_{nt} z_t}{2(2w + i)}\right), \tag{4}$$

$$E_s(w, w_s, z) = E_{0s} \exp\left(-\frac{\mu_{es} z}{2}\right) \exp\left(-i \frac{\mu_{ns} z}{2(2(w - w_s) + i)}\right). \tag{5}$$

Here, z_t (z_s) denotes the thickness of the transmitter (scatterer); μ_{nt} (μ_{ns}), the linear absorption coefficient of radiation by the nuclei at resonance in the transmitter (scatterer); μ_{et} (μ_{es}), the electronic absorption coefficient of the transmitter (scatterer); E_t (E_s), the propagating coherent field amplitudes of the transmitter (scatterer); and E_{0t} (E_{0s}), the amplitude of the radiation field at the entrance of the transmitter (scatterer). C_A and C_C are constants, and τ_1 and τ_2 are the starting and ending times, respectively, of the measured time window in units of the lifetime of the nuclear excited state. In order to calculate the measured spectra correctly, we must take into account the exact experimental conditions because of the absorption of fluorescent X-rays, internal conversion electrons, and γ-rays in the scatterer.

It is known that exact line shapes can be obtained in the Mössbauer spectrum by the transmission integral method [13], and that the Lorentzian line shape is observed only in the case of an ideally thin sample. Furthermore, the time-window effect, which is known in delayed coincidence Mössbauer spectroscopy [14],[15], yields a narrower and modified line shape of Mössbauer spectra obtained using radioactive sources. These attributes can be clearly seen in the developed method because of the pulsed nature of synchrotron radiation. If the measured time range can be approximated to the time range from $\tau_1 = 0$ to $\tau_2 = \infty$, then a Lorentzian line shape can be obtained using thin-sample approximation by ignoring the contribution of I_C [16].

3. Experiments

Experiments were performed at the nuclear resonant scattering beamline (BL09XU) at SPring-8. The storage ring was operated in the 203-bunch mode, yielding a bunch distance of 23.6 ns. We used a powder sample of EuF_3 (not enriched, 40-μm thick) as the transmitter. A disk of enriched EuF_3 powder (not enriched, 30-μm thick) was used as the scatterer. Delayed emission from the irradiated surface of the scatterer was measured using a multielement Si-APD detector (4 × 2 APDs, each with a detection area of 3 × 5 mm^2).

4. Results and discussion

We studied the time-window effect in the Mössbauer spectra of the 21.54 keV excited state of ^{151}Eu; the measured spectra are shown in figure 2. In this measurement, we used EuF_3 samples, which are

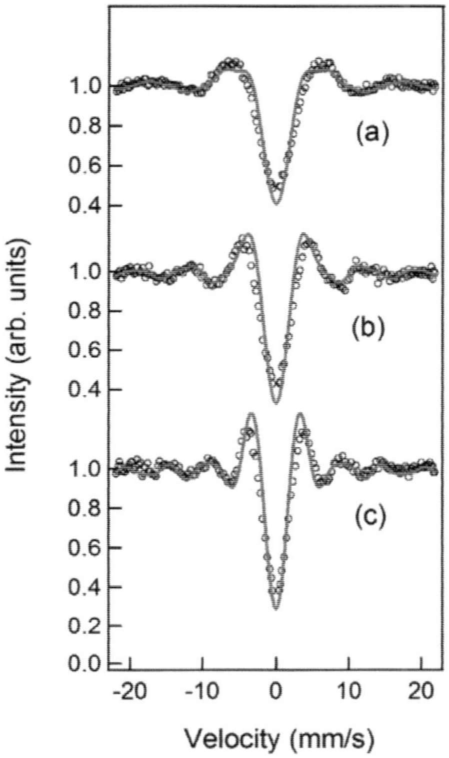

Figure 2. Mössbauer spectra of 151Eu in EuF3 used as a transmitter and a scatterer with time windows of (a) 5.7–17.0 ns, (b) 8.1–17.0 ns, and (c) 10.5–17.0 ns. Lines represent the spectra calculated for corresponding time windows using the developed method.

single-line materials, for both the transmitter and the scatterer. In figure 2, the values of τ_1 for the three measured spectra are (a) 5.7 ns, (b) 8.1 ns, and (c) 10.5 ns, and τ_2 = 17.0 ns is common to all the measurements. The changes in the spectral lineshape can be clearly seen as being dependent on τ_1. Because the half-life of the first excited state of ^{151}Eu is 9.6 ns, the used values of τ_1 cannot be approximated as zero. We observe that as τ_1 increases, the linewidth becomes narrower, the background becomes more distinctly wavy, and the cycle becomes larger. As discussed, these effects were observed in the Mössbauer spectra measured by the delayed coincidence method [14],[15]. Moreover, similar spectral changes were observed in the Mössbauer spectra obtained by the nuclear resonant Bragg scattering of synchrotron radiation [17]. We have calculated the spectral lines using the previously developed theory for corresponding time windows; the results are shown in figure 2. It was confirmed that the developed method reproduced the measured spectra well. This result shows that the theoretical and experimental results are consistent with each other.

In conclusion, we have successfully measured the time-window effect and demonstrated that the developed theory works well. In our method, we use two samples as the transmitter and the scatterer. By using only one single-line reference material with a sufficient recoilless fraction, we can carry out good-quality measurements for the samples under investigation. This situation is similar to the selection of a Mössbauer RI source; however, this method does not require radioactive material. Moreover, one of the features of this method is the flexibility of the experimental setup in that the measured sample can be used as a transmitter or as a scatterer, depending on the sample conditions.

The transmission experiments are suitable for high-pressure measurements, measurements of small samples, measurements of natural (not-enriched) samples, etc. On the other hand, scattering experiments are suitable for surface measurements, films on a substrate, measurements of very thick samples, etc. Mössbauer nuclides are widely available; therefore, this method offers the potential for unique measurement capabilities for various scientific requirements such as microscopic measurements, measurements under extreme conditions for ultrasmall samples, and measurements of complex nanostructured materials.

Acknowledgements
We are thankful to the members of the Accelerator Group of SPring-8 for their efforts in obtaining pure bunches in the top-up operation and to the entire staff of SPring-8 for their support. This study was conducted under the Power User Priority Program of SPring-8.

References

[1] Ruby S L 1974 *J. Phys. (Paris). Colloq.* **35** C6-209
[2] Gerdau E, Rüffer R, Winkler R H, Tolksdorf W, Klages C P and Hannon J P 1985 *Phys. Rev. Lett.* **54** 835
[3] Hastings J B, Siddons D P, van Bürck U, Hollatz R and Bergmann U 1991 *Phys. Rev. Lett.* **66** 770
[4] Sergueev I, Chumakov A I, Deschaux Beaume-Dang T H, Rüffer R, Strohm C and van Bürck U 2007 *Phys. Rev. Lett.* **99** 097601
[5] Röhlsberger R, Toellner T S, Sturhahn W, Quast K W, Alp E E, Bernhard A, Burkel E, Leupold O and Gerdau E 2000 *Phys. Rev. Lett.* **84**, 1007
[6] Coussement R, Cottenier S and L'abbé C 1996 *Phys. Rev.* B **54** 16003
[7] Callens R, Coussement R, Kawakami T, Ladriére J, Nasu S, Ono T, Serdons I, Vyvey K, Yamada T, Yoda Y and Odeurs J 2003 *Phys. Rev.* B **67** 104423
[8] Smirnov G V, Van Bürck U, Chumakov A I, Baron A Q R and Rüffer R 1997 *Phys. Rev.* B **55** 5811
[9] Smirnov G V, Sklyarevskii V V, Voskanyan R A and Artem'ev A N 1969 *Pis'ma Zh. Eksp. Teor. Fiz.* **9** 123
[10] Mitsui T, Seto M, Masuda R, Kiriyama K and Kobayashi Y 2007 *Jpn. J. Appl. Phys.* **46** L703
[11] Seto M, Yoda Y, Kikuta S, Zhang X W and Ando M 1995 *Phys. Rev. Lett.* **74** 3828
[12] Seto M, Masuda R, Higashitaniguchi S, Kitao S, Kobayashi Y, Inaba C, Mitsui T and Yoda Y 2009 *Phys. Rev. Lett.* **102** 217602
[13] Margulies S and Ehrman J R 1963 *Nucl. Instrum. Methods* **21** 217
[14] Lynch F J, Holland R E and Hamermesh M 1960 *Phys. Rev.* **120** 513
[15] Hamill D W and Hoy G R 1968 *Phys. Rev. Lett.* **21** 724
[16] Smirnov G V, van Bürck U, Arthur J, Brown G S, Chumakov A I, Baron A Q R, Petry W and Ruby S L 2007 *Phys. Rev.* A **76** 043811
[17] Tischler J Z, Larson B C, Boatner L A, Alp E E and Mooney T 1996 *J. Appl. Phys.* **79** 3686

Mössbauer spectroscopic microscope

Y Yoshida[1], T Kamimura[1], M Ichino[1], K Hayakawa[1], K Yukihira[1], and H Soejima[1,2]

[1]Shizuoka Institute of Science and Technology,
Fukuroi-city, 437-8555 Shizuoka, Japan
[2] Shimadzu Corporation, Nakagyo-ku, Kyoto 604-8511, Japan

E-mail: yoshida@ ms.sist.ac.jp

Abstract. A "Mössbauer spectroscopic microscope" has been developed using a Multi-Capillary-X-ray (MCX) lens, operating in a laboratory combined with a scanning electron microscope (SEM). This microscope yields a two-dimensional mapping image of ^{57}Fe probes with a space resolution of about 50μm in comparison with the microstructure observed by SEM.

1. Introduction

In materials science different measuring techniques are usually combined to characterize the microstructure of materials. Mössbauer spectroscopy provides atomistic information at a nuclear probe in a matrix through the hyperfine interactions, while the microstructure is studied, for instance, by scanning electron microscope and/or X-ray diffraction. The information obtained by Mössbauer spectroscopy is limited to a few nearest neighbors from the probe. In the case of state-of-art materials for electronic and magnetic devices, however, there seems to be strong demands to clarify the atomistic nature of the nuclear probes and the microstructure down to submicron meter scale simultaneously.

Recently, X-ray optical devices for focusing X-rays down to several 10 nm have been widely developed for X-ray microanalysis at synchrotron facilities as well as at laboratories. We use one of such technique, i.e., a Multi-Capillary-X-ray (MCX) lens [1], which makes us possible to focus the γ-rays of 14.4keV emitted from a standard ^{57}Co-in-Rh source in order to build a Mössbauer spectroscopic microscope. This MCX lens consists of thousands of small capillaries with a diameter of about 2μm. The γ-rays are introduced and totally reflected inside of the capillaries, changing the propagation directions, so that the γ-rays can be focused down to a diameter of 250μm. In this communication, we will describe the Mössbauer microscope and will show its typical applications to a contamination problem of multi-crystalline silicon wafers for solar cells. A preliminary result was already reported in the last international conference of the applications of the Mössbauer effect (ICAME2007) at Kanpur [2].

2. Mössbauer Microscope

A new MCX lens is specially designed for 14.4keV γ-rays, which provides a spot size of 256 μm in diameter at a focal distance of 58 mm from the outlet of lens for γ-rays. The 35% fluence of the parallel incoming γ-rays is transmitted through the MCX lens. The sample is fixed on a X-Y stage in a vacuum chamber. The conversion electrons are measured as a function of the position of the γ-ray spot on the sample using a three-stage micro channel plate (MCP) assembly, which is produced by Hamamatsu photonics Japan. The number of the mapping positions is typically 40×40 = 1600. The position is varied with a step size between 25μm and 250 μm. The mapping image is produced from a contour plot based on the counts of conversion electrons detected in the whole area, for instance, 1×1 and $10\times10mm^2$. One complete mapping will take 10 hours typically. Both the step size and the spot size of the focused γ-ray determine the space resolution of the mapping images. Furthermore, we have installed a field-emission type scanning electron microscope (Mini-EOC, APCO Ltd.) to our system (figure 1) with a space resolution of 20 nm in order to observe a microstructure of a sample which can be compared with the Fe mapping image measured by the Mössbauer spectroscopic microscope.

Figure 1. Mössbauer spectroscopic microscope combined with SEM.

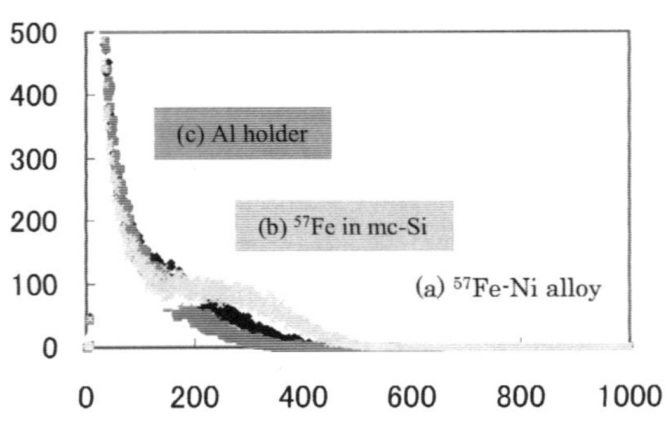

Figure 2. Pulse height spectra of MCP corresponding to the γ-rays' spot positions of (a) ^{57}Fe enriched Fe-Ni alloy, (b) ^{57}Fe deposited multi-crystalline Si, (c) Al-sample holder.

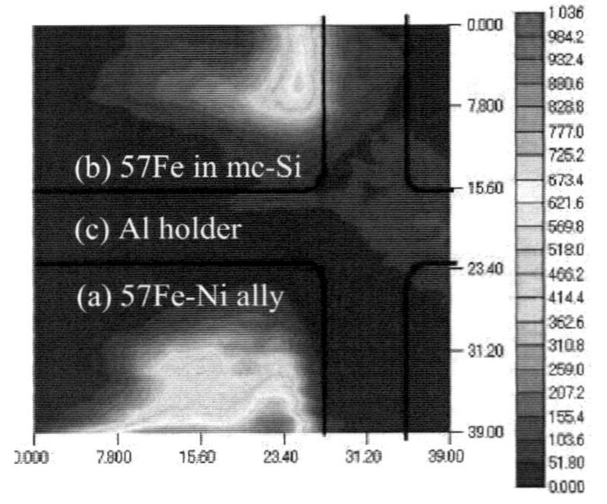

Figure 3. Mapping image of conversion electrons emitted from ^{57}Fe in mc-Si, FeNi alloy and Al holder for an area of $10\times10mm^2$ by setting the PHA-window between 200 and 1000 channel.

The electrons emitted from each focused position of the γ-rays are detected by the MCP which is sensitive not only to the conversion electrons, but also to the photo electrons and other radiation like X-rays and γ-rays. It is shown in figure 2 and 3 that these electrons can be well distinguished by setting a proper measuring window for the output signals from the MCP, because the pulse heights of the signals from the ^{57}Fe containing samples (figure 2 (a) and (b)) are much higher than those from the aluminum sample holder (figure 2 (c)). The measuring window of the MCA output was set between 200 and 1000 channels of the PHA spectrum in figure 2. Figure 3 shows a mapping image from an area of 10×10 mm^2 measured by a vibrating ^{57}Co source within a Doppler velocity range of ±8 mms^{-1}. Two different samples of ^{57}Fe in mc-Si and ^{57}Fe-Ni alloy are fixed by an Al sample holder. The black lines drawn in Fig.3 show the boundaries between the Al holder and the samples. Inhomogeneous distributions of ^{57}Fe atoms can be seen in both samples. Note that the highest counts zone is between 984 and 1036 in figure 3, while the lowest zone is close to zero, indicating an extremely high signal to noise ratio of about 20 for detecting ^{57}Fe atoms in a sample.

3. A mapping image of ^{57}Fe deposited multi-crystal Si

Silicon materials in solar cells are known to be contaminated by Fe impurities through the cell production process, which degrade the photovoltaic efficiency of the cell. We have installed a FE-SEM into our Mössbauer microscope, in order to observe an area using both microscopes. A SEM-picture of two crystal grains of multi-crystalline Si is shown in figure 4. The ^{57}Fe deposited sample with a thickness of 3 nm was annealed at 1000 °C for one week. We performed to map roughly the same area of the sample as is shown in figure 4, and the mapping intensity, which is presented in figure 5, changes considerably from one grain (left) to another (right). This appears to be due to either different concentrations of ^{57}Fe in the grains, or to different depth distributions of ^{57}Fe from the surface to 100nm, i.e., a measuring depth possible for the conversion electrons. Furthermore, the intensity depends on the local concentration of ^{57}Fe as well as on the resonance conditions corresponding to the chemical states and the lattice sites of ^{57}Fe atoms in the sample. The latter conditions provide us a possibility to get a mapping image of a certain spectral component selectively. A result obtained from a solar cell of multi-crystalline Si, which is not shown in this report because of

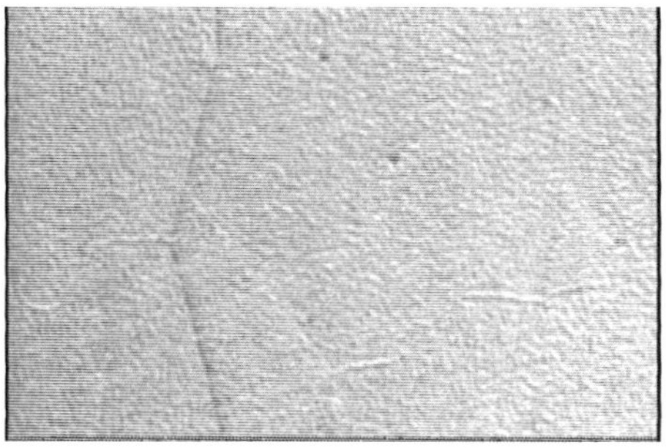

Figure 4. SEM picture of ^{57}Fe deposited multi-crystalline Si, showing two different grains.

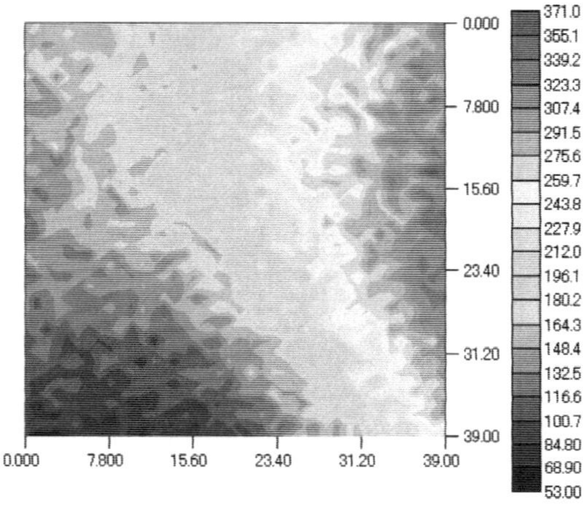

Figure 5. Mapping image of ^{57}Fe deposited multi-crystalline Si for an area of 1 × 1mm^2.

the lack of the space, suggests that substitutional and interstitial Fe atoms distribute differently in different grains. This result may be due to an inhomogeneous distribution of vacancies in different grains, which was produced during the solidification process of the sample. In Si crystal, on the other hand, Fe impurities are thought to diffuse dominantly via interstitial mechanism. Substitutional Fe may be formed by a defect reaction between an interstitial Fe and a vacancy during the annealing. Accordingly, such reaction is considered to be one of the reasons for the inhomogeneous distribution of substitutional and interstitial impurities. The details will be published elsewhere.

4. Summary

A new instrument "Mössbauer spectroscopic microscope" is presented in this paper. This microscope is sensitive only to ^{57}Fe atoms, because we detect the conversion electrons emitted from a ^{57}Fe containing sample after recoil-free absorptions of γ-rays, i.e., Mössbauer effect. Subsequently, we map the electron counts as a function of the sample position, where the γ-rays are focused by a multi-capillary X-ray lens (MCX) [1]. The microscope can be applied for materials research on Fe containing electronic and magnetic devices to investigate the microstructure and the electronic and magnetic states simultaneously.

Acknowledgments

Mrs. Akio Suzuki and Naoki Uchimura, Hamamatsu Photonics K. K., is greatly acknowledged for technical supports with the MCP. This work was supported by the research grants of Japan Science and Technology Agency (JST) through the development of systems and technology for advanced measurement and analysis (Program T) as well as the Research for Promoting Technological Seeds.

References

[1] Soejima H 1986 *Japan Patent* 2014379
Soejima H 1988 *Japan Patent* 2001797
Soejima H and Narusawa T 2001 *Advances in X-ray Analysis* **44** 320-324
[2] Yoshida Y, Suzuki K, Hayakawa K, Yukihira K and Soejima H 2009 *Hyperfine Int.* **188** 121- 126

Installation of an IR/Raman measuring station at the ESRF for simultaneous detection of vibrational and nuclear resonant scattering spectra

K Muffler[1], JA Wolny[1], HP Hersleth[2], KK Andersson[2], K Achterhold[3], R Rüffer[4] and V Schünemann[1]

[1] Department of Physics, University of Kaiserslautern, D-67663 Kaiserslautern, Germany

[2] Department of Molecular Biosciences, University of Oslo, N-0316 Oslo, Norway

[3] Department of Physics, Technische Universität München, D-85748 Garching, Germany

[4] ESRF, F-38043 Grenoble Cedex, France

schuene@physik.uni-kl.de

Abstract. Research on micro-structured iron-containing samples can be efficiently carried out by means of nuclear resonant scattering techniques at synchrotron facilities, because the high brilliance and low divergence of the beam facilitates the characterization of micrometer-sized samples containing Mössbauer active isotopes. Additional information can be obtained if the nuclear scattering techniques are combined with classical vibrational methods such as Raman and IR microspectroscopy. The sample environment presented within this paper provides the possibility to measure Raman spectra and to obtain optical (visible) on-line information about possible changes/damages of the sample during synchrotron beam exposition.

1. Introduction

Analytical methods like vibrational spectroscopy, and nuclear resonance scattering techniques have become independently evolved and established. Raman and infrared spectrometers are applied as standard instruments for solving analytical problems in academia and industry, furthermore, they are actually applied for routine analytics concerning quality processes as well as for quantitative measurements in large scale industrial process analytics by now. However, the fundamental aspect within the material and biophysical sciences is to enlighten the structure-property relationships. Due to the high brilliance of 3rd generation synchrotron radiation sources even small samples with a low content of $\sim 10^{12}$ atoms of ^{57}Fe can be probed.

Thus, the nuclear inelastic scattering (NIS-) and nuclear forward scattering (NFS-) techniques can be used as an efficient tool for analyzing metabolic steps, which involve iron-containing enzymes, due to obtain information about the iron's surrounding as well as information about the iron's central atom itself. These techniques are meanwhile well established for characterization of iron-containing organic and inorganic samples [1,2].

There is an increasing demand for the investigation of micro-structured iron-containing samples not only from biogenous material but also from synthetic inorganic and organic compounds by application of synchrotron radiation simultaneously with vibrational spectroscopic methods such as infrared spectroscopy and Raman scattering, respectively. Davies et al. already described an experimental set-up at ESRF ID 13, which enables combined microfocus Raman and microfocus XRD to probe deformation of high performance fibers [3].

This contribution describes the set-up of the first combined Raman-microscope and NFS/NIS measuring station at a synchrotron radiation facility enabling the simultaneous measurement of vibrational nuclear scattering techniques while the sample is additionally probed by optical methods with an applied microscope. As an example first results obtained from this sample environment on the reaction of a metmyoglobin single crystal with hydrogen peroxide are presented.

2. Experimental Set-up of the Measuring station

The installed measuring station currently allows the simultaneous probing of iron containing micro-structured samples by Raman spectroscopy - via a SENTERRA® Raman microscope - and NFS/NIS techniques, provided and developed as in-house methods at ESRF / ID-18. It is planned to extend the workplace with the IR microscope HYPERION® in 2010. The key features of the already provided and installed techniques are presented within this subchapter.

2.1. Raman microsope SENTERRA®

The SENTERRA® dispersive Raman microscope (figure 1, Bruker Optik GmbH, Germany) is a full featured confocal system which is capable to accommodate different excitation wavelengths with a high spatial resolution. The microscope features three different lasers with excitation wavelengths of 488 nm, 532 nm and 785 nm with a corresponding power of 20 mW, 20 mW, and 100 mW, respectively. Since the microscope was primarily developed as stand-alone system the integration into an existing NFS/NIS workplace requires small modifications in particular those affecting general safety regulations. Usually the Raman-experiments were carried out with closed safety doors of the measurement compartment, thus the system fulfill the laser safety class I. If the microscope is put into the synchrotron beam it is beneficial with respect to the intensity of nuclear forward scattering signals to endue the safety doors with pin holes enabling the lossless input and output of the beam or the scattered signals, respectively. During operation at the synchrotron the microscope therefore fulfills the laser safety class IIIb.

The optical device of the Raman microscope is based on the Olympus BX51 microscope, which provides the necessary tools concerning sample visualization and characterization. Visualization can be achieved using the standard objectives with a magnification of 20x (NA: 0.4, working distance is 1.3 mm) or 50x (NA: 0.75, working distance is 0.38 mm). Larger working distances can be realized by the additionally provided long distance working objective (20x, NA: 0.4, working distance is 12 mm).

Spectral resolution of the system is smaller than 3 cm^{-1} and the spatial resolution is 1 µm depending on the excitation laser wavelength. Confocal depth resolution is higher than 2 µm and depends on the utilized objective. For achieving a higher flexibility concerning the probing of larger samples or if the examination requires a higher working distance (e.g. investigation of samples in cooling or heating stages), the system is equipped with an optical Raman probe, which is connected to the excitation laser (only 785 nm) and the detector via fiber optic cables. The working distance of the current system is 15 mm and 60 mm, but other lenses with smaller or larger working distances are available.

Figure 1. Set-up of the Raman microscope (Bruker Optik GmbH, Germany) inside experimental hutch of beamline ID 18 during experiment CH 2839; the microscope is mounted on a goniometer with an adequate adapter plate. The figure shows the microscope with opened safety doors.

2.2. NFS/NIS technique at ESRF/ID 18

For the protein single crystal NIS measurements presented below the incident beam was monochromatized by a Si(111) double-crystal premonochromator to a bandwidth of 2 eV. A further decrease of bandwidth down to 1 meV (8 cm^{-1}) was obtained with a refractive beryllium collimator and a high-resolution monochromator. This beam was focused by a Kirkpatrick Baez mirror optic to a spot size of 8*15 μm. The NIS data were collected during several energy scans with 0.25 meV step size. The metmyoglobin single crystal was oxidized by treatment with hydrogen peroxide according to a procedure described in [4], frozen in a loop, and placed into the focused beam.

Cooling during the NIS measurements was performed by a cryostream from Oxford Cryostream Ltd. (figure 2). The examined sperm whale metmyoglobin was enriched in the Mössbauer isotope ^{57}Fe according to [5].

3. Initial experiments using the installed technical equipments

Figure 3 shows the NIS spectra obtained at 110 K of a ^{57}Fe-Myoglobin single crystal (700*150*150 μm^3) after addition of $H_2{}^{18}O_2$ and subsequent freezing. The single crystal data identify a localized iron mode of 18 cm^{-1}. Modes in this region have also been observed near 25 cm^{-1} in MbCO and deoxyMb and have been assigned to translational modes in response to torsional oscillations of the polypeptide backbone and side chains [6]. This assignment was confirmed by a NIS study of millimeter-sized metmyoglobin single crystals which reports haem sliding motions in the region of 32 - 40 cm^{-1} [7]. The reason why in the case of $H_2{}^{18}O_2$ treated myoglobin a localized mode is seen at lower energy is unclear and subject to further investigations by combined quantum mechanical and molecular mechanics calculations. Unfortunately the obtained Raman spectra do only show a significant band at 975 cm^{-1} using an excitation wavelength of 785 nm which corresponds to the ammonium sulfate (usually found at 983 cm^{-1}) within the crystal. Clearly more measurements also at 488 and 532 nm excitation wavelength as well as directly in the Soret band e.g. at 407 or 413 nm have to be performed to increase the up to now none significant scattering signals which are related to the protein vibrations. However, these data might be correlated with recently published synchrotron-derived results concerning investigations on different protonation states of oxidized myoglobin [8].

Figure 2. Experimental set-up of the cryostream and APD detector for probing the myoglobin crystal. The pin of the crystal mount has a diameter of 0.65 mm.

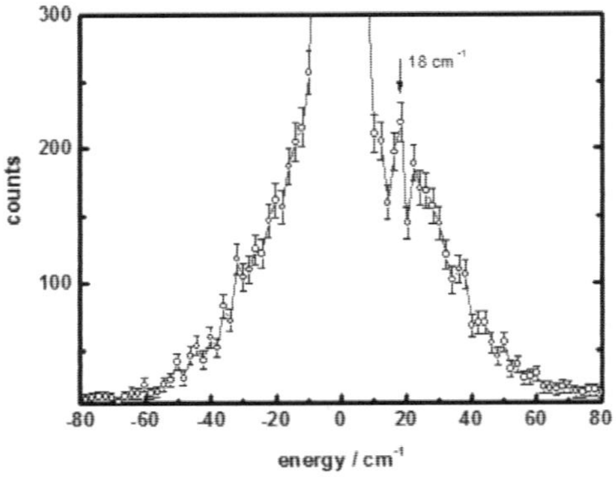

Figure 3. Nuclear Inelastic Scattering spectra obtained at 110 K of a ^{57}Fe-Myoglobin single crystal (dimension of crystal: 700*150*150 μm^3) after addition of $H_2{}^{18}O_2$ and subsequent freezing.

4. Conclusion and Outlook

The installed unique measuring station at ID 18 (ESRF, Grenoble) provides analytical information by complementary vibrational techniques and nuclear resonant methods. Thus, proteins, organic films, and micro-structures can be investigated spatially resolved, whereby the resolution depends on the selected vibrational technique, which is in the µm range. Since the molecular vibrational techniques are probed via corresponding microscopes, possible structural changes and potential damages of the sample during synchrotron beam exposition can be detected online with this equipment. New users can use the systems with their full potential within short adaptational training phases. This world-wide unique workplace will be available for the whole Mössbauer community.

Acknowledgement

This work was supported by the German Federal Ministry of Education and Research (BMBF) under 05KS7UK2, the Research Council of Norway grant 177661/V30 to K. K. Andersson, and the ESRF (exp. CH 2839).

References

[1] Schünemann V, Winkler H 2000 *Rep. Prog. Phys.* **63** 263
[2] Paulsen H, Schünemann V, Trautwein AX, Winkler H 2005 *Coord. Chem. Rev.* **249** 255
[3] Davies R J, Burghammer M, Riekel C 2006 *Macromol.* **39** 4834
[4] Hersleth H P, Uchida T, Rohr A K, Teschner T, Schünemann V, Kitagawa T, Trautwein A X, Gorbitz C H, Andersson K K 2007 *J. Biol. Chem.* 282 (32) 23372
[5] Teale F W J 1959 *Biochim. Biophys. Acta* **35** 543
[6] Sage J T, Durbin S M, Sturhahn W, Wharton D C, Champion P M, Hession P, Sutter J, Alp E E 2001 *Phys. Rev. Lett.* **86** 4966
[7] Achterhold K, Parak F G 2003 *J. Phys. Condens. Matter* **15** S1683
[8] Zeng W, Barabanschikov A, Zhang Y, Zhao J, Sturhahn W, Alp E E, Sage J T 2008 *J. AM. Chem. Soc.* **130** 1816

A closed cycle-cryostat for high-field Mössbauer spectroscopy

A Janoschka[1], G Svenconis[2] and V Schünemann[1]

[1] Department of Physics, University of Kaiserslautern, Erwin-Schrödinger-Str. 56, 67663 Kaiserslautern, Germany
[2] CRYO Industries, 11124 South Willow Street, Manchester, NH 03103, USA

E-mail: schuene@physik.uni-kl.de

Abstract. A closed cycle-cryostat coupled to a Mössbauer spectrometer has been installed at the University of Kaiserslautern and is in full operation since march 2007. The setup is equipped with a low vibrating two-stage pulse tube cooler and has a cool down time of 48 h. The sample can be top loaded without the need to shut off the refrigerator. With the static helium exchange gas in the variable temperature insert the sample may be cooled down from room temperature to 50 K within several hours. Dynamic exchange gas with external supply of gaseous helium is used to cool the sample down to 2 K. The superconducting self-shielding split-coil generates a magnetic field of up to 5 Tesla and a stray field of ca. 60 mT at the outer cryostat walls. Mössbauer measurements can be performed in perpendicular or parallel field orientations. The sample holder and the Mössbauer drive are rigidly connected to the cryostat. In this way a line width of the two inner α-Fe lines of 0.32 mm/s has been currently achieved.

1. Introduction

Mössbauer Spectroscopy at liquid helium temperature and high field conditions is indispensable for the characterization of paramagnetic iron sites in biomolecules and related chemical model complexes as well as nanometer-sized iron particles [1, 2]. Conventional cryostats for Mössbauer spectroscopy use considerable amounts of liquid helium. This is not only expensive if no recovery equipment is at hand but helium is also a limited natural resource on earth.

2. Instrumentation

2.1. Closed Cycle-Cryostat

A closed cycle-cryostat manufactured by Cryo Industries of America, Inc. [3] coupled to a Mössbauer spectrometer from WissEL GmbH [4] has been installed at the University of Kaiserslautern and is in full operation since march 2007 (figure 1 and 3). It provides a quick and easy way to reach low temperatures and high magnetic fields. The setup is equipped with a Cryomech, Inc. low vibrating two-stage pulse tube cooler and has a cool down time of 48 h. The sample can be top loaded at low temperatures without the need to shut off the refrigerator. The helium exchange gas in the variable temperature insert serves as an adjustable link to the refrigerator. Using static exchange gas the sample may be cooled down from room temperature to 50 K within several hours. Dynamic exchange gas with external supply of gaseous helium is used to cool the sample down to 2 K.

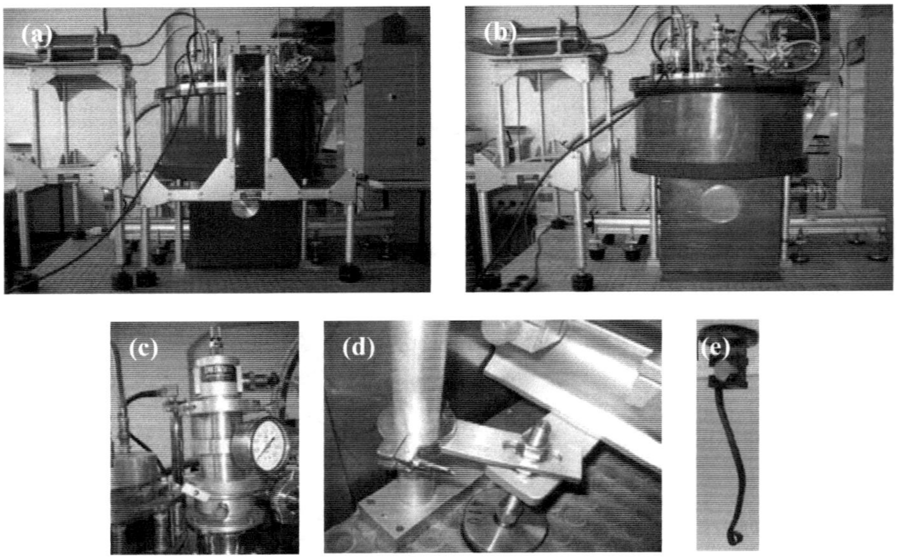

Figure 1. Top-loading closed cycle-cryostat for Mössbauer spectroscopy at temperatures of 2-300 K and fields of up to 5T. (a) Cryostat with vibration isolation assembly, (b) the vibration isolation assembly has been removed, (c), (d) and (e) rigid connections to the cryostat.

Vibrations caused by the cooling devices can lead to severe line broadening in the Mössbauer spectra. Therefore the vibrations from the refrigerator are often isolated from the sample by a vibration isolation assembly connecting the sample holder with the cryostat. Because line width broadening could not be minimized to satisfaction ($\Gamma \approx 0.48$ mm/s) with the construction shown in figure 1a it was removed (Figure 1b). Now the Mössbauer drive (Figure 1c) and sample holder (Figure 1d) are rigidly connected to the cryostat. The end of the sample holder was coupled to the bottom of the sample chamber by a copper spring (Figure 1e). In this way a line width of the two inner α-Fe lines of 0.32 ± 0.04 mm/s has been currently achieved.

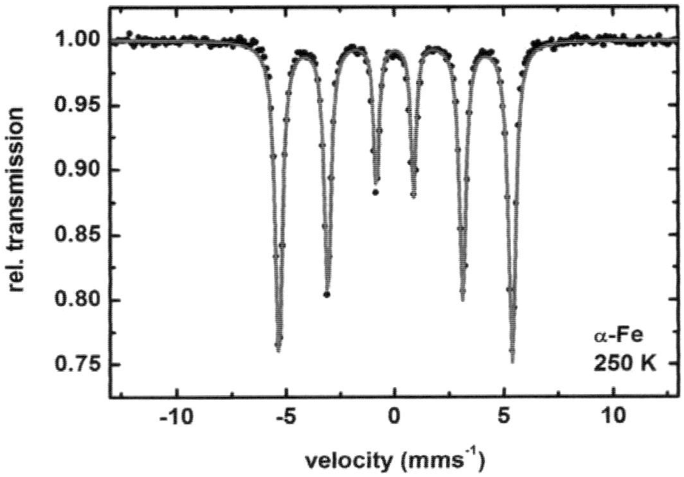

Figure 2. Mössbauer-spectrum of α-Fe taken at 250 K with inner line width of 0.32 ± 0.04 mm/s.

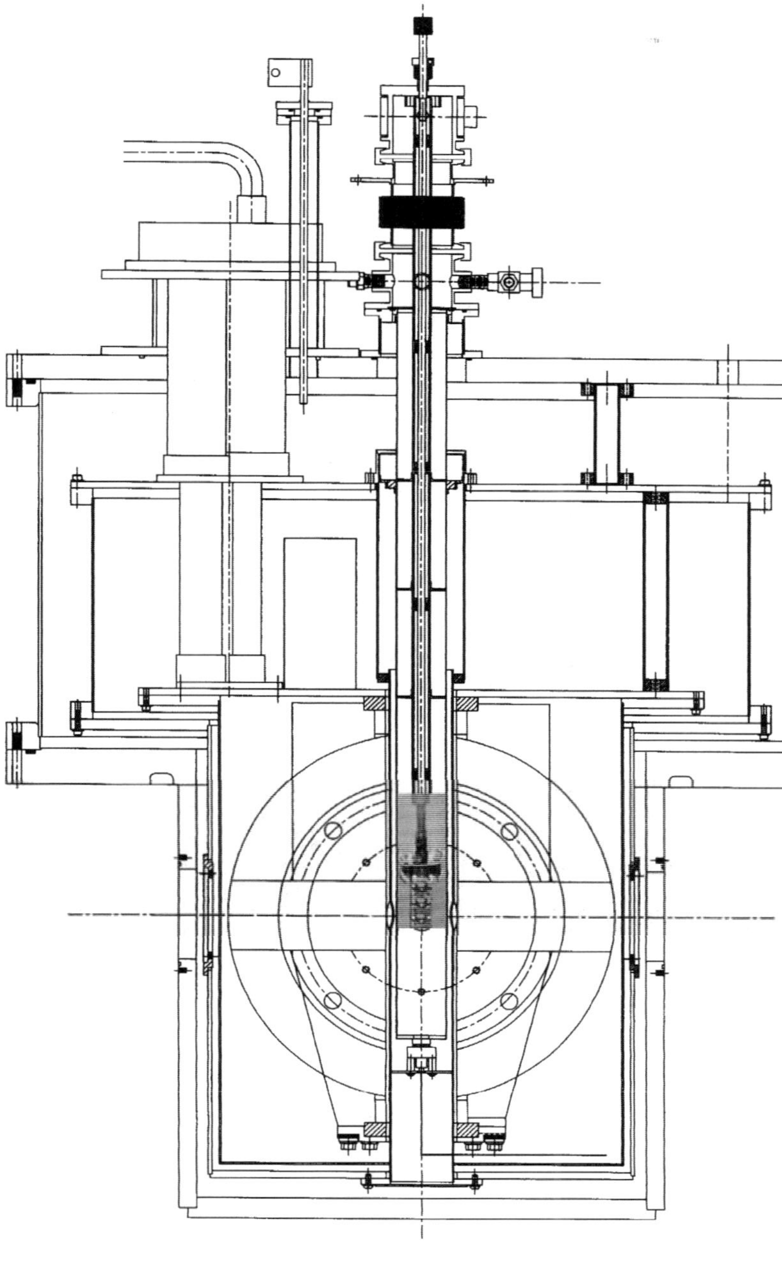

Figure 3. Drawing of the closed cycle-cryostat [3] with the second generation sample holder and damping springs.

A second generation sample holder is coupled to the wall of the sample chamber by a copper and a Teflon spring (figure 3). It offers place for three samples and allows the change of the sample by shifting its position without the need of opening the sample chamber. The new damping system did not lead to significant improvements of the line width (figure 2).

2.2. Superconducting Coil

The magnet is thermally coupled directly to the 2nd stage of the Cryomech pulse tube cooler and has a temperature of ~3 K during normal operation. The current is supplied directly via a LakeShore Model 625 Superconducting Magnet Power Supply. No superconducting switch is fitted, because the power supply will discharge the magnet in the case of a power failure. The superconducting self-shielded split-coil (figure 4) generates a magnetic field of up to 5 Tesla and a stray field of ca. 60 mT at the outer cryostat walls. Mössbauer measurements can be performed in perpendicular or parallel field orientations.

Figure 4. Superconducting coil.

3. Conclusion

The Cryogen Free Superconducting Magnet/ Closed Cycle-Cryostat System is a versatile tool for Mössbauer spectroscopy. It is easy and safe to use. It allows not only a quick and uncomplicated sample change but also provides an easy way to adjust the sample temperature in a wide temperature range. The magnetic field can be applied in perpendicular or parallel orientation to the γ-beam and can be sweeped to 5 T in less than three hours. The system may be left unattended for several hours or even days if needed. Due to the closed cycle the helium consumption is decreased and even no helium is needed if temperatures below 50 K are not required. The line widths of the spectra may still be decreased by an improved vibration damping system. The development of the next generation damping systems is in progress.

References

[1] Trautwein A X et. al 1991 *Struct. Bonding* **78** 1

[2] Mørup S, Dumesic J-A and Topsøe H 1980 *Applications of Mössbauer spectroscopy* Vol. II, ed R.L. Cohen, Academic Press, New York, pp. 1-53

[3] http://www.cryoindustries.com

[4] http://www.wissel-instruments.de

Universal LabVIEW-powered Mössbauer spectrometer based on USB, PCI or PXI devices

J Pechousek, R Prochazka, D Jancik, J Frydrych and M Mashlan

Centre for Nanomaterial Research, Palacky University, Slechtitelu 11, 783 71, Olomouc, Czech Republic

E-mail: pechous@prfnw.upol.cz

Abstract. A new design of the universal Mössbauer spectrometer is presented. Hardware solution is based on commercial-available data acquisition devices working on the USB, PCI or PXI platform controlled by the main application running on the personal computer. Final application allows, in addition to Mössbauer spectra accumulation, the detailed analysis of the acquired detector signal in energy and time domains, and also to tune the velocity driving system separately. The experimental results show a high flexibility in various detectors and velocity transducers usage. It is easy to change the way of operation according to the different experimental requirements. This concept can be used with all common spectrometric benches with different velocity transducers, radioactive sources and gamma-ray detectors. This is a new approach in the Mössbauer spectrometer construction.

1. Introduction

Different programming techniques and instrument solutions were used in the development of Mössbauer spectrometers [1-3]. The development of the original measurement electronic devices can be much demanding and expensive. The better approach can utilize verified and fully serviced commercial devices and develop only the software to its control. With this design, the end-user can alter the whole application fulfilling the user's requirements.

A new design of Mössbauer spectrometer which uses the virtual instrumentation techniques [4] is presented. This design is based on National Instruments (NI) commercially available USB, PCI or PXI devices. The NI LabVIEW software flexibility allows a wide number of hardware combinations to make which can provide the required function when used for building the application. The developed spectrometer application is nearly "hardware platform independent" [3,5,6].

2. LabVIEW powered Mössbauer spectrometer

The most classical electronic parts of the standard Mössbauer spectrometer are now replaced by the digital signal processing (DSP) algorithms. Here, virtual instrumentation technique is implemented in the first Mössbauer spectrometer powered by LabVIEW. The LabVIEW programming environment allows to realize such a system with minimum single purpose electronic devices, and final product can be used with all the common spectrometric benches.

2.1. γ-ray detection unit

The γ-ray detection unit consists of the detector and the multichannel amplitude analyzer based on a high-speed digital oscilloscope (digitizer). The unprocessed or amplified detector signal is recorded by the digitizer for consecutive processing. The acquired signal is sampled with a limited sampling rate given by the used digitizer. This device is selected with respect to its maximum sampling rate which is chosen by the time-length of the detector impulses for correct acquisition. The most common NaI:Tl scintillator impulses are sampled by 5 MS/s sampling rate. The much shorter impulses from the faster detectors are suitable for sampling with higher rates, i.e. 100 MS/s. Consequently, the sampling rate of the detector output signal differs with the detector type used. In figure 1, the detector signal analysis in the multichannel analysis (MCA) application is shown.

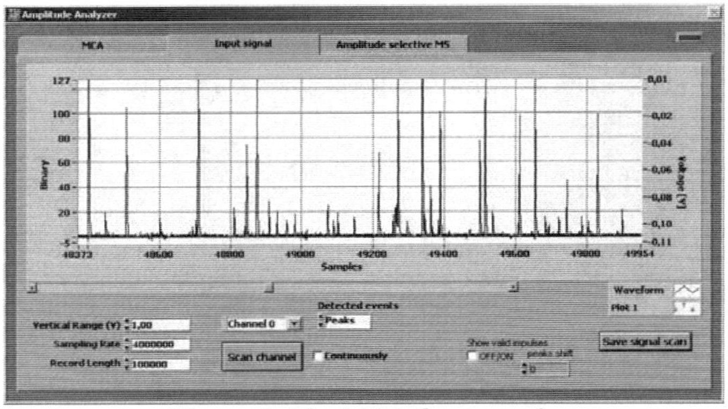

Figure 1. The MCA front panel.

In this window, the user can set proper sampling rate, vertical range of the analog input, detected events with positive or negative impulses, and detector channel while the digitizer device has two input channels. This sub-application allows detailed analysis of the acquired signal in energy and time domains. After this process, the MCA could be performed. In this step, the user can perform the MCA analysis of the sampled detector signal acquired with parameters applied in the previous step. After MCA, the user can set new discrimination window for selecting the events to be accumulated in the Mössbauer spectrum. In figure 2, the MCA-window is shown.

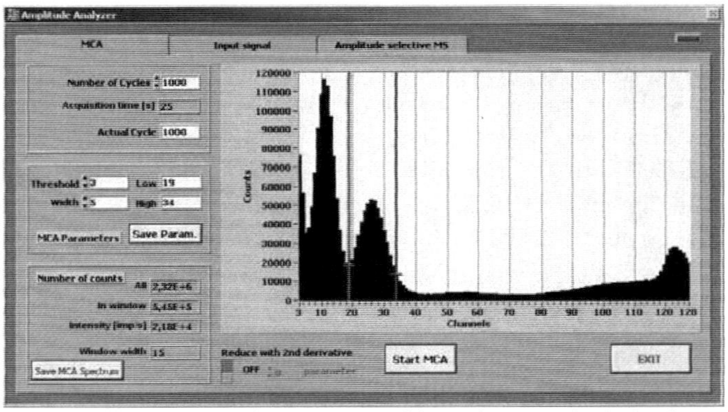

Figure 2. The detector signal processing setup and impulses analysis in the MCA application.

2.2. Velocity driving system

A common velocity-driving system for the Mössbauer spectrometer consists of the velocity signal generator, feedback control system (proportional-integral-derivative - PID) with power input/output buffers, and electromechanical linear transducer coupled with a Mössbauer source.

Velocity signal generator is based on the function generator which allows to generate user-defined waveforms with all standard ones. In figure 3, the set-up window for a velocity generation is shown.

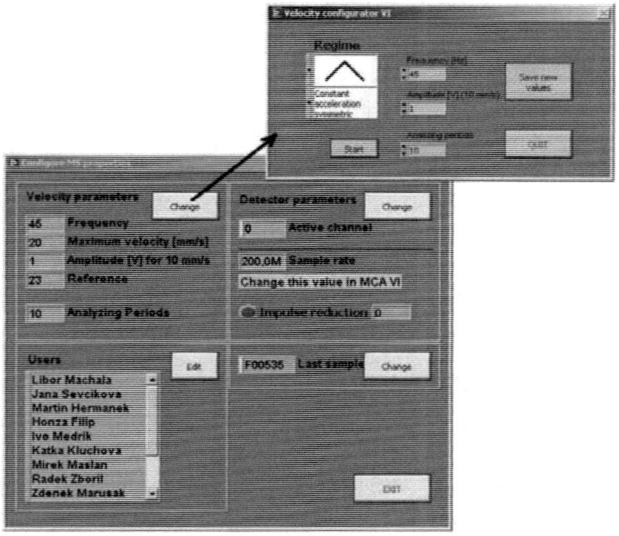

Figure 3. Spectrometer configuration-window.

During the further improvement, the commonly used analog PID controller was here replaced by the digital one [7,8]. This controller is implemented on Field Programmable Gate Array (FPGA) in the NI CompactRIO system [9]. The control algorithm is based on a Discrete PID LabVIEW function and operates in real-time conditions. The reference velocity signal can be also generated in the CompactRIO system or the external one (above mentioned) can be used.

The proposed velocity driving system is flexible enough to cover the wide range of disturbances (including vibrations coming from vacuum and cryogenic equipment, external magnetic forces, etc.) and increases system reliability. In addition, the CompactRIO system works as a remote system controlled via Ethernet, i.e. the PID controller parameters can be changed in the safe distance from a radioactive source and a high magnetic field.

2.3. Hardware solutions

The main unit provides spectra accumulation with the amplitude analysis of the data acquired by the detection unit, and reference velocity signal generation. All the tasks use DSP abilities of LabVIEW system running in the main computer. Several hardware implementations for each unit were tested. The detection unit with NI PXI 5102 (8-bit, 20 MS/s) digitizer [3,5], NI PCI-5124 (12-bit, 200 MS/s) [6] and NI USB-5133 (8-bit, 100 MS/s) were used. Higher sampling rates open the possibility to use the faster detectors and the high-activity radiation sources.

At the place of velocity driving unit, different analog input/output devices were used. Early presented spectrometer used NI PXI 5401 (12-bit, 40 MS/s update rate) function generator [3,5,6] coupled to the analog PID circuit. This generator can be replaced by other multifunction card with a proper analog and digital output on the USB, PCI or PXI platform. Limiting parameters are 12-bit resolution and the 150 kS/s update rate at the analog output as the minimum. Two USB devices, NI USB-6221 (16-bit, 833 kS/s) and NI USB-6215 (16-bit, 250 kS/s), with analog outputs were used. The trigger signal has to be produced by those devices to synchronize the spectra accumulation process.

The second available generation of the velocity unit takes advantage of CompactRIO system building a digital PID controller and the reference velocity generator on the joint device. The analog and digital input/output cRIO modules are used in this driving system where the FPGA provides the digital PID algorithm [7,8].

3. Conclusion

The accumulation unit has a major influence on the performance of the whole system. Not the accumulation process itself but rather the detector data DSP analysis can lead to the overloading of the main computer and data acquisition parameters have to be optimized. The performance tests showed a significant dependence of the measurement time on the actual computer configuration.

Several Mössbauer spectrometers are daily used in the Centre for Nanomaterial Research at the Palacky University in Olomouc. The part of these spectrometers use the principles described above. The advantages of the digital PID control come into effect for the high magnetic field system used in in-field Mössbauer spectroscopy. In figure 4, the main application window of the spectrometer is shown, where the presented spectrum for the standard α-Fe sample has a background level of 115 000 counts and a resonance effect of 28%.

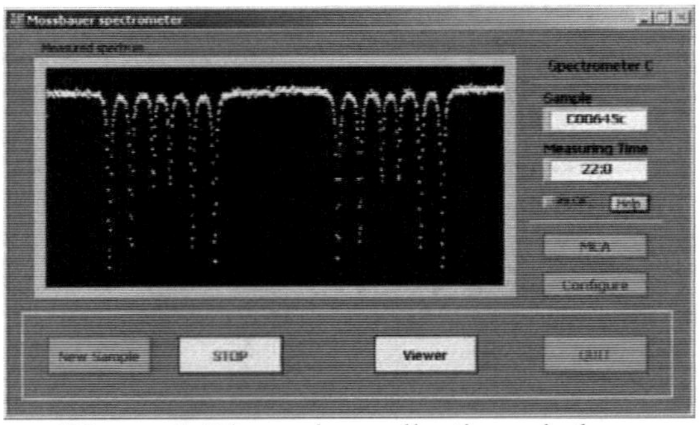

Figure 4. The main application window.

The measurement system described in this article represents a new way of Mössbauer spectrometers construction. The former single-purpose spectrometer units were replaced by universal data acquisition modules. LabVIEW software flexibility allows a wide number of hardware combinations to make which can thus provide the required function. The spectrometer is becoming to be nearly "hardware platform independent" and is prepared to be a commercial product.

Moreover, the commercial devices can be used independently in other measurement processes besides Mössbauer spectroscopy.

The Ministry of Education, Youth and Sports under the project of the research centre 1M6198959201 and research program 106/08/1440 are gratefully acknowledged.

References

[1] Stevens J G, Bowen L H and Whatley K M 1990 *Anal. Chem.* **62** 12 125R-39R
[2] Gibb T C 1976 *Principles of Mössbauer Spectroscopy* (New York: Wiley)
[3] Pechousek J and Mashlan M 2005 *Czech. J. Phys.* **55** 7 853-63.
[4] http://www.ni.com/labview/
[5] Pechousek J and Mashlan M 2005 *iJOE* **1** 2
[6] Pechousek J, Mashlan M, Frydrych J, Jancik D and Prochazka R 2008 *Hyperfine Interact.* **175** 1-8
[7] Pechousek J, Prochazka R, Mashlan M, Jancik D and Frydrych J 2009 *Meas. Sci. Technol.* **20** 017001
[8] Pechousek J, Jancik D, Evdokimov V and Prochazka R 2009 *Nucl. Instrum. Meth.* B **267** 846-8
[9] http://www.ni.com/compactrio/

Journal of Physics: Conference Series **217** (2010) 012007 doi:10.1088/1742-6596/217/1/012007

Investigation on New Scintillators for Subnanosecond Time-Resolved X-Ray Measurements

R Haruki[1,2], K Shibuya[2,3], F Nishikido[2,4], M Koshimizu[2,5], Y Yoda[2,6] and S Kishimoto[2,7]

[1]Advanced Science Research Center, Japan Atomic Energy Agency, 2-4 Shirane, Shirakata, Tokai, Ibaraki 319-1195, Japan

[2]CREST, Japan Science and Technology Agency, 5 Sanbancho, Chiyoda-ku, Tokyo 102-0075, Japan

[3]Institute of Physics, Graduate of Arts and Sciences, University of Tokyo, 3-8-1 Komaba, Meguro-ku, Tokyo 153-8902, Japan

[4]Molecular Imaging Center, National Institute of Radiological Sciences, 4-9-1 Anagawa, Inage-ku, Chiba 263-8555, Japan

[5]Department of Applied Chemistry, Tohoku University, 6-6-7 Aramaki Aza Aoba, Aoba-ku, Sendai 980-8579, Japan

[6]Japan Synchrotron Radiation Research Institute, 1-1-1 Kouto, Sayo, Hyogo 679-5198, Japan

[7]Institute of Materials Structure Science, High Energy Accelerator Research Organization, 1-1 Oho, Tsukuba, Ibaraki 305-0801, Japan

E-mail:haruki.rie@jaea.go.jp

Abstract. We investigated new x-ray detectors for nuclear resonant scattering measurements with high energy x rays. The organic-inorganic perovskite scintillator of phenethylamine lead halide ($(C_6H_5C_2H_4NH_3)_2PbX_4$) (X:Br, I) was used. These compounds have fast light emission due to an exciton. They include heavy atoms, which make the detector to have high efficiency to high energy x rays. The merit of these scintillators is that we can make a thick crystal compared to a Si wafer which is used in an avalanche photo-diode detector. We successfully measured 67.41 keV x ray signals, the energy of ^{61}Ni nuclear resonant scattering, with high detection efficiency of 42.5 % by the scintillator.

1. Introduction

In nuclear resonant scattering measurements, a large dynamic range is required, since large prompt electronic scattering signals must be masked to detect a single nuclear resonant quantum delayed by some nanoseconds. Silicon avalanche photo-diode (Si-APD) detectors were mainly used in these measurements since Si-APD detectors have good time resolution. However the nuclear resonant scattering measurements are spreading to high energy region and detectors are also required to have high efficiency to high energy x-ray signals. Si-APD detectors are made by very thin Si wafers up to 150 μ m, and have low efficiency to high energy x rays (lower than 15 % to 20 keV x rays). High efficiency detectors for these measurements are necessary for further development in this field.

Recently, organic-inorganic perovskite crystals are developed as scintillators[1]. These compounds have fast light emission caused by excitons[2]. Inorganic layers consist of metal halide (PbX_4, X:Br, I) are separated by organic layers includes alkyl chains. Inorganic layers are semiconductors and organic layers are insulators. Both layers make multiple quantum well structure, which keeps a strong photoluminescence in the material even at room temperature (RT). The merit of this material is heavy Pb atoms, since it shows the good absorption of x rays compared to Si-APD with light Si atoms. But thickness of 0.1mm was not enough to the detection of x rays with energy higher than 20 keV. To obtain larger crystal, alkyl chains are changed to phenetyl groups. A first one is an organic-inorganic perovskite scintillator of phenethylamine lead bromide ($(C_6H_5C_2H_4NH_3)_2PbBr_4$, abbreviated as PhE-$PbBr_4$). We have succeeded to measure ^{61}Ni nuclear resonant scattering (67.41 keV, lifetime: 7.6 ns)[3]. Second one is the same kind of scintillator of phenethylamine lead halide ($(C_6H_5C_2H_4NH_3)_2PbBr_{4-z}I_z$, abbreviated as PhE-$PbBr_{4-z}I_z$), in which bromide and iodide is mixed. The iodide sample showed faster scintillation decay in alkyl organic-inorganic crystals than scintillation decay of the bromide sample[2]. We have investigated optical properties of these crystals.

2. Experimental

The crystals we investigated here are PhE-$PbBr_4$, PhE-$PbBr_{4-z}I_z$ (z:low), PhE-$PbBr_{4-z}I_z$ (z:high). In fabrication of bromide and iodide mixture crystal, 20% of $PbBr_2$ was alternated with PbI_2 to make a crystal of PhE-$PbBr_{4-z}I_z$ (z:low), and 40% to make a crystal of PhE-$PbBr_{4-z}I_z$ (z:high). The structure and the fabrication of the first crystal is reported in ref. 4. The scintillation properties were investigated at beamline BL-14A of the Photon Factory, KEK, Tsukuba. The storage ring was operated in single bunch mode, which was suitable to investigate time domain properties of the scintillation. To obtain scintillation decay curve, crystal was set 38mm away from photo-multiplier tube (PMT, Hamamatsu R7400P) window in a vacuum chamber, to detect 1 photon or lesser in each x-ray pulse. The energy spectra are obtained by the crystals mounted on the window of PMT using optical grease and covered with Teflon tape.

3. Results and Discussion

The Scintillation decay curves from the samples are shown in Figure 1. The decay of iodide and bromide mixture crystals is faster than that of bromide crystals.

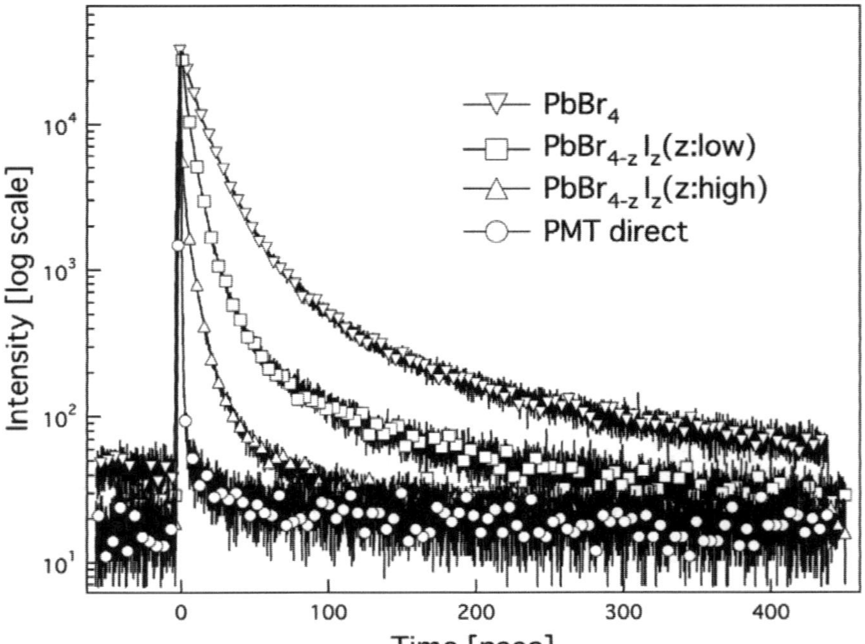

Figure 1. Scintillation decay curves of samples.

Table 1 shows the efficiency to 67.4 keV x ray and the decay components of each scintillator. Decay components values are obtained by fitting experimental data by four exponential functions. The energy spectra and energy resolution are shown in Figure 2. The values of energy resolution are calculated from FWHM. Compared to bromide crystals, bromide and iodide mixture crystals showed lower light emission, and bad energy resolution. The PhE-$PbBr_4$ crystal is transparent, and the iodide mixture crystals are transparent with yellow color. The color may cause the self absorption in the crystals. Since higher ratio of iodine retards the growth of the crystals, the thicknesses of iodide mixture samples are under 1 mm. Efficiency is mainly determined by the thicknesses of the crystals.

Table 1. Efficiency to 67.4 keV x ray and decay components of scintillators.

Material	Efficiency (%)	Decay components (ns (%))
PhE-$PbBr_4$	42.5	5.3(10.3), 15(62.2), 54(20.8), 213(6.7)
PhE-$PbBr_{4-z}I_z$ (z:low)	34.3	2.6(22.2), 8.1(58.0), 36(11.8), 168(7.9)
PhE-$PbBr_{4-z}I_z$ (z:high)	18.4	1.3(12.4), 5.1(49.5), 15(23.9), 142(14.2)

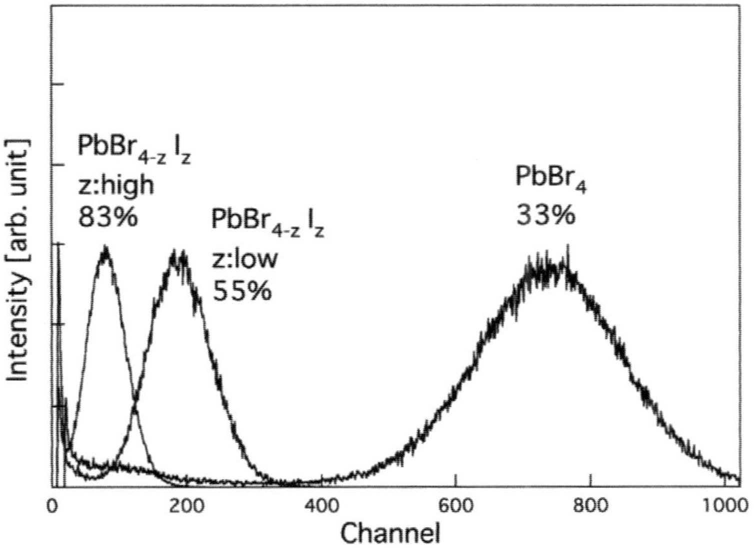

Figure 2. Energy spectra and energy resolution (FWHM) of scintillators for 67.4 keV x rays.

4. Conclusion
In practical use in nuclear resonant scattering, bromide crystals showed best performance. Although bromide and iodide mixture crystals showed fast scintillation decay, the light emission and the energy resolution were not enough for practical application. It is supposed to be caused by self-absorption of the crystals.

Acknowledgments
This study was performed with the approval of Photon Factory Advisory Committee (Proposal No. 2008G104). This study was supported by the Japan Science and Technology Agency, CREST.

References

[1] Shibuya K, Koshimizu M, Asai K and Shibata H 2004 *Appl. Phys. Lett.* **84** 4370

[2] Shibuya K, Koshimizu M, Murakami H, Muroya Y, Katsumura Y and Asai K 2004 *Jpn. J. Appl. Phys. Part 2* **43** L1333

[3] Kishimoto S, Shibuya K, Nishikido F, Koshimizu M, Haruki R and Yoda Y 2008 *Appl. Phys. Lett.* **93** 261901

[4] Shibuya K, Koshimizu M, Nishikido F, Saito H and Kishimoto S 2009 *Acta Crystallogr. Sect. E***65** m1323

Nuclear Resonant Scattering at PETRA III : Brillant opportunities for nano – and extreme condition science

H.-C. Wille, H. Franz, R. Röhlsberger, W.A. Caliebe, F.-U. Dill

Deutsches Elektronen-Synchrotron DESY, Notkestr. 85, D-22603 Hamburg

hans.christian.wille@desy.de

Abstract. The beamline P01, actually under construction at the PETRA III synchrotron in Hamburg, will be dedicated to nuclear resonant scattering (NRS) and to inelastic x-ray scattering (IXS). It will profit from a 20m long straight section with four 5m long undulator-segments. Due to the resulting outstanding brilliance in combination with micron and sub-micron focussing of the beam P01 will be especially well suited to study small samples in nano- and extreme condition science. In this paper the beamline P01 is introduced and future possibilities applying nuclear resonant spectroscopy at P01 are discussed.

1. Beamline layout

P01 will consist of a total of five hutches and three control cabins (CC). A three dimensional view of the beamline layout is shown in figure 1.

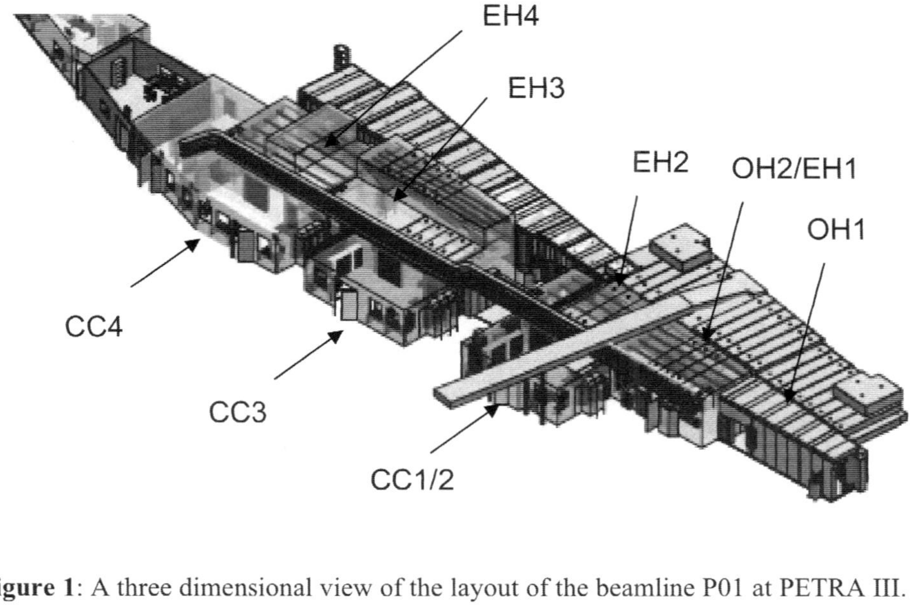

Figure 1: A three dimensional view of the layout of the beamline P01 at PETRA III.

First the beam will pass the optics hutch one (OH1) where basic optics like the high heat load monochromator and x-ray lenses are installed. The second hutch (OH2 / EH1) is reserved for high resolution monochromators and for experiments which do not require any special sample environment and not much space. EH2 is the first pure experimental hutch for nuclear forward scattering and nuclear inelastic scattering experiments in rather simple setups without helium cooling requirements. The experimental hutch EH3 will provide space for larger experiments and it will also host the inelastic x-ray scattering spectrometer. EH4 is an experimental hutch dedicated to experiments with samples at low temperature and/or high magnetic field using a helium cryostat with a supra conductive coil for magnetic fields up to about 10 Tesla. To host the cryostat the roof of the hutch is extended in height with respect to the other hutches.

2. Research topics envisaged at P01

Nuclear resonant scattering can nowadays be separated into four techniques: nuclear forward scattering (NFS), nuclear inelastic scattering (NFS), synchrotron radiation based perturbed angular correlation spectroscopy (SRPAC) and synchrotron radiation based Mössbauer spectroscopy (SRMS). We give some examples of future applications on the most common techniques NFS and NIS in the following (for an overview see [1]) .

2.1. Applications using nuclear forward scattering (NFS)

NFS is an excellent tool to study magnetism as the value and the direction of the magnetic field at the nucleus position is determined. Examples of applications using NFS are: magnetic properties of nanostructures and multilayers, magnetic properties and phase transitions under high pressure and/or high temperature, diffusion in layered systems.

2.2. Applications using nuclear inelastic scattering (NIS)

NIS provides isotope specific information on the atomic motions in a system. From the phonon energy spectra the vibrational density of states is directly calculated without any further models or assumptions. Examples of applications using NIS are: isotope specific dynamics in thermoelectric (nano-) materials, dynamics in phase change materials, dynamics in artificially materials and nano structures, protein dynamics, glass dynamics, dynamics in materials at high pressure and or high temperature like geological samples.

References

[1] *Hyperfine Interact. 123-125* edited by H. de Waard and E. Gerdau

Calculations of Mössbauer parameters in solids by DFT bandstructure calculations

Peter Blaha
Institute of Materials Chemistry, TU Vienna, Vienna, Austria
E-mail: pblaha@theochem.tuwien.ac.at

Abstract. A short introduction to periodic bandstructure methods suitable for the calculation of Mössbauer parameters is presented. These methods are based on density functional theory (because we want to treat big and complicated systems) and must be accurate not only in the bonding region (like pseudopotential approaches), but also near the atomic nuclei since Mössbauer parameters depend crucially on the wave functions at/near the nucleus. A numerical basis set as in *augmented plane wave* (APW) based methods is very well suited for this purpose and the APW method is briefly sketched. The results for $YBaFe_2O_5$ are discussed in more detail and compared with experiment.

1. Introduction

The development of new and advanced materials requires both, experimental and theoretical support to gain deeper knowledge about the relationship between the desired materials properties and its atomic and electronic structure. Among the many experimental techniques, Mössbauer spectroscopy is rather outstanding because it can yield detailed *local* information about some atoms in the sample, even when no long range order exists or the concentration of that atom is very low. On the other hand, the interpretation of the obtained results (mainly the Mössbauer parameters for isomer shift (IS), quadrupole splitting (QS) and magnetic hyperfine field (HFF)) are not always straight forward and comparison with theory is highly desirable. The Mössbauer parameters are all linked to the electron (or spin-) density, which can be obtained from the wave functions after solving the quantum mechanical problem self-consistently.

Modern materials pose quite some challenges to theoretical approaches because of the more and more complex nature of these materials. Very often, the complexity comes from two facts: i) the composition of the material gets more and more complex (nanostructures with large unit cells and many atoms; disorder, non-stoichiometry, ...) ii) some specific atoms and their electronic structure is not so simple to describe by standard theoretical methods, since two or more electronic configurations may be very close in energy (low-spin vs. high-spin states, different ionicities, ...) and theory may not yield the correct ground state in all cases.

Theoretical methods can be classified in various ways. I will deal here only with periodic band structure approaches (contrary to finite cluster models), where the solid is described by a unit cell, which is repeated periodically in all three dimensions. In order to simulate surfaces, nanostructures, disorder or interfaces one has to create a *supercell* model, i.e. a larger unit cell which includes the relevant structural features and describes the real material as close as possible. On modern computers such model structures can contain up to about 1000 atoms which is usually sufficient to build a reliable model of the specific material under investigation.

All theoretical methods considered here, try to describe the material by quantum mechanics and are based on *Density Functional Theory* (DFT)[1], which in principle is an exact theory, but the underlying functional is unknown and one has to make approximations. The most common approximations are the *Local Density Approximation* (LDA)[2] or the *Generalized Gradient Approximation* (GGA)[3], where just the local electron density (and it's gradient) enters the equations for the exchange-correlation energy and potential. While these methods work pretty well for the vast majority of solids, they may fail severely for so-called *correlated electrons* in 3d transition metal oxides or 4f and 5f compounds. Unfortunately, the most important Mössbauer element, Fe, belongs often to this class of materials and the theoretical description of such Fe-compounds may not be so reliable as, for example, the electronic structure of Si in standard silicates. One has to go beyond LDA/GGA and use methods like hybrid-DFT[4], LDA+U[5] or DMFT[6], but this requires quite some advanced knowledge of both, the expected materials properties and these more complicated methods (and their applicability).

Besides the decision how to treat the electron-electron interaction, one also has to consider which particular magnetic order should be investigated (non-magnetic, ferromagnetic, various antiferromagnetic configurations, a disordered local moment approximation or even non-collinear magnetism[7]), and whether relativistic effects should be included (clearly, spin-orbit coupling is very important for 5f compounds, but even for Fe orbital contributions to the magnetic HFF can be significant in certain cases).

Most importantly, the specific electronic structure method must be suitable to calculate Mössbauer parameters, which requires an accurate description of the wave functions also in the core region of an atom. Thus the widely used plane wave (PW) pseudopotential methods[8] cannot be used here at all and also the popular Gauss basis functions of traditional quantum chemistry (LCAO methods) may not provide a very accurate description near/at the nucleus, in particular for heavier elements. On the other hand, methods with an adaptive numerical all-electron basis set are very well suited and I'll describe the *augmented plane wave* methods in more detail. Once the electronic structure has been solved accurately, the Mössbauer parameters can be easily obtained from the corresponding wave functions as described below.

2. The augmented plane wave (APW) methods

One among the most accurate schemes for solving the DFT problem in solids is based on the APW method[9]. Much more information can be found in the book by Singh and Nordström[10] with many references and details. In APW based methods a basis set is introduced that is especially adapted to the problems in solids, where atomic like regions alternate with large empty space. Thus we divide the unit cell into non-overlapping atomic spheres (centered at the atomic sites) and an interstitial region and use different basis sets in the two types of regions. In the interstitial region a PW expansion is used

$$\Phi_{k_n} = \frac{1}{\sqrt{\omega}} e^{ik_n r} \tag{1}$$

where $k_n = k + K_n; K_n$ are the reciprocal lattice vectors, k is the wave vector inside the first Brillouin zone and ω is the unit cell volume. While PWs are in principle a complete basis set, in praxis a PW expansion does not converge for the wave functions near the nucleus and we must augment them by atomic-like functions in every atomic sphere (of radius R_t). Thus we use inside the atomic spheres a linear combination of numerical radial functions $u_l(r, E)$ times spherical harmonics $Y_{lm}(r)$ (we omit the index t).

$$\Phi_{k_n} = \sum_{lm} A_{lm} u_l(r, E) Y_{lm}(r) \tag{2}$$

where $u_l(r, E)$ is the (at the origin) regular solution of the radial Schrödinger equation for energy E. This function depends on the energy (eigenvalues) leading to an expensive non-linear eigenvalue problem, but it can be linearized in various ways, either by using $\dot{u}_l(r, E_l)$, the energy derivative of u_l (leading to the standard LAPW method introduced by Anderson[11]) evaluated at a fixed energy E_l (usually chosen at the center of bands with the corresponding l-like character), or by means of additional local orbitals[12] (APW+lo method, see below). The coefficients A_{lm} are functions of k_n and are determined by requiring that this basis function matches at R_t the corresponding PW (labeled with k_n) of the interstitial region. The accuracy of this method comes from the fact that u_l is the (numerical) exact solution for a given (spherical) potential.

In order to solve the problem of the energy dependence of u_l, overcome linearization errors and to enhance the efficiency, Singh[12] introduced additional basis functions called "local orbitals" (LO), which look very similar to normal APWs, but the coefficients A_{lm} are not matched to plane waves, but set such that this additional basis function goes to zero at the sphere boundary (thus the name "local orbital").

Such an optimized mixed basis set[13] forms the basis of the WIEN2k code[14], which has been developed in our group during the last two decades and is used worldwide by more than 1600 groups at universities and industrial laboratories.

The convergence of this basis set is controlled by a single cutoff parameter $R_t K_{max} = 6 - 9$, where R_t is the smallest atomic sphere radius in the unit cell and K_{max} is the magnitude of the largest K_n vector. Relativistic effects for core states are fully included, while by default the valence states are treated in a so called "scalar-relativistic" approximation[10], which includes mass velocity and Darwin s-shift terms, but neglects spin-orbit coupling. The latter can optionally be included in a second variational step and also radial basis functions for $p_{1/2}$ states (which differ significantly from non-relativistic p or relativistic $p_{3/2}$ functions) can be added[15]. WIEN2k comes with an (optional) user-friendly graphical user interface ("w2web") and proved to be efficient and rather robust against user errors. For bigger systems an efficient parallelization on various levels (k-points or fine-grain mpi-based) is available and WIEN2k has been executed on 1000 cores in parallel, treating unit cells with more than 1000 atoms[16].

3. Calculation of Mössbauer parameters

The isomer shift of a given atom can be obtained by comparing its electron density at the nucleus (ρ_0) with a reference. For the most common Mössbauer element this is body centered cubic (bcc) α-Fe. For this purpose bcc-Fe needs to be calculated with parameters as close as possible as the ones for the specific material under investigation. In particular the same R_t, a comparable basis set cutoff ($R_t K_{max}$) and the same DFT functional (however, the GGA+U method cannot be used because it would yield wrong results for an itinerant metal) should be used. A magnetic metal requires a dense k-mesh (like 10000 k-points in the full Brillouin zone) and spin-orbit coupling should be enabled/disabled exactly as in the compound. The ρ_0 values are directly available from the calculations. The isomer shifts δ, which are always related to the 4s electron density and thus to ionicity, can be calculated using the following expression:

$$\delta = \alpha(\rho_0^{Sample} - \rho_0^{Reference}), \tag{3}$$

where α is a proportionality constant that can be estimated through calibration. In recent DFT calculations by Wdowik et al.[17] α was found to be $-0.291\, a.u.^3\, mm/s$, which is slightly larger than the commonly used value of -0.27 from the work of Duff[18].

The quadrupole splitting $\Delta = eQV_{zz}$ is proportional to the product of the nuclear quadrupole moment Q times the electric field gradient (EFG) V_{zz}. In order to compare theory with experiment one needs a value for the nuclear quadrupole moment Q, which was determined by Dufek et al.[19] to be about $0.16\,b$ (barn; $1\,b = 10^{-28}\,m^2$) by comparing theoretical and experimental data for a large number of Fe compounds. With this value the relation between EFG and quadrupole splitting is approximately $EFG\,[10^{21}\,V/m^2] \approx 3\Delta\,[mm/s]$[20]. (Note that Q has limited accuracy and could also be about 10 % larger, which one should consider when comparing theory and experiment.) The EFG can be calculated directly from the electronic and nuclear charge distribution $\rho(r)$[21] by

$$V_{zz} = \int \rho(r)\frac{2P_2(cos\theta)}{r^3}dr, \tag{4}$$

where P_2 is the second-order Legendre polynomial which projects out the $l = 2$ component of the charge density. In addition, the $1/r^3$ factor weights non-spherical contributions near the nucleus much stronger. For further analysis, the integral can be decomposed spatially (into a sphere around the nucleus and the rest of space) or the density ρ can be decomposed into contributions from $3p-3p$, $4p-4p$ and $3d-3d$ orbitals. The semicore ($3p-3p$) contribution can be large if these states get polarized by anions at short distances and a very anisotropic environment; the $4p-4p$ contribution originates mainly from the tails of neighboring atoms (eg. oxygens) and is almost always very important, while the $3d-3d$ contribution is large when eg. in high-spin Fe^{2+} ions the single Fe-3d spin-down electron occupies *one* particular orbital (like d_{z^2} or d_{xy}), but is small if all d-orbitals have similar occupation (like eg. in high-spin Fe^{3+} ions). It should be noted that the EFG can be very sensitive to structural details and even small relaxations of some atomic positions may lead to very different EFGs[22].

In magnetic materials one can also calculate a magnetic hyperfine field B_{tot}, which can be decomposed theoretically into the Fermi contact term $B_{contact}$, the dipolar field B_{dip} (due to the anisotropic spin distribution) and the field associated with the orbital moment B_{orb} (this requires to include spin-orbit interactions in the Hamiltonian). Usually $B_{contact}$ dominates and can be obtained from the spin-density at the nucleus

$$B_{contact} = 8\pi/3[\rho_{\uparrow}(0) - \rho_{\downarrow}(0)], \tag{5}$$

where the polarization of the core-orbitals due to the finite 3d moment accounts for a large fraction of $B_{contact}$. Unfortunately, in many cases the contact term is underestimated by about 10 % in DFT.

4. Charge order and mixed valence state of $YBaFe_2O_5$

$YBaFe_2O_5$ shows a Verwey transition slightly above room temperature from a low-temperature charge-ordered (CO) phase with two distinct Fe sites (with formal valencies of +2 and +3, respectively) into a high-temperature valence-mixed (VM) modification with just one nonequivalent Fe site (with a formal valency of +2.5 [23]. In addition the structure changes from a strongly distorted orthorhombic to a nearly tetragonal symmetry and also the type of antiferromagnetic ordering changes. The changes in symmetry and magnetic order can be seen in Figs. 1 and 2.

Details of the structural, magnetic and electronic properties above and below the Verwey transition have been presented elsewhere[24] and here I concentrate only on the changes of Mössbauer parameters. The results are summarized in tables 1 and 2. Comparison between experiment and theory immediately reveals that LDA or GGA calculations fail to describe the Mössbauer parameters. For instance the differences for the IS (δ), quadrupole splitting (eQV_{zz}) or HFF (B_{tot}) between Fe^{2+} and Fe^{3+} sites in the CO phase are much too small. We need to consider on-site correlation effects as included in GGA+U and with a reasonable value of U (7 eV) rather good results are obtained in agreement with experiment. In particular for the CO-Fe^{2+} site significant dipolar and orbital contributions to the HFF lead to the relatively

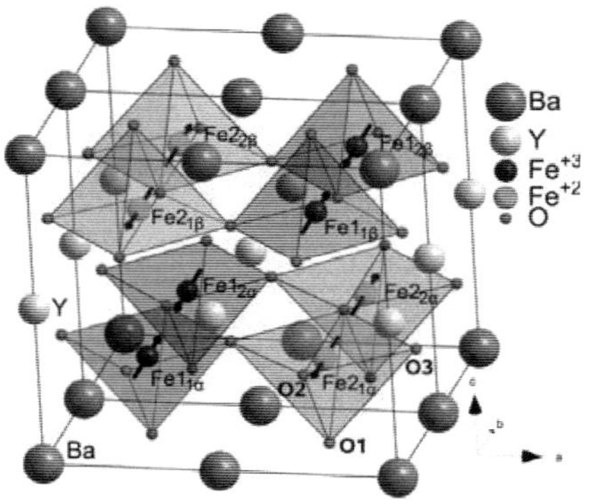

Figure 1. The magnetic 1 × 2 × 1 CO supercell. The arrows depict the direction of the magnetic moments.

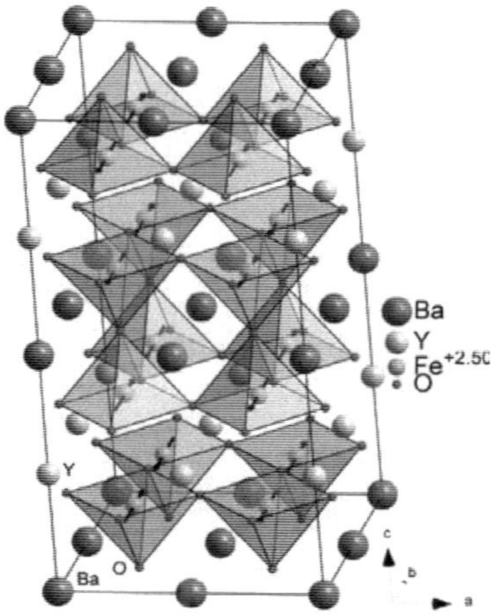

Figure 2. The magnetic 2 × 2 × 2 VM supercell. The arrows depict the direction of the magnetic moments.

Table 1. Theoretical and experimental[25, 26, 27] hyperfine fields B (in Tesla), isomer shifts δ (mm/s) and quadrupole splittings eQV_{zz} (mm/s) for the CO phase.

		exp.	GGA+U				LDA	GGA
	U_{eff} [eV]	—	5	6	7	8	—	—
$Fe2^{2+}$	B_{dip}	—	-16.29	-16.49	-16.66	-16.83	-6.68	-12.67
	B_{orb}	—	-6.73	-6.90	-8.26	-7.65	-9.57	-6.34
	$B_{contact}$	—	32.25	32.23	32.58	32.60	32.21	31.58
	B_{tot}	~ 8	9.23	8.83	7.66	8.13	15.96	12.57
	δ	~ 1	0.92	0.94	0.96	0.99	0.74	0.79
	eQV_{zz}	3.6 – 4	3.66	3.74	3.81	3.89	-0.82	2.60
$Fe1^{3+}$	B_{dip}	—	-0.67	-0.60	-0.52	-0.45	1.29	0.39
	B_{orb}	—	-0.52	-0.45	-0.37	-0.28	-7.96	-2.65
	$B_{contact}$	—	37.65	38.28	38.15	37.86	29.64	31.63
	B_{tot}	~ 50	36.46	37.24	37.26	37.12	22.97	29.37
	δ	~ 0.4	0.33	0.30	0.28	0.25	0.50	0.47
	eQV_{zz}	1 – 1.5	1.46	1.50	1.51	1.52	1.04	-0.30

small observed value. Of course, for the CO-Fe^{3+} ion with a full $3d^{\uparrow}$ and empty $3d^{\downarrow}$ shell these contributions are very small. The EFG in the VM phase is very small, because the EFG contributions from the tails of the O-neighbors and the on-site $3d$ electrons cancel partially. For the CO-Fe^{3+} ion V_{zz}^{3d} is very small (the five 3d electrons lead to nearly spherical symmetry), while for the CO-Fe^{2+} site a single $3d^{\downarrow}$ electron in the $3d_{xz}$ orbital leads to a large EFG.

5. Conclusions

Periodic bandstructure calculations based on DFT provide in most cases a very accurate description of Mössbauer parameters, provided one uses a method that is also accurate near the nucleus and the electron-electron interaction is described accurately. For so-called *highly*

Table 2. Theoretical and experimental[25, 26, 27] hyperfine fields B (in Tesla), isomer shifts δ (mm/s) and quadrupole splittings eQV_{zz} (mm/s) for the VM phase.

		exp.	GGA+U				LDA	GGA
	U_{eff} [eV]	—	5	6	7	8	—	—
$Fe^{2.5+}$	B_{dip}	—	-3.00	-2.98	-2.95	-2.87	-2.13	-2.83
	B_{orb}	—	-3.11	-2.99	-2.84	-2.74	-5.47	-4.56
	$B_{contact}$	—	41.17	40.96	41.45	41.17	33.10	36.36
	B_{tot}	~ 30	35.06	34.98	35.67	35.56	25.50	28.98
	δ	~ 0.5	0.53	0.52	0.51	0.49	0.60	0.60
	eQV_{zz}	~ 0.1	0.12	0.13	0.13	0.13	0.19	-0.27

correlated electrons it is necessary to go beyond standard DFT and unfortunately in that case the *ab initio* character of the calculations is partially lost, but one can still learn a lot about the proper physics in such a system. It should be stressed, that such calculations not only produce a few numbers which can be compared with experiment, but provide the possibility to gain deeper understanding of the specific compound. Theory can easily test various geometric or magnetic structures (which are not accessible in experiment), calculate various spectra (from infrared to X-ray absorption or emission spectroscopy) and explain the electronic properties.

References

[1] Dreizler R M and Gross E K U 1990 *Density Functional Theory* (Berlin:Springer)
[2] Kohn W and Sham L J 1995 *Phys. Rev.* **140** A1133
[3] Perdew J P, Burke K and Ernzerhof M 1996 *Phys. Rev. Lett.* **77** 3865
[4] Paier J, Marsman M, Hummer K, Kresse G, Gerber I C and Ángyán J G 2006 *J. Chem. Phys.* **125** 249901
[5] Anisimov V I, Zaanen J and Andersen O K 1991 *Phys. Rev.* B **44** 943
[6] Held K 2007 *Adv. Phys.* **56** 829
[7] Laskowski R, Madsen G K H, Blaha P and Schwarz K 2004 *Phys. Rev.* B **69** 140408
[8] Martin R M 2008 *Electronic Structure: Basic Theory and Practical Methods* (Cambridge:Cambridge University Press)
[9] Slater J C 1937 *Phys. Rev.* **51** 151
[10] Singh D J and Nordström L 2006 *Planewaves, Pseudopotentials and the LAPW Method* (New York:Springer)
[11] Andersen O K 1975 *Phys. Rev.* B **12** 3060
[12] Singh D J 1991 *Phys. Rev.* B **43** 6388
[13] Madsen G K H, Blaha P, Schwarz K, Sjöstedt E and Nordström L 2001 *Phys. Rev.* B **64** 195134
[14] Blaha P, Schwarz K, Madsen G K H, Kvasnicka D and Luitz J 2001 *WIEN2K, An Augmented Plane Wave and Local Orbitals Program for Calculating Crystal Properties*, (Vienna:Vienna University of Technology)
[15] Kuneš J, Novák P, Schmid R, Blaha P and Schwarz K 2001 *Phys. Rev. B* **64** 153102
[16] Laskowski R, Blaha P, Gallauner T and Schwarz K 2007 *Phys. Rev. Lett.* **98** 106802
[17] Wdowik U D and Ruebenbauer K 2007 *Phys. Rev. B* **76** 155118
[18] Duff K J 1974 *Phys. Rev. B* **9** 66
[19] Dufek P, Blaha P and Schwarz K 1995 *Phys. Rev. Lett.* **75** 3545
[20] Petrilli H, Blöchl P, Blaha P and Schwarz K 1998 *Phys. Rev. B* **57** 14690
[21] Blaha P, Schwarz K, Faber W and Luitz J 2000 *Hyperfine Interactions* **126** 389
[22] Body M, Legein C, Buzare J, Silly G, Blaha P, Martineau C and Calvayrac F 2007 *J. Phys. Chem. A* **111** 11873
[23] Woodward P M and Karen P 2003 *Inorganic Chemistry* **42** 1121
[24] Spiel C, Blaha P and Schwarz K 2009 *Phys. Rev. B* **79** 115123
[25] Lindén J, Karen P, Kjekshus A, Miettinen J, Pietari T and Karppinen M 1999 *Phys. Rev.* B **60** 15251
[26] Karen P, Woodward P M, Lindén J, Vogt T, Studer A and Fischer P 2001 *Phys. Rev.* B **66** 214405
[27] Lindén J, Karen P, Yamauchi H and Karppinen M 2004 *Hyperfine Interactions* **156-157** 321

Journal of Physics: Conference Series **217** (2010) 012010 doi:10.1088/1742-6596/217/1/012010

The method of invariants in ^{57}Fe Mössbauer spectroscopy on selected examples

K Szymański[1], D Satuła[1], L Dobrzyński[2], K Rećko[1], W Olszewski[1], K Brzózka[3] and J Jankowska-Kisielińska[4]

[1]University of Białystok, Faculty of Physics, Lipowa 41 15-424 Białystok, Poland

[2]also: The Andrzej Sołtan Institute for Nuclear Studies, 05-400 Otwock-Świerk, Poland

[3]Department of Physics, Technical University of Radom, Malczewskiego 29, Radom 26-600, Poland

[4]Institute of Atomic Energy, 05-400 Otwock-Świerk, Poland

corresponding author's e-mail address: kszym@alpha.uwb.edu.pl

Abstract. The method of invariants is an effective tool for decomposition of Mössbauer spectra into components resulting from electric quadrupole and magnetic dipole interactions. The invariants can be next analysed to obtain all possible values of hyperfine interaction parameters. In some cases it is possible to determine the parameters with satisfactory precision. The method was applied to not fully ordered UFe_4Al_8. Obtained results indicate that the dominant component of the electric field gradient is negative and is oriented almost perpendicularly to the direction of the hyperfine magnetic field. Another example shows that the method of invariants results in substantial reduction of the number of calculated components in modeling of hyperfine interactions in γ-FeMn.

1. Introduction

Mixed hyperfine interactions, e.g. magnetic dipole and electric quadrupole ones, in case of 1/2-3/2 spin transition cause difficulties in the interpretation of Mössbauer spectra because of the ambiguity problem [1]. Derivation of complete set of hyperfine parameters corresponding to specified values of invariants was considered in [2]. The proposed treatment was, however, not widely used. Explicit formulas for the intensity tensor [3] allow one to use the invariants (instead of hyperfine fields) for description of the absorption line positions and intensities as well. The method of Mössbauer spectra analysis making use of the invariants for proper description of the spectra components was recently proposed in [4]. Although the method described in [4] does not allow determination of the complete set of hyperfine interactions parameters, of invariants, and the analysis proposed in [2], may lead to satisfactory characterisation of the system. In particular the method is sensitive to the changes of local direction of hyperfine magnetic field (*HMF*) with respect to the electric field gradient (*EFG*) diagonal directions, see [4] for details.

UFe_4Al_8 intermetallic compound attracted considerable interest and the magnetic structure was recently determined [5,6]. Although the unit cell of UFe_4Al_8 is tetragonal, local symmetry of Fe is low [5], see Fig. 1a. The orientation of the *EFG* axes can not be simply deduced from the structure. Also experiments with ^{181}Ta local probe in UFe_4Al_8 [7] were not conclusive and to our best knowledge *EFG* at Fe(8f) in UFe_4Al_8 is not known. This motivated us to analyze the Mössbauer data of UFe_4Al_8 by the method of invariants. Unfortunately, probably because of the sample quality, the spectra collected close to the transition temperature could not be characterized by single site static behaviour. We thus restrict the analysis to the spectra measured at low temperatures only. The spectrum measured in paramagnetic phase will be analysed as well.

Another selected example was disordered γ-FeMn antiferromagnetic alloy, interesting for spintronics. The Mössbauer spectra of this alloy have been already measured in sixties [8], however orientation of hyperfine fields and strength of the *EFG* is still not exactly known [9,10]. The invariants offer strict description of the spectra components at temperatures close to the transition temperature and thus spectra measured in the wide temperature range can simultaneously be analysed. This results in information on the hyperfine interactions in poorly resolved spectra.

2. Example of UFe_4Al_8

Mössbauer measurements were performed on the powdered sample of UFe_4Al_8 (reported in [11]) at different temperatures. The spectra shown in Fig. 1b were simultaneously analysed using the set of four components. The line positions and intensities can be given explicitly by invariants S_0, S_1, S_2 constructed from the *EFG* tensor $\boldsymbol{V}$ and the *HMF* pseudovector $\boldsymbol{B}$ [3]:

$$S_0 = a\sqrt{\mathrm{Tr}\boldsymbol{V}^2}\,,\quad S_1 = a\,\boldsymbol{m}^T\cdot\boldsymbol{V}\cdot\boldsymbol{m}\,,\quad S_2 = a\sqrt{\boldsymbol{m}^T\cdot\boldsymbol{V}^2\cdot\boldsymbol{m}}\,, \tag{1}$$

where

$$a = eQc / E_\gamma \tag{2}$$

is the proportionality constant between the *EFG* components (in [V/m^2] in SI units) and the invariants expressed conveniently for applications in [mm/s]. Q is the nuclear quadrupole moment, e is the elementary charge, c is the speed of light, E_γ is the energy between the ground and the excited state, $\boldsymbol{m}=\boldsymbol{B}/B$. Because of presence of weak intensity components, constraints were applied to their center shifts according to the Debye model. The isomer shift at T=0 was a free parameter for each component. Constraints were also applied to the areas under subspectra and $T^{3/2}$ dependence of the *EFG* components on temperature. The invariants S_0, S_1, S_2 extracted from the spectra at temperatures 12, 35 and 80 K are shown in Fig. 1c for the main component (Fe(8*f*)). The measurements were also performed at higher temperatures close to the magnetic transition temperature (not shown). However, observed shape of the spectra indicate that simple approximation which uses constant number of subspectra is not sufficient.

All possible values of hyperfine parameters B, V_{zz}, η, θ, φ, corresponding to a point in Fig. 1c, were next extracted by a method similar to those presented in [2]. We note that the treatment described in [2] does not include situation when invariants are closely related to condition:

$$2S_0^4 + 18S_2^4 - 3S_0^2\left(S_1^2 + 4S_2^2\right) = 0 \tag{3}$$

shown by the lines in Fig. 1c. This case corresponds to the axially symmetric EFG and full discussion of the data treatment will be published elsewhere. The results obtained for all possible values of V_{zz}, η, θ, φ are shown in Fig. 1d

Low temperature measurements indicate presence of six line patterns, in agreement with earlier reports [5,6,11]. However, room temperature spectra consist of a slightly asymmetric doublet. It is hard to expect that Goldanski-Karyagin effect can be responsible for the asymmetry because a tetragonal axis of symmetry is perpendicular to the local two-fold axis of iron. More likely explanation can be searched in distribution of *EFG* originated due to chemical disorder in the crystal. One cannot also exclude presence of small amount of other phases, for example FeAl. That is why four components were used for simultaneous fit shown in Fig. 1b.

The method of invariants and constrained simultaneous fit allowed us to decompose the spectra into the main and other components. Although invariants are not single valued functions of hyperfine parameters, the results obtained for the main component in UFe_4Al_8 (Fig. 1d) are located within relatively narrow ranges. The V_{zz} component of the *EFG* is negative and the angle between the V_{zz} axis and the hyperfine magnetic field is close to $\pi/2$. No systematic changes of the orientation of hyperfine field with respect to the *EFG* axes were detected in the temperature range 13-80K.

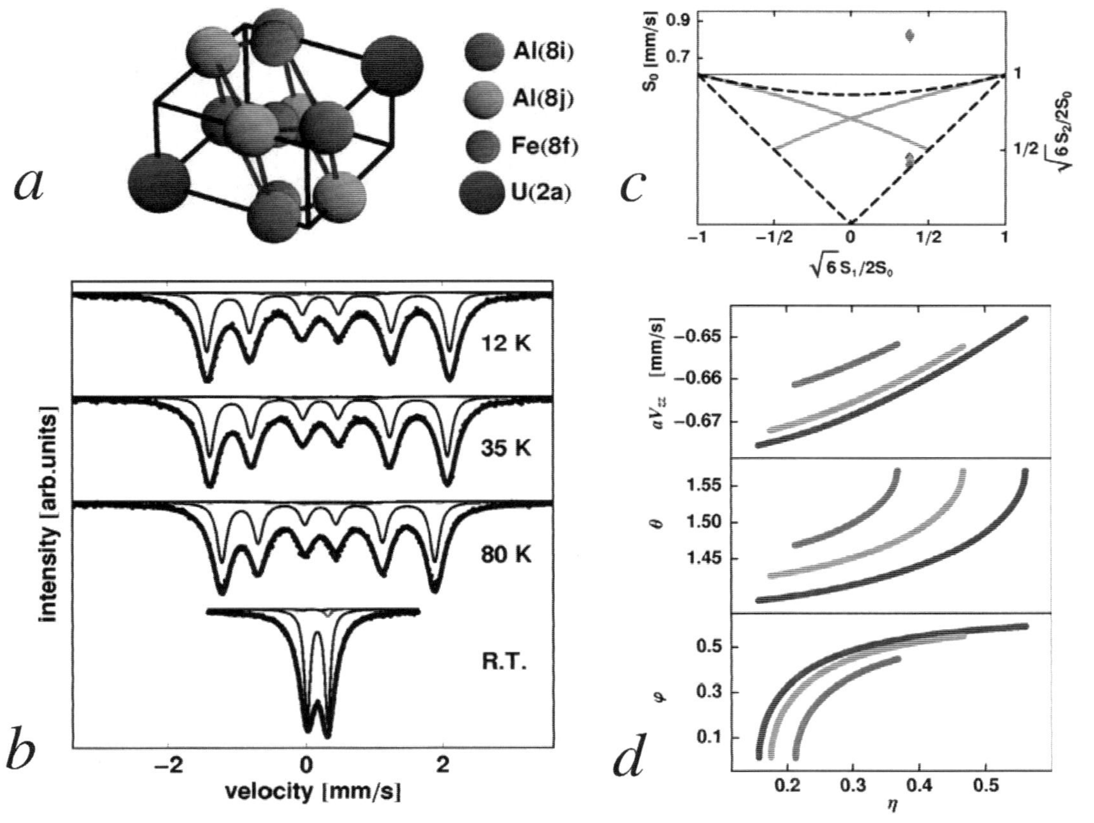

Figure 1 a) Local structure of Fe at 8*f* site in UFe_4Al_8, b) Mössbauer spectra of Fe in UFe_4Al_8 and results of simultaneous fit (weak-area components are reduce practically to horizontal lines in the figure c). Invariants S_0, S_1, S_2 of the main component (different temperatures are indicated by different colors, however the symbols overlap). Dashed lines indicate the region of all physically possible values of the invariants. Orange lines correspond to values of axially symmetric *EFG*, see eq. (3), d) All possible values of V_{zz}, η, θ, φ parameters for the S_0, S_1, S_2 invariants shown in c) (θ and φ in radians). Blue, green and red colours correspond to T=12, 35 and 80 K, respectively.

3. Example of γ-FeMn

Chemical disorder in cubic alloys causes breaking of local symmetry and creation of nonzero *EFG*. This was demonstrated in the case of bcc Fe-V alloys [12]. Similar situation may be present in disordered fcc Fe-Mn alloys. In case of magnetic order, number of different local environments is large. For exemplary estimation let us assume that local magnetization in fcc Fe-Mn is oriented along one of (111) directions. If the *EFG* gradient is next assumed to depend on the chemical configuration of the nearest neighbours (screened Coulombic pseudopotential [13]), one finds 2^{12} different chemical configurations and 8 different orientations of the local hyperfine fields (see example in Fig 2a) which makes $8{\cdot}2^{12}$=32768 different local environments for which diagonalisation has to be performed for correct calculation of the mixed hyperfine interactions. The problem, however, may be greatly simplified if the invariants are calculated. For all considered configurations only 160 different sets of invariants (S_0, S_1, S_2) exist. These invariants, once calculated, may be used in the fitting procedures. Efficiency of this procedure is shown in Fig.2 which displays excellent fit performed for the spectra measured in a wide temperature range (for details see [14]).

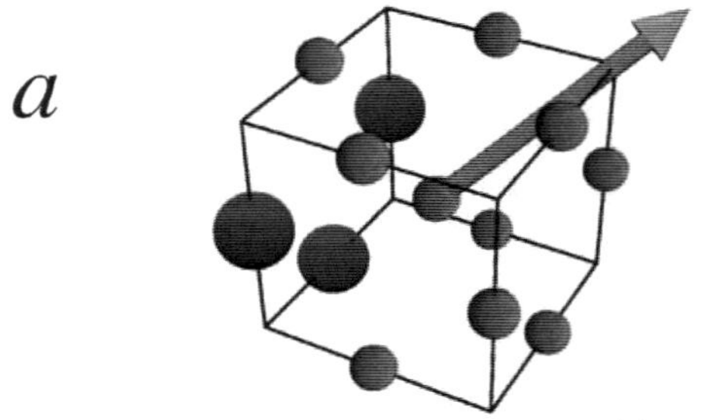

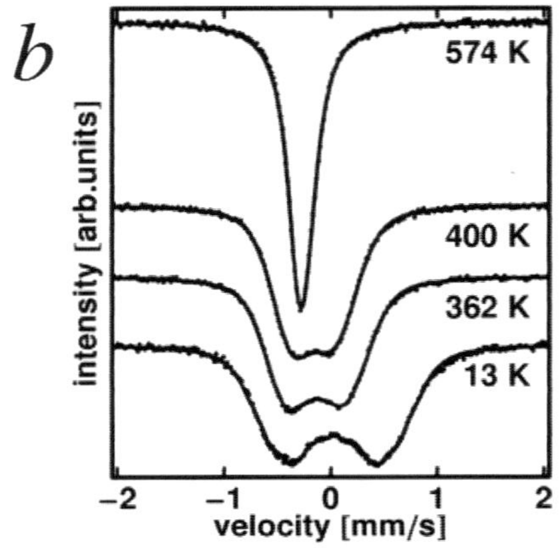

Figure 2 a) Example of local environment of iron at fcc binary alloy and *HMF* vector along (111) direction. b) Selected Mössbauer spectra of γ-FeMn fitted simultaneously at wide temperature range.

Acknowledgments:
The work was supported by a research project NN202172335.

References

[1] Karyagin S W 1966 *Fiz. Tverd. Tela* (Leningrad) **8** 493, 1966 *Sov. Phys. Solid State* **8** 391
[2] Dąbrowski L J, Piekoszewski J and Suwalski J 1972 *Nucl. Instrum. Methods* **103** 545
[3] Szymański K 2000 *J. Phys.: Condens. Matter* **12** 7495
[4] Satuła D Szymański K, Dobrzyński L, Tran V H and Troć R 2008 *Phys. Rev.* **B 78** 014411
[5] Paixao J A Lebech B, Goncalves A P, Brown P J, Lander G H, Burlet P, Delapalme A, and Spirlet J C 1997 *Phys.Rev.* **B 55** 14370
[6] Waerenborgh J C, Goncalves A P, Bonfait G, Godinho M and Almeida M 1999 *Phys. Rev.* **B 60** 4074
[7] Marques J, Barradas N, Alves E, Ramos A, Gonçalves A P, da Silva M and Soares J 2001 *Hyp. Int.* **136** 333
[8] Ishikawa Y and Endoh Y 1967 *J. Phys. Soc. Jpn.* **23** 205
[9] Kennedy S J and Hicks T J 1987 *J Phys. F: Met. Phys.* **17** 1599
[10] Szymański K, Olszewski W Dobrzyński L, Satuła D and Jankowska-Kisielinska J 2004 *Nukleonika* **49** S75.
[11] Szymański K, Rećko K, Dobrzyński L and Satuła D 1999 *J.Phys.Condens. Matter*, **11** 6451
[12] Preston R S, Lam D J, Nevitt M V and Ostenburg D O 1966 *Phys. Rev.* **149** 440
[13] Scholte P M L O, Tegze M, van der Woude F, Buschow K H J and Vincze I 1986 *Hyp. Int.* **27** 437
[14] Szymański K, Olszewski W, Satuła D, Dobrzyński L, Brzózka K and Jankowska-Kisielińska J 2009 *J. Phys. Soc. Jap.* **78** in press

Journal of Physics: Conference Series **217** (2010) 012011 doi:10.1088/1742-6596/217/1/012011

Mössbauer spectra and magnetization curves of nanoparticles in a weak magnetic field

M A Chuev[1]
Institute of Physics and Technology, Russian Academy of Sciences, Nakhimovskii pr. 36-1, 117218 Moscow, Russia

E-mail: chuev@ftian.ru

Abstract. Magnetic relaxation effects revealed in Mössbauer spectra and magnetization measurements of nanoparticles are discussed in the framework of specific models for magnetic dynamics of an ensemble of single-domain particles, which allow one to treat both the temperature and magnetic field-dependent magnetization curves and Mössbauer spectra in a self-consistent way within the same set of physical parameters inherent to the system studied. Based on the stochastic approaches, a simplified three-level stochastic model taking into account the magnetic anisotropy, precession and diffusion of uniform magnetization of single-domain particles is developed in order to describe the Mössbauer absorption spectra of nanoparticles in a weak magnetic field.

1. Introduction

A number of techniques are applied to study the non-equilibrium magnetism of nanoparticles, among which the most informative seem to be the conventional magnetization measurements and Mössbauer spectroscopy, which obviously supply one with a large amount of information about physical characteristics inherent to the systems studied. The principal difference between these two complementary techniques is that they can probe magnetic properties of the same material in different frequency ranges: the magnetization measurements are carried out at lower frequencies (of about 1-1000 Hz) while Mössbauer spectroscopy can reveal the magnetic dynamics of nanoparticles at higher frequencies due to the Mössbauer time window (10^{-11}–10^{-6} s for ^{57}Fe nuclei).

The only way to extract the reach information from the experimental data is to define a model of the magnetic dynamics in order to describe the whole set of the experimental data for the sample studied. Since a long time two fundamental ideas are used to describe qualitatively the magnetic properties of single-domain fine particles: (i) the Néel's equation for the relaxation rate between local energy minima of a ferromagnetic particle with the uniaxial magnetic anisotropy energy density K [1]:

$$p = p_0 \exp(-KV / k_B T) \qquad (1)$$

where p_0 is the fluctuation rate slightly dependent on temperature T, V is the particle's volume, k_B is the Boltzmann constant and (ii) the Stoner-Wohlfarth (SW) model for particles with the uniform magnetization M_0 and the energy density in an external magnetic field H [2]:

[1] To whom any correspondence should be addressed.

$$E = -K\cos^2(\Theta - \varphi) - HM_0\cos\varphi \tag{2}$$

where Θ and φ are the angles composed by the external field direction with the easiest magnetization axis and the magnetization vector, respectively.

Both the models are widely used in analyzing experimental magnetization curves and Mössbauer spectra of materials with magnetic nanoparticles, however such an analysis actually reduces to estimates of a finite number of empirical parameters such as the blocking temperature T_b and the coercivity H_c which are determined not only by the properties (e.g., K, V, M_0, and p_0) of the sample under investigation, but also by the external parameters of the measurement method (external field strength, temperature and their scanning rate, prehistory of the sample, etc.). In such a situation, it is necessary to search for a compromise solution, i.e., to develop a phenomenological model applicable for analyzing the experimental data and estimating the aforementioned parameters of single-domain particles.

2. Generalized Stoner-Wohlfarth model

The main limitation of the original SW model is a very approximate description of the relaxation process that transitions between the states with different energy minima are forbidden until the applied field strength does not exceed the critical value $H_C(\Theta)$ [2]. In accordance with the Néel's idea and equation (1), a more accurate description of the relaxation process within the SW model has been suggested under the assumption that the relaxation process is specified by only two parameters [3, 4]:

$$p_{12,21}(H,\Theta,T) = p_0\exp\left[-\left(E_{max}(H,\Theta) - E_{1,2}(H,\Theta)\right)/k_BT\right] \tag{3}$$

where $E_{max}(H, \Theta)$ and $E_i(H,\Theta)$ are the energies of the local maxima and minima, which are evaluated by simple numerical calculations from equation (2).

When temperature or applied field is changed in time, the actual populations of the local states $w_1(t)$ and $w_2(t)$ of each particle with a given Θ are non-equilibrium and can be determined using the differential equation for their difference $\widetilde{w}(t) = w_1(t) - w_2(t)$ [4]:

$$d\widetilde{w}(t)/dt = -p(t)[\widetilde{w}(t) - \widetilde{w}_0(t)] \tag{4}$$

where $p(t) = p_{12}(t) + p_{21}(t)$ and

$$\widetilde{w}_0(t) = \tanh\left([E_2(H,\Theta) - E_1(H,\Theta)]V/2k_BT\right) \tag{5}$$

is the difference of equilibrium populations at instantaneous values of H and T depending on time [5]. The time evolution of the magnetization of particles with the given angle θ is defined by [4]

$$M(t,\Theta) = M_0\left[(1+\widetilde{w}(t,\Theta))\cos\varphi_1(\Theta) + (1-\widetilde{w}(t,\Theta))\cos\varphi_2(\Theta)\right]/2\,. \tag{6}$$

Then, the total magnetic moment of an ensemble of the randomly oriented and non-interacting SW particles can be calculated by averaging equation (6) over Θ. One can see that the simple differential equation (4) appears to be universal in describing temperature- and field-dependent magnetization curves measured by any experimental technique (either conventional magnetization measurements or Mössbauer spectroscopy) at arbitrary measuring frequency, field strength and temperature [3, 4].

3. General theory of stochastic reorientation of particle's magnetization

The main limitation of the original and generalized SW models is that thermal excitations at local energy minima are not taken into account. The general theory of the stochastic reorientation of the vector **M** for the statistical ensemble of ferromagnetic single-domain particles in the presence of a rapidly fluctuating chaotic field $\mathbf{h}(t)$ has been developed under the assumption that the stochastic process $\mathbf{h}(t)$ is stationary and isotropic [6] followed by a differential equation for the probability density (population) $W(\theta,\phi)$ of states with a given **M** direction and the relaxation operator [6, 7]:

$$P = -D\left[\Delta - \frac{V}{k_{\mathrm{B}}T}\nabla\left(\nabla E(\theta,\phi)\right)\right] - \frac{\gamma}{M_0}\nabla\left(\frac{\mathbf{M}}{M_0}\times\nabla E(\theta,\phi)\right). \tag{7}$$

where D is the diffusion constant and γ is the gyromagnetic ratio. However, in spite of a number of simulations based on this equation, a real numerical analysis of the experimental data on the temperature- and field-dependent magnetization has not been yet performed within this approach due to first of all computational problems.

Recently, an alternative approach for describing the magnetic dynamics of an ensemble of single-domain particles in a magnetic field has been proposed [5], in which the magnetization precession orbits are considered as stochastic states of each particle, the states (orbits) with a given energy being characterized by the mean magnetization value determined by the curved integrals along the corresponding trajectory C_E:

$$\overline{M}_{z,x}(\Theta,E) = \int M_{z,x}dm/\widetilde{\Omega}_E(\theta,\varphi)\left(\int dm/\widetilde{\Omega}_E(\theta,\varphi)\right)^{-1} \tag{8}$$

where $\widetilde{\Omega}_E^2(\theta,\varphi) = \Omega_E^2(\theta,\varphi) + p_E^2(\theta,\varphi)$, $\Omega_E(\theta,\varphi)$ is the instantaneous angular velocity and $p_E(\theta,\varphi)$ is the predetermined probability of transition per unit time from the given trajectory C_E at the point (θ,φ) to the adjacent trajectories $C_{E\pm\Delta E}$ [5]. Based on this approach, a general model of magnetic dynamics can be determined for calculating the magnetic characteristics in various measurement methods and analyzing numerically the experimental data.

One can write down also the general expression for the Mössbauer absorption spectrum in this representation using the conventional stochastic approach [3, 8-10]:

$$\sigma(\omega,\Theta) \propto \mathrm{Im}\sum_{\eta}\mathrm{Sp}\left(V_\eta\langle W|\mathbf{A}^{-1}(\omega,\Theta)|1\rangle V_\eta^{+}\right) \tag{9}$$

where V_η is the operator for the interaction of the gamma-quantum with a given polarization η and the nucleus, W is the row vector of the equilibrium occupation probabilities of the stochastic states, $|1\rangle$ is a column vector with all components equal to unity. The superoperator $\mathbf{A}(\omega,\Theta)$ is defined by the Liouville operators of hyperfine interaction specified by the Hamiltonians in the ground and excited states for each stochastic state i [9]:

$$H_i^{(\mathrm{g,e})} = -g_{\mathrm{g,e}}\mu_{\mathrm{N}}H_{\mathrm{hf}}^{(0)}\overline{\mathbf{M}}_i\mathbf{I}^{(\mathrm{g,e})}/M_0 \tag{10}$$

and the relaxation matrix specified by the probabilities $p_E(\theta,\varphi)$ [5].

4. Mössbauer spectra of nanoparticles in a magnetic field within three-level relaxation model

However, the main problems of such an analysis within the multi-level relaxation model are associated with the optimization of a calculation procedure. This is why a simplified three-level stochastic model taking into account that the vector **M** for a given energy E describes a conic surface around one of the three poles corresponding to two local minima and the absolute maximum of energy (2) has been introduced in order to describe the Mössbauer absorption spectra of nanoparticles in a weak magnetic field [9]. Indeed, a similar three-level relaxation model has been already introduced as a formal intermediate step in passing from the Neel's two-level model to a multi-level model in the absence of a field [10]. Moreover, the physical reason for correctness of the three-level model results actually from the fact that the kinetic equation with the relaxation operator (7) is reduced to the Schrödinger equation with just a three-well potential even in the absence of a field [7].

The three-level relaxation model is valid under the condition $\Gamma_0/\hbar << D << \Omega_0 = 2\gamma K/M_0$ (Γ_0 is the width of the excited nuclear level) and takes into account physical mechanisms of formation of the magnetic hyperfine structure for nanoparticles in a weak (as compared to the hyperfine field) magnetic field [9]. A number of typical Mössbauer absorption spectra calculated within this model for the field direction composing the 'magic' angle with the gamma-beam (in order not to mix weak field and polarization effects) can be found in [9]. Here, figure 1 shows ^{57}Fe Mössbauer absorption spectra calculated within the three-level relaxation model for various values of the transverse and longitudinal external field as well as of the efficient energy barrier KV/k_BT and just demonstrates the polarization effects.

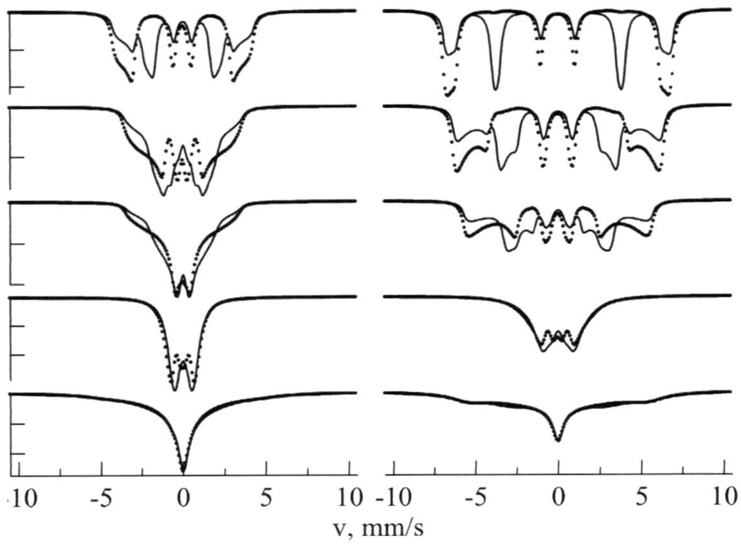

Figure 1. ^{57}Fe Mössbauer spectra of an ensemble of nanoparticles ($\mu_0 H_{hf}^{(0)}$ = 50 T, p_0 = 3 GHz) in a transverse (solid lies) and longitudinal (points) magnetic field, which calculated in the three-level relaxation model as a function of the effective energy barrier KV/k_BT=0.5 (left column), 2 (right column) and the normalized field strength $h=HM_0/2K$=0.01, 0.1, 0.5, 1, 2 (from bottom to top).

Thus, the three-level relaxation model taking into account the magnetic anisotropy, precession and diffusion of uniform magnetization can be efficiently used to describe the Mössbauer absorption spectra of an ensemble of magnetic nanoparticles in a weak magnetic field, e.g., measured long ago [11, 12], but not simulated yet. Applications of the model for analyzing the experimental Mössbauer absorption spectra of magnetic nanoparticles are presented in the contributions of Polikarpov et al. in this issue.

Acknowledgments
This work is financially supported by the Russian Foundation for Basic Research (projects No. 08-02-00388 and 09-02-12195).

References
[1] Néel L 1949 *Ann. Géophys.* **5** 99
[2] Stoner E C and Wohlfarth E P 1948 *Phil. Trans. Royal Soc. London* **A 240** 599
[3] Afanas'ev A M, Chuev M A and Hesse J 1999 *JETP* **89** 533
[4] Chuev M A and Hesse J 2007 *J. Phys.: Condens. Matter* **19** 506201
[5] Chuev M A 2009 *JETP* **108** 249
[6] Brown W F Jr. 1963 *Phys. Rev.* **130** 1677
[7] Afanas'ev A M and Sedov V E 1986 *Sov. Phys. Dokl.* **31** 651
[8] Chuev M A 2006 *JETP* **103** 243
[9] Chuev M A 2008 *J. Phys.: Condens. Matter* **20** 505201
[10] Jones D H and Srivastava K K P 1986 *Phys. Rev.* **B 34** 7542
[11] Eibschütz M and Shtrikman S 1968 *J. Appl. Phys.* **39** 997
[12] Lindquist R H, Constabaris G, Kündig W and Portis A M 1968 *J. Appl. Phys.* **39** 1001

Journal of Physics: Conference Series **217** (2010) 012012 doi:10.1088/1742-6596/217/1/012012

Theoretical investigation on the presence of different iron sites in $RbMn[Fe(CN)_6]\cdot H_2O$

Reshmi Kurian and Michael Filatov

Theoretical Chemistry, Zernike Institute for Advanced Materials,Rijksuniversiteit Groningen,Nijenborgh 4, 9747 AG Groningen, The Netherlands

E-mail: m.filatov@rug.nl

Abstract. The ^{57}Fe electron contact densities and the corresponding Mössbauer isomer shifts of $RbMn[Fe(CN)_6]\cdot H_2O$ are calculated with the use of *ab initio* wavefunction methods and double hybrid density functional method. The theoretical analysis of possible origin of the two slightly different signals in the Mössbauer spectra observed at 293 K and 50 K (Vertelman *et al.* 2008 *Chem. Mater.* **20** 1236-38) is carried out. The measured differences of 0.083 - 0.105 mm/s between the high-temperature and low-temperature phases can be attributed to the different oxidation states of iron: the iron is in Fe^{III} oxidation state in the high temperature phase and Fe^{II} state in the low temperature phase. The smaller isomer shift differences of 0.012 - 0.034 mm/s occurring in both high temperature and low temperature structures can be attributed to different distributions of the Rb^{I} ions within the unit cell of $RbMn[Fe(CN)_6]\cdot H_2O$.

1. Introduction

In the present work, theoretical analysis of ^{57}Fe Mössbauer isomer shifts in a Prussian blue analogue $RbMn[Fe(CN)_6]\cdot H_2O$ is carried out. Prussian blue analogues are compounds with the general molecular formula $A_xM^a[M^b(CN)_6]\cdot zH_2O$ (where, A = alkali cation and M^a/M^b = metal ion), which attract considerable interest due to their peculiar magneto-optical properties [1]. Thus, $RbMn[Fe(CN)_6]\cdot H_2O$ shows light and temperature induced switching of magnetization [2]. Upon cooling from 300K to approximately 150K, $RbMn[Fe(CN)_6]\cdot H_2O$ undergoes transition to a state with higher molar magnetization. A reversed transition to a low magnetization state is observed upon heating to *ca.* 250K. It has been hypothesized that the switching of magnetization occurs due to electron transfer from the Mn^{II} cations to the Fe^{III} ions (and reversed transition) as a result of structural phase transition upon cooling/heating the samples [3].

The Mössbauer spectra of the high-temperature (HT) and the low-temperature (LT) structures

Table 1. ^{57}Fe Mössbauer isomer shifts observed in $RbMn[Fe(CN)_6]\cdot H_2O$ at different temperatures [3] where number in the parentheses shows the experimental uncertainty in the last digit .

T, K	δ, mm/s	
50	-0.033(4)	-0.067(5)
293	-0.138(2)	-0.150(2)

were recorded whereby it was observed that, in both structures, two signals with the isomer shifts reported in Table 1 were present [3]. The presence of two different Fe^{III} sites in the HT phase of

RbMn[Fe(CN)$_6$]·H$_2$O with the shifts $\delta = -0.138(2)$ mm/s and $\delta = -0.150(2)$ mm/s was interpreted as a consequence of different distributions of the Rb cations over the interstitial sites $4c$ and $4d$ (see Figure 1) in the space group F$\bar{4}3m$ of the crystal. Although it was not possible to obtain reliable crystal structures of the LT phase of RbMn[Fe(CN)$_6$]·H$_2$O, it is clear that iron is present in the form of low-spin Fe^{II} ions octahedrally coordinated by the cyano ligands. Similar to the HT phase, a disorder in the occupation of the Rb sites around the Fe^{II} ions leads to emergence of two different iron sites as can be seen from the isomer shifts reported in Table 1.

To obtain insight into the origin of different iron sites in the HT and LT structures of RbMn[Fe(CN)$_6$]·H$_2$O we undertake quantum chemical calculations of ^{57}Fe Mössbauer isomer shifts in cluster models of these sites. More specifically, our goal is to elucidate whether the observed isomer shifts originate from variations in the oxidation state of iron ions (Fe^{II} *vs.* Fe^{III} isomer shifts) and from differences in the coordination sphere of iron ions due to different distribution of the Rb^{I} ions (see Figure 1).

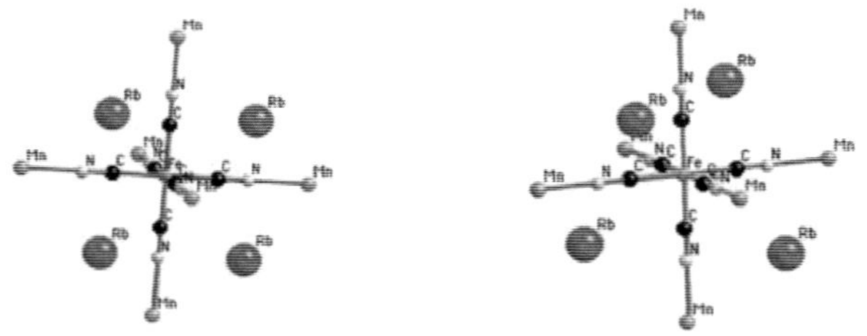

Figure 1. Two different iron sites in RbMn[Fe(CN)$_6$]·H$_2$O originating from different distribution of the Rb ions, the cluster shown first is referred as Type I and second as Type II respectively [3].

2. Details of calculations

The calculations are carried out in the embedded cluster approach, where the geometries of clusters shown in Fig. 1 are obtained from crystal structure of the HT phase reported in Ref. [3]. The iron is in octahedral coordination with six cyano ligands bridging Fe and Mn. The disorder in the distribution of Rb^{I} ions is modeled by selecting two possible distribution patterns of Rb^{I} around iron atoms as shown in Fig. 1. The Fe-C and Mn-N distances are 1.929 Å and 2.205 Å respectively. The same geometry of clusters is used for the LT phase calculations. The Madelung field is modeled by the field of a large array of point charges placed at the appropriate crystallographic positions.

The calculations are carried out using Hartree-Fock (HF), second order Møller-Plesset perturbation theory (MP2) [4], Coupled Cluster Singles and Doubles (CCSD) [4] and double hybrid density functional (B2-PLYP [5]) level. The 24s12p9d basis set of Partridge with a set of polarization functions taken from TZVpp basis set of Ahlrichs and May [6] is used for iron and 6-31G* [7] basis set is used for carbon and nitrogen. All basis sets were used in uncontracted form. The Rb^{I} core is modeled by the Stuttgart effective core potential (ECP) [8], whereas Mn$^{II/III}$ cores are modeled by the representative ECPs of Ca^{II} and Al^{III}, respectively. The closed shell systems are treated with spin restricted formalism and the open-shell systems with spin unrestricted formalism. The relativistic calculations are carried out within the one-electron approximation [9] and using the normalized elimination of the small component (NESC) [10] method which was implemented according to Ref. [11].

The ^{57}Fe isomer shifts are calculated using a formalism in which the isomer shift is treated as a response of the total electronic energy with respect to the variation of the nuclear charge radius [12, 13, 14]. The isomer shift is related to the contact density as $\delta = \alpha(\bar{\rho}_e^a - \bar{\rho}_e^s)$ where the calibration constant $\alpha(^{57}\mathrm{Fe})$ = -0.306 ± 0.009 mm/s is taken from theoretical calibration of ^{57}Fe isomer shift [15] and $\bar{\rho}$ is obtained from Eq. (1).

$$\bar{\rho}_e^{a/s} = \frac{5}{4\pi Z R}\frac{\delta E_e^{a/s}}{\delta R} \tag{1}$$

The derivatives $\delta E_e(R)/\delta R$ are calculated numerically using the increment of 10^{-6} Bohr for the root mean square (r.m.s.) nuclear charge radius ($R_{\mathrm{Fe}} = 0.70213 \times 10^{-4}$ bohr). Throughout this work, the

Gaussian nucleus model [16] is used in the calculations.

3. Results and Discussion

In this section, the influence of different factors, both electronic and geometric, on the ^{57}Fe contact densities of RbMn[Fe(CN)$_6$]·H$_2$O crystal are presented and discussed in detail. The main goal is to determine the origin of the different Fe sites observed in the HT and LT phases (see Table. 1)[3]. In

Table 2. ^{57}Fe electron contact densities and isomer shift differences obtained at different theory levels.

	Fe^{II}	Fe^{III}	$\Delta\rho$	$\Delta\delta$
HF	14938.17	14940.12	-1.95	0.60
MP2	14940.16	14939.82	0.35	-0.11
B2-PLYP	15140.18	15140.89	-0.71	0.22
CCSD	14940.38	14940.70	-0.52	0.16

RbMn[Fe(CN)$_6$]·H$_2$O, iron can be present either in the Fe^{II} or Fe^{III} oxidation states depending on the charge transfer between iron and manganese. In this work, we have carried out calculations on both oxidation states of iron incorporated in clusters described in the previous section. In both oxidation states, iron is assumed to be present in a low-spin state: spin 0 in Fe^{II} clusters and spin 1/2 in Fe^{III} clusters.

Table 2 gives the ^{57}Fe electron contact densities and isomer shifts obtained at different levels of theory (type I coordination sphere shown in Fig. 1 is assumed in these calculations). From the earlier theoretical works on K$_4$Fe(CN)$_6$ and K$_3$Fe(CN)$_6$, $\Delta\delta$ (Fe^{II} - Fe^{III}) is expected to be positive [18]. The Hartree-Fock, B2-PLYP and CCSD yield positive $\Delta\delta$ values while MP2 yields a negative value (see Table 2). From Table 2, it is obvious that the $\Delta\delta$ values obtained with the use of the methods including electron correlation are in much better agreement with the experimental observations than the HF value ($\Delta\delta_{\mathrm{CCSD}}$ = 0.16 mm/s and $\Delta\delta_{\mathrm{B2-PLYP}}$ = 0.22 mm/s *vs.* $\Delta\delta_{\mathrm{HF}}$ = 0.60 mm/s). The values obtained from the CCSD and B2-PLYP calculations are very close to one another. Therefore B2-PLYP is employed in the further calculations as a cost effective alternative to the coupled cluster method. Table 3 gives the contact densities and isomer shifts on iron in Fe^{II} and Fe^{III} oxidation

Table 3. ^{57}Fe electron contact densities and isomer shifts at different oxidation states incorporated in type I and type II clusters (see Fig. 1).

	Type I	Type II	$\Delta\rho$	$\Delta\delta$
Fe^{II}	15140.18	15139.98	-0.20	0.06
Fe^{III}	15140.89	15140.72	-0.18	0.05
$\Delta\rho$	-0.71	-0.74		
$\Delta\delta$	0.22	0.23		

states, for two different distribution patterns of Rb around Fe (see Fig. 1). In both types of Rb distribution patterns, change in the oxidation state of iron leads to a large variation of the isomer shift, of the order of $\Delta\delta$ (Fe^{II} - Fe^{III}) $\sim$ 0.2 mm/s. These large variations in δ of iron sites with iron in Fe^{II} and Fe^{III} oxidation states agree qualitatively with the experimentally observed variations of the isomer shift in HT and LT phases. Thus, it can be concluded that iron in the HT phase is present predominantly in the form of Fe^{III}, whereas in the LT phase it is in the form of Fe^{II}.

Changing the distribution pattern of Rb^{I} ions (see Fig. 1) leads to much smaller variations of the isomer shift for both oxidation states of iron (see Table 3). The $\Delta\delta$ values reported in the last column of Table 3 are consistent with the isomer shift differences of $\Delta\delta_{\mathrm{HT}} = 0.012$ mm/s and $\Delta\delta_{\mathrm{LT}} = 0.034$ mm/s observed for different iron sites within the HT phase and the LT phase respectively (see Table 1). Thus, these differences can be interpreted as originating from differences in the $\mathrm{Rb^I}$ ions distribution patterns around the $\mathrm{Fe^{III}}$ sites in the HT phase and around the $\mathrm{Fe^{II}}$ sites in the LT phase.

In the calculations reported above, it was assumed that the geometry of the coordination sphere of iron ions is the same in the HT and in the LT phases. In the HT phase, crystal structure of which is known experimentally, the $\mathrm{Fe^{III}}$ ions are surrounded by six cyano ligands at a distance $R_{\mathrm{Fe-C}} = 1.929$ Å. Because changing the oxidation state of iron to $\mathrm{Fe^{II}}$ should lead to contraction of the Fe-C bond length [17, 18], we have studied the effect of the bond length contraction on the contact densities and isomer shifts. In this study it was found that the bond length contraction in $\mathrm{Fe^{II}}$ coordination sites leads to a decrease in the isomer shift difference. Therefore, the value of $\Delta\delta(\mathrm{Fe^{II}} - \mathrm{Fe^{III}}) = 0.22$ mm/s obtained for the type I coordination at the HT experimental geometry should be corrected to a lower value (*e. g.* $\Delta\delta \approx 0.09$ mm/s for $\mathrm{Fe^{II}} - \mathrm{C}$ bond length of 1.89 Å), thus improving the agreement with the experiment (see Table 1) [3].

4. Conclusion

The origin of the different iron sites observed experimentally in the ^{57}Fe Mössbauer spectra of Prussian blue analogue $\mathrm{RbMn[Fe(CN)_6]{\cdot}H_2O}$ is analyzed with the help of theoretical calculations of the isomer shift on ^{57}Fe. In these calculations, a new formalism in which the isomer shift is treated as a response of the total electronic energy with respect to the variation of the nuclear charge radius (see Eq. (1)) is employed [12, 13, 14]. The relativistic calculations have been carried out at different levels of description of electron correlation varying from the Hartree-Fock method to the coupled cluster method (CCSD) and to the double hybrid density functional method (B2-PLYP) of Grimme[5].

The theoretically obtained ^{57}Fe isomer shift values enable one to interpret the experimentally observed $\Delta\delta_{\mathrm{HT-LT}}$ isomer shift differences (*ca.* 0.1 mm/s, see Table 1) between the HT phase and the LT phase of the $\mathrm{RbMn[Fe(CN)_6]{\cdot}H_2O}$ compound as originating from the change of the oxidation state of iron from $\mathrm{Fe^{III}}$ to $\mathrm{Fe^{II}}$. Two different iron sites observed within the HT and LT phases separately, which lead to $\Delta\delta$ values of the order of 0.01-0.03 mm/s (see Table 1), can be interpreted as originating from different distribution patterns of the $\mathrm{Rb^I}$ ions around the iron sites as shown in Fig. 1. The theoretical calculations yield the $\Delta\delta$ values of the order of 0.05 mm/s for the two types of $\mathrm{Rb^I}$ coordination thus corroborating the analysis of the experimental measurements presented by Vertelman *et al.* [3].

References

[1] Sato O, Iyoda T, Fujishima A and Hashimoto K 1996 *Science* **272** 704.
[2] Tokoro H, Ohkoshi S -L and Hashimoto K 2003 *Appl. Phys. Lett.* **82** 1245.
[3] Vertelman E J M, Lummen T T A, Meetsma A, Bouwkamp M W, Molnar G, van Loosdrecht P H M and van Koningsbruggen P J 2008 *Chem. Mater.* **20** 1236-38.
[4] Bartlett R J and Musial M 2007 *Rev. Mod. Phys.* **79** 291-352 and references cited therein.
[5] Grimme S 2006 *J. Chem. Phys.* **124** 034108.
[6] Ahlrichs R and May K *Phys. Chem. Chem. Phys.* **2000**, *2*, 943.
[7] Petersson G A and Al-Laham M A 1991 *J. Chem. Phys.* **94** 6081-90 and references cited therein.
[8] Fuentealba P, von Szentpaly L, Preuss H and Stoll H 1985 *J. Phys. B: At. Mol. Phys.* **18**, 1287.
[9] a) Dyall K G 2001 *J. Chem. Phys.* **115** 9136-43 b) Dyall K G 2002 *J. Comput. Chem.* *23* 786-93.
[10] Dyall K G 1997 *J. Chem. Phys.* **106** 9618-26.
[11] Filatov M and Dyall K G 2007 *Theor. Chem. Acc.* **117** 333-38.
[12] Filatov M 2007 *J. Chem. Phys.* **127** 084101.
[13] Kurian R and Filatov M 2008 *J. Chem. Theory. Comput.* **4** 278-85.
[14] Kurian R and Filatov M 2009 *J. Chem. Phys.* **130** 124121.
[15] Filatov M and Kurian R *(manuscript in preparation)*.
[16] a) Visscher L and Dyall KG 1997 *At. Data Nuc. Data Tables* **67**, 207. b) Visser O, Aerts PJ C, Hegarty D and Nieuwpoort W C 1987 *Chem. Phys. Lett.* **134**, 34-38.
[17] Neese F 2002 *Inorg. Chim. Acta* **337** 181-92.
[18] Nieuwpoort W C, Post D and van Duijnen P Th 1978 *Phys. Rev. B* **17** 91-98.

Journal of Physics: Conference Series **217** (2010) 012013 doi:10.1088/1742-6596/217/1/012013

T-MOKE for nuclear resonant reflectivity

M.A. Andreeva

Faculty of Physics, M.V. Lomonosov Moscow State University, Leninskie Gory, 119991, Moscow, Russia

marina@trtk.ru

Abstract. It has been shown that transverse magneto-optical Kerr effect (T-MOKE) can be essentially enhanced in multilayers with antiferromagnetic interlayer coupling and observable even at grazing angles. The description of this effects need the application of the exact theory of reflectivity. The performed calculations show the presence of the antiferromagnetic Bragg maximum on the delayed reflectivity curve for transverse geometry which is absent in the used at present theory of Mössbauer reflectivity. The energy or time spectra of reflectivity, calculated on the basis of the exact and approximate theory, also reveal the essential difference.

1. Introduction

Mössbauer reflectivity is observed at grazing angles because for wavelengths ~ 0.1 nm the medium susceptibility $\hat{\chi}$ is very small ~ 10^{-5}. So the general theory of reflectivity [1-2] had been simplified for this case [3-4]. This approximation was equivalent to the approach used in the further investigations of the synchrotron Mössbauer reflectivity [5-6]. However, sometimes the exact theory of reflectivity can give the different results from the used approximation. Here we show that even for Mössbauer wavelengths and grazing angles some new effects can be predicted on the basis of the general theory of reflectivity.

2. Exact theory

Transverse magneto-optical Kerr effect (T-MOKE) is characterized by the difference in reflectivity for π-polarized light from the sample, magnetized perpendicular to the scattering plane, when the direction of the magnetization changes sign. It has maximum value for the angles θ close to 45°, so for Mössbauer reflectivity at grazing angles it is supposed negligible.

The origin of T-MOKE is the specific relationship between the tangential components of the electric $\mathbf{E}_t$ and magnetic $\mathbf{H}_t$ fields of radiation. In the used at present theory of the nuclear resonant reflectivity [3-6] the tangential components of the radiation field in the scattering plane at grazing angles are ignored, thus T-MOKE could not appear.

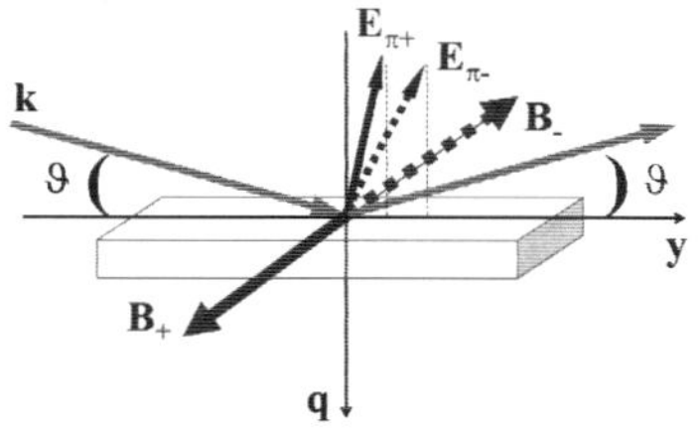

Figure 1. For the explanation of the T-MOKE. The tangential component of the π-polarized electric field of radiation in the sample are different for two opposite directions of the magnetization.

In the correct theory the exact 4x4-propagation matrices [1-2] should be used, for description of the depth evolution of the tangential components $\mathbf{H}_t = \mathbf{I}\,\mathbf{H}$ and $\mathbf{q}^{\times}\mathbf{E}$:

$$\hat{M} = \begin{pmatrix} \dfrac{\mathbf{q}^{\times}\hat{\chi}\mathbf{q}\circ\mathbf{x}\cos\theta}{1+\mathbf{q}\hat{\chi}\mathbf{q}} & \dfrac{\mathbf{I}\,\tilde{\hat{\varepsilon}}\,\tilde{\hat{\varepsilon}}^{-1}\mathbf{I}}{1+\mathbf{q}\hat{\chi}\mathbf{q}} - \cos^2\theta\,\mathbf{y}\circ\mathbf{y} \\ \mathbf{I} - \dfrac{\cos^2\theta\,\mathbf{x}\circ\mathbf{x}}{1+\mathbf{q}\hat{\chi}\mathbf{q}} & \dfrac{-\mathbf{x}\circ\mathbf{q}\hat{\chi}\mathbf{q}^{\times}\cos\theta}{1+\mathbf{q}\hat{\chi}\mathbf{q}} \end{pmatrix}, \tag{1}$$

where $\mathbf{x},\mathbf{y},\mathbf{q}$ is the unit vector basis (Fig. 1), a symbol $\circ$ means the outer product of vectors, θ is the glancing angle, $\mathbf{I} = 1 - \mathbf{q}\circ\mathbf{q}$, $\hat{\varepsilon} = 1 + \hat{\chi}$, $,\tilde{\hat{\varepsilon}}$ means a transposed tensor, $\mathbf{q}^{\times}$ is a dual tensor, i.e. $\mathbf{q}^{\times}\mathbf{a} = \mathbf{q}\times\mathbf{a}$, which is a vector product. For dipole transitions different hyperfine transitions with the change of the magnetic quantum number $\Delta m = \pm 1, 0$ have a specific contribution to the resonant susceptibility $\sim \mathbf{h}_{\Delta m}\circ\mathbf{h}^*_{\Delta m}$ ($\mathbf{h}_{\Delta m}$ are the spherical orts of the hyperfine field reference system), so the permittivity tensor has a form:

$$\hat{\varepsilon} = 1 + \hat{\chi} = 1 + \chi^{el} + A_0\,\mathbf{h}_0\circ\mathbf{h}^*_0 + A_{+1}\,\mathbf{h}_{+1}\circ\mathbf{h}^*_{+1} + A_{-1}\,\mathbf{h}_{-1}\circ\mathbf{h}^*_{-1}. \tag{2}$$

When the hyperfine magnetic field directions are perpendicular to the scattering plane (transverse geometry) we have (in $\mathbf{x},\mathbf{y},\mathbf{q}$ basis):

$$\hat{\varepsilon} = 1 + \hat{\chi} = \begin{pmatrix} 1+A+C & 0 & 0 \\ 0 & 1+A & iB \\ 0 & -iB & 1+A \end{pmatrix}, \tag{3}$$

where the notations are inserted

$$A = \chi^{el} + \frac{A_{+1}+A_{-1}}{2}, \quad C = A_0 - \frac{A_{+1}+A_{-1}}{2}, \quad B = \frac{A_{+1}-A_{-1}}{2}. \tag{4}$$

The propagation matrix (1) for the transformation of the vector-column $\{H_x, H_y, -E_y, E_x\}$ takes a form

$$\hat{M} = \begin{pmatrix} -\dfrac{iB\cos\theta}{1+A} & 0 & B_{11} = 1 + A - \dfrac{B^2}{1+A} & 0 \\ 0 & 0 & 0 & \sin^2\theta + A + C \\ \dfrac{\sin^2\theta + A}{1+A} & 0 & \dfrac{iB\cos\theta}{1+A} & 0 \\ 0 & 1 & 0 & 0 \end{pmatrix}, \tag{5}$$

In this geometry σ- and π–polarizations become the eigen polarizations of the exact propagation matrix (5), as well as that of the approximate propagation matrix. The normal component of the wave vector k_z^{π} (in ω/c units) for π–polarisation equals to

$$k_z^{\pi} \cong \sqrt{\sin^2\theta + A} = \sqrt{\sin^2\theta + \chi^{el} + \frac{A_{+1}+A_{-1}}{2}} = \sqrt{\sin^2\theta + \chi_{33}}, \tag{6}$$

which is the same as that obtained from the approximate theory. If the transverse external field changes sign than $A_+ \to A_-$ and $A_- \to A_+$ and (6) is not changed. However, for calculation of reflectivity we should solve the boundary task which for π–polarization takes a form

$$\begin{cases} H_x^0 + H_x^R = H_x^T, \\ \sin\theta(-H_x^0 + H_x^R) = \gamma^{\pi} H_x^T \end{cases}, \tag{7}$$

and one needs the exact relation γ^{π} between $\mathbf{E}_t$ and $\mathbf{H}_t$:

$$E_y = \gamma^{\pi} H_x \cong -\left(\frac{k_z^{\pi}}{1+A} \pm i\frac{(A_{+1} - A_{-1})}{2}\cos\theta\right) H_x \,. \tag{8}$$

In approximate theory we have only

$$E_y = -k_z^{\pi} H_x \,. \tag{9}$$

Comparing (8) and (9), we see that in the exact theory the relationship between $\mathbf{E}_t$ and $\mathbf{H}_t$ contains the additional term, which changes sign when the direction of the hyperfine field changes sign. That is the origin of the T-MOKE effect. But the absolute value of the additional term is quite small (the first term in (8) ~ $\sqrt{\chi}$ ~10^{-3}, the second term ~ χ ~10^{-5}), so in the most cases it can be neglected for Mössbauer reflectivity. Notice also that the resonant 14.4 keV transition in ^{57}Fe is M1 type and we should determine the polarization with respect to the magnetic field of radiation, so in (8,9) we should change $\mathbf{H} \leftrightarrow \mathbf{E}$. It occurs also that "$\pi$–polarization" is exactly the polarization of the synchrotron beam.

3. Some example calculations

Since T-MOKE is a consequence of the boundary conditions (the continuity of $\mathbf{E}_t$ and $\mathbf{H}_t$) we can enhance the effect at grazing angles by considering the model with many boundaries. For calculations we have chosen the periodic multilayer with antiferromagnetic interlayer coupling $[^{57}Fe\uparrow(6\,Å)/V(6\,Å)/^{57}Fe\downarrow(6\,Å)/V(6\,Å)]_{30}/Si$ (Fig. 2). Hyperfine magnetic field in ^{57}Fe layers (B_{hf}=33 T with broadening 1T) has been put perpendicular to the scattering plane and antiferromagnetically aligned in sequent ^{57}Fe layers.

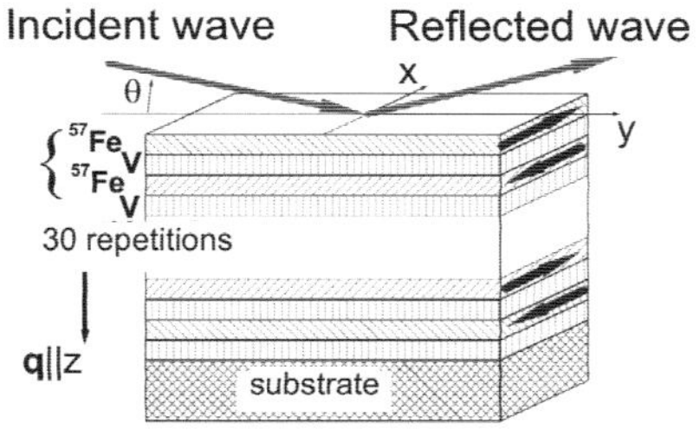

Figure 2. Model structure.

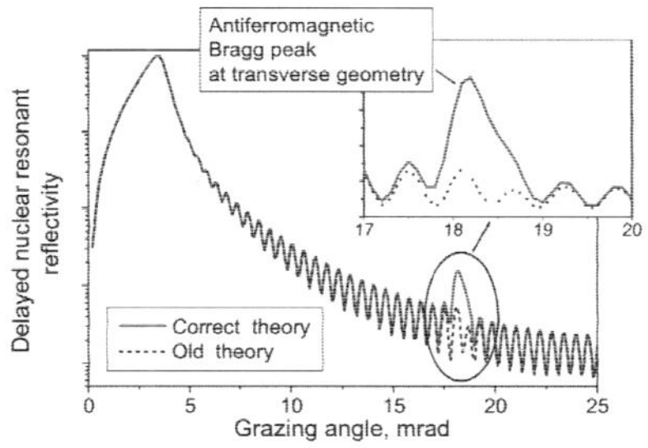

Figure. 3. Delayed nuclear resonant reflectivity, calculated on the basis of the correct and approximate theory

Reflections from the two ^{57}Fe layers in one magnetic period, magnetized perpendicular to the scattering plane but in opposite directions, are different due to T-MOKE, so the half-order magnetic Bragg reflection appears in the delayed nuclear resonant reflectivity curve at ~ 18 mrad (Fig. 3).

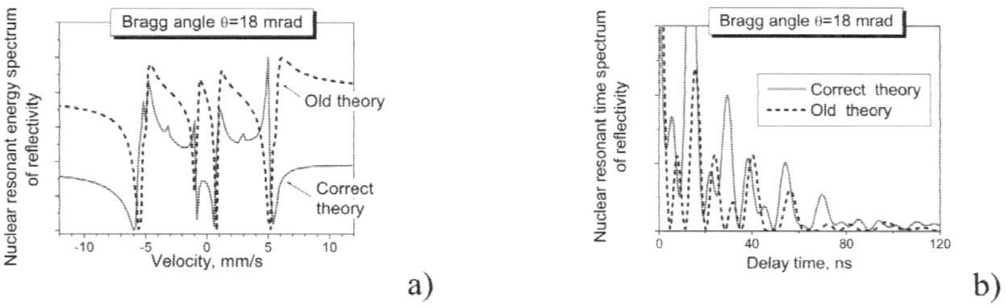

Figure 4. Mössbauer (a) and time (b) spectra of reflectivity at the antiferromagnetic Bragg maximum in transverse geometry, calculated on the basis of the exact and approximate theory.

Notice that calculations on the basis of the approximate theory are perfectly coincided with the results of the exact theory with exception of the small angular region in vicinity of the antiferromagnetic Bragg peak. The energy or time reflectivity spectra for the magnetic Bragg peak, calculated on the basis of the correct theory, are different from that calculated by the used at present theory (Fig. 4).

It is known that in longitudinal geometry when the hyperfine magnetic fields are aligned in the scattering plane the antiferromagnetic Bragg maximum perfectly described on the basis of the approximate theory and influence of the additional term in (8) disappears. However, at the other azimuth angles the exact theory can give the different results from the approximate theory. The most noticeable difference is presented on the shape of the energy or time spectra (Fig. 5). For this geometry σ- and π- polarizations are not the eigen polarizations so the calculations need the application of the general propagation matrix [1-2].

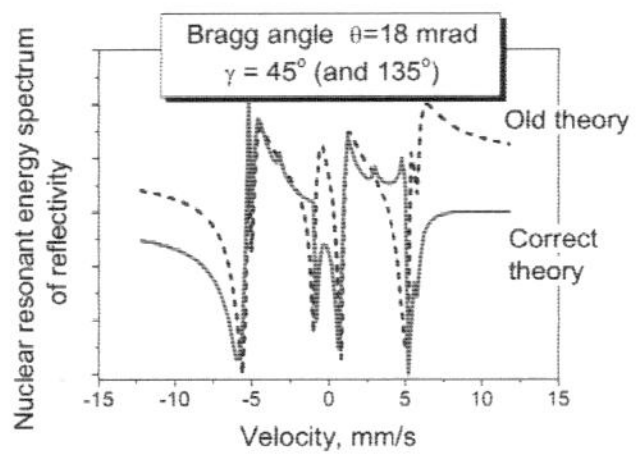

Figure 5. Mössbauer spectra of reflectivity at the antiferromagnetic Bragg maximum at the azimuth angle 45°, calculated on the basis of the exact and approximate theory.

5. Conclusions

The magnetic/nonmagnetic superlattices with antiferromagnetic interlayer coupling play an important role in modern technology. The detailed investigations of the peculiarities of the magnetic ordering in such systems are very interesting from the physical point of view. Resonant reflectivity can supply us with the very specific information. So the correct interpretation of the reflectivity data is desirable. Here we show that the exact theory of reflectivity leads sometimes to the essential difference in the shapes of the energy or time spectra of reflectivity in vicinity of the antiferromagnetic Bragg peak, existence of which has been here predicted by the exact theory of reflectivity in the transverse geometry.

We hope that this new effect will be observed for the nuclear resonant reflectivity in the nearest future. Notice that the antiferromagnetic Bragg reflections in T-MOKE geometry were successfully used for resonant X-rays (e.g. near L_3 edge of Ni in superlattice $[Ag(1.1\ nm)/Ni(1.75\ nm)]_n$ [7]).

References

[1] Azzam R M A and Bashara N M 1977 *Ellipsometry and polarized light* (North-Holland publishing company)
[2] Borzdov G N, Barkovskii L M and Lavrukovich V I 1976 *Zhurnal Prikladnoi Spectrosk.* **25** 526
[3] Andreeva M A and Rosete K 1986 *Vestnik Moscovskogo Universiteta, Fizika* **41** 65
Andreeva M A and Rosete K 1986 *Poverkhnost'* (In Russian) No.**9** 145
[4] Irkaev S M, Andreeva M A, Semenov V G, Belozerskii G N and Grishin O V 1993 *Nuclear Instrum. and Methods in Phys. Res. B* **74** 554
[5] Deák L, Bottyán L, Nagy D L and Spiering H 1996 *Phys. Rev.* B **53** 6158
[6] Röhlsberger R 1999 *Hyperfine Interactions* **123/124** 301
[7] Tonnerre J M, Sève L, Raoux D, Soullié G, Rodmacq B, and Wolfers P 1995 *Phys. Rev. Lett.* **75** 740

Acknowledgments

The work is supported by RFBR No. 09-02-01293_a, 09-02-12207-ofi_m.

IMSG09: A ^{57}Fe-^{119}Sn Mössbauer spectra computer fitting program with novel interactive user interface

A P Douvalis, A Polymeros and T Bakas

Department of Physics, University of Ioannina, 45110 Ioannina, Greece

adouval@cc.uoi.gr

Abstract. A recently developed computer based ^{57}Fe-^{119}Sn Mössbauer spectra fitting program is described. The program uses full Hamiltonian matrices to generate theoretical Mössbauer spectral contributions with which fits the experimental data based on a least-squares minimization procedure. The user of the program can exploit its interactive user interface to built and modify on-the-fly a starting theoretical model of proposed contributions without the need to write or alter the initial Mössbauer parameter values to or from any file. Changes in the Mössbauer parameters values for the proposed model can be made on the screen and the theoretical contributions can be displayed during those changes in real time together with the experimental spectrum in order to give the user the ability to approximate with the best and fast way the desired solution before launching the minimization procedure. Single as well as multiple sub-component contributions with Gaussian-type spreading of quadrupole splitting/shift or hyperfine magnetic field values can be realized to fit transmission, conversion electron or X-ray Mössbauer spectra of randomly or non-randomly oriented powder or single crystal samples. The program is currently written for 3/2→1/2 nuclear spin transitions but can be modified to include higher spins, as well as distributions of other types and of other Mössbauer parameters values.

1. Introduction

Mössbauer spectroscopy throughout its 50 years history has been established as a unique method for the investigation of the physical and chemical properties of materials. From its birth, this technique had gone through important developments both theoretically and experimentally [1]. One of the most important tools which follow these developments is the computer based programs utilized to fit Mössbauer spectra (MS) and determine quantitatively the Mössbauer parameters (MPs) in order to lead to the best interpretation of the experimental results.

In the past many useful programs based on several models of reproducing theoretical MS contributions have been released. Regardless of the specific program, its user has to propose a set of contributions with initial MPs, which will be used as the starting theoretical model to fit a given experimental spectrum, following usually a least-squares minimization procedure. It is well known that the success of this venture and the extraction of the best set of MPs with physical significance carrying important information, depend on the complexity of the given MS and on the proximity of the initial proposed model to the pursued solution [2]. However, in order to approximate this solution, the procedure of building the initial theoretical model in most of the programs requires to write, usually to a file, a group of starting MPs values for the set of proposed contributions, and then make several test runs by changing these values before the least-square minimization procedure is launched. Especially in complex spectra with many different contributions, this procedure is quite difficult, time-consuming and it doesn't always leads to the desired approximation to the experimental data.

In this work we present, for the first time, a recently developed computer program, IMSG09 (Ioannina Mössbauer Spectroscopy Group 2009), which fits MS with a least-squares minimization procedure. Among other characteristics, its novelty is based on its interactive-user-interface, which gives the ability to build and modify on-the-fly a starting theoretical model with a very quick and easy way, in order to produce the best first approximation to the experimental spectrum, without writing initial values to any file.

2. Structure and main procedures of the program

IMSG09 is written in Delphi and an outline of the structure of its main components is given schematically in figure 1. It combines a number of visual (displayed on screen) and non-visual classes, which are properly connected to provide the final optical and numerical result of the fitting. An example of a snapshot of such a result after the successful termination of the minimization procedure for the ^{57}Fe Mössbauer spectrum of bulk Fe_3BO_5 at 4.2 K [3] is shown in figure 2.

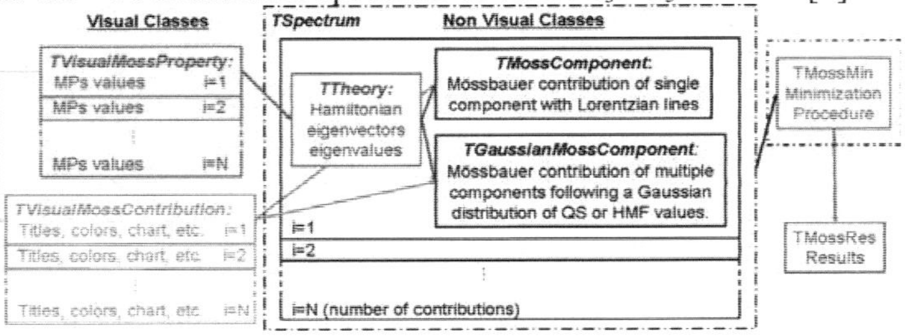

Figure 1. Structure of the main classes of the IMSG09 program.

The user can import an experimental spectrum as a sequence of numbers in a column representing the counts in each channel (the maximum channel number is user-defined), collected from a Mössbauer spectrometer operated in multi-channel scaling mode, accompanied by the calibration constants for the particular spectrum. The TVisualMossProperty class is used to insert the proposed MPs from which the theoretical contributions are produced in order to build a model to fit the given spectrum. The theoretical points of each contribution, which are displayed on the screen together with the points of the experimental spectrum, are always calculated using full Hamiltonian matrices. At present, these matrices correspond to the ground and excited nuclear states with nuclear spins of $I_g=1/2$ and $I_e=3/2$ respectively and are expressed following the formalism of Kunding [4] and Hoy & Chandra [5]. This means that they can generate theoretical MS for the cases of ^{57}Fe and ^{119}Sn. However the structure of the program allows higher spins to be added in following versions.

The TTheory class uses the provided MPs to calculate the elements of the ground and excited Hamiltonian matrices, solves the Hermitian eigenvector-eigenvalue problem with the QR decomposition method, using the LAPACK 3.0 library [6] as realised in ALGLIB [7] and extracts the line positions of the theoretical contribution. The lines are generated as Lorentzians, but the program can be easily modified to include other types of functions in future versions. In this formalism [4, 5] the MPs entering the Hamiltonian matrices include: i) $QS=eV_{zz}Q/2$ in mms^{-1}, where e is the electron's charge, V_{zz} is the main component of the electric field gradient (EFG) tensor defining the z direction of the Kartesian system and Q is the nuclear electric quadrupole moment of the excited state, ii) $\eta=(V_{xx}-V_{yy})/V_{zz}$, the asymmetry parameter, iii) HMF in kG, the magnitude of the hyperfine magnetic field, iv) theta (θ) and phi (φ) in degrees, the polar and azimuthial angles defining the orientation of the HMF vector with respect to V_{zz} and v) alpha (α) and beta (β) in degrees, the polar and azimuthial angles defining the orientation of the γ-ray direction with respect to the V_{zz} direction. Apart from these, the global MPs of: vi) IS in mms^{-1}, the isomer shift, vii) HWHM in mms^{-1}, the half line-width ($\Gamma/2$) of the generated Lorentzian lines, viii) Int. in counts, the intensity of the far most left line of the generated multiplet, are also used to produce the sub-spectrum of the inserted contribution. The relative intensities of the resulting resonant lines are also determined by the TTheory. The user can choose between contributions corresponding to polycrystalline powder (arbitrary orientation), which is the default case, and single crystal or oriented powder samples by ticking on the corresponding box. In each case the relative intensities of the absorption lines are calculated through the appropriate eigenvectors, the Glebsch-Gordan coefficients and the vector spherical harmonic matrices [4, 5].

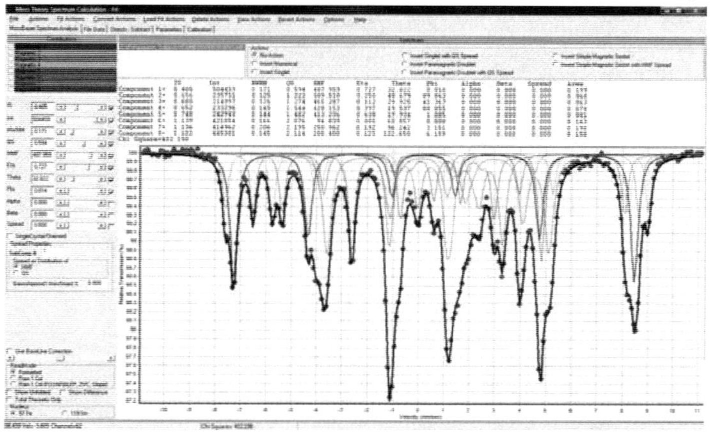

Figure 2. A snapshot of the user interface of the IMSG09 program.

The user can also choose if the contribution will be inserted as a single component (in this case the TMossComponent class is used), or if it will allow a spreading of HMF (for the magnetic contributions) or QS (for the paramagnetic or magnetic contributions) values around a central HMF_C or QS_C value. In the second case the TGaussianMossComponent class is used in order to produce the contribution which is composed of the superimposition of a set of N_C+1 interlinked sub-components (the number N_C in the set is user-defined with a current upper limit of 900 components). All sub-components in the set are generated using the Hamiltonian matrices through TTheory and posses the same $\Gamma/2$. The intensity I_i and the HMF_i/QS_i values for the i sub-component of the set are calculated as a function of the maximum intensity I_C of the set (parameter Int. of the contribution), which corresponds to the central HMF_C/QS_C value (parameter HMF/QS), and the spreading σ (parameter Spread), following a Gaussian distribution function, which in the case of HMF values spreading can be written as:

$I_i=I_C\exp\{-(HMF_i-HMF_C)^2/2\sigma^2\}$, where $HMF_i=HMF_C+S(2i-N_C)/2N_C$, $i=0, 1, \ldots, N_C$, $S=2\sigma[2\ln(1/r)]^{1/2}$

The user must choose the quantity (1-r), where r is the minimum to maximum intensity ratio $r=I_0/I_C$, for which the minimum (HMF_0) and maximum (HMF_{N_C}) values of the set will be calculated. In addition case-dependent boundary conditions eliminating the intensities of those generated sub-components of the set having $HMF_i<0$ also apply. The same distribution is also applied for QS values spreading, however with no boundary conditions in this case. Although currently only the Gaussian distribution is used for the calculation of the spreading of HMF/QS MPs values, the structure of the program allows any other distribution function to be added formally, which could also involve other MPs (like IS, η, θ etc.) as well.

While the experimental spectrum is displayed on the screen, the user can choose between two ways to insert a theoretical contribution: (i) insert a contribution by giving all the proper starting MPs numerically from the keyboard, or (ii) clicking with the mouse on the screen at the positions of those main resonant lines which he/she thinks that correspond to a certain contribution of the spectrum. In case (ii), the starting MPs of IS, QS, Int. and HMF are calculated and inserted automatically in the appropriate insertion MPs boxes using the velocities v_i (in mms^{-1}) detected from the 'clicked' positions on the screen (at the x-velocity axis) and the relations: a) $IS=(v_1+v_2)/2$, $QS=(v_2-v_1)$, with two (v_1, v_2) clicks in the case of paramagnetic doublet contributions (for a paramagnetic singlet only one click for v_1 is necessary, and then $IS=v_1$ and $QS=0$) and b) $IS=(v_1+v_2+v_3+v_4)/4$, $QS=(v_1-v_2-v_3+v_4)/4$ and $HMF=\{(v_4-v_1)\cdot(330/10.6248\ kG/mms^{-1})\}$ (for ^{57}Fe spectra) [8], with four (v_1, v_2, v_3, v_4) clicks at the outer lines of the sextet in the case of magnetic contributions; the Int. parameter in each case is calculated as the length in the y axis (in counts) between the first clicked position (for v_1) and the baseline of the spectrum. It is obvious that these values, calculated from the clicked positions, are good approximations to the expected corresponding real MPs of the contribution in the simple cases of having either diagonalized Hamiltonian matrices, or small perturbations in main diagonalized Hamiltonian matrices, i.e. when the electric quadrupole interaction can be taken as a small perturbation upon the main magnetic interaction [8]. In these cases, when the EFG is not considered to be axially symmetric ($\eta \neq 0$) and η, θ and φ cannot be uniquely determined, QS includes also η, $QS=\Delta E_Q=eV_{zz}Q\cdot(1+\eta^2/3)^{1/2}/2$ in the case of paramagnetic contributions and θ and φ as well in the case of magnetic contributions, $QS=2\varepsilon=eV_{zz}Q\cdot(3\cos^2\theta-1+\eta\sin^2\theta\cos2\varphi)/4$ [9]. QS then represents the combined parameters ΔE_Q or 2ε and the individual η, θ, φ values are inserted as equal to 0. Depending on the nature of the contribution to be inserted, the user can choose between a number of six different kinds displayed in the appropriate box, including paramagnetic singlets and doublets as also magnetic sextets, with or without QS or HMF spreading. For all other cases the user must choose the first numerical insertion method (i). Regardless of the insertion method, the displayed characteristics (title, color, chart etc.) of each contribution are controlled by the TVisualMossContribution class.

After insertion, all the MPs of the contribution (IS, Int., $\Gamma/2$, QS, HMF, η, θ, φ, α, β, σ) can be modified just by using the appropriate sliders provided for each parameter, or the arrows from the keyboard. For every change in its value through the movement of the slider or arrow, the contribution is recalculated by TTheory and redisplayed on the screen instantly. The limiting values of the slider, as well as other display and chart parameters are user-defined from a special window. The user can insert limitless number of contributions (practically restricted by the amount of the available memory of the system) in order to build the initial theoretical model to fit the spectrum. After insertion of all desired contributions and appropriate manipulation of their MPs targeting the best -user determined- initial approximation to the experimental data, the minimization procedure can be lunched. The user must define those MPs for each contribution which will be free to vary in the minimization procedure by checking their appropriate boxes. Currently all MPs are varied independent of each other, but correlations between MPs variation (within the same contribution or between different ones) are scheduled to be included in following versions of the program.

IMSG09 fits MS based on least squares (minimum χ^2) minimization procedure. The program currently follows the well known Levenberg-Marquardt algorithm [10] for the minimum χ^2 convergence, through the use of the LAPACK 3.0 library [6] as realized in ALGLIB [7], but it is designed to adapt other minimization methods too. It can be compiled to run in all major computer operating systems including Microsoft Windows™ (98, ME, 2000, XP, Vista), Unix-Linux, Macintosh™ and others.

3. Conclusions

IMSG09 introduces a new way to fit MS, using an interactive user interface to approximate on-the-fly the best proposed theoretical model to the experimental data before the final minimization procedure. We hope that the current inclusion of different cases of theoretical contributions, as well as the ability to expand in additional cases will give the 'Mössbauer community' a useful tool for MS analyses.

References

[1] Bandyopadhyay D 2006 *International Materials Reviews* **51** 171
[2] Longworth G 1984 *Mössbauer Spectroscopy Applied to Inorganic Chemistry* vol 1 ed G J Long (New York: Plenum Press) p 43
[3] Douvalis A P et al. 2002 *J. Phys.: Cond. Matter* **14** 3303
[4] Kündig W 1967 *Nucl. Instr. Meth.* **48** 219
[5] Hoy G R and Chandra S 1967 *J. Chem. Physics* **47** 961
[6] www.netlib.org/lapack
[7] www.alglib.net
[8] Greenwood N N and Gibb T C 1971 *Mössbauer Spectroscopy* (London:Chapman & Hall)
[9] Le Caer G et al. 1978 *Nucl. Instr. Meth.* **157** 127
[10] Marquardt D W 1963 *SIAM J. Appl. Math.* **11** 431

Journal of Physics: Conference Series **217** (2010) 012015 doi:10.1088/1742-6596/217/1/012015

Evaluation of the $\frac{f_s}{1+B}$ Factor in the Study of Dynamical Properties of Hyperfine Interactions

G Spina[1], E Pugliese[1], L Cianchi[2], F Del Giallo[2], M Lantieri[2] and P Moretti[2]

[1] INFM–CNISM and Dipartimento di Fisica, Università di Firenze, Via G. Sansone 1, 59100 Sesto Fiorentino (FI), Italy

[2] ISC–CNR, Via Madonna del Piano 10, 59100 Sesto Fiorentino (FI), Italy

E-mail: gabriele.spina@unifi.it

Abstract. The interpretation of Mössbauer spectra often requires the use of the Transmission Integral function (TI). It contains the quantity $\frac{f_s}{1+B}$, where f_s is the source Mössbauer factor and B depends on the background radiation rate and on the inhomogeneity of the target: both of them usually obtained by ancillary measurements and tests. We present a simple method, that avoids ancillary measurements in evaluating the quantity $\frac{f_s}{1+B}$. It is useful when the Mössbauer line-shape rapidly evolves in narrow temperature ranges: a real application for single molecule magnets spectra and an analytical example are reported, too

1. Introduction

Mössbauer spectroscopy is an excellent local probe for the study of magnetic solids containing iron. In particular, relevant information about dynamical parameters are obtained from the thermal evolution of the spectrum shape, which usually evolves, as a function of the temperature, from a set of well resolved lines to a single narrow collapsed structure. As a result of this evolution, the sample thickness, giving rise to a good signal/noise ratio at the bottom of the temperature region, will produce strongly saturated spectra at the top of the range. Consequently, in order to prevent erroneous evaluation of the temperature trend of the dynamical parameters, we should use the Transmission Integral (TI), whose expression is:

$$I(v,T) = F(v)\left\{1 - \overline{f_s^r}\int_{-\infty}^{+\infty} L(\omega - v, \Gamma_s)\left[1 - \exp\left(-\overline{t_a}\sigma(\omega, w(T))\right)\right]\mathrm{d}\omega\right\} \tag{1}$$

where $F(v)$ is the background function, L denotes the line shape of the source characterised by the line width Γ_s, t_a is the absorber effective thickness and $\sigma(\omega, w)$ is the Mössbauer absorption cross section expressed in terms of a set of temperature dependent physical parameters $w(T)$ determining the actual shape. Moreover, (1) contains the term $f_s^r = \frac{f_s}{1+B}$, i.e. the source Mössbauer factor f_s reduced by $(1+B)$ due to both background radiation rate and target inhomogeneity effects. The fitting procedure requires the knowledge of f_s^r, usually obtained by ancillary measurements and tests on the targets, as described in literature [1, 2]. On the contrary, the following method does not require additional measurements and consists of a preliminary

fitting procedure. It is useful for all those cases characterised by radical changes of the cross section belonging to small changes of the temperature and/or other external parameters.
The only approximation we use consists of the omission of the temperature dependence of the sample thickness t_a. The method can be applied in place of a simultaneous fit of the whole set of spectra and consists of a preliminary fitting procedure followed by independent fits of the single spectra. It allows to use straightforward computer code and procedures, limits the number of free parameters and reduces the correlation between t_a and $w(T)$ parameters.

2. Theory

Let us consider the TI expression of (1). It does not have Lorentzian shape for two reasons: firstly, from a physical point of view, the Mössbauer absorption cross section $\sigma(\omega, w)$ cannot be generally described as a sum of Lorentzian lines; secondly, the cross section appears as argument of an exponential function giving rise to distortion effects depending on the experimental t_a value. In the analysis of Mössbauer measurements, physical and experimental sources for the actual spectrum shapes are correlated to each other and erroneous evaluations of the physical parameters w arise.

For very low t_a values, the spectrum is not saturated, $I(v)$ and $\sigma(\omega, w)$ show identical line shape, apart from the values of the respective line–widths; consequently TI is a linear function of the $f_s^r t_a$ product but it is independent of the assumed f_s^r value. On the contrary, for real spectra, $I(v)$ separately depends on $f_s^r t_a$ product and f_s^r. Observing (1), it is easy to deduce that values for the $f_s^r t_a$ product obtained by a fitting procedure – as a function of assumed values for f_s^r – show a decreasing monotone trend, which passes through $\left(\overline{f_s^r}, \overline{f_s^r t_a}\right)$, representing the actual values. The trend depends both on the specific t_a and w values.

The *prefitting* procedure consists of suitable reproductions of each spectrum, or at least of its apparent area and peak intensities, by using a not-Lorentzian $\sigma(\omega, w)$ cross section shape, for a set of predefined f_s^r values. From these reproductions the trend of $f_s^r t_a$ vs. f_s^r will be determined for each spectrum.

The line-shapes depend on temperature, but if the T-range is fairly narrow, as we will suppose, $\overline{t_a}$ can be assumed to be constant. In this case, for small changes of f_s^r around $\overline{f_s^r}$, all the series of $(f_s^r, f_s^r t_a)_i$ points, where i defines the specific spectrum, can be fitted by straight lines

$$\left(f_s^r t_a\right)_i = \overline{f_s^r t_a} - \alpha_i \left(f_s^r - \overline{f_s^r}\right) \tag{2}$$

which obviously pass through the common point $\left(\overline{f_s^r}, \overline{f_s^r t_a}\right)$.

Now, If we report the zero intercepts obtained for the various spectra as a function of the angular coefficient α_i, all the points lie on a new straight line, whose angular coefficient and zero intercept are respectively $\overline{f_s^r}$ and $\overline{f_s^r t_a}$. The obtained estimates for t_a and $f_s^r = \frac{f_s}{1+B}$ will be then used in independent fitting procedures of each spectrum.

3. An Experimental Example

We will describe an application of the method for the heterometallic $Cr_7Fe(II)$ wheel. Six Mössbauer spectra were collected between 1.8 and 14.8 K. They show the hyperfine structure to completely collapse in this narrow range of temperature, where t_a can be reasonably assumed to be constant. Each spectrum was fitted by assuming a set of defined values for f_s^r in order to obtain corresponding values of $f_s^r t_a$. The strong correlation between sample thickness and dynamical parameters, affecting the sample absorption cross section, gives rise to very good reproductions of the spectra independently of the assumed f_s^r values. Figure 1 shows results along with the fit of corresponding linear trends for the lowest temperature and the highest one. The final step of the *prefitting* procedure consists in plotting the Zero Intercept (ZI) values vs. the angular coefficient α of the respective interpolative line. We obtain the series

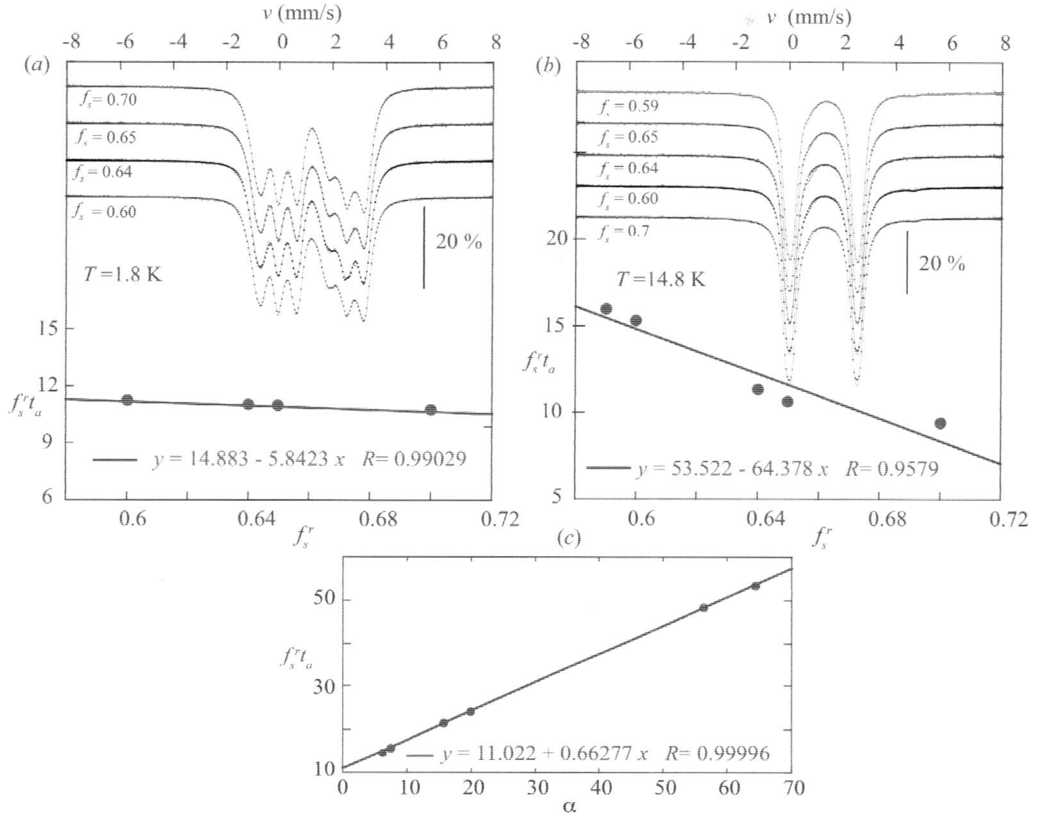

Figure 1. Mössbauer spectra of $Cr_7Fe(II)$ collected at the lowest (a) and the highest (b) temperatures and fitted for a series of f_s^r values. The obtained $f_s^r t_a$ values vs. f_s^r and their respective linear fits are displayed at the bottom of each spectrum. Zero Intercept values are reported in (c) as a function of corresponding angular coefficient α for the six spectra together with the linear fit that provides values for f_s^r and t_a.

of points reported in Figure 1 for the whole set of spectra along with the subsequent linear interpolation. As a result of the procedure, we have $f_s^r = \frac{f_s}{1+B} = 0.663(2)$ and $t_a = 16.6(1)$, which corresponds to $f_a = 0.78(1)$ calculated applying the well–known formula $t_a = n_a \sigma_0 f_a$[1] and in agreement with the typical values for magnetic molecules. The values will then be used to interpret the Mössbauer spectra in terms of dynamical interactions affecting the hyperfine fields of the heterometallic $Cr_7Fe(II)$ wheel.

4. An Analytical Example

A simple situation that permits analytical evaluations is represented by a cross section described by a single Lorentzian function, whose line width Γ_a is proportional to the natural one: $\Gamma_a(T) = \beta(T)\, 2\Gamma_n$, where $\beta(T)$ depends on the temperature monotonically.

The Mössbauer cross section is given by $\sigma(\omega, \beta) = \frac{\beta \Gamma_n^2}{\omega^2 + \beta^2 \Gamma_n^2}$ and has its maximum value equal to 1 for $\beta = 1$. Adapting standard evaluation procedures [3, 4] to the previous equation, we obtain the following expressions of spectrum area and peak intensity:

$$A\left(f_s^r,\, t_a,\, \beta\right) = \pi\,\Gamma_n\, f_s^r t_a \,\exp\left(-\frac{t_a}{2\beta}\right)\left\{I_0\left(\frac{t_a}{2\beta}\right) + I_1\left(\frac{t_a}{2\beta}\right)\right\} \tag{3}$$

$$\varepsilon(0;\, f_s^r,\, t_a,\, \beta) = f_s^r\left\{1 - \exp\left(-\frac{t_a}{2\beta}\right)\left[I_0\left(\frac{t_a}{2\beta}\right) + 2\sum_{n=1}^{\infty}\left(\frac{1-\beta}{1+\beta}\right)^n I_n\left(\frac{t_a}{2\beta}\right)\right]\right\} \tag{4}$$

which reduce to the usual expressions [1] for $\beta \to 1$.

If the right value $\overline{f_s^r}$ is not used, the spectrum shape will be reproduced by incorrect values for t_a and β and the fitting procedure will give rise to erroneus evaluation of β. From the series development of $A\left(f_s^r,\, t_a,\, \beta\right) - A\left(\overline{f_s^r},\, \overline{t_a},\, \overline{\beta}\right)$ around $\left(\overline{f_s^r},\, \overline{t_a},\, \overline{\beta}\right)$ we have:

$$I_0\left(\frac{\overline{t_a}}{2\overline{\beta}}\right)\left(f_s^r t_a - \overline{f_s^r t_a}\right) + \overline{t_a}\, I_1\left(\frac{\overline{t_a}}{2\overline{\beta}}\right)\left(f_s^r - \overline{f_s^r}\right) + \frac{\overline{f_s^r t_a}}{\overline{\beta}}\, I_1\left(\frac{\overline{t_a}}{2\overline{\beta}}\right)\left(\beta - \overline{\beta}\right) \simeq \frac{\exp\left(\frac{\overline{t_a}}{2\overline{\beta}}\right)}{\pi\,\Gamma_n}\,\sigma_{f_s^r t_a} \tag{5}$$

where $\sigma_{f_s^r t_a}$ is the standard deviation for the $f_s^r t_a$ product and the factor term $\frac{\exp\left(\frac{\overline{t_a}}{2\overline{\beta}}\right)}{\pi\,\Gamma_n}$ is connected with the experimental sensitivity, which exponentially decreases with $\frac{1}{\beta}$ and gives rise to point dispertion shown in Figure 1 for the spectrum at 14.8 K.

By assuming $\frac{\exp\left(\frac{\overline{t_a}}{2\overline{\beta}}\right)}{\pi\,\Gamma_n}\,\sigma_{f_s^r t_a} \simeq 0$, we obtain

$$f_s^r t_a \simeq \overline{f_s^r t_a} - \frac{\overline{t_a}}{\overline{\beta}}\frac{I_1\left(\frac{\overline{t_a}}{2\overline{\beta}}\right)}{I_0\left(\frac{\overline{t_a}}{2\overline{\beta}}\right)}\left(\overline{f_s^r}\left(\beta - \overline{\beta}\right) + \overline{\beta}\left(f_s^r - \overline{f_s^r}\right)\right). \tag{6}$$

Similarly, the peak value reproduction gives rise to the following second expression for $f_s^r t_a$

$$f_s^r t_a \simeq \overline{f_s^r t_a} + \frac{C_\beta(\overline{f_s^r},\, \overline{t_a},\, \overline{\beta})}{\overline{\beta}}\left(\beta - \overline{\beta}\right) + \frac{C_f(\overline{f_s^r},\, \overline{t_a},\, \overline{\beta})}{\overline{f_s^r}}\left(f_s^r - \overline{f_s^r}\right) \tag{7}$$

where $C_\beta(\overline{f_s^r},\, \overline{t_a},\, \overline{\beta})$ and $C_f(\overline{f_s^r},\, \overline{t_a},\, \overline{\beta})$ are algebrically evaluated coefficients. From (6) and (7) we obtain

$$\alpha_i = -\overline{t_a}\frac{I_1\left(\frac{\overline{t_a}}{2\overline{\beta}}\right)}{I_0\left(\frac{\overline{t_a}}{2\overline{\beta}}\right)}\left(1 - \frac{\overline{f_s^r t_a} I_1\left(\frac{\overline{t_a}}{2\overline{\beta}}\right) + C_f I_0\left(\frac{\overline{t_a}}{2\overline{\beta}}\right)}{\overline{f_s^r t_a} I_1\left(\frac{\overline{t_a}}{2\overline{\beta}}\right) + C_\beta I_0\left(\frac{\overline{t_a}}{2\overline{\beta}}\right)}\right) \tag{8}$$

for the α_i coefficient of (2) and an analytical expression for the error propagation on β due to erroneous estimations of f_s^r.

As numerical examples, by assuming $\overline{f_s^r t_a} = 5.5$, $\overline{f_s^r} = 0.66$ and using $\beta = 1.55$, $\beta = 5.0$ we obtain $\alpha_i \simeq 64$ and 5.8, respectively. These values are close to the experimentally derived ones for the 14.8 K and 1.8 K spectra. Moreover we respectively obtain $8.7\frac{(f_s^r - \overline{f_s^r})}{\overline{f_s^r}}$ and $0.8\frac{(f_s^r - \overline{f_s^r})}{\overline{f_s^r}}$ for the relative error on β.

The last results suggest that, in the Cr_7Fe(II) case, the relative systematic error on dynamical parameters is smaller than 5% and show that the described method prevents from erroneous evaluations of the temperature trend of electronic physical features.

References

[1] Kolk B 1984 Studies of dynamical properties of solids with the Mössbauer effect *Dynamical Properties of Solids* ed Horton G and Maradudin A (Amsterdam: North–Holland)

[2] Spina G, Cianchi L and DelGiallo F 2002 *Nucl. Instrum. Methods Phys. Res., Sect. B* **192** 415

[3] Capaccioli M, Cianchi L, DelGiallo F, Moretti P, Pieralli F and Spina G 1995 *Nucl. Instrum. Methods Phys. Res., Sect. B* **101** 280

[4] Chen Y L and Yang D P 2007 *Mössbauer Effect in Lattice Dynamics* (Weinheim: WILLEY-VCH)

Transparency of a thin absorber in Moessbauer optics: effect of electron relaxation

E K Sadykov[1], F G Vagizov[1,2], V V Arinin[1], O A Kocharovskaya[2]
[1] Kazan State University, Kazan 420008, Kremlevskaya 18, RUSSIA
[2] Physics Department, Texas A&M University, College Station, TX 77843, USA
E-mail: esad@ksu.ru

Abstract. A model of Moessbauer absorption taking into account the effects of the electron relaxation and the nuclear levels anticrossing simultaneously has been proposed. The dependence of the absorption deficit (transparency) on the relaxation and mixing parameters has been obtained using the stochastic theory of Moessbauer relaxation spectra. The role of quantum interference in occurrence of a partial transparency on Moessbauer transitions has been explored under these conditions.

1. Introduction

The problem of Moessbauer radiation transparency has took growing attention after the experiment [1] on the observation of the absorption deficit in $FeCO_3$ at 30.5 K under nuclear levels anticrossing conditions [2]. The theory of this phenomenon is discussed in references [3, 4, 5, 6] and recently a progress in understanding of possible mechanisms of the absorption deficit formation is achieved. As it was for the first time shown in [4], the electromagnetically induced transparency (EIT) [7] analog for Moessbauer radiation is possible if the crossing nuclear levels have different decay rates (see [8] also). In this report we consider this problem concerning to the case of thin absorber [5] in order to investigate conditions which lead to the realization of this mechanism. A stochastic relaxation model of the Moessbauer absorption [9] is used for this. One enables to determine the conditions leading to different relaxation widths of the crossing nuclear levels and establishes a quantitative relation between degree of the transparency and the electron relaxation parameters. It should be noted, this model provides an adequate explanation of the transparency observed earlier [1,4] and in some extent evolves a model proposed in [4].

2. The partial transparency at Moessbauer transitions due to electron relaxation

The main feature of the model used below is taking into account by the explicit way a fluctuation of the electron magnetic moment when the absorption cross section under a nuclear level anticrossing condition is calculated. The Hamiltonians of a nucleus in the ground (g) and excited (e) states have the form [9]:

$$\mathrm{H}^g = H^g f(t)\,, \quad \mathrm{H}^e = H^e f(t) + H^Q + H^{Q\perp}, \tag{1}$$

where $H^{g,e} = Z_{e,g}\hat{I}_z^{e,g}$ and $H^Q = Q[(\hat{I}_z^e)^2 - I^e(I^e+1)/3]$ are Zeeman and coaxial to them quadrupole interactions, respectively. $H^{Q\perp} = Q^{\perp}[(\hat{I}_+^e)^2 + (\hat{I}_-^e)^2]$ is quadrupole interaction,

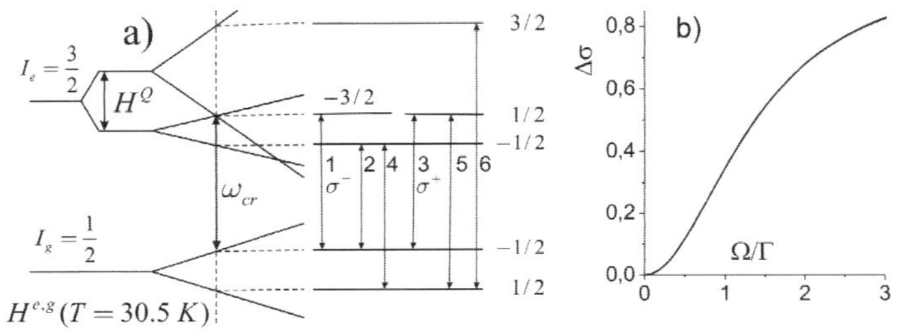

Figure 1. a) The Zeeman structure of Moessbauer transitions under levels crossing condition ($FeCO_3$). b) The absorption deficit $\Delta\sigma\,(p,\Omega)$ at crossing point, $\omega_\gamma = \omega_{cr}$; the dependence on Ω the same for polarizations $\sigma^{\mp}$ and for unpolarized photons. The longitudinal geometry and the values of ν_1, ν_2 as for Fig. 2 are used.

leading to a nuclear level mixing, where $Q^{\perp} = Q\eta/6$ and η is the asymmetry parameter. The crossing condition has the form $Z_e = Q$. The stochastic function $f(t)$ in Eq. (1) reflects stepwise changes in the hyperfine magnetic field due to the fluctuations of the electron spin. The function $f(t)$ in the simplest case has two values $a = \pm 1$ (two-level relaxation). The random process is characterized by the matrix of the stochastic transitions W_{ab}, where $W_{+-} = \nu_1$ is the probability of the transition from the state characterized by the value $f(t) = +1$ to the state corresponding to $f(t) = -1$. It should be noted, that $W_{++} = -W_{+-} = -\nu_1$, and $W_{-+} = \nu_2 \neq \nu_1$ in the general case. The cases $\nu_2 = \nu_1$ and $\nu_2 \neq \nu_1$ are called symmetric and asymmetric relaxation processes, respectively. According to Blume [9] the absorption cross section for gamma photons is given by the known expression:

$$\sigma\,(p) \sim \mathrm{Re} \sum_{MmM'm'\sigma,ab} p_a \left\langle m\left|k_1\sigma\right| M\right\rangle U^{ab}_{MmM'm'}\,(p) \left\langle M'\left|k_1\sigma\right| m'\right\rangle. \quad (2)$$

Here, $\langle M'\,|k_1\sigma|\,m'\rangle$ is the matrix element of the absorption of a photon with the wave vector k_1 and polarization σ (the transition of the nucleus from the ground state I_g, m' to the excited state I_e, M'). $p = -i(\omega_\gamma - \omega_0) + \Gamma/2$, where ω_γ is the energy of the photon and ω_0 and Γ are the energy and width of the Moessbauer level, respectively. $p_a = p(a)$ is the probability that the stochastic function $f(t)$ takes the value $a = \pm 1$. The condition $p\,(+)\,W_{+-} = p\,(-)\,W_{-+}$ is assumed. $U^{ab}_{MmM'm'}\,(p) = \left\{[pE - W - iL_1]^{-1}\right\}_{aMm,bM'm'}$ is the matrix of the Laplace transform of the evolution superoperator. This matrix has the dimension $(2I_g+1)(2I_e+1)2$, where a factor of 2 corresponds to the dimension of the stochastic space (± 1). The matrices W, E, and L_1 have the form:

$$W = \begin{pmatrix} -W_{ab}E' & W_{ab}E' \\ W_{ba}E' & -W_{ba}E' \end{pmatrix}, \quad E = \begin{pmatrix} E' & 0 \\ 0 & E' \end{pmatrix}, \quad L_1 = \begin{pmatrix} L^Q + L & 0 \\ 0 & L^Q - L \end{pmatrix}, \quad (3)$$

where E' is the identity matrix and L and L^Q are the Liouville operator matrices; all of these matrices have the dimension $(2I_g + 1)(2I_e + 1)$. For the conditions of anticrossing described above:

$$L_{MmM'm'} = H^e_{MM'}\delta_{mm'} - H^g_{m'm}\delta_{MM'}, L^Q_{MmM'm'} = \left(H^Q_{MM'} + H^{Q\perp}_{MM'}\right)\delta_{mm'}. \quad (4)$$

If $H^{Q\perp}$ is neglected, the matrix $[pE - W - iL_1]$, which should be inversed, consists of second-order block matrices Mm. The inversion of matrix Mm enables one to obtain the effect of the stepwise change in the hyperfine field on the line shape of the Moessbauer transition Mm. A small perturbation $H^{Q\perp}$ becomes significant when two transitions, Mm and $M'm'$, have the same energy $\left(L + L^Q\right)_{MmMm} = \left(L + L^Q\right)_{M'm'M'm'}$. Now we should to inverse a fourth-order matrix composed of second-order matrices Mm and $(M \pm 2)m$ joined with each other by the perturbation Ω; $L^{Q\perp}_{MmM'm'} = H^{Q\perp}_{MM'}\delta_{mm'} = 2\sqrt{3}Q^{\perp}\delta_{MM\pm2} = \Omega\delta_{MM\pm2}$. The expression (2) becomes as sum of the terms represented by a fourth-order matrices (MM'):

$$\sigma(p, \Omega) \sim \mathrm{Re} \sum_{(MM');\ M,M' \in (MM'), m, \sigma, ab} p(a) \langle m |k_1\sigma| M' \rangle U^{ab}_{M'mMm}(p) \langle M |k_1\sigma| m \rangle . \quad (5)$$

Using expression (5) it was shown [5] the outer lines of Zeeman sextet are broader than the inner lines $\Gamma_1 > \Gamma_3$ (see transitions presented in fig.1) for the asymmetric electron relaxation ($\nu_1 \neq \nu_2$). The expression (5) enables us to introduce an absorption deficit factor $\Delta\sigma(p, \Omega) = (\sigma(p, 0) - \sigma(p, \Omega))/\sigma(p, 0)$, which defines the degree of the transparency at the crossing point. It should be noted, the $\Delta\sigma$ factor depends on mixing parameter and on relaxation parameters ν_1, ν_2 (or the values Γ_1, Γ_3 corresponding to them). The dependence of $\Delta\sigma(p, \Omega)$ versus Ω at the crossing point, $\omega_\gamma = \omega_{cr}$, is shown in Fig. 1b. Let us discuss possible mechanisms leading to the absorption deficit.

If the widths of crossing levels are equal, $\Gamma_1 = \Gamma_3$ (it means $\nu_1 = \nu_2$), the deficit $\Delta\sigma(p, \Omega)$ is a consequence of Rabi splitting (the repulsion) of nuclear levels [2]. One can be called as Rabi deficit. A different result we have if $\Gamma_1 \neq \Gamma_3$. Now the decrease in the absorption cross section cannot be attributed only to Rabi effect, the explicit mechanism of deficit depends essentially on the mixing parameter value Ω. While $\Omega < \Omega_{sp} = (\Gamma_1 - \Gamma_3)/4$ the nuclear levels anticrossing leads to two mixed states with the same energy. The decay constants of these states vary from the values Γ_1 and Γ_3 to the common value $(\Gamma_1 + \Gamma_3)/2$ as Ω increases from 0 to Ω_{sp}. On the contrary, if $\Omega > \Omega_{sp}$, the mixed states have the same decay constants but their energies are different: $\omega_{cr} \pm \sqrt{\Omega^2 - (\Gamma_1 - \Gamma_3)^2/16}$. There is no Rabi splitting in first case, so the interference of the quantum amplitudes (QI) is responsible for the deficit in this case. If $\Omega > \Omega_{sp}$ two mechanisms (the Rabi and QI ones) contribute to the total absorption deficit. The last one is proportional to $\Gamma_1 - \Gamma_3$ and it vanishes when $\Gamma_1 = \Gamma_3$. The figure 2 illustrates aforesaid for the longitudinal geometry, $\vec{k} \parallel \vec{H}_z$ (Fig. 1a). In this case the transitions 1 and 3 have the pure polarizations σ^- and σ^+. The spectra $\sigma(p, \Omega)$ and $\Delta\sigma(p, 0)$ are calculated (thin and thick solid curves respectively) using (5) for $\Omega < \Omega_{sp}$ and $\Omega > \Omega_{sp}$. Let us note, in cases 2 a) and 2 b) the $\sigma(p, \Omega)$ (thin curve) consists of two interference parts (dotted lines). In both cases however the entire result of interference is destructive one. It results in decreasing of the absorption cross section in the vicinity $\omega_\gamma = \omega_{cr}$. Let us now pay attention to the cases 2 c) and 2 d). Now the thin curves are a sum of four terms. There are two Rabi contributions (dashed lines) and two interference terms (dotted lines) in each case. It is easy to see, the interference contribution is destructive in case 2 c) and constructive in case 2 d). In other words for the transition to the level with a larger width (absorption of the σ^- polarization photon) QI leads to the additional decrease of $\sigma(p, \Omega)$, or increase of the Rabi deficit. That is similar to phenomenon of EIT discussed firstly in [7]. For σ^+ photon the effect is opposite, now the interference contribution causes the decreasing of available value of Rabi deficit.

Separate contributions of two mechanisms to the absorption cross section $\sigma(p, \Omega)$ (presented by dashed and dotted lines in a Fig. 2) have been calculated by us using the method of the optical theorem. For this purpose the scattering amplitudes were constructed on the basis of nuclear spin states having different decay rates and mixed in anticrossing regime. Such calculations enable

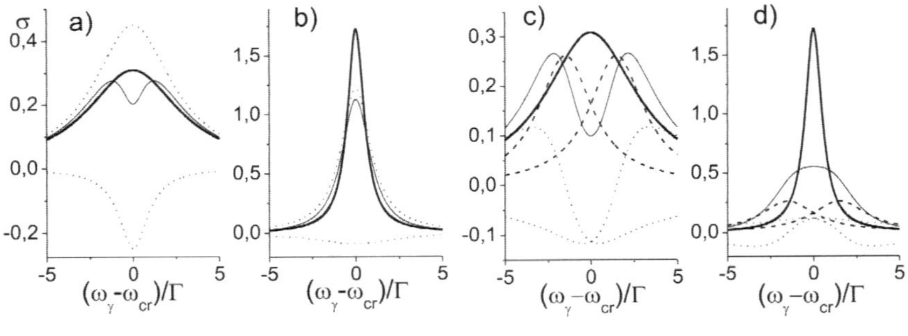

Figure 2. Cross section $\sigma(p,\Omega)$ in the vicinity $\omega_\gamma \simeq \omega_{cr}$. The values of ν_1, ν_2 used correspond to $\Gamma_1 = 6.49\Gamma$, $\Gamma_3 = 1.16\Gamma$. Thick lines represent $\sigma(p,0)$, thin lines - $\sigma(p,\Omega)$. The interference contributions to $\sigma(p,\Omega)$ are indicated by dotted lines, Rabi splitting terms - by dashed lines. The mixing parameters $\Omega = \Gamma < \Omega_{sp}$ (a,b) and $\Omega = 2\Gamma > \Omega_{sp}$ (c,d) are used. Longitudinal scheme of experiment is supposed, a) and c) correspond to polarization σ^-, b) and d) - to σ^+.

us to obtain simple analytical expressions for $\sigma(p,\Omega)$ in the case of longitudinal geometry. The mixing of levels having different decay rates was discussed earlier in [10] and partly in [4].

3. Concluding remarks

The asymmetric relaxation of the electron spin essentially changes the Moessbauer sublevels anticrossing mechamism. It leads also to the absorption deficit (partial transparency) of the Moessbauer spectrum at the crossing point due QI mechanism. Both of these are a result of the difference between the relaxation widths of the crossing nuclear levels. The absorption deficit model proposed recently [5] describes adequately the results of the experiments [1,4]. It serves also as a tool of the quantitative relation between the partial transparency on the Moessbauer transitions and the dynamical (relaxation) parameters of the system. In weak relaxation limit this model yields a results which are very close to a results obtained by methods [4,7] using an empirical widths Γ_1, Γ_3 for Moessbauer sublevels.

The same effect of partial transparency due QI is expected if the mixing of Moessbauer sublevels with different relaxation widths is induced by strong (RF and optical) electromagnetic field.

Acknowledgments

This work was supported in part by the RBRF (project no. 08-0200903-a) and the U.S. NSF.

References

[1] Coussement R, Rostovtsev Y and Odeurs J 2002 *Phys. Rev. Lett.* **89** 107601-1
[2] Coussement R, Put P, Scheveneels G and Hardeman F 1985 *Hyp. Int.* **23** 273
[3] Shakhmuratov R N, Odeurs J and Gheysen S 2005 *Appl. Phys.* B **81** 883
[4] Anisimov P, Vagizov F and Rostovtsev Y 2007 *J. Mod. Opt.* **54** 2595
[5] Sadykov E K, Arinin V V, Vagizov F G and Kocharovskaya O A 2008 *JETP Lett.* **88** 380
[6] Gheysen S and Odeurs J 2008 *J. Phys.: Condens. Matter* **20** 485214
[7] Scully M O and Zubairi M S 1997 *Quantum Optics* (Cambridge: Cambridge Univ. Press)
[8] Ok H N 1969 *Phys. Rev.* **185** 742
[9] Blume M 1968 *Phys. Rev.* **174** 351
[10] Rautian S G and Sobel'man I I 1961 *JETPh* **41** 456

Journal of Physics: Conference Series **217** (2010) 012017 doi:10.1088/1742-6596/217/1/012017

Mössbauer thermal scan study of a spin crossover system

P Mendoza Zélis[1,2], G A Pasquevich[1,2], F H Sánchez[1,2], A Veiga[1], M Ceolin[3], A F Cabrera[1,2], E Coronado-Miralles[4], M Monrabal-Capilla[4], J R Galan-Mascaros[4].
[1] Departamento de Física, FCE-UNLP, La Plata, Argentina
[2] Instituto de Física La Plata, CONICET,La Plata, Argentina
[3] Instituto de Investigaciones Físico-Químicas Teóricas y Aplicadas (UNLP-CONICET), La Plata, Argentina
[4] Instituto de Ciencias Moleculares, Universidad de Valencia, Valencia, España

E-mail: pmendoza@fisica.unlp.edu.ar

Abstract. Programmable Velocity equipment was used to perform a Mössbauer Thermal Scans to allow a quasi-continuous temperature study of the magnetic transition between the low-spin and a high-spin configurations in $[Fe(Htrz)_2(trz)](BF4)$ system. The material was studied both in bulk as in nanoparticles sample forms.

1. Introduction

Mössbauer Thermal Scans (MTS) were usually used to determine order temperatures [1,2]. Recently the technique was applied to obtain information on the temperature dependence of relative phase abundances, hyperfine and other spectrum parameters [3,4]. In these works different scans at one selected Doppler energy were done. The recent development of Programmable Velocity Scaler equipments [5], which allow to measure at arbitrary energies and for arbitrary times, highly improved the versatility of MTS techniques[6].

The experimental system developed in our laboratory allows the selection of Regions Of Interest (ROIs) in any spectral region, with a very high flexibility[7], while it undergoes changes due to the variation of an external parameter such as temperature, magnetic field, etc. The ROI can be composed of any number of points (up to 4096) with a freely selectable energy and live time for each point. We called the working mode of this equipment the Programmable Velocity Mode (PVM), to differentiate it from the conventional constant acceleration mode and constant velocity mode.

In this work we show that PVM can be used to perform Mössbauer Thermal Scans allowing a quasi-continuous temperature study of a thermally activated transition. In this case we study a spin crossover between a low-spin (LS) and a high-spin (HS) configuration.

Spin Crossover systems [8] are very interesting materials because of the ability to change their electronic state, between LS and HS configuration, under the influence of external stimuli (temperature, electromagnetic irradiation, pressure, etc.) being potential candidates e.g. for memory devices.

Among them, the $[Fe(Htrz)_n(trz)_{3-n}]X_m$ family (trz: triazole derivative) is a promising candidate for exhibiting bistability [9]. For the specific case of $[Fe(Htrz)_2(trz)](BF_4)$, the Fe(II) ions are coordinated by the ligands forming a 1-D polymeric structure exhibiting a cooperative spin transition from LS (S=0) to HS (S=2). Our study is performed with this material.

2. Experimental procedure

[Fe(Htrz)$_2$(trz)](BF$_4$) was prepared as bulk material and as 11 nm nanoparticles following the protocols described in reference [10] for compound 1a and using conventional reverse micelle techniques [11].

The experimental setup for a MTS consists of a conventional pulse height selection branch, a commercial driving system and two recently developed NIM modules serially interfaced with a computer. The first one is a previously introduced programmable constant-velocity scaler [5] that replaces the usual multiscaler. This module allows the independent acquisition of every spectrum channel based on a constant-velocity strategy and consequently the acquisition of partial Mössbauer spectra in selected energy regions. The second module is an analog input/output interface used to record the temperature and to command a linear DC power source that is used to modify the temperature of the sample. Both NIM modules communicate serially with the PC that runs a high-level technical computing language that can interact directly with hardware. A previously developed interactive environment for data visualization and data analysis was used in the workstation. In this way the programmable scaler and the temperature control could be fully operated and re-programmed as required.

The MTS and the conventional Mössbauer spectroscopy were obtained with a 25 mCi ^{57}Co*Rh*. Both calibrations were performed recording a complete α-Fe spectrum at RT.

3. Mössbauer spectroscopy

The LS and HS states of iron(II) can clearly be distinguished by Mössbauer spectroscopy by means of its hyperfine parameters (see figure 1). Both states are represented by a quadrupolar interaction but with a very different hyperfine parameters, while the LS state is characterized by an isomer shift of 0.42 mm/s and quadrupolar splitting of 0.27 mm/s at 298K, the HS state is characterized by an isomer shift of 0.97mm/s and quadrupolar splitting of 2.3 mm/s at 400K. Then these states can be clearly differentiated in the Mössbauer spectrum. It was noted that at temperatures beyond transition about 10% of the spectral area corresponds to LS phase.

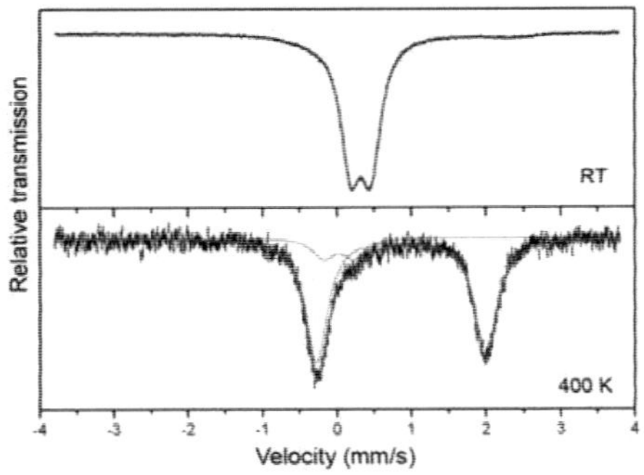

Figure 1. Mössbauer spectra of [Fe(Htrz)$_2$(trz)](BF$_4$) bulk sample. The LS and HS states of iron(II) are clearly distinguished. Each state is represented by a quadrupolar interaction.

4. Mössbauer Thermal Scans

We performed a Mössbauer thermal scan using a fixed ROI that was selected in order to include the patterns of the LS and HS phases, based on information previously obtained from Mössbauer Spectra taken at RT and 400 K. For the present case a ROI of 51 equally spaced points in the range from -1 to +3 mm/s, approximately centered at the isomer shift of the high temperature phase was chosen; and one additional point at an energy corresponding to background transmission (-8 mm/s) for control and normalization purposes was also incorporated. The ROI was periodically swept at 2 s/point while the sample was warmed up from RT to 411 K, and then cooled back to RT at rates of about ± 0.23 K/min. The whole experimental time was 16 h. Both scans of a [Fe(Htrz)$_2$(trz)](BF$_4$) bulk material absorber (6% effect at RT) are shown in figures 2-a and 2-b. This approach provided a detailed temperature and Doppler energy observation of the transition leading to a wealth of experimental information.

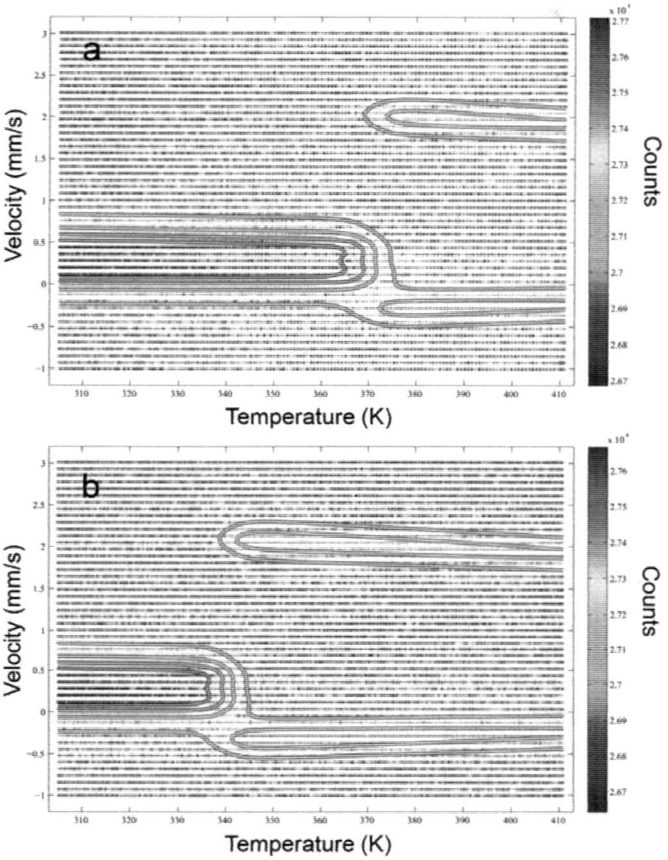

Figure 2. Mössbauer thermal scan (dots) for the $[Fe(Htrz)_2(trz)](BF_4)$ bulk material absorber using a fixed ROI (51 points) in the range from -1 to +3 mm/s and one additional point at an energy corresponding to background transmission (-8 mm/s) not shown in the figure. The ROI was periodically swept at 2 s/point while the sample was warmed up (a) to 411 K from RT, and then cooled back (b) to RT at rates of about $\pm$ 0.23 K/min. The continuous lines are the level curves of the fitted surface.

The whole set of MTS data was analyzed as an absorption surface $F(T,v)$ against velocity and temperature using an appropriate function. This consists of two quadrupolar interactions, each one with a fixed line width, but with temperature dependent isomer shift ($\delta(T) = \delta_0 + T\,\delta_1$), quadrupolar splitting ($\Delta Q(T) = \Delta Q_0\ (1+T^{3/2})$) and spectral areas. The spectra area temperature dependence was modeled considering the product of the Mössbauer–Lamb factor temperature dependence (assumed linear within the small scanned temperature interval) and the temperature evolution of the phases relative abundances $f(T)$.

The results obtained in figures 2-a and 2-b confirm the presence of thermal hysteresis in the transition. This hysteresis is due to the metastability of the high temperature configuration. For the time being we have decided to ignore metastability and treat the spin crossover transition within an equilibrium approach. A two level system with energy difference E and multiplicities 1 and Ω is proposed for the polymeric chain. The low spin configuration (S=0) has multiplicity 1, while the high spin one has multiplicity $\Omega = \omega^N$, where $\omega = 3x(2S+1) = 15$ (assuming that t_{2g} group is degenerate from where factor 3 arises[12]). N is the number of Fe atoms in the polymeric chain. Then

$$f_{HS}(T) = \frac{\Omega e^{(-E/T)}}{1+\Omega e^{(-E/T)}} \tag{1}$$

We define the transition temperature T_{05} through the condition $f\,(T_{05})=1/2$ and the temperature transition width ΔT by $f'\,(T_{05})\,\Delta T=1$. From these, it follows that

$$E = \frac{4(T_{05})^2}{\Delta T} \qquad \log(\Omega) = \frac{4T_{05}}{\Delta T} \tag{2}$$

Since the transition presents hysteresis, at least two different T_{05} values will be required to reproduce the experimental data. However, the E and Ω values on the heating up process are not very different from those of the cooling down one. Owing this, we may take them as representative of the physics of the problem. The fitting parameters corresponding to the temperature evolution of the two phases relative abundances are E_{up} (heating up) , E_{down} (cooling down) and Ω. The obtained values are: E_{up} = 4.1 eV, E_{down} = 4.5 eV , and Ω = $7\text{x}10^{60}$ corresponding to 371K (ΔT_{up}=10K) and 341K (ΔT_{down}=7K) for T_{05} in the heating up and cooling down respectively, and $N \sim 50$.

The experimental result and the level curves of the fitted surface corresponding to the warming and cooling process are shown in figure 2-a and 2-b. The level curves demonstrate the goodness of the fit.

In addition, in figure 3 we present temperature intersects of the previous figure at different temperature to show the agreement between experimental data and the fitted surface. In each graph are shown: the values measured in an interval of ±2K around the specified temperature (open blue circles), the measure average in this temperature interval (filled red points) and the intersect of the fitted surface (black line). In these intersects the presence at high temperature of the previously mentioned remnant LS phase fraction was clearly observed (see for example the 382 K intersect). As this fraction was not observed to be temperature dependent it was not considered in the analysis of the absorption surface $F(T,v)$ evolution.

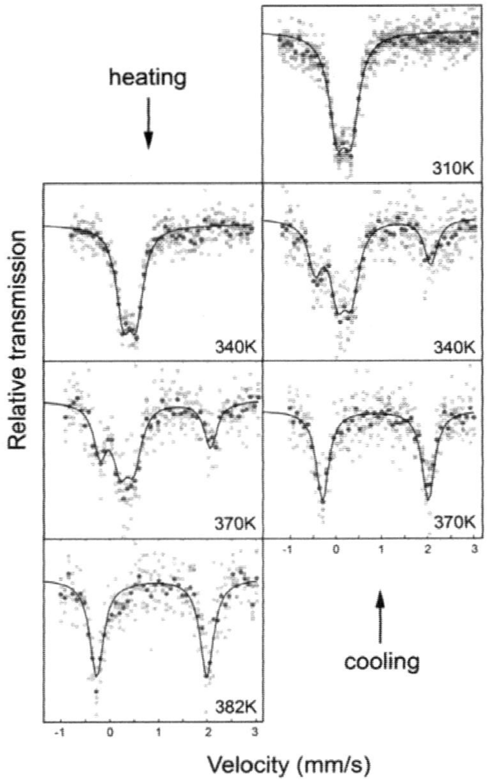

Figure 3. Temperature intersects of the MTS at different temperature. In each graph are shown the values measured in an interval of ±2K around the specified temperature (blue circles), the measure average in this temperature interval (red points) and the intersect of the fitted surface (black line).

The Mössbauer Spectra of the nanoparticles sample (not shown) indicate that the hyperfine parameters of both samples are similar (isomer shift: 0.44 mm/s, quadrupolar splitting: 0.35 mm/s at 298K for the LS phase and isomer shift: 1.0 mm/s, quadrupolar splitting: 2.6 mm/s at 411K for the HS phase). Mössbauer thermal scan of the nanoparticles warmed up to 415 K from RT is shown in figure 4 (the cooling back to RT is not shown). The whole experimental time was 80 hs. A 47 channels ROI plus background, and a time/channel ratio of 8 seconds were chosen. The data are analyzed as a absorption surface against velocity and temperature using the mentioned function. The level curves of the fitted surface corresponding to the warming process are shown in figure 4. The experimental result confirm the presence of magnetic bistability even in nanoparticles samples. The values of T_{05} determined in the fitting procedure are 372 K and 341 K corresponding to the heating up and cooling down respectively.

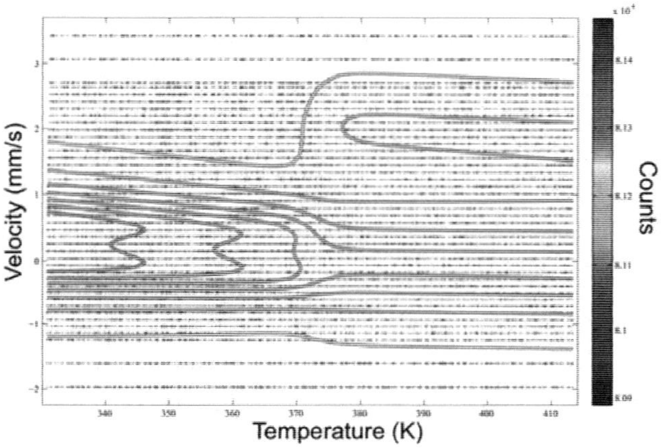

Figure 4. Mössbauer thermal scan of the nanoparticles warmed up from RT to 415K (dots). The continuous line are the level curves of the fitted surface.

5. Conclusions

Programmable Velocity Scaler equipment was used to perform Mössbauer Thermal Scans to follow up the magnetic transition through a quasi-continuous temperature study of the low-to-high spin transition of the [Fe(Htrz)2(trz)] (BF4) system. The study was made in bulk material and 11 nm nanoparticles. With information retrieved from conventional Mössbauer spectra, the ROI was chosen. Fitting the experimental data with an appropriate function, an absorption vs. energy⊗temperature surface was reconstructed. In this way a detailed quasi continuous observation of the spin thermal transition vs. temperature in terms of phases absorption spectral line intensities and positions was accomplished.

References

[1] Chien C L 1963 *Phys. Rev. B* **18** pp1003.
[2] Gonser U, Meechan C J, Muir C J, Wiedersich H 1963 *J. Appl. Phys.* **34** pp2373.
[3] Mendoza Zélis P, Pasquevich G A, Sánchez F H, Martínez N, Veiga A 2002 Phys. Lett. A 298 pp 55.
[4] Pasquevich G A, Mendoza Zélis P, Fernandez van Raap M B, Sánchez F H 2004 *Physica B* **31** pp369.
[5] Veiga A, Martínez N, Mendoza Zélis P, Pasquevich G A, Sánchez F H 2006 *Hyperfine Interact.* **167** pp905.
[6] Veiga A, Martínez N, Mendoza Zélis P, Pasquevich G A, Sánchez F H 2009 *Hyperfine Interact.* **188** pp137.
[7] Mendoza Zélis P, Pasquevich G A, Veiga A, Fernández van Raap M B and Sánchez F H 2009 *Hyperfine Interact.* **195** pp161.
[8] Gütlich P, van Koningsbruggen P J, Renz F 2004 *Struct. Bonding* **107** pp27.
[9] Haasnoot J G 2000 *Coord. Chem. Rev.*, **200** pp131.
[10] Krober J, Audiere J P, Claude R, Kahn O, Codjovi E, Haasnoot JG, Groliere F, Jay C, Bosseksou, A, Linares J, Varret F, Gonthier-Vasal A 1994 *Chem.Matter.* **6** pp1404.
[11] Coronado E, Galán-Mascarós J R, Monrabal-Capilla M, García-Martínez J, Pardo-Ibáñez P 2007 *Adv.Mater.* **19** pp1359.
[12] Koudriavtsev A B, Jameson R F, Linert W 2001 *The Law of Mass Action* (Berlin: Springer)

Journal of Physics: Conference Series **217** (2010) 012018 doi:10.1088/1742-6596/217/1/012018

Magnetic fluid: Comparative study of nanosized Fe_3O_4 and Fe_3O_4 suspended in Copaiba oil using Mössbauer spectroscopy with a high velocity resolution

M I Oshtrakh[1,7], A F R Rodriguez[2], V A Semionkin[1,3], J G Santos[4], O B Milder[1,5], L B Silveira[4], E M Marmolejo[4], M V Ushakov[1,3], M de Souza-Parise[6], P C Morais[6]

[1] Faculty of Physical Techniques and Devices for Quality Control, Ural State Technical University – UPI, Ekaterinburg, 620002, Russian Federation;
[2] Universidade Federal do Acre, Rio Branco AC 79900-000, Brazil;
[3] Faculty of Experimental Physics, Ural State Technical University – UPI, Ekaterinburg, 620002, Russian Federation;
[4] Fundação Universidade Federal de Rondònia, Departamento de Física, Ji-Paraná RO 78961-970, Brazil;
[5] Radio-Technical Department, Ural State Technical University – UPI, Ekaterinburg, 620002, Russian Federation;
[6] Universidade de Brasília, Instituto de Física, Núcleo de Física Aplicada, Brasília DF 70910-900, Brazil

E-mail: oshtrakh@mail.utnet.ru

Abstract. Comparative study of nanosized magnetite and magnetite suspended in Copaiba oil (biocompatible magnetic fluid) was made using Mössbauer spectroscopy with a high velocity resolution (spectra were measured in 4096 channels). The better fit of room temperature spectra was done using 15 sextets and 1 doublet employing different parameters while spectra measured at 90 K were better fitted using 15 sextets with different parameters. These component numbers were related to multi-domain structure and non-stoichiometry of magnetite. Observed differences of magnetic hyperfine fields and relative areas of spectral components for nanosized Fe_3O_4 and Fe_3O_4 suspended in Copaiba oil may be related to the effect of surface interactions of Fe_3O_4 and polar molecules of Copaiba oil.

1. Introduction

Biocompatible magnetic fluids are used and developed for instance as contrast agents for magnetic resonance imaging, as drug carriers and for drug delivery, for local hyperthermia effect in tumor therapy [1]. Typical magnetic fluids consist of iron-containing magnetic nanoparticles and may be well studied using Mössbauer spectroscopy [2]. Recently, new capabilities of Mössbauer spectroscopy with high velocity resolution were demonstrated in various biomedical researches [3]. Therefore, in this work we applied this technique to study native nanosized magnetite and nanosized magnetite suspended in Copaiba oil as a biocompatible magnetic fluid sample.

[7] To whom any correspondence should be addressed.

2. Materials and methods

Magnetite nanoparticles were synthesized by co-precipitation of a heated mixture of ferrous and ferric chloride aqueous solution with concentrated ammonia (25 % v.v.), under vigorous stirring. After precipitation of nanosized Fe_3O_4 and repeated washing using hydrochloric acid a stable sol at pH 2, with nanoparticle volume fraction around 1%, was produced. Copaiba oil was collected from Copaifera tree (*Copaifera Langsdorffii Desf.*) in the Ji-Paraná region, Rondônia State, Brazil. Copaiba oil dispersed in cyclohexane was then added to the as-produced sol under stirring. The resulting suspension was dried to remove out the organic solvent whereas the precipitated sample (CO–Fe_3O_4) as well as native Fe_3O_4 were used for Mössbauer measurement. Both samples had effective thicknesses about 5 mg Fe/cm^2 and about 18 mg Fe/cm^2 for 295 and 90 K measurements, respectively.

Mössbauer spectrometric complex with a high velocity resolution and its characteristics were described in details in [3, 4]. Spectra were measured at 295 and 90 K, in 4096 channels, and computer fitted with the least squares procedure using UNIVEM-MS program (from Research Institute of Physics, Southern Federal University, Rostov-on-Don, Russian Federation) with Lorentzian line shape. This program realizes two methods: direct search by method of configurations (the Hook–Jives method) and gradient search by the Newton–Gauss algorithm, both for minimization of χ^2. Therefore, minimization of χ^2 can be done using direct search firstly, then by gradient search. It is possible to use any number of spectral components and varied parameters for fitting spectra contained up to 4096 points. The Mössbauer parameters isomer shift δ, quadrupole splitting (quadrupole shift for magnetically split component) ΔE_Q, magnetic hyperfine field H_{eff}, line width Γ, subspectrum relative area S and statistical criterion χ^2 were determined. Magnetic sextets were fitted using the ratio $S_{16}:S_{25}:S_{34}=3:2:1$. Criteria of spectra fitting quality were χ^2, differential spectrum and physical meaning of parameters. Hyperfine field distribution program was additionally used for comparison. The values of isomer shift are given relative to α–Fe at 295 K.

3. Results and discussion

Mössbauer spectra of nanosized native Fe_3O_4 and CO–Fe_3O_4 samples are shown in Figure 1. These spectra were fitted using various numbers of components up to 18 sextets. For instance, fitting of the native Fe_3O_4 spectrum at 295 K using 1 paramagnetic doublet and 13, 14, 15 and 16 sextets demonstrated the following variation of χ^2 values: 3.434, 3.392, 3.366 and 3.366, respectively. Fitting of the CO–Fe_3O_4 spectrum at 295 K using the same number of components showed variation of χ^2 values: 1.629, 1.621, 1.608 and 1.611. Therefore, comparison of χ^2 values, differential spectra and physical meaning of parameters permitted us to suggest the better fit of Mössbauer spectra using 15 magnetic sextets and 1 paramagnetic doublet for the room temperature and 15 magnetic sextets for 90 K. However, hyperfine parameters of these spectral components and its relative areas were different. Parameters of paramagnetic doublet for native Fe_3O_4 and CO–Fe_3O_4 were also different: δ=0.47±0.02 mm/s, ΔE_Q=1.19±0.06 mm/s and δ=0.52±0.02 mm/s, ΔE_Q=1.03±0.04 mm/s, respectively. Differences of the plots of magnetic hyperfine field versus isomer shift for spectral components are shown in Figure 2. Unfortunately, the reason of large number of components in Mössbauer spectra of nanosized magnetite samples is not clear yet. These magnetic components may be a result of particle structure variations from surface to the core region (multi-domain structure) or layers as well as the presence of A and B sites in magnetite and non-stoichiometry of magnetite. Moreover, it is possible that some variations of the δ values (see Figure 2) may be related to the A and B sites in magnetite that is in agreement with published data (see, for instance, [5]). However, the δ value of about 1.2 mm/s cannot be explained. The room temperature data for nanosized Fe_3O_4 and CO–Fe_3O_4 samples demonstrated differences which may be a result of the effect of surface interactions of Fe_3O_4 and polar molecules in Copaiba oil. These interactions may lead to reduction of spin off–orientations of Fe^{3+} and increase of internal magnetic field. Then surface changes of the internal field may affect deep layers or domains in nanoparticles. Differences of relative areas of sextets and the values of H_{eff} are shown in Figure 3, ***a*** and ***b***. It should be noted that the hyperfine field distribution was not informative in comparison with

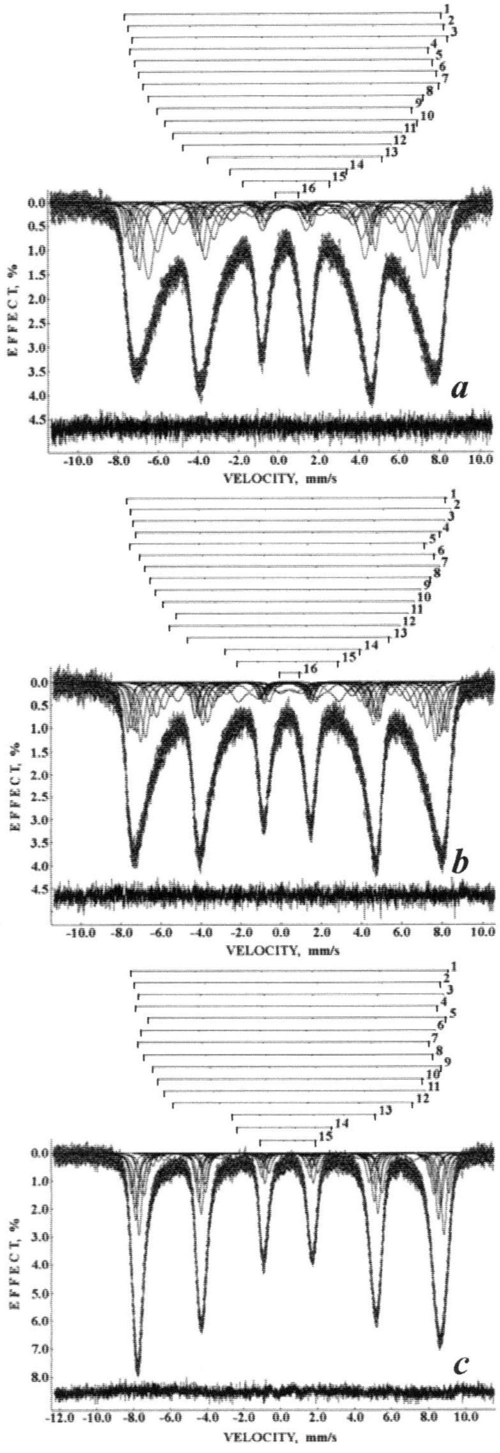

Figure 1. Mössbauer spectra of native Fe_3O_4 nanoparticles (***a***) and CO–Fe_3O_4 (***b***, ***c***) measured in 4096 channels. Components are the results of the better fit. T=295 K (***a***, ***b***), T=90 K (***c***).

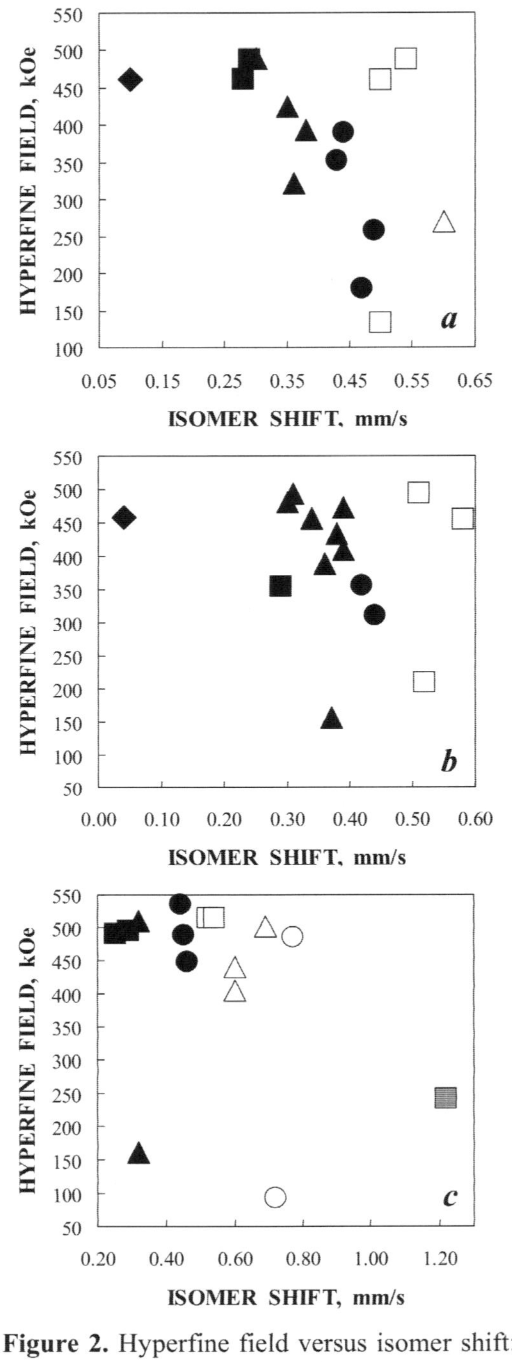

Figure 2. Hyperfine field versus isomer shift: ◆ – for δ=0.0–0.1 mm/s, ■ – for δ=0.2–0.3 mm/s, ▲ – for δ=0.3–0.4 mm/s, ● – for δ=0.4–0.5 mm/s, □ – for δ=0.5–0.6 mm/s, △ – for δ=0.6–0.7 mm/s, ○ – for δ=0.7–0.8 mm/s, ▦ – for δ=1.2–1.3 mm/s. ***a*** – native Fe_3O_4, T=295 K; ***b*** – CO–Fe_3O_4, T=295 K; ***c*** – CO–Fe_3O_4, T=90 K.

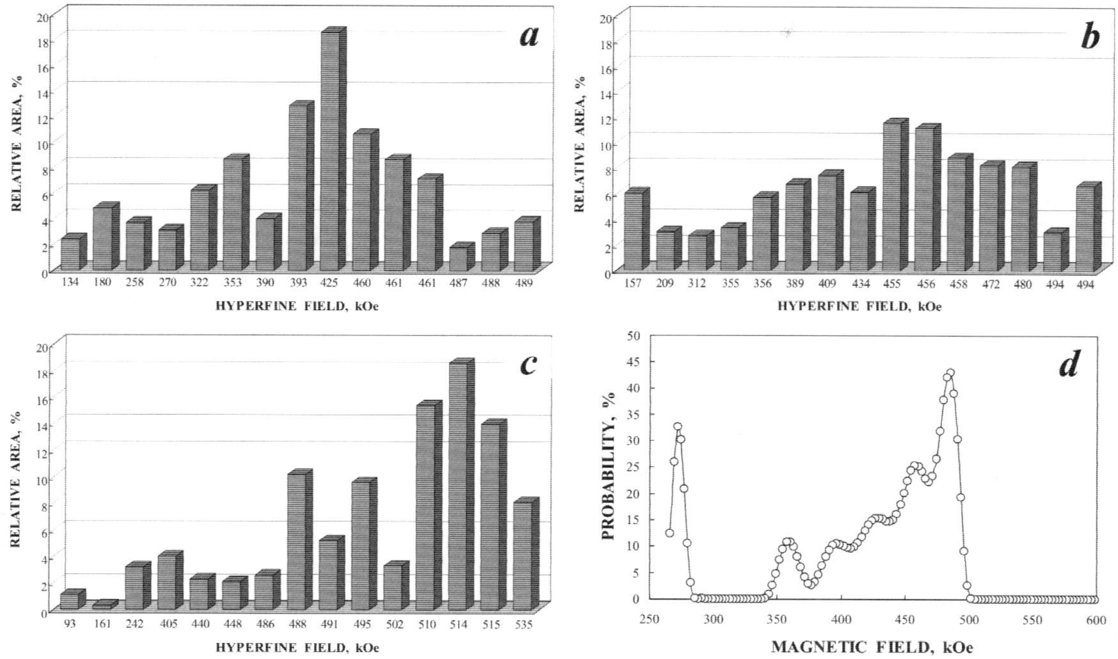

Figure 3. Differences of magnetic components relative areas versus hyperfine fields obtained from the better fit of Mössbauer spectra of native Fe_3O_4 nanoparticles at T=295 K (***a***) and CO–Fe_3O_4 at T=295 K (***b***) and T=90 K (***c***) measured in 4096 channels as well as result of hyperfine field distribution for Mössbauer spectrum of CO–Fe_3O_4 at T=295 K (***d***).

discrete fitting (Figure 3, ***b*** and ***d***) and demonstrated less components in the spectrum. Comparison of room and low temperature data for nanosized Fe_3O_4 and CO–Fe_3O_4 samples demonstrated disappearance of paramagnetic doublet, changes of hyperfine parameters (compare Figure 2, ***b*** and ***c***) and redistribution of relative areas of magnetic sextets (compare Figure 3, ***b*** and ***c***). This fact may be a result of temperature dependent variation of nanoparticles' internal structure and multi-domain magnetic structure. Moreover, differences of Mössbauer parameters for nanosized Fe_3O_4 and CO–Fe_3O_4 were also found at 90 K.

4. Conclusion

Study of native nanosized Fe_3O_4 and biocompatible Fe_3O_4 nanoparticles suspended in Copaiba oil using Mössbauer spectroscopy with a high velocity resolution (in 4096 channels) at 295 and 90 K demonstrated complicated structure of nanoparticles. This may be a result of structural variations from surface to core region (multi-domain structure) or layers as well as the presence of A and B sites in magnetite and non-stoichiometry of magnetite. Differences between native nanosized Fe_3O_4 and CO–Fe_3O_4 may reflect the effects of uncoated nanoparticle surface–carrier liquid interaction. Temperature dependent variation of Mössbauer parameters for CO–Fe_3O_4 was also observed.

References

[1] Pankhurst Q A, Connolly J, Jones S K and Dobson J 2003 Applications of magnetic nanoparticles in biomedicine *J. Phys. D: Appl. Phys.* **36**, R167–81

[2] Morais P C, Santos R L, Pimenta A C M, Azevedo R B and Lima E C D 2006 Preparation and characterization of ultra-stable biocompatible magnetic fluids using citrate-coated cobalt ferrite nanoparticles *Thin Solid Films* **515**, 266–70

Morais P C 2008 Using Mössbauer spectroscopy as key technique in the investigation of nanosized magnetic particles for drug delivery *Hyperfine Interact.* **181**, 1–12

[3] Oshtrakh M I, Milder O B and Semionkin V A 2008 Mössbauer spectroscopy with high velocity resolution in the study of ferritin and Imferon: preliminary results *Hyperfine Interact.* **185**, 39–46

Oshtrakh M I, Semionkin V A, Grokhovsky V I, Milder O B and Novikov E G 2009 Mössbauer spectroscopy with high velocity resolution: New possibilities of chemical analysis in material science and biomedical research. *J. Radioanal. Nucl. Chem.* **279**, 833–46

Oshtrakh M I, Semionkin V A, Milder O B and Novikov E G 2009 Mössbauer spectroscopy with high velocity resolution: New possibilities in biomedical research *J. Mol. Struct.* **924–926**, 20–6

[4] Oshtrakh M I, Semionkin V A, Milder O B and Novikov E G 2009 Mössbauer spectroscopy with high velocity resolution: An increase of analytical possibilities in biomedical research. *J. Radioanal. Nucl. Chem.* **281**, 63–67

Semionkin V A, Oshtrakh M I, Milder O B and Novikov E G 2010 Mössbauer spectrometric system with a high velocity resolution for biomedical research. *Bull. Rus. Acad. Sci., Phys.* **74**, 475–479

[5] Panda R N, Gajbhiye N S and Balaji G 2001 Magnetic properties of interacting single domain Fe_3O_4 particles. *J. Alloys Comp.* **326**, 50–53

Zheng H, Yang Y, Zhou M and Li F 2009 Microwave absorption and Mössbauer studies of Fe_3O_4 nanoparticles. *Hyperfine Interact.* **189**, 131–136

Low temperature ^{57}Fe Mössbauer study of cucumber root

K Kovács[1], E Kuzmann[1,2], F Fodor[3], Z Homonnay[1], L Machala[4] and A Vértes[1,2]

[1]Institute of Chemistry, Eötvös Loránd University, P.O. Box 32, 1512 Budapest, Hungary

[2]Laboratory of Nuclear Chemistry, Chemical Research Center , HAS, P.O. Box 32, 1512 Budapest, Hungary

[3]Department of Plant Physiology and Molecular Plant Biology, Eötvös Loránd University, P.O. Box 32, 1512 Budapest, Hungary

[4]Centre for Nanomaterial Research, Palacky University, Svobody 26, Olomouc 771 46, Czech Republic

E-mail: kkriszti@chem.elte.hu

Abstract. Iron uptake and distribution in cucumber root were studied with the help of ^{57}Fe Mössbauer spectroscopy at low temperature applying external magnetic field. Cucumber was grown in iron sufficient modified Hoagland nutrient solution. Mössbauer spectra of the frozen roots taken at 4.2 and 1.5 K, at 5 T external magnetic field support the identification of the main iron species (Fe^{III}-carboxylates, hydrous ferric oxides, Fe^{III}-sulfate-hydroxide) suggested according to its Mössbauer spectra taken between 35-200 K [1]. The magnetic ordering temperature of the hydrous ferric oxide and Fe^{III}-sulfate-hydroxide was found to be in the range of 4.2-1.5 K, which suggests the incorporation of H_3O^+, PO_4^{3-} and citrate into these minerals.

1. Introduction

The mechanism of iron uptake and the storage of iron in plants have great scientific interest since they can help to improve remediation techniques for iron chlorosis which is an old worldwide problem occurring in areas of calcareous and/or alkaline soils [2].

Iron is nor readily available for plants due to the low solubility of iron oxyhydroxides at the pH of the soil. For this reason, plants have evolved different strategies for iron uptake. Non-graminaceous plants, just like cucumber, follow Strategy I that starts with reduction of iron chelates at the outer surface of the plasma membrane [3]. The ferrous chelate releases iron due to its much lower stability and the Fe^{II} ions are taken up by a divalent cation transporter which is located in the plasma membrane. For this iron uptake mechanism, we gave direct evidence by the help of Mössbauer spectroscopy [1]. Furthermore, three main iron(III) components (Fe^{III}-carboxylate complexes, a ferritin-like hydrous ferric oxide and a jarosite-like Fe^{III}-sulfate-hydroxide containing species) were

Journal of Physics: Conference Series **217** (2010) 012019 doi:10.1088/1742-6596/217/1/012019

suggested according to its Mössbauer spectra taken between 35 – 200 K, exhibiting only three doublet subspectra [1].

The aim of the present work was to support the previously proposed iron bearing chemical species occurring in the root of cucumber applying Mössbauer spectroscopy below 4.2 K in external magnetic field.

2. Experimental

Cucumber (*Cucumis sativus* L. cv. Joker) was grown in modified Hoagland nutrient solution with 10 μM Fe^{III}-citrate (Fe was enriched in 90% with ^{57}Fe). After 3 weeks growth the plants were harvested and the roots were frozen and stored in liquid nitrogen. The Mössbauer spectra of the cucumber roots were recorded in a temperature controlled cryostat (Oxford) at 4.2 and 1.5 K without and with applied 5 T external magnetic field (perpendicular to the direction of γ-rays).

^{57}Fe Mössbauer spectra were recorded in the transmission geometry with a conventional Mössbauer spectrometer. γ-rays were provided by a 3×10^{9} Bq $^{57}Co/Rh$ source. Isomer shifts are given relative to α-iron. The Mössbauer spectra were analysed by the least-squares fitting using Lorentzian lines using the MOSSWINN code [4].

3. Results and discussion

The Mössbauer spectra of the cucumber root taken at low temperature are shown in figure 1.

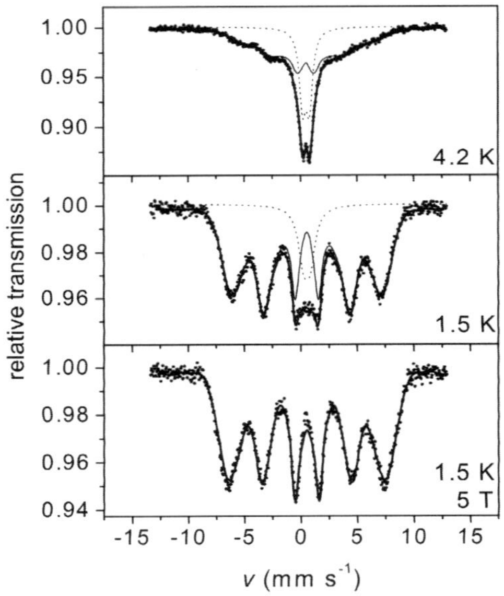

Figure 1. Low temperature Mössbauer spectra of cucumber roots grown in 10 μM $^{57}Fe^{III}$-citrate containing nutrient solution

The Mössbauer spectrum of the cucumber root recorded at 4.2 K exhibits a ~70 % magnetically split subspectrum while one of the three doublet components, existing at higher temperatures and identified as Fe^{III}-sulfate-hydroxides, disappears. The remaining doublet with parameters of δ=0.5 mms^{-1} and Δ=0.6-0.7 mms^{-1} can represent the hydous ferric oxides and paramagnetic Fe^{III}-complexes. At 1.5 K, which is below the magnetic transition temperature of Fe^{III}-sulfate-hydroxides and ferrihydrites, almost all species are magnetically ordered [5], [6]. The remaining paramagnetic species represented by the doublet can be easily transformed into magnetically split spectrum by applying 5 T external magnetic field.

The results can be interpreted in terms of the variation of the characteristic magnetic transition temperatures with the composition of the species. Namely, in the case of Fe^{III}-sulfate-hydroxides (jarosites with the composition of $MFe_3(SO_4)_2(OH)_6$ ($M=K^+$, NH_4^+, H_3O^+, ...)) the temperature of the paramagnetic-antiferromagnetic transition strongly depends on the stoichiometry, especially on the M^+ cation. It was shown in synthetic jarosites, where H_3O^+ units are built in the structure, that the transition took place even below 20 K and in a wide temperature range [7]. Taking into account that the iron uptake in Strategy I type plants is accompanied by significant proton release [8] and that high amount of sulfate ions are present in the nutrient solution, the formation of jarosites with various stoichiometry can be favored in the cell wall.

The magnetic properties of hydrous ferric oxides formed via biomineralization processes (as in the case of the iron storage protein ferritin) also strongly depend on its composition [6]. The high PO_4^{3-} ratio and the incorporation of organic ions, such as citrate can lead to amorphous structure and to the decrease of the antiferromagnetic ordering temperature (< 4 K). These findings support that the magnetically split subspectra at 4.2 and 1.5 K can be attributed to Fe^{III}-sulfate-hydroxides and hydrous ferric oxides, respectively.

The paramagnetic component, which can be transformed to magnetically split spectrum by applying 5 T external magnetic field at 1.5 K, can be identified as ferric-carboxylates, since the parameters of the doublet (δ=0.5 mms^{-1}, Δ=0.6 mms^{-1} present in figure 2 at 4.2 and 1.5 K) correspond with those of high spin Fe^{3+} in octahedral O environment. The Fe^{III}-carboxylate component can be present both in the cell wall and in the cytoplasm, since citric acid was shown to be secreted into the rhizosphere and also to take part in the iron transport inside the cell.

4. Conclusions

Low temperature Mössbauer measurements supported the identification — based on the measurements between 35-200 K — of the iron species found in cucumber root. These main iron containing chemical species were found to be Fe^{III}-carboxylate complexes, a ferritin-like hydrous ferric oxide and a jarosite-like Fe^{III}-sulfate-hydroxide.

The paramagnetic-antiferromagnetic ordering temperature of Fe^{III}-sulfate-hydroxides and hydrous ferric oxides was found to be in the range of 4.2-1.5 K, which suggests the incorporation of H_3O^+ ions into jarosite and high PO_4^{3-} and citrate content in the ferrihydrite species.

5. Acknowledgement

Support by the Hungarian Science Foundation OTKA (K67835, K68135, NN74045) and by the NKTH (CZ 11/2007) is gratefully appreciated.

References

[1] Kovács K, Kuzmann E, Tatár E, Vértes A and Fodor F 2009 *Planta* **229** 271
[2] Lucena J J In: Barton J and Abadia J 2007 *Iron Nutrition in Plants and Rhizospheric Miscroorganisms* (Springer)
[3] Römheld V 1987 *Physiol. Plant.* 70 231
[4] Klencsár Z, Kuzmann E and Vértes A 1996 *J. Radioanal. Nucl. Chem.* **210** 105
[5] Stevens J G 1958-2002 *Mössbauer Effect Reference and Data Index* (Interscience, N. Y)
[6] Chasteen N D and Harrison P M 1999 *J. Struct. Biol.* **126** 182
[7] Kovács K, Kuzmann E, Homonnay Z, Vértes A, Gunneriusson L and Sandström A 2008 *Hyp. Int.* **186** 69
[8] Schmidt W, Michalke W and Schikora A 2003 *Plant Cell Environ.* **26** 361

Magnetite formation via membrane-bound ferritin and an iron(II) species at the cytoplasmic membrane and in magnetosomes of *Magnetospirillum gryphiswaldense*

L H Böttger[1], D Faivre[2], D Schüler[3], A X Trautwein[1] and B F Matzanke[4]

[1]Institut für Physik, Universität zu Lübeck, Ratzeburger Alle 160, 23538 Lübeck, Germany
[2]Max-Planck-Institut für Kolloid- und Grenzflächenforschung, Abteilung Biomaterialien, Wissenschaftspark Golm, 14424 Potsdam, Germany
[3]LMU München, Mikrobiologie, Institut für Biologie, Maria-Ward-Straße 1a, 80638 München, Germany
[4]Isotopenlaboratorium TNF, Universität zu Lübeck, Ratzeburger Allee 160, 23538 Lübeck, Germany

E-mail: boettger@physik.uni-luebeck.de; matzanke@isolab.uni-luebeck.de

Abstract. Growth and cell fractions of the magnetic bacterium *Magnetospirillum gryphiswaldense* were studied by Mössbauer spectroscopy. In isolated magnetosomes only magnetite particles were observed. The membrane fraction of *Magnetospirillum gryphiswaldense* contains a ferritin-like component and a Fe^{2+} species and also magnetite particles smaller than those observed in the magnetosomes fraction. In the cytosol only ferritin was identified.

1. Introduction

Magnetotactic bacteria (MTB) were discovered in 1975 by Blakemore [1] and exhibit a unique cellular compartments referred to as magnetosomes (MS). MTB accomplish biosynthesis of magnetite in MS. *Magnetospirillum gryphiswaldense* biomineralizes single domain crystals of cubooctahedral magnetite (Fe_3O_4) with up to 4% of the total dry weight [2]. MS vesicles originate from invaginations (IV) of the cytoplasmic membrane forming a cavity at periplasmic side of the cytoplasnic membrane. The IV detach individually as vesicles (MS) from the cytoplasmic membrane [3-8]. The assembly of highly ordered MS chains is under genetic control and involves several specific proteins [9]. This structure is essential for MTB-to navigate along the lines of the earth's magnetic field in order to find growth-favouring microoxic zones in their aquatic environment. MTB were studied intensively in the last three decades [3-5, 10-12].These MS have recently gathered multidisciplinary interest due to their crystalline and magnetic properties, e.g. as distinct biomarkers for life on Mars [13,14], or as material for biotechnological applications[15].

. Following questions were addressed in this study: (i) Which iron metabolites are formed after iron uptake and in which compartments are they detectable? (ii) Is magnetite detectable in the membrane fraction? (ii) Are there any inorganic phases or precursors in addition to magnetite in the magnetosomes?

2. Materials

Cells were grown in an oxystat bioreactor, which allows control of temperature, pressure and oxygen concentration using a modified protocol for large scale cultivation [16]. Throughout all experiments the strain MSR-1 (DSMZ 6361) of *Magnetospirillum gryphiswaldense* was used.

Cells in the mid-logarithmic growth phase were transferred to a low carbon medium [11]. Cells kept under these conditions are still viable, however display significant cell growth and allow, therefore, growth-independent experiments. Iron uptake (induction) was achieved via $^{57}Fe^{3+}$- dicitrate added to the cell suspension. For separation of cell fractions, mature cells with fully formed magnetic chains were disrupted and spun down at different g-values. In addition, the soluble fraction was passed through a 100 kDa filter.

For analysis, cells were removed from a bioreactor sequently at different time periods after ^{57}Fe induction and were analyzed by magnetic light scattering and bright-field TEM (Zeiss EM10 transmission electron microscope operating at 60kV, data shown elsewhere [11,12]) For Mössbauer spectroscopy, cells were washed, spun down into Delrin Mössbauer sample holders, frozen in liquid nitrogen , and kept at this temperature until measurement in bath and continuous flow cryostats from Oxford Instruments (constant acceleration mode). Spectra were calibrated against α-iron foil at room temperature and were analyzed by least-squares fits of Lorentzian line shapes to the experimental data.

3. Results

3.1. Growth analysis

Mössbauer spectra of whole cells are shown in Figure 1. Already 20min after ^{57}Fe induction first traces of magnetite, ferritin (see chapter 3.3) and ferrous iron are detectable. In all Mössbauer spectra (taken at 130 K) two magnetically split sextets resembling magnetite sites A and B (δ_A=0.38 mms^{-1}, $B_{hf,A}$=49.5 T; δ_B=0.76 mms^{-1}, $B_{hf,B}$=47.4 T), a high-spin Fe^{3+}-doublet (δ=0.47 mms^{-1}, ΔE_q=0.77 mms^{-1}) and a high-spin Fe^{2+}-doublet (δ=1.27 mms^{-1}, ΔE_q=2.81 mms^{-1}) were detected. The amount of all components increases until 40min after ^{57}Fe induction. At later times after ^{57}Fe induction only the growth of the magnetite components continues further.

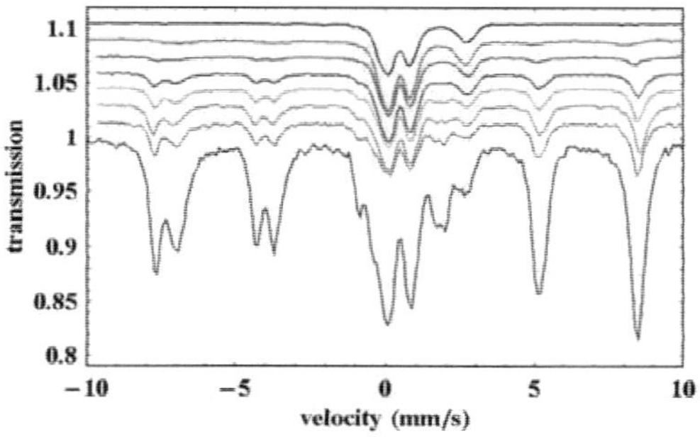

Figure 1. Mössbauer spectra at different times after ^{57}Fe induction: From top to bottom: 20min, 40min, 60min, 95min, 125min, 155min, 215min and 1230min.

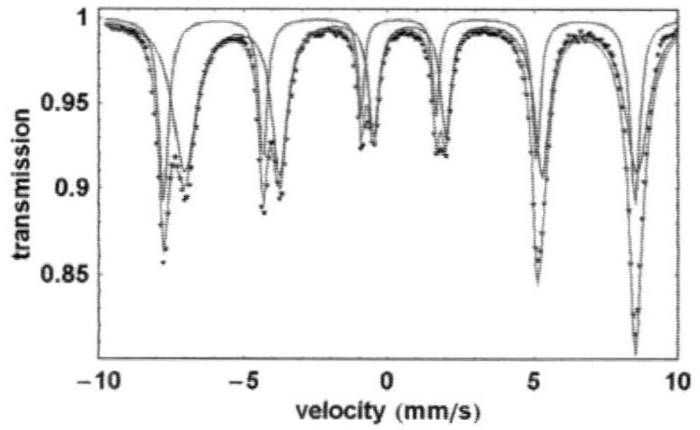

Figure 2. Mössbauer spectrum of isolated magnetosomes

3.2. Isolated magnetosomes

Mössbauer spectra of isolated magnetosomes (Figure 2) exhibit two magnetically split sextets attributed to sites A and B of magnetite (δ_A=0.38 mms^{-1}, $B_{hf,A}$=49.9 T; δ_B=0.76 mms^{-1}, $B_{hf,B}$=47.9 T). The magnetic hyperfine field of the isolated magnetosomes is slightly higher than that in the whole-cell fraction.

3.3. Membrane fraction

A Fe^{3+} species exhibiting parameters similar to bacterioferritin (δ=0.47 mms^{-1}, ΔE_q=0.77 mms^{-1}, see also next paragraph) and a membrane-bound Fe^{2+} species (δ=1.26 mms^{-1}, ΔE_q=2.86 mms^{-1}) were identified (Figure 3). These species were also detected in the cell spectra (Fig. 1) and were found previously in the cytoplasm of other Gram negative bacteria [17]. Surprisingly, the Mössbauer spectrum of the membrane fraction (Figure 3) also exhibits the two sextets of magnetite. However, the magnetic hyperfine field is significantly smaller than magnetite in isolated MS (δ_A=0.38 mms^{-1}, $B_{hf,A}$=46.8 T; δ_B=0.76 mms^{-1}, $B_{hf,B}$=44.8 T). This provides evidence for the presence of very small magnetite crystallites (<6nm) located in IV of the membrane fraction. This indicates in addition that (i) these IV are still well integrated in the cytoplasmic membrane and (ii) initial magnetite formation occurs at these IV.

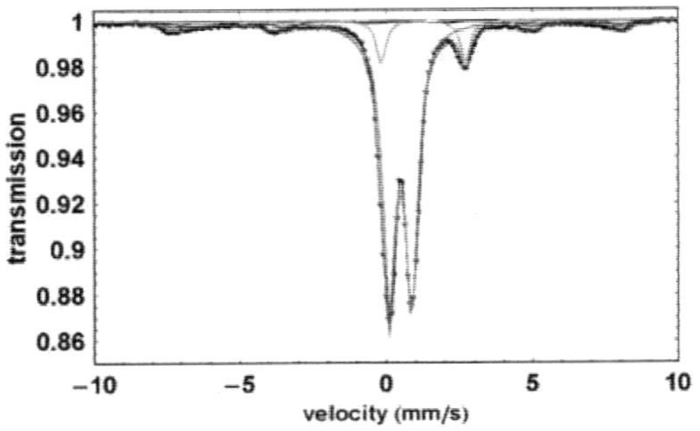

Figure 3. Mössbauer spectrum of the membrane fraction at T=130 K: The spectrum is dominated by the ferritin (green) component, accompanied by a high-spin Fe^{2+} species (cyan). Minor amounts of small magnetite particles (red and blue) were also identified at the high-velocity sides of the spectrum.

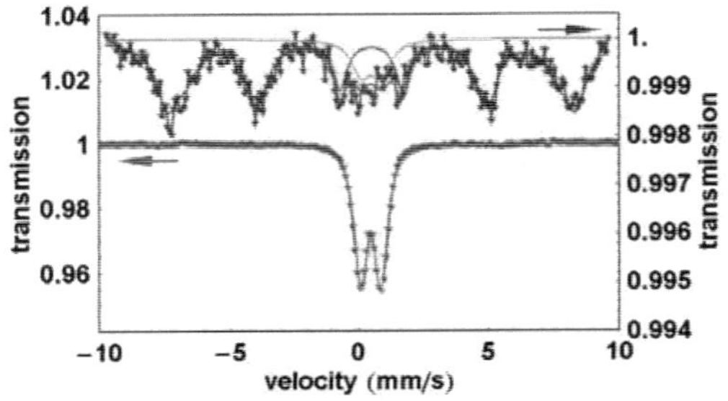

Figure 4. Mössbauer spectra of the cytoplasmic fraction (molecular mass >100 kDa) containing ferritin as sole iron species at T=4.2 K (top, blue). The spectrum is fitted with a split ferritin sextet and a non-split ferritin doublet. The spectrum recorded at T=77 K (bottom, red) is fitted with one ferritin doublet alone.

3.4. Fraction above 100 kDa

The Mössbauer spectrum of the 100 kDa supernatant obtained from the cytoplasmic fraction was recorded at 77K (Figure 4). It exhibits a single doublet the parameters of which (δ=0.47 mms^{-1}, ΔE_q=0.77 mms^{-1}) are similar to typical bacterioferritins (e.g. *Azotobacter*: δ=0.48 mms^{-1}, ΔE_q=0.78 mms^{-1}). However, at 4.2 K the majority (86.8 %) of the doublet (observed at 77 K) has transformed into a magnetically split sextet typical for superparamagnetic transitions found in ferritins (δ=0.49 mms^{-1}, B_{hf}=47.4 T). The remaining 13.2 % are fitted with a doublet (δ=0.49 mms^{-1}, ΔE_q=0.77 mms^{-1}). Gel electrophoresis (native & SDS-PAGE) of the ferritin verifies the existence of a high-molecular mass protein (> 100 kDa) that carries large amounts of iron and decays under SDS into 20-40 kDa subunits.

4. Conclusion

With the exception of ferritin-like iron mineral precursors of magnetite were not detected in cell fractions. Therefore it is tempting to assume that magnetite formation in *Magnetospirillum gryphiswaldense* proceeds during subsequent steps by fast coprecipitation of ferrous and ferric iron within the invagination pit, which is likely alkaline to enable magnetite thermodynamic stability [18]:

$$\text{(i) } [Fe^{2+}R_a]^0 + 2\,[Fe^{3+}R_b]^0 \text{ (ferritin)} + (2x + y)\,H_2O$$
$$\rightarrow 2\,Fe(OH)_x^{3-x} + Fe(OH)_y^{2-y} + (2x + y)\,H^+ + R_a^{2-} + 2R_b^{3-}$$

$$\text{(ii) } 2\,Fe(OH)_x^{3-x} + Fe(OH)_y^{2-y}$$
$$\rightarrow Fe_3O_4 \text{ (crystal in statu nascendi)} + (2x + y - 4)\,H_2O$$

At the CM level Fe^{2+} and Fe^{3+} ions are ligated to organic carriersR_a and R_b. These carriers release iron-hydroxo species directly at the membrane-invagination pit interface (step (i)). The presence of ferrous hexaquo and ferric hydroxo complexes in the complete membrane compartment is very unlikely because of their toxic nature (Fenton chemistry) and chemistry. According to this scenario, a mechanism (step (ii)) is assumed in which nucleation of magnetite occurs at the CM/pit interface. by coprecipitation of two equivalents of $Fe(OH)_x^{3-x}$ and one equivalent of $Fe(OH)_y^{2-y}$ to form crystalline Fe_3O_4 and H_2O. How magnetite grows after release of the fully formed vesicle from their membrane sites remains yet unclear.

In summary, we propose a mechanism for initial magnetite formation, where the iron required for magnetite biomineralization is released directly at the membrane invagination pit interface without iron flux *via* the cytoplasm. This might point to distinct transport pathways for magnetite formation and for biochemical utilization iron. Magnetite formation occurs *via* membrane-associated crystallites, whereas the final step of magnetite crystal growth is possibly spatially separated from the CM.

Acknowledgments
This research was supported by the Max Planck Society and the Biofuture program of the BMBF. D.F. acknowledges support from the Max Planck Society

References

[1] Blakemore R 1975 *Science* **190** 377
[2] Schüler D and Baeuerlein E 1998 *J. of Bacteriology* **180** 159
[3] Balkwill D L, Maratea D and Blakemore R P 1980 *J. of Bacteriology* **141** 1399
[4] Schüler D. 2004 *Arch. Microbiol.* **181** 1
[5] Bazylinski D A and Frankel R B 2004 *Nat. Rev. Microbiol.* **2** 217
[6] Komeili A, Li Z, Newman D K and Jensen G J 2006 *Science* **311** 242.
[7] Scheffel A, Gruska M, Faivre D, Linaroudis A, Plitzko J M and Schüler D 2006 *Nature* **440** 110
[8] Grünberg K, Müller E C, Otto A, Reszka R, Linder D, Kube M, Reinhardt R and Schüler D 2004 *Appl. Eviron. Microbiol.* **70** 1040
[9] Jogler C and Schüler D 2009 *Ann. Rev. Microbiol.* **63** 501
[10] Simmons S L, Bazylinski D A and Edwards K J 2006 *Science* **311** 371
[11] Faivre D, Böttger L H, Matzanke B F and Schüler D 2007 *Angew. Chem. Int. Ed.* **46** 8495
[12] Böttger L H 2008 PhD-Thesis, Universität zu Lübeck
[13] McKay D S, Gibson Jr. E K, Thomas-Keprta K L, Vali H, Romanek C S, Clemett S J, Chilier X D F, Maechling C R and Zare R N 1996 *Science* **273** 924
[14] Faivre D and Zuddas P 2006 *Planet. Sci. Lett.* **243** 53
[15] Lang C, Schüler D and Faivre D 2007 *Macromol. Biosci.* **7** 144
[16] Heyen U and Schüler D 2003 *Appl. Microbiol. Biotechnol.* **61** 536.
[17] Matzanke B **:** Iron Transport: Siderophores. In: Encyclopedia of Inorganic Chemistry, R. B. King, ed; Wiley publishers, 2nd edition September 2005, Vol IV,2619-2646
[18] Schüler D 2006 *Magnetoreception and magnetosomes in bacteria* Vol. 3 (Heidelberg: Springer)

Journal of Physics: Conference Series **217** (2010) 012021 doi:10.1088/1742-6596/217/1/012021

Influence of Cd^{2+} on the spin state of non-heme iron and on protein local motions in reactions centers from purple photosynthetic bacterium *Rhodospirillum rubrum.*

M Lipińska[1], A Orzechowska[2], J Fiedor[3], A I Chumakov[4], T Ślęzak[3], M Zając[3], K Matlak[3], J Korecki[3], A.Hałas[3], K Strzałka[5], L Fiedor[5], and K Burda[3]

[1]Institute of Nuclear Physics PAN, 31-342 Kraków, Radzikowskiego 152, Poland

[2]Institute of Physics, Jagiellonian University, 30-059 Kraków, Reymonta 4, Poland

[3]Faculty of Physics and Applied Computer Science, AGH University of Science and Technology, 30-059 Kraków, Reymonta 19, Poland

[4]European Synchrotron Radiation Facility, BP220, F-38043 Grenoble Cedex, France

[5]Faculty of Biochemistry, Biophysics and Biotechnology, Jagiellonian University, 30-387 Kraków, Gronostajowa 7, Poland

E-mail: burda@novell.ftj.agh.edu.pl

Abstract. Non-heme Fe is a conservative component of the Q-type photosynthetic reaction centers but its function remains unknown. Applying Mössbauer spectroscopy we show that in *Rhodospirillum rubrum* the non-heme Fe exists mostly in a ferrous low spin state. The binding of Cd^{2+} ions in the vicinity of the quinone-Fe complex changes the high spin state of the non-heme Fe into a low spin one characterized by hyperfine parameters similar to those obtained for the non-heme Fe low spin state in untreated reaction centers, as confirmed by Mössbauer measurements. The nuclear inelastic scattering of synchrotron radiation experiments show that the contribution of vibrations at low energies, between 3-15 meV, activated at 240 K are damped in the bacterial reaction centers treated with $CdCl_2$. No influence of Cd^{2+} ions is observed on the soft vibrational states at 60 K. These results suggest that binding of cadmium cations within the reaction centers may enhance decoupling of the non-heme Fe from the surrounding protein matrix at temperatures higher than 200 K, what can explain the slowing down of electron transfer between the Q_A and Q_B quinones by Cd^{2+}.

1. Introduction

Photosynthesis is the ultimate source of energy for all living organisms on our planet, which relies on the conversion of solar energy into biologically-useful chemical energy by specialized pigment-protein complexes called photosynthetic reaction centers (RCs). Depending on the type of the final electron acceptor in the electron transport (ET) chain, RCs are classified into two types: Q-and Fe-S type. Bacterial Q-type RCs belong to the best recognized photosynthetic structures. Due to their structural simplicity they serve as a model system in studies of intramolecular electron transfer (ET) within proteins [1, 2]. The core of bacterial RC is composed of two peptides L and M binding several cofactors arranged in two symmetric branches: a dimer of two bacteriochlophylls called a special pair (BP870A and BP870B), two accessory bacteriochlorophylls (B_A and B_B) in proximity to the special pair, two bacteriopheophytins ($BPhe_A$ and $BPhe_B$), a pair of ubiquinones (UQ_A and UQ_B) with a non-heme iron (NHFe) located between them (Fig.1A). It is believed that in native systems only left branch is active in linear electron transfer and the direction of electron flow from BP_{870} to the final quinone acceptor at site Q_B is marked by arrows in Fig.1A. When ubiquinone at the Q_B site is doubly reduced it accepts two H^+. Then, the ubiquinol form, UQH_2, is exchanged by the external oxidized UQ. The whole process is light driven by excitation of BP_{870}, which is a primary electron donor. Its oxidized form is reduced by cytochrome c_2.

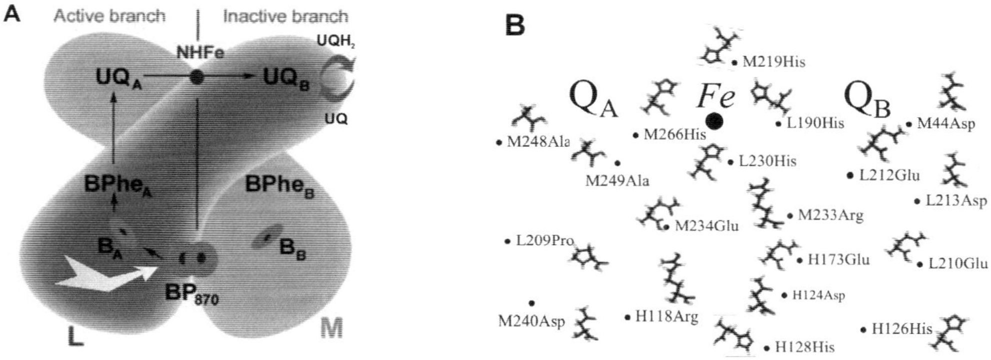

Figure 1. Scheme of the core proteins M and L of bacterial reaction center with redox active cofactors participating in linear ET (A). The black arrows show the subsequent steps of ET and the white arrow indicates the site excited by light. In (B), an arrangement of aminoacids from the proteins: M, L and H in the vicinity of the NHFe binding site for *Rs. rubrum* RC is shown [19, 20]. The aminoacids of the H protein are taken from Rb. sphaeroides. Symbols: BP_{870} – bacteriochlorophyll dimer (special pair); $B_{A,B}$ – bacteriochlorophyll monomers; $BPhe_{A,B}$ – bacteriopheophytin; $UQ_{A,B}$ – ubiquinone, UQ – oxidized form of ubiquinone, UQH_2 – ubiquinol form, NHFe – non-heme iron.

The efficiency, the rate of primary charge separation and its pathways in RCs critically depend on the distances, specific arrangement and electronic properties of the ET cofactors, which in turn are strongly determined by protein matrix. The molecular mechanisms regulating photosynthetic ET have long been a subject of intensive studies but because of their complexicity it is still unclear how exactly the protein matrix exerts its control over the ET cofactors [3-6]. In particular, the role of NHFe, a very conservative component of the Q-type RCs, in photosynthetic charge separation and the temperature effects on ET between the two external quinone acceptors are among the most challenging issues [7, 8]. Intriguingly, ET from reduced Q_A to Q_B is inhibited at temperatures below 200±20 K whereas Q_A in RCs from various organisms remains fully active at cryogenic temperatures [9, 10]. This is in line with the theoretical calculations which suggest that some intrinsic flexibility of the protein matrix in the vicinity of the Q_A–Fe–Q_B complex is required for efficient ET on the RCs acceptor side [11]. The collective motions of the protein matrix in RCs, activated above 180 K seem to be involved in the activation of ET within the Q_A–Fe–Q_B complex, along with the protonation and deprotonation events on the acceptor side of RCs [12-14]. However, the role of NHFe in stabilization of the Q_A and Q_B binding sites and in the primary ET is still under debate because a change of the

NHFe valence state has never been detected in native systems [15]. Such changes were observed only in RCs in algae PSI⁻ mutant [16, 17].

In this paper we show that NHFe in RC isolated from *Rhodospirillum* (*Rs.*) *rubrum* occurs almost exclusively in a low spin state, while in RC from *Rhodobacter sphaeroides* it was observed in a high spin state [7, 18]. At the same time, the rate of ET in the former RC is slower [19]. We suggest that this effects may be related to the differences of the amino acid sequences in the NHFe surrounding between these two types of bacterial RCs (see Fig. 1B, [20,21]). These intriguing observations prompted us to investigate in detail NHFe in *Rs. rubrum*, in particular, how its low spin state is affected by fluctuations of protein matrix and by Cd^{2+} ions, which are known to bind in the vicinity of the iron-quinone complex [22, 23]. To this end, Mössbauer spectroscopy was applied to compare the valence and spin states of Fe atoms in native RC from *Rs. rubrum* and in RC treated with Cd^{2+} salt. Further, nuclear inelastic scattering (NIS) was used to monitor the collective motions in the NHFe-binding sites in both the native and the Cd^{2+}-treated RCs.

2. Materials and Methods

Chromatophores preparation and isolation of reaction centers. The wild type bacterium, *Rs. rubrum* S1 was grown under anaerobic conditions in the white light at 27°C in a modified Hutner medium [24] supplemented with ^{57}Fe. The enrichment of the chromatophores in ^{57}Fe was 30-50% as estimated by the use of atomic absorption (ICP-MS ELAN 6100 spectrometer, Perkin Elmer) and Mössbauer spectroscopy. The RCs were prepared according to a described method with some modifications [25, 26]. The final purification step was done by ion-exchange chromatography on DEAE-Cellulose (DE52 Whatman) in the presence of 0.08% LDAO. The purified RCs were stored at -40°C. The quality of the RC preparation was assessed by absorption spectroscopy (Fig. 2; Cary 50 Bio spectrophotometer, Varian USA) and confirmed in the test for the photochemical activity (data not shown).

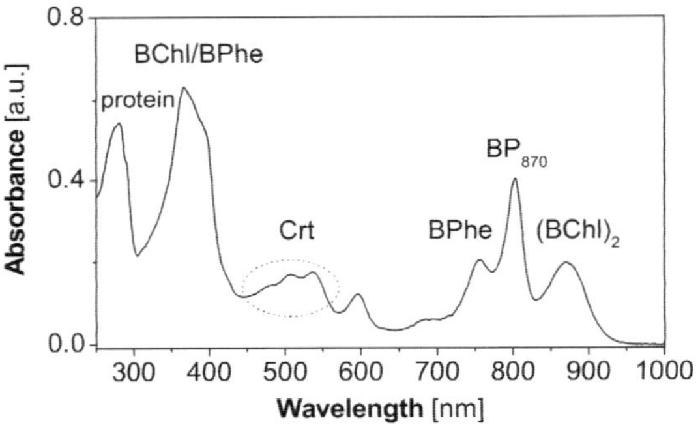

Figure 2. Absorption spectrum of *Rs. rubrum* reaction center. Symbols: BP_{870} – bacteriochlorophyll dimer (special pair); $BChl_{A,B}$ – bacteriochlorophyll monomers; $BPhe_{A,B}$ – bacteriopheophytin; Crt- carotenoid.

Mössbauer spectroscopy. **The** Mössbauer ^{57}Fe spectra were recorded in a home made cryostat using 50 mCi $^{57}Co/Rh$ as a source of 14.4 keV γ radiation and a proportional counter to detect the radiation. The temperature stabilization was within 0.1 K. The isomer shifts are given vs. metallic Fe at room temperature. The recorded spectra were fitted using a Recoil program [27].

Nuclear inelastic scattering of synchrotron radiation. The nuclear inelastic scattering of synchrotron radiation experiment was performed at the Nuclear Resonance beamline ID18 [28] at the European Synchrotron Radiation Facility in Grenoble, France. The storage ring was run in hybrid mode, providing 24 groups of 8 radiation pulses with the period of 88 ns. X rays were monochromatized to the energy bandwidth of 0.5 meV. The energy of incident radiation was tuned around the 14.4 keV energy of ^{57}Fe nuclear transition within the range from -40 meV to 100 meV for 60 K and from -80 meV to 100 meV for 240 K. More details on the experimental method and setup are described elsewhere [28, 29]. A statistically meaningful spectrum of iron vibration modes in RC was obtained after 10-12 h of data collection, depending on the ^{57}Fe concentration in the sample.

3. Results and Discussion

Typical Mössbauer spectra of the untreated and exposed to approximately a 200-fold excess of $CdCl_2$ chromatophors and RCs isolated from *Rs. rubrum* are presented in Fig. 3. The spectra were collected at 85 K.

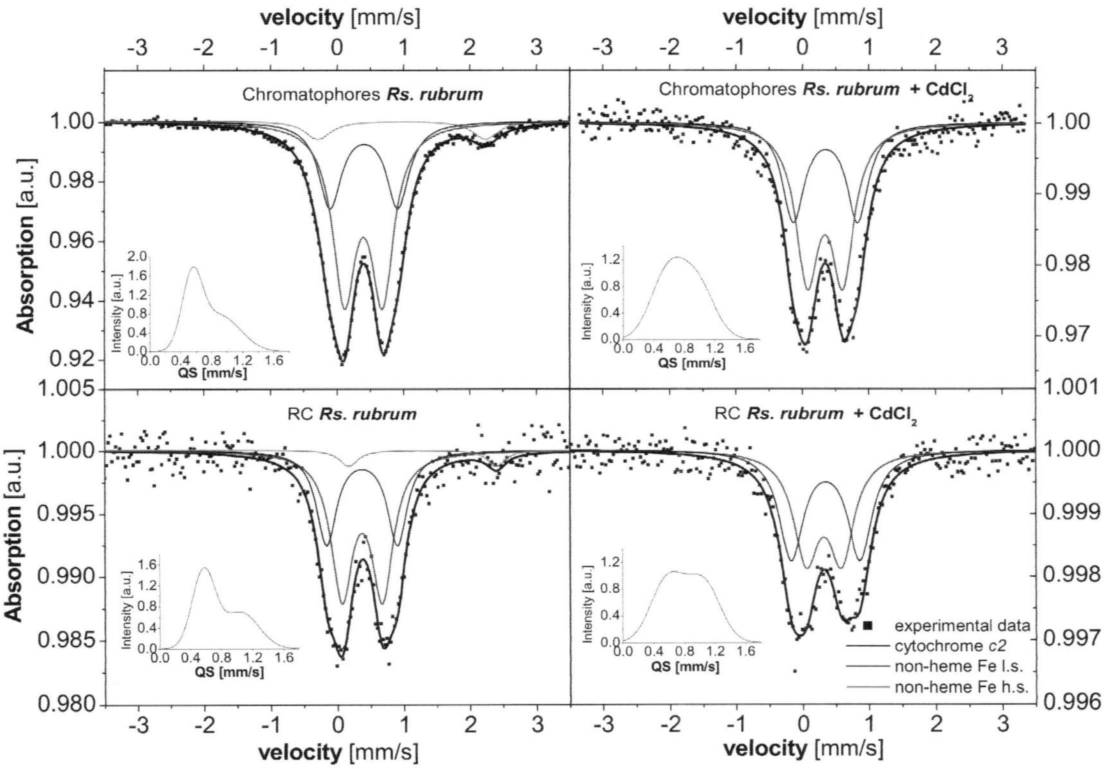

Figure 3 The ^{57}Fe Mössbauer spectra of chromatophores and reaction centers isolated from *Rs. rubrum*, left panel - control and right panel - treated with $CdCl_2$. The spectra were measured at 85 K. The lines represent fits assuming symmetrical doublets. Subspectra correspond to NHFe and HFe in cytochrome c_2. Quadrupole splitting distributions in the subsequent cases are presented in the insets.

In the case of samples treated with Cd^{2+} the signal is weaker because a lower amount of RCs was available due to losses during the purification procedure following the treatment. In the theoretical evaluations of the Mössbauer spectra we applied two approaches: (a) assuming quadrupole splitting distribution for the central doublet or (b) superposition of two quadrupole doublets with different quadrupole splittings and similar isomer shifts. From the distributions of the quadrupole splitting (see insets in Fig.3) we found that the central doublet detected for chromatophores and RCs can be approximated by two doublets with the line-width of 0.21 ± 0.02 mm/s and 0.19 ± 0.02 mm/s, respectively. The subspectrum with IS (isomer shift) = 0.41 ± 0.01 mm/s and QS (quadrupole splitting) = 1.02 ± 0.07 mm/s for chromatophores and IS = 0.38 ± 0.02 mm/s and QS = 1.05 ± 0.05 mm/s for RC's is assigned to the heme iron (HFe) from the cytochromes. From gel electrophoresis we learnt that in the purified RC a contamination of cytochrome c_2 is present in addition to the RC core proteins binding NHFe (data not shown). This HFe is in a low spin ferrous state [30]. The second component with IS = 0.40 ± 0.01 mm/s for chromatophores and IS = 0.37 ± 0.02 mm/s for RCs and characterized by QS = 0.58 ± 0.10 mm/s comes from NHFe (in chromatophores one expects also a contribution of some Fe-S proteins to this subspectrum). Such hyperfine parameters are characteristic for low spin Fe^{2+} [16]. Finally, the spectrum obtained for the control sample of chromatophores contains a third component with IS = 0.97 ± 0.02 mm/s and QS = 2.53 ± 0.14 mm/s, typical hyperfine parameters for the high spin Fe^{2+} state. This component is usually ascribed to NHFe, which high spin ferrous state in photosynthetic RC of type Q is well known [10, 18]. In order to improve the fit of the spectrum collected for control RCs one can also add a minor contribution of this high spin ferrous state but IS in this case is shifted to 1.18 ± 0.10 mm/s. The discussed above components are presented in Fig.3.

The most interesting results concern NHFe from the quinone-iron complex (Q_A–Fe–Q_B). Surprisingly, NHFe in *Rs. rubrum* RCs exists predominantly in a low spin ferrous state. It is believed that NHFe should exist solely in a high spin ferrous state [18, 31] in photosynthetic RCs although this diamagnetic state of NHFe has been already observed in mutated BBY PSII isolated from algae [16, 17] and a similar Mössbauer spectrum of *Rs. rubrum* chromatophores with only a small contribution of the high spin NHFe was detected [9]. Moreover, some EPR results can be interpreted as a signature for the presence of LS NHFe in *Rs. rubrum* chromatophores since otherwise it would not be possible to observe an electron paramagnetic resonance (EPR) signal having a g-value of 2.0050 ± 0.0003 characteristic for a semiquinone molecule in the whole cells of *Rs. rubrum* [32]. The same EPR signal of the primary quinone acceptor Q_A in its semiquinone form was monitored in the absence of this iron atom [33, 34, 35].

Here, we also show that Cd^{2+} at molar ratio to RC as high as 200:1 does not remove the NHFe. Nevertheless, as seen in the Mössbauer spectrum (Fig. 3), after the $CdCl_2$ treatment of chromatophores (RC) NHFe exists only in the low spin state. Only two components were found in these samples, described by the following hyperfine parameters, (i) for NHFe in chromatophoes: IS = 0.34 ± 0.02 mm/s and QS = 0.54 ± 0.05 mm/s, and in RCs: IS = 0.32 ± 0.01 mm/s, and QS = 0.55 ± 0.07 mm/s (ii) for HFe from cytochrome in chromatophoes: IS = 0.35 ± 0.01 mm/s and QS = 0.97 ± 0.04 mm/s, and in RCs: IS = 0.35 ± 0.02 mm/s, and QS = 1.08 ± 0.05 mm/s.

For chromatophores we could observe that cadmium ions caused the transfer of NHFe high spin state into a low one. In the case of the low spin ferrous state of NHFe, the tendency of the isomer shift decrease can be noticed in the presence of the cadmium cations. These effects suggest that Cd^{2+} is directly bound in the vicinity of the iron-quinone complex affecting the arrangement of the iron – ligands. No direct interaction of Cd^{2+} ions with HFe of the cytochrome is observed.

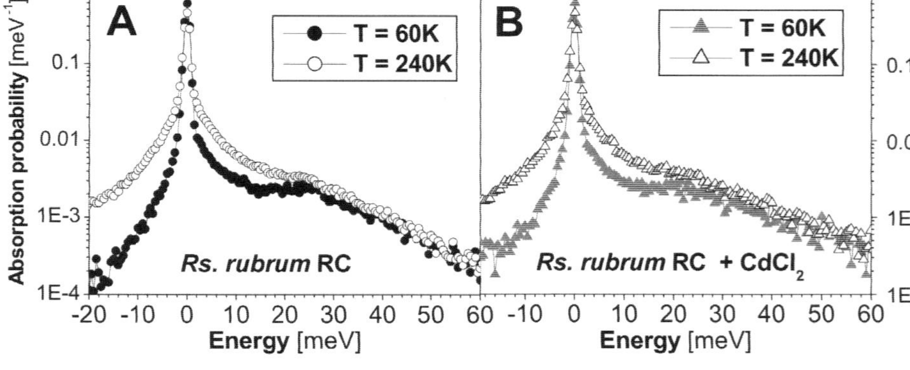

Figure 4 The energy dependencies of nuclear inelastic absorption in RCs isolated from *Rs. rubrum* enriched in ^{57}Fe: native (A) and incubated with $CdCl_2$ (B) measured at 60 K and at 240 K. The spectra are normalized so that their first energy moments are equal to the recoil energy of a free iron nucleus. In this presentation, they show the density of absorption probability.

The experimental data of nuclear inelastic scattering indicates that the presence of Cd^{2+} in the vicinity of the iron-quinone complex may influence the transfer of vibrations of the surrounding protein matrix on the NHFe bonds. The measurements were performed at two temperatures, 60 K and 240 K because of expectations of some differences in fluctuations above 200 ± 20 K (the characteristic temperature above which the electron transfer between Q_A and Q_B within the iron-quinone complex is activated). Fig. 4 shows that the spectra of the control and incubated with $CdCl_2$ RCs, measured at 60 K, differ from those collected at 240 K. As one expects at the higher temperature a significant enhancement of the low energy vibrations is observed which is caused by the increase of the population of these vibrational states with temperature. However, the comparison between the recorded spectra of the control and incubated with $CdCl_2$ RCs measured at the same temperature are much more interesting. The differences in the inelastic fluctuations are more visible in the density of vibrational states (DOS) calculated from the spectra after subtraction of the elastic contribution using the instrumental function measured in parallel [28]. The comparison of DOS obtained for control RCs and those incubated with $CdCl_2$ is presented in Fig.5.

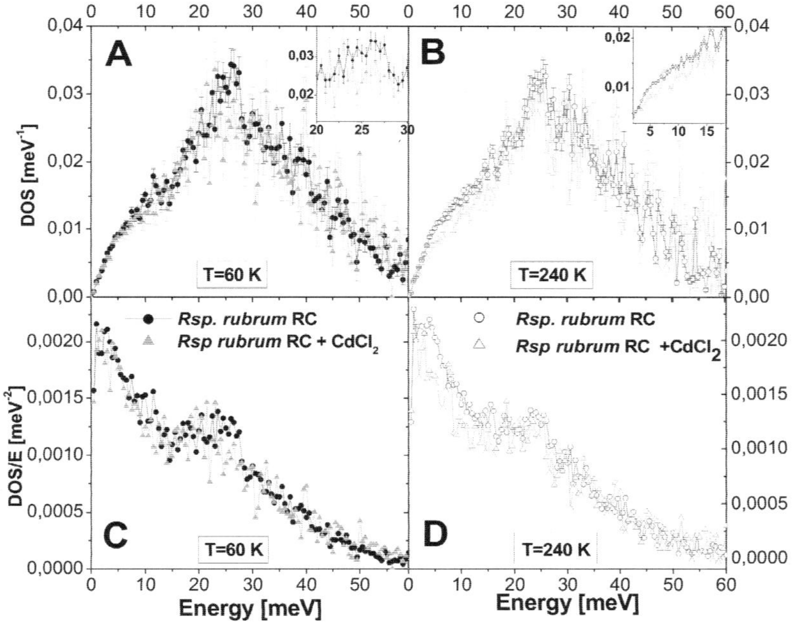

Figure 5 The iron-partial density of vibrational states, DOS, of the native and treated with $CdCl_2$ RCs isolated from *Rs. rubrum* measured at 60 K (A) and at 240 K (B). The reduced DOS of the native and treated with $CdCl_2$ RCs measured at 60 K (C) and at 240 K (D). The error bars along the horizontal line (A and B) evaluate the statistical reliability of the data. The insets show the mentioned in the text differences in the DOS spectra.

The low frequency vibrations originate from the resonant ^{57}Fe bound in the quinone-Fe complex due to the efficient transfer of protein fluctuations to NHFe bonds because only this atom is directly bound to the protein matrix. Thus, NHFe is more strongly connected to a big fragment of the protein core of RC than it is observed in the case of heme-proteins. It participates in the fast collective motions of the surrounding protein matrix fragment. The increase of the coupled mass of the protein matrix to NHFe (the increase of the moment of inertia) results in the decrease of the energy of the vibrational states [36, 37]. Therefore, in RCs from *Rs. rubrum,* NHFe vibrations dominate at low energies similarly to Fe-S clusters which have several bonds to the protein matrix [38]. It was shown in [39, 40] that the heme-Fe vibrations, regardless of whether heme is bound to protein or not, contribute significantly to the DOS at energies higher than 30 meV (in cytochrome c even higher than 35 meV) and they show only a very low contribution in the range of low vibrational frequencies (< 30 meV). Thus, the differences observed for energies < 30 meV (discussed below) come from the NHFe binding site modified by Cd^{2+}. At 60 K Cd^{2+} cations cause damping of fluctuations within the range of 23 - 27.5 meV (Fig.5A and C), which for our statistics is slightly larger than the statistical error. At 240 K the main statistically significant differences in DOS between the control and treated with $CdCl_2$

RCs occur at energies from 3.5 to 16.5 meV. Within this range of energies the contributions of the fluctuations detected for the control RCs are damped by Cd^{2+}. This effect is much more pronounced in Fig. 5D which shows the densities of vibrational states divided by the square of the energy, i.e., the so-called reduced DOS. Such presentation enhances differences in the vibrational states between the control and $CdCl_2$ treated RCs within the low energetic spectrum, observed at 240 K. The data shows that NHFe in RCs in the presence of Cd^{2+} has smaller contribution in the low-energy vibrational states and, consequently, feels less massive fragment of the RC protein core. This may result from the increased force constants of bonds between NHFe and its ligands from its first coordination sphere in comparison to the force constants of bonds from the more distant protein matrix. This decreasing contribution of vibrations at low energies due to the action of Cd^{2+} ions is especially interesting because it is important for understanding of the molecular mechanism related to the down regulation of ET between the two external quinone acceptors Q_A and Q_B by cadmium cations. It was observed that Cd^{2+} ions may slow down the electron and proton transfer within the iron-quinone complex [19, 35].

4. Conclusions

The presented results shed new light on the role of NHFe in the Q-type RCs. The Mössbauer experiments demonstrate that NHFe in *Rs. rubrum* is mainly in the low spin ferrous state while usually it occurs in a ferrous high spin state. Exposure of RCs to Cd^{2+} ions causes transfer of either the high or low spin state into a new low spin state. This indicates that Cd^{2+} binds in the vicinity of the iron-quinone complex and it does not remove NHFe from this complex (at least at the applied concentrations).

The NIS experiments provide a first direct evidence on the regulatory role of the NHFe on the activation of ET within the Q_A – Fe – Q_B complex. These studies performed on the RCs treated with $CdCl_2$ confirmed that Cd^{2+} ions are bound in the vicinity of the quinone-Fe complex and that they can influence the low energy protein fluctuations causing their damping, especially at 240 K. Because ET between theses two quinones is down regulated by Cd^{2+} and is initiated only at temperatures above 200 K, we suggest that the collective fluctuations damped by cadmium cations at about 3-16 meV are responsible for a proper activation of the coupling between Q_A and Q_B and efficient ET. This means that a certain flexibility of the RC core is required for the efficient action of the Q-type photosystem. This process can additionally be accompanied/regulated by protonation and deprotonation of residues located near NHFe as well as near the Q_A and Q_B ubiquinones binding sites (see Fig. 1B). This, in turn, may not only affect the strength of coordination bonds to NHFe but also regulate its redox potential.

Acknowledgements: This work was supported partially by grant No 2 P04A 04427 (2004-2007), N N302 2225 33 (2007-2008), N N302 454434 (2007-2008) and N N302 195035 (2008-2011) from Polish Ministry of Science and Higher Education and ESRF SC264 (155/ESR/2006/03). AH acknowledged the financial support from the project operated within the Foundation for Polish Science MPD Programme co-financed by the EU European Regional Development Fund.

References

[1] Deisenhofer J, Epp O, Miki R, Huber R and Michel H 1985 *Nature* **318** 618
[2] Feher G, Allen J P, Okamura M Y and Rees D C 1989 *Nature* **339** 111
[3] Hopfield J J 1974 *Proc. Natl. Acad. Sci. USA* **71** 3640
[4] Bixon M and Jortner J 1986 *J. Phys. Chem.* **90** 3795
[5] Balabin I A and Omuchnic J N 2000 *Science* **290** 114
[6] Lee H, Cheng Y-C and Fleming G R 2007 *Science* **316** 1462
[7] Kleinfeld D, Okamura M Y and Feher G 1984 *Biochemistry* **23** 5780
[8] Burda K 2007 *Cell Biochem. Biophys.* **47** 271

[9] Parak F, Frolov E N, Kononenko A A, Mössbauer R L, Goldanskii V I and Rubin A B 1980 *FEBS Lett.* **117** 368
[10] Garbers A, Reifarth F, Kurreck J, Renger G and Parak F 1998 *Biochemistry* **37** 11399
[11] Beroza P, Fredkin D R, Okamura M Y and Feher G 1995 *Biophys. J.* **68** 2233
[12] Kleinfeld D, Okamura M Y and Feher G 1985 *Biophys. J.* **48** 849
[13] Stowell M H B, McPhillips T M, Rees D C, Soltis S M, Abresch E and Feher G 1997 *Science* **276** 812
[14] Remy A and Gerwert K 2003 *Nature Struct. Biol.* **10** 637
[15] Hermes S, Bremm O, Garczarek F, Derrien V, Liebisch P, Loja P, Sebban P, Gerwert K and Haumann M 2006 *Biochemistry* **45** 353
[16] Burda K, Kruk J, Borgstadt R, Stanek J, Strzalka K, Schmid G H and Kruse O 2003 *FEBS Lett.* **535** 159
[17] Burda K, Kruk J, Strzałka K, Stanek J, Schmid GH and Kruse O 2006 *Acta Phys. Polon. A* **109** 237
[18] Boso B, Debrunner P, Okamura M Y and Feher G 1981 *Biochim. Biophys. Acta* **638** 173
[19] Lipińska M 2009 Studies of photosynthetic reaction centers activity under stress conditions *Department of Applied Spectroscopy* (Cracow: Institute of Nuclear Physics PAN) pp 1-237
[20] Belanger G, Berard J, Corriveau and Gingras G 1988 *J.Biol.Chem.* **263** 7632
[21] Komiya H, Yeates TO, Rees DC, Allen JP and Feher G 1988 *Proc. Natl. Acad. Sci. USA* **85** 9012
[22] Paddock ML, Graige MS, Feher G and Okamura MY 1999 *Proc. Natl. Acad. Sci. USA* **96** 6183
[23] Ishikita H and Knapp E-W 2005 *Proc. Natl. Acad. Sci. USA* **102** 16215
[24] Cohen- Bazire G, Sistrom W R and Stanier R Y 1957 *J. Cell. Comp. Physiol.* **49** 25
[25] Feher G and Okamura M Y 1978 *The Photosynthetic Bacteria*, ed R K Clayton and W R Sistrom (Plenum Press, New York) p 349
[26] Noel H, van der Rest M and Gingras G 1972 *Biochim. Biophys. Acta* **275** 219
[27] Rancourt DG, and Ping J-Y 1991 *Nucl. Instr. and Meth. B* ***58*** 87
[28] Rüffer R and Chumakov A I 1996 *Hyperfine Interactions*, **97/98** 589
[29] Chumakov A I and Sturhahn W 1999 *Hyperfine Interactions* **123/124** 781
[30] Moss T H, Bearden A J, Bartsch R G and Cusanovich M A 1968 *Biochemistry* **7** 158
[31] Debus R J, Feher G and Okamura M Y 1986 *Biochemistry* **25** 2276
[32] Kohl DH, Townsend J, Commoner B, Crespi H, Dougherty R C and Katz J J (1965) *Nature* **206** 1105
[33] His E S P and Bolton J B 1974 *Biochim. Biophys. Acta* **347** 126
[34] Breton J, Boullais C, Burie J R, Nabedryk E and Mioskowski C 1994 *Biochemistry* **33** 14378
[35] Axelrod H L, Abresch E C, Paddock M L, Okamura M Y and Feler G 2000 *Proc. Natl. Acad. Sci. USA* **97** 1542
[36] Sage J T, Durbin S M, Sturhahn W, Wharton D C, Champion P M, Hession P, Sutter J and Alp E E 2001 *Phys. Rev. Lett.* **86** 4966
[37] Achterhold K, Keppler C, Ostermann A, van Bürck U, Sturhahn W, Alp E E and Parak F G 2002 *Phys. Rev. E* **65** 051916
[38] Xiao Y, Tan M L, Ichiye T, Wang H, Guo Y, Smith M C, Meyer J, Sturhahn W, Alp E E, Zhao J, Yoda Y 2008 *Biochemistry* **47** 6612
[39] Adams K L, Tsoi S, Yan J, Durbin S M, Ramdas A K, Cramer W A, Sturhahn W, Alp E E and Shultz C 2006 *J. Phys. Chem. B* **110** 530
[40] Leu B M., Zhang Y, Bu L, Straub J E, Zhao J, Sturhahn W, Alp E E and Sage J T 2008 *Biophys. J.* **95** 5874

Physical and chemical induced spin crossover

F Renz[1]
Leibniz University Hannover, Institute of Inorganic Chemistry, Callinstr. 9,
D-30167 Hannover, Germany

E-mail: Franz.Renz@acd.uni-hannover.de

Abstract. Physical spin crossover (spin transition), one of the most fascinating dynamic electronic structure phenomena occurring in coordination compounds of third row transition metal ions, mostly of iron(II), iron(III) and cobalt(II), has attracted increasing attention by many research groups. Some examples for physical and chemical induced spin crossover are discussed. Among the physical is the temperature and pressure induced switching. Novel effects observed by irradiation and magnetic and electric field will be discussed. Some chemical induced switching effects caused by solvation, ligand exchange, ligand isomerisation and bond breaking will be briefly outlined.

1. Introduction

Molecular switching is one of the major targets in various disciplines, such as technology, biology and medicine. One approach is to develop chemically and physically induced transitions between electronic states. In some transition metal complexes such a switching has been discovered and labelled spin crossover (SC) or spin transition phenomenon [13, 14] (see Figure 1). The phenomenon of thermal spin crossover, though already well known in inorganic chemistry for more than seven decades, has developed in recent years to one of the most actively studied subjects in coordination chemistry [8, 14, 10, 12, 11, 15].

2. Physical Induced

2.1. Thermal Induced SC (TISC)

The thermal spin crossover was first observed in an Fe(III) complexes by Cambi, et al. in 1931. After 60 years the first Fe(II)-spin crossover complex, the $[Fe(II)(phen)_2(NCS)_2]$, was found by Baker, et al. in 1963. The Mössbauer effect was very important for the understanding of the nature of this switching phenomenon. The essential condition for TISC is that the zero-point-energy gap, the difference in zero-point vibrational energy between high-spin and low-spin state, is at the order of thermal energy kT [13, 14].

---

1 To whom any correspondence should be addressed.

2.2. Pressure Induced SC (PISC)

The application of pressure leads to a first order approximation to an increased zero-point-energy gap. This shifts the transition temperature in the TISC to higher values [16]. In case of structural changes due to the application of pressure the opposite may be the case [16].

2.3. Light Induced SC

2.3.1. Batch Light Induced SC

2.3.1.1. Light Induced Excited Spin-State Trapping (LIESST): The phenomena of LIESST (Light-Induced Excited Spin State Trapping) and reverse LIESST are well reviewed in the literature [14, 15]. For LIESST: green light irradiation on the Fe(II)-low-spin (S=0) state leads to an excited singlet state and decays via triplet states (S=1) to a trapped meta-stable quintet state, the Fe(II)-high-spin state (S=2) [14].

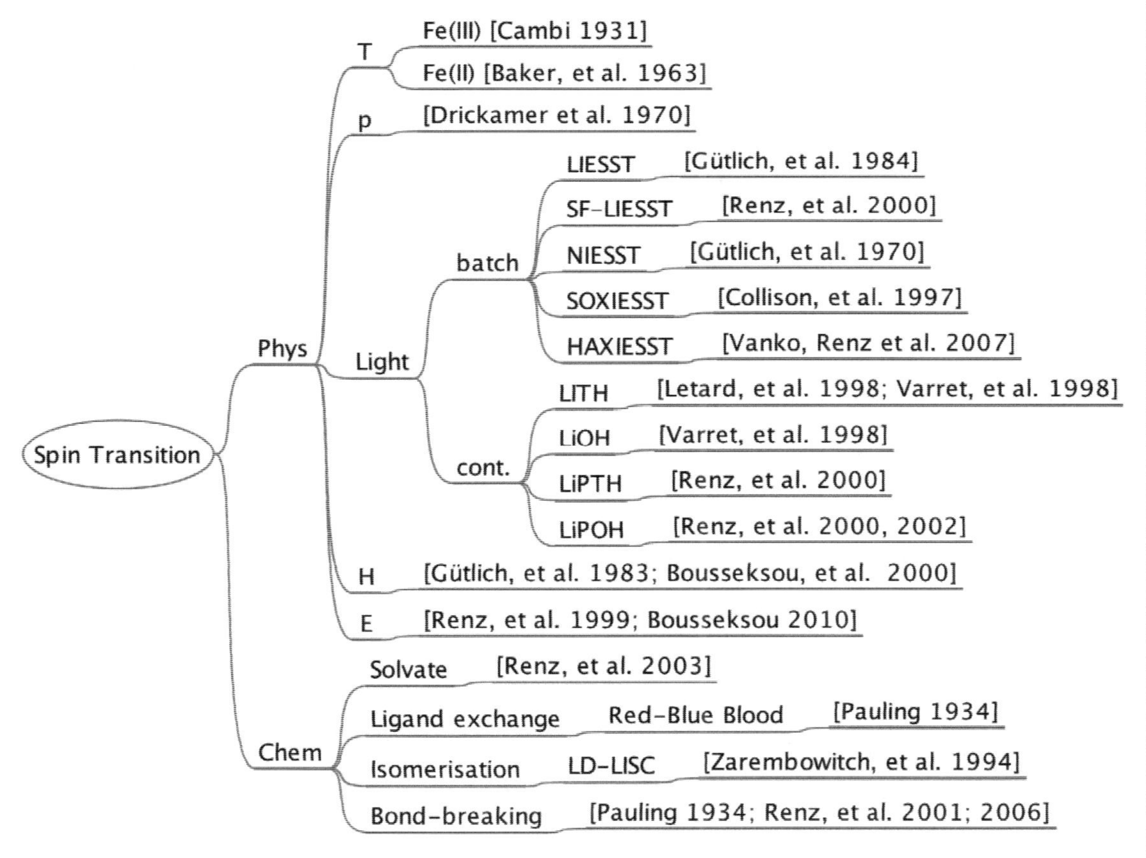

Figure 1. Spin Transition (or Spin Crossover) induced by physical and chemical changes

2.3.1.2. Strong-Field- Light Induced Excited Spin-State Trapping (SF-LIESST)[26]: Unexpected long life-time (mega-seconds) of the light-induced meta-stable state are observed for the strong-field complexes $[Fe(II)_xMn(II)_{1-x}(terpy)_2](ClO_4)_2$, $(x < 0.1)$. In opposite the non-dilluted $(x = 1)$ compound follows the expected LIESST mechanism at a life-time of sub-nanoseconds. The SF-LIESST effect seems to correlate with the tri-dentate inherent structure motive performing a double mode transition [22]. In comparison to that, usual mono-dentate perform single mode transitions. In such compounds the LIESST effect was first observed, e.g. $[Fe(II)(ptz)_6]^{2+}$.

2.3.1.3. Nuclear Induced Excited Spin-State Trapping (NIESST)[14]: The nuclear decay of radioactive atoms embedded in a compound is known to lead to various chemical and physical "after effects" such as bond rupture, redox processes, formation of metastable states and others. One ways of investigating such effects is the "Mössbauer Emission Spectroscopy" (MES). Using this technique some metastable HS states of 57-Fe(II) in strong and intermediate ligand fields are observed. Many examples are described in a recent review article [9]. In analogy to the LIESST effect (generating and trapping of metastable HS states by irradiation with an external light source) the mechanism of the "Nuclear Decay Induced Excited Spin State Trapping (NIESST)" with its nuclear decay process 57-Co(EC)57-Fe may be regarded as an "internal" molecular excitation source.

2.3.1.4. Soft-X-ray Induced Excited Spin-State Trapping (SOXIESST)[5]: The soft X-ray electronic L-shell excitation in iron(II) coordination complexes leads to trapped metastable HS states. This is similar to the LIESST effect. Unfortunately further "after" effects such as bond rupture are also observed. This limits the application of this effect.

2.3.1.5. Hard-X-ray Induced Excited Spin-State Trapping (HAXIESST)[30]: The hard X-ray electronic K-shell excitation in iron(II) coordination complexes leads to trapped metastable HS states. This is similar to NIESST, SOXIESST and LIESST effect. In comparison to the SOXIESST and NIESST the HAXIESST has a much smaller probability for irreversible bond breaking. While L-shell experiments are performed under vacuum a K-shell measurement is done at ambient pressure under atmospheric conditions. A further advantage of the HAXIESST is its ability of single molecule addressing and switching, due to the atomistic spatial resolution of the short wavelength. In opposite the LIESST effects has a resolution of around 500 nm.

2.3.2. Continuous Light Induced SC

2.3.2.1. Light Induced Thermal Hysteresis (LITH)[17, 7]: Monitoring the HS fraction under constant-intensity continuous irradiation and variable temperature allows to observe a light induced thermal hysteresis. The LITH indicates a measure for cooperativity among the switching centers. If the cooperativity between the switching centers vanishes the thermal hysteresis width collapses to zero, so that the heating and cooling curve become identical.

2.3.2.2. Light Induced Optical Hysteresis (LIOH)[7]: Monitoring the HS fraction under variable-intensity continuous irradiation and constant temperature allows to observe a light induced optical hysteresis. LITH and LIOH can be interpreted in mean approximation with a master equation including a non-linear relaxation term[7].

2.3.2.3. Light Perturbed Thermal Hysteresis (LiPTH)[27]: Monitoring the HS fraction under constant-intensity continuous irradiation and variable temperature within a thermal hysteresis allows to Light-perturb the thermal hysteresis (LiPTH). LiPTH allows to switch even at room temperature and in batch irradiations[28]. So far the maximum shift of a hysteresis was found to be around 50 K.

2.3.2.4. Light Perturbed Optical Hysteresis (LiPOH)[28, 27]: Monitoring the HS fraction under variable-intensity continuous irradiation and constant temperature within a thermal hysteresis allows to Light-perturb the thermal hysteresis seen as an optical hysteresis (LiPOH). LiPOH allows to switch even at room temperature and in batch irradiations[28].

2.4. Magnetic Field Induced SC (H)[4]

The effect of the magnetic field on the spin crossover transition temperature was weak and yielded a shift of ca. 2 K with a field of 5 Tesla as seen by Gütlich, et al. [21]. Almost two decades later Bousseksou, et al. used pulsed high-fields of up to 32 Tesla [3]. This turned out to be very effective[4].

2.5. Electric Field Induced SC (E)

The effect of the electric field on the spin crossover was seen on stretching a polar polymer matrix [25]. The matrix-incorporated spin crossover compound was exposed and affected by the electrostatic field of the polymer. More recent examples show the effect of the electrostatic field affects the bistability [6, 1, 19].

3. Chemical Induced

3.1. Solvate Induced SC

In solution the cooperativity is lowered and exhibit a Bolzmann-like behaviour. In opposite, non-coordinated solvates or adsorbats induce molecular switching (AIMS)[24] with potential room for applications.

3.2. Ligand Exchange Induced SC

Many compounds change their spin state on ligand exchange. E.g. hemoglobine reversibly coordinates oxygen. This causes the blue and red blood-color. Along with that a spin state change was observed by Pauling and Corvell [20].

3.3. Ligand Isomerisation Induced SC[29]

Roux et al. [29] have reported a method to obtain ligand-driven light-induced spin changes (LD-LISC) in transition metal compounds. Photo-sensitive ligands, such as cis-trans photo-isomerizable ligands, can alter the ligand field strength on irradiation and induce a spin state conversion at the metal center. LD-LISC was first observed in $[Fe(II)L_4(NCS)_2]$ with L = cis, trans 4-styrylpyridine. For further examples see[2].

3.4. Bond-Breaking Induced SC

A non-reversible spin state change after a coordination bond-breaking is quite normal for many such chemical reactions. However, reversibility of such a phenomena is rare. One example is a rod-type structure consisting of $[Fe(II)(acetate)_2]$ bridged by terpy-R-terpy (spacer R). Such systems switches pressure/digest and light-reversible [18, 23].

4. Outlook

This paper is a short compromise on some representative examples in an egocentric attempt to cover the vast development in the field and is far from complete. The intended compromise is to give a broad overview on selected effects and is not focused to systematically cover all the beautiful new compounds. Several very recent appealing new effects have not been included due to space restrictions.

Acknowledgments

Author wishes to acknowledge the assistance of his working group and his collaborators. This work war partly funded by the Leibniz University of Hanover (LUH), the ZFM/LUH, the LNQE/LUH, and the "Deutsche Forschungsgemeinschaft" (Priority Program SPP 1136 "Molecular Magnetism"; Re 1627/1-3), the "Fonds der chemischen Industrie".

References

[1] Baadji N, Piacenza M, Tugsuz T, Sala F D, Maruccio G and Sanvito S 2009 Electrostatic spin crossover effect in polar magnetic molecules *nature materials* **8** 813-7
[2] Boillot M-L, Zarembowitch J and Sour A 2004 *Spin Crossover in Transition Metal Compounds II,* ed P Gütlich and H Goodwin (Heidelberg: Springer) pp 261-76
[3] Bousseksou A, Negre N, Goiran M, Salmon L, Tuchagues J P, Boillot M L, Boukheddaden K and Varret F 2000 Dynamic triggering of a spin-transition by a pulsed magnetic field *Eur. Phys. J. B* **13** 451-6
[4] Bousseksou A, Varret F, Goiran M, Boukheddaden K and Tuchagues J P 2004 *Spin Crossover in Transition Metal Compounds III,* ed P Gütlich and H Goodwin (Heidelberg: Springer) pp 65-84
[5] Collison D, Garner C D, McGrath C M, Mosselmans J F W, Roper M D, Seddon J M W, Sinn E and Young N A 1997 Soft X-ray induced excited spin state trapping and soft X-ray photochemistry at the iron L2,3 edge in [Fe(phen)2(NCS)2] and [Fe(phen)2(NCSe)2] (phen = 1,10-phenanthroline) *J. Chem. Soc. Dalton Trans.* 4371–6
[6] D'Avino G, Grisanti L, Guasch J, Ratera I, Veciana J and Painelli A 2008 Bistability in Fc-PTM Crystals: The Role of Intermolecular Electrostatic Interactions *J. Am. Chem. Soc.* **130** 12064-72
[7] Desaix A, Roubeau O, Jeftic J, Haasnoot J G, Boukheddaden K, Codjovi E, Linares J, Nogues M and Varret F 1998 Light-induced bistability in spin transition solids leading to thermal and optical hysteresis *Eur. Phys. J. B* **6** 183-93
[8] Gütlich P 1981 Spin crossover in iron(II)-complexes *Struct Bonding* **44** 83-192
[9] Gütlich P 2004 *Spin Crossover in Transition Metal Compounds II,* ed P Gütlich and H Goodwin (Heidelberg: Springer) pp 231-60
[10] Gutlich P, Garcia Y and Goodwin H A 2000 Spin crossover phenomena in Fe(II) complexes *Chem. Soc. Rev.* **29** 419-27
[11] Gütlich P, Garcia Y and Spiering H 2002 *Magnetism: Molecules to Materials IV,* ed J S Miller and M Drillon (Weinheim: Wiley-VCH) pp 271-344
[12] Gutlich P, Garcia Y and Woike T 2001 Photoswitchable coordination compounds *Coordination Chemistry Reviews* **219** 839-79
[13] Gütlich P and Goodwin H A 2004 *Spin Crossover in Transition Metal Compounds I,* ed P Gütlich and H Goodwin (Heidelberg: Springer) pp 1-48

[14] Gütlich P, Hauser A and Spiering H 1994 Thermal and optical switching of iron(II) complexes *Angewandte Chemie - International Edition in English* **33** 2024-54

[15] Gütlich P, van Koningsbruggen P J and Renz F 2004 Recent advances of spin crossover research *Optical Spectra and Chemical Bonding in Transition Metal Complexes* **107** 27-75

[16] Ksenofontov V, Gaspar A B and Gütlich P 2004 *Spin Crossover in Transition Metal Compounds III,* ed P Gütlich and H Goodwin (Heidelberg: Springer) pp 23-64

[17] Letard J F, Guionneau P, Rabardel L, Howard J A K, Goeta A E, Chasseau D and Kahn O 1998 Structural, magnetic, and photomagnetic studies of a mononuclear iron(II) derivative exhibiting an exceptionally abrupt spin transition. Light-induced thermal hysteresis phenomenon *Inorg. Chem.* **37** 4432-41

[18] Lindner A, Menzel M, Renz F, Kurth D G and Thünemann A F 2006 Metallosupramolecular coordination polyelectrolytes investigated by Mössbauer spectroscopy *Hyperfine Interact.* **166** 465-8

[19] Mahfoud T, Molnar G, Bonhommeau S, Cobo S, Salmon L, Demont P, Tokoro H, Ohkoshi S-I, Boukheddaden K and Bousseksou A 2009 Electric-Field-Induced Charge-Transfer Phase Transition: A Promising Approach Toward Electrically Switchable Devices *J. Am. Chem. Soc.* **131** 15049-54

[20] Pauling L and Corvell C D 1936 The Magnetic Properties and Structure of the Hemochromogens and related Substances *PNAS* **22** 159-63

[21] Qi Y, Müller E W, Spiering H and Gütlich P 1983 The effect of a magnetic field on the high-spin α low-spin transition in bis(1,10-phenanthroline)iron thiocyanate ($[Fe(phen)_2(NCS)_2]$) *Chem. Phys. Lett.* **101** 503-5

[22] Renz F 1995 *Diploma Thesis,* (Vienna: Vienna University of Technology)

[23] Renz F 2001 Light Induced Excited States in Mono and Polynuclear Strong-Field Fe(II) Complex with Terpyridine-Derivatives *Molecular Photomagnetism Meeting*

[24] Renz F, De Souza P A, Klingelhofer G and Goodwin H A 2002 Molecular sensors for moisture detection by Mossbauer spectroscopy *Hyperfine Interact.* **139** 699-704

[25] Renz F, Hasegawa M, Hoshi T, El-Ayaan U, Linert W and Fukuda Y 1999 Anisotropic Effect after Stretching of the Spin-Crossover Compound [Fe(II)(2,6-bis-(benzimidazol-2'-yl)pyridine)(2)] (ClO4)(2) in Polyvinylalcohol Polymer Matrix *Mol. Cryst. Liq. Cryst. Sci. Technol. Sect. A-Mol. Cryst. Liq. Cryst.* **335** 1243-52

[26] Renz F, Oshio H, Ksenofontov V, Waldeck M, Spiering H and Gutlich P 2000 Strong field iron(II) complex converted by light into a long-lived high-spin state *Angew. Chem.-Int. Edit.* **39** 3699-700

[27] Renz F, Spiering H, Goodwin H A and Gütlich P 2000 Light-perturbed hysteresis in an iron(II) spin-crossover compound observed by the Mossbauer effect *Hyperfine Interact.* **126** 155-8

[28] Renz F, Spiering H, Gütlich P, Sugiyarto K H and Goodwin H 2002 Light Perturbed Thermal Hysteresis (LiPTH) in [Fe(bpp)2][Fe(CN)5NO]. In: *TOSS-Meeting,* (Seeheim, Germany)

[29] Roux C, Zarembowitch J, Gallois B, Granier T and Claude R 1994 Toward Ligand-Driven Light-Induced Spin Change - Influence of the Configuration of 4-Styrylpyridine (Stpy) on the Magnetic- Properties of Feii(Stpy)4(Ncs)2 Complexes - Crystal-Structures of the Spin-Crossover Species Fe(Trans-Stpy)4(Ncs)2 and of the High-Spin Species Fe(Cis-Stpy)4(Ncs)2 *Inorg. Chem.* **33** 2273-9

[30] Vanko G, Renz F, Molnar G, Neisius T and Karpati S 2007 Hard-X-ray-induced excited-spin-state trapping *Angew. Chem.-Int. Edit.* **46** 5306-9

^{99}Ru and ^{61}Ni Mössbauer Spectroscopic Studies Using the Accelerator at RIKEN

Y Kobayashi

Advanced Meson Science Laboratory, Nishina Center for Accelerator Based Science (RNC), RIKEN, Hirosawa, Wako, Saitama 351-0198, Japan

E-mail: kyoshio@riken.jp

Abstract. Ruthenium and nickel are important constituent elements of a wide variety of compounds, including oxides, catalysts, and functional and industrial materials for solar butteries and shape-memory alloys. ^{99}Ru and ^{61}Ni Mössbauer spectroscopy can reveal significant information about the physical and chemical properties of these compounds. However, there have not been many studies that use Mössbauer spectroscopy to investigate Ru and Ni compounds. This paper discusses Mössbauer source preparation methods using the RIKEN accelerator for conventional Mössbauer absorption spectra in addition to our recent results.

1. Introduction

The Mössbauer effect has been observed for nearly 100 nuclear transitions in about 80 nuclides distributed over 43 elements [1]. Of these elements, about 20 have yet to be investigated by Mössbauer spectroscopy, since not all transitions are amenable to normal experimental conditions. Mössbauer elements of Fe, Ni, Zn, Ru, Sn, Sb, Te, I, Xe, Eu, Hf, Ta, W, Re, Os, Ir, Pt, Au, Hg, U are used in chemistry, solid-state physics, biology, metallurgy, and earth science. Some Mössbauer nuclides including ^{57}Fe, ^{119}Sn and ^{151}Eu have been extensively investigated in a wide variety of fields, whereas other nuclides have not been studied greatly by Mössbauer spectroscopy despite being significant elements for materials science. There are several reasons for this state of affairs: the half-lives of Mössbauer mother nuclides are too short, inappropriate lifetimes of the first excited levels, few Mössbauer transitions, and complicated preparation procedures or their chemical treatments to produce appropriate Mössbauer mother nuclides.

In addition to Fe and Sn, Ru and Ni are important iron-group elements. Ruthenium is a major element of important oxides, catalysts utilized in organic synthesis, and solar energy technologies. An interesting feature of Ru chemistry is that Ru has a large variety of oxidation states ranging from 0 to 8+. For example, Ru ions have an oxidation state of 4+ ($4d^4$ configuration) in superconducting Sr_2RuO_4 (T_c=1.5 K) [2], and an oxidation state of 8+ ($4d^0$) in RuO_4. On the other hand, Ni is ferromagnetic near room temperature. It is used in many industrial and electrical devices including magnets, rechargeable butteries, shape-memory alloys, and catalysts. In addition, several enzymes and microorganisms contain nickel.

It is important to study the physicochemical properties of many Ru and Ni compounds using ^{99}Ru and ^{61}Ni Mössbauer spectroscopy. However, there have not been many Mössbauer spectroscopic studies of Ru and Ni compounds because of experimental difficulties. Specifically, ^{99}Ru and ^{61}Ni do not have long-lived nuclides that can be used as Mössbauer sources, and both the source and the absorber should be cooled to low temperatures because of the low Debye–Waller factors caused by the relatively high energies of the γ-ray transitions (see Fig. 1). Table 1 compares typical precursors, the nuclear reactions for source preparation, and nuclear data of ^{99}Ru and ^{61}Ni Mössbauer spectroscopy with those of ^{57}Fe Mössbauer spectroscopy. Facilities for the source preparation, including a heavy-ion accelerator and the liquid helium supply are necessary for continuous spectroscopic measurement. In this paper, the experimental procedures for ^{99}Ru and ^{61}Ni Mössbauer spectroscopy using the accelerator at RIKEN are introduced and recent results for $CaRuO_3$ [3], $Hg_2Ru_2O_7$ [4,5] and Ru-containing skutterudites [6,7] are presented.

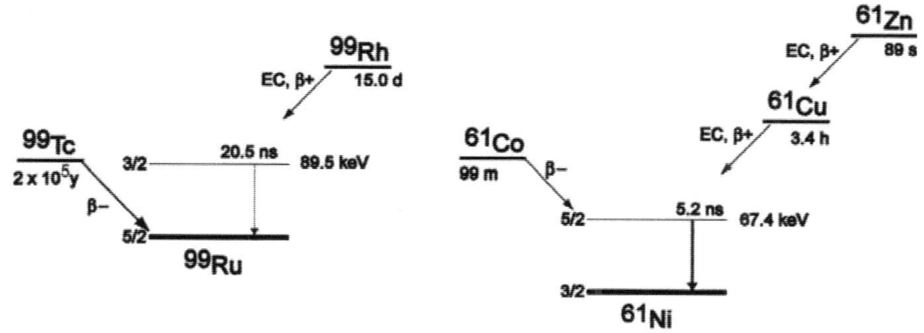

Figure 1. Simplified decay schemes of ^{99}Ru and ^{61}Ni. Only transitions relevant to Mössbauer studies are shown.

Table 1. Mössbauer precursor nuclides and simplified nuclear data of ^{57}Fe, ^{61}Ni, and ^{99}Ru.

nuclide	abundance	precursor (reaction)	$T_{1/2}$	E_γ (keV)	$t_{1/2}$ (ns)	Γ_0 (mm/s)
^{57}Fe	2.1 %	^{57}Co	270 d	14.4	97.8	0.1946
^{61}Ni	1.1 %	^{61}Co (^{62}Ni(γ,p)^{61}Co) ^{61}Cu (^{58}Ni(α,p)^{61}Cu)	99 m 3.3 h	67.4	5.27	0.7805
^{99}Ru	12.8 %	^{99}Rh (^{99}Ru(p,n)^{99}Rh)	16 d	89.8	20.5	0.1468

$T_{1/2}$: half-life of Mössbauer precursor nuclide, E_γ : energy of Mössbauer γ-ray transition, $t_{1/2}$: lifetime of excited nuclear state, Γ_0 : natural linewidth (FWHM).

2. ^{99}Ru Mössbauer spectroscopic studies at RIKEN

2. 1. Source preparation and ^{99}Ru Mössbauer measurement

First, the source nuclide, ^{99}Rh ($T_{1/2}$=15.0 d) of ^{99}Ru Mössbauer spectroscopy must be produced by the ^{99}Ru(p,n)^{99}Rh reaction. Figure 2 shows a schematic of the experimental layout at a beam-line connected with the AVF cyclotron (K70-MeV) in RIKEN Nishina Center (RNC) [8]. The AVF cyclotron typically provides protons with energies up to 14 MeV [9]. The proton beam for the (p,n) reactions was extracted from the AVF cyclotron and used to bombard the target of 96 %-enriched ^{99}Ru metallic powder. The energy and intensity of the p-beam were 12 MeV and 10 μA, respectively. The ^{99}Ru-enriched powder was pressed and packed into an Al holder. The holder was tightly fixed to the irradiation equipment and the target was cooled by He gas flow and water flow during the p-irradiation. The active area (*i.e.*, the spot size of the p-beam) was 8×8 mm^2. After p-irradiation for about 24 hours, the ^{99}Rh Mössbauer source was estimated to have an activity of 10 MBq. The Al holder was removed from the beam-line, and immediately transported to the Mössbauer laboratory in RNC for off-line measurement without being subjected to annealing or chemical treatment. The obtained ^{99}Rh Mössbauer source could be used for measurements up to about 7–8 weeks after *p*-irradiation.

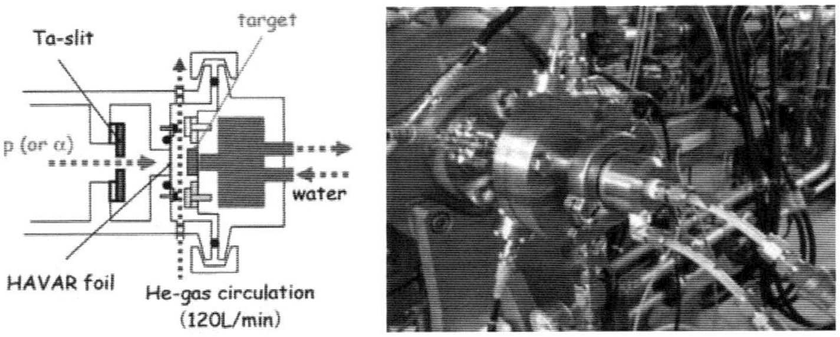

Figure 2. A schematic of the experimental setup for the Mössbauer source preparation. The beam-line is directly connected with the AVF cyclotron in RNC.

^{99}Ru Mössbauer spectra were obtained using a Mössbauer spectrometer (Wissel, MVT-1000) and a multichannel analyzer (EG&G, ORTEC MCS-plus). Due to the relatively high energy of the Mössbauer γ-ray (89.8 keV), both the source and the absorbers were maintained at low temperatures in a liquid-helium cryostat during the measurements. The Mössbauer γ-rays were detected by a 2-mm-thick NaI(Tl) scintillator. Velocity calibration of the system was performed by measuring the Mössbauer absorption lines of ^{57}Fe in an iron foil against a ^{57}Co/Rh source. In our experimental setup, the linewidth (FWHM) and the area intensity of the resonance singlet were 0.25(2) mm/s and 1.56(5) %, respectively, against an absorber of natural ruthenium metal.

We analyzed the ^{99}Ru Mössbauer spectrum by least-squares fitting, adopting the following values for the electric quadrupole (Q) and magnetic dipole moments (μ) of the ^{99}Ru nucleus: Q_e = 0.22 b and μ_e = −0.285 nm for the excited level (I_e = 3/2), and Q_g= 0.076 b and μ_g = −0.623 nm for the ground level (I_g = 5/2). The E2/M1 mixing ratio, δ^2, for the γ-transition was assumed to be 2.7 [10]. When the nuclei of ^{99}Ru are in a pure magnetic field with no electric field gradient, 18 absorption lines appear symmetrically in the spectrum. An asymmetric doublet, each branch consisting of three unresolved lines, appears in the presence of an electric field gradient of ordinary magnitude [11]. The chemical valence state dependence of the isomer shift of ^{99}Ru is similar to that of ^{57}Fe, although the sign is reversed because $\Delta R/R$ is positive for ^{99}Ru.

2. 2. Applications of ^{99}Ru Mössbauer spectroscopy

Since the 1990s, we have used ^{99}Ru Mössbauer spectroscopy to investigate ruthenium trichlorides [11], Fe_3Si-based Heusler alloys [12] and to perform nuclear probe studies in iron oxides [13,14]. We summarize our recent results below.

2. 2. 1. *Perovskite $CaRuO_3$* : Orthorhombic ruthenates MRuO$_3$ (M = Sr or Ca) have attracted renewed interest as a result of the superconductivity found in Sr_2RuO_4 [2]. $SrRuO_3$ and $CaRuO_3$ have orthorhombically distorted $GdFeO_3$-type structures and exhibit metallic conductivity. However, despite having similar crystallographic and electrical properties, they exhibit quite different magnetic properties. $SrRuO_3$ is ferromagnetic with a T_C of 165 K [15,16], whereas the magnetic ground state of $CaRuO_3$ has yet to be determined (possibilities include antiferromagnetic [17], weak ferromagnetic [18–20], a spin-glass-like state [21], exchange-enhanced paramagnetic [22]). We have studied the magnetism of $CaRuO_3$ by magnetization measurements and Mössbauer spectroscopy. Our previous

studies revealed that no spin glass-like state exists in $CaRuO_3$ and that its magnetism strongly depends on its purities. Single crystals of $CaRuO_3$ were prepared by a self-flux method using $CaCl_2$ as the flux. An off-stoichiometric mixture of $CaCO_3$, RuO_2, and $CaCl_2$ was charged into a Pt crucible and heated to 1350 °C in an electric furnace. It was then carefully cooled to 900 °C, and quenched to room temperature to prevent other phases from growing. A small-crystal sample of $CaRuO_3$ was found to be mostly composed of rough cubes of 0.5 mm^3. These crystals were characterized by XRD, EPMA, magnetization and electrical resistivity measurements [3], ^{99}Ru Mössbauer spectroscopy, and μSR method.

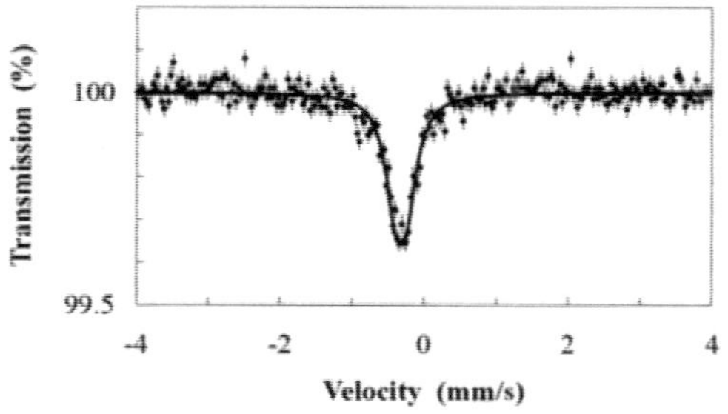

Figure 3. ^{99}Ru Mössbauer spectrum of the single crystals of $CaRuO_3$ at 4.2 K.

From the measurements of the temperature dependence of the magnetic susceptibility between 2 and 300 K, it was found that the magnetic susceptibility, $\chi(T)$, exhibited paramagnetic behavior and obeyed the Curie-Weiss law. The effective magnetic moment in the temperature range 90 K < T < 300 K was estimated to be P_{eff} = 2.92 μ_B. Figure 3 shows the ^{99}Ru Mössbauer spectrum obtained from single crystals of $CaRuO_3$ at 4.2 K. This spectrum could be satisfactorily analyzed by assuming an axially symmetric electric field gradient. The isomer shift was −0.31 mm/s and the quadrupole splitting was 0.22 mm/s. Based on these Mössbauer parameters, the Ru ion was assigned to be in a low-spin state of Ru^{4+}. The Mössbauer measurement results indicate that $CaRuO_3$ has no long-range magnetic order down to 4.2 K. In addition, the μSR measurements were performed down to 0.3 K at RIKEN RAL muon facility. The results obtained for the temperature dependences of ZF-, TF-, and LF-μSR measurements agree with those described above.

2. 2. 2. *Pyrochlore $Hg_2Ru_2O_7$*: $Hg_2Ru_2O_7$ is a new ruthenium oxide with pyrochlore structure, $A_2B_2O_7$, which was recently synthesized by Yamamoto *et al.* [4,5]. Of the wide variety of *4d*- and *5d*-pyrochlore oxides, pyrochlores with tetravalent ruthenium Ru^{4+} occupying the *B*-site, $A_2^{3+}Ru_2^{4+}O_7$, have been intensively studied. In contrast to Ru^{4+} pyrochlores, only two Ru^{5+} pyrochlores, $Cd_2Ru_2O_7$ and $Ca_2Ru_2O_7$, has been reported, partly due to the difficulty of oxidizing Ru to make it pentavalent. $Hg_2Ru_2O_7$ is a new pyrochlore oxide and it is the third Ru^{5+} compound to be synthesized. The properties of Cd and Ca analogues have not been completely explored. However, both have a relatively low resistivity (ρ) of the order of milliohm-centimeters but neither exhibits a metallic T dependence (*i.e.*, $d\rho/dT > 0$) [23,24]. In contrast to Cd and Ca analogues, Yamamoto *et al.* found a well defined a metal–insulator transition (MIT) at ~107 K in $Hg_2Ru_2O_7$. ^{99}Ru Mössbauer spectroscopy was used to determine the oxidation state of Ru ions in $Hg_2Ru_2O_7$. The $Hg_2Ru_2O_7$ sample was prepared in a small Au cell under 6 GPa using a cubic-anvil-type high-temperature and high-pressure apparatus. Consequently, it was very difficult to produce an adequate amount for ^{99}Ru Mössbauer measurements. Although the quality of the ^{99}Ru Mössbauer spectrum was low, preliminary analysis suggests that the oxidation state of Ru ions in $Hg_2Ru_2O_7$ is Ru^{5+}; however, further studies are needed to confirm this.

2. 2. 3. *Skutterudite $SmRu_4P_{12}$* : $SmRu_4P_{12}$ is an attractive compound among filled skutterudites. Its electric resistivity shows MIT at T_{MI} = 16 K. Specific heat measurements with an applied magnetic field suggest that successive transitions occur around T_{MI}. The magnetic susceptibility of $SmRu_4P_{12}$ has a cusp around 14 K, implying an antiferromagnetic ordering. The successive transitions were initially ascribed to antiferroquadrupole and antiferromagnetic orderings, since the entropy at T_{MI} reaches *R log* 4. The possibility of an octupole ordering has been proposed in $SmRu_4P_{12}$ based on a macroscopic point of view. The clear jump in the specific heat and antiferromagnetic-like temperature dependence of the magnetic susceptibility in $SmRu_4P_{12}$ resemble those in NpO_2, which is a typical system with octupole ordering. If the ordering in $SmRu_4P_{12}$ is caused by an octupole similar to NpO_2, the absence of the hyperfine field and/or magnetic moment should be clarified at the Sm site in addition to the observation of the hyperfine field at ligand sites.

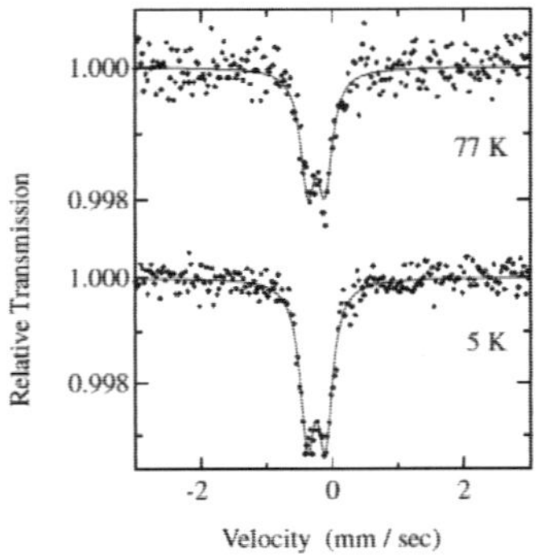

Figure 4. ^{99}Ru Mössbauer spectra of $SmRu_4P_{12}$ at 77 K and 5 K [7].

Figure 4 shows ^{99}Ru Mössbauer spectra of $SmRu_4P_{12}$. The spectra obtained at 77 and 5K could be satisfactorily analyzed using typical doublets with no obvious magnetic field. These doublets are caused by a sufficiently large electric field gradient near the ^{99}Ru nuclei that possess small asymmetries. The value of the isomer shift (*I.S.* = –0.24(2) mm/s at 5 K) suggests that the Ru ions are in the tetravalent state.

The Debye temperature at the Ru site in $SmRu_4P_{12}$ was estimated to be high because the relative area intensity of the spectrum exhibited little variation with increasing temperature. The Mössbauer spectroscopic results support the presence of a dipole moment, which was demonstrated by X-ray absorption spectroscopy and ^{149}Sm nuclear resonant forward scattering at low temperatures performed by Tsutsui *et al.* [7]. The presence of the magnetic dipole moments at the Sm ions is in a sharp contrast with their absence in NpO_2. The experimental results reported thus far suggest that a possible primary order parameter is a T^{α}-type octupole.

3. ^{61}Ni Mössbauer spectroscopic studies at RIKEN

Two parent nuclides of ^{61}Co and ^{61}Cu are used for ^{61}Ni Mössbauer spectroscopy (see Fig. 1). Although the half-life of ^{61}Co ($T_{1/2}$ = 99 m) is considerably shorter than that of ^{61}Cu ($T_{1/2}$ = 3.4 h), ^{61}Co is more suitable for Mössbauer measurements because it undergoes β^- decay direct to 67.4-keV Mössbauer levels. The Mainz group in Germany has performed extensive ^{61}Ni Mössbauer spectroscopic investigations using the ^{61}Co source. ^{61}Co source are typically produced by ^{62}Ni(γ,*p*)^{61}Co using bremsstrahlung (100 MeV) from an electron accelerator or by ^{64}Ni(*p*, α)^{61}Co with proton bombardment. These nuclear reactions require enriched ^{62}Ni or ^{64}Ni, but both isotopes have very low isotopic abundances (3.6 % for ^{62}Ni and 0.9 % for ^{64}Ni). On the other hand, we adopted ^{61}Cu as the source at RNC, although it decays in a complex manner with only 2.4 % populating the Mössbauer level. The nuclear reactions of $^{58}Ni(\alpha,p)^{61}Cu$ and $^{58}Ni(\alpha,n)^{61}Zn \rightarrow EC\ \beta^{+} \rightarrow ^{61}Cu$ are utilized to produced the source nuclide in the AVF cyclotron. When a paramagnetic $Ni_{0.84}V_{0.16}$ alloy was irradiated for 3 hours with 25 MeV α-particles, a ^{61}Cu Mössbauer source with an activity of 10 MBq was obtained. This preparation method has an advantage that it does not require an enriched isotope because ^{58}Ni has a high natural abundance of 68.1 %. The Mössbauer source preparation procedure was the same as that for ^{99}Rh. Recently, we have used ^{61}Ni Mössbauer spectroscopy to evaluate the catalytic performance of NiO samples; detailed results of this study will soon be published elsewhere.

4. Conclusion

Ru and Ni are important constituent elements of a wide variety of compounds, including oxides, catalysts, and functional materials in solar butteries and shape-memory alloys. ^{99}Ru and ^{61}Ni Mössbauer spectroscopy can reveal important information about the physicochemical properties of these compounds. We have developed irradiation equipment for these Mössbauer sources and an off-line measurement system at low temperatures at RNC. In addition, ^{61}Ni Mössbauer spectra have been obtained by a new technique using synchrotron radiation at SPring-8 and ESRF. Moreover, short-lived radioisotope-beam implantation methods will be available for studying of these elements. Conventional methods that use Mössbauer sources produced by nuclear reactions play a complementary role to these sophisticated Mössbauer techniques, but they remain the most basic Mössbauer measurement technique.

References

[1] Gütlich P, Link R and Trautwein A 1978 *Mössbauer Spectroscopy and Transition Chemistry* (Berlin: Springer-Verlag)
[2] Maeno Y, Hashimoto H, Yoshida K, Nishizaki S, Fujita T, Bednorz J G and Lichtenberg F 1994 *Nature,* **372** 532
[3] Koriyama A, Ishizaki M, Ozawa T C, Taniguchi T, Nagata Y, Samata H, Kobayashi Y and Noro Y 2004 *J. Alloys Comp.* **372** 58
[4] Yamamoto A, Sharma P A, Okamoto Y, Nakao A, Katori H A, Niitaka S, Hashizume D and Takagi H 2007 *J. Phys. Soc. Jpn.* **76** 0437031
[5] Takeshita N, Terakura C, Tokura Y, Yamamoto A and Takagi H 2007 *J. Phys. Soc. Jpn.* **76** 0637071
[6] Tsutsui S, Umemura J, Kobayashi H, Tazaki T, Nasu S, Kobayashi Y, Yoda Y, Onodera H, Sugawara H, Matsuda T D, Kikuchi D, Sato H, Sekine C and Shirotani I 2006 *Hyp. Int.* **168** 1073
[7] Tsutsui S, Kobayashi Y, Okada T, Haba H, Onodera H, Yoda Y, Mizumaki M, Tanida T, Uraga T, Sekine C, Shirotani I, Kikuchi D, Sugawara H and Sato H 2006 *J. Phys. Soc. Jpn.* **75** 0937031.
[8] Haba H and Enomoto S 2005 *RIKEN Prog. Accel. Prog. Rep.* **38** 105
[9] refer to WEB of RNC (http://www.nishina.riken.jp/UsersGuide/accelerator /overview.html).
[10] Kistner O C 1966 *Phys. Rev.* **144** 1022
[11] Kobayashi Y, Okada T, Asai K, Katada M, Sano H and Ambe F 1992 *Inorg. Chem.* **31** 4570
[12] Kobayashi Y, Asai K, Okada T and Ambe F 1994 *Hyp. Int.* **84** 131
[13] Ohkubo Y, Kobayashi Y, Asai K, Okada T and Ambe F 1993 *Phys. Rev.* B **47** 11954
[14] Ohkubo Y, Kobayashi Y, Harasawa K, Ambe S, Okada T, Ambe F, Asai K and Shibata S 1995 *J. Phys. Chem.* **99** 10629
[15] Callaghan A, Moeller C W and Ward R 1966 *Inorg. Chem.* **5** 1572
[16] Bouchard R J and Gillson J L 1972 *Mater. Res. Bull.* **7** 873
[17] Longo J M, Raccah P M and GoodenoughJ B 1968 *J. Appl. Phys.* **39** 1327
[18] Yoshimura K, Imai T, Kiyama T, Thurber K R, Hunt A W and Kosuge K 1999 *Phys. Rev. Lett.* **83** 4397
[19] Kiyama T, Yoshimura K and Kosuge K 1999 *J. Phys. Soc. Jpn.* **68** 3372
[20] Mukuda H, Ishida K, Kitaoka Y, Asayama K, Kanno R and Takano M 1999 *Phys. Rev.* B **60** 12279
[21] Felner I, Nowik I, Bradaric I and Gospodinov M 2000 *Phys. Rev.* B **62** 11332
[22] Kiyama T, Yoshimura K, Kosuge K, Ikeda Y and Bando Y 1996 *Phys. Rev.* B **54** R756
[23] Wang R and Sleight A W 1998 *Mater. Res. Bull.* **33** 1005
[24] Munenaka T and Sato H 2006 *J. Phys. Soc. Jpn.* **75** 103801

Journal of Physics: Conference Series **217** (2010) 012024 doi:10.1088/1742-6596/217/1/012024

Mössbauer spectroscopy on superparamagnetic nanoparticles used as catalyst supports for the production of fine chemicals

S Shylesh[1], J Schweizer[2], R Ulber[3], W Thiel[1] and V Schünemann[2]

[1]Department of Chemistry, University of Kaiserslautern, D-67663 Kaiserslautern, Germany

[2]Department of Physics, University of Kaiserslautern, D-67663 Kaiserslautern, Germany

[3]Institute of Bioprocess Engineering, University of Kaiserslautern, D-67663 Kaiserslautern, Germany

E-mail: schuene@physik.uni-kl.de

Abstract. Temperature and field dependent Mössbauer spectroscopy has been used to characterize silica coated iron oxide nanoparticles with a core size of ~ 10 nm. Covalent attachment of oxodiperoxo molybdenum complexes to the silica surface of these particles yields an organic-inorganic hybrid nanocatalyst which is an active, magnetically separable epoxidation catalyst. Mössbauer spectroscopy identifies superparamagnetic γ-Fe_2O_3 particles, whereas the presence of significant amounts of Fe_3O_4 (≥ 5 %) can be excluded.

1. Introduction

Grafting of catalytically active compounds on the surface of a support (heterogenization) allows to recover the catalyst from the reaction mixture by simple filtration and thus the reuse of the expensive noble metal species. Obviously supports with high specific surfaces – such as nanoparticles, or mesoporous materials - are of special interest in this context, since in these cases, a large number of active sites can be heterogenized. However, decreasing the diameter of the support particles down to the nanometer scale will give rise to problems in catalyst separation by classical filtration. Therefore methods like ultrafiltration, dialysis or zentrifugation have to be applied. This problem can be overcome by using magnetic nanoparticles as support, which can be separated from the product by simple application of a magnetic field. As long as these particles show superparamagnetic behavior, they can be resuspended without any agglomeration for the next catalytic transformation. Within this study silica coated iron oxide nanoparticles were used as support for catalytically active metal complexes. They have been characterized by temperature and field dependent Mössbauer spectroscopy. The goal of this project is to produce new nanoparticle based catalysts for the production of fine chemicals.

2. Experimental and methods

Silica coated nanoparticles with an inner iron oxide core diameter of ~10 nm and a silica layer thickness of ~50 nm have been prepared as described in [1]. Mössbauer spectra were recorded using a conventional spectrometer in the constant-acceleration mode. Isomer shifts are given relative to α-Fe at room temperature. Temperature and field dependent Mössbauer spectra were recorded with a closed-cycle cryostat from CRYO Industries of America Inc. and were analyzed by least-square fits using Lorentzian line shapes. Organic-inorganic hybrid heterogenous nanocatalyst systems were synthesized by covalent anchoring oxodiperoxo molybdenum complexes [(L-L)$MoO(O_2)_2$] (L-L = (3-triethoxysilyl propyl)[3-(2-pyridyl)-1-pyrazolyl] acetamide and N-(2-aminoethyl)-3-aminopropyl triethoxysilane) on the silica coated magnetic nanoprticles. The activity of the organic/inorganic hybrid catalysts was evaluated for the epoxidation of *cis*-cyclooctene [1].

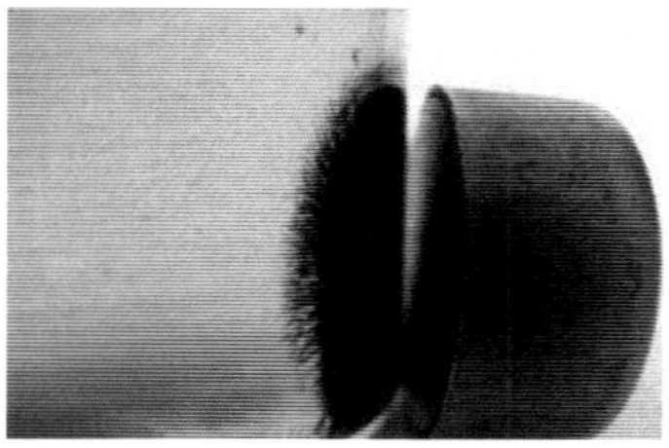

Figure 1. Magneto filtration of finely dispersed silica coated particles.

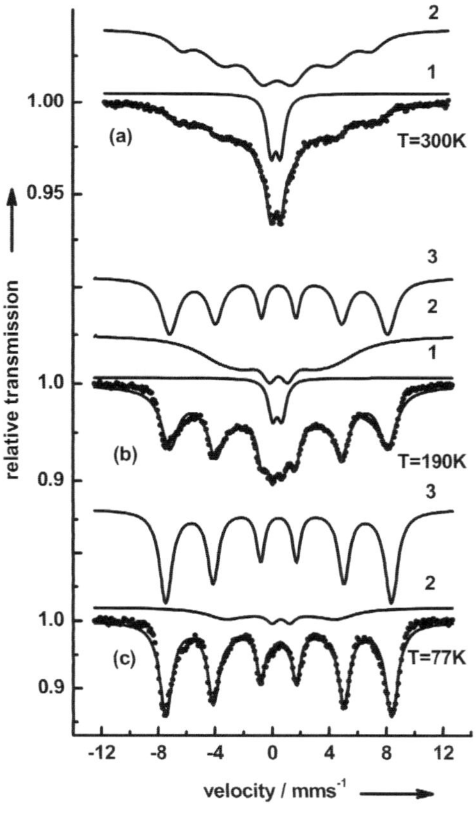

Figure 2. Temperature dependent Mössbauer spectra of silica coated particles. The solid lines represent simulations with parameters given in Table 1.

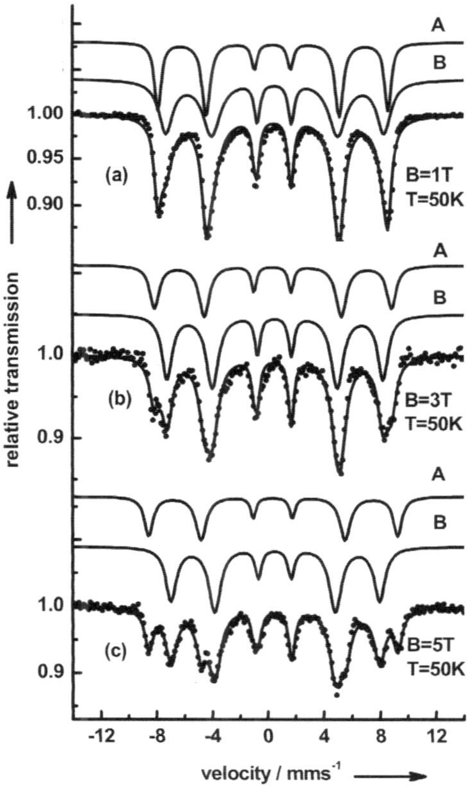

Figure 3. Field dependent Mössbauer spectra of silica coated particles obtained at T = 50 K. The solid lines represent simulations with parameters given in Table 2.

3. Results and Discussion

Figure 1 shows a magneto filtration of the finely dispersed nanocatalyst after 2 minutes magnetic field application. Due to its magnetic function it is possible to reuse the catalyst in the epoxidation of cyclooctene without a significant lost of catalytic activity [1].

Mössbauer spectroscopy at T = 300 K (figure 2a) of the silica coated particles exhibits two fractions of particles: (i) superparamagnetic particles the magnetic moments of which fluctuate with a relaxation time shorter than the time window of Mössbauer spectroscopy ($\sim 10^{-7}$ s) [2-4] and (ii) particles with magnetic moments which fluctuate with a relaxation time in the order of 10^{-7} s.

Table 1. Parameters of the simulations of the Mössbauer spectra shown in figure 2. The relative ratio of the magnetically split subspectra have been taken as 3:2:1, except at 300 K where the ratio of component 2 has been taken as 1.4:4.5:5.2 in order to phenomenological reproduce the intermediate relaxation regime.

T (K)	Component	Rel. area (%)	δ (mms^{-1})	ΔE_Q (mms^{-1})	$\Gamma_{1,2,3}$ (mms^{-1})	B_{hf} (T)
300	1	16.5	0.28	0.66	0.72	-
	2	83,5	0.36	0	1.9, 3.1, 2.5	40.9
190	1	7.5	0.38	0.66	0.70	-
	2	47.0	0.48	0	4.4, 3.4, 1.2	24.3
	3	45.5	0.48	0	1.4, 1.2, 0.7	46.8
77	2	20	0.48	0	3.0, 4.8, 0.9	24.2
	3	80	0.46	0	1.1, 0.9, 0.7	48.5

Field-dependent Mössbauer spectroscopy has been used to identify the iron phase of the superparamagnetic nanoparticles as γ-Fe_2O_3 (figure 3). The Mössbauer spectrum obtained at 5 T (Fig. 3c) shows two resolved magnetic sextets (component A and B) with δ_A= 0.36 mms^{-1} and δ_B=0.50 mms^{-1} and hyperfine fields of $B_{hfA} \approx 55$ T and $B_{hfB} \approx 46$ T. Component A represents high-spin Fe^{III}-ions located in the A-site of γ-Fe_2O_3 and component B represents high-spin Fe^{III}- ions in the B-site of the γ-Fe_2O_3 lattice [2].

Table 2. Parameters of the simulations of the field-dependent Mössbauer-spectra shown in figure 3. The relative ratio of the lines has been taken as 3:4:1.

B (T)	Component	Rel. area (%)	δ (mms^{-1})	ΔE_Q (mms^{-1})	$\Gamma_{1,2,3}$ (mms^{-1})	B_{hf} (T)
1	A	37.5	0.36	0	0.6, 0.7, 0.5	50.5
	B	62.5	0.50	0	1.2, 1.5, 0.5	47.8
3	A	37.5	0.36	0	0.8, 0.9, 0.4	52.0
	B	62.5	0.50	0	0.9, 1.1, 0.5	47.3
5	A	37.5	0.36	0	0.8, 0.9, 0.5	54.6
	B	62.5	0.50	0	0.9, 1.0, 0.6	45.8

Obviously the external field of 5 T decreases the magnetic hyperfine field of the ions in the B-sites to a value of ~ 46 T which shows that the sign of B_{hfB} is negative. In turn the external field leads to an increase of the hyperfine field of the ions in the A-sites to ~ 55 T, which shows that the sign of B_{hfA} is positive. Because B_{hf} is dominated by negative Fermi contact contribution, an external field B leads to a decrease in B_{hf} in the case of field parallel alignment of the magnetic moments. Thus figure 3c shows that the magnetic moments of the Fe^{III}-B-sites of the ferrimagnetic particles are oriented almost field parallel and those of the Fe^{III}-A-sites are oriented antiparallel to an external field of 5 T at T = 50 K.

Ferrimagnetic γ-Fe_2O_3 has a spinel (AB_2O_4) structure, but there are too few Fe^{III} cations to fill all the A and B sites. Assuming that all vacancies are located on the B lattice gives a ratio of A/B = 0.60 [2], which is in exact agreement with the ratio of the relative spectral areas of component A and B in the spectra displayed in figure 3. The fact, that the cores of the silica coated nanoparticles investigated here contain γ-Fe_2O_3, has been indeed corroborated by electron diffraction [1].

4. Conclusion

In conclusion Mössbauer spectroscopy identifies superparamagnetic γ-Fe_2O_3 particle cores of silica coated nanoparticles whereas significant amounts of Fe_3O_4 ($\geq$5%) can be excluded because there is no evidence of spectral components representing Fe^{II}-phases.

Acknowledgments

This work was supported by the University of Kaiserslautern within NANOKAT.

References

[1] Shylesh S P, Schweizer J, Demeshko S, Schünemann V, Ernst S, Thiel W R 2009 *Advanced Synthesis & Catalysis* in press
[2] Greenwood N N and Gibb TC 1971 *Mössbauer Spectroscopy* (London: Chapman and Hall)
[3] Mørup S, Dumesic J-A and Topsøe H 1980 *Applications of Mössbauer spectroscopy* (vol II) ed R L Cohen (New York: Academic Press) pp 1–53
[4] Schünemann V and Paulsen H 2007 *Applications of Physical Methods To Inorganic and Bioinorganic Chemistry,* eds. R A Scottt and C M Lukehart (Wiley Interscience) pp 243–270

^{121}Sb Mössbauer Spectra of M or A-Doped Antimonic Acid (M = Ta and Nb, A = Bi and Y) Fine Nanoparticles Prepared by Soft Chemical Solution Process

Junhu Wang[1], Tao Zhang[1], Masashi Takahashi[2] and Masuo Takeda[2]
1) State Key Laboratory of Catalysis, Dalian Institute of Chemical Physics, Chinese Academy of Sciences, 457 Zhongshan Road, Dalian, 116023 China
2) Department of Chemistry, Faculty of Science, Toho University, 2-2-1 Miyama, Funabashi, 274-8510 Japan

E-mail: wangjh@dicp.ac.cn, taozhang@dicp.ac.cn

Abstract. ^{121}Sb Mössbauer spectroscopic studies on the pyrochlore-type M or A-doped antimonic acids (M = Ta and Nb, A = Bi and Y) nanoparticles were carried out at 12 K in order to determine the oxidation state of Sb and also find more useful information for making clear the detail mechanisms of the photocatalytic property and the proton conductivity. The ^{121}Sb Mössbauer results indicated that all of Sb were Sb^{5+} in $HSb_{1-x}M_xO_3{\cdot}nH_2O$ (M = Ta and Nb) which Sb was partially substituted by M, however, small amount of Sb were Sb^{3+} in $H_{1-3x}A_xSbO_3{\cdot}nH_2O$ (A = Bi and Y) which H was partially substituted by A. The amounts of Sb^{3+} in the Bi-containing material were clearly larger than that in the Y-containing material. The photocatalytic properties and the proton conductivities of the pure and M or A-containing materials were discussed in relation to their ^{121}Sb Mössbauer parameters, especially isomer shift (δ) and quadrupole coupling constant (e^2qQ).

1. Introduction

We are interested in developing novel double oxide photocatalytic nano-materials with having not only strong adsorption capability but also high photocatalytic ability to water-splitting and decomposition of the organic pollutants [1-6]. In general, pyrochlore(P)-type compounds are presented by the formula $A_2M_2O_6O'$ and correspond to the space group symmetry of *Fd*-3*m* [7]. As shown in Figure 1, the P-type structure is characterized by a three-dimensional ($M_2O_6^{n-}$) framework built up of vertex-linked MO_6 octahedra. In this framework, there are interconnected tunnels with large cavities (8 per unit cell) [7]. Taking into account the compositional flexibility of the P-type structure, previously, we developed a novel series of P-type photocatalytic materials, Bi_2MTaO_7 (M = In, Ga and Fe), which were promisingly applied for water-splitting to produce the clean-energy of hydrogen [1,3,5]. Moreover, relationships between the photocatalytic activity under UV light irradiation and the P-type structural property were found. The photocatalytic activity became larger when the lattice distortion became larger and the closer the angle between the vertex-linked MO_6 octahedra is to 180° [5].

Recently, we also reported the photocatalytic properties and ^{121}Sb Mössbauer spectra of two kinds of P-type antimonic acids, $HSbO_3{\cdot}nH_2O$ (or $Sb_2O_5{\cdot}nH_2O$), fine nanoparticles prepared by using different starting materials [6]. For the P-type $HSbO_3{\cdot}nH_2O$, A = H or H_3O, M = Sb and O' can be replaced by H_2O. Its full structure has been determined by Slade et al. from the Rietveld refinement using neutron diffraction data [8]. The photocatalytic experiment indicated that the one prepared by using $Sb(O\text{-}i\text{-}C_3H_7)_3$ showed very strong adsorption capability and high photocatalytic activity for methylene blue degradation under UV light irradiation. The ^{121}Sb Mössbauer experiment confirmed that the one prepared by using $Sb(O\text{-}i\text{-}C_3H_7)_3$ only contained Sb^{5+} with $4d^{10}$ electronic configuration, however, the another one prepared by using Sb metal powder contained a small amount of Sb^{3+} with $4d^{10}5s^2$ electronic configuration. The high photocatalytic activity was pointed out due to the high dispersion, fine crystallinity and large relative surface area as well as only containing the octahedrally coordinated Sb^{5+} species with $4d^{10}$ electronic configuration. Taking into account the previous reports by Inoue's group [9], a small amount of Sb^{3+} species with $4d^{10}5s^2$ electronic configuration are considered to be the main reason why the photocatalytic performances of the two kinds of antimonic acids were definitely different to each other.

In addition, Ozawa et al. successfully prepared three novel series of P-type M or A-doped antimonic acid (M = Nb, A = Bi and Y) fine nanoparticles by modifying the soft chemical solution process and found that the proton conductivities of the prepared materials were significantly enhanced after the M or A-doping [10-12]. The full structures of the M or A-doped materials were also been performed from the Rietveld refinement using powder X-ray diffraction data. In the Rietveld refinement, Sb and O1 atoms, consisting of the SbO_6 octahedra, were placed on a 16*c* site (0, 0, 0) and a 48*f* site (*x*, 1/8, 1/8), respectively; the O2 atoms of the H_2O or H_3O^+ species were placed on a 32*e* site (*x*, *x*, *x*). Varieties of the proton conductivities of the M or A-doped materials were qualitatively interpreted in terms of the strength of the hydrogen bonding between the oxygen atoms of the ($Sb_2O_6^{2-}$) frameworks and H_2O or H_3O^+ species, which changes according to the M or A-doping amount [10-12].

In this paper, ^{121}Sb Mössbauer spectroscopic studies on the four kinds of M or A-doped antimonic acids (M = Ta, and Nb, A = Bi and Y, M or A/Sb = 0.1) were conducted in order to determine the oxidation state of Sb and also try to find more useful information for making clear the mechanism of the photocatalytic property as well as the proton conductivity in detail.

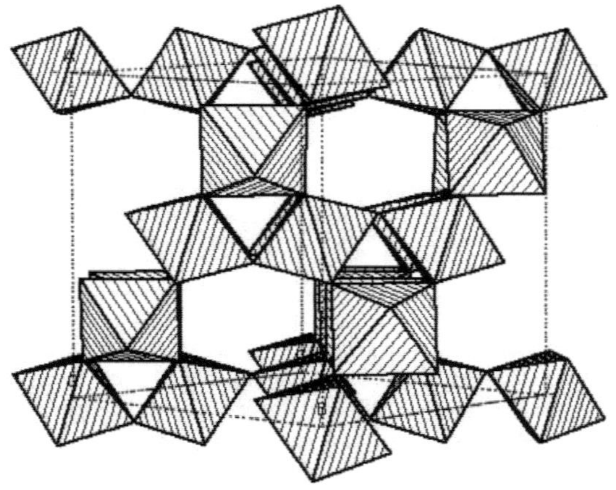

Figure 1. The [$Sb_2O_6^{2-}$] pyrochlore framework formed by vertex-shared SbO_6 octahedra for pyrochlore-type $HSbO_3 \cdot nH_2O$.

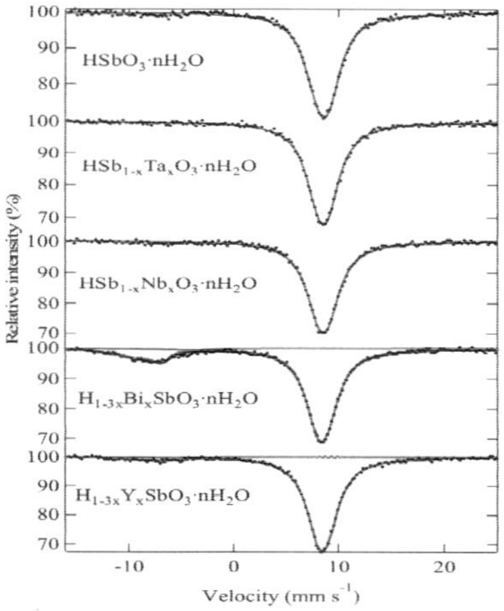

Figure 2. ^{121}Sb Mössbauer spectra of M or A-doped antimonic acids (M = Ta and Nb, A = Bi and Y, the molar ratios of M or A/Sb = 0.1).

2. Experimental procedures

The samples used in the present study were supplied by Ozawa of National Institute for Materials Science in Japan. The molecular formulas of the Ta and Nb-containing materials were confirmed as $HSb_{1-x}M_xO_3 \cdot nH_2O$, which M locates on the position of Sb. The molecular formulas of the Bi and Y-containing materials were confirmed as $H_{1-3x}A_xSbO_3 \cdot nH_2O$, which A locates on the position of H. ^{121}Sb Mössbauer spectra of the prepared materials were observed by a Wissel Mössbauer spectrometer system using a $Ca^{121m}SnO_3$ source (16 MBq) and a Ge detector; in these experiments both the source and the samples containing Sb were kept at 12 K in a cryostat incorporating a closed-cycle refrigerator [13].

3. Results and discussions

Figure 2 shows ^{121}Sb Mössbauer spectra of the pure and M or A-doped materials (M = Ta and Nb, A = Bi and Y, the molar ratios of M or A/Sb = 0.1) at 12 K obtained in the present study. The spectra of $HSb_{1-x}M_xO_3 \cdot nH_2O$ (M = Ta and Nb) exhibited one absorption peak which is typical for Sb^{5+} species as that of the reported pure material, $HSbO_3 \cdot nH_2O$, prepared by using the Sb alkoxide complex as a starting material [6]. However, the spectra of $H_{1-3x}A_xSbO_3 \cdot nH_2O$ (A = Bi and Y) were two distinct absorption peaks which are typical for Sb^{3+} and Sb^{5+} species, respectively, as that of the reported pure material, $HSbO_3{\cdot}nH_2O$, prepared by using the Sb metal powder as a starting material [6]. The amount of Sb^{3+} in the Bi-containing material was clearly larger than that in the Y-containing material. Therefore, the least square curve fitting was performed by assuming only one Sb^{5+} site in the Ta- and Nb-containing materials, one Sb^{5+} site and one Sb^{3+} site were assumed in the Bi- and Y-containing materials.

The spin numbers of ^{121}Sb Mössbauer transition are 5/2 for the ground state and 7/2 for the excited state, respectively. The ^{121}Sb Mössbauer spectra should be consist of 12 absorption lines in the case of the asymmetric parameter, $\eta \neq 0$ and 8 absorption lines in the case of $\eta = 0$. However, since the natural line-width of ^{121}Sb Mössbauer transition is broad to 2.1 mm s^{-1}, the splitting can not be completely observed [13]. In the present study, considering the symmetry of the cubic P-type structure, the ^{121}Sb Mössbauer spectra were analyzed by assuming 8 quadrupole-split lines corresponding to the asymmetric parameter, $\eta = 0$, from using a transmission integral method [14].

Table I. ^{121}Sb Mössbauer parameters of M or A-doped antimonic acids (M = Ta and Nb, A = Bi and Y, the molar ratios of M or A/Sb = 0.1) fine nanoparticles at 12 K (These parameters were obtained by assuming the asymmetric parameter, $\eta = 0$).

Sample	Site	δ (mm s^{-1})[a]	e^2qQ (mm s^{-1})[b]	$2\Gamma_{exp}$ (mm s^{-1})[c]	A (%)[d]
$HSbO_3{\cdot}nH_2O$[e]	Sb^{5+}	9.34	4.7	2.38	100
$HSb_{1-x}Ta_xO_3{\cdot}nH_2O$	Sb^{5+}	8.94	3.2	2.33	100
$HSb_{1-x}Nb_xO_3{\cdot}nH_2O$	Sb^{5+}	8.70	3.5	2.37	100
$H_{1-3x}Bi_xSbO_3{\cdot}nH_2O$	Sb^{5+}	8.31	2.8	2.29	89
	Sb^{3+}	-7.61	17.2		11
$H_{1-3x}Y_xSbO_3{\cdot}nH_2O$	Sb^{5+}	8.10	1.9	2.26	97
	Sb^{3+}	-7.49	16.8		3

a. Isomer shift, relative to a standard absorber of InSb at 12 K (δ = 7.91 ± 0.05 mm s^{-1} at 20 K [13]), the experimental error was estimated to be ± 0.05 mm s^{-1}. b. Quadrupole coupling constant, the experimental error was estimated to be ± 0.5 mm s^{-1}. c. Experimental line-width. d. Relative peak area. e. The water content was estimated to be 0.5.

The peak area ratio of Sb^{5+}/Sb^{3+} was 89:11 for the Bi-containing material. The result was clearly different to that measured by the $KMnO_4$ titration method [11]. It showed that the existence of Sb^{3+} ion should be taken into account when a structural model for the Bi-containing material was proposed to apply for the Rietveld structural refinement [11]. The peak area ratio of Sb^{5+}/Sb^{3+} was 97:3 for the Y-containing material. The result was almost consistent to that measured by the $KMnO_4$ titration method [10]. It showed that the structural model for the Y-containing material was almost reasonable to apply for the Rietveld structural refinement [10]. Table 1 lists the obtained ^{121}Sb Mössbauer parameters at 12 K, including isomer shift (δ), quadrupole coupling constant (e^2qQ), experimental line-width (2Γ) and relative area (A). The isomer shifts were determined relative to a standard compound InSb at 12 K (δ = -7.91± 0.05 mm s^{-1} at 20 K [13]).

As listed in Table 1, all of the δ values for the Sb^{5+} species were smaller than the pure material and were in order: $\delta_{pure} > \delta_{Ta} > \delta_{Nb} > \delta_{Bi} > \delta_{Y}$. The change trend of δ was likely to be in opposition to the change trend of the maximal value of the proton conductivity that can be obtained on the M or A-doped materials in the same temperature. For example, the proton conductivities at room temperature were 8.3 x 10^{-5}, 6.4 x 10^{-3}, 4.8 x 10^{-3} and 3.8 x 10^{-3} S cm^{-1}, respectively, for the pure (x = 0), Nb- (x = 0.09), Bi- (x = 0.081) and Y- (x = 0.12) containing materials. Here, the M or A/Sb mole ratios x were obtained by chemical analysis [10-12].

It is well-known that Mössbauer δ mainly depends on the s-electron density at the Mössbauer nucleus and the variation of the ^{121}Sb Mössbauer δ for the Sb^{5+} and Sb^{3+} compounds are mainly because of the effects of the Sb 5s electrons [15]. So, here the δ results for the Sb^{5+} species indicated that the 5s electron density at the Sb nucleus position decreased in order: $\rho_{s\,(pure)} < \rho_{s\,(Ta)} < \rho_{s\,(Nb)} < \rho_{s\,(Bi)} < \rho_{s\,(Y)}$ after the M or A-doping since $\Delta R/R < 0$ for ^{121}Sb. Thus, we can reasonably suppose that the interactions between the Sb 5s and O 2p orbitals became bigger and then the Sb-O bond became stronger in the SbO_6 octahedra after the M or A-doping. This should lead to that the strength of the hydrogen bonding between the oxygen atoms of the ($Sb_2O_6^{2-}$) frameworks and H_2O or H_3O^+ species became weaker after the M or A-doping since there is an equilibrium in the P-type antimonic acid: Sb-OH + $H_2O \leftrightarrow$ Sb-O- + H_3O^+ [8]. This should be considered to be the main reason for causing the increase of the proton conductivity since the proton conductivity of the P-type antimonic acid was known to be influenced by the strength of the hydrogen bonding between the oxygen atoms of the ($Sb_2O_6^{2-}$) frameworks and H_2O or H_3O^+ species. This consequence is consistent to that obtained by the Rietveld structural analysis reported by Ozawa et al. [10-12].

By the way, the ^{121}Sb Mössbauer parameters of the Sb^{3+} species were very different to the Sb^{5+} species. Comparing with the Sb^{5+} species, the Sb^{3+} species had very smaller δ values and very larger e^2qQ values. This was caused by the different electron configurations of $Sb^{3+}(4d^{10}5s^2)$ and $Sb^{5+}(4d^{10})$.

All of the e^2qQ values of the M or A-doped materials were smaller than the pure material. The e^2qQ is a measure on the magnitude of the electric field gradient at the Mössbauer nucleus position. So, the e^2qQ results indicated that the electric field gradient, namely the distortion of the SbO_6 octahedra existing in the P-type structure of $HSbO_3{\cdot}nH_2O$, became smaller due to the M or A-doping. The photocatalytic activities of the four M or A-doped materials (M = Ta and Nb, A = Bi and Y) [16] were investigated by using the same method and the same model micro-reaction as reported previously [6]. The results indicated that their photocatalytic activities were clearly lower than the pure material. That is to say, the photocatalytic ability of the P-type antimonic acid decreased when the distortion of the SbO_6 octahedra became smaller due to the M or A-doping. This was consistent to the trend that we observed in the novel series of photocatalysts, Bi_2TaMO_7 (M = In. Ga and Fe) [1, 3, 5]. As mentioned in the introduction section, we investigated the phtocatalytic abilities and the crystal structures of Bi_2TaMO_7 and found a trend: the larger the distortion of the MO_6 octohedra is, the higher the photocatalytic ability is. This is probably because that the electron-hole pairs produced by UV light irradiation have bigger drive-force for moving to the nanoparticle surface when the distortion of the MO_6 octohedra become larger in a similar series of P-type structure.

4. Summary

The ^{121}Sb Mössbauer spectroscopic studies on the four M or A-doped P-type antimonic acids were performed and the results indicated that the ^{121}Sb Mössbauer spectroscopy was very effective to determine the Sb oxidation sate and investigate the mechanism of the Sb-containing proton conductive and photocatalytic materials. All of Sb were confirmed in Sb^{5+} in $HSb_{1-x}M_xO_3{\cdot}nH_2O$ (M = Ta and Nb), however, small amount of Sb were confirmed in Sb^{3+} in $H_{1-3x}A_xSbO_3{\cdot}nH_2O$ M (A = Bi and Y). The amount of Sb^{3+} in the Bi-containing material was clearly larger than that in the Y-containing material. All of the δ values of the M or A-doped materials were smaller than the pure material and were in order: $\delta_{pure} > \delta_{Ta} > \delta_{Nb} > \delta_{Bi} > \delta_{Y}$. This implied that the Sb-O bond became stronger in the SbO_6 octahedra after the M or A-doping and then the strength of the hydrogen bonding between the oxygen atoms of the ($Sb_2O_6^{2-}$) frameworks and H_2O or H_3O^+ species became weaker after the M-doping. This should be related to the enhancement of the proton conductivity. All of the e^2qQ values of the M-doped materials were smaller than the pure material. This indicated that the distortion of the SbO_6 octahedra existing in the P-type antimonic acid became smaller due to the M-doping. It should be related to the decrease of the photocatalytic ability that had been demonstrated experimentally by photocatalytic degradation of model organic pollutant.

References

[1] Wang J, Zou Z and Ye J 2003 *Mater. Sci. Forum* **423-425** 485
[2] Wang J, Zou Z and Ye J, 2003 *Book of Abstracts, 3rd Asia-Pacific Congress on Catalysis **III** Dalian China* 460-461
[3] Wang J, Zou Z and Ye J 2004 *J. Alloys Compounds* **377** 248
[4] Wang J, Nonami T 2004 *J. Mater. Sci.* **39** 6376
[5] Wang J, Zou Z and Ye J 2005 *J. Phys. Chem. of Solids* **66** 349
[6] Wang J, Ozawa K, Takahashi M, Takeda M and Nonami T 2006 *Chem. Mater.* **18** 2261
[7] Subramanian M A, Aravamudan G and Subba G V 1983 *Prog. Solid State Chem.* **15** 55
[8] Slade R C T, Hall G P, Ramanan A and Prince E 1996 *Solid State Ionics* **92** 171
[9] Sato J, Saito N, Nishiyama H and Inoue Y 2002 *J. Photochem. Photobiol. A: Chem.* **148** 85
[10] Ozawa K, Wang J, Ye J, Sakka Y and Amano M 2003 *Chem. Mater.* **15** 928
[11] Ozawa K, Eguchi M, Nakamura H and Sakka Y 2004 *Solid State Ionics* **172** 109
[12] Ozawa K, Hase M, Fujii H, Eguchi M, Yamaguchi H and Sakka Y 2005 *Electrochim. Acta* **50** 3205
[13] Takeda M 1985 *Radioisotpes* **34** 628
[14] Shenoy G K, Friedt J M, Maletta H, Ruby S L 1975 *Mössbauer Methodology* Editors: Gruverman I J, Seidel C W, Dieterly D K New York: Plenum Press 277
[15] Lippens P E, 2000 *Solid State Commun.* **113** 399
[16] Our unpublished data

Electrical conductivity and local structure of iron-containing lithium barium vanadate glass

S Kubuki[1], Y Tomota[1], R Yoshimura[1], Z Homonnay[2], K Sinkó[3], E Kuzmann[4] and T Nishida[5]

[1]Department of Chemical and Biological Engineering, Ube National College of Technology, Tokiwadai 2-14-1, Ube, Yamaguchi, 755-8555, JAPAN
[2]Laboratory of Nuclear Chemistry, Eötvös Loránd University, Pázmany P. s., 1/A 1117 Budapest, HUNGARY
[3]Institute of Chemistry, Eötvös Loránd University, Pázmany P. s., 1/A 1117 Budapest, HUNGARY
[4]Laboratory of Nuclear Chemistry, HAS CRC, Eötvös Loránd University, 1518 Budapest, HUNGARY
[5]Department of Biological and Environmental Chemistry, Faculty of Humanity-Oriented Science and Engineering, Kinki University, Kayanomori 11-6, Iizuka, Fukuoka, 820-8555, JAPAN

E-mail: kubuki@ube-k.ac.jp

Abstract. A relationship between the local structure and electrical conductivity of lithium barium iron vanadate glass with a composition of xLi$_2$O·(20-x)BaO·10Fe$_2$O$_3$·70V$_2$O$_5$, abbreviated as xLBFV glass, was investigated by means of ^{57}Fe-Mössbauer spectroscopy, differential thermal analysis (DTA), as well as DC four-probe method. When the Li$_2$O content increased from 0 to 20 mol%, a constant isomer shift (δ) value of 0.35±0.01 mm s^{-1} and a decreasing quadrupole splitting (Δ) value from 0.71 to 0.67±0.01 mm s^{-1} were observed in the Mössbauer spectra. From the DTA curves, decrease in both glass transition temperature (T_g) and crystallization temperature (T_c) was confirmed from 317 to 236±5 °C and from 374 to 299±2 °C, respectively. A large slope value of 1560 °C/mm s^{-1} was estimated from 'T_g - Δ plot', showing that the Fe^{3+} acts as a network former (NWF). When the xLBFV glass with 'x' of 20 was heated up to 125 °C, a larger increase in the electric conductivity (σ) was observed from 3.2×10^{-5} to 1.3×10^{-3} S cm^{-1}. It was found that replacement of Li^+ for Ba^{2+} reduces the local distortion of FeO$_4$ (T_d), and increases the temperature dependence of electrical conductivity.

1. Introduction

Vanadate glass is known as a semiconducting oxide glass with an electrical conductivity (σ) of 10^{-7}-10^{-5} S cm^{-1}, caused by 3d electron hopping from V^{VI} to V^{V} [1]. Nishida *et al.* reported that a drastic increase in σ was observed after annealing of K$_2$O-V$_2$O$_5$-Fe$_2$O$_3$ and Li$_2$O-Fe$_2$O$_3$-V$_2$O$_5$-P$_2$O$_5$ glasses [2,

3]. Barium iron vanadate glass, $BaO\text{-}Fe_2O_3\text{-}V_2O_5$, has already been registered as a Japanese patent [4], with a trade mark of '*NTA glass*™'. Recently, Kubuki *et al.* revealed that '*NTA glass*™' has a σ of 1 S cm^{-1} after a prolonged heat treatment, owing to a structural relaxation [5].

In the present study, substitution of Li^+ for Ba^{2+} in *NTA* glass™ was investigated in order to achieve higher electrical conductivity. A relationship between the local structure and electrical conductivity of iron-containing lithium barium vanadate glasses was investigated by means of ^{57}Fe-Mössbauer spectroscopy, differential thermal analysis (DTA) and electrical conductivity measurement.

2. Experimental

New vanadate glass expressed by xLi$_2$O · (20-x)BaO · 10Fe$_2$O$_3$ · 70V$_2$O$_5$, hereafter abbreviated as xLBFV glass, was prepared by a conventional melt-quenching method. Li_2O content (x) was changed from 0 to 20 mol %. Weighed amounts of Li_2CO_3, $BaCO_3$, V_2O_5 and Fe_2O_3 of a reagent grade were well mixed in an agate mortar and heated at 1100 °C for 1 h with a platinum crucible. Dark brown glass samples were obtained by dipping the bottom of the crucible into ice-cold water. For Mössbauer measurement, enriched isotope of $^{57}Fe_2O_3$ (^{57}Fe = 95.54 %) was used. Mössbauer spectra were recorded by a constant acceleration method with a source of ^{57}Co(Rh). DTA was conducted from RT to 500 °C by changing the heating rate from 5 to 10, 15, 20 and 30 K min^{-1}. α-Al_2O_3 was used as a reference of the temperature. The electrical conductivity (σ) was measured by a dc-four probe method at temperatures ranging from RT to 125 °C.

3. Results and Discussion

Mössbauer spectra of xLBFV glass with 'x' of 0, 5, 10, 15 and 20 were shown in Figure 1. All the spectra were composed of paramagnetic doublet due to the tetrahedral Fe^{III}. When the Li_2O content increased, a constant isomer shift (δ) value was obtained for 0.35 (±0.01) mm s^{-1}, together with a small decrease in the quadrupole splitting (Δ) value from 0.71 to 0.69, 0.68, 0.67 and 0.66 (±0.01) mm s^{-1}. These results indicate that Fe-O bond strength remained constant, and that the local distortion of FeO_4 tetrahedra became smaller by substituting Li^+ for Ba^{2+}.

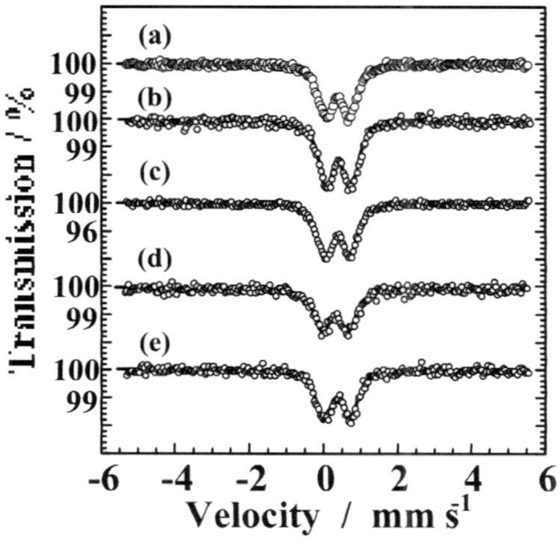

Figure 1. Mössbauer spectra of xLi$_2$O·(20-x) BaO·10Fe$_2$O$_3$·70V$_2$O$_5$ glass with 'x' of (a) 0, (b) 5, (c) 10, (d) 15 and (e) 20.

DTA curves of xLBFV glass are depicted in Figure 2. Both glass transition temperature (T_g) and crystallization temperature (T_c) decreased from 317 to 284, 274, 258 and 236 (±5) °C, and from 374 to 346, 331, 314 and 299 (±2) °C, respectively. These results show that the thermal durability was

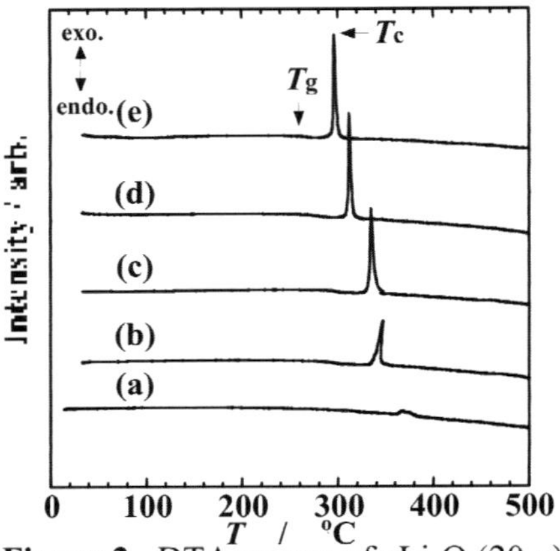

Figure 2. DTA curves of $xLi_2O{\cdot}(20\text{-}x)BaO{\cdot}10Fe_2O_3{\cdot}70V_2O_5$ glass with 'x' of (a) 0, (b) 5, (c)10, (d) 15 and (c) 20.

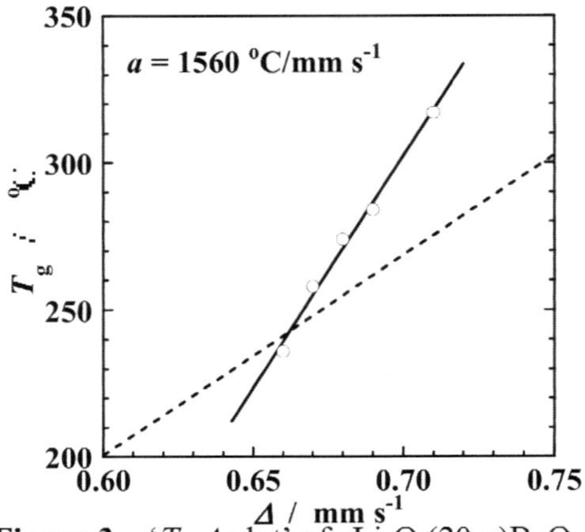

Figure 3. 'T_g-Δ plot' of $xLi_2O{\cdot}(20\text{-}x)BaO{\cdot}10Fe_2O_3{\cdot}70V_2O_5$ glass. A dashed line, having a slope value of 680 °C / mm s^{-1}, is a guide for eyes.

lowered by the introduction of Li_2O. A linear relationship between T_g and Δ for oxide glasses, discovered by Nishida [6], and denominated as 'T_g-Δ rule', is expressed by:

$$T_g = a\Delta + b \tag{1}$$

When Fe^{3+} ion is located at a network former (NWF) site, slope 'a' becomes more than 680 °C/mm s^{-1}. While a smaller 'a' value of 35 °C/mm s^{-1} is obtained when Fe^{3+} acts as a network modifier (NWM). As shown in Figure 3, a large 'a' value of 1560 °C/mm s^{-1}, obtained from the 'T_g-Δ plot', indicates that the Fe^{3+} plays a role of NWF. The activation energy for crystallization (E_a) can be calculated by Kissinger equation [7], *i. e.*,

$$\ln (T_c^2/\alpha) = E_a/RT_c + \text{const.} \tag{2}$$

where, α and R are heating rate and gas constant, respectively. The E_a value of xLBFV glass was varied from 2.3 to 1.7, 1.6, 1.9 and 2.3 (±0.2) eV with an increase of Li_2O content. Because the chemical bond energy of V-O is reported to be 3.9-4.9 eV [8]. It can be considered that the glass skeleton structure composed by VO_4 and VO_5 was not affected by the introduction of Li_2O.

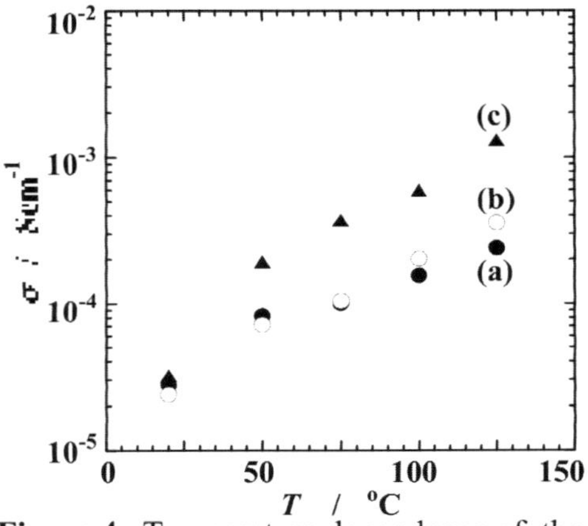

Figure 4. Temperature dependence of the electrical conductivity(σ) of $xLi_2O{\cdot}(20\text{-}x)BaO{\cdot}10Fe_2O_3{\cdot}70V_2O_5$ glass with 'x' of (a) 0, (b) 10 and (c) 20.

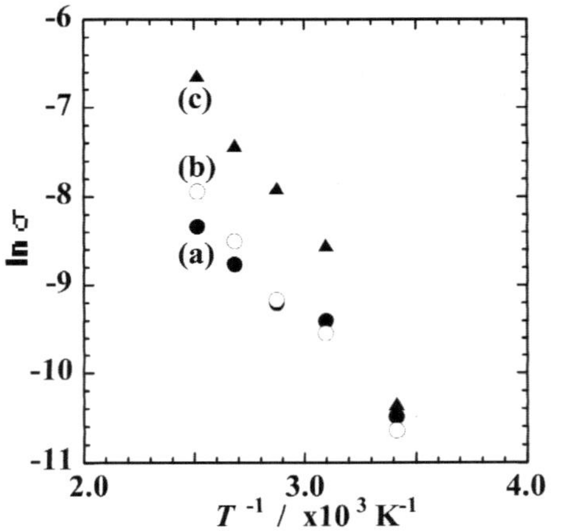

Figure 5. ln σ *vs.* $1/T$ plot of $xLi_2O{\cdot}(20\text{-}x)BaO{\cdot}10Fe_2O_3{\cdot}70V_2O_5$ glass with 'x' of (a) 0, (b) 10 and (c) 20.

A temperature dependence of the electric conductivity (σ) of the *x*LBFV glass is shown in Figure 4. Along with an increase in the measuring temperature from RT to 125 °C, σ value was increased from 3.2×10^{-5} to 1.3×10^{-3} S cm^{-1} in *x*LBFV glass with '*x*' of 20(figure 4(c)), while a smaller increase in σ value was observed from 2.8×10^{-5} to 2.4×10^{-4} S cm^{-1} when '*x*' was 0 (figure 4(a)). It is noteworthy that the temperature dependence of σ value is increased by substituting Li^+ for Ba^{2+} because of higher ionic mobility of Li^+. Temperature dependence of electric conductivity can be expressed as Arrehnius equation , *i. e.*,

$$\sigma = \sigma_o \exp(-W / kT) \tag{3}$$

where W and k are activation energy for electric conduction and Boltzmann constant, respectively. When the Li_2O content was increased from 0 to 10 and 20, W value gradually increased from 0.19 to 0.25 and 0.34 eV, as represented in Figure 5. This result shows that the temperature dependence in the σ value of *x*LBFV glass is enhanced by the introduction of Li^+. It can be concluded that the temperature dependence of the electrical conductivity is strongly affected by the decrease in local distortion of FeO_4 tetrahedra, as reflected in the as Δ value of Mössbauer spectra.

4. Summary

A relationship between the local structure and electric conductivity of lithium barium iron vanadate glass with a composition of xLi$_2$O·(20-x)BaO·10Fe$_2$O$_3$·70V$_2$O$_5$ (*x*LBFV) was investigated. When the Li_2O content was increased, local distortion of FeO_4 tetrahedra was decreased; a decrease of Δ value was observed from 0.71 to 0.67 mm s^{-1} in the Mössbauer spectra. In addition, both T_g and T_c values respectively decreased from 317 to 236 °C and from 374 to 299 °C, showing that the thermal durability was lowered by the introduction of Li_2O. From the 'T_g-Δ' plot, a large slope value of 1560 °C/mm s^{-1} was estimated, indicating that the Fe^{3+} acts as a NWF. A larger increase in the electrical conductivity from 3.2×10^{-5} to 1.3×10^{-3} S cm^{-1} was observed when *x*LBFV glass with '*x*' of 20 was heated from RT to 125 °C. This result indicates that the temperature dependence of the electrical conductivity was increased by substituting Li^+ for Ba^{2+}.

It is concluded that the electric conductivity is strongly affected by the local distortion of FeO_4, VO_4 and probably VO_5 units.

Acknowledgement
One of the authors (SK) expresses his great gratitude for the support by Electric Technology Research Foundation of Chugoku.

References

[1] Mott N F1967 *Adv. Phys.* **16**(No. 61) 49
[2] Nishida T, Kubota J, Maeda Y, Ichikawa F and Aomine T 1996 *J. Mater. Chem.* **6** 1889
[3] Nishida T, Yoshida Y, Takahashi Y, Okada S and Yamaki J 2008 *J. Radioanal. Nucl. Chem.* **275** 417
[4] Nishida T 2006 *Japanese Patent* No. 3854985
[5] Kubuki S, Sakka H, Tsuge K, Homonnay Z, Sinkó K, Kuzmann E, Yasumitsu H and Nishida T 2007, *J. Ceram. Soc. Jpn.* **115** 776
[6] Nishida T 1994 *J. Non-Cryst. Solids* **177** 257
[7] Kissinger H E 1957 *Anal. Chem.* **29** 1702
[8] Sun K H 1957 *J. Am. Ceram. Soc.* **30** 277

Journal of Physics: Conference Series **217** (2010) 012027 doi:10.1088/1742-6596/217/1/012027

Molecular switching in iron complexes bridged via tin-cyanides observed by Mössbauer and ESR spectroscopy

S Jung[1], F Renz[1*], M Klein[1], M Menzel[2], R Boča[3] and R Stößer[4]
[1]Leibniz University Hannover, Institute of Inorganic Chemistry, Callinstr. 9, D-30167 Hannover, Germany,
[2]Federal Institute for Materials Research and Testing (BAM) Richard-Willstätter-Str. 11, D-12489 Berlin, Germany
[3]Slovak Technical University, Institute for Inorganic Chemistry, Bratislava, Slovakia,
[4]Humbold University, Institute of Chemistry, Brook-Taylor-Str. 2, D-12489 Berlin, Germany

* corresponding author: E-mail: Franz.Renz@acd.uni-hannover.de

Abstract. The precursor $[Fe^{III}(L)Cl]$ (LH_2 = N,N'-bis(2'-hydroxy-3'-X-benzyliden)-1,6-diamino-3-N-hexane) is a high-spin (S = 5/2) complex (with X = $-CH_3$, $-O-CH_3$). This precursor is combined with the bridging unit $[Sn^{IV}(CN)_4]$ to yield star-shaped pentanuclear clusters, $[(L-X-Fe^{III})_4Sn(CN)_4]Cl_4$ ^{57}Fe-Mössbauer, ^{119m}Sn-Mössbauer, and ESR spectroscopy are used to study our samples. For X = $-CH_3$ the ^{57}Fe-Mössbauer data show a multiple spin transition between iron(III) in the high-spin and low-spin state. Changing the functional group from X = $-CH_3$ to X = $-O-CH_3$ turns the switchability off.

1. Introduction

Molecular switching is one of the major targets in various disciplines, such as technology, biology and medicine. One approach is to develop multistable molecules, i.e. chemically and physically induced transitions between electronic states. In some transition metal complexes such a multistability has been discovered in 1931 and labelled spin transition phenomenon [1]. Spin transitions are rare for multinuclear compounds [1].

Our systematic approach [2-13] enabled us to apply chemical tuning on pentadentate ligand Fe(III) complexes as building units $[LFe^{III}-X]$ (L= bis(3-salicylidene aminopropyl) amine) and show multistability, e.g. multiple thermally induced transitions between low-spin (LS) $^2T(O_h)$ and high-spin (HS) $^6A(O_h)$ in various star-shaped compounds: tetranuclear [4], heptanuclear [3, 5, 6, 7, 8, 9], nonanuclear [2, 3, 6, 7, 8, 9], and dodecanuclear [10].

In the present work we extend this approach and introduce a novel pentanuclear system. With tin we introduce a main-group element as star-shaped center [14] and coincidental we modify the salicylidene-component with other functional groups. Such pentanuclear Fe_4Sn compounds may be capable of performing 4 spin transitions in one cluster, e.g. between 4 and 20 unpaired electrons.

2. Experimental

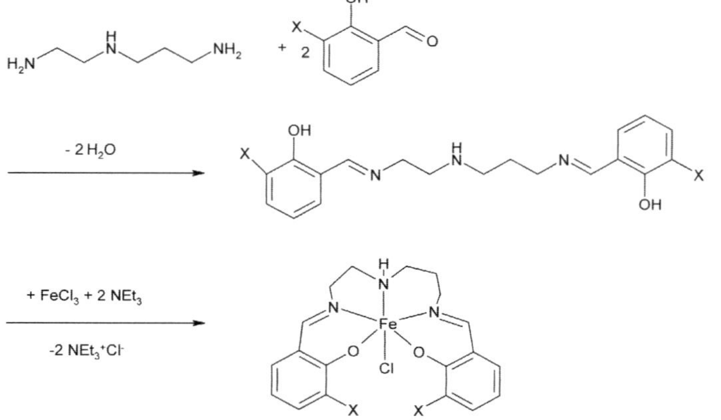

Figure 1. Synthesis of the [Fe(L)Cl] precursor (with X= $-CH_3$, $-OCH_3$)

The ligand LH_2 = N,N'-bis(2'-hydroxy-3'-X-benzyliden)-1,6-diamino-3-N-hexane) is prepared. Complexation with Fe(III) gives the used iron precursor [2,3/H/Sal-3X/Fe/Cl]. This building brick, $[Fe^{III}(L)Cl]$, is a high-spin (S = 5/2) complex (see Figure 1). This precursor is combined with the bridging unit $[Sn^{IV}(CN)_4]$ to yield a star-shaped pentanuclear cluster, $[(LFe^{III}\text{-}NC)_4Sn^{IV}]Cl_4$ (see Figure 2).

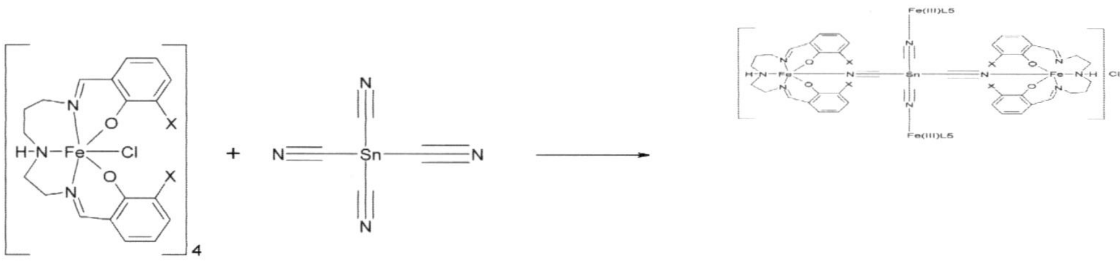

Figure 2. Synthesis of the $[(LFe^{III}\text{-}NC)_4Sn^{IV}]Cl_4$ cluster

2.1. Precursor
The precursor was prepared as described in literature [2]. The dark brown powder was characterized by IR.

2.2. Tin cyanides
The tin cyanides were prepared with their analogous compound of tin chloride and potassium cyanide in acetone under nitrogen atmosphere. The mixtures stirred for six days under reflux.

2.3. Complexes
The complexes were prepared in analogy to the literature [2]. The reaction solution stirred for 30 minutes and then was cooled for 5 days. The products were characterized by IR- and Mössbauer spectroscopy, ESI-MS and ESR-spectroscopy.

2.4. IR-spectroscopy
A Tensor 27 by Bruker Optics IR-spectrometer with a 5000 – 200 cm^{-1} visual appearance and with an ATR-supplement is used at ambient temperatures.

2.5. Mössbauer spectroscopy
The Mössbauer spectra were recorded using constant acceleration Mössbauer spectrometers from WissEl Electronics Starnberg (Germany) (295K and 78K) and from the Institute for Analytical Instrumentation St. Petersburg (Russia) (CM-2201, for 295K). The used Mössbauer source of ^{57}Co in chromium matrix had an average activity of about 0.4 GBq. The ^{119m}Sn source in $CaSnO_3$ matrix had an average activity of 0.35 GBq. Proportional counters (LND Inc., Ocean Side, New York, USA) were used with a voltage between 1800 V and 1950 V to detect the spectra. The sample of 33.5 mg was placed into a sealable cylindrical polyethylene sample compartment with an inner diameter of 19 mm. The ^{57}Fe Mössbauer spectra were recorded with a maximum velocity of +/- 3.20 mm/s (T=295 K) and +/- 3.6 mm/s (T=78 K) because no magnetic splitting could be observed. In the case of the ^{119m}Sn the Mössbauer spectra were recorded with a maximum velocity of +/- 7.95 mm/s (T=295 K) and +/- 6.15 mm/s (T=78 K). All ^{57}Fe spectra were referred to a standard α-Fe foil (Goodfellow). The ^{119m}Sn were referred to SnO_2 and a Sn foil.

2.6. ESR spectroscopy
The ESR spectra were measured with an ESR-spectrometer ESR 300 of the "Zentrum für wissenschaftlichen Gerätebau" (today: Magnettech GmbH), Berlin Adlershof with a test frequency of 9.426 GHz.

3. Results and Discussion
The structure of the [2,3/H/Sal-3X/Fe/Sn$(CN)_4$] complexes are shown in Figure 2. The MM+ force field structure of these clusters are shown in Figure 3. The complexes are characterized by IR-spectroscopy. On bridging the typical CN-bond stretching vibration switched from 2080 cm^{-1} to higher wavenumbers which are above 2100 cm^{-1}.

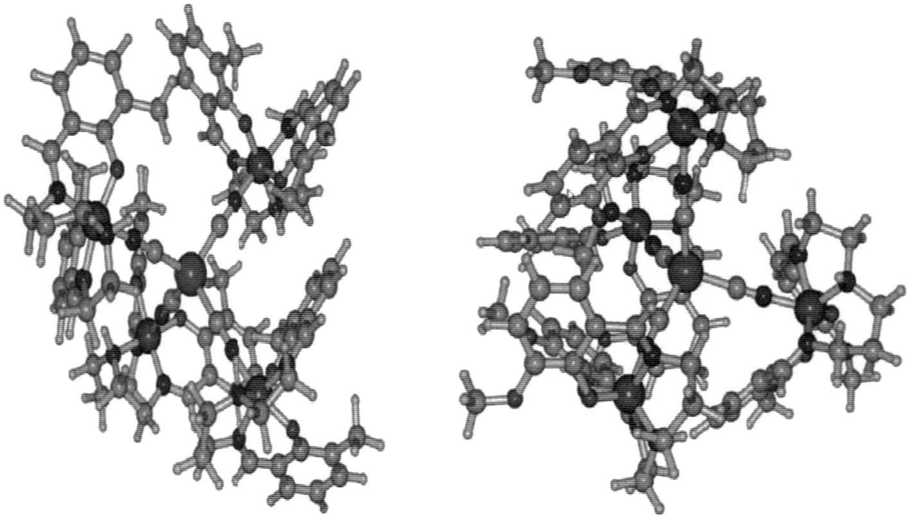

Figure 3. MM+ force field structure of [2,3/H/Sal-3Me/Fe/Sn$(CN)_4$](left) and of [2,3/H/Sal-3MeO/Fe/Sn$(CN)_4$] (right) optimized based on standard X-ray data

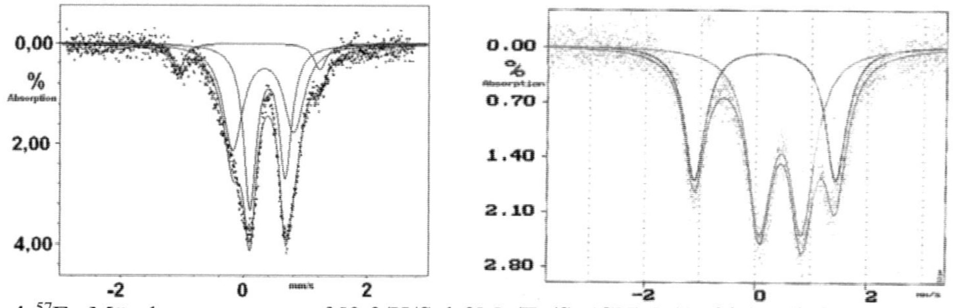

Figure 4. ^{57}Fe-Mössbauer-spectra of $[2,3/H/Sal\text{-}3Me/Fe/Sn(CN)_4]$ (T=295 K (left); T=78 K (right))

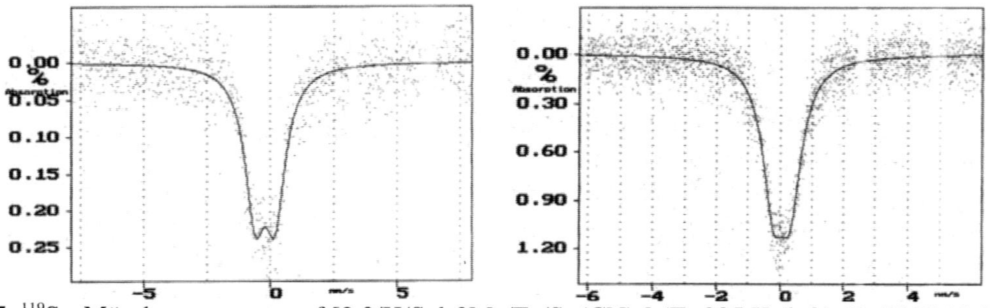

Figure 5. ^{119}Sn-Mössbauer-spectra of $[2,3/H/Sal\text{-}3Me/Fe/Sn(CN)_4]$ (T=295 K (left); T=78 K (right))

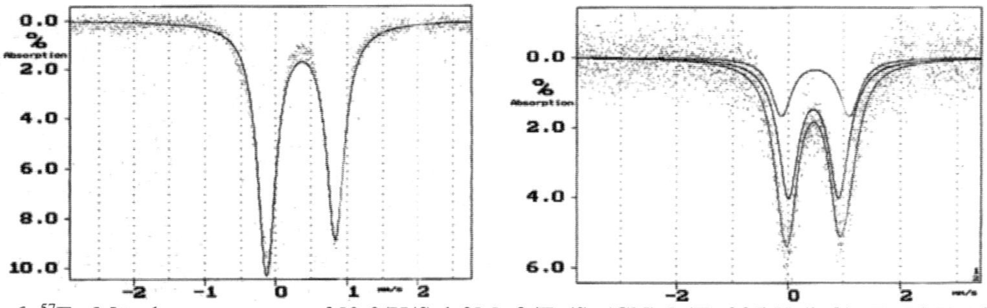

Figure 6. ^{57}Fe-Mössbauer-spectra of $[2,3/H/Sal\text{-}3MeO/Fe/Sn(CN)_4]$ (T=295 K (left); T=78 K (right))

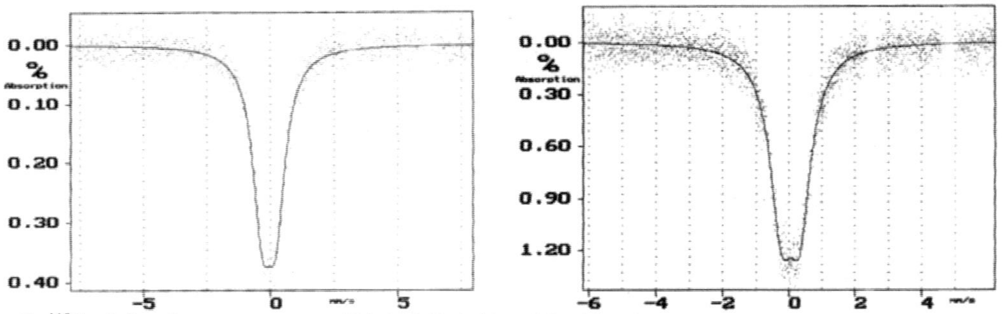

Figure 7. ^{119}Sn-Mössbauer-spectra of $[2,3/H/Sal\text{-}3MeO/Fe/Sn(CN)_4]$ (T=295 K (left); T=78 K (right))

Figure 4 and 5 show the ^{119}Sn and ^{57}Fe Mössbauer spectra for the compound $[2,3/H/Sal\text{-}3Me/Fe/Sn(CN)_4]$. The tin-spectra show that the oxidation state of tin is +4 at high and low temperature. For the iron-spectra there are three doublets at 295K. One with an isomeric shift (δ) of

0.388 mm/sec, a quadrupol-splitting (QS) of 0.568 mm/sec and a site population of 44.96 %, and the other with a δ of 0.315 mm/sec, a QS of 0.987 and a site population of 47.00 %. Both are characteristic for a high-spin state for Fe(III). The third one is typical for a Fe(III) low-spin state with a δ of 0.084 mm/sec, a QS of 2.288 mm/sec and a site population of 8.04%. At 78 K two doublets are presented, one with an isomeric shift of 0.177 mm/sec, a quadrupol-splitting of 2.550 mm/sec and a signal plot of 40.01 %, and the other one with an isomeric shift of 0.465 mm/sec, a QS of 0.776 mm/sec and a signal plot of 59.99 %. This indicates that the molar ratio shifts from the dominant high-spin state at room temperature towards the low-spin state at 78 K for Fe(III). Figure 6 and 7 show the ^{119}Sn and ^{57}Fe Mössbauer spectra for the compound [2,3/H/Sal-3MeO/Fe/Sn(CN)$_4$]. The tin-spectra show that the oxidation state of tin is +4 at high and low temperature like before. In the iron spectra just iron(III) in high-spin state can be seen. A low-spin state cannot be verified, neither at room temperature nor at low temperature. That means by changing the functional group at the salicylidene from methyl to methoxy turns the switchability off. It shows, that some iron(III)-molecules can reversibly be converted between high-spin and low-spin state, but the switchability is addicted to their functional groups.

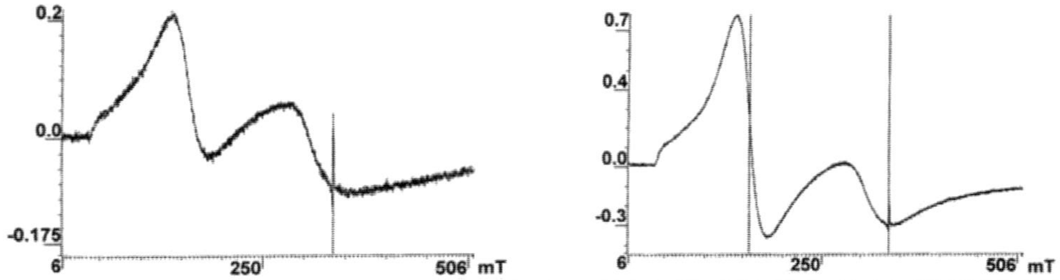

Figure 8. ESR-spectra of [2,3/H/Sal-3Me/Fe/Sn(CN)$_4$] (T=293 K (left); T=77 K (right))

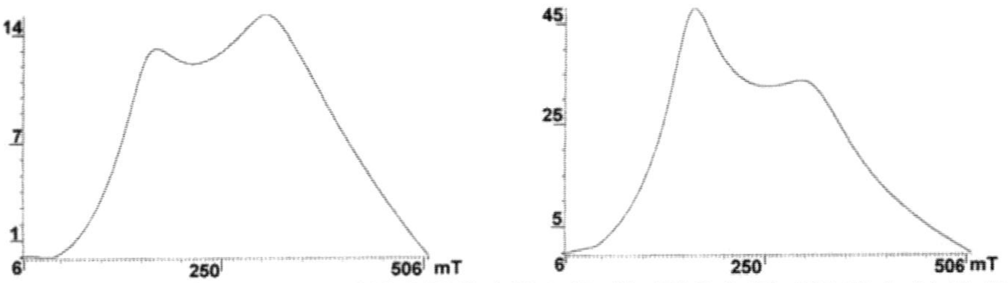

Figure 9. Integral of the ESR-spectra of [2,3/H/Sal-3Me/Fe/Sn(CN)$_4$] (T=293 K (left);T=77 K (right))

Figure 8 show the ESR-spectra of the compound [2,3/H/Sal-3Me/Fe/Sn(CN)$_4$]. For 77 K a broad transition at g'~4,3 (first step) and a broad segue with a lower intensity at g'~2 (second step) can be seen. For 293K the intensity-ratio switches to the segue at g'~2. These intensity effects are clearly verified by their corresponding integrals (see Figure 9). This confirms [2,3/H/Sal-3Me/Fe/Sn(CN)$_4$] as a switching molecular complex. The analysis for the corresponding ESR-data of the compound [2,3/H/Sal-3MeO/Fe/Sn(CN)$_4$] are not straight forward. A detailed ESR-analysis will be published somewhere else.

4. Conclusion

The ^{57}Fe-Mössbauer data indicate that the [2,3/H/Sal-3Me/Fe/Sn(CN)$_4$] show multiple spin transition between Fe(III) in the high-spin and low-spin state. Changing the functional group at the salicylidene from methyl to methoxy turns the switchability off. The multiple spin transition for the 3-methyl-derivate is confirmed by the ESR-data of this compound. The results show that the compound [2,3/H/Sal-3Me/Fe/Sn(CN)$_4$] is a switching molecular complex with a main group element as central atom. This proves the usage of main group elements such as tin as central atoms in switching molecules and a continuative investigation of the influence on this switchability from functional groups in complexes. The ESR-measurement shows that this method of spectroscopy is useful to verify switching complexes and further investigations are in progress.

References

[1] Gütlich P, Koningsbruggen P van, Renz F 2004 *In Optical Spectra and Chemical Bonding in Transition Metal Complexes, Springer-Verlag, Structure and Bonding* **107** 27-75
[2] Renz F, Kerep P 2004 *Hyperfine Interactions* **156/157** 371-377
[3] Renz F, Kerep P 2005 *Polyhedron* **24** 2849-2851
[4] Herchel R, Boča R, Gembický M, Kožišek J, Renz F 2004 *Inorg. Chem.* **43** 4103-4105
[5] Gembicky M, Boča R, Renz F 2000 *Inorg. Chem. Comm.* **3** 662-665
[6] Renz F, Hefner J, Hill D, Klein M 2007 *Polyhedron* **26** 2330-2334
[7] Renz F, Hill D, Kerep P, Klein M, Müller-Seipel R, Werner F 2006 *Hyperfine Interactions* **168** 1051-1056
[8] Renz F, Kerep P, Hill D, Klein M 2006 *Hyperfine Interactions* **168** 981-987
[9] Klein M, Renz F 2006 *Hyperfine Interactions* **168** 1001-1007
[10] Renz F, Hill D, Klein M, Hefner J 2007 *Polyhedron* **26** 2325-2329
[11] Boča R, Fukuda Y, Gembicky M, Herchel R, Jarosciak R, Linert W, Renz F, Yuzurihara J 2000 *Chemical Physics Letters* **325** 411-419
[12] Glöckle M, Kaim W, Klein A, Roduner E, Hubner G, Zalis S, van Slageren J, Renz F, Gütlich P 2001 *Inorganic Chemistry* **40** 2256-2262
[13] Boča M, Baran P, Boča R, Fuess H, Kickelbick G, Linert W, Renz F, Svoboda I 2000 *Inorganic Chemistry* **39** 3205-3212.
[14] Renz F, Jung S, Klein M, Menzel M, Thünemann A F 2009 *Polyhedron* **28** 1818-1821.

Competing redox reactions in Fe-containing AlO(OH) and Al_2O_3 matrices: A combined investigation by Mössbauer, ESR spectroscopy and thermal analysis

R Stößer[1], M Menzel[2*], M Feist[1], M Nofz[2] and F Renz[3]

[1]Humboldt University, Institute of Chemistry, D-12489 Berlin, Brook-Taylor-Str. 2, Germany,
[2]Federal Institute for Materials Research and Testing (BAM), D-12489 Berlin, Richard-Willstätter-Str. 11, Germany,
[3]Leibniz University, Institute of Inorganic Chemistry, D-30167 Hannover, Callinstr. 9, Germany,

* corresponding author: Michael.Menzel@bam.de

Abstract. The investigation of iron-doped AlO(OH)/Al_2O_3 systems revealed that the combined employment of Mössbauer and ESR spectroscopies together with thermal analysis yields meaningful data with complementary information. This mutual complementarity is based on the coexistence of Fe point defects with the corresponding aggregated FeO_x species which has been observed even for very low Fe concentrations. Competing redox processes between the dopant, the AlO_x matrix, and the gas atmosphere during the thermal treatment enable the generation of solid phases exhibiting specific chemical properties. The entire reaction process is influenced by a specific mechanical and thermal pre-treatment that affects predominantly oxydative processes in the matrix. A protecting influence of the matrix preventing further reductive attack of the Fe^{3+} and Fe^{2+} ions by hydrogen has been established.

1. Introduction

Crystalline and amorphous hydroxylic (AlOOH) and oxidic (Al_2O_3) precursors of α-Al_2O_3 have been doped with iron species for various reasons: (i) to reduce the temperature of the phase transition yielding α-Al_2O_3 when forming corundum protective layers, and (ii) to yield particular mechanical properties of the polycrystalline product phases via an enhanced grain boundary concentration of FeO_x or $(Fe_xAl_{1-x})_2O_3$. These processes are strongly influenced by mechanical activation both of the educts and the precursors for the Fe doping.

Both the mechanical activation and the thermal transformation to give α-Al_2O_3 have been investigated by employing the title methods. The activation depends on the kind of Fe compounds used for the doping and the gas atmosphere during the thermal analysis (TA) heating run. The competing redox processes occurring during the TA runs deliver unique information about the active role of the AlO_x matrix for the solid state reactions. Furthermore, they allow directly to influence certain properties of the resulting materials.

2. Results and Discussion

For the present study, the Fe concentration given as Fe_2O_3 was limited by the range $c \leq 0.1$mol-%. Under these conditions, the combination of Mössbauer, ESR, and TA measurements yields complementary information. As an example, a sample of α-Al_2O_3 calcined at 1500 °C can be regarded (Figure 1g). It contains only very few iron impurities and exhibits the typical ESR spectrum for the system $\{\alpha\text{-}Al_2O_3:Fe^{3+}\}$. The position of the fine-structure transitions is undoubtedly governed by the symmetry of the matrix [1] (Figure 2a). Shape and intensity of the transitions depend on the mode of the Fe^{3+} insertion into the matrix and, therefore, on the given structural perfectness of the α-Al_2O_3 lattice. Under these conditions, the concentration of the Fe^{3+} species placed on Al^{3+} positions is sufficient to yield a meaningful ESR spectrum such as displayed in Figure 2a. The integral of the spectrum, however, shows that these Fe^{3+} point defects affect only a comparably small part of the total ESR intensity (Figure 2b). The latter is predominantly located at g'~2 and originates from phase precursors of the approximative formula $(Fe_xAl_{1-x})_2O_3$ [2] which yields the Mössbauer spectrum according to Figure 1g. Taking into account the above mentioned aspect of the present study to obtain complementary information from different analytical methods, this result can be regarded as a first approach: the Mössbauer data primarily correspond to the aggregated species being in coexistence or even in equilibrium with the point defects.

Comparable effects could be observed for the common educt of the α-Al_2O_3 formation, i.e. the crystalline AlO(OH) which already contains iron impurities. Here, we are dealing with the weakly aggregated and rhombohedrally distorted Fe-O species which cause ESR signals at g'~ 4.3 whereas the aggregated species are indicated by broad lines at g'~2. In this case as well, the Mössbauer data correspond to aggregated species and yield the parameters $\delta = 0.27$ mm/s, $\Delta = 0.23$ mm/s and $\Gamma = 0.39$ mm/s (Figure 1a). After strong mechanical activation by 50 hours milling (cf. Figure 3) the Mössbauer parameters change as follows: $\delta = 0.22$ mm/s, $\Delta = 0.38$ mm/s and $\Gamma = 0.47$ mm/s. This means that the mechanochemical reaction which is accompanied by a water release [3], shifts the value of δ down to a 4 to 5 coordination of the Fe^{3+} ions, whereas the values of Δ and Γ increase due to the mechanically induced broader statistical distribution of the bond angles in the $\{FeO_x\}$ polyhedra. These findings can be related to the properties of a nominally undoped sample being prepared via a sol-gel route and showing a low Fe concentration [$c(Fe_2O_3) \sim 0{,}01$ mol%] (Figure 1d). The corresponding Mössbauer parameter at 293 K are as follows: $\delta = 0.24$ mm/s, $\Delta = 0.29$ mm/s and $\Gamma = 0.37$ mm/s and, respectively, at 78 K: $\delta = 0.34$ mm/s, $\Delta = 0.36$ mm/s and $\Gamma = 0.44$ mm/s. The increasing values for Δ and Γ can be explained by the stronger distortion of the matrix at the low temperature. In general, the values represent the typical ones for Fe^{3+} ions in $\{FeO_x\}$ polyhedra with $x = 4\ldots6$.

After co-milling of the originally crystalline AlO(OH) with 1 wt-% $Fe(NO_3)_3 \cdot 9H_2O$ and a subsequent TA heating run under $N_2/10\%H_2$ (Pt) (Figure 3) one obtains a doublet ($\delta = 1.10$ mm/s, $\Delta = 1.82$ mm/s, $\Gamma = 0.64$ mm/s) in coexistence with a sextet ($\delta = 0.01$ mm/s, $\Delta = 0.01$ mm/s, $\Gamma = 0.27$ mm/s, hf = -33.34 T) with ~8.2 % and 91.8 % relative intensity, respectively (Figure 1c). These parameters unambiguously represent $Fe^{(0)}$ being formed by reduction. Interestingly enough, the doublet of ~8% represents Fe^{2+} ions, which were not reducible due to a stabilization via reactive implementation into the matrix.

This idea is supported regarding a further sample doped by adding $Fe(NO_3)_3 \cdot 9H_2O$ in a sol-gel process. The sample contains 0.13 mol-% Fe_2O_3 and, after TA under N_2, yields a part of the total intensity of ~60% indicating Fe^{3+} ions ($\delta = 0.26$ mm/s and $\Delta = -0.02$ mm/s), and ~39% of a second phase with $\delta = 0.14$ mm/s, $\Delta = -0.66$ mm/s, hf = -33.35 T. It may represent $Fe^{(0)}$ or Fe_xN [4]. Preferably, however, the second phase is assigned to be Fe_xN due to an isomer shift larger than zero. Fe_xN should be not only a product of a reduction (e.g. by organic residues), but also of an internal nitridation of $Fe^{(0)}$ via intermedially formed NH_3. This would explain that the T_P decrease ends up at ~1110 °C, i.e. the number of appropriate FeO_x seeds [5] is not sufficient.

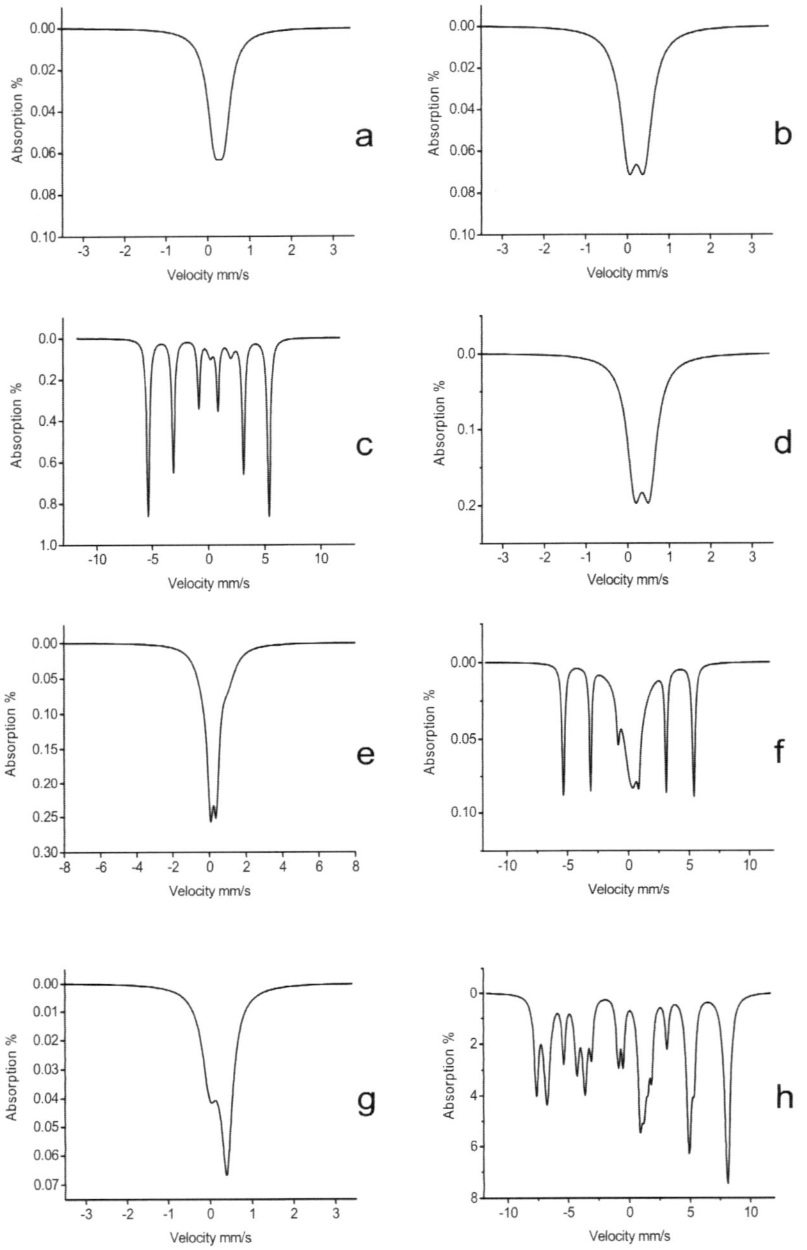

Figure 1. Mössbauer spectra of differently treated compounds:

(a) AlO(OH) crystalline (commercial, Nabaltec);
(b) as (a), 50 h milled, after TA under N_2/10%H_2 (1200 °C);
(c) as (a), 8h co-milled with 1 wt-% $Fe(NO_3)_3 \cdot 9H_2O$, after TA under N_2/10%H_2 (1200 °C);
(d) AlO(OH) xerogel, undoped, measured at 78 K;
(e) AlO(OH) xerogel, $Fe(NO_3)_3 \cdot 9H_2O$ (0.52 mol-% Fe_2O_3) added to the sol;
(f) as (e) after TA under N_2/10%H_2 (1200 °C);
(g) α-Al_2O_3 (commercial, VEB Laborchemie Apolda), calcined at 1500 °C;
(h) $Fe(NO_3)_3 \cdot 9H_2O$ after TA under N_2/10%H_2 (1200 °C);

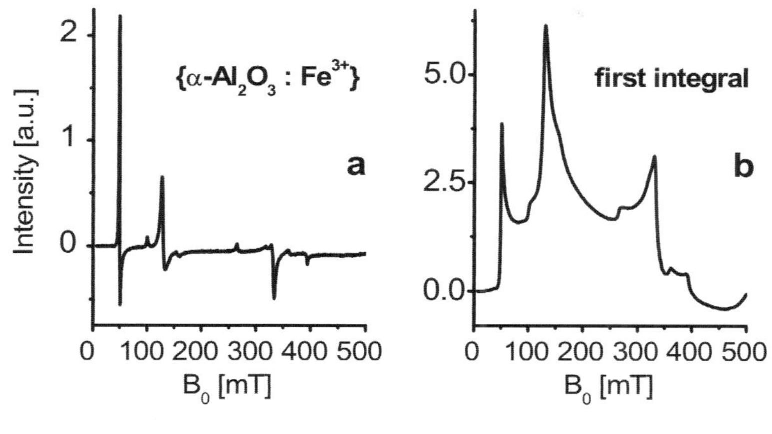

Figure 2. ESR spectra of {α-Al_2O_3:Fe^{3+}} powder measured at 78 K (a) and its first integral (b)

If the external reductive force is enhanced (e.g. by heating under N_2/10%H_2) an unexpected result is obtained. The sample yields a Mössbauer spectrum with a Fe^{3+} doublet (δ = 0.21 and Δ = -0.002 mm/s). Quite surprisingly, it amounts to ~50% of the total intensity, whereas the remaining 50 % are caused by $Fe^{(0)}$. Despite of the reductive conditions, a large part of Fe^{3+} is

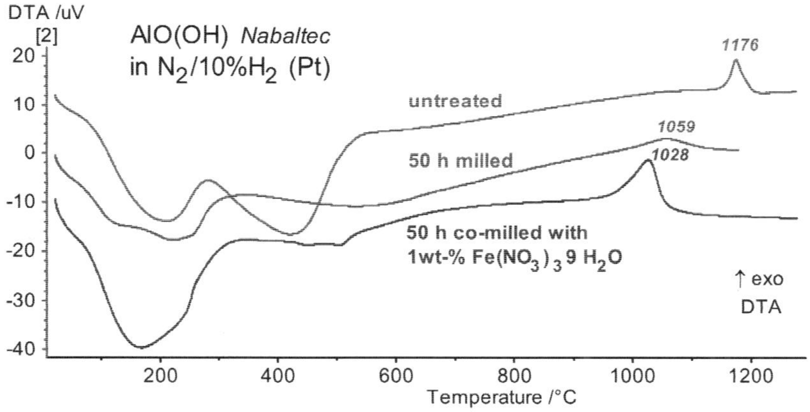

Figure 3. DTA curves of AlO(OH) in N_2/10%H_2 (Pt crucibles) for different sample pre-treatments

formed by chemical reaction with the matrix being under formation and structural reorganization. As a consequence, further reductive attack is prevented.

Competing redox processes *without* the participation of a matrix can be observed in the case of pure $Fe(NO_3)_3 \cdot 9H_2O$ being subjected to a TA heating run under N_2/10%H_2.
The characterization of the residue of the TA treatment by Mössbauer spectroscopy (Figure 1h) led to the following composition:
13.0% Fe^{2+} (δ = 1.08 mm/s, Δ = 0.31 mm/s and Γ = 0.49 mm/s);
15.6% $Fe^{(0)}$ (δ = 0.006 mm/s, Δ = - 0.02 mm/s and h_f = 33.27 T);
27.5% Magnetite A-sites δ = 0.27 mm/s, Δ = - 0.03 mm/s and h_f = 49.10 T;
43.7% Magnetite B-sites δ = 0.65 mm/s, Δ = - 0.02 mm/s and h_f = 46.09 T.

It is noteworthy that only 15.6% $Fe^{(0)}$ and a considerable amount of magnetite are formed. The oxydative atmosphere, formed during the thermal decomposition of $Fe(NO_3)_3{\cdot}9H_2O$, effectively competes with the hydrogen of the carrier gas in the TA run. However, at higher temperatures (TA up to 1200 °C) the fraction of reduced Fe species in the product mixture increases.

Finally, another interesting way to obtain Fe-doped α-Al_2O_3 should be mentioned here. It concerns the formation of alumina in an O_2/H_2 flame starting from a Fe-doped AlN precursor yielding a strongly distorted system (g'~ 4.3 in the ESR [6], not shown here). There is a coexistence of weakly aggregated point defects with precursors of $(Fe_xAl_{1-x})_2O_3$. This system is characterized by two doublets in the Mössbauer spectrum:

δ_1 = 0.23 mm/s, Δ_1 = 0.82 mm/s and Γ_1 = 0.78 mm/s (Fe^{3+} ions representing ~80 % of the spectral intensity);

δ_2 = 0.84 mm/s, Δ_2 = 1.77 mm/s and Γ_2 = 0.78 mm (Fe^{2+} ions in the aggregated part, ~20 %).

Obviously, this kind of a "pulse-like" pyrolysis of AlN, which yields hollow balls of α-Al_2O_3, creates a partly reductive atmosphere thus explaining the formation of Fe^{2+} species. A further reduction, however, which would lead to $Fe^{(0)}$, or even to a nitridation of iron could not be observed here. The applied temperature-time regime, obviously, prevents the reductive action.

3. Conclusions

For concentrations of $c(Fe_2O_3) \leq 0{,}01$ mol-% already, a distribution of Fe species could be established both in polycrystalline and amorphisized systems: (i) as point defects in the matrix lattice which can be followed by ESR spectroscopy, and (ii) as aggregated species acting as precursors of phases like $(Fe_xAl_{1-x})_2O_3$, Fe_2O_3, Fe_3O_4, and $Fe^{(0)}$ which can be successfully studied by Mössbauer spectroscopy. The actual phase composition together with the possible formation of further species such as $FeAlO_4$ precursors can be directed by redox processes of internal (residual carbon from the sol-gel process, NO_3^- etc.) or external (H_2; O_2) reaction partners during or prior (!) to the thermal matrix transformation. A comprehensive inspection of the spectroscopical and thermoanalytical findings shows that

(i) the FeO_x phases can significantly decrease the phase transformation temperature to yield α-Al_2O_3 provided the FeO_x phases exhibit both a suitable grain formation kinetics and particle size;

(ii) the transforming matrix does not exhibit only oxydative properties due to the *in situ* water release and the related formation of reactive surface sites, but is able as well to prevent the reduction of the matrix-incorporated Fe^{3+} ions up to ~50 % of the total amount by the external hydrogen gas flow covering the temperature range up to 1200 °C.

4. References

[1] Stösser R, Scholz G, Buzaré J Y, Silly G, Nofz M and Schultze D 2005, A magnetic resonance investigation of the process of corundum formation starting from sol-gel precursors, *J. Am. Ceram. Soc.* **88**, 2913-22

[2] Scholz G, Stößer R, Krossner M and Klein J 2001, Modelling of multifrequency ESR spectra of Fe^{3+} ions in crystalline and amorphous materials: A simplified approach to determine statistical distributions of spin-spin coupling parameters, *Appl. Magn. Res.* **21,** 105-23

[3] Stößer R and Feist M 2008, Exothermal water release from Pseudoboehmite gels and their mechanically treated analoga caused by activated hydrogen, *J. Phys. Chem. C* **112**, 16438-444

[4] Rissanen L, Neubauer M, Lieb K P and Schaaf P 1998, J The new cubic iron-nitride phase FeN prepared by reactive meagnetron sputtering, *J. Alloys Compounds* **274**, 74-82

[5] Nofz M, Stößer R, Dörfel G and Schultze D 2005, The thermally induced transformation of pseudoboehmite gels - a comparison of the effects of corundum seeding and iron doping, *J. Europ. Ceram. Soc.* **25**, 1095-107

[6] Brenneis R, personal communication.

Journal of Physics: Conference Series **217** (2010) 012029 doi:10.1088/1742-6596/217/1/012029

Mössbauer study of oxygen adducts in solid Fe(II) phthalocyanines

E Kuzmann[1,2], Z Homonnay[1], A Mylonakis[3], H Yin[3], Y Wei[3], K Kovács[1], S Kubuki[4], Z Klencsár[2], A Vértes[1,2] and A Nath[5]

[1] Institute of Chemistry, Eötvös Loránd University, P.O. Box 32, 1512 Budapest, Hungary

[2] Laboratory of Nuclear Chemistry, Chemical Research Center , HAS, P.O. Box 32, 1512 Budapest, Hungary

[3] Department of Chemistry, Drexel University, Philadelphia, PA 19104, USA

[4] Department of Chemical and Biological Engineering, Ube National College of Technology,Yamaguchi Japan

[5] Department of Chemistry, University of North Carolina, Asheville, NC 28804, USA

E-mail: kuzmann@chem.elte.hu

Abstract. Beta iron phthalocyanines (β-FePc) oxygenated at low temperatures were investigated with the help of ^{57}Fe Mössbauer spectroscopy and X-ray diffractometry (XRD). The oxygenation of FePc complex is the key step in the pyrolysis of the flat layered molecules that yields acicular iron carbide particles encaged in uniform carbon shells. Mössbauer spectroscopy revealed that upon oxygenation of β-FePc new species were formed which could be associated with Fe^{III}Pc oxygen adducts in high spin and low spin states. The temperature dependent Mössbauer measurements showed the interconversion of the low spin species into high spin species upon raising the temperature.

1. Introduction

Phthalocyanines and metallophthalocyanines have been known for many years as colorants and pigments. Nowadays their use as electrochemical catalysts, gas sensors, and regular heterogenous catalysts has widely stimulated their research [1]-[3]. Furthermore it was shown that controlled pyrolysis of flat layered iron phthalocyanine (FePc) can be used for preparing iron carbide nanocrystallites encapsulated in uniform 2.7 nm wide carbon coating [4]. The encapsulation of magnetic carbides by nanotubes enhances their applications since the carbon shells protect the encaged species from the ambient atmosphere where hydrolysis and oxidation could occur. Such carbon-coated magnetic particles have various industrial applications in magnetic data storage or as ferrofluids.

It is essential to know the mechanism of the formation of the nanocapsules in order to be able to control the diameter and thickness of the carbon coatings. Since the oxygenation of the FePc complex is the key step in the synthesis, there is a great interest in the interaction of oxygen with FePc.

Phthalocyanines can acquire several polymorphic forms of which α and β are fairly well characterized. The difference in the α- and β-polymorph lies essentially in the orientation of the flat molecules with respect to the crystallographic axes, where the perpendicular distance between the stacks of molecules is approximately the same. The inclination of the molecular stacks in the α-form is much less than in the β-form. In an FePc molecule, the metal atom is at the centre of a square of nitrogens. In the β-polymorph, there are additional nitrogen atoms of neighbouring molecules present axially above and below the central metal atom whereas in the α-form, the nitrogens of the nearest molecules are not in axial positions and therefore do not form an octahedron about the central metal atom.

In our previous works [5] Mössbauer spectroscopy and X-ray diffractometry were used to investigate the nature of oxygen bonding after the oxygenation of $Fe^{II}Pc$. The results indicated that considerable amount of oxygen can be introduced into the solid β-$Fe^{II}Pc$ at relatively low temperatures in the dry state and in aqueous suspensions. A variety of oxygen adducts with β-$Fe^{II}Pc$ were shown to be stabilized upon oxygenation, while only a very small amount of oxygen could be introduced into the α-$Fe^{II}Pc$ under the same conditions. Here we report cases when an oxygen adduct with β-$Fe^{II}Pc$ formed in aqueous suspension exhibits low-spin high-spin transition in the temperature range of 30 K – 293 K.

2. Experimental

Beta forms of $Fe^{II}Pc$ were oxygenated under O_2 flow in an aqueous suspension. For the wet oxygenation β-$Fe^{II}Pc$ was suspended in distilled water and the oxygenation was accomplished by bubbling O_2 through the aqueous suspension at 50 °C for 75 days.

^{57}Fe Mössbauer measurements of powdered compounds were carried out in transmission geometry between 20 K and 293 K in a closed circuit He refrigerator cryostate (SHI-APD Cryogenics). A 57Co/Ph source of 10^9 Bq activity was used. The Mössbauer spectra were evaluated using the MOSSWIN code [7].

X-ray diffractograms of the pristine and oxygenated β-$Fe^{II}Pc$ were recorded by a computer–assisted DRON-2 diffractometer using CoK_α radiation and β-filter.

3. Results and discussion

A detail of the powder X-ray diffractogram of the β-$Fe^{II}Pc$ oxygenated in aqueous suspension can be seen in figure 1. All lines in the diffractogram can be indexed according to the monoclinic standard β-$Fe^{II}Pc$ [5] with lattice parameters of a=14.8148 Å, b=4.7779 Å, c=19.6296 Å and β=121.36° . This reflects that the basic structure is preserved after the long term oxygenation. However, the lattice parameters *a* and *c* decreased considerable compared to those of the pristine compound [8], furthermore, significant changes in the relative intensities of the corresponding lines and line broadening can be observed in the diffractogram upon oxygenation. All these changes are well consistent with and indicate the incorporation of oxygen between the molecular layers. The volume of the elementary cell of the oxygenated β-$Fe^{II}Pc$ decreased to 1186 $Å^3$ from 1216 $Å^3$ characteristic of pristine compound. This strongly supports the oxygen uptake in good agreement with our earlier result [5].

The temperature dependent Mössbauer spectra of the β-$Fe^{II}Pc$ oxygenated in aqueous suspension are shown in figure 2. There are significant changes in the spectral envelope upon raising the temperature. The spectra of the β-$Fe^{II}Pc$ could be decomposed into three doublets (D1, D2, D3), the data obtained by the computer fit are shown in table 1. The Mössbauer parameters of D1 correspond well with those of the pristine β-$Fe^{II}Pc$ with large quadruple splitting [9]. The doublets D2 and D3 can be associated with Fe^{III} oxygen adducts in high spin (HS) and low spin (LS) states, respectively.

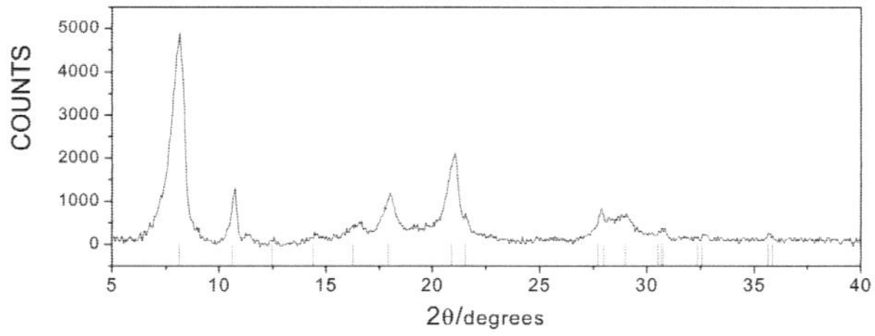

Figure 1. A detail of PXRD of β-$Fe^{II}Pc$ oxygenated in aqueous suspension. The bars indicate the calculated reflections.

Although in our previous publication we suggested an assignment of the observed doublets to real oxygen adduct species, in the light of our new experimental findings, it is worthwhile to return to this question and see if alternative models may be possible.

The most remarkable finding in these experiments compared to our previous work is the limited number of species, i.e., actually only one oxygen adduct species is found, which shows up in LS and HS state, depending on the temperature. This species is found now in a sample having the highest oxygen content ever synthesized by us, which is demonstrated by the very low spectral contribution of the doublet of the pristine β-$Fe^{II}Pc$ (7 % at room temperature, see table 1).

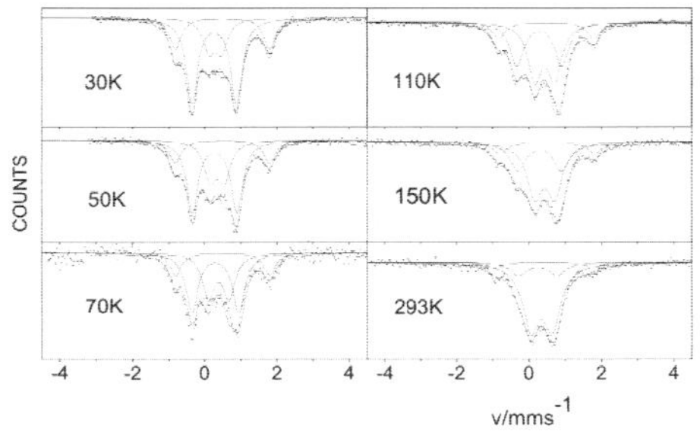

Figure 2. Temperature dependence of Mössbauer spectra of β-$Fe^{II}Pc$ oxygenated in aqueous suspension.

Table 1. Mössbauer parameters of the oxygen treated β-$Fe^{II}Pc$. δ = Chemical Shift[a], Δ = Quadrupole splitting, A = Abundance.

T (K)	**Doublet 1 (D1)**			**Doublet 2 (D2)**			**Doublet 3 (D3)**		
	δ (mms^{-1})	Δ (mms^{-1})	A (%)	δ (mms^{-1})	Δ (mms^{-1})	A (%)	δ (mms^{-1})	Δ (mms^{-1})	A (%)
30	0.52	2.57	23	0.41	0.43	24	0.31	1.24	53
50	0.50	2.59	19	0.40	0.44	31	0.30	1.24	50
70	0.48	2.61	18	0.36	0.46	35	0.29	1.26	48
110	0.47	2.62	14	0.40	0.52	46	0.28	1.24	40
150	0.45	2.67	11	0.40	0.54	54	0.27	1.25	34
293	0.40	2.70	7	0.34	0.58	77	0.24	1.07	16

[a]relative to α-iron

In our previous model [10], where the oxygenation resulted roughly in 60 % oxygen related Fe species, the "spin-crossover" species was assumed to be an Fe atom having a peroxo bridge on one side of the Pc plane, and a dihapto bound O_2 on the other side (figure 3 (a), species c, d). Those samples also contained another species (figure 3 (a), species b) in which Fe had one dihapto bound O_2 only. Now one can expect that when the oxygen content of the sample increases, the relative contribution of the species (c-d), and also of b would increase at the expense of species a (Fe in pristine $Fe^{II}Pc$). However, as presented in figure 2 and table 1, species b disappears while the pristine FePc is still there.

On the other hand, in case of O_2 type ligands placed in between β-FePc layers in order to form a bridge between iron atoms, the Fe–O distance is around 0.19 nm – almost ideal to form a bond. To reach the regular bond length, found in peroxo bridged complexes of similar type [11], roughly 5 % linear contraction is needed only. Now, considering species (c-d), it is very likely that if the dihapto bound O_2 transforms into end-on bound O_2 by thermal excitation, the rotation of this O_2 molecule would automatically result in the formation of another Fe–O bond with an Fe atom in the neighbouring Pc plane. The final result is the formation of long chains of peroxo bridged irons as presented in figure 3 (b). Note that iron has a strong tendency to form long …–Fe–O–Fe–O–Fe–O–… chains for example in aqueous Fe^{3+}-solutions [12]. In the lattice of FePc, μ-oxo bridges are not, only peroxo bridges are feasible due to the large Fe–Fe distance [5].

In this scenario, the reduction of the number of species is easier to understand: the dominating oxygen adduct species in the high oxygen content sample is the in-chain site (peroxo bridge on both sides of Fe). Chain-terminal species may be five-coordinate Fe atoms with one peroxo bridge only at one side or "quasy" seven-coordinate with an additional dihapto O_2 at the other side of iron.

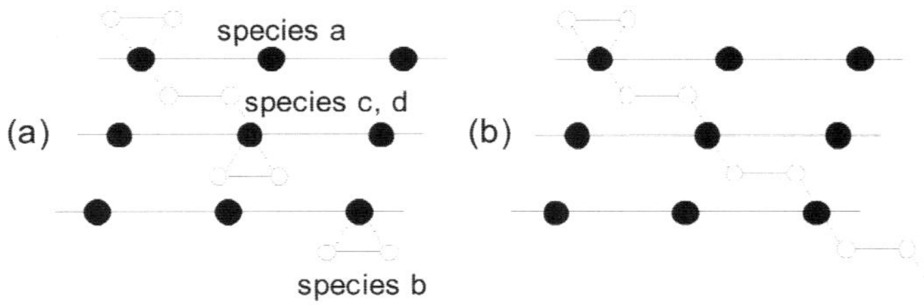

Figure 3. Possible distribution of dioxygen in the lattice of β-Fe^{II}Pc after oxygenation. (a) Low oxygen content (~60 %) (b) high oxygen content (~93 %)

According to our previous reasoning, a dihapto bound O_2 at the end of the chain can transform into a new bridge easily unless it is on the surface of the crystal lattice and thus another iron is not available. It is difficult to judge which species is favoured energetically, but higher coordination may be preferred by iron (and this is drawn in figure 3). Nevertheless, disregarding the type of the terminal species, the relative number of them is negligible if the chains are long enough, and therefore there is no other doublet found in the Mössbauer spectra, that would represent a single dihapto O_2–Fe oxygen adduct (species b).

The required 5 % bond length contraction upon O_2-bridge formation and the existence of long chains are in perfect agreement with the decreasing lattice parameters found by our XRD measurement. This can also explain why the Mössbauer parameters of doublets D2 and D3 are only close but not identical to those of species c and d found earlier in samples with lower oxygen content.

In our modified model, the species showing spin-crossover is therefore a six-fold coordinated species with two axial oxygens and four equatorial nitrogens, with the O–Fe–O axis being non-perpendicular to the Pc plane. The spin-crossover may be connected with electron-density shift from iron to the O_2 ligands.

Association of the electron density shift with a possible structural deformation is likely because of the non-perpendicular O–Fe–O axis. Taking into consideration that the O–O bond lengths are 0.121 nm, 0.128 nm, and 0.149 nm for neutral dioxygen (O=O), superoxide ($O=O^-$), and peroxide ($O=O^{2-}$), respectively, an electron density shift from iron to oxygen (e.g., shift form a formal Fe^{2+} state to a formal Fe^{3+} state with a concomitant shift from O_2 to O_2^-) can result in lower deformation of the ligand sphere of iron, i.e., the O–Fe–O axis closer to perpendicularity. This simultaneous electronic and structural change can be accompanied by a spin state change, and therefore it is manifested as a spin-crossover process.

4. Conclusions

With the help of temperature dependent Mössbauer spectra, a spin-crossover of the six coordinate in-chain Fe-oxygen adduct has been shown. Based on the XRD and Mössbauer spectroscopic data, the formation of long μ-peroxo bridged –Fe–O–O–Fe– chains are suggested upon the introduction of high amount of oxygen into the lattice of β-FePc. The driving force for the high spin to low spin transformation at low temperature is a decreasing distortion of the ligand sphere of iron (toward a regular octahedron) which assumes an electron density shift from iron to the bridging O_2 ligands.

5. Acknowledgement

This work was supported by OTKA grants No. K068135, K067835, NI67702.

References

[1] Zagel J H 1998 *Coord. Chem. Rev.* **119** 89
[2] Meunier B and Sorokin A 1991 *Acc. Chem. Res.* **30** 470
[3] Guillaud G, Simon J and Germain J P 1998 *Coord. Chem. Rev.* **178-180** 1443
[4] Kopelev N S, Chechersky V, Nath A, Wang Z L, Kuzmann E, Zhang B and Grayson V H 1995 *Chem. Mater.* **7** 1419
[5] Kuzmann E, Homonnay Z, Vértes A, Li S, Yin H, Wei Y, Nath A, Chen X and Li J 2003 *J. Solid State Chem.* **170** 118
[6] Kuzmann E, Nath A, Chechersky V, Li S, Wei Y, Chen X, Li J, Homonnay Z, Gál M, Garg V K, Klencsár Z and Vértes A 2002 *Hyp. Int.* **139-140** 631
[7] Klencsár Z, Kuzmann E and Vértes A 1996 *J. Radioanal. Nucl. Chem.* **210** 105
[8] Coppens P and Li L 1984 *J. Chem. Phys.* **81** 1983
[9] Srivastava T S, Przybylinski J L, Nath A 1974 *Inorg. Chem.* **13** 1562
[10] Yin H, Kubuki S, Homonnay Z, Kuzmann E, Vértes A, Wei Y and Nath A 2008 *The Open Inorganic Chemistry Journal* **2** 69
[11] Kurtz, D.M., Jr. 1990 *Chemical Reviews* **90** 585
[12] Cornell R M and Schwertmann U 2006 *The Iron Oxides* (Darmstadt: Wiley-VCH GmbH & Co. KGaA)

Journal of Physics: Conference Series **217** (2010) 012030 doi:10.1088/1742-6596/217/1/012030

Vibrational dynamics of a trinuclear oxo–bridged iron complex as studied by nuclear inelastic scattering, Mössbauer spectroscopy and DFT calculations

S Rajagopalan[1], **T Asthalter**[1,4], **V Rabe**[2], **S Laschat**[2], **G Rauhut**[3] **and E Roduner**[1,5]

[1] Institut für Physikalische Chemie, Universität Stuttgart, Pfaffenwaldring 55, D–70569 Stuttgart, Germany
[2] Institut für Organische Chemie, Universität Stuttgart, Pfaffenwaldring 55, D–70569 Stuttgart, Germany
[3] Institut für Theoretische Chemie, Universität Stuttgart, Pfaffenwaldring 55, D–70569 Stuttgart, Germany

E-mail: e.roduner@ipc.uni-stuttgart.de

Abstract. We have applied nuclear inelastic scattering (NIS) and computational predictions based on density functional theory (DFT) to explore the vibrational dynamics of ^{57}Fe in a trinuclear oxo-bridged iron(III) complex. Quantitative comparison of the experimentally measured vibrational dynamics of the Mössbauer nuclei ^{57}Fe, with normal mode analysis, calculated *via* quantum chemical calculation based on DFT methods generally yields a good overall agreement and enables the assignment of Fe vibrational modes. The oxidation and spin state of the complex were determined by ^{57}Fe Mössbauer spectroscopy.

1. Introduction

There is a continuous interest in the search for selective metal complexes which serve as catalysts for the oxygenation of unactivated hydrocarbon substrates *via* polynuclear non-heme iron catalysts. The understanding on how such activation reactions work relies on the vibrational dynamics of the metal centre in the catalysts. Vibrational spectroscopy provides a sharp and clear probe of the structure, dynamics and reactivity of biological molecules. Identifying the active metal–site vibrations and relating them to the functional dynamics is of particular interest in enzymes. So far the published works by several authors were conducted on the vibrational properties of iron heme porphyrins [1] and non-heme Fe-S hydrogenase [2]. Recently, the mechanisms governing the spin crossover in Fe compounds were also reported in detail [3]. In this work, we report the first information about the vibrational dynamics of a trinuclear oxo-bridged iron complex (figure 1), investigated by NIS and DFT methods. The inhomogeneity of the iron coordination is determined by Mössbauer spectroscopy.

[4] Present address: European Patent Office, Patentlaan 2, 2288 EE Rijswijk, The Netherlands.
[5] To whom any correspondence should be addressed.

2. Experiment

The trinuclear complex is prepared by the reaction of the deprotonated ligand **1** with iron(III) perchlorate (figure 1). On precipitation, the complex crystallizes out as red crystals [4]. An ^{57}Fe transmission Mössbauer spectrum was measured at room temperature (RT) using a conventional constant–acceleration spectrometer equipped with a gas proportional counter and a ^{57}Co/Rh source with a strength of 50 mCi. The resolution of the spectrometer was $\sim$0.24 mm s^{-1} (full–width at half–maximum of the inner lines when measured with a α–Fe foil at RT). The experimental data were fitted to Lorentzian profiles by the least–squares method, and the isomer shift is relative to the centroid (= zero mm s^{-1}) of the α–Fe foil spectrum at RT. NIS is a new

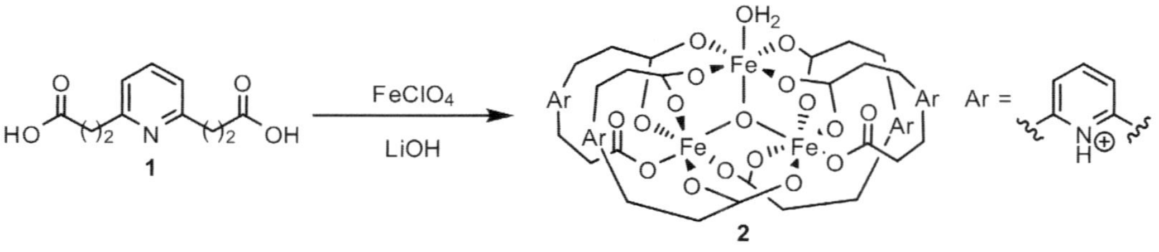

Figure 1. Formation and structure of the trinuclear oxo–bridged Fe complex.

synchrotron based spectroscopic technique, which is used in the present work to study the vibrational dynamics of the Mössbauer nuclei, ^{57}Fe. The method obeys very simple selection rules, as it senses only the ^{57}Fe vibrational motion and thus measures the complete vibrational density of states (VDOS) for the Fe atoms including the mean square displacement (msd) of the Fe atom. The NIS experiments [5] were carried out at the nuclear resonance beamline ID18 of the European Synchrotron Radiation Facilities (ESRF) in Grenoble, with an energy bandwidth of 1.1 meV. Details of NIS theory, experimental procedure and data processing are described elsewhere [6, 7]. The unenriched crystalline sample is scanned at 3 s per point with a total of 36 scans (500 points per scan). In order to minimize the multiphonon contributions measurements were performed at 26.5 K (obtained from the ratio of anti-Stokes to Stokes intensities).

3. Computational details

The geometry optimizations and subsequent vibrational frequency calculations were carried out using the DFT B3LYP hybrid exchange correlation functional [8], as implemented in the GAUSSIAN03 program [9]. A correlation consistent polarized valence double–ζ basis set has been used for all atoms except Fe [10]. For the metal Fe atoms the 10 core electron wavefunctions was replaced by the scalar relativistic Stuttgart/Dresden effective core potential, which was combined with the corresponding basis set for the Fe valence shells as described in reference [11]. DFT calculations were performed on the complete structure including all the H atoms, thus in total 119 (= N) atoms. In the calculation, the spin state of each Fe atom was taken as 5/2 with a total multiplicity of the cluster of 16. The VDOS is extracted from the DFT calculated normal coordinates of all modes assuming the low temperature approximation according to [12]

$$g(E) = \frac{1}{3}\sum_{j=1}^{3N-6} \frac{m_{\mathrm{Fe}} u_{\mathrm{Fe},j}^2}{\sum_{k=1}^{N} m_k u_{k,j}^2} \delta(E - E_j),$$

where m_{Fe} (56.94 a.u.) is the mass of the resonant atom (Mössbauer nuclei), $u_{\mathrm{Fe},j}^2$ and $u_{k,j}^2$ are the msd of the Fe and k^{th} atom, respectively, in the jth mode, E is the difference between the incoming photon energy and the recoil–free resonance energy, and E_j is the energy of the molecular vibration. The term $\delta(E - E_j)$ is a normalized Gaussian line shape function, which includes a width equal to 1.1 meV.

4. Results and discussion

The core structure can be well described as a central planar Fe_3O unit with three planar FeO_4 units perpendicular to the Fe–O–bonds of the Fe_3O (figure 1). Each unit of the Fe atoms is coordinated octahedrally by six O atoms, four of them are provided by the carboxylic acid groups of the ligand, the fifth O is the common central ion of the Fe_3O unit. For two of the Fe centers the sixth O is also provided by carboxylic ligand groups, but for the third Fe the sixth position is occupied by a H_2O molecule. This renders one of the three Fe centers slightly different from the other two. The Mössbauer spectrum of the complex at RT shown in figure 2 consists of an

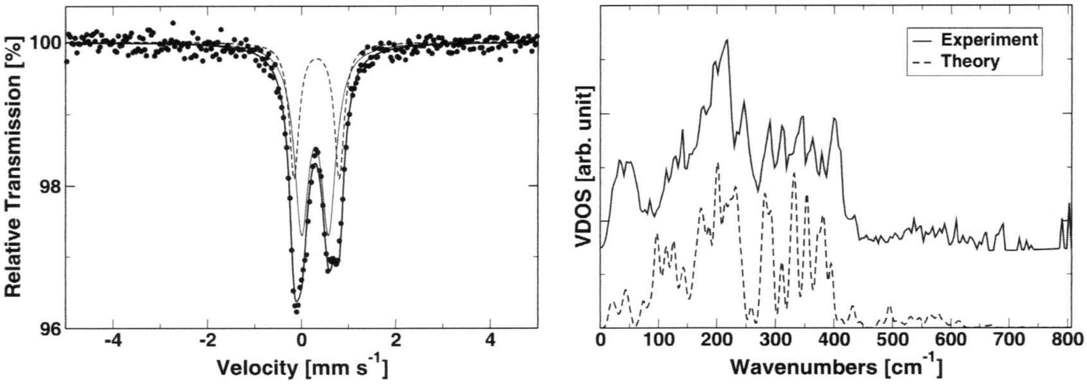

Figure 2. RT Mössbauer spectrum (left) and VDOS (right) at 27.5 K of the trinuclear complex.

asymmetric broadened quadrupole doublet indicating the existence of two quadrpole doublets. While the doublet structure is in agreement with the expectation for inequivalent Fe sites and the isomer shift is typical for Fe^{3+} the asymmetry is an immediate consequence of the inequivalence of the three Fe centers. Initial attempts to fit with two quadrupole doublets with isomer shift δ, quadrupole splitting ΔE_Q, line width Γ and area ratios as free variables gave an area ratio far from the theoretically expected value of 2:1 [13]. Hence we fixed the area ratio as 2:1 (area ratio of inner doublet to outer doublet). A good fit is shown in figure 2 with values: inner doublet $\delta = (0.40 \pm 0.01)\,\mathrm{mm\,s^{-1}}$, $\Delta E_Q = (0.58 \pm 0.01)\,\mathrm{mm\,s^{-1}}$ and $\Gamma = (0.37 \pm 0.03)\,\mathrm{mm\,s^{-1}}$; outer doublet $\delta = (0.43 \pm 0.01)\,\mathrm{mm\,s^{-1}}$, $\Delta E_Q = (0.96 \pm 0.01)\,\mathrm{mm\,s^{-1}}$ and $\Gamma = (0.25 \pm 0.03)\,\mathrm{mm\,s^{-1}}$. The outer doublet, which has the large ΔE_Q and lower weight by area is attributed to the single Fe, which is coordinated to a H_2O molecule. The other two Fe sites, which are equivalent by coordination, can then be collectively attributed to inner doublet with large area. It should be noted that Mössbauer is a local probe technique which provides information about the spin state of the individual Fe centers, while magnetization and EPR measurements as reported by Rabe et al. [4] provides the spin state of the coupled Fe_3O cluster, and this was reported to have a spin 1/2 doublet ground state.

The experimental together with the DFT predicted VDOS of the trinuclear iron complex is shown in figure 2. In overall the DFT simulated spectrum for the $S = 15/2$ state is in agreement with the measured one. The calculation of the same structure for the $S = 1/2$ ground state yielded a completely different spectrum (not shown) and suffers from spin contamination. This discrepancy has to be further investigated. The acoustic modes in the spectrum extends up to $\sim 100\,\mathrm{cm^{-1}}$, which was estimated by the procedure prescribed by Chumakov et al. [14]. In this region, the calculated spectrum exhibits less than the measured number of peaks. This is because the calculation has been performed for a free single molecule. The Fe–O vibrations i.e. the skeletal modes comprising of the central FeO_3 unit, are generally present in the low frequency region, typically $< 700\,\mathrm{cm^{-1}}$. The measured and simulated VDOS displays a dominant peak at about 210 cm^{-1} and a number of weaker peaks above 250 cm^{-1}. In a local D_{3h} symmetry the central metal Fe_3O unit has six fundamental modes of vibration: ν_s and ν_2 being nondegenerate while ν_3 and ν_4 are doubly degenerate [15]. By direct comparison of NIS with theory the stretching mode ν_s can be assigned to the peak at 244 cm^{-1}. Since this mode is Raman active we clarified by Raman analysis, where it appears at 245 cm^{-1} (not shown). This is in contrast to the previously reported value of 148 cm^{-1} [16]. One possible reason may a different spin state with larger force constant in our complex so that excitation of the stretching mode ν_s occurs at higher frequency, as the larger ligand mass would tend to the opposite effect. In the experimental spectrum we notice a broad dominant peak around 210 cm^{-1}. Since this peak is broad and intense, one may expect that it may be coupled with two or more unresolved modes. The complex has three planar FeO_4 units, D_{4h} symmetry sites perpendicular to the Fe–O–bonds of the Fe_3O unit. Johnson et al. observed the O–Fe–O bending vibration E_u of the planar FeO_4 units at 200 cm^{-1} [16]. A similar feature is observed in our spectrum at 202 cm^{-1} in the experimental as well as calculated VDOS and thus this mode should be assigned to the in-plane bend E_u mode. On considering D_{3h} symmetry of the Fe_3O unit interacting under the D_{4h} symmetry of three FeO_4 units, the E_u mode of the D_{4h} symmetry leads to a rotational symmetry A_2' of the Fe_3O unit. Thus E_u as well as A_2' contribute to the VDOS at 200 cm^{-1}.

5. Conclusion

The Mössbauer results show that the Fe_3O complex, which based on other methods was reported to have a spin doublet ground state, consists of high–spin Fe^{3+} ions with two differnet inequivalent Fe sites. The vibrational dynamics obtained by NIS and DFT calculations are presented. Some of the vibrational modes of the central metal unit that are potentially important in the functional dynamics of the catalyst are assigned.

Acknowledgments

We thank Drs. A.I. Chumakov, I. Sergeev and R. Rüffer of ID 18 at ESRF for the assistance with NIS experiments. Financial support for this work by the Deutsche Forschungsgemeinschaft within the framework of SFB 706 and the Ministerium für Wissenschaft, Forschung und Kunst des Landes Baden–Württenberg is gratefully acknowledged.

References

[1] Rai B K et al. 2002 *Phys. Rev.* E **66** 51904
Leu B M et al. 2004 *J. Am. Chem. Soc.* **126** 4211
[2] Cramer S P, Xiao Y, Wang H, Guo Y and Smith M C 2006 *Hyperfine Interact.* **170** 47
Guo Y et al. 2008 *Inorg. Chem.* **47** 3969
[3] Ronayne K L et al. 2006 *Phys. Chem. Chem. Phys.* **8** 4685
Winkler H, Chumakov A I and Trautwein A X 2004 *Top. Curr. Chem.* **235** 137
[4] Rabe V et al. Manuscript submitted to *Eur. J. Inorg. Chem.*
[5] Seto M, Yoda Y, Kikuta S, Zhang X W and Ando M 1995 *Phys. Rev. Lett.* **74** 3828
Sturhahn W et al. 1995 *Phys. Rev. Lett.* **74** 3832
[6] Chumakov A I and Sturhahn W 1999 *Hyperfine Interact.* **123/124** 781
[7] Kohn V G and Chumakov A I 2000 *Hyperfine Interact.* **125** 205
[8] Becke A D 1993 *J. Chem. Phys.* **98** 5648
[9] Frisch M J et al. Gaussian 03, Revision E.01 (Gaussian, Inc., Wallingford, CT, 2004)
[10] Dunning T H Jr 1989 *J. Chem. Phys.* **90** 1007
[11] Dolg M, Wedig U, Stoll H and Preuss H 1987 *J. Chem. Phys.* **86** 866
[12] Paulsen H, Winkler H, Trautwein A X, Grünsteudel H, Rosanov V and Toftlund 1999 *Phys. Rev.* B **59** 975
Asthalter T 2006 *Z. Phys. Chem.* **220** 979
[13] Boudalis A K et al. 2005 *Polyhedron* **24** 1540
[14] Chumakov A I, Rüffer R, Leupold O and Sergueev I 2003 *Struct. Chem.* **14** 109
[15] Bass C D, Lynds L, Wolfram T and DeWames R E 1964 *J. Chem. Phys.* **40** 3611
[16] Johnson M K, Powell D B and Cannon R D 1981 *Spectrochem. Acta.* **37A** 995

International Conference on the Applications of the Mössbauer Effect (ICAME 2009) IOP Publishing
Journal of Physics: Conference Series **217** (2010) 012031 doi:10.1088/1742-6596/217/1/012031

Influence of a thermochromic anion on the spin crossover of iron(II) trinuclear complexes probed by Mössbauer spectroscopy

François Robert, Anil D Naik and Yann Garcia

Institut de la Matière Condensée et des Nanosciences, Université Catholique de Louvain, Place L. Pasteur 1, 1348 Louvain-la-Neuve, Belgium.
E-mail: yann.garcia@uclouvain.be

Abstract. The insertion of thermochromic anions (**1** and **2**) into triazole based iron(II) trinuclear spin crossover complexes $[Fe^{II}_3(hyetrz)_6(H_2O)_6]$(**1** or **2**)$_4Cl_2$ paves the way to a large panel of multifunctional materials. In addition to a gradual thermo-induced spin conversion of the central metal ion on cooling below room temperature, an unexpected split of the Mössbauer signal of the external iron(II) ions as well as pronounced relaxation phenomenon for the central HS iron(II) ion was discovered. This phenomenon is suggested to be induced by the tautomeric equilibrium between enol and keto forms of the anion.

1. Introduction

Spin crossover (SCO) is a highly versatile phenomenon which involves a complete electronic reorganization between a diamagnetic low-spin (LS) and a paramagnetic high-spin (HS) state, in iron(II) coordination compounds [1]. Many applications are thought in sensors [2] and displays [3] but the most promising field is presumably the development of new molecular switches for molecular electronics [3]. The large majority of iron(II) SCO compounds are made of mononuclear or large polynuclear units such as linear chains [2]. Some dinuclear and tetranuclear self-assemblies [4,5] were obtained by using ligands with specific coordination sites although trinuclears were synthesised with suitable metal/ligand ratios with 1,2,4-triazole ligands [6]. In these later model SCO complexes, only the central iron(II) ion, which is isolated from the terminal HS iron(II) ions, is SCO capable [6] (Figure 1a). In addition, the use of an alcohol substituted triazole such as 4-(2'-hydroxyethyl)-1,2,4-triazole (hyetrz) can lead to dense supramolecular networks which can modulate the SCO phenomenon thanks to steric and electronic factors [6]. A strong influence of the anion nature was also nicely demonstrated because of the sensitivity of the external iron(II) ions on the SCO of the central ion in the case of an anion-directed rigid supramolecular network [7]. In this work, novel thermochromic anions derivated from *N*-salicylidene aniline were introduced in 1,2,4-triazole based trinuclear compounds. The sodium salt of *N*-salicylidene *p*-aminobenzene sulfonate anion (**1**) and its bromo derivative (**2**) display thermochromism thanks to a thermo-induced tautomeric equilibrium between an uncoloured enol form and a yellow *cis*-keto form (Figure 1b) [8]. An indirect influence of the anion nature activated by temperature is proposed to explain the unusual Mössbauer features of the trinuclear complexes (**3**,**4**).

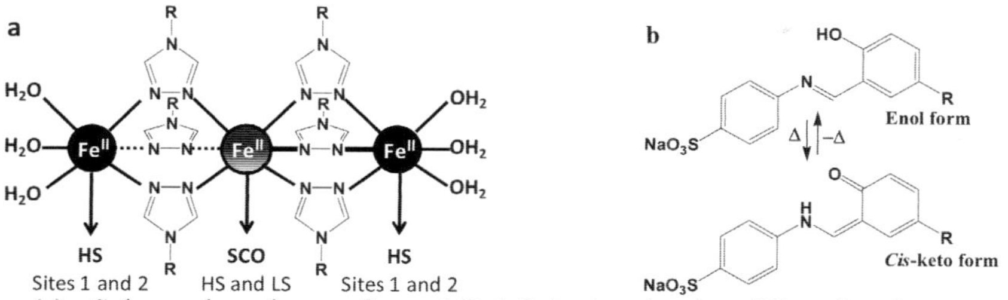

Figure 1. (a) Schematic view of a 4-R-1,2,4-triazole iron(II) trinuclear compound with R = CH_2CH_2OH. (b) Sodium salt of the *N*-salicylidene *p*-aminobenzene sulfonate with R = H (**1**), Br (**2**). The thermo-induced equilibrium between the uncoloured enol and the yellow *cis*-keto forms is shown.

2. Results

1 and **2** was synthesized following the reported method [9]. Interestingly, the trinuclear compound **3**, $[Fe^{II}_3(hyetrz)_6(H_2O)_6](\mathbf{1})_4Cl_2$, (Figure 1) was obtained although a ratio of one equivalent of iron(II) chloride salt for three equivalents of hyetrz and two equivalents of the sodium salt of the anions **1** was engaged in the reaction medium (methanol, technical grade). It strongly suggests that **1** prevents the formation of the usually reported 1D chain [2].

δ [mm/s]	T [K]	Fe^{II}_{LS}	$Fe^{II}_{HS\text{-}central}$	$Fe^{II}_{HS\text{-}Ext.\,1}$	$Fe^{II}_{HS\text{-}Ext.\,2}$	Fe^{III}_{HS}
3	145(1)	0.43(1)	1.71(2)	1.14(1)	1.17(1)	-
	175(1)	0.45(2)	1.50(9)	1.12(1)	1.16(1)	-
	225(1)	0.41(2)	1.32(1)	1.03(2)	1.17(1)	-
	275(1)	0.41(4)	1.14(2)	0.817*	1.18(2)	0.403*
4	145(1)	0.33(3)	1.88(2)	1.12(6)	1.22(5)	-
	175(1)	0.33(4)	1.83(4)	1.111*	1.17(1)	-
ΔE_Q [mm/s]						
3	145(1)	0.32(2)	2.66(4)	3.15(8)	3.39(4)	-
	175(1)	0.23(4)	2.6(2)	2.96(2)	3.38(2)	-
	225(1)	0.20(4)	2.28(2)	2.93(2)	3.01(2)	-
	275(1)	0.22(7)	2.52(5)	2.70(3)	2.94(3)	0.448*
4	145(1)	0	2.52(5)	3.35(1)	3.42(3)	-
	175(1)	0	2.49(7)	2.95(2)	3.35(2)	-
Γ/2 [mm/s]						
3	145(1)	0.14(1)	0.16(4)	0.16(2)	0.13(2)	-
	175(1)	0.18(4)	0.35*	0.15*	0.18(1)	-
	225(1)	0.12(3)	0.48(9)	0.145*	0.17(1)	-
	275(1)	0.15(7)	0.22(3)	0.16(3)	0.22(3)	0.36*
4	145(1)	0.27(3)	0.14(3)	0.15(6)	0.17(3)	-
	175(1)	0.25(5)	0.22*	0.20(9)	0.18(2)	-
Relative Area Fraction [%]						
3	145(1)	22(1)	12(3)	33(2)	32(2)	-
	175(1)	17(1)	17*	27*	39(2)	-
	225(1)	10(1)	23*	24*	43*	-
	275(1)	7(2)	29*	18(5)	43(7)	4*
4	145(1)	22(2)	12(3)	19(3)	47(4)	-
	175(1)	13(2)	16(3)	13(1)	58(9)	-

Table 1. Temperature dependent ^{57}Fe Mössbauer parameters for **3** and **4**. Isomer shift (δ) values are given respective to α-Fe. * Fixed parameters. ΔE_Q = quadrupole splitting. $\Gamma/2$ = half width at half maximum.

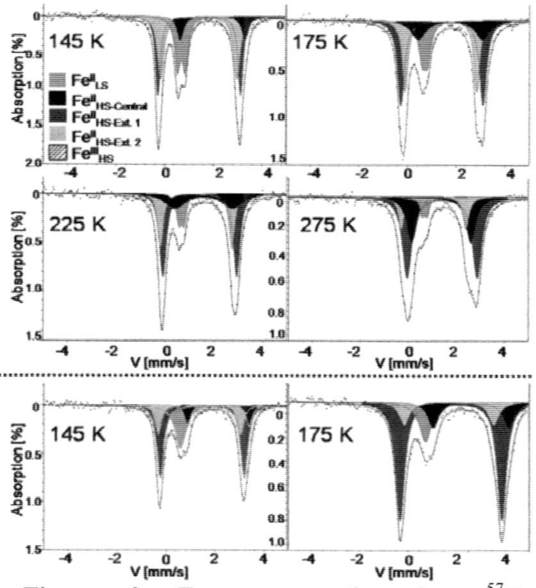

Figure 2. Temperature dependant ^{57}Fe Mössbauer spectra of **3** (top) and **4** (bottom).

3 was analysed by temperature dependent ^{57}Fe Mössbauer spectroscopy (Figure 2). At 145 K, the central metal ion gives rise to two ^{57}Fe Mössbauer signals: a narrow doublet (δ_{LS} = 0.43(1) mm/s and $\Delta E_{Q\text{-}LS}$ = 0.32(2) mm/s) and a large quadrupole doublet ($\Delta E_{Q\text{-HS-central}}$ = 2.66(4) mm/s) with a higher isomer shift ($\delta_{HS\text{-central}}$ = 1.71(2) mm/s) which appears to be unusually large and higher than reported values [7] (Table 1). These contributions are attributed to a LS iron(II) and a HS iron(II) produced by a thermally induced HS↔LS spin conversion [10]. This is consistent with the $Fe^{II}N_6$ coordination sphere which is very often encountered for SCO centres [1]. The external iron(II) ions have a N_3O_3 coordination sphere which is not suitable for SCO in the present case (Figure 1a). Surprisingly, for **3**, the external $Fe^{II}_{HS\text{-ext.}}$ Mössbauer contribution appears to be split into two slightly different doublets labelled $Fe^{II}_{HS\text{-}Ext.1}$ and $Fe^{II}_{HS\text{-}Ext.2}$ (Figure 2). These two signals are detected at a lower isomer shift ($\delta_{HS\text{-}ext.1}$ = 1.14(1) mm/s and $\delta_{HS\text{-}ext.2}$ = 1.17(1) mm/s) and a higher quadrupole splitting ($\Delta E_{Q\text{-}HS\text{-}ext.1}$ = 3.15(8) mm/s and $\Delta E_{Q\text{-}HS\text{-}ext.2}$ = 3.39(4) mm/s) than for the central iron(II) HS ion because of the strongly different coordination spheres of these ions (Table 1). The difference between the two $Fe^{II}_{HS\text{-}ext.}$ is clearly identified in the ΔE_Q values and translates the presence of two types of long-range molecular environments through a different H-bonding network at the end of the trinuclear entities.

The influence of temperature on **3** was analysed by ^{57}Fe Mössbauer spectroscopy on warming over the temperature range 145-275 K (Figure 2). Mössbauer spectra of **3** are fitted best by the combination of the four contributions described above although the presence of a small amount of Fe^{III}_{HS} impurity is identified in the 275 K spectrum thanks to the higher *f* factor of this iron site (Table 1). An optimization of Mössbauer parameters all together was successfully achieved for the 145(1) K spectrum but did not lead to a good fit for other spectra. Some parameters need indeed to be fixed during the optimization. A gradual and incomplete SCO phenomenon is clearly identified for the central metal ion because of the slow decrease of the $Fe^{II}_{LS\text{-}central}$ relative area fraction (from 22 to 7 %) compensated by a slow increase of the Fe^{II}_{HS} relative area fraction (from 12 to 29 %) (Figure 3a). The two $Fe^{II}_{HS\text{-}ext.}$ doublets seem to be involved in a thermal equilibrium although the error on their relative area fraction seems to be large. $Fe^{II}_{HS\text{-}ext.1}$ relative area fraction decreases from 33 to 18 % whereas the relative area fraction of $Fe^{II}_{HS\text{-}ext.2}$ increases from 32 to 43 % (Figure 3a). Interestingly, a large sensibility of the isomer shift values is noted for the $Fe^{II}_{HS\text{-}central}$ and $Fe^{II}_{HS\text{-}Ext.\ 1}$ (Table 1). Although this variation is supported by the presence of a second order Doppler effect, the modifications is higher than the expected ones. An influence of the thermo-sensibility of the anion could be invoked to explain this observation. Moreover, the half width at half maximum, $\Gamma/2$, of the central HS iron(II) ion increases on cooling and reaches a maximum at 225(1) K, before returning to a similar value on further cooling (Figure 3b). This dramatic line broadening reflects a pronounced relaxation phenomenon, that might be promoted by the tautomeric equilibrium between the enol and keto forms of the anion. Accurate population of respective species shall be derived using an appropriate relaxation model [11] to fit our spectra in order to take into account the present dynamics.

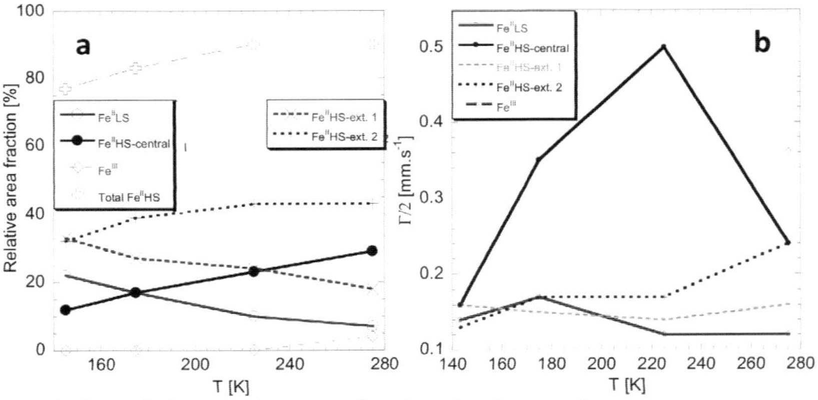

Figure 3. (a) Evolution of the relative area fraction for **3** over the temperature range 140-275 K. (b) Evolution of $\Gamma/2$ for **3** over the same temperature range.

A second trinuclear SCO compound $[Fe^{II}_3(hyetrz)_6(H_2O)_6]$(**2**)$_4Cl_2$ (**4**) was prepared by mixing one equivalent of iron(II) chloride salt, three equivalents of hyetrz and two equivalents of the sodium salt of the anions **2** in methanol and investigated by Mössbauer spectroscopy from 145 K to 175 K. The presence of four iron(II) signals as in **3** is confirmed (Figure 2). The change of the anion from **1** to **2** leads to a significant change into some Mössbauer parameters. The Fe^{II}_{LS} site is a singlet and has a lower isomer shift than for **3**. The isomer shift of $Fe^{II}_{HS\text{-}central}$ increases and shows a lower quadrupole splitting (See Table 1). The SCO phenomenon is still observed with a decrease of the Fe^{II}_{LS} relative area fraction: from ~ 22 to 13 % and an increase of the $Fe^{II}_{HS\text{-}central}$ relative area fraction: from ~ 12 to 16 %, which appears to be less influenced by the anion equilibrium (Table 1). Although the isomer shifts of $Fe^{II}_{HS\text{-}ext.}$ do not differ much comparing **4** and **3**, a large change in the relative $Fe^{II}_{HS\text{-}ext.1}/Fe^{II}_{HS\text{-}ext.2}$ ratio is observed (from 33/32 for **3** to 19/47 for **4** at 145(1) K). The variation of the relative area fraction of $Fe^{II}_{HS\text{-}ext.}$ sites with the temperature which was observed for **3** is preserved in **4**.

3. Conclusion

The influence of a thermochromic anion was probed by ^{57}Fe Mössbauer spectroscopy for this novel type of multifunctional material built with a triazole based trinuclear SCO compound, for the first time. The stabilization of the trinuclear pattern, revealed by a ratio $Fe^{II}_{central}/Fe^{II}_{ext.}$ of ~ 1/3, is the first consequence of the successful insertion of the organic anions **1** and **2** which is believed to prevent the formation of the 1D chain. A reaction mechanism involving a steric hindrance of anions must be considered. A second very interesting impact of the presence of the anions is highlighted by the splitting of the doublet corresponding to the $Fe^{II}_{HS\text{-}ext.}$ into two slightly different doublets ($Fe^{II}_{HS\text{-}ext.1}$ and $Fe^{II}_{HS\text{-}ext.2}$). This behaviour translates the presence of two slightly different molecular environments for the external metal ions. The electronic density of the organic anion has a strong influence on the relative area fraction of the two $Fe^{II}_{HS\text{-}ext.}$ doublets as shown by compound **4** where the anion was substituted by a bromide. Thus, a new type of multifunctional material which combines a thermo-sensitive anion and a SCO inorganic compound was synthesized, for the first time. A direct influence of the presence of the anion was revealed thanks to ^{57}Fe Mössbauer spectroscopy. The communication between the anion and the complexes is thought to be favoured by the expected dense supramolecular network developed by these molecules [6]. Further studies on the incorporation of thermochromic anions into SCO networks are in progress in our laboratories.

Acknowledgement

This work was partly funded by a Fonds Special de Recherche de l'Université Catholique de Louvain, the FNRS (FRFC project N°2.4508.08), a ARC project of the 'Communauté Française de Belgique' and the IAP-VI (P6/17) INANOMAT program. We thank the F.R.I.A. for a doctoral scholarship allocated to F.R. We also thank the GFSM for supporting F.R. for attending GFSM 2009 in Grenoble and A.D.N. for ICAME 2009 in Vienna.

References

[1] Gütlich P, Garcia Y and Spiering H 2003 *Magnetism: From Molecules to Materials IV* (New-York: Willey) pp 271-344
[2] Garcia Y, Ksenofontov V and Gütlich P 2002 *Hyperfine Interact.* **139/140** 543-51
[3] Kahn O, Kröber J and Jay C 1992 *Adv. Mater.* **4** 718-28
[4] Garcia Y, Grunert C M, Reiman S, van Campenhoudt O and Gütlich P 2006 *Eur. J. Inorg. Chem.* **17** 3333-9
[5] Breuning E, Ruben M, Lehn J-M, Renz F, Garcia Y, Ksenofontov V, Gütlich P, Wegelius E and Rissanen K 2000 *Angew. Chem. Int. Ed.* **39** 2504-7
[6] Garcia Y, Guionneau P, Bravic G, Chasseau D, Howard J A K, Kahn O, Ksenofontov V, Reiman S and Gütlich P 2000 *Eur. J. Inorg. Chem.* **7** 1531-8
[7] Kolnaar J J A, van Dijk G, Kooijman H, Spek A L, Ksenofontov V G, Gütlich P, Haasnoot J G and Reedijk J 1997 *Inorg. Chem.* **36** 2433-40
[8] Fujiwara T, Harada J and Ogawa K 2004 *J. Phys. Chem.* B. **108** 4035-38
[9] Robert F, Naik A D, Hidara F, Tinant B, Robiette R, Wouters J and Garcia Y 2010 *Eur. J. Org. Chem.* 621-37
[10] Greenwood N N and Gibb T C 1971 *Mössbauer Spectroscopy* (London: Chapman & Hall)
[11] Adler P, Spiering H and Gütlich P 1987 *Inorg. Chem.* **26** 3840-45

Effect of texture alteration by thin film fabrication on the spin crossover of $[Fe(3\text{-}Br\text{-}phen)_2(NCS)_2]\cdot 0.5CH_3OH$

Anil D Naik, Marinela M Dîrtu and Yann Garcia

Institut de la Matière Condensée et des Nanosciences, Université Catholique de Louvain, Place L. Pasteur 1, 1348, Louvain-la-Neuve, Belgium.
E-mail: yann.garcia@uclouvain.be

Abstract: A mononuclear complex of composition $[Fe(3\text{-}Br\text{-}phen)_2(NCS)_2]\cdot 0.5CH_3OH$ which exhibits a gradual, incomplete spin crossover (SCO) in bulk sample (**1**) was selected to study the effect of alteration of particle size and shape on SCO properties. Thin films (**2**) were fabricated using spin coating technique on quartz substrates and silicon wafers. Morphology analysis by scanning electron micrographs shows a drastic reduction in particles size and deformation in shape compared to the bulk sample. Nearly spherical particles of 200-500 nm diameter were obtained and further reduction in size down to 50-120 nm were obtained with increase in spin coating speed. A thin film on a quartz substrate was studied by ^{57}Fe Mössbauer spectroscopy in transmission mode which shows only a low-spin signal as opposed to a dominant high-spin signal in bulk sample thus hinting out the effect of texture modification on increasing the ligand field strength.

1. Introduction

Molecular-level information technology, including domains like molecular electronics, optical switches and data storage are the future scenario towards the miniaturisation of microelectronics components [1-3]. The principle that governs such devices is the molecular bistability and in particular the stimulating phenomenon of SCO in iron(II) compounds. External stimuli such as light, temperature, and pressure can be used to induce entropy driven transition from the low-spin state (LS, S = 0) to the high-spin state (HS, S = 2) in such bistable systems [2]. Tuneable properties in such bistable systems are of technological importance; in particular regarding the concept of cooperativity which is defined as a communication medium among metal centres during the SCO process. The distortions around iron centres during a spin transition are elastically propagated through the lattice or through meticulously designed rigid networks. The magnitude of such cooperative phenomena drastically affects the properties of the spin transition. In many instances, synthetic methods, suitable ligand field, presence of solvent governs this process. Since the era of 'miniaturisation' is constantly developing, there is a surge in the development of nanoparticles or thin films of such SCO materials due to demand in advanced technology. Either 'top-down' or 'bottom-up' approaches were the choice in hand to design such systems. But adverse effects of size reduction are also known. Thus, reducing particle size results in reducing cooperative effects as predicted by a recent theoretical model for SCO nanoparticles [4] and noted for $[Fe(pz)Pt(CN)_4]$ (pz = pyrazine) [5]. This result is particularly attractive in the frame of the current miniaturization of SCO molecular units at the nanometre scale [5-9], in view of their potential use in information storage devices thanks to micro and nanopatterning [10,11]. Nevertheless, exact role of effect of size reduction or shape deterioration on SCO properties on nanoparticles or thin films is still under debate. With this context, we focus here on fabricating a thin film of a SCO system which was previously designed and studied in bulk sample by us [12a].

2. Results and discussion

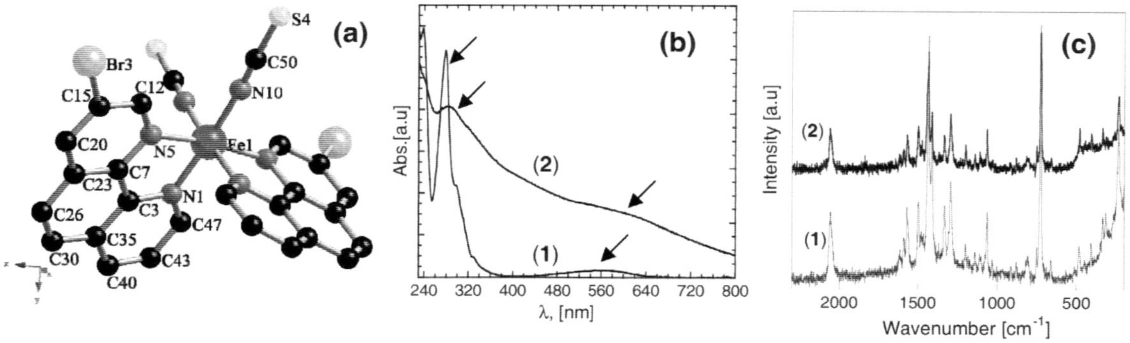

Figure 1. (**a**) Perspective view of the mononuclear complex [Fe(3-Br-phen)$_2$(NCS)$_2$][12a] (**b**) Uv-Vis spectra of (**1**) in CH_2Cl_2 [12a] and thin film (**2**) on quartz at room temperature (Arrow indicates the position of bands) (**c**) Raman spectra of (**1**) and thin film (**2**) on quartz, at room temperature (100-4000 cm^{-1}).

The complex was prepared following our reported procedure [12a], obtained either by precipitation (**1**) or solvent extraction methods (**1a**). A perspective view of the mononuclear complex is shown in Figure 1a. It consists of two molecules of 3-Br-phenanthroline and two molecules of thiocyanato anions bonded to iron(II). Only **1** shows temperature dependent spin crossover. The complex is dark bluish pink and shows no visible thermochromism due to the strong charge transfer (CT) band tailing into the visible region of the Uv-Visible spectrum (Figure 1b). Since CT bands strongly dominate the visible region no bands due to d-d transition were observed (Figure 1b). The shoulder band observed in **2** at 604 nm (compared to 555 nm in **1**) was assigned to CT transition and the higher energy band at 284 nm (compared to 280 nm in **1**) to π-π^* intra-ligand transition. The thin film was also characterized by Raman microscopy (Figure 1c, λ = 785 nm). Spectra for (**1**) and (**2**) are identical, thus confirming that the deposited thin film retains the structural organization observed in the bulk sample. The diagnostic feature for identifying the compound is the presence of a prominent band near 2061 cm^{-1} (in **2**) associated to the C-N stretching mode of the NCS^- ligand [12b]. This signal is slightly shifted to higher frequency, compared to **1** (2054 cm^{-1}) as expected for the LS state [12b]. The possible splitting of this signal could not be detected due the broadness of the signal.

This system presents a few advantages, it is resistant to aerial oxidation, easily soluble in common solvent like MeOH and halogenated solvents and their SCO properties in bulk are well understood [12a]. Film fabrication was carried out using the spin-coating technique on both quartz substrate and silicon wafer (100 orientation) with different spinning speed from a saturated solution of CH_2Cl_2 and MeOH mixture. The colour of the obtained film was pale pink. There are few studies on thin films of mononuclear SCO reported in the literature [13]. We recall here the first SCO compound in a Langmuir-Blodgett film by Kahn et al [13a] who used an amphiphilic compound based on tri-substituted 1,10 phenanthroline and studied the SCO by variable temperature IR spectroscopy on CaF_2 slides. A more exhaustive investigation was done by Delhaes et al. [13b] on substituted bipyridine in an amphiphilic iron(II) complex with variable temperature IR spectroscopy as the prime tool. More recently Matsuda and Tajima [13c] reported a study on thin film of a SCO system based on 2,6-di(pyrazolyl)pyridine and used absorption spectroscopy and electrical resistance change to investigate SCO. The reported film thickness obtained by spin coating method was 30 nm and AFM confirms smoothness of the film. Mössbauer spectroscopy was never or rarely used to investigate thin films probably due to the low amount of sample deposition, which imply a long time of acquisition. Thus we have increased the sample amount by performing multiple coating steps improving Mössbauer spectroscopy studies of the film, sidelining the fact about film thickness and smoothness. Although we could not observe any signals in the powder diffraction due to low sample deposition and loss of crystallinity, optical spectra in absorbance mode gave bands which are comparable with that of bulk sample thus confirming the identity of the complex upon film formation. ^{57}Fe Mössbauer spectroscopy in transmission mode shows a signal fitted for the LS state (δ^{LS} = 0.24(1) mm/s). We recall that in (**1**) the 293 K spectrum was dominated by HS species with 75% population. Two quadrupole doublets were observed with an isomer shift characteristic of the HS (δ^{HS} = 1.09(1) mm/s) and LS states (δ^{LS} = 0.25(3) mm/s). Interestingly, the isomer shifts for the LS state for **1** and **2** and even for **1a** are identical (Table 1). This indicates that there is no appreciable change in the electronic distribution around each individual iron atom for these two compounds. It is to be noted that the value of ΔE_Q^{HS} for compound **1** is elevated compared to un-substituted compounds, [Fe(phen)$_2$(NCS)$_2$] [14]. When the FeN_6 coordination polyhedron deviates from the O_h symmetry a large ΔE_Q^{HS} is observed. The local distortion shown by the quadrupole splitting, ΔE_Q^{LS} ~ 0.3 mm/s, for **1** and **1a**, are identical, and indeed a distortion is identified in the LS state for **1** by X-ray single crystal studies, which shows that the N-Fe-N angles deviates from perfect octahedron [12a]. Table 1 gives an overview of some physical properties like size and shape obtained from Scanning electron micrographs (SEM) and Mössbauer parameters. SEM helps us to establish a logical relationship among the property and morphology relationships.

Table 1. Comparison of physical properties among bulk sample and thin film.

Method	**Shape**	**Size [nm]** Thickness / **Ø**	**Property**	**^{57}Fe Mössbauer parameter at 293 K**				
				δ [mm/s]	**ΔE_Q [mm/s]**	**$\Gamma/2$ [mm/s]**	**$A_{HS}/A_{tot.}$**	**Spin state**
Precipitation,	Crystalline;	450-790	Gradual	1.09(1)	2.81(2)	0.23(1)	74.0	HS
1	Blocks		SCO	0.25(3)	0.3*	0.20(1)	26.0	LS
Extraction,	Fibrous	260	No SCO	1.14(3)	2.5*	0.21(1)	16.0	HS
1a				0.24(3)	0.30(1)	0.16(1)	84.0	LS
Thin film	Spherical	200-500	No SCO	0.24(1)	0.29(1)	0.17(1)	100	LS
on quartz, **2**	grains							

δ: isomer shift (with respect to α-Fe at 293 K); ΔE_Q: quadrupole splitting; $\Gamma/2$: half width at half maximum; A_{HS}: Area of HS doublet; A_{tot}: total area. *Fixed parameter; Ø: diameter.

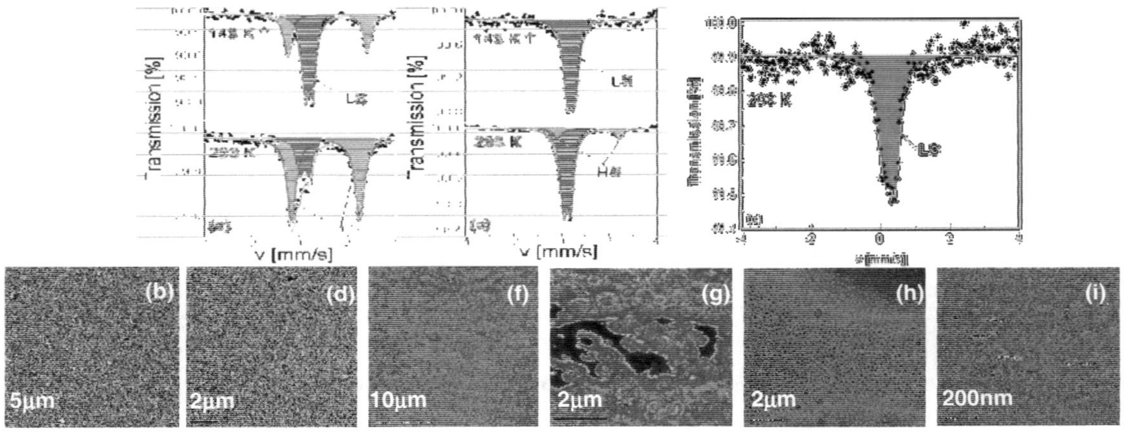

Figure 2. (**a**) Mössbauer spectra of **1** showing SCO; (**b**) SEM of **1** showing large crystals [12a]; (**c**) No SCO in **1a**; (**d**) fibrous morphology as seen from SEM in **1a**; (**e**) Mössbauer on thin film for **2**; (**f** and **g**) SEM under different magnification of thin film on quartz; (**h** and **i**) SEM of **2** on silicon wafers.

Thus **1,** which is large crystalline material (Figure 2) shows SCO while **1a** which was subjected to continuous solvent bleaching during synthesis is highly fibrous and shows no SCO despite both sharing the same composition. Spin coating with a speed of 2000 rpm and 3000 rpm on quartz plate or silicon wafers does not differ much with respect to particle shape, they are near to spherical but there is a drastic reduction in size from 200-500 nm diameters to 50-120 nm. The present investigation was restricted to the batch of 200-500 nm size particles. Grains are found distributed unequally over the initial deposited layer but no aggregations of particles were observed.

3. Concluding remarks

Molecular systems unlike coordination polymers form isolated clusters of iron centres and therefore there is no strong connectivity between clusters. It is understood that secondary interactions like H-bonding, π-π interactions, and presence of solvent which are prime modulators for tuning SCO in bulk sample with through space interactions particularly in mononuclear complexes will be drastically altered during thin film fabrication, eventually minimizing or destroying the cooperativity or modifying the ligand field strength. In the supramolecular material under study a change in complex surrounding or reorganization around metal centres during thin film fabrication was ruled out because of identical isomer shifts. Since films are deposited quickly from solution where reactants may be still in equilibrium with each other, a unique preferential orientation of constituents other than bulk is feasible due to steric requirements in the context of crystal packing. The increase in ligand field strength which is observed may be due to drastic change in crystal shape from well developed blocks to spherical grains that affects crystal packing and defects.

Acknowledgments

We acknowledge financial support from FRFC (No 2.4508308), IAP-VI (P6/17) INANOMAT and ARC. We thank the GFSM for supporting A. D. N. and M. M. D. to attend ICAME 2009 in Vienna.

References

[1] Kahn O, Kröber J and Jay C 1992 *Adv. Mater.* **4** 718
[2] Garcia Y, Ksenofontov V and Gütlich P 2002 *Hyperfine Interact.* **139/140** 543
[3] Garcia Y, Ksenofontov V, Mentior S, Dîrtu M M, Gieck C, Bhatthacharjee A and Gütlich P 2008 *Chem. Eur. J.* **14** 3745
[4] Kawamoto T and Abe S 2005 *Chem. Comm.* 3933
[5] Boldog I, Gaspar A B, Martínez V, Pardo-Ibañez P, Ksenofontov V, Battacharjee A, Gütlich P and Real J A 2008 *Angew. Chem. Int. Ed.* **47** 6433
[6] Coronado E, Galan-Mascaros J, Monrabal-Capilla M, Garcia-Martinez J and Pardo-Ibanez P 2007 *Adv. Mater.* **19** 1359
[7] Volatron F, Catala L, Rivière E, Gloter A, Stéphan O and Mallah T 2008 *Inorg. Chem.* **47** 6584
[8] Forestier T, Mornet S, Daro N, Nishihara T, Mouri S I, Tanaka K, Fouché O, Freysz E and Létard J F 2008 *Chem. Comm.* 4327
[9] Larionova J, Salmon L, Guari Y, Tokarev A, Molvinger K, Molnár G and Bousseksou A 2008 *Angew. Chem. Int. Ed.* **47** 8236
[10] Molnár G, Cobo S, Real J A, Carcenac F, Daran E, Vieu C, and Bousseksou A 2007 *Adv. Mater.* **19** 2163
[11] Cavallini M, Bergenti O, Milita S, Ruani G, Salitros I, Qu Z R, Chandrasekar R and Ruben M 2008 *Angew. Chem. Int. Ed.* **47** 8596
[12] (a) Naik A D, Tinant B, Muffler K, Wolny J A, Schünemann V and Garcia Y 2009 *J. Solid State Chem.* **182** 1365 (b) König E and Madeja K 1967 *Spectrochim. Acta.* **23A** 45
[13] (a) Coronel P, Barraud A, Claude R, Kahn O, Teixier A R and Zarembowitch J 1989 *J. Chem. Soc. Chem.Commun.* 193 (b) Soyer H, Mingotaud C, Boillot M L and Delhaes P *Langmuir* 1998 **14** 5890 (c) Matsuda M and Tajima H 2007 *Chem. Lett.* **36** 700
[14] Ganguli P, Gütlich P, Müller E W and Irler W *J* 1981 *J. Chem. Soc. Dalton Trans.* 441

Mössbauer spectroscopy, Raman- and IR- microscopy studies of hysteretic behaviour in spincrossover systems.

A Ahrens-Botzong[1], K Muffler[1], S M Stuppy[1], S Rackwitz[1], R Rüffer[2], V Schünemann[1] and J A Wolny[1]

[1]Department of Physics, University of Kaiserslautern, D-67663 Kaiserslautern, Germany

[2]ESRF, F-38043 Grenoble Cedex, France

E-mail: wolny@physik.uni-kl.de, schuene@physik.uni-kl.de

Abstract. The polymeric spin-crossover systems $[Fe(atz)(X)_2]_2$ ($X = Cl^-$, $CH_3SO_3^-$) have been investigated by complementary Mössbauer, and Raman and IR-microscopy as well as nuclear inelastic scattering (NIS). The systems reveal SCO transition temperatures around room temperature including hysteresis. Within the hysteresis loop the areas of LS and HS phases could be detected.

1. Introduction

Iron (II) spin crossover (SCO) complexes can be switched reversibly from the low-spin state (LS, S=0) to a high-spin state (HS, S=2) [1-5] by variation of temperature, pressure or by irradiation with light [6-8]. A feature important for the practical applications of SCO materials [9-11] is their hysteretic behaviour and the resulting bistability of the systems. One of the central questions is whether a domain structure is present during the spin transition [5,12]. The recently introduced techniques of Raman and IR-microimaging may shed more light on the above problem. To the best of our knowledge there is only one published study describing the use of Raman microscopy to study the SCO systems. Molnar et. al.[12] showed the absence of domains smaller than 1μ in the studied Fe(pz)[Ni(CN)4] system.

The polymeric material investigated in this study is based on 4-aminotriazole ligand $[Fe(atz)_3(X)_2]_n$ ($X=Cl^-$, $CH_3SO_3^-$) [13]. The complexes have SCO transition temperature at ca. 300 and 323 K, respectively and reveal a hysteresis. In order to elucidate whether spin domain structures are present in these systems the complementary techniques like Mössbauer spectroscopy, nuclear inelastic scattering (NIS) as well as Raman- and IR-microscopy have been employed.

2. Experimental and methods

The polymer $[Fe(atz)_3(CH_3SO_3)_2]_n$ **(1)** has been prepared by means of reaction of water solution of $Fe(CH_3SO_3)_2*7H_2O$ with ethanol solution of the ligand in a 1:3 ratio. $[Fe(atz)(Cl)_2]_n*(H_2O)_2$ **(2)** has been prepared according to the method of Lavrenova et al.[14]. The Mössbauer spectra have been measured in a closed cycle-cryostat manufactured by Cryo Industries of America, Inc. coupled to a Mössbauer spectrometer from WissEL GmbH. Raman mapping has been performed with a Senterra Raman microscope (Bruker AG) equipped with a cryostat (Linkam Ltd) at an excitation wavelength of 785 nm. The IR spectra were collected using a Hyperion IR microscope (Bruker AG) also equipped with a cryostat (Linkam Ltd). The spectra were detected in reflection with the sample mounted on a copper plate. For the nuclear inelastic scattering (NIS) experiments the Senterra Raman microscope was coupled to the synchrotron beam at ID18 of the ESRF during exp. CH 3829. The assignment of the vibrational modes presented here is based on DFT calculations for a model trimeric system $[Fe_3(atz)_6(H_2O)_6]$ [15] which were performed with Gaussian 2003 package, using the B3LYP functional and the CEP-31G basis set as described previously [16].

3. Results and Discussion

Mössbauer spectroscopy of **(1)** reveals a spin transition temperature of ~ 300 K as well as a strong hysteresis effect (figure 1). As a representative example a Mössbauer spectrum at the spin crossover temperature T_c is shown in figure 1a. The LS component exhibits $\delta = (0.48 \pm 0.02)$ mm/s and $\Delta E_Q = (0.38 \pm 0.01)$ mm/s while the HS component shows $\delta = (1.03 \pm 0.02)$ mm/s and $\Delta E_Q = (2.90 \pm 0.01)$ mm/s. Figure 1b shows a strong the hysteresis effect which obviously depends on the history of the sample: During heating from T = 260 K, a temperature where ~ 10 % of the HS isomer is still present, the HS component stays relatively constant but increases very fast at 295 K to almost 100 %. Subsequent cooling to 240 K leads to a slow decrease of the HS component starting at 290 K.

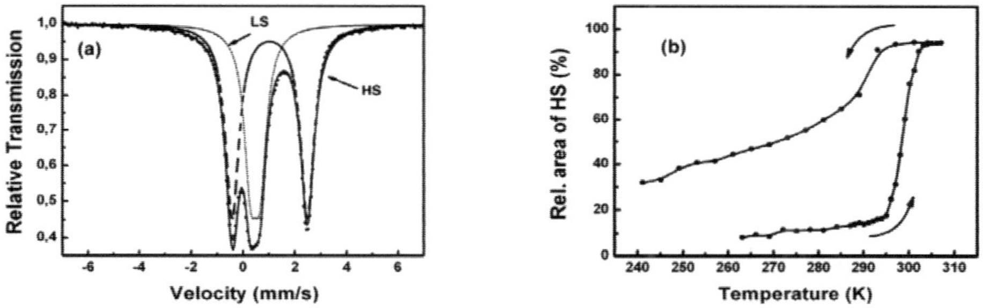

Figure 1. Mössbauer spectrum of the polymeric SCO system (1) taken at 299 K with parameters given in the text (a) and temperature dependence of the HS component which shows a strong hysteretic behaviour of the SCO system (b).

Raman spectroscopy shows the presence of two distinct LS marker bands (see figure 2). Preliminary DFT calculations identify the marker bands for LS (ca. 250 cm^{-1}) and HS (ca. 135 cm^{-1}) as vibrations with mainly Fe-N bending character.

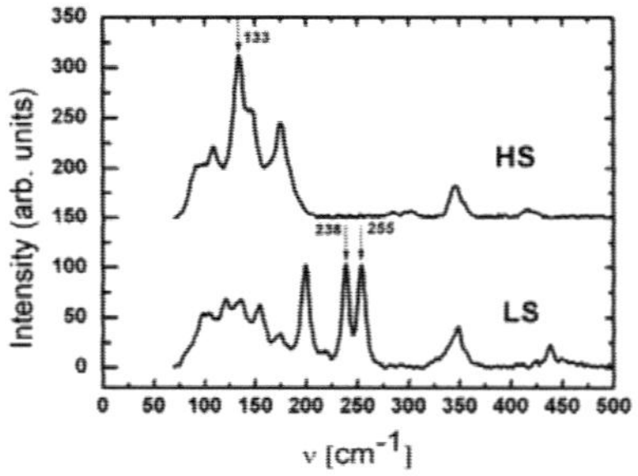

Figure 2. Raman spectra of (1) obtained at 273 K (LS, bottom) and 293 K (HS, top).

The temperature dependence of the Raman spectra shows a hysteretic behaviour of the spin transition (see figure 3). In addition and the SCO temperature T_c increases from ~15 °C at 10 mW laser power to ~24 °C at 1 mW laser power. Therefore Raman mapping experiments have been performed with a laser power of 1 mW. Figure 3b shows the presence of LS as well as HS domains with domain sizes of ca. 20*30 μm. Surprisingly, thermal cycling back to 10 °C after each mapping experiment leads to a consecutive build up of HS domains instead of an expected statistical arrangement of the spin domains with fixed proportions of HS and LS.

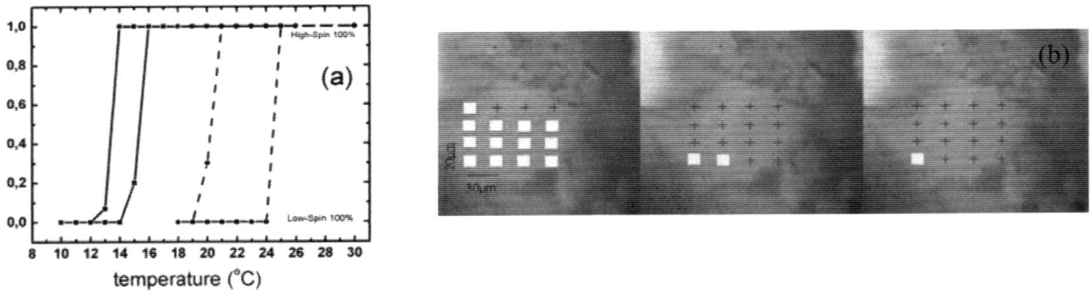

Figure 3. (a) Temperature dependence of relative intensity of LS marker band (254 cm^{-1}) on temperature for laser power of 1mw (●) and 10 mw (■) for **(1).** (b) Raman mapping for **(1)** at 25 °C. (laser power 1 mW). After each measurement the sample was cooled back to 10 °C. The areas of the LS state are denoted with white squares, while those of the HS state have red crosses.

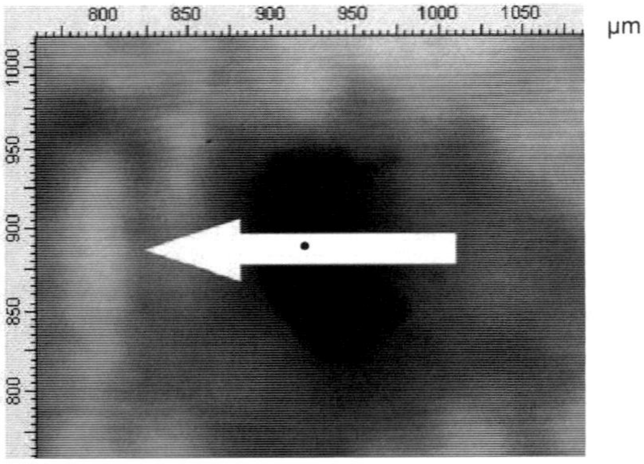

Figure 4. Micrograph of the sample grain of (2) which has been investigated by NIS and Raman spectroscopy in parallel. The arrow depicts the synchrotron beam and the black dot the focus point of the Raman excitation.

The question arises whether the observed spin distribution is due to a real spin domain structure at the SCO temperature or whether this effect is caused by local heating of the sample by means of the Raman excitation laser. In order to answer this question we have performed infrared microimaging of the IR-active HS and LS marker bands at 880/920 cm-1 and 865/910 cm-1 with a spatial resolution of 10 µm at the SCO temperature of 25 °C. These experiments (not shown here) show indeed a different space distribution of HS and LS phases in two adjacent areas (20 µm * 10 µm).

Since the spatial resolution of infrared microimaging has a diffraction limit of ~10 µm nuclear inelastic scattering (NIS) on 57Fe samples is an attractive alternative to Raman and IR micro spectrometry. Therefore we have performed first experiments in order to measure simultaneous NIS and Raman microspectrometry on the same sample spot. The chosen sample grain of compound (2) had a diameter in the order of 100 µm and is shown in figure 5. With the focus of the Senterra Raman microscope (less than 1 µm diameter spot size) in the middle of the sample we oriented the whole microscope with respect to the focussed synchrotron beam (area 15 * 8 µm) in such a way that the beam was passing the short axis of the sample grain (figure 4). The corresponding Raman spectrum reveals a signature typical for the LS form as shown in figure 5 together with the corresponding NIS spectrum. These preliminary experiments show that NIS measurement with a beam size of less than 1 µ can answer the question whether spin domains are really present in SCO systems during spin crossover.

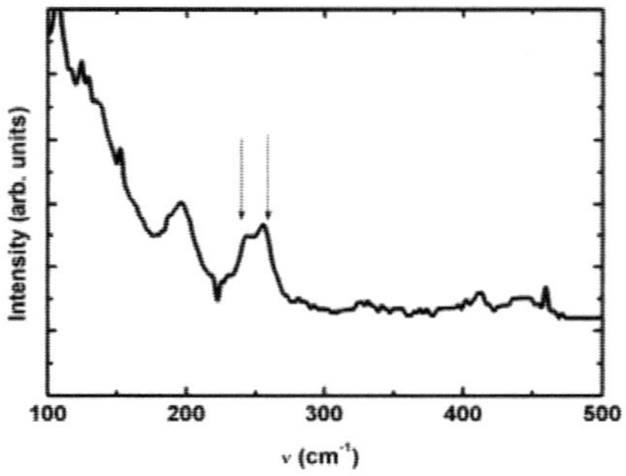

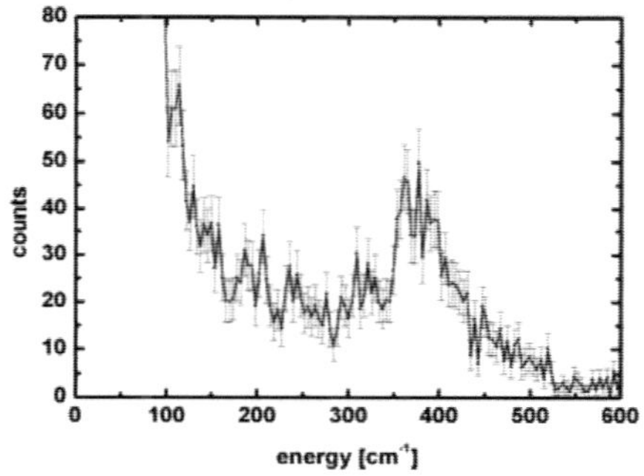

Figure 5. Left: Raman spectra of LS phase of **(2)** measured on the sample position given in figure 4. Right: Corresponding NIS spectrum. Because of different selection rules the intensity of the Raman bands do not correspond to those in the NIS spectrum.

4. Conclusion

The hysteretic behaviour of an iron (II) polymeric SCO system has been studied by combining the Mössbauer spectroscopy and Raman and IR microimaging. The Raman mapping at the spin crossover temperature revealed a complicated behaviour: Obviously there is a spatial distribution of HS and LS phases, but the distribution is strongly dependent on laser power, time and thermal cycling.

Acknowledgments
Financial support of BMBF (05-KS7UK2) and the DFG (Schu1251/9-1) and the ESRF (exp. CH 2839) is gratefully acknowledged.

References

[1] Gütlich P 1981 *Struct. and Bond.* **44** 83
[2] Toftlund H 2001 *Monatshefte für Chemie* **1321** 1269
[3] König E 1991 *Struct. and Bond.* **76** 51
[4] Toftlund H 1989 *Coord. Chem. Rev.* **94** 67
[5] Gütlich P, H.A. Goodwin 2004 In P. Gütlich, H. Goodwin (eds.), Spin Crossover in Transition Metal Complexes. *Top. Curr. Chem.* **233** 1
[6] McGarvey J.J, Lawthers I 1982 *J. Chem. Soc. Chem. Commun.* 906
[7] Gütlich P, Hauser A, Spiering H 1994 *Angew. Chem.* **106** 2109
[8] Decurtins S, Gütlich P, Köhler CP, Spiering H, Hauser A, 1984 *Chem. Phys. Lett.* **105** 1
[9] Kahn O, Martinez CJ 1998 *Science* **279** 44
[10] Gütlich P, Hauser A, Spiering H 1994 *Angew. Chem. Int. Edition in English* **33** 2024
[11] Létard JF, Guionneau P, Goux-Capes L 2004 In P. Gütlich, H. Goodwin (eds.), Spin Crossover in Transition Metal Complexes. *Top. Curr. Chem.* **235** 221
[12] Molnar G, Bousseksou A, Zwick A, McGarvey JJ 2003 *Chem. Phys. Lett* **367** 593
[13] van Koningsbruggen PJ 2004 In P. Gütlich, H. Goodwin (eds.), Spin Crossover in Transition Metal Complexes. *Top. Curr. Chem.* **233** 123
[14] Lavrenova LG, Shakirova OG, Ikorskii VN, Varnek VA, Sheludyakova LA, and Larionov SV 2003 *Russ. J. Coord. Chem.* **29(1)** 24
[15] Garcia Y, Ksenofontov V, Levchenko G, Gütlich P 2000, **J. Mater. Chem. 10** 2274
[16] Wolf M M N, Groß R, Schumann C, Wolny JA, Schünemann V, Døssing A, Paulsen H, McGarvey JJ, Diller R 2008 *Phys. Chem. Chem. Phys.* **10** 4264

Linkage-isomerization and charge-transfer in the formation of iron mixed valence complexes, $(n\text{-}C_3H_7)_4N[Fe^{II}Fe^{III}(dto)_3]$ (dto = $C_2O_2S_2$) and $(n\text{-}C_4H_9)_4N[Fe^{II}Fe^{III}(mto)_3]$ (mto = C_2O_3S)

K Kagesawa[1]**, N Kida**[1]**, Y Ono**[1]**, M Enomoto**[1] **and N Kojima**[1]

[1]Graduate School of Arts and Sciences, The University of Tokyo, Komaba 3-8-1, Meguro-ku, Tokyo 153-8902, Japan

E-mail: cnori@ mail.ecc.u-tokyo.ac.jp

Abstract. In iron mixed-valence complex, $(n\text{-}C_3H_7)_4N[Fe^{II}Fe^{III}(dto)_3]$ (dto = $C_2O_2S_2$), the linkage isomerisation and charge-transfer between Fe^{II} and Fe^{III} occur in the precipitation process. We succeeded the suppression of linkage-isomerization and charge-transfer in the formation process of $(n\text{-}C_3H_7)_4N[Fe^{II}Fe^{III}(dto)_3]$ by the synthesis on low temperature condition and the selection of solvent. Moreover, we have found that the charge transfer occurs in the formation of $(n\text{-}C_4H_9)_4N[Fe^{II}Fe^{III}(mto)_3]$ (mto = C_2O_3S) without linkage-isomerization.

1. Introduction

In the case of iron mixed-valence complexes whose spin states are situated in the spin-crossover region, conjugated phenomena coupled with spin and charge are expected. From this viewpoint, recently, Kojima, et al. discovered a new type of phase transition for a dithiooxalato-bridged iron mixed-valence complex, $(n\text{-}C_3H_7)_4N[Fe^{II}Fe^{III}(dto)_3]$ (dto = $C_2O_2S_2$) (**1**), where the thermally induced charge transfer phase transition (CTPT) reversibly occurs around 120 K in order to minimize the free energy in the whole system [1]. From the analysis of ^{57}Fe Mössbauer spectra of **1**, it is clearly observed that the iron spin-states are $Fe^{III}S_6$ (S = 1/2)-$Fe^{II}O_6$ (S = 2) in the high temperature phase (HTP), above 120 K, and $Fe^{III}O_6$ (S = 5/2)-$Fe^{II}S_6$ (S = 0) in the low temperature phase (LTP), below 120 K, respectively. The CTPT occurs because of the difference in spin entropy between the HTP and the LTP. Moreover, we have also synthesized and investigated physical properties for an monothiooxalato-bridged iron mixed-valence complex, $(n\text{-}C_4H_9)_4N[Fe^{II}Fe^{III}(mto)_3]$ (mto = C_2O_3S) (**2**), consisting of $Fe^{III}O_3S_3$ and $Fe^{II}O_6$ octahedra, which behaves as a ferrimagnet with its magnetic transition temperature of T_N = 38 K and Weiss temperature of $\theta = -93$ K [2]. From the analysis of ^{57}Fe Mössbauer spectra and the dielectric constant of **2**, we have found the valence fluctuation between Fe^{II} and Fe^{III} in the paramagnetic phase [3].

In the case of **1**, we have observed the linkage-isomerization in the precipitation process of **1**. According to Hard and Soft Acids and Bases (HSAB) principle [4], the flipping of dto by 180° takes place here and there, which induces a linkage-isomerization from -Fe^{III}-$S_2C_2O_2$-Fe^{II}- to -Fe^{III}-$O_2C_2S_2$-Fe^{II}-. From the analysis of ^{57}Fe Mössbauer spectra of **1**, although only one of the two Fe sites was enriched with ^{57}Fe isotope in preparation and the Mössbauer spectra of **1** were measured at room temperature where the CTPT does not occur, the observed spectra of **1** show a considerable amount of the other Fe site. In the first step, we assumed that the iron spin-state is in the LTP at the formation of linkage-isomerization. Following this, the charge-transfer induced by linkage-isomerization occurs between Fe^{II} and Fe^{III} to stabilize the HTP in the precipitation process of **1**. Since the formation of linkage-isomerization strongly depends on the temperature in the precipitation process, it is possible to suppress the linkage-isomerization by the control of temperature conditions in the formation process of **1** [5]. On the other hand, from the analysis of ^{57}Fe Mössbauer spectra of **2** in which only one of the two Fe sites was enriched with ^{57}Fe isotope in preparation at –20 °C, we have confirmed that the linkage-isomerization does not occur.

In this paper, we employed ^{57}Fe enriched samples, $(n\text{-}C_3H_7)_4N[Fe^{II}\,{}^{57}Fe^{III}(dto)_3]$ and $(n\text{-}C_4H_9)_4N[Fe^{II}\,{}^{57}Fe^{III}(mto)_3]$. They were prepared at –20 and 25 °C in which only Fe^{III} was substituted with ^{57}Fe in preparation to investigate the temperature dependence of the linkage-isomerization of **1** and **2** by means of ^{57}Fe Mössbauer spectroscopy. Moreover, we used methanol or methanol-water mixture as a solvent for the preparation of $(n\text{-}C_3H_7)_4N[Fe^{II}\,{}^{57}Fe^{III}(dto)_3]$ to investigate the solvent dependence on the formation of linkage-isomerization. We also confirmed the CTPT and ferromagnetic phase transition for $(n\text{-}C_3H_7)_4N[Fe^{II}\,{}^{57}Fe^{III}(dto)_3]$ from the analysis of magnetic susceptibility.

2. Experimental

2.1. Preparation

2.1.1. $(n\text{-}C_3H_7)_4N[Fe^{II}\,{}^{57}Fe^{III}(dto)_3]$. $^{57}Fe(NO_3)_3{\cdot}9H_2O$ was prepared by dissolving ^{57}Fe (19.5 mg, 0.35 mmol) in nitric acid under inert atmosphere. A methanol solution (20 ml) of $^{57}Fe(NO_3)_3{\cdot}9H_2O$ (141.4 mg g, 0.35 mmol) was added to a methanol solution (100 ml) of $K_2(dto)$ (208.2 mg, 1.05 mmol) and 18-crown-6 (555.0 mg, 2.1 mmol). After stirring for 5 minutes, $[^{57}Fe^{III}(dto)_3]^{3-}$ solution was given. This solution was filtered once to remove solid impurities. The filtrate was added to a methanol solution (20 ml) of iron(II) chloride (68.9 mg, 0.35 mmol), $(n\text{-}C_3H_7)_4NBr$(93.1 mg, 0.35 mmol) and ascorbic acid (17.6 mg, 0.1 mmol) at –20 (**1a**) and 25 °C (**1b**), respectively. The mixture was stirred for an hour. The black product was collected by suction filtration and dried in vacuo. Similarly, we also synthesized $(n\text{-}C_3H_7)_4N[Fe^{II}\,{}^{57}Fe^{III}(dto)_3]$ in a methanol-water mixture of 3: 2 ratio at 25 °C (**1c**).

2.1.2. $(n\text{-}C_4H_9)_4N[Fe^{II}\,{}^{57}Fe^{III}(mto)_3]$. $(n\text{-}C_4H_9)_4N[Fe^{II}\,{}^{57}Fe^{III}(mto)_3]$ was prepared according to the previous paper [3]. We synthesized $(n\text{-}C_4H_9)_4N[Fe^{II}\,{}^{57}Fe^{III}(mto)_3]$ in the temperature conditions at –20 (**2a**) and 25 °C (**2b**).

2.2. Measurements

The ^{57}Fe Mössbauer spectra were measured for all the samples. The spectra were calibrated by using the six lines of a body-centered cubic iron foil (α-Fe), the center of which was taken as zero isomer shift. The static magnetic susceptibility was measured by a Quantum Design, MPMS5 SQUID susceptometer between 2 and 300 K under 5000 Oe.

3. Results and discussions

Figures 1(a), (b) and (c) show the ^{57}Fe Mössbauer spectra at 300 K for **1a**, **1b** and **1c**, respectively. The Mössbauer parameters of **1a**, **1b** and **1c** are summarized in Table 1. In all of the Mössbauer spectra of **1a**, **1b** and **1c**, they clearly indicate the presence of three quadrupole doublets, which are assigned to Fe^{II} (S = 2), Fe^{III} (S = 1/2) and Fe^{II} coordinated by S_2O_4. Although only Fe^{III} site was enriched with ^{57}Fe isotope in preparation, the observed spectra show the existence of Fe^{II} site. This result suggests that the charge transfer induced by the linkage-isomerization between Fe^{II} and Fe^{III} takes place in the formation of **1**. As shown in Figure 1(a), Fe^{III} (S = 1/2) site in the **1a** occupies the great majority of the spectra and the linkage-isomerization is almost suppressed. As a result of preparing **1** at –20 °C, namely **1a**, we succeeded in the suppression of linkage-isomerization. Comparing the Mössbauer spectra of **1b** and that of **1c**, Fe^{III} (S = 1/2) component of **1b** is larger than that of **1c**, although the

Table 1. The Mössbauer parameters

sample	site	δ [mms^{-1}]	ΔE_Q [mms^{-1}]	Γ [mms^{-1}]	Area [%]
1a	Fe^{II} (S = 2)	0.92	1.82	0.83	8.0
	Fe^{III} (S = 1/2)	0.38	0.71	0.62	87.5
	Fe^{II} (S_2O_4)	0.87	2.84	0.86	4.5
1b	Fe^{II} (S = 2)	1.00	1.47	0.41	15.8
	Fe^{III} (S = 1/2)	0.35	0.55	0.66	78.0
	Fe^{II} (S_2O_4)	0.87	2.84	0.80	6.2
1c	Fe^{II} (S = 2)	1.05	1.35	0.56	45.5
	Fe^{III} (S = 1/2)	0.34	0.46	0.50	44.5
	Fe^{II} (S_2O_4)	0.83	2.72	0.89	10.0
2a	Fe^{II} (S = 2)	1.10	1.63	0.65	52.2
	Fe^{III} (S = 1/2, 5/2)	0.37	0.62	0.35	47.8
2b	Fe^{II} (S = 2)	1.09	1.65	0.67	50.4
	Fe^{III} (S = 1/2, 5/2)	0.37	0.59	0.36	49.6

preparing temperature condition is almost the same as each other. As shown in Figure 1(c), the observed spectra of Fe^{II} (S = 2) site show a comparable amount of Fe^{III} (S = 1/2) site. These results imply that the linkage-isomerization depends on the solvent and is enhanced by the presence of water in solvent. Judging from these results, we consider the following formation process of **1c**. The linkage-isomerizarion is enhanced because $^{57}Fe^{III}$ mononuclear complex coordinated by six O atoms is stabilized by the presence of water in solvent. Consequently, Fe^{II} is coordinated by six S atoms. Then, the charge-transfer occurs between Fe^{II} and Fe^{III} because of the difference in spin entropy between the HTP and LTP. Figure 2 shows the molar magnetic susceptibility as a function of temperature, $\chi_M T$, for **1a**. As shown in the inset of Figure 2, the thermal hysteresis derived from the CTPT appears around 120 K. Moreover, we confirmed that **1a** behaves as a ferromagnet below 7 K. These behavior of **1a** closely resemble those of **1c** having a considerable amount of the linkage-isomerization component.

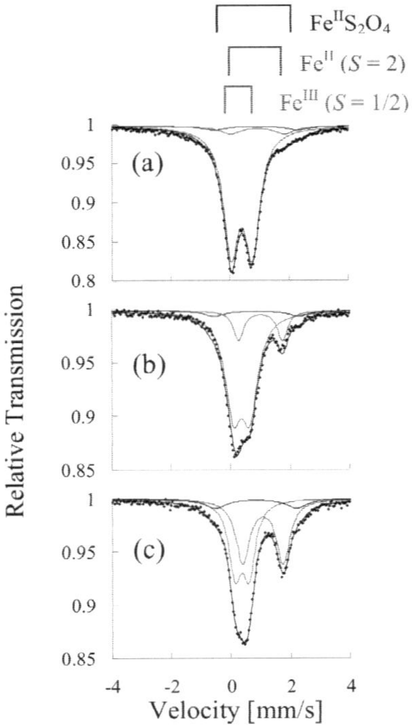

Figure 1. ^{57}Fe Mössbauer spectra at 300 K. (a): **1a**, (b): **1b**, (c): **1c**.

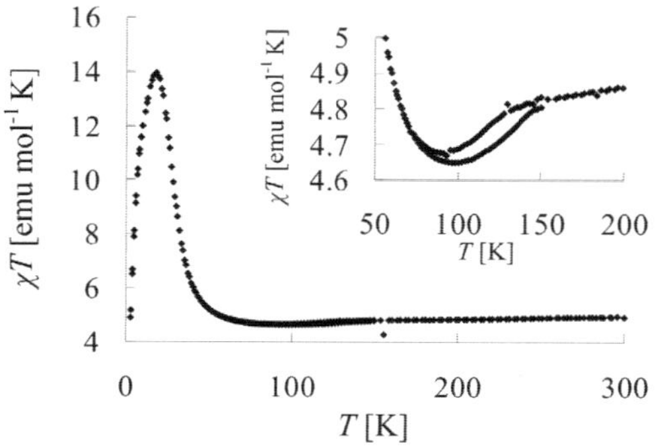

Figure 2. Temperature dependence of the effective magnetic moment for **1a**. Inset is magnified figure of dependence of the effective magnetic moment.

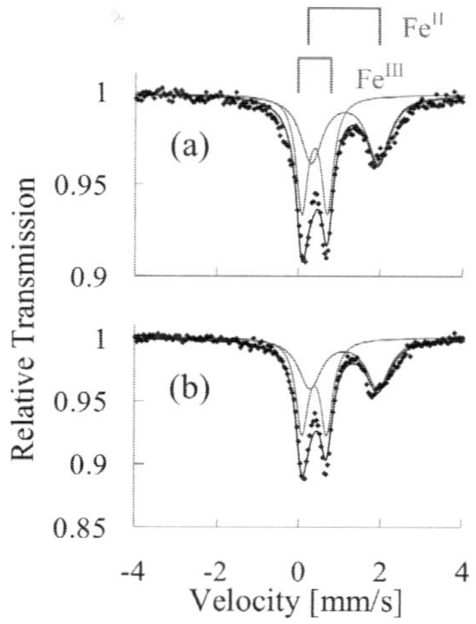

Figure 3. ^{57}Fe Mössbauer spectra at 300 K. (a): **2a**, (b): **2b**.

Figures 3(a) and (b) show the Mössbauer spectra at 300 K for **2a** and **2b**, respectively. The Mössbauer parameters of **2a** and **2b** are summarized in Table 1. In both of the Mössbauer spectra of **2a** and **2b**, they clearly indicate the presence of two quadrupole doublets of Fe^{II} ($S = 2$) and averaged single spectrum of Fe^{III} between the high-spin state and low-spin state [6]. Although only Fe^{III} site was enriched in preparation, the observed spectra show a considerable amount of the Fe^{II} site. This result suggests that the charge transfer occurs between Fe^{II} and Fe^{III} regardless of temperature condition for the formation of **2**. As mentioned in the previous paper, from the analysis of Mössbauer spectra, $Fe^{II}O_6/Fe^{III}O_6$ sites and $Fe^{II}S_3O_3/Fe^{III}S_3O_3$ sites were observed for $(n\text{-}C_4H_9)_4N[^{57}Fe^{II}Fe^{III}(mto)_3]$ and $(n\text{-}C_4H_9)_4N[Fe^{II}\,^{57}Fe^{III}(mto)_3]$, respectively, which implies that the linkage-isomerization does not occur in the formation process of **2** at –20 °C, namely **2a**. Therefore, from the result of that the Mössbauer parameters of **2a** are very close to those of **2b**, the linkage-isomerization also does not occur in the formation process of **2** at 25 °C. Since mto ligand having coordination sites of SO and O_2 has less effect of HSAB principle than dto ligand, the linkage-isomerization does not occur in the formation of **2** on the temperature conditions between −20 and 25 °C.

4. Conclusion

In the preparation of **1** in methanol at –20 °C, we could suppress the linkage-isomerization and charge-transfer in the formation process of **1**. Furthermore, the linkage-isomerization does not affect the CTPT and ferromagnetism of **1**. In contrast, the linkage-isomerization does not occur in the formation of **2** between −20 and 25 °C. However, the charge transfer between Fe^{II} and Fe^{III} occurs regardless of temperature condition for the formation of **2**. In conclusion, mto ligand having coordination sites of SO and O_2 has less effect of HSAB principle than dto ligand.

References

[1] Kojima N, Aoki W, Itoi M, Ono Y, Seto M, Kobayashi Y, Maeda Y, *Solid State Commun*, 2001, **120**, 165.
[2] Ono Y, *Doctor thesis* (The University of Tokyo, 2005).
[3] Kagesawa K, Ono Y, Enomoto M and Kojima N, *Polyhedron*, 2009, **28**, 1822.
[4] Pearson R G, *J. Am. Chem. Soc*, 1963, **85**, 3533.
[5] Coucouvanis D, Piltingsurd D, *J. Am. Chem. Soc*, 1973, **95**, 5556.
[6] Kagesawa K, to be published

Spin state of mixed crystals of iron with zinc or cobalt for the assembled complexes bridged by 1,3–bis(4–pyridyl)propanes

S Nakashima[1], T Dote[2], M Atsuchi[2] and K Inoue[2]

1 Natural Science Center for Basic Research and Development, Hiroshima University, Kagamiyama, Higashi-Hiroshima 739-8526, Japan

2 Graduate School of Science, Hiroshima University, Kagamiyama, Higashi-Hiroshima 739-8526, Japan

E-mail: snaka@hiroshima-u.ac.jp

Abstract. Mixed crystals, $[Fe_{1-x}M_x(NCX)_2(bpp)_2]$ (M=Zn, Co; X=S, Se; bpp= 1,3–bis(4–pyridyl)propane), had a similar structure with 2D interpenetrated structure of $[Fe(NCX)_2(bpp)_2]$. The proportion of Fe^{II} low-spin state in the mixed crystals of NCSe complex increased compared with that of the corresponding Fe complex, while such change in the Mössbauer spectra was not observed in the NCS complex.

1. Introduction

The design and construction of various structures for assembled complexes have attracted great interest to many chemists because of their potential abilities for selective inclusion and transformation of ions and molecules [1–3]. Especially, the assembled complexes including the transition metal ions with the $3d^n$ ($4 \le n \le 7$) electron configuration attract attention to know the relation between assembled structures and their spin states [4–10].

Recently, we have synthesized assembled the iron complex $[Fe(NCS)_2(bpp)_2]$ (bpp = 1,3–bis(4–pyridyl)propane) [11]. This bridging–ligand bpp has three methylenes, and this is a flexible ligand. We revealed that $[Fe(NCS)_2(bpp)_2]$ consisted of the heap of 2D interpenetrated sheets. The sheet consisted of two separated grids. And $[Fe(NCBH_3)_2(bpp)_2]$, having 2D interpenetrated structure, shows the spin-crossover phenomenon.

In the present study, we synthesized mixed crystals of $[Fe_{1-x}M_x(NCX)_2(bpp)_2]$ (M=Zn, Co; X=S, Se; bpp=1,3–bis(4–pyridyl)propane). The aim of the present study is to know the dilution effect by Zn^{2+} or Co^{2+} ion which has a smaller ion radius compared with Fe^{2+}(HS) ion. We studied the spin state of the mixed crystals in the 2D interpenetrated structure and we discussed the effect of the formation of mixed crystals on the spin state of iron ion.

2. Experimental

All complexes were prepared by the diffusion method under the argon atmosphere in order to prevent oxidation of Fe^{II}. The same amounts of iron ions and cobalt or zinc ions were dissolved with the anion into water, while the bpp ligand was dissolved in EtOH. By diffusing the solutions slowly, mixed crystals were obtained for several days. The structures of the mixed crystals were checked by comparing the powder X-ray diffraction patterns with those of the corresponding iron, zinc, or cobalt complex.

A ^{57}Co(Rh) source in a constant acceleration mode was used for the ^{57}Fe Mössbauer spectroscopic measurements. ^{57}Fe Mössbauer spectra were obtained by using a Wissel Mössbauer spectrometer and a proportional counter. The isomer shift values are referred to metallic iron foil. The Mössbauer parameters were obtained by a least-squares fitting to Lorentzian peaks.

The magnetic susceptibilities were obtained on a SQUID magnetometer.

3. Results and discussion

$[Fe(NCS)_2(bpp)_2]$ has a novel 2D interpenetrated structure [11]. Powder X-ray diffraction patterns showed that $[Co(NCS)_2(bpp)_2]$ has a similar structure with the corresponding Fe complex, while $[Zn(NCS)_2(bpp)_2]$ has a different structure from the corresponding Fe complex. The powder X-ray diffraction pattern of the mixed crystals, $[Fe_{1-x}Zn_x(NCS)_2(bpp)_2]$, is not a superposition of those of the corresponding Fe and Zn complexes, but similar to that of the Fe complex. The color of the crystals was different from the corresponding Fe complex, suggesting the formation of mixed crystals. The powder X-ray diffraction pattern of the mixed crystals, $[Fe_{1-x}Co_x(NCS)_2(bpp)_2]$, is similar to those of the corresponding Fe and Co complexes. In this case, it is not easy to judge whether the crystals are mixed crystals or not. But, we judged that the obtained crystals are mixed crystals because of the color change from the corresponding Fe or Co complex. Although it is not easy to determine the value of x, we judged from the results of magnetic susceptibilities that the x does not differ much from 0.5.

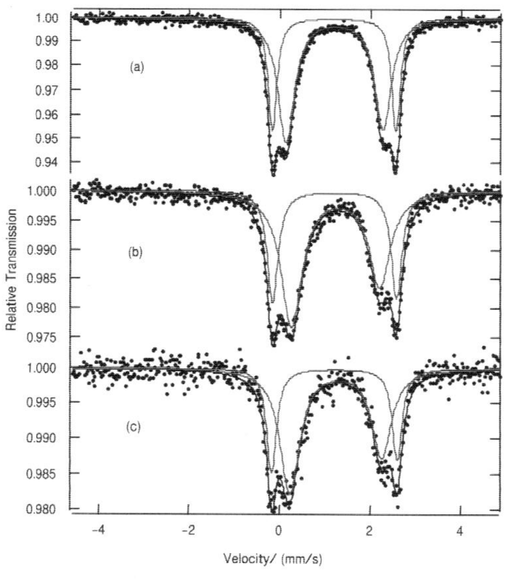

Figure 1. Mössbauer spectra of (a) $[Fe(NCS)_2(bpp)_2]$, (b) $[Fe_{1-x}Zn_x(NCS)_2(bpp)_2]$, and (c) $[Fe_{1-x}Co_x(NCS)_2(bpp)_2]$ at 78 K.

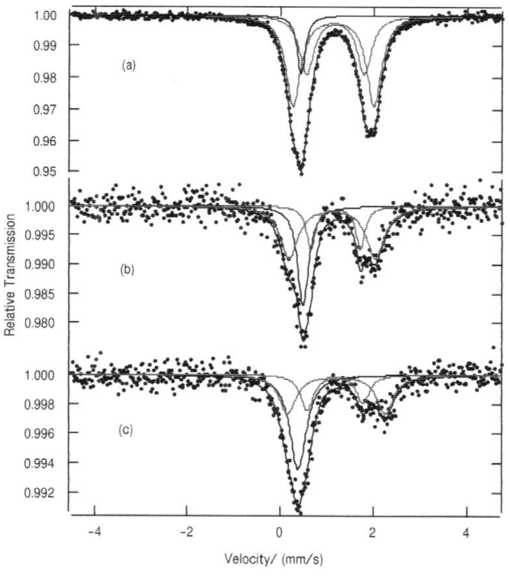

Figure 2. Mössbauer spectra of (a) $[Fe(NCSe)_2(bpp)_2]$, (b) $[Fe_{1-x}Zn_x(NCSe)_2(bpp)_2]$, and (c) $[Fe_{1-x}Co_x(NCSe)_2(bpp)_2]$ at 78 K.

Figure 1 shows the Mössbauer spectra of $[Fe(NCS)_2(bpp)_2]$, $[Fe_{1-x}Zn_x(NCS)_2(bpp)_2]$, and $[Fe_{1-x}Co_x(NCS)_2(bpp)_2]$ at 78 K. The Mössbauer parameters are summarized in Table 1. Two doublets were observed in $Fe(NCS)_2(bpp)_2$. About 1.1 mm/s of IS reveals the Fe^{II} high-spin (HS) state. The appearance of the two doublets is due to the two independent iron atoms in the X-ray structural analysis [11]. $[Fe_{1-x}Zn_x(NCS)_2(bpp)_2]$ and $[Fe_{1-x}Co_x(NCS)_2(bpp)_2]$ also showed the two Fe^{II} HS doublets. The results show that the cobalt or zinc atoms were introduced into the two iron sites randomly. Although the effect of mixed crystals were not observed on the spin state, the Mössbauer parameters slightly changed by forming mixed crystals. The QS value of the inner doublet decreased by forming mixed crystals, while the outer doublet did not change. This slight change might be an effect of the formation of mixed crystals.

Figure 2 shows the Mössbauer spectra of $[Fe(NCSe)_2(bpp)_2]$, $[Fe_{1-x}Zn_x(NCSe)_2(bpp)_2]$, and $[Fe_{1-x}Co_x(NCSe)_2(bpp)_2]$ at 78 K. The Mössbauer parameters are summarized in Table 1. $[Fe(NCSe)_2(bpp)_2]$ showed a slight Fe^{II} low-spin (LS) state in addition to the two Fe^{II} HS states. The small amount of Fe^{II} LS state agreed with the result of magnetic susceptibility. $[Fe_{1-x}Zn_x(NCSe)_2(bpp)_2]$ and $[Fe_{1-x}Co_x(NCSe)_2(bpp)_2]$ show the two doublets of the Fe^{II} HS state and one singlet of the Fe^{II} LS state. The difference was observed in the proportion of the LS state between mixed crystals and the corresponding Fe complex. The proportion of the LS state increased by forming the mixed crystals. In this case also, the QS values of the two doublets of the HS state slightly changed by forming mixed crystals. That is, the QS value of the outer doublet increased and the QS value of the inner doublet decreased by forming mixed crystals. SQUID measurement showed that the χT value of $[Fe_{1-x}Zn_x(NCSe)_2(bpp)_2]$ is less than half the χT value of the corresponding iron complex. The smaller value of $[Fe_{1-x}Zn_x(NCSe)_2(bpp)_2]$ was explained by the increase of the Fe^{II} LS state.

Table 1. Mössbauer parameters at 78 K

Sample	IS[a)]/ mm s^{-1}	QS/ mm s^{-1}	Area/ %
$Fe(NCS)_2(bpp)_2$	1.14	2.72	34
	1.15	2.13	66
$Fe_{1-x}Zn_x(NCS)_2(bpp)_2$	1.17	2.71	33
	1.20	1.99	67
$Fe_{1-x}Co_x(NCS)_2(bpp)_2$	1.17	2.71	32
	1.20	1.99	68
$Fe(NCSe)_2(bpp)_2$	1.16	1.73	55
	1.19	1.23	35
	0.46	0	10
$Fe_{1-x}Zn_x(NCSe)_2(bpp)_2$	1.10	1.84	50
	1.17	1.05	21
	0.48	0	29
$Fe_{1-x}Co_x(NCSe)_2(bpp)_2$	1.22	2.08	37
	1.19	1.19	22
	0.41	0	41

a) Relative to iron foil

The ligand field of $[Fe(NCX)_2(bpp)_2]$ in the 2D interpenetrated structure is close to the spin-crossover point, because $[Fe(NCBH_3)_2(bpp)_2]$ shows the spin-crossover phenomenon [11]. In this situation, such a small change by forming mixed crystals has affected the spin state. The ion radii for Zn^{2+} and Co^{2+}(HS) were 88pm and 88.5pm, respectively, which are smaller than that of Fe^{2+}(HS, 92pm). Zn or Co ions are connected with Fe ion by bpp in the present mixed crystals. Introduced small metal ions are thought to enforce the reduction of iron ion radius; i.e., the proportion of Fe^{II} LS state increases. This means that the chemical pressure effect worked by forming the mixed crystals with the metal ion having smaller ion radii than Fe^{2+}. Gütlich et al. already reported the dilution effect by using mixed crystals such as $[Fe_xCo_{1-x}(NCS)_2(phen)_2]$ [12]. They showed that the Fe^{2+}(HS) complex molecules embedded in the cobalt host lattice feels like being exposed to pressure. The present results do not contradict the results by Gütlich et al. The remarkable effect was observed in the NCSe complex compared with the NCS complex in the present study. This may be due to the softness of Se atom compared with S atom.

References

[1] Moulton B and Zaworotko M J 2001 *Chem. Rev.* **101** 1629
[2] Gütlich P, Garcia Y and Goodwin H A 2000 *Chem. Soc. Rev.* **29** 419
[3] Batten S R and Robson R 1998 *Angew. Chem. Int. Ed.* **37** 1460
[4] Real J A, Andrés E, Muñoz, M C, Julve M, Granier T, Bousseksou A and Varret F 1995 *Science* **268** 265
[5] Halder G J, Kepert C J, Moubaraki B, Murray K S and Cashion J D 2002 *Science* **298** 1762
[6] Kahn O and Martinez C J 1998 *Science* **297** 44
[7] Moliner N, Muñoz M C, Létard S, Salmon L, Tuchgues J–P, Bousseksou A and Real J A 2002 *Inorg. Chem.* **41** 6997
[8] Morita T, Nakashima S, Yamada K and Inoue K 2006 *Chem. Lett.* **35** 1042
[9] Agustí G, Muñoz M C, Gaspar A B and Real J A 2008 *Inorg. Chem.* **47** 2552
[10] Quesada M, Prins F, Roubeau O, Gomez P, Teat S J, Koningsbruggen P J, Haasnoot J G and Reedjik J 2007 *J. Inorg. Chem. Acta* 3787
[11] Atsuchi M, Higashikawa H, Yoshida Y, Nakashima S and Inoue K 2007 *Chem. Lett.* **36** 1064
[12] Ganguli P, Gütlich P and Müller E W 1982 *Inorg. Chem.* **21** 3429

Mössbauer study of iron perovskites of Er, Sm and Nd.

C Piña[1], H Arriola[1], N Nava[2]
[1]Division de Estudios de Posgrado, Facultad de Química, UNAM, 04510 México, D.F.
[2]IMP, México, D.F.

E-mail: has@servidor.unam.mx

Abstract. In solid state chemistry the perovskite structure is very important due to their interesting technological properties between this as ceramic pigments. In this work we present the synthesis of $ErFeO_3$, $SmFeO_3$ and $NdFeO_3$. It's characterization by X-ray diffraction, SEM, energy dispersive X-ray analysis and Mössbauer spectroscopy was performed. energy dispersive X-ray analysis on the compounds was also conducted to confirm the assumptions made from X-ray diffraction results as well as Mössbauer spectroscopy showing the iron as Fe (III).

1. Introduction

A number of $LnFeO_3$ perovskites have been reported in 1960 years in relation to their magnetic properties and Mössbauer studies and were obtained by the ceramic method from the oxides Ln_2O_3 and Fe_2O_3 and heated to 1380-1420°C for 16h in air [1-4]. In the last years are reported as chemical sensors, in particular $SmFeO_3$ [5], and the development of ceramic pigments as $LaFeO_3$ and $HoFeO_3$ because have high temperature stability and colorimetric properties en the reddish range color. [9]

In this work the ferrites of Erbium, Samarium and Neodymium were synthesized by precursor method from 800°C to 1300°C with the idea of using them as ceramic pigments. These perovskites present reddish colors and great stability and resistance under high temperature and corrosive atmosphere during firing. Their perovskite structures were established by powder X-ray diffraction, their morphology and size particles were determined by SEM, their elemental analysis by EDX and the iron oxidation state by Mössabuer spectroscopy.

2. Experimental

The $LnFeO_3$ (Ln = Er, Sm, Nd) were synthesized using Ln $(NO_3)_3$. H_2On. They were heated to dryness in a stove at 110°C; once dry they were weighed and dissolved in deionized water and gauged to 100ml. Aliquots of each were taken and the stoichiometric amounts of $FeCl_3.6H_2O$ were added in a 1:1 ratio each of them. These samples were put in a platinum crucible and heated to complete dryness in an electric plate. The dry samples were calcined in an electric furnace at 800°C, 1000°C and finally at 1300°C for 12 hours, every time with intermittent grinding in an agate mortar in each of the thermal process.

The X-ray diffraction patterns of these samples were taken with a Siemens D5000 diffractometer using Cu, Kα radiation (λ = 1.5406Å) and a Ni filter in a 5-85° 2θ range. Details of the microstructure and morphology were done by a (SEM) scanning electron microscope JEOL model JSM- 5900LV. energy dispersive X-ray analysis (EDX) shows the microanalysis of the elements in the samples. Mössbauer spectra were obtained at 77 K using a constant acceleration Austin Scientific Associates S-600 spectrometer and a 370 MBq (10 mCi) ^{57}Co/Rh source, calibration was carried out using a natural iron foil and the spectra were fitted by using the NORMOS program [8].

3. Results and Discussion

The examination of the X-ray diffractogram of $SmFeO_3$ (see figure 1) is representative of the perovskite structure of these phases: $ErFeO_3$, $NdFeO_3$ with traces of lanthanide oxides. These compounds are isomorphous, their structure is similar and the space group is *Pbnm* which is a distorted perovskite in where each iron remains essentially octahedral; however the axes of four octahedral sites, equivalent at four iron ions, are in different directions. [1]

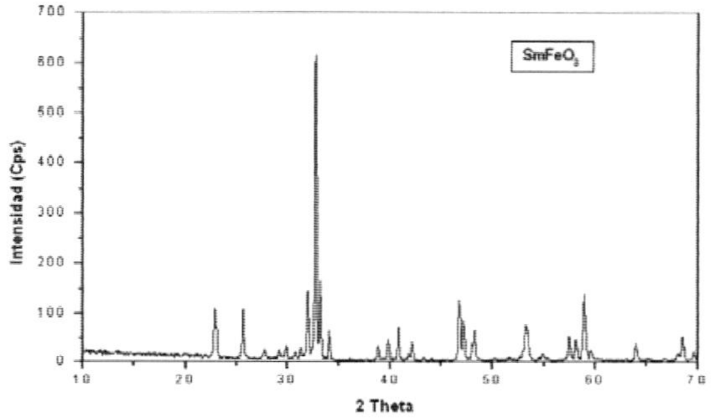

Figure 1. X-ray diffractogram of $SmFeO_3$.

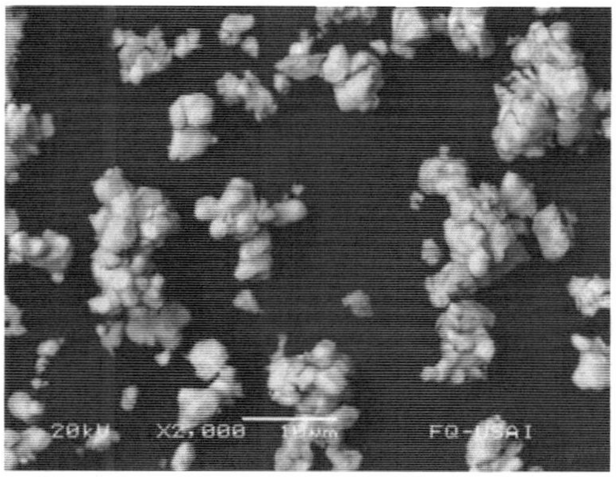

Figure 2. Microphotograph of $SmFeO_3$.

Figure 2 shows a representative SEM micrograph. All have the same morphology and similar particle sizes in the range of 1-2 µm.

Figure 3 shows a typical energy dispersive X-ray spectrum of the erbium compound to confirm the assumptions made from the X-ray diffraction powder results.

The Mössbauer results of Fe^{57} in $LnFeO_3$ (Ln = Sm, Er and Nd) at 77 K, have a characteristic sextet pattern for the three compounds, meaning that all the iron atoms are in equivalent sites, in agreement with the crystallographic structure of these compounds. Figure 4 shows the spectrum for $SmFeO_3$.

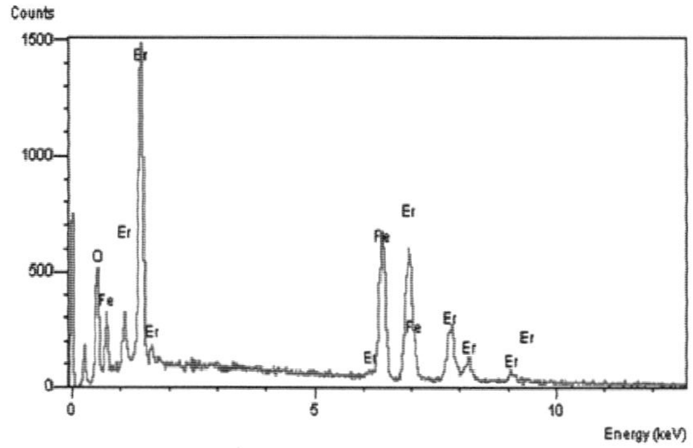

Figure 3. X-ray energy dispersive analysis.

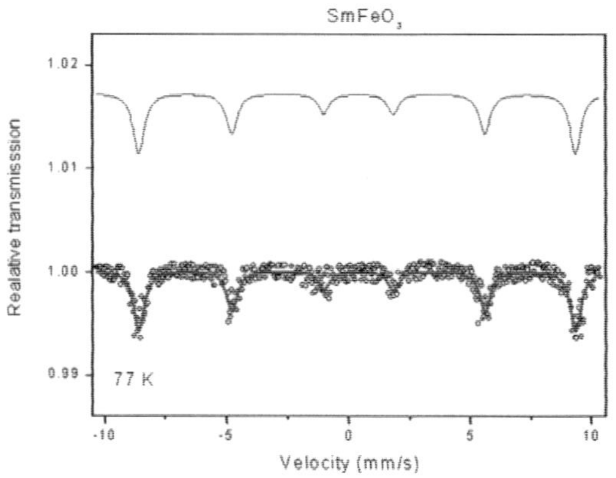

Figure 4. Mössbauer spectroscopy of $SmFeO_3$.

A comprehensive study of the magnetic properties of orthoferrites by M. Eibschutz [3] shows among other things that the internal magnetic field at the iron nucleus does not result only from the iron ion, but it is also influenced by the environment, this effect can be seen in the values given for the magnetic field and the values for the isomeric shift are very close for the three compounds and they are in agreement with the values expected for a $3d^5$ configuration of the Fe^{+3} ion (see table 1).

The quadrupole sppliting is proportional to the electric field at the site of the Fe^{57} nucleus and reflects the crystalline structure; bellow the Néel temperature, which is our case (77 K), the quadrupole interaction, is a small perturbation on the magnetic nuclear field and the values for QS are given.

4. Conclusions

The ferrites of erbium, samarium and neodymium were synthesized by precursor method to 1300°C, different to the ceramic method realized to 1420°C, may be was necessary to rise the temperature because the X-ray diffractograms show traces of lanthanide oxides. Nevertheless, these perovskites of Er, Nd and Sm present reddish dark colors increasing the darkness respectively. The micrographics of these pigments show fine particle size and present thermal stability and resistance at high temperature. These qualities give good possibilities of using as ceramic pigments.

Table 1. Mössbauer parameters of the iron perovskites

Sample	IS (mm/s)	QS (mm/s)	H/eff (T)	Oxidation state
$SmFeO_3$	0.430±.02	0.034±.01	55.6	Fe^{3+}
$ErFeO_3$	0.392±.02	0.011±.01	53.5	Fe^{3+}
$NdFeO_3$	0.422±.02	0.101±.01	54.4	Fe^{3+}

5. References

[1] Eibschütz M, Cohen R and Uitert G 1971 *Journal de Physique* **32** 922-3
[2] Eibschütz M 1965 *Acta Cryst.* **19** 337-9.
[3] Eibschütz M, Shtrikman S, and Treves D 1967 Physical Review **156** 2 562-77.
[4] Traversa E ,Villanti S, Gusmano G, Aono H and Sadaoka Y 1999 *J of the Am. Cer. Soc.* **82** 9 2442-50.
[5] Hosoya Y, Itagaki Y, Aono H and Sadaoka Y 2005 Sensors and Actuators B Chem.**108** 1 198-201.
[6] Bittler K and Ostertag W 1980 *Angew Chem. Int.* **19** 190-6.
[7] Piña C, Buentello R, Arriola H and Nava N 2008 *Hyperfine Interactions* **185** 173-7.
[8] Brand R A 1987 *Nucl. Instr. and Meth.* B **28** 398

Journal of Physics: Conference Series **217** (2010) 012037 doi:10.1088/1742-6596/217/1/012037

Mössbauer spectroscopy of iron phosphate $Fe_3 (PO_4)_2(H_2O)_n$ pigments

C Piña[1], H Arriola[1] and N Nava[2]
[1]Division de Estudios de Posgrado, Facultad de Química, UNAM, 04510 México, D.F.
[2]IMP, México, D.F.

E. mail: cppqic@unam.mx

Abstract. In this work the $Fe_3 (PO_4)_2.(H_2O)_n$ pigment was synthesized from a $FeSO_4.7H_2O$ and digested during some time in a solution of hydrogen ammonium phosphate at 75°C. This compound of a light blue colour was heated at 200°C, 300°C and 800°C, changing their colour and characterized by X-ray diffraction, scanning electron microscopy (SEM), energy dispersive X-ray (EDX), and Mössbauer spectroscopy. The Mössbauer results of the compound subjected to thermal treatment show only the occurrence of Fe (III).

1. Introduction

Commercial phosphates based pigments are widely used by art painters. The iron phosphate $Fe_3 (PO_4)_2.(H_2O)_8$, is used as oil, acrylic, temple and aquarelle pigment. It is a mineral that occurs naturally in the form of light blue-green crystals [1,2] There are different hydrated compounds of iron (II) phosphate. The blue pigment $Fe_3(PO_4)_2.(H_2O)_8$ occurs as the vivianite mineral but, due to a progressive loss of water, $Fe_3(PO_4)_{20}.(H_2O)_4$ ludlamite, $Fe_3 (PO_4)_2.(H_2O)_3$ phosphoferrite and a monohydrate compound $Fe_3 (PO_4)_2.H_2O$ are usually also formed[2, 3].

Alternatively, the vivianite may change to various other mineral species namely hydroxi phosphates, by an oxidation process, metavivianite or kertschenite $Fe_2^{2+}Fe^{3+}(PO_4)_2(OH)(H_2O)_7$, ferrostrunzite $Fe^{2+}Fe_2^{3+}(PO_4)_2(OH)_2(H_2O)_6$ and allanpringite $Fe_3^{3+}(PO_4)_2(OH)_3(H_2O)_5$ [2]. At this stage of chemical change all divalent iron is oxidized. Some of these chemical species have been early prepared and characterized by Mössbauer spectroscopy [4 – 6].

In this work a vivianite pigment was prepared and processed at 200°C, 300°C and 800°C and the following studies to the samples: structure X-ray diffraction, (EDX), microstructure and morphology, with scanning electron microscopy (SEM) and Mössbauer spectroscopy

2. Experimental

The laboratory synthesis of vivianite was from a soluble iron compound, $FeSO_4 .7H_2O$, and hydrogen ammonium phosphate $(NH_4)_2HPO_4$, made by starling digested in a solution at 75°C, for 30 minutes. The compound so obtained was a light blue precipitate in the form of crystals, formed under stirring. Crystals were washed with distilled water and dried at room temperature. The purity of this is vivianite $Fe_3 (PO_4)_2(H_2O)_8$ was checked by X-ray diffraction and elemental chemical analysis. The sample, then was heated at 200° C, 300° C and 800° C. The initial product obtained was of a light blue colour and during the thermal process there was a progressive change of colour in the following sequence: green, dark orange and light yellow.

The X-ray diffraction shows intermediate amorphous products at 200°C and 300°C due to a chemical alteration until a crystalline phase of ferric phosphate at 800°C was obtained.
X-ray diffraction patterns were recorded using a Siemens D500 diffractometer and Cu Kα radiation. The study of the microstructure and morphology as well as the energy dispersive X-ray (EDX) analysis of the composition elements of these pigments was done with a scanning electron microscope (SEM) Jeol model JSM-5900LV.

Mössbauer spectra were obtained at room temperature using a constant acceleration Austin Scientific Associates S-600 spectrometer and a 370 MBq (10 mCi) ^{57}Co/Rh source; calibration was carried out using a natural iron foil and the spectra were fitted by using the NORMOS program [7].

3. Results and Discussion

The examination of the X-ray diffractograms revels the following evidences: vivianite is synthesized at 75°C, as a blue pigment, (Figure 1) and corresponding chemical formula is $Fe_3(PO_4)_2.(H_2O)_8$ indicates only the Fe^{2+} presence. However, Mössbauer spectroscopy shows two oxidation states due to the iron partial oxidation. This situation agrees with its chemical change by exposing the material to air or light, meaning the occurrence of in the chemical structure. [1].

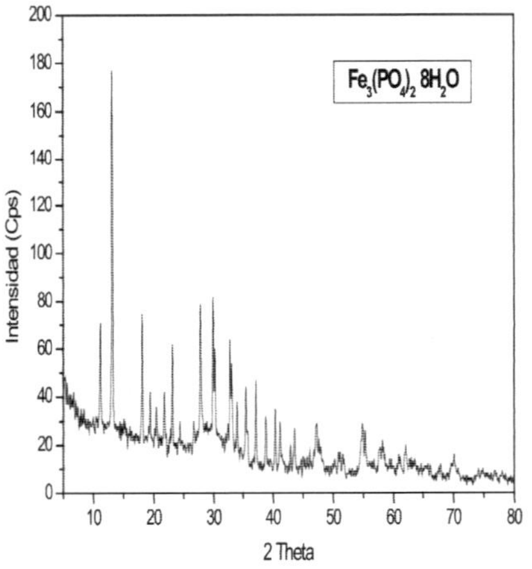

Figure 1. X-ray diffractogram of vivianite.

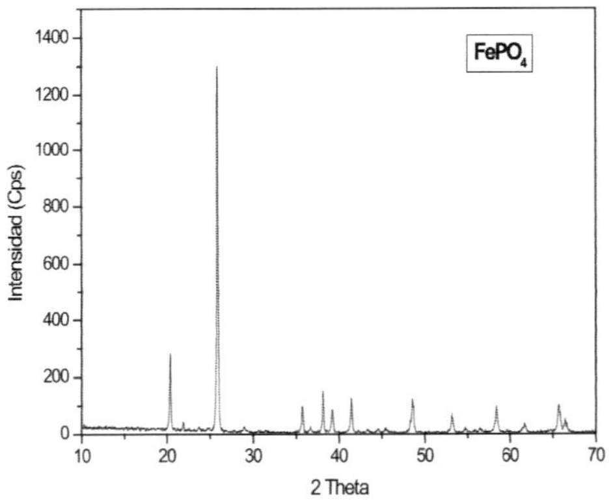

Figure 2. X-ray diffractogram of ferric phosphate.

At 200°C vivianite completely disappears and the pigment becomes green at 300°C it changes to a dark orange colour and the corresponding X-ray pattern shows evidences of amorphous products with short range order and finally the sample processed at 800°C shows only a light yellow ferric phosphate, $FePO_4$, see figure 2. This means a chemical alteration during the thermal process. In these samples the divalent iron has been oxidized to trivalent iron. This is confirmed by Mössbauer spectroscopy, see figure 3. Some changes in vivianite are due to the partial oxidation of Fe^{2+} to Fe^{3+} by being exposed to air and light but with the thermal treatment total oxidation occurs.

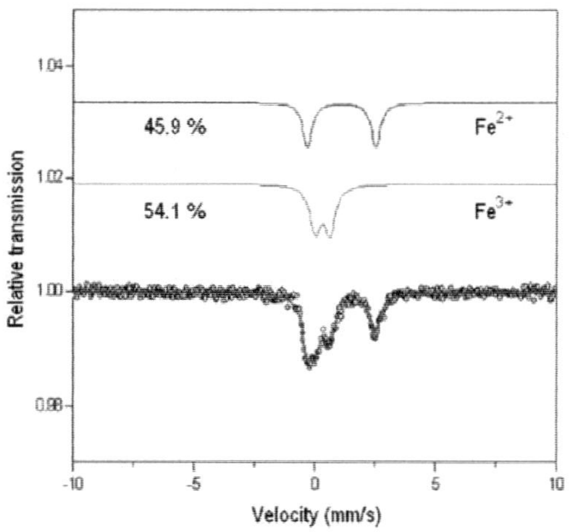

Figure 3. Mössbauer spectrum at R.T. of $Fe_3(PO_4)_2.8H_20$.

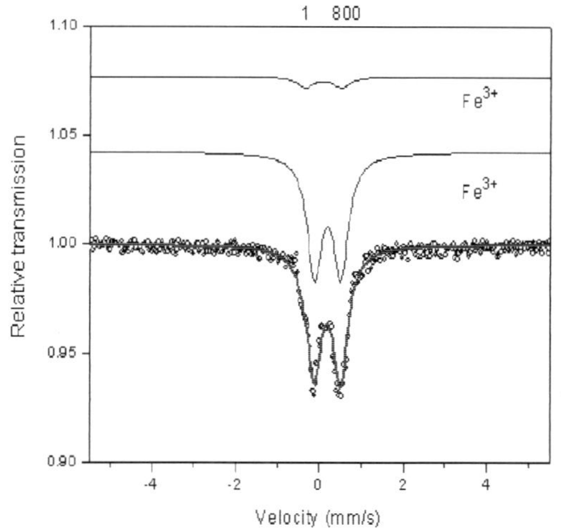

Figure 4. Mössbauer spectroscopy of ferric phosphate.

Previous studies report that the laboratory synthesis of vivianite is complicated by the large number of Fe^{2+} hydrates phosphates phases: ludlamite (n = 4), phosphoferrite (n = 3), two anhydrous forms, sarcopside and graphtonite; one acidic ferrous trihydrate and two basic ferrous species [1, 3] with the tendency of solutions and powdered products to be promptly oxidize in air [2-6]. According to Mathevich, it is relatively easy to prepare vivianite by various methods, particularly by the Evans [8] method, in which preparation usually avoids any contact with air. Nevertheless our synthesis, similar to the Evans method, was done in contact with air. In opinion of Rouzies and Millet [6] the concentration of approximately 50% of Fe^{3+} corresponds to the limit of stability of the vivianite structure after 375 days.

Table 1. Mössbauer parameters at room temperature of vivianite synthesized and samples of thermal process.

Compound	IS (mm/s)	QS (mm/s)	Spectral area (%)	Oxidation state
$Fe_3(PO_4)_2.(H_2O)_8$	1.102±0.05	2.856±0.1	45.9	Fe^{2+}
	0.317±0.01	0.607±0.03	54.1	Fe^{3+}
200°C	0.010±0.003	0.622±0.03	17.0	Fe^{3+}
	0.351±0.01	0.751±0.04	83.0	
300°C	0.292±0.01	1.146±0.06	50.0	Fe^{3+}
	0.310±0.01	0.728±0.04	50.0	
Fe PO_4 800° C	0.011±0.003	0.861±0.04	9.0	Fe^{3+}
	0.191±0.01	0.623±0.03	91.0	

The Mössbauer parameters at room temperature, from table 1 show that the vivianite synthesized at 75°C contains 54% of high spin Fe^{3+} and 46% of Fe^{2+}, the values for IS and ΔQ indicate that the Fe^{3+} and Fe^{2+} ions are in an octahedral or distorted octahedral coordination, also not having an internal magnetic field, vivianite is considered paramagnetic. In opinion of Mattievich [3] a precise structure determination revels an extraordinarily complex atomic arrangement and the spectra differ notably from those of all the other ferrous phosphates an indicate the presence of Fe^{2+} occupying three non equivalent sites in the crystal, the environment of two sites are distorted octahedral and the one is a very distorted tetragonal pyramid. The Mössbauer results at room temperature, from table 1, after the thermal process at 200°C, 300°C and 800°C show only Fe^{3+} figure 4.

The blue coloration of partially oxidized vivianite is attributed to the $Fe^{2+} \rightarrow Fe^{3+}$ electron transition [1-3]. The vivianite blue pigment has been used by dipping it in to oil or sealing it with a thin coat of lacquer. It has been reported that the bulk of vivianite is more resistant to oxidation than the surface region. Figure 5, shows SEM microstructure of the blue pigment as agglomerates of fine particles and finally Energy dispersive X-ray (EDX) analysis on the vivianite, see figure 6, confirms the assumptions made from the X-ray diffraction pigment results.

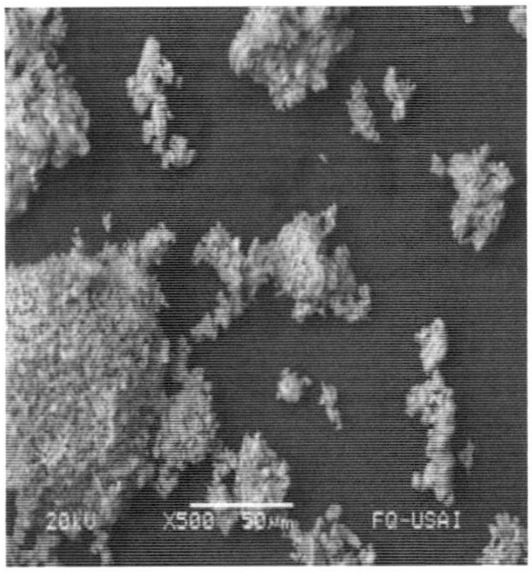

Figure 5. Microstructure of blue pigment (vivianite).

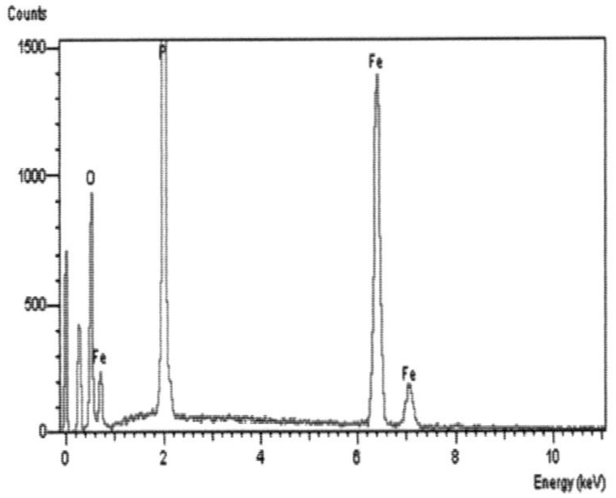

Figure 6. X-ray energy dispersive analysis of vivianite.

4. Conclusions.

X-ray diffraction pattern corresponding to a blue vivianite pigment structure. Results by Mössbauer spectroscopy show partial oxidation, 54% of Fe^{3+} and 46% Fe^{2+}.In the thermal process, the total oxidation occurs from 200°C to 800°C. Loss of water and reorganization of the chemical elements at 800°C results in a ferric phosphate structure. The microstructure of the blue pigment show fine particle size appropriate to be used by art painter.

References

[1] Gudel H U 1983 *Inorg.Chem.* **22** 3812-15
[2] Hanzel D, Meisel W, Hanzel D and Gütlich P 1990 *Hyperfine Interactions* **57** 2201-08
[3] Mattievich E and Danon J 1997 *J. of Inorg. and Nuc. Chem.* **39** 4 560-80
[4] Kolitsch U, Bernhardt H J, Lengauer C L, Blass G and Tillmanns E 2006 *European Journal Mineralogy* **18** 6 793-801
[5] Vochten R and De Grave E 1981*Phys. Chem. Minerals* **7** 197-203
[6] Rouzies Dand and Millet J M M 1993 *Hyperfine Interactions* **77** 19-28
[7] Brand R A Nucl. 1987 *Instr. and Meth.* B **28** 398
[8] Pascal P 1959 *Nouveau Traité de Chimie Minérale* XVIII Paris

Journal of Physics: Conference Series **217** (2010) 012038 doi:10.1088/1742-6596/217/1/012038

Mössbauer study of isomorphous substitutions in $Cu_2Fe_{1-x}Cu_xSnS_4$ and $Cu_2Fe_{1-x}Zn_xSnS_4$ series

V S Rusakov[1], **N I Chistyakova**[1,4], **I A Burkovsky**[1] **A M Gapochka**[1], **T L Evstigneeva**[2] **and S Schorr**[3]

[1] Faculty of Physics, M.V. Lomonosov Moscow State University, Leninskie Gory, Moscow 119991, Russia
[2] IGEM RAS, Staromonetny per. 35, Moscow, 119017, Russia
[3] Institute of Mineralogy, Crystallography and Materials Science, University Leipzig, Scharnhorststr. 20, 04275 Leipzig, Germany

E-mail: nchistyakova@yandex.ru

Abstract. The investigation of isomorphous substitutions in minerals of stannite group was carried out for two series $Cu_2FeSnS_4 - Cu_3SnS_4$ (stannite – kuramite) and $Cu_2FeSnS_4 - Cu_2ZnSnS_4$ (stannite – kesterite) by ^{57}Fe and ^{119}Sn Mössbauer spectroscopy. The calculation of the lattice contribution to the electric field gradient tensor at the ^{57}Fe nuclei and an estimation of the quadrupole shift for the determination of Fe^{3+} atom positions in the structure were carried out. The charge and structural states of Fe and Sn atoms in compounds of the two systems were determined. The schemes of isomorphous substitutions were found for both series in all range of iron concentrations. The changes in the covalence degree of the bonds $Fe^{2+} - S$, $Fe^{3+} - S$ and $Sn^{4+} - S$ in the process of the substitution were found.

1. Introduction

The compounds of stannite group (Cu_2FeSnS_4) are important among natural and synthetic sulfides. The interest to study sulfide minerals is generated by their wide abundance in nature. Many isomorphous elements are typical for minerals of the stannite family. The most known is Fe in kuramite (Cu_3SnS_4) and kesterite (Cu_2ZnSnS_4). Stannite and kesterite have newly attracted attention as possible photovoltaic materials. The compounds of stannite family are the fine objects for the solution of different crystal chemistry problems. The knowledge about details of sulfide structure is necessary for development of modern technology of natural raw material processing.

Mössbauer spectroscopy is the powerful and effective method, which allows one to determine the valence state and atom coordination in a structure. Therefore these two isomorphous series $Cu_2FeSnS_4 - Cu_3SnS_4$ (stannite – kuramite) and $Cu_2FeSnS_4 - Cu_2ZnFeSnS_4$ (stannite – kesterite) have been studied by ^{57}Fe and ^{119}Sn Mössbauer spectroscopy.

2. Objects and methods

Synthesis of the series $Cu_2Fe_{1-x}Cu_xSnS_4$ was carried out using standard techniques described in [1]. The synthesis of compounds of the series $Cu_2Fe_{1-x}Zn_xSnS_4$ was described in [2]. Three-stage process

[4] To whom any correspondence should be addressed

of re-ordering of the Cu^{+}, Zn^{2+} and Fe^{2+} cations among tetrahedral positions of the structure under $Fe^{2+} \leftrightarrow Zn^{2+}$ substitution using simultaneous Rietveld analysis of X-ray and neutron powder diffraction data were found [2].

Mössbauer investigations of all samples were carried out at room temperature with the spectrometer operated in constant acceleration mode and equipped with ^{57}Co and ^{119}Sn sources. Mössbauer spectra were processed by model fitting (for ^{57}Fe Mössbauer spectra) and extracting hyperfine parameter distributions (for ^{119}Sn Mössbauer spectra). The isomer shifts values are reported relative to α-Fe and $BaSnO_3$ for ^{57}Fe and ^{119}Sn Mössbauer spectra accordingly. The lattice contribution to the electric field gradient tensor at the ^{57}Fe nuclei was calculated for Fe^{3+} cations using the formal point charge model to estimate the quadrupole shifts of the Mössbauer spectrum components. The programs SPECTR, DISTRI and LATTICE from MSTools Complex were used [3].

3. Results and discussion

^{57}Fe Mössbauer spectra for the samples of series $Cu_2Fe_{1-x}Cu_xSnS_4$ are shown in figure 1. It was found that below the boundary concentration $x_0 \sim 0.5$ well-resolved quadrupole doublet corresponding to Fe^{2+} atoms and poorly resolved quadrupole doublet corresponding to Fe^{3+} atoms were observed in the spectra. The relative intensity of the second doublet rapidly increases with increase in x [1, 4]. For ensure the charge balance the increase in Fe^{3+} concentration should be accompanied with the increase in Cu^{1+} ions. The isomorphous substitution is realized in this range $0 \le x \le x_0$ by the scheme $2Fe^{2+} \rightarrow Cu^{1+}+Fe^{3+}$ and the formula of compounds is the following $Cu_2^{1+}(Cu_x^{1+}Fe_x^{3+}Fe_{1-2x}^{2+})Sn^{4+}S_4^{2-}$. When x exceeds x_0 the quadrupole doublet corresponding to Fe^{2+} atoms disappears in the spectra. This allows to propose for this range $x_0 \le x \le 1$ the following scheme of isomorphous substitution $Cu^{1+}+Fe^{3+} \rightarrow 2Cu^{2+}$ and corresponding formula is $Cu_2^{1+}(Cu_{2x-1}^{2+}Fe_{1-x}^{3+}Cu_{1-x}^{1+})Sn^{4+}S_4^{2-}$.

^{57}Fe Mössbauer spectra for all compounds of series $Cu_2Fe_{1-x}Zn_xSnS_4$ are a superposition of two quadrupole doublets corresponding to Fe^{2+} and Fe^{3+} atoms (figure 2). The analysis of spectra showed that about 90% of iron atoms were divalent. The relative intensities of subspectra corresponding to Fe^{2+} and Fe^{3+} atoms remain virtually constant for all range of Zn concentrations in contrast to the data of [5]. It could be caused by the synthesis conditions.

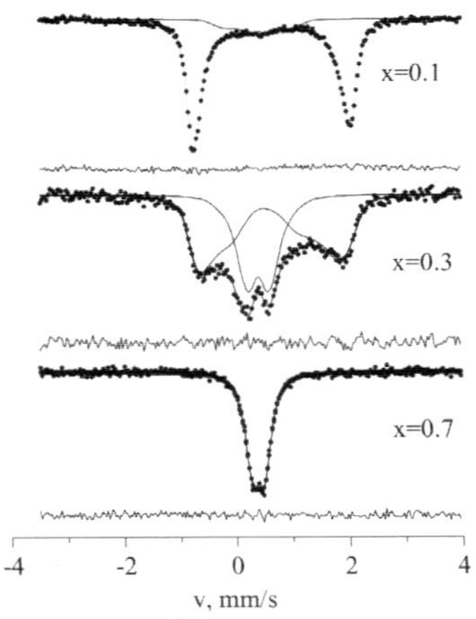

Figure 1. Typical ^{57}Fe Mössbauer spectra of $Cu_2Fe_{1-x}Cu_xSnS_4$ compounds for $x = 0.1$, $x = 0.3$ and $x = 0.7$.

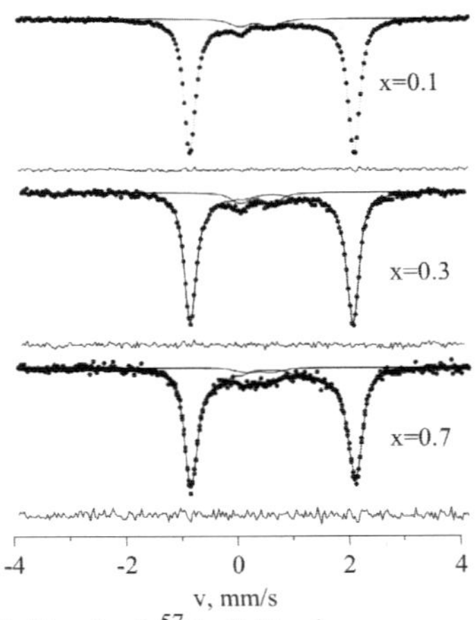

Figure 2. Typical ^{57}Fe Mössbauer spectra of $Cu_2Fe_{1-x}Zn_xSnS_4$ compounds for $x = 0.1$, $x = 0.3$ and $x = 0.7$.

^{119}Sn Mössbauer spectra of both series are poorly resolved paramagnetic type spectra. More intensive doublet with the parameters $\delta \sim 1.5$ mm/s and $\varepsilon \sim 0.15 \div 0.35$ mm/s is related to tin atoms for strong covalent bonds Sn^{4+} - S. Small amount of SnO_2 are presented in all samples.

The dependencies of hyperfine parameters of ^{57}Fe and ^{119}Sn Mössbauer spectra on x for both series are shown in figures 3 and 4. The average isomer shift δ_{mid} corresponding to Fe^{3+} ions increases in both series with an increase in x (figure 3a). This behavior of the isomer shift for the nuclei ^{57}Fe corresponds to a decrease in charge density of electrons at nucleus by decreasing the effective number of 4s-electrons. Hence the covalence degree of the bonds Fe^{3+} - S increases with an increase in iron atom concentration. Observed increase in δ_{mid} for Fe^{2+} in $Cu_2Fe_{1-x}Cu_xSnS_4$ corresponds to a decrease in the covalence degree of Fe^{2+} - S bonds with increasing iron concentration. A small decrease in ε_{mid} for Fe^{3+} under Fe substitution by Cu in $Cu_2Fe_{1-x}Cu_xSnS_4$ (figure 3b) can be attributed to a slight decrease in the distortion degree of local symmetry of Fe atoms. For Fe^{2+}ions in $Cu_2Fe_{1-x}Cu_xSnS_4$ the average quadrupole shift ε_{mid} rapidly decreases with an increase in x. This behavior can be explained by a decrease in number of Jahn-Teller ions Fe^{2+}, which cause a distortion of local structures along the axis $\bar{4}$ of local symmetry. The average values δ_{mid} and ε_{mid} for Fe^{2+} atoms in $Cu_2Fe_{1-x}Zn_xSnS_4$ virtually do not change with increase in Zn concentration (figure 3). This behavior is due to the isovalent substitution $Fe^{2+}(T_d) \rightarrow Zn^{2+}(T_d)$ and the small difference between the radii of the cations Zn^{2+} and Fe^{2+}. Thus, the covalence degree of Fe^{2+} – S bond does not change with increasing x.

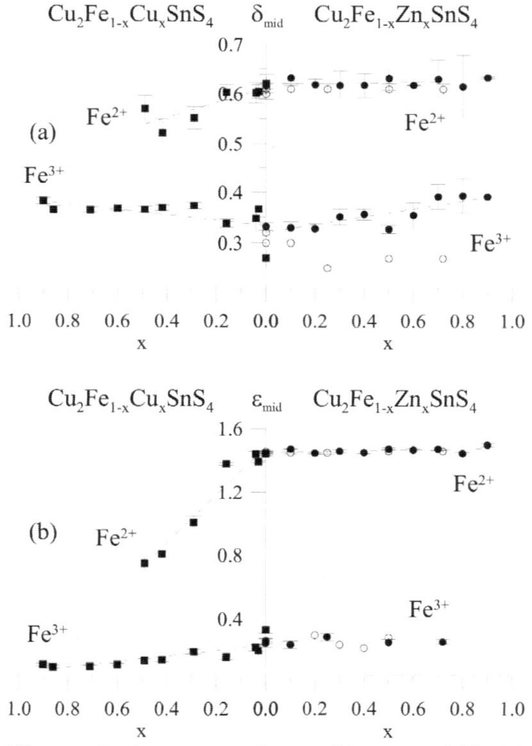

Figure 3. Average values of isomer shift (a) and quadrupole shift (b) of ^{57}Fe Mössbauer spectra of $Cu_2Fe_{1-x}Cu_xSnS_4$ and $Cu_2Fe_{1-x}Zn_xSnS_4$ compounds. Points (○) are the data obtained by Di Benedetto F et.al. [5].

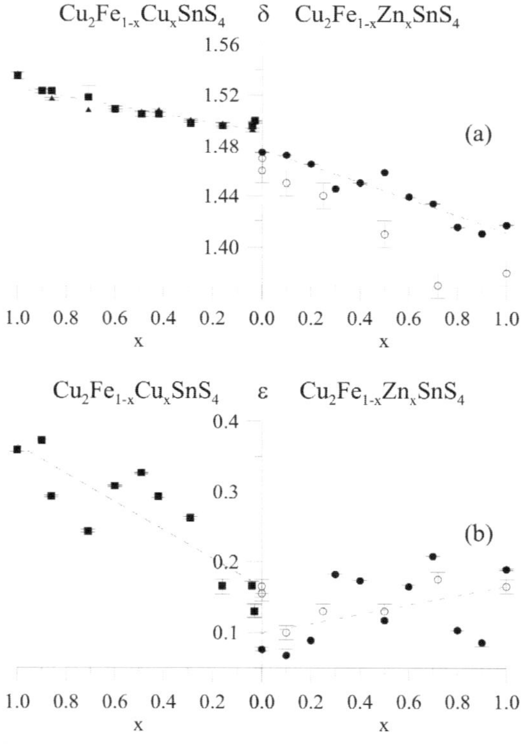

Figure 4. Isomer shift (a) and quadrupole shift (b) of ^{119}Sn Mössbauer spectra of $Cu_2Fe_{1-x}Cu_xSnS_4$ and $Cu_2Fe_{1-x}Zn_xSnS_4$ compounds. Points (○) are the data obtained by Di Benedetto F et.al. [5].

The values of isomer shift of subspectra corresponding to Fe^{2+} ions are typical for Fe^{2+} ions in the tetrahedral environment of sulfur atoms [6]. The values of δ for subspectra corresponding to ions Fe^{3+} are more typical for octahedral surrounding of Fe^{3+} cations than tetrahedral one [6]. Therefore, it was previously suggested that the Fe^{3+} atoms occupied the octahedral positions 8g formed by sulfur atoms [1, 4]. The best value of R-factor during the structure refinement by Rietveld method was obtained for some compounds of series $Cu_{3-x}Fe_xSnS_4$ based on this assumption [1]. However new data of XAFS analysis showed that iron atoms occupied only tetrahedral sites in the structure [7].

To determine what type of structural positions are occupied by Fe^{3+} atoms in the structure of $Cu_2Fe_{1-x}Cu_xSnS_4$ compounds we have calculated the lattice contribution to the electric field gradient tensor at the ^{57}Fe nuclei using the formal point charge model. The bond valences of Sn atoms in different structure positions were calculated using the known values of bond lengths [8]. These calculated bond valence values for all positions except b-positions significantly exceed 4+. The quadrupole shifts for Fe^{3+} cations in the compounds with a low concentration of iron atoms for different cations distribution among nonequivalent 2a, 2c and 2d- crystallographic positions of the structure were estimated. The average experimental value of quadrupole shift is close to ε for Fe^{3+} atoms in tetrahedral site (especially for 2d-) but substantially less than ε for Fe^{3+} atoms in octahedral 8g-site. Therefore the copper atoms replaces Fe^{3+}atoms in sulfur tetrahedra seems be preferable, and more likely there are the 2d-positions. Thus the isomorphous substitution in $Cu_2Fe_{1-x}Cu_xSnS_4$ is realized by the two schemes: $2Fe^{2+}(T_d) \rightarrow Cu^{1+}(T_d)+Fe^{3+}(T_d)$ for $0 \le x \le x_0$, and $Cu^{1+}(T_d)+Fe^{3+}(T_d) \rightarrow 2Cu^{2+}(T_d)$ in the range $x_0 \le x \le 1$.

The observed increase in the ^{119}Sn isomer shift in $Cu_2Fe_{1-x}Cu_xSnS_4$ series (figure 4a) under the substitution of Fe atoms for Cu atoms corresponds to an increase in the charge electron density at nucleus due to increasing the effective number of 5s-electrons. Thus, the degree of covalent bond $Sn^{4+} - S$ increases with the iron concentration decrease. In case of $Cu_2Fe_{1-x}Zn_xSnS_4$ compounds a decrease in the covalence degree of bond $Sn^{4+} - S$ is observed with the increase in Zn concentration. We estimated the effective charge of Sn atoms using the dependence $\delta(x)$: $Q_{Sn} = 3.32 \pm 0.11$ for $Cu_2Fe_{1-x}Cu_xSnS_4$ and $Q_{Sn} = 3.35 \pm 0.11$ for $Cu_2Fe_{1-x}Zn_xSnS_4$ series. The increase in quadrupole shift ε for the ^{119}Sn subspectrum is observed with the increase in Cu concentration in case of $Cu_2Fe_{1-x}Cu_xSnS_4$ series and Zn concentration in case of $Cu_2Fe_{1-x}Zn_xSnS_4$ series (figure 4b) indicating an increase in the distortion of tetrahedrons with Sn atoms in their centers.

It should be noted that our results did not reveal the three-stage process of cation re-ordering among tetrahedral positions of the structure for $Cu_2Fe_{1-x}Zn_xSnS_4$ series found in [2]. This is apparently due to the fact that such re-ordering process of cations does not change the number of Cu^+, Fe^{2+} and Zn^{2+} atoms in the second coordination sphere for ^{119}Sn and ^{57}Fe atoms and hence does not lead to a noticeable changes in the hyperfine parameters of Mössbauer spectra.

Acknowledgments
This work was supported by RFBR (Grant 08-05-01016)

References

[1] Evstigneeva T L, Rusakov V S and Kabalov Yu K 2003 *New data of minerals* **38** 65
[2] Schorr S, Hoebler H –J and Tovar M 2007 *Eur. J. Mineral* **19** 1
[3] Rusakov V S 2000 *Mössbauer Spectroscopy of Locally Inhomogeneous Systems* (Almaty: Institute of Nuclear Physics of the NNC of the Republic of Kazakhstan) p 431
[4] Rusakov V S, Chistyakova N I, Burkovsky I A, Gapochka A M, Evstigneeva T L and Schorr S 2008 *Aip Conference Proceedings* **1070** 96
[5] Di Benedetto F, Bernardini G P, Borrini D, Lottermoser W, Tippelt G and Amthauer G 2005 *Phys. Chem Minerals* **31** 683
[6] Reiff W M 1973 *Mossbauer Effect Methodology* (NY: L. Prenum Press) vol. 8 pp 89–105. Goodenough J B, Fatseas G F 1982 *J. Solid State Chem.* **41** 1
[7] Zalewski W, Bacewicz R, Antonowicz J, Evstigneeva T L and Schorr S 2008 Book of Abstracts of 16th International Conference on Ternary and Multinary Compounds (ICTMC16), Berlin ID 211
[8] Brese N E and O'Keeffe M 1991 *Acta Cryst.* **B47** 192

Journal of Physics: Conference Series **217** (2010) 012039 doi:10.1088/1742-6596/217/1/012039

^{197}Au Mössbauer study of gold selenide, AuSe

F E Wagner[1], P Palade[2], J Friedl[1], G Filoti[2], N Wang[3]

[1]Physics Department E15, Technical University of Munich, 85747 Garching, Germany
[2]National Instiute of Physics of Materials, 77125 Bucharest-Magurele, Romania
[3]Institute of Mineralogy, University of Heidelberg, 69120 Heidelberg, Germany

E-Mail: `fwagner@ph.tum.de`

Abstract. The two crystallographic modifications of AuSe, α-AuSe and β-AuSe, were studied by ^{197}Au Mössbauer spectroscopy. Despite the different crystal structures, the Mössbauer spectra of the two phases are nearly the same, and reveal the presence of two gold sites with Mössbauer parameters typical for Au^{I} and Au^{III}. The Mössbauer-Lamb factor of the Au^{I} is less than half of that of the Au^{III}. Calculations using the WIEN2k program reproduce the experimental results quite well and also yield the signs of the electric field gradients.

1. Introduction

In the gold-selenium system, only the intermetallic phase AuSe exists. Of this phase, however, two different crystallographic modifications are known, α-AuSe and β-AuSe [1-3]. Both crystallize in the monoclinic space group C2/m, but with different lattice constants and with 4 (β-AuSe) or 8 (α-AuSe) formula units in the unit cell.

In both modifications there are two different gold lattice sites: Half of the gold atoms occupy sites with a square-planar coordination to 4 nearest selenium neighbours at a distance of about 0.25 nm, while the other half occupies sites with a linear coordination to two nearest selenium neighbours at a distance of about 0.24 nm [3]. This prompted the idea that the gold on the former sites is Au^{III}, while that on the latter ones is Au^{I}. We have used Mössbauer spectroscopy with the 77 keV γ rays of ^{197}Au to confirm this assumption.

2. Experimental details

The Mössbauer experiments were performed at 4.2 K in a liquid He bath cryostat with sources of ^{197}Pt ($T_{1/2}$ = 19 h) produced by neutron irradiation of isotopically enriched ^{196}Pt metal in the Munich Research Reactor. The spectra were fitted with transmission integral lineshapes based on Lorentzian lines grouped into quadrupole doublets.

The gold selenides were prepared by reacting pellets of appropriate amounts of elemental gold and selenium powder in sealed quartz tubes. The materials thus obtained were analyzed by X-ray diffraction and 3 samples were selected for Mössbauer spectroscopy. A specimen with a composition of $AuSe_{1.1}$ kept for 120 days at 350°C yielded nearly pure α-AuSe with only a minor β-phase impurity (Sample 1). By reacting a $AuSe_{\sim 1.3}$ mixture at a temperature decreasing from 400 to 375°C within a day and subsequent annealing at 350°C for 120 days yielded a mixture of comparable amounts of α-AuSe and β-AuSe (Sample 2). The best result of our efforts to prepare pure β-AuSe was obtained by keeping an AuSe mixture at 280°C for 18 days, which yielded almost pure β-AuSe with only a minor α-phase impurity and some metallic gold (Sample 3).

3. Mössbauer Results and Discussion

The Mössbauer spectra of the 3 samples are shown in figure 1. Those of Samples 1 and 2 could be fitted with two overlapping symmetric quadrupole doublets. Since the right components of the two doublets nearly coincide, both a nested and a side-by-side arrangement of the doublets yield good fits. Figure 1 shows the latter fits, which yielded marginally better χ^2 values. The results of both ways of fitting the spectra are compiled in table 1. The spectrum of Sample 3 exhibits the two quadrupole doublets attributable to gold selenide and an additional component of metallic gold, whose isomer shift was fixed to the value of −1.22 mm/s [4] in the least squares fits.

For Sample 1 (α-AuSe) the doublet with the larger quadrupole splitting and smaller isomer shift can be attributed to Au^{I} and the other one to Au^{III}. On a plot of the quadrupole splittings versus the isomer shifts of Au^{I} and Au^{III} compounds [5], the values of both types of fit agree reasonably well with the notion that the gold on the two sites is Au^{I} and Au^{III}, respectively. One should keep in mind, however, that the electric quadrupole interactions are expected to be negative for Au^{I} and positive for Au^{III} [5], though the sign cannot be determined from the Mössbauer spectra of powder samples.

The fact that the spectrum of the mixture of comparable amounts of α-AuSe and β-AuSe (Sample 2) is virtually identical to that of the nearly pure α-phase (Sample 1) shows that the two phases yield nearly the same Mössbauer patterns. This is borne out by the Mössbauer parameters for Sample 3, which contains but very little of the α-phase. We thus conclude that the hyperfine parameters for α-AuSe and β-AuSe are the same within the accuracy of the Mössbauer measurements, and that two sets of parameters are possible because of the ambiguity in the fitting of the spectra.

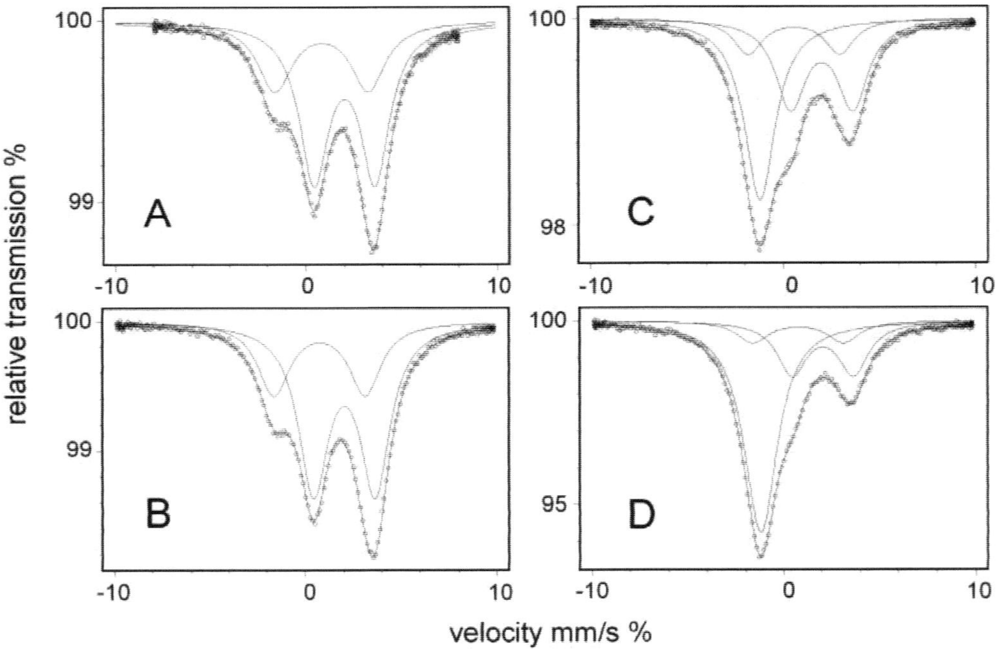

Figure 1. ^{197}Au Mössbauer spectra of gold selenides: (A) nearly pure α-AuSe (Sample 1), (B) a roughly equal mixture of α-AuSe and β-AuSe (Sample 2) and (C) nearly pure β-AuSe containing a minor amount of α-AuSe and some metallic gold (Sample 3). A spectrum of Sample 2 with an additional absorber of metallic gold is also shown (D)

Although the amounts of gold on the two lattice sites of the gold selenides is the same, the Au^{I} component in the Mössbauer spectra exhibits less than half of the intensity of the Au^{III} component. This shows that the Lamb-Mössbauer factor *f* is much smaller for Au^{I} than for Au^{III}. For α-AuSe

(Sample 1) one obtains $f(Au^{I})/f(Au^{III})$ = 0.40(3) from the intensities given in Table 1. The f-factor ratio for the β-phase is similar, but the experimental uncertainty is larger because of the presence of the Au metal impurity phase in Sample 3.

Table 1. Quadrupole splittings (QS), isomer shifts (IS) with respect to the Pt-metal source and relative spectral areas (A) of the two components attributable to Au^{I} and Au^{III} in gold selenides. Last digit errors are given in parentheses. The two sets of data given for each sample represent the two possible ways of fitting the spectra described in the text. When the relative intensities of the Au^{I} and Au^{III} components do not add up to 100%, the remainder represents the single line of metallic gold at –1.22 mm/s with respect to the Pt metal source. D is the absorber thickness.

Sample	D (mg/cm2)	QS_I (mm/s)	IS_I (mm/s)	A_I (%)	QS_{III} (mm/s)	IS_{III} (mm/s)	A_{III} (%)
Sample 1	53	4.87(5)	0.82(3)	33.7(13)	3.19(3)	2.01(2)	66.3(13)
		5.57(7)	1.06(2)	27.5(9)	3.00(2)	1.89(2)	72.5(9)
Sample 2	82	4.76(5)	0.74(2)	31.4(8)	3.22(2)	2.03(2)	68.6(8)
		5.50(5)	1.09(2)	29.3(7)	2.94(3)	1.89(2)	70.7(7)
Sample 3	84	4.78(9)	0.55(6)	15.5(13)	3.24(3)	2.00(2)	40.3(13)
		5.89(7)	0.96(3)	12.2(7)	2.92(3)	1.82(2)	43.4(7)
2 + Au-metal	82 + 35	4.76*	0.74*	11.0(5)	3.22*	2.03*	25.2(5)
		5.89*	0.96*	8.3(3)	2.92*	1.82*	28.2(5)

*Values fixed to those obtained for Sample 2 alone.

In order to obtain the f-factors for the two gold species relative to the f-factor of metallic gold (f = 0.189 [7]), the absorber of Sample 2 (82 mg/cm^2) was also measured together with an absorber of metallic gold (35 mg/cm^2) [4]. From the mean intensity ratios observed in the resulting spectrum (figure 1 and table 1), and taking into account that Sample 2 has a nominal composition of $AuSe_{1.3}$, one obtains $f(Au^{I})/f(Au_{Metal})$ = 0.20(2) and $f(Au^{III})/f(Au_{Metal})$ = 0.55(2). The small value of the f-factor for the Au^{I} species can be attributed to the large vibrational amplitudes of the gold perpendicular to the direction of the Se-Au-Se linear bonds. A similar situation has been observed for other Au^{I} compounds (e. g. [6, 8]).

4. Calculations of the hyperfine parameters of α-AuSe and β-AuSe

For comparison with the experimental results, the electric field gradients and electron densities at the Au nuclei in the gold selenides were calculated using the WIEN2k computer code [10, 11]. The Au 6s and 5d atomic states and the Se 4p states, which lie less than 1 Ry below the Fermi level, were treated as true valence states characterized by forming the (APW + lo) basis set. The Au 5p and Se 4s and Se 3d atomic states with binding energies of several Ry were described as local orbitals plus APW + lo wave functions. The atomic states with higher binding energies were considered as core states. About 100 APW + lo per atom in the unit cell and a k-mesh with 260–300 k-points in the irreducible mesh of the first Brillouin zone were used. The Fermi level was determined by the modified tetrahedron method [12] and the GGA parameterization [13] was used to calculate the exchange correlation energy. Muffin-tin radii of 2.50 a.u. and 2.16 a.u. were used for Au and Se, respectively. The optimization of the atomic positions was achieved via the PORT minimization facility of the WIEN2k program.

Values of the electron densities at the gold nuclei in a number of intermetallic gold compounds have previously [14] been obtained by calculations with the WIEN97 program, the predecessor of WIEN2k. These calculations were now repeated with the WIEN2k code. The results confirm the $\Delta\langle r^2\rangle$ given in [14] and were used together with the electron densities calculated for the AuSe phases to obtain the theoretical isomer shift values given in table 2. The isomer shift values for Au^{III} agree well with the experimental data, while those for Au^{I} are somewhat too small.

The V_{zz} value for the Au^{I} site is large and negative, as is expected for Au^{I} [4, 5], the Au^{III} value is smaller and positive (table 2). The EFG values decrease but slightly when the atomic positions are allowed to vary. The asymmetry parameter η is nearly zero for Au^{I} and small for Au^{III}. The conversion of the EFG data into electric quadrupole splittings is subject to the uncertainty in the value of the nuclear quadrupole moment of the ground state of ^{197}Au [15]. We have used the value of Q = 0.56 barn [16] to obtain the values of table 2, which are in good agreement with the experimental results.

Table 2. Calculated values for the EFG (V_{zz}, η), the electric quadrupole splitting (QS) and the isomer shift with respect to a Pt-metal source (IS) for the Au^{I} and Au^{III} sites in AuSe. Both the values calculated for the atomic positions given in [3] and for optimised positions are listed.

Phase	V_{zz}^{I} (10^{21} /cm^2)	η^{I}	QS^{I} (mm/s)	IS^{I} (mm/s)	V_{zz}^{III} (10^{21} /cm^2)	η^{III}	QS^{III} (mm/s)	IS^{III} (mm/s)
α-AuSe	–48.1	0.068	–5.29	1.66	32.1	0.335	3.59	1.85
α-AuSe opt.	–43.3	0.068	–4.77	1.27	30.1	0.316	3.37	1.68
β-AuSe	–50.1	0.020	–5.51	1.87	32.4	0.260	3.60	1.91
β-AuSe opt.	–47.4	0.025	–5.22	1.67	30.4	0.233	3.37	1.65

5. Conclusions

The Mössbauer study confirms that the two gold positions in α-AuSe and β-AuSe can be attributed to Au^{I} and Au^{III} species. The hyperfine parameters are nearly the same for α-AuSe and β-AuSe, as can be expected from the similar nearest-neighbour distances [3]. There is good agreement between the experimental results and calculations of the hyperfine parameters by the APO + lo method using the WIEN2k computer code.

Acknowledgements
Support by the grant IDEI 1382/2008 of the Romanian Ministry of Education and Research is gratefully acknowledged.

References

[1] Cranton G E, Heyding, R D, 1968 *Can. J. Chem.* **46** 2637
[2] Rabenau A, Rau H, Rosenstein G, 1971 *J. Less-Common Metals* **24** 291
[3] Rabenau A, Schulz H, 1976 *J. Less-Common Metals* **48** 89
[4] Thomson J O, Obershain F E, Huray P G, Love J C, Burton J, *Phys. Rev B* **11** 1835
[5] Parish R V,1985 in: *Mössbauer Spctroscopy in Chemistry* Vol. 1 (ed. G J Long) 577 (Plenum Press New York)
[6] Prosser H, Wagner F E, Wortmann G, Kalvius G M, Wäppling R, 1975 *Hyperfine Interactions* **1** 25
[7] Erickson D J, Roberts L D, Burton J W, Thomson J O, 1971 *Phys. Rev. B* **3** 2180
[8] Schmidbaur H, Hartmann Ch, Wagner F E, *1988 Hyperfine Interactions* **40** 335
[9] Palade P, Wagner F E, Filoti G, 2002 *Hyperfine Interactions C* **5** 185
[10] Blaha P, Schwarz K, Madsen G K H, Kvasnicka D, Luitz J, 2001 *WIEN2k, An Augmented Plane Wave + Local Orbitals Program for Calculating Crystal Properties* (Schwarz K, Technische Universität Wien, Austria) ISBN 3-9501031-1-2
[11] Madsen G K H, Blaha P, Schwarz K, Sjöstedt E, Nordström L, 2001 *Phys. Rev. B* **64** 195134
[12] Blöchl P E, Jepsen O, Andersen O K, 1994 *Phys. Rev. B* **49** 16223
[13] Perdew J P, Burke K, Ernzerhof M, 1996 *Phys. Rev. Lett.* **77** 3865
[14] Palade P, Wagner F E, Filoti,G, 2002 *Hyperfine Interactions C* **5** 185
[15] Schwerdtfeger P, Bast R, Gerry M C L, Jacob Ch R, Jansen M, Kellö V, Mudring A V, Sadlej A J, Saue T, Söhnel T, Wagner F E, 2005 *J. Chem. Phys.* **122** 124317
[16] Palade P, Wagner F E, Jianu A D, Filoti G, 2003 *J. Alloys and Compounds* **353** 2

Moessbauer study of tris-(methyldietoxysiloxy)iron and its derivatives

N A Tebeneva[1], A M Muzafarov[1], E S Trofimchuk[2], M A Chuev[3], S K Dedushenko[2], L A Kulikov[2] and Yu D Perfiliev[2]

[1]Institute of the Synthetic Polymer Materials RAS, Moscow, Russia
[2]Department of Chemistry, Moscow State University, Russia
[3]Institute of Physics and Technology RAS, Moscow, Russia

E-mail: dedushenko@mail.ru

Abstract. Moessbauer study of tris-(methyldiethoxysiloxy)iron, $[CH_3(C_2H_5O)_2SiO]_3Fe$, tris(polymethylsiloxy)iron, $[\{CH_3[(CH_3)_2SiO]_4\}_2CH_3SiO_3]_3Fe$, and its derivatives with 1-decanthiol showed, that these compounds are the complexes of high-spin trivalent iron. The analysis of the spectra by means of Afanas'ev-Chuev method allowed us to suggest different ways for their interpretation.

1. Introduction

Immobilized molecular systems containing iron-centres are very promising for present-day catalysis. The long history of the metallosiloxanes demonstrates evident progress in the molecular design of these prospective compounds [1-3]. But the use of the cage-like metallosiloxanes has limitations due to their insolubility in organics and also due to the crowded coordination sphere of metals' ions. The new approach based on the sodium salts of organoalkoxysilanes allows us to obtain the spider-like oligomers with a single iron atom as a branching centre. Good solubility in organic solvents combined with the absence of strong ligands in oligomers' composition makes them prospective intermediates in catalytic applications.

All these substances are maroon liquids being in a viscous-flow state at room temperature. They transform into glass state at lower temperatures with no crystallization taking place. The synthesized samples can contain the impurities of the reactants and/or by-products. They can be sensitive to moisture if containing functional groups. These substances can also exist as a mixture of stereoisomers and intent to oligomerization during the chemical transformations of their functional groups. That is why the number of the techniques which are suitable for the investigation of these compounds is very limited.

The state and the behavior of iron centers within oligomeric structure during their chemical transformations are very important for further applications. Moessbauer spectroscopy is a suitable technique for their investigation. It can allow us to determine the oxidation state of iron, magnetic properties, coordination number, to estimate the purity of the product etc. But for the oligomer samples which have no definite structure and in which the environment of iron atom could be considerably disordered it is not easy to propose correct models for the spectra observed. The model should provide a description of the distributions of quadrupole splittings and isomer shifts, as the existence of these distributions is an inherent property of the compounds with disordered structure. Moreover, during the freezing of the liquid sample, which is necessary to record the Moessbauer spectrum, the distortion of iron coordination polyhedron can also take place and should be taken into account. Finally, it is necessary to keep in mind the influence of the impurities, because they commonly have similar chemical nature and their lines could be in the same spectral range.

In this report the results of a Moessbauer study of some iron organosiloxanes are presented.

2. Experimental

To synthesize tris-(methyldiethoxysiloxy)iron $[CH_3(C_2H_5O)_2SiO]_3Fe$ we used the interaction of $FeCl_3$ with sodium oxy(methyl)dietoxysilan $(CH_3)(C_2H_5O)_2SiONa$:

$$(CH_3)(C_2H_5O)_2SiONa + FeCl_3 \rightarrow [(CH_3)(C_2H_5O)_2SiO]_3Fe + NaCl$$

This organosiloxan was further treated with nonamethyltetrasiloxan-1-ol $CH_3[(CH_3)_2SiO]_4H$ to produce tris(polymethylsiloxy)iron $[\{CH_3[(CH_3)_2SiO]_4\}_2CH_3SiO]_3Fe$:

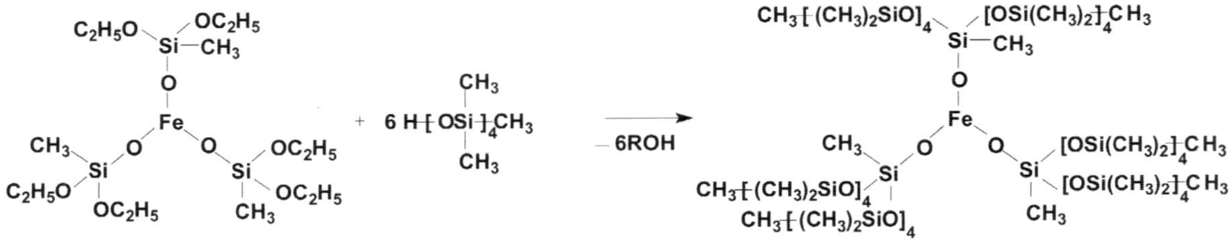

Finally, this oligomer was treated with different quantities of 1-decanthiol $CH_3(CH_2)_9SH$ to introduce sulfur to the first coordination sphere of iron. These products as well as two intermediate oligomers were studied by Moessbauer spectroscopy.

The ^{57}Fe Moessbauer absorption spectra were measured at 77K on a Perseus spectrometer working at constant velocities. A standard γ-source of ^{57}Co in metallic rhodium matrix with the activity of 0.8 GBk (a product of Cyclotron, Co., Ltd., Obninsk, Russia) was employed. Isomer shifts in this paper are presented relative to α-Fe.

To fit the Moessbauer spectra we used the DISCVER program based on the Afanas'ev-Chuev method [4] which uses the Voigt profile to fit the line shape. At the first step this program calculates the maximal possible number of well-defined lines in a spectrum with the intended statistical level. It also takes into account the source line-shape and the thickness of an absorber. Then an operator can apply the binding (e.g. doublets etc.) according to the chemical nature of the sample under study.

3. Results and discussion

The Moessbauer spectra obtained for all samples (Figure 1) are slightly asymmetric doublets which can be reliably related to high-spin trivalent iron (Table 1).

The DISCVER program allowed us to fit each of these spectra with one doublet (parameters presented in Table 1). The shapes of the doublets' lines are described by the Voigt profile

$$V(v,\Gamma,\gamma) = \frac{\Gamma}{2\pi}\frac{1}{\gamma\sqrt{2\pi}}\int\frac{1}{(v-x)^2+(\Gamma/2)^2}\exp\left(-x^2/2\gamma^2\right)dx$$

where Γ is a sum of double natural linewidth and the additional Lorentzian line broadening in the Moessbauer source ($\Gamma = 2\Gamma_0 + \Gamma_s$), γ is the Gaussian width of the line. According to the suggested model each iron ion in the substance is characterized by the symmetric doublet of two Lorentzians. For all iron atoms in the substance the isomer shifts and quadrupole splittings of the doublets are distributed according to the Gaussian probability function. The scatter of both isomer shift and quadrupole splitting for the substance causes the asymmetry of the doublet, so that two parameters γ_1 and γ_2 characterize the degree of the disorder. Assuming the linear correlation between scattered values of the isomer shift and quadrupole splitting ($\gamma_\delta = \alpha\gamma_\Delta$), one can evaluate the widths of their Gaussian distributions γ_δ and γ_Δ.

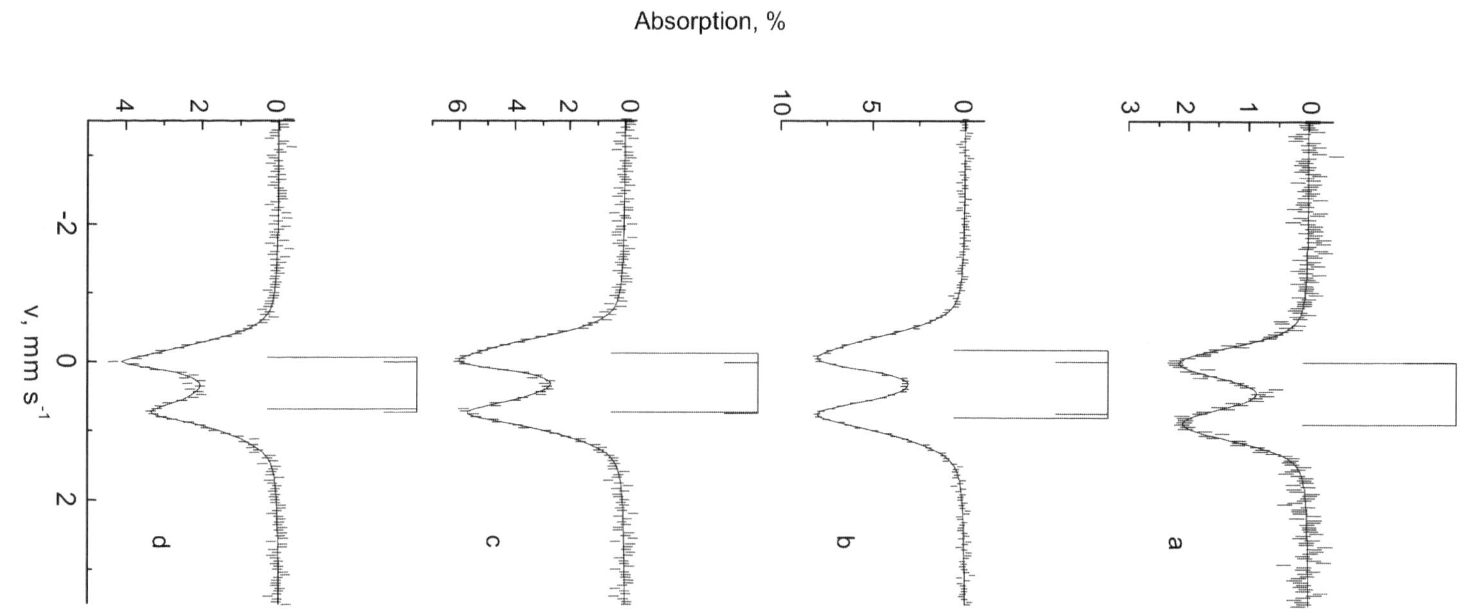

Figure 1. Moessbauer spectra of four iron organosiloxanes at 77 K.

Table 1. Moessbauer parameter of four iron organosiloxanes at 77 K.

Sample	Fit with one doublet					Fit with two doublets					
	δ, mm·s^{-1}	Δ, mm·s^{-1}	γ_Δ, mm·s^{-1}	α	χ^2	δ, mm·s^{-1}	Δ, mm·s^{-1}	S, %	γ_Δ, mm·s^{-1}	α	χ^2
$[CH_3(C_2H_5O)_2SiO]_3Fe$	0.453(4)	0.881(8)	0.36	0.01	1.17						
$[\{CH_3[(CH_3)_2SiO]_4\}_2CH_3SiO]_3Fe$	0.449(2)	0.897(3)	0.38	0	1.64	0.427(9) 0.486(5)	0.98(2) 0.74(1)	75(5) 25(5)	0.44 0.18	0.04 0.02	0.92
$[\{CH_3[(CH_3)_2SiO]_4\}_2CH_3SiO]_3Fe$ + $CH_3(CH_2)_9SH$ (Fe:S=1:0,92)	0.443(3)	0.823(5)	0.37	0.01	1.23	0.415(6) 0.49(1)	0.86(1) 0.72(2)	81(5) 19(5)	0.42 0.2	0.05 0.05	0.96
$[\{CH_3[(CH_3)_2SiO]_4\}_2CH_3SiO]_3Fe$ + $CH_3(CH_2)_9SH$ (Fe:S=1:1,40)	0.448(4)	0.746(6)	0.37	0.08	1.27	0.424(6) 0.487(7)	0.75(1) 0.72(2)	82(3) 18(3)	0.45 0.5	0.08 0.05	1.00

The better fit for the spectra can be obtained using two doublets. The respective results are presented in Tables 1 and in Figure 1(b-d). It is clearly seen that all the parameters of smaller doublet including relative intensity are similar for each sample. For the doublets with the lager splitting the isomer shifts are also very similar, but quadrupole splitting is decreased with the rise of the sulfur content in the sample.

Regarding the values of the isomer shifts of siloxyiron it is reasonable to take into account the known values for iron silicates. Iron has an octahedral arrangement in $Ca_3Fe_2Si_3O_{12}$ [5,6] with garnet structure (δ=0.41 mm·s^{-1}) and in $NaFeSi_2O_6$ [7,8] with the structure of pyroxene (δ=0.41 mm·s^{-1}). These values are slightly different from the isomer shifts of α-Fe_2O_3 [9,10], which consists of regular oxygen octahedra of corundum structure (δ=0.36 mm·s^{-1}). Tetrahedral polyhedron is in $KFeSi_2O_6$ [11] with leucite structure (δ=0.25 mm·s^{-1}) and $KFeSi_3O_8$ [11,8] with pollucite structure (δ=0.23 mm·s^{-1})*.

First, to make a comparison we have to refer our calculated isomer shifts to room temperature. In our case it is possible to apply an increment of appox. -0.1 mm·s^{-1}. Now it could be seen that all calculated isomer shifts are between the above mentioned values for iron in tetrahedral and octahedral arrangements. It is possible to suggest that the coordination polyhedron of iron in our compound is distorted. Such an interpretation is correlated with Moessbauer data for $C_{120}H_{100}Si_{20}O_{43}Fe_6Na_8{\cdot}9C_4H_9OH{\cdot}C_7H_8$ which contains penta- and hexacoordinated iron (δ=0.33±0.05 mm·s^{-1}) [12].

The treatment of $[\{CH_3[(CH_3)_2SiO]_4\}_2CH_3SiO_3]Fe$ with 1-decanthiol can lead to substitution of σ-unbonded oxygens in iron polyhedron into sulfur. This can explain the difference in quadrupole splitting for the doublets with the larger splitting. Increase of sulfur content in the samples results in decreasing of the splitting.

What is remarkable is that doublets with smaller splitting have thin lines. The isomer shift is very close to the shift of α-Fe_2O_3. The doublet can be related to a by-product, which can be formed during the synthesis of $[\{CH_3[(CH_3)_2SiO]_4\}_2CH_3SiO]_3Fe$ as a result of hydrolysis of Fe-O-Si groups followed by the condensation. It is important to emphasize that this constituent remains unchanged during the treatment of the oligomer with the thiol.

4. References

[1] Schmidbaur H and Schmidt M 1962 *JACS* **84** 3600
[2] Blagodatskikh I V, Pozdnyakova Yu A, Shchegolikhina O I and Zhdanov A A 1996 *Russ.Chem.Bull.* **45** 964
[3] Levitsky M M, Zavin B G and BilyachenkoA N 2007 *Russ.Chem.Rev.* **76** 847
[4] Afanas'ev A M and ChuevM A 1995 *JETF* **80** 550
[5] Amthauer G, Annersten H and Hafner S S 1997 *Z.Krystallogr.* **143** 14
[6] Nishizawa H, Shimada M, Matsuoka K and Koizumi M 1997 *Bull.Chem.Soc.Jap.* **50** 3186
[7] Kurkjian C R and Sigey E A 1968 *Phys.Chem.Glasses* **9** 73
[8] Kurkjian C R 1970 *J.Non-Crystalline Solids* **3** 157
[9] van der Woude F 1966 *Phys.Stat.Sol.* **17** 417
[10] Artman J O, Muir A H Jr and Wiedersich H 1968 *Phys.Rev.* **173** 337
[11] Hirao K, Soga N and Kunigi M 1979 *J.Am.Ceramic Soc.* **62** 109
[12] Levitskiy M M, Zavin B G, Bilyachenko A N, Morguniv P B, Kurganova E V, Ovanesyan N S, and Polshin E V 2008 *Russ.Chem.Bull.* **77** 1600

* Room-temperature isomer shifts are presented in this paragraph.

Estimations of the isomer shifts for tetraoxoferrates

O Poleshchuk[1], N Kruchkova[2], Yu Perfiliev[3] and S Dedushenko[3]
[1]Tomsk State Pedagogical University, Kievskaya 60, Tomsk 634061, Russia
[2]Institute Inorganic Chemistry RAS, Lavrentieva 3, Novosibirsk 630090, Russia
[3]Department of Chemistry, Moscow State University, Lenin Hills 1, Moscow 119992, Russia

E-mail: poleshch@tspu.edu.ru

Abstract. ^{57}Fe isomer shifts and the electron density on iron nucleus for a range of tetraoxoferrate ions have been analyzed by means of DFT calculations. The bond lengths and the electron density on the nucleus obtained by our calculations substantially corresponded to the known structural and Mössbauer data. For ferrates the isomer shift depends on the 4s-orbital population. The results obtained allowed us to estimate the isomer shifts for tetraoxoferrate(VII), $KFeO_4$, and iron(VIII) oxide, FeO_4.

1. Introduction

There is constantly growing interest in iron in high oxidation states primarily due to its high potential in water treatment [1, 2]. Ferrates of alkali metals contain complex tetraoxoferrate anions $[FeO_4]^{n-}$ (n=4÷2), which give ferrate(VI)-anion, FeO_4^{2-} , when come into water. FeO_4^{2-}-ion is a very powerful oxidant. They are able to neutralize extremely toxic substances, whereas common oxidizing agents used in water treatment, such as chlorine and ozone, often result in more toxic products. Ferrates (VII) and (VIII) are exotic part of modern inorganic chemistry. But these compounds could be much stronger oxidizers than ferrate (VI). That is why it is very important to know their Mössbauer parameters. This could allow us to detect iron (VII) and (VIII) in the reaction mixture betimes and to optimize the synthetic way.

The focus of this study is the estimation of the isomer shift values on the base of non-empirical calculations for ferrate-anions in the highest oxidation degrees.

2. Calculations

To calculate the geometrical parameters of the isolated ions $[Fe^{m}O_4]^{m-8}$ (m>3) the optimization procedure at the gradient-corrected DFT level by using the three-parameter fit of the exchange-correlation potential suggested by Becke in conjunction with the LYP exchange potential [3] (B3LYP) within the GAUSSIAN'03 program [4] was applied. The use of all-electron basis set DGDZVP for all atoms is better with respect to accuracy and efficiency. The analysis of the atomic charge has been done using the NBO approach [5].

Isomer shifts in this paper are related to α-iron and are presented in [$mm \cdot s^{-1}$].

3. Results and discussion

First, the geometries of ions $[FeO_4]^{n-}$ (n=0÷4*)* were optimized by the above mentioned procedure. The calculated Fe-O bond lengths are presented in Table 1. To check the accuracy of our quantum-chemical calculations it is necessary to compare the experimental and calculated geometries of molecules. For the comparison we used the structural data for the following known ferrates: Na_4FeO_4 [6], K_3FeO_4 [8], K_2FeO_4 [10, 11]. In these compounds iron is in practically regular tetrahedral oxygen

Table 1. Calculated and experimental parameters for tetraoxoferrates (IV) – (VIII)

Oxidation state of iron	$R_{Fe\text{-}O}^{cal.}$, Å	$R_{Fe\text{-}O}^{exp.}$, Å	$\delta_{\alpha\text{-}Fe}^{cal.}$, $mm \cdot s^{-1}$	$\delta_{\alpha\text{-}Fe}^{exp.}$, $mm \cdot s^{-1}$	ρ_0, a_0^{-3}	N_s, e
+4	1.848	1.804 for Na_4FeO_4	-0.18	-0.22 for Na_4FeO_4	11617.4	0.223
+5	1.736	1.695 for K_3FeO_4	-0.55	-0.55 for K_3FeO_4	11618.6	0.241
+6	1.668	1.640 for K_2FeO_4	-0.87	-0.90 for K_2FeO_4	11619.6	0.255
+7	1.619	-	-1.18	-	11620.6	0.268
+8	1.586	-	-1.40	-	11621.3	0.277

arrangement. The average Fe-O distances used for comparison are presented in Table 1. The correlation diagram is presented in Figure 1 and gives us very good results:

$$R_{fe}\text{-}O^{exp.}\,[\text{Å}] = -0.2 + 1.1 R_{Fe\text{-}O}^{cal.}\,[\text{Å}] \qquad (r=0.987;\ s=0.02;\ n=4)$$

In this and subsequent correlation equations r is a correlation coefficient, s is the standard curve fit error, and n is the number of compounds.

The calculations showed that the increasing of the iron oxidation state (OS) lead to decreasing both natural charge on the oxygen atom (Q_O) and natural charge of the iron atom (Q_{Fe}). The respective dependencies are described by the following correlation equations:

$$Q_O\,[e] = -2.5 + 0.27OS \qquad (r=0.9999;\ s=0.007;\ n=5)$$

$$Q_{fe}\,[e] = 1.9 - 0.08OS \qquad (r=0.981;\ s=0.03;\ n=5)$$

Rising of the oxidation state is also accompanied by the increase of the electron density on the iron nucleus (ρ_0) (see Figure 4):

$$\rho_0\,[a_0^{-3}] = 11613.6 + 0.98OS \qquad (r=0.996;\ s=0.16;\ n=5)$$

Finally ρ_0 has been compared with experimental isomer shifts of known ferrates (Figure 5).
This dependency has also good correlation parameters:

$$\delta\,[mm \cdot s^{-1}] = 3635.023 - 0.3129\rho_0\,[a_0^{-3}] \qquad (r=0.999;\ s=0.02;\ n=3)$$

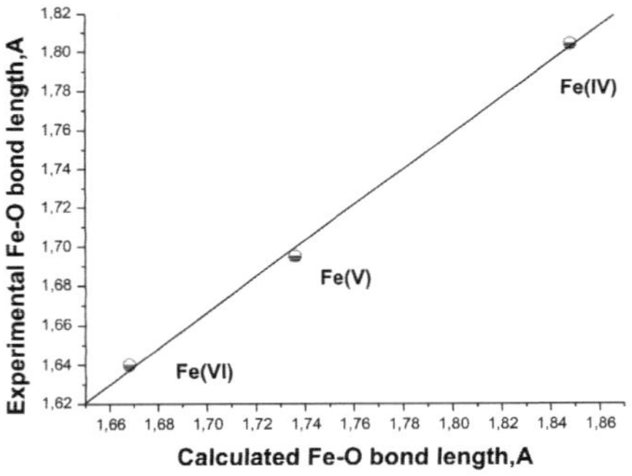

Figure 1. Dependence between experimental and calculated Fe-O bond length

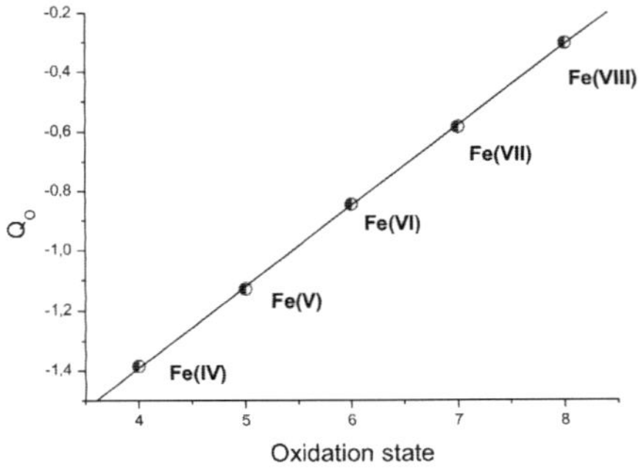

Figure 2. Dependence between Q_O and oxidation state in FeO_4 ions

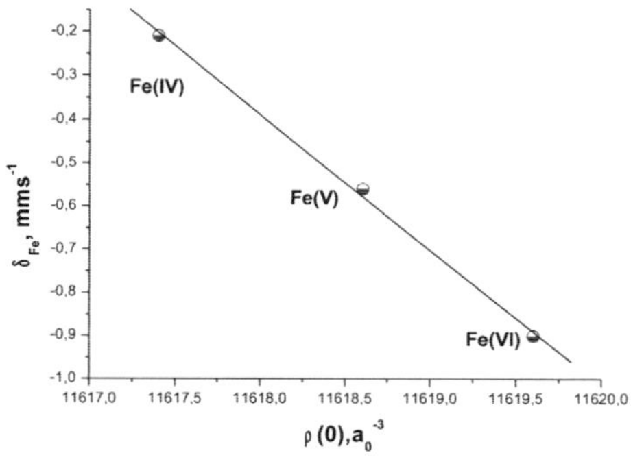

Figure 3. Experimental isomer shift versus calculated relativistic electronic charge densities at the ^{57}Fe nucleus

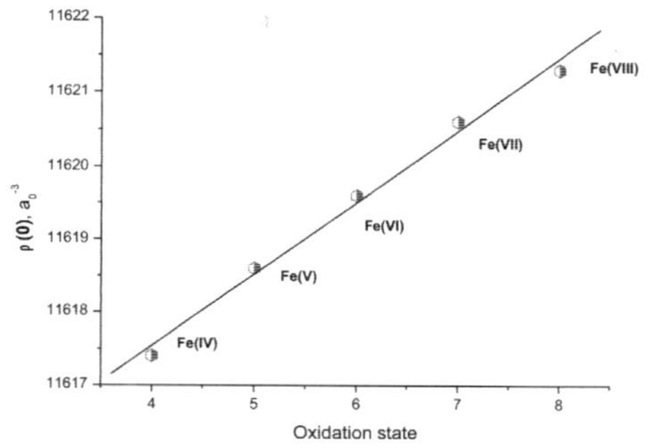

Figure 4. Dependence between electron density on Fe nucleus and oxidation state

Based on the calculated values of the electron density on the iron nucleus the obtained correlation allows us to estimate the isomer shifts for tetraoxoferrate (VII) and (VIII) (-1.18 and -1.40 mm·s^{-1} relative to α-Fe respectively).

It will be noted that absorption lines in the respective range were already observed formerly at the spectra of Na_4FeO_4 [7]. This compound decomposes at room temperature by disproportionation mechanism. During this multistage process several higher oxidation states of iron can be observed. Interaction of Na_4FeO_4 with Na_2O_2 also allowed us to synthesized iron derivatives showing absorption lines with extremely low shifts [7]. The line with isomer shift -1.52 mm·s^{-1} (at 78 K) was also observed in the frozen solutions, obtained by anodic dissolution of metallic iron in concentrated NaOH at high current densities [17]. This line was assigned to Fe(VIII).

It is well known that for Mössbauer atoms the magnitude of the isomer shift depends simultaneously on the s-, p- and d-orbital populations of these atoms [13]. Earlier for various so-called Mössbauer atoms very good correlations between isomer shifts and orbital populations calculated at B3LYP/3-21G(d) by similar procedure have been found [14-16]. For iodine compounds the main contribution to isomer shift comes from the 5s-orbital population, but for tin and antimony compounds a considerable contribution comes from the shielding by 5p-orbitals. For the Au(I) and Au(III) compounds the dependences between the isomer shift and the orbital populations include the direct effect of the valence-shell s-electrons and their shielding of the d-electrons [11]. According to these results it was possible to confirm the conclusion about the greater contribution of the 6s-orbital than 5d-orbital of a gold atom to the isomer shift. If in Au(I) compounds the chemical bonding is determined basically by s- and to a lower extent by d-orbitals of the central atom, in Au(III) compounds the contribution of d-orbital is considerably increased, that is coordinated by a natural image with the increase of the number gold atom bonds. The similar results were obtained by us at B3LYP/LanL2DZ level of Pt(II) and Pt(IV) compounds [16]. For Pt(II) compounds the main contribution to isomer shift comes from the 6s-orbital population, but for Pt(IV) compounds the shielding of the nucleus by 5d-orbitals brings in the greater contribution.

The main observed trends in the variations of the isomer shifts for ferrates can now be interpreted in terms of the valence electronic populations, which depends on iron oxidation state. Our calculation leads to the conclusion that in tetraoxoferrate ions the main contribution to isomer shift brings the 4s-electrons, which is confirmed by the excellent correlation equation:

$$\delta\ [\mathrm{mm\,s^{-1}}] = -0.214 - 0.045 N_S\,[e] \qquad (r=0.999;\ s=0.001;\ n=5)$$

This correlation is valid for both calculated and experimental isomer shifts of ferrates (IV)-(VIII).

An addition of the N_d population increment to the last equation impairs the quality of the correlation, the standard curve fit error being 0.01 mm·s^{-1}. Thus, our calculations show that the shielding of the iron nucleus by d-electrons does not influence on the isomer shift.

This work was supported by the Russian Foundation for Basic Research (project #09-03-01041a) and NATO Collaborative Linkage Grant # 983119.

Journal of Physics: Conference Series **217** (2010) 012041 doi:10.1088/1742-6596/217/1/012041

4. References

[1] 2004 *Innovative Ferrate (VI) Technology in Water and Wastewater Treatment* ed V K Sharma, Q Jiang and K Bouzek (Prague: ICT Press)
[2] 2006 *Ferrates: Properties and Applications in Water and Wastewater Treatment: Preprints of Extended Abstracts, Division of Environmental Chemistry of the American Chemical Society* **46-2** 540
[3] Lee C, Yang W, Parr R G 1988 *Phys. Rev.* **B 37** 785
[4] Frisch M J, Trucks G W, Schlegel H B, Gill P M W, Johnson B G, Robb M A, Cheeseman J R, Keith T, Petersson G A, Montgomery J A, Raghavachari K, Al-Laham M A, Zakrzewski V, Ortiz J V, Foresman J B, Closlowski J, Stefanov B B, Nanayakkara A, Challacombe M, Peng C Y, Ayala P Y, Chen W, Wong M W, Andress J L, Replogle E S, Gomperts R, Martin R L, Fox D J, Binkley J S, Defress D J, Baker J, Stewart J P, Head-Gordon Gonzales C and Pople J A 2003 *Gaussian 03, Revision B03* (Pittsburg: Gaussian Inc PA)
[5] Reed A E, Curtiss L A and Weinhold F 1988*Chem. Rev.* **88** 899
[6] Jeannot C, Mulaman B, Gerardin R and Oullodiaf B 2002 *J. Solid State Chem.* **165** 266
[7] Dedushenko S K, Kholodkovskaya L N, Perfiliev Yu D, et al 1997 *J.Alloys Comp.*, **262-263** 78
[8] Hoppe R, Mader K 1990 *Z. anorg. allg. Chem.* **586** 115
[9] Dedushenko S K, Perfiliev Yu D and Saprykin A A 2008 *Hyp. Int.* **185** 197
[10] Audette R J, Quall J W, Black W H and Robertson B E 1973 *J. Solid State Chem.* **8** 43
[11] Hope M L, Schlemper E O and Murmann R K *Acta Cryst.* 1982 **B38** 2237
[12] Dedushenko S K, Perfiliev Yu D, Goldfeld M G and Tsapin A I 2001 *Hyp. Int.* **136/137** 373
[13] Parish R V 1982 *Coord. Chem. Rev.* **42** 1
[14] Poleshchuk O Kh, Latosinska J N and Yakimov V G 2000 *Phys. Chem. Chem. Phys.* **2** 1877
[15] Poleshchuk O Kh, Shevchenko E L, Branchadell V and Schulz A 2004 *Hyp. Int.* **159** 293
[16] Poleshchuk O K, Branchadell V, Ritter R A and Fateev A V 2008 *Hyp. Int.* **181** 27
[17] Kopelev N S, Perfiliev Yu D, Kiselev Yu M 1992 *J. Radioanal. Nucl. Chem.* **162** 239

Mössbauer Spectroscopic Studies of ^{237}Np in Frozen Nitric Acid Solution

M. Nakada[1], J. Wang[1,2], T. Kitazawa[1], M. Takahashi[1], M. Takeda[1], S. Tsutsui[3], N. M. Masaki[1] and T. Yamashita[1]
1) Department of Materials Science, Japan Atomic Energy Research Institute(JAERI, present JAEA: Japan Atomic Energy Agency), Tokai, Naka, Ibaraki 319-1195, Japan
2) Department of Chemistry, Faculty of Science, Toho University, Miyama, Funabashi, Chiba 274-8510, Japan
3) Advanced Science Research Center, JAERI, Tokai, Naka, Ibaraki 319-1195, Japan

E-mail: wangjh@dicp.ac.cn

Abstract. We applied Mössbauer spectroscopy to ^{237}Np in nitric acid solutions to study chemical states. Sample solutions were adjusted to 50 mg/ml of Np(V) and Np(VI) concentration in 0.6 M nitric acid solutions. These solutions were put into original designed containers for samples and mounted to sample holders. The values of isomer shift (IS, relative to $NpAl_2$) of Np(V) and Np(VI) in 0.6 M nitric acid were -17.4(1) and –37.2(3) mm/s at about 10 K, respectively. It was found that Np(V) had 7 coordination number and Np(VI) had 8 in nitric acid solutions from values of IS.

1. Introduction

It is difficult to study actinide elements. ^{237}Np Mössbauer spectroscopy is a useful tool to investigate the physical and chemical properties of Np in solid state compounds. ^{237}Np sample must be packed and sealed in sample containers for measurement since ^{237}Np is an α emitter. ^{241}Am metal is the standard source for ^{237}Np Mössbauer spectroscopy. Mössbauer γ-ray is 59.5 keV. The sample containers can be made of metals. Because the 59.5 keV γ-ray pass through a thin metal wall of container. Then, ^{237}Np Mössbauer spectroscopy is the most convenient for handling and measurement in actinides.

We have been investigated neptunyl compounds[1-4]. In the present work, we applied ^{237}Np Mössbauer spectroscopy to ^{237}Np in nitric acid solutions to study chemical states.

2. Experimental

Np(V) nitric acid solution: Np(V) hydroxide was precipitated from Np(V) nitric acid solution with sodium hydroxide. The precipitate was weighed and solved by dilute nitric acid solution. Sample solution of Np(V) was adjusted this solution to 50 mg/ml of Np concentration in 0.6 M nitric acid

[1,2] To whom correspondence should be addressed. Present address: State Key Laboratory of Catalysis, Dalian Institute of Chemical Physics, Chinese Academy of Sciences, 457 Zhongshan Road, Dalian, 116023 China

solution. Np(VI) nitric acid solution: Np(VI) hydroxide was precipitated from Np(VI) nitric acid solution with sodium hydroxide. Weighed precipitate was solved by concentrated nitric acid. Sample solution of Np(VI) was adjusted this solution to 50 mg/ml of Np concentration in 0.6 M nitric acid solution. All sample preparations were done in air.

The Sample solution containing Np was put into original designed containers for measurement of ^{237}Np Mössbauer spectra. Containers were described in Figure 1. Containers were constructed with three containers. The inner container was made of graphite as to keep acid and alkali solutions and high transmittance of γ-rays. Other two containers were made of aluminum as high transmittance of γ-rays and metal material. Outer container was sealed by indium metal ring driven turn screws.

Experimental arrangement was shown in Figure 2. 1) Containers were mounted to a sample holder and soaked into liquid nitrogen to freeze quickly; 2) The sample holder was inserted in a cryostat cooled by refrigerator of which cold head was about 6 K; 3) Mössbauer transducer was mounted on the sample holder and attached a rod with source. Mössbauer spectra were measured in the transmission geometry at about 10 K. Mössbauer γ-ray was detected by pure Ge detector or NaI(Tl) scintillation detector. Each spectrum was measured for a month.

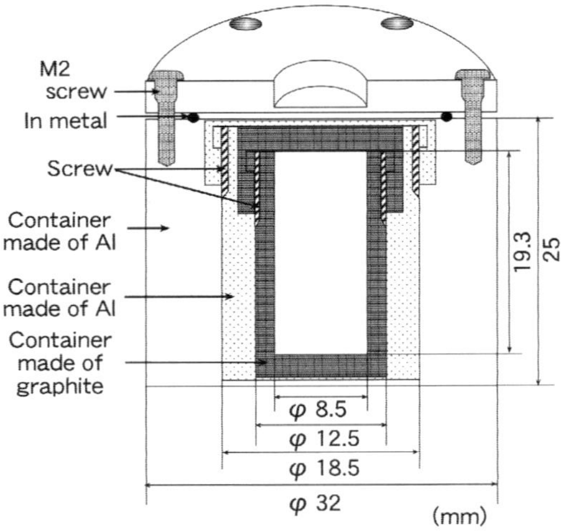

Figure 1. Structure of containers for sample solution containing Np.

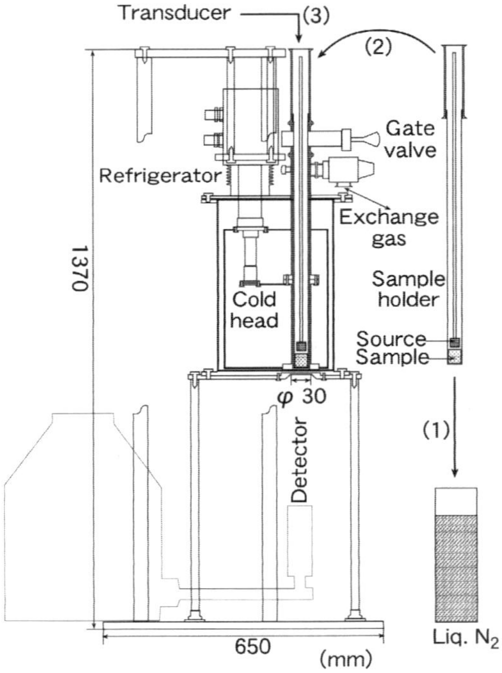

Figure 2 Experimental arrangement for ^{237}Np Mössbauer spectrum measurement of solution sample.

3. Results and discussion

Mössbauer spectra of Np(V) and Np(VI) species in 0.6 M nitric acid solution were shown in Figure 3. Each spectrum was measured for a month as small absorption intensity by small recoilless fractions. Mössbauer parameters were listed in Table 1.

Np(V) species in aqueous solution is known as NpO_2^+. The spectrum of Np(V) was observed 16 absorption lines. Jové et al. reported similar pattern's Mössbauer spectra for frozen solution of $HClO_4$ systems [5]. This pattern was also the same of some neptunyl(V) compounds. The magnitudes of magnetic splitting (MS) and quadruple splitting (QS) were near to some neptunyl(V) compounds that were measured Mössbauer spectra [2-4]. These facts showed that Np(V) was coordinated by some NO_3^- and/or H_2O molecules in nitric acid solution. Isomer shift (IS) value was −17.4(1) mm/s relative to $NpAl_2$. The coordination number of Np(CNN) was 7 (see Figure 4). The value of the full-width at half maximum(FWHM) was larger than other neptunyl(V) compounds. It seems that the broad line width was caused by the paramagnetic relaxation phenomena.

Np(VI) species in aqueous solution is known as NpO_2^{2+}. The observed spectrum of Np(VI) was consisted with mainly 16 absorption lines. Vasili'ev et al. reported existence of three neptunyl complex species, $[NpO_2(H_2O)_6]^{2+}$, $[NpO_2(NO_3)_2(H_2O)_2]$ and $[NpO_2(NO_3)_3]^-$ [6]. CNN of all species was 8. For neptunyl(V) malonate, Np atoms had two sites in the crystal, but two species could not be distinguished by Mössbauer spectroscopy [3]. Because CNN of both sites were 7. Whereas Np atoms of neptunyl(V) nitrate had two sites in the crystal, two species could be distinguished by Mössbauer spectroscopy [7]. Because CNN of one site was 7, that of the other site was 8. CNN of Np(VI) in 0.6 M nitric acid solution was 8 (see Figure 4). This result was agreed with ref. 6. The FWHM value was larger than Np(V) solution. The broad line width disturbed corrective fitting for the spectrum. The better fitting will be taken by the program with relaxation effect [8]. Small amount of Np(V) species was also found on the spectra of Np(VI) in 0.6 M nitric acid solution. Mössbauer parameters were almost same that of Np(V) in 0.6 M nitric acid solution.

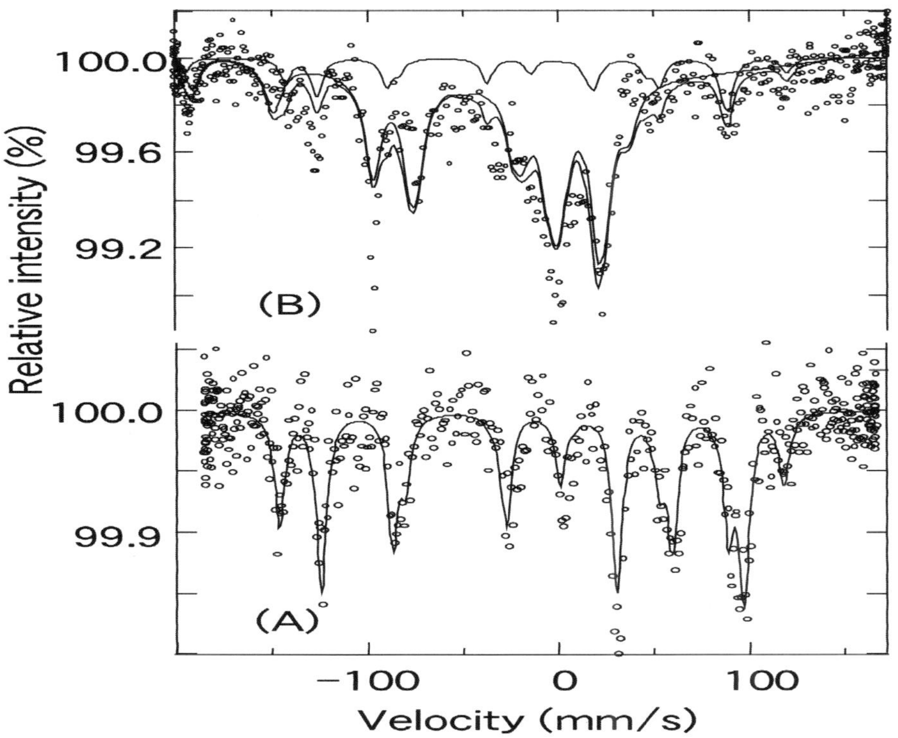

Figure 3. ^{237}Np Mössbauer spectra of (A) Np(V) at 10 K and (B) Np(VI) at 11 K in 0.6 M nitric acid solutions.

Journal of Physics: Conference Series **217** (2010) 012042 doi:10.1088/1742-6596/217/1/012042

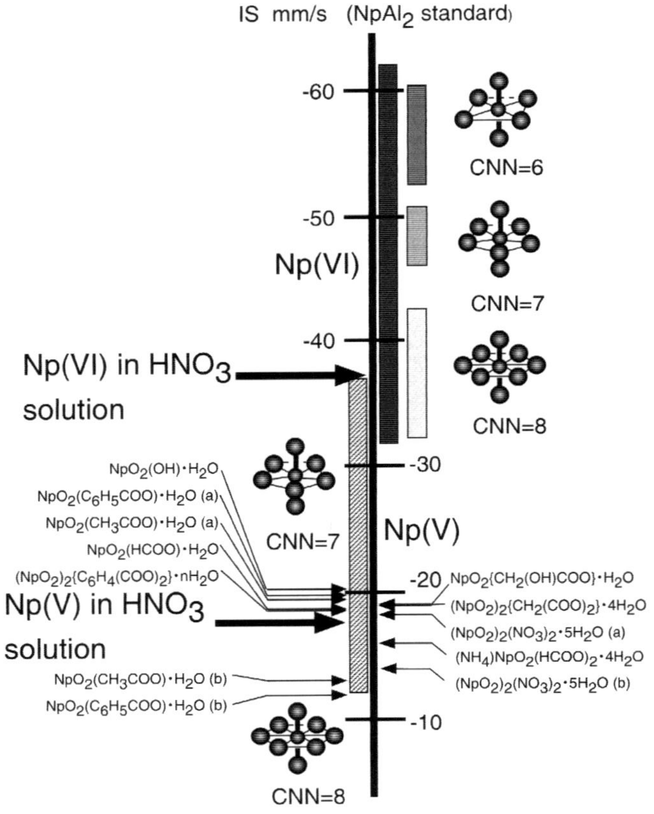

Figure 4. Relationship between isomer shift and coordination number of Np(V) and Np(VI).

Table 1 Mössbauer parameters of Np(V) and Np(VI) in 0.6 M nitric acid solutions

	Np(V) in 0.6 M nitric acid at 10 K	Np(VI) in 0.6 M nitric acid at 11 K
IS (mm/s) *	-17.4(2)	-37.2(3)
Intensity (%)**	0.49(2)	0.36(2)
FWHM (mm/s)	6.6(4)	10.8(9)
QS (mm/s)	100(1)	237(4)
MS (T)	539(1)	291(5)

*relative to $NpAl_2$; ** was sum of 16 absorption lines.

References

[1] Nakada M, Saeki M, Masaki N M and Tsutsui S 1998 *J. Radioanal. Nuc. Chem.* **232** 201

[2] Saito T, Wang J, Kitazawa T, Takahashi M, Takeda M, Nakada M, Nakamoto T, Masaki N M, Yamashita T and Saeki M 1999 *J. Radioanal. Nuc. Chem.* **239** 319

[3] Saeki M, Nakada M, Nakamoto T, Yamashita T, Masaki N M and Krot N N 1999 *J. Radioanal. Nuc. Chem.* **239** 221

[4] Nakada M, Nakamoto T, Masaki N M, Saeki M, Yamashita T and Krot N N 1998 *Hyperfine Interactions* (C) **3** 129

[5] Jové J, He L, Proust J, Pagès M and Pyykkö P 1991 *J. Alloy. Comp.* **177** 285

[6] Ya Vasili'ev V, Andereichuk N N and Rykov A G 1975 *Radiokhimiya* **17** 21.

[7] Nakada M, Saeki M, Nakamoto T, Masaki N M, Yamashita T and Krot N N, to be published.

[8] Nakamoto T, Nakada M, Masaki N M, Saeki M and Yamashita T, 1998 *J. Radioanal. Nuc. Chem.* **227** 137

Mössbauer study of iron-based perovskite-type materials as potential catalysts for ethyl acetate oxidation

D Paneva [1], M Dimitrov[2], N Velinov[3], H Kolev[1], V Kozhukharov[3], T Tsoncheva[2] and I Mitov[1]

[1]Institute of Catalysis, Bulgarian Academy of Sciences, Bulgaria

[2]Institute of Organic Chemistry, Bulgarian Academy of Sciences, Sofia, Bulgaria

[3]University of Chemical Technology and Metallurgy, Sofia, Bulgaria

E-mail:daniela@ic.bas.bg

Abstract. La-Sr-Fe perovskite-type oxides were prepared by the nitrate-citrate method. The basic object of this study is layered Ruddlesden-Popper phase $LaSr_3Fe_3O_{10}$. The phase composition and structural properties of the obtained materials are investigated by Mössbauer spectroscopy, X-ray diffraction (XRD), X-ray Photoelectron Spectroscopy (XPS) and temperature programmed reduction (TPR). The preliminary catalytic tests show a high potential of these materials for volatile organic compounds (VOCs) elimination as they possess high conversion ability and selectivity to total oxidation of ethyl acetate. Catalytic performance of $LaSr_3Fe_3O_{10}$ is depended on the stability of structure and Fe^{4+}-oxidation state.

1.Introduction

Perovskite-type oxides (ABO_3) are promising catalytic materials for the total oxidation of hydrocarbons and oxygenates [1] due to their much lower cost in comparison to noble metals. The Ruddlesden-Popper (RP) type crystal structure is closely related to the perovskite structure and is inherent for composition with cations ratio A/M>1. This structure type is generally described by the formula $A_{n+1}M_nO_{3n+1}$ denoting a series of compounds with n=1,2,3,…∞. The iron-based perovskites with RP structure and their related structures have recently drawn attention as they can provide further insight into some other perovskite systems. This interest is related to the potential of these oxides for applications as catalysts, high-temperature fuel cell electrodes and membrane materials, as well. Perovskite-related oxides with iron in the high oxidation state +4 are an interesting class of materials for studying correlations between chemical composition, crystal structure, electronic states, magnetism and electrical transport behavior. Merino et al. [2] point that Fe^{3+}-Fe^{4+} couple play important role in VOC catalytic oxidation. The appearance of Fe^{4+} would modify the redox ion couple responsible for controlling the electronic mobility in the perovskite [2] and affects the catalytic properties. As mixed valence of Fe^{3+} and Fe^{4+} is accompanied of oxygen non-stoichiometry in perovskite-type compounds,

[1] To whom any correspondence should be addressed.

a study of the effect of oxygen vacancy and mixed valence in different chemical transformations will be interesting to study. The Mössbauer spectroscopy is suitable technique for detailed investigation of iron state and study transformation of Fe^{3+}-Fe^{4+} couple in different reaction media.

The aim of present study was to synthesize several iron-based perovskite-type oxides, characterize the iron state and test the performance of these materials as catalysts.

2. Experimental

The basic object of this study Ruddlesden-Popper (RP) phase $LaSr_3Fe_3O_{10}$ was synthesized by the nitrate-citrate method as described in [3] and was characterized by XRD, SEM and Neutron diffraction (ND) [4]. The sample was characterized by Mössbauer spectroscopy, XPS, TPR with hydrogen and catalytic experiments of methanol decomposition (MD) as described in [5] and ethylacetate oxidation (EAO) as described in [6].

3. Results and discussion

The preliminary ND results [4] indicated that the crystalline structure of $LaSr_3Fe_3O_{10}$ is tetragonal *I*4/mmm with the lattice parameters a=3.87 Å and c=28.06 Å, and Fe-ions occupy two different crystallographic sites. One third of the iron atoms – Fe(1) occupy (2a) Wickoff positions and form the inner perovskite layer of FeO_6 octahedra; the remaining Fe(2)-ions are on (4e) positions and form FeO6 layers, interconnecting the Fe(1)-layer with La(Sr)–O rock salt structure. Due to random distribution of La^{3+} and Sr^{2+} ions, higher oxidation mixed-valence of Fe^{3+} and Fe^{4+} states appear. Starting from charge neutrality requirement, the chemical formulae of studied composition can be written as $La^{3+}Sr_3^{2+}Fe^{3+}Fe_2^{4+}O_{10}^{2-}$. Assuming this model we expect similar ratio of Fe^{3+} to Fe^{4+} ions in Mössbauer spectrum. But calculated ratio between the two coordination states of iron ions Fe^{3+} and Fe^{4+} is 0.2 for the initial sample. Its Mössbauer spectrum is shown in Figure 1a and calculated parameters are listed in Table 1.

Table 1. Mössbauer parameters of investigated samples.

Sample	Components	δ, mm/s	Δ, mm/s	B, T	Γ, mm/s	G, %
$LaSr_3Fe_3O_{10}$ - ini	Sn - Fe^{4+}	0.13	0.00	-	0.57	83
	Db - Fe^{3+}	0.23	0.64	-	0.50	17
$LaSr_3Fe_3O_{10}$ cat EAO	Sn - Fe^{4+}	0.12	0.00	-	0.58	36
	Db - Fe^{3+}	0.33	0.76	-	0.58	64
$LaSr_3Fe_3O_{10}$ cat MD	Sx1 - Fe^{3+}	0.35	-0.09	47.9	0.68	31
	Sx2 - Fe^{3+}	0.35	-0.21	45.7	0.50	26
	Sx3 - Fe^{3+}	0.35	-0.10	43.2	0.94	33
	Sx4 – Fe^{3+}	0.33	-0.21	30.7	0.51	10
$LaSr_3Fe_3O_{10}$ - red	Sx1 - Fe^{3+}	0.35	-0.15	46.7	0.66	34
	Sx2 - Fe^{3+}	0.35	-0.15	44.3	0.73	40
	Sx3 - Fe^{3+}	0.35	-0.15	40.4	1.00	26

δ - isomer shift value relative to α-Fe; Δ - quadrupole splitting; B - hyperfine magnetic field; G – relative area of components

The Mössbauer spectrum of the initial sample ($LaSr_3Fe_3O_{10}$ ini) can be interpreted as one sharp singlet (Sn) with a weak, low-intensity component in the right handside of the velocity scale. The low-intensity component – quadrupole doublet (Db) has hyperfine parameters different from those of the octahedrally coordinated iron. In accordance with previous investigations [7,8] and ND data of oxygen nonstoichiometry - 9.8 (O/per formula unit), this component can be attributed to one with oxygen deficiency. The broad line width (0.50-0.58 mm/s) is a result of the crystallographic disorder both at

Fe(1) and Fe(2) sites and also is a possible cause of the discrepancy between XRD and Mössbauer determined Fe^{3+}/Fe^{4+} ratios.

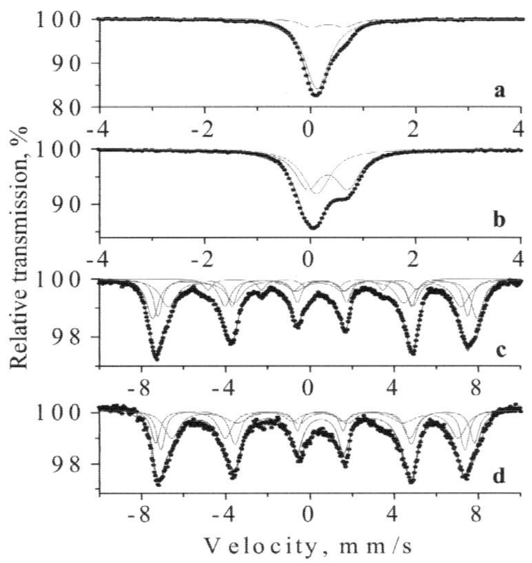

Figure 1. Mössbauer spectra of sample $LaSr_3Fe_3O_{10}$ after different treatment: initial sample (a); sample after EAO (b); sample after MD (c); sample after reduction with H_2 (d).

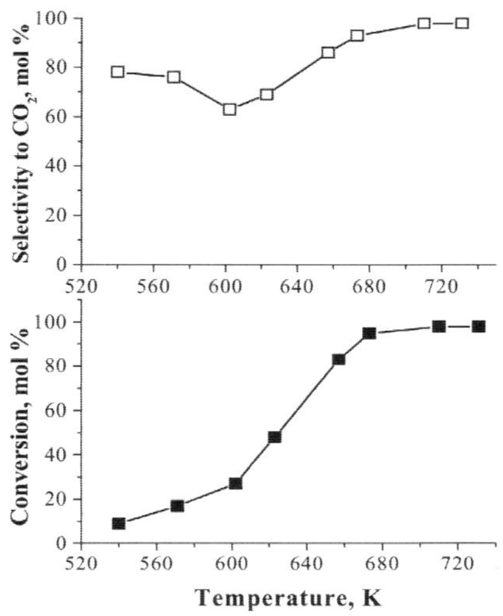

Figure 2. Conversion of ethyl acetate and selectivity to CO_2 as a function of the reaction temperature

In order to test the catalytic properties and stability of $LaSr_3Fe_3O_{10}$ we carried out 2 catalytic tests (EAO – important for VOCs elimination and MD - for H_2 production) and TPR experiments. Three types of reactions characterized with increasing reduction degree of the respective medium, were used: EAO, MD and reduction with H_2.The results of the catalytic evaluation in ethyl acetate oxidation are shown in Figure 2. The studied sample exhibits noticeable catalytic activity in ethyl acetate oxidation above 550 K. At lower temperatures ethanol, acetaldehyde and acetic acid are also registered as by-products. Total oxidation to CO_2 is achieved above 670 K. At the same time, the catalytic activity in methanol decomposition is much lower and the conversion is about 20% at the same temperatures. The main observed carbon containing products are CO and methane with about 80% selectivity to the later (not shown). A single peak is registered in the TPR profile up to 773K (not shown). The TG and DTG profiles show that the reduction proceeds in one step with maximum at 578K. The observed weight loss (about 1.8 wt. %) is indicative of the reductive phase transformations within the sample as result a reduction of Fe^{4+} to Fe^{3+} [9]. In order to obtain information of iron state we acquired Mössbauer spectra after treatment of initial sample under these reaction mediums. They are shown in Figure 1b-1d and corresponding parameters are included in Table 1. Changes in the ratio of the singlet and doublet component are obtained in sample after EAO ($LaSr_3Fe_3O_{10}$ cat EAO). The new doublet component has parameters close to RP with n=1 - $LaSrFeO_4$. [10] Therefore, partial transformation of triple layered perovskite to single layered one realizes in oxidation type of catalytic reaction (EAO). X-ray photoelectron spectroscopy data are used to characterize the surface chemical composition and the oxidation states of the elements presented. The Fe 2p XPS spectrum is split by the Fe 2p spin–orbit effect into the $2p_{3/2}$ and $2p_{1/2}$ regions. The main $2p_{3/2}$ peak around 710 eV is followed by a distinct satellite around 718 eV. The positions of the main and satellite peaks for the Fe $2p_{1/2}$ and Fe $2p_{3/2}$

peaks are also very sensitive to the oxidation states and these peaks were used for qualitatively determination of the ionic states of iron. An analysis of the $Sr3_d$ peak also shows that it can be resolved into two spin-orbital doublets, i.e. Sr ion has two valence states: (133.5 and 135.0 eV) [11] confirming multivalence of iron state. After curve fitting of Fe2p3/2 peak of initial sample and sample after catalytic test EAO we obtain Fe^{3+}/Fe^{4+} ratio of 2.4 and 3.2, respectively. The increase of Fe^{3+} on the surface corresponds to the Mössbauer data, i.e. parameter G changes from 17 to 64%. On the other hand, the results obtained by Mössbauer spectroscopy confirm the existence of Fe^{4+}. The singlet components in both Mössbauer spectra - initial and after EAO have close parameters. Therefore the initial $LaSr_3Fe_3O_{10}$ compound is relatively stable in reaction conditions of EAO. After catalytic test of MD and TPR (Figures 1c,d) the paramagnetic components disappear and new broad magnetically split component appears. The hyperfine parameters deduced after least squares minimization computer processing (Table 1) show that these magnetic components correspond only to Fe^{3+}. They are located at lattice positions in perovskite related phases with different degrees of distortion and/or disorder and different ratio La to Sr [12,13]. The collecting of complex magnetically split hyperfine patterns from the samples after catalytic test - MD and TPR reflects the interaction between Fe^{3+} ions which is not possible in Fe^{4+} phases. Therefore, $LaSr_3Fe_3O_{10}$ transforms into series of $La_{1-x}Sr_xFeO_3$ with PR structure with $n=\infty$ during TPR and catalytic reaction of reductive type.

4. Conclusions

The main idea behind this work is to test reducibility and stability of Ruddlesden-Popper type material $LaSr_3Fe_3O_{10}$ in catalytic reactions. The carried out investigations showed that the studied sample has potential application as catalyst system for VOCs elimination in air where the sample is much more stable. On contrary, it transforms easily under reductive media (MD reaction and TPR). To be used in practice as a catalyst a sample with Ruddlesden-Popper (RP) structure needs additional modification aiming chemical stabilization.

Acknowledgments: The authors D. Paneva, H. Kolev, T. Tsoncheva and I. Mitov are grateful for the financial support to the Bulgarian Science Fund at the Ministry of Education and Science - Project (DO 02-295/2008).

5. References

[1] Szabo V, Bassir M, Van Neste A and Kaliaguine S 2002 *Appl. Catal. B* **37** 175-80.
[2] Merino N A, Barbero B P, Ruiz P and Cadus L E 2006 *J. Catalysis*, **240** 245-57.
[3] Velinov N, Brashkova N and Kozhukharov V 2005 *Ceramics-Silicaty* **49** 29-33.
[4] Neov S, Prokhnenko O, Velinov N, Kozhukharov V, Neov D and Dabrowski L 2007 *Bulg. J. Phys.* **34** 219-25.
[5] Kohn R, Paneva D, Dimitrov M, Tsoncheva T, Mitov I, Minchev Ch and Froba M 2003 *Micropor. Mesopor. Mater.* **63** 125-37.
[6] Tsoncheva T, Linden M, Areva S, Minchev C 2006 *Catal. Commun.*, **7** 357.
[7] Kuzishita K, Morimoto S, Nasu S, Kawasaki S,. Takano M, 2002 *Solid State Communication*, **123** 107.
[8] Reiff W M, Lee J Y and H Steinfink 1992 *Z. Anorg. Allgem. Chem.* **616** 172.
[9] Hirabayashi D, Yoshikawa T, Mochizuki K, Suzuki K and Sakai Y 2006 *Catal. Lett.* **110** 155-60.
[10] Omata T, Ueda K, Hosono H, Katada M, Ueda N and Kawazoe H, 1994 *Phys rev* **49** 10194-99.
[11] Simsa Z, Simsova, Zemek J, Vigen P E and Pardavi-Horvath M 1988, *J. de Physique, Colloque* **C8,** C-975.
[12] Leontiou A A, Ladavos A K, Bakas T V, Vaimakis T C, Pomonis P J 2003 *Appl. Catal. A*:**241** 143–54.
[13] Leontiou A A, Ladavos A K, Giannakas A E, Bakas T V, Pomonis P J 2007 *J.Catal.* **251** 103–12.

Journal of Physics: Conference Series **217** (2010) 012044 doi:10.1088/1742-6596/217/1/012044

In situ Mössbauer investigation of iron oxide catalyst in Water Gas Shift Reaction - Impact of oxyreduction potential and temperature

Z Cherkezova-Zheleva[1] **and I Mitov**

Institute of Catalysis, Bulgarian Academy of Sciences, Acad. G. Bonchev St., Block 11, 1113 Sofia, Bulgaria

E-mail: zzhel@ic.bas.bg

Abstract. The aim of the study is to obtain the exact state of iron oxide catalyst active phase in reaction conditions, as well as the correlation between the active phase and catalytic properties of iron-containing catalysts. *In situ* Mössbauer spectroscopy is the major investigation technique. It is established that the change of reaction conditions (temperature and gas reaction mixture) lead to redistribution of the relative weight of spectra components and influence mainly tetrahedrally and octahedrally coordinated cations in Fe_3O_4 phase. It was concluded, that the active sites of the catalyst in studied reaction are probably pairs of Fe^{3+}+Fe^{2+}-($Fe^{2.5+}$) ions, i.e. the mixed valance iron ions. The obtained catalytic activity can be explained with combination of the natural thermo-activated and catalytically induced electron exchange and better synchronizing of oxidation and reduction steps of the catalytic reaction.

1. Introduction

Nowadays, the water gas shift reaction (WGSR) has attracted an increasing interest mainly because of its application in fuel cells [1, 2]. By WGSR, the hydrogen production from steam reforming is increased and, most importantly, the hydrogen stream is purified by the removal of carbon monoxide. Moreover, the WGSR is a key step in automobile exhaust processes, since the hydrogen produced is a very effective reductant for NO_x removal [3]. The WGSR is one of the basic processes in chemical fertilizer industry, also [4]. Iron based samples are mainly used for catalyzing the WGSR at medium- and high temperatures (MHT) (340 - 500 °C). The industrial MHT catalyst is usually hematite (α-Fe_2O_3), which is converted to magnetite (Fe_3O_4) during the activation process, which is the active phase in the MHT reaction [3, 4]. A detail characterization of the target catalyst under operational conditions is then required to succeed in its optimization. The main part of investigations considers the study of structural properties and catalytic performance using different promoters in order to avoid the sintering process. But the morphological/textural properties can be insufficient in explaining the catalytic behavior due to the major impact of the electronic state of the iron ions. On the other hand the active phases form and change their structure and oxidation state depending on in catalytic reaction conditions (temperature and reaction gas mixture oxidation potential). This requires the use of *in situ* spectroscopies. As one of a handful tools capable of giving chemical and structural information on working catalysts at reaction conditions, Mössbauer spectroscopy is in an elite group of *in situ*

[1] To whom any correspondence should be addressed.

characterization methods. The high energy photons "see" inside chemical reactors to reveal catalyst chemistry and structure under *in situ* conditions [5, 6].

2. Experimental

2.1. Catalyst preparation

The studied catalysts α-Fe_2O_3 were prepared by thermal dissociation of different iron oxy-hydroxide precursors (α-FeOOH and $Fe_5OH_8.4H_2O$) [7], which allows the investigation of the catalyst optimal precursor. There are presented obtained results with the first precursor, only.

2.2. *In situ* Mössbauer spectroscopy

The *in situ* experiments were carried out in a Mössbauer reactor-cell, which permits spectra recording under reaction conditions (high temperature and reaction gas mixture). Sample of 20 mg/cm^2 was placed in a chemically inert holder situated in the cell. The Mössbauer spectrometer used worked in a constant acceleration mode. A ^{57}Co/Cr source with an activity of about 50 mCi was used. The spectra were collected over 1024 channels in mirror image format. Data analysis was performed using a least squares fitting routine that models the spectra as a combination of quadruple doublets and magnetic sextuplets based on a Lorentzian line shape profile. Identification of the spectral components was based on their isomeric shift (δ), quadruple splitting (Δ) and hyperfine magnetic field (B) values. All isomer shift values are reported relative to metallic iron (α-Fe). The line widths (Γ) were determined also. The content of the spectral components (G) was determined from the relative area of each component, which was corrected by value of the Debye-Waller factor at this temperature according to literature data [8]. On the basis of different fitting procedures (giving fixed or non-fixed ratio of magnetite line intensities) were obtained firstly the number of different phases (shown on the Figures) and secondly was calculated the non-stoichiometry parameter of magnetite phase in each spectrum (see the Table).

2.3. Catalytic measurements

The investigation was performed with two different gas mixtures: CO/Ar (14% CO+86% Ar) – *Reaction media 1* and CO/H_2 (15% CO+10% CO_2+63% H_2+12% N_2) – *Reaction media 2*. Purity of all used gases was 99.995%. The gas mixtures were saturated with water vapour up to the chosen pressure: P_{H_2O} = 0, 5.6, 15.7 or 31.2 kPa. The reaction products were determined with gas an analyzer "Infralit-2000". The catalytic activity was expressed by the CO moles converted over a 1 m^2 of the catalyst for 1 h (W, mol/m^{2}h). The experiments were performed according to the following procedure: blowing through Ar at a temperature T = 472 K for an one hour. The catalyst was reduced for 7 h with the gas mixture CO/H_2 under P_{H_2O} = 5.6 kPa, a flow rate V = 4 ml/min and a temperature was increased from T = 470 K to T = 620 K. A steady state of catalyst was attained at the given T and P_{H_2O} with a flow V = 2 ml/min. Recording of the spectrum and determination of the reaction mixture concentration at the outlet and the inlet of the reactor-cell under the steady state conditions was attained.

3. Results and discussion

The samples were studied using Mössbauer spectroscopy before and after catalytic test (*ex situ*) and in catalytic reaction conditions (*in situ*), also. The registered change of catalyst steady state depends on the temperature (293-673K) and oxyreduction potential of reaction gases (P_{CO}/P_{H_2O}), studying two catalytic reaction medias, which can be seen on Figures 1 and 2. The calculated parameters are listed in Table 1. It is well known, that studied WGSR is catalytic reaction of oxyreduction type, having both oxidation and reduction steps. Thus, the catalyst must be characterized by the facile change of the oxidation state of its cations. The catalytic cycle involves two adsorbed redox couples: oxidation of molecule in which the first injects electrons into the oxide catalyst and reduction step, in which the second molecule extracts them from the oxide. In order to study the impact of oxidation potential of reaction media on the steady state of catalyst, two different gas mixtures were investigated, having predominantly oxidation (*Reaction media 1*) or predominantly reduction potential (*Reaction media 2*), respectively. The main attention was paid to the Fe^{+2}/Fe^{+3} ratio in magnetite phase, as it has been suggested that it is responsible for the catalytic activity in HT WGSR [9, 10].

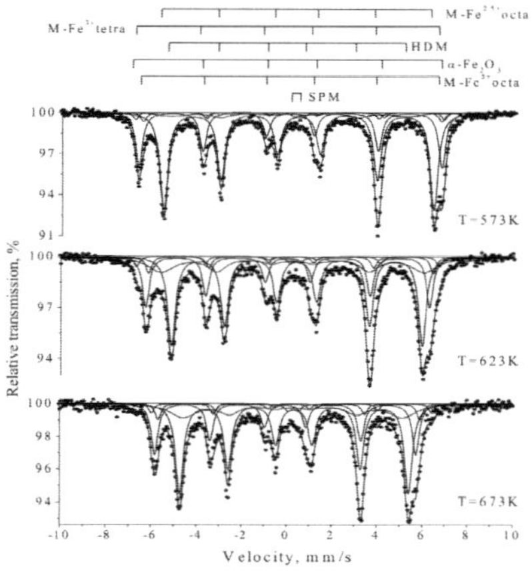

Figure 1. *In situ* Mössbauer spectra at different temperatures and constant P_{H_2O} = 31.2 kPa in *Reaction media 1*.

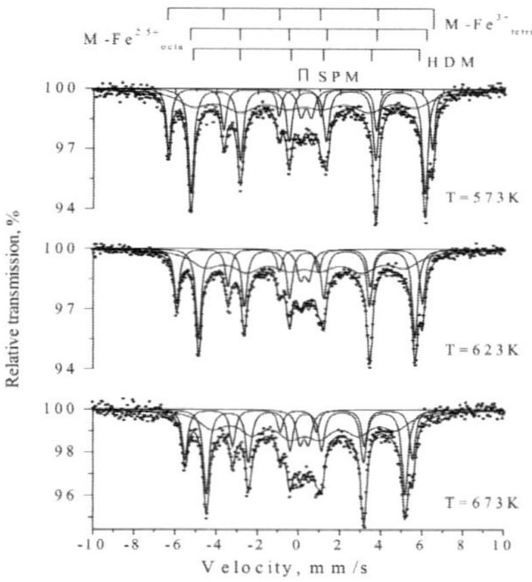

Figure 2. *In situ* Mössbauer spectra at different temperatures and constant P_{H_2O} = 31.2 kPa in *Reaction media 2*.

The obtained *in situ* Mössbauer spectra with the increase of temperature in reaction conditions of *Reaction media 1*, i.e. in predominantly oxidation media are shown on Figure 1. It can be seen the steady state of iron oxide catalyst, i.e. phase and dispersity transformations performing with the catalyst during catalytic reaction. Formation of $(Fe^{3+})_{tetra}[Fe^{2+}_{(1-3x)}Fe^{3+}_{(1+2x)}\square_x]_{octa}O_4$; $\square$ - vacancies; (x=0.012 –0.023), i. e. nonstoichiometric magnetite, as catalyst active phase was registered. The value of nonstoichiometry parameter (*x*) ($Fe_{3-x}O_4$) was obtained according to relative arias of the octahedrally and tetrahedrally coordinated iron ions (S_B/S_A). Depending on T and P_{H_2O} the amount of vacancies passes through a maximum (0.023) at T = 623 K, when the catalytic activity have maximal value (24.4 mol/m^{2}h). It was obtained a small amount of α-Fe_2O_3, which grow up, when the P_{H_2O} increases. High dispersed (HDM) and superparamagnetic (SPM) components were registered also.

The change of reaction conditions (temperature and gas mixture) lead to formation of different catalyst steady state during the *Reaction media 2* (Figure 2). The predominantly reduction gas mixture lead to a presence of other type catalyst active phase - non-stoichiometric magnetite (Fe_3O_{4-x}) - $(Fe^{2+}_xFe^{3+}_{1-x})_{tetra}[Fe^{2+}_{(1+x)}Fe^{3+}_{(1-2x)}]_{octa}O_{4-x}$; x=0.004 – 0.005. An effect of change of catalyst dispersity was registered as a result of catalyst performance. It was obtained SPM component in the spectra. Its partial weight increases with an increasing of number of consecutive oxidation/reduction cycles of catalytic process (i.e. with an increasing of residence time in reaction conditions) and it is higher for the *Reaction media 2*, then for the *Reaction media 1*.

The absence of catalytic reaction, i.e. the *in situ* Mössbauer spectra obtained in Ar media didn't show neither phase composition of catalyst (which is stoichiometric Fe_3O_4), nor catalyst dispersity.

Comparison between the change of vacation quantity (x) and catalyst performance with the temperature increase gives a correlation between the catalysts steady state and structure, and the catalytic activity (Table 1). It was obtained a parallel behavior. On the other hand study of the change of all spectra components showed that the line width of octahedrally coordinated $Fe^{2.5+}$ ions in this non-stoichiometric $Fe_{3-x}O_4$ phase have the same change with temperature increase as catalytic activity (Table 1).

Table 1. Calculated *in situ* Mössbauer parameters in *Reaction media 1 (RM-1)* and *2 (RM-2)*. Experimental errors were ±0.01 mm/s for IS, QS and G; ±0.1 T for B, respectively.

Conditions	Spectra components	δ, mm/s	Δ, mm/s	B, T	Γ mm/s	G, %	S_B/S_A; Parameter of non-stoichiometry (*x*)	Catalytic activity, W, mol/m²h
573K,	α-Fe_2O_3	0.15	-0.11	42.5	0.28	2	-	
RM-1	Fe_3O_4 - Fe^{3+}_{tetra}	0.07	0	41.8	0.36	98	1.817	20.0
	$Fe^{2.5+}_{octa}$	0.44	0	37.3	0.42		0.012	
623K,	α-Fe_2O_3	0.12	-0.11	41.7	0.28	3	-	
RM-1	Fe_3O_4 - Fe^{3+}_{tetra}	0.05	0	39.1	0.50	97	1.667	24.4
	$Fe^{2.5+}_{octa}$	0.41	0	34.6	0.52		0.023	
673K,	α-Fe_2O_3	0.12	-0.11	39.4	0.28	7	-	
RM-1	Fe_3O_4 - Fe^{3+}_{tetra}	0.02	0	36.0	0.49	93	1.717	21.2
	$Fe^{2.5+}_{octa}$	0.37	0	31.5	0.52		0.017	
573K,	Fe_3O_4 - Fe^{3+}_{tetra}	0.07	0	40.8	0.33	100	2.085	18.9
RM-2	$Fe^{2.5+}_{octa}$	0.43	0	36.6	0.39		-0.005	
623K,	Fe_3O_4 - Fe^{3+}_{tetra}	0.03	0	38.2	0.36	100	2.075	19.5
RM-2	$Fe^{2.5+}_{octa}$	0.39	0	33.6	0.41		-0.004	
673K,	Fe_3O_4 - Fe^{3+}_{tetra}	0.01	0	35.4	0.40	100	2.085	18.4
RM-2	$Fe^{2.5+}_{octa}$	0.36	0	30.9	0.44		-0.005	

Then it can be concluded, that under the catalytic reaction conditions the natural electron exchange, which is an intrinsic property of solid state magnetite catalyst is probably combined with electron exchange between reactants in the course of redox catalytic reaction, i.e. with catalytically induced electron exchange. This leads to the observed redistribution of electron density and valence state of the iron ions. Therefore the active sites of this catalyst might be related with pairs of Fe^{3+}+Fe^{2+}-($Fe^{2.5+}$) ions. These mixed valance pairs have an improved possibilities of exchange interactions. The formation of such active complex allow synchronizing of electron exchange leading to simultaneous participation of these ions in the oxidation and reduction steps of the catalytic reaction. Its activation energy decreases due to the more appropriate initial mobility of the catalyst electrons.

Acknowledgments: The authors are grateful for the financial support to the Bulgarian National Science Fund at the Ministry of Education and Science - Project DO 02-295/2008.

4. References

[1] Pereira A L, Berrocal G J, Marchetti S, Albornoz A, de Souza A and Rangel M C 2008 *J. Mol. Catal. A: Chemical* **281** 66–72
[2] Scariot M, Francisco M S, Jordao M, Zanchet D, Logli M and Vicentini V 2008 *Catal. Today* **133–135** 174–80
[3] Kappen P, Grunwaldt J D, Hammersho B, Tröger L and Clausen B 2001 *J. Catal.* **198** 56–65
[4] Byun I S, Choi O L, Choi J G and Lee S H 2002 *Bull. Korean Chem. Soc.* **23/11** 1513-8
[5] Niemantsverdriet J W 2007 *Spectroscopy in catalysis: an introduction* Third revised edition, (Weinheim: Wiley-VCH) pp 121-46
[6] Niemantsverdriet J W and Delgass W N 1999 *Topics Catal* **8** 133–40
[7] Mitov I, Paneva D and Kunev B 2002 *Thermochim. Acta* **386** 179-88
[8] Sawatzky G, Van der Woude F and Morish A 1986 *Phys. Rev.* **183** 383-6
[9] Cherkezova-Zheleva Z, Tsoncheva T, Tyuliev G and Mitov I 2006 *Appl. Catal. A* **298** 24-31
[10] Haber J and Witko M 2003 *J. Catalysis* **216** 416–24

International Conference on the Applications of the Mössbauer Effect (ICAME 2009) IOP Publishing
Journal of Physics: Conference Series **217** (2010) 012045 doi:10.1088/1742-6596/217/1/012045

Investigation by Mössbauer spectroscopy of Sn redox transformations during H_2/O_2 titration on Pt/Al_2O_3Sn-Cl and Pt/Al_2O_3SnIn-Cl naphtha reforming catalysts

Ali Jahel[1,2], **Priscilla Avenier**[1], **Sylvie Lacombe**[1], **Josette Olivier-Fourcade**[2] **and Jean-Claude Jumas**[2]

[1] IFP-Lyon, Rond-point de l'échangeur de Solaize, B.P.3, 69360 Solaize, France
[2] ICG/AIME (UMR 5253 CNRS), Université Montpellier II CC 15-02, Place E. Bataillon, 34095 Montpellier Cedex 5, France

E-mail: priscilla.avenier@ifp.fr, sylvie.lacombe@ifp.fr

Abstract. Bimetallic Pt/Al_2O_3Sn-Cl and trimetallic Pt/Al_2O_3SnIn-Cl catalysts were characterized by H_2/O_2 double titration and ^{119}Sn Mössbauer spectroscopy. The addition of Sn or both Sn and In to the monometallic Pt/Al_2O_3-Cl catalyst leads to an increase in the volume of chemisorbed O_2 at each titration. For bimetallic and trimetallic catalysts the increase in the Pt dispersion, calculated from the second oxygen titre, abnormally surpasses the barrier of 100%. Characterization of samples obtained after each step of the H_2/O_2 titration sequence using ^{119}Sn Mössbauer spectroscopy allowed to follow the redox transformations of Sn species suspected of contributing to O_2 consumption. ^{119}Sn Mössbauer spectra show that after reduction unalloyed metallic Sn and PtSn alloy phases are formed in bi- and trimetallic catalysts respectively. These metallic phases are oxidized upon the first oxygen titre into the so-called $Pt_xSn(O)$ oxometallic phases. The index calculated from this titre thus expresses the Pt-Sn bimetallic character and is called the bimetallicity index (BMI). These latter phases further undergo reduction and regeneration (oxidation) upon the second O_2 titre and thus contribute to the calculated Pt dispersion as well.

1. Introduction

Bimetallic Pt-Sn catalysts supported on chlorinated γ-alumina are extensively studied and used in the naphtha reforming process for the production of high-octane fuels. Recently, trimetallic naphtha reforming catalysts have also received close attention [1]. We have been concentrating our efforts on studying the effect of indium addition on the bimetallic Pt-Sn interactions in this system. The effect of metallic promoters on Pt dispersion (accessible Pt fraction) and the consequent Pt-promoter interaction are important subjects being linked to the catalyst's performance. H_2/O_2 double titration is a common technique which permits the determination of Pt-promoter interactions and the accessible fraction of Pt (dispersion) [2]. We have found that addition of Sn or both Sn and In to the Pt/Al_2O_3-Cl catalyst leads to a significant gain in the Pt dispersion value, abnormally exceeding the barrier of 100%. In order to investigate the possibility that certain Sn species could contribute to the oxygen titres and thus to the bimetallicity indices and the abnormal dispersion values (>100%) as well, we have used ^{119}Sn Mössbauer spectroscopy to determine the various Sn species present at each step of the titration sequence. Characterization and classification of Sn species in Sn-doped naphtha reforming catalysts according to their hyperfine parameters has been previously carried out by Jumas *et al.*[3].

2. Experimental

2.1. Catalyst preparation

Catalysts were prepared by introducing proper amounts of H_2PtCl_6 and HCl by the incipient wetness impregnation technique to a bare γ-alumina support (monometallic catalyst) or to a support already doped with either Sn (bimetallic catalyst) or both Sn and In (trimetallic catalyst). Sn and/or In doped supports were prepared by co-precipitating known amounts of $In(NO_3)_3.5H_2O$ and/or $SnCl_2.2H_2O$ with aluminum hydroxide using NH_4NO_3 as a precipitation agent (sol-gel method). The precipitate (spherical pellets) was calcined in dry air (740°C, 4 hr). Supports possessed a specific surface area (S_{BET}) and porous volume close to 200 m^2/g and 0.55 ml/g respectively. Catalysts obtained were dried at 120°C overnight and calcined (520°C, 2 hr) in a dry air stream (1 $L.g^{-1}.h^{-1}$). Metal and chlorine loadings were determined by X-ray fluorescence spectroscopy. The monometallic, bimetallic and trimetallic catalysts are referred to as Pt/Al_2O_3-Cl, Pt/Al_2O_3Sn-Cl and Pt/Al_2O_3SnIn-Cl respectively.

2.2. H_2/O_2 double titration

H_2/O_2 double titration measurements were performed on a χ-sorb apparatus (IFP license) using a conventional pulse technique. Titrations were carried out on approximately 2 g of catalyst. Samples were calcined (2 hr, 520°C, 20 ml/hr) in dry air and reduced (2 hr, 500°C) in hydrogen flow (hydrogen U 99.995% AIR GAZ). After cooling to room temperature and flushing with He (Helium C 99.995% Air Liquide), pulses (0.24 ml) of O_2 (Oxygen N48 99.998%) were given until saturation occurred as indicated by the constant peak surface of the last 10 pulses given by an online catharometer device. The first oxygen titre VO_1 corresponds to hydrogen chemisorbed on Pt as well as the oxidation reduced metal promoter(s) at high temperature '[2]'. This allowed us to calculate the bimetallicity index BMI. After flushing with He, a second reduction under H_2 flow was performed (1 hr, 30°C). This should reduce only Pt-O into Pt-H. After flushing physisorbed H_2 with He, a second titration gives the second oxygen titre VO_2 which normally corresponds to the oxidation of only Pt-H into Pt-O and which is thus expected to allow the calculation of Pt dispersion. Calculations were based on the Pt loading of each catalyst considering a stoichiometry of one oxygen atom per Pt-H and an assumed chemisorption molar ratio H/Pt of 1 and using already published equations '[2]'.

2.3. ^{119}Sn Mössbauer spectroscopy

Characterization procedure of naphtha reforming catalysts was described elsewhere '[3]'. Samples were characterized after each oxidation/reduction step of the H_2/O_2 sequence: after calcination in air at 520°C, after reduction in H_2 at 500°C, after first O_2 titre VO_1, after second reduction in H_2 at 30°C and finally after second O_2 titre VO_2. The fraction of each phase is given herein using its peak area.

3. Results and discussion

Table 1 lists the Pt dispersion and BMI indices of the characterized catalysts and their metal/chlorine loadings. The monometallic catalyst presents expected equal first and second O_2 titres. Indices calculated from both O_2 titres are thus equal to the platinum dispersion (92%). In the case of the Pt/Al_2O_3Sn-Cl catalyst the two indices are now different. Metallic Sn^0 formed after reduction at 500°C contributes to VO_1 and causes the difference between the two titres. Surprisingly, the dispersion value derived from VO_2 is superior to 100%. One possible explanation is that the H/Pt ratio is superior to 1. Another explanation is that some Sn species might also contribute to this titre. Addition of In to the bimetallic system induces the same effect, with a yet bigger difference between the two indices. It is likely that some In is found in the metal state and contributes in turn to the first oxygen titre increasing the BMI index to 130%. Figures 1 and 2 show ^{119}Sn Mössbauer spectra obtained for the bimetallic and trimetallic catalysts respectively at different stages of the H_2/O_2 double titration sequence.

Table 1. Characterized catalysts and their BMI and Pt dispersion values.

Catalyst	Pt wt.%	Sn wt.%	In wt.%	Cl wt.%	BMI %[a]	Pt Disp. %[b]
Pt/Al2O$_3$-Cl	0.29	/	/	0.98	92	92
Pt/Al_2O_3Sn-Cl	0.30	0.32	/	1.02	122	106
Pt/Al_2O_3SnIn-Cl	0.30	0.30	0.39	1.02	130	109

[a] Index calculated from VO_1 relative to Pt loading.
[b] Index calculated from VO_2 relative to Pt loading.

In both catalysts, only Sn^{IV} is present after calcination. Two forms are distinguished depending on their hyperfine parameters δ and Δ (mm/s): Sn^{IV} 2 (Sn^{IV}-O-M, M= Al, Pt or In) and SnO_2 1 (Sn^{IV}-O-Sn^{IV}) 'figures 1a and 2a'. Reduction of Pt/Al_2O_3Sn-Cl at 500°C forms a metallic unalloyed Sn phase (δ= 2.22 mm/s and Δ= 0.70 mm/s) 'figure 1b'. This phase most likely consists of a solid solution of Sn in Pt, with its isomer shift at the upper limit of the range of Sn-rich Pt_xSn alloys '[3]'. In the presence of In [4], a singlet (δ= 1.93 mm/s, Δ= 0) is observed after reduction 'figure 2b', characteristic of PtSn alloys '[3]'. One Sn^{II} phase (Sn^{II} 2b: Sn^{II}-O-Sn^{II}) '[3]' is also observed for both reduced catalysts.

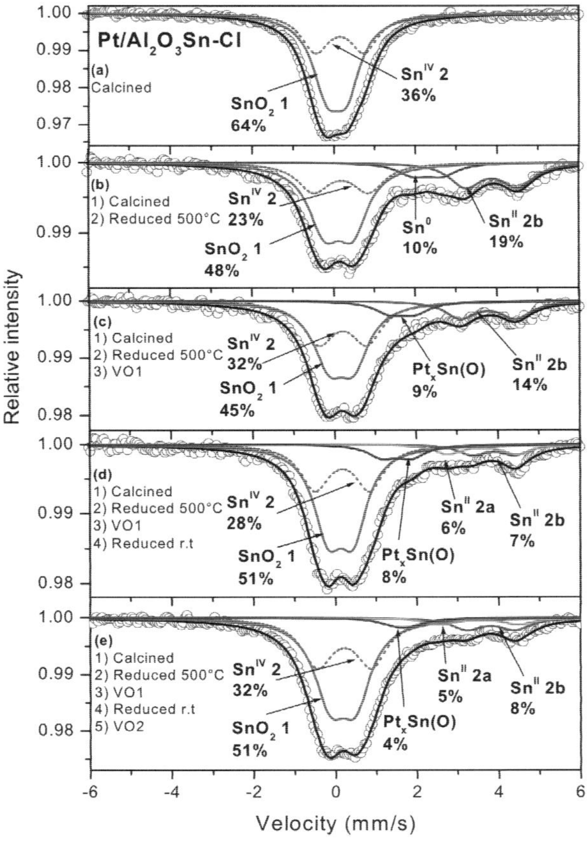

Figure 1. ^{119}Sn Mössbauer spectra of Pt/Al_2O_3Sn-Cl catalyst recorded at room temperature after different steps of the H_2/O_2 titration sequence.

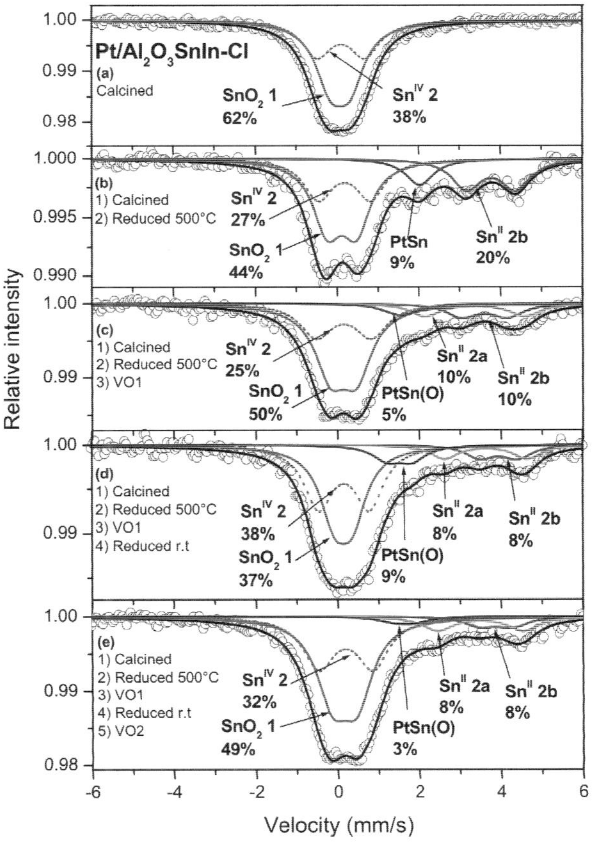

Figure 2. ^{119}Sn Mössbauer spectra of Pt/Al_2O_3SnIn-Cl catalyst recorded at room temperature after different steps of the H_2/O_2 titration sequence.

^{119}Sn Mössbauer spectra recorded after first O_2 titration (VO_1) show that these metallic Sn phases in both catalysts are active to oxidation, whereby they are transformed into $Pt_xSn(O)$ oxometallic phases observed at δ= 1.57 and δ= 1.46 mm/s respectively 'figures 1c and 2c respectively'. Such phases have been observed in previous studies [5]. Their structure is probably similar to that of alloys but they are characterized by a non-zero Δ value, indicating the presence of oxygen bonds, mostly on surface layers of the phase. This is in agreement with the proposition that during the first O_2 titration, reduced promoters contribute to VO_1 '[2]' thus allowing to calculate the index BMI. An enhanced oxidation capacity of the reduced Sn phase is also observed in the trimetallic catalyst, with a fraction of the phase being oxidized into Sn^{II} or/and Sn^{IV} phases. This partially counts, along with the oxidation of other reduced In phases, for the bigger BMI value obtained for this catalyst and is probably resulting from the intimacy between Sn and Pt in the PtSn alloy. Hydrogen atoms would not be necessarily localized on the metallic Pt atoms but could also migrate by a spill-over type mechanism on the nearby Sn atoms [6]. Moreover, a second Sn^{II} phase, Sn^{II} 2a (Sn-O-M) with δ= 3.30 mm/s and Δ= 2.43 mm/s, is formed after VO_1 in the case of Pt/Al_2O_3SnIn-Cl 'figure 2c', probably by rearrangement of a Sn^{II} 2b fraction via oxygen bridging with the nearest Pt particles. The absence of this phase at this stage 'figure 1c' in the absence of In suggests a lower Pt-Sn proximity.

Reduction under H_2 flow at 30°C slightly affects the oxometallic phase contribution in Pt/Al_2O_3Sn-Cl but its peak position is shifted from 1.57 to 1.43 mm/s and its quadrupole splitting increases from 0.61 to 0.73 mm/s. This is most likely due to hydrogen chemisorption on the nearest Pt atoms in the $Pt_xSn(O)$ structure, which creates a bigger partial positive charge on proximate Sn atoms and explains the shifts in hyperfine parameters 'figures 1c and 1d'. A Sn^{II} 2a phase also appears 'figure 1d' (δ= 3.52 mm/s, Δ= 1.69 mm/s), rather due to reduction of some of the Sn^{IV} 2 phase. Reduction of Pt/Al_2O_3SnIn-Cl regenerates the PtSn(O) phase from Sn^{II} and Sn^{IV} phases. Similar shifts are observed for the newly formed phase (δ= 1.43 mm/s and Δ= 0.61 mm/s) relative to the phase formed after VO_1 (δ= 1.46 mm/s and Δ= 0.30 mm/s) 'figures 2c and 2d'. This apparently reactivates these phases to the next oxygen titration (VO_2) whereby a similar mechanism to that upon first titration would take place. The second O_2 titration decreases oxometallic fractions in both catalysts, and a part of tin is oxidized into Sn^{II} or Sn^{IV} 'figures 1e and 2e'. Therefore, Sn oxidation in these phases directly contributes to the second oxygen titre VO_2, which was usually considered to result from the oxidation of only Pt-H into Pt-O oxides '[2]'. The remaining $Pt_xSn(O)$ phases are observed at (δ= 1.56 mm/s, Δ= 0.34 mm/s) and (δ= 1.54 mm/s, Δ= 0.53 mm/s) for Pt/Al_2O_3Sn-Cl and Pt/Al_2O_3SnIn-Cl respectively.

4. Conclusions

^{119}Sn Mössbauer spectroscopy results revealed a continuous phenomenon of bimetallic interplay during H_2/O_2 treatments in multimetallic naphtha reforming catalysts. Metallic Sn in proximity to Pt is sensitive to O_2 treatment and contributes to both VO_1 and VO_2 oxygen titres in the H_2/O_2 titration sequence with a higher apparent oxidation/reduction sensitivity when Sn is in the form of a PtSn alloy.

Acknowledgements

Financial support by the Région Languedoc Roussillon for the X-ray and γ-ray platform is acknowledged. Ali Jahel is also grateful to IFP and ANRT for research grants.

References

[1] Carnevillier C, Epron F, Marecot P, 2004 *Appl. Catal.* A: General **275** 25.
[2] Sharma L D, Kumar M, Saxena A K, Rawat D S, Prasada Rao T S R, 1998 *Appl. Catal.* A: General **168** (2) 251.
[3] Olivier-Fourcade J, Womes M, Jumas J-C, Le Peltier F, Morin S, Didillon B, 2004 *ChemPhysChem* **5** 1734.
[4] A detailed study on the effect of In will be published elsewhere.
[5] Womes M, Le Peltier F, Morin S, Didillon B, Olivier-Fourcade J, Jumas J-C, 2007 *J. Mol. Catal.* A: Chemical **266** 55.
[6] Muller A C, Engelhard P A, Weisang J E, 1979 *J. Catal.* **56** 65.

Journal of Physics: Conference Series **217** (2010) 012046 doi:10.1088/1742-6596/217/1/012046

Mössbauer study of modified iron-molybdenum catalysts for methanol oxidation

K I Ivanov[1], I G Mitov[2], St V Krustev[1], B S Boyanov[3]
[1]Agricultural University, Department of Chemistry, 12 Mendeleev St., 4000 Plovdiv, Bulgaria
[2]Institute of Catalysis, Bulgarian Academy of Science, 1113 Sofia, Bulgaria
[3]Plovdiv University "Paisii Hilendarski", 24 Tsar Assen St., 4000 Plovdiv, Bulgaria

E-mail: kivanov1@abv.bg

Abstract. The preparation and catalytic properties of mixed Fe-Mo-W catalysts toward methanol oxidation are investigated. Mössbauer spectroscopy, X-ray diffraction and chemical studies revealed the formation of two types of solid solutions with compositions $Fe_2(Mo_xW_{1-x}O_4)_3$ and $(Mo_xW_{1-x})O_3$. The solid solutions formed are characterized by high activity and selectivity upon methanol oxidation and are of interest in view of their practical application. Sodium-doped iron-molybdenum catalysts are also investigated and the $NaFe(MoO_4)_2$ formation was established.

1. Introduction

The most important results of the investigations of oxide catalysts for selective oxidation of methanol, carried out during the last two decades, lead to the conclusion that the system $Fe_2(MoO_4)_3$-MoO_3 has no alternative [1, 2]. Mössbauer spectroscopy is an excellent technique for investigation of these catalysts [1, 3]. According to Pernicone [4] and Ivanov [5], attention should be focused on the enhancement of the activity, selectivity and stability of the catalysts by the introduction of suitable additives. The present work deals with a Mössbauer study of tungsten and sodium modified iron-molybdenum catalysts.

2. Experimental

The tungsten modified catalysts were obtained by precipitation from 5 mass % aqueous solutions of $FeCl_3$ and of mixtures of ammonium heptamolybdate and Na_2WO_4 in various ratios, at pH values by the end of the process ranging from 1.6 to 2.0. The precipitates obtained were washed, dried and annealed at 500°C for 5 hours.

The preparation of the sodium-doped catalysts was being carried out in two ways: (i) by precipitation from 5 mass % aqueous solutions of iron nitrate and sodium molybdate and (ii) different amounts of sodium were incorporated into preliminary synthesized catalysts by impregnation with aqueous solutions of $NaNO_3$, vaporizing, drying and calcination at 500°C in the course of 5 hours.

Chemical analysis of the samples was performed by atomic absorption (AA) and atomic emission (ICP) measurments. X-ray and Mössbauer spectra were obtained to establish the phase composition of the catalysts. The X-ray analysis of the annealed specimens was performed by a TUR-M 62 diffractometer with Co-K_α radiation. The exact positions of the lines and their breadths were determined according to a program for decomposition and profile analysis of the diffractograms [6].

The Mössbauer spectra were recorded on an electromechanical spectrometer Wissenschaftliche Electronic GmbH, (Germany) operating at a constant acceleration mode at room temperature. A $^{57}Co/Cr$ (activity of about 50 mCi) source and an α-Fe standard were used. The experimentally obtained spectra were mathematically treated by the least squares method. The parameters of the isomer shift (IS), quadrupole splitting (QS) and effective internal magnetic field (H_{eff}) were determined. The relative weight of the partial components (G) of the spectra was calculated on the basis of the intensity of the respective lines.

3. Results and discussion

3.1. Tungsten modified catalysts

The results of X-ray investigation suggest that the addition of tungsten leads to considerable changes in the phase composition and structure of the precipitated specimens: (i) the increase of the WO_3 content causes changes in the position and intensity of the basic lines in the diffractograms; (ii) the lines characteristic of $Fe_2(MoO_4)_3$ are slightly shifted to higher values of d (nm); (iii) the most considerable change is observed in the intensity of the lines belonging to MoO_3 (interplanar distances d = 0.685, 0.386, 0.347, 0.327 and 0.230 nm) - some of them almost disappear at high tungsten content. These changes in the X-ray diffractograms suggest that the replacement of some of the molybdenum atoms in the $Fe_2(MoO_4)_3.MoO_3$ catalyst by tungsten leads to the formation of two types of solid solutions. The drastic decrease of the intensity of the lines characteristic of MoO_3 and the absence of lines related to the formation of WO_3 lead to the conclusion that during precipitation and subsequent thermal treatment, compounds of the type $(Mo_xW_{1-x})O_3$ are formed, the latter being described in powder diffraction files PDF 9-0224, PDF 9-0225, PDF 28-0667 and PDF 28-0669. At x < 0.5, the diffraction lines correspond in the best way to a compound with the composition $(Mo_{0.68}W_{0.32})_{12}O_{35}$ [7].

On the other hand, the interplanar distances, characteristic of $Fe_2(MoO_4)_3$, increase with the rise of the WO_3 content. This indicates that a portion of tungsten participates in the formation of solid solution having a $Fe_2(Mo_xW_{1-x}O_4)_3$ composition. The results described above lead to the conclusion that some of the Mo^{6+} ions in $Fe_2(MoO_4)_3$ and the free MoO_3 are isomorphically substituted by W^{6+} ions. This is confirmed by the Mössbauer spectra of the specimens.

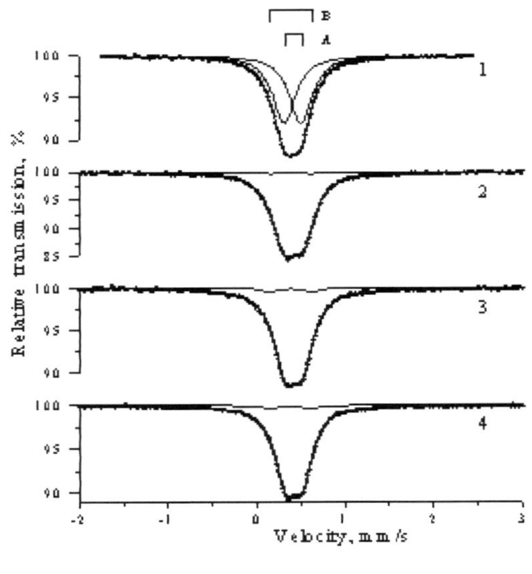

Figure 1. Mössbauer spectra of $Fe_2(MO_4)_3$+xWO_3, where x=0.0(1), 1.9(2), 8.0(3) and 15.9(4)

The spectra of the specimens containing tungsten are processed as composed of two doublets having the same IS values and differing in the QS values (Fig. 1). The parameters obtained and the relative weights of the areas corresponding to the doublets for $Fe_2(MoO_4)_3.MoO_3 + xWO_3$ catalysts, where x = 0.0, 1.9, 8.0 and 15.9 are given in Table 1. The parameters shown in Table 1 suggest that the Fe^{3+} ion are high-spin ones and are situated in a slightly deformed octahedral field. The higher QS values of the lover-intensity doublets (Fig.1, component B) are attributed to the ferric ions having a different ligand surrounding, composed of both Mo^{6+} and W^{6+} ions.

Table 1. Isomeric shift (IS), quadrupole splitting (QS), line width (FWHM) and relative weight of the components (G) of specimens with composition $Fe_2(MoO_4)_3.MoO_3$ and $Fe_2(MoO_4)_3.MoO_3 + xWO_3$, where x = 0, 1.9, 8.0 and 15.9

Samples		IS, mm/s	QS, mm/s	FWHM, mm/s	G, %
$Fe_2(MoO_4)_3.MoO_3$		0.41	0.187	0.32	100
$Fe_2(MoO_4)_3.MoO_3$ +1.9 % WO_3	1 doublet	0.416	0.198	0.315	98.2
	2 doublet	0.395	0.463	0.320	1.8
$Fe_2(MoO_4)_3.MoO_3$+8.0 % WO_3	1 doublet	0.415	0.190	0.313	96.3
	2 doublet	0.392	0.469	0.316	3.7
$Fe_2(MoO_4)_3.MoO_3$ +15.9 % WO_3	1 doublet	0.416	0.189	0.306	96.3
	2 doublet	0.392	0.460	0.320	3.7

The fact that the ratio of the relative weights of the doublets remains unchanged with the rise of the WO_3 content over 5% allows the assumption that the components of the system considered here have restricted solubility.

3.2. Sodium modified catalysts

The results from the chemical analyses of the precipitates indicate that washing with distilled water has a significant effect on their composition. The content of sodium decreases sharply from 6.5 to about 0.3 wt. % (presented as Na_2O).

The calcination of the precipitates prepared in the described way leads to the formation of well defined phases with composition $Fe_2(MoO_4)_3$ and MoO_3. The presence of a considerable amount of sodium in some of the samples, however, could lead to the formation of a new phase with composition $NaFe(MoO_4)_2$. This expectation was clearly confirmed by the results presented in Fig. 2. The catalytic activity and selectivity (S) for oxidation of methanol were determined on standard flow equipment at a volumetric rate of 1200 h^{-1}, methanol concentration in the alcohol-air mixture of 6% and temperature of the catalyst layer of 200 - 350°C. The amount of catalyst was 3 cm^3 and a fraction of 1.0-1.6 mm was used in all cases. The formaldehyde content in the off-gases was determined by means of the bisulfite method, while the amounts of CO, CO_2, $(CH_3)_2O$ and unreacted methanol were defined by gas chromatography.

The results for tungsten modified catalysts confirmed the expectation that solid solutions formed are characterized by high activity and selectivity upon methanol oxidation and are of interest in view of their practical application.

No negative effect up to 0.2 wt. % sodium content was indicated at the precipitated catalysts. On the contrary, a dramatic change in the catalytic activity of the impregnated catalysts was established at sodium content even under 0.05 wt. %.

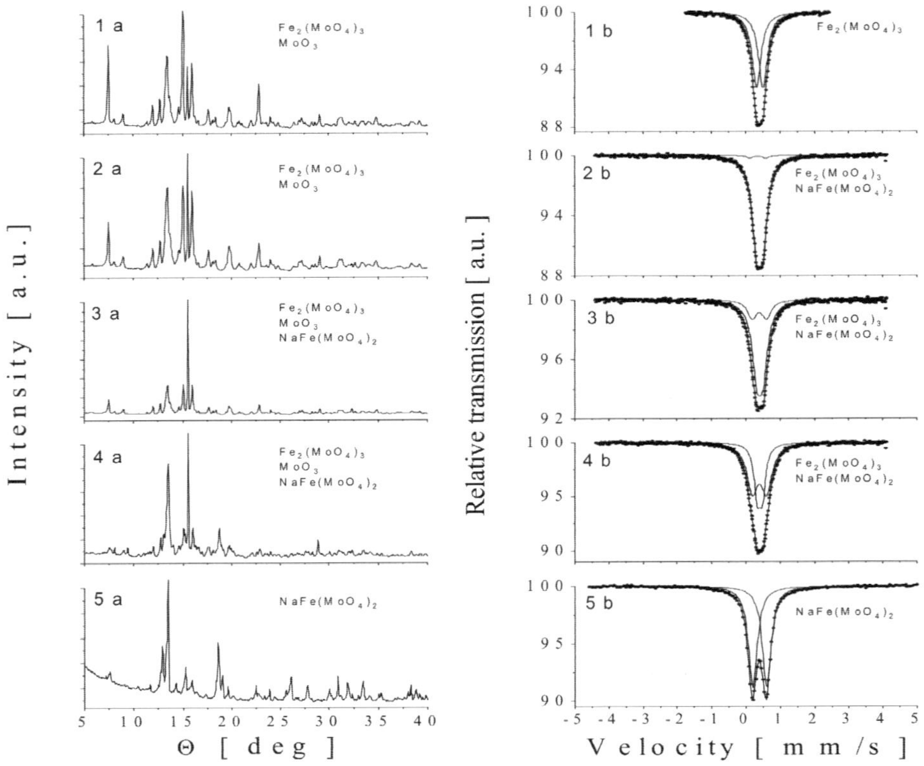

Figure 2. X-ray diffraction patterns (a) and Mössbauer spectra (b) of Na-doped iron-molybdenum catalysts: $Fe_2(MoO_4)_3$ (1a and 1b), $NaFe(MoO_4)_2$ (5a and 5b) and samples containing 0.3 % Na_2O (2a and 2b), 1.4 % Na_2O (3a and 3b) and 6.5 % Na_2O (4a and 4b).

4. Conclusions

1. The coprecipitation of aqueous solutions of $FeCl_3$ and of a mixture of ammonium heptamolybdate and Na_2WO_4 leads to the formation of a mixture of two types of solid solutions with compositions $Fe_2(Mo_xW_{1-x}O_4)_3$ and $(Mo_xW_{1-x})O_3$. The solid solutions formed are characterized by high activity and selectivity upon methanol oxidation and are of interest in view of their practical application.
2. A strong nucleation effect of sodium to form new compounds takes place. Regardless of the way of its introduction, $NaFe(MoO_4)_2$ formation after calcination was proved.

Acknowledgement: This work is supported by Bulgarian Ministry of Education, Project VU- X-305/07.

References

[1] Soares A P V, Portela M F, Kiennemann A 2004 Methanol Selective Oxidation to Formaldehyde over Iron-Molybdate Catalysts *Catal. Rev.* **47** pp 125-174
[2] Ivanov K, Dimitrov D 2009 Methanol Oxidation over Iron-Molybdate Catalysts. Optimization of the Process *Topics in Catalysi* **52** (6) pp 813-822
[3] Carbucicchio M, Trefiro F 1976 Surface and Bulk Redox Processes in Iron-Molybdate-Based Catalysts *J. Catal.* **45** pp 77-85

[4] Pernicone N, Advances and trends in the catalysis of formaldehyde production over Fe-Mo catalysts, Meeting on Industrial and Environmental Catalysis, Lisbon, May 15-19, 1995, in: *Actas do 38 Encontro Nacional de Cata´ lise Ba´ sica e Aplicada (Industrial e Ambiental); Farinha Portela, M., Gama Freire, F. and Pilar Castilho, M., Eds.; Lisboa*

[5] Ivanov K 1997 Oxide catalyst for methanol oxidation *Dissertation* Plovdiv

[6] Petkov V, Bakaltchev N1990 FIT, a computer program for decomposition of powder diffraction patterns and profile analysis of pair correlation functions *J. Appl. Crist.* **23** pp 138 -140

[7] Powder Diffraction Files (Joint Committee on Powder Diffraction Standards), Philadelphia PA, USA, 1997

Effect of extent of natural subsurface bioreduction on Fe-mineralogy of subsurface sediments

Ravi K Kukkadapu*, Nikolla P Qafoku, Bruce W Arey, Charles T Resch and Philip E Long

Pacific Northwest National Laboratory, P O Box 999, Richland, WA 993552, USA

*E-mail: ravi.kukkadapu@pnl.gov

Abstract. Naturally bioreduced zones with considerable sorbed U were recently identified at a former U mining and processing site at Rifle, CO, USA. Most of the sorbed U appears to be associated with Fe minerals. Variably reduced sediment samples were analyzed by suite of techniques, primarily by room temperature Mössbauer spectroscopy. Fe-oxides of different types and crystallinity, and Fe(II)/Fe(III)-containing clays are dominant in all the sediments. The amounts of poorly crystalline Fe(III)-oxide, however, was lower in the reduced samples. In addition, framboidal pyrites with sorbed U were common in the highly reduced sediments. Overall, the information gained from this work may help develop design field strategies for immobilization and stabilization of U(VI) in contaminated subsurface environments.

1. Introduction

The U.S. Department of Energy (DOE) faces the challenge of cleaning up and/or monitoring large, dilute groundwater plumes contaminated by radioactive and hazardous metals, such as U, Tc, Hg, Pu, Cr etc. These contaminants resulted from the legacy of nuclear weapons production at various US-DOE (e.g., Hanford, WA) and metal-mining and ore-processing sites (e.g., Rifle, CO). Various techniques are underway to remediate the contaminated sites with a combination of physical, chemical, and biological methods [1 and references therein]. For example, stimulation of groundwater dissimilatory iron-reducing bacteria (DMRB) is being explored as a potential strategy to reduce soluble U(VI) to insoluble U(IV) at a number of DOE sites including the Rifle site. At Rifle, the addition of acetate (an electron donor) to the subsurface resulted in an initial decrease in U(VI) [2]. The initial loss of U was associated with enrichment of *Geobacter sp* (a known Fe- and U-reducing bacterium) and production of aqueous Fe(II). Prolonged acetate injection resulted in the longer term stabilization of U, sulfate reduction, and blackening of the sediment, presumably due to precipitation of Fe-sulfide minerals [3].

Iron bearing minerals (both oxides and clays) of the sediment matrix and groundwater chemical composition play an important role in controlling the mobility of aqueous U(VI) [e.g., 4]. Iron oxides such as ferrihydrite, goethite, hematite, and magnetite, and Fe(II)/Fe(III)-containing clays (e.g., smectites, illites) that are ubiquitous in sediments, are good U(VI) sorbents [e.g., 5, 6]. In addition, Fe-sulfides, such as Fe monosulfide $(Fe,Ni)S_{0.9}$ (mackinawite), greigite (Fe_3S_4), and pyrite (FeS_2) that also exist in the subsurface may be good hosts for U, as well. Laboratory batch studies with pyrite demonstrated that U(VI) adsorption and partial reduction occur on the surfaces of pyrites [e.g., 7, 8].

Naturally reduced zones that are relatively rich in reduced Fe [Fe(II)], sulfide, U(IV), and organic C (OC), compared to oxidized zones, are common at the Rifle site [9]. An understanding of the geo-microbiological aspects of these samples (oxidized and reduced sediments) may thus provide insights into biotic [direct reduction of U(VI) by bacteria] and/or coupled biotic-abiotic processes [bioreduction of Fe(III) followed by abiotic reduction of U by Fe(II)] that may occur under the natural conditions. The objective of this study is to gain insights into the relationship between the extent of natural reduction and associated Fe mineralogical changes on groundwater U attenuation. A series of sediment samples obtained at different locations that reflect varying aquifer conditions was analyzed by various microscopic techniques, ^{57}Fe-Mössbauer spectroscopy, and chemical extraction methods. The results, for example, biotransformation of native Fe-minerals to Fe-sulfides and their role in U removal, will enhance our understanding of the long-term behavior of U in the groundwater.

2. Materials and Methods

2.1. Sediment Materials

The cores represented a cross section of sediment conditions that ranged from typical aquifer (minimally reduced) to highly bioreduced (within the saturated zone) conditions. The oxidized sample was obtained from a minimally contaminated zone by use of a backhoe (~3 m depth) from an area with relatively low U groundwater contamination (Figure 1a). This sample is designated as RABS (Rifle aquifer background sample). Samples with higher levels of contamination/reduction were obtained from cores drilled in 2007 (Figure 1; expanded scale of the injection gallery is shown in Figure 1b).

Figure 1: a) Aerial view of Rifle site, CO, with experimental plot highlighted in circle (reduced samples) and square (RABS sample); b) location of samples in the well lay out [(G60 to G-51 are acetate injection wells); U01 and U02 are up gradient monitoring wells; and down gradient wells (D-series)], and c) a core from D08 well (a highly reduced zone) showing the boundary between reduced and oxidized sediments.

The minimally reduced sediment was designated as D05-16 (16 ft sample from well D-05), while the highly bioreduced sample was designated as D08-16 (16 ft sample from well D-08). Transition between reduced zone and unreduced is highlighted in Figure 1c. We focused on the

<53-micron size fractions that contained relatively more sorbed U [6]. The details of size separation and chemical extraction are described elsewhere [6].

2.2. Scanning Electron Microscopy (SEM)

For the SEM measurements, polished sections of sediments were carbon coated to make them electrically conductive. The sections were examined using a JEOL 6340f instrument. Images were then collected using a backscattered electron detector for atomic number contrast. A JEOL model 8200 EMPA, with a detection limit of ~100 μg g^{-1} and optical resolution of 50 nm, was used to take additional measurements including the quantification of U concentration in pyrite.

2.3. ^{57}Fe-Mössbauer spectroscopy

Mössbauer spectra were collected at room temperature (RT) using a 50-mCi (initial strength) ^{57}Co/Rh source. The velocity transducer MVT-1000 (WissEL) was operated in a constant acceleration mode (23 Hz, ±12 mm/s). An Ar-Kr proportional counter was used to detect the radiation transmitted through the holder, and the counts were stored in a multichannel scalar (MCS) as a function of energy (transducer velocity) using a 1024-channel analyzer. Data were folded to 512 channels to give a flat background and a zero-velocity position corresponding to the center shift (CS) of a metal Fe foil at RT. Calibration spectra were obtained with a 25-μm-thick Fe(m) foil (Amersham, England) placed in the same position as the samples to minimize any errors due to changes in geometry. The Mössbauer data were modeled with Recoil software (University of Ottawa, Canada) using a Voigt-based structural fitting routine [10]. The coefficient of variation of the spectral areas of the individual sites generally ranged between 1% and 2% of the fitted values.

3. Results and Discussion

3.1. Chemical analysis

The 0.5N HNO_3 acid extractable Fe(II) contents (a measure of extent of reduction) of the oxidized sample/background (RABS) was lower than the reduced samples (D05-16 and D08-16) (Figure 2a); the Fe(II) content of D08-16 is 1.3x D05-16 and 2.5x RABS. It's acid extractable S and U contents (Figure 2b-c), however, was significantly lower than the reduced samples.

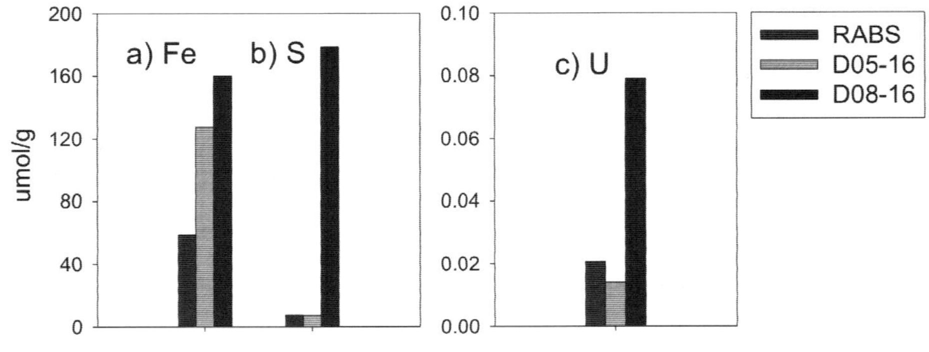

Figure 2: 0.5 N HNO_3 acid extractable a) Fe, b) S, and c) U of each of the three sediments.

The S and U contents of the D08-16 sample are significantly higher (and 10xS and 4xU) than those of the RABS and D05-16 sediments. The D05-16 and RABS sediments have similar U and S compositions. This trend is similar in all of the size fractions (<53-mm, in the present study; 53 to 106 μm and 106 to149 μm [6]; and >2-mm [9]). These data imply that the D08-16 zone may have experienced relatively more sulfate-reducing conditions but somewhat similar Fe-reducing conditions. The organic carbon (OC) content of D08-16 is 2-3x higher than that of D05-16, based on analysis of whole sediment [9]. The relatively higher OC content of the zone may have played a role in the extent of reduction.

3.2. Scanning Electron Microscopy

The backscattered electron image of the S-rich D08-16 sediment (highly reduced) is significantly different than the S-poor RABS (oxidized) and D05-16 (minimally reduced) sediments (Figure 3c). Framboidal pyrites are common in the S-rich sediment, unlike in the other two (Figures 3a,b).

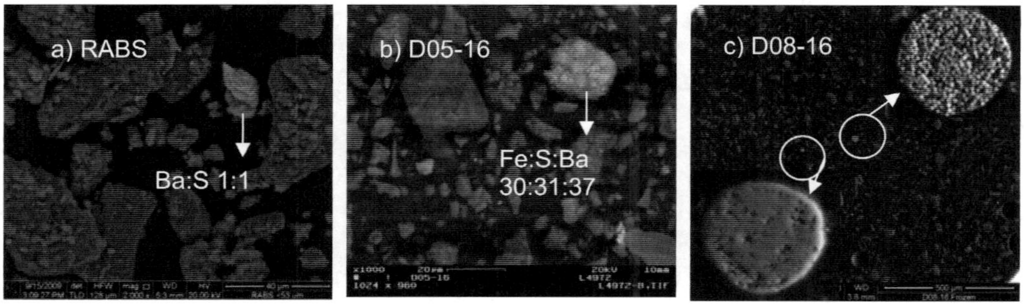

Figure 3. Low magnification SEM images of a) oxidized (RABS) showing Ba and S rich particles, b) minimally reduced (D05-16) with Fe, Ba, and S associations (as coatings) framboids), and c) highly reduced (D08-16) sediments showing frequent framboids (within white circles). The approximate atomic elemental composition (major elements) of the bright spots was obtained by energy dispersive x-ray analysis method. High magnification images of two representative framboidal pyrites of different microcrystal morphology and aggregation are shown in Figure 3c (inserts) to emphasize diverse biogeochemical conditions that exist in the zone and their probable effect on framboidal pyrite morphologies.

Framboidal pyrites are spherical or sub spherical aggregates of distinct microcrystals with distinct internal architecture (Figures 3c inserts). The major difference between the RABS and D05-16 sediments, despite similar extractable Fe(II) contents, appears to be in the residence of S; Fe-S-Ba aggregates as coatings (Figure 3b) and incipient Fe-sulfide framboids (not shown) in the D05-16 sample, although not common; while almost all of the S appears to be associated with Ba in the RABS (Figure 3a). Almost the entire sulfide in D08-16 is associated with Fe. The brightness of the spots in the micrographs is due to association of U with framboidal pyrite in D08-16 (Figure 3c), probable association of U with Fe-sulfide in D05-16 (Figure 3b), and Ba alone with or without little U in RABS (Figure 3a).

3.3. ^{57}Fe-Mössbauer spectroscopy

3.3.1. Oxidized/background (RABS) sediment. Mössbauer spectra of various size fractions were analyzed at different temperatures, but this paper focuses only on the <53-μm-fraction spectrum obtained at RT (Figure 4). The dominant Fe minerals in the sediment are different types of Fe-oxides and clays. From the Mössbauer parameters alone, it is not possible to distinguish the type and relative composition of various clays; muscovite, illite, and clinochlore are identified by x-ray diffraction (XRD) (Figure not included). We couldn't detect any XRD peaks due to crystalline Fe-oxide(s). Various oxides, such as, hematite (10%), near stoichiometric "well-crystalline" magnetite (11%), and RT-blocked oxide (Al-substituted Fe-oxides); 31%, contribute to the spectrum. In addition to these Fe-oxide phases, substantial amounts of amorphous/superparamagnetic (SP) is present in this sample, based on a 12-K spectrum of <2-mm size fraction [5]. Thus, the Fe(III) doublet (30%) centered around 0-1.5 mm/sec is a mix of clay Fe(III) and SP Fe-oxide. Finally, magnetite's Mössbauer features are similar to that of a well-crystalline/large particle (>50 nm sample [11]). Biogenic magnetites, derived from transformation of Fe-oxides by DMRB, are small

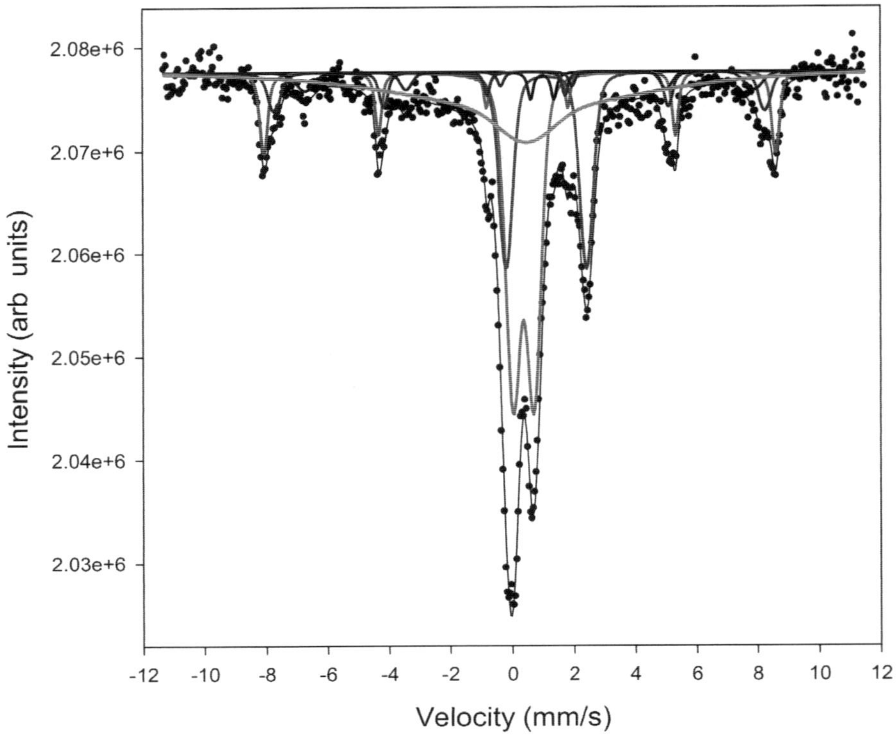

Figure 4: RT Mössbauer spectrum of RABS: Legend: green = hematite; brown = magnetite Fe(III) from octahedral/tetrahedral sites (assuming non-stoichiometry); dark blue = magnetite Fe2.5+ from octahedral sites; red = combination peaks of clay Fe(III) + SP Fe-oxide; blue = clay Fe(II); and grey = RT-blocked Fe-oxide.

(5-10 nm) and display a significantly different RT Mössbauer spectrum than that of pedogenic magnetites [12].

3.3.2. Minimally reduced sediment (D05-16). The RT Mössbauer spectrum of the D05-16 sample (Figure 5a) is significantly different from the background (RABS) sample, which is in agreement with the chemical data; it shows a higher extractable Fe/S ratio. The main differences are lower amounts of RT-blocked Fe (22% *vs.* 31%) and higher central doublet (0 to 1.5-mm/s region) intensity (36% *vs.* 30%). Thus, the central doublet is a combination peak of clay Fe(III), SP Fe-oxide, and low-spin, diamagnetic Fe-sulfide compounds [pyrite (FeS_2) and mackinawite $(Fe,Ni)S_{0.9}$)]. This is because Fe(II) in these Fe-sulfides and Fe(III) in clay and SP Fe-oxide exhibit slightly different RT Mössbauer parameters, which make it rather difficult to model unambiguously their relative contributions to the spectrum [13]. Partial assignment of the central doublet feature to these Fe-sulfides is in agreement with 1) the presence of Fe-sulfide particles (Figure 3b) and higher

Fe/S mole ratio than the oxidized (Figure 2a,b), and 2) indirectly from the derived clay Fe(II)/central doublet intensity ratio. The derived clay Fe(II)/central doublet ratio is lower than the oxidized (RABS), 0.5 *vs.* 0.6, which is not expected for a reduced clay. Clay Fe(III) is readily bioavailable under field biostimulated conditions [14,15]. Low-temperature applied field Mössbauer studies that are in progress may be useful if they distinguish peaks of these Fe-sulfides from clay Fe(III).

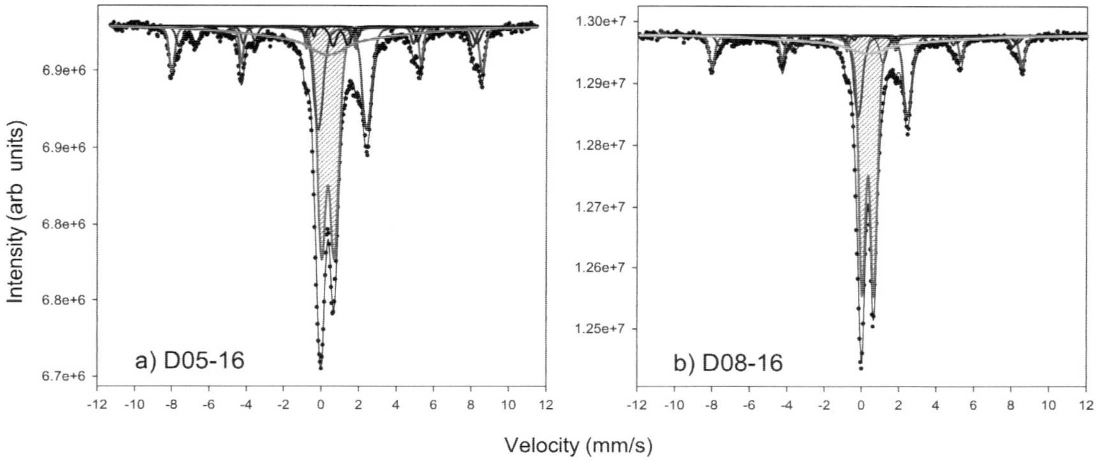

Figure 5: Room temperature Mössbauer spectra of a) minimally reduced (DO5-16) and b) highly reduced (D08-16). Legends are identical to Figure 1 expect the red pattern with lines is a combination peak of clay Fe(III) + SP Fe-oxide + Pyrite.

3.3.3. Highly reduced sediment (D08-16). This sample was extensively studied [9]. Fe-oxides (particularly magnetite) and pyrites, which are known U sorbents, are common in this sediment. Extensive reduction affected the Mössbauer spectrum in two ways (Figure 5b). It differs from the RABS and D05-16 sediments mainly in its lower RT-blocked Fe(III)-oxide content (18% *vs.* 22% of D05-16 and 31% of RABS), and its higher central doublet peak intensity (45% *vs.* 36% and 30% in D05-16 and RABS, respectively). Lower RT-blocked Fe-oxide content is in agreement with its redox status. Komlos *et al.*, [5] have shown the gradual loss of Fe-oxide [with simultaneous reduction of clay Fe(III)] upon biostimulation. As argued earlier, the lower clay Fe(II)/central doublet ratio is due to accumulation of Fe-sulfides (0.3 *vs.* 0.5 in D05-16). Framboidal pyrites (and occasional euhedral pyrites) and near stoichiometric magnetites (not included in this study) with surface sorbed U are common in this sediment; U-pyrite interactions are reported in [6]. Fe-S particles with stoichiometry similar to mackinwaite are also noted in some micrographs, *albeit* not common. Thus, the doublet centered around 1-1.5 mm/sec (Figure 5b) is a combination peak of Fe(II) clay, SP Fe-oxide, and predominantly pyrite. The tabulated mole ratio of Fe/S (Figure 2) is 1.8 is significantly higher than that expected if the extracted Fe(II) is coming from Fe-sulfides alone (0.5 in pyrite, and 1 in mackinawite). This discrepancy may be due to extraction of Fe(II) from clays; acid readily dissolves Fe(II) from the reduced clays [14]. Finally, the Mössbauer parameters of the high-spin Fe(II) doublet (assigned to clay in Figure 4) are significantly different from siderite ($FeCO_3$), a phase anticipated under the described conditions. This assignment is also in agreement with its 12 K Mössbauer spectrum (not included), which is free of siderite sextet; siderite magnetically orders below <35 K [13], and peaks due to griegite (Fe_3S_4) sextet, a metastable intermediate product of mackinwaite transformation to pyrite [16], are absent.

4. Conclusions

An understanding of field biogeochemical process is required to engineer long-term immobilization of groundwater contaminants. Bioavailability/biotransformation of Fe(III)-oxides and Fe(III) in clays and their subsequent role in reduction/attenuation of redox sensitive contaminants, e.g., U(VI), is a key element of this process. The role of the Fe mineral composition of the sediment, groundwater chemistry, and native microbial community composition and dynamics on such transformations would be significant [17]. The comparison of Fe mineralogy of the sediments obtained from various zones—oxidized and from reduced zones that appear to be reduced under Fe-only and/or Fe-and sulfate-reducing conditions—provided insights into the reduction/removal of Fe-oxides and precipitation of pyrite. Sulfate reducing conditions appear to be present in the naturally bioreduced zone and have resulted in loss of Fe-oxide abundance. Precipitation of sulfide apparently plays an important role in sequestering U under these conditions [6] but we also note the possible importance of mixed-valence Fe oxides (e.g. magnetite) in sequestering U. Detailed examination of Fe-U interactions of sediments from column experiments with little or no sulfate in pore water [5] and where reduction of phyllosilicate Fe(III) was noted, may provide additional information on the role of Fe minerals in U attenuation.

Acknowledgments
Ravi Kukkadapu and Nikolla Qafoku equally worked on this research and it was funded as part of the Integrated Field Research Challenge site at Rifle, CO, USA, supported by the U.S. DOE, Office of Science's Environmental Remediation Science Program (ERSP). Pacific Northwest National Laboratory (PNNL) is operated for the DOE by Battelle Memorial Institute under the Contract DE-AC06-76RLO 1830. Scanning electron microscopy and Mössbauer spectroscopy measurements were conducted at Environmental Molecular Sciences Laboratory (EMSL), a national scientific user facility sponsored by DOE's Office of Biological and Environmental Research program. EMSL is located at the PNNL in Richland, WA, USA.

References

[1] N'Guessan A, Vrionis H A, Resh C T, Long P E and Lovley D R 2008 Sustained removal of uranium from contaminated groundwater following stimulation of dissimilatory metal reduction *Environ. Sci. Technol.* **42** 2999-3004

[2] Anderson R T, Vrionis H A, Ortiz-Bernad I, Resch C T, Long P E, Dayvault R, Karp K, Marutzky S, Metzler D R, Peacock A, White D C, Lowe M and Lovley D R 2003Stimulating the in situ activity of geobacter species to remove uranium from the groundwater of a uranium-contaminated aquifer *Appl. Envir. Microbiol.* **69** 5884-91

[3] Abdelouas A, Lutze W and Nuttall E 1998 Chemical reactions of uranium in ground water at a mill tailings site *J. Contam. Hydrol.* **34** 343-61

[4] Mohanty S R, Kollah, B, Hedrick D B, Peacock A D, Kukkadapu, R K and Roden E E 2008 Biogeochemical processes in ethanol stimulated uranium-contaminated subsurface sediments *Environ. Sci. Technol.* **42** 4384-90

[5] Komlos J, Peacock A, Kukkadapu, R K and Jaffe P R 2008 Long-term dynamics of uranium reduction/reoxidation under low sulfate conditions *Gechim.et Cosmochim. Acta* **72** 3603-15

[6] Qafoku N P, Kukkadapu R K, McKinley J P, Arey B W, Kelly S D, Wang C, Resch C T and Long P E 2009 Uranium in framboidal pyrite from a naturally bioreduced alluvial sediment *Environ. Sci. Technol.* **43** 8528-8534

[7] Scott T B, Riba Tort, O and Allen G C 2007 Aqueous uptake of uranium onto pyrite surfaces; reactivity of fresh versus weathered material *Gechim.et Cosmochim. Acta* **71** 5044-53

[8] Hua B. and Deng B 2008 Reductive Immobilization of Uranium(VI) by amorphous iron sulfide *Environ. Sci. Technol.* **41** 8703-8

[9] Cambell K, Qafoku N P, Kukkadapu R K, Williams K H, Lesher E, Figueror L, Peacock A, Wilkins M J, Davis J, A, Icenhwoer and Long P. E. Characterizing the extent and role of natural subsurface bioreduction in a uranium-contaminated aquifer *Goldschmid*t 2008, July 13-13 (2008).

[10] Rancourt D J and Ping J Y 1991 Voigt-based methods for arbitrary-shape static hyperfine parameter distribution of mossbaur spectroscopy *Nucl. Instrum. Meth. Phys. Res. B* **58** 85-7

[11] Goya G F, Berquo T S, Fonseca F and Morales M P 2003 Static and dynamic magnetic properties of spherical magnetite nanoparticles *J. Appl. Phys.* **94** 3520

[12] Kukkadapu R K, Zachara J M, Fredrickson J K and Kennedy D W 2004 Biotransformation of two-line silica-ferrihydrite by a dissimilatory Fe(III)-reducing bacterium: formation of carbonate green rust in the presence of phosphate *Gechim.et Cosmochim. Acta* **68** 2799-8114

[13] Murad E and Cashion J 2004 Mossbauer Spectroscopy of Environmental Materials and Their Industrial Utilization Kluwer Academic Publishers.

[14] Kukkadapu R K, Zachara J M, Fredrickson J K, McKinley J P, Kennedy D W, Smith S C and Dong H 2006 Reductive biotransformation of Fe in shale-limestone saprolite containing Fe(III) oxides and Fe(II)/Fe(III) phyllosilicates *Gechim.et Cosmochim. Acta* **70** 3662-76

[15] Stucki J W, Lee K, Goodman B A and Kostka J E 2007 Effects of in situ biostimulation on iron mineral speciation in a sub-surface soil *Gechim.et Cosmochim. Acta* **71** 835-43

[16] Posfai M, Buseck P R, Bazylinski D A and Frankel R B 1998 Reaction sequence of iron sulfide minerals in bacteria and their use as biomarkers *Science* **280** 880-3

[17] Fang Y, Yabusaki S B, Morrison S J, Amonette J P and Long P E 2009 Multicomponent reactive transport modeling of uranium bioremediation field experiments *Geochim.et Cosmochim. Acta* **73** 6029-51

Iron speciation in bleached rocks by hydrocarbon leaching in Dushanzi Mud Volcano, NW China

G D Zheng[1,2*], B H Fu[2], A Kuno[3] and M Matsuo[2]

[1] Key Laboratory of Petroleum Resources Research, Institute of Geology and Geophysics, Chinese Academy of Sciences, Lanzhou 730000, PR China
[2] Key Laboratory of Earth's Deep Interior, Institute of Geology and Geography, Chinese Academy of Sciences, Beijing 100029, PR China
[3] Graduate School of Arts and Sciences, The University of Tokyo, Komaba 3-8-1, Meguro-ku, Tokyo 153-8902, Japan

E-mail: gdzhuk@hotmail.com

Abstract. Mud volcano is a typical petroleum leaching system, which normally result in bleaching effect to surrounding rocks. The result of Mössbauer spectroscopy of rock samples collected from the Dushanzi mud volcano system revealed clear variations in iron species. Compared to the unbleached reddish sedimentary rocks, the bleached rocks are dominated by reducing iron species such as paramagnetic ferrous iron (*para*-Fe^{2+}), ferrous iron in siderite (*sid*-Fe^{2+}), sulphide and pyrite (*pyr*-Fe^{2+}) whereas the original reddish rock is enriched in ferric iron including iron in hematite (*hem*-Fe^{3+}) and paramagnetic ferric iron (*para*-Fe^{3+}). A reduction of ferric iron species and hydrolysis of iron along with oxidation of hydrocarbons should be one of the main processes along bleaching by hydrocarbons to rocks.

1. Introduction

The occurrence of oil and gas related mud volcanoes normally result in sharp variations of rocks properties such as surface colours and structures, especially the colour change of rocks from bright reddish to green, deep gray or black known as bleaching effect. Mud volcanoes that produce bleaching effect on rocks in and around the mudflow craters and vents, can serve as a direct indicator of the characteristics of petroleum reservoirs. The changes of rocks mineral composition and chemical transformation of elements, especially the redox sensitive elements such as iron, caused by the migrating hydrocarbons can be utilized in surface exploitation and environmental issues. However, only limited information is available on the chemical speciation of iron-bearing minerals, probably due to limitations of technology. It is not so easy to get information on chemical species of iron either qualitatively or quantitatively in solid materials such as soils, sedimentary rocks by routine or typically chemical techniques. As one of few available methods, Mössbauer Effect offers a potential for obtaining information about the proportion of ferrous and ferric iron species in both crystalline and amorphous materials and has been applied in a wide range of scientific fields, including the iron chemistry of sediments (Manning. and Ash, 1979; Mørup et al., 1985; Hilton et al., 1986; Zheng et al., 2001). In this paper, iron species of bleached rock samples collected from the Dushanzi mud volcano were investigated for their iron speciation in order to understand the bleaching effect on the reddish sedimentary beds by released hydrocarbons in an inland basin under a very arid climate.

* To whom any correspondence should be addressed.

2. Samples and experiment

The Dushanzi mud volcano is located on the northern piedmont of the Tianshan Mountains, which are standing as the southern margin of the Junggar Basin in Xinjiang Uighur Autonomous Region of northwest China. The Junggar Basin is one of the largest Mesozoic-Cenozoic sedimentary basins enriched in fossil fuels including coal, crude oil and natural gas in China. Because of continuous northward thrusting of the Tianshan Mountains, there occurred several zones of tectonic folds in front of the mountains. Dushanzi anticline is one part of the fourth zone lied from west to east. The Dushanzi mud volcano occurs in the northern side of the Dushanzi anticline, which appears as a single yellowish hill with an elevation of about 940 meters above sea level. A normal fault was found along the crest of the Dushanzi anticline, which provides pathways for eruption of fluids and mud. The Tertiary formation of red beds at surface was bleached because of hydrocarbon-induced alteration (Fu et al., 2007).

Seven samples (DSZ-01 to DSZ-07) with various colours were collected from the bleached rocks around the mud craters. A reddish rock sample, as numbered as DSZ-00, was also collected from the original sedimentary sequence far from the hydrocarbon craters for comparison. The sample DSZ-07 with strong gasoline smell was freeze dried whereas other samples are dry. All samples were crushed into powder with an agate mortar and pestle in the laboratory.

For ^{57}Fe Mössbauer measurement, 250-335 mg of the powdered sample without any chemical pretreatment was gently pressed into a brass sample holder (16 mm in diameter, 1 mm thick). The sample holder was closed at both ends with iron-free plastic tap. The Mössbauer spectra were obtained with an Austin Science S-600 Mössbauer spectrometer using a γ-ray source of 1.11 GBq ^{57}Co/Rh at a constant temperature of either 293 K(room temperature) or 78 K (liquid nitrogen temperature). The obtained spectra were fitted to Lorentzian lineshapes using standard line shape fitting routines. Half-width (HW) and peak intensity of each quadruple doublet was constrained to be equal. Isomer shifts (IS) were expressed with respect to the centroid of the spectrum of metallic iron foil.

3. Results and discussion

3.1 Characterization of iron speciation

The ^{57}Fe Mössbauer spectra of all the samples at room temperature (RT, constant at 293 K) and liquid nitrogen (LN) temperature (78 K) were shown in Fig. 1 and Fig. 2, respectively. Several overlapped doublets and sextets were observed in these spectra. The curve fitting with iron components was robust because the Chi-squared values were sufficiently small. For the spectra obtained at RT (Fig 1), two doublets with a smaller quadruple splitting were ascribed to either paramagnetic high-spin ferric iron (*para*-Fe^{3+}) or iron in pyrite (*pyr*-Fe^{2+}). The former, *para*-Fe^{3+}, probably originated from clay minerals and/or hydrated Fe(III) oxides (Manning and Ash, 1979). The *para*-Fe^{3+} in the samples DSZ-01 through DSZ-07 was most likely in the form of hydroxides whereas the *para*-Fe^{3+} in the sampleDSZ-00 was mainly due to ferric iron in clay minerals such as smectite based on the Mössbauer parameters and the XRD pattern (data are not shown).

The ferrous iron in pyrite was fitted using the following Mössbauer parameters constrained by values of isomer shift (IS) 0.307 mm/s and quadruple splitting (QS) 0.610 mm/s for the Mössbauer spectra at 293 K. Pyrite was considered absent when the calculated peak area of *pyr*-Fe^{2+} was lower than the detection limit, and calculated as 3 times larger than the standard deviation of the baseline count (Kuno et al., 1998). The doublets with a larger quadruple splitting were attributable to paramagnetic low-spin and high-spin ferrous iron, *para*-Fe^{2+}(inner) and *para*-Fe^{2+}(outer), respectively. According to their Mössbauer parameters, the *para*-Fe^{2+}(outer) with higher IS, QS, and HW might be the iron in clay minerals and/or oxide whereas the *para*-Fe^{2+}(inner) with lower IS, QS and HW might be the ferrous iron in siderite (Ram, et al., 1998; John et al., 2004; Medina et al., 2006).

For the sextets, at least two types of iron minerals could be identified as magnetite and pyrrhotite according to their RT Mössbauer parameters. Compared with the Mössbauer spectra at RT(Fig 1) and LN(Fig 2), there was some variation in the samples DSZ-01, 03 and 05 that displayed as a doublet at RT but a sextet with about 48.3-4 Tesla at LN temperature, indicating a special iron species other than very fine particles of hematite. This iron species may be the super-paramagnetic ferric iron, because the magnetic field of the sextet in these three samples appeared to be small. Existence of an admixture of ferric hydroxides or ferric oxides with various degree of hydration was estimated based on the difference of the parameters between RT and LN Mössbauer spectra.

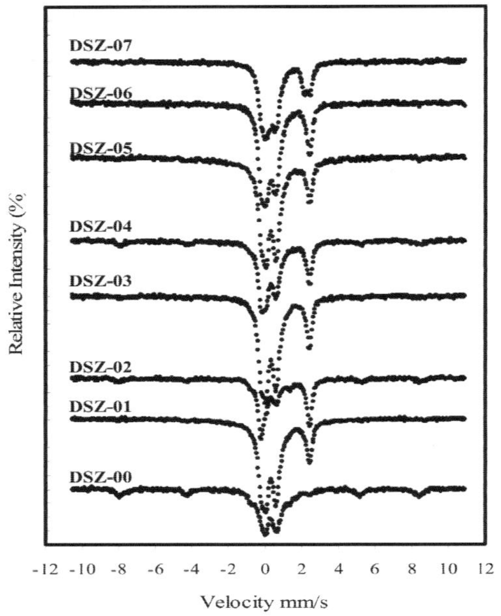

Figure 1. Mössbauer spectra of samples from Dushanzi mud volcano, measured at room temperature (298 K)

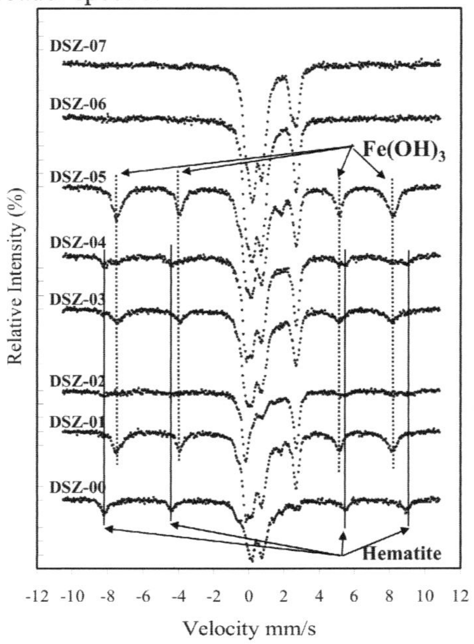

Figure 2. Mössbauer spectra of samples from Dushanzi mud volcano measured at liquid nitrogen temperature (78 K)

With the presence of various iron species among the bleached and unbleached rocks, reduction of ferric iron species into ferrous ones appeared to be the major geochemical process of iron in the rocks with petroleum bleaching. In addition, iron reduction processes appeared to follow certain sequences; probably the reduction of hematite was the earliest reaction, and then goethite and magnetite followed. At the same time, the reduction of ferric iron in smectite to ferrous iron in chlorite would be the main mechanism in the conversion of clay minerals from smectite to chlorite and illite. In addition, the hydrolysis of ferric iron compounds could be also one important step of geochemical reactions during the bleaching processes.

3.2 Significance to reservoirs properties

The inorganic-organic interactions have been considered to be important geochemical reactions for the change of petroleum reservoirs (Spiro, 1984; Mayer, 1994; Larter and Aplin, 1995) because the mineralogical and geochemical changes may give strong impact on the properties of reservoir rocks, including porosity and permeability. Reservoir properties such as the porosity and permeability may change with the formation of secondary minerals as well as the dissolution of original minerals such as quartz, feldspar and other silicate minerals. These interactions could lead to the enlargement of fractures and pores of the oil-gas reservoirs along with the depletion of minerals, especially the dissolution of felsic minerals such as quartz and feldspar. On the other hand, the decreasing of the porosity and permeability of reservoir rocks could be caused by the precipitation of secondary minerals and newly formed cements. Siderite, pyrite and other ferrous iron-bearing compounds as revealed in this study may contribute much to the properties change of petroleum reservoirs, and also the cover layers as the same.

The newly formed minerals during the bleaching process, including clay, silicate and various carbonate minerals, could change the porosity and permeability of rocks. Bleaching process could significantly affect the reservoir and capping layers with low porosity and low permeability, such as the Triassic rocks in the Erdos Basin in northwest China. In some cases, the porosity of a reservoir might be reduced until their fractures and holes were closed, and the reservoir layer became a cap layer under some specific conditions. On the other hand, increase in the porosity and permeability might occur in some reservoirs by the dissolution of minerals due to the hydrocarbons-mineral interactions during oil and gas migration and accumulation. In such cases, the petrologic features of the reservoir could be changed and even converted to different types of formations such as the dolomitized carbonate reservoirs for sour gas deposits in the Sichuan Basin, SW China (Ma et al., 2008). All those changes could be investigated by Mössbauer Effect as ME is effective to almost all iron species in solid materials and iron speciation is one of the major process along with hydrocarbons bleaching.

4. Conclusions

Chemical speciation of iron was greatly influenced by the releasing of hydrocarbons in the Dushanzi mud volcano system. Along with bleaching, a reduction of iron was the primary process in oil-gas bleaching to the rocks. Almost all ferric iron species in the original reddish rocks were reduced into ferrous species such as the iron in siderite, chlorite, even sulphide and pyrite associated with reduced sulphur species.

Acknowledgements

G.Z. gratefully acknowledged the support of K.C. Wong Education Foundation, Hong Kong to attend the ICAME2009 in Vienna. This study was financially supported as the "100-Talent" program from Chinese Academy of Sciences in 2006-2009.

References

[1] Fu BH, Zheng GD, Ninomiya Y, Wang CY and Sun G 2007 *Terra Nova* **19** 225
[2] Hilton J, Long GJ, Chapman JS and Lishman JP 1986 *Geochim. Cosmochim. Acta* **50** 2147
[3] Kuno A, Matsuo M and Takano B 1998 *Hyperine Interatcions* **C3** 328
[4] Larter SR and Aplin AC 1995 In: Cubitt JM and England WA, eds, *The Geochemistry of Reservoirs* **86**, Geological Society Special Publication, pp. 5–32
[5] Ma YS, Zhang SC, Guo TL, Zhu GY, Cai XY and Li MW 2008 *Marine and Petroleum Geology* **25** 357
[6] Manning PG and Ash LA 1970 *Canadian Mineralogist* **17** 111
[7] Mayer LM 1994 *Chemical Geology* **114** 347
[8] Medina G, Tabares JA, Alcázar PGA and Barraza JM 2006 *Fuel* **85** 871
[9] Mørup S, Franck J, van Wonterghem J, RoyPoulsen H and Larsen L 1985 *Fuel* **64** 528
[10] Ram S, Patel KR, Sharma SK and Tripathi RP 1998 *Fuel* **77** 1507
[11] Spiro B 1984 *Organic Geochemistry* **6** 543
[12] Zachara JM, Kukkadapu RK, Gassman PL, Dohnalkova A, Fredrickson JK and Anderson T 2004 *Geochim Cosmochim Acta* **68** 1791
[13] Zheng GD, Takano B, Kuno A and Matsuo M 2001 *Applied Geochemistry* **16** 1201

Journal of Physics: Conference Series **217** (2010) 012049 doi:10.1088/1742-6596/217/1/012049

Mössbauer spectroscopic study of some iron and antimony – containing minerals

R Bayliss[1], F J Berry[1,2,4], A Bowden[2], C Greaves[1] and M F Thomas[3]

[1] School of Chemistry, The University of Birmingham, Edgbaston, Birmingham, B15 2TT, United Kingdom

[2] Department of Chemistry, The Open University, Walton Hall, Milton Keynes, MK7 6AA, United Kingdom

[3]Department of Physics, The University of Liverpool, Liverpool L69 3BX

[4]To whom any correspondence should be addressed : f.j.berry@open.ac.uk

Abstract Iron-57 Mössbauer spectra have been recorded from three minerals containing both iron and antimony. Schafarzikite of composition $FeSb_2O_4$ contains Fe^{2+}. The ^{121}Sb Mössbauer spectrum shows only the presence of Sb^{3+}. The ^{57}Fe Mössbauer spectrum corresponds with that recorded from a material of identical composition synthesised by a solid state reaction during the course of this work. Apuanite of formulation $Fe_{20}Sb_{16}O_{48}S_4$ contains both Fe^{2+} and Fe^{3+} in the ratio 1:3.35 The result is consistent with crystal structure determinations and the formulation of apuanite as $Fe_4^{2+}Fe_{16}^{3+}Sb_{16}O_{48}S_4$. Versiliaite of composition $Fe_{12}Sb_{12}O_{32}S_2$ contains $Fe^{2+}$and Fe^{3+} in the ratio 1:2.12 and, also consistent with structural characterisations, can be formulated $Fe_4^{2+}Fe_8^{3+}Sb_{12}^{3+}O_{32}S_2$.

1. Introduction

The application of Mössbauer spectroscopy for the characterisation of minerals has been known for many years [1]. We are beginning a programme of synthesis and characterisation of inorganic solids containing iron and antimony and have become aware of minerals containing these elements which appear to have received sparse attention in the past. As a preliminary to the main body of work we have examined, for the first time, the ^{57}Fe Mössbauer spectra from three minerals containing iron and antimony.

Schafarzikite was first identified in the Slovak Republic in 1921 [2] and was subsequently shown to adopt the composition $FeSb_2O_4$[3]. The mineral has been characterised by X-ray diffraction and described [4] in terms of columns of edge–sharing iron-containing octahedra connected by corner sharing to parallel chains of three-coordinate pyramidal antimony. The mineral is rare but after no findings for over eighty years has recently been found near Pernek also in the Slovak Republic [5]. Schafarzikite is the parent phase of apuanite, $Fe_{20}Sb_{16}O_{48}S_4$, and versiliate, $Fe_{12}Sb_{12}O_{32}S_2$, found in the Apuan Alps in Italy. Apuanite and versiliaite are derived from schafarzikite by substitution on some antimony sites by iron [6,7] and the insertion of sulphide bridges between these iron ions. We report here on the characteristion of the iron oxidation states in schafarzikite and a synthetic analogue made during the course of this work and in apuanite and versiliaite by Mössbauer spectroscopy.

2. Experimental

Schafarzikite from the Kriznica mine, Pernek, Slovak Republic [5] was obtained from Systematic Mineralogy, Czech Republic. Apuanite and versiliaite from the Bucca della Vena Mine, Alpuan Alps, Italy [6,7] were obtained from Roger's Minerals, Canada. A sample of $FeSb_2O_4$ was prepared by heating a stoichiometric mixture of α– iron(III) oxide and antimony(III) oxide in air at 925^{0}C for 24 hours to produce iron antimonate ($FeSbO_4$). The iron antimonate was heated with an appropriate amount of elemental antimony in an evacuated sealed silica tube at 600^{0}C for 24 hours. The product was ground and heated again in an evacuated sealed silica tube at 600^{0}C for 24 hours to give single phase $FeSb_2O_4$.

Mössbauer spectra were recorded at 298K (^{57}Fe) and 77K (^{121}Sb) with a constant acceleration spectrometer in transmission geometry using a 400 MBq $^{57}Co/Rh$ and a 0.6 mCi $Ca^{121m}SnO_3$ source. The drive velocity was calibrated with the $^{57}Co/Rh$ source and a natural iron foil. All the chemical isomer shift data are reported relative to that of metallic iron at 298K.

3. Results and discussion

3.1 Schafarzikite

The ^{57}Fe Mössbauer spectra recorded from schafarzikite, apuanite, versiliate and the synthesised $FeSb_2O_4$ are shown in Figures 1-3. The ^{57}Fe Mössbauer parameters are collected in Table 1.

The ^{57}Fe Mössbauer spectrum recorded here from the recently discovered sample of schafarzikite [5] (Figure 1a) was dominated by a quadrupole split absorption characteristic of Fe^{2+}. The ^{121}Sb Mössbauer spectrum showed a resonance, δ= -11.84 mms^{-1}, charactaristic of Sb^{3+}. The result is consistent with the structural description of earlier–[4] and recently–[5] found samples of schafarzikite from which chains of FeO_6 octahedra, held together by chains of 3-coordinate antimony in which the lone pairs of electrons on Sb^{3+} point into the channels, can be envisaged. A doublet corresponding to Fe^{3+} was observed in the spectrum but, given that schafarzkite is found with other iron-containing minerals [2,5] ,we associate this with a minority impurity phase. Indeed X-ray powder diffraction showed weak reflections not attributable to schafarzikite but which could not be unequivocally associated with another Fe^{3+}-bearing mineral.

The spectrum recorded from the synthesised material of composition $FeSb_2O_4$ (Figure 1b) showed a single quadrupole split absorption with a chemical isomer shift characteristic of Fe^{2+}.

3.2 Apuanite

The ^{57}Fe Mössbauer spectrum is shown in Figure 2. The spectrum was best fitted to two components characteristic of Fe^{2+} and Fe^{3+}. The peak area ratio of the Fe^{2+} to Fe^{3+} components was *ca.* 1: 3.35 (Table 1). The result validates the structural description [6,7] in terms of the substitution of every third antimony cation in the pyramidal chains of schafarzikite by an Fe^{3+} ion with pairs of these ions being linked by the sulphide anions; charge balance is achieved by oxidation of some Fe^{2+} within the chains. The results recorded here are consistent with the formulation of apuanite as $Fe_4^{2+}Fe_{16}^{3+}Sb_{16}^{3+}O_{48}S_4$

Journal of Physics: Conference Series **217** (2010) 012049 doi:10.1088/1742-6596/217/1/012049

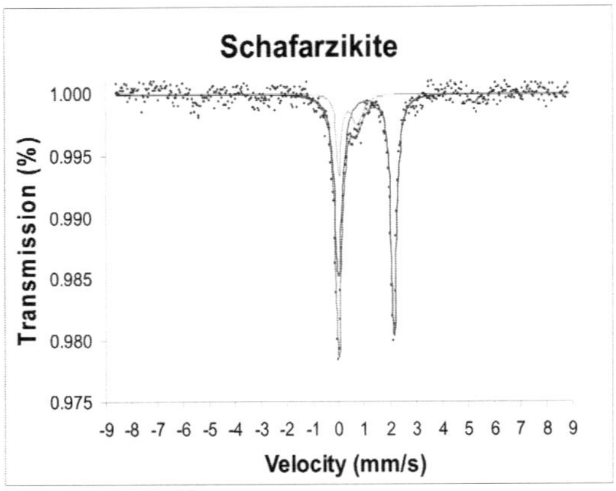

Figure 1a. ^{57}Fe Mössbauer spectrum recorded from schafarzikite

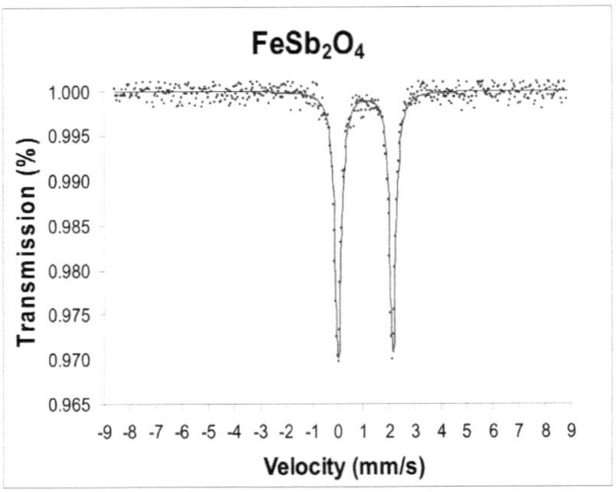

Figure 1b. ^{57}Fe Mössbauer spectrum recorded from $FeSb_2O_4$

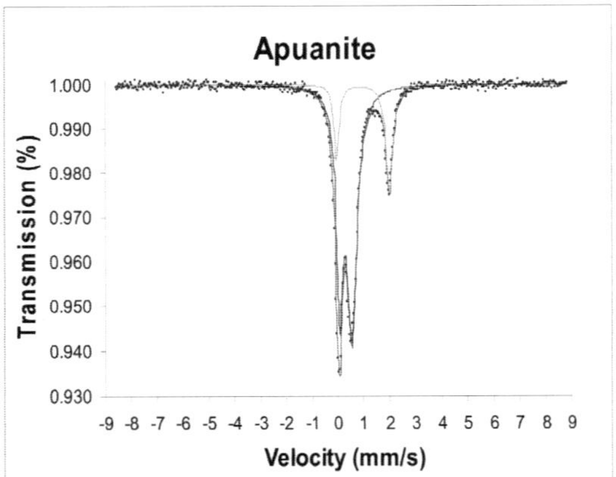

Figure 2. ^{57}Fe Mössbauer spectrum recorded from apuanite

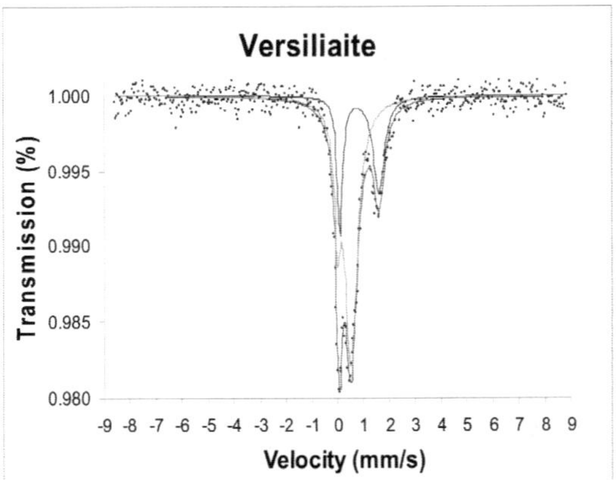

Figure 3. ^{57}Fe Mössbauer spectrum recorded from versiliaite

3.3 Versiliaite

The ^{57}Fe Mössbauer spectrum is shown in Figure 3. The spectrum was again fitted to two components characteristic of Fe^{2+} and Fe^{3+} but with a peak area ratio of 1:2.12 (Table 1). This result is also consistent with the structural description of versiliate [6, 7] involving the substitution of every fourth Sb^{3+} cation in the pyramidal chains of schafarzikite by an iron ion and the insertion of sulphide anions with corresponding oxidation of Fe^{2+}. The results recorded here are consistent with the formulation of versiliaite as $Fe_4^{2+}Fe_8^{3+}Sb_{12}^{3+}O_{32}S_2$.

Table 1. ^{57}Fe Mössbauer parameters recorded at 298K from schafarzikite-related structures.

Sample		$\delta \pm 0.02/mms^{-1}$	$\Delta \pm 0.02/mms^{-1}$	Area ±2 %
Schafarzikite	Fe^{2+}	1.08	2.13	80
	Fe^{3+}	0.38	0.69	20
$FeSb_2O_4$	Fe^{2+}	1.08	2.11	100
Apuanite	Fe^{2+}	0.99	2.03	23
	Fe^{3+}	0.32	0.47	77
Versiliaite	Fe^{2+}	0.85	1.50	32
	Fe^{3+}	0.25	0.51	68

4. Conclusions

A material of composition $FeSb_2O_4$ containing only Fe^{2+} identical to that of the mineral schafarzikite has been prepared. ^{57}Fe Mössbauer spectroscopy shows the minerals apuanite and versiliaite to contain both Fe^{2+} and Fe^{3+} consistent with predictions from crystal structure determinations.

5. References

[1] Bancroft G M 1973 *Mössbauer Spectroscopy: An Introduction for Inorganic Chemists and Geochemisty* Maidenhead: McGraw-Hill
[2] Krenner J A 1921 *Z. Kristallogr.* **56** 198
[3] Zemann J 1951 *Tschermaks Mineral. Petrog. Mitt.* **2** 166
[4] Fischer R and Pertlik F 1975 *Tschermaks Mineral. Petrog.Mitt.* 22, 236
[5] Sejkora J, Ozdin D, Vitalos J, Tucek P, Cejka J and Duda R. 2007 *Eur. J. Mineral.* **19** 419
[6] Melini M, Merlino S and Orlandi P *Amer. Min.* **64** 1230
[7] Mellini M and Merlino S 1979 *Amer. Min.* **64** 1235

Journal of Physics: Conference Series **217** (2010) 012050 doi:10.1088/1742-6596/217/1/012050

Study of oxygen fugacity influence on redox state of iron in granitoidic melts

M V Volovetsky[1,3], V S Rusakov[2], O A Lukanin[1] and A A Kargaltsev[1]

[1] Vernadsky Institute of Geochemistry and Analytical Chemistry, Russian Academy of Sciences 19, Kosygin Str, 119991 Moscow, Russia
[2] Lomonosov Moscow State University, Department of Physics, Vorobyovy Gory, 119899, Moscow, Russia

E-mail: volovetsky@gmail.com

Abstract. A series of experiments has been carried out on two granitoid melt compositions under the wide range of oxygen fugacities. Room temperature ^{57}Fe Mössbauer spectroscopy has been used to study the structural and oxidation state of Fe in granitoid composition glasses. Samples were produced in a series of melting experiments at temperatures from 1120 to 1420 °C and oxygen fugacities from $10^{-0.7}$ (air) to 10^{-12} (CCO buffer) bars. The analysis has shown that Fe^{3+}/Fe^{2+} ratio in the melt considerably decreases with fO_2 decreasing and/or temperature increasing. Dependence of Fe^{3+}/Fe^{2+} ratio on fO_2 is linear in logarithmic scale with slope 0.18-0.24. Ferrous iron occupies octahedral and probably five-fold positions and is regarded as a network-modifier. Ferric iron coordination depends on iron redox state. Its structural role changes from network-modifier to network-former.

1. Introduction

Iron is an important natural probe that is present in more than one oxidation state in silicate minerals and glasses. The redox state of iron influences differentiation trends in crystallizing silicate melts. Besides, ferric and ferrous cations play different role in the structure of melt, so ferric-ferrous ratio influences the physical properties of melts.

The redox state of iron in melts depends on some external and internal parameters: temperature, oxygen fugacity and composition. The iron oxidation reaction in silicate melts can be expressed by the reaction:

$$4FeO + O_2 = 2Fe_2O_3 \quad (1)$$

with the equilibrium constant

$$K = [a_{FeO}](fO_2)^{1/4}/[a_{FeO_{1.5}}], \quad (2)$$

where a_i are the activities of the species and fO_2 – oxygen fugacity.

If we suppose activity coefficients to be equal and take logarithms of both sides of the equation (2) we find

$$\log(FeO/FeO_{1.5}) = -0.25\log(fO_2) + \log(K) \quad (3)$$

[3] To whom any correspondence should be addressed.

Thus in the case of ideal mixing ferric and ferrous components relation between logarithms of Fe redox ratio and oxygen fugacity is expressed by the straight line with the slope of 0.25.

The knowledge of influence of external and internal parameters on ferric-ferrous equilibrium in silicate melts could give an opportunity to reconstruct the conditions of acid rocks forming. This question has been investigated by many authors. However no one of the proposed equations describes experimental results well in a wide range of compositions, temperatures and fO_2 [1]. Moreover most of them did not consider acid melts in a wide range of temperatures and oxygen fugacities [2].

2. Samples and experimental methods

Two series of samples were prepared for melting experiments: 1) pantellerite glass – P9 (71% SiO_2); 2) melted granite – I2 (73% SiO_2) (table 1). P9 series have larger alkali and Fe content. Such compositions were chosen because of their resemblance to acid volcanic glasses and tektites compositions. Tektites are natural silicate glasses which were formed under very high temperature (≥2000 °C) and probably low oxygen fugacity [3].

Table 1. Chemical composition of some experimental samples: P9- and I2-series.

Conditions (T, t, log(fO_2))	Sample	SiO2	TiO2	Al2O3	FeO	MnO	MgO	CaO	Na2O	K2O	Cl	ZrO2	Total
Initial	P9	68.28	0.36	7.39	7.82	0.31	0.07	0.34	6.61	4.17	0.81	0.33	96.48
1102 °C, 46h, -0.7	P9-N2	70.00	0.36	7.71	7.60	0.28	0.09	0.37	6.05	4.13	0.70	0.36	97.64
1420 °C, 15h, -5.7	P9-37	63.69	0.35	20.77	4.05	0.36	0.22	0.35	3.52	3.85	0.12	0.31	97.81
Initial	I2	70.46	0.06	15.28	3.41		0.14	2.06	4.07	3.95			99.48
1240 °C, 20h, -7.0	I2-N15	73.53	0.06	14.15	2.98		0.14	1.62	3.76	3.87			100.13
1420 °C, 7h, -5.7	I3R-29	74.22	0.06	13.03	2.84		0.13	1.55	2.06	3.77			97.70

Starting materials were melted in a vertical muffle furnace under different T-fO_2 conditions: T = 1120-1420 °C and -log(fO_2) = 0.7-12. Oxygen fugacity was imposed by different gas mixings: 1) pure air; 2) air+N_2; 3) CO_2+H_2. About 500 mg of ground starting materials was used for one experiment. Alumina crucibles were used as a sample containers. Run durations were 2-15 hours at temperature about 1420 C° and up to 50 hours at lower temperatures. Samples were quenched by immersing them into water after they have been drawn from the furnace.

Starting materials and obtained samples were studied using electron microprobe to investigate chemical changes during the experiments.

Mössbauer samples were prepared by grinding glasses under ethanol to avoid some oxidation prior to analisys, then were mounted in plexiglas holders. Room temperature Mössbauer spectra show broadened, slightly asymmetric quadrupole split doublets. Some spectra show contribution of magnetic phases. Spectra were analyzed in terms of two independent quadrupole splitting distributions corresponding to Fe^{3+} and Fe^{2+} sites. It was supposed that there was the positive correlation of hyperfine parameters (δ and ε) for Fe^{2+}. When the contribution of magnetic phases was essential additional distribution function of hyperfine magnetic field was calculated.

3. Results and discussion

Microprobe analysis has shown that some quantity of Na volatilizes during melting. For more viscous I2-series the volatilization was revealed at 1420 °C. But for the P9-series the volatilization of Na has been revealed at any used temperatures. There was a strong influence of alumina crucible on the melts at 1420 °C for these glasses. It led to the enrichment of Al and the non-homogeneity of glass (table 1).

Mössbauer spectra of the most of studied glasses are paramagnetic ones with broad lines and at least two subspectra. Experimental spectra have been analyzed in terms of two quadrupole splitting distributions corresponding to Fe^{3+} and Fe^{2+} sites. Each distribution shows one peak (figure 1).

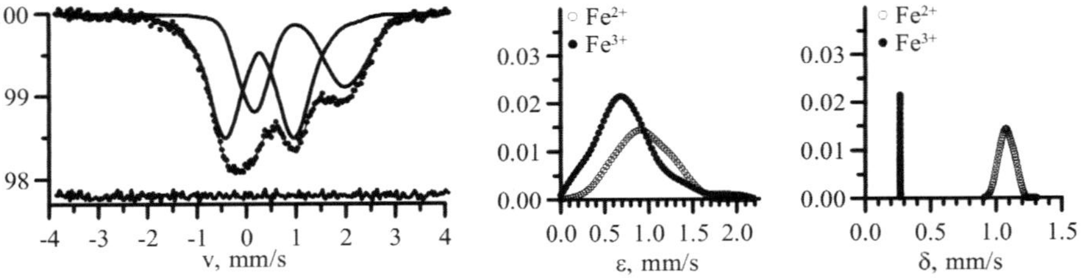

Figure 1. Typical Mössbauer spectra of glass and distribution funcions of hyperfine parameters (sample from I2-series; T=1140 °C, $fO_2=10^{-6.1}$, 10 hours).

At the lower temperature and oxidative atmosphere (T=1140 °C, $fO_2=10^{-0.7}$ bar) there was observed crystallization of oxide magnetic phases in the samples of I2 series. At the same time there were not any oxide phases in the series P9 in remarkable quantities. The series I2 is more viscous and its liquidus probably lies above 1140 °C.

The rates of redox reactions in investigated melts are strongly dependent on temperature. At T = 1420 °C about 5 hours is enough to attain eqilibrium between glass and furnace atmosphere. At the same time at T = 1220 °C we could not reach the eqilibrium for 50 hours even for less viscous melt of P9-series.

Oxygen fugacity was varied in the range from $10^{-0.7}$ (air) to 10^{-12} (near CCO buffer: graphite-CO-CO_2) bar. Even in the most oxidative atmosphere some quantity of Fe^{2+} remained. The Fe^{3+}/Fe^{2+} ratio in the melt considerably decreases with fO_2 decreasing and temperature increasing according to the following equation [4] (figure 2).

$$\log(Fe^{3+}/Fe^{2+}) = a\cdot\log(fO_2) + b/T + c \tag{4}$$

In the figure there are given approximation equations for experimental data. It should be noted that the slope of lines is in the range 0.17-0.24 and is smaller than theoretical value for ideal ferric-ferrous mix (0.25).

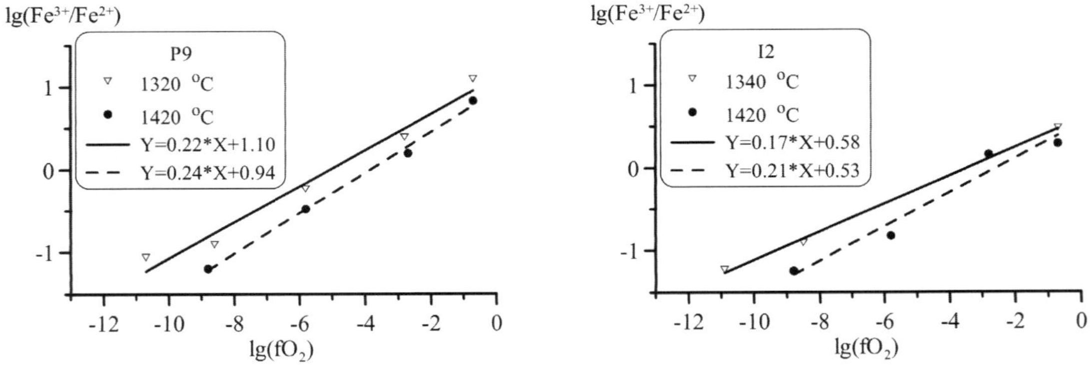

Figure 2. Relationship between logarithms of Fe^{3+}/Fe^{2+} ratio and fO_2. In legends there are given equations of approximation lines.

It is difficult to obtain temperature dependence in a wider range due to kinetic limitations at the bottom side and instrumental limitations at the upper side. Thus we can not reach temperatures of the formation of tektites and volcanic glasses.

To reach extremely reducing conditions we used closed graphite crucible. So fO_2 in this system was near CCO buffer. Under these very reducing conditions (about 10^{-12} bar) at temperature 1420 °C some quantity of metallic iron segregated in visible spherules and almost all iron in glass was divalent (some quantity of Fe^{3+} remained).

Mössbauer line shifts of ferric iron change from values typical for octahedral oxygen coordination to those for tetrahedral coordination with increasing in its content. The transitional border lies at Fe^{3+} content near 50 at.%Fe (figure 3). Probably there is a coordination transition from octahedral to tetrahedral positions with increasing in ferric iron content. Such transition was noted by Mysen et al. for Na-, Ca-, Mg-alumosilicate glasses [5].

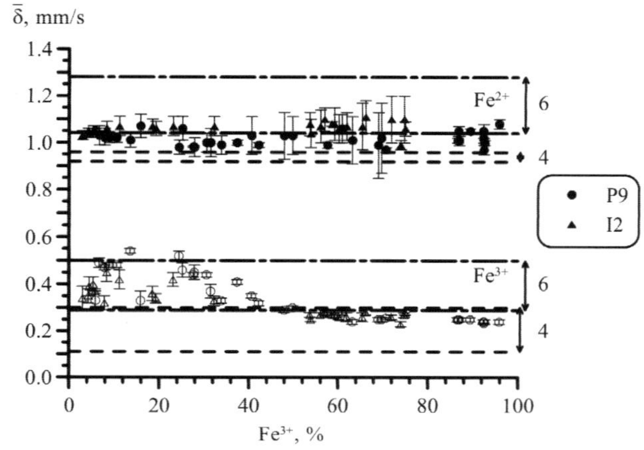

Figure 3. Relation between δ and Fe^{3+} content. Typical shift ranges for crystalline samples are taken from Menyl [6].

It is worth that shifts of Fe^{3+} in reduced granitoid glasses are the same as in the case of tektites [3]. It points to similar structure and probably similar thermal history.

Mössbauer line shifts of ferrous ions lie in relatively narrow regions. Values for I2 series lie in regions typical for octahedral and five-fold sites. Values for P9 are between regions typical for octahedral and tetrahedral sites. These values are independent of Fe^{3+}/Fe^{2+} ratio (figure 3). We assigned these values to distorted octahedral sites. Ferrous iron in silicate is usually regarded as network-modifier and assigned to octahedral positions.

4. Conclusions

Using Mössbauer spectroscopy it has been shown that relation between Fe^{3+}/Fe^{2+} ratio and oxygen fugacity is linear in logarithm scales according to the equation (4). However slope of the lines are 0.17-0.21 for I2-series and 0.22-0.24 for P9-series. They deviate from theoretically predicted value 0.25. This deviation may be caused by non-ideality of the mixing ferric and ferrous components.

Ferrous iron occupies octahedral and probably five-fold positions and is regarded as network-modifier. Ferric iron coordination changes from octahedral to tetrahedral with increasing Fe^{3+}/Fe^{2+} ratio. This transition occurs at Fe^{3+} content about 50 at.%Fe.

References

[1] Partzsch G M, Lattard D and McCammon C 2004 *Contrib. Mineral. Petrol.* **147** 565
[2] Borisov A A and Shapkin A I 1990 *Geochem. Int.* **27** 111
[3] Volovetsky M V, Rusakov V S, Chistyakova N I and Lukanin O A 2008 *Hyperfine Interactions* **186** 83
[4] Sack R O, Carmichael I S E, Rivers M and Ghiorso M S 1980 *Contrib. Mineral. Petrol.* **75** 369
[5] Virgo D, Mysen B O 1985 *Phys. Chem. Minerals.* **12** 65
[6] Menil F 1985 *J. Phys. Chem. Solids.* **46** 763

Journal of Physics: Conference Series **217** (2010) 012051 doi:10.1088/1742-6596/217/1/012051

Changes of pyrite and pyrrhotite in coal upon microwave treatment

F B Waanders[1], W Mohamed[2] and N J Wagner[2]

[1]North-West University, Potchefstroom, 2520, South Africa
[2]University of the Witwatersrand, Johannesburg, South Africa
frans.waanders@nwu.ac.za

Abstract. More stringent environmental regulations and the demand for higher quality final products require optimum sulphur removal from sulphur-containing ores. Microwave treatment is an emerging technique in minerals processing and may be used to reduce sulphur-containing minerals. In the present study two coal samples containing low and high amounts of sulphur, respectively, were subjected to microwave and caustic soda treatment with a view to desulphurization. In order to assess the microwave treatment, the same procedures were applied to a pyrite/pyrrhotite rock obtained from a coal mine. XRD and Mössbauer analyses were performed, before and after treatment in order to determine changes brought about in the structure and bonding. Following the treatment, both the low and high-sulphur coal samples showed a decrease in sulphur content of ≈ 40%, whilst the XRD and Mössbauer analyses did not detect the formation of new minerals or changes in structure. In the case of the pyrite / pyrrhotite rock sample the Mössbauer investigations showed the presence of pyrite (26%) and pyrrhotite (74%) After treatment the concentrations changed to 8% and 92% respectively, indicating a change in sulphur mineral composition during treatment, but no structural changes.

1. Introduction

Coal consists mainly of hydrocarbons and also contains inorganic constituents, which are mainly minerals, incorporated in various ways in the coal. The amounts of these minerals, which are the inert solid material, vary considerably in coals of diverse origin and remain behind in an altered form as ash after coal combustion or gasification [1]. Clays, quartz and carbonate minerals and the iron-containing minerals pyrite (FeS_2), jarosite ($KFe_3(SO_4)_2(OH)_6$), troilite and pyrrhotite (both FeS) and the non-sulphur-containing iron minerals ankerite ($CaFe(CO_3)_2$), illite (clay mineral) and siderite ($FeCO_3$) [2] can occur in varying amounts in South African coals. The pyritic, organic and sulphatic sulphur minerals in the coal are difficult, and sometimes impossible, to remove [3]. The sulphur content is of great concern owing to its harmful effect on boilers where sulphates cause corrosion, its lowering of the quality of metallurgical coal and atmospheric pollution. The removal of sulphur from coal prior to utilization has, therefore, become a primary concern to industry.

Microwave applications have been extensively used in minerals processing, due to its intense energy flow and penetration depth. The primary advantage of microwave processing is the specificity of microwave energy absorption, allowing volumetric heating of materials [4]. Desulphurization of coal using high power microwave energy was previously investigated [5-7]. The addition of strong solutions of NaOH or KOH, and repeated microwave irradiation, yielded high sulphur removal values [8-12].

In the present work, two South African coals containing different amounts of sulphur were desulphurized by exposure to a strong alkali solution. Subsequently the samples were subjected to both XRD and Mössbauer analyses.

2. Experimental

Two high-volatile bituminous coals, with sulphur contents of 0.5% and 3.6%, respectively, were used in this study. The as-received coal samples were in the form of large lumps, requiring crushing to a size fraction of > 212μm. Proximate analyses were performed and the calorific values were determined. Typically, 15g of coal was mixed with a 300g/L of NaOH solution, with a solid to liquid ratio of 25:75. The solution was then microwave-treated at a microwave energy input of 650W for 10 minutes. After treatment the mixture was filtered and washed until all traces of the alkali were removed (red litmus paper). The extent of desulphurization was determined by sulphur analysis using a Leco analyzer. The changes in structure and bonding were determined by means of XRD and Mössbauer analyses, before and after treatment.

A rock sample containing pyrite and pyrrhotite was also milled to a desirable size fraction and further investigated to be able to gauge the reaction to microwave processing after NaOH solution treatment. The possible changes in structure and bonding were determined by means of XRD and Mössbauer analyses.

3. Results and Discussion

The results for the coals, containing high or low amounts of sulphur, are discussed in with reference to the "pyrite / pyrrhotite" containing rock sample.

3.1. High-sulphur coal

Milling liberated the pyrite particles, which were found to occur randomly in the coal matrix as nodules with a size > 25μm. The sulphur in this sample was mainly (97%) in a pyritic form with the balance occurring in the organic form. Following microwave treatment and exposure to the NaOH solution, a decrease in the sulphur content of ≈ 40% (see Table 1) was found. XRD (see Figure 1) and Mössbauer analysis (see Table 2 and Figure 2) did not show the formation of new minerals or a change in the crystallographic structure following the NaOH treatment. After combustion, negligible changes were noted in the ash content. The volatile matter and calorific value of the treated coal decreased slightly; therefore, the benefit of lower sulphur content needs to be weighed against the loss of calorific value. The doublet fitted in the Mössbauer spectrum is indicative of the pyrite in the sample, with no formation of pyrrhotite visible. The parameters are presented in Table 2.

3.2. Low-sulphur coal

The low-sulphur coal contained predominantly fine nodules of pyrite embedded in the coal structure making this coal more difficult to desulphurize. A decrease in sulphur of ≈ 40% was possible after the NaOH treatment (see Table 1), even though the sulphur content of the coal was low initially. XRD and Mössbauer analyses did not show the formation of new minerals or a change in the crystallographic structure following the NaOH treatment. Again, only a doublet was fitted in the Mössbauer spectrum, indicative of the pyrite in the sample, with no formation of pyrrhotite visible (see Table 2).

3.3. Pyrite/pyrrhotite rock sample

In the case of the pyrite / pyrrhotite rock sample the Mössbauer investigations showed the occurrence of pyrite (26%) and pyrrhotite (74%) as the sulphur-containing minerals. After NaOH and microwave treatment the intensities changed to 8% and 92% respectively, indicating a change in sulphur mineral composition during treatment. The XRD and Mössbauer spectra (see Figure 2) did not show new mineral formation and only a decrease in the amounts of each constituent was observed.

Table 1. Analysis of the > 212μm coal particle samples before and after treatment

	Proximate Analysis (wt%)			
	High S (≈ 3.3%) Before treatment	High S (≈ 3.3%) After treatment	Low S (≈ 0.5%) Before treatment	Low S (≈ 0.5%) After treatment
Moisture	4.0	7.8	4.2	6.8
Ash	18.5	17.9	24.9	24.8
Volatile Matter	32.0	27.6	22.1	19.7
Fixed Carbon	45.5	46.6	48.8	48.7
Total Sulphur	3.3	1.8	0.5	0.3
Calorific Value (MJ/kg)	24.6	22.4	21.1	20.5

The Mössbauer parameters are presented in Table 2 and are in agreement with literature [13].

Table 2. Mössbauer parameters of Fe components found in the samples in this investigation

Sample	Component	IS mms^{-1} (±0.001)	QS mms^{-1} (±0.001)	H Tesla (±0.3)	Relative intensity (%)
High S-coal untreated	Pyrite	0.27	0.59	-	100
High S-coal microwave and NaOH treated	Pyrite	0.26	0.59	-	100
Low S-coal untreated	Pyrite	0.27	0.59	-	100
Low S-coal microwave and NaOH treated	Pyrite	0.26	0.59	-	100
Pyrite / pyrrhotite untreated	Pyrite	0.30	0.59	-	26
	Pyrrhotite	0.68	-0.07	30.6	25
		0.72	0.15	23.8	49
Pyrite / pyrrhotite microwave and NaOH	Pyrite	0.27	0.60	-	8
	Pyrrhotite	0.65	-0.09	29.9	29
		0.63	0.13	23.7	63

Note: IS = Isomer shift relative to α-Fe, QS = Quadrupole splitting, H = Magnetic hyperfine field

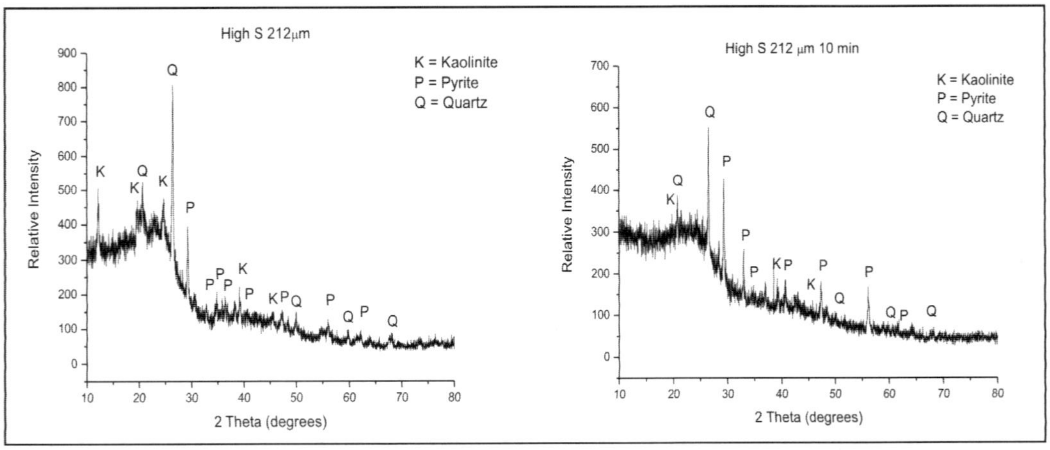

Figure 1: On the left the XRD scan of the >212μm particles before treatment and on the right the XRD scan after treatment

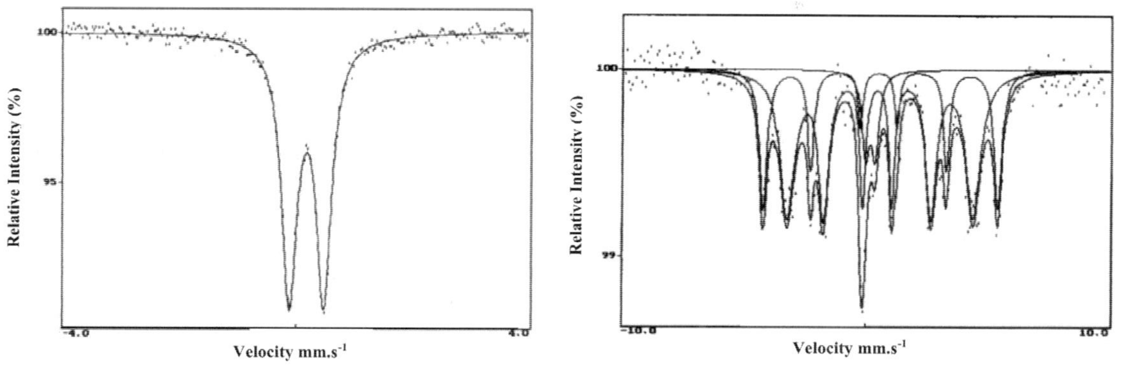

Figure 2. On the left is the Mössbauer spectrum of the pyrite in the coal, where the spectrum before and after treatment remained the same, whilst on the right the Mössbauer spectrum of the pyrite / pyrrhotite rock, after treatment, is shown. Only a decrease in the amounts of each constituent was observed.

4. Conclusion

The study was undertaken to determine whether desulphurization of two South African coals is possible using microwave power and an alkali solution treatment. It forms part of a larger investigation into the desulphurization methods of coals. Both the high and low-sulphur coals showed an approximate sulphur removal of 40%, with the volatile matter content decreasing by 14% and the calorific value decreasing slightly, impacting negatively on the coal quality. The uniform sulphur removal for the high and low-sulphur coals indicates that the form of pyrite in the coal does not affect desulphurization, as was also observed for the pyrite / pyrrhotite sample. Mössbauer spectroscopy showed no change in the Fe-S structure.

5. Acknowledgement

Special thanks to Sasol through the Sasol Hub-and-Spoke initiative and the National Research Foundation for financial support.

References

[1] Waanders F B, Vinken E, Mans A and Mulaba-Bafubiandi A 2003 *Hyp. Int.* **148-149**, 21
[2] Clark T M, Evans B J, Wynter C, Pollak H, Taole, S and Radcliffe D 1989 *Hyp. Int.* **112**, 227
[3] Mehliss A T M 1987, *Department of Mineral and Energy Affairs Report* **9/87** 2
[4] Bykov Y V, Rybakov K I, Semenov V E 2000 *J. Phys. D: Applied Physics* **34** R55
[5] Zavitsanos P D and Golden J A 1982 *J. Appl. Phys.* **53** 2730
[6] Uslu T and Atalay Ü 2003 *Fuel Proc. Tech.* **85** 21
[7] Rowson N A and Rice N M 1990 *Min. Eng.***3 No** ¾ 363
[8] Hayashi J, Oku K, Kusakabe K, Morooka S 1990 *Fuel* **69** 739
[9] Araya P E, Badilla-Ohlbaum R, Droguett S E 1981 *Fuel* **60** 1127
[10] Mukherjee S and Borthakur P C 2001 *Fuel* **80** 2037
[11] Yürüm Y, Elsamak G G, Oztas N A 2002 *Fuel* **82** 531
[12] Andrés J M, Ferrando A C, Membrado L 1995 *Energy and Fuels* **10** 425
[13] Stevens J G, Khasanov A M, Miller J W, Pollak H and Li Z (eds.) *Mössbauer Mineral Handbook* Mössbauer Effect Data Centre University of North Carolina Baltimore Press Asheville North Carolina ISSN 0272-2755

Journal of Physics: Conference Series **217** (2010) 012052 doi:10.1088/1742-6596/217/1/012052

Comparison between the quasi-continuous quadrupole splitting distributions (QSD) for Mössbauer spectra of glauconites and the QSD-profiles simulated on the basis of crystal-chemical model

L G Dainyak[1], V S Rusakov[2], I A Sukhorukov[2], B B Zviagina[1], N I Chistiakova[2] and V A Drits[1]

[1] Geological Institute of the Russian Academy of Science, Pyzhevsky Street 7, 119017, Moscow, Russia

[2] Lomonosov Moscow State University, Department of Physics, Vorobyovy Gory, 119899, Moscow, Russia

E-mail: dainyak@ginras.ru

Abstract. Structural peculiarities of Fe^{3+}-rich dioctahedral mica, glauconite, such as a wide variety of isomorphous octahedral cation substitutions (Al, Fe^{3+}, Fe^{2+}, Mg) and the occurrence of these cations in *cis*-sites only, make it possible to consider glauconites as model locally inhomogeneous objects. To fit room and liquid-nitrogen Mössbauer spectra of two glauconite samples of close compositions with unusually low Fe^{3+}/Fe^{2+} ratios, reconstruction of quadrupole splitting distributions (QSD) in terms of quasi-continuous model-independent approach was applied. Interpretation of the obtained QSD^{exp}s became possible owing to application of a crystal-chemical model (CCM) taking account of the effects of the local structural and chemical heterogeneity of layer silicates. The main parameters of the CCM are individual quadrupole splittings, Δ_i, for Fe^{3+} and Fe^{2+} in different local cationic arrangements (LCAs) having the occurrence probabilities, w_i, provided by two-dimensional simulation of the cation distribution (CD). The agreement between CD reconstructions, the QSD^{sim} profiles simulated on the basis of the CCM parameters and the QSD^{exp}s provided, first, the corroboration of the Δ_i^{pred} values for Fe^{3+}; second, the refinement of the Δ_i^{ref} values for Fe^{2+}; and third, revealing the difference in the octahedral cation distributions for the studied glauconites.

1. Introduction

Glauconite belongs to Fe^{3+}-rich dioctahedral micas. Distribution of isomorphous octahedral cations (Fe^{3+}, Fe^{2+}, Al, Mg) is of great importance in studying diagenetical transformation and in geochronological dating. According to X-ray and oblique-texture electron diffractions, and EXAFS spectroscopy [1 and references therein], of the available structurally independent octahedral sites (two *cis*- and one *trans*-) only the *cis*-sites are occupied by cations. Taking account of Fe *cis*-occurrence in the local cationic arrangement (LCA) consisting of three nearest cations, also in *cis*-positions, broadened Mössbauer spectra of glauconites display an immediate effect of the LCAs around Fe on the electric field gradient.

The crystal-chemical model (CCM) for interpreting the traditional Lorentzian fittings of Mössbauer spectra of 2:1 dioctahedral *trans*-vacant micas has been developed in [2 and references therein]. The main parameters of the CCM are individual quadrupole splittings, Δ_i, for Fe^{3+} and Fe^{2+} in the different LCAs with their occurrence probabilities, w_i, provided by two-dimensional simulation of the octahedral cation distribution [3]. The Δ_i^{pred} values for Fe^{3+} in the range of 0.0 – 1.40 mm/s are obtained from the empirical equation [4]. The assignment of the tentative Δ_i^{tent} values for Fe^{2+} in the range of 1.20 – 2.96 mm/s to specific LCAs results from the correlation between the octahedral cation composition and the maximum quadrupole splitting Fe^{2+} extracted from Mössbauer spectra for a representative mineral collection [5].

The aim of the work is to apply quasi-continuous model-independent quadrupole splitting distribution analysis [6] for Mössbauer spectra of two glauconites, BSH and KUL, and to interpret the obtained QSD^{exp}s using the crystal-chemical model.

2. Samples and experimental results

Two samples of glauconites, BSH and KUL (mineralogical description is in [7]) with the crystal-chemical formulae $K_{0.74}Ca_{0.02}Mg_{0.04}[Si_{3.71}Al_{0.29}](Al_{1.06}Fe^{3+}_{0.36}Fe^{2+}_{0.25}Mg_{0.35})O_{10}(OH)_2$ and $K_{0.78}Ca_{0.01}Mg_{0.04}[Si_{3.71}Al_{0.29}](Al_{0.99}Fe^{3+}_{0.41}Fe^{2+}_{0.27}Mg_{0.33})O_{10}(OH)_2$, respectively, were investigated.

2.1. IR spectroscopy

The OH-stretching regions of the IR spectra recorded on a Nicolet ESP-260 FTIR spectrometer were decomposed using the Galactic PeakSolve ™ program. The assignment of the individual OH-stretching bands corresponding to specific types of OH-bonded cation pairs follows the band attributions suggested in [8]. The octahedral cation compositions of the samples calculated using relative integrated intensities of individual OH bands are in good agreement with crystal-chemical formulae.

2.2. Mössbauer spectroscopy

Room (figure 1) and liquid-nitrogen (RT and NT, respectively) spectra were measured on a constant acceleration spectrometer using a Co^{57}(Rh) source (source half-width Γ=0.105 mm/s) and calibrated with reference to α-Fe. Orientation effects were eliminated preparing the absorbers in a hollow cone form with 55° half-cone angle. The absorber thickness was less than 5 mg Fe/cm^2.

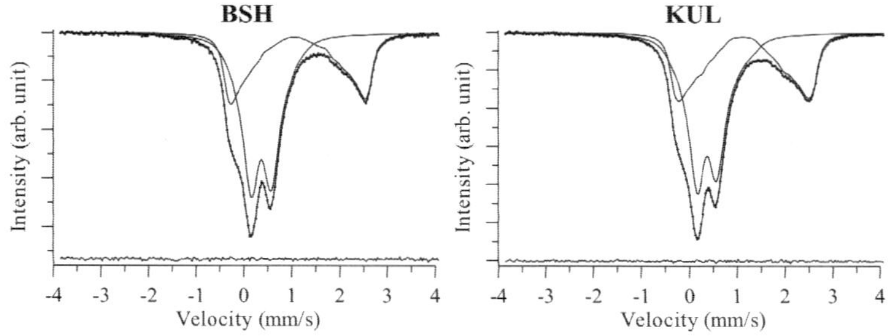

Figure 1. Quasi-continuous QSD fits of the RT spectra of glauconites BSH and KUL.

2.2.1. Quasi-continuous model-independent QSD analysis. The QSD^{exp}s, which were obtained with the DISTRI code from the software MSTools [6], are shown for Fe^{2+} in figure 2 (solid circles). Because the source line width Γ_s = 0.105 mm/s is close to the natural line width Γ_τ = 0.097 mm/s, a Lorentzian line shape for the resonance line has been used in the QSD reconstruction. The linear correlation coefficients for the main Fe^{3+} and Fe^{2+} contributions to the QSD^{exp}s correspond to the narrow variation ranges for isomer shifts: $\Delta\delta^{main}(Fe^{3+})\leq 0.03$ mm/s and $\Delta\delta^{main}(Fe^{2+})\leq 0.05$ mm/s.

A comparative analysis of the RT and NT data made it possible to choose the Δ range for the simulation of the RT QSD^{sim} profiles, and to identify the minor unsimulable $P(\Delta)$ values at ~1.6 mm/s for Fe^{3+} QSD as the effect of the boundary octahedra sharing only two edges with the nearest two octahedra [9] and at ~1.0 – 1.2 mm/s for Fe^{2+} QSD as the presence of some Fe-sulphate [10].

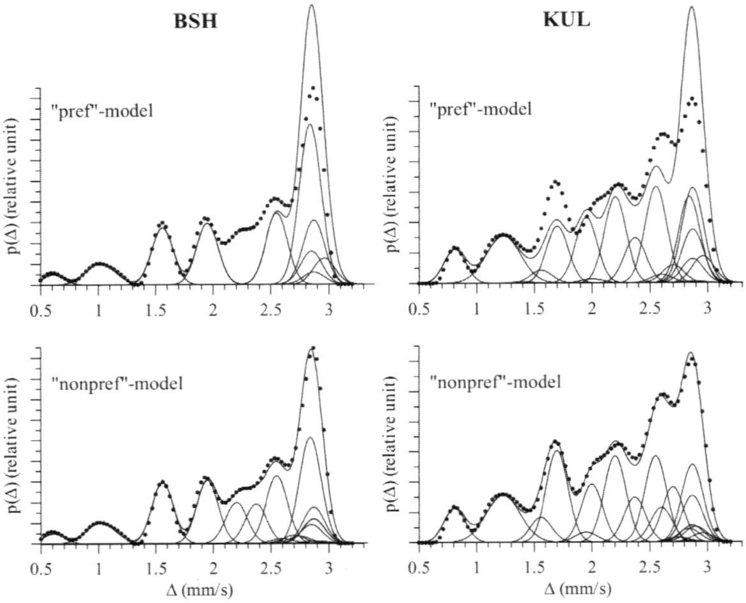

Figure 2. Comparison between experimental Fe^{2+} QSD^{exp}s (solid circles) and Fe^{2+} QSD^{sim} profiles (solid envelope) simulated using the refined Δ_i^{ref} values and the occurrence probabilities, w_i, for the LCAs corresponding to the "pref"- and "nonpref"-models used in CD reconstruction.

3. Comparison between QSD^{exp}s and QSD^{sim} profiles

3.1. Two-dimensional CD simulation

The CD simulation uses the integrated optical densities of individual bands, w_{ik}, where i and k are cations (Al, Fe^{3+}, Fe^{2+}, Mg) bound to the OH groups, as initial parameters. Such a technique yields a solution that satisfies the two spectroscopic methods. Because the w_{ik} values are proportional to the sum of occurrence probabilities, w_{ik} and w_{ki}, any cation may equally well occupy the M2 and M2' *cis*-sites or with variable preference to one of these sites. This feature, combined with the set of optional limitations along the directions running at ±120° or ±60° with respect to the OH-bonded cationic pair (*b*-direction), allows analyzing a wide variety of CD reconstructions in terms of the obtained IR data for a given sample.

3.2. CD models satisfying QSD^{exp}s for glauconites BSH and KUL

Using successive approximations during CD simulation, both for glauconites BSH and KUL, the preset individual occurrence probabilities, w_i, for Fe^{3+} and Fe^{2+} in different LCAs, as well as the Δ_i^{tent} values, were gradually changed so as to co-ordinate the CD reconstructions, the simulated QSD^{sim} profiles and the experimental QSD^{exp}. Table 1 shows Δ_i^{pred} for Fe^{3+}, the refined Δ_i^{ref} for Fe^{2+} and the sets of w_i values for Fe^{3+} and Fe^{2+} in specific LCAs corresponding to the preferable Fe^{2+},Mg-occurrence ("pref"-model) in one of *cis*-sites (columns 1) and to equally probable occupancy ("nonpref"-model) of *cis*-sites by di- and tri-valent cations (columns 2). The differences between the refined, Δ_i^{ref}, and tentative, Δ_i^{tent}, values do not exceed 0.1 mm/s, and Δ_i^{ref} sequence for Fe^{2+} corresponding to specific LCAs has not changed.

For Fe^{3+}, the contributions of cluster-forming LCAs ($2Fe^{3+}Al$ and $2Fe^{2+}Fe^{3+}$), which include three or four nearest neighboring Fe cations, are approximately the same for both models for the two samples providing good agreement between QSD^{exp}s and QSD^{sim} profiles (not shown). For Fe^{2+}, such agreement is only achieved with "nonpref"-model (figure 2, solid envelope). Table 1 shows that the main w_i values satisfying the QSD^{exp} for sample BSH, in contrast to sample KUL, belong to the LCAs which are homogeneous with respect to charge distribution ($3Fe^{3+}$, $2Fe^{3+}Al$, $2AlFe^{3+}$ and 3Al).

Table 1. Individual occurrence probabilities, w_i, for Fe^{3+} and Fe^{2+} in specific LCAs with preferable Fe^{2+},Mg-occupancy in one of two *cis*-sites (1) and with equally probable occupancy of *cis*-sites by di- and tri-valent cations (2).

LCAs	Δ_i^{pred},	For Fe^{3+} QSD^{sim} profile				Δ_i^{ref}	For Fe^{2+} QSD^{sim} profile			
		KUL		BSH			KUL		BSH	
	mm/s	1	2	1	2	mm/s	1	2	1	2
$3Fe^{2+}$	0.	-	0.025	0.003	-	1.20	-	-	-	-
$2Fe^{2+}Mg$	0.13	-	-	0.002	-	1.40	-	-	-	-
$3Fe^{3+}$	0.17	0.029	-	0.003	0.006	1.56	0.019	0.037	0.117	0.127
$2MgFe^{2+}$	0.26	-	-	0.009	0.012	1.76	0.083	0.135	-	-
$2Fe^{3+}Al$	0.33	0.338	0.314	0.323	0.301	1.95	0.096	0.015	0.126	0.131
$2Fe^{2+}Fe^{3+}$	0.34	0.141	0.160	0.112	0.12	2.00	-	0.085	-	-
3Mg	0.39	-	-	0.041	0.041	2.00	-	-	-	-
$MgFe^{2+}Fe^{3+}$	0.48	0.066	0.121	0.131	0.213	2.20	0.126	0.127	-	0.085
$2Fe^{2+}Al$	0.49	0.023	0.037	0.076	0.045	2.37	0.066	0.066	-	0.082
$2AlFe^{3+}$	0.51	0.172	0.128	0.123	0.060	2.55	0.141	0.127	0.147	0.140
$2MgFe^{3+}$	0.61	0.004	0.002	0.017	0.025	2.60	0.012	0.051	-	0.010
$AlMgFe^{2+}$	0.63	0.015	0.018	0.028	0.030	2.70	0.026	0.081	-	0.017
$2Fe^{3+}Fe^{2+}$	0.65	0.041	0.020	0.011	0.018	2.75	-	0.007	-	0.015
3Al	0.74	0.005	0.010	-	-	2.84	0.126	0.020	0.331	0.220
2MgAl	0.78	-	-	0.014	0.010	2.85	-	0.002	0.068	0.039
$2Fe^{3+}Mg$	0.79	0.022	0.013	0.008	0.010	2.87	0.035	0.025	0.025	0.004
$AlFe^{2+}Fe^{3+}$	0.86	0.059	0.047	-	0.015	2.87	0.078	0.114	-	0.051
$AlMgFe^{3+}$	1.02	0.042	0.035	0.040	0.035	2.87	0.139	0.068	0.132	0.075
$2AlFe^{2+}$	1.06	0.033	0.030	0.058	0.045	2.90	-	0.023	-	0.005
2alMg	1.40	0.009	0.004	-	0.029	2.96	0.039	0.013	0.054	-

4. Conclusion

- The CCM proved to be consistent with QSDs for Mössbauer spectra of two investigated glauconites, thus filling these QSDs with the crystal-chemical meaning.
- The high resolution of the quasi-continuous model-independent QSD analysis provided, first, corroboration of the Δ^{pred} = 0.0 mm/s for Fe^{3+} in the arrangement $3Fe^{2+}$ previously assumed as the minimum value in the Δ_i^{pred} sequence; and second, refinement of the Δ_i^{ref} set for Fe^{2+} in specific LCAs.
- For the CD reconstruction corresponding to the absence of preference of di-valent cations to one of the two *cis*-sites, the studied glauconites obey Pauling's principle of homogeneous dispersion of charge to different extents.

5. References

[1] Drits V A, McCarty D K and Zviagina B V 2006 *Clays Clay Minerals* **54** 132
[2] Dainyak L G, Drits V A and Lindgreen H 2004 *Eur. J. Mineral.* **16** 451
[3] Dainyak L G, Drits V A and Heifits L M 1992 *Clays Clay Minerals* **40** 470. Dainyak L G, Drits V A, Zviagina B B and Lindgreen H 2006 *American Mineralogist* **91** 589
[4] Drits V A, Dainyak L G, Muller F, Besson G and Manceau A 1997 *Clay Minerals* **32** 153
[5] Dainyak L G and Drits V A 2009 *Eur. J. Mineral.* **21** 99
[6] Rusakov V S 1999 *Bulletin Russian Academy of Science Physics* **7** 1093; Rusakov V S and Kadyrzhanov K K 2005 *Hyperfine Interactions* **164** 87. Rusakov V S 2000 Mössbauer spectroscopy of locally inhomogeneous compounds INPNNCRK, Almaty, 431
[7] Zaitseva T S, Gorokhov I M, Ivanovskaya T A, Semikhatov M A, Kuznetsov A B, Mel'nikov N N and Arakelyants M M 2008 *Stratigraphy and Geological Correlation* **16** 227
[8] Besson G and Drits V A 1997 *Clay Clay Minerals* **45** 158
[9] Dainyak L G, Zviagina B B, Rusakov V S and Drits V A 2006 *Eur. J. Mineral.* **18** 753
[10] Hikita T, Iton K, Okada N, Ohkubo N and Kodera E 1996 *J. Phys. Soc. Japan* **65** 1293

Various redox conditions in Boda Claystone as reflected in the change of Fe^{2+}/Fe^{3+} ratio in clay minerals

K Lázár[1], Z Máthé[2] and M Földvári[3]

[1] Institute of Isotopes, Budapest, P.O.B 77, H-1525, Hungary,
[2] Mecsekérc Environmental, Esztergár L. 19, Pécs, H-7633, Hungary
[3] Geological Institute of Hungary, Stefánia 14, Budapest, H-1143, Hungary

E-mail: lazar@iki.kfki.hu

Abstract. Fe^{2+} and Fe^{3+} species in layered clay minerals are analysed in samples obtained from two boreholes in Boda Siltstone Formation. In the first series occurrence of Fe^{2+} components and Fe^{2+} => Fe^{3+} oxidation is demonstrated in upper layers of a strata near to the surface. In the second series minerals are identified with iron present exclusively in ferrous form in a thin strata found in a borehole at 1050 m depth.

1. Introduction

Redox processes may influence the migration of certain radionuclides which are able to change their valency, and as a consequence, to change the charge of the migrating species. For example, reduction may proceed in the following processes: UO_2^{2+} <=> UO_2, $^{79}SeO_3^{2-}$ <=> ^{79}Se (0), $^{237}NpO_2^{+}$ <=> $^{237}NpO(OH)_2$, $^{99}TcO_4^{-}$ <=> $^{99}TcO_2$.

In turn, the opposite process of the previous reduction should also proceed in neighbouring media, first in pore water, then, in second stage, in host rock. In several cases the Fe^{2+} => Fe^{3+} oxidation is considered as the possible counterpart of the previously mentioned reductions.

Boda Siltstone Formation (BSF) is considered as a perspective media for disposal of nuclear waste in Hungary. Thus, the ability of the constituting mineral components to participate in the redox processes may also be evaluated. In general, overwhelming part of the Boda Formation was formed under strong oxidation conditions in strong alkaline media at arid/semi-arid conditions. In correspondence, dominant part of the iron is present as hematite [1]. Thus, further oxidation can hardly be expected.

In the present study Fe^{2+} and Fe^{3+} components are studied with Mössbauer spectroscopy in BSF samples. Two examples are discussed in detail. In the first one the secondary oxidation of Fe^{2+} to Fe^{3+} is demonstrated in layered minerals in regions close to the upper boundary of the Boda Formation in a borehole driven from the ground. In the second example samples obtained from a deep drill crossing a reductive zone below the surface at 1050 m are analysed, and the iron containing minerals are identified.

With the present study we would like to demonstrate that minor amounts of iron can be present in ferrous form in spite of the predominance of the oxidative conditions governing the genesis of most of the BSF. Furthermore, it is also shown that Fe^{2+} ions in layered clay minerals are accessible to $Fe^{2+} \rightarrow Fe^{3+}$ oxidation in natural environment. In addition, it is shown with the second case that regions may occur in the formation where iron exists exclusively in ferrous state.

2. Experimental

The studied samples were obtained from two different drills of the BSF. The first one, the vertical BAT-14 drill was bored from the ground. The claystone strata commences in 22.3 m depth. It is covered with Quaternary sediments (loess, redeposited loess and marl). The second set of samples was collected from another drill, Delta-9, bored in horizontal direction in a roadway at 1050 m depth below ground level. The mineral compositions of samples were analysed by X-ray diffraction or by thermogravimetry (TG) and differential thermal analysis. Only those minerals can be analysed with the latter methods which exhibit weight changes or thermal effects upon heating. The respective amounts of these thermally responding minerals can also be estimated as well. Iron-bearing minerals were analysed by Mössbauer spectroscopy as well.

3. Results

3.1. Fe^{2+} and Fe^{3+} bearing species in samples from top layers of a BSF strata

The relative amounts of thermally active minerals found in the studied samples are shown in table 1.

Mössbauer spectra were collected in two velocity ranges on the samples. The spectra recorded in the wider velocity range (± 12 mm/s) contain the contributions of all the iron bearing components, namely the whole sextet of hematite. The spectra recorded in the narrower range (± 4 mm/s) provide better resolution for identifying the respective components, however they show only the two inner lines of the hematite sextet. The corresponding spectra are shown in figure 1, and data are collected in table 1.

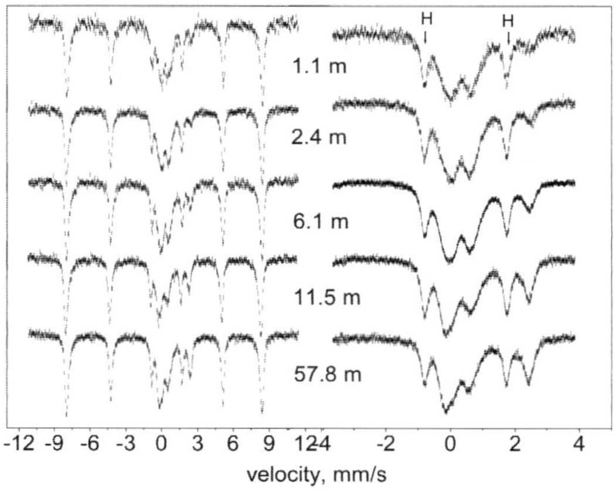

Figure 1. Mössbauer spectra of samples taken from different depths of BAT-14 borehole. Spectra are collected in two velocity ranges. (H marks the positions of the two central peaks of the sextet of hematite)

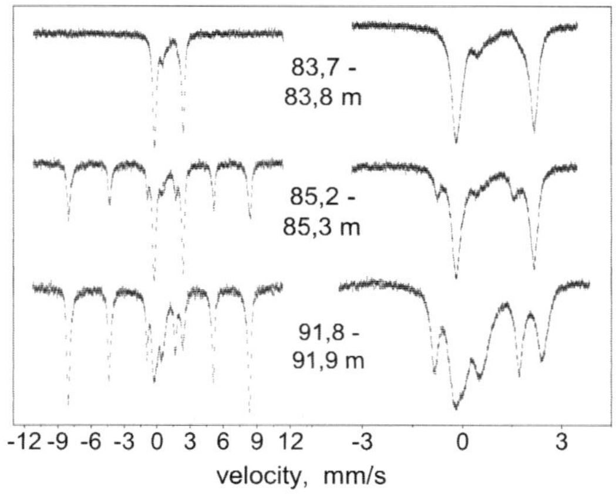

Figure 2. Mössbauer spectra of samples taken from different distances in the Delta-9 borehole drilled horizontally in 1050 m depth. Spectra are collected in two velocity ranges.

Assignment of hematite, amounting to 64 – 66 % of iron, is unambigous. Data published in the Mössbauer Mineral Handbook [2] can be considered for the identification of the components filling in the remaining 34 – 36 % spectral area. The IS = 1.1 and QS = 2.6 mm/s pair of data is reported for the characteristic Fe^{2+} positions in chlorite in most cases. However, in several studies IS = 1.2 and QS = 2.4 pair of data is also specified. These data are also characteristic for some other layered clay minerals (illite, montmorillonite). Thus, adopting these values, the Fe^{2+} doublet can probably be attributed to ferrous ions located in chlorite or in other clay minerals. Several positions are available for iron in chlorite – both Fe^{2+} and Fe^{3+} may be located in it [3]. The assignment of the Fe^{3+} components is similar – the Fe^{3+} (a) and Fe^{3+} (b) doublets may attest either the presence of chlorite or illite-muscovite as well [2].

Thus, in short, it can be concluded from the relevant RI data that the Fe^{2+} portion decreases in the clay minerals from 14 % to c.a. 4 % as approaching the top few meters of the Boda Formation. This can probably be attributed to a secondary oxidation in the layered clay mineral (the amount of the other main constituing mineral, hematite, practically does not vary). This phenomena demonstrates that $Fe^{2+} \rightarrow Fe^{3+}$ oxidation may proceed under natural conditions in Boda Claystone which may influence occasional migration processes as well.

Table 1. Mineral composition and the corresponding Mössbauer data obtained on samples from different depths of BAT-14 drill.

Depth[a]	Mineral compsition [b] (%)				Mössbauer data [c]				
(m)	Chl.	Ill/mus.	Mont.	Calc.	Component	IS	QS	MHF	RI
1.1	0.5	24	7	35	Fe^{2+}	1.26	2.39		4.0
					Fe^{3+} (a)	0.19	0.85		3.2
					Fe^{3+} (b)	0.33	0.73		27.0
					Fe^{3+} (hematite)	0.40	0.22	50.7	65.7
2.4	33	10	15	8	Fe^{2+}	1.30	2.37		3.9
					Fe^{3+} (a)	0.12	0.81		2.3
					Fe^{3+} (b)	0.32	0.72		28.0
					Fe^{3+} (hematite)	0.37	0.22	50.7	65.6
6.1	8	10	10	12	Fe^{2+}	1.27	2.36		7.7
					Fe^{3+} (a)	0.21	0.79		15.1
					Fe^{3+} (b)	0.43	0.78		13.6
					Fe^{3+} (hematite)	0.37	0.23	50.6	63.6
11.5	13	13	5	7	Fe^{2+}	1.26	2.35		10.5
					Fe^{3+} (a)	0.19	0.76		8.5
					Fe^{3+} (b)	0.33	0.98		12.5
					Fe^{3+} (hematite)	0.38	0.22	50.6	68.5
57.8	27	25	1	5	Fe^{2+}	1.27	2.32		14.1
					Fe^{3+} (a)	0.18	0.76		12.9
					Fe^{3+} (b)	0.32	0.95		8.5
					Fe^{3+} (hematite)	0.38	0.22	50.7	63.6

[a] The depth is measured from the upper boundary of Boda Claystone, which is covered with loess and marl in thickness amounting together to 22.3 m.

[b] The listed minerals are: Chl. – chlorite, Ill/m. – illite/muscovite, Mont. – montmorillonite, Calc. – calcite

[c] The respective data are: IS: isomer shift, relative to metallic iron, mm/s; QS: quadrupole splitting, mm/s; MHF: magnetic hyperfine field, Tesla; RI: relative intensity, %

3.2. Crossing a strata formed under reductive conditions

Reductive conditions were prevailing during the genesis of the formation in exceptional periods. This is clearly reflected in the state of the iron in the respective samples collected from a horizontal bore drilled in a depth at 1050 m. The occurrence of the reductive zone is estimated in a layer positioned in between the 83 – 85 m distance from the commencement of the drill.

Table 2. Mössbauer data extracted from spectra of figure 2.

Distance[a]	Component	IS	QS	MHF	RI
83.7 – 83.8	Fe^{2+} (chlorite)	1.13	2.62	-	70
	Fe^{2+}	0.87	2.34	-	7
	Fe^{2+} (pyrite)	0.34	0.63	-	23
85.2 – 85.3	Fe^{2+} (chlorite)	1.13	2.62	-	40
	Fe^{2+} (pyrite ?)[b]	0.37	0.59	-	14
	Fe^{3+} (hematite)	0.38	0.22	50.9	45
91.8 – 91.9	Fe^{2+} (chlorite)	1.12	2.67	-	15
	Fe^{2+}/Fe^{3+} [b]	0.35	0.58	-	24
	Fe^{3+} (hematite)	0.38	0.22	50.9	60

[a] The distance is measured from the commencement of the horizontal drill (in meters)
[b] The IS, QS parameters of Fe^{2+} in the pyrite (low spin state) and Fe^{3+}(high spin state) in silicate minerals can hardly be distinguished.

Data shown in table 2 and spectra of figure 2 attest that the Delta-9 boring crossed a zone which was primarily formed under reducing conditions. Hematite is not found by XRD at all in the sample taken from 83.2 m, and in correspondence, only ferrous iron is detected in the spectra of the sample taken from 83.7 m. (Reduction of hematite to form chlorite and pyrite in secondary diagenetic processes could hardly be assumed.) The IS, QS data in this sample are exactly the same as reported for chlorite [2], thus in this sample the presence of chlorite is evidenced. The overall Fe^{3+} / Fe^{2+} ratio characterizing the Boda Claystone is reversed in average by departing from this reductive layer located at 83 – 85 m, since the 60 % portion of ferric iron in hematite is restored in the sample taken at the ~ 92 m distance.

4. Conclusions

Two examples are presented to prove and analyse the occurence of Fe^{2+} ions in minerals of Boda Claystone. In the first case presence of the Fe^{2+} is demonstrated in a vertical borehole driven near to the ground level. In this instance, Fe^{2+} is present in minor amounts beside the preponderance of the hematite. $Fe^{2+} \rightarrow Fe^{3+}$ oxidation in the topmost region is also evidenced. In the second case an example is shown to demonstrate the occasional occurrence of a strata formed under reducing conditions. In this region – similarly to the previous example - most of the ferrous iron is present in layered clay mineral (chlorite). Thus, it can be assumed that the $Fe^{2+} \rightarrow Fe^{3+}$ oxidation may provide the counterpart process for reduction of certain migrating radionuclides, provided the redox equilibrium is mediated by the pore water.

Acknowledgements

This study was partly supported from the EU FP7 ReCosy (CP-212287) project. The courtesy of PURAM for providing the samples for characterisation is acknowledged.

References

[1] Árkai P, Balogh K, Demény A, Fórizs I, Nagy G and Máthé Z 2000 *Acta Geologica Hungarica* **43** 351
[2] Stevens J G, Khasanov A M, Miller J W, Pollak H and Li Z 1998 *Mössbauer Mineral Handbook* (Mössbauer Effect Data Center, Asheville, North Carolina)
[3] Zazzi A, Hirsch T K, Leonova E, Kaikkonen A, Grins J, Annersten H and Edén M 2006 *Clays and Clay Minerals* **54** 252

Journal of Physics: Conference Series **217** (2010) 012054 doi:10.1088/1742-6596/217/1/012054

Properties of iron sulphides from a copper mine in southern Brazil

Wagner N Mussel[1], Enver Murad[1], Natalie C Magalhães[1], Walter A P Abrahão[2], Jaime W V Mello[2] and José D Fabris[1]

[1]Departamento de Química, ICEx, Universidade Federal de Minas Gerais Campus – Pampulha, 31270-901 Belo Horizonte, Minas Gerais, Brazil

[2]Departamento de Ciência do Solo, Universidade Federal de Viçosa, Avenida Peter Henry Rolfs, Campus Universitário, 36570-000 Viçosa, Minas Gerais, Brazil

E-mail: wdmussel@ufmg.br

Abstract. Chemical analysis, X-ray diffraction and Mössbauer spectroscopy showed two iron sulphide samples from a copper mine in Camaquã, Rio Grande do Sul, Brazil, to consist of essentially pure pyrite and chalcopyrite associated with a minor amount of pyrite and possibly some cubanite. While the pyrite was well crystallized and of simple mineralogy, Mössbauer data indicated the chalcopyrite contained in the second sample to consist of coexisting tetragonal and (remnant) cubic modifications, as has been previously described for another sample from this locality.

1. Introduction

Pyrite, (FeS_2) is a very common mineral that can be found in virtually all geological environments, whereas marcasite, the polymorph of pyrite, occurs mainly in sedimentary settings. Marcasite is orthorhombic and often has a tabular morphology that is distinctly different from that of (cubic) pyrite. Chalcopyrite, $CuFeS_2$, is another common iron-bearing sulphide that is often associated with pyrite and marcasite. When exposed to the atmosphere, these minerals can rapidly oxidize, leading to the formation of iron sulphates and abundant sulphuric acid. Mining wastes containing iron sulphides may thus cause "acid mine waters" and associated precipitates, and thereby severe environmental problems.

Because of its sulphur content, pyrite is of commercial interest and is currently used mainly as a starting material for the production of sulphur dioxide and sulphuric acid. High-quality pyrite powder is also used in lithium metal sulphide batteries, moulding compounds (resin-bonded grind wheels, brake linings etc), as well as in battery recycling discharge processes. Bioleaching of pyrite concentrates has been successfully used to liberate gold from the sulphide matrix for subsequent cyanidation [1], but cannot be recommended for leaching chalcopyrite concentrates for copper recovery since copper is a biocide. A somewhat exotic use of pyrite that has been suggested is the application as an amendment for calcareous soils, to which it can supply sulphur and iron as fertilizers and affect phosphorus sorption and dissolution [2]. Pyrite nanocrystals and nanorods also show promise for solar energy conversion devices, solid-state batteries and catalysis [3].

In the context of the industrial uses of pyrite and other iron sulphides, ^{57}Fe Mössbauer spectroscopy, being one of the most efficient methods for the characterization of Fe-bearing minerals, can play an important role as an analytical tool to characterize the principal iron-bearing phases and to help identify impurities. To improve our knowledge of the physical properties of these minerals, we used X-ray diffraction with Rietveld analysis and Mössbauer spectroscopy in the present paper to characterize pyrite and chalcopyrite pyrite and chalcopyrite from a mine in southern Brazil that display slightly different spectral features from those reported earlier for samples from this deposit [4].

2. Materials and Methods

Two sulphide-rich samples from a copper mine in Camaquã (30° 30' 44" S 53° 29' 29" W), Rio Grande do Sul, Brazil [5], were selected with the purpose of identifying the main iron sulphide phases in samples from two stages of the industrial mining process. The samples were manually ground under acetone to prevent sulphide oxidation. X-ray diffraction was carried out by scanning from 5 to 80 °2θ at 0.025 °2θ per step on a Siemens D5000 diffractometer equipped with a Cu tube and a graphite diffracted-beam monochromator. Data evaluation was performed by Rietveld analysis. Total major and minor-element analyses were carried out by X-ray fluorescence on samples fused with a 2:1 mixture of $Li_2B_4O_7$ and $LiBO_2$ using a Rigaku spectrometer equipped with a Rh tube. Mössbauer spectra of both samples were taken at room temperature (298 K) and an additional spectrum of the chalcopyrite was taken at 110 K using a constant acceleration drive with a ^{57}Co/Rh source. The experimental data were fitted using Lorentzian functions with a least-squares procedure based on the NORMOS program and calibration was effected relative to α-Fe.

3. Results

Chemical analyses given in table 1 show the pyrite (sample 1) to be associated with a minute amount of silica that could not be manually removed (possibly encapsulated quartz), whereas the other sample (sample 2) consists mainly of chalcopyrite with a subordinate admixture (6 wt%) of pyrite.

Table 1. Chemical analyses of samples 1 (pure pyrite) and 2 (chalcopyrite and pyrite) and theoretical (idealized) compositions of pyrite and chalcopyrite.

Element/ wt%	Fe	Cu	S	Si
Sample 1	46.4(2)	–	53.6(2)	0.002(1)
Theory	46.55	–	53.45	–
Sample 2	33.60(3)	27.53(3)	41.87(5)	–
Theory	30.43	34.63	34.94	–

X-ray diffraction indicates sample 1 to consist exclusively of pyrite and sample 2 to consist predominantly of chalcopyrite. On the basis of line widths and lattice parameters, sample 1 is well crystallized, whereas sample 2 has an average particle size of 42 nm.

The room-temperature Mössbauer spectrum of sample 1 consists of the low-spin Fe^{2+} doublet that is characteristic for pyrite (table 2 and figure 1a). At room temperature the chalcopyrite sample displays a broad sextet indicating a wide distribution of magnetic hyperfine fields and a poorly-defined central component (figure 1b). At 110 K (figure 1c), the spectrum of this sample still has rather high line widths, whereas the central component now is less pronounced than at room temperature. A quadrupole-distribution fit (figure 2a) to the central portion of the spectrum of sample 2 at 110 K has a complex appearance with a major maximum at 0.61 mm/s that can be attributed to pyrite and a subordinate maximum at ~ 1.36 mm/s that may result from an associated superparamagnetic iron sulphide, a superparamagnetic iron oxide, or a paramagnetic iron-bearing silicate that was not detected by X-ray diffraction, or it could be a fitting artefact. This spectrum was additionally fit with a magnetic field distribution to look into the possibility of coexisting tetragonal and cubic chalcopyrite phases. The resultant hyperfine distribution (figure 2b) is dominated by a broad maximum that can be broken up into a major maximum at 36.9(1) T and a subordinate maximum at 35.5(1) T. Another minor maximum, at 33.2(2) T, agrees well with published values on cubanite, $CuFe_2S_3$, at liquid-nitrogen temperature [6,7].

Table 2. Mössbauer data. The chalcopyrite spectrum (sample 2) at 110 K was fit with hyperfine field and quadrupole splitting distributions.

Sample	Room Temperature			110 K			
	B_{hf}	δ	Δ	B_{hf}	δ	Δ	A
Pyrite	–	0.32(1)	0.59(1)				
Chalcopyrite				35.5(2)	0.36(1)	0.00(1)	0.69(3)
				36.9(1)	0.31(2)	0.00(0)	0.27(1)
				–	0.33(1)	1.10(4)	0.04(1)

Magnetic hyperfine fields (B_{hf}) in Tesla; isomer shifts (δ) relative to α-Fe, quadrupole splitting and quadrupole shifts (Δ) in mm/s, fractional areas (A). Errors on the last digit are given in parentheses.

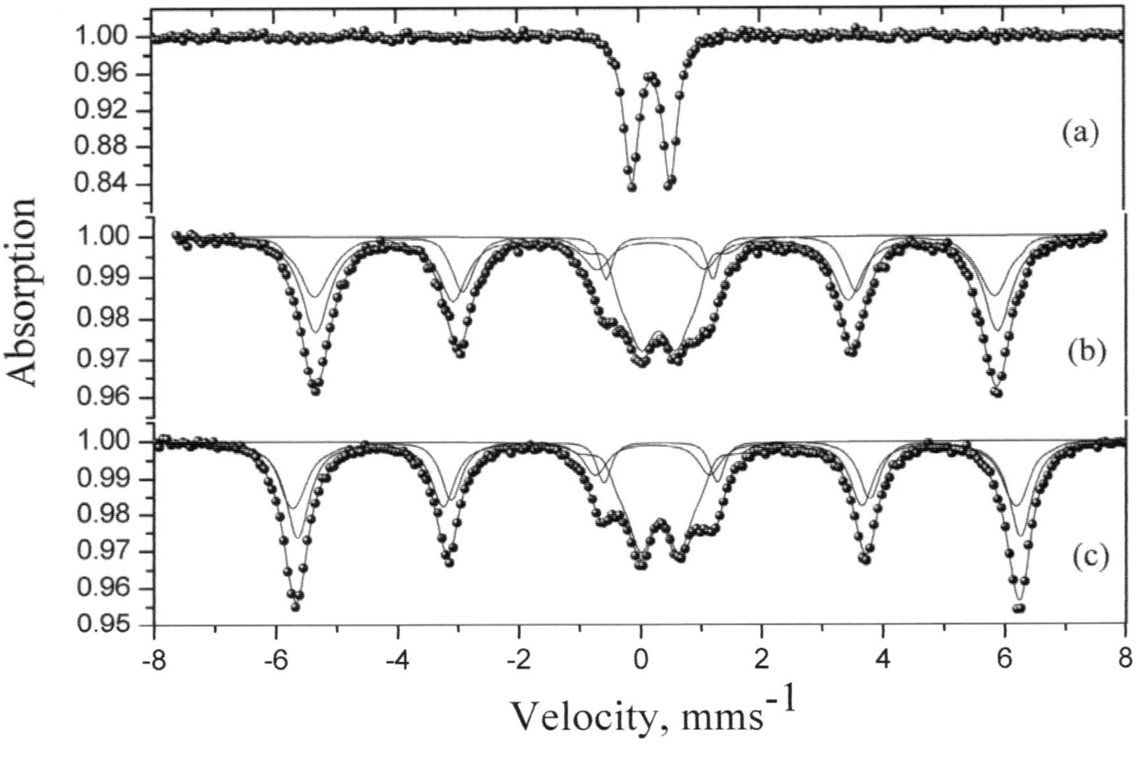

Figure 1. Mössbauer spectra at 298 K of samples 1 (a) and 2 (b) and of sample 2 at 110 K (c).

4. Discussion

Chemical analysis, quantitative X-ray analysis and room-temperature Mössbauer spectrum of sample 1 (table 2 and figure 1a) are consistent in showing the sample to consist almost exclusively of pyrite. The minor admixture of a silicate revealed by chemical analysis has no influence on these data. Similarly, the X-ray diffraction data and Mössbauer spectra of sample 2 (figures 1b,c) are dominated by chalcopyrite, and an apparent "deficit" in Cu shown by the chemical analysis (table 1) can be attributed to the presence of pyrite.

A narrowing of the sextet lines and concurrent reduction of the paramagnetic component between room temperature and 110 K indicates some effect of particle size at room temperature. The relatively broad hyperfine field distribution of the principal component at 110 K is attributed to the presence of two chalcopyrite phases that differ somewhat in their physical properties. We associate these with two

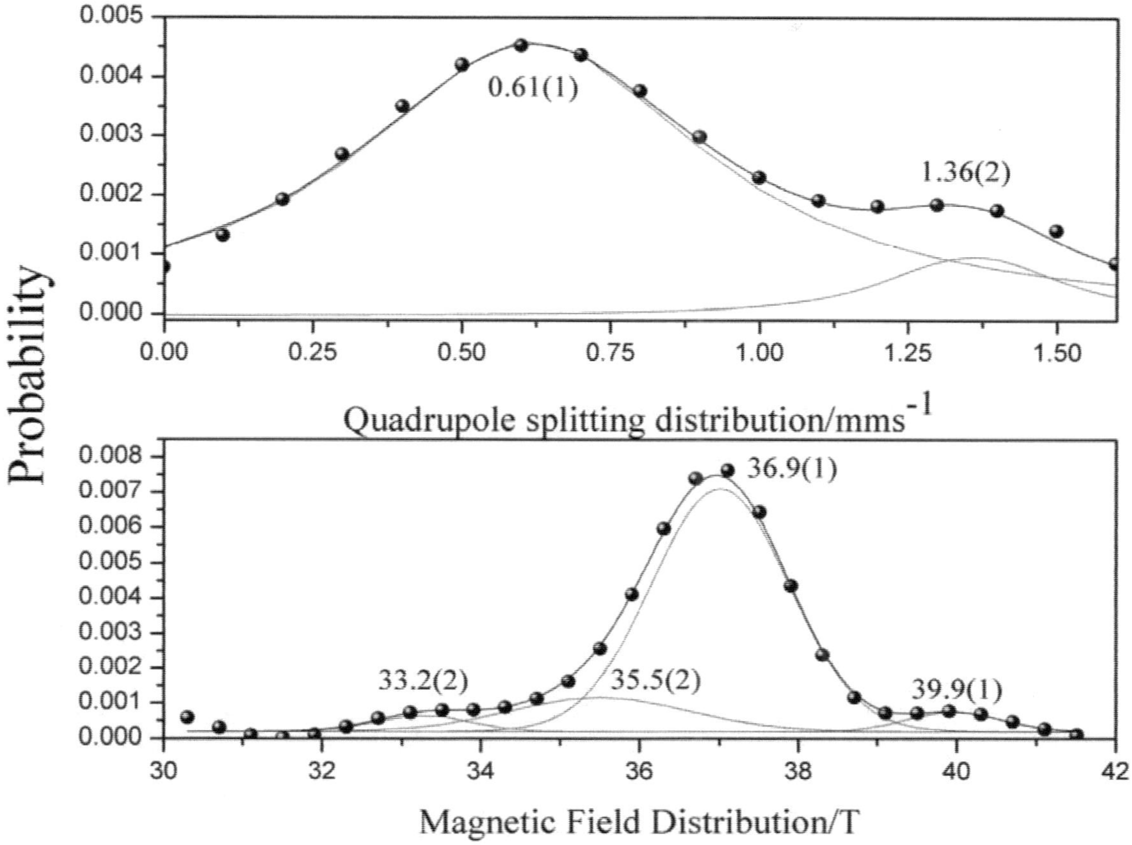

Figure 2. Quadrupole splitting (a) and hyperfine field distributions of sample 2 at 110 K (b).

chalcopyrite (tetragonal and cubic) modifications having hyperfine fields of 36.5 and 36.9 T (Table 2). The tetragonal to cubic ratios determined by XRD (68:32) agree adequately well with the relative Mössbauer areas (72:28) The coexistence of two chalcopyrite modifications may be due to an inhibited transformation of the original cubic constituent to the low-temperature tetragonal modification [4]. The presence of a further component with a hyperfine field of 33.2(2) T indicates sample 2 to contain cubanite, while a minor bulge near 40 T could either be a fitting artefact or result from a small amount of poorly-ordered goethite as a sulphide oxidation product.

Acknowledgments
This study has been supported by FAPEMIG and CNPq (Brazil).

References

[1] Habashi F 2006 In: *Metallurgy of the XXI Century*, International Conference Proceedings, Institute of Metallurgy and Ore Beneficiation, Almaty, Kazakhstan **1** 17
[2] Castelo-Branco M A, Santos J, Moreira J O, Oliveira A, Pires F P, Magalhaes I, Dias S, Fernandes L M, Gama J, Vieira e Silva J M and Ribeiro J R 1999 *J. Geochem. Expl.* **66** 363
[3] Barnard A S and Russo S P 2007 *J. Phys. Chem. C* **111** 11742
[4] Mussel W N, Murad E, Fabris J D, Moreira S, Barbosa J, Murta C, Abrahão W, Mello J and Garg V K 2007 *Phys. Chem. Miner.* **34** 383
[5] Laux J H, Lindenmayer Z G, Teixeira J B G and Neto A B 2005 *Ore Geology Rev.* **26** 71
[6] Marfunin A S and Mkrtchyan A R 1967 *Geochem. Int.* **4** 980
[7] Greenwood N N and Whitfield H J 1968 *J. Chem. Soc. A* 1697

Mössbauer study of formation of iron oxides and carbonate by dissimilatory alkaliphilic bacterium

N I Chistyakova[1,3], V S Rusakov[1], K A Nazarova[1], A A Shapkin[1], T N Zhilina[2] and D G Zavarzina[2]

[1] Faculty of Physics, M.V. Lomonosov Moscow State University, Leninskie Gory, Moscow 119992, Russia
[2] Winogradsky Institute of Microbiology, Russian Academy of Sciences, Prospect 60-letiya Oktyabrya 7/2, 117312 Moscow, Russia

E-mail: nchistyakova@yandex.ru

Abstract. The process of amorphous Fe(III)-hydroxide (AFH) reduction by anaerobic alkaliphilic bacterium – *Geoalkalibacter ferrihydriticus* (strain Z-0531) was investigated by Mössbauer spectroscopy methods. Strain Z-0531 was isolated from the sediments of soda Lake Khadyn (Tuva, Russian Federation). The influence of AFH concentration and the concentration of anthraquinone-2,6-disulfonate added to the cultivation medium as well as the incubation time influence on the reducing process were studied. It was found that increase in the time of cultivation of bacteria led to an increase in the relative content of phases containing ferrous atoms that could be explained by the adaptation of bacterium to the cultivation medium.

1. Introduction

One of the possible ways of iron mineral formation is an extracellular reduction of amorphous Fe (III) oxides and hydroxides by the group of microorganisms that has been found in the 1980's and named dissimilatory iron-reducing bacteria [1]. A microbiological pathway of a mineral formation has been intensively studied during recent decades. Investigations of the kinetics of iron mineral formation by bacterium *Thermincola ferriacetica* (strain Z-0001) and a study of different physical and chemical conditions of the formation of biogenic minerals were carried out [2]. The possibility of the reduction of amorphous and weakly crystalline iron oxides in alkaline environments has been subject to doubt [3] due to the low mobility of Fe(III) under these conditions. However, the iron-reducing process with magnetite and siderite formation may occur in alkaline medium for anaerobic bacterium *Geoalkalibacter ferrihydriticus* (strain Z-0531) [4].

The goal of this work is to investigate the iron mineral formation process by alkaliphilic dissimilatory iron-reducing bacterium (strain Z-0531) by ^{57}Fe Mössbauer spectroscopy.

2. Experimental

Anaerobic alkaliphilic bacteria *Geoalkalibacter ferrihydriticus* (strain Z-0531) was isolated from a bottom sediment sample from the weakly mineralized soda Lake Khadyn (Tuva, Russian Federation) [5]. It is a mesophilic representative of alkaliphilic microbial community. Optimal conditions for strain

[3] To whom any correspondence should be addressed

growth are pH 8.6, T = 35°C. The strain uses the amorphous Fe(III)-hydroxide as an electron acceptor and acetate CH_3COO^- as an electron donor. The bacterium can also reduce sulfur S^0 resulted in the formation of H_2S.

Amorphous Fe(III) hydroxide (AFH) was prepared by titration of a $FeCl_3$ solution with 10% NaOH. AFH and acetate were added to a mineral medium prepared under anaerobic conditions. The medium was poured into flasks, and then inoculated flasks were incubated at 35°C. In order to reduce errors we prepared two series for each experiment, and the second series (b) was the repetition of the first one (a).

Mössbauer investigations of all samples were carried out at 300K on a spectrometer operated in constant acceleration mode and equipped with a ^{57}Co-source into an Rh-matrix.

3. Results and discussion

Mössbauer investigations carried out at room temperature showed that the addition of anthraquinone-2,6-disulfonate (quinone) to the cultivation medium of the strain Z-0531 leads to a more intensive reduction of AFH and an increase in the relative content of magnetite and siderite unlike the strain growth with citrate addition and without any additional electron acceptors [4].

Therefore we studied the iron mineral formation by strain Z-0531 with quinone in the cultivation medium. The samples were obtained during the process of bacterium growth with different concentrations of AFH and quinine. The influence of incubation time and time of bacterium cultivation on biomineralization process was also studied.

3.1. AFH concentration influence

We investigated the influence of AFH concentration $n_{Fe(III)}$ on iron reducing process. The content of initial AFH varied from 5 to 150 mM, quinone concentration was 0.1 g/l, and incubation time was 1 month. Typical Mössbauer spectra of obtained samples are shown in figure 1. It was found that the formation of siderite $FeCO_3$ was observed at a small concentration of AFH. Sufficiently large particles of magnetically ordered phases were formed under the content of AFH above 70 mM. Two hyperfine field distributions were extracted from these spectra. The analysis has shown that distribution with a lower isomer shift (δ ~0.3 mm/s) could be attributed to cations Fe^{3+} located in A-positions of magnetite Fe_3O_4 and in both positions of maghemite γ-Fe_2O_3. Distributions with a large isomer shift value (δ ~0.6 mm/s) could be attributed to cations $Fe^{2.5+}$, located in the B-positions of magnetite. The relative intensity of the subspectrum corresponding to B-position of magnetite was noticeably larger for b-series than for a-series. Thus, the more intensive reduction of AFH was observed in the case of b-series.

The monotonic change of relative subspectrum intensities corresponding to the initial AFH, siderite and magnetically ordered phases was not observed within the range of 30-70 mM of AFH concentration for both series (figure 1). This behavior can be explained by the following considerations. In applying the method of series, when one flask corresponds to one experimental point, the fluctuations in bacterium cell number are inevitable. We suspect that biomineralization process becomes sensitive to the ratio between AFH concentration and the bacterium cell numbers.

3.2. Quinone concentration influence

We are interested in the influence of the quinone concentration on iron reducing process. The content of initial AFH was 100 Mm and quinine concentration was changed from 0 to 1 g/l. The typical Mössbauer spectra of obtained samples of both series for different quinine concentration are shown in figure 2. In general Mössbauer spectra are a superposition of three subspectra. Two quadrupole doublets correspond to AFH and siderite. The sextet with broad asymmetrical lines corresponds to magnetically ordered phases.

The increase in the sextet intensity is observed with an increase in n_q in the range of quinone concentration $n_q = 0$ g/l ÷ 0.05 g/l; further this intensity remains virtually unchanged up to $n_q = 0.3$ g/l. The relaxation character of the Mössbauer spectra of the samples for both series (a and b) becomes more pronounced for $n_q \geq 0.4$ g/l, reflecting an increase in the relaxation rate of magnetic moments of iron atoms due to reducing of superparamagnetic particle size.

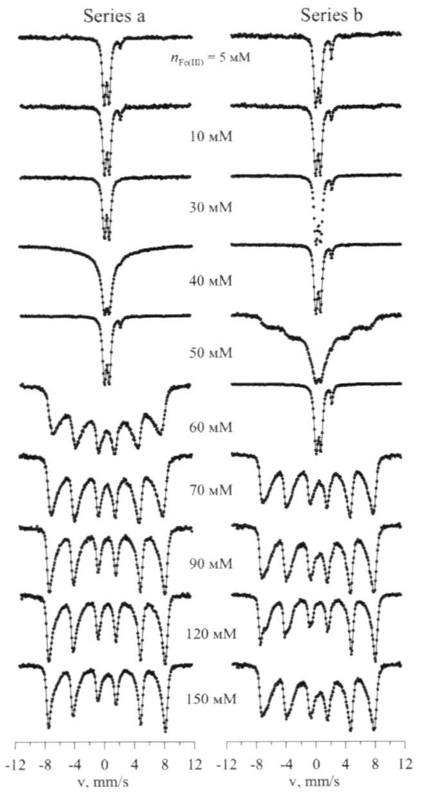

Figure 1. Mössbauer spectra of samples obtained as a result of bacterium Z-0531 growth for the different concentration of AFH in the cultivation medium (the concentration of quinone $n_q = 0.1$ g/l).

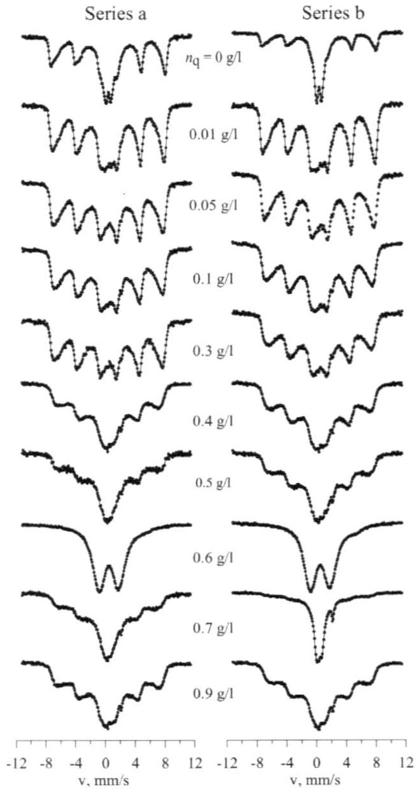

Figure 2. Mössbauer spectra of samples obtained as a result of bacterium Z-0531 growth for the different concentration of quinone in the cultivation medium (the concentration of AFH $n_{Fe(III)} = 100$ mM).

Thus, the quinone concentration affects the size of particles of magnetically ordered phases formed in the process of the AFH reduction by strain Z-0531. The sharp changes in the relative intensities of subspectra corresponding to obtained mineral phases were observed for b-series in the concentration range $n_q = 0.7$ g/l ÷ 1.0 g/l. Apparently, the biomineralization process in this concentration range is sensitive to fluctuations in bacterium cell number as it was observed in the previous series.

3.3. Incubation time influence

We have studied how the incubation time would affect the process of mineral phase formation. Samples were synthesized by the strain Z-0531 in a mineral medium with the addition of AFH (100 mM) and quinone (0.05 g/l). Incubation time was varied from 1 month to 30 months. This experiment is still in the progress.

Analysis of the Mössbauer spectra showed that the relative intensity of the quadrupole doublet corresponding to siderite virtually had not been changed with the increase in incubation time (figure 3). Herewith the increase in the relative intensity of the subspectrum corresponding to cations $Fe^{2.5+}$ located in the B-positions of magnetite and the decrease – to cations Fe^{3+} located in A-positions of magnetite and in both positions of maghemite were observed. Thus, an increase in the degree of initial AFH reduction occurs with the increase in the incubation time.

3.4. The influence of bacterium cultivation time

We found that the different times of bacterium cultivation connected with the number of bacterium inoculations and number of consecutive transfers of the culture affected the ratio of the mineral phases formed in the process of iron reduction under the same conditions. Thus, the relative amount of magnetically ordered phase was 75% for our first experiment with the addition of AFH (90 mM) and quinone (0.1 g/l) and 100% for the last one. Similar results were also obtained for another strain Z-0001. An increase in cultivation time led to a more intensive reduction of AFH which could be explained by the adaptation of bacteria to the culture medium.

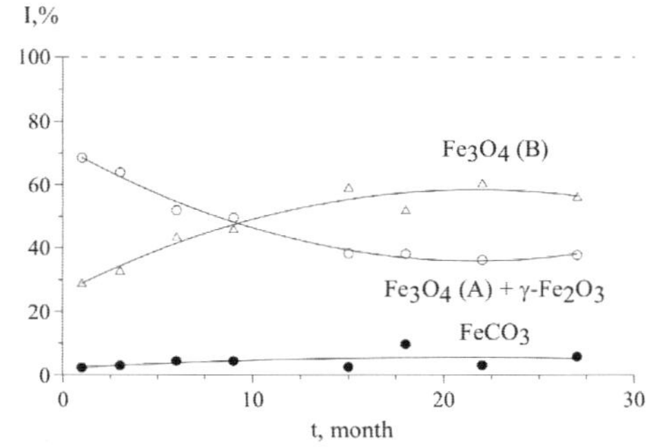

Figure 3. The dependences of relative intensities of subspectra corresponding to iron atoms in siderite, A-positions of magnetite and both positions of maghemite and the B-positions of magnetite on incubation time.

4. Conclusion

The Mössbauer investigations of the process of amorphous Fe(III)-hydroxide reduction by the strain Z-0531 have shown that the increase in AFH concentration leads to a decrease in siderite content and to an increase in the content of the magnetically ordered phase. The phase is not a well-crystallized magnetite, or a mixture of magnetite and maghemite. The concentration range of AFH $n_{Fe(III)} = 30$ mM ÷ 70 mM and quinone $n_q = 0.7$ g/l ÷ 1.0 g/l is the region of unstable phase formations by the strain Z-0531. For this strain the process of biomineralization is sensitive to fluctuations in bacterium cell number. Quinone concentration has a significant influence on the size of the particles of magnetically ordered phases. The increase in the degree of initial AFH reduction is observed with the increase in the incubation time and the time of cultivation in iron-containing mineral medium.

Acknowledgments

This work was supported by the programs no. 15 of the Presidium of the Russian Academy of Sciences "Origin of the Biosphere and Evolution of Geobiological Systems" and Russian Federal Agency of Science and Innovation (Rosnauka), state contract no. 02.740.11.0023.

References

[1] Balashova V V and Zavarzin G A 1979 *Mikrobiologiya* **48** 773. Lovley D R and Phillips E J P 1986 *Appl. Environ. Microbiol* **51** 683. Lovley D R, Giovannoni S J, White D S, Champine J E, Phillips E J P, Gorby Y A and Goodwin S 1993 *Arch. Microbiol.* **159**, 336.

[2] Chistyakova N I, Rusakov V S, Zavarzina D G and Kozerenko S V 2001 *The Physics of Metals and Metallography* **92** S138. Chystyakova N I, Rusakov V S, Zavarzina D G, Slobodkin A I and Gorohova T V 2005 *Czechoslovak J. of Physics* **50** 781. Chystyakova N I, Rusakov V S, Koksharov Yu A, Zavarzina D G and Greneche J -M 2009 *Solid State Phenomena* **152-153** 431.

[3] Ye Q, Roh Y, Carroll S L, Blair B, Zhou J, Zhang C L, Fields M W 2004 *Arch. Microbiol.* **70** 5595.

[4] Chystyakova N I, Rusakov V S, Nazarova K A, Koksharov Yu A, Zavarzina D G and Greneche J -M *Hyperfine Interact.* 2008 **182** 55.

[5] Zavarzina D G, Kolganova T V, Boulygina E S, Kostrikina N A, Tourova T P and Zavarzin G A 2006 *Microbiology* **75** 673.

NGR, XRD and TEM/SAED investigations on waste dumps materials with a view to recover precious metals

S S Udubasa[1,3], S Constantinescu[2], N Popescu-Pogrion[2], M Feder[2] and G Udubasa[1]

[1] University of Bucharest, Faculty of Geology and Geophysics, 1 N.Balcescu Blv., Bucharest, Romania
[2] National Institute for Materials Physics, Bucharest-Magurele, Romania

E-mail: udubasa@geo.edu.ro

Abstract. Two types of ores were selected for the investigation of the fate of the ore minerals during relatively long time of residence in the waste dumps (active mining: 25 years ago in the Bădeanca Valley and some 50 years ago in the Valea lui Stan area). The pentametallic ores (Co-Ni-Bi-Ag-U) in the Leaota Mts. contain a great number of primary minerals of Cu, As etc. Although the waste dumps materials have whitish-yellowish colors their NGR spectra show the presence of iron minerals very finely dispersed. In the gold ores of Valea lui Stan deposit numerous minerals were identified, such as arsenopyrite, pyrite, pyrrhotite, etc., as major gold bearing sulfides. The waste dumps materials naturally show different mineral constituents, with clay minerals as major phases. Detailed NGR investigations show however Mössbauer spectra pointing out the presence of finely dispersed iron or iron-bearing minerals. Under supergene conditions gold is commonly sequestrated by iron hydroxydes; further TEM/SAED and XRD investigations are contributing to localize the gold. In some samples Mössbauer spectra resembling those of greigite have been obtained. Greigite is also a principal concentrator of gold under supergene conditions.

1. Aim of the paper

Waste dumps are quite widespread in Romania, a country with a long history of mining activity. Preliminary investigations showed that such waste dumps material still contains significant amounts of metals which cannot be recovered by conventional methods. Detailed investigation of the waste dump material is the first step of an integrated research aiming to localize the metals (especially precious one) and then to find out methods to recover them using chemical and phytoextraction procedures, as shown by [1] and [2].

The methods used involve a combination of optical microscopy, EMPA, XRD, EDX, TEM/SAED and NGR (Mössbauer), in order to obtain as many as possible data at different observation scales.

2. Geological setting and petrography of the Valea lui Stan area

The Valea lui Stan area is located in the South Carpathians (Căpăţânii Mts.) near the Olt river. The dominant rocks are micaschists, red, ortochlase-dominated gneisses, amphibolites, migmatites, limestones and scarcely developed ultramafics. The dominant sedimentary formations are represented

[3] Corresponding author.

by cretaceous conglomerates. The ores are shear-zone related and display a typical association of Au-As-Cu. The gold ores have been mined during the XXth century yielding significant waste dumps. The mines are now closed. More details are given in [3] and [4].

3. Geological setting and petrography of the Bădeanca Valley area

The Bădeanca Valley belongs to the Leaota Mts. in the South Carpathians. The rocks occurring in the area are mainly micaschists in which intercalations of quartzites, amphibolites, gneisses and scarcely developed eclogites and granites occur in places. Regional retromorphic changes are widespread, on which retrograde transformation on shear planes overlap. The ore occurrences contain a great number of metals, Au and U included. (For details see [5] and [4]).

4. Optical microscopy

The fine grained materials from waste dumps have been embedded in epoxy and thin and polished sections have been realized. The mineral observed in these thin sections are generally the same with those observed mainly in the host rocks from the two occurrences. Photos taken from the thin sections are presented in figures 1 and 2.

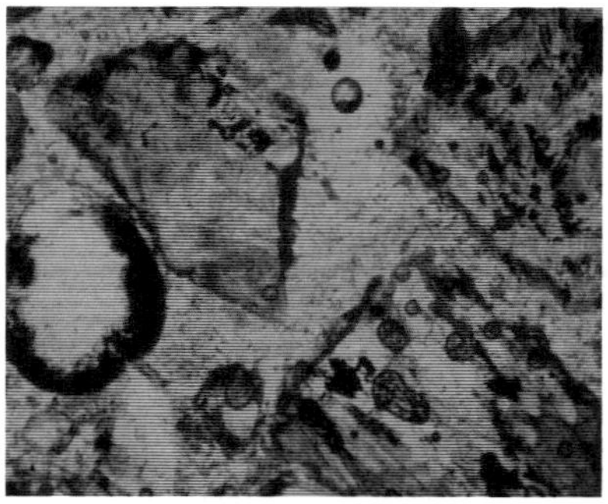

Figure 1. Rock fragments with euhedral opaque minerals (black), muscovite (white) and chlorite (green). Left side: a "ghost" of garnet (spessartine?) with black rims, presumably consisting of a Mn-spinel. Badeanca Valey area. Transmitted light (TL), N //, x40.

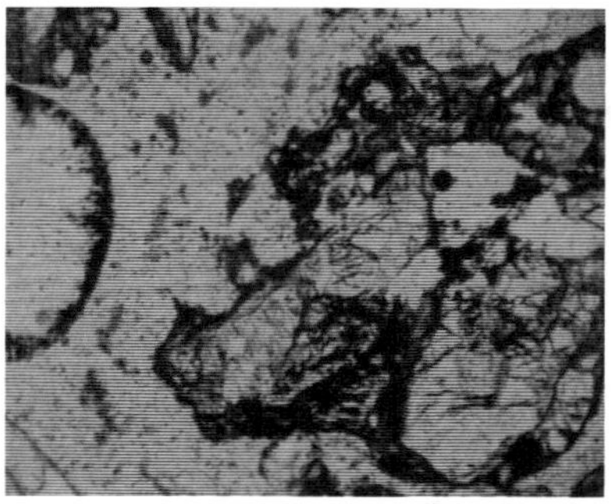

Figure 2. Fine opaque blades (ilmenite), euhedral grains (magnetite) and Fe-hydroxides (yellowish-brown spots). Left: a garnet (spessartine) "ghost" with opaque rims (similar to figure 1). Valea lui Stan area, TL, N //, x40.

5. TEM/SAED investigations

Electron microscopy investigations (transmission electron microscopy and selected area electron diffraction) on powder samples (2B1 – Badeanca Valley) have shown the followings:

- nanometric "precipitates" (average size ~ 9.8 nm) of Au, Ag or (Au, Ag) alloy in presence of maghemite (γ-Fe_2O_3) (figures 3, 4)
- nanometric particles of Au, Ag or (Au, Ag) alloy near β-cristobalite (agglomerates or monocrystals); it can be observed the mode of growth of the particles starting from the isolate spherical crystal $\rightarrow$ agglomerates (2 – 4 spheres) $\rightarrow$ monocrystals;
- the sample reveals a great diversity of morphologies: isolated spherical gold nanoparticles, between 24 – 75 nm in size, with average value of 42 - 0.412 nm up to monocrystals (average value 300 nm);

During the microstructural investigations, some minerals have been identified:

- α-quartz and β-quartz hexagonal - very small monocrystals about 140 – 350 nm in size (figure 4);
- maghemite, hematite (γ-Fe_2O_3, α-Fe_2O_3), (figure 4).
- pyrrhotite ($Fe_{1-x}S$) – rarely

6. XRD and EDX investigations

The XRD spectrum has been obtained using a *Bruker D8 Advance* diffractometer and CuKα_1 radiation (λ=1.540598 Å). The dominant (> 50%÷80%) contribution is given by SiO_2 phases, different Si-Al – oxides and – oxihydrites (with alkali, alkaline earth metals, 3d-metals, etc.).

Generally, due to the high XRD intensity on *011* planes of quartz, other contributions are hardly to detect. One can remark low amount (<3%÷5%) of Fe, Ti, Ni, Co oxides/silicates, in some of samples.

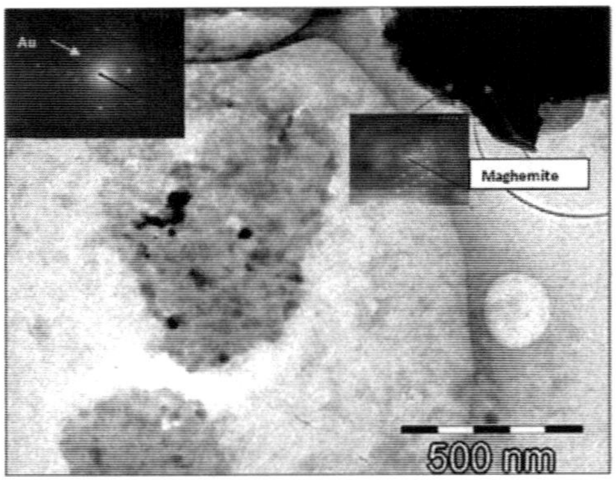

Figure 3. Gold and maghemite

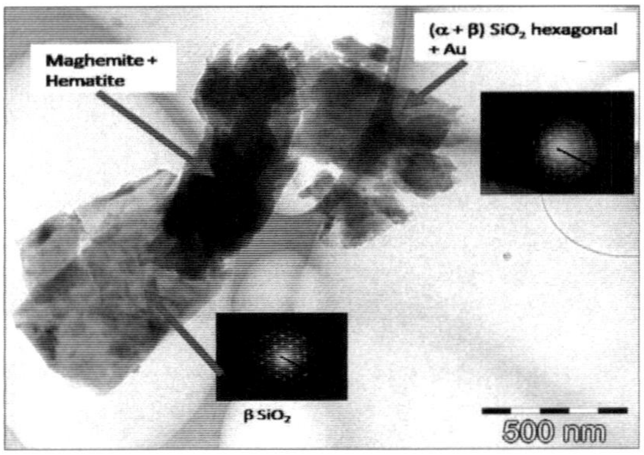

Figure 4. Cristobalite, hematite, maghemite and Au

It is notable that low detectable and/or no evidence of the sulfides, arsenides and/or arseno-sulfides are revealed by the XRD spectra. So, in the case of 2B-3m sample, the main contributions of SiO_2 phases and the non-Fe silicates, oxides and oxihydrites are about 76.63% and the minor (19.31%) one is given by sulfides and arsenides. The Fe-compound contribution is up 5%. The XRD data of 2B-3m sample are shown in the table 1.

The energy dispersive X-ray analysis (EDX) has been used, in order to identify the elemental composition of the specimens. The peak spectrum of the bombarded sample by electrons permits to identify the containing element and the type of X-ray. EDX analysis reveals the notable presence of iron.

Table 1. The XRD-data of 2B-3m sample

Sample	Phase	XRD Line Intensity [%]	$2\theta_{hkl}$ [grd]	d_{hkl} [A]	Phase Contribution [%]	Miller Indices		
						h	k	l
2B-3m	Quartz	100.0	26.64	3.343	21.971	1	0	1
		16.4	20.86	4.255	3.603	1	0	0
		10.4	50.14	1.818	2.285	1	1	2
	Lavendulan	100.0	8.884	9.946	15.999	1	0	0
		43.3	27.96	3.188	6.928	1	2	0
		41.1	26.96	3.304	6.576	3	1	0
	Jacobsite	100.0	35	2.562	4.631	3	1	1
		41.0	42.52	2.124	1.899	4	0	0
		38.4	61.7	1.502	1.778	4	4	0

7. NGR / Mössbauer investigations

The Mössbauer spectra were obtained in transmission geometry on the velocity scale ± 15mm/s, using a 57Co:Rh at room temperature (RT) [6]. 2B-3 and a spectrum at 77K (LNT) has been also recorded. All 2B spectra (from Badeanca Valley) are showing an intense magnetic-phase contribution and in their centres a non-magnetic one. The less intense magnetic contribution is shown by 2B-9 sample. The contribution of the paramagnetic signal (characterizing the ferrous and ferric valence states) has both different line shapes and intensities. Generally the sublattices of more/less intense magnetic fields are corresponding to ferric/ferrous ions in samples [7].

The fit of 2B RT-spectra reveals that some spinelic and pyrrhotite phases are responsible for the magnetic contribution and biotites, nonmagnetic spinel and ferrihydrites phases are responsible for the central signals in spectra. LNT-spectra are more complex (see table 2 and figure 5) due to paramagnetic to magnetic-ordering phase transition supported by some chemical Fe-compounds vs. decrease of the temperature [8]. The RT indexed phases are evidenced at LNT too, but with changed parameters.

Table 2. The RT and LNT Mössbauer parameters of 2B-3m sample spectrum

	RT				LNT			
Sample	δ_{Fe} [mm/s]	Δ/ε^1 [mm/s]	B [T]	A [%]	δ_{Fe} [mm/s]	Δ/ε^1 [mm/s]	B [T]	A [%]
	0.24	-0.13	48.09	36.5	0.09	0.16	44.93	8.0
	0.97	2.74	-	4.6	0.48	0.60	35.22	3.8
	0.89	2.26	-	9.2	0.37	-0.09	50.58	42.3
	0.25	0.72	-	14.4	0.86	1.64	-	1.2
	0.38	0.32	-	8.7	0.22	0.60	-	10.9
2B-3m	0.21	0.42	-	5.3	0.16	0.46	31.09	6.4
	0.34	-0.06	47.93	9.2	0.50	0.30	-	8.3
	0.74	2.06	-	3.7	0.29	2.12	-	1.0
	0.64	-0.42	27.69	8.4	1.16	1.94	-	5.0
					0.94	2.52	-	6.5
					1.11	2.90		6.4
Errors	±0.05	±0.07	±0.05	±2.5	±0.05	±0.07	±0.05	±2.5

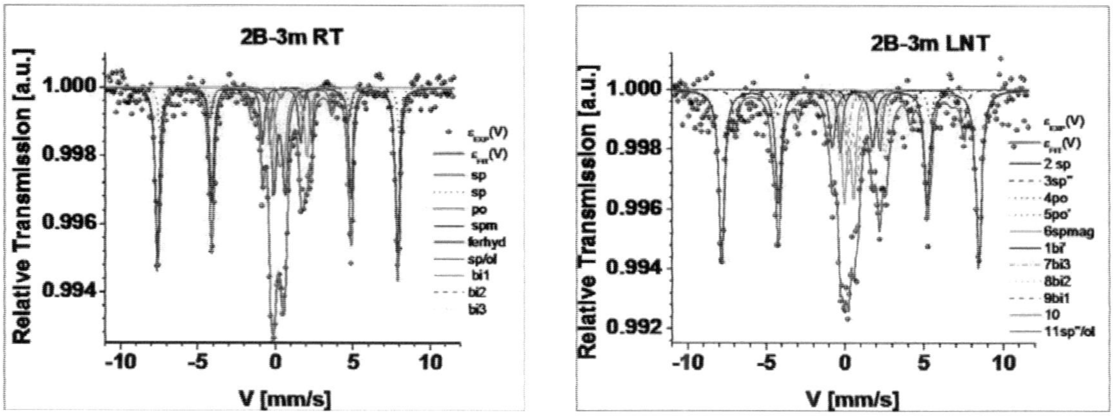

(a) (b)

Figure 5. The fit of the B-3m RT-spectra (a) and LNT-spectra (b) and the corresponding hyperfine parameters. Abbreviations: *sp* = spinel; *po* = pyrrhotite; *spm* = magnetic spinel; *ferhyd* = ferrihydrite; *sp/ol* = spinel-olivine; *bi* = biotite; *spmag* = superparamagnetic.

8. Conclusions

The investigation of waste dump material can produce valuable results, provided that several methods are applied. Combination of optical and structural methods is in a way the key of obtaining information about the fate of the minerals down to the nanometric scale. The special properties of the nanominerals such as high(er) reactivity and probably the availability for the plants uptake are of interest in both theoretical and practical aspect.

The waste dump materials used in this study have been collected from two different occurrences located in metamorphic rocks of the South Carpathians: Valea lui Stan gold ores (Au-As-Cu) in the Capatana Mts. and Badeanca Valley pentametallic occurrence (Co-Ni-Bi-Ag-U) in the Leaota Mts.

The study material has been first investigated by using optical microscopy in order to locate the opaque minerals, some of them having magnetic properties and/or producing reasonably good Mössbauer spectra. Macroscopic examination of the material revealed whitish-yellowish colours, however at the microscopic scale a number of opaque minerals do occur, either euhedral (magnetite) or elongated (ilmenite), as well as irregular patches within silicates (Fe-hydroxides) and rutile and "leucoxene"; rounded grains showing opaque rims are reminiscent of intensely transformed spessartine garnet; the rims might host the Mn-bearing spinel evidenced by Mössbauer spectra.

In addition to the clay minerals which might adsorb gold or other metals, a lot of possibly gold-bearing minerals have been identified: magnetite, ferrihydrite, greigite, even "leucoxene" loose-aggregates.

The complementarity of the methods used was a good choice to follow up the distribution of minerals and metals (especially gold) down to the nanometric scale; the next step will be the investigation of specially grown plants on metal-enriched substratum and the role of nanominerals availability for plants.

9. Acknowledgements

The financial support of the Ministry of Education, Research and Innovation of Romania is acknowledged through the PNCD-II project No. 31-081/2007.

References

[1] Anderson C W N, Brooks R R, Stewart R B and Simcock R 1999 Gold uptake by plants *Gold Bulletin* **32** 48-51
[2] Durães N, Bobos I and Ferreira da Silva E 2008 Chemistry and FT-IR spectroscopic studies of plants from contaminated mining sites in the Iberian Pyrite Belt, Portugal *Min. Magazine* **72** 405-409
[3] Udubasa G and Hann H P 1988 A shear-zone related Cu-Au ore occurrence: Valea lui Stan, South Carpathians *Dari Seama Inst. Geol. Geofiz.* **72-73/2** 259-282
[4] Udubasa S S, Udubasa G, Hirtopanu P, Constantinescu S, Popescu-Pogrion N, Popescu I V, Stihi C and Petrescu L 2008 How to detect submicroscopic minerals and their bearing on the extension of the mineral parageneses *Rom. J. Mineral Deposits* & *Rom. J. Mineralogy* **83** (joint volume) 148-152
[5] Udubasa G 2000 Brookite from Badeanca Valley, Leaota Mts. *Rom. J. Mineralogy* **80** 51-56
[6] Barb D 1980 *Grundlagen und Anwendungen der Mössbauer-Spektroskopie* (Berlin: Akademie Verlag) p. 305-325
[7] Greenwood N N and Gibb T C 1971 *Mössbauer Spectroscopy* (London: Chapman and Hall Ltd.) p. 283-305
[8] Constantinescu S, Udubasa G and Calogero S 1999 Mössbauer spectroscopy in mineralogy and geochemistry *Rom.J. Mineralogy* **79** 35-84

Mössbauer spectra of white micas from the Central Western Carpathians Mountains

J Sitek[1], M Sulák[2], M Putiš[2] and I Tóth[1]
[1]Department of Nuclear Physics and Technology, Faculty of Electrical Engineering and Information Technology, Slovak University of Technology, Ilkovicova 3, Bratislava, Slovakia
[2]Department of of Mineralogy and Petrology, Faculty of Natural Sciences, Comenius University Bratislava, Mlynská dolina G, 842 15 Bratislava, Slovakia

E-mail: jozef.sitek@stuba.sk

Abstract. Potassium white micas from the rocks included into Cretaceous deformation zones (ca. 100-70 Ma in age) of the Central Western Carpathians were investigated by Mössbauer spectroscopy. White micas formed during a polystage evolution and changing P-T conditions of their crystallization in crustal-scale shear zones. We found criteria for distinguishing generations of celadonite-poor (muscovitic) and celadonite-rich (phengitic) white micas using Mössbauer spectroscopy. This method revealed contrasting spectra characterized by typical quadrupole doublets corresponding to Fe^{2+} and Fe^{3+} contents in white micas. They are in the range of 2.9-3.0 mm/s for phengite, and 2.6-2.7 mm/s for muscovite. Mössbauer spectra reflect well the chemical changes in white mica aggregates, especially of those close to the end-member muscovite and (alumino-)celadonite compositions.

1. Introduction

Tectonic shear zones of the central Western Carpathians are located in basement/cover structural complexes subjected to Mesozoic and/or Tertiary metamorphism and deformations. These shear zones are analogous to those within the Austroalpine complexes of the Eastern Alps. The Central Western Carpathian are dissected by a few major shear zones. Here, a crustal shortening by underthrusting of some basement/cover complexes was accompanied by crystallization of celadonite-rich muscovite (phengite). This kind of phyllosilicates, besides the other associating minerals like pumpellyite, zoisite, grossular-rich garnet, chloritoid and kyanite, belongs to the group of white micas which often occur in higher-pressure zones. The studied zones fall into varying pressure interval of 300 – 1000 MPa. We specified chemical and spectroscopic characteristics [1],[2],[3], which allow distinction between celadonite-poor (muscovitic) and celadonite-rich (phengitic) white micas. White mica generations formed during a polystage evolution and changing P-T conditions from the very low to medium temperatures at medium pressure within the Alpidic Central Western Carpathians orogenic wedge.

The main goal of this study was to document the correlation between chemical composition of a natural white mica [1],[4] and its Mössbauer characteristics, respectively. These data should enhance identification of different white mica aggregates in blastomylonitic rocks. Mössbauer spectroscopy is a very efficient tool for the white mica investigation, because it provides information on relative abundance of Fe^{2+} and Fe^{3+}, which are not distinguishable by electron microprobe. Moreover, it also allows identification of all iron containing phases, when characterized by different hyperfine parameters. Celadonite poor muscovite and celadonite rich muscovite (phengite) are distinguished in presented Mössbauer spectra by the superposition of several quadrupole doublets corresponding to Fe^{2+} or Fe^{3+} contents.

2. Sample preparation and experimental technique

We separated original grain-size aggregates to avoid crushing of larger muscovite or phengite flakes during separation procedure. Ms porphyroclasts were separated from roughly crushed samples, nearly in original grain-size. White mica aggregates were separated from representative samples, containing more than 90% of newly-formed white mica aggregates, concentrated in foliation planes. The samples containing fine-grained micas were gently crushed after removal of the weathered parts. Subsequently, they were ground with pestle and mortar, sieved under 0.16 mm, soaked in distilled water and ultrasonically disaggregated. For the Mössbauer spectroscopy, samples were powdered and measured at room temperature by a standard transmission geometry Mössbauer spectrometer using a Co(Rh) source. Isomer shifts are given relative to the source. Spectra were evaluated by NORMOS program. Polished sections of samples were analysed by using a CAMECA SX100 and JEOL JXA-8100. Beam diameter varied from 2-10 μm and current from 20 to 10 nA according to size of the white mica flakes.

3. Results and discussion

Mössbauer spectra of celadonite-poor muscovite are shown in Fig.1. The spectra consist of two quadrupole doublets. The higher one corresponds to Fe^{2+} with QS = 2.63 – 2.73 mm/s and IS = 1.01-1.20 mm/s and the lower one to Fe^{3+} with QS = 0.66 – 0.73 mm/s and IS = 0.24 mm/s. Experimental error of QS values in all investigated samples is 0.01 mm/s. The relative amount of each component was determined from the area of the subspectra and is given in Tab. 1. The chemical composition of studied samples is documented in Tab. 2. Our results indicate that part of measured samples (MF-84, MF-KU) represent celadonite-poor muscovite with different ratio of Fe^{2+}/Fe^{3+}. For a comparison, the Mössbauer spectra of a typical celadonite-rich muscovite (phengite) are reported in Fig. 2 and Tab.1 (MK-237, PI-17HZ, MK-P1) with parameters Fe^{2+} (QS = 2.95 – 3.00 mm/s, IS=1.01 mm/s) and Fe^{3+} (QS = 0.71 mm/s, IS = 0.23 mm/s).

Table. 1 Mössbauer parameters of white micas from the Tatric tectonic unit.

sample	Fe^{2+} (%)	QS Fe^{2+} (mm/s)	Fe^{3+} (%)	QS Fe^{3+} (mm/s)
MK-237	21	2.95	79	0.73
PI-17HZ	48	3.00	52	0.71
MK-P1	57	2.99	43	0.66
MF-84	90	2.63	10	0.70
MF-KU	77	2.66	23	0.71

Table. 2 EMPA chemical analysis of white micas from Tatric tectonic unit.

mass % ox.	SiO_2	TiO_2	Al_2O_3	Cr_2O_3	FeO	MnO	MgO	CaO	Na_2O	K_2O	F	Cl	NiO	Total
MK-237	48.65	0.39	27.60	0.00	5.68	0.02	1.56	0.05	0.09	10.70		0.04	0.00	94.78
MK-P1	50.67	0.37	26.96	0.00	4.06	0.06	1.60	0.02	0.06	10.90	0.00	0.01	0.02	94.71
MF-84	49.31	0.08	31.80	0.00	1.43	0.03	1.98	0.01	0.21	10.98	0.00		0.00	95.82
MF-KU	46.66	0.36	36.20	0.05	0.93	0.00	0.74	0.01	2.02	8.37	0.00		0.00	95.35

Mol. Frac. (11 O)	Si	Al 3+ (IV)	Al 3+ (VI)	Ti	Cr	Fe	Mn	Mg	Ca	Na	K	Total
MK-237	3.35	0.65	1.59	0.02	0.00	0.33	0.00	0.16	0.00	0.01	0.94	7.10
MK-P1	3.46	0.54	1.62	0.02	0.00	0.23	0.00	0.16	0.00	0.01	0.95	7.04
MF-84	3.26	0.74	1.74	0.00	0.00	0.08	0.00	0.19	0.00	0.03	0.93	7.02
MF-KU	3.09	0.91	1.91	0.02	0.00	0.05	0.07	2.05	0.00	0.26	0.71	7.05

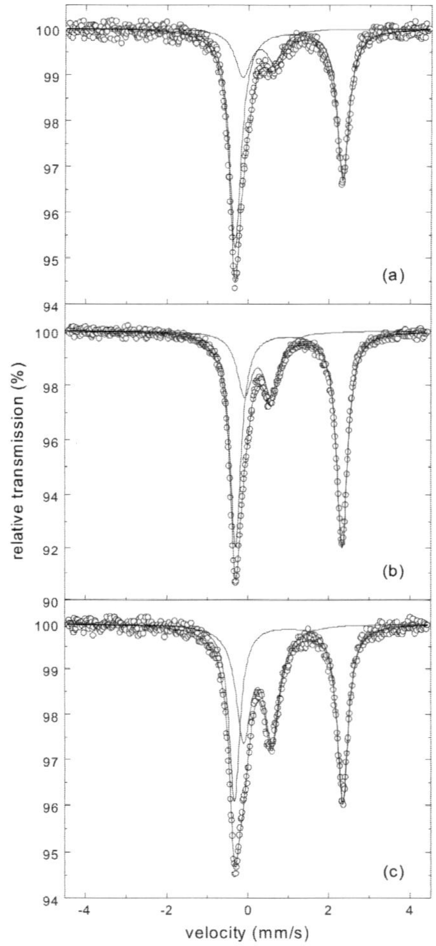

Figure. 1. Mössbauer spectra of celadonite-poor muscovite; a (MF-KU), b (MF-84), c (MF-P1)

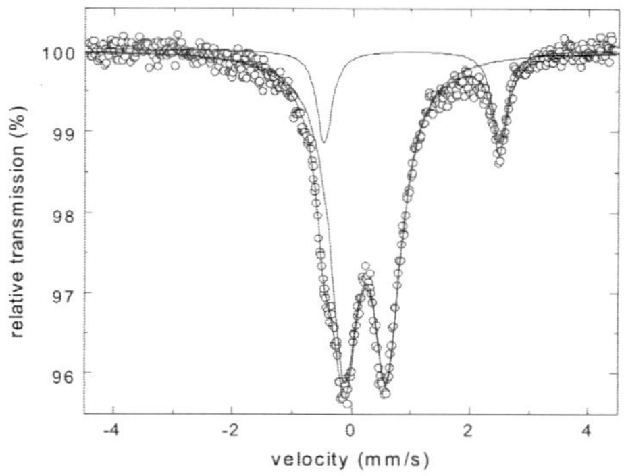

Figure. 2. Mössbauer spectrum of celadonite-rich muscovite MK-237

Mössbauer spectra well reflect the chemical changes in white mica aggregates, especially close to the end-member muscovite and (alumino-)celadonite composition, respectively.
For a practical aspect and for distinguishing of muscovite from phengite we utilize Fe^{2+} parameter, grouped into two different, but quite narrow intervals, with a "gap" of relevant values between them (see Tab. 1).

Similarly, the micro-Raman spectra indicate clear differences between muscovite porphyroclasts and newly formed phengite aggregate [3].

Acknowledgements
This work was supported by grants APVT 20-016104, APVV-0279-07 , VEGA 1/4038/07 and VEGA 1/0606/09.

References

[1] Rieder M, Cavazzini GD, Yakonov Y, Frank-Kamanetskii V, Gottardi G, Guggenheim S, Koval PV, Mülle, G, Neiva AMR, Radoslovich EW, Robert JL, Sassi FP, Takeda H, Weiss Z, Wones D.R. 1998. Nomenclature of micas *American Mineralogist* **36** (3), 905-912.

[2] Sitek J., Sulák M, Putiš M, Tóth I 2007. Chemical changes of Wmca across the West-Carpathian tectonic shear zones, registered by electron microprobe and Mössbauer spectroscopy. In: Pudiš D, Martinček I, Jamnický I, (Eds.), APCOM 2007, Conference Proceedings, Bystrá, Slovakia.

[3] Sulák M., Kaindl R., Putiš M., Sitek J., Krenn K., Tóth I., 2009: Chemical and spectroscopic characteristics of potassium white micas related to polystage evolution of the Central Western Carpathians orogenic wedge. Lithos, 113, 709-730.

[4] Tischendorf G, Rieder M, Förster H-J, Gottesmann B, Guidotti CV, 2004. A new graphical presentation and subdivision of potassium micas. *Mineralogical Magazine* **68** (4), 649-667.

Hyperfine and radiological characterization of soils of the province of Buenos Aires, Argentina

M L Montes, M A Taylor, R C Mercader, F R Sives and J Desimoni
Departamento de Física, Facultad de Ciencias Exactas, Universidad Nacional de La Plata, Instituto de Física La Plata – CONICET, CC 67, 1900 La Plata.Argentina.

E-mail: taylor@fisica.unlp.edu.ar

Abstract. The depth profile concentration of both natural and anthropogenic gamma-ray-emitter nuclides were determined in soil samples collected in an area located at 34° 54.452′ S, 58° 8.365′ W, down to 50 cm in depth, using an hyper-pure Ge spectrometer. The soil samples were also characterized by means of Mössbauer spectrometry and X-ray diffraction. The activities of ^{238}U and ^{232}Th natural chains remain constant in depth at 41 Bq/kg and 46 Bq/kg, respectively, while the ^{40}K activity increases from 531 Bq/kg to 618 Bq/kg between 2.5 cm y 25.5 cm of depth. The only anthropogenic detected nuclide is ^{137}Cs, whose activity changes form 1.4 Bq/kg to values lower than the detection limit (L_D) for depths below 25 cm, exhibiting a maximum at 10 cm beneath the surface. The Mössbauer spectra show two magnetic sextets associated with α-Fe_2O_3 and Fe_3O_4, as well as two Fe^{+3} and Fe^{+2} doublets, probably originated in octahedral and tetrahedral sites of paramagnetic phases. The Fe^{3+} paramagnetic signal relative fraction increases up to 82% at the expense of the α-Fe_2O_3 one when de depth increases. No correlation between Fe_3O_4 and the ^{137}Cs was identificated.

1. Introduction

The concentration of metals and radionuclides in uncontaminated soil is primarily related to its origin and to pedogenetic processes. However, human activities can strongly modify the natural concentrations due to the presence of residues or accumulation of elements caused by the release of effluents to the environment. On a long-term scale, soil is the major source of nuclides entering the food chain or contaminating the ground water [1]. Consequently, the knowledge of the distribution of radionuclides in soils becomes extremely important. Once the distribution and profiles of nuclides are determined, to predict the nuclide dispersion in the soil, it is necessary either to modify existing models [2, 3] or to develop new ones to include the local conditions and to account for the very complex system that soil is. Indeed, nuclide transport in soil is strongly affected by the soil composition, pH, concentration of inorganic ions, redox potential, concentration of organic mater, physical and chemical properties, etc. [2, 3]. In addition, it has been suggested [1] that magnetite plays an important role for the radionucleide migration in soils.

Within this frame, we have started to study of soils belonging to a densely populated area of the Province of Buenos Aires, Argentina. The northeastern plain of Buenos Aires province exhibits an undulating topography which results from the erosion by tributary streams of the Paraná and La Plata rivers. This is an important agricultural and industrial area where nuclear power plants and research nuclear reactors are located, adding to the natural inventory small amounts of activities of anthropogenic nuclides. Despite the long term nuclear activity in the region, no systematic study has been carried out until now to establish the natural conditions of the soils. Hence, it is essential to determine the radioactive baseline for undisturbed soils and their chemical and physical properties to be able later to understand the degree of pollution that the anthropic activities may have produced and to establish reliable models that predict the possible dissemination of radionuclides. To this end, the natural radioactive element contents of gamma-emitters in soil profiles at different depths have analyzed and, at the same time, the soil properties have been characterized by x-ray diffraction and Mössbauer spectrometry. In this contribution, the first findings in an area close to the La Plata district located at about 50 km from the Argentine Federal Capital are reported.

2. Experimental

The samples for radiological analysis were collected with a special spade of 3 cm high and 30 cm wide, while the corresponding ones for Mössbauer spectroscopy were taken with small plastic tubes. After collection, the samples were dried at 363 K, crushed and sieved through a 2 mm mesh without previous separation. The sampling was performed at depths down to 50 cm from the surface.

For gamma spectroscopy (GS), the soil was properly accommodated in a Marinelli-type box, sealed and kept for nearly three weeks before analysis to achieve the secular equilibrium. The spectra were taken inside an EG&G Ortec low background chamber, using a high-resolution hyper-pure Ge gamma EG&G Ortec spectrometer in the range of 30 keV to 1.8 MeV with a standard electronic chain and 8192 channels multichannel. Energy calibration was performed with ^{60}Co, ^{133}Ba, ^{137}Cs and ^{152}Eu sources. The efficiency calibration was carried out using an admixture of known amounts of naturally occurring ^{176}Lu and ^{138}La long-lived isotopes, dispersed in a soil substrate [4]. The absolute efficiency and the energy resolution of the 1332 keV ^{60}Co line were 0.5 % and 1.9 keV, respectively. The room background was determined and peak corrections were done to the soil spectrum. The spectra were recorded during four days and analyzed with a commercial program.

Mössbauer spectra (MS) were recorded between 30 K and room temperature in a 512 channels conventional constant acceleration spectrometer with a ^{57}Co*Rh* source of approximately 10 mCi of nominal activity. Velocity calibration was performed with a 12 μm thick α-Fe foil at room temperature. Isomer shifts are referred to this standard. The spectra were numerically analyzed using hyperfine field and quadrupole distributions [5].

The structural analyses were done by X-ray diffraction (XRD) with a Philips PW1710 diffractometer (CuKα radiation) in the Bragg-Brentano geometry, with a step mode collection of 0.02°, 1 s by step, with 2θ ranging from 10° to 100°. The patterns were analyzed using the PowederCell program [6] and the structural data were obtained from ref. 7.

3. Results and discussion

The XRD patterns showed a broad background and many peaks among which illite, feldspar and quartz could be identified, in good agreement with previous studies carried on in a close area [8]. The fractions of the different detected phases – being illite the most abundant– remain nearly constant within the monitored depth. Because of the statistical uncertainty of the X-ray data, the existence of other phyllosilicates different from illite cannot be ruled out.

Figure 1 shows some selected Mössbauer spectra from the collected samples recorded at different temperatures. Two main contributions to the spectral area are observed; a magnetically split signal and a central set of doublets. The fitting procedure was carried out using hyperfine distributions with Voigt-shaped line profiles.Table 1 displays the resulting fitting parameters, which have been assigned as described below, and whose values remain almost constant for all depths.

While the parameters of the sextets belong unequivocally to α-Fe_2O_3 and Fe_3O_4, the assignment of the Mössbauer parameters of the doublets is more uncertain because different Fe species have similar parameters that overlap in the same region of the spectra [9-12] –e.g. hydroxides, finely divided Fe oxides in superparamagnetic relaxation regimes, or Fe ions partially substituted for other ions, like Al or Mn. The magnetically split signal of Fe_3O_4 is noticeable only down to 25.5 cm. The Fe^{3+} doublets may arise from paramagnetic clay minerals and other silicates or super-paramagnetic Fe oxides and related minerals. Because of the small particle size, the low crystallinity degree and the low iron concentration in paleosols and soils of Buenos Aires Province [13], the detection of iron oxides or oxyhydroxides by means of XRD is very difficult [14]. For example, the X-ray data did not reveal the presence of hematite or magnetite even after searching for their most intense peaks. The lack of noticeable changes at low temperatures (see Figure 1) inhibits the unambiguous assignment of the contribution of the doublets to the existence of superparamagnetic phases. Therefore, the determined interactions can only be related to the Fe^{3+} and the Fe^{2+} octahedral and tetrahedral sites in paramagnetic species [9-12]. It is worth mentioning that illite is the species that can take the higher contents of Fe, and hence, it is likely that the main Fe^{3+} doublet area be related to this phase. The Fe^{2+} signal cannot be originated in Fe oxides found in the soil and therefore can only be associated to the clay minerals existent in the samples.

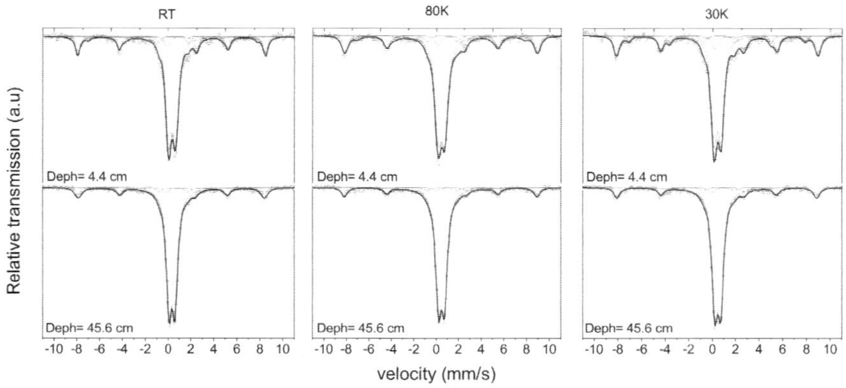

Figure 1: Mössbauer spectra of two different analyzed depth at RT (left), 80K (center) and 30 K (right). The thick solid line is the result of the fitting as described in the text. The thinner lines are the components of parameters shown in Table I.

The evolution of the RT relative fractions of the different Fe environments with depth is shown in Figure 2a. It is observed that the main component to the MS corresponds to the Fe^{3+} doublets, followed in importance by the α-Fe_2O_3 sextet. It is worth mentioning that the Fe^{3+} fraction increases up to 82% at the expense of the α-Fe_2O_3 one.

The determined gamma activities for the natural nuclides of the ^{238}U and ^{232}Th resulted 41 Bq/kg and 46 Bq/kg, respectively, remaining constant in depth. Regarding the activity of the ^{235}U chain was always lower than the L_D (0.02 Bq/kg). The ^{40}K activity varies from 531 Bq/kg to 618 Bq/kg between 2.5 cm and 25.5 cm of depth, as shown in Figure 2b. The only anthropogenic nuclide detected was ^{137}Cs, with a decreasing activity from 1.4 Bq/kg (2.5 cm depth) up to values lower than the corresponding L_D (0.2 Bq/kg) at a depth of 25.5 cm (see Figure 2b). The surface activity is compatible with the fallout of the nuclear weapons tests in the South Pacific [15]. Regarding the ^{137}Cs depth profile and its correlation with Fe_3O_4, controversial results on their sorption by Fe_3O_4 have been reported [1, 16, 17]. By comparing the present results displayed in Figures 2 it can be seen that there is not an apparent correlation between the ^{137}Cs activity and the MS relative fraction of Fe_3O_4. Further experiments and theoretical developments are in progress to fully describe the local soil characteristics entering in the existing models to really account for the diffusion coefficient and convection velocity of ^{137}Cs in the plain of Buenos Aires province. The soil characterization by structural, hyperfine and magnetic techniques will allow developing a model to understand the diffusion of the elements produced by the nuclear and industrial activities produced in the area.

Table 1: Hyperfine parameters obtained after fitting the Mössbauer spectra at room temperature using an extended Voigt-based fitting method. For doublets: average values of isomer shift <δ> (mm/s) with respect to α-Fe, quadrupole splitting <Δ> (mm/s), and relative fractions rf (%). For sextets: average magnetic field, <H> (T), quadrupole shift ε (mm/s) and relative fractions rf (%). The numbers between parentheses denote the Gaussian widths, σ (mm/s).

Depth (cm)		2.3	4.4	9.4	18.3	25.5	29.8	37.3	45.6	48.9
Fe^{3+}	<δ>	0.37(0.15)	0.37 (0.16)	0.38 (0.14)	0.37 (0.14)	0.38 (0.14)	0.37 (0.15)	0.36 (0.14)	0.36 (0.14)	0.36 (0.15)
	<Δ>	0.60 0.12)	0.60 (0.09)	0.59 (0.16)	0.59(0.14)	0.58 (0.13)	0.53 (0.12)	0.53 (0.14)	0.53 (0.13)	0.53 (0.10)
	rf(%)	58 ± 3	63 ± 3	61 ± 5	59 ± 2	61 ± 1	73 ± 2	73 ± 2	73 ± 2	73 ± 2
Fe^{2+} (SiteI)	<δ>	0.54 (0.22)	0.55 (0.10)	0.60 (0.16)	0.55(0.17)	0.56 (0.14)	0.53 (0.25)	0.51(0.20*)	0.51 (0.22)	0.52 (0.00)
	<Δ>	1.49 (0.20)	1.63 (0.15)	1.57 (0.16)	1.49(0.15)	1.55 (0.19)	1.47(0.10*)	1.48 (0.20)	1.50 (0.20)	1.51 (0.47)
	rf(%)	11 ± 2	9 ± 2	12 ± 3	12 ± 1	12 ± 1	12 ± 2	11 ± 1	11 ± 1	11 ± 2
Fe^{2+} (Site II)	<δ>	0.93 (0.20)	0.88(0.00*)	0.82 (0.20)	0.88(0.25)	0.88(0.15*)*	0.91(0.10*)	0.90 (0.10*)	0.90(0.10*)	0.84 (0.25)
	<Δ>	2.8 (0.0*)	3.1 (0.0*)	3.1 (0.0)	3.1 (0.2)	3.0 (0.2)	3.0* (0.3*)	3.0*(3*)	3.0* (0.1*)	3.0 (0.3*)
	rf(%)	6 ± 1	6 ± 1	7 ± 2	7 ± 1	7 ± 1	5 ± 1	5 ± 1	4 ± 1	5 ±1
αFe_2O_3	<δ>	0.38 (0.00)	0.38 (0.04)	0.38 (0.07)	0.38(0.02)	0.38 (0.00*)	0.38(0.00*)	0.36 (0.00*)	0.38 (0.00*)	0.38 (0.01)
	<ε>	-0.10 (0.00)	-0.11(0.00)	-0.12(0.01)	-0.11(0.01)	-0.10 (0.00)	-0.10(0.01)	-0.10(0.00*)	-0.11 (0.03)	-0.11(0.02)
	<H>(T)	51(1)	51 (1)	50 (0)	51 (1)	51(1)	50 (1)	51 (1)	51 (1)	51 (1)
	rf(%)	17 ± 2	17 ± 2	18 ± 4	17 ± 1	21 ± 1	11 ± 1	12 ± 1	12 ± 1	11 ± 1
Fe_3O_4 (Site A)	<δ>	0.3 (0.0)	0.3 (0.0)	0.3 (0.0)	0.3 (0.0)					
	<ε>	0.00	0.00	0.00	0.00					
	<H>(T)	48 (1)	48 (0)	49 (1)	49 (1)					
	rf(%)	6 ± 2	3 ± 3	1 ± 3	2 ± 1					
Fe_3O_4 (Site B)	<δ>	0.7 (0.3)	0.6 (0.0)	0.6 (0.0)	0.7 (0.0)					
	<ε>	0	0	0	0					
	<H>(T)	44.0 (0.2)	46.0 (0.1)	45.8 (0.2)	44.0 (0.2)					
	rf(%)	2 ± 1	2 ± 2	3 ± 2	3 ± 1					

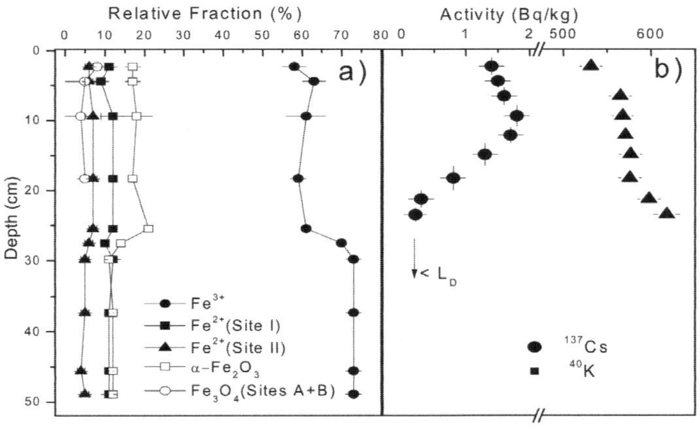

Figure 2: a) Evolution of the Mössbauer relative fractions. b) ^{137}Cs and ^{40}K profiles. (LD is the detection limit and the horizontal and vertical bars represent the activity errors and analyzed soil depth, respectively.

4. Conclusions

The results reported in this contribution –for samples taken from the surface down to a depth of 50 cm– are the first ones to establish the baseline activity for the naturally occurring gamma emitter radionuclides in an area around Buenos Aires. The ^{238}U and ^{232}Th chains were detected in the soil samples. The ^{40}K activities are in agreement with previous reported data. The only anthropogenic detected nuclide was ^{137}Cs arising from the fallout of the South Pacific weapon test. No correlation has been found yet between the magnetite fraction and the ^{137}Cs activity. More studies are in progress.

References

[1] Singh B K, Jain A, Kumar S, Tomar B S, Tomar R Manchanda V K, and Ramanathan S 2009 *J. Contam. Hydrol.* **106** 144

[2] Bossen P and Kircher G 2004 *J. Environ. Radioactiv.* **73** 127

[3] Almgren S and Isaksson M 2006 *J. Environ. Radioactiv.* **91** 90

[4] Perillo Isaac M C, Hurley D, McDonald R J, Norman E B, and Smith A R 1997 *Nucl. Instrum. Meth. A* **397** 310

[5] Lagarec K, and Rancourt D G 1998 *University of Ottawa- Department of Physics* Version 1.0

[6] Kraus W, and Nolze G 1999 *Federal Institute for Mater* PowderCell for Windows Version 2.3

[7] Downs R T, and Hall-Wallace M 2003 *Am. Mineral.* **88** 247

[8] Zabala S M, Conconi, M S. Alconada M, and Torres Sanchez R M 2007 *Ciencia de Suelos* **25** 65

[9] Fabris J D, and Coey J M D 2002 Topics em Ciencia do Solo, *Soc. Brasileira de Ciencia do Solo* **2** 47

[10] Murad E 1998 *Hypefine Interact.* **117** 39

[11] Paduani C, Samudio Pérez C A, Gobbi D, and Ardisson J D 2009 *Appl. Clay Sci.* **42**, 559

[12] MacCannon C A, 1995 Mössbauer spectroscopy of minerals in *Mineral Physics & Crystallography: Hanndbook of Physical Constants,* American Geophysical Union, Washington DC, pp. 332–347

[13] Rico Y, Bidegain J C, and Elsner C I 2009 Geofísica Internacional **48** (2), 221

[14] Bidegain J C, 1998 Quaternary of South America and Antartic Peninsula, V12, 207

[15] UNSCEAR 1982 *Report Annex E*, http://www.unscear.org/unscear/en/publications/1982.html

[16] Catallette H, Dumonceau J, and Ollar P 1998 *J. Contam. Hydrol.* **35** 151

[17] Marnier N, and Fromage F 2000 *J. Colloid Interf. Sci.* **223** 83

Iron Oxides of Fazendão Deposit, East Border of Quadrilátero Ferrífero, Minas Gerais, Brazil

A R P Pereira[1,2], F J Rios[2], C A Rosière[3], J D Fabris[1]

[1]Departamento de Química – Universidade Federal de Minas Gerais, Brasil
[2]Centro de Desenvolvimento da Tecnologia Nuclear – CDTN, Minas Gerais, Brasil
[3]Instituto de Geociências – Universidade Federal de Minas Gerais, Brasil

E-mail: arpp@qui.dout.ufmg.br

Abstract. The iron oxides highly influence the soil structure and aggregation of mineral particles in soil. They also play an important role in some economical variables, as those related to the use of land for agriculture practices or mineral exploitation, in ore mining activities. About 60 % of all industrial activities on iron processing in Brazil is based on ores mined in the geodomain of Quadrilátero Ferrífero, state of Minas Gerais, Brazil. Samples of a mining area for iron, the Fazendão Deposit in the east border of Quadrilátero Ferrífero have been studied in an attempt to contribute to a better understanding of the key chemical and mineralogical pathways related to the formation and transformation of iron oxides, involving hematite. From powder diffraction patterns, it is confirmed that hematite is the main mineralogical phase in all samples. The total iron contents were found to range between 65.15 and 70.00 mass%. The saturation magnetization values, $\sigma = 6.9$ and $2.1\ \mathrm{J\,T^{-1}\,kg^{-1}}$, are significantly measurable only for the two samples showing some evidences of magnetite. 298 K-Mössbauer data confirmed the dominant occurrence of hematite in all samples, along with magnetite for the two samples showing non-zero magnetization. 110 K-Mössbauer data indicate that the hematite in all samples undergoes the Morin transition ($T_M \approx 260$ K), as expected for the relatively pure oxide. Maghemite could not be detected in any sample. New numerical analysis are now being performed, using Rietveld refinement of XRD data, in an attempt to obtain crystallographic results that could indicate more reliable evidences about the mechanisms of formation of hematite, particularly in the magnetic samples, for which the precursor is presumably magnetite.

1. Introduction

Iron oxides are found in large proportions in nature, being the most abundant metal oxides in soils. They rank second in the mineralogical occurrence order in the clay fraction of tropical and sub-tropical pedosystems, following the kaolinite group [1].

Iron oxides play an important influence on some critical properties of the geo-superficial layer, such those related to the use of land for agriculture practices or involved in the genesis of iron ore deposits [2]. About 60 % of all starting material used in industrial processing of iron in Brazil comes from exploitation of mines in the geodomain broadly known as Quadrilátero Ferrífero, in the state of Minas Gerais, Brazil. Itabirite is the dominant metamorphical hematite-rich rock of that area [3]. The name "itabirite" has more commonly been applied in this sense to the metamorphosed banded iron formations (BIFs) of the Cauê Formation [4], but its use also comprehends similar BIFs of other parts of Brazil and western Africa. This denomination is also used to generically describe [5] any massive rock with a granular to schistose fabric, composed primarily of specularite, granoblastic hematite and, more locally, magnetite (Fe_3O_4), a primary oxide, which tends to be converted to hematite presumably via two main broad chemical mechanisms [6]: (i) by direct oxidation or (ii) through an intermediate formation of maghemite (γFe_2O_3).

The main objective of this work was to contribute to a better understanding of the key chemical and mineralogical pathways related to the formation and transformation of iron oxides, involving hematite, in a mining area for iron, the Fazendão Deposit, in the Quadrilátero Ferrífero geo-province.

2. Experimental

Nine samples were collected from sampling sites in that mine (geographical coordinates of the mineral deposit 20° 07' 51.71" S 43° 25' 00.09" W), in the east border of Quadrilátero Ferrífero, state of Minas Gerais, Brazil. The samples were analyzed with conventional chemical analytical methods, as redox titration (dichromatometry), for total iron contents, and with physical techniques, namely powder X-ray diffraction, ^{57}Fe Mössbauer spectroscopy at 298 K and 110 K and saturation magnetization (σ) measurements, with a portable magnetometer [7] under a fixed magnetic field of ~0.3 tesla.

3. Results and Discussion

From powder X-ray diffraction patterns (radiation CoKα), hematite appears to be the main mineralogical phase in all samples (Figure 1). The corresponding rhombohedral-hexagonal cell parameters for hematite [8], as obtained from Rietveld refinement of powder X-ray diffratometry patterns, are presented in table 1.

Table 1. Rhombohedral-hexagonal cell parameters for hematite

Parameters / Samples	*a*/Å	*b*/ Å	*c*/ Å
FZ-01	5.038(1)	5.038(1)	13.750(1)
FZ-02	5.035(6)	5.035(6)	13.748(5)
FZ-03	5.039(2)	5.039(2)	13.750(5)
FZ-04	5.034(7)	5.034(7)	13.75(3)
FZ-05	5.029(5)	5.029(5)	13.74(3)
FZ-06	5.03(3)	5.03(3)	13.74(7)
FZ-10	5.032(8)	5.032(8)	13.74(3)
FZ-11*	5.035(9)	5.035(9)	13.752(6)
FZ-12*	5.035(9)	5.035(9)	13.749(4)

*Samples with magnetite

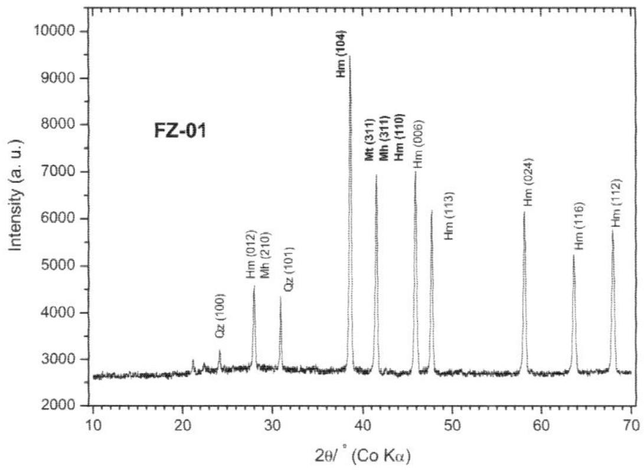

Figure 1 - XRD pattern from FZ-01 sample [Hm = hematite; Mt = magnetite; Qt = quartz]

From volumetric determinations (dichromatometry), the total iron contents were found to range between 65.15 and 70.00 mass%, as shown in Table 2. These results agree with previously reported data on related materials of this same area [9,10].

Table 2. Total iron content from volumetric determinations (σ = standard deviation over the mean, estimated from results of three determinations)

Sample	Iron content (% Fe $\pm\ \sigma$)
FZ – 01	65.15 ± 0.04
FZ – 02	68.69 ± 0.04
FZ – 03	70.00 ± 0.01
FZ – 04	68.51 ± 0.05
FZ – 05	68.74 ± 0.04
FZ – 06	67.69 ± 0.04
FZ – 10	68.07 ± 0.05
FZ – 11	67.93 ± 0.02
FZ – 12	65.93 ± 0.05

The saturation magnetization values observed are consistent with the values for natural hematite, $\sigma \sim 0.1\ T^{-1}\ kg^{-1}$, except for two samples, for which $\sigma = 6.9$ and $2.1\ J\ T^{-1}\ kg^{-1}$ (samples FZ-11 and FZ-12, respectively). In these two samples, crystals of magnetite could be optically observed from microscope examination of thin petrographic sections; results were confirmed from data of XRD and ^{57}Fe Mössbauer spectroscopy at 298 K. They were then chosen to be treated with sodium dithionite-citrate-bicarbonate (DCB) [11], in an attempt to selectively remove hematite and concentrate the magnetic phase.

298 K-Mössbauer data also confirmed the dominant occurrence of hematite in all samples (Figure 2, table 3), along with some magnetite for the two samples showing non-zero magnetization (Figure 3).

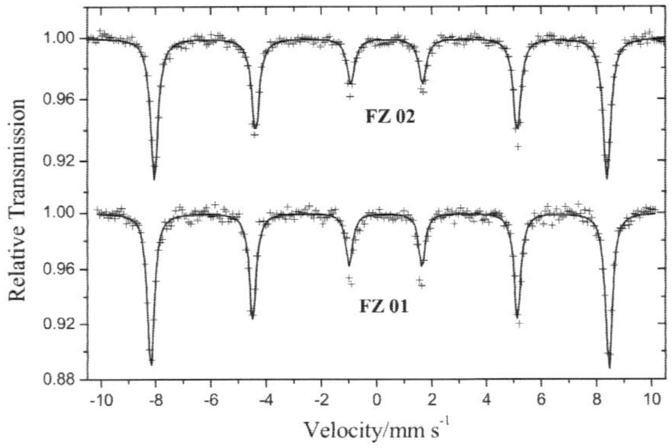

Figure 2 - 298 K-Mössbauer spectrum for sample FZ-01 and FZ-02

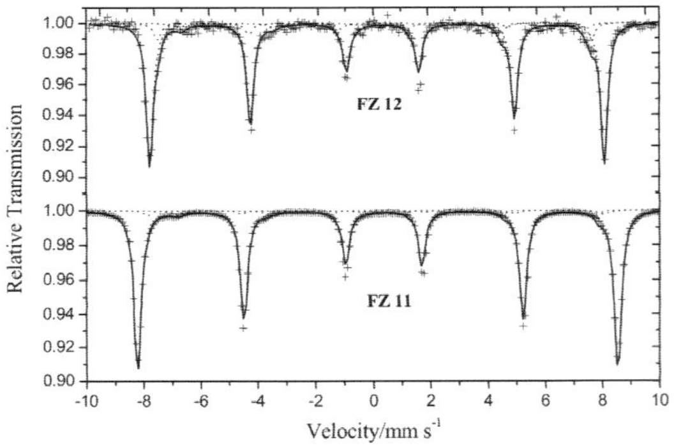

Figure 3 - 298 K-Mössbauer spectrum for sample FZ-11 and FZ-12 (magnetic samples)

Table 3. Mössbauer Parameters for samples FZ-01, FZ-02, FZ-11 and FZ-12 (δ = isomer shift relative to α-Fe; ε = quadrupole shift; B_{hf} = hyperfine field; Γ = line width; AR = relative area of the subspectrum)

Sample	Temperature/K	δ/mm s^{-1}	ε/mm s^{-1}	B_{hf}/T	Γ/mm s^{-1}	AR/%
FZ-01	**298**	0.3(7)	-0.21(1)	51.5(2)	0.32(9)	100.(2)
	110	0.4(8)	0.41(1)	54.7(6)	0.37(7)	100.(1)
FZ-02	**298**	0.36(7)	-0.2(1)	50.9(2)	0.33(8)	100.(1)
	110	0.52(7)	0.41(9)	54.1(6)	0.31(1)	100.(1)
FZ-11	**298**	0.375(3)	-0.20(6)	51.9(8)	0.30(6)	94.2(5)
		0.144(1)	0^{a}	49.3(1)	0.31(1)	2.0(0)
		0.715(7)	0^{a}	46.5(7)	0.4(2)	3.7(5)
	110	0.468(6)	0.37(5)	53.6(9)	0.33(2)	83.(1)
		0.451(1)	-0.27(2)	50.8(0)	0.6(8)	16.(9)
FZ-12	**298**	0.36(8)	-0.20(0)	51.6(3)	0.31^{a}	85.1(5)
		0.18(8)	-0.02^{a}	49.3(3)	0.2(9)	9.8(4)
		0.63(5)	-0.2(3)	46.0(1)	0.45^{a}	5.0(1)
	110	0.48(2)	0.4(1)	54.2(2)	0.37(3)	96.(3)
		0.42(1)	-0.2(5)	50.(9)	0.3(6)	3.(7)

^{a} Values fixed.

110 K-Mössbauer data indicate that the hematite in all samples undergoes the Morin transition ($T_M \approx 260$ K), as expected for the relatively pure oxide (Figure 4 and Figure 5). The 110 K-spectrum for sample FZ-01 (Figure 4) evidences a magnetically ordered component with exceptionally high hyperfine field, B_{hf} = 54.7(6) tesla, which is a somewhat unusual value for hematite at this temperature. However similar values may be found in the literature, as those reported by Spier at al. [12], for hematites at 50 K, in samples from the same banded formation of Quadrilátero Ferrífero, in Minas Gerais, of sites geographically located only about a hundred kilometers apart from our sampling area.

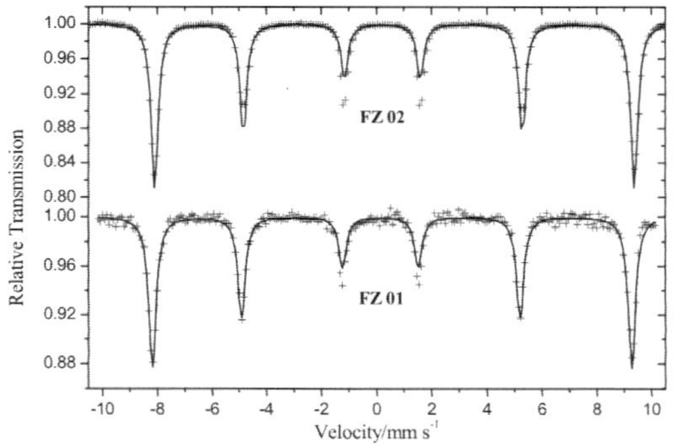

Figure 4 - 110 K-Mössbauer spectrum for sample FZ-01 and FZ-02

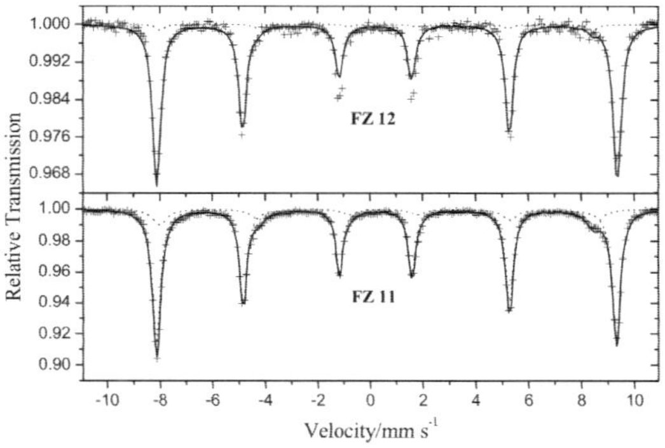

Figure 5 - 110 K-Mössbauer spectrum for sample FZ-11 and FZ-12 (magnetic samples)

4. Conclusions

Maghemite could not be detected, either from X-ray or from Mössbauer analysis, in any sample [13]. Rietveld refinement of XRD data indicates only slight, but significant, differences on dimensions of the rhombohedral-hexagonal cell, as for sample FZ-01 (which is, dominantly, specularite). This sample corresponds also to an exceptionally high hyperfine field for hematite at 110 K. New numerical analysis Rietveld refinements are now in progress, in an attempt to obtain a better crystallographic description that could indicate more reliable evidences about the mechanisms of hematite formation, particularly as they could be drawn from data for sample FZ-01 and for the magnetic samples FZ-11 and FZ-12. Results from those calculations are expected to point also to a more accurate interpretation about subtle differences on crystallographic structures, allowing us to propose consistent models for the mineralogical transformation involving hematite in this geosystem.

Acknowledgments
Work supported by Brazilian Council for the Scientific and Tecnological Development (CNPq) and Minas Gerais State Foundation for the Science and Technology Development (FAPEMIG) (Brazil).

References

[1] Costa A C S and Bigham J M In: Melo V F and Alleoni L R F (eds) 2009 *Quimica e Mineralogia do Solo – Parte I-Conceitos Básicos*. Soc. Bras. de Ciência do Solo, (Viçosa) 505-572
[2] Fabris J D and Coey J M D. In: Alvarez V V H, Schaefer C E G R, Barros N F de, Mello J W V de and Costa L M da (eds) 2002 *Tópicos em Ciências do Solo*. Soc. Bras. de Ciência do Solo v.2 (Viçosa) 47-102
[3] Rosière C A et al. 2001 *J. Struct. Geology* **23**, 1429-1440
[4] Rosière C A and Rios F J 2004 *Econ.Geology* **99**, 611-624
[5] Rosière C A et al. 2008 *The Itabirites of the Quadrilátero Ferrífero and related high-grade Iron Ore Deposits: an Overview*. SEG Reviews vol. 15, 223-254
[6] Lagoeiro L E 1998 *J. Methamorphic Geol* **16**, 415-423
[7] Coey J M D, Cugat O, McCauley J, Fabris J D 1992 *Rev.Fís. Aplic. Instrumentação* **7**(1), 25-30
[8] Cornell, R.M. and Schwertmann,U. *The Iron Oxides: Structure, Properties, Reactions, Occurences and Uses*.2003 Wiley-VCH Verlag GmbH & Co. KGaA, Weinheim pp 5
[9] Silva A M, Gibotti Jr M. *Geologia das Minas de Fazendão e Adjacências*. 1989 Monografia. UFOP (Ouro Preto) 147 p
[10] Rosière C A, Chemale Jr F, Guimarães M L V 1993 *Geonomos* **1** (1), 65-84
[11] Mehra O P, Jackson M L 1960 *Proc. Seventh Nat. (U.S.) Conj.Clay and Clays Minerals*, 1958. Pergamon Press. Lond
[12] Spier C A, Oliveira S M B, Rosière C A, Ardisson J D 2008 *Minerallium Deposita* **43**, 229-254
[13] Pereira A R P, Fabris J D, Rios F J, Rosière C A, Couceiro P R C, Furlan F F, Menezes L M 2010 *Hyperfine Interactions* **195**, 69-76

Mössbauer study of Haltern 70 amphora sherds from Castro do Vieito, North of Portugal

B F O Costa[1,5], G Pereira[2], A J M Silva[3], M Ramos Silva[1] and S J C do Carmo[4]

[1] CEMDRX, Department of Physics, University of Coimbra, P-3004-516, Coimbra, Portugal
[2]Science Museum of the University of Coimbra, P-3000-272, Coimbra Portugal
[3]Centro de Estudos Arqueológicos das Universidades de Coimbra e do Porto/CAM, Palácio de Sub-Ripas P-3004-395 Coimbra, Portugal
[4]GIAN, Department of Physics, University of Coimbra, P-3004-516, Coimbra, Portugal

Email: benilde@ci.uc.pt

Abstract. Haltern 70 amphora sherds obtained from Castro do Vieito were studied by Mössbauer spectroscopy, X-ray diffraction and X-ray fluorescence. In all our samples non magnetic Fe^{3+} and haematite species were found. In some samples, also Fe^{2+} was found. We conclude that the samples were fired under changing atmosphere, air being admitted at the end of the firing cycle following firing in a reducing atmosphere.

1. Introduction

In many areas of the world, sherds from broken pottery vessels are the most common artefacts found during the excavation of archaeological sites, and indeed sometimes the only evidence of past cultures.

Haltern 70 is the name of a particular type of amphora (a ceramic vase used in the Antiquity to transport commodities by the sea), that had been produced in the Hispania Baetica Roman province during the second half of the first century BC and the first century AC. A typical Haltern 70 amphora is shown in Figure 1. Recently, the excavation of the Castro do Vieito (Northwest of Portugal, Rio Lima estuary), an indigene settlement from the beginning of the Roman occupation of the region in the early empire, provided the most extended collection of Haltern 70 amphorae of all the roman world.

The signification of the massive presence of this type of artifact in the norhwest of the Roman province of Hispania is actually the object of an intense debate. Traditionally, the presence of the amphorae was explained by the importance of the Atlantic coast of Hispania for the ancient commercial rotes in the beginning of the Roman Empire. Today, the most consensual theory explains this fact mostly by the role of the Haltern 70 amphorae in the supply system of the roman troops who controlled the region after its definitive conquest by the emperor Augustus [1].

Haltern 70 sherds were subjected to Mössbauer spectroscopy (MS) and X-ray diffraction (XRD) studies. The combination of XRD and MS has a considerable potential in the study of the chemical and physical transformations occurring in pottery clays during firing, as non iron-bearing minerals are

[5] To whom any correspondence should be addressed.

only accessible by XRD, while iron-containing oxidic and amorphous phases may be difficult to detect by XRD due to poor crystallinity and small particle size, but can readily be detected by MS.

2. Experimental

Twelve Haltern 70 sherds obtained from the excavation of the Castro do Vieito were studied. The specimens were red in color, were no more than 10 cm long and all exhibited a slight curvature. The archeological codes of the items are used in this paper. Small quantities of powder were taken from the sherds and subjected to MS and XRD analysis.

XRD was performed at room temperature (RT) using Cu K_α radiation (λ = 0.154184 nm). Powder diffractograms were collected in a X-ray powder diffractometer equipped with a CPS120 curved position sensitive detector, from 0 to 120° 2Θ, in Debye-Scherrer geometry. ^{57}Fe Mössbauer spectra were recorded at RT in transmission geometry using a standard constant acceleration spectrometer. A ^{57}Co source in Rh matrix with strength of ≈ 20 mCi was used. The least squares fitting of the spectra were carried out with a set of Lorentzian lines. The isomer shifts are given with respect to α-Fe at RT.

X-ray fluorescence (XRF) spectrometry has been used as a qualitative chemical elemental analysis method. A gas proportional scintillation counter filled with pure Xe at atmospheric pressure and with a mylar film window [2] was used to detect the X-ray lines emitted by the specimens under study, when excited with α-particles from a ^{244}Cm source placed under vacuum. The spot area incident at the sherds had about 3 cm diameter. X-ray energy spectra were acquired for up to 15 min time.

3. Results and Discussion

Figure 2 shows only the XRF spectra obtained for one specific sherd because it is similar to all the other ones obtained for the different sherds. The elements found are Si, Mg and/or Al, Ca, Fe and small amounts of Ti or V. In our conditions it is not possible to distinguish clearly between Si, Mg and Al as all the peaks lie in the same energy range [2].

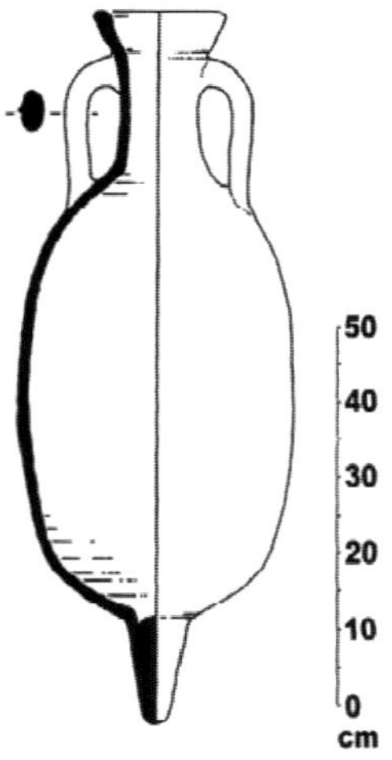

Figure 1. Characteristic form of Haltern 70 amphora [3].

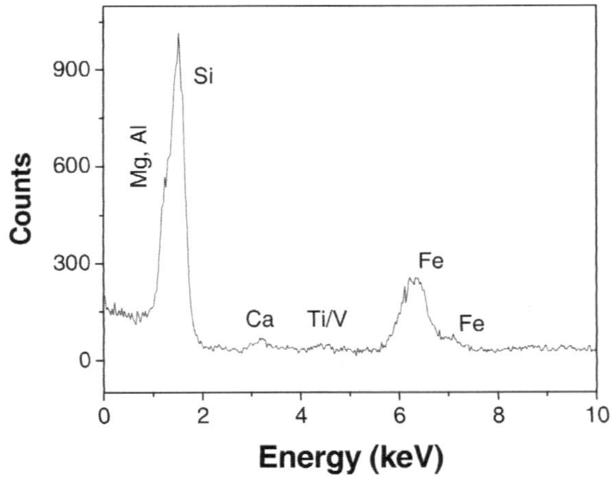

Figure 2. Representative XRF spectrum and identification of elements.

Figure 3 shows some photographs of representative sherds. The most significant feature is the sandy texture, sometimes with big inclusions of silicates. In figure 4 we present some ^{57}Fe RT Mössbauer spectra of Haltern 70 sherds. Solid lines are fits to the spectra. Mössbauer parameters obtained from representative spectra are shown in table 1.

Figure 3. Photographs and respective amplification (5x) of sherds C121 (top) and C112 (bottom). The most obvious feature is the sandy texture, sometimes fine (bottom) but in other cases with rather large inclusions (top).

The conclusions we can draw from Mossbauer spectra of clay-based ceramics depend on the chemical variability of iron, which may occur as Fe^{2+} or Fe^{3+} depending on the redox conditions in the kiln, and which may occur in a variety of silicate and oxide phases, also depending on the firing conditions.

We can derive the relative amounts of iron in individual phases from the fractional area of the respective pattern in the spectrum. Doing so, one often assumes that the Lamb-Mossbauer f-factors are the same for all components, which may not be strictly the case, but appears to be a reasonable guess for ceramic materials as long as one does not need to obtain a very high accuracy.

In raw clays (not heated) an intense Fe^{3+} doublet with a small QS of about 0.70 mm/s, is characteristic of iron contained in the structure of the clay minerals. Oxidising firing between 400 and 600ºC causes the splitting increases to 1.2 mm/s or even higher. It decreases again to below 0.8 mm/s firing above 800ºC there is an increase of the splitting until about 0.80 mm/s [4]. The small quadrupole splitting of the Fe^{3+} doublet thus indicates either a firing temperature above 800ºC or a very low one of less than 600 ºC.

It is also known that haematite forms from iron present in the clay minerals when the layer structure collapses above 800ºC and that another frequent source of haematite in fired clays are the iron oxyhydroxides, mainly goethite, that may be present in the clay and decompose to haematite between 200 and 400ºC [4].

In all our samples non magnetic Fe^{3+} and haematite species were found. Haematite was found in a small percentage, indicating that the iron in the clay matrix was not liberated during firing, which rules out firing above 800ºC in an oxidising atmosphere.

In most of the samples also Fe^{2+} was found. Firing under reducing conditions would show a fraction of Fe^{2+} higher than that obtained. It is possible that the Fe^{2+} oxide is $FeAl_2O_4$ as the Mössbauer parameters are close to this oxide [5].

The Mössbauer data for the Haltern 70 amphorae thus seem to be incompatible with a purely oxidising or a strongly reducing firing. They rather seem to have been fired under changing conditions, most probably being fired under reducing conditions and then oxidised towards the end of the firing cycle.

Recently it has been reported that similar parameters were obtained at RT for Celtic pottery. Mössbauer measurements at 4.2 K have been helpful to understand the firing conditions the sherds were subjected to [6].

XRD (figure not shown) identified SiO_2 (quartz) as the main crystalline phase common to all samples.

This indicates that sand had been added to the clay [7], as also observed in photographs of Figure 3. Other less crystalline phases, changing slightly from sample to sample, are also present in small percentages, and are mostly identified as mineral silicates. Iron oxides were not found in crystalline form; however the shape of the background indicates the presence of amorphous phases.

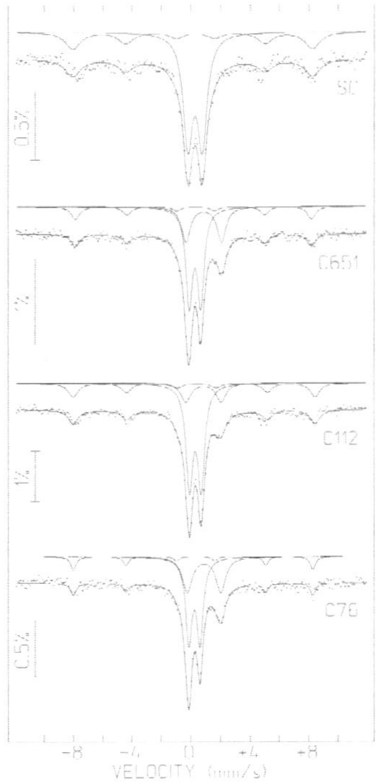

Figure 4. RT Mössbauer spectra of some of the samples.

Table 1. RT ^{57}Fe Mössbauer data for representative samples.

	QS (mm/s)	IS (mm/s)	Γ (mm/s)	B (T)	%
SC					
Fe3+	0.98(1)	0.41(2)	0.79	-	66.8
haematite	-0.20(1)	0.32(2)	1.42	51.2(2)	33.2
C112					
Fe3+	0.79(2)	0.42(2)	0.56	-	60.4
Fe2+	2.4(1)	0.97(2)	0.76	-	16.0
haematite	-0.21(1)	0.42(2)	0.95	51.5(2)	23.7
C651					
Fe3+	0.78(2)	0.40(2)	0.55	-	55.1
Fe2+	2.42(1)	1.00(1)	0.72	-	27.0
haematite	-0.18(1)	0.37(1)	0.93	50.5(2)	17.9
C76					
Fe3+	0.75(1)	0.37(2)	0.52	-	52.6
Fe2+	2.28(1)	1.03(1)	0.78	-	33.8
haematite	-0.20(1)	0.36(1)	0.67	51.0(2)	13.6
C121					
Fe3+	0.84(2)	0.37(2)	0.64	-	59.9
Fe2+	1.14(1)	2.36(1)	0.60	-	7.1
haematite	-0.19(1)	0.40(1)	0.81	50.5(1)	33.0
C499					
Fe3+	0.82(2)	0.33(2)	0.87	-	75.7
Fe2+	2.17(1)	1.07(2)	0.50	-	7.0
haematite	-0.18(2)	0.40(1)	0.83	51.0(2)	17.3

In conclusion, from the above observations it is inferred that the samples were fired under a changing atmosphere, air being admitted at the end of the firing cycle following the reducing firing, and that is the reason for red colour of the amphorae. In order to confirm this conclusion we intend to perform in the future Mössbauer measurements at 4.2K

References

[1] Silva A J M 2008 "PhD thesis in Archaeology", University of Coimbra.
[2] Do Carmo S J C, Borges F I G M and Conde C A N 2009 *IEEE Trans. Nucl. Sci.* **56** 437–40
[3] (http://www.potsherd.uklinux.net/atlas/Ware/H70)
[4] Wagner F E and Wagner U 2004 *Hyperfine Interact.* **154** 35–82
[5] McCammon C 1995 *Mineral Physics and Crystallography*, (American Geophysical Union) p 332
[6] Friederich J, Gebhard R, Krause R, Wagner F E and Wagner U 2009 *ICAME09, abstract book* p 195
[7] Gonzalez J, Gonzalez C and Gonzalez M (2004), pp. 159-164, in Carreras Monfort C et al, *Culip VIII i les àmfores Haltern 70, Monografias del CASC 5*, Museu d'Arqueologia de Catalunya – Centre d'Arqueologia Subaquàtica de Catalunya, Girona.

International Conference on the Applications of the Mössbauer Effect (ICAME 2009) IOP Publishing
Journal of Physics: Conference Series **217** (2010) 012061 doi:10.1088/1742-6596/217/1/012061

Study of geo-chemical environment in organic rich sub-surface sediments of Jaisalmer petroliferous basin of Rajasthan, India, using Mössbauer spectroscopy

K R Patel[1], S Ram, A Tripathi, R Sharma and R P Tripathi
Department of Physics, Jai Narain Vyas University, Jodhpur, India

E-mail: sr_panwar@yahoo.co.in

Abstract. A ^{57}Fe Mössbauer Spectroscopic study was carried out on brown red coloured samples collected from different depth intervals of two wells (LNR-1 & DND-1) of Jaisalmer petroliferous basin, India. All these samples show a sextet corresponding to hematite along with two quadrupole doublets one correspond to iron in siderite and other correspond to Fe^{2+} in clay minerals. Relative presence of hematite is seem to be correlated with nature of gaseous hydrocarbon discovered in this basin. Our earlier studies also pointed out such correlation between redox environment and presence/absence of hydrocarbon in a basin. We suggest that such correlation is not surprising because rate of sedimentation determines both hydrocarbon prospecting and redox environment in the sediments.

1. Introduction

Sediments rich in organic matter are known as source rocks. These rocks are tiny generators of oil and gas in a basin . As the time passes the organic matter present in sediments is modified by bacteria and undergoes thermal alteration, commonly termed as maturation. The quality and maturity of the organic matter present in source rocks have a decisive effect on the hydrocarbon prospect and accumulation pattern in the basin. A systematic and excellent description on petroleum geology and its occurrence available in literature where the importance of study of inorganic phases for characterization of source rocks is highlighted [1] .

Several workers have reported the Mössbauer spectroscopic studies on samples collected from wells drilled in different petroliferous basins. These studies show that, iron in sediments is mainly present only in four variety of minerals i.e in sulphur containing minerals, in carbonate minerals, in clay minerals (both Fe^{2+} and Fe^{3+} in octahedral site of clay minerals) and in oxides/oxyhydroxide but their relative amount varies from sample to sample and also from basin to basin.

As far as Jaisalmer Basin is concerned number of wells have been drilled in several parts of this basin but economically viable oil has not been explored so far. However many structures have yielded hydrocarbons; mainly methane accompanied with appreciable amount of nitrogen and carbon dioxide. For example in Lunar structure (LNR) nitrogen forms more than 80 percent of the gases discovered while in Dandewala structure (DND) it is less than 20 percent. In earlier studies our group have reported Mössbauer spectroscopic study on samples collected from several wells located in various structures of the Jaisalmer basin [4-10]. From Table 1, it can be seen that presence of hydrocarbons in North sea offshore wells and in Jaisalmer basin are markedly different. Main aim of earlier studies was to correlate this difference with relative distribution of iron bearing minerals in these two basins. It was observed that pyrite was the dominating phase in Offshore wells while siderite was dominating phase in Jaisalmer basin samples. This difference clearly suggest that sediments in these two basins were deposited in different geochemical (redox) conditions. Number of wells in Jaisalmer basin were drilled from top to basement. In these wells brownish red coloured samples were encountered at larger depth (only in middle Jurassic and Lower Jurassic Sedimentary Sequence). Sample at such depth were not studied earlier. None of the samples from cretaceous rocks have shown this characteristic colour In present investigation, we have carried the Mössbauer spectroscopic study on these brownish red coloured samples. Interestingly in Jaisalmer basin the Jurassic rocks also contain potential source rocks with adequate maturity. Here studies are confined to two typical wells LNR-1 and DND-1 as a representative wells.

Table 1. A Most general trend of authegenic iron-bearing minerals observed in different basins.

Basin	Iron minerals which are dominantly present in majority of samples	Status of hydrocarbon discovery	Reference	Reducing environment on the basis of authegenic minerals present in basin
Danish North Sea (Offshore)	Pyrite, Fe2+ in octahedral site in clay minerals	Oil and gas are present	2, 3	Good
Danish North Sea (Onshore)	Pyrite, appreciable presence of carbonate minerals	Mainly dry	2, 3	Poor
Eastern Krishna-Godavary	Pyrite, iron in sulphate minerals	Oil and gas are present (relatively it is gaseous field)	9	Good
Cambay	Pyrite, iron in sulphate minerals	Oil and gas are present	10	Good
Jaisalmer	Iron in Carbonates, Fe3+ in octahedral site of clay minerals	No oil but gaseous hydrocarbons mainly accompanied by large nitrogen and methane	4-8	Good

2. Experimental

Mössbauer spectra were recorded at room temperature with a conventional constant acceleration Mössbauer spectrometer using a 10 mCi ^{57}Co source (initial strength). The Isomer shift has been reported with respect to centroid of a 25 μm thick α-iron foil spectrum. Details of experimental setup, method of sample preparation and fitting programme have been already reported earlier work [4-6].

3. Results and Discussions

In Figure 1, we display some representative Mössbauer spectra recorded at room temperature for samples collected from different depth intervals of wells LNR-1 and DND-1 . Depth interval and name of the well from which the sample has been collected is marked in the Figure itself. It can be seen from Fig 1 that all the samples show presence of intense sextet with HMF value (hyperfine magnetic splitting) 516 kOe, and IS value 0.34 mms^{-1} and QS value -0.15 mms^{-1} characteristic of hematite(α-Fe_2O_3). Apart from this sextet, two quadrupole doublets are also present. One corresponding to iron in siderite marked as BB$'$ and other corresponding to Fe^{2+} in octahedral site of clay minerals marked DD$'$. In LNR-1 the thickness of sedimentary band containing hematite is

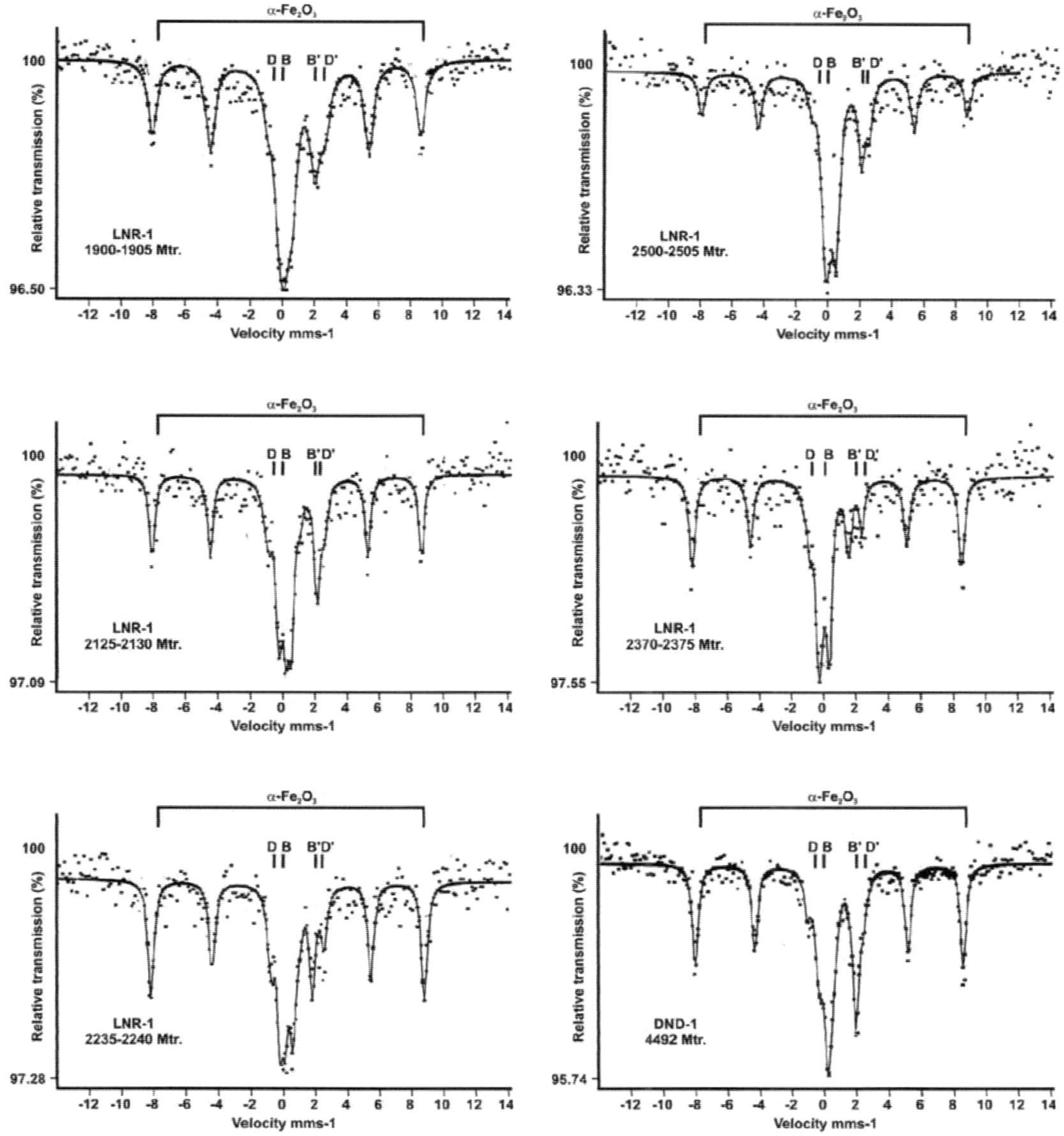

Figure 1. Room temperature Mössbauer spectra of samples collected from wells LNR-1 and DND-1. The depth from which the sample is collected is mentioned in figure itself.

600m (1900 m to 2500 m) while in well DND-1 only one sample colleted from depth 4492 m has shown presence of hematite. The sample below this was devoid of hematite. In well DND-1 (Dandewala structure) the thickness of sedimentary sequence in which hematite is present is certainly very small in comparison of LNR-1 (Lunar structure). It is worth mentioning that hematite is formed in oxidising environment.

The most important finding is that in the structure (Lunar) where rigorous oxidizing conditions are present poor quality of hydrocarbon is found in comparison of where oxidizing condition are not so rigorous (Dandewala structure). Similar trend was also observed in other part of basin (Ghotaru Structure, Bhakri-Tibba Structure etc.). On the basis of present study and studies reported earlier (see Table 1) it can be suggested that redox condition may be one of the significant marker for hydrocarbon presence/absence provided other geochemical and geophysical parameters are same.

This correlation can be explained on the basis of following facts; as it is well known that hydrocarbon generation is highly correlated with rate of sedimentation because this affects the maturity of rocks and also controls early and late digenesis[1,11]. Simultaneously, the rate of sedimentation also controls the stability of authegenic minerals. If the rate of sedimentation is fast, it quickly cuts off sediment from environment and this favours formation of minerals like pyrite which are digenetically stable in reducing environment. Apart from this organic matter also escapes oxidation due to quick sedimentation. On the other hand if deposition rate is slow, sediments remain with contact of atmosphere relatively more time. This favour formation of minerals like carbonates, goethite, hematite etc. and the organic matter will also suffer oxidation. If we put these two facts together, the correlation between hydrocarbon prospecting and redox environment is not surprising. However more studies are needed to establish correlation conclusively. The ability of Mössbauer spectroscopy to provide information of geochemical environment in sediments can be utilised in future for discovery of oil and gas in a basin.

4. References

[1] Tissot B P and Welte D H, 1984 *in Petroleum Formation and Occurrence, 2nd ed. Springer, Berlin,* p 699

[2] Mørup S and Lindgreen H B, 1981 *in Proceedings of the International Conference on the Application of Mössbauer Effect, Jaipur, India,* Ind. Nat. Sci. Acad., New Delhi, 1982, p 290.

[3] Mørup S, Frank J, Wontergham J, Poulsen R H and Larsen L, 1985 *Fuel,* Vol 64, p 528

[4] Nigam A N, Tripathi R P, Singh H S, Gambhir R S and Lukose N G, 1989 *Fuel,* Vol 68, p 209

[5] Nigam A N, Tripathi R P, Singh H S and Gambhir R S, 1991 *Fuel,* Vol 70, p 262

[6] Nigam A N, Tripathi R P, Singh H S and Gambhir R S, 1987 *Ind. J. Pure and Appl. Phys,* Vol 25, p 188

[7] Ram S, Patel K R, Sharma S K and Tripathi R P , 1997 *Fuel,* Vol 76, p 1369

[8] Ram S, Patel K R, Sharma S K and Tripathi R P , 1998 *Fuel,* Vol 77, p 1507

[9] Kulshreshtha A, Tripathi A, Agarwal T N, Patel K R, Sisodia M S, and Tripathi R P, 2004 *Fuel,* Vol 83, p 1333-1339

[10] Tripathi A, Kulshrestha A, Patel K R, Sharma Y K and Tripathi R P, 2008 *Hyperfine Interact,,* Vol 186, p 23-29

[11] Datta A K, 1983 *Petroleum Asia J.,* p 1993

In-situ Mössbauer Spectroscopy with MIMOS II at Rio Tinto, Spain

I. Fleischer[1], G. Klingelhöfer[1], F. Rull[2], S. Wehrheim[1], S. Ebert[1], M. Panthöfer[1], M. Blumers[1], D. Schmanke[1], J. Maul[1], and C. Schröder[1]

[1]Institut für Anorganische und analytische Chemie, Johannes-Gutenberg-Universität Mainz, Staudinger Weg 9, 55128 Mainz, Germany. [2]Unidad Asociada UVA-CSIC, al Centro de Astrobiología, Universidad de Valladolid, 47006-Valladolid, Spain

fleischi@uni-mainz.de

Abstract. The Rio Tinto, located in southwest Spain, exhibits a nearly constant, acidic pH-value along its course. Due to the formation of sulfate minerals, Rio Tinto is considered a potential analogue site for sulfate-rich regions on Mars, in particular at the landing site of the Mars Exploration Rover Opportunity, where the ferric sulfate mineral jarosite was identified with Opportunity's Mössbauer spectrometer. Primary and secondary mineralogy was investigated in situ with portable Raman and Mössbauer spectrometers at four different Rio Tinto sampling sites. The two techniques analyse different sample portions due to their specific field of view and sampling depth and provide complementary mineralogical information.

1. Introduction

The Rio Tinto, located in southwest Spain, exhibits a sulfate-rich mineralogy, linked to significant acidophilic biogenic activity. An important characteristic of the river is its nearly constant acidic pH value of ~2.3 along its approximately 100 km long course. Sulfates mainly form through aqueous alteration of iron-rich sulfide minerals of the Iberian Pyrite Belt [1]. The MER rovers identified sulfate minerals on the Martian surface. In Meridiani Planum, along the 14 km traverse of the rover Opportunity, jarosite was detected by the Miniaturised Mössbauer spectrometer MIMOS II [2-4]. The identification of sulfates and in particular jarosite provides clear mineralogical evidence for past water activity at Meridiani Planum and acidic conditions. The Rio Tinto region is thus considered a potential analogue site for Meridiani Planum. Micro-Raman and Mössbauer spectroscopy are among the techniques available for in-situ analyses on the Martian surface. Both instruments can be mounted on the robotic arm of a rover, constituting the contact-analytical forefront for surface mineral analysis. On a Mars mission, the combination of these techniques will provide definitive mineralogical and chemical identification of a great number of surface minerals and in particular of sulfates on Mars. To develop the operation methodology associated with mission objectives, in situ analyses were performed during a field campaign at Rio Tinto in June 2008.

For the in situ Mössbauer setup, two MIMOS II sensorheads were equipped with ^{57}Co-sources (~30 mCi). Spectra of good quality were obtained within less than two hours and at ambient temperatures of ~35 °C. The spot diameter for a measurement is 15 mm. MIMOS II sensorheads were also used for laboratory measurements. The enhanced follow-up version of the spectrometer, MIMOS IIA, will be equipped with silicon-drift-detectors (SDD) instead of Si-PIN detectors, thus providing a significantly better signal-to-noise ratio. The instrument will also be able to measure X-ray fluorescence to determine elemental composition.

Raman spectra were obtained in-situ with a portable i-Raman from B&W TEC Inc, adapted to work under field conditions. The optical head was positioned in front of the samples using a mechanical device which allows mapping the surface at near the mineral grain scale. A baffle was used to minimise the solar light background. The excitation used was a 532nm wavelength laser with ~15 mW power on the sample and a spot diameter of 100μm. Spectral resolution was 5 cm^{-1}.

To characterize the samples with an independent laboratory method, XRD measurements were recorded on finely ground samples using a Bruker AXS D8 Advance diffractometer with MoKα radiation and a SolX energy-dispersive detector in Bragg-Brentano mode.

2. Results

Four different sampling sites were studied with the aim of investigating primary volcanic rocks and different stages of weathering due to hydrological activity and evaporite minerals as the direct product of the acidic water evaporation mediated by acidophilic activity. Measurements were performed in-situ on evaporative precipitates and bedrock. Due to time constraints, Raman and Mössbauer spectra were not always taken on the exact same spot, but in close proximity to each other. In addition to the in-situ measurements, representative samples from all sampling sites were analysed in detail after the field trip with the laboratory instrument setups.

Outcrops of primary volcanic rock were investigated a few kilometers from the river. These show some signs of weathering on the surface and, due to water circulation, along minor cracks. Spectra were obtained on a drill core, which included surface material and a crack. Mössbauer spectra show a ferrous doublet, a ferric doublet and a sextet (Figure 1a). The ferrous doublet, with an isomer shift of 1.14 mm/s and a quadrupole splitting of 2.73 mm/s stems most likely from chlorite crystals in the sample. The ferric doublet, with an isomer shift of 0.36 mm/s and a quadrupole splitting of 0.69 mm/s, can be attributed to superparamagnetic goethite. The sextet, with an isomer shift of 0.39 mm/s and a magnetic hyperfine field of 51.9 T, can be attributed to hematite. Raman measurements confirmed the presence of goethite and hematite and revealed the silicate phases quartz and muscovite, which do not contain iron and are therefore not detectable in Mössbauer spectra.

Porous volcanosedimentary materials originally constituted the rock complex at the Pena de Hierro site (e.g., Figure 1 from [1]) before they were affected by hydrothermal fluids, forming rocks rich in quartz, other silicates and pyrite. Acidic weathering driven by microbial activity then led to the replacement of the sulfide ore body by gossan [5]. The resulting rocks contain black, hard material intermixed with red, softer material. A representative spectrum is shown in Figure 1b, with contributions from a ferric doublet and magnetic components. The ferric doublet with an average isomer shift of 0.37mm/s and an average quadrupole splitting of 0.64 mm/s can be attributed to superparamagnetic goethite. XRD analyses confirmed the presence of goethite in these samples instead of other iron oxides which would be consistent with the Mössbauer parameters. Crystalline goethite is also present in most of the samples, characterized by Mössbauer sextets with a distribution of magnetic hyperfine fields ranging from 26T to 38T. Most spectra show hematite with an average isomer shift of 0.37mm/s and an average magnetic hyperfine field of 51T. Some samples contain hematite with different degrees of crystallinity. Well-crystalline hematite shows a larger magnetic hyperfine field and smaller line widths. Silicate phases were detected in some samples based on a ferrous doublet in their Mössbauer spectra. The presence of hematite with different degrees of crystallinity and goethite were confirmed by Raman analyses. In addition to iron-bearing minerals, epsomite ($Mg(SO_4)\cdot 7(H_2O)$) and quartz were detected through Raman and XRD measurements, respectively.

The lithified sedimentary deposits at the "Huerto del Loco" area (e.g., Figure 1 from [1]) were found to be conglomeratic, cemented materials with clast sizes in the range up to ~3cm. The surface of the rocks is in many cases visibly affected by the acidic water and exhibits a dark crust. Several samples from the river bedrock and a drill core were taken for laboratory analyses. Mössbauer spectra show a prominent ferric doublet with an average isomer shift of 0.37 mm/s and an average quadrupole splitting of 0.54 mm/s, attributed to superparamagnetic goethite. Again, XRD analyses confirmed the presence of goethite instead of other iron oxides. Minor amounts of hematite are also present. Some clasts show ferrous doublets due to silicate phases. Raman spectra confirm the presence of goethite and hematite and identify quartz as the main silicate phase in the matrix. Two spectra obtained from the drill core are shown in Figure 1c and d.

Efflorescent precipitates and evaporite crusts can be observed as seasonal deposits in a number of locations close to the stream margin. During our field campaign, precipitates with "popcorn"-like texture a few cm in diameter were most common, their main colors were white and different shades of yellow, sometimes mixed with small amounts of red or grey. Loose crusts up to ~5mm thick and up to a few cm in diameter were encountered on some rocks close to the water, exhibiting a variety of colors including white and different shades of green and blue. A representative amount of crusts was sampled for laboratory analyses. As evaporite minerals may easily be altered by a change of their environment (e.g., temperature and humidity), representative samples were analysed in situ. Typical integration times in the field were between 30 minutes and two hours. The samples studied in situ were again analysed with laboratory equipment after ~ 8 weeks, mineralogical changes could not be detected.

The Mössbauer spectra of "Popcorn-texture" evaporites exhibit single ferric doublet, with an average isomer shift of 0.42 mm/s and an average quadrupole splitting of 0.55 mm/s (Figure 1e). This doublet results from a mixture of the ferric sulfates copiapite ($Fe^{2+}Fe_4^{3+}(SO_4)(OH)_2{\cdot}20H_2O$) and coquimbite ($Fe_2(SO_4)_3{\cdot}9H_2O$), which can be detected separately in Raman spectra. In addition to copiapite and coquimbite, minor amounts of rhomboclase ($(H_5O_2)^+Fe^{3+}(SO_4)_2{\cdot}2(H_2O)$), were detected in XRD measurements. Mössbauer spectra obtained on white, green and blue crusts show a single ferrous doublet with an average isomer shift of 1.25 mm/s and an average quadruple splitting of 3.1 9mm/s (Figure 1f), resulting from the ferrous sulfates melanterite ($Fe^{2+}(SO_4){\cdot}7(H_2O)$) and rozenite ($Fe^{2+}(SO_4){\cdot}4(H_2O)$), which can be distinguished in Raman and XRD analyses.

Portions of the river bedrock were covered with a red crust, in which jarosite ($KFe_3(SO_4)_2(OH)_6$) was detected with the Mössbauer and the Raman spectrometer. A spectrum was obtained on red crust on top of black substrate. It shows pyrite from the substrate layer, with an isomer shift of 0.33mm/s and a quadrupole splitting of 0.59 mm/s, and jarosite from the red crust with an isomer shift of 0.37mm/s and a quadrupole splitting of 1.18 mm/s (Figure 1g). Sometimes, an additional, thin, white or yellow crust covers the red crust. A spectrum obtained on light crust on top of red crust is shown in Figure 1h. Jarosite from the red crust and copiapite with coquimbite from the light crust were clearly detected. In this spectrum, jarosite and copiapite have similar isomer shifts of ~0.38 mm/s, but different quadrupole splittings of 1.15 mm/s and 0.55 mm/s, respectively. MIMOS II is able to measure two layers simultaneously. Different subspectral areas of mineral phases from both layers result from their thickness and Fe content. Raman spectra obtained on this sample show copiapite or coquimbite from the light colored top layer, jarosite was detected at some spots.

A stream with a slightly higher pH-value (~3.5) runs parallel to the Rio Tinto close to its source. Orange sediment was sampled and dried for laboratory analyses. With an isomer shift of 0.36 mm/s and a quadrupole splitting of 0.61 mm/s, it was identified as schwertmannite ($Fe^{3+}{}_{16}O_{16}(OH)_{12}(SO_4)_2$), The spectrum (shown in Figure 1i) shows a slight asymmetry, as has been reported for schwertmannite previously [6].

3. Conclusions

Mössbauer- and Raman- spectroscopy provide complementary information about a sample: Raman spectroscopy is sensitive to the very surface of a sample, and the Raman field of view is small enough (~100 μm) to perform measurements at the scale of single mineral grains. With its field of view of ~1.5 cm and sampling depth of ~200 μm, MIMOS II provides mineralogical context information about a larger volume and depth selective information about a sample by detecting backscattered 14.4 keV γ- and 6.4 keV X-rays simultaneously [7]. The two instruments also provide complementary mineralogical information: While Raman spectroscopy is sensitive to a large number of minerals and organic phases, Mössbauer spectroscopy provides detailed information about Fe-bearing mineral phases, Fe oxidation states and the distribution of Fe among them. Redox cycling of Fe can provide energy for microbial metabolism [e.g. 8]. Raman and Mössbauer spectroscopy may prove to be a valuable combination to determine organic phases and Fe oxidation states to search for past or present traces of life on Mars.

Acknowledgements
Development of the MIMOS II Mössbauer spectrometer was funded by the German Space Agency under contract 50QM99022 and supported by the Technical University of Darmstadt and the University of Mainz. The support of the Russian space agency is acknowledged.

References

[1] Fernandez-Remolar C, Morris R V, Gruener J E, Amils R and Knoll H 2005 *Earth Planet. Sci. Lett.* **240**, 149-167.
[2] Klingelhöfer G et al. 2003 *J. Geophys. Res.* **108**, 8067.
[3] Klingelhöfer G et al. 2004 *Science* **306**, 1740-1745.
[4] Morris R V et al. 2006, *J. Geophys. Res.* **111**, E12S15.
[5] Fernandez-Remolar D, personal communication.
[6] Murad E and Cashion J 2004, Mössbauer Spectroscopy of Environmental Materials and their Industrial Utilization (Kluwer Academic Publishers, Boston), 237-238.
[7] Fleischer I, Klingelhöfer G, Schröder C, Morris R V, Hahn M, Rodionov D, Gellert R and de Souza P A 2008 *J. Geophys. Res.* **113**, E06S21.
[8] Schröder C, Bailey B, Klingelhöfer G and Staudigel H 2006 *Planet. Space Sci.* **54**, 1622-1634.

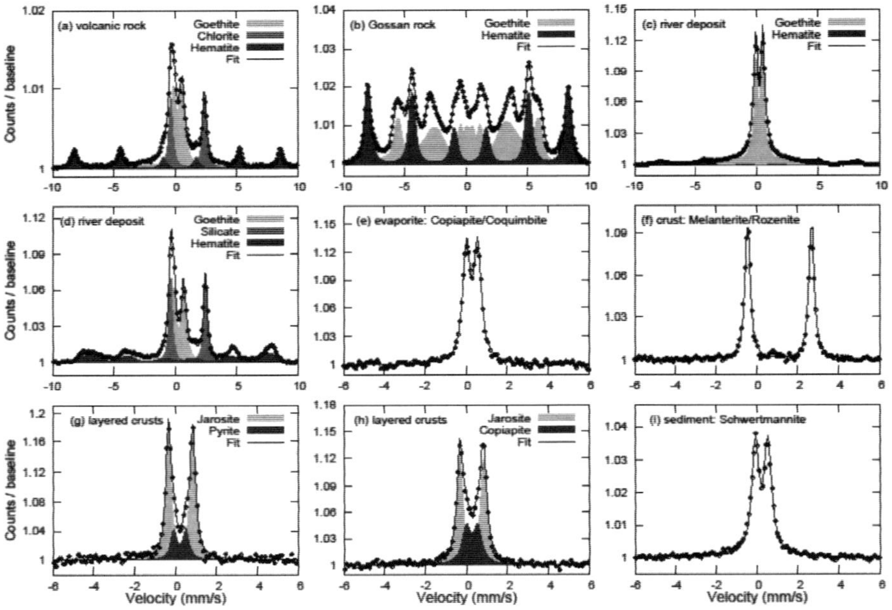

Figure 1: representative spectra from Rio Tinto samples: (a) Volcanic outcrops show minor amounts of weathering phases such as hematite and goethite. (b) Gossan rocks show mixtures of hematite and goethite. (c, d) conglomeratic river deposits are dominated by nanophase goethite and hematite. (d) Ferrous doublets due to silicate clasts are also observed. (e) The common popcorn texture evaporites are dominated by copiapite and coquimbite. (f) Ferrous sulfates (melanterite, rozenite) were identified in crust precipitates. (g) A layered sample of red crust on top of black substrate shows jarosite from the crust and pyrite from the substrate. (h) Light crust on top of red crust shows copiapite from the light crust and jarosite from the red crust. (i) Schwertmannite was identified in river sediment.

Journal of Physics: Conference Series **217** (2010) 012063 doi:10.1088/1742-6596/217/1/012063

On simfitting MER Mössbauer data to characterize Martian hematite

David G Agresti[1], Iris Fleischer[2], Göstar Klingelhöfer[2] and Richard V Morris[3]

[1] University of Alabama at Birmingham, Birmingham, AL 35294-1170, USA
[2] Institut für Anorganische und Analytische Chemie, Universität Mainz, Germany
[3] NASA Johnson Space Center, Houston, Texas 77058, USA

E-mail : agresti@uab.edu ; fleischi@uni-mainz.de

Abstract. Mössbauer spectra of Eagle Crater outcrop rocks in Meridiani Planum were acquired by the Mars Exploration Rover (MER) Opportunity. Sixty spectra, containing ~20 to 60% hematite by area, were simultultaneously fit (simfit) in a self-consistent manner to a single chi-squared minimum, where relations among parameters from different spectra were defined for both sol (Martian day) and acquisition temperature (200–280 K). Different spectral models were compared, hematite being modeled optimally with two sextets. Sextet S1 (~35% of total sextet area) has narrower linewidths, a larger magnetic hyperfine field, and a quadrupole shift that changes smoothly from positive to negative values as the temperature increases through the bulk Morin transition temperature. Sextet S2 has broader linewidths, a likely skewed line shape, a smaller hyperfine field, and a quadrupole shift that remains negative at all temperatures, implying the S2 phase is weakly ferromagnetic at all temperatures.

1. Introduction

Hematite (α-Fe_2O_3) is paramagnetic above the Curie temperature (T_C = 956 K) and weakly ferromagnetic (wfm) below T_C, down to the Morin temperature T_M, where hematite undergoes a phase transition to an antiferromagnetic (afm) state [1, 2]. For well crystalline, chemically pure hematite, T_M = 264 K [3]. Wfm and afm phases can coexist over a range of temperatures ΔT_M (e.g. [4]). Small particle sizes, poor crystallinity, and cation substitution can reduce T_M and broaden the temperature interval ΔT_M over which the transition occurs.

On Mars, backscatter Mössbauer spectra were acquired in 10-K-wide temperature windows between 180 K and 290 K with the Mössbauer spectrometers (MIMOS II) on board the two Mars Exploration Rovers (MERs) [5]. Here, we focus on the hematite component identified in spectra of outcrop rocks from Eagle Crater acquired by the MER rover Opportunity at Meridiani Planum during sols 16–51 (Martian days after landing).

Analyses of Mössbauer spectra of Eagle Crater outcrop were reported in [6], for which single-sol spectra were summed over several temperature windows and fit individually, with the hematite component modeled as a single sextet. Given the possibility that afm and wfm phases may coexist in the temperature range over which these spectra were acquired, we investigated two-sextet models for hematite, fitting unsummed spectra from many sols and temperatures simultaneously (simfitting).

Compared with single-spectrum fits, multi-spectrum fits result in reduced correlations among parameters because parameters in different spectra are held in fixed relations (e.g. equal) during a fit. As a result, fitted parameter values are automatically consistent among all spectra in the simfit set, whereas repeated single-spectrum fits might otherwise be required to arrive at compatible parameter values. Simfits were performed with *MERFit* [7] after spectrum extraction, velocity calibration, and fold-summing to ~256 data values with *MERView* [8]. Written in Lahey-Fujitsu Fortran, both programs are Windows-based, enabling data entry and control through specially designed dialog boxes, with graphical display of spectra that can change as parameter values are changed.

2. Two-dimensional simfitting

In a preliminary application [9], Mössbauer spectra from Eagle Crater outcrop were organized into equal-temperature groups. Each group was simfit separately, and sextet hyperfine parameters were held equal for all sols within the group, resulting in unique hyperfine values at each temperature. However, because a given sol appeared in several different simfit groups, there was no way to establish sol-related equalities. Consequently, temperature-independent quantities, such as most doublet hyperfine parameters and relative doublet areas for a given sol, were not uniquely determined in this "one-dimensional" simfitting procedure.

Here we use "two-dimensional" simfitting capability, where parameter relations are established for both environmental dimensions, sol and temperature, in the same simfit. We analyzed 60 Eagle Crater outcrop spectra with a single two-dimensional simfit. We assume all samples contain the same hematite and use the simfit results to compare four distinct temperature-dependent, two-sextet models for this hematite. The other Fe-bearing phases are modeled identically for all 60 spectra. *Exactly one completed simfit is required to establish all parameter values for a given spectral model for all sols and temperatures of the entire 60-spectrum set.*

3. The 60-spectrum simfit

The 60 spectra included in the simfit set are listed in Table 1 by sol (row) and mid-window temperature (column). Also tabulated are the total number of spectra (N) for a given sol (right column) and temperature (bottom row). Spectrum numbers (1-60) are used by *MERFit* to identify spectra over which particular parameters are related.

Table 1. 60 Eagle Crater outcrop spectra.

	Temperature (K)								
	205	215	225	235	245	255	265	275	N
B16	1	2	3	4	5	6			6
B29		7	8	9	10	11	12	13	7
B30	14	15	16	17	18	19	20	21	8
B32	22	23	24	25	26	27	28		7
B35	29	30	31	32	33	34	35	36	8
B43							37	38	2
B45	39	40	41	42	43	44	45		7
B46					46	47	48	49	4
B49					50	51	52		3
B51	53	54	55	56	57	58	59	60	8
N	6	7	7	7	9	9	9	6	60

Table 2. Models used for the simfits.

	Sextet S1		Sextet S2	
Model	Skew	Widths	Skew	Widths
S1e2e	no	equal	no	equal
[a] S1e2d	no	equal	no	[b] w+d
S1e2ke	no	equal	[c] yes	equal
S1e2kd	no	equal	yes	w+d

[a] Model used by [9]
[b] Sextet widths are broadened outward from the center as w+2d, w+d, w, w, w+d, w+2d.
[c] Peaks are composed of two half-Lorentzians joined at their midline, the left one with half-width = ½W (1–Skew) and the right one with half-width = ½W (1+Skew). W is full width at half maximum for the skewed Lorentzian.

3.1. Spectral models

Spectra were modeled with four symmetric Lorentzian doublets, one each for olivine, pyroxene, an unidentified ferric doublet phase Fe3D3 [6], and jarosite, and two Lorentzian sextets (skewed for some models) for hematite. Four models were defined, as listed in Table 2, differing in the treatment of line shape (skew) and width of the second sextet S2. Separate two-dimensional simfits were carried out for each model.

An example from a simfit with model S1e2kd (see Table 2) is shown in Figure 1, which also illustrates the effect of skew. The absolute value of the skew parameter is the same for all six peaks of the sextet, being positive for peaks 1–3 and negative for peaks 4–6.

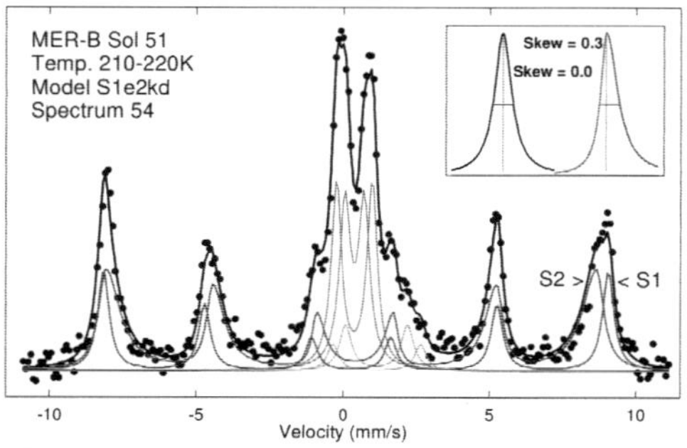

Figure 1. Example fit with model S1e2kd. Skew parameter is illustrated at top (see text).

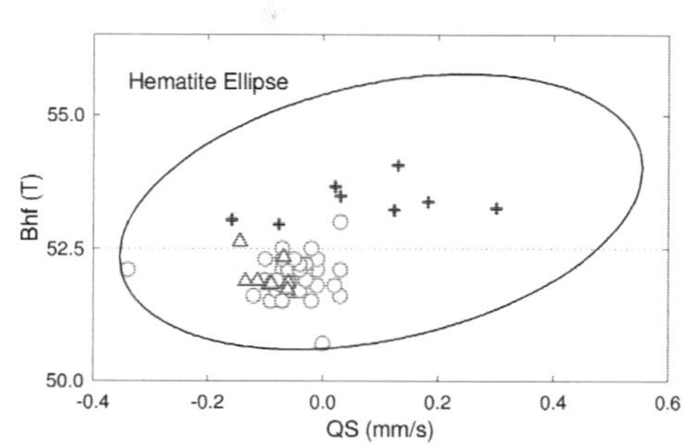

Figure 2. Phase plot for hematite. Values for Burns Outcrop Class (o) are from [6]. S1(+) and S2(Δ) are from model S1e2kd.

3.2. Parameter constraints

Simfitting employs single-spectrum constraints to define the details of a spectral model, the model and single-spectrum constraints being the same for all spectra of the simfit set. Multi-spectrum constraints are used to establish relationships among same-named parameters in different spectra.

Single-spectrum constraints used here: Doublet widths and areas were paired equal. Center shift CS (referenced to α-Fe) and quadrupole splitting QS for the weaker olivine and pyroxene were fixed to values given in [6]. Center shift CS, quadrupole shift QS, and hyperfine field B_{hf} were varied independently for each sextet, and peak areas were fixed in the ratio 3:2:1:1:2:3. The nature of sextet width constraints is shown in Table 2. When present, a single skew parameter refers to all six peaks (see example, Figure 1). We used $g(3/2^-)/g(1/2^-) = -0.572$ [10].

Multi-spectrum constraints used here: Doublet parameters except areas were held equal for all spectra. Doublet area ratios were held equal for all spectra of a given sol. Sextet parameters and area ratios were held equal for a given temperature.

4. Simfitting results

In Figure 2 we compare sextet hyperfine parameters QS and B_{hf} for hematite derived by two-sextet simfitting of Eagle Crater outcrop spectra with the one-sextet model of Burn Outcrop Class rocks [6]. The one-sextet model leads to a cluster of points (open circles) whose spread in values results in part from the different temperature ranges for each spectrum [6] and, as discussed next, the presence of a second hematite sextet. Using simultaneous fitting we obtained unique values of QS and B_{hf} at each temperature for each sextet of the two-sextet model. The data points for sextet S2 (triangles) overlap those for the single-sextet model, which is expected because sextet S2 is ~65% of total sextet area. For sextet S1 (~35% of total sextet area) the points (crosses) are clearly distinct from those for S2. As shown in Figure 3, S1 has a diffuse Morin transition because there are both positive and negative

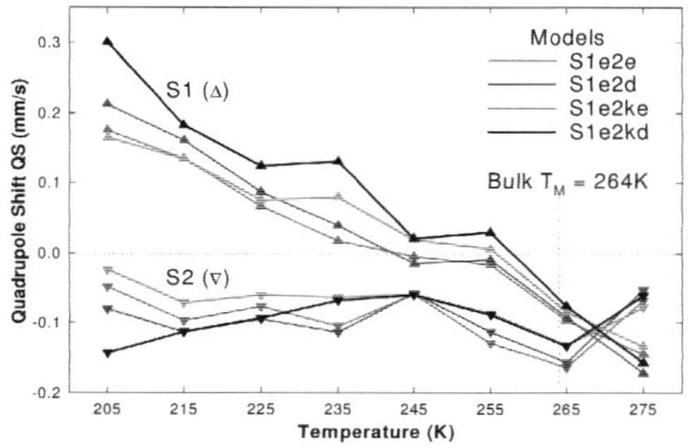

Figure 3. Fitted quadrupole shifts QS(S1) and QS(S2) (mm/s) for all 4 simfit models.

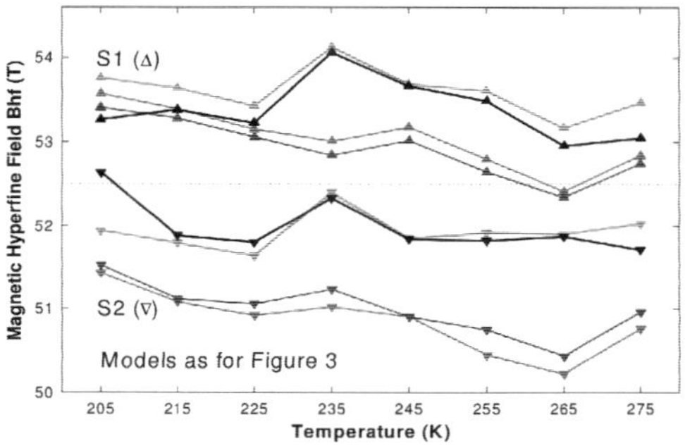

Figure 4. Fitted hyperfine fields B_{hf}(S1) and B_{hf}(S2) (T) for all 4 simfit models.

values of QS, while S2 lacks a Morin transition because its QS is always negative. Figure 4 shows values for the magnetic hyperfine field B_{hf} as a function of temperature for the two-sextet model.

Generally, for all models: **1**) QS(S1) decreases quasi-linearly with temperature, changing sign at T ~ 255 K (Figure 3); **2**) QS(S2) < 0 at all temperatures, and is nearly constant; **3**) B_{hf}(S1) > ~52.5 T > B_{hf}(S2) (Figures 2 and 4), both decreasing slightly with temperature; **4**) B_{hf}(S2) is ~1 T larger for skew models, likely reflecting outward shifting of peak area; **5**) CS(S1) ≈ CS(S2) at all temperatures; **6**) S2 linewidths > S1 linewidths, with no clear temperature dependence; **7**) models with skew converge to smaller chi-squared values than models without skew; **8**) Skew increases strongly with temperature, from ~0.34 to 0.60 for S1e2ke and from ~0.32 to 0.46 for S1e2kd (based on fitted linear trend lines); and **9**) sextets S1 and S2 are ~35% and 65% of total sextet area, respectively.

It is instructive to contrast single-spectrum and multi-spectrum fitting from a statistical point of view. Fitting one spectrum with model S1e2kd, and the given single-spectrum constraints, implies 24 free parameters for the 256-value data matrix. Simultaneous fitting of 60 such spectra with no multi-spectrum constraints would use 1,440 independent parameters to fit 15,360 values. Multi-spectrum constraints leave 98 doublet (90 area + 8 non-area) and 148 sextet parameters (68 area + 80 non-area), for a total of 246 independent parameters to fit the same 15,360 data values.

5. Conclusions

Simultaneous fitting is an effective method to fit a large number of MER spectra in a systematic way because summing of spectra to enhance statistics is avoided, fewer parameters are held fixed, sol-related quantities are unique, and deriving parameter values as a function of temperature is enabled. The method applied to 60 Mössbauer spectra from Eagle Crater outcrop demonstrates that modeling the hematite component with two independent hematite sextets provides additional information and statistically better fits compared to a one-sextet model [6] or a two-sextet model with identical linewidths for both sextets [11]. Sextet S1 has narrower lines, larger B_{hf}, and a QS that changes sign at T ~ 255 K, resulting from a diffuse Morin transition (bulk T_M ~264 K). Sextet S2 has broader, likely skewed lines, a smaller B_{hf}, and QS < 0 at all temperatures (i.e. weakly ferromagnetic with no Morin transition). The results are consistent, for example, with sextet S1 corresponding to a population of hematite particles with larger mean particle diameter than the population corresponding to sextet S2.

Acknowledgments

MERFit development was supported by NASA grant NNX06AD93G to DGA, who has benefited from discussions on hematite with R. Vandenberghe. IF and GK acknowledge support from the German Space Agency (DLR) and the University of Mainz. RVM acknowledges support from the NASA Mars Exploration Program.

References

[1] Cornell R M and Schwertmann U 1996, The Iron Oxides (VCH Publishers), 120–123.
[2] de Grave E and Vandenberghe R E 1990, *Phys Chem Minerals* **17,** 344–352.
[3] Amin N and Arajs S 1987, *Phys. Rev. B* **35,** 4810–4811.
[4] Murad E and Cashion J 2004, Mössbauer Spectroscopy of Environmental Materials and their Industrial Utilization (Kluwer Academic Publishers, Boston), 237–238.
[5] Klingelhöfer G et al. 2003 *J. Geophys. Res.* **108**, 8067.
[6] Morris R V et al. 2006 *J. Geophys. Res.* **111**, E12S15.
[7] Agresti D G and Gerakines P A 2009 *Hyp. Interact.* **188**, 113–120.
[8] Agresti D G, Dyar M D and Schaefer M W 2006, *Hyp. Interact.* **170**, 67–74.
[9] Fleischer I, Agresti D G and Klingelhöfer G 2009 *Lunar Planet. Sci.* **40**, Abstract #1803.
[10] Guetlich P, Link R F and Trautwein A 1978, Mössbauer Spectroscopy and Transition Metal Chemistry (Springer-Verlag, New York).
[11] Van Cromphaut C, de Resende V G, de Grave E, van Alboom A, Vandenberghe R E and Klingelhöfer G 2007, *Geochim. Cosmochim. Acta* **71**, 4814–4822.

Mössbauer study of Celtic pottery from Bopfingen, Baden-Württemberg, Germany

J Friederich[1], R Gebhard[2], R Krause[1], J Riederer[3], F E Wagner[4], U Wagner[4]

[1]Institut für Archäologische Wissenschaften, Johann Wolfgang Goethe-Universität, 60323 Frankfurt, Germany
[2]Archäologische Staatssammlung München, 80538 München, Germany
[3]Rathgen-Forschungslabor, Schloßstraße 1a, 14059 Berlin
[4]Physik-Department E15, Technische Universität München, 85747 Garching, Germany

E-Mail: uwagner@ph.tum.de

Abstract. In the course of a study of pottery from Celtic Central Europe, we report on an investigation of pottery found in the context of a Celtic ditched square enclosure near the modern town of Bopfingen in Baden-Württemberg, Germany. The studied pottery may be visually subdivided into wheel-turned ware, graphite ware, and coarse ware. The Mössbauer spectra are surprisingly uniform, indicating that all types of pottery were fired in a similar manner. Firing in a reducing environment at 800 °C was usually followed by re-oxidation during cooling, which leads to unique features in the low temperature Mössbauer patterns.

1. Introduction

The pottery of the Celtic civilisation in Central Europe exhibits a considerable degree of uniformity. Apparently it was made locally, but the style and methods of production were similar throughout the Celtic world during several centuries before the beginning of the Christian era. An outstanding feature of the Celtic ceramics is the abundance of graphite pottery, which was made by mixing natural graphite into the clay and hence required graphite deposits at a viable distance to the production site. This is probably the reason why graphite pottery is abundant only east of the Rhine.

In the course of an ongoing study of Celtic pottery [1], we studied about 60 pottery fragments found in or near a ditched square enclosure from the 2nd century BC, with features from the preceding 3rd century, excavated near the modern town of Bopfingen in Baden-Württemberg, Germany, during the years 1989-1992 [2]. Bopfingen is situated about 100 km west of the Celtic oppidum of Manching, one of the biggest Celtic settlements in Central Europe. By their appearance, the finds from Bopfingen were classified as graphite ware, wheel-turned ware and coarse ware. The main purpose of this study is to gain information on the material properties of the sherds.

The methods used to study the material are Mössbauer spectroscopy [3, 4], on which this paper will focus, as well as neutron activation analysis (NAA) [5] and thin-section microscopy [6].

2. Samples and Experiments

The sherds were ground in a mortar and absorbers of about 100 mg/cm^2 thickness were prepared.

Mössbauer spectra of all samples were recorded at room temperature (RT) and 4.2 K with sources of ^{57}Co in rhodium that were always at the same temperature as the absorber. Isomer shifts are always given with respect to the source. In some cases, Mössbauer spectra were also taken at temperatures between 4.2 K and RT. Moreover, some samples were re-fired in the laboratory before being subjected to Mössbauer spectroscopy.

3. Results and discussion

About 60 samples were studied in order to obtain a set of representative data. The sherds are mainly dark brown or dark grey; only a few are brick-coloured. NAA showed a rather wide spread of element contents overlapping in part with that of pottery from the oppidum of Manching [7, 8]. Thin section microscopy revealed a rather varied composition of the ceramic material in terms of mineral content, texture and particle size of the constituents.

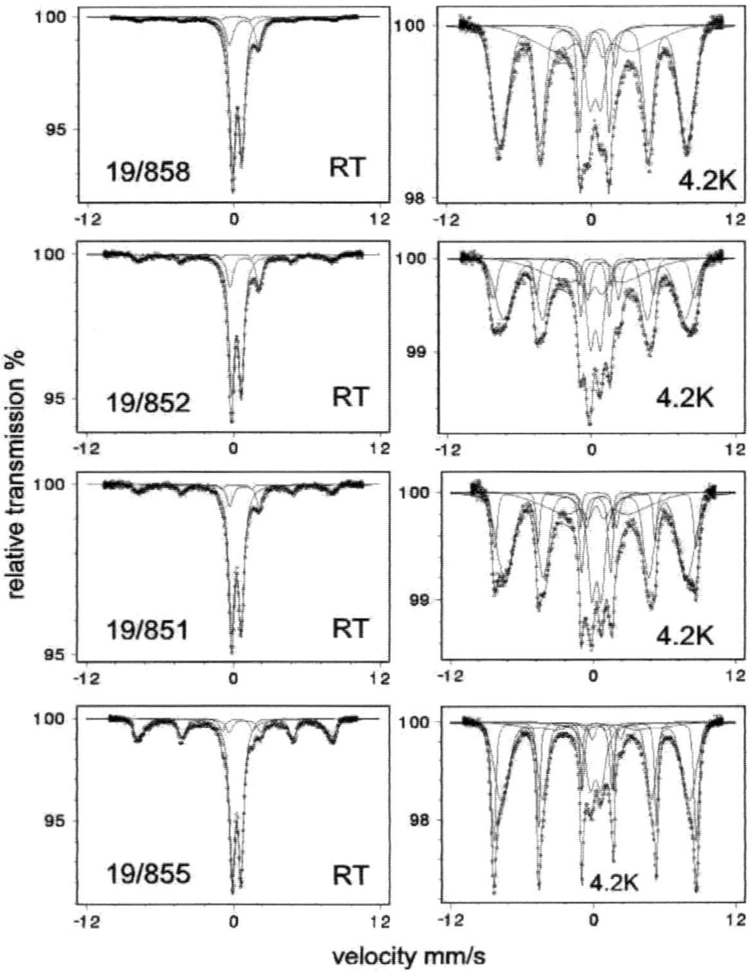

Figure 1. ^{57}Fe Mössbauer spectra of pottery from Bopfingen. RT spectra are shown on the left, 4.2 K spectra on the right. The numbers given on the left identify the individual sherds.

The Mössbauer spectra are, however, remarkably uniform. With very few exceptions, they are all like those shown in figure 1. RT spectra consist of a dominant Fe^{3+} quadrupole doublet with an isomer shift (IS) of 0.33 mm/s and a quadrupole splitting (QS) between 0.7 and 0.8 mm/s. An additional weak Fe^{2+} doublet (IS=1.02 mm/s; QS=2.6 mm/s) has intensities between 10 and 20%. Only in a few cases, more than one Fe^{2+} doublet is present. In all spectra, one also finds a weak (5–15%) magnetic sextet pattern with hyperfine fields between 48.0 and 49.5 T. In some cases this can be identified as hematite by its quadrupole shift near −0.21 mm/s, but in most cases the quadrupole shift is near zero, which is typical for maghemite. Intermediate values may be due to a mixture of both.

In the 4.2 K spectra one observes mainly a broad magnetic hyperfine pattern that can be fitted assuming a Gaussian distribution of hyperfine fields with mean values of 48 to 49 T and a variance of 3 to 4 T. The quadrupole shift is close to zero, and the IS around 0.22 mm/s. In some spectra, nearly all the Fe^{3+} yielding the doublet at RT splits in this manner, with only a weak Fe^{3+} quadrupole doublet remaining (figure 1, sample 19/858). In most cases, a more substantial fraction of the iron is still paramagnetic at 4.2 K. The temperature dependence of the spectra was studied in detail in two cases, which exhibited a very similar behavior: The intensity of the broad magnetic pattern begins to decrease already at about 25 K; at 40 K its intensity is only about half of that at 4.2 K, and at 70 K it has vanished altogether. While its intensity decreases, the hyperfine field of the remaining magnetically split component is virtually unchanged.

This behaviour is reminiscent of that of 2-line ferrihydrite, which becomes superparamagnetic in the same temperature range [9]. The presence of ferrihydrite in ceramics is, however, difficult to understand, since ferrihydrite converts into hematite at about 300°C. In pottery it could therefore only be formed during burial by contact with iron-bearing groundwater.

However, the broad magnetic component cannot be ferrihydrite, since a sample that was heated in the laboratory in air to temperatures of 300, 500, 700 and 800 °C for 24 hours still showed it even on heating to 800 °C, though with reduced intensity. The sample that was subjected to these re-firing experiments contained 15 % of maghemite according to its RT spectrum. This was found to be still present on heating to 700 °C; on heating to 800 °C it had converted to hematite.

The broad magnetic component also occurs in the Mössbauer spectra of pottery from other locations (e.g., [10]), but it is particularly prominent in the pottery from Bopfingen. It thus seems to form easily during firing, but the firing conditions or the clay composition of the ceramics from Bopfingen appear to have been particularly favorable to its formation. It does not seem to be a common iron oxide phase. The distribution of hyperfine fields indicates that it is a spin-glass like magnetic system. In fact, it probably also is a structurally amorphous phase.

When clays are fired at around 800°C, the sheet silicate structure of the de-hydroxylated clays will collapse into an amorphous state, from which new crystalline phases (pyroxenes, spinels) may crystallize at somewhat higher temperatures. The vitreous state is particularly favored during reducing firing; Fe^{2+} readily dissolves in it. The Mössbauer spectra of this amorphous phase exhibit mainly a Fe^{2+} quadrupole doublet at both RT (QS= 2.6 mm/s; IS=1.03 mm/s) and 4.2 K (QS =2.8; IS = 1.03 mm/s) [7, 11]. When this phase is oxidized at or above 800 °C, the iron exsolves as hematite. A subsequent reduction at 800 °C has been found to result in a re-dissolution of the hematite in the amorphous phase. In fact, this process can be repeated many times [11].

When the oxidation occurs at a temperature that is too low for diffusive processes to take place, the iron will remain in the vitreous phase and, when its concentration is sufficiently high, it may order magnetically at low temperatures, probably as a spin glass. The appearance of the amorphous, Fe^{3+}-bearing phase would thus be connected with firing the pottery near 800°C under reducing conditions and then, at the end of the firing cycle, re-oxidising the material at a lower temperature. This explains the mainly trivalent state of the iron in the spectra of the Bopfingen pottery. The gradual disappearance of the magnetic pattern of the vitreous phase when the temperature is raised above 4.2 K can be understood in terms of a distribution of ordering temperatures, though superparamagnetism of nano-scale ordered regions may also play a role.

In the case of the Bopfingen pottery, the oxidation seems to have taken place at a temperature where only a minor exsolution of iron into oxides would occur. When the oxide formation occurred at a sufficiently high temperature, hematite would form, while maghemite may have formed at lower temperatures. The temperature limit of maghemite formation probably lies somewhere between 700 and 800°C, since we found maghemite in the Bopfingen ceramics to transform into hematite between 700 and 800°C. This explains that some of the sherds contain hematite and some contain maghemite.

The described phenomena seem to have occurred in a similar manner in all types of pottery from Bopfingen (graphite ware, wheel-turned ware and coarse ware). Thus, the firing seems to have followed the same procedure for all these types of pottery. The predominance of the vitreous phase containing Fe^{3+} may also be due to the special composition of the clay used to make this pottery. The initially reducing firing was probably chosen because the potters wanted to avoid oxidation of graphite in graphite ware, and the pottery not containing graphite was fired in the same way. The graphite as such does not noticeably influence the Mössbauer patterns of the pottery, which also indicates that the graphite did not contain noticeable amounts of iron.

One should also note that the described process would explain the frequent presence of maghemite in graphite pottery from other locations, like Manching or sites in Bohemia [7, 11]. If the cooling is sufficiently slow, the exsolution of iron may be nearly complete, and maghemite may form if this happens at a sufficiently low temperature. Graphite ware from Manching and Bohemia is, however, often found to contain mainly divalent iron in the amorphous phase [7, 11]. In these cases, the cooling of the pottery in the kiln apparently occurred under persistently reducing conditions.

Laboratory firing experiments intended to test the firing sequence described above are presently under way. First results show that Mössbauer spectra like those observed for the Bopfingen pottery are, indeed, obtained for samples fired in the described manner.

4. Conclusions

One may thus conclude that the firing conditions of the graphite ware, the wheel-turned ware and the coarse ware were quite similar over a wide region of Celtic Europe. Firing apparently occurred at 800 to 850 °C, and the differences observed in the Mössbauer spectra of the pottery are mainly due to differences in the cooling procedure, i.e., the atmosphere during cooling and the speed of cooling.

For the pottery found at the ditched square enclosure near Bopfingen, even the cooling conditions seem to have been remarkably uniform and favourable for the preservation of the Fe^{3+}-bearing vitreous phase that gives rise to the broad magnetic pattern at 4.2 K and becomes paramagnetic between 20 and 50 K.

References

[1] Gebhard R and Wagner U, 2002 *Kolloqu. z. Vor- und Frühgeschichte* **7** 243-252
[2] Krause R and Wieland G, 1993 *Germania* **71** 59-112
[3] Wagner F E and Kyek A, 2004 *Hyp. Int.* **154** 5-33
[4] Wagner F E and Wagner U, 2004 *Hyp. Int.* **154** 35-82
[5] Glascock M D, Neff H and Vaughn K J, 2004 *Hyp. Int.* **154** 95-105
[6] Riederer J, 2004 *Hyp. Int.* **154** 143-158
[7] Gebhard R, Bott R, Distler N, Michàlek J, Riederer J, Wagner F E and Wagner U, 2004 *Hyp. Int.* **154** 199-214
[8] Bott R D, Grosse G, Wagner F E, Wagner U, Gebhard R and Riederer J, 1994 *Naturwissenschaften* **81** 560-562
[9] Schwertmann U, Friedl J and Kyek A, 2004 *Clays and Clay Min.* **52** 221-226
[10] Lumbreras L G, Gebhard R, Häusler W, Kauffmann-Doig F, Riederer J, Sieben G and Wagner U, 2003 *Hyp. Int.* **150** 51-72
[11] Bott R D, Gebhard R, Wagner F E and Wagner U, 1994 *Hyp. Int.* **91** 639-644

Origin of the two quadrupole doublets in NAu-1 nontronite

J D Cashion[1], W P Gates[2] and G M Riley[1]

[1] School of Physics, Monash University, Melbourne, Vic 3800, Australia
[2] Department of Civil Engineering, Monash University, Melbourne, Vic 3800, Australia

john.cashion@sci.monash.edu.au

Abstract. Mössbauer spectra of octahedral Fe^{3+} in NAu-1 nontronite, saturated with Na^+, K^+, NH_4^+, Ca^{2+} and La^{3+} have been analysed as two quadrupole doublets. The interlayer cations were found to affect the mean isomer shift, the quadrupole splitting and the spectral area. By considering the statistically significant nearest cation neighbour coordination environments, the inner doublet has been assigned to an octahedral Fe^{3+} surrounded by three Fe^{3+} in the octahedral sheet and eight Si^{4+} in the tetrahedral sheets. The outer doublet is due to the same octahedral coordination, but combined with [$7Si^{4+}$, $1Al^{3+}$] and [$6Si^{4+}$, $2Al^{3+}$].

1. Introduction

The true end-member nontronite contains ferric iron [1] in the octahedral sheet, although Fe^{3+} is partly replaced by Al^{3+} (and to a lesser extent Mg^{2+}) in most nontronites. They also typically have Al^{3+} and/or Fe^{3+} substitutions (~1 in 8 atom) in the tetrahedral sheet [2]. Substituted Al^{3+} often occupies preferentially tetrahedral sites so that some well-characterised nontronites (e.g. Garfield) contain Fe almost entirely in the octahedral layers [3, 4, 5]. However, since both Fe^{3+} and Al^{3+} can be in both tetrahedral and octahedral sites, variations in nontronite chemistry can produce distributions in Fe^{3+} occupancy within sites of both the octahedral and tetrahedral sheets [6].

Considerable disagreement exists in the literature regarding fitting Mössbauer spectra of nontronite and other smectites and in the interpretation of the origin of doublets. Pitfalls associated with fitting nontronite spectra are well covered previously [7]. Analysis of the spectra is complicated by overlapping lines and because a distribution of quadrupole splittings is often used, but the fitted parameters depend markedly on the quality of the data and ambiguous fits are frequent. For example, the two most common doublets (quadrupole splitting, QS, of ~0.25 and ~0.65 mm/s, and with similar isomer shift, (IS) have been assigned in the past to ferric iron in octahedral layers occupying *cis* and *trans* sites [8, 9], but, perhaps more correctly, also to ferric iron in octahedral layers occupying *cis* sites having different ions in the neighbouring octahedral and tetrahedral sites [3, 10, 11]. The existence of a tetrahedral site in many nontronites has been controversial, and fitted parameters are inconsistent; IS = 0.16 mm/s and QS = 0.50 mm/s being most common. Gates et al. [6] observed that the results of some fittings were incompatible with the actual chemistries of the nontronites under study.

We have developed [11] a systematic means of identifying potential nearest neighbour cation arrangements within the octahedral sheet, based on combining chemistry and infrared data [12]. Each Fe atom has three nearest octahedral cation neighbours, Fe, Al or Mg, and an array can be produced which details the distribution of octahedral neighbours to Fe, namely Fe(xyz), where x, y and z are integers (which must sum to 3) representing, respectively, octahedral Fe, Al and Mg atoms. Thus, Fe(210) represents a central Fe atom surrounded by 2 Fe and 1 Al atoms in the trigonal array of the octahedral sheet. An important advance was to find real examples in which a doublet can be identified by other means, such as spectra of several closely related clay minerals. In the case of *cis*-vacant ferruginous smectites, we identified [11] that the doublet with the largest splitting was due to a *cis*-Fe^{3+} ion which contained at least one Mg^{2+} ion as an octahedral neighbour. For nontronites, however, substitutions of Mg in the octahedral sheet are negligible. Thus, another explanation for the occurrence of two doublets in nontronites is needed.

The array[11] can be expanded to incorporate the tetrahedral sheet. In the *trans*-vacant nontronites [13], tetrahedral Si^{4+} is substituted by either Fe^{3+} or Al^{3+}, so a given octahedral Fe centre could have as nearest neighbour environments eight Si, Fe or Al in the tetrahedral sheets, in addition to Fe, Al and Mg in the octahedral sheet. Thus the predominant sites of a central octahedral Fe in a representative nontronite would be: Fe(300)[Si_8], Fe(300)[Si_7Fe], and Fe(300)[Si_7Al] where the [Si], [Fe] and [Al] represent tetrahedral substitutions. With increasing aluminium content the above sites could be mixed with Fe(210)[Si_8], Fe(210)[Si_7Al] and possibly Fe(210)[Si_7Fe]. For the highly Al-enriched nontronites, Fe(120)[Si_8], or Fe(120)[Si_7Al] sites will also become common.

We contrast the present approach to the crystallographically intensive approach of Ballet and Amthauer [14] and other work reviewed there. We believe that the concepts of point group symmetry of the cation sites are not useful in these mixed substitution systems and our approach is very similar to that of Dainyak, Drits and co-workers [e.g. 3, 15, 16] who have done detailed Mössbauer effect, IR and EXAFS studies of various layer silicates. We are concentrating on selecting simple systems in order to avoid ambiguities involved even in careful fitting of "sites", as emphasized by Dyar et al. [17].

2. Experimental

2.1. Sample preparation

The nontronites used here were the same as studied by Gates et al. [6]. As detailed in that study, all samples were carefully purified by saturating initially with either Na^+ or Li^+ and washing by centrifugation to obtain dispersed fine fractions, generally of <0.2, 0.15 or 0.1 μm. Random and orientated powder X-ray diffraction and infrared spectroscopy were used to check the purity. To prepare samples with different counter cations, a portion of each sample was reacted and washed three times with 0.1M chloride salts of the following cations NH_4^+, Na^+, K^+, Ca^{2+} and La^{3+}. Pure fine fractions were then dialyzed to remove excess salts. All materials were finally oven dried at 105 °C. The mean structural formula [6] of NAu-1 is $M^+_{1.05}[Si_{7.00}Al_{1.00}](Al_{0.29}Fe_{3.68}Mg_{0.04})$. Thus, the octahedral layer is strongly ferric, with $Fe^{3+} : Al^{3+} : Mg^{2+}$ as 1 : 0.08 : 0.01. However, on average, one Al^{3+} in the eight tetrahedral positions is coordinated to octahedral Fe^{3+}.

2.2. Mössbauer spectroscopy

Mössbauer spectra were taken at room temperature of the samples described above. The spectra were collected using a ^{57}Co in Rh source mounted on a conventional, symmetrical waveform, constant acceleration drive. The spectra were fitted with Voigtian profiles using the program Recoil. All IS are quoted relative to α-Fe at room temperature.

3. Results and fitting

The spectra obtained are shown in figure 1. It can be seen that the top four spectra all have the same basic shape of two nested quadrupole doublets, while the La^{3+} spectrum has some additional structure. The first simple analysis that we can do is to determine the centre of the spectral intensity by simply treating the spectra as histograms, which is completely independent of any fitting models. This gave the centres as: NH_4^+: 0.380 mm/s, K^+: 0.378 mm/s, Ca^{2+}: 0.370 mm/s, Na^+: 0.369 mm/s and La^{3+}: 0.349 mm/s. The second model-independent parameter which we can obtain is the spectral area normalized to equal sample masses, which is the effectively the relative recoilless fraction. These gave a variation of nearly 40%, with the order $NH_4^+ > Na^+ > Ca^{2+} > K^+ \sim La^{3+}$.

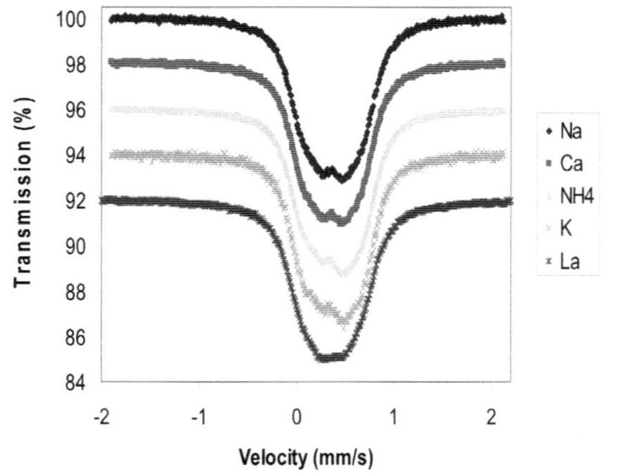

Figure 1. ^{57}Fe Mössbauer spectra of NAu-1 nontronite having exchange cations listed. Spectra have been displaced vertically by 2% for clarity.

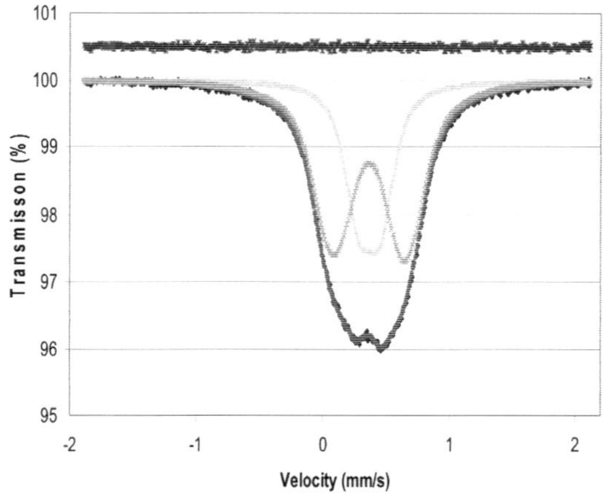

Figure 2. Fitted spectrum of Ca^{2+}saturated NAu-1 nontronite, with the error in the fit as the top line.

The spectra, except for La^{3+}, were all fitted with two nested Voigtians. No attempt was made to include a possible tetrahedral site as its existence amounts to only ~2% [6] and such a small contribution would not affect the larger-picture conclusions which we were trying to draw about the octahedral sites. The fitted spectrum of the Ca^{2+} sample is shown in figure 2. The mean parameters for the inner doublet of the four spectra (excluding the La^{3+} spectrum for which we do not yet have a definite fit) were IS = 0.38(1) mm/s, QS = 0.18(2) mm/s, area = 34(4) %, and for the outer doublet, IS = 0.38(1) mm/s, QS = 0.56(3) mm/s, area = 66(5) %. There was some evidence for additional intensity of a few percent with a larger quadrupole splitting which was fitted as an additional site in the outer doublet, but this has been ignored for the present interpretation.

4. Discussion

The hydrated radii of the cations used are, in increasing order, $NH_4^+ \sim K^+ < Na^+ < Ca^{2+} < La^{3+}$; which correlates well with the model-independent mean isomer shift, $(NH_4^+, K^+) < (Na^+, Ca^{2+}) < La^{3+}$. The electron density at octahedral Fe therefore appears to be influenced by the hydrated interlayer cations according to the relative localization of tetrahedral layer charge neutralization.

From the atom concentrations, it is clear that the dominant octahedral coordination is Fe(300). Assuming that tetrahedral substitution is random we can use the binomial theorem to estimate the proportions for the tetrahedral coordinations as Fe(300)[Si_8] : 34%, Fe(300)[Si_7Al] : 39%, Fe(300)[Si_6Al_2] : 20%. Since we would expect the symmetrical [Si_8] to produce the smallest quadrupole splitting, we see that its expected area agrees perfectly with the experimental area of the inner doublet. The outer doublet is then due to the sum of, principally, the other two configurations. This interpretation is supported by comparison with the spectrum of NAu-2 which has the composition $M^+_{.83}[Si_{7.52}Al_{.06}Fe_{.42}](Al_{.42}Fe_{3.43}Mg_{.04})$. Note that the silicon content of the tetrahedral layer for NAu-2 is 94% compared to 87.5% for NAu-1, so one would expect the contribution of the narrow central doublet due to [Si_8] to increase accordingly. This is the case as can be seen from the spectrum of Jaisi et al. [18] (inset in their figure 2A), which agrees with our spectrum (not shown). They have fitted it to a single doublet with IS = 0.38 mm/s, QS = 0.44 mm/s. Our detailed interpretation of the spectra of NAu-2, including in applied magnetic field, will be published separately.

There have been two previous very detailed papers on the Mössbauer spectra of nontronites. NAu-1 has a very similar composition to Garfield nontronite [6] and the spectra obtained by Johnston and

Cardile [8] are very similar to ours. They strongly favoured a two octahedral site model, distinguished electrostatically as suggested by Goodman [10], but no definitive configurations were suggested. Their fitted quadrupole splittings are larger by about 0.8mm/s than ours and the relative intensity of their inner doublet is slightly larger than their outer doublet. We believe that these changes are due mainly to their use of Lorentzians, whereas we have used Voigtians for which the quoted QS is <QS> of the Gaussian distribution, and they also included a tetrahedral site. We note that they observed the outer doublet to be broader than the inner one, compatible with our conclusion that is it due to two different configurations.

Luca [9] studied the effect of different exchangeable cations and of dehydration in SWa-1 nontronite. However, the octahedral layer of SWa-1 [6] has Fe^{3+} : Al^{3+} : Mg^{2+} as 1 : 0.42 : 0.10, making the major configuration Fe(210)[Si_8], followed by Fe(210)[Si_7Al] and Fe(300)[Si_8], with other configurations, including Mg^{2+} contributions, unable to be neglected. Such a combination resulted in a smooth envelope with no definite bumps to aid the analysis. We note that the large QS of ~ 0.7-1.0 mm/s in the ferruginous smectites assigned by Cashion et al. [11] to *cis*-Fe^{3+} having at least one Mg^{2+} octahedral neighbour, is not observed here, helping to confirm the correctness of the assignment.

5. Conclusions

By considering the most likely nearest cation neighbour configurations for octahedral Fe^{3+} in NAu-1 nontronite, we have identified the two major quadrupole split doublets as being due to Fe(300)[Si_8] for the narrower doublet, and to Fe(300)[Si_7Al] and Fe(300)[Si_6Al_2] for the broader doublet. These assignments are also compatible with the spectra taken by other workers on Garfield, SWa-1 and NAu-2 nontronites, and therefore we expect them to be universal.

Acknowledgements

We are grateful for the support of the Australian Research Council, Monash University and SmecTech Research Consulting for parts of this work.

References

[1] Eggleton R A 1977 *Clay Miner.* **12** 181
[2] Goodman B A, Russell J D, Fraser A R and Woodhams F W D 1976 *Clays and Clay Miner.* **24** 53
[3] Besson G, Bookin A S, Daynyak L G, Rautureau M, Tsipursky S I, Tchoubar C, Drits V A 1983 *J. Appl. Cryst.* **16**, 374
[4] Bonnin D, Calas, G, Suquet H and Pexerat H 1985 *Phys. Chem. Miner.* **12** 55
[5] Manceau A, Chateigner D and Gates W P 1998 *Phys. Chem. Miner.* **25** 347
[6] Gates W P, Slade P G, Manceau A and Lanson B 2002 *Clays and Clay Miner.* **50** 223
[7] Murad E 1987 *Z. Pflanz. Boden.* **150** 279
[8] Johnston J H and Cardile C M 1985 *Clays and Clay Miner.* **33** 21
[9] Luca V 1991 *Clays and Clay Miner.* **39** 478
[10] Goodman B A 1978 *Clays and Clay Miner.* **26** 176
[11] Cashion J D, Gates W P and Thomson A 2008 *Clay Miner.* **43** 83
[12] Gates W P 2008 *Clays and Clay Miner.* **56** 10
[13] Manceau A, Lanson B, Drits V A, Chateigner D, Gates W P, Wu J, Huo D and Stucki J W 2000 *Amer. Mineral.* **85** 133
[14] Ballet O and Amthauer G 1986 *J. Phys. C: Solid State Phys.* **19** 7099
[15] Drits, V A, Dainyak L G, Muller F, Besson G and Manceau A 1997 *Clay Miner.* **32** 153
[16] Dainyak L G and Drits, V A 2009 *Eur. J. Mineral.* **21** 99
[17] Dyar M D, Schaefer M W, Sklute E C and Bishop J L 2008 *Clay Miner.* **43** 3
[18] Jaisi D P, Kukkadapu R K, Eberl D D, Dong H 2005 *Geochim. Cosmochim. Acta.* **69** 5429

International Conference on the Applications of the Mössbauer Effect (ICAME 2009) IOP Publishing
Journal of Physics: Conference Series **217** (2010) 012066 doi:10.1088/1742-6596/217/1/012066

Kaolin mining and beneficiation: the role of iron

Enver Murad and José D Fabris
Departamento de Química, ICEx, Universidade Federal de Minas Gerais
Campus – Pampulha, 31270-901 Belo Horizonte, Minas Gerais, Brazil

E-mail: emurad@yahoo.com

Abstract. Mössbauer spectra of raw kaolins generally comprise components from iron bound in the structures of kaolinite minerals and iron in ancillary minerals. Of the latter, iron oxides are among the most detrimental to industrial uses. Although a distinction of structural and oxidic iron is generally possible if the oxides are magnetically ordered, Mössbauer spectra of kaolins are generally complicated as a result of slow paramagnetic relaxation of iron bound in the kaolinite structure. A distinction of iron bound in oxides and the kaolinite structure may therefore require the application of selective dissolution procedures in addition to Mössbauer spectroscopy.

Thus Mössbauer spectroscopy, where necessary in conjunction with chemical analyses, enables a quantitative assessment of the iron mineralogy of kaolins to be made at all stages of industrial beneficiation. This allows the efficiency of individual stages of processing and the quality of the final product to be monitored, problems during processing to be identified, and individual processing stages to be fine-tuned.

1. Introduction

Kaolinite is a layer silicate with a structure that consists of two-dimensional tetrahedral silica (SiO_4) sheets linked to octahedral alumina ($Al(O,OH)_6$) sheets to form 1:1 layers (figure 1). The formula of kaolinite is $Al_2Si_2O_5(OH)_4$, generally with little to no cation substitution. Dickite, nacrite and halloysite

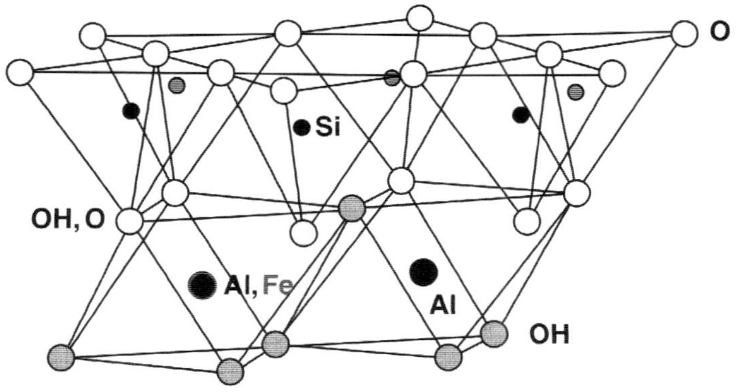

Figure 1. Structure of kaolinite. Note the SiO_4 tetrahedra that are linked to $(Al,Fe)O_6$ octahedra by sharing of oxygen atoms (open circles). Modified from [1].

are polymorphs of kaolinite, i.e. they have the same chemistry (halloysite contains additional interlayer water) but different structures.

The kaolinite structure can accommodate moderate amounts of iron that substitute for Al^{3+} on the octahedral sites. Such substitutional iron allows an assessment of the octahedral site geometry to be made using ^{57}Fe Mössbauer spectroscopy. The Mössbauer parameters given by [2] and [3] probably constitute the most reliable data on a wide range of kaolinites to date. The average Fe^{3+} isomer shift δ/Fe of 0.35 mm/s confirms the octahedral coordination of iron. The relatively low, temperature-independent "best" value for the Fe^{3+} quadrupole splitting of Fe^{2+}-free kaolinite of about 0.51 mm/s is in good agreement with the moderate octahedral flattening angle Ψ_{oct} of 57.5° (ideally 54.73° for undistorted octahedral sites) observed by X-ray and neutron diffraction [4]. The Fe^{3+} quadrupole splitting increases from the value given above with the Fe^{2+}/Fe^{3+} ratio, indicating the incorporation of a relatively large ion such as Fe^{2+} in the kaolinite structure to increase the average octahedral site distortion.

Kaolin, as opposed to kaolinite, is a kaolinite-rich product of intense weathering or hydrothermal alteration of aluminosilicate minerals such as feldspar and mica. The Glossary of Geology [5] defines kaolin as "A soft, relatively nonplastic but dispersible, usually white or nearly white claystone comprised mainly of minerals of the kaolin group, principally kaolinite ...". In reference to porcelain that originated in China, "China Clay" is a common synonym for high-grade white (i.e. iron-poor) kaolin.

Kaolin has a variety of industrial uses, the most important of which is the production of paper, where it is used as a filler and coating (requiring the kaolin to have a high whiteness). Other major uses of kaolin include the production of ceramic products such as porcelain and refractories, of rubbers, to which it adds strength, abrasion resistance and rigidity, and of paints and inks, in which it can act as a filler and thickening agent or substitute for titanium-based pigments. More exotic uses include cosmetics, soaps, toothpastes and medicines (for example as antidiarrheal adsorbents or inactive excipients).

Genetically, there is a twofold classification of kaolin deposits: those that have formed *in situ* ("primary" deposits) and others that have been eroded, transported and redeposited after formation ("secondary" or "sedimentary" deposits). Although numerous kaolin deposits exist worldwide, the market is currently dominated by three regions: the southeastern U.S.A., the lower Amazon Basin in Brazil and the Cornwall area of southwestern England [6]. Current mining is almost exclusively performed in open pits. Besides kaolinite, kaolins often contain other mineral alteration products, the most deleterious of which for industrial uses are iron oxides and oxyhydroxides (here generically termed "iron oxides"), quartz, and the TiO_2 polymorphs anatase and rutile. Raw kaolin ores must therefore be subjected to processing ("beneficiation") following mining to remove associated minerals that would decrease the commercial value of the final product or even render it unsuitable for specific uses. Kaolin processing involves a variety of stages to homogenize the raw ore and concentrate desired particle size fractions while removing unwanted ancillary components. Depending on the properties of the raw material and the desired properties of the final product, the individual steps and their sequence can vary significantly.

Like kaolinite, hematite (α-Fe_2O_3) and goethite (α-FeOOH) are common products of intense weathering and therefore also frequent constituents of kaolins. Because of their intense colour, iron oxides (in particular hematite) must be removed from kaolins for most industrial applications. The removal of iron oxides is generally effected by either physical (magnetic separation, flotation) or chemical (selective dissolution) means, or both. A more modern approach involves biological leaching using Fe^{3+}-reducing bacteria or fungi. The particular value of Mössbauer spectroscopy for the current issue lies in the possibility of enabling a distinction to be made between iron in ancillary iron oxides and iron that is bound in the kaolinite structure, and to quantify these forms of iron.

2. Mössbauer spectroscopy

Iron oxides formed in the weathering environment are generally of small particle size, and thus often superparamagnetic at room temperature, and furthermore generally aluminium substituted. Thus, rather than being pure iron oxides, natural hematites and goethites have compositions given by α-$(Al_xFe_{1-x})_2O_3$ and α-$(Al_xFe_{1-x})OOH$, where x can amount to as much as 0.16 and 0.36, respectively [7]. In the (super)paramagnetic state, the Mössbauer parameters of these iron oxides resemble those of Fe^{3+} in kaolinite and cannot be distinguished from these. A discrimination between iron bound in the kaolinite structure and associated iron oxides therefore requires Mössbauer spectra to be taken at low temperatures, usually in the liquid-nitrogen range (~ 77 K). A further complication can arise because the generally low structural iron content of kaolinite may result in slow paramagnetic relaxation. For Fe^{3+}, this leads to a six-line spectrum with a magnetic field of ~ 55 T at low temperatures. Although the lines are considerably broader than for a typical magnetically ordered hematite, the spectrum "mimics" a magnetically ordered iron oxide. Such a component therefore may be (and has been) mistaken for hematite. The lower magnetic field that is developed at room temperature, in contrast, is more suggestive of a non-Lorentzian broadening of the paramagnetic doublet. As an example, figure 2 shows Mössbauer spectra of an iron oxide-free kaolin from Brazil at room temperature and 4.2 K. The broken subspectra show Lorentzian doublet fits to both spectra. Note the broadening of the observed Fe^{3+} doublet relative to the Lorentzian fit at room temperature and the sextet component at 4.2 K. Both result from slow paramagnetic relaxation and not from the presence of associated iron oxides [2].

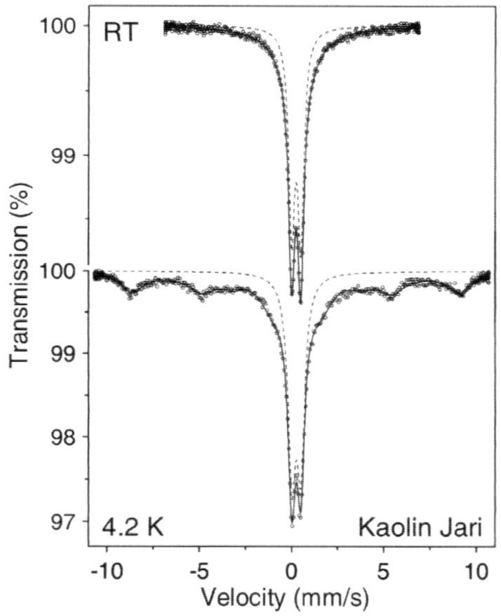

Figure 2. Mössbauer spectra of an iron-oxide free kaolin from Brazil taken at room temperature and 4.2 K. The broken subspectra indicate Lorentzian Fe^{3+} doublets that were fit to the spectra. Modified from [2].

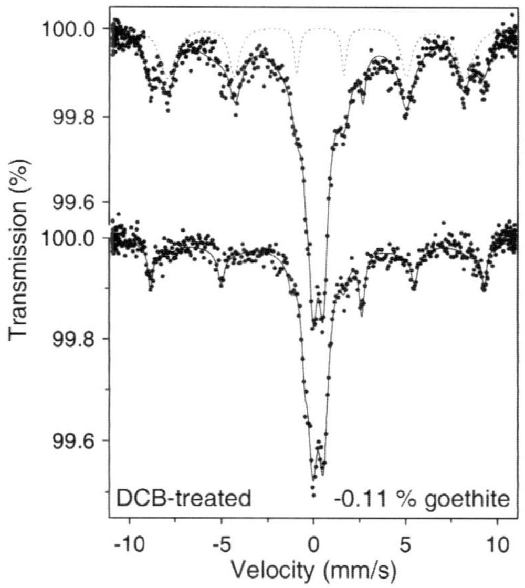

Figure 3. Mössbauer spectra of a kaolin taken at 4.2 K before (top) and after (bottom) iron oxide removal with Na dithionite ("DCB"). The dotted subspectrum indicates a sextet fit to the iron oxide (here goethite) component of the original sample. Modified from [2].

Figure 3 shows Mössbauer spectra of a commercial kaolin from Germany taken at 4.2 K before and after selective iron oxide removal with Na dithionite [2]. The parameters of the inner sextet in the spectrum of the original sample show this to result from goethite. The slight amount of iron dissolved by the dithionite treatment (0.07 mass %), corresponding to a mere 0.11 % goethite in the original kaolin, attests to the sensitivity of Mössbauer spectroscopy towards magnetically ordered phases. The magnetic component remaining after goethite removal (bottom spectrum) can again be attributed to paramagnetic relaxation of iron in the kaolinite structure. It must, however, be pointed out that the significant difference in magnetic fields between that generated by paramagnetic relaxation (~ 56 T) and goethite (~ 50 T) renders this case particularly favourable for a direct observation of the two magnetic components in the untreated sample. A distinction of the magnetic field due to paramagnetic relaxation in kaolinite from the hyperfine field of hematite at 4.2 K (≤ 54.2 T or ≤ 53.5 T for hematite that has and has not passed through a Morin transition, respectively [8]), in contrast, would hardly be possible, and such a case would require spectra to be taken before and after the selective removal of hematite for a definitive discrimination of the two contributions.

3. Mining and beneficiation: an example

The possibilities of Mössbauer spectroscopy as a probe for the physical and chemical changes of iron during the industrial production of kaolin are best exemplified by presenting a specific example. For this purpose we select here the raw ore and beneficiated products of a kaolin mined by Pará Pigmentos S.A. in the Ipixuna area of the State of Pará, Brazil. A detailed description of the deposit, its mineralogy and the mining process has been recently published [9] and the constitutions and selected properties of the raw ore and the intermediate and final products of the industrial beneficiation have been described by [10]. Here we restrict ourselves to the original (raw) product extracted from the deposit, the reject of magnetic separation in the course of beneficiation, and the final product.

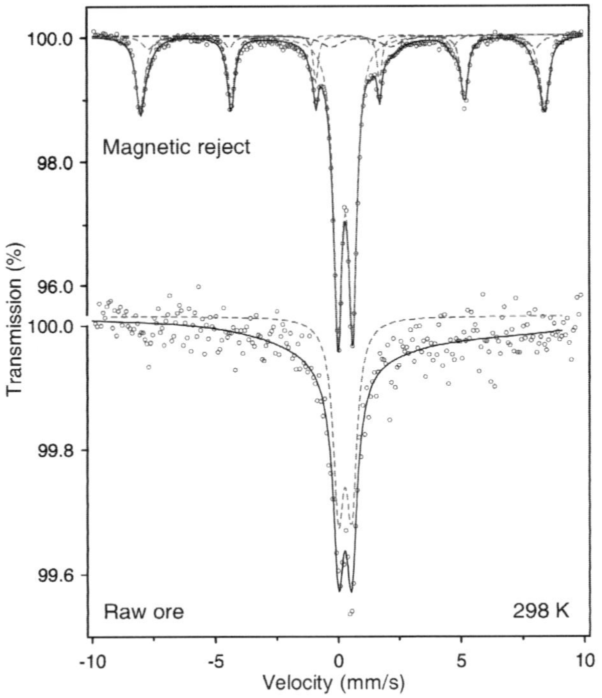

Figure 4. Room-temperature Mössbauer spectra of the raw ore and the reject of magnetic separation of kaolin mined at the Ipixuna deposit of Pará Pigmenos S.A., Brazil. The broken subspectrum in the bottom diagram indicates a Lorentzian Fe^{3+} doublet that was fit to spectrum. Modified from [10].

Figure 4 shows room-temperature Mössbauer spectra of the raw product and the magnetic reject. The spectrum of the raw ore shows – in spite of considerable data scatter because of the low iron content of the sample – the already familiar discrepancy between an Fe^{3+} doublet fitted to the spectrum and the actual data. In contrast to Mössbauer spectra of other samples from the same region [11,12], the sample we studied showed no evidence of iron oxides that are magnetically ordered at room temperature, indicating at most very low iron oxide contents. The spectrum of the magnetic reject, in contrast, displayed a well-developed sextet in addition to the paramagnetic components, and could be fit with four components: a major Fe^{3+} doublet, a subordinate Fe^{2+} doublet, and two sextets. The magnetic hyperfine fields of 51.1 and 49.5 T of the two sextets that were fit to the magnetic component (at an average quadrupole shift of ~ -0.18 mm/s) can, rather than indicating two discrete mineral components, be taken as ensuing from a distribution of hyperfine fields in hematite of moderate crystallinity and/or significant Al substitution. A Mössbauer spectrum of this sample taken at 110 K could be fit with a major Fe^{3+} doublet, a subordinate Fe^{2+} doublet, and three sextets (figure 5). In addition to hematite, the two other sextets evidence the presence of goethite and (tentatively) indicate maghemite. The latter can also be observed (though not quantified) in the magnetic reject by X-ray diffraction, thereby validating this assignment in the Mössbauer spectrum, whereas overlap of the most intense X-ray peaks of goethite with peaks of kaolinite prevents a definitive identification of goethite by X-ray diffraction.

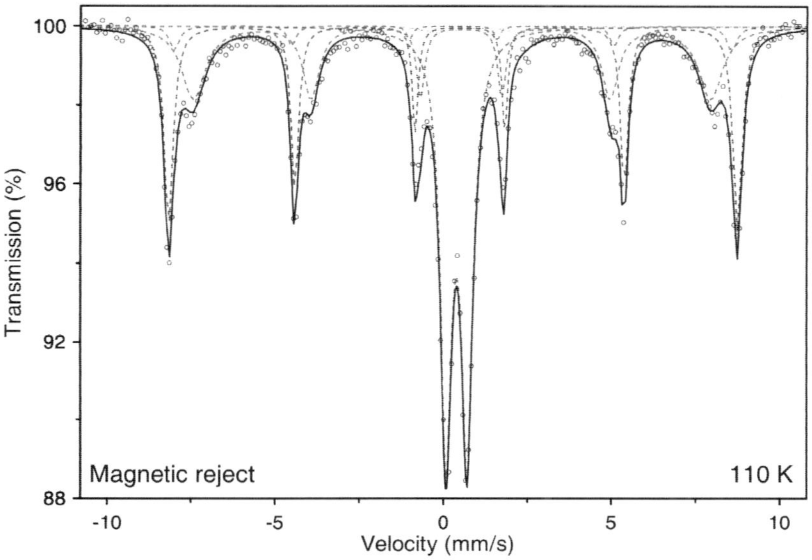

Figure 5. Mössbauer spectrum at 110 K of the magnetic reject of the Pará Pigmentos kaolin. Modified from [10].

The iron mineralogy of the magnetic reject deduced from the spectrum taken at 110 K is summed up in table 1. Chemical analysis shows this sample to have a total iron content of 10.89 % Fe_2O_3 [10]. Breaking this up into the components indicated by the Mössbauer spectrum at 110 K allows an assessment of the relative amounts of iron oxides to be made. The Fe^{3+} doublet is attributed to iron that is bound in the structure of the final kaolin (0.69 mass %), although contributions from iron oxides that are still superparamagnetic at 110 K cannot be excluded.

Table 1. Iron mineralogy of the magnetic reject of the Pará Pigmentos kaolin derived from the Mössbauer spectrum at 110 K (Figure 5).

Mineral	Relative area	Mass % [a]
Hematite	0.29	3.74
Maghemite	0.04	0.59
Goethite	0.28	4.22
Fe^{2+} [paramagnetic] [b]	0.02	
Fe^{3+} [paramagnetic] [c]	0.38	

[a] Corrected for recoil-free fractions [13,14].

[b] Paramagnetic Fe^{2+} in kaolin and ancillary minerals.

[c] Includes contributions from paramagnetic Fe^{3+} in kaolin and ancillary minerals and possibly superparamagnetic Fe^{3+} oxides.

4. Conclusions

^{57}Fe Mössbauer spectra of kaolinite are generally complicated by slow paramagnetic relaxation of iron bound in the kaolinite structure. Because kaolins tend to be mixtures of minerals of the kaolin group (kaolinite with possible admixtures with dickite, nacrite and/or halloysite) and other minerals formed during intense weathering, their Mössbauer spectra are correspondingly complex, comprising components from iron bound in the structures of the kaolinite minerals and iron in ancillary minerals. During industrial processing the proportions of kaolinite and other kaolin minerals are increased relative to those of the ancillary minerals, and the resultant changes in iron mineralogy can be monitored by Mössbauer spectroscopy. A comprehensive characterization of the kaolin and ancillary mineral mineralogy will nevertheless usually require a combination of Mössbauer spectroscopy with other analytical procedures such as X-ray diffraction, Raman spectroscopy and selective chemical dissolution.

Acknowledgments

This study has been supported by FAPEMIG (grants CEX21/08-32 and APQ 00217-08) and CNPq (grant 303447/2007-9) (Brazil).

References

[1] Gruner J W 1932 *Z. Krist.* **83** 75
[2] Murad E and Wagner U 1991 *N. Jahrb. Min. Abh.* **162** 281
[3] St. Pierre T G, Singh B, Webb J and Gilkes B 1992 *Clays Clay Min.* **40** 341
[4] Giese R F Jr. 1988 In: *Reviews in Mineralogy* **19** 29 (Washington, DC: Mineralogical Society of America), ISBN 0-939950-23-5
[5] Neuendorf K K E, Mehl J P and Jackson J A 2005 *Glossary of Geology*, 5th edition (Alexandria, VA: American Geological Institute), ISBN 0-922152-76-4
[6] Murray H H 2002 *Mining, Minerals and Sustainable Development (MMSD)*, International Institute for Development and Environment, No. 64
[7] Cornell R M and Schwertmann U 2003 *The Iron Oxides*, 2nd edition (Weinheim: Wiley-VCH), ISBN3-527-30274-3
[8] Murad E and Cashion J 2004 *Mössbauer Spectroscopy of Environmental Materials and their Industrial Utilization* (Norwell, MA: Kluwer), ISBN 1-4020-7726-2
[9] Murray H H, Alves C A and Bastos C H 2007 *Clay Min.* **42** 145
[10] Mussel W N, Murad E, Criscuolo P S R, Pinheiro P G and Fabris J D 2008 *Clay Min.* **43** 381
[11] Souza D J L, Varajão A D F C, Yvon J, and da Costa G M 2007 *Clay Min.* **42** 69
[12] Scorzelli R B, Bertolino L C, Luz A B, Duttine M, Silva F A N G and Munayco P 2008 *Clay Min.* **43** 129
[13] Meisel W and Kreysa G 1973 *Z. anorg. allg. Chem.* **395** 31
[14] De Grave E and Van Alboom A 1991 *Phys. Chem. Min.* **18** 337

Mechanical strength and local structure of 'new' Hagi porcelain investigated by ^{57}Fe-Mössbauer spectroscopy

S Kubuki[1], H Uehara[1], K Akagi[1], A Mikuni[2], N Isobe[3], Z Homonnay[4], K Sinkó[5], E Kuzmann[6] and T Nishida[7]

[1]Department of Chemical and Biological Engineering, Ube National College of Technology, Tokiwadai 2-14-1, Ube, Yamaguchi, 755-8555, JAPAN
[2] Environmental material technology division, Yamaguchi prefectural industrial technology institute, Astopia 4-1-1, Ube, Yamaguchi, 755-0195, JAPAN
[3]Shuto-en Inc., Oazatubaki3775, Hagi, Yamaguchi, 758-0061, JAPAN
[4]Laboratory of Nuclear Chemistry, Eötvös Loránd University, Pázmany P. s., 1/A 1117 Budapest, HUNGARY
[5]Institute of Chemistry, Eötvös Loránd University, Pázmany P. s., 1/A 1117 Budapest, HUNGARY
[6]Laboratory of Nuclear Chemistry, HAS CRC, Eötvös Loránd University, 1518 Budapest, HUNGARY
[7]Department of Biological and Environmental Chemistry, Faculty of Humanity-Oriented Science and Engineering, Kinki University, Kayanomori 11-6, Iizuka, Fukuoka, 820-8555, JAPAN

E-mail : kubuki@ube-k.ac.jp

Abstract. A relationship between the local structure and mechanical strength of 'new' Hagi porcelain '*A*' and '*B*' prepared by sintering two types of iron containing aluminosilicate soils under oxidizing and reducing atmospheres were investigated by means of ^{57}Fe-Mössbauer spectroscopy, X-ray diffraction (XRD), scanning electron microscopy (SEM), as well as three point bending test. The largest mechanical strength (σ) value of 64.3±3.1 MPa was estimated from the three-point bending test of reductively sintered Hagi porcelain '*A*'. The Mössbauer spectrum was composed of two paramagnetic doublets due to tetrahedral Fe^{II} and Fe^{III} with the isomer shift (δ) values of 1.13±0.02 and 0.31±0.01 mm s^{-1}. On the other hand, a paramagnetic doublet and a magnetic sextet with the δ values of 0.30±0.01 and 0.35±0.02 mm s^{-1} were observed from the Mössbauer spectra of other samples. It can be concluded that the mechanically strengthened Hagi porcelain was successfully fabricated by choosing soil '*A*' and by sintering under a reducing atmosphere.

1. Introduction

Hagi porcelain is known as one of the most famous Japanese traditional porcelains. It originates from the end of the 16th century; a Korean craftsman started their fabrication at Hagi district, Yamaguchi prefecture, west Japan. Hagi porcelain has a characteristic cracking structure, which causes a change of color due to tea component, being known as a popular 'Seven-color' change of Hagi porcelain.

However, the cracking structure observed in Hagi porcelain is not always favourable for common use at schools, restaurants, *etc*., because the crack causes a weak mechanical strength.

In order to overcome this problem, new composition and new fabrication condition have been studied. A relationship between the local structure and mechanical strength was investigated by means of ^{57}Fe-Mössbauer spectroscopy, X-ray diffraction (XRD), scanning electron microscopy (SEM) and three-point bending test.

2. Experimental

Hagi porcelain *'A'* and *'B'* were prepared by a conventional sintering method. Weighed amount of porcelain clay *'A'* or *'B'*, 30-40 vol.% of water and 0.3-0.4 mass% of dispersing agent were well mixed and poured into a gypsum mold. When the dried mixture was sintered at around 1230 °C under a reducing or an oxidizing atmosphere, opaque white cylindrical samples were obtained. In order to achieve a reducing atmosphere, propane gas was flown in the electric muffle furnace. For Mössbauer measurement, enriched isotope of $^{57}Fe_2O_3$ (^{57}Fe = 95.54 %) was used in the sample preparation. Mössbauer spectra were recorded by a constant acceleration method with a source of ^{57}Co(Rh) and with a reference of α-Fe. X-ray fluorescence (XRF) analysis was carried out by a conventional glass-bead method. A transparent medal-like bead was prepared by melting the pulverized mixture of Hagi porcelain (0.4 g) and lithium tetraborate (4.0 g) at 1100 °C for 10 min. The XRD pattern was conducted from $2\Theta = 10$ to 90° at a scanning rate of 6 degree min^{-1}, using the Cu-K_α X-rays generated by setting the tube voltage and the current to 40 kV and 40 mA, respectively. SEM image was taken by energizing 10 kV of electric voltage under the magnification of 30 to 10^4 using gold-coated samples. Three-point bending test was carried out for 20 pieces under the head speed of 1 mm min^{-1} and 50 mm span.

3. Results and Discussion

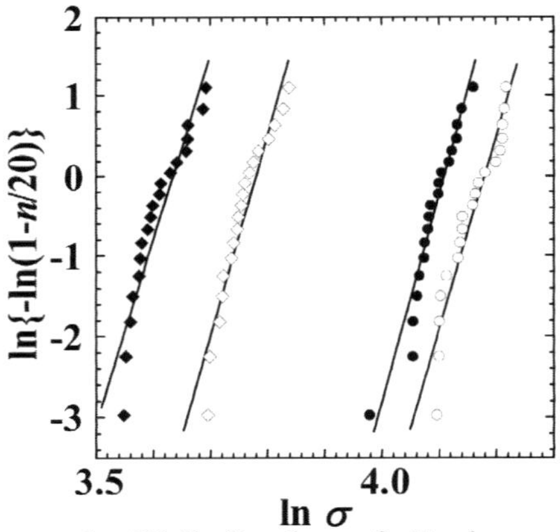

Figure 1. Weibull plot of Hagi porcelain '*A*'(circle) and '*B*' (diamond) prepared under a reducing (open symbols) and an oxidizing (closed symbols) atmospheres.

The mass percent of Hagi porcelain '*A*' was determined to be SiO_2 (57.9 %), Al_2O_3 (26.1 %), Fe_2O_3 (8.75 %) and others (7.25 %), while that of Hagi porcelain '*B*' to be SiO_2 (63.1 %), Al_2O_3 (28.6 %), Fe_2O_3 (2.2 %) and others (6.1 %) from the XRF analysis. This result shows that the both Hagi porcelains were mainly composed of iron containing aluminosilicates.

The mechanical strength (σ in Pa) was estimated by the following equation;

$$\sigma = 8PL/\pi d^3 \qquad (1)$$

where P, L and d are breaking load (in N), span (in m) and diameter (in m), respectively. The σ values of Hagi porcelain '*A*' prepared under a reducing and an oxidizing atmospheres were estimated to be 64.3±3.1 and 60.1±2.8 MPa, whereas those for Hagi porcelain '*B*' were 43.1±1.9 and 37.2±1.9 MPa. These results show that the mechanically strengthened Hagi porcelains were successfully fabricated because the σ value of conventional Hagi porcelain was reported to be 25 MPa [1]. It should be noted that the σ value of the porcelain '*A*' was larger than that of the porcelain '*B*' by 20 MPa, and that the σ value of reductively prepared porcelains were larger than that of oxidatively prepared ones by 5 MPa. It is considered that the Hagi porcelain with the highest mechanical strength can be fabricated by sintering soil '*A*' under a reducing atmosphere.

The obtained σ value was utilized for Weibull analysis, *i. e.*,

$$\ln \{- \ln\{1-(n/20)\}\} = m \ln\sigma + \text{const.} \tag{2}$$

where n and m respectively show ranking number and Weibull index. $\ln\{-\ln(1-n/20)\}$ *vs.* $\ln\sigma$ plot of Hagi porcelains were shown in Figure 1. Weibull index, a slope value of each fitted line, for Hagi porcelain '*A*' and '*B*' prepared under a reducing and an oxidizing atmospheres were respectively calculated to be 23.7, 25.6, 25.1 and 22.5. These values indicate that the homogeneous porcelains were fabricated because the m values of traditional ceramics are reported to be 5-20 [2].

SEM photographs of Hagi porcelains were shown in Figure 2. Particles with the size of about 1 μm can be seen in the photographs of Hagi porcelain prepared under a reducing atmosphere (figure 2 (A-a) and (B-a)), while those with the size of around 2-3 μm can be observed in the Hagi porcelains prepared under an oxidizing atmosphere (figure 2 (A-b) and (B-b)). These results shows that the enhancement of mechanical strength was occurred due to the compaction of the porcelain surface when the Hagi porcelain was reductively sintered.

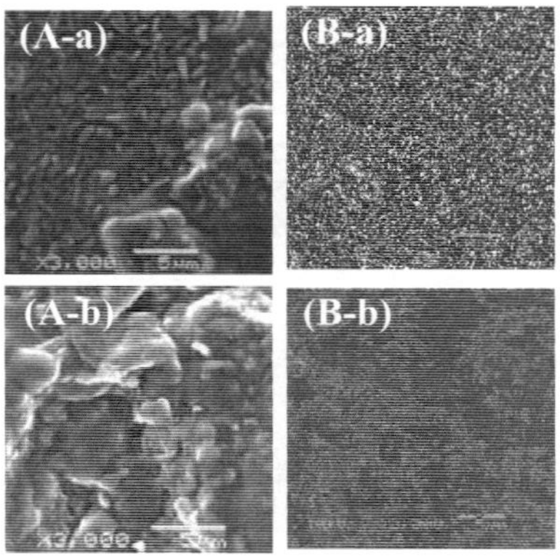

Figure 2. SEM images taken under the magnitude of 3×10^3 of Hagi porcelain (A) '*A*' and (B) '*B*' prepared under (a) a reducing and (b) an oxidizing atmospheres.

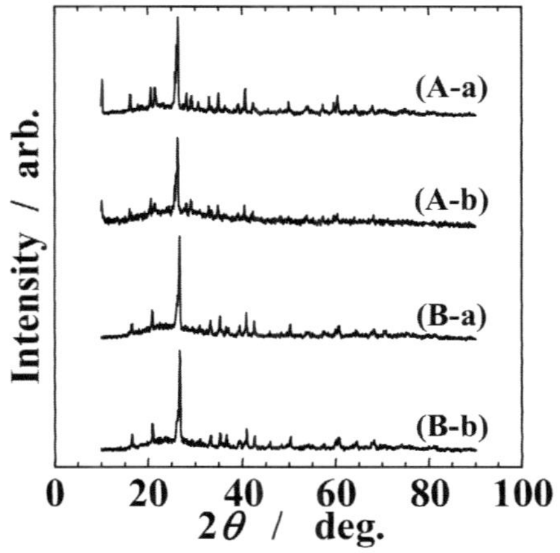

Figure 3. XRD patterns of Hagi porcelain (A) '*A*' and (B) '*B*' prepared under (a) a reducing and (b) an oxidizing atmospheres.

XRD patterns of the Hagi porcelains were represented in Figure 3. A precipitation of Mullite ($Al_6Si_2O_{13}$) phase was confirmed from JCPDS card for all Hagi porcelains. In addition, specific peaks attributed to Cordierite ($Mg_2Al_4Si_5O_{18}$) and Sekaninaite ($Fe_2Al_4Si_5O_{18}$) were detected at 2θ = 10.1, 21.6, 28.2 and 29.2 ° in the XRD patterns of Hagi porcelain '*A*' (figure 3 (A-a) and (A-b)). On the other hand, peaks for Cordierite and Sekaninaite were absent from those of Hagi porcelain '*B*'. The mechanical strength of Hagi porcelain '*A*' is about 1.5 times larger than that of Hagi porcelain '*B*' from the three-point bending test; thus the precipitation of Cordierite and/or Sekaninaite enhances the mechanical strength of the Hagi porcelain.

Mössbauer spectra of Hagi Porcelain '*A*' sintered under a reducing or an oxidizing atmosphere were shown in Figure 4. Due to tetrahedral Fe^{II} and Fe^{III} with the isomer shift (δ) values of 1.13±0.02 and 0.31±0.01 mm s^{-1}, two paramagnetic doublets were observed in the spectra of the reductively sintered porcelain '*A*' (figure 4(a)). While a paramagnetic and a magnetic Fe^{III} species with δ values of 0.30±0.01 and 0.35±0.02 mm s^{-1} were observed for the Hagi porcelains prepared under other conditions. These results indicate that the precipitation of magnetic Fe^{III} resulted in the decrease of mechanical strength of Hagi porcelain prepared under an oxidizing atmosphere. Because of less precipitation of magnetic Fe^{III}, the weakening of mechanical strength was restrained when sintered under a reducing atmosphere.

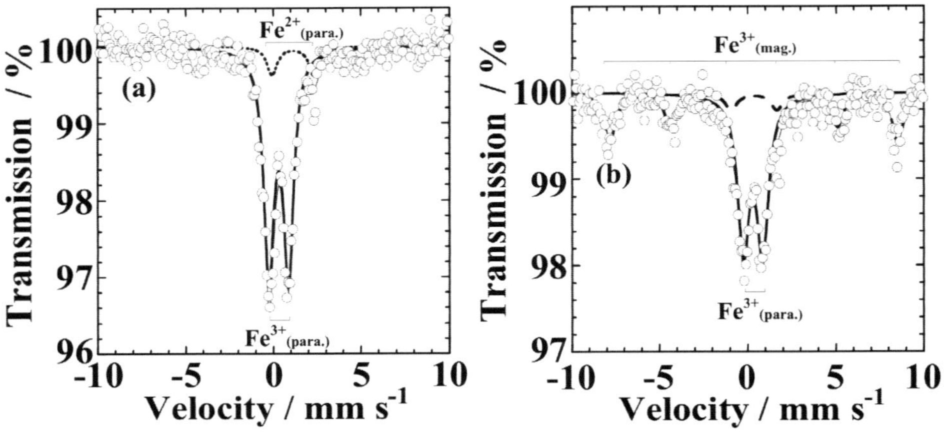

Figure 4. Mössbauer spectra of Hagi porcelain '*A*' prepared under (a) a reducing and (b) an oxidizing atmospheres.

4. Summary

A relationship between mechanical strength and local structure of the newly fabricated Hagi porcelain was investigated. The largest mechanical strength of 64.3±3.1 MPa was estimated from the three-point bending test of Hagi porcelain '*A*' prepared under a reducing atmosphere. Precipitation of Cordierite ($Mg_2Al_4Si_5O_{18}$) and Sekaninaite ($Fe_2Al_4Si_5O_{18}$), were confirmed by the XRD pattern. Due to tetrahedral Fe^{II} and Fe^{III} with the isomer shift (δ) values of 1.13±0.02 and 0.31±0.01 mm s^{-1}, two paramagnetic doublets were observed from the Mössbauer spectra. It is concluded that less precipitation of magnetic Fe^{III} in the matrix restrain the decrease of mechanical strength.

Acknowledgement
One of the authors (SK) expresses his great gratitude for the support given by the Ministry of Economy, Trade and Industry, Japan.

References

[1] METI Chugoku 2008 *Hagiyaki no Huai wo ikasita jituyouteki na ko-kyodo tei-kyusuisei touki no kaihatsu (a fabrication of Hagi porcelain with high mechanical strength and low water absorption ability)* A report of the research project on regional resources the Ministry of Economy, Trade and Industry, Japan (in Japanese)
[2] Okada A 1998 *Ceramics no hakai-gaku* (Tokyo: Uchidarokakuho) p 45 (in Japanese)

Study of irradiated Hadfield steel using transmission Mössbauer spectroscopy with high velocity resolution and conversion electron Mössbauer spectroscopy

V A Semionkin[1,2], F G Neshev[1], V A Tsurin[3], O B Milder[2,4], M I Oshtrakh[2,5]

[1] Faculty of Experimental Physics, Ural State Technical University – UPI, Ekaterinburg, 620002, Russian Federation
[2] Faculty of Physical Techniques and Devices for Quality Control, Ural State Technical University – UPI, Ekaterinburg, 620002, Russian Federation
[3] Institute of Metal Physics, Ural Division of the Russian Academy of Sciences, Ekaterinburg, 620041, Russian Federation
[4] Radio-Technical Department, Ural State Technical University – UPI, Ekaterinburg, 620002, Russian Federation

E-mail: oshtrakh@mail.utnet.ru

Abstract. Proton irradiated Hadfield steel foil was studied using transmission Mössbauer spectroscopy with high velocity resolution and conversion electron Mössbauer spectroscopy. It was shown that proton irradiation leads to structural changes in the foil as well as to surface oxidation with ferric hydrous oxide formation (ferrihydrite). Moreover, oxidation on the foil underside was higher than on the foil right side.

1. Introduction

Radiation effects on iron alloys are subject to various studies including Mössbauer spectroscopy [1]. This is very important for analysis of alloys aging and corrosion in nuclear reactors. Mössbauer spectroscopy is useful for analysis of structural damages and oxidation processes in irradiated alloys. Therefore in the present work we discuss results of the study of the effect of proton irradiation in Hadfield steel foil by Mössbauer spectroscopy using transmission and conversion electrons techniques.

2. Materials and methods

Hadfield steel (19 wt% Mn, 0.8 wt% C and Fe as a balance) foil with thickness of 20 μm (rolling with high temperature annealing) was irradiated with protons (energy of 5.7 MeV, fluence of 5×10^{15} particles/cm^2). Non-irradiated and irradiated foils were studied at room temperature using transmission Mössbauer spectroscopy with high velocity resolution and conversion electron Mössbauer spectroscopy (CEMS) with back scattering counter with He using high precision and sensitive spectrometer SM–2201.

[5] To whom correspondence should be addressed.

Characteristics of Mössbauer spectrometer SM–2201 with high velocity resolution were described in details in [2]. Spectra were measured in 4096 channels and then presented in 1024 channels for transmission spectra and in 512 channels for CEMS spectra. Mössbauer spectra were computer fitted with the least squares procedure using UNIVEM-MS program with Lorentzian line shape. Mössbauer parameters isomer shift δ, quadrupole splitting (quadrupole shift for magnetically split component) ΔE_Q, magnetic hyperfine field H_{eff}, line width Γ, subspectrum relative area S and statistical criterion χ^2 were determined. Magnetic sextets were fitted using the ratio $S_{16}:S_{25}:S_{34}=3:2:1$ for transmission spectra and equal $S_{16}:S_{25}:S_{34}$ ratio for sextets in CEMS spectra. The values of isomer shift are given relative to α–Fe at 295 K.

3. Results and Discussion

Mössbauer spectra of irradiated foil obtained using transmission spectroscopy with high velocity resolution and CEMS are shown in Figures 1 and 2. Transmission spectra of non-irradiated and irradiated foil were better fitted using two sextets (1 and 2), one doublet (3) and two singlets (4 and 5). Mössbauer hyperfine parameters for these components were similar for non-irradiated and irradiated foils and related to iron environment correspondent to α–Fe (H_{eff} were 329.9±0.5 and 330.7±0.5 kOe, respectively), Fe–Mn (H_{eff} were 305.7±0.5 and 305.6±0.5 kOe, respectively), Fe–C (ΔE_Q and δ were 0.521±0.015 mm/s, –0.025±0.015 mm/s and 0.558±0.015 mm/s, –0.024±0.015 mm/s, respectively), ε–Fe (δ were –0.120±0.015 and –0.132±0.015 mm/s, respectively) and γ–Fe (δ were 0.029±0.015 and 0.017± 0.015 mm/s, respectively). Therefore, these components and iron environments were further denoted as α–Fe, Fe–Mn, Fe–C, ε–Fe and γ–Fe, respectively. However, relative areas of these components were changed after irradiation (Figure 3). Relative area of α–Fe component increased from ~19 to ~24 %, while that of Fe–C component decreased from ~19 to ~13 % as well as relative area of ε–Fe decreased from ~25 to ~20 % while that of γ–Fe component increased from ~21 to ~25 %. Relative area of Fe–Mn component remained the same. These results demonstrated that proton irradiation of the foil led to removing of C from iron environment and, therefore, increasing of α–Fe environment as well as transformation of ε–Fe environment into γ–Fe one.

The results of the fitting of conversion electron Mössbauer spectra showed that non-irradiated foil spectrum consisted of three components which may be related to α–Fe (H_{eff} value was 328±1 kOe), Fe–Mn (H_{eff} value was 315±1 kOe) and ε–Fe (δ value was –0.05±0.05 mm/s). Spectra of the right side and underside of irradiated foil consisted of two sextets which were related to α–Fe (H_{eff} values were 328±1 kOe and 329±1 kOe) and to Fe–Mn (H_{eff} values were 311±1 kOe and 311±1 kOe), one quadrupole doublet with parameters δ=0.31±0.05 mm/s, ΔE_Q=0.83±0.05 mm/s and δ=0.35±0.05 mm/s, ΔE_Q=0.77±0.05 mm/s, and one singlet which was related to γ–Fe (δ values were 0.14±0.05 mm/s and 0.23±0.05 mm/s). Parameters of revealed doublet were similar to ferric hydrous oxides such as akaganite (β–FeOOH) or ferrihydrite (see data for comparison in [3]). β–FeOOH was considered as a corrosion product of steel corrosion including corrosion in the nuclear reactor [4]. However, we can consider ferrihydrite as a result of foil oxidation in the absence of Cl^- ions. It was interesting to observe twice increase of relative area of ferrihydrite component in the spectrum of the underside of irradiated foil (Figure 4). It is possible that energy lost with protons passing foil raises protons interaction cross-section with indoor air molecules (O_2, H_2, H_2O) with formation of radicals, ozone, oxygen atoms, etc. Therefore, radiation induced corrosion in the foil increases with increase of products of radiolysis in the air.

4. Conclusion

Comparative study of the Hadfield steel foil non-irradiated and irradiated with protons using Mössbauer spectroscopy demonstrated structural variations in the foil after irradiation as well as surface oxidation with ferrihydrite formation. It was also shown that radiation induced oxidation was higher on the underside of the foil.

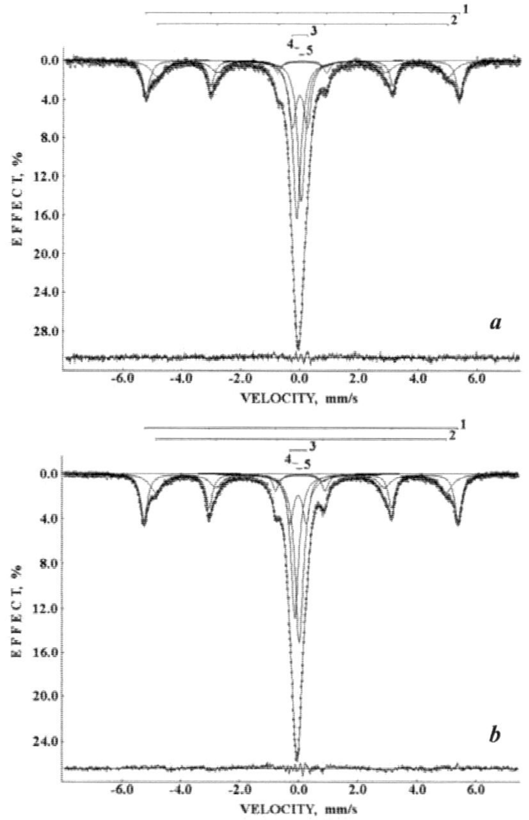

Figure 1. Transmission Mössbauer spectra of Hadfield steel foils presented in 1024 channels: non-irradiated (***a***) and irradiated with protons (***b***). 1 – α–Fe, 2 – Fe–Mn, 3 – Fe–C, 4 – ε–Fe, 5 – γ–Fe. T=295 K.

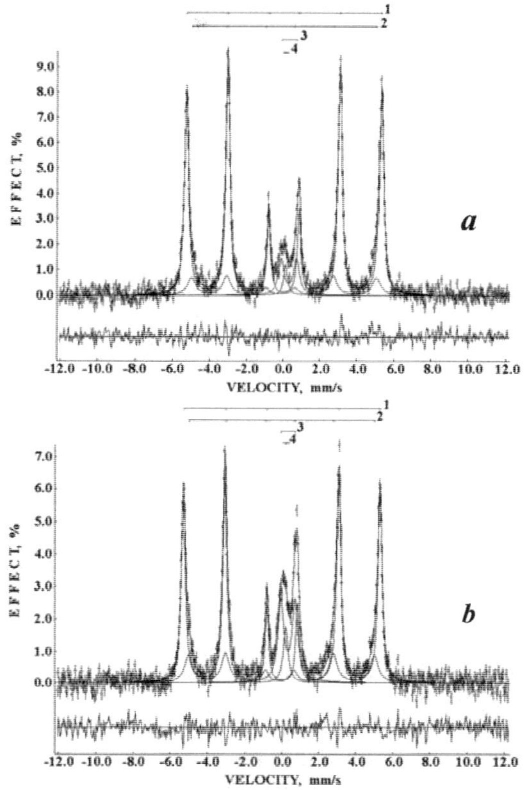

Figure 2. Conversion electron Mössbauer spectra of Hadfield steel foil irradiated with protons presented in 512 channels: the right side of the foil (***a***) and the underside of the foil (***b***). 1 – α–Fe, 2 – Fe–Mn, 3 – ferrihydrite, 4 – γ–Fe. T=295 K.

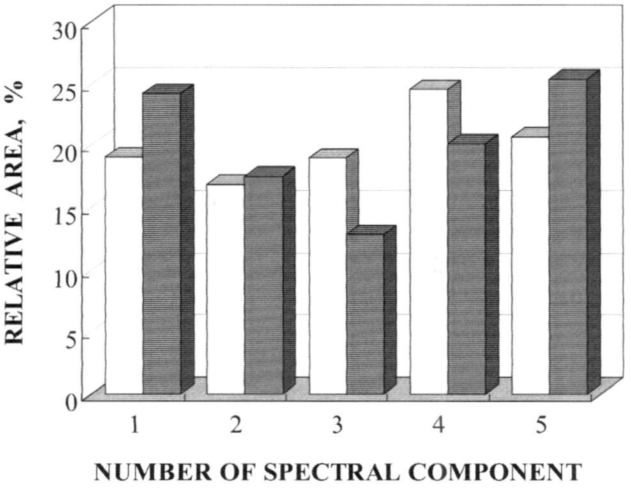

Figure 3. Relative areas of spectral components for transmission Mössbauer spectra of Hadfield steel foils: non-irradiated (white) and irradiated with protons (grey). 1 – α–Fe; 2 – Fe–Mn; 3 – Fe–C; 4 – ε–Fe; 5 – γ–Fe. Error for relative areas was 10 % or less.

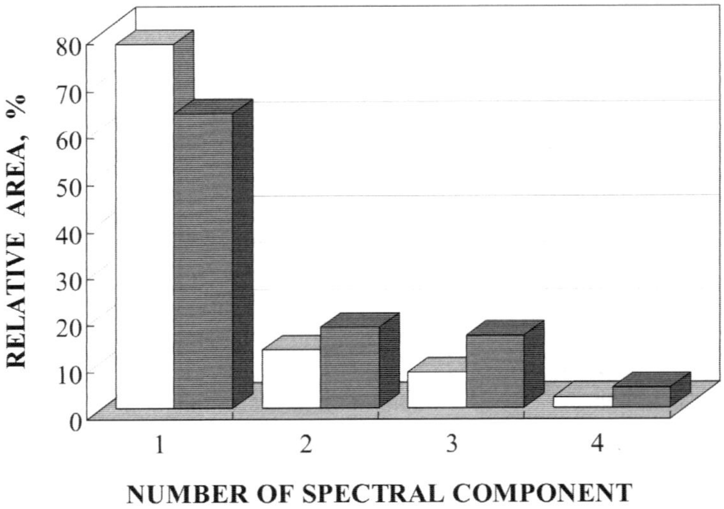

Figure 4. Relative areas of spectral components for conversion electron Mössbauer spectra of Hadfield steel foil irradiated with protons: the right side of the foil (white) and the underside of the foil (grey). 1 – α–Fe, 2 – Fe–Mn, 3 – ferrihydrite, 4 – γ–Fe. Error for relative areas was 10 % or less.

References

[1] Merkel D G, Tanczikó F Sajti Sz, Major M, Ńmeth A, Bottyán L, Horváth Z E, Waizinger J, Stankov S and Kovács A 2008 A Modification of local order in FePd films by low energy He^+ irradiation *J. Appl. Phys.* **104**, No 013901

Zeman A, Debarberis L, Kupča L, Acosta B, Kytka M and Degmová J 2007 Study of radiation-induced degradation of RPV steels and model alloys by positron annihilation and Mössbauer spectroscopy *J. Nucl. Mater.* **360**, 272–281

Koloskov V M, Kozlov A V, Semionkin V A, Milder O B and Portnykh I 2004 A Mössbauer study of 0Kh16N15M3T1 structural steel after neutron irradiation *Phys. Metals Metall.* **97**, 259–265

[2] Oshtrakh M I, Grokhovsky V I, Abramova N V, Semionkin V A and Milder O B 2009 Iron–nickel alloy from iron meteorite chinga studied using Mössbauer spectroscopy with high velocity resolution *Hyperfine Interact.* **190**, 135–142

Oshtrakh M I, Semionkin V A, Grokhovsky V I, Milder O B and Novikov E G 2009 Mössbauer spectroscopy with high velocity resolution: New possibilities of chemical analysis in material science and biomedical research *J. Radioanal. Nucl. Chem.* **279**, 833–46

[3] Bowen L H and Weed S B 1984 Mössbauer spectroscopy of soils and sediments *Chemical Mössbauer Spectroscopy*, ed. R.H. Herber (Plenum Publishing Corporation) p. 217–242

Murad E and Johnston J H 1987 Iron oxides and oxyhidroxides *Mössbauer Spectroscopy Applied to Inorganic Chemistry*, vol. 2, ed. G.J. Long, (Plenum Publishing Corporation) p. 507–582

Murad E, Bowen L H, Long G J and Quin T G 1988 The influence of crystallinity on magnetic ordering in natural ferrihidrites *Clay Minerals* **23**, 161–173

Oshtrakh M I, Semionkin V A, Prokopenko P G, Milder O B, Livshits A B and Kozlov A A 2001 Hyperfine interactions in the iron cores from various pharmaceutically important iron–dextran complexes and human ferritin: A comparative study by Mössbauer spectroscopy *Int. J. Biol. Macromol.* **29**, 303–314

[4] Slugn V, Lipka J, Tóth I, Haščík J, Hinca and Lehota M 2005 Corrosion of steam generator pipelines analysed using Mössbauer spectroscopy *Nucl Engin Design* **235**, 1969–1976

Long-term corrosion study at nuclear power plant Bohunice (Slovakia)

V Slugen[1], J Lipka[1,3], J Dekan[1], I Tóth[1] and I Smieško[2]

[1]Department of Nuclear Physics and Technology, Slovak University of Technology Bratislava, Ilkovicova 3, 812 19 Bratislava, Slovakia; Vladimir.Slugen@stuba.sk

[2]NPP Jaslovské Bohunice, SE, a.s., Slovakia.

[3]Slovak Institute of Metrology, Bratislava, Slovakia

Abstract. Steam generators of four VVER-440 units at nuclear power plants V-1 and V-2 in Jaslovske Bohunice (Slovakia) were gradually changed by new original "Bohunice" design in period 1994-1998. Corrosion processes before and after these design and material changes in Bohunice secondary circuit were studied using Mössbauer spectroscopy during last 25 years. Innovations in the feed water pipeline design as well as material composition improvements were evaluated positively. Mössbauer spectroscopy studies of phase composition of corrosion products were performed on real specimens scrapped from water pipelines or in form of filter deposits. The corrosion of new feed water pipelines system (from austenitic steel) in combination to innovated operation regimes goes dominantly to magnetite. The hematite presence is mostly on the internal surface of steam generator body and its concentration increases towards the top of the body. In the results interpretation it is necessary to consider also erosion as well as scope and type of maintenance activities. The long-term study of phase composition of corrosion products at VVER reactors is one of precondition for the safe operation over the projected NPP lifetime.

1. Introduction

Steam generator (SG) is generally one of the most important components at all nuclear power plants (NPP) with close impact to safe and long-term operation. The Russian water cooled and water moderated reactors (VVER) are unique because of the horizontal position of SGs. It takes several advantages (large amount of cooling water in case of loss of coolant accident, good accessibility, large heat exchange surface, etc. ...) but also some disadvantages, which are important to take into account during the operation and maintenance. Material degradation and corrosion/erosion processes are serious risks for long-term reliable operation. In the period of about 10-15 years ago, the feed water pipelines were changed at all SGs in all 4 Bohunice units (V-1 and V-2, in total at 24 SGs). Also, a new design of this pipeline system was performed. Actually, there is a time to evaluate the benefit of these changes.

Application of Mössbauer spectroscopy for precise analysis of phase composition of corrosion products was performed from selected areas of secondary circuit and SGs. The interpretation of measured results, having in vision the long-term operation and nuclear safety, is not easy, nor straightforward. Thanks to our more than 25 years of experiences in this area, there exists already a base for the relevant evaluation of results. Optimisation of operating chemical regimes as well as regimes at decontamination and passivation seems to be an excellent output.

Based on operational experiences, the mitigation of damages and leak tightness defects in pipelines or collectors require much more time and money, than prevention measures. It is necessary to keep in mind the actual development in nuclear industry towards NPP lifetime prolongation and power increase (one of essential goals of European Commission 7FP-NULIFE). Fortunately, VVER-440 steam generators were designed with the huge power reserve (possible overloading of 20%). Beside several leakages in primary pipelines (ø16 mm), which can be in case of VVER-440 SGs relatively easy solved (blended), the corrosion deposits in feed water pipeline system occurred at many VVER-440 units [1]. The identified damages were caused mostly due to corrosion/erosion processes attacking materials familiarly called "black steels" with insufficient resistance against corrosion.

2. Actual results

In the period 2007-2008 six samples for Mössbauer effect experiments collected from different parts of the NPP Bohunice unit were prepared by crushing to powder pieces (Table 1 and Fig.1).

All measured spectra contained iron in magnetic and many times also in paramagnetic phases. Magnetic phases contained iron in nonstoichiometric magnetite $Fe_{3-x}M_xO_4$ where M_x are impurities and vacancies which substitute iron in octahedral (B) sites. Another magnetic fraction is hematite, α-Fe_2O_3. In one sample also the magnetic hydroxide (goethite α-FeOOH) was identified.

Paramagnetic fractions are presented in the spectra by quadrupole doublets (QS). Their parameters are close to those of hydroxides e.g. lepidocrocite γ–FeOOH or to small, so called superparamagnetic particles of iron oxides or hydroxides with the mean diameter of about 10 nm. It should be noted that there is no problem to distinguish among different magnetically ordered phases when they are present in a well crystalline form with low degree (or without) substitution. Both the substitutions and the presence of small superparamagnetic particles make the situation more complicated [2]. In such cases, it is necessary to perform other supplementary measurements at different temperatures down to liquid nitrogen or liquid helium temperatures without and with external magnetic field [3].

The major fraction in all samples consists of magnetically ordered iron oxides, mainly magnetite (apart from the sample No. 1 and 2, where also goethite and hematite has appeared, respectively). Magnetite crystallizes in the cubic inverse spinel structure. The oxygen ions form a closed packed cubic structure with Fe ions localized in two different sites, octahedral and tetrahedral. The tetrahedral sites (A) are occupied by trivalent Fe ions. Tri- and divalent Fe ions occupying the octahedral sites (B) are randomly arranged at room temperature because of electron hopping. At room temperature, when the electron hopping process is fast, the Mössbauer spectrum is characterized by two sextets. The one with the hyperfine magnetic field B_{hf} = 48.8 T and the isomer shift IS = 0.27 mm/s relative to α-Fe corresponds to the Fe^{3+} ions at the tetrahedral A - sites. The second one with B_{hf} = 45.7 T and IS = 0.65 mm/s is the $Fe^{2.5+}$ - like average signal from the cations at octahedral B sites. Fe^{2+} and Fe^{3+} are indistinguishable due to fast electron transfer (electron hopping), which is faster (~1 ns) than the ^{57}Fe excited state lifetime (98 ns). The magnetite unit cell contains eight Fe^{3+} ions and eight Fe^{2+} and Fe^{3+} ions, 16 in total at the B sites, therefore, the intensity ratio β = I(B)/I(A) of the two spectral components is a sensitive measure of the stoichiometry. Assuming that the room temperature ratio of the recoil-free fractions f_B/f_A for the B and A sites is 0.97, the intensity ratio β for a perfect stoichiometry should be 1.94 [4,5]. In non-stoichiometric magnetite, under an excess of oxygen, cation vacancies and substitutions at the B sites are created. The vacancies screen the charge transfer and isolate the hopping process. For each vacancy, five Fe^{3+} ions in octahedral sites become trapped. In the Mössbauer spectrum these trapped Fe^{3+} ions at the octahedral sites and Fe^{3+} ions at tetrahedral sites are indistinguishable without applying an external magnetic field. Therefore, in the spectrum of non-stoichiometric magnetite, intensity transfer from the $Fe^{2.5+}$ to Fe^{3+}-like components is observed. Thus, the intensity ratio β decreases markedly with the oxidation process, until the stoichiometry reaches the

γ-Fe_2O_3 phase. It should be noted that in our samples the intensity ratio β is far from 1.94 (for perfect stoichiometry), it varies from 0.97 up to 1.85.

Table 1. Spectral parameters of individual components including spectral area (A_{rel}), isomer shift (IS), quadrupole splitting (QS), as well as hyperfine magnetic field (B_{hf}) for each sample with according components.

Sample	Component	Area [%]	Isomer shift [mm/s]	Quadrupole shift/splitting [mm/s]	Hyperfine field [T]
Sample no.1	Magnetite A-site	36.3	0.28	0.00	48.90
Surface deposit at small coolant circuit of pumps	Magnetite B-site	37.2	0.64	0.00	45.60
17. 10. 2007	Goethite	14.4	0.36	-0.25	35.80
	Hydroxide	12.1	0.36	0.70	-
Sample no. 2.	Hematite	15.8	0.38	-0.23	51.56
Deposites scraped from filters after filtration of	Magnetite A-site	32.6	0.28	0.00	49.14
SG - feed water during	Magnetite B-site	41.8	0.65	0.00	45.91
operation	Hydroxide	9.7	0.38	0.56	-
Sample no. 3.	Magnetite A-site	34.6	0.28	0.00	49.14
SG42 pipelines - low level	Magnetite B-site	65.4	0.65	0.00	45.83
Sample no. 4.	Hematite	9.2	0.38	-0.22	51.29
Mixture of corrosion products, ionex, sand	Magnetite A-site	45.4	0.28	0.00	49.20
taken from filter of	Magnetite B-site	40.7	0.66	0.00	45.87
condenser to TG 42	Hydroxide	4.7	0.37	0.56	-
Sample no. 5.	Hematite	8.3	0.36	-0.22	51.33
Deposit from filters after 340 l of feed water of SG	Magnetite A-site	49.3	0.30	0.00	49.11
S3-09 during pasivation	Magnetite B-site	38.5	0.61	0.00	45.51
27. and 28. 5. 08	Hydroxide	3.9	0.37	0.55	-
Sample no. 6.	Hematite	6.4	0.38	-0.25	51.26
Deposit from filters after 367 l of feed water of SG	Magnetite A-site	50.3	0.29	0.00	49.14
S4-09 during pasivation	Magnetite B-site	40.7	0.66	0.00	45.61
27. and 28. 5. 08	Hydroxide	2.6	0.37	0.54	-

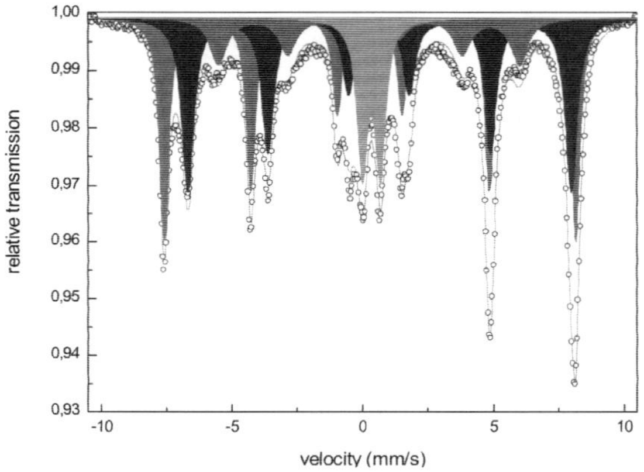

Figure 1. Mössbauer spectrum of sample no. 1. A-site (red), B-site (dark red) magnetite, goethite (pink) and hydroxide (green) were identified.

3. Conclusion

Newest results in our long-term corrosion study confirm good operational experiences and suitable chemical regimes (reduction environment) which results mostly in creation of magnetite (on the level 70% or higher) and small portions of hematite, goethite or hydrooxides.

Regular observation of corrosion/erosion processes is essential for keeping NPP operation on a high safety level. The output from performed material analyses influences the optimisation of operating chemical regimes and it can be used in optimisation of regimes at decontamination and passivation of pipelines or secondary circuit components. It can be concluded that a longer passivation time leads to a higher magnetite fraction in the corrosion products composition.

The long-term study of the phase composition of corrosion products at VVER reactors is one of the precondition to the safe operation over the projected NPP lifetime. The long-term observation of the corrosion situation by Mössbauer spectroscopy is in favour of utility and is not costly. Based on the achieved results, the following points could be established as an outlook for the next period:

A) In collaboration with NPP-Bohunice experts for operation as well as for chemical regimes, several new additional samples from places not studied up to now should be extracted and measured by Mössbauer spectroscopy with the aim to complete the existing "corrosion map" of the VVER-440 secondary circuit.

B) Optimisation of chemical regimes (having in mind the measured phase composition of measured corrosion specimens from the past) could be discussed and perhaps improved.

C) Optimisation and re-evaluation of chemical solutions used in cleaning and/or decommissioning processes during the NPP operation can be considered.

D) In connection to the planned NPP Mochovce 3, 4 commissioning (announced officially at 3.10.2008) it is recommended that all feed water pipelines and water distribution systems in steam generators should be replaced immediately before putting in operation by new ones constructed from austenitic steels. The Bohunice design with feed water distribution boxes is highly recommended and it seems to be accepted from the utility side.

Acknowledgement

This work was supported by company ENEL Produzione, Pisa and by VEGA 1/0129/09.

References

[1] Slugen V 1999 *Mössbauer spectroscopy in material science* ed. M Miglierini and D Petridis, (Amsterdam: Kluwer) 119-130.
[2] Lipka J and Miglierini M 1994 *J. of El. Eng.* **45** 15
[3] Morup S, Topsoe H and Lipka J 1976 *J. de Phys.* **37** 287
[4] Korecki J et al. 2002 *Thin Solid Films* **412** 14
[5] Cornell RM, Schwertmann U 1996 *The Iron Oxides* ISBN 3-527-28576-8

Journal of Physics: Conference Series **217** (2010) 012070 doi:10.1088/1742-6596/217/1/012070

The influence of surface layer nitriding on phase composition and tribological properties of cast steel

K Brzózka[1], T W Budzynowski[2], B Górka[1] and M Gawroński[1]

[1]Department of Physics, Faculty of Mechanical Engineering, Technical University of Radom, Krasickiego 54, 26-600 Radom, Poland
[2]Institute of Machines Building, Faculty of Mechanical Engineering, Technical University of Radom, Krasickiego 54, 26-600 Radom, Poland

E-mail: k.brzozka@pr.radom.pl

Abstract. The effect of two-stage low-temperature nitriding on atomic structure and mechanical properties of selected cast steels is investigated. Conversion electron Mössbauer spectroscopy has been used to investigate nitrides formation. In order to study tribological characteristics, tests of friction and reflecting electron microscopy measurements have been performed. It has been found that thin nitrides layer (composed mainly of γ'-Fe_4N) arises in the course of the nitriding procedure in most of investigated cast steels, what considerably affects their microstructure and tribological properties.

1. Introduction

Nitriding consists in saturation of the surface layer of metals or alloys with nitrogen during chemical and thermal treatment. It is known as method of improving some mechanical and chemical properties of the host material [1-3]. This is particularly important in case of cast steel, because of the possibility of inexpensive and precise production of metal details of machines and devices. In contrast to steel [3-8], there are not many reports related to properties of nitrided cast steel up to now [9-11]. In this paper, an influence of nitriding on phase composition and tribological properties of surface layer of highly-alloyed cast steel was examined. In order to study phase composition of the surface layer, conversion electron Mössbauer spectroscopy (CEMS) was used that yields information from the surface layer about 100 nm thick.

2. Experimental

Six kinds of highly-alloyed cast steel (denominated: A, ..., F) were investigated. The materials differed mainly in respect of chromium as well as carbon percentage. The concentration of alloying constituents of investigated cast steel samples is presented in Table I.

Two-stage low-temperature nitriding was performed in Institute of Precision Mechanics in Warsaw. In the first stage the cast steel was subjected to four hours long treatment in pure ammonia at 480°C, while in the second one the mixture of ammonia (40%) and previously dissociated ammonia (60%) was applied at 530°C during 16 hours. Mössbauer measurements were carried out in a backscattering geometry, using $^{57}Co(Rh)$ source of gamma radiation and a gas-flow detector supplied with He+4%CH_4 mixture (for electron detection). CEMS spectra were numerically analyzed by means of NORMOS program. Investigations of abrasive wear resistance were realized according to the PN-

83/H-04302 Polish Standard, performing test of friction in system: three rollers - cone. Besides, tracks of wiping were analysed by means of reflecting electron microscope.

Table 1. Real concentration of alloying constituents (in weight %) in investigated cast steel samples; *p* means the total alloying percentage. The small amount of phosphorus (<0.03%) and sulphur (<0.015%) was neglected.

sample	C	Si	Mn	Ni	Cr	Mo	V	Cu	*p*
A	0.21	0.55	0.46	0.69	2.80	0.27	0.27	1.46	6.50
B	0.64	0.48	0.54	0.75	2.63	0.23	0.54	1.22	6.40
C	0.11	0.31	0.54	0.73	7.40	0.48	0.25	1.80	11,5
D	0.70	0.37	0.39	0.68	7.49	0.43	0.26	1.90	11.5
E	0.18	0.54	0.42	0.65	13.70	0.76	0.24	1.58	17.9
F	0.55	0.53	0.42	0.60	13.70	0.35	0.26	1.82	17.7

3. Results

Mössbauer spectra collected before and after the nitriding process (Fig. 1) show significant changes in the phase structure of investigated specimens. While components characteristic of ferrite, martensite and (in a much smaller degree) austenite are mainly present in the spectra of the as-cast samples (Table 2), a significant contribution of subspectra attributed to iron nitrides has been found for most cast steels after nitriding.

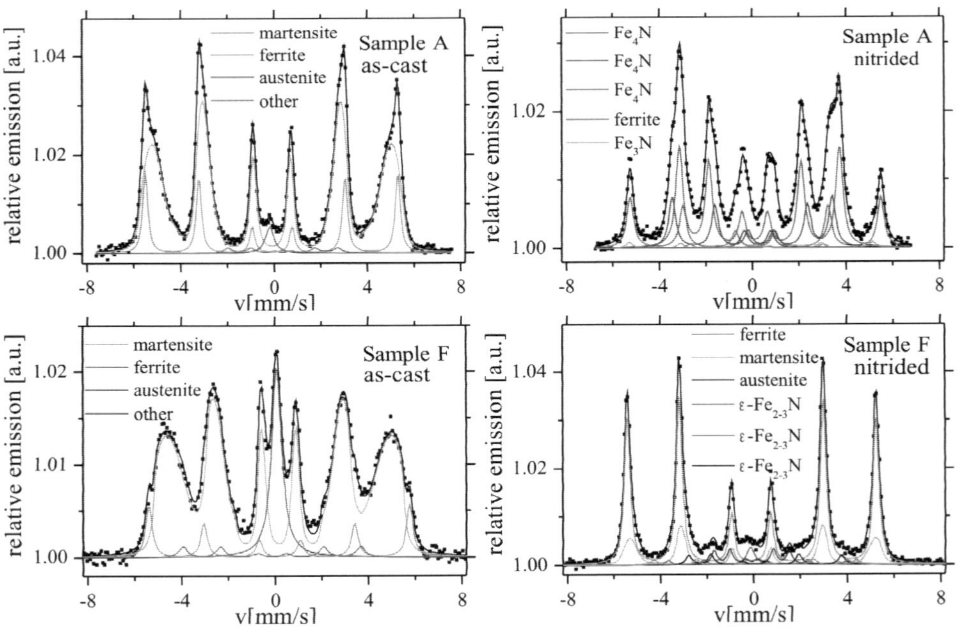

Figure 1. Example Mössbauer spectra of selected samples (see Table I).

In iron – nitride system, γ`-Fe_4N, ε-$Fe_{3.2}N$, ε-$Fe_{3-x}N$ and ς-Fe_2N are well known stable phases at room temperature. Unfortunately, their Mössbauer spectra, mostly composed of several Zeeman sextets, are strongly overlapped. Components related to other phases containing iron, for instance carbon nitrides or carbides should be also taken into account. Therefore, in order to separate the phases credibly, fitting the spectra with some imposed restrictions was realized. Since it was found that γ`-$Fe_4N$ dominated among phases comprising nitrogen in most nitrided samples, the main constraint was a fixed intensities ratio (1:2:1) of three subspectra composing $\gamma$`-Fe_4N pattern – with magnetic hyperfine field values equal to about 34 T, 21.5 T, 21.9 T, isomer shift values (relative to pure iron) equal to 0.24 mm/s, 0.31 mm/s, 0.30 mm/s and quadrupole splitting values: 0, 0.19 mm/s, –0.39 mm/s, respectively. Assuming similar *f*-factor, the relative fraction of iron atoms contributed to individual phases was estimated.

Table 2. Relative fraction of iron atoms (in %) contributing to individual phases, estimated from Mössbauer spectra (a.c. – as cast samples, nitr. – nitrided samples)

sample	A		B		C		D		E		F	
	a.c.	nitr.	a.c.	nitr.	a.c.	nitr.	a.c.	nitr.	a.c.	nitr.	a.c.	nitr.
martensite	77	-	73	-	74	-	70	-	87	-	74	20
ferrite	19	3	17	1	22	4	13	3	8	71	9	64
austenite	2	2	6	2	2	2	14	2	3	-	13	2
γ`-Fe_4N	-	94	-	94	-	92	-	92	-	26	-	-
ε-$Fe_{2-3}N$	-	1	-	3	-	2	-	3	-	-	-	<10
other	2	-	4	-	2	-	3	-	3	3	3	4

The results derived by means of numerical analysis of spectra obtained for as-cast specimens as well as for nitrided ones are collected in Table 2. The outcomes prove that applied nitriding procedure was very efficient in all cases apart from samples F and E, which show much smaller amount of iron-nitrogen phases. These samples are characterized by maximal content of chromium and specimen F – also relatively high carbon concentration, which probably prevents creating iron nitrides. In other samples, γ`-$Fe_4N$ predominates in the surface layer. A small amount of phases $\varepsilon$ is also present, which is probably the effect of oxidation of the perovskite $\gamma$`-Fe_4N phase towards the hexagonal one [12].

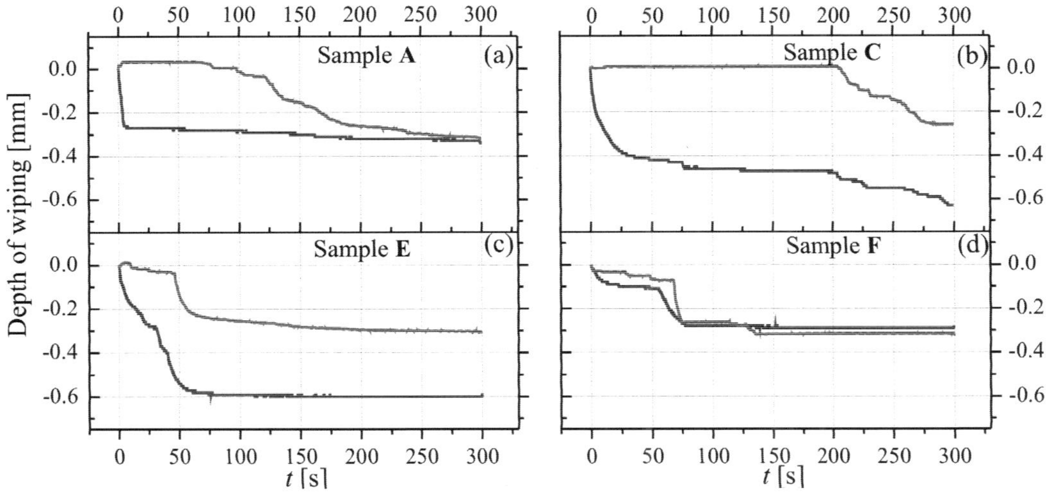

Figure 2. Selected results of the test of friction in system: three rollers – cone (according to the PN-83/H-04302 Polish Standard). The blue lines and the red ones are related to the as-cast and nitrided specimens, respectively.

The results obtained by means of Mössbauer spectroscopy are well correlated with the outcomes of tribological examinations, presented in Figure 2a, b, c, d. Although it has been found that all investigated samples show enlarged abrasive wear resistance after nitriding, the value of this effect is different. The minimal reduction of depth of wiping (close to zero) is observed for sample F, the maximal for samples C, B, A. In sample D the relative improvement is only just of the order of 25%, but the abrasive wear resistance is good both before and after nitriding (the depth of wiping is of the order of 0.1÷-0.15 mm). Time evolution of wear suggests that nitrides layer is relatively thin.

The analysis of microscopic outcomes proves that in all the samples the nitriding procedure causes comminution and unification of surface layer structure, independently on the chemical composition of the cast steel.

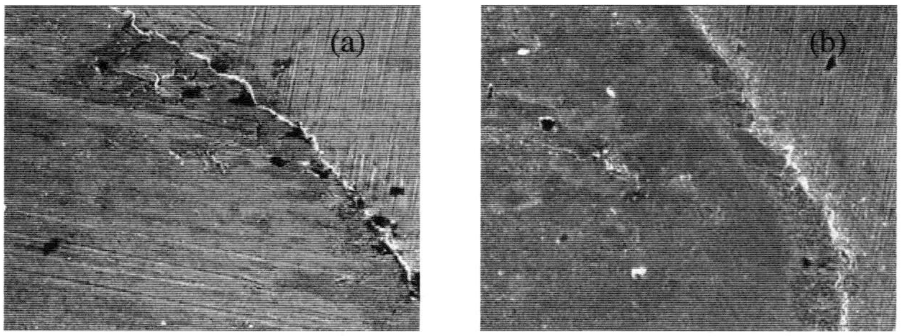

Figure 3. Tracks of wiping recorded by means of reflecting electron microscope (100x) for: (a) as cast sample A; (b) nitrided sample A.

4. Summary

An influence of nitriding on phase composition and tribological properties of surface layer of highly-alloyed cast steel was examined. The Mössbauer spectroscopy outcomes proved that low-temperature nitriding was efficient in all cases apart from the sample comprising maximal amount of chromium together with relatively high carbon concentration, which probably prevented creating iron nitrides. In other samples, γ'-Fe_4N predominated in the surface layer. The results of tribological measurements (test of friction) evidenced enlarged abrasive wear resistance of the specimens submitted to the nitriding.

References

[1] Davis Joseph R 2005 *Gear Materials, Properties and Manufacture* (Technology & Engineering) chapter 10 pp 227–244
[2] Kamminga J-D, Janssen G C A M 2006 *Surf. Coat. Tech.* **200** 5896
[3] Tarcikowski J, Senatorski J and Panasiuk W, 1995 *Met. Sci. Heat Treat.* **37** 48
[4] Kochmański P and Nowacki J 2006 *Surf. Coat. Tech.* **200** 6558
[5] Lin J F, Chen K W, Wei C C and Ai C-F 2005 *Surf. Coat. Tech.* **197** 28
[6] Piekoszewski J, Waliś L, Langner J, Werner Z, Białoskórski J, Nowicki L, Kopcewicz M and Grabias A 1996 *Nucl. Instr. Meth.* B **114** 263
[7] Budzyński P, Tarkowski P, Jartych E and Kobzev A P 2001 *Vacuum* **63** 737
[8] Nicoletto G, Tucci A and Esposito L 1996 *Wear* **197** 38
[9] Kabała-Trzaskowska A, Budzynowski T W 2005 *Inż. Mat.* **147** 403
[10] Nie X, Wang L, Yao Z C, Zhang L, Cheng F 2005 *Surf. Coat. Tech.* **200** 1745
[11] Starostin V N, Pytyaeva E I and Pershina V V 2006 *Met. Sci. Heat Treat.* **40** 421
[12] Garg V K, Oliveira A C, Azevedo R B, Wagener M, Buske N and Morais P C 2004 *J. Magn. Magn. Mater.* **272-276** 2326

Journal of Physics: Conference Series **217** (2010) 012071 doi:10.1088/1742-6596/217/1/012071

Characterization of the crystal field interaction in rare earth intermetallics using Mössbauer spectroscopy

G A Stewart

School of Physical, Environmental and Mathematical Sciences, University of New South Wales, Australian Defence Force Academy, Canberra, Australia 2600

g.stewart@adfa.edu.au

Abstract. Mössbauer spectroscopy is a valuable tool for the characterization of the crystal field interaction at rare earth sites in a wide range of compounds. This overview focuses on the complementary role of the ^{1169}Tm- and ^{55}Gd-Mössbauer resonances when investigating compounds with low magnetic ordering temperatures and highlights special issues that need to be kept in mind when working with intermetallics.

1. Introduction

The electrostatic charge distribution external to the 4f shell of a rare earth ion is responsible for the so-called crystal field (CF) interaction which removes the degeneracy of the 4f multiplet levels (as shown schematically in figure 1 for the Tm^{3+} ion). Given that the 4f shell is well shielded by the outer closed $5s^2$ $5p^6$ shells, the CF interaction is small in comparison with the spin-orbit interaction and can be treated as a perturbation. For temperatures that are low (≤ 2000 K) it is usual to express the CF Hamiltonian in terms of a multipole expansion of Stevens operator equivalents, $O_n^m(J)$,

$$\mathcal{H}_{CF} = \sum_{m=-2}^{+2} B_2^m\, O_2^m(J) + \sum_{m=-4}^{+4} B_4^m\, O_4^m(J) + \sum_{m=-6}^{+6} B_6^m\, O_6^m(J) \tag{1}$$

acting on the ion's (2J+1)-degenerate ground state multiplet. The characterization of the CF interaction is then reduced to a process of determining the relevant set of CF parameters, $B_n^m = \theta_n(1 - \sigma_n)\langle r^n\rangle_{4f} A_n^m$. In this expression, the A_n^m represent a bare estimate of the electrostatic interaction, the θ_n are reduced matrix elements, the σ_n represent the shielding effect of the outer closed shells, and the $\langle r^n\rangle_{4f}$ are radial expectation values for the 4f shell [1-3].

Inelastic neutron scattering (INS) is the most direct experimental approach to the determination of the CF parameters. This is because the positions of the spectrum's absorption lines are related to the CF-split energy levels and the intensities are related to their eigenfunctions [1]. However, characterization can also be achieved via the analysis of temperature-dependent data recorded for physical properties that depend on the thermal population of the CF levels. These include the bulk properties of specific heat and magnetic susceptibility as well as the quadrupole interaction at the rare earth nucleus, which can be monitored using Mössbauer spectroscopy. In practice, it has been found that it is advantageous to test proposed sets of CF parameters against data from two or more experimental techniques. This is particularly important for rare earth sites with low point symmetries. As the degree of symmetry decreases, the number of CF parameters needing to be determined increases from 2 (cubic symmetry) to 27 (triclinic). It is often the case that more than one non-

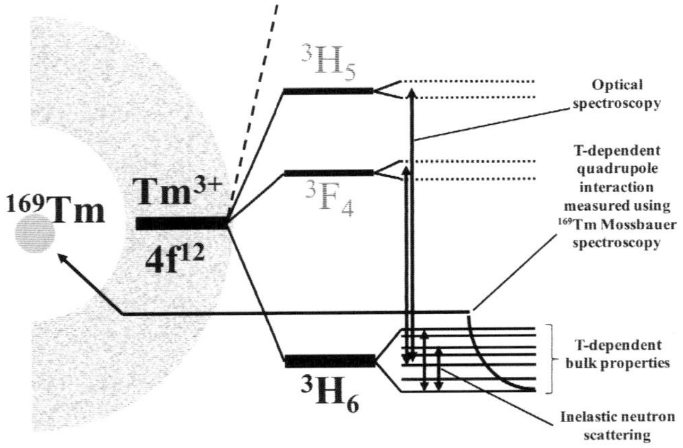

Figure 1. Schematic diagram showing the relationship between alternative experimental CF characterization techniques for the special case of a pure CF interaction.

equivalent set of CF parameters will adequately describe the same set of experimental data. This brief overview considers the application of rare earth Mössbauer spectroscopy to the characterization of the CF interaction via monitoring the nuclear quadrupole interaction as a function of temperature. Mössbauer resonances commonly employed for this purpose include ^{169}Tm(8.4 keV), ^{161}Dy (25.7 keV), ^{166}Er(80.6 keV) and ^{170}Yb(84.3 keV). The first two have the advantage that their gamma energies are small and the recoil-free fraction is large over a wide temperature range. ^{169}Tm has the additional advantage that Tm^{3+} is a non-Kramers ion and is therefore less susceptible to paramagnetic relaxation effects that are often associated with Kramers ground state doublets. There will be a focus on the complementary roles of the ^{169}Tm- and ^{155}Gd-Mössbauer resonances. Most of the considerations apply equally well to the other resonances listed above.

2. The nuclear quadrupole interaction approach to CF characterization.

The electric field gradient (EFG) tensor acting at the rare earth nucleus can be expressed in its most general (lowest symmetry) form as

$$\begin{bmatrix} V_{xx} & V_{xy} & V_{xz} \\ V_{xy} & V_{yy} & V_{yz} \\ V_{xz} & V_{yz} & V_{zz} \end{bmatrix} = k_1 \left\langle \begin{matrix} 3O_2^{+2} & \frac{3}{2}O_2^{-2} & 3O_2^{+1} \\ & & 3O_2^{-1} \\ & & O_2^0 \end{matrix} \right\rangle_{4f} + k_2 \begin{bmatrix} B_2^{+2} & \frac{1}{2}B_2^{-2} & \frac{1}{4}B_2^{+1} \\ & & \frac{1}{4}B_2^{-1} \\ & & B_2^0 \end{bmatrix}_{latt} \quad (2)$$

where the temperature-dependent 4f contributions are due to the CF-induced distortion of the 4f shell and are proportional to the thermally averaged expectation values of the rank n = 2 Stevens operators. The temperature-independent, lattice contributions are due to the direct lattice EFG and are proportional to the rank 2 CF parameters. Note that some liberty has been taken with the highlighted elements of the tensors, in order to show that $\langle O_2^{+2}\rangle_{4f}$ and B_2^{+2}contribute to the difference $V_{xx} - V_{yy}$. The factors k_1 and k_2 involve the Sternheimer shielding and antishielding parameters (R_Q and γ_∞, respectively), as well as the 4f-shell radial expectation values, $\langle r^{-3}\rangle_{4f}$ and $\langle r^2\rangle_{4f}$. It is useful to combine these with the quadrupole moment, Q, to form the experimental "shielding parameters"

$$\rho_{4f} = Q(1 - R_Q)\langle r^{-3}\rangle_{4f} \quad \text{and} \quad \rho_{latt} = Q(1-\gamma_\infty)/\{(1-\sigma_2)\langle r^2\rangle_{4f}\} \quad (3)$$

that are convenient for the purpose of analyzing temperature-dependent, quadrupole interaction data.

2.1. TmES and Tm_2O_3

The relationship between the CF interaction and the nuclear quadrupole interaction was first demonstrated in the seminal paper of Barnes *et al.* [2] that considered two non-metallic compounds.

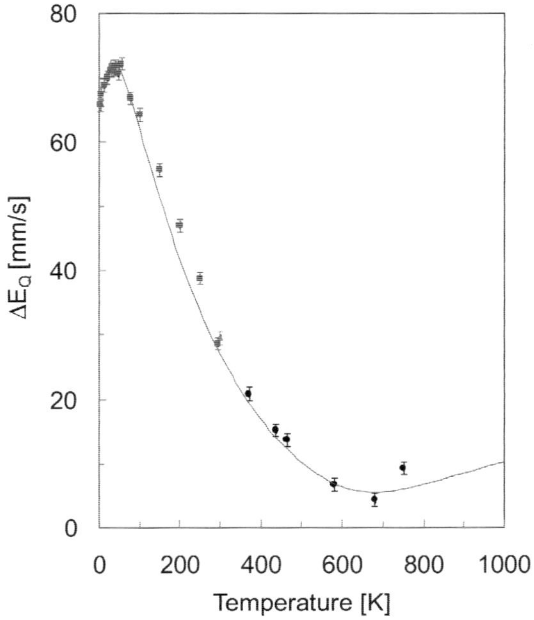

Figure 2. Quadrupole splitting, ΔE_Q, versus temperature for ^{169}Tm (I=3/2 level) in Tm_2O_3. The solid squares [3] and diamonds [4] (red and green, respectively, on line) are data for the resolved monoclinic site doublet and the solid circles at higher temperatures are the original unresolved data of Barnes *et al.* [2]. The solid curve represents the shielding parameter fit using the CF parameter set due to Leavitt *et al.* [5].

The authors used ^{169}Tm-Mössbauer spectroscopy to record the quadrupole splitting, ΔE_Q, over an extensive temperature range for the excited I = 3/2 state of ^{169}Tm in both thulium ethylsulphate (TmES) and Tm_2O_3. In each case, CF parameters had already been determined using optical spectroscopy. The impressive outcome was that they needed only to adjust the shielding parameters to match theoretical curves to the respective sets of experimental data. However, the resultant values of ρ_{4f} were -15.1 b a_0^{-3} for single hexagonal symmetry site of TmES and -17.1 b a_0^{-3} for the dominant unresolved monoclinic symmetry site of Tm_2O_3. This discrepancy arose out of an assumption that the EFG tensor was diagonal [6] for both cases. For monoclinic symmetry the CF Hamiltonian involves 15 CF parameters.

$$\mathcal{H}_{CF} = B_2^0O_2^0 + B_2^2O_2^2 + B_2^{-2}O_2^{-2} + B_4^0O_4^0 + B_4^2O_4^2 + B_4^{-2}O_4^{-2} + B_4^4O_4^4 + B_4^{-4}O_4^{-4}$$
$$+B_6^0O_6^0 + B_6^2O_6^2 + B_6^{-2}O_6^{-2} + B_6^4O_6^4 + B_6^{-4}O_6^{-4} + B_6^6O_6^6 + B_6^{-6}O_6^{-6} \quad (4)$$

In order to simplify the optical spectroscopy analysis process, it is commonly assumed that an appropriate rotation about the CF z-axis has been used to set B_2^{-2}=0. Although this reduces the lattice contribution of equation 2 to diagonal form, this is not so for the 4f contribution where the $\langle O_2^{-2}\rangle_{4f}$ term remains non-zero. A subsequent reanalysis [3] using revised CF parameters [5], resolved site data (figure 2) and a full EFG tensor diagonalization yielded a shielding parameter of ρ_{4f} = -15.3 b a_0^{-3} for the monoclinic site of Tm_2O_3. Based on close agreement with the earlier value for TmES, it was recommended that ρ_{4f} = -15.2 b a_0^{-3} and the theoretical value of $Q(I = 3/2)$ = -1.5 b [7] be adopted for the ^{169}Tm-Mössbauer resonance [3].

2.2. The special case of the ^{155}Gd-Mössbauer resonance

For S-state (L = 0) ions, such as Gd^{3+}, the local CF has relatively little influence on the spherical charge distribution associated with the ground state multiplet. This means that there will be no 4f shell contribution to the EFG tensor at the ^{155}Gd nucleus (equation 2). Hence the ^{155}Gd-Mössbauer determination of the EFG tensor components can be exploited to estimate the rank 2 CF parameters, B_2^m, for another trivalent rare earth ion in a compound of the same isostructural series.

3. $TmCu_2Si_2$ as an instructive intermetallic case study

The system of ternary rare earth intermetallics RT_2X_2 (R = rare earth, T = transition metal, X = Si or Ge) continues to be studied because of the extensive array of compounds that form with the same body-centred tetragonal structure and their diverse range of physical properties. There is a single R site whose tetragonal D_{4h} (4/mmm) symmetry requires the following CF Hamiltonian with just 5 CF parameters to be determined.

$$\mathcal{H}_{CF} = B_2^0 O_2^0 + B_4^0 O_4^0 + B_4^4 O_4^4 + B_6^0 O_6^0 + B_6^4 O_6^4 \tag{5}$$

$TmCu_2Si_2$ orders at 2.8 K so that the temperature-dependent quadrupole splitting shown in figure 3 was able to be recorded over a wide temperature range using the ^{169}Tm-Mössbauer resonance [8]. At the time, it was observed that the low temperature maximum could be described using a small positive B_2^0 and relatively large negative values for B_4^0 and B_4^4. The data were then interpreted in terms of the CF parameters given in the first row of table 1. For this initial analysis, the lattice shielding parameter was set at $\rho_{latt} = -555$ b a_0^{-2}, based on the early work of Barnes *et al.* [2].

The CF parameters presented in row 2 of table 1 were fitted to the same data set using starting CF parameters converted from those determined for $ErCu_2Si_2$ by Gromychkin *et al.* [10] via INS. The conversion was made by adjustment of the R-specific components of the CF parameters according to

$$B_n^m(Tm^{3+}) = \frac{\{\theta_n(1-\sigma_n)\langle r^n\rangle_{4f}\}_{Tm^{3+}}}{\{\theta_n(1-\sigma_n)\langle r^n\rangle_{4f}\}_{Er^{3+}}} B_n^m(Er^{3+}) \tag{6}$$

Note that the new fitted B_2^0 is opposite in sign and larger in magnitude, while the rank 4 terms are significantly reduced in magnitude. This is a very different set of CF parameters but it provides an equally good description of the experimental ^{169}Tm-Mössbauer data. The final two sets of CF parameters in table 3 were taken, respectively, from a combined INS/single-crystal magnetism investigation of $TmCu_2Si_2$ due to Kosaka *et al.* [11] and from a recent re-appraisal by Chatterji *et al.* [12] of all available $TmCu_2Si_2$ experimental data. They were then employed without modification to provide theory curves that are indistinguishable from the others in figure 3. The last three theory curves discussed were all calculated with the lattice V_{zz} contribution fixed at the ^{155}Gd-Mössbauer

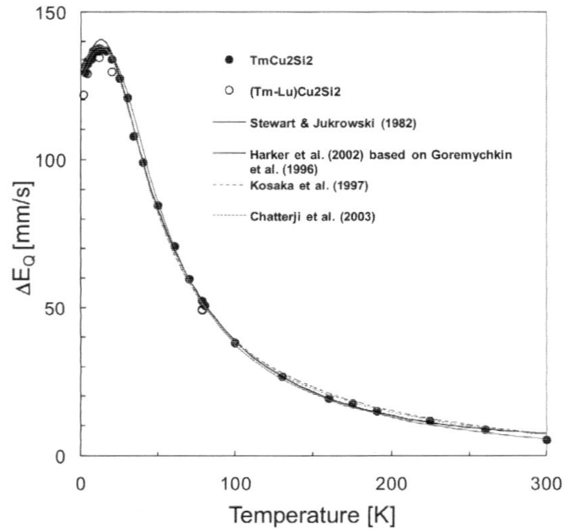

Figure 3. Theory curves for the $TmCu_2Si_2$ quadrupole splitting data (based on the CFparameters presented in table 1)

Table 1. Sets of CF parameters employed for the description of the $TmCu_2Si_2$ temperature-dependent ^{169}Tm quadrupole splitting data.

B_2^0 [K]	B_4^0 [K]	B_4^4 [mK]	B_6^0 [mK]	B_6^4 [mK]
Stewart & Jukrowski 1982 [8]				
+.40	-39.7	-19.8	-.104	+1.78
Harker *et al.* 2002[9] (based on [10])				
-.98	-4.81	-11.4	-.020	+1.14
Kosaka *et al.* 1997 [11]				
-1.18	+.285	-.553	-.078	+1.05
Chatterji *et al.* 2003 [12]				
-1.15	-0.29	-.565	-.073	+1.05

value determined for isostructural $GdCu_2Si_2$ [13, 14] rather than setting it proportional to B_2^0.

The clear message from this sequence of measurements and analyses for $TmCu_2Si_2$ is that a satisfactory fit of CF parameters to good quality quadrupole splitting data need not represent the only solution. To arrive at an unambiguous solution, it will almost certainly be necessary to test the outcome against other experimental techniques. It is also evident from the breakthrough brought about by the $ErCu_2Si_2$ INS work that CF parameters determined for one rare earth intermetallic can be converted to provide useful starting parameters for the CF analysis of a second isostructural compound.

4. $ErNiAl_4$ and a semi-empirical approach for dealing with low symmetry

The orthorhombic $RNiAl_4$ series is another intermetallic series that orders magnetically at low temperatures. It is of interest because of the intermediate magnetic phases that are observed both as a function of temperature and as a function of external magnetic field. For the orthorhombic C_{2v} (mm) point symmetry of the single R site, the CF Hamiltonian requires 9 CF parameters to be determined. The CF Hamiltonian is given by

$$\mathcal{H}_{CF} = B_2^0O_2^0 + B_2^2O_2^2 + B_4^0O_4^0 + B_4^2O_4^2 + B_4^4O_4^4 + B_6^0O_6^0 + B_6^2O_6^2 + B_6^4O_6^4 + B_6^6O_6^6 \qquad (7)$$

In an effort to reduce the number of parameters that needed to be fitted, a semi-empirical approach was recently employed for a preliminary CF analysis of an INS spectrum that had been recorded for $ErNiAl_4$ using an incident neutron energy of 20.5 meV [15, 16]. The 8.6 K INS spectrum is shown in figure 4. Until further INS spectra are recorded using higher neutron energy, only 3 of the 7 excited CF levels are known with confidence.

The approach taken was to use the simple point charge model for the near-neighbour ions (figure 5) to estimate the within-rank ratios $r_n^m = B_n^m/B_n^0$ for the rank n = 4 and 6 CF parameters. For this purpose the CF axes $(x,y,z)_{CF}$ were chosen to align with the crystallographic axes *a*, *b* and *c*, respectively. Next, the rank 2 CF parameters B_2^0 and B_2^2 were estimated using the EFG tensor values determined earlier from a ^{155}Gd-Mössbauer spectroscopy investigation of isostructural $GdNiAl_4$ [17] For orthorhombic point symmetry, the principal EFG axes are also aligned with the crystallographic axes but the particular orientation of the axes is unknown. This leads to 6 possible sets of B_2^0 and B_2^2 as shown in table 2 [16]. Finally, a grid search was performed over the values of the CF parameters B_4^0 and B_6^0 for each of the six possible B_2^0 - B_2^2combinations. INS spectra were then simulated for the CF parameter sets that matched the known excited CF energies. The only close fit to the spectrum without further refinement (solid curve in figure 4) was obtained for the $(x,y,z)_{EFG}$ // $(x,-z,y)_{CF}$ combination of B_2^0 and B_2^2 values with B_4^0 = -6.83 mK and B_6^0 = -27.28 μK.

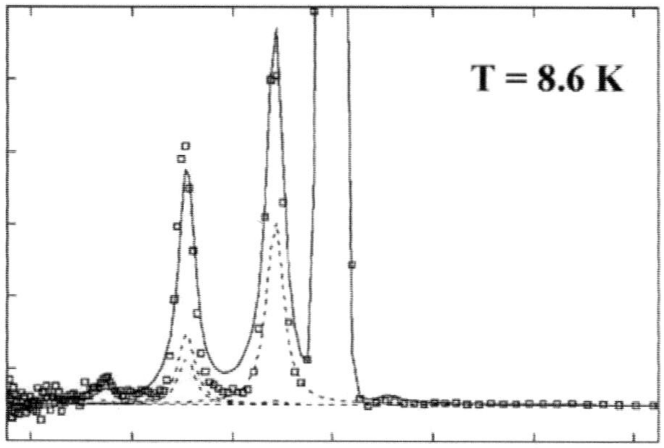

Figure 4. CF theory fit (solid line) to the INS neutron energy loss spectrum for $ErNiAl_4$ (taken from reference [15])

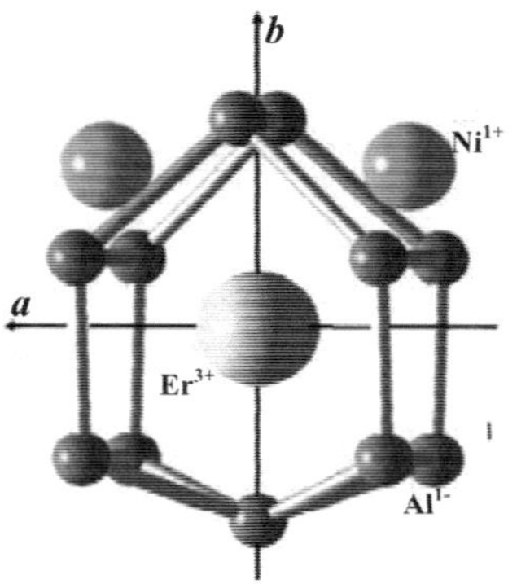

Figure 5. Local near-neighbour environment for the Er^{3+} site in orthorhombic $ErNiAl_4$ (from [15]).

Table 2. The possible combinations of B_2^0 and B_2^2 for Er^{3+} in $ErNiAl_4$ that are derived from the ^{155}Gd-Mössbauer investigation of isostructural $GdNiAl_4$ (taken from reference [16].

$(x,y,z)_{EFG}$ //	$(y,z,x)_{CF}$	$(x,-z,y)_{CF}$	$(z,x,y)_{CF}$	$(z,-y,x)_{CF}$	$(-y,x,z)_{CF}$	$(x,y,z)_{CF}$
$B_2^0[K]$	+0.432	+0.432	+0.199	+0.199	-0.631	-0.631
$B_2^2[K]$	-0.829	+0.829	+1.065	-1.065	+0.233	-0.233

This semi-empirical approach can also be applied to the CF analysis of ^{169}Tm quadrupole splitting data where the ^{155}Gd- and ^{169}Tm-Mössbauer resonances then play complementary roles. However, for intermetallics the point-charge model calculations of the within-rank CF parameter ratios are, at best, an approximate indication of the true situation. These types of calculations are much more reliable for ceramic compounds where the ligands are generally of a single type with the same charge and, in the context of the superposition model, the within-rank ratios are determined largely by the angular arrangement of the ligands.

5. Concluding remarks

In conclusion, the conversion of CF parameters between isostructural rare earth intermetallics is a useful tool and the ^{155}Gd- and ^{169}Tm-Mössbauer resonances complement one another as a means of characterising the CF interaction. However, it is always wise to test the outcome against data obtained using other experimental techniques. Finally, the semi-empirical approach using theoretical calculations of within-rank CF parameters to reduce the number of free parameters is a useful first approximation, although it is likely to be more reliable for non-metallic compounds than for the rare earth intermetallics.

Acknowledgments

The author wishes to thank Professors Günter Kaindl and Gerhard Wortmann for introducing him to rare earth Mössbauer spectroscopy, in particular, the ^{155}Gd- and ^{169}Tm-Mössbauer resonances. The $ErNiAl_4$ results presented here were part of the PhD thesis work of Mr Banchachit Saensunon and the Australian Institute of Nuclear Science and Engineering is acknowledged for its support over many years.

References

[1] Moze O 1998 *Handbook on the Physics and Chemistry of Rare Earths* vol 11, ed K A Gschneidner Jr and L Erwin (Amsterdam: Elsevier) p 493
[2] Barnes R G, Mössbauer R L, Kankeleit E and Poindexter J M 1964 *Phys. Rev.* **A136** 175
[3] Stewart G A, Day R K, Dunlop J B and Price D C 1988 *Hyp. Interact.* **40** 339
[4] Wortmann (private communication)
[5] Leavitt R P, Gruber J B, Chang N C and Morrison C A 1982 *J. Chem. Phys.* **76** 4775
[6] Stewart G A, 1985 *Hyp. Interact.* **23** 1
[7] Olesen M C and Elbek B 1960 *Nucl. Phys.* 15 123 and equation 21 of Bleaney B, Bowden G J, Cadogan J M, Day R K and Dunlop J B 1982 J. *Phys. F: Met. Phys.* 12 795 (Q = -1.5 b)
[8] Stewart G A and Zukrowski J 1982 *Crystalline Electric Field Effects in f-Electron Magnetism*, ed P Guertin *et al.* (New York: Plenum) pp 319-325
[9] Harker S J, van Dijk B D, Mulders A M, Gubbens P C M, Stewart G A, de Vroege C F and Buschow K H J 2002 *J. Phys.: Condens. Matter* **14** 2705
[10] Goremychkin E A, Muzychka Yu A and Osborn R 1996 *JETP* **83** 738
[11] Kosaka M, Onodera H, Ohoyama K, Ohashi M, Yamaguchi Y, Nakamura S and Goto T 1997 *J. Phys. Soc. Jp* **66** 2844
[12] Chatterji A, Chatterjee S and Chatterjee R 2003 *phys. stat. sol. (b)* **238** 180

[13] Dirken M W 1991 PhD Thesis (Leiden)
[14] Nowik I, Felner I and She M 1980 *J. Magn. Magn. Mater.* **15-18** 1215
[15] Saensunon B, Stewart G A, Gubbens P C M, Hutchison W D and Buchsteiner A 2009 *J. Phys.: Cond. Matter* **21** 124215
[16] Saensunon B, Stewart G A, Gubbens P C M, Hutchison W D and Buchsteiner A 2009 *J. Phys.: Cond. Matter* (Corrigendum, in press)
[17] Stewart G A, Hutchison W D, Edge A V J, Rupprecht K, Wortmann G, Nishimura K and Isikawa Y 2005 *J. Magn. Magn. Mater.* **292** 72

Journal of Physics: Conference Series **217** (2010) 012072 doi:10.1088/1742-6596/217/1/012072

Concerning the use of standards for identifying coordination environments in glasses

S D Forder[1], O M Hannant[2], P A Bingham[2] and R J Hand[2]

[1] Materials and Engineering Research Institute, Sheffield Hallam University, Sheffield, S1 1WB, UK

[2] Immobilisation Science Laboratory, University of Sheffield, Mappin Street, Sheffield S1 3JD, UK

Email: s.d.forder@shu.ac.uk

Abstract. It is an established methodology to use crystallographically well-defined standard materials for understanding site geometries in glasses. Here we discuss the benefits and the limitations of this approach for the investigation by Mössbauer and EXAFS of Fe^{2+} and Fe^{3+} in aluminosilicate glasses. As a case study we specifically consider $^{[5]}Fe^{2+}$ and $^{[5]}Fe^{3+}$ sites and whether these exist in our glasses; and if so, whether they have defined site geometries or occur simply as a consequence of the site distortions and Fe-O bond length distributions. Results are consistent with the existence of $^{[5]}Fe^{2+}$ and $^{[5]}Fe^{3+}$ but do not *prove* this because site distortion and a mixture of 4- and 6-coordinated sites can produce comparable results. This exemplifies the need for caution when interpreting glass data based on standards.

1. Introduction

Understanding the relationships between oxidation state and structural environment of Fe in glasses is important because they affect processing and performance, as discussed in our recent studies of vitrified sewage sludge ashes (SSA) [1, 2]. ^{57}Fe Mössbauer spectroscopy and Fe K-edge X-ray absorption spectroscopy (XAS) have been widely used in similar studies and both rely, to an extent, on data for crystallographically well-defined materials in which the oxidation state and local environment of Fe is known. For Mössbauer data this involves comparing fitted centre shift (CS), quadrupole splitting (QS), linewidth (LW) and area ratio between samples and standards and thereby determining Fe environments and the Fe^{2+}/Fe^{3+} ratio. Further information may be obtained from XAS, with the pre-edge peak and XANES providing a measure of coordination and redox ratio whilst the EXAFS region provides Fe-O bond lengths and (average) Fe coordination. Some authors [3, 4] have suggested the existence of $^{[5]}Fe$ in oxide glasses. As a case study on the benefits and limitations of using standards to identify coordination in glasses we present research into the possible existence of $^{[5]}Fe^{2+}$and/or $^{[5]}Fe^{3+}$ in aluminosilicate silicate glasses using Mössbauer spectroscopy and Fe K-edge EXAFS.

2. Experimental procedures

Six mineral standards were analysed. Staurolite $(Fe^{2+},Mg,Zn)_{1.5\text{-}2}Al_9[O_6|(OH,O)_2|(SiO_4)_4]$, grandidierite $(Mg,Fe^{2+})Al_3(BO_4)(SiO_4)O$ and siderite $Fe^{2+}CO_3$ represent Fe^{2+} in 4-, 5- and 6-fold symmetries and Fe-Berlinite $Fe^{3+}PO_4$, yoderite $Mg_2(Al,Fe^{3+})_6Si_4O_{18}(OH)_2$ and aegerine $NaFe^{3+}(Si_2O_6)$ represent Fe^{3+} in 4-, 5- and 6-fold symmetries. Several mixed-redox glasses modelling vitrified SSA were studied (see [1, 2] for compositional details); some (Table 1) were melted under conditions ensuring that > 95 % of Fe was present in one oxidation state, i.e. redox-pure. Glasses were reduced by melting in a H_2/N_2 atmosphere for 48h or were oxidized by adding CeO_2 at a Ce:Fe ratio of 3:2.

Sample	SiO_2	Al_2O_3	Fe_2O_3	CaO	P_2O_5	SO_3	MgO	Na_2O	K_2O
SSA + Ca[a]	48.7	12.2	5.4	14.3	8.1	0.0	5.9	1.8	1.8
DoE 39	44.4	13.1	8.3	19.0	1.4	5.0	5.8	1.1	1.9

Table 1. Nominal and ([a] analysed) compositions of redox-pure glasses in mol %

^{57}Fe Mössbauer spectroscopy was performed at room temperature using a Wissel spectrometer in transmission with a $^{57}Co/Rh$ source and calibrated relative to α-Fe. Extended Voigt-based fitting (*x*VBF) was performed using RECOIL software [5]. It has been proven useful to adjust the hyperfine parameters of glasses by Voigt profiles. A sigma value of 1 describes a large distribution width and 0 represents a pure Lorentzian line with no distribution of parameters. Two sites were fitted to represent Fe^{3+} and one for Fe^{2+} except in the redox pure glasses, in which one site for each oxidation state gave the best fit, consistent with previous results [1, 2]. Fe K-edge EXAFS was performed in air at room temperature in transmission at Station 16.5 SRS, KMC-2 BESSY II and BM29 ESRF. Energy calibrations used Fe foil. Spectra were fitted using FEFF software packages [6].

3. Results and Discussion

Fitted Mössbauer parameters are shown in Table 2 and spectra in Figure 1. Spectral quality varies due to relative abundances of Fe and distributions of lattice sites. Some minerals also contain multiple sites, especially Staurolite and Grandidierite. Dyar *et al.* [7] associated "zones" with Fe oxidation state and coordination in a plot of CS vs. QS. Our glass CS and QS (Figure 2a) correspond with these zones and are consistent with our standards and Dyar's [7]. Two Fe^{3+} sites occur in all mixed-redox glasses [1, 2] but in redox-pure glasses only one Fe^{3+} site consistent with distorted $^{[4]}Fe^{3+}$ or distorted $^{[5]}Fe^{3+}$ occurs. CS and QS values for mixed-redox glasses suggest distorted ($^{[4]}Fe^{3+}$ or $^{[5]}Fe^{3+}$) with distorted $^{[6]}Fe^{3+}$ (since site distortion increases QS) [7]. The Fe^{2+} CS and QS exhibit two distinct groupings consistent with $^{[4]}Fe^{2+}$ and $^{[5]}Fe^{2+}$. Only one doublet was fitted to Fe^{2+} components (two doublets produced unsatisfactory results) so it is equally possible that (CS, QS) values suggesting $^{[5]}Fe^{2+}$ are produced by a combination of $^{[4]}Fe^{2+}$ and $^{[6]}Fe^{2+}$; a point noted by others [4]. Linewidths indicated by σ are broader for the glasses than most standards owing to the range of site distortions typical of glasses.

We have studied (not shown here) the integrated pre-edge peak area as a function of centroid position for samples and standards [8] using Fe K-edge XANES. Average CN of ~5 (Fe^{2+}) and ~4.5 (Fe^{3+}) are obtained for the samples. As with the Mössbauer results, no distinction can be made between $^{[5]}Fe$ sites and a distribution of distorted $^{[4]}Fe$ and $^{[6]}Fe$ sites [9]. EXAFS (Figure 3) provides average CN of ~4.5 (Fe^{2+}) and ~4.2 (Fe^{3+}) for the redox-pure glasses. Whilst there is broad agreement between the Mössbauer and EXAFS results, they do not fully agree. However, all results are consistent with the existence of $^{[5]}Fe^{2+}$ and $^{[5]}Fe^{3+}$. The main difficulty in definitively answering the $^{[5]}Fe$ question is that CS and QS for $^{[5]}Fe$ lie between those for $^{[4]}Fe$ and $^{[6]}Fe$ and both XANES and EXAFS provide only average CN and Fe-O distances. Therefore any data points found to lie between $^{[4]}Fe$ and $^{[6]}Fe$ can be attributed to $^{[5]}Fe$ sites or, equally justifiably, to a mixture of distorted $^{[4]}Fe$ and $^{[6]}Fe$ sites.

The differences between CS, QS and fitted EXAFS data for the glasses and for the standards are consistent with a distribution of site geometries and distortions found in glasses but they also lead us to question the use of conventional site geometries to describe local environments in glasses. Based on the results of this study a limited "continuum of sites" may be a more suitable description. This allows for the existence of exotic site geometries as a consequence of the range of site distortions and Fe-O bond lengths occurring in glasses.

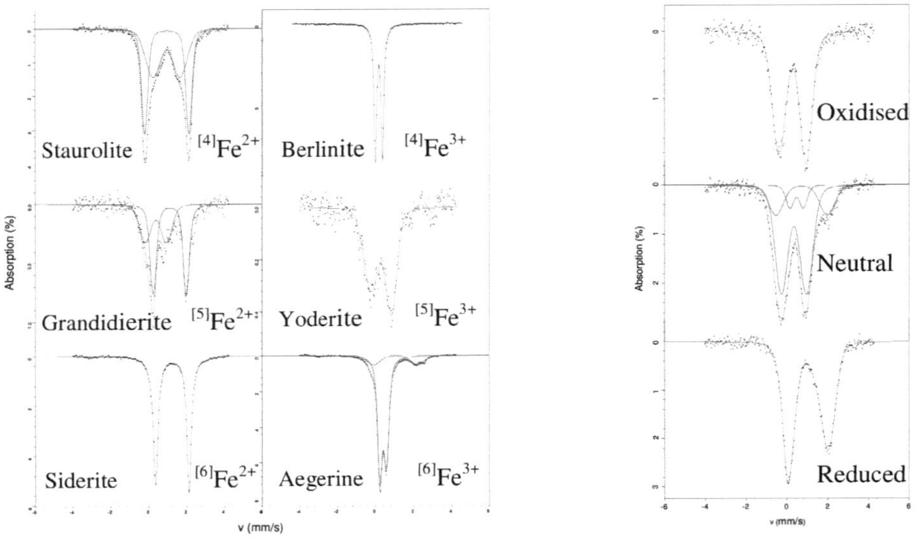

Figure 1. Fitted Mössbauer spectra for mineral standards (left) and sample glasses (right)

Material	Fe Valence	Nominal CN	CS ±0.02 / mm s^{-1}	σ (CS) / mm s^{-1}	QS ±0.02 / mm s^{-1}	σ (CS) / mm s^{-1}
Staurolite	II	4	0.90	0.12	2.35	0.00
Grandidierite	II	5	1.10	0.07	1.77	0.20
Siderite	II	6	1.24	0.08	1.80	0.12
Fe-Berlinite	III	4	0.29	0.13	0.62	0.01
Yoderite	III	5	0.30	0.22	1.07	0.42
Aegerine	III	6	0.40	0.01	0.34	0.16
SSA + Ca (Red.)	II	~4	0.99	0.29	1.84	0.48
DoE 39 (Red.)	II	~4	1.00	0.33	1.82	0.47
SSA + Ca (Oxi.)	III	~4	0.27	0.18	1.33	0.42
DoE 39 (Oxi.)	III	~4	0.29	0.23	1.29	0.35

Table 2: Fitted hyperfine Mössbauer parameters for the mineral standards and the redox-pure glasses

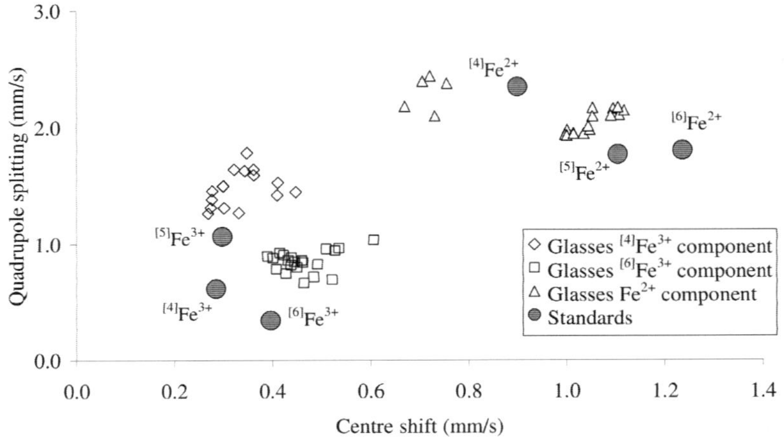

Figure 2. Mössbauer CS vs. QS for standards, redox-pure glasses and mixed redox glasses from [1, 2].

Glass Sample	CN	d(Fe-O) / Å	$\Delta\sigma^2$
SSA+Ca (Oxi.)	4.2	1.89	0.0081
SSA+Ca (Red.)	4.5	1.98	0.0121
DoE 39 (Oxi.)	4.2	1.88	0.0086
DoE 39 (Red.)	4.5	1.98	0.0143

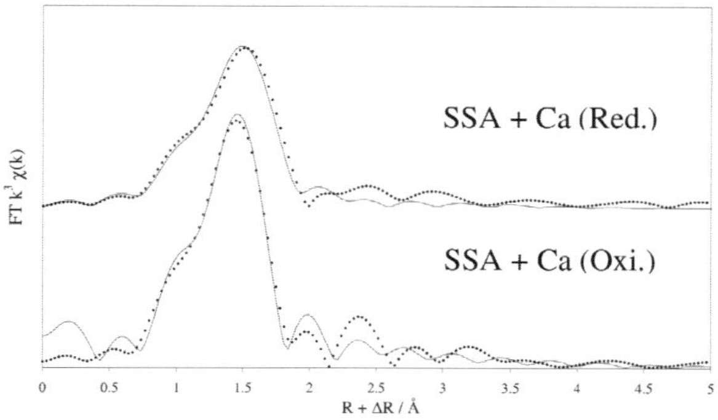

Figure 3. Fourier transforms of Fe K-edge EXAFS for oxidised and reduced glass samples.

Conclusions

Measurable differences have been observed between the fitted parameters of the glass spectra and those of the mineral standards for the Mössbauer and XAS techniques employed. This confirms that comparisons between glasses and crystalline standards by either technique should be made with caution. The apparent identification of "exotic" coordination geometries such as pentahedral may indeed be correct, but it may also be due to the presence of a combination of distorted tetrahedral and octahedral sites. Neither our study nor others have provided conclusive evidence either for or against the presence of distinct $^{[5]}$Fe sites in glasses and this serves to highlight the limitations of crystalline analogues for modelling vitreous coordination sites. Structural differences in terms of Fe-O bond lengths and site distortions make direct comparisons between glasses and crystalline materials difficult and crystalline standards are best used to ascertain qualitative trends in coordination states.

Acknowledgements

The authors acknowledge travel grants from RAEng and WCAB. We acknowledge BESSY II and ESRF for provision of synchrotron radiation facilities and thank A. Erko and O. Mathon for assistance in using beamlines KMC-2 and BM29, respectively.

References

[1] Hannant O M, Forder S D, Bingham P A and Hand R J 2009 *Hyp. Interact.* **192** 37-42.
[2] Hannant O M, Bingham P A, Hand R J and Forder S D 2008 *Glass Technol.: Eur. J. Glass Sci. Technol. A* **49** 27-32.
[3] Farges F, Lefrère Y, Rossano S, Berthereau A, Calas G and Brown G E Jr 2004 *J. Non-Cryst. Solids* **344** 176-188.
[4] Rossano S, Behrens H and Wilke M 2008 *Phys. Chem. Minerals* **35** 77-93.
[5] K. Lagarec and D. G. Rancourt, Recoil: Mössbauer spectral analysis software for windows, http://www.isapps.ca/recoil/ (1998).
[6] B. Ravel and M. Newville 2005 *J. Synchrotron Rad.* **12** 537-541.
[7] Dyar M D, Agresti D G, Schaefer M G, Grant C A and Sklute E C 2006 *Annu. Rev. Earth Planet Sci.* **34** 83-125.
[8] Wilke M, Farges F, Petit P E, Brown Jr. G E and Martin F 2001 *Am. Mineral.* **86** 714-730
[9] Wilke M, Partzsch G M, Bernhardt R and Lattard D 2004 *Chem. Geol.* **213** 71-87

Journal of Physics: Conference Series **217** (2010) 012073 doi:10.1088/1742-6596/217/1/012073

Characterization of Ag-Ge-Se bulk glasses by means of Mössbauer effect on ^{57}Fe and ^{119}Sn atomic nuclei.

B Arcondo[1], M A Ureña, J M Conde Garrido, J A Rocca and M Fontana

Laboratorio de Sólidos Amorfos, INTECIN, Facultad de Ingeniería, Universidad de Buenos Aires-CONICET, Argentina.

E-mail: barcond@fi.uba.ar

Abstract. In this work, the structure of Fe and Sn doped $Ag_x(Ge_{0.25}Se_{0.75})_{100-x}$ (x=0 to 25 at.%) intrinsically inhomogeneous glasses is analyzed employing ^{119m}Sn and ^{57}Fe Mössbauer spectroscopy, X-ray diffractometry and scanning electron microscopy. ^{119m}Sn enters in the glass as a substitutional impurity for Ge whereas ^{57}Fe enters as an interstitial impurity. Mössbauer spectra obtained with $^{119m}SnO_3Ca$ source, from samples containing about 1% ^{119}Sn for Ge, reveal that the local order of Ge in both amorphous phases is basically the same whereas Mössbauer spectra obtained with ^{57}Fe(Rh) source, from samples containing about 0.5% ^{57}Fe, evidence the differences between both phases.

1. Introduction

The $GeSe_2$-Se system presents an extensive glass forming composition range by the melt quenching technique. These glasses can be easily obtained either as massive glasses or as films and present great stability and resistance to water and corrosive media.

The fundamental structural unit of pure chalcogen glasses, e.g. Se, is based on a single atom, therefore the short range order is rather straightforward and Se coordination number is n=2. [1] For binary glasses the situation is more complex. Their structure consists of tetrahedral units centered on the tathogen, e.g. $GeSe_{4/2}$. Ge coordination number is n=4. In $GeSe_2$-Se system glasses, Ge centered tetrahedra are connected by Se bridges whereas in glasses with higher Ge content the tetrahedra are either corner sharing or edge sharing. [2] Crystalline $GeSe_2$ has a layered structure formed by chains of tetrahedra whereas crystalline Se is formed by Se chains. [3] The binary glasses, as well as the corresponding crystalline phases, are semiconductors with a band gap above 2 eV. [4] However, it is possible to control their physicochemical properties by adding a third element even in a low concentration. When Ag is incorporated in the $GeSe_2$-Se system above certain threshold concentration (~8at.%) ternary glasses result in fast ionic conductors. [5]

The Ag-Ge-Se glass forming system [6] is dominated by liquid immiscibility. Glasses in the Se-$GeSe_2$-Ag_2Se triangle may have different morphologies according to their composition but, in general, they are intrinsically inhomogeneous [7]. The physical properties of these glasses are dependent on their composition and can be easily correlated to their morphology. Accordingly, the electric conductivity evolves from semiconducting to ionic conductivity whereas a percolation transition occurs in a narrow Ag concentration range. [5]

[1] To whom any correspondence should be addressed.

In this work, the structure of Fe and Sn doped $Ag_x(Ge_{0.25}Se_{0.75})_{100-x}$ (x=0 to 25 at.%) is characterized employing X-ray diffractometry. Their morphology is analyzed by means of scanning electron microscopy, and the local order about the ^{57}Fe and ^{119}Sn, analyzed by means of Mössbauer spectroscopy, is employed to infer the physical characteristics of these glasses.

In section 2 is described the synthesis procedure of the samples as well as the characterization techniques. In section 3 are described and discussed the results obtained and, finally, the conclusions are stated in the last section.

2. Experimental Procedure

$Ag_x(Ge_{0.25}Se_{0.75})_{100-x}$ samples (x=0 to 25 at.%), named Ag_x, Ag_2Se, Ag_8GeSe_6 and $GeSe_2$, were synthesized in 10 mm diameter quartz tubes evacuated to 3.10^{-5} mbar and sealed. 1at.% Sn (>90% ^{119}Sn) or 0.5 at.% Fe (>90% ^{57}Fe) were added to the components (which purity is 99.99% for Ge, Se and 99.9% for Ag) in the desired stoichiometry in the synthesis step. The batches were heated in a furnace at 910 °C for 8 hrs and then, either quenched in a mixture of ice-water or slowly cooled. Either glasses or crystalline samples were obtained with these procedures.

The amorphous or crystalline nature of the samples was checked by X-ray diffractometry (XRD) using a θ-θ diffractometer with monochromatised Cu (Kα) radiation. Samples morphology was analyzed by scanning electron microscopy (SEM), employing an acceleration voltage of 25 kV, and energy dispersive x-ray analysis (EDX). Microscopy images were obtained by means of a back scattering electron detector (BSED) to reveal the chemistry differences on the sample surface.

Mössbauer measurements (MS) were performed in transmission geometry at room temperature, employing $^{119m}SnO_3Ca$ or ^{57}Co(Rh) sources. The spectra were fitted employing sites or quadrupole splitting (Δ) distributions with the Normos programs (Site or Dist) [8]. Isomer Shift (δ) is reported either relative to SnO_3Ca or to α-Fe.

3. Results and Discussion

3.1. XRD characterization

Rapidly quenched $Ag_x(Ge_{0.25}Se_{0.75})_{100-x}$ samples (x=0 to 25 at.%) doped either with ^{57}Fe or with ^{119}Sn resulted in completely amorphous samples. FeSe was segregated in Ag_2Se doped with ^{57}Fe slowly cooled. Segregation of FeSe in other Fe doped crystalline samples cannot be ignored. However, no segregation was detected in samples cooled in air. The $Ag_8(Ge_{0.85}{}^{119}Sn_{0.15})Se_6$ pattern evidenced the unlimited solubility of Sn in $Ag_8(Ge_{1-x}Sn_x)Se_6$.[9]

3.2. SEM and EDX analyzes

Bulk glasses resulted in homogeneous glasses for x<7. On the other hand, bulk glasses for $7 \leq x \leq 20$ present notorious inhomogeneity that denotes a spinodal decomposition for $8 \leq x \leq 10$ outflanked by two bi-nodal composition ranges (figure 1). For 5<x<8 bright dots appear isolated embedded in a dark matrix. For 10<x<25 dark dots appear embedded is a Ag rich matrix. According to EDX analyzes on glasses with $8 \leq x \leq 20$, bright zones composition are characterized by larger Ag and smaller Se concentration. Ge composition of both zones is very similar in agreement with the liquid miscibility gap reported in the ternary phase diagram.[6] Sample with x=25, in spite of being non-homogeneous as observed with a magnification of 48000x, does not present the same morphology as the above described glasses.

3.3. MS analyzes

Mössbauer spectra of ^{57}Fe doped $Ag_x(Ge_{0.25}Se_{0.75})_{100-x}$ glasses for x=5, 10 and 25 at. % are shown in figure 2. The best fit was obtained with two Gaussian quadrupolar splitting (Δ) distributions plus a single environment with δ~0.69 mm s^{-1} and Δ~1.40 mm s^{-1}. Two types of Fe environments are present: high spin (HS) and low spin (LS) Fe^{2+} in octahedral environments. The tendency of <Δ> in the

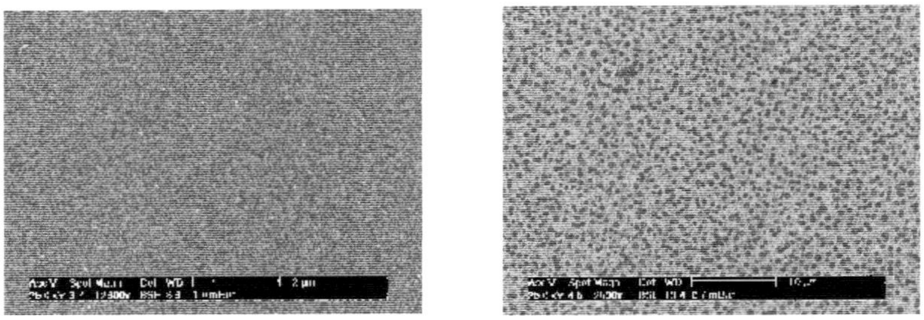

Figure 1. SEM images of $Ag_7(Ge_{0.25}Se_{0.75})_{93}$ (left) and $Ag_{20}(Ge_{0.25}Se_{0.75})_{80}$ (right) glasses.

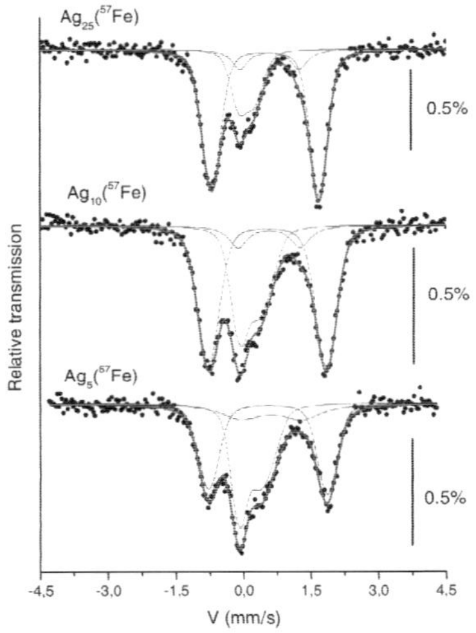

Figure 2. MS of ^{57}Fe doped $Ag_x(Ge_{0.25}Se_{0.75})_{100-x}$ glasses. The experimental points, the fitting (solid line) and each interaction are depicted.

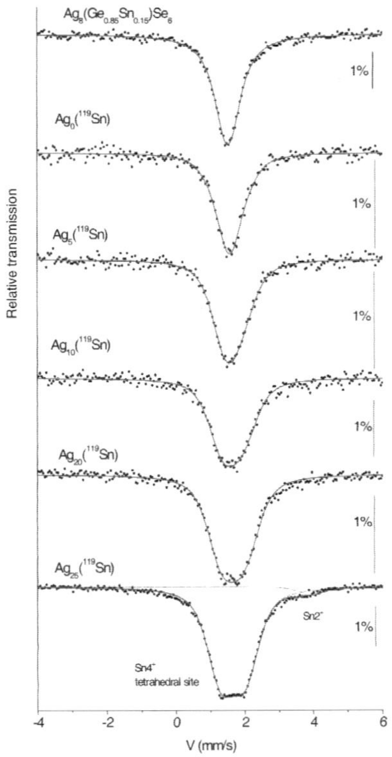

Figure 3. MS of ^{119}Sn doped Ag_8GeSe_6 and $Ag_x(Ge_{0.25}Se_{0.75})_{100-x}$ glasses. The experimental points and the fitting (solid line) are depicted.

Gaussian distribution corresponding to HS environments is that it decreases as x grows. These environments have already [7] been identified as corresponding to Ag rich zones and Ag depleted zones respectively and the Debye temperature (Θ_D) for each phase had been determined being Θ_D (bright zone) ~ $0.78\Theta_D$ (dark zone). This fact can be correlated with a different conducting regime in Ag rich and Ag depleted zones.

The local diffusion of Ag^+ in the neighborhoods of the impurity Fe atom in Ag rich zones modifies the electric field gradient (EFG) on the Fe atom, strongly dependent on the distortions of its octahedral environment. As a consequence, the effective value of the EFG is its average value calculated in the mean life of the 14.4 keV ^{57}Fe excited level and, consequently, the effective value of $<\Delta>$ would be smaller as the ionic conductivity increases.

Figure 3 shows Mössbauer spectra of samples doped with ^{119}Sn. The fittings were performed with a Gaussian Δ distribution. However fittings with sites are not less satisfactory denoting a strong short range order around Sn (that is around Ge). Sn environments in the studied glasses are similar to those of Sn doped Ag_8GeSe_6

Isomer shift for Sn doped Ag_x (~1.65 mm s^{-1}) corresponds to $Sn4^+$ tetrahedrally coordinated by 4 Se. This δ is larger than that of Ge (Sn) in Sn doped Ag_8GeSe_6. This fact could be attributed to an octahedral coordination [10]. However, as Sn is tetrahedrally coordinated by 4 Se in this phase this decrease in δ may be attributed to a decrease in Ge (Sn)-Se distances for Ag_8GeSe_6 (0.234 nm) [11] as compared to Sn-Se distances in glasses (0.236-0.239 nm) [12]. Sn doped Ag_{25} presents in addition a Sn^{2+} contribution.

A general trend is observed in the spectra fittings with sites mode, Δ and δ values increase with x as well as the linewidth (Γ) ($0 \leq x < 25$ at. %). This fact together with the linear dependence of the diffusion coefficient of Ag^+ ions reported in [5] let us correlate the line widths with the diffusion coefficient (D) in agreement with the case described in [13] for the motion, governed by the diffusion equation, of the diffusive atom in the absorbent. In addition, the increase in δ can be attributed to the increase in Ge-Se average distances with x reported in [12]. In a similar manner the increase in Ag-Se correlation with x will be responsible for the increase of Δ as the EFG on ^{119}Sn atoms is strongly dependent on the charge state of the near neighbor Se atoms and consequently on the ionicity of their bonds.

4. Conclusions

^{57}Fe and ^{119}Sn doped Ag-Ge-Se glasses in the Ag_2Se-Se-$GeSe_2$ are intrinsically inhomogeneous in a wide range of composition. They are formed by a Ag rich phase (bright) and a Ag depleted one (dark). The balance of these phases volume controls the conductivity regime.

Se concentration in the bright phase is smaller than in the dark phase whereas Ge concentration is similar in both phases.

From MS on ^{57}Fe a decreasing of the average value in the Δ distribution corresponding to HS Fe^{2+} environment in the Ag rich phase, was found and explained as a relaxation effect, due to the diffusion of Ag ions, that distorts the EFG on Fe.

An increase in linewidths in MS on ^{119}Sn with x is observed and attributed to the increase in the diffusion coefficient of Ag^+ in the samples. In addition, the increase of δ and Δ are attributed respectively to the increasing of Ge-Se distances and Ag-Se correlation with x.

References

[1] Koslowski T 1997 *Journal of Physics: Condensed Matter* **9** pp 613-26.
[2] Armand P, Ibanez A, Dexpert H, Bittencourt D, Raoux D, Philippot E 1992 *J. Phys. IV Colloq.* C2 **2** pp C2–189.
[3] Gaskel P H 1999 *Physics of glasses: Structure and dynamics. AIP Conf. Proc.* **489** p 38.
[4] Trodahl H J and Viña L 1983 *Phys. Rev. B* **27** pp 6498-501.
[5] Ureña M A, Piarristeguy A, Fontana M, Arcondo B 2005 *Solid State Ionics* **176** pp 505-12.
[6] Prince A 1988 *Ternary Alloys*, ed. G Petzow and G Effemberg (New York: VCH) p 195.
[7] Arcondo B, Ureña M A, Piarristeguy A, Pradel A and Fontana M 2007 *Physica B* **389** pp 77-83.
[8] Brand R A 1990 *NORMOS Programs*.
[9] Gulay L D, Olekseyuk I D, Parasyuk O V 2002 *Journal of Alloys and Compounds* **339** pp 113-17.
[10] Lippens P E 1999 *Phys. Rev. B* **60** pp 4576-4586.
[11] Conde Garrido J M, Rocca J A, Erazú M, Ureña M A, Fontana M, Arcondo B 2009 *Physica B* doi:10.1016/ j.physb.2009.06.083
[12] Piarristeguy A, Mirandou M, Fontana M, Arcondo B 2000 *Journal of Non-Crystalline Solids* **273** pp 30-5.
[13] Singwi K S and Sjölander A 1960 *Phys. Rev. B* **120** pp 1093-102.

Mössbauer study of steels cooled to dry ice temperature

B.S.Boyanov[1], D.G.Paneva[2] and K.I.Ivanov[3]

[1]Plovdiv University "Paisii Hilendarski", 24 Tsar Assen St., 4000 Plovdiv, Bulgaria
[2]Institute of Catalysis, Bulgarian Academy of Science, 1113 Sofia, Bulgaria
[3]Agricultural University, Department of Chemistry, 12 Mendeleev St., 4000 Plovdiv, Bulgaria

E-mail:boyanb@uni-plovdiv.bg

Abstract. Based on the change of hardness H_B, the parameter β, the microstructure and Mössbauer spectra of 7 kinds of steels the conclusion is made that in the conditions of CO_2 (dry ice) cleaning, using dry ice for repair and modification of pipes and dry snow for cooling (-78.3 °C) no significant change in the mechanical properties of the cleaned surfaces, mechanically processed steels and the steel pipes can be expected.

1. Introduction

CO_2/dry ice cleaning is effectively used in wide ranges of application – from slag removal to cleaning delicate semi-conductors and complicated electrical chips. This method allows for direct use (on-line) with no damage to the cleaned equipment [1].

The essence of the method for quick repair and modification of pipes with "Qwikfreezer" is in obtaining two ice "plugs" on both sides of the section of the pipe that will be repaired or modified. This can be done by quick freezing using carbon dioxide and using the developed system for quick freezing [2].

A new technology is developed to use solid CO_2 to cool and grease the processed metal surfaces as well as removal of the cut waste elements. This technology is called "snow machining" and one day it could eliminate the use of petrol or synthetic chemical fluids for metal cutting and cleaning metal parts in industry [3]. Very important in terms of using CO_2/dry ice particles are the consequences of thermal shock on the cleaned or cooled material. In this connection the main objective of the current study is to check the influence of cooling to the temperature of dry ice (-78.3 °C) on the microstructure and Mössbauer spectra of various types of steels and their hardness.

2. Experimental

Objects of study are 7 kinds of steels - C10, C45, C20Ch, 65G, U8A, Ch12, SchCh15 - (Table 1).

Based on the measurement of hardness an assessment can be made for the resistance to cold of the steels using a comparatively simple and repeatable methodic. It consists of measuring hardness at a room temperature Hr and comparing with the hardness at the low temperature Ht, during which the parameter β is calculated [4]:

$$\beta = Ht/Hr \qquad (1)$$

The higher the value of the parameter β, the more susceptible the metal is to crumbling under the effect of cold.

Mössbauer spectra (MS) were obtained at room temperature (RT) with a Wissel electromechanical spectrometer (Wissenschaftliche Elektronik GmbH, Germany) working at a constant acceleration mode. A $^{57}Co/Cr$ (activity ≅ 45 mCi) source and an α-Fe foil standard were used. The experimentally obtained spectra were treated using the least squares method. The parameters of hyperfine interaction such as isomer shift (δ), quadrupole splitting (Δ) and effective internal magnetic field (B) as well as the line widths (Γ) and the relative area (G) of the partial components of the spectra were determined. The areas of the observed signals have been used to evaluate the relative populations of the different iron species, assuming an equal free recoil fraction for all species.

Table 1. Chemical composition of the studied steels, %

Type of steel	C	Mn	Si	Cr	Ni	P	S
C10	0.07 – 0.14	0.17 – 0.37	0.35 – 0.65	0.035	0.04	-	-
C45	0.42 – 0.50	0.50 – 0.80	0.17 – 0.37	0.25	0.25	0.040	0.040
C20Ch	0.17 – 0.23	0.60 – 0.90	0.17 – 0.37	0.70 – 1.0	-	-	-
65G	0.62 – 0.70	0.90 – 1.20	0.17 – 0.37	0.25	0.25	0.040	0.040
U8A	0.76 – 0.83	0.17 – 0.28	0.17 – 0.33	0.2	0.2	0.025	0.018
Ch12	2.00 – 2.20	0.15 – 0.45	0.10 – 0.40	11.5 – 13.0	0.35	0.030	0.030
SchCh15	0.95 – 1.05	0.20 – 0.40	0.17 – 0.37	1.3 – 1.65	0.3	0.027	0.020

3. Results and discussion

The obtained results about the phases and the structural phases of the samples are shown in Table 2.

After cooling for the duration of 24 h at the temperature of dry ice the observed microstructure, shown in Table 2, is preserved. Based on the obtained results and the microphotographs it is confirmed that the cooling at the temperature of dry ice for a duration significantly greater than the duration of influence of the blasting stream from the blasting machine upon the cleaned material, and the time necessary for the repair and modification of pipes does not influence to a measurable degree the microstructure of the studied steels. An insignificant decrease of the size of grains is observed with C10, C45 and C20Ch.

Table 2. Results from the microstructure analysis of the studied steels

Samples	Microstructure
C10	Ferrite + perlite, with ferrite grain size 7-8 (6) grade and Widmanstätten pattern of 1.5 grade
C45	Perlite + broken ferrite net. Size of perlite grain ~7-6-8, single 5 grade and Widmanstätten pattern up to 1.5 grade
C20Ch	Ferrite + perlite, with ferrite grain size ~8-7 (6) grade with axial disalignment and Widmanstätten pattern of 1 grade
65G	Perlite with different dispersity + ferrite
U8A	Troostite
Ch12	Sorbitic perlite + carbides
SchCh15	Sorbite, with small amount of fine carbides

The steels studied for the influence of low temperature treatment on their strength are also subjected to a study of their hardness with the Brinell method. For the studied steel samples the parameter β is calculated [4]. The obtained results show that the cooling of up to -78.3 °C influences the hardness of samples insignificantly. The observed increase is measured with values that do not give reason to reach the conclusion that at the processing of pieces and items with dry ice their mechanical properties can be changed to such a degree that irreversible changes take place in their hardness and entirety.

The model of the Mössbauer spectra processing (Figure 1) for each sample before and after cooling 24 h at -78.3 °C is the same.

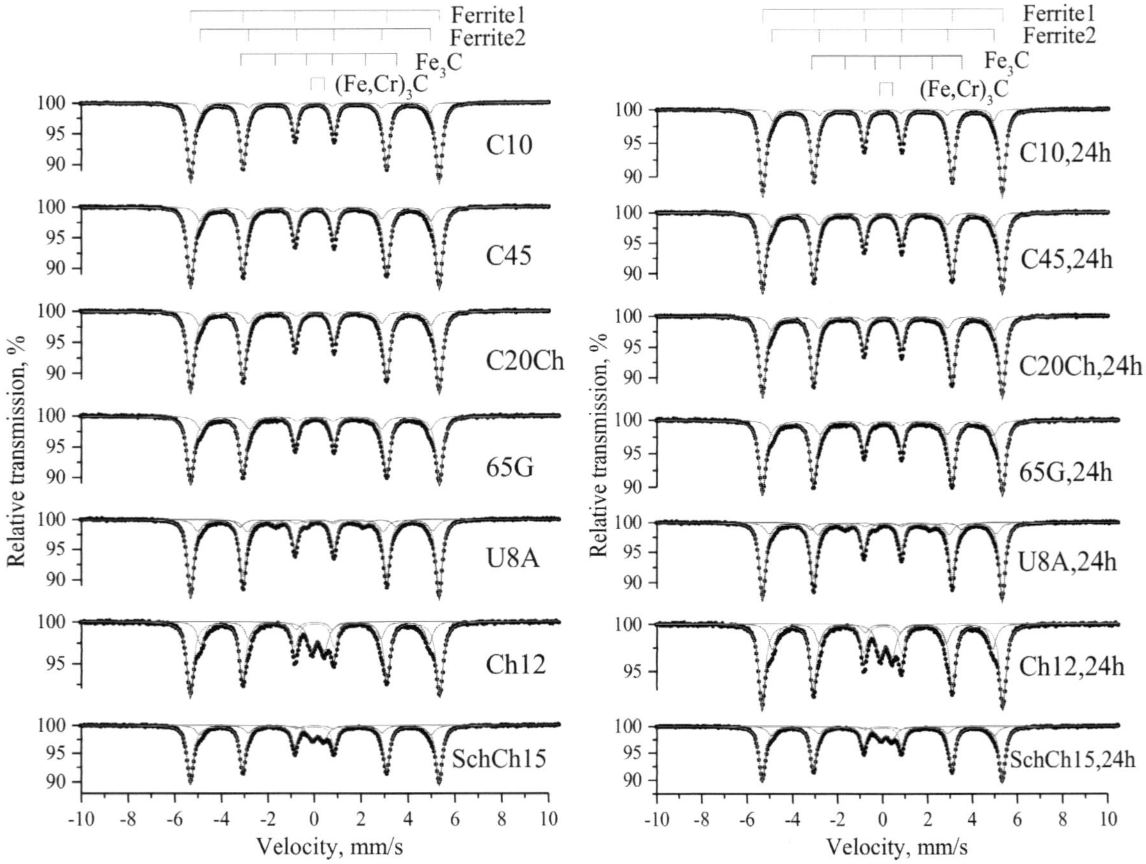

Figure 1. Mössbauer spectra of initial steel samples and cooled ones at -78.3 °C for 24h

The accepted in literature [5] designations of ferrite 1 and ferrite 2 are used in the processing of analogical Mössbauer spectra with more than 1 sextet. According to the degree of deviation of the components spectral lines with superfine structure from the theoretical Lorentians either 2 or 3 sextets are introduced. All mathematical processing of the spectra includes components designated as Ferrite 1 and Ferrite 2 (see bar-diagram above the spectra in Figure 1) with different ratio for the different samples. The component Ferrite 1 has approximately the same hyperfine parameters as the α-Fe and it is result of the presence of rich-iron ferrite phase. The value of B is in the range 33-33.3T for all 14 samples. The hyperfine field in some samples is 1% larger than that of the iron field (B = 33T) and is probably related to the plastically-deformed structure accompanying rolling. The second component Ferrite 2 has lower values of B=30.5 - 31.5T compared to the α-Fe, which can be explained with the inclusion of solute atoms on substitutional sites in iron lattice such as Cr, Ni and Mn. Samples C10,

C45, C20Ch, 65G consist of 2 magnetic components only - Ferrite 1 and Ferrite 2. Relative content of these components are 90/10, 83/17, 80/20, 78/22%, respectively. The analysis of Mössbauer spectra of the pair of samples (before and after cooling at -78.3°C) shows the equal values in range of error 1%.

A new sextet component (9%) is introduced in the processing of the spectra of steel U8A and U8A, 24h samples except components Ferrite 1 and Ferrite 2. The parameters (δ=0.18 mm/s, Δ=0.04 B=20.8T) of new component allow identification of carbide phase Fe_3C – cementite. Ratio of Ferrite 1 and Ferrite 2 is 71%/20% of sample before cooling and it changes to 73%/18%.

A doublet component with the designation of $(Fe,Cr)_3C$ is included for SchCh15 (SchCh15, 24h) and Ch12 (Ch12, 24h). The main components are magnetic - Ferrite 1 and Ferrite 2. The doublet parameters (δ=0.15-0.17 mm/s Δ=0.52- 0.54 mm/s) are close in the two spectra, and its content is different in the two pairs of samples – it is 14 and 13% respectively for the pair Ch12/(Ch12, 24h). Relative content of paramagnetic components in the pair SchCh15/(SchCh15, 24h) is 11 and 10%. It is visible that these samples have the greatest chrome content (Table 1). The determined isomer shift is close to that of the cementite, and can be said that the doublet is connected with a carbide phase or other similar to carbide. It is presented by a doublet [6] because of the inclusion of ions other than iron (for example chrome) and collapse of hyperfine magnetic interactions. As much as this component is not registered in the spectra of the other samples where the chrome content is far lower, we connect this component with phase $(Fe,Cr)_3C$. This designation must be considered as conditional and as one possibility to explain the registered doublet component in these two spectra. It is possible that other admixture atoms are included in this carbide phase.

The comparison of the parameters for the samples before and after 24 h treatment at -78.3 °C and the presented data show insignificant deviation in their values.

All obtained results from the detailed study of steels with the help of Mössbauer spectroscopy give reason for the conclusion that treatment at low temperatures (-78.3 °C) for the duration of up to 24 h does not lead to essential changes in the microstructure of the samples.

4. Conclusions

1. The data from the Mössbauer spectroscopy for the initial steels and the steels cooled for 24 hours at the temperature of dry ice (-78.3 °C) show that the microstructure of the separate specimens (initial and cooled) does not change measurably due to the cooling.
2. Based on the change of Mössbauer spectra, H_B, the parameter β, and the microstructure of the steels, it can be concluded that at conditions of cleaning with CO_2 (dry ice), using dry ice for repair and modification of pipes and for replacement of lubricants (during cutting, grinding, drilling, etc.), a significant change in the mechanical properties of the cleaned surfaces, the steel pipes and mechanically processed steels cannot be expected.

Acknowledgement: The studies are carried out through project IF-02-1402/2006, financed by the National Innovation Fund, in which the Plovdiv University Paisii Hilendarski is a Partner.

References

[1]. Spur G, Uhlmann E, Elbing F 1999 *Dry-ice blasting for cleaning: process, optimization and application, Wear,* **233-235**, pp 402-411
[2] http://www.QwikFreezer.com
[3]. *Process news: Snow-machining cuts out lubricants* 2005 *Chemical Engineer* **767**, 17
[4]. Matyunin V M *Operative diagnostics of mechanical properties of constructive materials*, Publishing House MEI, Moscow, 2006, p.214 (in Russian).
[5] Hanc A, Binczyk F 2008 *Archives of Materials Science and Engineering*, **31** 101-104
[6] Oztürk Ö, Onmuş O, Williamson D L 2005 *Surf. Coat. Technol.* **196** 341– 348

Mechanically and thermally induced structural transformations in Fe-Pt-B alloys

A Grabias[1], M Kopcewicz[1], D Oleszak[2] and J Latuch[2]

[1] Inst. of Electronic Materials Technology, Wólczyńska 133, 01-919 Warsaw, Poland
[2] Faculty of Materials Science and Engineering, Warsaw University of Technology, Wołoska 141, 02-507 Warsaw, Poland
E-mail: agrabias@itme.edu.pl

Abstract. Mechanically and thermally induced structural transformations in the $Fe_{56}Pt_{24}B_{20}$ and $Fe_{59.5}Pt_{25.5}B_{15}$ alloys are compared. Both alloys were prepared as ribbons by the melt spinning and were subsequently exposed to high-energy ball milling. Thermal behaviour of the samples was studied by differential scanning calorimetry measurements. The structure of the samples was characterised by Mössbauer spectroscopy and X-ray diffraction. The as-quenched alloys have shown disordered structures with a major contribution of the disordered cubic FePt solid solution. Ball milling of the as-quenched alloys induced a separation of the Fe-Pt and Fe-B regions. The heating of the final powders up to 993 K led to a transformation of the fcc FePt solid solution to the ordered tetragonal FePt phase, and to a formation of the Fe_2B phase.

1. Introduction

Nanocomposite Fe-Pt-B alloys consisting of hard and soft magnetic phases have revealed excellent hard magnetic properties with a potential for fabrication of efficient exchange-coupled spring magnets [e.g. 1,2]. For this system the most popular preparation technique is a rapid quenching and subsequent annealing in order to produce the required two-phase nanocrystalline structure consisting of the hard magnetic ordered tetragonal $L1_0$-FePt and soft magnetic Fe_2B phases [1,2].

In the present study, structural properties of the $Fe_{56}Pt_{24}B_{20}$ and $Fe_{59.5}Pt_{25.5}B_{15}$ alloys are compared. Both alloys were prepared by high-energy ball milling of the ribbons obtained by the rapid quenching technique in order to study the influence of milling on the structure of the alloys. The nanocrystalline ribbons are usually very brittle. The nanocrystalline powders obtained by a combination of the rapid quenching and ball milling techniques and subsequent annealing may serve as precursors used for the preparation of bulk magnets. A similar procedure has been recently used in the study of the nanocrystalline $Fe_{48.75}Pt_{26.25}B_{25}$ alloy [3]. The structures of the as-quenched ribbons and the powders collected after milling for 1, 5 and 10 h as well as of the final powders after their heating up to 993 K were characterised by Mössbauer spectroscopy and X-ray diffraction.

2. Experimental details

The $Fe_{56}Pt_{24}B_{20}$ and $Fe_{59.5}Pt_{25.5}B_{15}$ alloys were prepared in the form of ribbons by melt-spinning. The ribbons were subjected to high-energy ball milling for up to 10 h under Ar atmosphere. The Fritsch P5 planetary ball mill operating at 250 rpm was employed. Steel vial and balls with 10 mm in diameter were used. The ball-to-ribbon weight ratio was about 10:1. Differential scanning calorimetry (DSC) measurements were performed up to 993 K with the heating rate of 40 K min^{-1}. Structural properties of the ribbon and powder samples were investigated by Mössbauer spectroscopy in a transmission geometry and by X-ray diffraction (XRD). A Rigaku MiniFlex II diffractometer with CuK_{α} radiation was employed. All Mössbauer spectra were measured at room temperature using a constant accelaration spectrometer with the ^{57}Co-in-Rh source. Relative spectral fractions of the phases identified in the Mössbauer spectra were calculated as a ratio of the area of the relevant subspectrum to the total spectral area. Isomer shifts are related to α-Fe standard.

3. Results

The XRD patterns of the rapidly quenched ribbons and as-milled powders are presented in figure 1 and figure 2 for the $Fe_{56}Pt_{24}B_{20}$ and $Fe_{59.5}Pt_{25.5}B_{15}$ alloys, respectively. The XRD patterns of both as-quenched alloys reveal diffraction peaks related to the disordered cubic (fcc) FePt solid solution. Significantly narrower peaks are observed for the $Fe_{59.5}Pt_{25.5}B_{15}$ alloy than for the $Fe_{56}Pt_{24}B_{20}$ one, indicating a better defined crystalline structure. Mechanical milling of the ribbons causes broadening of the XRD peaks due to the structure refinment. A broad unresolved diffraction pattern observed in the range from 38 to 48° is assigned to disordered Fe-B regions.

The Mössbauer spectra of the $Fe_{56}Pt_{24}B_{20}$ as-quenched and milled samples are shown in figure 3. The spectrum of the as-quenched ribbon consisting of six broad overlapped lines resembles the ones that are characteristic for amorphous ferromagnetic alloys. It is fitted by a distribution of hyperfine magnetic fields, $P(B_{hf})$, using a histogram method [4,5]. The average value of the hyperfine field is about 24 T. The broad unresolved shape of the $P(B_{hf})$ distribution indicates a variety of local atomic environments. The major peak at about 28 T can be related to the disordered Fe-Pt regions, in agreement with the XRD result. The minor peak at about 11 T originates most probably from FeB environments. The Mössbauer spectra of the $Fe_{56}Pt_{24}B_{20}$ samples milled for 1-10 h were analyzed with two magnetic components: (i) a broadened sextet fitted with the hyperfine field distribution with the average value of B_{hf} of about 27 T, isomer shift $\delta = 0.19$ mm/s and quadrupole shift QS = 0.00 mm/s, and (ii) a discrete sextet with $B_{hf} = 10.1$ T, $\delta = 0.26$ mm/s and QS = 0.15 mm/s. The dominating $P(B_{hf})$ component is related to the disordered cubic FePt solid solution [6] whereas the low field sextet (10 T) can be attributed to the crystalline FeB phase [7]. A small fraction of the 10 T subspectrum is contained also in the $P(B_{hf})$ distribution. Relative fractions calculated from the Mössbauer spectra of the samples milled for 5-10 h are 76 and 24% for the FePt and FeB components, respectively.

Figure 4 presents the Mössbauer spectra of the $Fe_{59.5}Pt_{25.5}B_{15}$ alloy in the as-quenched and as-milled states. The spectrum of the as-quenched ribbon is fitted using the $P(B_{hf})$ distribution with the average value of B_{hf} of about 26 T. The major peak at the $P(B_{hf})$ with maximum at about 30 T is significantly narrower than in the case of the $Fe_{56}Pt_{24}B_{20}$ as-quenched alloy indicating a better defined crystalline nature of the $Fe_{59.5}Pt_{25.5}B_{15}$ alloy, in agreement with the XRD results. The minor peak at about 11 T is assigned to FeB environments. The Mössbauer spectra of the samples milled for 1-10 h were fitted with three magnetic components: (i) the broad sextet fitted with the $P(B_{hf})$ distribution (the average value of B_{hf} of about 30 T) representing the fcc FePt solid solution, (ii) the FeB sextet as observed

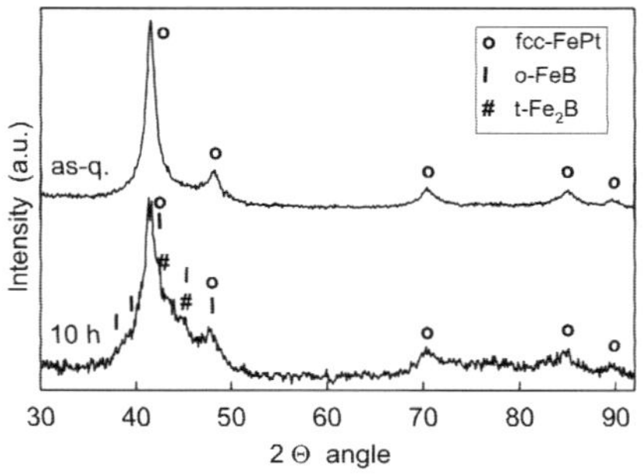

Figure 1. XRD patterns of the $Fe_{56}Pt_{24}B_{20}$ as-quenched ribbon and the powder milled for 10 h.

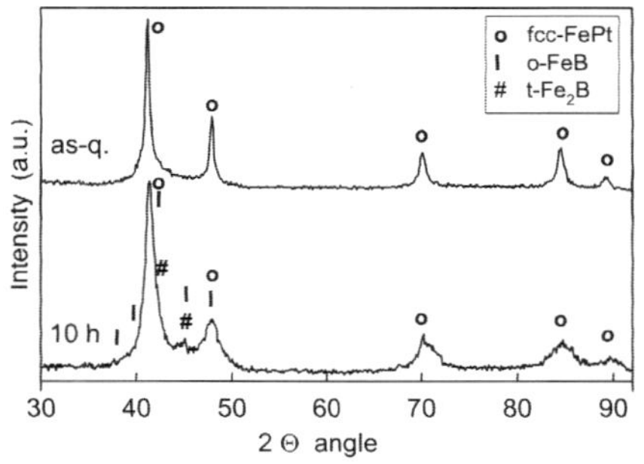

Figure 2. XRD patterns of the $Fe_{59.5}Pt_{25.5}B_{15}$ as-quenched ribbon and the powder milled for 10 h.

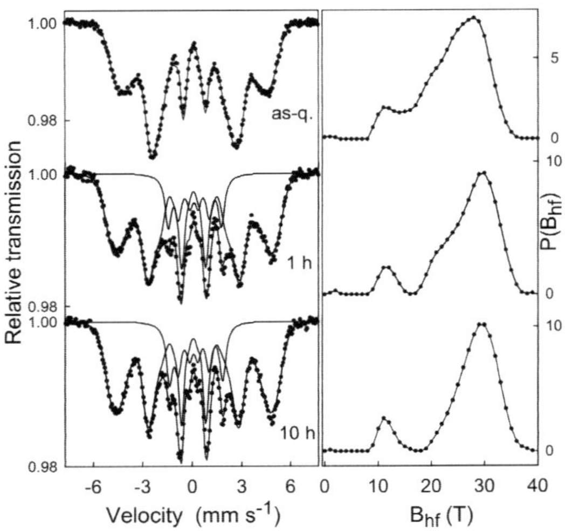

Figure 3. Mössbauer spectra of the $Fe_{56}Pt_{24}B_{20}$ as-quenched ribbon and the powders milled for 1 and 10 h.

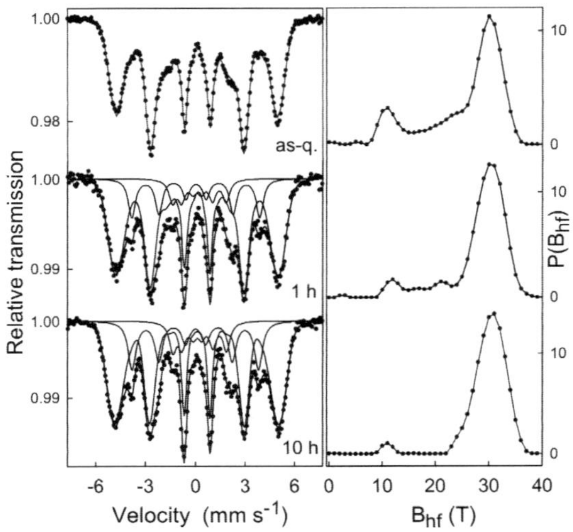

Figure 4. Mössbauer spectra of the $Fe_{59.5}Pt_{25.5}B_{15}$ as-quenched ribbon and the powders milled for 1 and 10 h.

before for $Fe_{56}Pt_{24}B_{20}$, and (iii) an additional sextet with B_{hf} = 23.6 T, δ = 0.11 mm/s and QS = 0.00 mm/s. The latter component is related to the formation of the Fe_2B regions [8]. After 10 h milling the relative fractions of the FePt, Fe_2B and FeB spectral components are 68, 21 and 11%, respectively.

Summarizing, the changes in the shape of the $P(B_{hf})$ distributions extracted from the Mössbauer spectra of both ball milled alloys clearly evidence mechanically induced separation of the Fe-B and Fe-Pt atomic environments from the starting disordered precursors.

Thermal behaviour of the powders milled for 10 h was studied by the DSC method (figure 5). The DSC curve of the $Fe_{56}Pt_{24}B_{20}$ powder shows two broad exothermal peaks with the maximal effects at T_1 = 668 K and T_2 = 831 K. The first one is related to atomic rearrangements in the amorphous and solid solution phases leading to a formation of iron-rich Fe-B environments. For $Fe_{59.5}Pt_{25.5}B_{15}$ a broad exothermal peak with maximum at T_3 = 840 K is observed. The XRD and Mössbauer results obtained for the powders after their DSC exposure are shown in figure 6 and figure 7, respectively. The XRD pattern of the $Fe_{56}Pt_{24}B_{20}$ alloy contains a set of well-defined narrow peaks that is assigned to the ordered tetragonal (fct) FePt phase. The broadened diffraction peak at 45° can be attributed to both Fe_2B and FeB crystalline phases. The corresponding Mössbauer spectrum consists of two dominating sextets: (i) with B_{hf}=27.7 T, δ=0.27 mm/s and QS=0.29 mm/s assigned to the tetragonal FePt phase [9], and (ii) with B_{hf}=23.4 T, δ=0.11 mm/s and QS=0.00 mm/s related to Fe_2B. In addition to these

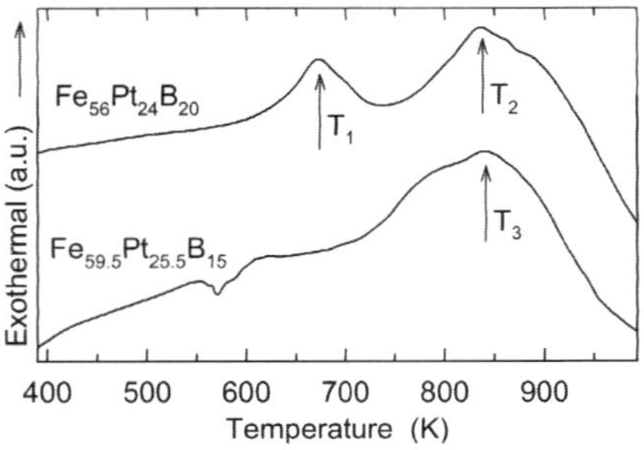

Figure 5. DSC curves recorded for the alloys milled for 10 h.

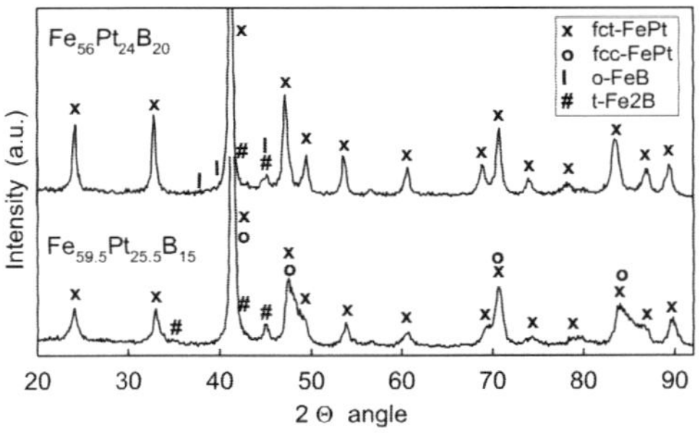

Figure 6. XRD patterns of the alloys milled for 10 h and heated up to 993 K.

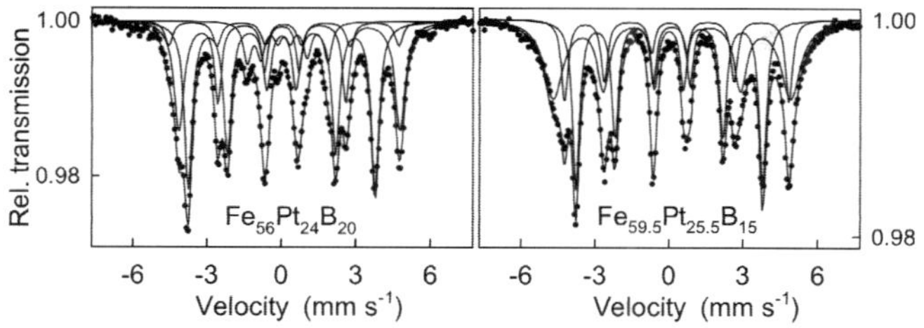

Figure 7. Mössbauer spectra of the alloys milled for 10 h and heated up to 993 K.

components small contributions of the subspectra of the remained fcc FePt and FeB phases are observed. Similar result was obtained for the ribbon sample. In the case of the $Fe_{59.5}Pt_{25.5}B_{15}$ alloy the XRD peaks related to the ordered fct FePt phase are superimposed on the peaks of the disordered fcc solid solution. The Fe_2B phase can be also identified in the XRD pattern. The corresponding Mössbauer spectrum is fitted with three sextets with B_{hf} of 29.8, 28.2 and 23.5 T representing the fcc FePt, fct FePt and Fe_2B phases, respectively. For the ribbon sample, however, only the fct FePt and Fe_2B phases are observed, what suggests that the disordered fcc FePt obtained due to the milling is more stable than after the melt spinning. Summarizing, the exothermal effects at T_2 and T_3 are related to a transformation of the disordered fcc solid solution to the ordered tetragonal FePt phase accompanied by the formation of the Fe_2B phase. The relative fractions of the fct FePt and Fe_2B components calculated for the samples after DSC exposure are 32% and 47% for $Fe_{56}Pt_{24}B_{20}$, and 19% and 39% for $Fe_{59.5}Pt_{25.5}B_{15}$, respectively. For the latter alloy a significant fraction (about 42%) of the fcc FePt phase remained untransformed. Average sizes of the fct FePt crystallites, estimated by the Sherrer formula, are 25 and 20 nm for $Fe_{56}Pt_{24}B_{20}$ and $Fe_{59.5}Pt_{25.5}B_{15}$, respectively.

4. Conclusions

The $Fe_{56}Pt_{24}B_{20}$ and $Fe_{59.5}Pt_{25.5}B_{15}$ as-quenched ribbons have shown disordered structures with a contribution of the fcc FePt solid solution. High-energy ball milling of the ribbons led to the separation of the Fe-B and Fe-Pt atomic environments. For the $Fe_{59.5}Pt_{25.5}B_{15}$ alloy the crystalline Fe_2B phase was observed already after the milling. The thermally induced transformation of the disordered fcc FePt solid solution to the ordered tetragonal FePt phase was almost complete for the $Fe_{56}Pt_{24}B_{20}$ alloy. In the case of the $Fe_{59.5}Pt_{25.5}B_{15}$ alloy a significant part of the fcc phase remained untransformed unlike in the case of the ribbon sample. The formation of the Fe_2B phase was observed for both alloys after their DSC exposure. The crystalline FeB phase remained only in the case of the $Fe_{56}Pt_{24}B_{20}$ alloy. In conclusion, the $Fe_{56}Pt_{24}B_{20}$ nanocrystalline powder seems to be a promising material as concerns good hard magnetic properties. A study of the relation between the structure and hard magnetic properties of the prepared nanocrystalline powders will be published elsewhere.

Acknowledgement

This work was financially supported by the Ministry of Science and Higher Education (Grant No. N507 140 32/4048).

References

[1] Zhang W, Sharma P, Shin K, Louzguine D V and Inoue A 2006 *Scripta Mater.* **54** 431
[2] Zhang W, Yubuta K, Sharma P and Inoue A 2006 *J. Appl. Phys.* **99** 08E914
[1] Grabias A, Kopcewicz M, Oleszak D, Latuch J, Kowalczyk M and Pękała M 2009 *J. Phys.: Conf. Series* **144** 012077
[4] Hesse J and Rübartsch A 1974 *J. Phys.* **E 7** 526
[5] LeCaer G and Dubois J M 1979 *J. Phys.* **E 12** 1083
[6] Hesse J, Nölle G and Körner H 1983 *Solid State Commun.* **46** 721
[7] Shinjo T, Itoh F, Takaki H, Nakamura Y and Shinkazono N 1964 *J. Phys. Soc. Jpn.* **19** 1252
[8] Fernandez van Raap M B, Sanchez F H and Zhand Y D 1995 *J. Mater. Res.* **10** 1917
[9] Goto T, Utsugi H and Kashiwakura A 1992 *J. Magn. Magn. Mater.* **104-107** 2051

Features of structure transformations in $La_{1-x}Sr_xMn_{0.98}Fe_{0.02}O_{3+\delta}$ (x = 0.05 – 0.30)

V D Sedykh, I I Zverkova and A V Dubovitskii
Institute of Solid State Physics, Russian Academy of Sciences, Chernogolovka, Moscow Region, 142432 Russia

E-mail: sedykh@issp.ac.ru

Abstract. The change in the structure properties of lanthanum manganite $La_{1-x}Sr_xMn_{0.98}Fe_{0.02}O_{3+\delta}$ (x = 0.05--0.30) depending on the Sr concentration and heat treatment conditions has been systematically investigated by Mössbauer spectroscopy and X-ray diffraction analysis. The features of structure transformations in the given compound have been compared to those of the undoped compound $LaMnO_{3+\delta}$.

1. Introduction

The lanthanum manganite compounds doped with bivalent impurities are the materials with colossal magnetoresistance and widely used in different fields of modern engineering.

To understand the behavior of doped systems, it is necessary to systemize data of the parent $LaMnO_{3+\delta}$ compound. According to the literature data a series of structural modifications can be obtained in the undoped $LaMnO_{3+\delta}$ compound depending on the synthesis and heat treatment conditions, the main of which are two orthorhombic *PnmaI* and *PnmaII* (total space group *Pnma*) and rhombohedral (space group *R-3c*) modifications. Manganese ions in lanthanum manganites have a mixed valence, Mn^{3+} and Mn^{4+}. Each phase has the definite Mn^{4+} concentration (less than 10% for *PnmaII*, less than 20% for *PnmaI*, and 20-30% for *R-3c*) [1,2]. A portion of Mn^{3+} transfers into Mn^{4+} under oxidation and structural phase transitions occur which are reversible under heat treatment [3]. All positions in the lattice are occupied and additional oxygen coming with Mn^{4+} will be placed in interstitial positions and distort the local environment of Fe [3]. In *R-3c* a quantity of Mn^{4+} and, consequently oxygen is maximal that results in greater distortion of the local environment as compared to *PnmaI* [3].

The presence of Jahn-Teller Mn^{3+} ions coexisting with not Jahn-Teller Mn^{4+} ions is one of the features of the compounds based on lanthanum manganite [4]. The doping of cation positions actively affects the structure and, correspondingly, the fundamental physical properties of these compounds. The change in the structure leads to significant changes in physical properties (conductivity, magnetization) since there is a strong correlation between lattice and electron subsystems in lanthanum manganites.

2. Experimental

Polycrystalline powder samples $La_{1-x}Sr_xMn_{0.98}Fe_{0.02}O_{3+\delta}$ were prepared by sol-gel method. Preparation of the samples was described in detail in [5].

X-ray diffraction (XRD) patterns of powder samples were measured using a diffractometer Siemens D-500 (Cu K_α and Fe K_α radiation). The lattice parameters were calculated using the program PowderCell (Werner Kraus & Gert Nolze, BAM Berlin).

The ^{57}Fe Mössbauer measurements were carried out at room temperature using a conventional Mössbauer spectrometer operating in the constant acceleration mode.

3. Results and discussion

La and Sr ion radii are close, therefore they should not induce notable distortions in the lattice. The synthesized $La_{1-x}Sr_xMn_{0.98}Fe_{0.02}O_{3+\delta}$ samples with x = 0.05, 0.10 и 0.20 were single phase and had the rhombohedral structure. It follows from the XRD data (Fig.1) that all the mentioned phases *PnmaI*, *PnmaII* and *R-3c* are formed for the small Sr concentration up to 10% with polymorphous reversible transformations under heat treatment. The lattice parameters change insignificantly for *PnmaI* and *R-3c* with an increase in Sr concentration. For *PnmaII* the lattice parameters *b* and *c* change insignificantly with an increase in Sr concentration, whereas the *a* parameter and cell volume notably decrease (Fig.2). For 10%Sr the *a* and *c* parameters approach.

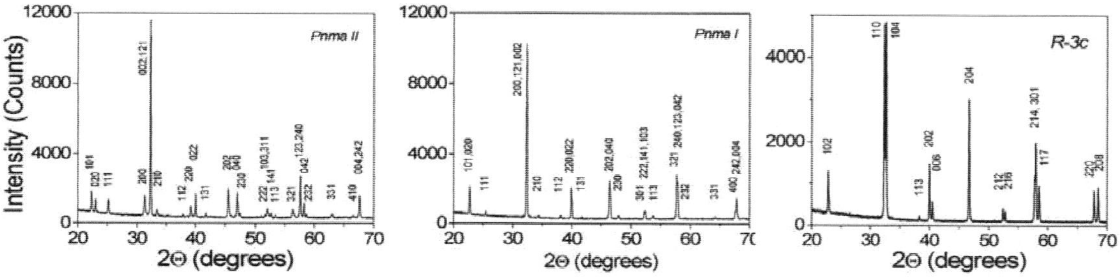

Figure 1. Standard X-ray diffraction patterns for *R-3c, PnmaI,* and *PnmaII* phases ($CuK_{\alpha 1}$ radiation).

When Sr concentration is higher than 10%, *PnmaII* is not formed. When Sr concentration is 20%, only the rhombohedral phase is formed which is stable at any heat treatment.

There are some substantial differences as compared to the undoped compound. It follows from XRD data that the monoclinic phase (distorted cube) appears in addition to *R-3c* for 30%Sr.

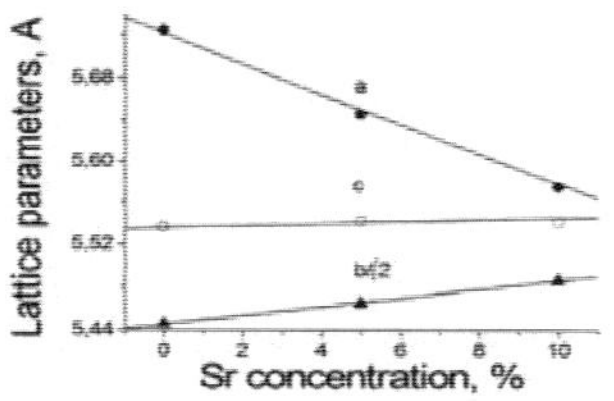

Figure 2. Dependence of the lattice parameters for *PnmaII* in $La_{1-x}Sr_xMn_{0.98}Fe_{0.02}O_{3+\delta}$ on the Sr concentration.

The obtained Mössbauer data are in a good agreement with the XRD results.

The linewidth of Mössbauer spectra for *PnmaI* and *R-3c* of the polycrystalline Sr-doped samples is rather narrow (Γ = 0.33 mm/s) and the spectra were fitted by single doublet. The lines of Mössbauer spectra for the *PnmaII* phase represent strongly broadened quadrupole-split doublets (Fig.3). It can be due to the presence of several subspectra. In [5] we tried to fit qualitatively similar spectra by three doublets. Since the spectra are smooth, some uncertainty in a choice of linewidths appears. In addition, the lattice distortions can be due to several reasons. (1) We can assume that not Jaht-Teller Mn^{4+}ions have a concentration dispersion within the *PnmaII* phase which results in the formation of the regions with a different degree of distortion and, consequently in the Δ distribution. (2) We have previously shown [6] that particles of *PnmaII* are nanosized, therefore a surface of nanoparticles will also contribute. (3) Interstitial oxygen will distort the local environment. It is quite difficult to separate all these contributions. Therefore it is more correctly to determine the quadrupole splitting ($\Delta = 2\varepsilon$) distribution probability (Fig.3). For this purpose the DISTRI code from the software MSTools [7] was used. Note that with an increase in Sr concentration, i.e. with an increase in the Mn^{4+} ion portion, within the *PnmaII* phase the maximum probability of distribution is displaced towards a decrease in the Δ value. It means a decrease in distortion of the local symmetry.

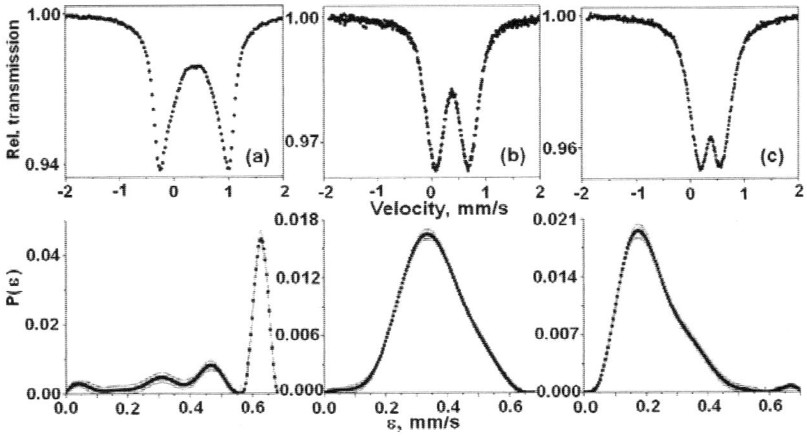

Figure 3. Mössbauer spectra (at the top) and distribution probability of quadrupole displacement P(ε) (at the bottom) for the *PnmaII* phase in $La_{1-x}Sr_xMn_{0.98}Fe_{0.02}O_{3+\delta}$ for x = 0 (a), 0.05 (b), and 0.1 (c).

In the Sr-doped compound an increase in the Sr^{2+} concentration leads to an increase in the Mn^{4+} concentration and to a decrease in lattice distortions. For *R-3c* the Δ value decreases with an increase in Sr concentration and has the minimal value for 20%Sr (Δ = 0.23 mm/s) (Fig.4, Table 1). It means that for this Sr concentration the quantity of interstitial oxygen is small and the lattice distortions are insignificant.

For the sample with 30%Sr the additional subspectrum appears in the Mössbauer spectrum in addition to the doublet corresponding to the *R-3c* phase which can be attributed to the magnetic component with H = 3.41 T (Fig.4). It is known from literature that Tc increases with an increase in impurity concentration, T_c = 375 K for 33%Sr [8] and Tc is lower than room temperature for 18%Sr [9]. As for the monoclinic phase observed in the XRD patterns, it is most probable that the Δ values for rhombohedral and monoclinic phases are close and it is not possible to determine the contribution of each of them in the Mössbauer spectrum. A small asymmetry of the spectra can be related to the sample texture [2].

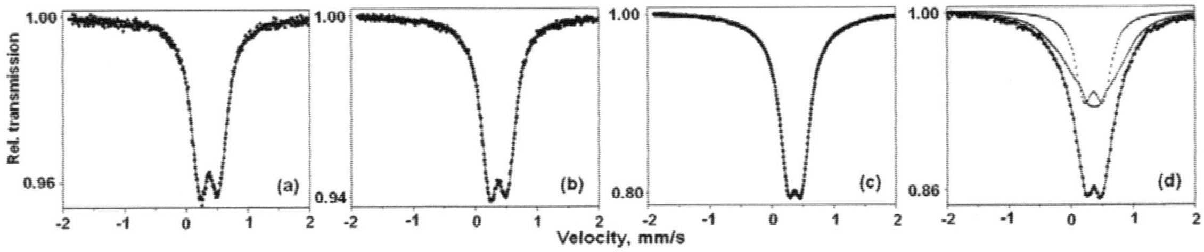

Figure 4. Mössbauer spectra for the rhombohedral phase in $La_{1-x}Sr_xMn_{0.98}Fe_{0.02}O_{3+\delta}$ for x = 0.05 (a), 0.1 (b), 0.2 (c), and 0.3 (d).

Comparing the experimental data for the Sr-doped and undoped compounds we can conclude the following. Lattice distortions in the undoped $LaMnO_{3+\delta}$ compound with a stoichiometric composition and a small Mn^{4+} portion (*PnmaII*) are due to the cooperative Jahn-Teller effect. With an increase in Mn^{4+} ion concentration the breaking of orbital order occurs and lattice distortions related to them decrease. However lattice distortions related to oxygen coming with Mn^{4+} ions and occupying only interstitial positions in the undoped compound appear which are maximal for the rhombohedral phase.

Table 1. Parameters of the ^{57}Fe room-temperature Mössbauer spectra for the *R-3c* phase of $La_{1-x}Sr_xMn_{0.98}Fe_{0.02}O_{3+\delta}$. Δ and IS are quadrupole splitting and isomer shift which is relative to bcc Fe at room temperature; A is relative intensity.

%Sr	Δ, mm/s	IS, mm/s	Γ, mm/s	A, %	Sextet
5	0.31±0.01	0.37±0.01	0.33±0.01		
10	0.29±0.01	0.37±0.01	0.33±0.01		
20	0.23±0.01	0.37±0.01	0.33±0.01		
30	0.24±0.01	0.37±0.01	0.33±0.01	47	H=3.41 T

The introduction of bivalent impurity occupying trivalent La positions leads to the appearance of vacant oxygen sites in the lattice. Therefore a part of Mn^{3+} ions transfers to Mn^{4+} already during synthesis. When the Mn^{4+} and Sr^{2+} quantity is the same, the oxygen stoichiometric phase is formed and all extra oxygen first of all occupies vacant oxygen lattice sites. In this case lattice distortions are minimal. However, for the fixed Sr concentration in the doped sample the Mn^{4+} ion concentration can increase up to the maximum value under oxidation, the rest extra oxygen will now occupy interstitial sites and the lattice is notably distorted. Thus in the Sr-doped sample one can obtain both the minimum Mn^{4+} concentration equal to the bivalent impurity concentration and the maximum one under oxidation. While oxidized, the Mn^{4+} portion increases and the phase transitions from *PnmaII* to *PnmaI* and to *R-3c* occurs for small Sr concentration. An increase in the Sr^{2+} concentration, on the one hand, increases the Mn^{4+} ion concentration, and on the other hand, decreases the quantity of extra oxygen occupying interstitial positions, which results in a decrease in lattice distortions and, respectively, in the Δ value.

The work was supported by the Russian Foundation for Basic Research, project no. 09--02--00767 and the Russian Academy of Sciences (program Properties of Condensed Matter).

4. References

[1] Huang Q, Santoro A, Lynn J W, et al.1997 *Phys. Rev.* B **55** 14987
[2] Sedykh V D, Smirnova I S, Bagautdinov B Sh, et al.2002 *J. Surf. Inv.* **12** 9
[3] Sedykh V, Shekhtman V Sh, Zverkova I I, et al.2006 *Physica* C **433** 189
[4] Kugel K I, Khomskii D I 1982 *UFN* **136** 621 (in Russian)
[5] Sedykh V D, Shekhtman V Sh, Dubovitskii A V, et al.2009 *Phys. Sol. State* **51** 373
[6] Aristova I M, Sedykh V D, Shekhtman V Sh, et al.2008 *Mat. Lett.* 62, 1036
[7] Rusakov V S 1999 *Bulletin Russian Academy of Science Physics* **7** 1093; Rusakov V S and Kadyrzhanov K K 2005 *Hyperfine Interactions* **164** 87. Rusakov V S 2000 Mössbauer spectroscopy of locally inhomogeneous compounds INPNNCRK, Almaty, 431
[8] Kim Ch S, Shim I-B, Kim S B 2003 *JMMM* **254-255** 568
[9] Tkachuk A, Rogacki K, Brown D E, et al. 1998 *Phys. Rev.* B **57** 8509

Study of a sigma-phase formation in an equiatomic Fe-V alloy

B F O Costa[1, 3], J Cieślak[2] and S M Dubiel[2]

[1] CEMDRX, Department of Physics, University of Coimbra, P-3004-516, Coimbra, Portugal

[2] Faculty of Physics and Computer Science, AGH University of Science and Technology, 30-059 Kraków, Poland

E-mail: benilde@ci.uc.pt

Abstract. Kinetics of the α- to σ-phase transformation in an equiatomic Fe-V alloy has been studied by Mössbauer Spectroscopy and X-ray diffraction. The *B2* superstructure seems to form prior to the precipitation of the σ-phase. The results obtained in this study give evidence that in the investigated interval, the transformation kinetics strongly depends on temperature. The kinetics of the α- to σ-phase transformation, as determined from the temperature dependence of the average hyperfine field derived from the Mössbauer spectra, has been quantitatively described in terms of the Johnson-Mehl-Avrami equation, from which the kinetics parameters n and k have been determined.

1. Introduction

According to the Fe-V phase diagram, a disordered bcc solid solution (α-phase) can be formed over an entire composition range. However, for certain compositions around 50%, the solution is not stable, and, upon annealing it transforms, depending on the temperature, either into the σ-phase with a complex tetragonal structure, or into the ordered metastable *B2* phase. Precipitation of the σ-phase severely deteriorates Fe-based steel properties, and, in particular, decreases ductility and impact toughness, lowers the corrosion resistance and increases brittleness and cracking [1]. For as-melted alloys, the transformation into the σ-phase is rather slow and it requires a prolonged annealing time whose exact duration strongly depends on temperature [2].

In this study we were interested in the effect of temperature on the α- to σ-phase transformation in plastically deformed samples. It is known that a plastic deformation can significantly accelerate the formation of the σ-phase [3, 4].

2. Experimental

An equiatomic Fe-V alloy has been prepared in an induction furnace by melting together appropriate amounts of Fe (99.9 % purity) and V (99.7% purity) elements. The ingots were then cold-rolled down to form platelets of about 200 μm thickness, which were next isothermally annealed at various temperatures, T_a, namely 600°C, 700°C and 800°C for different periods.

X-ray diffraction (XRD) was performed at room temperature (RT) using Cu K_α radiation (λ = 0.154184 nm) to characterize the microstructural changes. ^{57}Fe Mössbauer spectra were recorded at

[3] Author to whom any correspondence should be addressed.

RT in a transmission geometry using a standard constant acceleration spectrometer. A ^{57}Co source in Rh matrix with the activity of about 20 mCi was used. The spectra were analysed by a constrained Hesse-Rübartsch method [5] which yields a hyperfine magnetic field distribution, $P(B)$.

3. Results and Discussion

Figure 1 shows the X-ray diffraction patterns taken on the Fe-V cold rolled sample annealed at 700°C. It is observed that the full transformation from the α- to σ-phase has taken place after 30 min. When analyzing the diffractogram obtained on the sample annealed at 700°C for 2 min, it is evident that some structural change has taken place – see the shape of the peak at $2\theta \approx 135°$. For the other temperatures of annealing the same transformation is observed, but it is slower at 600°C and quicker at 800°C. From the XRD patterns, it can be also concluded that while the structure had remained bcc, the lattice constant has decreased relative to the initial value which is indicative of the formation of *B2* [2]. The Mössbauer spectroscopy also gives evidence that the transformation is fast, and that its rate grows with increasing isothermal annealing temperature. The Mössbauer spectra have dramatically changed their shape, as illustrated in Figure 2 for the annealing temperature of 700°C, giving thereby an evidence that the hyperfine field has dramatically decreased. The quick decrease both of the lattice parameter of the bcc phase and that of the mean hyperfine field (Figure 2b) are characteristic features of the formation of the *B2* superstructure [2]. This seems to be in agreement with a previous study according to which the *B2* superstructure forms prior to the σ-phase in the equiatomic Fe-V system [6].

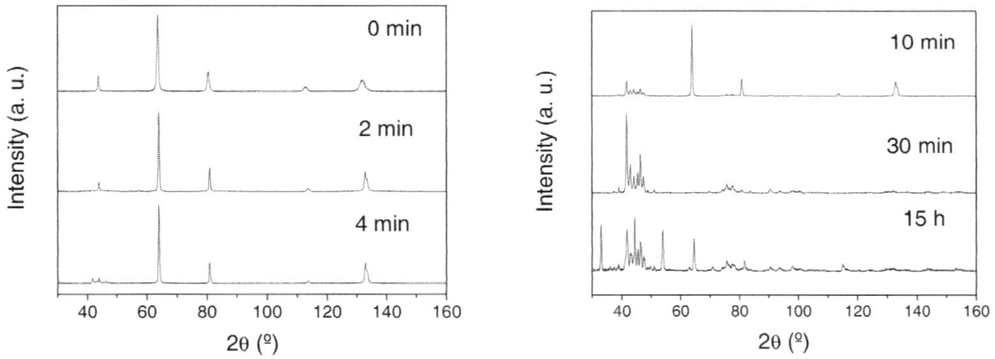

Figure 1. X-ray diffraction patterns of the Fe-V cold rolled sample annealed at 700°C for the indicated periods.

Upon further annealing all the samples have eventually transformed into the σ-phase which was evident both from the XRD pattern as well as from the Mössbauer spectrum – see Fig. 2c.

Assuming that the shape of the Mössbauer spectrum corresponding to the *B2* superstructure is not changing with annealing time, the kinetics of the σ-phase formation was followed by studying the temperature dependence of the average hyperfine field, $\langle B \rangle$, which, to our knowledge, is a procedure that has been for the first time used in the study of the kinetics of the σ-phase formation. Figure 2 shows also the hyperfine field distribution curves obtained as a function of the annealing time for the three different annealing temperatures.

The time dependence of the isothermal decrease of $\langle B \rangle$ has been fitted using Johnson-Mehl-Avrami (JMA) type equation, of the form:

$$F = 1 - \exp[-(kt)^n] \qquad (1)$$

where F is the relative fraction of the σ-phase precipitated after the isothermal annealing time t, k is the rate of formation and n is the Avrami form factor.

The relative fraction F has been determined by normalizing the $\langle B \rangle(t)$-values to the initial value of the average hyperfine field of the pure α-phase, $\langle B \rangle_{\alpha}$, corresponding to 100% α-phase. The plots based on F-quantity are shown in Figure 3b, while the fitted values of n and k are given in Table 1.

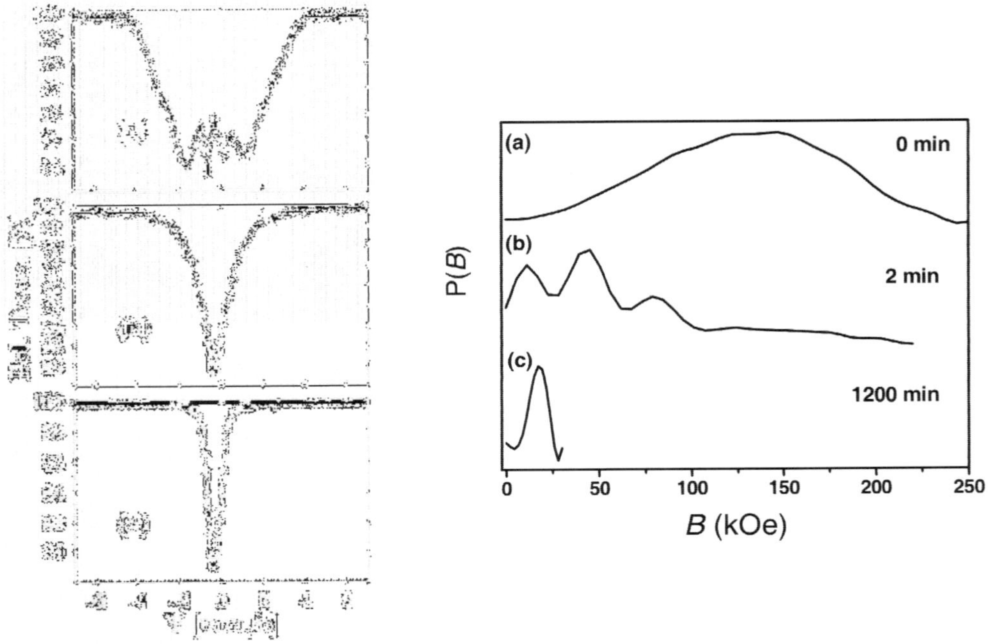

Figure 2. Room temperature ^{57}Fe Mössbauer spectra, and the corresponding hyperfine field distribution curves (right-hand side), for Fe-V cold rolled sample annealed at 700°C and for different periods shown (a) α-phase; (b) $B2$ phase and (c) σ-phase.

From Figure 3b, it can be deduced that at 800°C the transformation from α- to σ-phase is the quickest one, and it takes about 2 min to be completed. At 700°C the full transformation takes about 25 min and there is one process prior to the formation of σ, while at 600°C the full transformation takes about 1500 min and there are two processes preceding the formation of σ. The first one can be ascribed to the ordering of α into $B2$ (it takes ca. 1000 min), and the second one is related to the transformation of $B2$ into σ (it takes ca. 6 min).

The transformation rate to the σ-phase (3rd step in Table 1) increases with increasing temperature, as already mentioned above.

In order to interpret the third regime, ascribed to the precipitation of the σ-phase, the n values can be compared to theoretical ones [7]. The obtained values indicate a diffusion controlled growth mechanism with zero or decreasing nucleation rate, which agrees with the literature data on the transformation mechanism in other systems [6-8].

In summary, the kinetics of the α- to σ-phase transformation in an equiatomic Fe-V alloy that was determined from the temperature dependence of the average hyperfine field measured by means of the Mössbauer Spectroscopy, has been quantitatively described in terms of the Johnson-Mehl-Avrami equation.

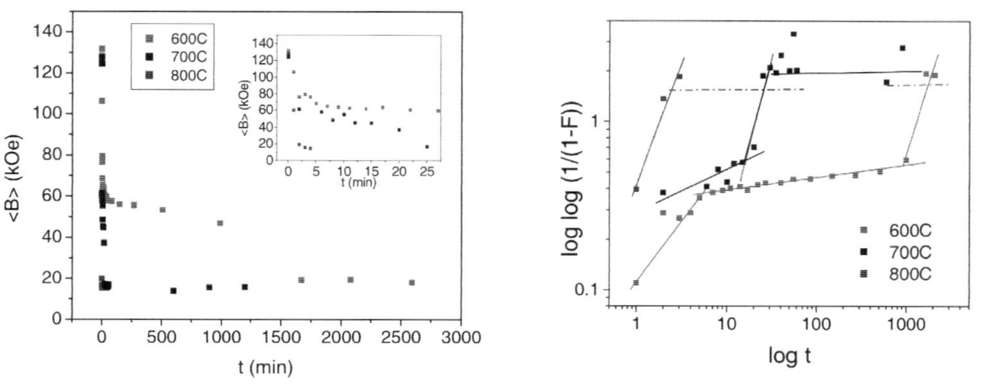

(a) (b)

Figure 3. (a) Hyperfine field distribution as a function of annealing time for different annealing temperatures; (b) Johnson-Mehl-Avrami plots from the data in figure (a). Dotted lines in (b) are guides to the eye, indicating stabilization of final stage of transformation.

Table 1. Values of *n* and *k* obtained from the fits to the data in Figure 3b to the Johnson-Mehl-Avrami equation.

Annealing Temperature	1st step	2nd step	3rd step	
600°C	0.72	0.075	1.70	*n*
	9.04	1.57	0.07	*k* (h^{-1})
700°C		0.25	2.1	*n*
		12.42	4.33	*k* (h^{-1})
800°C			1.45	*n*
			59.5	*k* (h^{-1})

From the JMA-equation the kinetics parameters *n* and *k* have been calculated. The results obtained give evidence that there are different steps in the transformation depending on the annealing temperature. For T_a = 600°C and 700° the ordered *B2* superstructure forms prior to the precipitation of the σ-phase.

References

[1] Dubiel S M 2009 *Hyperfine Interact.* **189** 53–61; and references therein
[2] Ziller T, Le Caer G, Isnard O, Cenedese P and Fultz B 2001 *Phys. Rev.* B **65** 024204
[3] Costa B F O, Le Caër G and Ayres de Campos N 2001 *Phys. Stat. Sol.* (a) **183** 235–50
[4] Cieslak J Reissner M Dubiel S M Wernisch J and Steiner W 2008 *J. Alloys Compd.* **460** 20–25
[5] Le Caër G and Dubois J-M 1979 *J. Phys.*: E **12** 1083-1090
[6] Ustinovshikov Y Pushkarev B and Sapegina I 2005 *J. Alloys Compd.* **398** 133–8
[7] Christian J W 1981 *The theory of transformations in metals and alloys*, Part I, *Equilibrium and General Kinetis Theory*, 2nd edition, (New York: Pergamon Press)
[8] Costa B F O, Le Caër G and Ayres de Campos N 1997 *Phys. Stat. Sol.* (a) **164** 687–97
[9] Costa B F O, Amado M M, Le Caër G, Amaral V S, Sá M A, Sousa J B and Ayres de Campos N 1997 *Phys. Stat. Sol.* (a), **161** 349-60.

Journal of Physics: Conference Series **217** (2010) 012078 doi:10.1088/1742-6596/217/1/012078

Mösbauer spectroscopy study of the influence of Si addition to ordered and disordered $Fe_{60}Al_{40}$ alloys

E Legarra[1], E Apiñaniz[2] and F Plazaola[1]

[1]Elektrizitatea eta Elektronika Saila, UPV-EHU, 644, 48080 Bilbao
[2]Fisika Aplikatua I Saila, UPV-EHU, Alameda Urquijo s/n, 48013, Bilbao

fernando.plazaola@ehu.es

Abstract. We have used Mössbauer spectroscopy to study the influence of Si addition to $Fe_{60}Al_{40}$ composition without changing the iron content of the alloy. The study has been performed in ordered and mechanically deformed samples. This work shows that the B2 structure present in ordered $Fe_{60}Al_{40}$ alloy does not change with Si introduction in the studied ternary alloys. However, the binary $Fe_{60}Si_{40}$ alloy shows D03 and B20 structures. Mechanical deformation of the ordered samples induces a clear change of the Mössbauer spectra of binary alloys with the presence of new sextets or variation of the previous ones; however, the Mössbauer spectra of the studied ternary alloys are very similar to the ordered ones.

1. Introduction

The magnetic behavior of FeAl alloys has been a subject of several theoretical and experimental studies because they often display unique magnetic properties. Hernando et al. [1] showed that a paramagnetic to ferromagnetic transition linked to an order-disorder transition between ordered B2 structure (paramagnetic) and disordered A2 structure (ferromagnetic) can be observed after mechanical deformation in $Fe_{60}Al_{40}$. In $Fe_{70}Al_{30}$ an enhancement of the magnetism of the intermetallic alloy after mechanical deformation has been observed too [2].

In this work we study the influence of Si addition to $Fe_{60}Al_{40}$ binary alloy in ordered and mechanically deformed samples by means of Mössbauer spectroscopy and X-ray diffraction.

2. Experimental

Five $Fe_{60}Al_{1-x}Si_x$ alloys with x= 0, 12, 20, 28 and 40 were prepared by induction melting in alumina crucibles and cast into ingots under a 40 kPa pressure of high purity argon. Iron, Aluminum and Silicon with a purity of 99.99, 99.99 and 99.999%, respectively were used. To obtain the ordered samples, they were annealed for 2 h at 1043 K, then they were kept at 843 K for 10 h and afterwards they were cooled down to room temperature at a rate of 3 K/min. The mechanical deformation has been performed crushing the ordered samples to obtain homogeneous powders with a diameter of about 10 μm. ^{57}Fe Mössbauer spectroscopy measurements were carried out at room temperature in transmission geometry using a conventional spectrometer with a ^{57}Co-Rh source. The measured spectra were fitted using the NORMOS program developed by Brand et al. [3]. All the isomer shift values presented in this work are relative to metallic iron at room temperature. XRD measurements were carried out in a Philips PW 1710 powder diffractometer using CuKα radiation.

3. Results and discussion

From the structural point of view the largest difference in the binary phase diagrams of Fe rich FeAl(Si) system corresponds to alloys with composition 60 at.%Fe. Indeed, the alloy $Fe_{60}Al_{40}$ belongs to a region of the phase diagram where the equilibrium structure corresponds to the B2 structure; however, $Fe_{60}Si_{40}$ alloy is located in a biphase zone of B20 and D03 structures. The B20 structure (ε-FeSi) is of distorted NaCl type. It consists of a simple cubic Bravais lattice having four Fe atoms and four Si atoms in the unit cell, with $P2_13$ spatial group, where all Fe positions are equivalent. Fe local symmetry is trigonal, with a unique Si atom as nearest neighbor, located at 2.28 Å, three Si atoms as second nearest neighbors located at 2.35 Å, and three Si atoms as third nearest neighbors located at 2.52 Å [4]. In such a structure all Fe and Si atoms are equivalent and each atom is surrounded by 7 atoms of different kind (the distances up to the third nearest neighborhood are very similar).

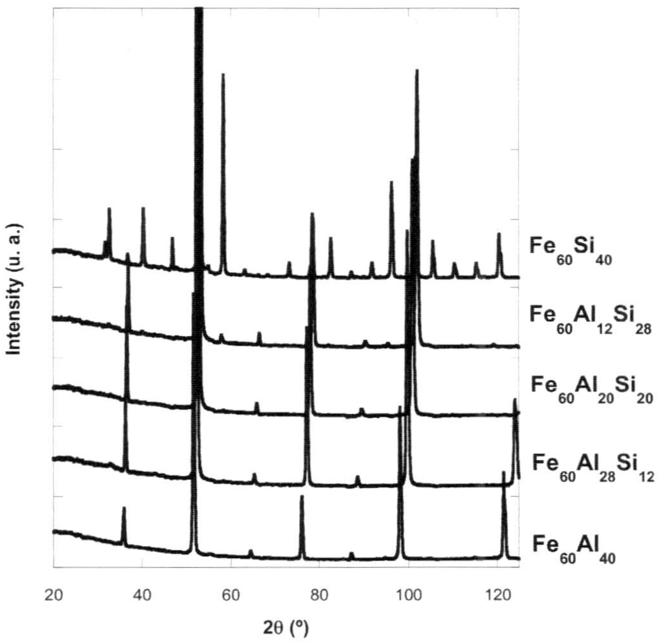

Figure 1. X-ray diffraction patterns of the ordered samples.

3.1. Ordered $Fe_{60}Al_{40-x}Si_x$ alloys

Figure 1 shows the XRD pattern of the alloy series $Fe_{60}Al_{40-x}Si_x$. The pattern of binary alloys correspond to phase diagram equilibrium ones, that is to say, $Fe_{60}Si_{40}$ presents a mixture of phases D03 and B20. On the other hand, $Fe_{60}Al_{40}$ alloy shows a perfect B2 structure. As far as ternary alloys are concerned Figure 1 shows that Al addition causes the disappearance of the B20 structure. Indeed, no characteristic reflection corresponding to B20 structure is observed. Neither can any characteristic (111) reflection of the D03 structure be observed in the diffractograms of the alloys containing Si. Therefore, all the studied ternary alloys present the B2 structure, alike in $Fe_{60}Al_{40}$ alloy, with no hints of the D03 structure. So, the minimum Al addition in this study, 12 at. %Al, is enough to destroy completely the phases present in the binary $Fe_{60}Si_{40}$ alloy. Figure 1 shows clearly that the lattice parameter increases with Al addition. It increases quite linearly from 2.82 Å (12 at. %Al) to 2.90 Å (40 at. %Al).

Figure 2 shows the Mössbauer spectra of the ordered alloy series studied. The alloys that present B2 phase at room temperature in the XRD do not present any sextet at all. Indeed, $Fe_{60}Al_{40}$ alloy presents a unique singlet. On the other hand, $Fe_{60}Si_{40}$ alloy, which presents D03 characteristic reflections, has been fitted by means of five sub-spectra (see figure 2 a, bottom). Four of the five sub-spectra are sextets corresponding to the D03 structure (two sextets for 8Fe position, and one for positions 4Fe and 3Fe). The fifth sub-spectrum is a doublet corresponding to the B20 structure. It is worth noting that this results show the expected pattern of FeAl binary alloys, where the D03 structure gives sextets in Mössbauer spectra taken at room temperature and the B2 structure does not give any sextet [5]. From the area relative to the sub-spectra it is concluded that 58 % of the sample has D03 structure and 42 % B20 one. Moreover, analyzing the relative areas of the sub-spectra corresponding to the D03 structure by means of the binomial distribution of atoms in it, we obtain that such structure is highly ordered, only a 4% of disorder is found on it.

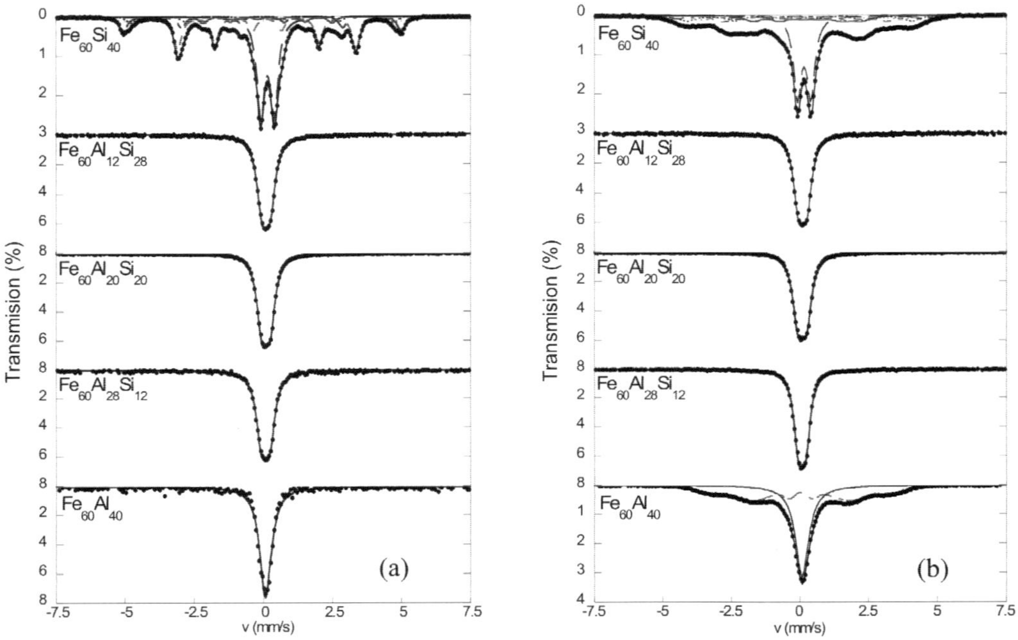

Figure 2. Fitted Mössbauer spectra. a) Ordered samples. b) Mechanically deformed samples.

The ternary alloys do not follow in a straightforward way neither of the Mössbauer patterns shown by the two binary alloys of this series. For a proper fit of the spectra of the ternary alloys it is necessary to use a doublet, instead of a singlet, like in $Fe_{60}Al_{40}$ alloy. The appearance of a doublet instead, is due to the different configurations of Al/Si atoms as nearest neighbors of Fe atoms. In the B2 structure the probability of finding a Fe atom surrounded by 8 Fe atoms as nearest neighbors is 16,7%; so a large number of Al/Si atoms surround Fe atoms in this structure. Moreover, each nearest neighbors configuration around one Fe atom gives place to a singlet, but taking into account that adding Si in these alloys the isomer shift increases, the singlets will have an isomer shift that depends on the particular configuration and therefore the peak of the spectrum is formed by singlets with an isomer shift that changes slightly between configurations. Due to the large number of configurations the Mössbauer peak becomes a doublet with not very large quadrupolar momentum. Table 1 shows clearly that increasing the Si content in the alloy the isomer shift increases but on the other hand the quadrupolar splitting decreases.

Table 1. Hyperfine parameters IS (isomer shift), Δ (quadrupolar splitting) and Width (line width) obtained from the the Mössbauer spectra of the alloys $Fe_{60}Al(Si)$

% Al	Width (mm/s)	IS (mm/s)	Δ (mm/s)
12	0.42 ± 0.01	0.0989 ± 0.0005	0.281 ± 0.002
20	0.43 ± 0.01	0.0903 ± 0.0003	0.288 ± 0.001
28	0.48 ± 0.01	0.077 ± 0.001	0.295 ± 0.003
40	0.50 ± 0.01	0.074 ± 0.001	

3.2. Mechanically deformed $Fe_{60}Al_{40-x}Si_x$ alloys

Figure 3 shows the Mössbauer spectra of the alloys after mechanical deformation. It is interesting to compare the results obtained after deformation with the ones obtained in the annealed samples. In the case of $Fe_{60}Si_{40}$ alloy the width of the doublet obtained in the deformed alloy is very similar to the one obtained in the ordered one. However, there is a clear difference in the sextets corresponding to the D03 structure, the ones of the deformed sample are much wider than the ones of the ordered alloy. So, this result indicates clearly that at the deformation level presented in this work most of the deformation of the alloy occurs in the D03 structure. In the case, of the binary alloy $Fe_{60}Al_{40}$, it is clear that the mechanical deformation induces the appearance of a sextet, indicating that mechanical deformation of the binary $Fe_{60}Al_{40}$ induces the appearance of a magnetic contribution. This result is in agreement with the results reported in Reference 1 and there have been different studies about the causes of the induction (or enhancement) of magnetism in FeAl alloys [1, 6]. On the other hand, it is interesting to notice that unlike the binary alloys the ternary ones present very similar Mössbauer spectra to the ordered ones. Therefore, this result shows that mechanical deformation does not induce any sextet in the ternary alloys, indicating that in opposition to what happens in FeAl binary alloys mechanical deformation does not cause any magnetic contribution in the studied ternary alloys.

4. Conclusions

The Mössbauer study performed in this work shows clearly that in ordered alloys Al addition to $Fe_{60}Si_{40}$ changes the structural order and destroys the magnetic contribution present in it. In mechanically deformed alloys the deformation produces a variation of the magnetic contribution present in ordered $Fe_{60}Si_{40}$ alloy and gives place to a new magnetic contribution in $Fe_{60}Al_{40}$ alloy. However, no magnetic contribution is observed after mechanical deformation in the studied ternary alloys.

This work has been undertaken under projects IT-382-07 (Eusko Jaurlaritza) and MAT2006-12743 (Spanish government).

References

[1] Hernando A, Amils X, Nogués J, Suriñach S, Baró MD, Ibarra MR 1998 *Phys Rev. B* **58** 11864
[2] Apiñaniz E, Plazaola F, Garitaonandia JS, Martín D, Jiménez JA 2003 *J. Appl. Phys.* **93** 7649
[3] Brand RA et al. 1983 *J. Phys. F: Met. Phys.* **13** 675
[4] Iasonna A and Magini M 1996 *Acta Mater.* **44** 1109
[5] Apiñaniz E, Martín-Rodriguez D, Legarra E, Garitaonandia JS, Plazaola F 2009 *Mater. Res. Soc. Symp. Proc.* **1128** 65
[6] Apiñaniz E, Plazaola F, Garitaonandia JS 2004 *J. Magn. Magn. Mat.* **272-276** 794

Structural and hyperfine properties of Cr-doped SnO_2 nanoparticles

F H Aragón[1], J A H Coaquira[1], D S Candela[2], E Baggio Saitovitch[2], P Hidalgo[3], D Gouvêa[3] and P C Morais[1]

[1] Universidade de Brasília, Instituto de Física, Núcleo de Física Aplicada, Brasília DF 70910-900, Brazil.
[2] Centro Brasileiro de Pesquisas Físicas, Rua Dr. Xavier Sigaud 150, Rio de Janeiro RJ 22290-180, Brazil.
[3] Departmento de Metalurgia e Engenharia de Materiais, Escola Politécnica, Universidade de São Paulo, São Paulo SP 05508-900, Brazil.

E-mail: coaquira@unb.br

Abstract. The structural and hyperfine properties of Cr-doped SnO_2 nanoparticles are report in this work. X-ray diffraction patterns indicate the formation of only the tetragonal phase in samples with Cr content up to 10mol%. The presence of Cr ions drives to the particle size reduction. Mössbauer spectra are well-resolved by a distribution of quadrupole doublets. The solubility limit of the dopant is found to be at ~3mol% Cr base on the quadrupole splitting evolution with the Cr content. The linear increase of the isomer shift as the Cr content is increased is assigned to a linear increase in the concentration of oxygen vacancies.

1. Introduction

Doped wide band-gap semiconducting oxides, such as SnO_2, TiO_2, ZnO are very interesting materials for the technological and fundamental point of view [1]. The doping of those semiconducting oxides with transition metals (Mn, Fe, Co, Cr, Ni) are reported to provide oxide dilute magnetic semiconductors (ODMS) with ferromagnetic temperature (T_C) above the room temperature [2].

Unfortunately, due to the complex solid-state chemistry involved in the preparation, the resulting properties depend on the conditions and preparation technique. Reports indicate that the doping of SnO_2 with Co concentrations below 10% is enough to obtain ferromagnetic order at room temperature [3]. Thin films of Cr-doped SnO_2 show ferromagnetism well above room temperature, which depends on the kind of substrate they are grown [4]. Despite the intense research in ODMS systems there are controversies mainly related to the origin of the magnetism observed in these materials.

In this work, the structural and hyperfine properties of Cr-doped SnO_2 nanoparticles with Cr content up to 10mol% are reported.

2. Experimental details

SnO_2 and Cr-doped SnO_2 nanoparticles with Cr content up to 10mol% were produced by a polymer precursor method. Details of the preparation are published elsewhere [1]. The phase formation and mean particle size has been determined from X-ray diffraction and corroborated by Transmission Electron Microscopy (TEM) data. Mössbauer spectroscopy measurements were carried out using a

$Ca^{119m}SnO_3$ as the radiation source. The analysis of spectra were obtained with a least-square fitting routine assuming a Lorentzian lineshape and considering the hyperfine splitting of I=3/2, 1/2 nuclear levels in a usual way.

3. Results and discussion

Some XRD patterns of the Cr-doped SnO_2 nanoparticles are shown in figure 1a. The Rietveld refinement of the XRD pattern of the undoped tin oxide powder indicates the formation of only the tetragonal phase (space group, $P4_2/mnm$) with lattice constants a=4.737nm and c=3.187nm.

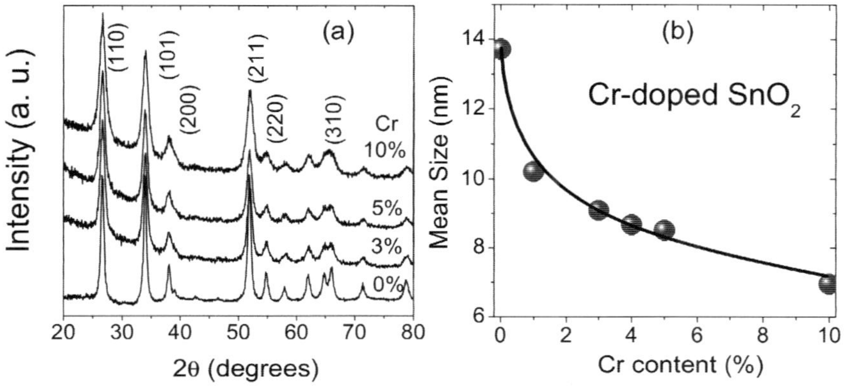

Figure 1. (a) X-ray diffraction patterns of the Cr-doped SnO_2 nanoparticles. The numbers between parentheses *(h k l)* represent the corresponding Bragg reflections. (b) Evolution of the mean particle size with the Cr content.

It is found that the Cr doping does not modify the crystal structure since the diffraction patterns of Cr-doped SnO_2 particles are also consistent with the tetragonal phase and no signal of a second phase formation is determined even at doping levels up to 10 mol% Cr. The Bragg reflections remain essentially at the same position and the linewidth (full width at the half maximum) increases steadily (see figure 1a). The latter indicates that the particle size decreases as the Cr concentration is increased. The evaluation of the mean particle size using the Scherrer's relation [5] provides a mean crystalline size of ~13nm for the un-doped sample. This size decreases to ~7nm for the 10 mol% Cr-doped sample. These results are in agreement with TEM data. The particle size evolution with the Cr content is shown in figure 1b.

In figure 2 are shown the Mössbauer spectra for three samples obtained at room temperature. Within the experimental resolution, all spectra seem to show no magnetic components and the preliminary analysis indicates that the whole set is well resolved by fitting with a single doublet. The linewidth (Γ) of the spectrum of the undoped sample is Γ~1.2 mm/s. That linewidth is larger than that expected for the bulk SnO_2 and suggests a distribution of quadrupole splitting (QS) in our samples. The fit considering the histogram distribution of QS is shown in figure 2. For each sample, the QS′s distribution (see right side of figure 2) shows a broad and strong symmetric maximum in the low-QS region. An isomer shift (IS) of −0.014 mm/s and a median value of the QS distribution of +0.65 mm/s are determined for the undoped sample. These values are slightly larger than those reported for bulk SnO_2 [6] and the differences could be associated with the size effect (*i.e.* structural distortions and symmetry breaking on the particle surface) which is expected to affect mainly the surrounding of Sn ions located on the particle surface.

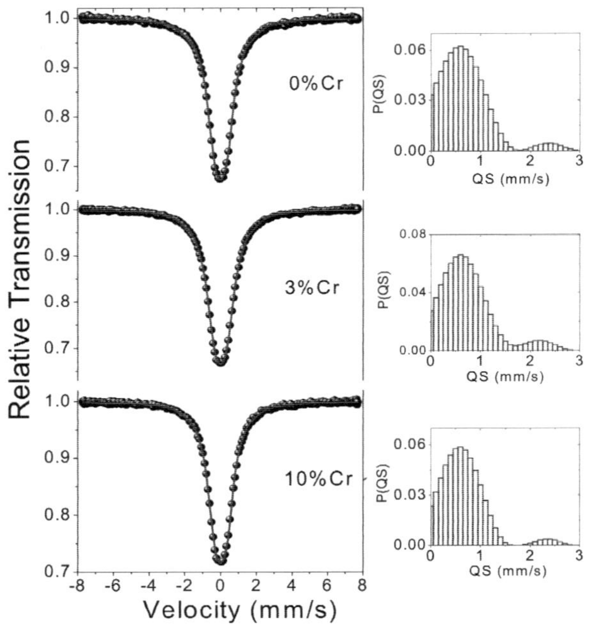

Figure 2. Mössbauer spectra obtained at room temperature. The points are the experimental data and the lines represent the fit.

When the Cr content is increased the median distribution value of QS shows a smooth increase reaching the value of +0.68 mm/s for the sample doped with 3mol% Cr content (data not shown here). That increase suggests the occurrence of additional distortions introduced by the substitution of Sn by Cr ions (ions with different size and chemical properties) and remains at around the same value for samples with higher Cr content. However, the width of the QS distribution (σ) shows clear tendency to decrease when the Cr content is above 3mol% (see figure 3a). That tendency of σ suggests the homogenization on the structural distortions which drives to narrow distribution of QS. Based on previous reports [1,7], the ~3mol% Cr is assigned as the region where the regime change from bulk solubility of Cr ions to their segregation onto the particle surface occurs, and the behaviour of σ should reflect that change.

As observed in figure 3b, the IS values shows a linear increase (*dIS/dx~0.14 mm/s*) when the Cr content is increased. The increase in IS with respect to the Cr-free sample suggests that the s-electronic density [8] reaching the nucleus of Sn ions is larger in Cr-doped sample and that electronic density is enhanced when Cr content increases. It is known that the occurrence of oxygen vacancies provides conduction electrons to the SnO_2 system (n-type doping) [6]. Therefore, if one correlates the linear increasing in the isomer shift with the linear increase in conduction electron density, then it is expected to have a linear increase in the concentration of oxygen vacancies as the Cr content is increased, which is perfectly plausible since the substitution of mainly Cr^{3+} ions by Sn^{4+} ions has been determined from magnetic measurements. Those magnetic results will be published elsewhere soon.

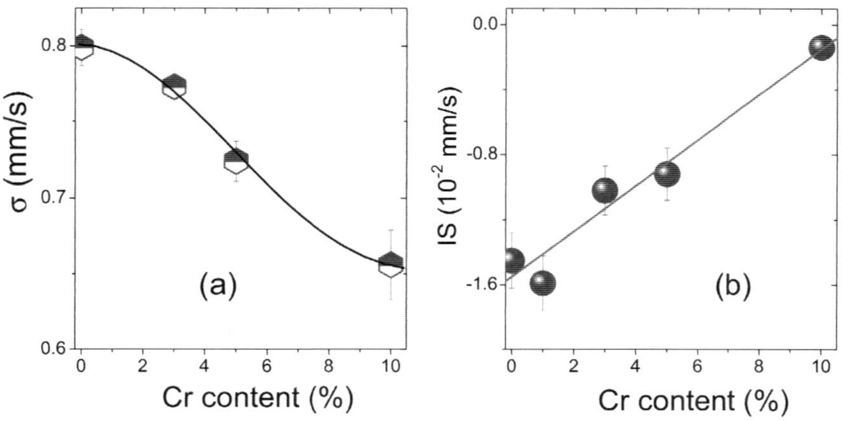

Figure 3. (a) Width of the quadrupole splitting (QS) distribution as a function of the Cr content. The line is drawn just for a guide to the eyes. (b) Evolution of the isomer shift (IS) with the Cr content. The solid line represents the fit.

4. Conclusions

The presence of Cr ions drives to the reduction of the particle size. Mössbauer spectra are well-resolved by a distribution of non-magnetic components. The 3mol% Cr is assigned as the concentration where the regime change from the solubility to surface segregation of Cr ions occurs. The linear increase of the isomer shift with the Cr content is assigned to a linear increase in the concentration of oxygen vacancies.

Acknowledgments
The authors thank the Brazilian agencies CNPq, CAPES and FAP/DF for the financial support.

References

[1] Hidalgo P, Castro R H R, Coelho A C V and Gouvêa D 2005 *Chem. Mater.* **17** 4149
[2] Wang X L, Daí Z X and Zeng Z 2008 *J. Phys.: Condens. Matter* **20** 045214
[3] Punnoose A and Hays J 2005 *J. Appl. Phys.* **97** 10D321
[4] Hong N H, Sakai J, Prellier W and Hassini A 2005 *J. Phys.: Condens. Matter* **17** 1697
[5] Dutta P, Vanimannan A, Seehra M S, Shah N and Huffman G P 2004 *Phys. Rev. B* **70** 174428
[6] Stjerna B, Granqvist C G, Seidel A and Häggströn L 1990 *J. Appl. Phys.* **68** 6241
[7] Castro R H R, Hidalgo P, Coaquira J A H, Bettini J, Zanchet D and Gouvêa D 2005 *Eur. J. Inorg. Chem.* **11** 2134
[8] Solis J L, Frantti J, Lantto V, Häggström L and Wikner M 1998 *Phys. Rev. B* **57** 13491

Phase composition and morphology as a function of depth for boride coatings grown on Fe-Ni alloys

M Carbucicchio[1], R Ciprian[1], L Micele[2] and G Palombarini[2]

[1]Department of Physics, University of Parma, Parma, Italy
[2]Department of Metals Science, Electrochemistry and Chemical Techniques, University of Bologna, Bologna, Italy

E-mail: massimo.carbucicchio@fis.unipr.it

Abstract. Two Fe-Ni alloys very differing in Ni content, $Fe_{64}Ni_{36}$ and $Fe_{20}Ni_{80}$, were borided at 850°C for 8 h using a high boron potential powder medium. On both alloys ~150 μm thick boride coatings grew in the form of two stratified sublayers of different thickness depending on the Ni content of the metal substrate. The outer sublayer mainly consisted of a ternary (Fe,Ni)B compound with small amounts of Ni borides in its outermost region, the inner by a ternary $(Fe,Ni)_2B$ compound. It is shown and explained how the same (Fe,Ni)B compound could form on both high-Ni alloys in spite of the considerable difference in Ni content.

1. Introduction

A considerable interest is addressed to iron-nickel alloys as candidate materials for the production of structural components displaying a satisfactory resistance to chemical corrosion, as well as to surface treatments suitable to improve their performance to stress corrosion cracking, fatigue and wear [1]. In this regard, coatings consisting of iron borides display hardness values in excess of 20 GPa, as well as good resistance to wear under both sliding and abrasive conditions [1-2]. Borided steel components display excellent performance in several tribological applications in the mechanical engineering and automotive industries. The composition, growth rate and properties of the boride coatings are dependent on type and behaviour of the alloying elements to an extent requiring further investigation and clarification.

In the present work attention has been addressed to the role of Ni. Two binary Fe-Ni alloys with Ni contents of 36 at.% and 80 at.%, respectively, were borided by means of a pack-cementation process under conditions (boron potential of the medium, exposure time) selected in order to produce protective coatings with a thickness of practical interest.

2. Experimental

1 mm thick sheets of two commercial Fe-Ni alloys, $Fe_{64}Ni_{36}$ and $Fe_{20}Ni_{80}$, received in an annealed state, were surface finished with SiC emery papers up to 600 grit and then thermochemically borided at 850°C for 8 h using a powder mixture consisting of amorphous boron (boronizing agent) and KBF_4 (10 wt.%, activator). The treatments were carried out in crucibles of refractory steel filled with the boriding medium and tightly sealed in order to avoid undesired oxidizing reactions. The analyses were carried out on as-grown coatings as well as at different depths after layer-by-layer abrasions performed parallel to the external surface. X-ray diffraction (XRD) was performed using a computer-controlled

goniometer and Co$K\alpha$ radiation with 0.02 2θ degrees step size and 1 s scan step time. The Mössbauer measurements were carried out by detecting the 6.4 keV X-rays in scattering geometry (CXMS) using a 40 mCi ^{57}Co(Rh) source. A least squares minimization routine with a combination of linear and non linear regressions was used to fit the spectra as a superposition of Lorentzian lines. The isomer shifts are referred to α-Fe. The metallographic analyses were performed on cross sections of borided samples by means of optical microscopy (OM) and scanning electron microscopy (SEM) and, for elemental compositions, by energy dispersive spectrometry (EDS).

3. Results and Discussion

Figure 1(a) shows a metallographic cross section of the borided $Fe_{64}Ni_{36}$ alloy. A coating ~150 μm thick grew on the substrate in the form of two sublayers (A and B in figure) almost equal in thickness. EDS analyses carried out on the cross section at different depths from the external surface show that the reaction products contain Fe and Ni in an atomic ratio which everywhere in the coating is only slightly lower than that measured for the base alloy (Ni/Fe ≅ 0.53). In addition, the coating contained boron in amounts decreasing on going towards the coating-alloy interface.

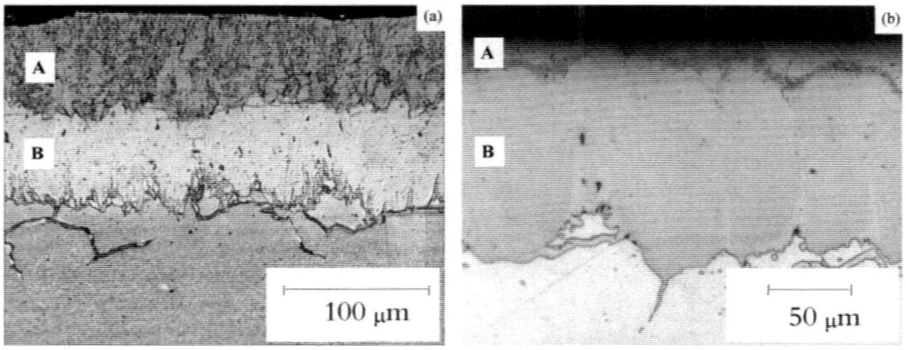

Figure 1. Cross sections of (a) $Fe_{64}Ni_{36}$ and (b) $Fe_{20}Ni_{80}$ samples borided at 850°C for 8 h with a powerful boriding medium: A: outer sublayer; B: inner sublayer;

The XRD pattern measured on the external surface of the coating [figure 2(a)] shows contributions of FeB phase displaying a strong (002) preferred crystallographic orientation, Ni_2B and Fe_2B, the last only contributing with its (002) reflection. A diffraction peak at 46.80 2θ might be attributed to the (501) reflection of orthorhombic Ni_4B_3, other peaks of this compound being masked by overlapping to peaks of the above phases in the coating, or missing at all as the consequence of crystallographic texturing effects.

The CXMS spectrum measured for the same sample [figure 3(a)] shows a quadrupole doublet, (Table 1) which can be attributed to an (Fe,Ni)B compound [3]. It is to be noted that the diffraction peaks for this ternary compounds are almost coincident with those of pure FeB because of the isomorphism existing between orthorhombic FeB and NiB compounds. (Fe,Ni)B could form due to the powerful reactivity between the base alloy and the adopted boriding medium, whose boronizing potential is very high. The effects of the boron reactivity, which are different for iron and nickel when lower boron potential media such as a B_4C-base powder mixtures are used, vanished and the active boron reacted without selectivity between Fe and Ni giving rise to a ternary (Fe,Ni)B compound. On the basis of the Ni/Fe atomic ratio evaluated by means of the above mentioned EDS analysis, the formula of this compound might be $(Fe_{1-x}Ni_x)B$ with x slightly lower than the Ni at.% in the base alloy, taking into account that a small amount of nickel borides is present in the outermost part of the coating.

After removing from the coating an ~11 μm thick external layer of reaction products, the XRD pattern only shows the contribution of FeB, together with the (002) reflection of Fe_2B [figure 2(b)].

This result indicates that nickel borides only formed in the outermost part of the coating. The CXMS spectrum for the same sample only shows the quadrupole doublet attributed before to the (Fe,Ni)B ternary compound. Similar XRD and CXMS results were found after layer-by-layer analyses carried out at different depths within the outer sublayer of the boride coating [figure 1(a), zone A].

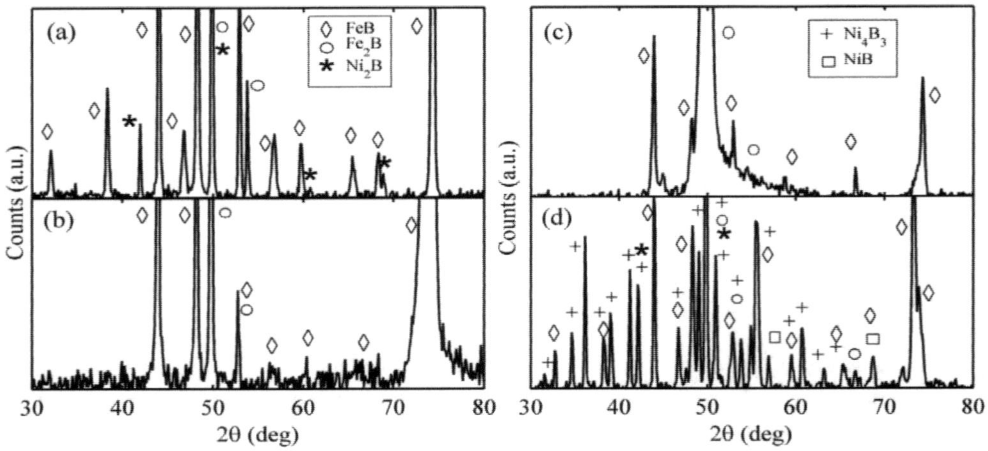

Figure 2. XRD patterns for borided samples: $Fe_{64}Ni_{36}$ (a) as-borided and after removal of (b) 11 μm and (c) 75 μm outer layers; (d) $Fe_{20}Ni_{80}$ as-borided.

After removing practically the whole outer sublayer of the coating, i.e. the zone A in figure 1(a), the XRD pattern shows a very intense (002)-Fe_2B reflection, together with some relatively low intensity peaks due to both FeB and Fe_2B [figure 2(c)]. The CXMS spectrum for the same sample [figure 3(b)] shows a prevailing ferromagnetic contribution (β_1 and β_2 sextets), whose parameters are reported in Table 1, which can be attributed to a $(Fe,Ni)_2B$ ternary compound. The spectrum also shows the contribution of the δ doublet indicating the presence of small amounts of the (Fe,Ni)B ternary compound.

Table 1. Hyperfine parameters of the contributions to the CXMS spectra in figure 3.

Contribution	IS (mm/s)	ΔE_Q (mm/s)	H_{hf} (kOe)
δ	0.21 ± 0.01	0.50 ± 0.01	-
β_1	0.17 ± 0.02	-0.14 ± 0.03	184
β_2	0.07 ± 0.01	-0.04 ± 0.02	210

In conclusion, the coating grown on the $Fe_{64}Ni_{36}$ alloy using a very powerful boriding medium consists of an outermost thin region of nickel borides and two main sublayers [zones A and B in figure 1(a)] containing respectively the (Fe,Ni)B and $(Fe,Ni)_2B$ ternary compounds.

The hypothesis on the stoichiometry of the $(Fe_{1-x}Ni_x)B$ compound with $x \approx 0.36$, was proposed considering that no preferred reaction could occur between boron and iron or nickel under the adopted boriding conditions. Under this hypothesis, (Fe,Ni)B ternary compounds differing in stoichiometry might form in boride coatings grown under the same conditions on binary alloys differing in the Ni content.

Figure 1(b) shows a metallographic cross-section of a ~150 μm thick boride coating formed on the $Fe_{20}Ni_{80}$ alloy. It consists of an ~5 ÷ 20 μm thick fine-grained sublayer grown on an inner ~130 μm thick coarse-grained sublayer of reaction products. The EDS analysis shows that, also in this case, the Ni/Fe ratio in the coating was everywhere similar to that in the base alloy. XRD measurements confirmed that a FeB phase is the main component of the outer sublayer whose more external part, as shown by layer-by-layer analysis, consists of nickel borides Ni_2B and Ni_4B_3 [figure 2(d)].

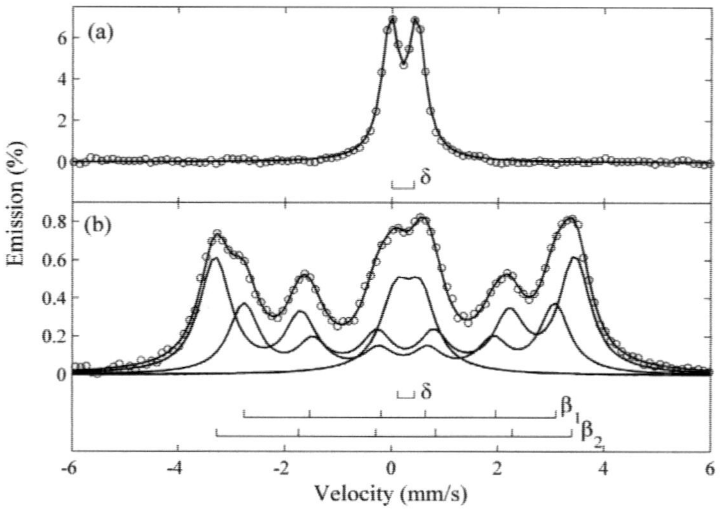

Figure 3. CEMS spectra for $Fe_{64}Ni_{36}$ (a) as borided and (b) after the removal of 75 μm outer layer.

The CXMS spectrum measured for this sublayer shows the δ quadrupole doublet due to the same (Fe,Ni)B ternary compound already found in the outer part of the boride coating grown on the $Fe_{64}Ni_{36}$ alloy. This fact suggests that the same (Fe,Ni)B ternary compound can form in the boride coatings growing on high-Ni binary alloys irrespective of the Ni content when a medium with a powerful boriding potential is used.

4. Conclusions

The boriding treatments carried out at 850°C on two Fe-Ni alloys considerably differing in Ni content using a very high boron potential medium for an exposure time of 8 h allowed the following conclusions to be drawn:

- The boriding reactions strongly involved both iron and nickel leading to the formation of coatings consisting of Fe-Ni-B ternary compounds, together with nickel borides located in the outermost regions of the coatings.
- The coatings grew in the form of two stratified sublayers. The outer composed of an (Fe,Ni)B ternary compound containing a small amount of nickel borides in a few μm thick region of its outer border, while the inner consists of $(Fe,Ni)_2B$ compound. The thickness of the (Fe,Ni)B-base sublayer was considerably lower for the alloy with the highest Ni content, i.e. the $Fe_{20}Ni_{80}$, while the total thickness of the boride coating was practically the same on both alloys.
- The Mössbauer spectra measured for the (Fe,Ni)B formed on both Fe-Ni alloys are characterized by the same hyperfine parameters, a result indicating that the same ternary compound formed in the coatings irrespective of the Ni content in the base alloy. According to EDS analysis and considering the presence of Ni borides in the coatings, the Ni fraction in (Fe,Ni)B was evaluated to be slightly lower that 0.36.

References

[1] Sinha A K, ASM Handbook Online, vol. 4 Heat Treating. ASM International, Materials Park, Ohio, USA (2002)
[2] Carbucicchio M, Casagrande A. and Palombarini G., 1990 *Hyper. Interact.* **57** 1769-74
[3] Ciprian R, Palombarini G, Carbucicchio M and Sambogna G 2009 *Hyper Interact.* **192** 43-9

Step-by-step powder composite mechanosynthesis for functional nanoceramics

A O Polyakov[1], T Yu Kiseleva[1], A A Novakova[1], T F Grigoryeva[2] and A P Barinova[2]

[1]Moscow M.V.Lomonosov State University, Department of Physics, Vorob'evy Gory, Russia
[2]Institute of solid state chemistry and mechanochemistry SO RAS, Novosibirsk, Kutateladze st.

E-mail: alexpol.msu@gmail.com

To study the possibility of Fe_2O_3 mechanochemical reduction by preliminary mechanically alloyed Fe+20%Al compound their powder mixture was subjected to high-energy ball-milling in Ar atmosphere, with the milling time varying between 2 and 12 minutes. The milled samples obtained at various times of milling were characterized by X-ray diffraction and Mössbauer spectroscopy. As a result gradual α-Fe_2O_3 reduction via formation of intermediate Fe-Al-O oxides was observed. The presence of the intermediate Fe_2AlO_4 spinel phases stable over long milling time is stated. Mechanocomposite Fe+20%Al transformation to α-Fe(Al) solid solutions which evolve peculiarly with the milling time, was observed also. The kinetics of α-Fe_2O_3 reduction process was analyzed in comparison with the same processes in the systems: α-Fe_2O_3 + Al and α-Fe_2O_3 + Al + Fe.

1. Introduction

The possibility of taking the advantage of particular properties of the constituent materials to meet specific demands is the most important motivation for the development of composites. In this way we are working on varying the mechanosynthesised composites final structure for different purposes. Recently, we have studied the processes of Al_2O_3/Fe-Al [1], Fe_2O_3/Fe_3O_4/Fe [2] nanocomposites mechanochemical formations. The thermite solid state reduction reaction of α-Fe_2O_3 with metallic reductant (Al, Fe) was realized in those processes. α-Fe_2O was consistently reduced to Fe through intermediate binary and ternary oxides. Reduced Fe subsequent alloying with reductant surplus resulted in Al_2O_3/intermetallic phases composites formation. The metal-reductant surplus in the mixture leads to deceleration of Fe_2O_3 reduction and incapsulated structures formation. Fe as additional reductant caused some shift of the activation barrier of Fe_2O_3 reduction. The ordering-disordering phenomena led to amorphous iron-based magnetic phase formation in grain-boundary regions. This phase is stable for a long milling time and after heating up to 500°C [2]. Its formation creates the additional competitive process to Fe_2O_3 destruction. To study the possibility of Fe_2O_3 mechanochemical reduction, alloyed Fe-Al intermetallic compounds in high energy planetary ball mill activation of pre-milled Fe+20%Al mixture with Fe_2O_3 have been performed.

2. Experimental

Ball milling of 8 g Fe powder with 2 g of Al powder was performed in AGO-2 planetary ball mill for 20 minutes in a vial sealed under Ar. Then 3 g of α- Fe_2O_3 powder was added into 5 g of pre-milled Fe-Al mixture and again subjected to 2, 6 and 12 minute ball milling. Vial volume was 250 cm^3. Balls diameter and mass were 5 mm and 200 g respectively. The speed of drums rotation was ~1000 rpm.

The [57]Fe Mössbauer spectra (MS) were recorded at room temperature with [57]Co(Rh). X-ray diffraction (XRD) of as-milled and annealed powder samples was performed at Rigaku DX/Max diffractometer with Cu Kα radiation.

3. Results and discussion

X-ray diffraction image of Fe+20%Al powder mixture milled for 20 min is shown in Fig1 (a). The

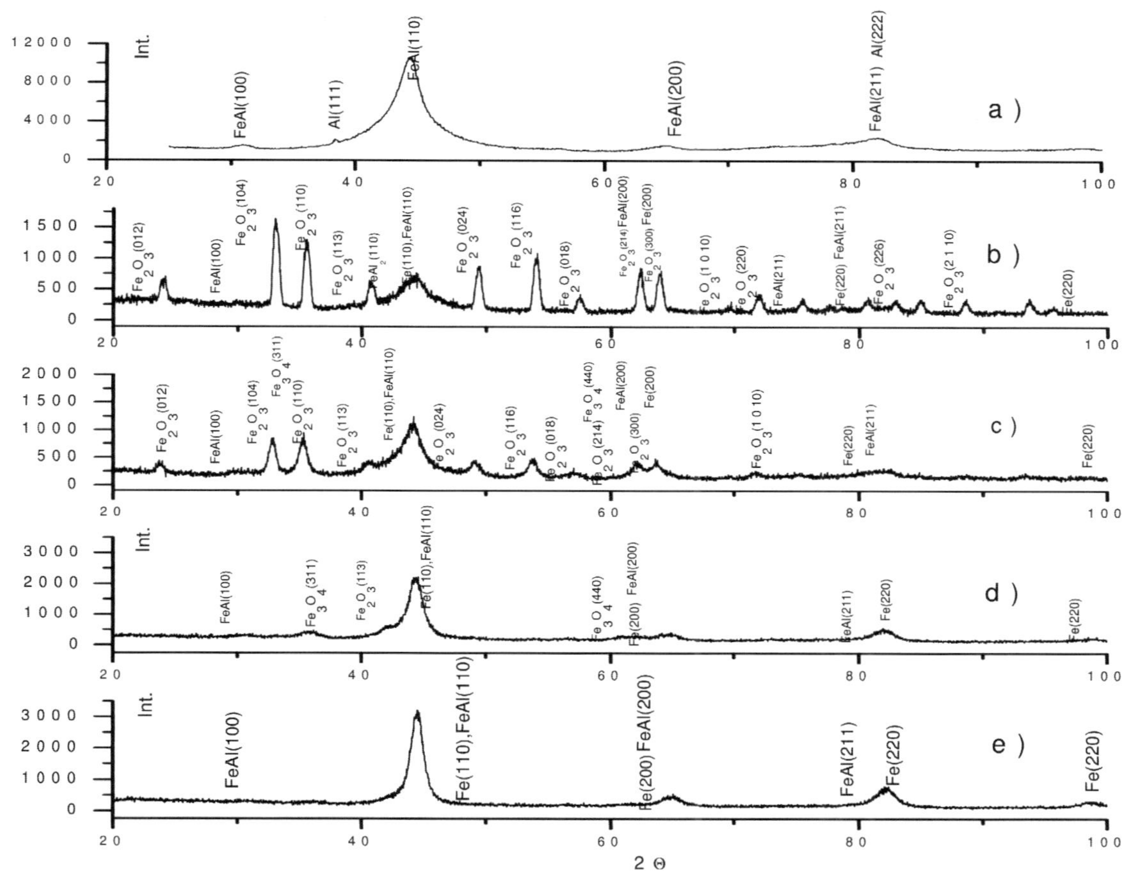

Figure 1. XRD pattern for 20 minutes ball milling Fe+20%Al (a); mechanoactivated Fe+20%Al composite and Fe_2O_3 simply mixed (b) and mechanoactivated during 2(c), 6(d) and 12(e) minutes.

broadened peaks with maxima at the angles of FeAl may correspond not only to disordered (or even amorphous) FeAl phase and its solid solution but also to structural reflections of $FeAl_2$ and Fe_2Al_5 intermetallics also. The presence of well resolved narrow Al peaks on the diffraction pattern indicate unreacted Al.

Mössbauer (Fig. 2a) analysis of pre-milled Fe+20%Al presented in bar diagram shows that after 20 min of milling interaction between Fe and Al leads to the formation of highly disordered composite consisted of FeAl, Fe_2Al_5 and $FeAl_2$ intermetallic phases and disordered Fe(Al) solid solution. We observed isomer shifts and quadrupole splittings change to some extent, increased linewidths of subspectra and even additional subspectra appearence in the case of FeAl phase. As was observed [4] FeAl disordering leads to the magnetically split component being resolved in addition to the non-magnetic singlet. Thus, after 20 minutes of high energy ball milling of Fe+20%Al we obtain intermetallic phase composite.

The XRD pattern (Fig. 1b) and Mössbauer spectrum (Fig. 2b) of simple mixture: Fe_2O_3 and pre-activated Fe-Al composite correspond to mixed components contributions in the ratio 1:2. The percent of intermetallics component is about 60% that indicate metallic component prevalence.

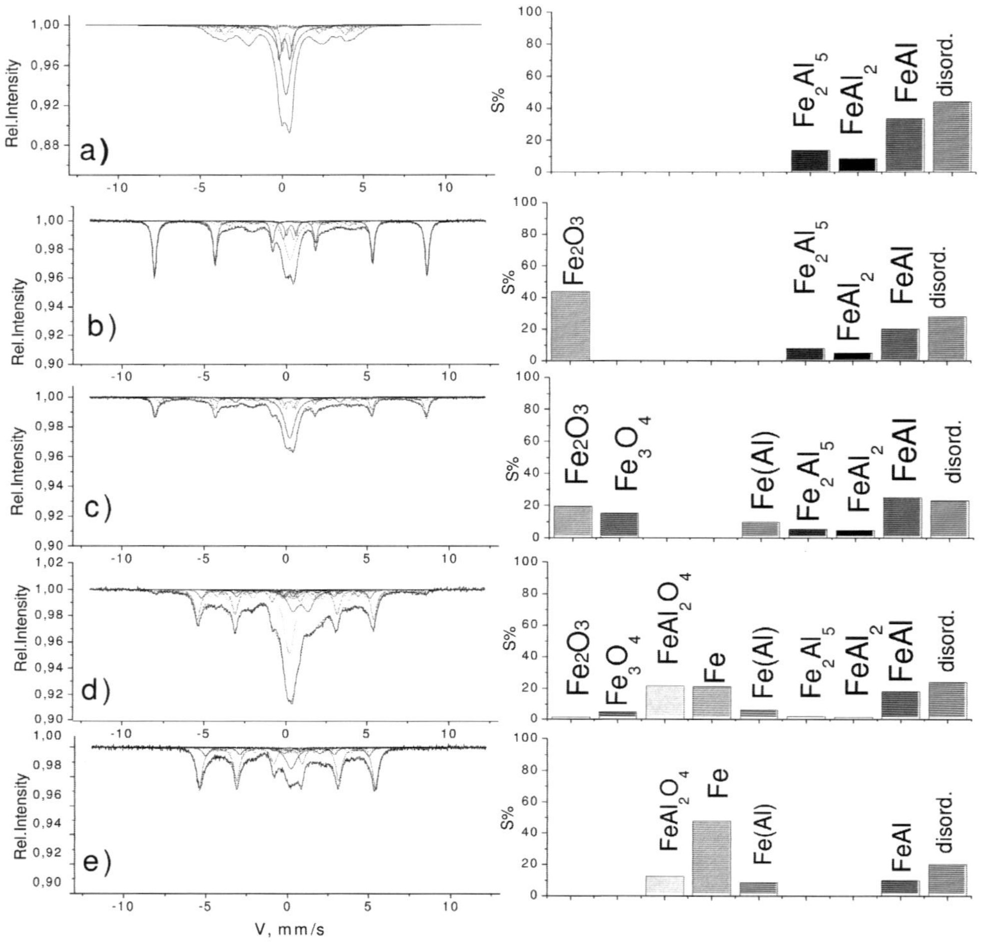

Figure 2. Mössbauer spectra and phase diagrams of pre-milled Fe+20%Al (a), pre-milled Fe+20%Al simply mixed with Fe_2O_3(b), mechanoactivated during 2(c), 6(d) and 12(e) minutes.

As is seen from XRD pictures (Fig. 1 c,d,e) the subsequent milling of this mixture results in phase transformations arising from the Fe_2O_3 and Fe-Al composite interaction. Fe_2O_3 destruction after 2 min. is reflected in its intensity decrease and structure peaks widening is observed. The main wide composite intermetallic peak becomes narrow. Al peaks disappeared fully. According to Mössbauer data (Fig. 2c) the mixture after 2 min contains Fe_2O_3, Fe_3O_4, Fe and 42 % of a disordered intermetallics mixture. Milling for 6 min (Fig. 1 d; Fig. 2) reveals Fe_2O_3 peaks disappearance on XRD pattern and wide Fe_3O_4 maxima rise. Mössbauer data confirm that the mixture is composed of Fe_2O_3, Fe_3O_4 and disordered intermetallics mixture. It is interesting that at this time of milling traces of ternary compound $FeAl_2O_4$ were detected. Finally, after 12 min. (Fig. 1 e) the XRD pattern consists of only wide and asymmetric maxima which reflect the bcc α-Fe(Al) solid solution formation. Mössbauer spectrum (Fig. 2) analysis reveal not only α-Fe(Al) formation but also disordered intermetallics and ternary oxide. Generally, analyzing Mössbauer data of different milling time gradually Al rich intermetallic phases decrease in accordance with hematite reduction and bcc α-Fe(Al) solid solution formation is clearly detected. It should be noted that no oxide phases of aluminum were detected on XRD pattern which is assumed due to possible X-ray amorphous modification of it. Evaluation of aluminum concentration in a-Fe(Al) solid solution was performed by analysis of hyperfine fields distribution of the corresponding Mössbauer spectrum component (Fig. 2 (e)) and reveals the value 7 at% .

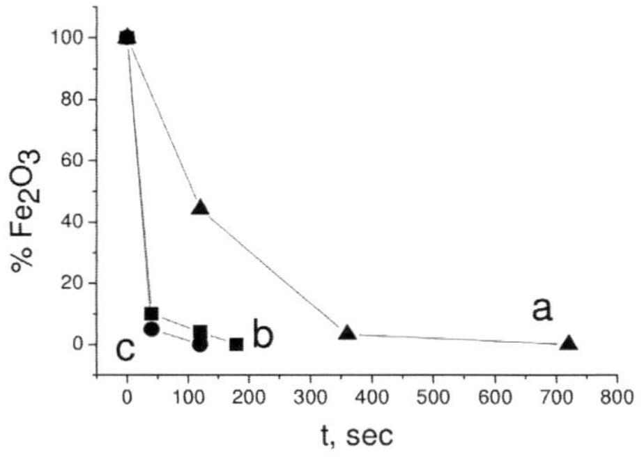

Figure 3. The % fraction of Fe_2O_3 depending on milling time t in (a) the system under study, (b) in comparison with analogous systems α-Fe_2O_3 + Al and (c) α-Fe_2O_3 + Al + Fe.

Fig. 3 demonstrates the kinetics of Fe_2O_3 destruction in different powder mixture with equal surplus of metal reductant component subjected to high energy ball milling. Amorphous phase formation in the presence of iron compounds leads to the slowing down the oxide reduction process.

4. Conclusions

Mössbauer spectroscopy and X-ray diffraction study of subsequent interaction of α-Fe_2O_3 with pre-mechanoactivated Fe+20%Al composite powder mixture reveal gradual α-Fe_2O_3 reduction via formation of iron binary and ternary oxides up to Fe. On the other hand Fe-Al mechanocomposite transformations lead to disordered α-Fe(Al) solid solution and Fe-based amorphous phase formation. Complete amorphous phase formation on the iron surface decreases its reductant ability and slows down the process of α-Fe_2O_3 reduction in comparison with the analogous systems.

References

[1] A.A.Novakova, T.Yu.Kiseleva, et al 14th *Int.Symp. on Metastable and Nano-Mater.ISMANAM-2007*, Corfu, Greece, Proceed., 179, (2007)

[2] T.Yu.Kiseleva, A.A.Novakova, M.I.Chystyakova, A.O.Polyakov, et al. *Solid State Phenomena* **152-153**, 25-28, (2009)

[3] T.Yu.Kiseleva, A.A.Novakova, T.F.Grigorieva et al. *Advanced materials* **6**, 1-10 (2008)

[4] S.Gialanella, X.Amils et al. *Acta mater.* **46**, 3305 (1998).

Hyperfine interactions in mechanosynthesized and thermally treated Co-Fe-Ni alloys

T Pikula[1] , D Oleszak[2] and E Jartych[1]

[1]Department of Experimental Physics, Institute of Physics, Technical University of Lublin, Nadbystrzycka str. 38, PL-20-618 Lublin, Poland

[2]Faculty of Materials Science and Engineering, Warsaw University of Technology, Wołoska str. 141, PL-02-507 Warsaw, Poland

E-mail: t.pikula@pollub.pl

Abstract. A series of Co-rich $Co_xFe_yNi_z$ alloys was successfully produced using the mechanical alloying method. To check the stability of their structure thermal treatment was applied subsequently. Mössbauer spectroscopy and X-ray diffraction were used to investigate some structural and micro-magnetic properties of mechanosynthesized as well as thermally treated alloys. X-ray diffraction measurements proved that during mechanical alloying solid solutions with b.c.c. or f.c.c. lattices were formed, while thermal processing in some cases provided to decomposition into the mixture of b.c.c. and f.c.c. phases. Mössbauer spectra were fitted using hyperfine magnetic field distribution method. The average value of hyperfine magnetic field induction ranged from 31.5 to 35.5 T. In the case of some thermally treated samples, shapes of distributions reflect two-phase character of the alloys. For this reason curves of distribution were numerically fitted using two Gaussian functions, which were attributed to the suitable phases. The area of individual components was calculated.

1. Introduction

Co-Fe - based alloys have been extensively studied due to their soft magnetic properties. Beside permalloy they may form competitive material for production of write-read heads in hard-disk drives [1-3]. While the structure and macroscopic magnetic properties of these alloys have been studied quite widely, the literature reports focused on hyperfine interactions are scant.

Mechanical alloying (MA) represents a simple way of fabricating a variety of materials e.g., crystalline, nanocrystalline, amorphous materials. MA is a fundamentally different approach to alloy-manufacture because it relies on deformation processes to mix materials [4]. Here alloying is a result of repeated cold welding and crushing of particles. However, different alloy systems react to milling in different ways depending on the mutual solubility and reactivity of the components, their mechanical properties and the type of used milling equipment [4].

This work follows our systematic investigations [5-6] of the mechanically synthesized $Co_xFe_yNi_z$ alloys in the context of their structure and hyperfine interactions. Products of milling reach saturation magnetization of the order of 2T and coercive field about 15-67Oe [6], therefore MA may represent potential technology of production of soft magnetic powders. After milling the obtained structures of alloys are often in the non-equilibrium state. Heat treatment may cause structural changes, i.e. an increase in the grain size accompanied by a decrease in the level of internal strains, the change of the crystalline lattice type or alloy decomposition into a mixture of two phases [6]. The aim of this work was: (1) to group and compare hyperfine magnetic field (HMF) distributions of the alloys after MA processes as well as after thermal processing and (2) to estimate the contributions of the suitable solid solutions in the decomposed alloys on the basis of the HMF distributions.

2. Experiment

Co, Fe and Ni powders were subjected to the MA process to obtain a series of $Co_xFe_yNi_z$ alloys. The compositions of alloys were chosen on the basis of the phase diagram reported for bulk, melted Co-Fe-Ni alloys in [2]. We concentrated on Co-rich compositions mainly, because of their location in a transitional area of the phase diagram. Milling was performed in a Fritsch P5 planetary ball mill. All MA processes were conducted up to 100h under an argon atmosphere.

X-ray diffraction (XRD) measurements were carried out using a Philips PW1830 diffractometer working in a continuous scanning mode with CuK_α or CoK_α radiation. The lattice constants were determined from the shift of the diffraction lines. The Williamson–Hall approach was used for determination of the average grain sizes, D, and the mean level of internal strains, ε [7].

Mössbauer measurements were performed at room temperature in standard transmission geometry using a source of ^{57}Co in a rhodium or chromium matrix.

Thermal treatment of the mechanosynthesized $Co_xFe_yNi_z$ alloys was performed in two ways: (1) heating from the room temperature up to 993 K in a calorimeter under an argon atmosphere with the rate of 20 K per min and (2) isothermal annealing in a furnace at 1173 K for 1 h in vacuum.

3. Results and discussion

Detailed analysis of XRD patterns allowed stating that during milling process of $Co_{65}Fe_{23}Ni_{12}$, $Co_{52}Fe_{26}Ni_{22}$ and $Co_{40}Fe_{35}Ni_{25}$ alloys the disordered solid solutions with f.c.c. lattice were formed, while for remnant samples solid solutions with b.c.c. crystalline lattice were obtained. In comparison with results for bulk, melted alloys reported in [3] after the MA processes two-phase samples were not formed. The final products of MA were alloys in powder state with D about tens of nanometers and ε of the order of 0.6 - 1.3 %. The lattice constants for b.c.c. alloys were from 0.2839 to 0.2865 nm and for f.c.c. alloys 0.3575 to 0.3581 nm.

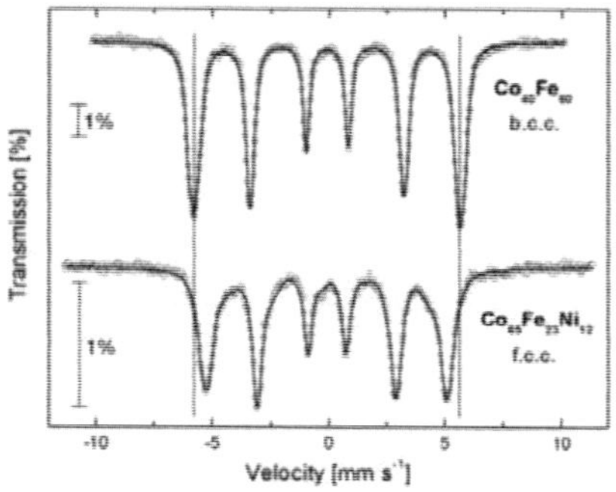

Figure 1. Room-temperature Mössbauer spectra registered for $Co_{40}Fe_{60}$ and $Co_{65}Fe_{23}Ni_{12}$ after 100h of MA.

The products of milling were often in non-equilibrium state. It may be stated that $Co_{40}Fe_{60}$, $Co_{52}Fe_{26}Ni_{22}$, $Co_{50}Fe_{45}Ni_5$ which lie relatively far from the phase boundary region predicted in [2] were the most stable. The other alloys were decomposed into the mixture of b.c.c. and f.c.c. phases during heating. After annealing in most of alloys the f.c.c. phase was dominating. Only $Co_{40}Fe_{45}Ni_{15}$, $Co_{50}Fe_{40}Ni_{10}$ and $Co_{60}Fe_{35}Ni_5$ became two-phase samples, while in the case of $Co_{40}Fe_{40}Ni_{20}$, $Co_{50}Fe_{35}Ni_{15}$ and $Co_{60}Fe_{30}Ni_{10}$ the type of crystalline lattice was changed from b.c.c. to f.c.c. In this context e/a ratio (number of valence electrons per atom) seems to be an important parameter. Alloys with the same e/a value present a lot of similarities in the case of structural properties.

Mössbauer studies allowed determining the hyperfine interactions in the mechanosynthesized and thermally treated alloys. The spectra after MA as well as after thermal treatment were six-line patterns. Fig. 1, as an example, presents spectra for the final products of alloying with f.c.c. and b.c.c. lattice.

Mössbauer measurements were performed after 5, 20, 50, 100 hours of milling, which also allowed monitoring the alloy formation process. For the longer milled samples, the spectral lines were slightly broadened up to 0.15 – 0.17 mms^{-1}. This broadening was the result of decreasing of the grain sizes as well as the alloy formation, i.e. different atomic configurations in the nearest neighbourhood of ^{57}Fe atoms appear. Taking this fact into consideration the HMF distribution seems to be the best numerical fitting of the spectra.

From the HMF distribution, the average ($\langle B_{hf} \rangle$) and the most probable (B_{max}) values of magnetic field induction were determined and in the case of some $Co_xFe_yNi_z$ alloys they are listed in Table 1. Generally, for mechanosynthesized alloys values of $\langle B_{hf} \rangle$ ranged from about 31.5 to 35.5 T, and they depended on two main factors: (1) the number of valence electrons per atom, e/a, and (2) concentration of iron in the alloy. In Fig. 2 the HMF distribution curves are arranged in the direction of increasing e/a value from the top. It may be stated that with an increase of e/a the value of $\langle B_{hf} \rangle$ decreases. For alloys with the same e/a ratio with the increasing of iron concentration in the sample, the value of $\langle B_{hf} \rangle$ increases.

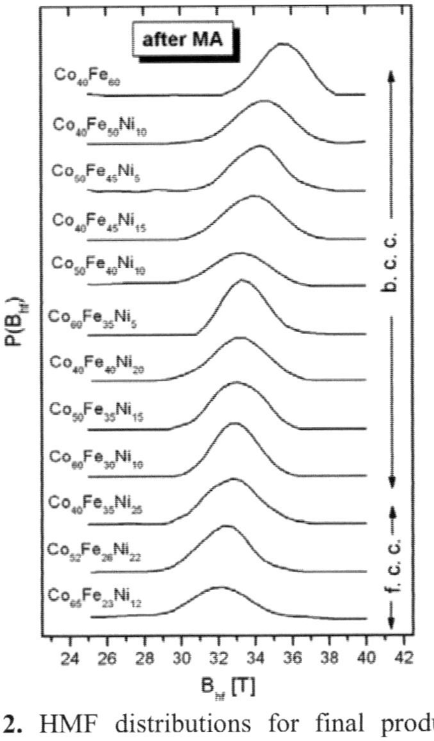

Figure 2. HMF distributions for final products of milling; $P(B_{hf})$ – probability in arbitrary units.

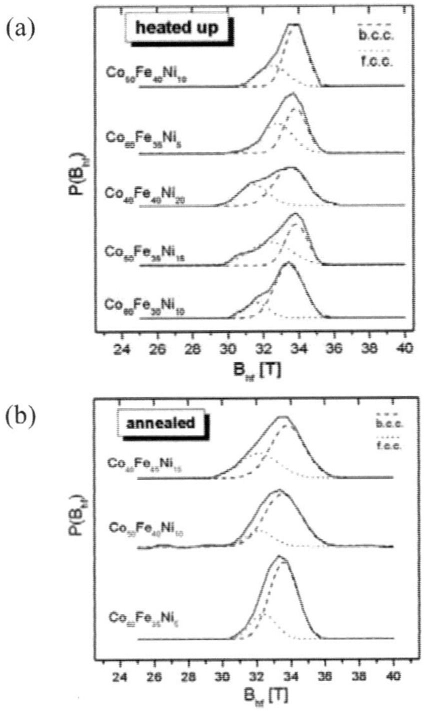

Figure 3. HMF distributions fitting for (a) heated alloys, (b) annealed alloys.

As mentioned, in some cases thermal treatment caused the formation of the mixture of b.c.c. and f.c.c. solid solutions. It is also reflected in HMF distributions. It may be noted that the shape of distributions obtained for mechanosynthesized samples is quite regular and similar to simple Gaussian function (Fig. 2), while for the thermally treated two-phase alloys, curve of distribution seems to be a combination of two components (Fig. 3a,b). However, in not all cases it is clearly visible probably because of a low amount of the second phase.

In order to estimate the contributions of the suitable phases in the decomposed alloys the experimental HMF distributions were numerically fitted using two Gaussian functions, which were attributed to the individual phases (Fig. 3a,b). Some starting parameters, i.e. center position of Gaussian curve were restricted to avoid the ambiguity of such procedure. The area of individual Gaussian components (in %) was calculated and attributed to the content of f.c.c. and b.c.c. phases in the alloy. Moreover, the average value of HMF, <B>, for each component was calculated. Both values mentioned above are listed in Table 1. Generally, it may be noted that values of <B> for components with b.c.c. lattice are larger than those with f.c.c. structure by about 1-2 T.

The isomer shift (IS) values for studied samples were ranged from 0.01 to 0.04 mms^{-1} and they were slightly different for fcc and bcc components in two-phase alloys. Unfortunately, these differences did not exceed the measuring errors and it was impossible to notice any trends. Furthermore, larger influence on the IS values in this case may have number of Co and Ni atoms in the nearest neighborhood of ^{57}Fe atoms than the type of crystalline lattice.

Table 1. Hyperfine magnetic fields and contributions of phases for mechanosynthesized $Co_xFe_yNi_z$ alloys before and after heat treatment; A – area of the component.

Alloy	State	e/a	$\langle B_{hf}\rangle$ [T]	B_{max} [T]	Phase			
					f.c.c. A	<B> [T]	b.c.c. A	<B> [T]
$Co_{50}Fe_{40}Ni_{10}$	after MA	26.7	32.03	33.41	-	-	100 %	-
	heated up		33.12	34.03	30 %	32.5	70 %	33.8
	annealed		33.06	33.38	20 %	32.1	80 %	33.5
$Co_{60}Fe_{35}Ni_5$	after MA	26.7	33.50	33.24	-	-	100 %	-
	heated up		33.30	33.72	48 %	32.8	52 %	33.8
	annealed		33.25	33.24	24 %	32.4	76 %	33.6
$Co_{40}Fe_{40}Ni_{20}$	after MA	26.8	33.09	33.48	-	-	100 %	-
	heated up		32.87	33.79	32 %	31.5	68 %	33.5
	annealed		32.80	32.76	100 %	-	-	-
$Co_{50}Fe_{35}Ni_{15}$	after MA	26.8	33.02	33.07	-	-	100 %	-
	heated up		33.02	33.83	50 %	32.4	50 %	33.9
	annealed		32.93	33.07	100 %	-	-	-

4. Conclusions

This work was extension of our previous studies concerned on ternary $Co_xFe_yNi_z$ alloys obtained during mechanical synthesis and thermally treated subsequently. Performed investigations allowed to state that structure of mechanosynthesized alloys was often different comparing with the structure of the same alloys obtained by melting. The main difference was that after milling processes the two-phase samples were not formed. Furthermore, the structure of such alloys was unstable and thermal treatment often provided to decomposition into two-phases.

Final products of MA processes were characterized by the relatively broad but smooth and regular HMF distributions similar to Gaussian functions. In the case of the thermally treated two-phase alloys, curve of distribution seems to be a combination of two Gaussian components. Moreover, it was possible to estimate contribution of the suitable phase in the alloy using HMF distribution fitting but only as a complementary method. It turned out that the value of e/a ratio plays important role in the context of similarities in structure properties of the alloys as well as in the case of the values of average hyperfine magnetic field induction.

References

[1] Osaka T 2000 *Electrochim. Acta* **45** 3311
[2] Jen S U, Chiang H P, Chung C M, Kao M N 2001 *J. Magn. Magn. Mater.* **236** 312
[3] Khomenko E V, Shalyguina E E, Chechenin N G 2007 *J. Magn. Magn. Mater* **316** 451
[4] Harris J R 2002 PhD thesis submited to The University of Nothingam
[5] Pikula T, Oleszak D, Pękała M, Mazurek M, Żurawicz J K, Jartych E 2008 *Acta Phys. Pol.* A **114** 1545
[6] Pikula T, Oleszak D, Pękała M, Jartych E 2008 *J. Magn. Magn. Mater* **320** 413
[7] Williamson G K, Hall W H, 1953 *Acta Metallurg.* **1** 22

Thermal stability of a supersaturated Fe-Ge-Nb solid solution produced by ball milling

J J Ipus, J S Blázquez, A Conde[1]
Dpto. Física de la Materia Condensada, ICMSE-CSIC, Universidad de Sevilla, P.O. Box 1065, 41080, Sevilla, Spain.

E-mail: conde@us.es

Abstract. Thermal evolution of Fe neighbourhood in a supersaturated bcc Fe(Ge,Nb) solid solution, obtained as the final product of mechanical alloying of $Fe_{75}Ge_{20}Nb_5$, was studied. No changes in Fe neighbourhood were detected after heating up to 473 K, although differential scanning calorimetry shows a clear deviation of the baseline at 400 K. After heating up to 723 K, a similar nanocrystalline microstructure is derived from X-ray diffraction. However, Mössbauer spectra evidence changes in the Fe neighbourhood. A proposed deconvolution of the hyperfine field distribution yields a Ge content of ~10 at. %, in agreement with the maximum solubility of Ge in bcc Fe in thermodynamical equilibrium.

1. Introduction

Mössbauer spectroscopy is found to be a useful tool in order to study the local Fe environment in Fe containing alloys. In this work, the evolution of microstructure and local Fe environments of the supersaturated bcc Fe(Ge,Nb) solid solution obtained as the final product of mechanical alloying of $Fe_{75}Ge_{20}Nb_5$ [1] is reported. X-ray diffraction (XRD) shows that, at the end of milling process (150 h at 350 r.p.m), powders show a single phase: nanocrystalline bcc Fe(Ge,Nb) supersaturated solid solution with a lattice parameter a =2.894 ±0.003 Å and a minimum crystal size of ~6 nm. The as-milled sample was heated up to 473 and 723 K in order to study its thermal stability.

2. Experimental

$Fe_{75}Ge_{20}Nb_5$ alloy was prepared by mechanical alloying from pure powders in a planetary ball mill Fritsch Pulverisette 5. Details of milling parameters can be found elsewhere [1]. Crystalline structure was studied by XRD using Cu Kα line and the local Fe environment was analyzed by Mössbauer spectroscopy (MS) at room temperature in a transmission geometry using a ^{57}Co(Rh) source. The values of hyperfine parameters were obtained by fitting with NORMOS program [2]. Thermal characterization of the samples was studied by differential scanning calorimetry (DSC) using a Perkin-Elmer DSC7.

3. Results and Discussion

MS spectra were fitted using a hyperfine field distribution (0 $\leq HF \leq$36 T). In figure 1(a) the MS spectrum along with *HF* distribution for as-milled and heated samples are shown. The average hyperfine field $<HF>$ ~27 T and the average isomer shift $<IS>$ ~0.15 mm/s (slightly increasing with heating temperature) are similar for all samples and different to those of pure bcc Fe, which agrees with the existence of the supersaturated bcc Fe(Ge,Nb) solid solution observed by XRD.

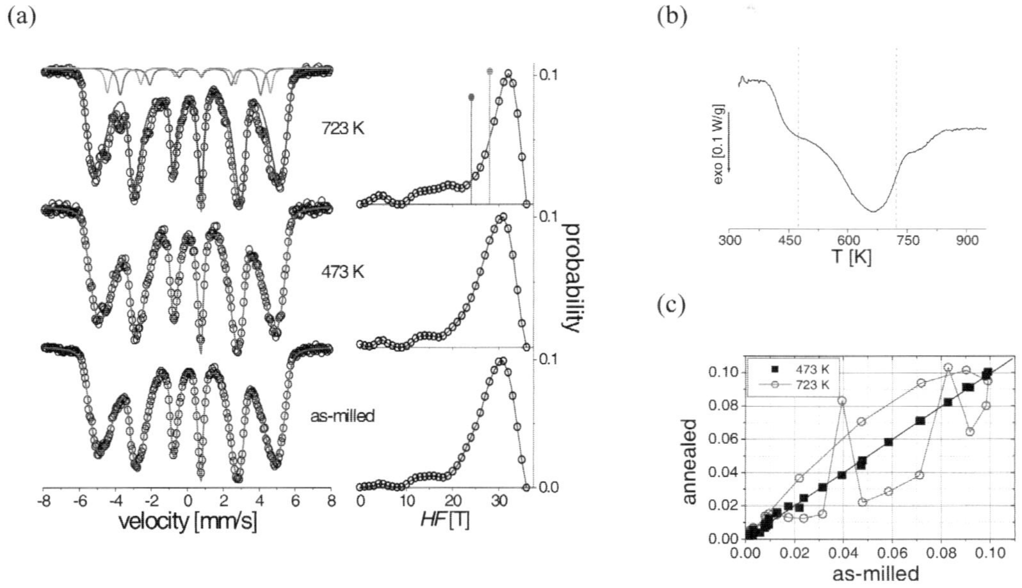

Figure 1. (a) Mössbauer spectra and *HF* distributions at 300 (as-milled), 473 and 723 K; (b) DSC scan of as-milled sample; (c) correlation between *HF* distributions of as-milled and heated samples.

The broad exothermic process (from 400 up to 800 K) shown in figure 1(b) ascribed to structural relaxation does not affect the microstructure or *HF* distribution after heating up to 473 K (first dotted line in figure). However, for the sample heated up to 723 K, two new site contributions (*HF* ~28 and ~24 T) are necessary to fit the spectrum, indicating an evident evolution of Fe environments in the solid solution. In fact, figure 1(c) shows a good correlation of the probabilities of each *HF* contribution between the samples heated up to 473 K and the as-milled one. In the case of the sample heated up to 723 K the lost of correlation is clearly observed.

Taken into account the effect of Ge [3] and Nb [4] on *HF* for Fe-Fe interactions, it is possible to estimate the impurity content in the bcc Fe phase [1]. In order to reduce the huge amount of possible configurations for Ge and Nb as nearest neighbours, NN, (n and p, respectively) and next nearest neighbours, NNN, (m and q, respectively) of a central Fe atom, average reductions on *HF* can be considered. In the case of Nb, both NN and NNN has been averaged to $<HF_{pq}>$, and for Ge as NNN, $<HF_m>$ was calculated. Finally, *HF* can be expressed as a function of Ge atoms as NN:

$$HF = 34.54 - 2.36n^{1.5} - <\Delta HF_{pq}> - <\Delta HF_m>. \qquad (1)$$

Table 1. Average reduction due to Nb as NN and NNN and Ge as NN.

% at. Nb	1	2	3	% at. Ge	10	15	20
$<\Delta HF_{pq}>$ [T]	0.45	0.90	1.35	$<\Delta HF_m>$ [T]	1.07	1.61	2.15

Table 1 shows the reduction in *HF* of Fe atoms due to Nb as NN and NNN and Ge as NN for different concentrations of these elements in the crystals. Fixing $HF(n)$ to the corresponding values from expression (1) and table 1, experimental *HF* distributions were fitted with a set of 5 contributions (n= 0, 1, 2, 3, 4 and >4). Figure 2(a) shows an example of the deconvolution of *HF* distribution for the spectrum of as-milled sample.

From deconvolution results, it is possible to obtain an experimental probability for each atomic environment and compare its value with the theoretical probability calculated for a totally disordered bcc Fe(Ge,Nb) phase, for known Ge and Nb contents, as is shown in figure 2 (b). It can be observed that for a composition of ~15 at.% Ge and between 1 and 2 at.% Nb, the experimental probability agrees with the theoretical one, and a $Fe_{83.5}Ge_{15}Nb_{1.5}$ composition can be estimated for the solid solution of as-milled sample (with error bar of 2.5 at.% Ge and 0.5 at.% Nb), for which a lattice parameter between 2.896 < a < 2.900 Å can be calculated considering an increase of $1.7 \cdot 10^{-3}$ Å per at.% Ge [5] and $\sim 4.4 \times 10^{-3}$ per at.% Nb [6] in the bcc Fe lattice, in agreement with XRD, a =2.894 Å.

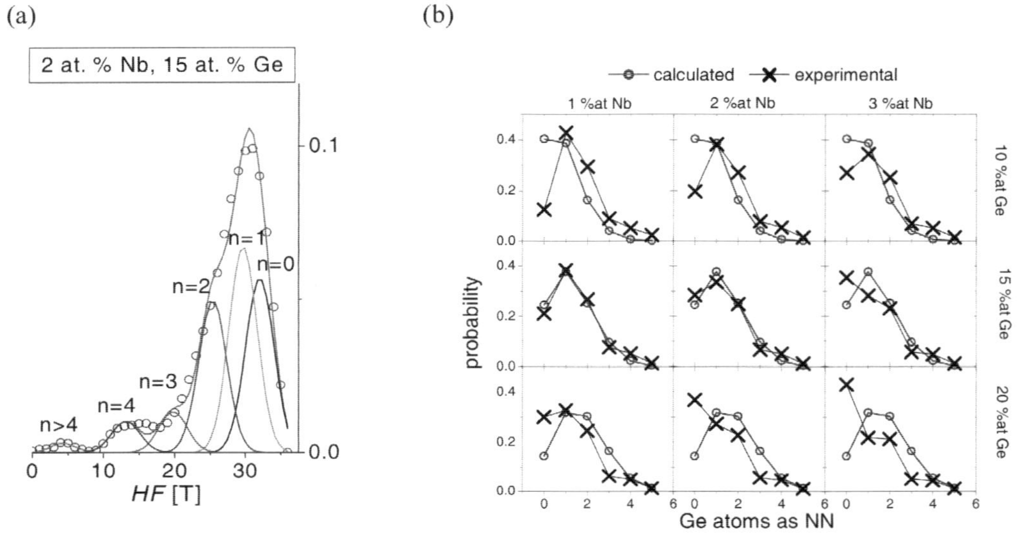

Figure 2. (a) Deconvolution of *HF* distribution at room temperature for 2 at% Nb, 15 at.% Ge in bcc phase; (b) Experimental and calculated probabilities for different Ge and Nb contents in crystals.

In the case of sample heated up to 723 K, the microstructure evolves mainly due to the rejection of Nb atoms to the grain boundaries. Therefore, the *HF* distribution was fitted considering 0 at.% Nb in the crystals and keeping free the different $HF(n)$ contributions (due to the numbers of Ge atoms as NN) as is shown in figure 3(a). Considering $HF(0,0)$ =34.54 T [3] as the hyperfine magnetic field of a Fe atoms with no Ge in the first and second shells and plotting $HF(0,0)–HF(n)$ as a function of $n^{3/2}$, a linear trend is observed in figure 3(b). The slope is 2.3 ±0.2 T being in good agreement with the dependence found for Fe-Ge system [3] and supporting the deconvolution performed. The value at n =0 is bigger than for crystalline Fe-Ge [3] and could be ascribed to the nanometric size of the crystals and to the presence of Cr from contamination [1].

From $HF(0,0)–HF(0)$ =2.4 ±0.1 T, the average number of Ge atoms in the second coordination shell is 1.44 atoms. Assuming a maximum ordered DO_3 phase, the concentration of Ge in the solid solution would be C_{Ge}~25(1.44/6) =6 at.%. However, considering a totally disordered phase, the presence of 1.44 atoms in the set of 14 atoms surrounding a central Fe, C_{Ge} ~10 at.%, which agrees with the solubility limit of Ge into bcc Fe in thermodynamic equilibrium [7]. The latter result is also in better agreement than the former one with the lattice parameter derived from XRD, a =2.888 ±0.003 Å, which implies a concentration of 13.2 at% Ge in bcc crystals.

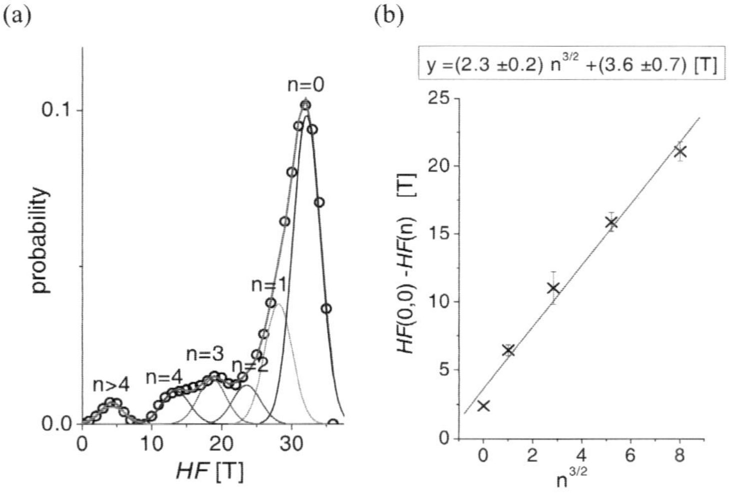

Figure 3. (a) deconvolution of *HF* distribution at 723 K; (b) *HF* reduction as a function of number of Ge atoms as near neighbour.

4. Conclusions

Thermal evolution of Fe environments of the supersaturated bcc Fe(Ge,Nb) solid solution produced by mechanical alloying of $Fe_{75}Ge_{20}Nb_5$ was studied. Sample heated up to 473 K shows no evolution and a $Fe_{73.5}Ge_{15}Nb_{1.5}$ composition can be estimated but after heated up to 723 K, the composition is $Fe_{90}Ge_{10}$, close to the limit of thermodynamical equilibrium.

Acknowledgements

This work was supported by ministry of Science and Innovation (MICINN) and EU FEDER (Project MAT2007-65227 and the PAI of the Regional Government of Andalucía (Project P06-FQM-01823). J.J.I. acknowledges a fellowship from MICINN. J.S.B. acknowledges a research contract from Junta de Andalucía.

References

[1] Blázquez J S, Ipus J J, Millán M, Franco V, Conde A, Oleszak D, Kulik T 2009 *J. All. Comp.* **169** 2469.
[2] Brand R A, Lauer J, Herlach D M 1983 *J. Phys. F: Met. Phys.* **13** 675.
[3] Cabrera A F, Sánchez F H 2002 *Phys. Rev. B* **65** 094202.
[4] Blachowski A, Ruebenbauer K, Zukrowski J 2005 *Phys. Stat. Sol. (b)* **15** 3201.
[5] Pearson W B 1967 *Handbook of lattice spacings and structures of metals and alloys* vol 2 (Oxford: Pergamon Press) p 911.
[6] JCPDS card numbers 6-696 and 35-789.
[7] Massalski T B, Okamoto H, Subramanian P R, Kacprzak L 1992 *Binary alloys phase diagrams* (Materials Park (OH):ASM International) pp 1187,1706.

Reflexion Mössbauer analysis of the *in situ* oxidation products of hydroxycarbonate green rust

S Naille, M Abdelmoula, D Louber, J-P Moulin and C Ruby
Laboratoire de Chimie Physique et Microbiologie pour l'Environnement
LCPME UMR 7564 CNRS - Nancy-Université
405 rue de Vandœuvre, 54600 Villers-lès-Nancy, France

E-mail: sebastien.naille@lcpme.cnrs-nancy.fr

Abstract. The purpose of this study is to determine the nature of the oxidation products of Fe^{II-III} hydroxycarbonate $Fe^{II}_4Fe^{III}_2(OH)_{12}CO_3{\cdot}3H_2O$ (green rust GR(CO_3^{2-})) by using the miniaturised Mössbauer spectrometer MIMOS II. Two Mössbauer measurements methods are used: method (i) with green rust pastes coated with glycerol and spread into Plexiglas sample holders, and method (ii) with green rust pastes in the same sample holders but introduced into a gas-tight cell with a beryllium window under a continuous nitrogen flow. Method (ii) allows us to follow the continuous deprotonation of GR(CO_3^{2-}) into the fully ferric deprotonated form $Fe^{III}_6O_4(OH)_8CO_3{\cdot}3H_2O$ by adding the correct amount of H_2O_2, without any further oxidation or degradation of the samples.

1. Introduction

The mixed-valence Fe^{II}-Fe^{III} hydroxysalt compounds, named green rusts (GRs) due to their "bluish-green" colour [1], play an important role in environmental science due to their high redox flexibility and reactivity [2,3]. They were first identified as intermediate products in the corrosion process of iron-based materials [4] but are now known to reduce many anionic pollutants [5,6]. Considered for a long time as laboratory synthetic products, it is only ten years ago that GRs were discovered as minerals in hydromorphic gleysoil [7].

They belong to the layered double hydroxide (LDH) family with the general formula $[Fe^{II}_{(1-x)}Fe^{III}_x(OH)_2]^{x+}{\cdot}[(x/n)A^{n-}{\cdot}mH_2O]^{x-}$ (with x = $[Fe^{III}]$ / $\{[Fe^{II}] + [Fe^{III}]\}$) which are made up of alternate brucite-type iron-hydroxide layers and interlayers consisting of water molecules and anions A^{n-} such as Cl^-, SO_4^{2-}, CO_3^{2-}, … Anions allow to establish the electroneutrality of these compounds as the iron-hydroxide layers are positively charged due to the presence of trivalent iron. Moreover, the crystal structure depends on the type of intercalated anions. When spherical (halogens) or planar (e.g. CO_3^{2-}) anions are intercalated, GRs crystallise in the trigonal system, space group $R\bar{3}m$, and are named green rusts one (GR1) [8,9] whereas green rusts two (GR2) crystallise in the hexagonal system, space group $P\bar{3}1m$, when tetragonal anions (e.g. SO_4^{2-}) are intercalated [10].

GRs are involved in the formation of iron oxides and (oxy)hydroxides in the natural environment. Since these discoveries, several works have been done on their oxidation processes [11,12]. The oxidation products depend on several factors such as oxidation rate and kinetic, nature of the GR, pH, temperature or Fe^{II} concentration. The most common products formed by GRs oxidation are lepidocrocite, goethite or ferrihydrite. However, recent studies have shown that Fe^{II}-Fe^{III}

hydroxycarbonate $Fe^{II}_4Fe^{III}_2(OH)_{12}CO_3 \cdot 3H_2O$ (stoichiometric green rust $GR(CO_3^{2-})$, $x = 0.33$) leads to the formation of Fe^{III} oxyhydroxycarbonate $Fe^{III}_6O_4(OH)_8CO_3 \cdot 3H_2O$ ($GR(CO_3^{2-})^*$, $x = 1$) by fast and/or violent oxidation [12,13]. The proposed mechanism of the $GR(CO_3^{2-})$ oxidation mode with H_2O_2 is a solid-state reaction due to the *in situ* progressive deprotonation of the hydroxyl sheets within the layered structure. The ferric molar ratio x is continuously rising with the increasing amount of H_2O_2 for the Fe^{II}-Fe^{III} oxyhydroxycarbonate intermediates $Fe^{II}_{6(1-x)}Fe^{III}_{6x}O_{12}H_{2(7-3x)}CO_3 \cdot 3H_2O$ ($GR(CO_3^{2-})^*$, $x \in]0.33\text{-}1[$) [13]. Nevertheless, except for the fully ferric form ($x = 1$), Fe^{II}-Fe^{III} hydroxycarbonate green rusts are air sensitive due to the divalent iron content. The aim of this study is to show how it is possible to monitor at room temperature the oxidation products of $GR(CO_3^{2-})$ with H_2O_2 by means of the Mössbauer spectrometer MIMOS II [14] without any further oxidation.

2. Experimental

$GR(CO_3^{2-})$ was synthesised by coprecipitation method. First, $Fe^{II\text{-}III}$ solution ($[Fe^{II}] + [Fe^{III}] = 0.4$ M and $x = 0.33$) was prepared by dissolving weighted amounts of $FeSO_4 \cdot 7H_2O$ (Aldrich, 99+%) and $Fe_2(SO_4)_3 \cdot 5H_2O$ (Aldrich, 97%) with 100 mL of demineralised water in an Erlenmeyer flask. Moreover, an additional amount of NaH_2PO_4 (Aldrich, 99%) was dissolved in the mixture in order to stabilise the $GR(CO_3^{2-})$ by the mechanism of phosphate adsorption [15]. Following this, a 100 mL basic solution of 0.8 M NaOH (Aldrich, 97+%) containing dissolved Na_2CO_3 (Sigma-Aldrich, 99.5+%) with $[CO_3^{2-}] = 0.466$ M, was rapidly transferred to the first mixture and a "bluish-green" precipitate appeared immediately. CO_3^{2-} ions were added in excess compared to SO_4^{2-} ions to prevent the formation of hydroxysulphate green rust $GR(SO_4^{2-})$. Finally, the erlenmeyer flask was filled with demineralised water and sealed off to avoid any air oxidation.

$GR(CO_3^{2-})^*$, with different x values, were obtained by violent oxidation of $GR(CO_3^{2-})$ with a H_2O_2 solution (Sigma-Aldrich, 30wt.%). Calculated volumes of H_2O_2 were dropped under vigorous stirring to reach four different compositions ($x = 0.5$, 0.67, 0.83 and 1) according to the following equation:

$$Fe_4^{II}Fe_2^{III}(OH)_{12}CO_3 \cdot 3H_2O + (3x-1)H_2O_2 \rightarrow Fe^{II}_{6(1-x)}Fe^{III}_{6x}O_{12}H_{2(7-3x)}CO_3 \cdot 3H_2O + 2(3x-1)H_2O$$

The increasing oxidation of divalent iron can be followed by the precipitates changing colour, which gradually transforms from "bluish-green" ($x = 0.33$) to "brownish-orange" ($x = 1$).

Figure 1. MIMOS II against the gas-tight cell supported by a three-legged stand.

The room temperature (RT) ^{57}Fe Mössbauer spectra were recorded in backscattered measurement geometry in the constant acceleration mode using the miniaturised Mössbauer spectrometer MIMOS II, which was designed and fabricated at the University of Mainz (Germany) [14]. The source used for experiments was ^{57}Co (50 mCi) embedded in a Rh matrix and the velocity scale was calibrated with

Journal of Physics: Conference Series **217** (2010) 012084 doi:10.1088/1742-6596/217/1/012084

the magnetic sextet of a high purity iron foil as the reference absorber. The spectra were fitted to either Lorentzian shape lines or pseudo-Voigt profile analysis with the Recoil© program [16] and the quality of the fits were controlled by the usual χ^2 test. Two methods of sample preparation were used to avoid any air oxidation, but in both cases the precipitates were first filtered and the wet green rusts pastes were spread on the Mylar window of a machined Plexiglas cell as sample holder. Method (i) consisted to coat every paste with glycerol and the sample holder was placed on the spectrometer head right above the four Si-PIN-diode detectors. For method (ii), the sample holders were incorporated into a home-made gas-tight cell with a beryllium window and the data recording was performed under a continuous nitrogen flow (figure 1).

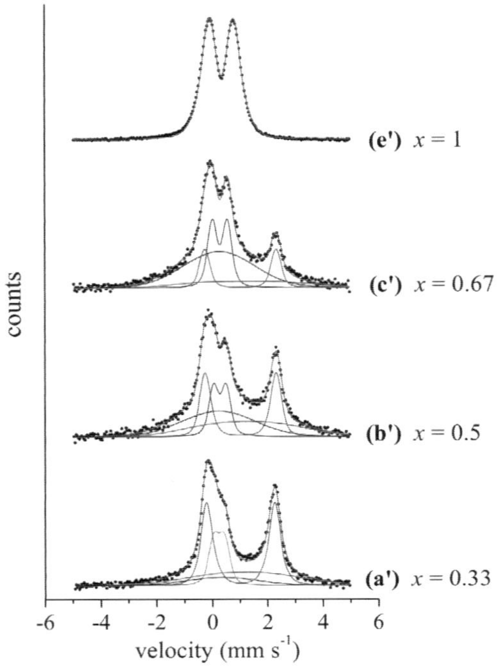

Figure 2. Backscattered Mössbauer spectra of $GR(CO_3^{2-})$ at $x = 0.33$ and $GR(CO_3^{2-})^*$ oxidised at different values of the ferric molar fraction x. The spectra are recorded with method (i) at room temperature.

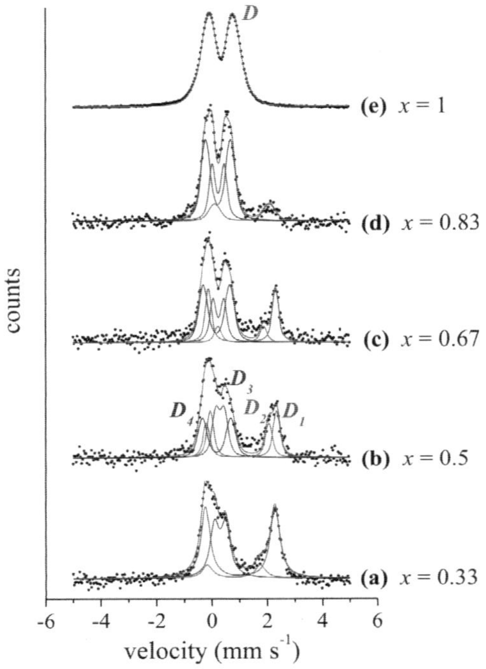

Figure 3. Backscattered Mössbauer spectra of $GR(CO_3^{2-})$ at $x = 0.33$ and $GR(CO_3^{2-})^*$ oxidised at different values of the ferric molar fraction x. The spectra are recorded with method (ii) at room temperature.

3. Results and discussion

The Mössbauer spectra of $GR(CO_3^{2-})$ and its oxidation products using method (i), are shown in figure 2. This sampling method, consisting of coating the pastes to avoid air oxidation, is often used for the X-ray diffraction (XRD) characterisation of LDHs [12,15]. However, the Mössbauer spectra exhibit a huge broadening, leading to a very large line widths as a result of the Mössbauer fitting procedure. Although the Fe^{II} components intensities decrease with the increasing added amount of H_2O_2, these spectra are not exploitable. This is due to glycerol which interacts chemically with LDHs by acting as a swelling agent [17]. Since glycerol is simply spread out on the GRs pastes, the reaction between glycerol and GRs takes time and thus explain why there is no evidence in the XRD patterns which are generally recorded in one hour. In the case of Mössbauer spectrometry, and especially in backscattering geometry, a spectrum can be recorded within four days. The reaction between GRs and glycerol can then occur and may lead to a delamination process which falsifies the Mössbauer spectra results.

In order to tackle this problem, another method, method (ii), has been used to characterise the oxidation products of GR(CO_3^{2-}). A gas-tight cell with a beryllium window has been designed (figure 1) at the laboratory to record spectra at room temperature under a continuous nitrogen flow (figure 3). The spectrum of GR(CO_3^{2-}) (figure 3(a)) is fitted with two ferrous doublets (D_1 and D_2) and one ferric doublet (D_3); the hyperfine parameters are summarised in table 1. The use of two doublets for ferrous cations and only one for ferric cations is consistent with the environment surrounding these cations. Indeed, Fe^{II} cations can be close to either water molecules ($Fe^{II}(H_2O)$) or CO_3^{2-} anions ($Fe^{II}(CO_3^{2-})$) whereas Fe^{III} cations environment cannot be easily distinguished [18]. The experimental abundances observed for the three different sites by Mössbauer spectroscopy (table 1) are in good agreement with the theoretical ones which are 1/2, 1/6 and 1/3 for $Fe^{II}(CO_3^{2-})$, $Fe^{II}(H_2O)$ and Fe^{III} respectively. The centre shifts (*CS*) values of the different doublets confirm that Fe–O bonds in the brucite-type layers, where each cation lies at the centre of an octahedron with OH^- ions at its six vertices, are highly ionic due to the inductive effect. The non-zero values of the quadrupole splittings (Δ) are characteristic of high-spin iron species in a distorted octahedral configuration as explained in [19].

Table 1. Mössbauer hyperfine parameters of the spectra presented in figure 3: the different components used in the Mössbauer fitting procedure and their relative area (*R.A.*), the ferric molar fraction (x = $[Fe^{III}]$ / $[Fe_{total}]$) either measured by Mössbauer spectroscopy or targeted from the added quantity of H_2O_2, the centre shift (*CS*) (Lorentzian shape lines, figure 2(a)) or the centre shift ($\langle CS \rangle$) at the maximum of the Gaussian distribution (Voigt profiles, figures 2(b-e)) with respect to metallic α-Fe at room temperature and the quadrupole splitting (Δ) or the quadrupole splitting at the maximum of the Gaussian distribution ($\langle\Delta\rangle$).

Fig. 3	Component	*R.A.* (%)	x (measured)	x (targeted)	*CS* or $\langle CS \rangle$ (mm s^{-1})	Δ or $\langle\Delta\rangle$ (mm s^{-1})
(a)	D_1 (Fe^{II})	54	0.35	0.33	1.13	2.51
	D_2 (Fe^{II})	11			0.90	1.91
	D_3 (Fe^{III})	35			0.40	0.40
(b)	D_1 (Fe^{II})	26	0.54	0.5	1.25	2.39
	D_2 (Fe^{II})	18			1.04	2.26
	D_3 (Fe^{III})	26			0.41	0.31
	D_4 (Fe^{III})	28			0.29	1.03
(c)	D_1 (Fe^{II})	28	0.63	0.67	1.20	2.42
	D_2 (Fe^{II})	9			1.16	1.65
	D_3 (Fe^{III})	22			0.36	0.39
	D_4 (Fe^{III})	41			0.29	0.96
(d)	D_1 (Fe^{II})	15	0.85	0.83	1.21	1.97
	D_3 (Fe^{III})	28			0.35	0.43
	D_4 (Fe^{III})	57			0.35	0.90
(e)	D (Fe^{III})	100	1	1	0.45	0.98

The formation of Fe^{II}-Fe^{III} oxyhydroxycarbonate green rusts GR(CO_3^{2-})* as intermediates with $x \in]0.33\text{-}1[$ by violent oxidation of GR(CO_3^{2-}) is confirmed by the MIMOS spectra (figures 3(b-d)) and the hyperfine parameters (table 1). Voigt profile analyses are used to fit every spectrum due to a global broadening which has already been observed by using transmission Mössbauer spectrometry [18]. A second ferric doublet (D_4) with a high quadrupole splitting appears and is in good agreement with the progressive deprotonation of the brucite-type layers, which was observed by XPS study [13], with a distortion of the octahedral sites that results from the mixture of OH^- and O_2^- ions at the octahedra vertices. Furthermore, the relative area of D_4 compared to D_3 is continuously increasing and ferrous doublets, D_1 and D_2, are gradually vanishing. The Mössbauer spectrometry at room temperature makes it thus possible to validate the *in situ* deprotonation process of the GR(CO_3^{2-}) hydroxide layers with H_2O_2 since there is a good agreement between the targeted x values and those measured (table 1). The fully ferric sample $GR(CO_3^{2-})^*$ ($x = 1$) is fitted with a broad Fe^{III} paramagnetic doublet (*D*), covering every Fe^{III} sites that can exist in the cationic layers (figure 3(e)). This shows that $GR(CO_3^{2-})$ can be fully oxidised in accordance with a solid-state reaction without any dissolution processes taking place.

4. Conclusion

Depending on the preparation method, significant differences were observed on the MIMOS spectra for $GR(CO_3^{2-})$ and its *in situ* oxidation products with H_2O_2. These oxidation products are Fe^{II}-Fe^{III} oxyhydroxycarbonate intermediates and include the fully ferric deprotonated salt at the final stage of the oxidation process. Method (i) leads to a significant broadening of the spectra that could be attributed to a chemical interaction between GRs and glycerol. Such a broadening decreases significantly when the experiments are done under a controlled atmosphere. The ferric molar ratio $x = [Fe^{III}] / \{[Fe^{II}] + [Fe^{III}]\}$, determined by using the miniaturised MIMOS(II) spectrometer, are then very close to the expected x values that take into account the amount of added H_2O_2.

Acknowledgments

The authors would like to acknowledge Nadia Sabri (McGill University, Montreal, Quebec, Canada) for her constructive comments, which improved the quality of the manuscript.

References

[1] Bernal J D, Dasgupta D and Mackay A L 1959 *Clay Min. Bull.* **4** 15–30
[2] Hansen H C B 2001 Environmental Chemistry of Iron(II)-Iron(III) LDHs (Green Rusts) *Layered Double Hydroxides: Present and Future* ed V. Rives (Huntington, N.Y.: Nova Science Publishers) chapter 13 pp 469–93
[3] Ruby C, Upadhyay C, Géhin A, Ona-Nguema G and Génin J-M R 2006 *Environ. Sci. Technol.* **40** 4696–702
[4] Boucherit N, Hugot-Le Goff A and Joiret S 1991 *Corros. Sci.* **32** 497–507
[5] Loyaux-Lawniczak S, Refait Ph, Ehrhardt J-J, Lecomte P and Génin J-M R 2000 *Environ. Sci. Technol.* **34** 438–43
[6] Hansen H C B, Guldberg S, Erbs M and Koch C B 2001 *Appl. Clay Sci.* **18** 81–91
[7] Trolard F, Génin J-M R, Abdelmoula M, Bourrié G, Humbert B, Herbillon A, 1997 *Geochim. Cosmochim. Acta* **61** 1107–11
[8] Refait Ph, Abdelmoula M and Génin J-M R 1998 *Corros. Sci.* **40** 1547–60
[9] Aissa R, Francois M, Ruby C, Fauth F, Medjahdi G, Abdelmoula M and Génin J-M R 2006 *J. Phys. Chem. Solids* **67** 1016–9
[10] Ruby C, Abdelmoula M, Aissa R, Medjahdi G, Brunelli M and François M 2008 *J. Solid State Chem.* **181** 2285–91
[11] Benali O, Abdelmoula M, Refait Ph and Génin J-M R 2001 *Geochim. Cosmochim. Acta* **65** 1715–26
[12] Legrand L, Mazerolles L and Chaussé A 2004 *Geochim. Cosmochim. Acta* **68** 3497–507
[13] Mullet M, Guillemin Y and Ruby C 2008 *J. Solid State Chem.* **181** 81–9
[14] Klingelhöfer G, Fegley B J, Morris RV, Kankeleit E, Held P, Evlanov E and Priloutskii O 1996 *Planet. Space Sci.* **44** 1277–81
[15] Bocher F, Géhin A, Ruby C, Ghanbaja J, Abdelmoula M and Génin J-M R 2004 *Solid State Sci.* **6** 117–24
[16] Lagarec K, Rancourt D G 1997 *Nucl. Instrum. Methods Phys. Res. Sect. B - Beam Interact. Mater. Atoms* **129** 266–280
[17] Auxilio A R, Andrews P C , Junk P C, Spiccia L, Neumann D, Raverty W, Vanderhoek N 2007 *Polyhedron* **26** 3479–90
[18] Génin J-M R, Abdelmoula M, Ruby C and Upadhyay C 2006 *C. R. Geosci.* **338** 402–19
[19] Raouche S, Naille S, Dobenesque M, Bot A, Jumas J-C, Cuq J-L and Marchesseau S 2009 *Int. Dairy J.* **19** 56–63

Journal of Physics: Conference Series **217** (2010) 012085 doi:10.1088/1742-6596/217/1/012085

Room temperature hysteretic spin transition in 1D iron(II) coordination polymers

Marinela M Dîrtu, Anil D Naik, Jacqueline Marchand-Brynaert and Yann Garcia*

Institut de la Matière Condensée et des Nanosciences, Université Catholique de Louvain, Place L. Pasteur 1, 1348, Louvain-la-Neuve, Belgium.
E-mail: yann.garcia@uclouvain.be

Abstract: The 1D spin transition compound $[Fe(L1)_3](ClO_4)_2$ (**1**) with L1 = ethyl-4H-1,2,4-triazol-4-yl-acetate, a novel neutral bidentate ligand, has been synthesised. The temperature dependence of the high-spin molar fraction derived from ^{57}Fe Mossbauer spectroscopy reveals an exceptionally abrupt single step transition between low-spin and high-spin states with a hysteresis loop of width 5 K ($T_c^{\uparrow}$ = 298 K and $T_c^{\downarrow}$ = 293 K). This spin transition operating around room temperature presents striking reversible thermochromism from white at 295 K to pink at ice temperature, behaving as an optical alert towards temperature variations. This first order phase transition was additionally followed by differential scanning calorimetry in good agreement with Mössbauer spectroscopy data. The spin transition properties of a freshly prepared sample of **1**, dried under a $N_{2(g)}$ stream which was formulated as $[Fe(L1)_3](ClO_4)_2 \cdot MeOH$ (**2**), are shifted below room temperature ($T_{1/2}^{\uparrow}$ = 273 K and $T_{1/2}^{\downarrow}$ = 263 K), thus showing a remarkable influence of solvent inclusion on the spin state of these chain compounds.

1. Introduction

Bistable molecular systems particularly materials exhibiting captivating scenario of spin crossover (SCO) are versatile section of the thriving field of molecular electronics.[1,2] Although SCO is well understood in $3d^4$-$3d^7$ transition metal ions in octahedral surrounding, the $3d^6$ iron(II) complexes received much more attention.[1] In these compounds, the reversible intra-ionic electron transfer from a diamagnetic low-spin (LS, $^1A_{1g}$) state to a thermally populated paramagnetic high-spin (HS, $^5T_{2g}$) state is mainly an entropy driven process. In the solid state, the presence of short and long range interactions acts as communication media between iron centres promoting cooperative first order spin transitions leading to a large memory domain that can be suitable for potential applications. Indeed, a SCO compound meeting display and data processing requirements would have in addition a good shelf life and an easily detectable optical response, and would operate near room temperature.[2] The family of thermochromic 1D iron(II) 1,2,4-triazole SCO coordination polymers have attracted our attention in recent years.[3] As a continuation of our work on amino acid functionalization,[4] we now introduce an ester of glycine on the 4 position of the triazole by designing L1 = ethyl-4H-1,2,4-triazol-4-yl-acetate (Figure 1) and studied the spin transition (ST) properties of two iron(II) complexes, $[Fe(L1)_3](ClO_4)_2$ (**1**) and as $[Fe(L1)_3](ClO_4)_2 \cdot MeOH$ (**2**).

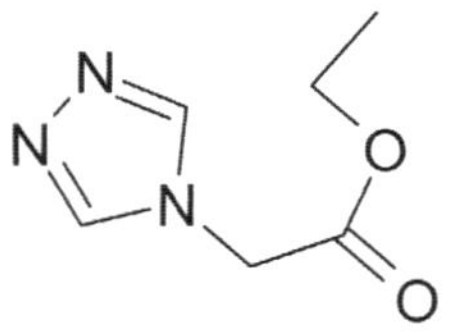

Figure 1. Ethyl-4H-1,2,4-triazol-4-yl-acetate (L1).

2. Results and discussion

The complexes were prepared by reacting L1 and $Fe(ClO_4)_2{\cdot}6H_2O$ in a 1:3 ratio in MeOH using Schlenk line and cannula techniques. These were filtered at room temperature, dried under vacuum (for **1**) or under a slow stream of $N_{2(g)}$ (for **2**). **1** is a white powder with a tinge of pale pink colour whereas **2** is a white powder. Composition examination by elemental analysis, atomic absorption and TGA-MS analysis of **1** leads to the formulation of the complex as $[Fe(L1)_3](ClO_4)_2$. For **2**, initial mass loss in TGA-MS analysis confirms the presence of one methanol molecule in the crystal lattice. **1** exhibits reversible pronounced thermochromism to pink on quenching of the sample in liquid nitrogen. Quench cooling in ice is also sufficient to reveal a strong colour change precluding a ST just below room temperature. Such optical properties are indeed a signature of a thermally induced SCO phenomenon. The pink colour results from two d-d transitions of LS Fe^{II} sites: ${}^1A_{1g} \rightarrow {}^1T_{1g}$ and ${}^1A_{1g} \rightarrow {}^1T_{2g}$, the ${}^1A_{1g} \rightarrow {}^1$MLCT transition being not located in the visible region. The HS Fe^{II} compounds are white because neither the d-d nor the MLCT transitions fall in the visible range. Compound **1** was investigated by Mössbauer spectroscopy over the temperature range 78 – 300 K to confirm this behavior (Figure 2). Mössbauer parameters are listed in Table 1. At 78 K the spectrum consists of one quadrupole doublet, with an isomer shift $\delta = 0.49(1)$ mm/s and a low quadrupole splitting of $\Delta E_Q = 0.19(1)$ mm/s corresponding to the LS state of Fe^{II}, indicating a complete ST. The presence of a quadrupole splitting is characteristic for a lattice contribution to the electric field gradient which implies a non-negligible distortion of the LS octahedron, as expected for a constrained chain.[5] Upon warming further, no change is observed in the spectrum until 298 K where a new quadrupole doublet characteristics of a HS state is detected. On cooling, a hysteretic behavior is clearly revealed by comparing Mössbauer spectra recorded at the same temperature (Figure 2). δ and ΔE_Q of both LS and HS species are typical of a $Fe^{II}N_6$ chain compound in which metallic centers are bonded by triple *N*1,*N*2-triazole bridges.[6]

Table 1. Overview of the ^{57}Fe Mössbauer parameters for (**1**) on warming and cooling modes over the temperature range 78-308 K.

T [K]	δ^a [mm/s]	ΔE_Q^b [mm/s]	$\Gamma/2^c$ [mm/s]	Relative Area [%]	Sites
78(1) ↑	0.49(1)	0.19(1)	0.15(1)	100	LS
288(1) ↑	0.43(1)	0.19(2)	0.15(1)	100	LS
293(1) ↑	0.46(2)	0.20(1)	0.12(2)	100	LS
298(1) ↑	1.04(1)	2.78(2)	0.16(1)	35.8	HS
	0.42(2)	0.19(1)	0.15(1)	64.2	LS
300(1) ↑	1.03(1)	2.78(1)	0.15(1)	91	HS
	0.34(5)	-	0.18(2)	9	LS
308(1)	1.03(1)	2.74(4)	0.16(1)	100	HS
298(1) ↓	1.03(1)	2.77(1)	0.16(1)	92.6	HS
	0.31(3)	0.21(2)	0.17*	7.4	LS
293(1) ↓	1.04(1)	2.80(1)	0.16(1)	48	HS
	0.43(1)	0.20(1)	0.13(1)	52	LS
288(1) ↓	1.00(1)	2.73(2)	0.15(1)	5.7	HS
	0.43(2)	0.19(2)	0.14(1)	94.3	LS
283(1) ↓	0.43(1)	0.19(1)	0.15(1)	100	LS

${}^a\delta$ = isomer shift relative to α-iron; ${}^b\Delta E_Q$ = quadrupole splitting;
${}^c\Gamma/2$ = half width at half maximum; * fixed parameters; ↓ cooling; ↑ warming.

Table 2. Overview of the ^{57}Fe Mössbauer parameters for **2** on warming and cooling modes over the temperature range 78-300 K.

T [K]	δ^a [mm/s]	ΔE_Q^b [mm/s]	$\Gamma/2^c$ [mm/s]	Relative Area [%]	Sites
78(1) ↑	0.48(1)	0.20(1)	0.16(1)	100	LS
240(1) ↑	0.47(1)	0.20(1)	0.15(1)	100	LS
260(1) ↑	1.06(1)	2.93(2)	0.17(1)	22.4	HS
	0.49(3)	0.20(1)	0.14(1)	77.6	LS
270(1) ↑	1.04(1)	2.92(2)	0.15(1)	38.0	HS
	0.45(3)	0.20(1)	0.15(1)	62.0	LS
280(1) ↑	1.04(1)	2.88(2)	0.15(1)	70.2	HS
	0.44(3)	0.21(1)	0.12(1)	29.8	LS
300(1)	1.03(1)	2.78(1)	0.15(1)	100	HS
280(1) ↓	1.05(1)	2.87(1)	0.15(1)	100	HS
					LS
270(1) ↓	1.06(1)	2.91(1)	0.15(1)	87.4	HS
	0.47(1)	0.22(1)	0.12(1)	12.6	LS
250(1) ↓	1.05(1)	2.91(1)	0.20(1)	83.2	HS
	0.46(1)	0.21(1)	0.14(1)	16.8	LS
240(1) ↓	0.46(1)	0.20(1)	0.15(1)	100	LS

${}^a\delta$ = isomer shift relative to α-iron; ${}^b\Delta E_Q$ = quadrupole splitting;
${}^c\Gamma/2$ = half width at half maximum; * fixed parameters; ↓ cooling; ↑ warming.

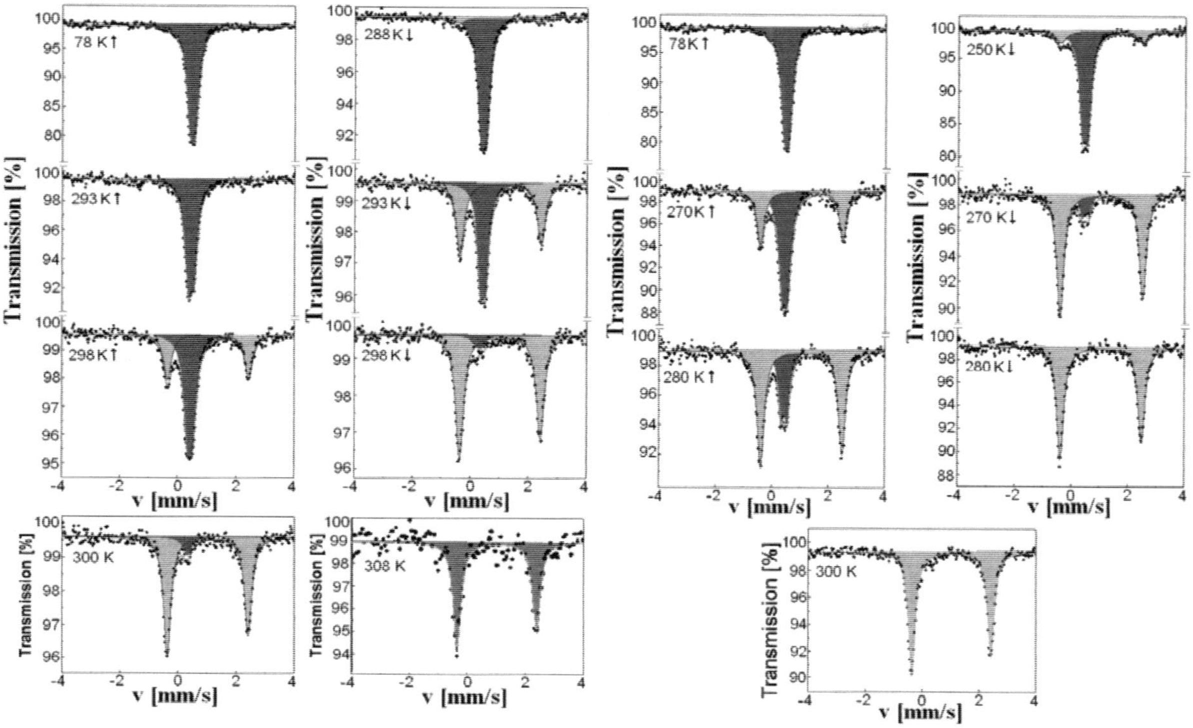

Figure 2. Selected Mössbauer spectra of **1** (left) and **2** (right). The signals in grey, dark grey refer to HS Fe(II) and LS Fe(II) ions respectively.

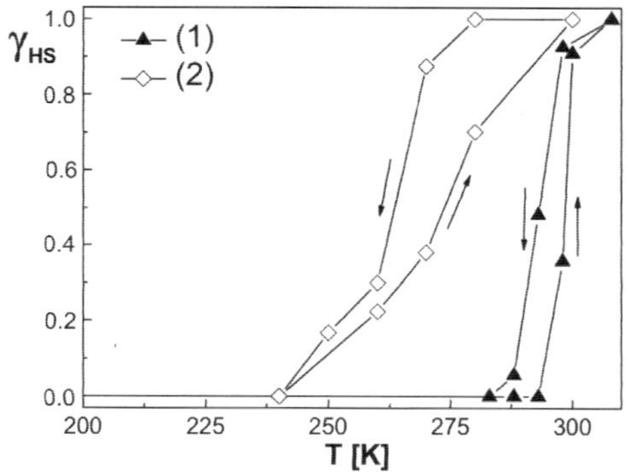

Figure 3. γ_{HS} *vs.* T derived by Mössbauer spectroscopy for **1** and **2**. The line is a guide to the eyes.

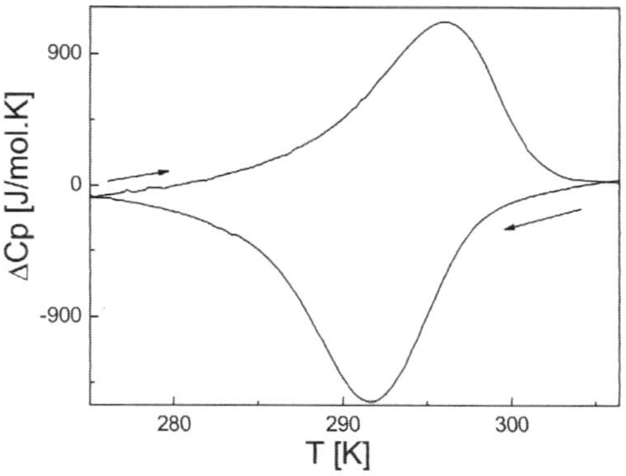

Figure 4. DSC profile for **1** in the cooling and warming modes.

9% of LS iron(II) ions are present at 300 K, in agreement with the pale pink colour observed at room temperature but the compound becomes fully HS at 308(1) K. HS molar fraction, γ_{HS}, was plotted, assuming equal Debye Waller factors for HS and LS sites (Figure 3). The critical temperatures, $T_c^{\uparrow}$ = 298 K and $T_c^{\downarrow}$ = 293 K, reveal an abrupt ST occurring near room temperature with an hysteresis width of 5 K.

Differential scanning calorimetric measurements were recorded over the T range 100 – 310 K at a rate of 10 K min^{-1}. Representative heat and cooling capacity profile are depicted in Figure 4 for **1**. An endothermic peak is observed on warming at $T_{max}^{\uparrow}$ = 296 K and an exothermic peak is recorded at $T_{max}^{\downarrow}$ = 291 K, on cooling, thus confirming the same hysteresis width identified by Mössbauer spectroscopy. The slight shift in the hysteresis loop is related to the higher heating and cooling rates of the calorimetric measurement. These peaks correspond to a first-order phase transition associated to the ST behaviour. The enthalpy and entropy values were derived as follows: ΔH_{HL} = 14.4 kJ mol^{-1} and ΔS_{HL} = 49 J. mol^{-1} K^{-1}. Compound **2** was also investigated by Mössbauer spectroscopy. At 78 K, the compound is fully in the LS state (Figure 2). The Mössbauer parameters are identical as for **1** suggesting that methanol inclusion has no effect on the local environment of the iron(II) ions. This strongly suggests that these non coordinated species are not located in the neighbouring of the chains. Upon warming further, another quadrupole doublet is observed which is typical of the HS state. The ST profile derived in Figure 3 clearly reveals an hysteretic ST at $T_{1/2}^{\uparrow}$ = 273 K and $T_{1/2}^{\downarrow}$ = 263 K, that is shifted to lower temperatures. Interestingly, the hysteresis loop of 10 K width is dissymmetric that may preclude a structural phase transition. The ST is here complete in both LS and HS state.

3. Concluding remarks

We have presented two novel iron(II) 1D ST chain compounds presenting a ST around room temperature, which has been observed only for rare cases.[7-9] Inclusion of a non coordinated methanol molecule has a paramount importance on the ST properties.

Acknowledgments

We acknowledge financial support from FRFC (No 2.4508308), IAP-VI (P6/17) INANOMAT and ARC. J. Marchand-Brynaert is a senior research associate of the FNRS (Belgium). We thank the GFSM for supporting A. D. N. and M. M. D. for attending ICAME 2009 in Vienna.

References

[1] Gütlich P, Garcia Y and Spiering H *In Magnetism: From Molecules to Materials* (Miller J S and Drillon M Eds, Wiley-VCH: Verlag GmbH) 2003 **IV** 271

[2] (a) Kahn O, Kröber J and Jay C 1992 *Adv. Mater.* **4** 718 (b) Kahn O and Martinez C J 1998 *Science* **279** 44

[3] (a) Garcia Y, Ksenofontov V and Gütlich P 2002 *Hyperfine Interact.* **139/140** 543 (b) Garcia Y, Moscovici J, Michalowicz A, Ksenofontov V, Levchenko G, Bravic G, Chasseau D and Gütlich P 2002 *Chem. Eur. J.* **8** 4992 (c) Garcia Y, Niel V, Muñoz M C and Real J A 2004 *Top. Curr. Chem.* **233** 229

[4] Naik A D, Marchand-Brynaert J and Garcia Y 2008 *Synthesis* **1** 149

[5] Dîrtu M M, Garcia Y, Nica M, Rotaru A, Linares J and Varret F 2007 *Polyhedron* **26** 2259

[6] Garcia Y, Ksenofontov V, Mentior S, Dîrtu M M, Gieck C, Bhatthacharjee A and Gütlich P 2008 *Chem. Eur. J.* **14** 3745

[7] Van Koningsbruggen P J, Garcia Y, Codjovi E, Lapouyade R, Kahn O, Fournes L, and Rabardel L 1997 *J. Mater. Chem.* **7** 2069

[8] (a) Cobo S, Molnár G, Real J A, Bousseksou A 2006 *Angew. Chem. Int. Ed.* **45** 5786 (b) Šalitroš I, Madhu N T, Boča R, Pavlik J and Ruben M 2009 *Monatsch. Chem.* **140** 695

[9] Garcia Y, van Koningsbruggen P J, Lapouyade R, Rabardel L, Kahn O, Wierczorek M, Bronisz R, Ciunik Z, Rudolf M F 1998 *C. R. Acad.Sc.* **IIc** 523

Mössbauer spectroscopy study of the high energy ball milling (HEBM) synthesis & kinetics of $CoFe_2O_4$ from $Co(OH)_2$ and α-Fe_2O_3

Oswald N C Uwakweh, Richard Pérez Moyet

Department of Engineering Science and Materials, University of Puerto Rico-Mayagüez, P. O. Box 9044, Mayagüez, Puerto Rico 00681-9044

uwakweh@ece.uprm.edu

Abstract.

Room temperature mechanochemical synthesis of single phase nanosized particles of $CoFe_2O_4$ by HEBM of stoichiometric mixtures of $Co(OH)_2$ and α-Fe_2O_3 under wet acetone was achieved after 25 hours of continuous ball milling. The particle size and lattice constant were 98.13 nm, and 8.398 Å after 10 hours, while 31.96 nm and 8.343 Å after 25 hours. The coercivity (25 hours state) was 12.78 mT with saturation and remanent magnetizations of 27.52 Am^2/kg and 3.94 Am^2/kg respectively due to small amount of a magnetic phase. Mössbauer spectroscopy measurements showed very limited reaction in 5 hours, with the single phase forming directly to superparamagnetic particles. The quadrupole splitting of the doublets was in the range of 0.24-1.14 mm/s, with corresponding isomer shift range of 0.80 -0.67 mm/s. The DSC measurements of the single phase material showed that it was metastable with characteristic stages whose activation energies were determined by the modified Kissinger kinetic analyses to be in the range of 3.51 kJ/mol to 11.89 kJ/mol.

1. Introduction

The oxides called spinel ferrites or ferrospinels are designated as MFe_2O_4 with "M" as the divalent cation different from the Fe trivalent cation based on their location within the crystal lattice. Néel [1], suggested that small antiferromagnetic particles can exhibit superparamagnetism and weak ferromagnetism due to uncompensated spins in the two sublattices. They are used in a host of scientific and technological applications such as in magnetic and switching devices [2-4] where their performance can be related to their compositions, microstructure, cationic distributions, and synthesis/processing histories [5].

$CoFe_2O_4$ is a well known hard magnetic material with very high cubic magnetocrystalline anisotropy, relatively high coercivity combined with moderate saturation magnetization. This explains why the material has a high potential for magnetic storage applications [6], with varied synthesis techniques such the modified coprecipitation method [7] etc. The synthesis based on high energy ball milling or mechanical alloying [8, 9] is among those used to obtain structures and properties that would have been otherwise impossible. Further, the mechanochemical synthesis route in obtaining a host of ferrite materials [10-11] is known to be valid alternatives to the conventional ceramic synthesis. Based on these, we undertook mechanochemical synthesis starting from solid precursors of

Journal of Physics: Conference Series **217** (2010) 012086 doi:10.1088/1742-6596/217/1/012086

$Co(OH)_2$ and α-Fe_2O_3 to investigate both its formation into the single phase, subsequent evolution with HEBM, and its thermal stability behaviour [12, 13].

2. Experimental

The solid precursors used for this study were Alfar Aesar supplied $Co(OH)_2$ and α-Fe_2O_3, weighed out in stoichiometric amounts to yield the single phase cobalt ferrite via the reaction scheme: $Co(OH)_2$ + α-$Fe_2O_3 \rightarrow CoFe_2O_4 + H_2O$, by ball milling 8 g of powder in a Fritsch Pulverisette-4 mill. The setting was 1400 rpm for the main disc, with a R-ratio – 3.0. 17 tempered steel grinding balls (10 mm diameter), a ball-to-powder ratio (BPR) of 40:1, and addition of 0.6 ml of acetone to the initial material mixture was used. The grinding jar was WC lined. Milling operations were stopped intermittently to retrieve powder samples for analyses.

The X-ray diffraction (XRD) was performed with a Siemens D500 diffractometer with CuK_α radiation. Debye Scherrer formula was used to determine the lattice parameters, while Williamson-Hall [14] analysis was used to determine the particle size and strain values. The magnetic properties were determined with a Lakeshore 7400 Series VSM. Mössbauer spectroscopic measurements were done with a WEBRES spectrometer (SEECO Company) operating at constant acceleration mode with a 50 mCi ^{57}Co gamma ray source in Rh matrix, and calibrated with α-Fe metal. Data was analyzed with WMOSS available at www.SEECo.us.

The thermal stability study was done with a differential scanning calorimeter (TA instrument (DSC-Q20) operated in nitrogen atmosphere (99.99 %) at 50 mL/min flow rate. Temperature and heat flow calibrations was done with a 10.5 mg indium, while the base calibration was carried out with an Al crucible.

3. Result and discussion

3.1 X-ray diffraction (XRD) measurements

Figure 1 shows XRD spectra of the manual mixture of $Co(OH)_2$, and α-Fe_2O_3 and its evolution with ball milling. As shown, new peaks emerged, with the disappearance of some corresponding to the precursor materials. The spectra showed both shifts in peaks positions, with their broadening as well. The progressive formation of the single phase $CoFe_2O_4$ commenced after 5 hours of ball milling, while structural changes including increase in structural defects resulted with ball milling time.

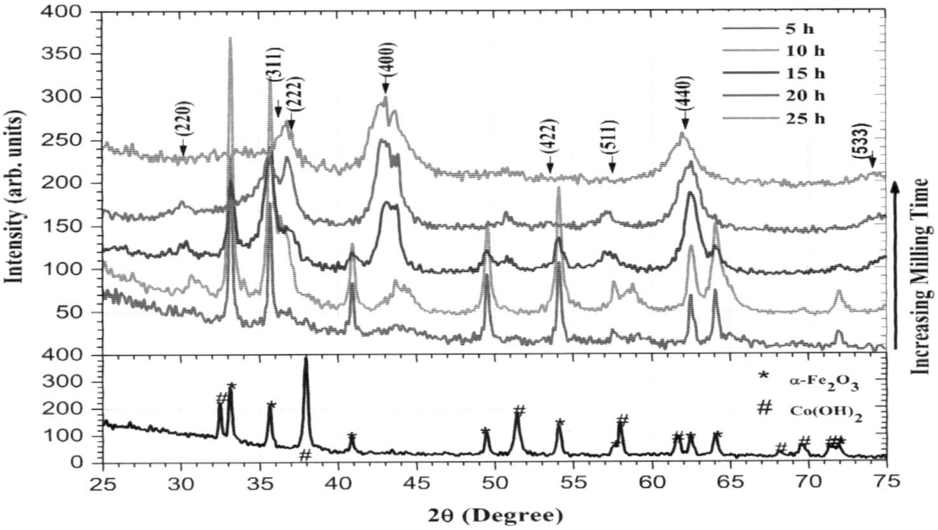

Figure 1: XRD spectra of the mechanochemical synthesis of $CoFe_2O_4$ as a function of time

The lattice constant, average internal strain, and particle size for the 10 hours milled state were, 8.398 Å, 0.00419 mm/mm, and 98.13 nm, respectively, in contrast to the corresponding values for the 25 hours milled state which were, 8.343 Å, 0.00004 mm/mm, and 31.96 nm.

3.2 Vibrating Sample Magnetometer (VSM) and Mössbauer spectroscopy (MS) measurements

The magnetization of spinel ferrites originates from the difference in the magnetic moments of the cations distributed in the tetrahedral and octahedral interstitial sites, and as a consequence directly points to the distribution of the magnetic Fe^{3+} and Co^{2+} ions in the two sublattices. Since ball milling caused a disruption of the crystallinity of the lattice, therefore the magnetization behaviors would change during mechanochemical synthesis, as shown in Figure 2. Based on XRD measurements, complete synthesis occurred after 25 hours of ball milling, therefore the coercivity, saturation and remanent magnetization values corresponding to the VSM hysteresis plot were due to the single phase nanosized $CoFe_2O_4$ material. The coercivity was 12.78 mT, with the corresponding saturation and remanent magnetization 27.52 Am^2/kg and 3.94 Am^2/kg respectively.

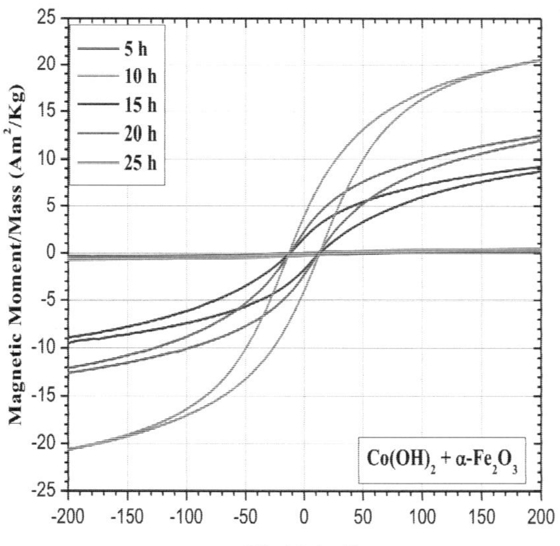

Figure 2: VSM measurements of the synthesis of $CoFe_2O_4$ by HEBM

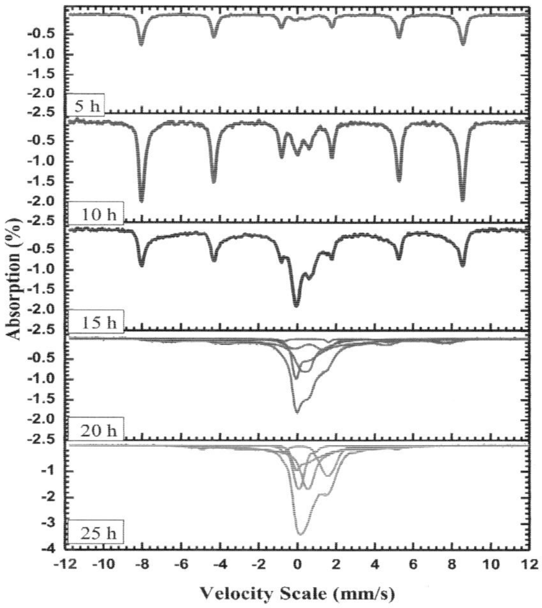

..... **Figure 3: Mössbauer spectra of $CoFe_2O_4$ by mechanochemical synthesis from $Co(OH)_2$ and α-Fe_2O_3**

The Mössbauer spectra of Figure 3 showed the progressive increase in intensity of the central portion with milling. The spectrum for the 25 hours milled material was fitted with a sextet, two doublets, and a singlet using the model by Rancourt and Ping [15]. It should be noted that the hyperfine distribution model of WMOSS (the program used in the spectral analyses) is an implementation of the Voigt based fitting method. In this model, each Fe site does not represent a single homogeneous iron species, but an ensemble of iron atoms with a Gaussian distribution in their internal field. Their relative abundances were 6.3%, 35.7%, 21.9%, and 36.1% correspondingly. The quadrupole splitting and isomer shifts of the doublets were: 0.24 mm/s and 0.81mm/s; and 1.14 mm/s and 0.67mm/s respectively. The degree of inversion λ referring to the sites occupancy based on the designation $(M^{2+}_{1-\lambda}Fe^{3+}_{\lambda})[M^{2+}_{\lambda}Fe^{3+}_{2-\lambda}]O_4$, where, M = Co, Mg, etc., varies as a function of processing history, and was not determined at the temperature of measurement.

The singlet isomer shift was 0.52 mm/s. The internal magnetic field of sextet was 311 kOe, and accounted for the weak coercivity recorded in the VSM measurements. Possible particle size distribution with varying cationic arrangements may have contributed to doublets and singlet observed associated with Fe^{3+} and Fe^{2+} in the octahedral and tetrahedral sites. With the corroborating XRD results, the paramagnetic peaks were associated with superparamagentism due to the nanosized particles of the mechanosynthesized phase, as the particle agglomeration leading to magnetic ordering did not occur based on prior SEM observation.

3.3 Thermal stability via differential scanning calorimetry (DSC) measurements

The thermal evolution of the single phase metastable $CoFe_2O_4$ corresponding to the 25 hours of ball milled state are shown in Figure 4, with the analyzed five characteristic stages shown on the 35°C/min trace only. The corresponding values for the other traces were used in the modified Kissinger method for non-isothermal kinetic analyses (similar to Lui and Uwakweh [12, 13]) to obtain the activation energies corresponding to the physical phenomena associated with them. These activation energies were obtained by plotting the following expression: $Ln(T_{f*}^2/\phi) = (E/R)1/T_{f*} + Ln(E/Rk_0)$, where T_{f*} = characteristic temperature, ϕ = heating rate, E, R, and k_0, are activation energy, gas constant, and kinetic parameter respectively. Since it is known that HEBM process in general induces physical phenomena such as inter/intra-particles pores on ceramic materials among other structural defects, therefore, the characteristic stages identified on the DSC traces could be associated with the elimination of, and/or rearrangements of surface atoms.

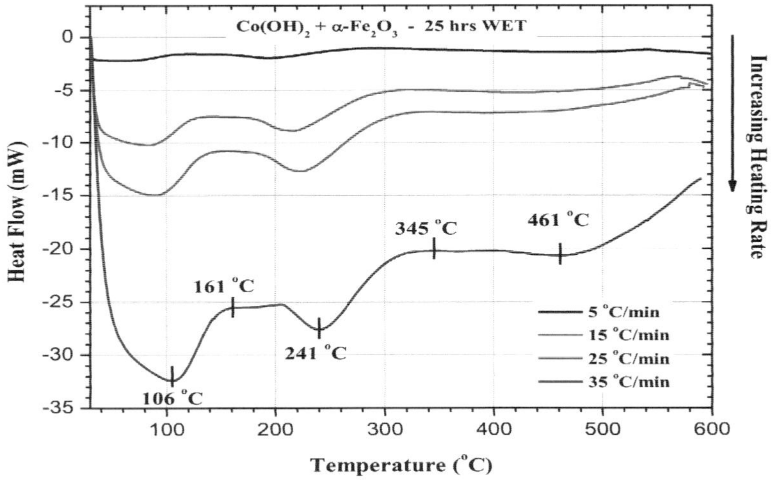

Figure 4: DSC traces of the mechanochemical synthesized $CoFe_2O_4$ from $Co(OH)_2$ and α-Fe_2O_3 solid precursors

The activation energies were 3.51 kJ/mol, 3.38 kJ/mol, 3.72 kJ/mol, 5.41 kJ/mol and 11.89 kJ/mol respectively, and are compatible with non-energetically tasking processes since the DSC measurements were done over temperature ranges where volume atomic diffusion in the material would be low. Further, it was not possible to associate the activation energies determined to the elimination or rearrangements of structural defects such as the zero order types (e.g., vacancies, or interstitials types, etc.) because of their relatively small values, and also, because they were observed at low temperatures were they should be substantially immobile. This showed that tailor-made materials could be realized after HEBM based on these measurements.

4. Conclusions

1. The mechanochemical synthesis of single phase $CoFe_2O_4$ by HEBM of stoichiometric amounts of $Co(OH)_2$ and α-Fe_2O_3 whose physical properties depended on milling time was demonstrated. The average particle size of 31.96 nm, and corresponding lattice constant of 8.343 Å determined for the 25 hours milled material.
2. The Mössbauer measurements showed non-magnetic Fe sites corresponding to two doublets and a singlet at room temperature that were associated with superparamagnetism arising from the nanosized particles. The origin of the two doublets and singlet Fe sites could be due to particle size distribution and presence of other effects such as structural defects associated with HEBM. The Fe site associated with the magnetically split portion of the spectrum with an internal field of 311 kOe accounted for the hysteresis observed in the VSM measurements. The corresponding coercivity in the 25 hours milled material was 12.78 mT, with saturation magnetization of 27.52 Am^2/kg, and remanent magnetization of 3.94 Am^2/kg.
3. The DSC measurements showed kinetically related stages with activation energies of 3.5 kJ/mol, 3.38 kJ/mol, 3.72 kJ/mol, 5.41 kJ/mol, and 11.89 kJ/mol. All the results point to the metastable state of the HEBM material.

The support of NSF-DMR PREM at UPRM on grant No. 0351449 is hereby acknowledged.

References

[1] Néel A L 1961 *Compt. Rend* **252** 4075
[2] Chantrell R W and O'Grady K 1994 *Appl. Magnetism* (Kluwer Academic, The Netherlands 113
[3] Nakamura T, Tsutaoka T and Hatakeyama H 1994 *J. Magn. Magn. Mater* **138** 319
[4] Tsutaoka T, Ueshima Y, Tokunaga T, Nakamura T and Hatakeyama K 1995 *J. Appl. Phys.***78** 3983
[5] Clark T M and Evans B J 1997 *IEEE Trans. Magn.***33** 3745
[6] Goldman A 1990 *A Modern ferrite Technology* (New York,Van Nostrand Reinhold)
[7] Cedeño-Mattei Y, Perales-Pérez O, Osorio-Cantillo C and Uwakweh O N C 2009 *NSTI-Nanotech-2009 ISBN 978-1-4398-1782-7* **1** 109-12
[8] Ding J, Miao W M, McCormick P G and Street R 1995 *Solid State Commun* **95** 31
[9] Ding J, Miao W M, McCormick P G and Street R 1997 *J. Magn. Magn. Mater.* **171** 309
[10] Druska P, Steinike U and Šepelák V 1999 *J. solid State Chem.* **146** 13
[11] Šepelák V, Baabe D, Mienert D, Litterst F J and Becker K D 2003 *Script. Mater.* **48** 961
[12] Liu Z T and Uwakweh O N C 1996 *J. Mater. Research* **11** 1665-72
[13] Zhentong Lui and Oswald N C Uwakweh 1997 *Metallurgical and Materials Trans A* **28,** 743
[14] Williamson G K and Hall W H 1953 *Acta Metall.* **1** 22
[15] Rancourt D G and Ping J Y 1991 *Nucl. Instr. Meth* **B58** 85-97

Journal of Physics: Conference Series **217** (2010) 012087 doi:10.1088/1742-6596/217/1/012087

Mössbauer spectroscopy study of the synthesis of $SnFe_2O_4$ by high energy ball milling (HEBM) of SnO and α-Fe_2O_3

Oswald N C Uwakweh[1], Richard Pérez Moyet[1], Rita Mas[2], Carolyn Morales[2], Pedro Vargas[3], Josué Silva[3], Ángel Rossa[3] and Neshma López[4].

[1]Department of Engineering Science and Material, [2]Department of Industrial Engineering, [3]Department of Mechanical Engineering, [4]Department of Chemical Engineering, University of Puerto Rico-Mayagüez

uwakweh@ece.uprm.edu

Abstract.

The formation of single phase nanoparticles of spinel structured ferrite, $SnFe_2O_4$, by mechanochemical syntheses using HEBM of stoichiometric amounts of solid SnO and α-Fe_2O_3 with acetone as surfactant was achieved progressively as function of ball milling time. Single phase $SnFe_2O_4$ formation commenced from five hours of continuous ball milling, and reached completion after 22 hours, thereby yielding a material with a lattice parameter of 8.543 Å, and particle size of 10.91 nm. The coercivity was 4.44 mT, magnetic saturation value of 17.75 Am^2/kg, and remanent magnetizations of 1.50 Am^2/kg, correspondingly. The nanosized particles exhibited superparamagnetic behavior phenomenon based on Mössbauer spectroscopy measurements. The kinetic analyses based on the modified Kissinger method yielded four characteristic stages during the thermal evolution of the 22 hours milled state with activation energies of 0.23 kJ/mol, 2.52 kJ/mol, 0.024 kJ/mol, and 1.57 kJ/mol respectively.

1. Introduction

Most chemical reactions in the solid state are slow and complex and are generally characterized by product formation at the interfaces of the reactants. Since atomic exchanges are required for the formation of a new phase, therefore diffusion of the reactants or precursor solids are inevitable with or without charge transfer. Mechanochemical processes by HEBM entail chemical transformations that occur in finite region of the induced particles in the range of 10^{-5} to 10^{-6} m [1]. The changes associated with or induced by ball milling include the alteration or generation of specific surface areas of the solid precursor particles by dynamic deformation, fracturing, etc. [1, 2]. All of these occur together with rapid solid particle reduction, while intimate mixings are enhanced with the use of surfactants such alcohol, acetone, etc. The spinel structured ferrites generally designated as $MeFe_2O_4$ are important class of materials that were generally synthesized via traditional ceramic methods [3-4], and have in recent times been synthesized with increased versatility from solid compounds. Ferrite materials having the spinel structure are important magnetic materials used in a wide variety of applications. Recently, Li and co-workers [5] reported the successful synthesis of $SnFe_2O_4$ nanoparticles through the precipitation exchange. Due to their technological importance, we carried out the synthesis of $SnFe_2O_4$ nanostructured materials via mechanochemical synthesis process entailing the room temperature HEBM of stoichiometric compositions of the SnO and α-Fe_2O_3 solid precursors with controlled amounts of acetone as facilitating agent.

2. Experimental

Alfar Aesar supplied SnO and α-Fe_2O_3, were weighed out stoichiometrically to yield the single phase tin ferrite: $SnO + \alpha\text{-}Fe_2O_3 \rightarrow SnFe_2O_4$ via mechanochemical synthesis through ball milling of 8 g of powder in a Spex-8000D mill. Tungsten carbide (WC) lined jars and 10 mm diameter tempered steel balls totalling 17 to achieve a BPR of 40:1 was used, with 0.6 ml of acetone. Milling was stopped intermittently to retrieve samples for analysis to reveal extent of the mechanosynthesis process.
Scanning electron microscopy (SEM) was done with a Joel 5410-LV, while X-ray diffraction (XRD) measurements were carried out with a Siemens D500 diffractometer with a Cu K_α source.
Magnetic properties were carried out with a Lakeshore 7400 Series VSM vibrating sample magnetometer (VSM), with the applied field up to 22000 gauss. Room temperature Mössbauer spectroscopic (MS) measurements were done with a WEBRES spectrometer [now SEECO Company] operating at constant acceleration mode with a 50 mCi ^{57}Co gamma ray source in Rh matrix, while the velocity calibration was carried out using α-Fe metal. Data analysis was done with WMOSS program, available at www.SEECo.us .
Lastly, the thermal behaviour of the single phase metastable materials was investigated with the aid of a differential scanning calorimeter (DSC), with a TA instrument, DSC-Q20 operated in nitrogen atmosphere 49.998% at 50 mL/min flow rate. The scan temperature range was 30 – 600°C, with ramps of 5, 15, 25, and 35°C /min.

3. Results and discussions

3.1. Scanning Electron Microscope (SEM) and X-ray diffraction (XRD) measurements

The SEM image of the 22 hours ball milled state revealed aggregated particles with inter/intra-agglomerated pores as shown in Figure 1. This is common with milled materials [6] with large volume of internal surfaces

Figure 1: SEM of synthesized $SnFe_2O_4$ for 22 hours of HEBM

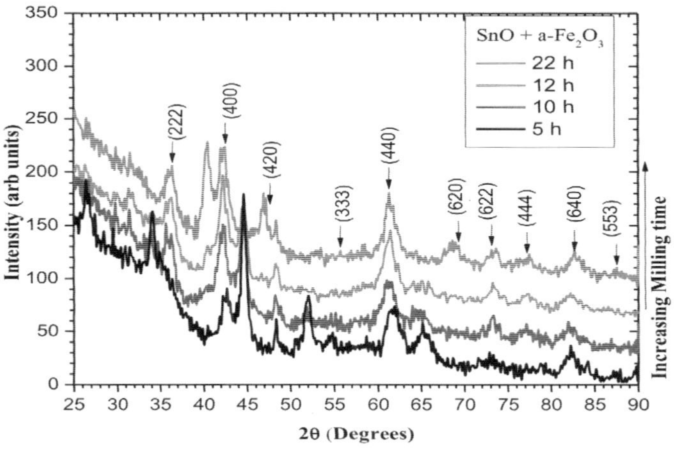

Figure 2: XRD of mechanochemical synthesis $SnFe_2O_4$

The XRD spectra corresponding to the different ball milled states are shown in Figure 2, which revealed the progressive formation of the single phase $SnFe_2O_4$ completed after 22 hours. This was confirmed with the disappearance of the peaks of precursor materials phases, and emergence of peaks of the tin ferrite. The lattice parameter of $SnFe_2O_4$ from Debye Scherrer formula was 0.8543 nm for the 22 hours case. With the Williamson-Hall method, the particle size and average internal strain were equally determined. The particle size for the 22 hours ball milled state was 10.91 nm. The value for internal strain for the 22 hours ball milled states was 0.00430 mm/mm. Altogether, the XRD measurements proved that synthesis of the desired $SnFe_2O_4$ single phase took place progressively with ball milling, thereby justifying the mechanochemical synthesis as a valid method for obtaining ferrospinel materials.

3.2 Vibrating Sample Magnetometer (VSM) and Mössbauer Spectroscopy (MS) measurements
The VSM plots of Figure 3, showed the decrease in the hysteresis with milling time, and are consistent with the general trend of the difficulty in the attainment of saturation, with decreasing saturation as a function of ball milling time due to the nanosized nature of the particles, or due to the presence of anisotropy of Fe spins inside nanoparticles and in their local environments, as with the substitution of Sn^{4+} in Fe_3O_4 which introduces simultaneously an equivalent increase of the Fe^{2+} concentration with modified magnetic properties. According to Néel [7], Sn enters magnetite exclusively as Sn^{4+} into the octahedral sublattice thereby causing the disappearance of two Fe^{3+} ions via (i): direct replacement, and (ii): valence change from three into two fold. The magnetic saturation of the 22 hours ball milled state was 17.753 Am^2/Kg, with magnetic saturation and remanent ratio, M_r/M_s, equal to 0.0848, and is consistent with the fact that $SnFe_2O_4$ phase formed progressively with ball milling time. Liu and Li [8] observed that the coercivity of $SnFe_2O_4$ decreased with particle size reduction. Though they attributed this behavior to the presence of noncolinear arrangements of Fe spins inside clusters and also in their local environments, however, the drastic loss of coercivity in their sample (particles size of 4 nm at 350 K when particle agglomeration should have taken place) suggested possibility of other factors to be responsible. The coercivity was 4.441 mT for the 22 hours milled material. Particle size reduction in the nanometer scale can result in unusual magnetic properties originating from changes in the A- and B-cationic populations in comparison to the bulk materials [9, 10].

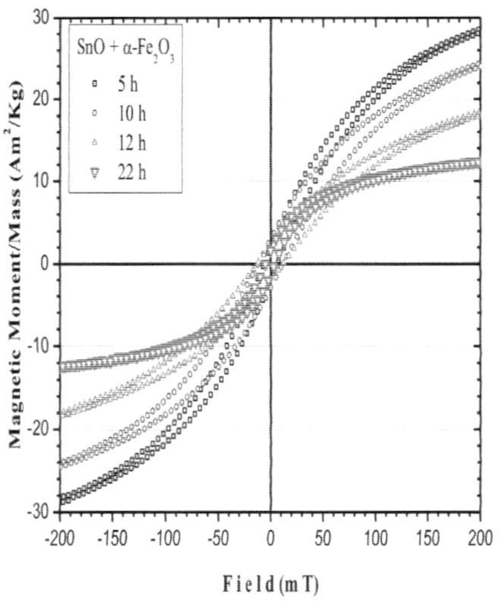

Figure 3: VSM Measurements of $SnFe_2O_4$ synthesis

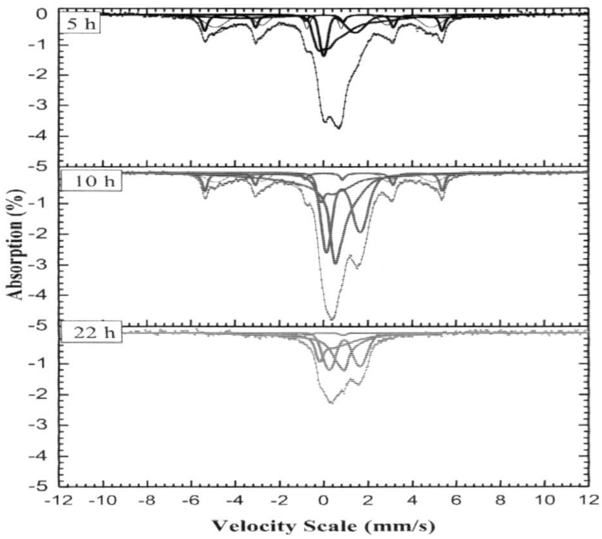

Figure 4: Mössbauer spectra of $SnFe_2O_4$ mechanochemical synthesis

The room temperature MS measurements shown in Figures 4, for the different mechanosynthesis states showed the presence of sextet peaks within the ± 6 mm/s, and a central region within the ± 2 mm/s with non-magnetic peaks. The sextets continuously decreased in intensity with ball milling time, while the non-magnetic portions of the spectra progressively increased.

The fitting of the spectra were carried out with the aid of the hyperfine field distribution model according to Rancourt and Ping [11]. Thus, the hyperfine field distribution model in WMOSS (the program used in spectral analyses) is an implementation of the Voigt based fitting method. In this model, each Fe site does not represent a single homogeneous iron species, but an ensemble of iron atoms with a Gaussian distribution in their internal field. The sub spectra in the accompanying Figure 4, showed the evolution of the various Fe sites which were associated with the mechanochemical synthesis of the single phase $SnFe_2O_4$ in intermediate or metastable stages. The magnetically split peaks were resolved into two subspectra with internal magnetic fields of 326 kOe, and 310 kOe respectively. Given the relatively small amounts of magnetically ordered component of the 22 hours state in comparison to the 5 and 10 hours HEBM states, it is possible that solid solutions of the sort $Sn_{1-x}Fe_xO$ formed, probably similar to the case of $Mg_{1-x}Fe_xO$ during the mechanosynthesis process according to Sepelák et al. [12]. They showed that prolonged milling resulted to the formation of nano sized particles and chemical reduction of $MgFe_2O_4$ leading to the formation of a solid solution of FeO and MgO, together with metallic Fe found as a by-product. Lastly, particle size reduction leading to the phenomenon of superparamagnetism was evidenced by doublet or singlet peaks in MS measurements. Typical isomer shift values of 0.24(6) mm/s is characteristic of octahedrally coordinated Fe^{3+}, while the quadrupole splitting of the doublet is typically in the range of 0.78(6) to 0.81(7) mm/s.

3.3 Differential scanning calorimetric (DSC) measurements

The thermal evolution traces of the metastable synthesized $SnFe_2O_4$ corresponding to the 22 hours of ball milled state are shown in Figure 5, with the characteristic stages identified as points “A”, “B”, “C”, and “D” for each heating rate.

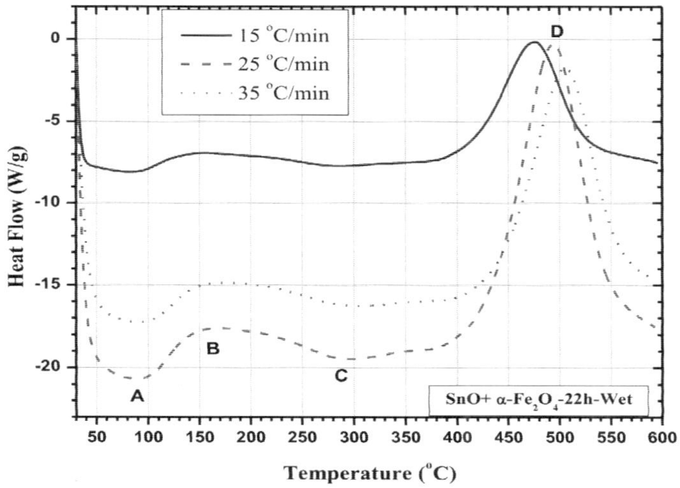

Figure 5: DSC traces of the mechanochemical synthesized $SnFe_2O_4$ corresponding to 22 hours of HEBM

Similar approach was adopted by Lui and Uwakweh [13] in the kinetic study of HEBM of Fe-Zn materials followed by thermal stability investigations aided by the use of the modified Kissinger method for non-isothermal kinetic analyses. With the kinetic analyses [13], the activation energies corresponding to the physical phenomena associated with these regions were determined to be 23 kJ/mol, 2.52 kJ/mol, 0.024 kJ/mol, and 1.57 kJ/mol respectively.

According to Heitjans and Indris [6], highly agglomerated powders of $LiBO_2$ resulting from high energy ball milling with a Spex8000, displayed a faster diffusion in comparison to the microcrystalline counterparts. They found activation energy values of up to three times higher than that of nanocrystalline BaF_2 compared to the microcrystalline one. These stages could be associated with surface phenomena, e.g., intercrystallite/intra-agglomerated pore elimination since smaller sized pores are eliminated faster during sintering operations. It was not possible to associate these characteristic stages with point defects movements due to the relatively low temperature ranges where they were detected. Nanoparticles can be viewed as a core that is isostructural with the bulk, having a large volume fraction surface layer with a different cationic distribution. This "core-surface" notion has been used to explain the magnetization reduction (or enhancement) in spinel ferrite nanoparticles compared to that bulk [12].

4. Conclusion

1. Mechanochemical synthesis by HEBM of solid SnO and α-Fe_2O_3 oxides yielded single phase $SnFe_2O_4$ particles, whose lattice parameter changed with ball time, from 0.8434 to 0.8543 nm or (8.434 to 8.543 Å), with corresponding particle sizes of 19.11 nm, and 10.91 nm.
2. The coercivity and magnetic saturation decreased with the progressive formation of $SnFe_2O_4$, and are for the 5 hours and 22 hours sates, 6.27 mT, and 4.44 mT respectively.
3. MS measurements confirmed superparamagnetism for the 22 hours milled state.. The magnetic parts had internal magnetic fields of 326 kOe and 310 kOe respectively.

Acknowledgement

The support of NSF-DMR PREM at UPRM based on grant No. 0351449 for the authors is hereby acknowledged.

References

[1] Butyagin P 1999 *Coll. Surf. A* **160** 107
[2] Kajdas C K 2005 *Tribology Intern* **38** 337
[3] James Reed S 1995 *Principles of Ceramics Processing 2nd ed.* (New York: Wiley-Interscience)
[4] Chiang Y M, Birnie D III and Kingery W D 1997 *Physical Ceramics, Principles for Ceramic Science and Engineering* (The MIT Series in Materials Science & Engineering- John Wiley)
[5] Fashen L, Wang H, Wang L and Wang J 2007 *J. Magn. Magn. Mater.* **309** 295-99
[6] Heitjans P and Indris S 2004 *Journal of Materials Science* **39** 5091-96
[7] Néel L 1995 Some theoretical aspects of rock magnetism *Adv. Phys* **4(2)** 91
[8] Liu F X and Li T Z 2005 *Mat. Letters* **59** 194-6
[9] Jiang J Z, Goya G F and Rechenberg H R 1999 *Journal of Physics: Condens. Matter* **11** 4063
[10] Sepelak V, Feldhoff A, Heijans P, Krumeich F, Menzel D, Litterst F J, Bergmann I and Becker K D 2006 *Chem. Mater* **18** 3057
[11] Rancourt D G and Ping J Y 1991 *Nucl. Instr. Meth* **B58** 85-97
[12] Sepelák V, Menzel M, Beckar D and Krumeich F: 2002 *J. Phys. Chem. B*, **106** 6672-8
[13] Liu Z and Uwakweh O N C 1997 *J. Metall. Mater. Trans. A* **28A, #3A** 743-7

Journal of Physics: Conference Series **217** (2010) 012088 doi:10.1088/1742-6596/217/1/012088

The effect of temperature on the structural and magnetic behaviour of Fe-Ni Invar alloys

M Kądziołka-Gaweł, W Zarek and E Popiel
Institute of Physics, University of Silesia, 4 Uniwersytecka Str., 40-407 Katowice, Poland

E-mail: mariolakadz@op.pl

The crystal structure and magnetic properties of polycrystalline $Fe_{1-x}Ni_x$ alloys with x = 0.30, 0.325 and 0.375 were investigated in the temperature range 20K<T<900K with the use of X-ray diffraction, magnetostatic and Mössbauer effect methods. Two structural transitions were observed above 600K. In the first one bcc (α) $\rightarrow$ fcc (γ) and in the second one fcc (γ) $\rightarrow$ fcc (γ') with a lower lattice constant. It was found that below 700K the lattice constant of the γ phase does not depend on temperature. The transition $\alpha \rightarrow \gamma$ changes the local environment of the Fe atoms and affects the magnetic properties of the Fe-Ni alloys. The disappearance of the ferromagnetic order in γ-$Fe_{0.70}Ni_{0.30}$ alloy can be explained by the presence of Fe atoms in the low ($0\mu_B$) and high ($2.8\mu_B$) spin state. The Fe atoms with zero magnetic moment can destroy the long range magnetic order in the investigated Fe-Ni alloys.

1. Introduction

The Fe-Ni alloys in the concentration range of „Invar" (~30$\pm$40% Ni) have a zero thermal expansion coefficient at room temperature. Due to this behavior they have been used for various practical applications since their discovery by Guillaume in 1897 [1]. Fe-Ni alloys in the Invar region of composition exhibit anomalous magnetic properties such as a sudden deviation of the saturation magnetization from the Slater-Pauling curve, strong forced volume magnetostriction and low ferromagnetic ordering temperature T_c in comparison to the T_c of α-Fe, pure Ni and non Invar γ-FeNi [2 - 4].

In this paper we present the results of investigations of the crystal and of the magnetic properties of $Fe_{1-x}Ni_x$ (x = 0.30, 0.325, 0.375) alloys in the temperature range 20K<T<900K with the use of X-ray diffraction, magnetostatic and Mössbauer effect methods.

2. Experimental

The polycrystalline $Fe_{1-x}Ni_x$ (x = 0.30, 0.325, 0.375) alloys were prepared by arc-melting, using stoichiometric quantities of the starting metals in argon atmosphere. Ingots were re-melted several times to ensure homogeneity. The investigations were performed on powdered samples with the granule sizes less than 30μm made by grinding on a diamond file. This procedure neither caused oxidation nor produced stresses in the alloys.

The XRD measurements were performed in the temperature range 20K<T<900K using a Siemens D5000 X-ray diffractometer and Fe K_α radiation. From these measurements the type of crystal structure and of the lattice constants were determined.

The Curie temperature and magnetic moments of the studied alloys were determined from magnetic measurements performed at temperatures up to 600K in the magnetic fields of up to 9T using the Faraday method and SQID magnetometer.

The Fe^{57} Mössbauer spectra were recorded in a temperature range from 300K to 843K, using a constant acceleration spectrometer with Co^{57}:Pd source. The absorbents were annealed at 673K, 773K and 843K for 3 h and cooled at a rate of 400K/h. Metallic iron powder was used for velocity calibrations of the Mössbauer spectrometer. All spectra were fitted by means of a hyperfine field distribution using the Hesse- Rübartsch procedure [5].

3. Results and discussion

The X-ray diffraction measurements showed that before annealing all samples at room temperature exhibited two cubic phases bcc (α) and fcc (γ). After this process only the fcc cubic phase was observed. The X-ray diffraction patterns at selected temperatures and with a temperature variation of the average lattice constant for $Fe_{0.625}Ni_{0.375}$ alloy are shown in Figure 1. The lattice constants for the investigated Fe-Ni alloys before and after annealing are presented in Table 1 and errors for these lattice constants are smaller then 2%.

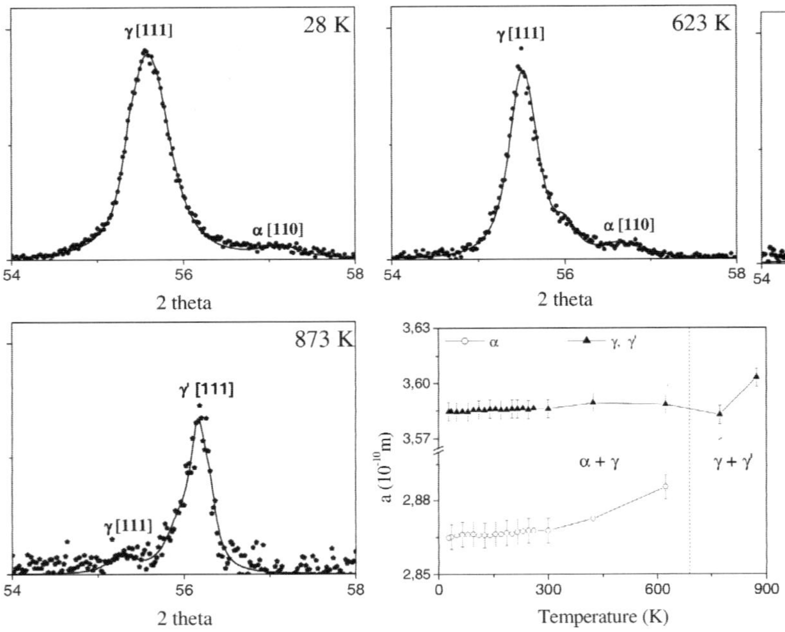

Figure 1. The XRD diffraction patterns at selected temperatures and with temperature variations of the lattice parameter for alloy with x = 0.375. Above 700K γ and γ' are paramagnetic.

Table 1. Lattice constant, contribution of crystal structure and XRD line width at room temperature for the investigated Fe-Ni alloys.

	Before annealing						After annealing		
Compound	0.30		0.325		0.375		0.30	0.325	0.375
Structure	fcc	bcc	fcc	bcc	fcc	bcc		fcc	
Lattice constant (Å)	3,583	2,862	3,586	2,859	3,586	2,852	3,589	3,587	3,598
Contribution (%)	10	90	60	40	95	5	100	100	100
XRD line width	0.42	0.72	0.4	0.75	0.4	0.5	0,1	0,15	0,08

The XRD diffraction patterns for each structure and sample before annealing up to 900K were broad. This can be the result of the composition fluctuation due to the presence of regions rich in iron or nickel. In the temperature range from 20K to about 600K the phases α and γ are present in the

Journal of Physics: Conference Series **217** (2010) 012088 doi:10.1088/1742-6596/217/1/012088

investigated alloys with a constant contribution whereas no structural transition was observed, as suggested in reference [6]. Above 600K the transition $\alpha \rightarrow \gamma$ was noticed. Above 700K the second transition fcc (γ) with a large lattice constant to fcc (γ') with a smaller one appears. Above 850K only the γ' phase with narrow Bragg lines was observed. The lattice constant of the α phase increases with temperature, however, the lattice constant of the γ phase exhibits a certain anomaly. In the temperature range 20K<T<700K this parameter, within experimental error, is independent of temperature and has an almost constant value (Figure 1). Above 900K the lattice constant of this phase increases with increasing temperature. Such an anomalous temperature dependence of the lattice constant leads to the anomalous thermal expansion of γ-Fe-Ni alloys.

Magnetic studies showed that all investigated alloys behaved like ferromagnets. Before annealing they are ferromagnetic with the Curie temperature at about 420K for the γ phase and above 650K for the α phase. Structural transition at high temperatures makes impossible the determination of the Curie temperature of the α phase. The mean magnetic moment was determined to be at 2K for $Fe_{1-x}Ni_x$ alloys with x = 0.30, 0.325 and 0.375 was 2.272, 2.082 and 1.914μ_B/f.u., respectively.

The Mössbauer studies of the investigated alloys confirm the results of structural and magnetic measurements. Figure 2 shows room temperature (RT) Mössbauer absorption spectra and diagrams of the hyperfine magnetic field (HMF) distribution before and after annealing for alloys with x = 0.30 and 0.375. The shape of the Mössbauer spectra at RT, before annealing, have a similar shape to those presented in work [7, 8]. These spectra indicate the existence of magnetic ordering as well as a very small contribution of the paramagnetic phase which is manifested by a single line in the centre of the spectra. The magnetically ordered phase is revealed as a superposition of several Zeeman sextets giving a wide HMF distribution with a high mean <HMF>. The α phase in the studied alloys can be seen as a narrow peak concentrated at the HMF at about 335kOe (x = 0.30). The γ phase gives a broad HMF distribution concentrated at the HMF at about 290kOe and a smooth tail extending towards lower fields.

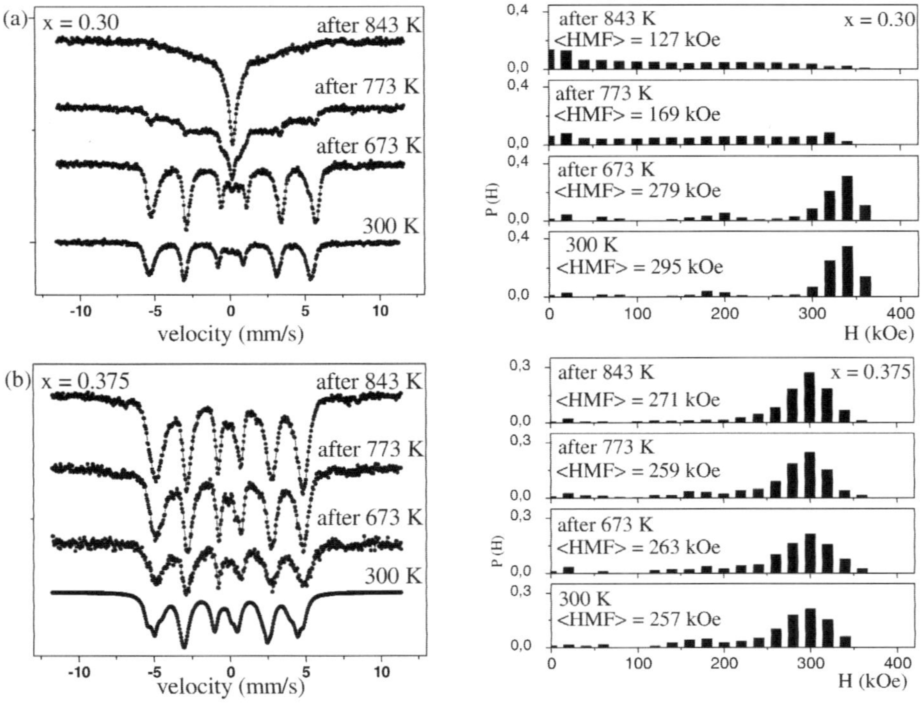

Figure 2. The room temperature Mössbauer absorption spectra and the hyperfine field distribution for $Fe_{1-x}Ni_x$ alloys, with x = 0.30 (a) and x = 0.375 (b), before annealing and after annealing.

With the increase of the annealing temperature the structural transition $\alpha \rightarrow \gamma$ and compositional homogenization of the γ phase take place and leads to a change in the HMF distribution. The Mössbauer spectra of $Fe_{0.70}Ni_{0.30}$ obtained at room and high temperatures are shown in Figure 3. Marked changes of the shape of the Mössbauer spectra for alloy with x = 0.30 were caused by the structural transition $\alpha \rightarrow \gamma$ as well as by the change of magnetic properties of the alloy. On the basis of [3, 8] we assume that in the case the Fe atom is surrounded by less than eight Fe atoms as its nearest neighbors (nn) it has a magnetic moment similar to that of α-Fe ($2.2\mu_B$). When the Fe atom has more then eight Fe atoms as its nn then the magnetic moment equals zero. This speculation has been confirmed by XPS studies of $Fe_{1-x}Ni_x$ alloys with fcc structure [9], which showed that Fe atoms can be both in the low-spin ($0\mu_B$) and high-spin ($2.8\mu_B$) state. At room temperature the $Fe_{0.70}Ni_{0.30}$ alloy was ferromagnetic and above 700K it behaves paramagnetic. After cooling to room temperature this alloy is also a paramagnetic but exhibits the fcc structure. Hence a marked contribution of lower magnetic fields to the HMF distribution is also observed. In γ-$Fe_{0.70}Ni_{0.30}$ the probability that the Fe atom has from nine to twelve other Fe atoms as nn is about 15%. In this case the ferromagnetic ordering is destroyed. With the increase of the nickel concentration in γ-$Fe_{1-x}Ni_x$ alloys this probability falls to zero (x > 0.4). These alloys exhibit a homogenous ferromagnetic order and therefore the annealing of $Fe_{0.625}Ni_{0.375}$ alloy leads to the increase of the Curie temperature by about 30K.

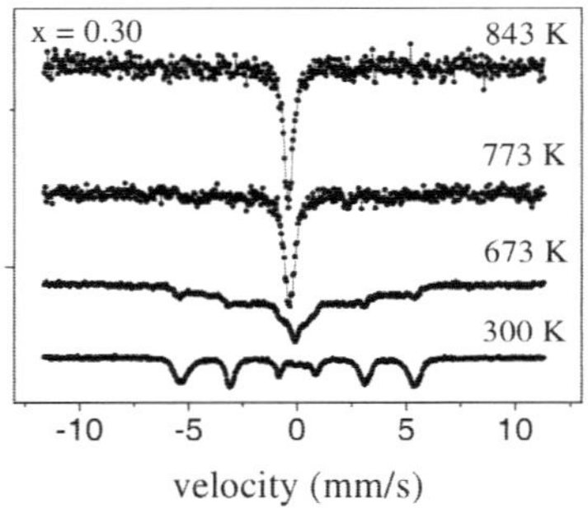

Figure 3. Mössbauer spectra for $Fe_{0.70}Ni_{0.30}$ alloy obtained at room temperature and high temperatures. Above 700K only a single line is present. The structural transition $\alpha \rightarrow \gamma$ at about 700K temperature makes it impossible to determine the Curie temperature of the α phase in Fe-Ni alloys. The narrow single line at 843K refers to the paramagnetic fcc phase only.

4. Summary

The structural and magnetic study of $Fe_{1-x}Ni_x$ (x = 0.30, 0.325, 0.375) alloys showed that above 600K there exist two structural transitions: bcc (α) → fcc (γ) and fcc (γ) → fcc (γ′). The lattice constant of γ-$Fe_{1-x}Ni_x$ alloys exhibit non-anomalous behavior up to 700K. The structural transition α → γ changes the local environment of the Fe atoms as well as the magnetic properties of the alloys. In the γ phase Fe atoms with nine to twelve other Fe atoms as nn appear. These atoms have a zero magnetic moment and can destroy the magnetic order. In γ-$Fe_{1-x}Ni_x$ alloys with x = 0.30 and 0.325 the probability of finding such atoms is higher than 10% and hence these alloys are magnetically inhomogeneous. If this is smaller than 5% alloys are ferromagnetic with T_C higher than room temperature.

References

[1] Guillaume C E 1897 *CR Acad. Sci.* **125** 235
[2] Rancourt D G, Smit H A and Thiel R C 1987 *J. Mag. Magn. Mat.* **60** 121 and references therein
[3] Muller J B, Hesse J 1983 *Z. Phys. B – Condensed Matter* **54** 35
[4] Frąckowiek J 1990 *Hyperfine Interactions* **54** 793
[5] Hesse J and Rübartsch A 1974 *J. Phys. E* **7** 526
[6] Rancourt D G, Hargaves P and Lamarche G 1990 *J. Mag. Magn. Mat.* **87** 71
[7] Hesse J 1989 *Hyperfine Interactions* **47** 357
[8] Hamada N 1979 *J. Phys. Soc. Japan* **46** 1759
[9] Kądziołka-Gaweł M, Zarek W, Talik E and Popiel E 2008 *Acta Phys. Polonica A* **114** 1493

Interfaces in Sequence Permutated Multilayers

J Balogh[1], L Bujdosó, D Kaptás, L F Kiss, T Kemény and I Vincze

Research Institute for Solid State Physics and Optics, 1525 Budapest P.O.Box 49., Hungary

E-mail: baloghj@szfki.hu

Abstract. Sequence permutation of three building block multilayers was recently suggested as a new approach in studying bottom and top interfaces formed of a given layer with either of the other two elements. It was applied to Fe-B-Ag multilayers with 5 nm Ag layers separating the Fe and the B layers. Now we examine the dependence of the chemical mixing and the consequent amorphous phase formation on the nominal thickness of the Ag layers in [2 nm B / 2nm Fe / x nm Ag]$_4$, $0.2 \leq x \leq 10$, multilayers. The ratio of the non-alloyed Fe layer and the amorphous Fe-B interface compound changes only below x=5 nm. It is attributed to discontinuities of the Ag layer due to its three dimensional island growth over the bcc-Fe layer. The results obtained on the variation of the hyperfine field distribution of the amorhous Fe-B layers also confirm that the top interfaces of Fe with B are more B-rich than the bottom ones.

1. Introduction

Mössbauer spectroscopy (MS) being sensitive to the local neighborhood of the Mössbauer isotope is widely used to study nanoscale multilayers. The ^{57}Fe marker method, when the position of a thin ^{57}Fe layer is varied across the natural Fe layer, detected differences of the bottom and the top interface [1] in several multilayer structures. The application of the above method is limited by possible mixing and diffusion between the ^{57}Fe and the natural Fe layers. A new approach, sequence permutation of three building block multilayers [2], was suggested recently. In order to distinguish the bottom and top interfaces of the Fe layer a third element is interleaved, and the multilayers composed of three different elemental layers are deposited by sequence permutation in two different forms, as it is illustrated in figure 1. The Fe/B multilayer system was chosen [2] to be studied this way because a significant mixing of the elements accompanied by solid state amorphization occurs at the interfaces [3] already during the sample deposition and Ag was interleaved since it does not mix with either of the two elements: both Fe-Ag and B-Ag are non-mixing element pairs. The sequence permutated sample pairs can be prepared simultaneously in such a way that two of the building blocks (e.g. Fe and B, as indicated in figure 1) are evaporated together with the exception of a bottom and a top layer in the two samples, respectively. This way the accuracy of the interface analysis is greatly enhanced, since the experimental uncertainties of the layer thickness measurement can be excluded for the selected layers. Analysis of the results obtained by MS for sequence permutated sample pairs had shown [2] that the composition of the amorphous Fe/B and B/Fe interfaces are different, chemical mixing is larger at the interface where B is on top of the Fe layer.

[1]Corresponding author

The layer permutation, however, influenced the sample morphology to a large extent, as it was revealed by transmission electron microscopy investigations [2]. Large interface roughness and waviness of the layers was observed when Ag was grown on top of the B layer, but B was found to grow on Ag smoothly. The Fe-Ag and Fe-B interfaces did not show such a large difference. As a consequence of the above interface properties, the variance of the Ag layer thickness was larger for the Fe/B/Ag than for the B/Fe/Ag sequence. From comparison of Mössbauer measurements on trilayer and multilayer samples the following conclusions have been drawn. In case of the B/Fe/Ag sequence the 5 nm nominal thickness Ag layers applied could be supposed to act as barriers to the diffusion between the Fe and B layers separated by them, but in case of the Fe/B/Ag sequence the variance of the Ag layer thickness allowed the mixing between them to a measurable extent. In this paper we study in more details how the thickness of the Ag layers interleaved into Fe-B multilayers influences the chemical mixing of Fe and B. In view of the above described morphological differences the B/Fe/Ag sequence was chosen for the study.

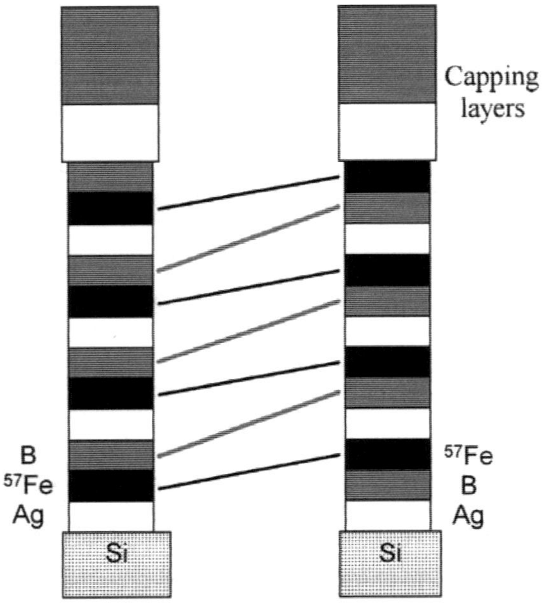

Figure 1. Schematic illustration of the sequence permutated sample pairs studied in Ref. 2. A possible scheme for the simultaneous preparation of two selected building blocks is also shown. All the Fe layers connected by black lines and all the B layers connected by gray lines can be deposited together.

2. Experimental

The samples were prepared by thermal evaporation in a high vacuum chamber with a base pressure of 10^{-7} Pa onto Si(111) wafers at room temperature. Ag and B were evaporated from Cu cold crucibles by electron guns, and ^{57}Fe was evaporated from a heated W boat. The layer thickness was measured by a quartz oscillator. The sample structures were similar to those investigated earlier [2] and illustrated in the right column of figure 1. The first layer was always 5 nm Ag and the subsequent 16 – 56 nm thick B/Fe/Ag multilayers were covered by 50 nm Ag and 100 nm B (cover layer in the followings) in order to prevent the samples from oxidation and to make them easily removable from the substrate. The layers removed were cut into smaller pieces and stacked to increase the measuring efficiency. The following samples were prepared:

Si / 5 nm Ag / [2 nm B / 2 nm ^{57}Fe / x nm Ag]$_4$ / cover layer, with x=0.2, 0.4, 1, 2, 4, 5 and 10.

The bracket and the subscript to it indicate the multilayer unit and the number of repetitions, respectively.

The nominal layer thickness measured by the quartz oscillator was checked by transmission electron microscopy investigations in case of the x=5 sample [2]. Due to the formation of the amorphous Fe-B alloy and the amorphous (nanocrystalline) nature of the non-alloyed B (Fe) layer, the results were not informative on the Fe and B layer thickness, but the presence of a continuous fcc-Ag layer with average thickness around 5 nm could be confirmed.

The transmission MS measurements were done with a 50 mCi ^{57}Co–Rh source in a standard constant acceleration spectrometer. The hyperfine field (HF) distributions were evaluated according to the Hesse–Rübartsch method [4].

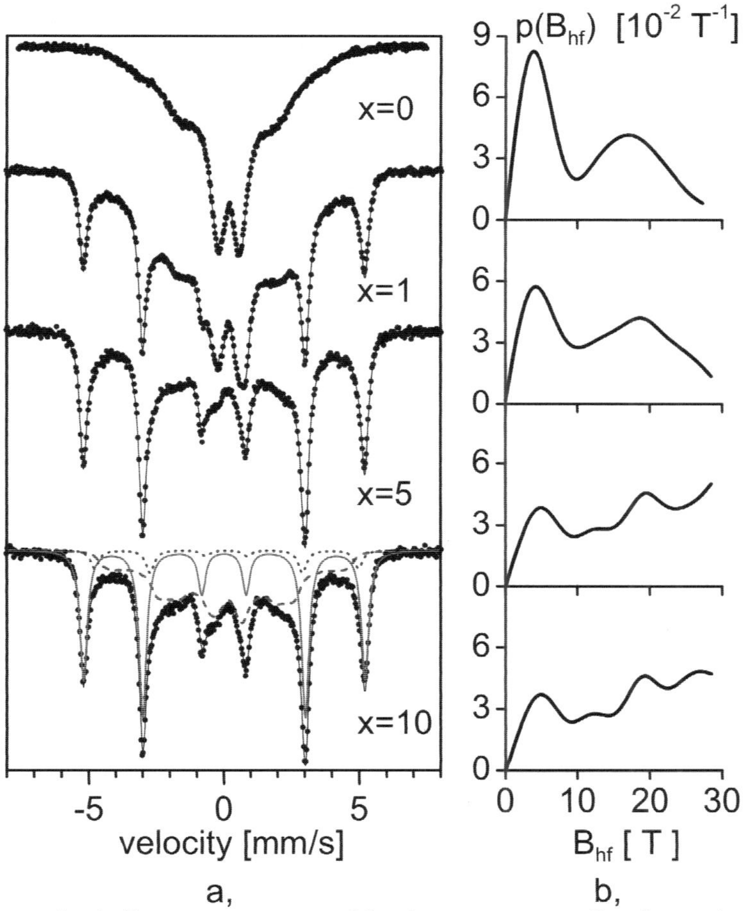

Figure 2. a) Room temperature Mössbauer spectra of a few selected samples of the [2 nm B / 2 nm ^{57}Fe / x nm Ag]$_4$ multilayer series. The spectra were fitted by two crystalline sextets and a distribution hyperfine fields, as indicated for the x=10 spectrum by full, dotted, and dashed lines, respectively. **b)** Hyperfine field distributions with normalized area evaluated from the measured spectra as described in the text.

3. Results and Discussion

Room temperature Mössbauer spectra of a few selected samples are shown in figure 2a. The spectrum belonging to x=0 is in accordance with the results obtained earlier for Fe-B multilayers [3]. The 2 nm Fe layers are fully intermixed with B and according to the fitted HF distribution, shown in Figure 2b, it forms an amorphous alloy with a broad concentration distribution. Since the layer thickness ratio of Fe and B belongs to the $Fe_{40}B_{60}$ average concentration, where the magnetic moment of amorphous Fe-B alloys disappears [5], the presence of magnetic components even at room temperature indicates that the 2 nm B layers are not fully mixed with Fe. The multilayer structure deduced from the MS results is depicted schematically in figure 3.

For x=5 and 10 the measured spectra agree within the experimental errors and can be described by two slightly broadened (Γ_1=0.326(4) and Γ_2=0.382(4) mm/s) sextets and a HF distribution. The two sextets (HF_1=32.25(1) and HF_2=30.24(10) T) can be attributed to the bulk and surface contributions of crystalline Fe layers in contact with Ag layers [6, 7], while the broad HF distribution belongs to the amorphous B/Fe interfaces [3]. According to the line intensities there are an equal number of Fe atoms that belong to the amorphous B/Fe interface and to the crystalline Fe layer. For 0<x<5 both the ratio of Fe atoms in the amorphous interface and the shape and average value of the HF distribution change continuously, while HF_1 and HF_2 remain practically constant. It is to be noted here that all the HF distributions were restricted to the range below 28.5 T in order to avoid strong correlations with the crystalline components. The variation of the spectral fraction and average HF value of the distribution is shown in figure 4. It is also worth noting that in the evaluations the close to 3:4:1:1:4:3 intensity ratio of the lines for both the crystalline and the amorphous components provided the best fit, indicating an in-plane magnetization of the samples.

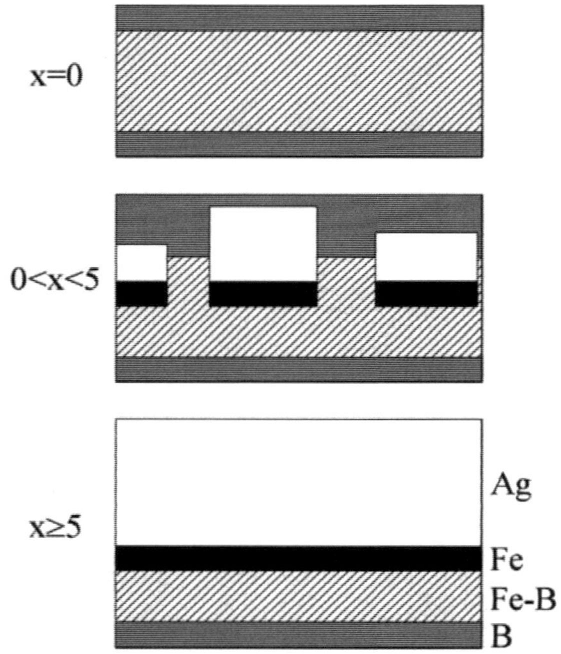

Figure 3. Schematic sample structures formed by interface mixing in case of different thickness of the Ag layer in the [2 nm B / 2 nm ^{57}Fe / x nm Ag]$_4$ multilayer series, as inferred from the Mössbauer measurements. In case of a continuous Ag layer (x≥5) mixing of Fe and B (dashed area) takes place only at the bottom side of the Fe layer. When there is no Ag layer interleaved (x=0), it occurs both at the bottom and the top sides and the 2 nm Fe layer is fully amorphized. For 0<x<5 Ag islands are formed with varying area and height and protect the top side of the underlying Fe area from amorphization.

The changes observed for varying Ag layer thickness can be understood if we suppose that below 5 nm nominal thickness the Ag layer is not continuous, as it is illustrated in figure 3. In case of a continuous Ag layer (x=5 and 10) mixing of Fe and B takes place only at the bottom side of the Fe layer. In the case, however, when there is no Ag layer interleaved (x=0), it occurs both at the bottom and the top sides and the bottom and top Fe-B interfaces overlap. In between these two extremes (0<x<5) Ag islands are formed with varying area and height and the area of the Ag covered Fe layer gradually increases with x. As a consequence the ratio of bcc-Fe increases and that of the amorphous interface decreases, as it can be seen in figure 4.

In parallel with the intensity change, the shape and the average HF value of the distribution also varies, as it can be seen in figures 2b and 4. These changes are in accordance with the result obtained from the sequence permutation studies [2], the average HF is larger when the ratio of the bottom

(B/Fe) interfaces is larger in the sample. The distributions cover a broad range of HFs for all the samples, the standard deviations vary between 7.4 and 8.3 T, and the increase of the average HF is mainly the result of a decrease in the respective ratio of the low field components. The increase of the average HF, accompanied also by a decrease of the average isomer shift, can be interpreted as a decrease of the average B concentration [5] of the amorphous Fe-B interface.

Of course, the validity of the above simple model is limited. In case of very small Ag islands the lateral diffusion of Fe and B should also be taken into account and non-equilibrium mixing of Ag with either Fe or B or with the amorphous Fe-B alloy might also take place. However, it is remarkable that both the fraction and the average HF of the amorphous phase keep varying in the small Ag thickness range, as well. Since the room temperature average values might also be influenced by the presence of superparamagnetic particles, low temperature and in-field measurements are under way to analyze in more details the role of the Ag atoms in the formation of the amorphous Fe-B interface. The possible effect of the Ag neighbors on the hyperfine fields of the different components is also to be studied. Our preliminary results indicate, however, that the changes observed in the room temperature data are indeed basically due to the variation of the amount and the composition of the amorphous phase.

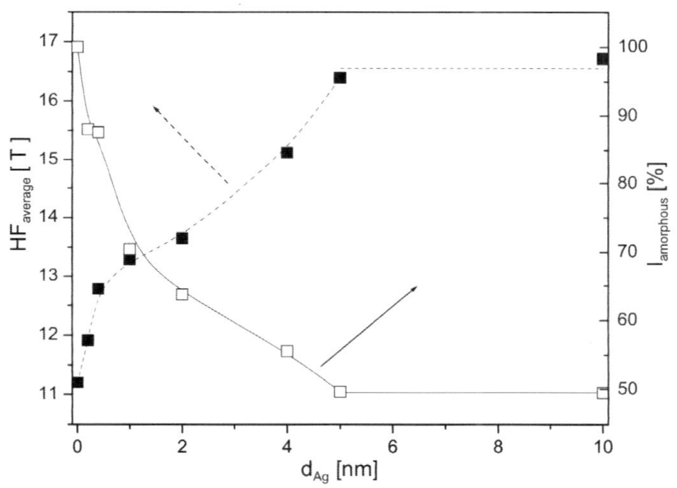

Figure 4. Spectral area (open symbols) and average HF (full symbols) of the fitted distribution as a function of the nominal Ag layer thickness. The lines are guide to the eye.

4. Conclusions

The hyperfine field distribution of the amorphous Fe-B interfaces are influenced by the Ag layer thickness in [2 nm B / 2nm Fe / x nm Ag]$_4$ multilayers. Since Ag does not mix with either Fe or B, the observed changes are attributed to the three dimensional island growth of Ag below 5 nm nominal thicknesses and the consequently different ratio of the B/Fe and Fe/B interfaces. The results also confirm the conclusion [2] that the bottom and the top interfaces of Fe are different in Fe-B multilayers, the top interfaces being richer in B than the bottom ones.

Acknowledgments

This work was supported by the Hungarian Scientific Research Fund OTKA K 68612.

References

[1] Shinjo T and Keune W 1999 *J. Magn. Magn. Mater.* **200** 598

[2] Balogh J, Bujdosó L, Kaptás D, Kemény T, Vincze I, Kovács A and Tóth L 2009 *J. Appl. Phys.* **105** 104303
[3] Balogh J, Bujdosó L, Kemény T, Pusztai T, Tóth L and Vincze I 1997 *Appl. Phys.* A: *Mater. Sci. Process.* **65** 23
[4] Hesse J and Rübartsch A 1974 *J. Phys.* E **7** 526
[5] Xiao G and Chien C L 1987 *Phys. Rev.* B **35** 8763
[6] Balogh J, Kaptás D, Vincze I, Temst K and Van Haesendonck C 2007 *Phys. Rev.* B **76** 052408
[7] Dézsi I, Fetzer Cs, Szűcs I, Degroote B, Vantomme A, Kobayashi T, Nakanishi A 2007 *Surface Science* **601** 2525

Magnetism of ultra-thin iron films seen by the nuclear resonant scattering of synchrotron radiation

T Ślęzak[1], K Freindl[2], A Kozioł-Rachwał[1], K Matlak[1], M Rennhofer[4], R Rüffer[3], B Sepioł[4], N Spiridis[2], S Stankov[3], M Ślęzak[1], D Wilgocka-Ślęzak[2], M Zając[1] and J Korecki[1,2]

[1] Faculty of Physics and Applied Computer Science, AGH University of Science and Technology, Al. Mickiewicza 30, 30-059 Kraków, Poland
[2] Institute of Catalysis and Surface Chemistry, Polish Academy of Sciences, ul. Niezapominajek 8, 30-239 Kraków, Poland
[3] European Synchrotron Radiation Facility, BP220, F-38043 Grenoble, France
[4] Fakultät für Physik, Universität Wien, Strudlhofgasse 4, A-1090 Wien, Austria

e-mail address: slezak@uci.agh.edu.pl

Abstract. Conversion electron Mössbauer spectroscopy proved in the past to be very useful in studying surface and ultrathin film magnetism with monolayer resolution. Twenty years later, its time-domain analogue, the nuclear resonant scattering (NRS) of synchrotron radiation, showed up to be by orders of magnitude faster and more efficient. The evolution of the spin structure in epitaxial ^{57}Fe films on a tungsten W(110) was studied via the accumulation of the NRS time spectra directly during Fe film deposition. In the 0.5 - 4 monolayers Fe thickness range, the complex non-collinear magnetic structure was derived from the NRS data, resulting from the deviation from the layer by layer growth mode. For thicker Fe films, the in-plane thickness induced spin reorientation transition could be clearly identified. Based on the NRS analysis it is shown that SRT process originates at the Fe/W(110) interface and proceeds through a transient fan-like magnetization structure.

1. Introduction

Magnetic properties of nanoscale materials are nowadays of great scientific and technological interest. Complex magnetic structures often found for low-dimensional systems, such as ultra-thin iron films, present a challenge for experimental methods. They have been studied by scanning probe microscopy (SPM) [1], spin sensitive electron scattering techniques [2], X-ray magnetic dichroism [3], magneto-optic Kerr effect (MOKE) [4], magnetometry [5], and ^{57}Fe Conversion Electron Mössbauer Spectroscopy (CEMS) [6]. Except the last one, all of these methods suffer from limited depth resolution: spin-polarized scanning tunneling is strictly surface sensitive, while the others integrate magnetic properties over the entire sample or over an element.In this respect, CEMS and, recently, the Grazing Incidence Nuclear Resonant Scattering (NRS) of synchrotron radiation are exceptional since, owing to their isotopic selectivity, they make it possible to study of magnetic properties with spatial resolution using isotopic probe-layers.

2. NRS technique

NRS is a synchrotron analogue of Mössbauer spectroscopy (MS), in the sense that recoilless excitation (induced by the resonant x-rays with energy 14.4 keV for ^{57}Fe) of the nuclear energy levels, split due to the hyperfine interactions, is involved. In this method, the hyperfine parameters can be obtained from a characteristic beat pattern seen in the time evolution of the intensity of nuclear resonant scattering (the so called time spectrum). In conventional MS methods, such as Conversion Electron Mössbauer Spectroscopy (CEMS), the hyperfine interactions are measured in an incoherent process, in which the decay of a nucleus resonantly excited by a γ-quant occurs via resonance fluorescence or internal conversion. The resulting spectrum is the incoherent sum of those single events. In contrast, the NRS signal, measured after the simultaneous excitation of an ensemble of nuclei by a synchrotron radiation pulse, results from the coherent superposition of the probability amplitude for scattering from all nuclei of the ensemble. This coherent character of the scattering process, in combination with the outstanding properties of the synchrotron radiation from third generation synchrotron sources (high brilliance, tunability in energy with monochromatization down to 0.5 meV, defined time structure, and polarization) opened new possibilities for time resolved nuclear spectroscopy based on hyperfine interactions. Magnetic properties in the nanoscale can be studied with sub-monolayer sensitivity. The isotopic sensitivity, in combination with the ^{57}Fe probe layer concept, gives local structural and magnetic information from a sample region pre-selected during the preparation process. Furthermore, the well defined polarization of the synchrotron x-rays allows for an enhanced (with respect to CEMS) sensitivity to the orientation of the hyperfine magnetic field and electric field gradient. Extensive description of the method and its application can be found in several review papers [7] and a comprehensive book [8].

NRS on thin films or surfaces is a nuclear diffraction technique performed under specular reflection geometry at grazing incidence (GI) as shown in Fig. 1a.

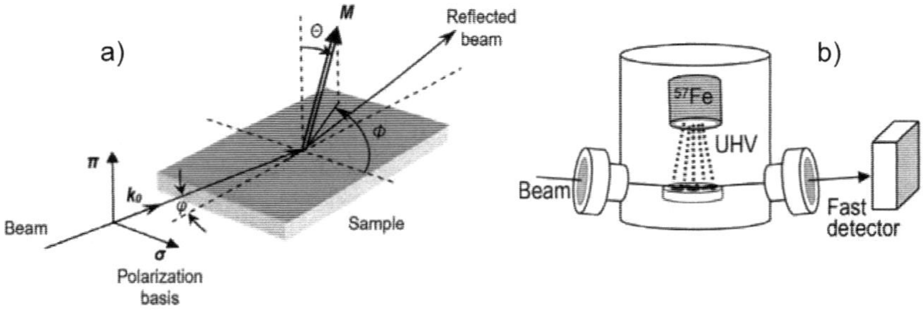

Figure 1. (a) Geometry of a grazing incidence nuclear resonance scattering experiment. Angles Θ and Φ give relative orientation of the incident synchrotron beam wave vector $\boldsymbol{k_0}$ to magnetization $\boldsymbol{M}$. $\boldsymbol{\sigma}$ and $\boldsymbol{\pi}$ are the linear polarization basis vectors. The grazing angle φ is typically a few milliradians (b) Schematics of the UHV NRS scattering chamber at ID 18 of the ESRF.

A newly constructed ultra high vacuum (UHV) system [9] at the beamline ID 18 at ESRF Grenoble [10] permits diverse GI-NRS experiments *in situ,* and also during or shortly after ^{57}Fe deposition. This multichamber UHV system ensures state-of-the-art preparation and characterization of single crystalline surfaces and epitaxial films, offering evaporation of several metals (including Fe isotopes). Unique features of the GI-NRS technique were for the first time demonstrated in the in-situ nuclear inelastic scattering [11] and GI-NRS experiments for epitaxial Fe nanostructures on a single crystalline tungsten substrate.

3. Results and Discussion

For the epitaxial Fe(110) film on W(110), there are two different thickness regimes where the most intriguing properties have been found: i) ultrathin Fe films with thickness in the range of 0.5 - 4 ML, where the onset of ferromagnetic behavior is expected [2], and ii) Fe films with thicknesses of a few

Journal of Physics: Conference Series **217** (2010) 012090 doi:10.1088/1742-6596/217/1/012090

tens of ML in the vicinity of the in plane spin reorientation transition (SRT) [12]. Using the UHV system at ID 18 these regimes could be studied with unprecedented spatial resolution.

As shown in Fig. 1b, ^{57}Fe could be deposited on a freshly cleaned and pre-aligned (to the X-ray beam) W(110) crystal. The first pseudomorphic monolayer was deposited at 600 K. The rest of the layer, beyond 1 ML, was deposited at room temperature, and all GI-NRS time measurements were also made at room temperature. Using a remote-operated shutter and a precision ^{57}Fe flux monitor, the whole preparation process could be operated on-line from the control hutch, without stopping the X-ray beam. Thus, the time of the experiment could be minimized, ensuring clean preparation and impurity-free films. It is also important that virgin magnetic states can be accessed, since, in contrast with most magnetic measurements, no magnetic field is necessary for the NRS. The measurement procedure that allows the acquisition of the GI-NRS time spectra along with the film deposition makes the method extremely fast. It was possible to complete the entire experimental run for the ^{57}Fe films in the thickness range 1.6 Å to 6.5 Å (step 0.4 Å) in 1.5 hours.

Figure 2 shows a selection of the fitted time spectra for $\boldsymbol{k_0}$ parallel to the [1 1 0] and [001] directions in the W(110) plane, as well as the evolution of the magnetic structure derived from the best

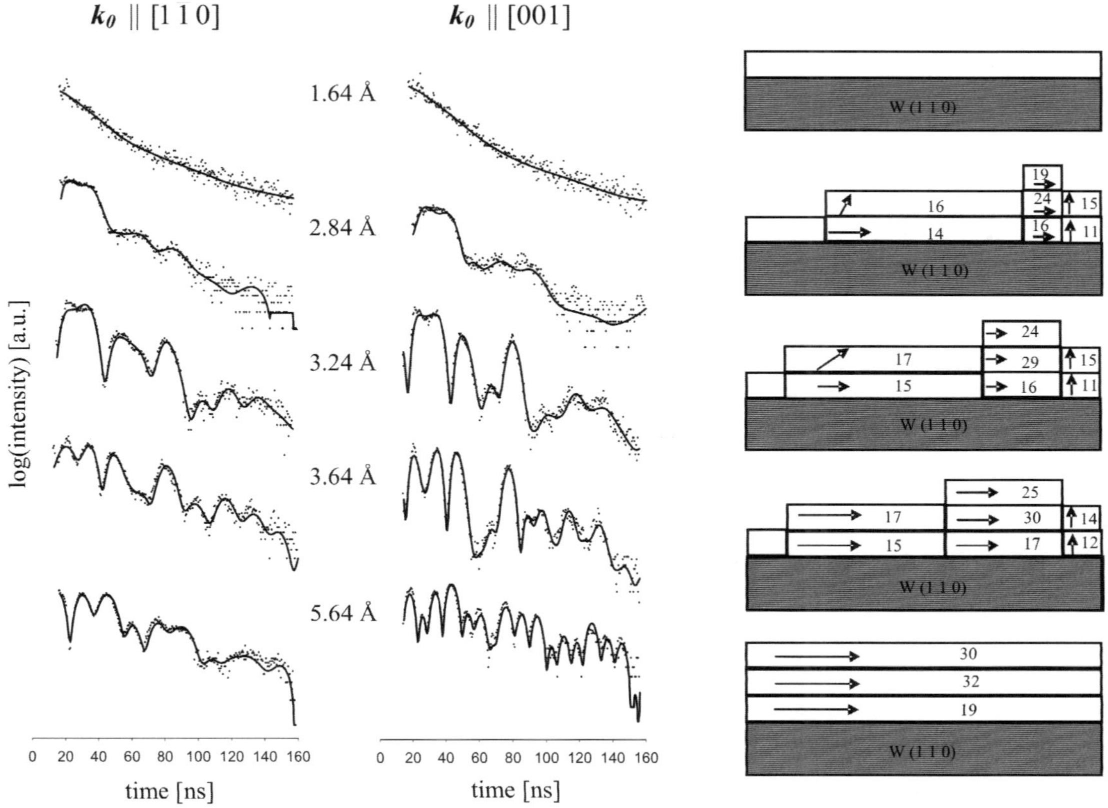

Figure 2. Fitted NRS time spectra of ultrathin ^{57}Fe films on W(110) for selected, nominal Fe coverage (in Å), for $\boldsymbol{k_0}$ parallel to [1 1 0] and [001] directions in the W(110) plane. The corresponding magnetic structures are shown. The arrows indicate directions of the hyperfine magnetic fields and numbers are their magnitudes in teslas.

fits obtained with the software package CONUSS [13]. For the given coverage, the fits for both the sample orientations were obtained assuming identical contributions of all components (represented by the areas of rectangles in the fig.2) as well as magnitudes of hyperfine parameters. The in-plane angles between hyperfine magnetic fields and $\boldsymbol{k_0}$ were fixed to 0 and 90 degree for $\boldsymbol{k_0}$ parallel to the $[1\bar{1}0]$ and [001] respectively, consistently with the $[1\bar{1}0]$ easy magnetization axis in this thickness range [2]. The spectrum of the 1.64 Å film [corresponding to 1.0 pseudomorphic Fe(110) monolayer (psML)] shows no quantum beat pattern, in agreement with the literature data that the monolayer Curie temperature is of about 240K [2]. A quantum beat structure that appears in the time spectra when the nominal Fe thickness reaches 2.84 Å (1.8 psML) is of a magnetic origin. For this coverage, formation of the magnetic double layer islands surrounded by the nonmagnetic monolayer areas is expected [1]. In fact, from the obtained spin structure more complicated film morphology could be deduced (see fig.2) yielding coexistence of the two types of the double layer patches (one with the homogenous out-of-plane magnetization and the second with non-collinear magnetic order) with an in-plane magnetized tri-layer areas. In addition, a nonmagnetic component corresponding to the uncoated monolayer had to be included. With increasing coverage the in-plane magnetic anisotropy becomes dominant as tri-layer areas expand. Finally, for coverage above 5Å (~3 psML), the out of plane magnetic anisotropy disappears and homogenous in-plane magnetization establishes as it can be concluded from the fit parameters and distinctly different spectra for the [001] and $[1\bar{1}0]$ directions. Our results reveal the complex character of this well known thickness induced SRT [2], from the out-of-plane to the in-plane magnetization direction. The non-collinear magnetic structure appears due to the exchange interaction between the mono-, double-, and tril-layer areas coexisting due to the deviation from the layer-by layer growth mode. The obtained profiles of the local magnetization across the film are unique and were not accessible with any other method.

As mentioned above, because of the reduced surface symmetry, the Fe(110) films provide an example where in-plane surface anisotropies occur in addition to the usual out-of plane anisotropies. These in plane anisotropies may induce an in plane SRT, as it was observed by Gradmann et al. for film thicknesses increasing 30 and 50 ML, where the magnetization switching from $[1\bar{1}0]$ to [001] was found [12]. The nature and mechanism of the transition is not fully explained. During the film growth, when approaching the critical film thickness, it can be considered either as continuous coherent magnetization rotation from $[1\bar{1}0]$ to [001] or as coexistence of the $[1\bar{1}0]$ and [001] oriented magnetic domains with different occupation. In-field measurements [14] or remanent measurements after saturating field pulses [15] point to the domain model. On the other hand, the early conversion electron Mössbauer study does not exclude the scenario of continuous rotation occurring without external field [12]. It is also plausible that the applied field could change the character of transition. The present GI-NRS experiment allowed us to get insight into the mechanism of the in plane SRT. In a single experimental run, multiple transition steps induced by the increase of the film thickness could be studied. Methodology of the thickness induced SRT studies was similar to that described above. ^{57}Fe was evaporated on the W(110) crystal held at 330 K with the rate of 0.3 ML/min to the final thickness of 30 ML. During the preparation, a set of NRS time spectra were collected (acquisition time per spectrum was only several seconds), thus probing the hyperfine parameters in thickness steps corresponding to the fraction of Fe monolayer. The fitted time spectra are shown in Fig.3 for the selected film thicknesses. A regular beat structure that is exemplified in Fig. 3a reflects, according to the theoretical fits, the uniform magnetization state with the easy axis along the $[1\bar{1}0]$ direction that persists up to the thickness $d \approx 50$ Å. Similarly, the spectra for the coverage above 56 Å (Fig.3f) can be simply fitted assuming homogeneous magnetization along [001]. It is clear that SRT extends over large thickness range $\delta \approx 6$ Å, corresponding to about 3ML Fe. The time spectra accumulated across the SRT process (Fig.3b-d) were unique and presented a challenge for the numerical analysis. Assuming a homogeneous magnetization depth profile across the Fe(110) films, the coherent magnetization and domain models, as described above were considered. Both models showed distinctly different spectra, however any combination of the magnetization configuration resulting from the models could give a satisfactory description of the NRS time spectra series. Moreover, various combinations of these models, such as for example simple coexistence of domains magnetized along [1 1 0], [001] and with magnetization continuously rotating with increasing thickness from [1 1 0] towards [001], were also excluded, as not reproducing the measured time spectra. Only when a depth-dependent magnetization structure was assumed, the perfect fits of the time spectra from Fig. 3 could be obtained (thin solid line).

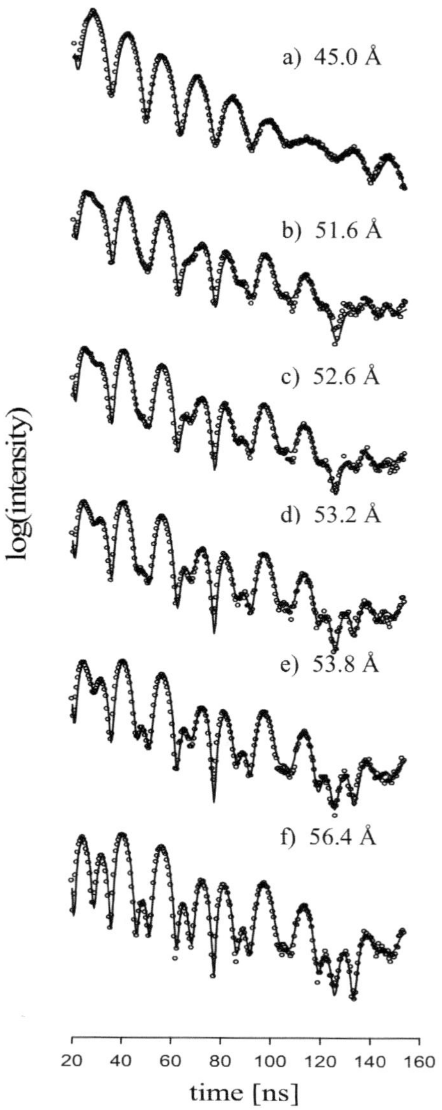

Figure 3. NRS time spectra measured with $\boldsymbol{k_0}$ parallel to $[1\bar{1}0]$ direction are shown for selected $^{57}Fe/W(110)$ films, demonstrating in-plane SRT.

To implement the depth variation of the magnetization direction, for each nominal coverage d_{nom} the iron film was divided into eight equivalent sub-layers, with the same thickness $d=d_1=d_2=...=d_8=d_{nom}/8$. Each sub-layer was characterized by the identical magnetic hyperfine field B_{HF} = 33T, while its in-plane orientation was defined by the angles φ_1, φ_1,...,φ_8 with respect to the [1 1 0] direction. In this way, the number of free parameters could be limited to the absolute minimum and, with increasing thickness, the subsequent spectra could be fitted assuming only the corresponding increase of d_{nom} and adjusting the set of the φ_1, φ_1,...,φ_8 values. From the best fits the evolution of the angular distribution of the magnetization along the film normal could be determined. We found that the spin structure evolves with increasing thickness in the smooth and continuous manner and that the SRT from $[1\bar{1}0]$ to [001] is initiated by the deepest layers (neighboring to the tungsten substrate). In the transition a non-collinear magnetization structure appears. With increasing thickness, the magnetization of the subsequent layers rotates, and finally the transition is completed by the top-most surface layers.

4. Conclusions

In conclusions, the magnetic and structural properties of epitaxial Fe films were studied with *in situ* NRS technique. In the low Fe thickness regime NRS provided unique information about the correlation between the spin structure evolving with the increasing thickness and film morphology resulting from the peculiar growth mode. For thicker films, details of the in-plane spin reorientation transition could be accessed revealing existence of the non-collinear magnetic structure at the thickness driven SRT. Combination of the NRS technique and standard UHV methodology resulted in the high perfection and surface cleanness of the investigated films eliminating unwanted effects of the residual gas adsorption. Finally, the described experiments were performed in the absence of the external magnetic field, which can strongly alter the intrinsic magnetism at the spin reorientation transition process.

5. Acknowledgements

This work was supported by the European Community under the Specific Targeted Research Project Contract No. NMP4-CT-2003-001516 (DYNASYNC), and in part by the Polish Ministry of Science and Higher Education and by the Team Program of the Foundation for Polish Science co-financed by the EU European Regional Development Fund.

6. References

[1] Pietzsch O, Kubetzka A, Bode M, Wiesendanger R 2000 *Phys. Rev. Lett.* **84** 5212.
[2] Elmers H. J., Hauschild J., Hoche H., Gradmann U., Bethge H., Heuer D., Kohler U., *Phys. Rev. Lett.* **73** (1994), 898.
[3] Lu L., Bansmann J., Meiwes-Broer K. H. *J. Phys. Cond. Matter* **10** (1998), 2873
[4] Ślęzak T., Karaś W., Krop K., Kubik M., Wilgocka–Ślęzak D., Spiridis N., Korecki J., *J.Magn. Magn. Mat.* **240** (2002), 362
[5] Elmers H. J., Hauschild J., Gradmann U., *Phys. Rev. B* **59** (1999), 3688
[6] Przybylski M., Korecki J., Gradmann U., *Appl. Phys. A* **52** (1991), 33
[7] Gerdau E., and de Waard H. (eds), *Hyperfine Interact.* **123/124** (1999), Baltzer Science Publishers; Rüffer R., *Hyperfine Interact.* **141/142** (2002), 83
[8] Röhlsberger R., *Nuclear Condensed Matter Physics with Synchrotron Radiation,* STMP 208, Springer-Verlag, Berlin, Heidelberg 2004.
[9] Stankov S., Rueffer R., Sladecek M., Rennhofer M., Sepiol B., Vogl B., Spiridis N., Ślęzak T., Korecki J. *Review of Scientific Instruments,* **79** (2008), 045108
[10] Rüffer R., Chumakov A. I., *Hyperfine Interact.* **97-98** (1996), 589 http://www.esrf.fr/exp_facilities/ID18/
[11] Stankov S., Röhlsberger R., Ślęzak T., Sladecek M., Sepioł B., Vogl G., Chumakov A.I., Rüffer R., Spiridis N., Łażewski J., Parliński K., Korecki J., *Phys.l Rev. Lett.* **99**, (2007) 185501
[12] Gradmann U., Korecki J., Waller C., *Appl. Phys. A* **39** (1986), 101
[13] Sturhahn W., *Hyperfine Interact.* **125** (2000), 149
[14] Elmers H. J. and Gradmann U., *Appl. Phys. A* **51** (1990), 255
[15] Baek I.-G., Lee H. G., Kim H.-J., Vescovo E., *Phys. Rev. B* **67** (2003) 075401

Nanoparticulated magnetic drug delivery systems: Preparation and magnetic characterization

P C Morais
Universidade de Brasília, Instituto de Física, Núcleo de Física Aplicada, Brasília DF 70910-900, Brazil

E-mail: pcmor@unb.br

Abstract. This paper describes how magnetic resonance can be successfully used as a tool to help customize and quantify nanosized magnetic particles while labeling cells and administered in animals for targeting different biological sites. Customization of magnetic nanoparticles is addressed here in terms of production of complex magnetic drug delivery systems whereas quantification of magnetic nanoparticle in different biological compartments emerges as a key experimental information to assess time-dependent magnetic nanoparticle biodistribution profiles. Examples of using magnetic resonance in unfolding information regarding the pharmacokinetics of intravenously-injected surface-functionalized magnetic nanoparticles in animals are included in the paper.

1. Introduction

The successful development of a nanosized magnetic drug delivery system (MDDS) requires full assess to different information regarding the material's properties and both *in vitro* and *in vivo* tests. Ultimately, the pre-clinical tests and clinical trials of a new MDDS is nowadays expected to be supported by the very knowledge of the computer-simulation of the interaction between the drug itself and the template material that carries, site-target, and delivers the drug. In between these two ends we found the synthesis and the physical characterization of the drug loaded and drug unloaded MDDS (l-MDDS and u-MDDS). Most of the currently developing MDDS based on the platform offered by precipitated and colloidal-suspended magnetic nanoparticles include iron-oxides, particularly cubic ferrites. Within this scenario magnetic resonance (MR) may play a key role in help in characterizing and understanding the very properties of the l-MDDS and u-MDDS. In addition to the use of magnetic resonance (MR) for the characterization of magnetic nanoparticulate-based materials the MR technique can also be successfully employed as an analytical tool to track these materials while in contact with biological systems, including time-dependent biodistribution of magnetic nanoparticulate-based materials incubated with cells or intravenously-injected in animals. This paper is focused on the description of the use of MR to assess important information regarding magnetic nanoparticulated materials while used to carry on *in vitro* and *in vivo* tests. The use of magnetic colloids as a material platform to produce l-MDDS and u-MDDS will be briefly described as well. Indeed, besides MR [1-3] other techniques have been used to characterize magnetic colloids, magnetic nanocomposites and MDDS, providing assess to morphological/crystalline aspects [4-9], magnetic properties [10, 11], and magneto-optical responses to DC/AC magnetic fields [12-15]. Applications of the l-MDDS discussed in this paper include cell-labeling [16-18], photodynamic therapy (PDT) [19-22], and magnetohyperthermia (MHT) of cancer cells and tissues [23-25].

2. The material platform provided by nanosized magnetic particles

Nanosized magnetic particles can be conveniently suspended as a very special colloid in which the suspended nanoparticle posseses a permanent magnetic moment. This special colloid, usually named magnetic fluid (MF), uses aqueous or organic carrier fluids, therefore providing an exceptional material platform to encapsulate the previously-synthesized magnetic nanoparticle in any hosting template in a very controllable way. Stable MF samples containing as much as 15% in particle volume fraction have been produced and shelved for long periods of time. Besides the Brownian motion electrostatic and steric repulsions are the main mechanisms supporting the MF colloidal stability. Electrostatic interaction is the dominant mechanism in ionic MFs whereas steric repulsion is the dominant mechanism supporting the MF colloidal stability in organic-based MFs. In more complex systems, such as in the biocompatible magnetic fluids (BMF), both electrostatic and steric repulsion mechanisms take place simultaneously, providing very stable magnetic colloids even at physiological conditions. Nevertheless, biocompatibility of magnetic fluids is assigned only after performing and evaluating in vitro and in vivo tests while is often considered a dose dependent issue.

The most extensively used synthesis route for MF samples described in the literature starts with the co-precipitation of the metal ions from their aqueous salt solution using weak or strong base addition under controlled chemical species (hydroxyl and metal-ions) concentration, stirring speed, and temperature. Peptization of the precipitated magnetic nanoparticle as ionic MF can be accomplished in low-pH as well as in high-pH values [26, 27]. Fresh precipitate containing magnetic nanoparticle can be suspended in both non-polar [28, 29] and polar-based media [30, 31]. Starting with MF samples magnetic nanosized particles can be encapsulated as anionic of cationic magnetoliposome (ML), trapped in magnetic nanoemulsion (MNE) or hosted in magnetic nanocapsule (MNC). The ex-situ incorporation of magnetic nanosized particles using the MF technological platform provides full control for nanoparticle concentration with excellent uniformity. MDDS consisting of biocompatible magnetic fluids (BMFs), magnetoliposomes (MLs), magnetic nanoemulsions (MNEs) and magnetic nanocapsules (MNCs) are presently under production and under biological test worldwide [19-23, 32-39].

Interesting aspects regarding MNE formulations are the enhanced drug solubilization, good thermodynamic stability and easily-achieved scaling up for mass production [40]. Whereas surface passivation of the suspended nanoparticle needs to be properly addressed to prevent chemical degradation and aging effect PDT and MHT may promote phase changes of the magnetic component in MDDS. In order to investigate the phase stability of the material and identify the thermally- and optically-induced production of new magnetic phases different spectroscopic techniques have been employed. Magnetic resonance, for instance, provides very useful information regarding magnetic nanoparticle core and surface engineering and particle-particle interaction. In addition, due to the strong MR signal provided by nanosized magnetic particles the technique can be used as an excellent spectroscopic tool to probe biodistribution of MDDS while introduced in animals or used to label cells.

Figure 1 presents a typical particle size (core diameter) histogram plot obtained from the analysis of transmission electron microscopy (TEM) micrographs. The vertical bars in Fig. 1 represent the experimental data whereas the solid line represents the best curve-fitting using the log-normal distribution function. Average particle diameter (D) and diameter dispersion (σ) obtained from the data shown in Fig. 1 was 5.6 nm and 0.22, respectively. The sample whose data are presented in Fig. 1 consists of spherical nanosized magnetite surface-coated with DMSA (dimercaptosuccinic acid). While suspended in physiological medium the DMSA-coated magnetite nanoparticle can be used to target cells (*in vitro* tests) and organs (*in vivo* tests). The next section will describe how MR can assist the *in vivo* and *in vitro* tests using surface-coated oxide-based magnetic nanoparticles suspended in physiological medium.

Journal of Physics: Conference Series **217** (2010) 012091 doi:10.1088/1742-6596/217/1/012091

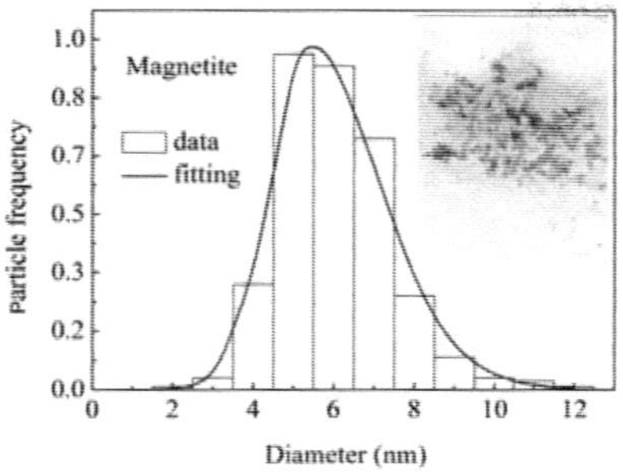

Figure 1. Typical particle size distribution of magnetite nanoparticles surface coated with DMSA. The inset is a representative TEM micrograph of the sample.

3. Magnetic resonance in cell labeling and organ targeting

Evaluation of the magnetic nanoparticle content in a given biological site is easily obtained via MR measurements as long as a calibration curve is provided. Figure 2 shows typical MR spectra recorded from two BMF samples, each one presenting distinct nanoparticle contents (2.5×10^{13} particle/mL and 5.0×10^{16} particle/mL). Resonance spectra recorded in a given range of nanoparticle concentration are used to build the calibration curve, as shown in Fig. 3. The vertical axis in Fig. 3 represents the area under the MR spectrum (absorption curve) whereas the horizontal axis represents the magnetic nanoparticle concentration (particle/mL). The samples whose MR spectra are shown in Fig. 2 are based on magnetite nanoparticles (9.4 nm in average diameter) surface-coated with Dextran and DMSA. Biocompatible magnetic fluid samples containing magnetite nanoparticle surface-coated with Dextran and DMSA were labeled DEX-MAG and DMSA-MAG, respectively. The room temperature MR data, as shown in Fig. 2, were recorded using a commercial spectrometer (Bruker ESP-300) tuned around 9.42 GHz.

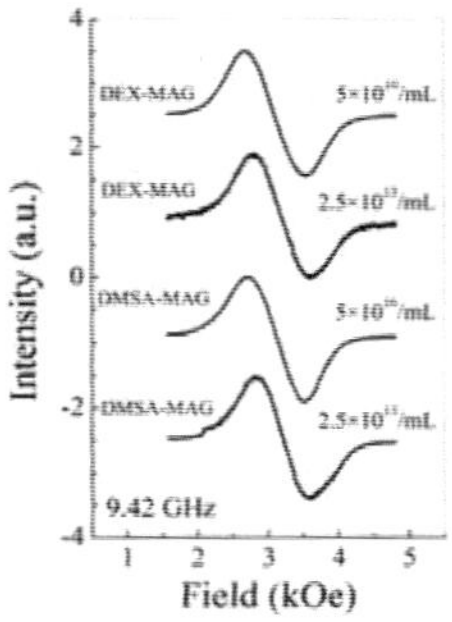

Figure 2. Typical magnetic resonance spectra of nanosized magnetite particles coated with Dextran and DMSA and suspended as biocompatible magnetic fluid samples (DEX-MAG and DMSA-MAG), at different particle concentration.

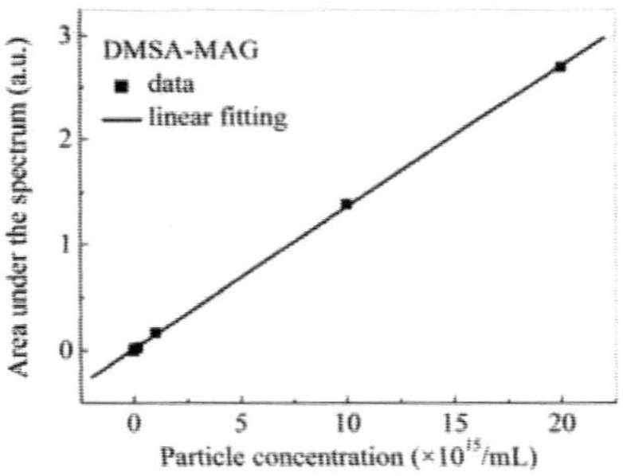

Figure 3. Typical magnetic resonance calibration curve. The sample used was the biocompatible DMSA-MAG, whose spectra were presented in Fig. 2.

Magnetic resonance was used to investigate the effects of incubating mouse Raw cells with DEX-MAG, with emphasis in finding the efficiency of magnetic nanoparticle (MNP) internalization by the cells. Symbols in Fig. 4 represent the time (t) dependence of the DEX-MAG concentration (C) in Raw cells [18]. The DEX-MAG concentration in the Raw cells versus time is described by $C(t) = C_o[1-\exp(-k_c t)]$, where $C_o = 53\times10^{14}$ particle/mL and $k_c = 0.23$ min^{-1}. The solid straight line in Fig. 4 (semi-log plot) represents the best curve-fitting of the experimental data (solid triangles) according to the previously described exponential behaviour. Above 12 hours we found the experimental data falling well below the straight solid line in Fig. 4, indicating a systematic reduction of the DEX-MAG concentration, probably due to cell metabolization, as suggested by independent analysis using light microscopy analysis. Indeed, the half-life associated to the MNP internalization by the cells is given by $t_{1/2} = 0.69/k_c = 3$ s.

Likewise, MR was used to investigate the time dependence of the MNP concentration in the blood circulation and in several organs following an intravenous bolus dose administration of sample DEX-MAG to female Swiss mice. Animals between 8 and 10 weeks of age, were used in this study. In order to perform the MR experiments 100 μL of the DEX-MAG sample containing about 10^{17} particle/mL were used. After injection the animals were sacrificed and the samples collected. Before the MR measurements were carried out the animals were submitted to a full perfusion procedure. One hour after the injection MNP were found mainly in liver and spleen, while small amounts of magnetic material were observed elsewhere. Blood samples from the control were mixed with the DEX-MAG sample in order to obtain the calibration curve. Open squares in Fig. 5 represent the time-decay (disposition) of the dextran-coated magnetite nanoparticle concentration in blood [41]. The data indicate that the bolus dose reaches the systemic circulation where, in a short period of time, it is homogenously distributed. The solid straight line going through the open squares in Fig. 5 represents the best fit of the MR data using one-exponential decay, thus signalling the presence of a single biological compartment [42]. Therefore, The DEX-MAG concentration in the blood versus time is described by $C(t) = C_o^B\exp(-k_B t)$, where C_o^B and k_B are the MNP concentration at $t = 0$ and the rate constant for elimination of the MNP from blood, respectively. The best fit of the MR data recorded from the blood samples using the equation presented above provides $C_o^B = 110\times10^{14}$ particle/mL and $k_B = 0.10$ min^{-1}. The half-life associated to the MNP disposition from the compartment is given by $t_{1/2} = 0.69/k_c = 6.9$ min. In addition, data in Fig. 5 allow determination of mouse blood clearance as $\mathrm{CL} = \mathrm{dose}/\int C(t)dt = \mathrm{dose}\times k_B/C_o^B = 90$ μL/min [43]. Finally, the volume of distribution (V) in the mouse systemic circulation was $V = \mathrm{CL}/k_B = 900$ μL. Figure 5 describes how fast MNP is transferred from the bloodstream compartment to the liver and spleen (LS) compartment. Open circles in Fig. 5 represent the time-dependence (absorption) of the dextran-coated magnetite nanoparticle concentration in liver and spleen. The time dependence of the MNP concentration in the organs (liver and spleen) is described by $C(t) = C_o^{LS}[1-\exp(-k_{LS}t)]$, where $C_o^{LS} = 35\times10^{14}$ particle/mL and $k_{LS} = 0.05$ min^{-1}. This finding allows us to estimate the effective half-life associated to the MNP absorption by the two compartments (liver plus spleen) as $t_{1/2} = 0.69/k_c = 13.8$ min. Mass-balance involving the two compartments (B and LS) indicates that about 8×10^{15} nanoparticles were found in liver and spleen 60 min after injection, while in the same time window about 16×10^{15} particles left the bloodstream. Blood, liver, and spleen volumes were quoted as 1.5, 2.3, and 0.2 mL, respectively. We argue that about 8×10^{15} particles were captured by the blood mononuclear phagocyte cells (neutrophyls and monocytes) and retained in the liver and spleen. The blood mononuclear phagocyte cells containing nanoparticles is certainly removed out by the perfusion procedure before the MR data were recorded using the liver and spleen samples. Based on the mass-balance the dashed line in Fig. 5 (with $k_{LS} = 0.10$ min^{-1}) would be the expected curve describing the time dependence of the MNP concentration in liver plus spleen if the organs were not submitted to the perfusion procedure. Consequently, due to the perfusion procedure, the open circles and the associated solid straight line (see Fig. 5), both representing the time-dependence of the dextran-coated magnetite nanoparticle concentration in liver and spleen, lie just below the dashed line. Therefore, semi-log plots as presented in Fig. 5 allow direct observation of the mass-balance scenario. Note in addition that all straight lines in Fig. 5 extrapolate to the unit as the time tends to zero.

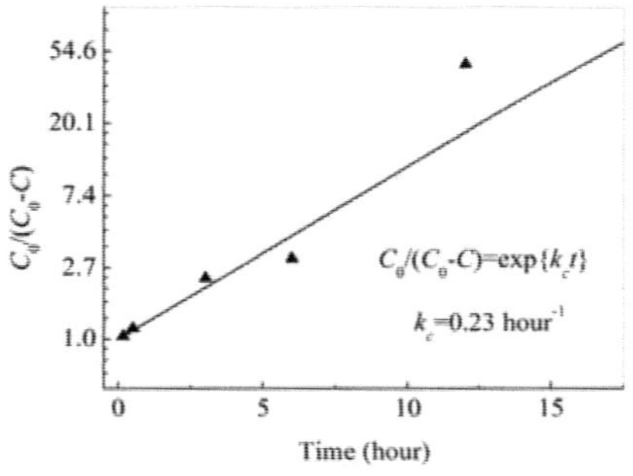

Figure 4. Time dependence of the magnetic nanoparticle concentration. The data were obtained from the MR spectra recorded from mouse Raw cells after incubation with DEX-MAG [18]. Note the natural logarithm vertical scale.

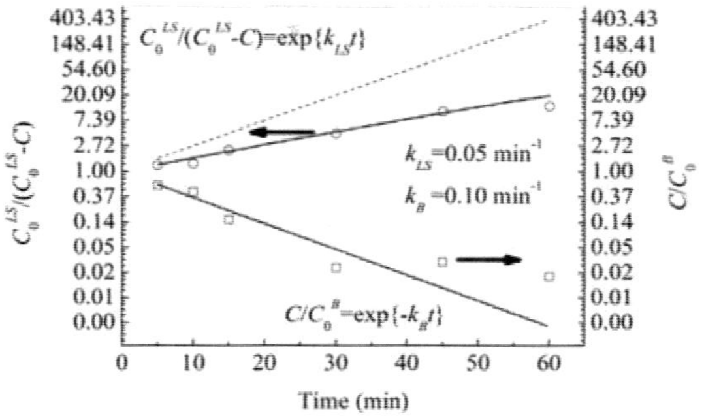

Figure 5. Time dependence of the magnetic nanoparticle concentration. The data were obtained from the MR spectra recorded from blood (open squares) and liver plus spleen (open circles) after intravenous injection of DEX-MAG as a bolus dose in female Swiss mice [41].

Acknowledgments
Data provided by colleagues and students from the Brazilian Network on Nanobiomagnetism (MCT/CNPq) are acknowledged. Support from FAP-DF is also acknowledged.

References

[1] Morais P C, Santos J G, Skeff Neto K, Pelegrini F, De Cuyper M 2005: *J. Magn. Magn. Mater.* **293** 526
[2] Pereira A R, Miranda K L C, Sartoratto P P C, Morais P C, Bakuzis A F 2006 *J. Appl. Phys.* **100** 086110
[3] Figueiredo L C, Lacava B M, Skeff Neto K, Pelegrini F, Morais P C 2008 *J. Magn. Magn. Mater* **320** E347
[4] Sjogren C E, Johansson C, Naevestad A, Sontum P C, Briley-Saebo K, Fahlvik A K 1997 *Magn. Res. Imaging* **15** 55
[5] Lacava B M, Azevedo R B, Silva L P, Lacava Z G M, Skeff Neto K, Buske N, Bakuzis A F, Morais P C 2000 *Appl. Phys. Lett.* **77** 1876
[6] Morais P C, Azevedo R B, Silva L P, Rabelo D, Lima E C D 2001 *Phys. Stat. Sol.* (a) **187** 203
[7] Kim D K, Mikhaylova M, Zhang Y, Muhammed M 2003 *Chem. Mater.* **15** 1617
[8] Mikhaylova M, Kim D K, Bobrysheva N, Osmolowsky M, Semenov V, Tsakalados T, Muhammed M 2004 *Langmuir* **20** 2472
[9] Klokkenburg M, Erne B H, Meeldijk J D, Wiedenmann A, Petukhov A V, Dullens R P A, Philipse A P 2006 *Phys. Rev. Lett.* **97** 185702
[10] Allen P D, St Pierre T G, Chua-anusorn W, Strom V, Rao KV (2000) *Biochim. Biophys. Acta* **1500** 186
[11] Morais P C, Lacava B M, Bakuzis A F, Lacava LM, Silva L P, Azevedo R B, Lacava Z G M, Buske N, Nunes W C, Novak M A 2001 *J. Magn. Magn. Mater.* **226** 1899
[12] Tartaj P, Gonzalez-Carreno T, Bomati-Miguel O, Serna, C J, Bonville P 2004 *Phys. Rev.* **B 69** 94401

[13] Xu M, Ridler P J 1997 *J. Appl. Phys.* **82** 326
[14] Morais PC, Gravina P P, Skeff Neto K, Lacava B M 2005 *Thin Sol. Films* **515** 226
[15] Oliveira F M L, Avelino S R, Eloi M T A, Gravina P P, Skeff Neto K, Lima E C D, Morais P C 006 *J. Non-Cryst. Sol.* **352** 3689
[16] Azevedo R B, Silva L P, Lemos A P C, Bao S N, Lacava Z G M, Safarik I, Safarikova M, Morais P C 2003 *IEEE Trans. Magn.* **39** 2660
[17] Godoy Morais J P, Azevedo R B, Silva L P, Lacava Z G M, Bao S N, Silva O, Pelegrini F, Gansau C, Buske N, Safarik I, Safarikova M, Morais P C 2004 *J. Magn. Magn. Mater.* **272/276** 2400
[18] Lacava Z G M, Lacava L M, Fonseca M J P, Souza T M M, Pereira L O, Silva O, Pelegrini F, Sabolovic D, Sestier C, Azevedo R B, Buske N, Morais P C 2006 *IEEE Trans. Magn.* **42** 3599
[19] Macaroff P P, Oliveira D M, Lacava Z G M, Lima E C D, Morais P C, Tedesco A C 2005 *J. Appl. Phys.* **97** 10Q906
[20] Macaroff P P, Oliveira D M, Ribeiro K F, Tedesco A C, Lacava Z G M, Lima E C D, Morais P C 2005 *IEEE Trans. Magn.* **41** 4105
[21] Oliveira D M, Lacava Z G M, Lima E C D, Morais P C, Tedesco A C 2006 *J. Nanosci. Nanotechnol.* **6** 2432
[22] Primo F L, Michieleto L, Rodrigues M M A, Macaroff P P, Morais P C, Lacava Z G M, Bentley M V L B, Tedesco A C 2007 *J. Magn. Magn. Mater.* **311**, 354
[23] Calderwood S K, Asea A 2002 *Int. Hyperthermia* **18** 597
[24] Guedes M H A, Guedes M E A, Morais P C, da Silva M F, Santos T S, Alves Jr J P, Bertelli C E, Azevedo R B, Lacava Z G M 2004 *J. Magn. Magn. Mater.* **272/276** 2406
[25] Simioni A R, Primo F L, Rodrigues M A, Lacava Z G M, Morais P C, Tedesco A C 2007 *IEEE Trans. Magn.* **43** 2459
[26] Massart R 1981 *IEEE Trans. Magn.* **17** 1247
[27] Morais P C, Garg VK, Oliveira AC, Silva LP, Azevedo RB, Silva A M L, Lima E C D 2001 *J. Magn. Magn. Mater.* **225** 37
[28] Muller R, Hiergeist R, Steinmetz H, Ayoub N, Fujisaki M, Schuppel W 1999 *J. Magn. Magn. Mater.* **201** 34
[29] Goossens A, de Jongh L J, Butter K, Philipse A P, Craje M W J, van der Kraan A M 2002 *Hyperfine Inter.* **141** 381
[30] Gotze T, Gansau C, Buske N, Roeder M, Görnert P, Bahr M 2002 *J. Magn. Magn. Mater.* **252** 399
[31] Morais P C, Santos R L, Pimenta A C M, Azevedo R B, Lima E C D 2006 *Thin Sol. Films* **515** 266
[32] Rocha F M, Pinho S C, Zollner R L, Santana M H A 2001 *J. Magn. Magn. Mater.* **225** 101
[33] De Cuyper M, Valtonen S 2001 *J. Magn. Magn. Mater.* **225** 89
[34] Hodenius M, De Cuyper M, Desender L, Muller-Schulte D, Steigel A, Lueken H 2002 *Chem. Phys. Lipids* **120** 75
[35] De Cuyper M, Hodenius M, Lacava Z G M, Azevedo R B, da Silva M F, Morais P C, Santana M H A 2002 *J. Colloid Interface Sci.* **245** 274
[36] Chen K Z, Luo W L, Rahman Z U, Guo Y, Schulte A 2007 *J. Dipersion Sci. Technol.* **28** 471
[37] Yu C H, Tam K Y, Lo C C H, Tsang S C 2007 *IEEE Trans. Magn.* **43** 2436
[38] Masuda M, Maeda K, Kobayashi T, Shiomi S, Fujiwara Y, Saito Y 2000 *Jpn. J. Appl. Phys.* **39** L733
[39] Chen C P, Chang T H, Wang T F 2002 *Ceramics International* **28** 925
[40] Baroli B, López-Quintela M A, Delgado-Charro M B, Fadda A M, Blanco-Mendez J J 2000 *Control. Release* **69** 209
[41] Lacava L M, Lacava Z G M, Azevedo R B, Chaves S B, Garcia V A P, Silva O, Pelegrini F, Buske N, Gansau C, Da Silva M F, Morais PC 2002 *J. Magn. Magn. Mater.* **252** 367
[42] Nony, P, Cucherat, M, Boissel, J-P 1998 *Trends Pharmacol. Sci.* **19** 49

[43] Wilkinson G R 1996 Pharmacokinetics: In: Hardman J G, Limbird L E, Molinoff P B, Ruddon R W, Gilman A G, editors. Goodman and Giman's The Pharmacological Basis of Therapeutics. 9th ed. New York: McGraw-Hill; p. 3-29

Hyperfine Interactions and Crystallization Kinetics of Co-Substituted NANOPERM-Type Alloys

Marcel Miglierini[1,2]

[1]Slovak University of Technology, Bratislava, Slovakia
[2]Centre for Nanomaterials Research, Palacky University, Olomouc, Czech Republic

E-mail: marcel.miglierini@stuba.sk

Abstract. Evolution of structural changes after (as well as during) heat treatment and their consequence upon hyperfine interactions is discussed for $(Fe_{1-x}Co_x)_{79}Mo_8Cu_1B_{12}$ and $(Fe_{1-x}Co_x)_{76}Mo_8Cu_1B_{15}$ ($0 \leq x \leq 0.75$) alloys which eventually exhibit nanocrystalline arrangement. Along with Mössbauer spectrometry, results obtained from AFM, and MFM are reported. The kinetics of crystallization is characterized by *in-situ* diffraction of synchrotron radiation. Substitution of Fe by Co has a remarkable effect on the average hyperfine field and average isomer shift values in the as-quenched alloys. This indicates structural modifications which lead to increase of Curie temperature with higher Co contents. In nanocrystalline state, presence of bcc FeCo crystalline phase is identified. The progress of crystallization is shown to depend upon the annealing conditions as well as on the Co contents. Differences in the onset of crystallization on both sides of the ribbon-shaped samples are revealed. They correlate with the degree of quenched-in surface crystallization.

1. Introduction

NANOPERM nanocrystalline alloys attract a lot of attention mainly due to their excellent soft magnetic properties provided by the interplay between the reduced dimensionality of the nanograins and the large fraction of grain boundaries [1]. Even though still some fundamental properties are not well understood presently these materials find increasing industrial applications [2]. Partial substitution of Fe by Co in NANOPERM alloys has introduced a new class of these materials, the so-called HITPERM alloys [3].

Investigations of Co-substituted NANOPERM-type alloys can be effectively performed namely by the help of Mössbauer spectrometry. In nanocrystalline $(Fe_{1-x}Co_x)_{90}Zr_7B_2Cu_1$ ($0<x<0.6$) alloys, Co was reported to be preferentially partitioned to the residual amorphous phase, and the bcc grains were accordingly enriched by Fe [4]. In the heat-treated amorphous $Fe_{42.7}Co_{42.7}Zr_{6.8-x}Nb_xB_{6.8}Cu_1$ ($x = 0$; 1) alloy an ordered nanocrystalline α'-FeCo phase was identified by the help of Mössbauer spectroscopy [5]. It has developed after 10 min annealing at 723 K. Neither of these works provides details on the fitting used to reconstruct the spectral contribution namely of the crystalline component.

While original HITPERM alloys contain Zr as one of its constitution elements, investigation of other types of materials where Zr is replaced by Hf [6], Nb [7, 8], and/or Mn [9] has recently started. Again, Mössbauer spectrometry was used as a principal method of characterization. In the $(Fe_{1-x}Co_x)_{86}Hf_7B_6Cu_1$ alloy [6], crystalline phase, interface, and residual amorphous matrix were found after nanocrystallization. It was concluded that Co addition increases the hyperfine field. In addition, the Fe concentration in bcc-FeCo crystals and the fraction of crystalline phase influenced by Co addition are opposite. The line profile used for reproduction of the crystalline phase is not mentioned. Room and high temperature (474 and 673 K) Mössbauer spectra of annealed $Fe_{78-x}Co_xNb_6B_{16-y}Cu_y$ ($x = 18, 39, 60$; $y = 0, 1$) alloys were decomposed into their spectral components by the help of up to 4 discrete sextets and one continuous distribution of hyperfine fields [7]. The former represented crystalline phase whereas the latter was ascribed to the amorphous + interface contribution. One continuous distribution of hyperfine fields was eventually decomposed into two separate distributions during the evaluation of room temperature Mössbauer spectra of a nanocrystalline $Fe_{60}Co_{18}Nb_6B_{15}Cu_1$ alloy [8]. In both cases, a correlation between magnetic parameters and the nanocrystallization process is discussed. Similar fitting model was applied to the study of kinetics of nanocrystallization and microstructure of the $Fe_{60-x}Co_{18}Mn_xNb_6B_{16}$ ($x = 0, 2, 4$) alloys [9]. Mössbauer spectra of amorphous and heat treated samples show that the effect of Mn addition is clear on the amorphous phase but crystalline sites are not affected.

In our earlier work [10], we have reported on the effect of Mo on the progress of crystallization in the $Fe_{91-x}Mo_8Cu_1B_x$ ($x = 12, 15, 17, 20$) NANOPERM-type alloys. Surface features were characterized by atomic force microscopy (AFM). Morphology exhibits different features at the air- and wheel-side of the ribbon. In the as-quenched state, presence of quenched-in nanocrystallites is detected. During annealing, formation of large agglomerates of nanocrystallites is observed. They are formed in the early stage of crystallization in order to minimize the surface energy of small nanocrystallites.

In this presentation, an evolution of structural changes after (as well as during) heat treatment of original as-quenched samples and their consequence upon hyperfine interactions will be discussed for (Fe,Co)-Mo-Cu-B-type alloys which eventually exhibit nanocrystalline arrangement. Along with Mössbauer spectrometry, results obtained from AFM and magnetic force microscopy (MFM) will be reported. The kinetics of crystallization is characterized by *in-situ* diffraction of synchrotron radiation.

2. Experimental Details

Two basic compositions of $(Fe_{1-x}Co_x)_{79}Mo_8Cu_1B_{12}$ and $(Fe_{1-x}Co_x)_{76}Mo_8Cu_1B_{15}$ alloys were substituted with Co ranging from $x = 0$ to $x = 0.75$. They were prepared in a form of 6 mm wide and about 20 μm thick ribbons by planar-flow casting at the Institute of Physics, Slovak Academy of Sciences in Bratislava. Samples with different contents of nanocrystallites were prepared by 30 min annealing of the as-quenched alloys at temperatures covering the first crystallization peak as determined by differential scanning calorimetry.

The acquisition of surface topographic images was carried out by scanning probe microscope NTEGRA Aura in a map topography mode (AFM) as well as magnetic properties mode (MFM).

Monochromatic synchrotron radiation of 7 keV ($\lambda = 0.178$ nm) provided in the KMC-2 beamline at BESSY Berlin was used for diffraction experiments. Selected as-quenched ribbons were placed in an evacuated furnace and continuously heated up to 800 °C with a temperature increase of 10 K/min. Diffraction patterns from a 2Θ region covering the main (110) bcc FeCo reflection were recorded *in situ* every 10 seconds using a 2D multiwire detector. The picture in figure 1 shows the furnace with a beryllium hemispherical dome and an evacuated conical tube attached to the 2D detector.

Figure 1. Evacuated furnace with hemispherical Be-dome.

Mössbauer effect experiments were carried out in transmission geometry at room temperature using a $^{57}Co/Rh$ source mounted on a constant acceleration driving system. The spectra are plotted with velocity scales calibrated by a 12.5 μm thick α-Fe foil at room temperature. Evaluation of Mössbauer spectra was accomplished with the Confit [11] fitting program using up to 5 discrete sextets to reconstruct the crystalline phase and continuous distribution of hyperfine fields to describe the amorphous residual and interface phase [7, 12].

With respect to the different surfaces of the ribbons we will use the following convention. The side of the ribbon, which was in direct contact with the quenching wheel, will be referred to as the wheel-side. The opposite side of the ribbon, *i.e.* the one which was exposed to the surrounding atmosphere, will be denoted as the air-side.

3. Results and Discussion

3.1. As-quenched alloys

Substitution of Fe by Co has a remarkable effect on the shape of Mössbauer spectra of the as-quenched amorphous precursors as demonstrated in figure 2. Depending on the B-contents, the Co-free alloys ($x = 0$) exhibit different magnetic arrangement. While for 12 at%. B the corresponding Mössbauer spectrum shows pure paramagnetic doublet, weak magnetic hyperfine interactions can be seen in the 15 at%. B alloy. Substitution of Fe by Co leads in both alloys to enhanced formation of magnetic regions which is demonstrated by a gradual formation of broadened sextuplet spectra.

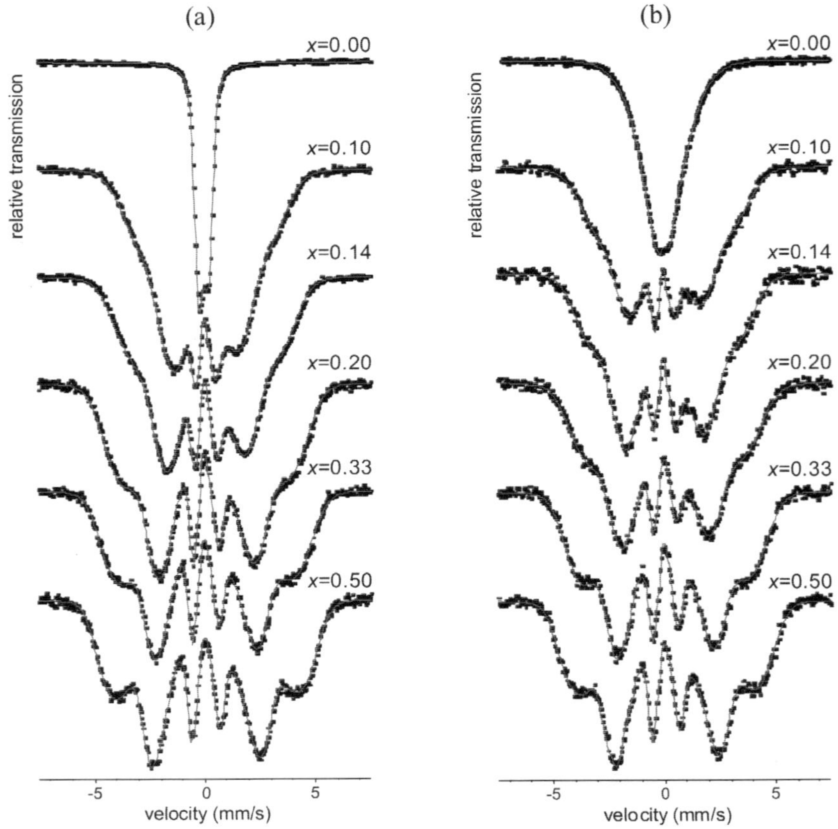

Figure 2. Room temperature Mössbauer spectra of as-quenched $(Fe_{1-x}Co_x)_{79}Mo_8Cu_1B_{12}$ (a) and $(Fe_{1-x}Co_x)_{76}Mo_8Cu_1B_{15}$ (b) alloys.

The Mössbauer spectra in figure 2 are characteristic for amorphous material and they were fitted using distributions of hyperfine magnetic fields P(B) except for the $x = 0$, 12 at% B sample where distribution of quadrupole splitting was used. A progressive increase in magnetic hyperfine interactions is observed with rising Co-substitution, x in both alloys. 3D mappings of P(B) distributions derived from the spectra are illustrated in figure 3. A bi-modal character of P(B) for low x in the 12 at% B alloy suggests presence of regions with low hyperfine fields which progressively vanish as the Co-contents increases. For Co-substitution $x > 0.2$ both alloys depict rather high magnetic fields. A tendency towards magnetic order in both alloys is demonstrated also by appearance of spontaneous formation of magnetic domains. They can be visualised by the help of MFM.

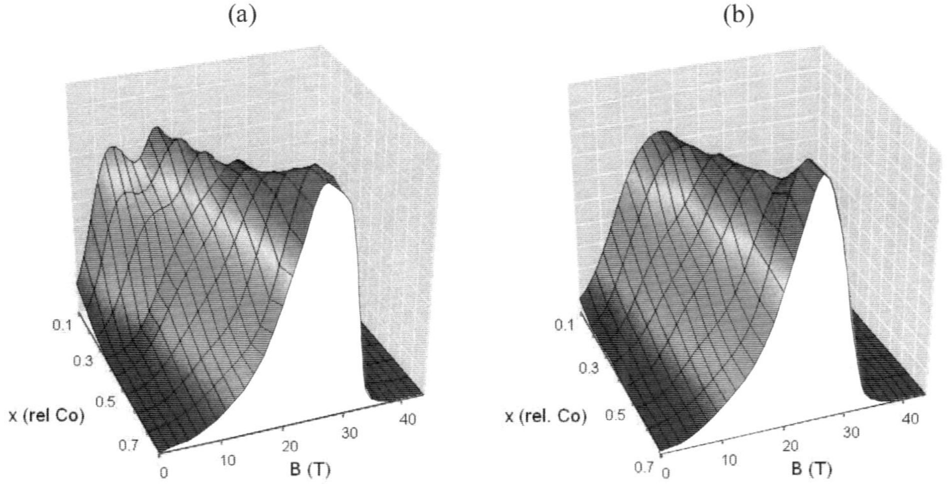

Figure 3. 3D mappings of P(B) distributions derived from room temperature Mössbauer spectra of as-quenched $(Fe_{1-x}Co_x)_{79}Mo_8Cu_1B_{12}$ (a) and $(Fe_{1-x}Co_x)_{76}Mo_8Cu_1B_{15}$ (b) alloys.

An example of magnetic domain formation in the $(Fe_{0.5}Co_{0.5})_{79}Mo_8Cu_1B_{12}$ as-quenched alloy is illustrated in figure 4. Presence of magnetic domains on the wheel and air side can be clearly seen from the MFM images in figure 4b and 4d. They are formed in the amorphous part of the alloy. In addition to the domains, dark spots can be also noticed. They are associated with magnetic moments of nanocrystallites which are present on both surfaces of the ribbon as documented also by the respective AFM images in figure 4a and 4c. It is noteworthy that the surface crystallization is more pronounced on the air side of the ribbon. Nevertheless, the total amount of crystallites is so small that they cannot be recognized by transmission Mössbauer spectroscopy which scans the entire bulk of the samples.

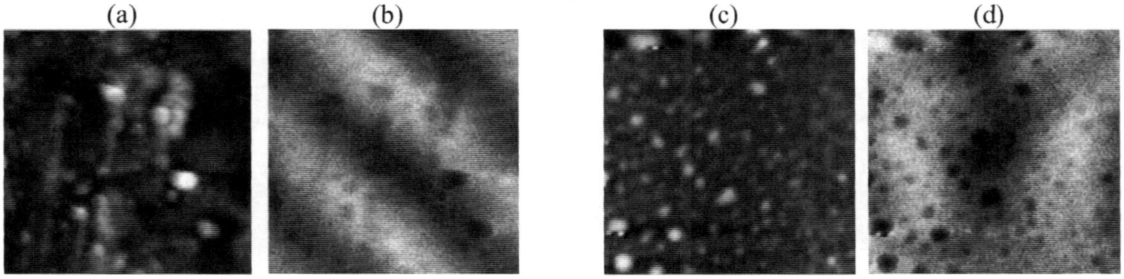

Figure 4. AFM (a, c) and MFM (b, d) images of the as-quenched $(Fe_{0.5}Co_{0.5})_{79}Mo_8Cu_1B_{12}$ alloy taken from the wheel side (a, b) and from the air side (c, d) of the ribbon. The size of scanned area is 3 x 3 μm and 5 x 5 μm, respectively.

Quantitative description of Mössbauer spectra is presented in figure 5 where average isomer shift <*IS*> and average hyperfine fields <B> are depicted as a function of Co-substitution, *x*. Changes in <*IS*>- and <*B*>-values indicate structural modifications which lead toward increase in Curie temperature with higher Co contents. Changes in <*IS*> are caused primarily by deviations in chemical composition around the Fe resonant atoms while the increase in <*B*> might be associated, along with the influence of composition, also by topology.

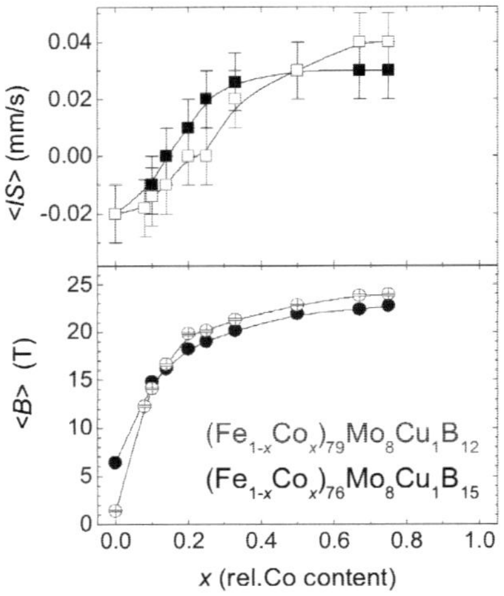

Figure 5. Average isomer shift, <*IS*> (a) and hyperfine field, <*B*> (b) plotted against the relative Co contents *x* in the as-quenched $(Fe_{1-x}Co_x)_{79}Mo_8Cu_1B_{12}$ (open symbols) and $(Fe_{1-x}Co_x)_{76}Mo_8Cu_1B_{15}$ (solid symbols) alloys.

3.2. Nanocrystalline alloys

The progress of crystallization in the studied alloys was followed by Mössbauer spectrometry and diffraction of synchrotron radiation. Mössbauer effect experiments were performed at room temperature upon samples that were first heat treated in a vacuum furnace at certain temperatures for 30 min and then cooled down to room temperature. In this way, different amount of crystalline phase was obtained and its influence on structural changes was investigated. In diffraction experiments, the as-quenched samples were exposed to temperature which was continuously increasing with the ramp of 10 K/min. Diffractograms were collected *in situ* during the temperature change every 10 s.

3.2.1. Mössbauer spectrometry. Typical examples of some Mössbauer spectra taken from the heat-treated $(Fe_{1-x}Co_x)_{76}Mo_8Cu_1B_{15}$ alloys are shown in figure 6 for $x = 0.33$, 0.5, and 0.67. They exhibit sharp lines, which belong to bcc-FeCo nanocrystals, that are superimposed upon broadened features assigned to the amorphous rest and interface phase. Crystalline components were fitted by up to five narrow sextets, the relative contributions of which were derived from binomial distribution describing the Co nearest neighbours of Fe resonant atoms in a bcc arrangement. The progress of crystallization (*i.e.*, relative fraction of the crystalline phase) depends upon the annealing conditions as well as upon the Co contents as illustrated in figure 7. With rising Co substitution, the decrease of Fe contents in the amorphous + interface regions is more rapid with respect to the temperature of annealing. A tendency of Fe atoms to leave the amorphous precursor and preferentially partitioned into bcc crystalline grains was reported in [4].

3.2.2. Diffraction of synchrotron radiation. Kinetics of crystallization process can be effectively followed by in situ diffraction of synchrotron radiation. Diffraction patterns have been obtained by integration of data from the 2D detector over angle χ. A typical example is illustrated in figure 8a and 8b as 3D mapping and contour plot, respectively, for the wheel side of the $(Fe_{0.5}Co_{0.5})_{76}Mo_8Cu_1B_{15}$ sample. The onset of the first and the second crystallization is marked by T_{x1}, and T_{x2}, respectively. Up to about 600 °C (T_{x2}), a presence of bcc-FeCo crystalline phase is detected.

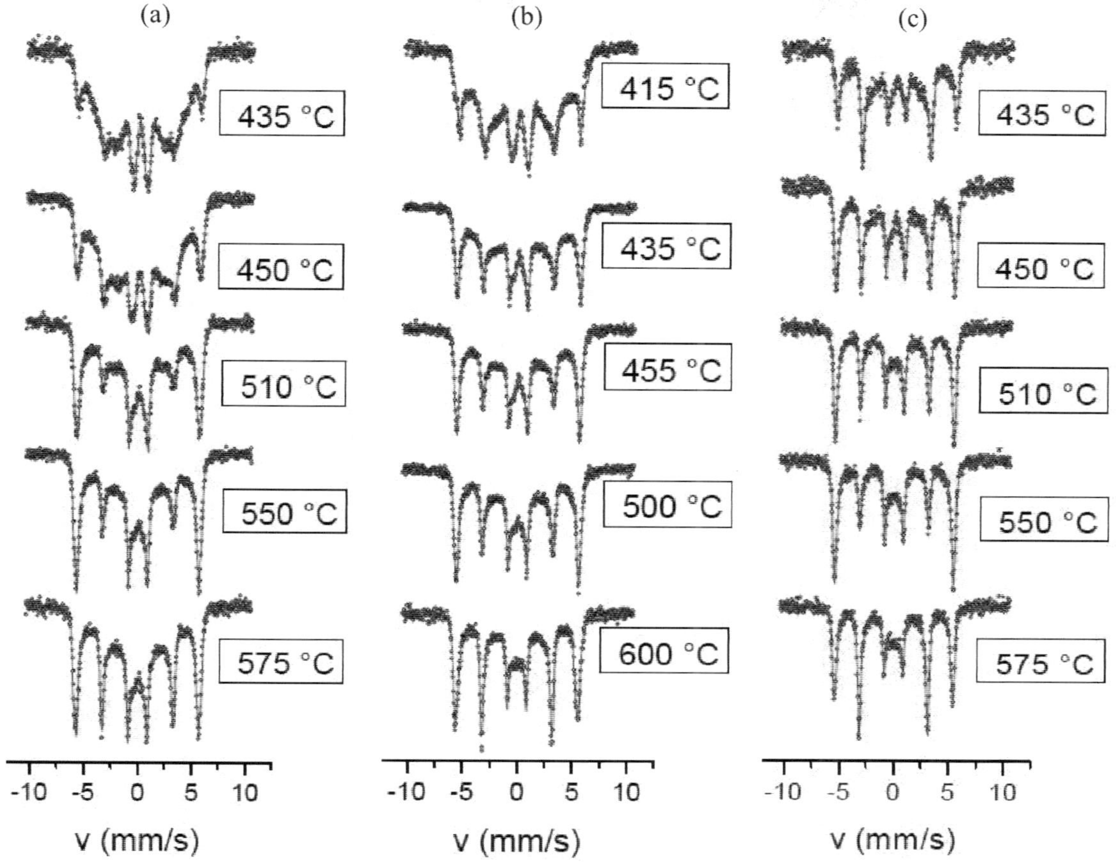

Figure 6. Room temperature Mössbauer spectra of the $(Fe_{1-x}Co_x)_{76}Mo_8Cu_1B_{15}$ alloys annealed at the indicated temperatures for $x = 0.33$ (a), $x = 0.5$ (b), and $x = 0.67$ (c).

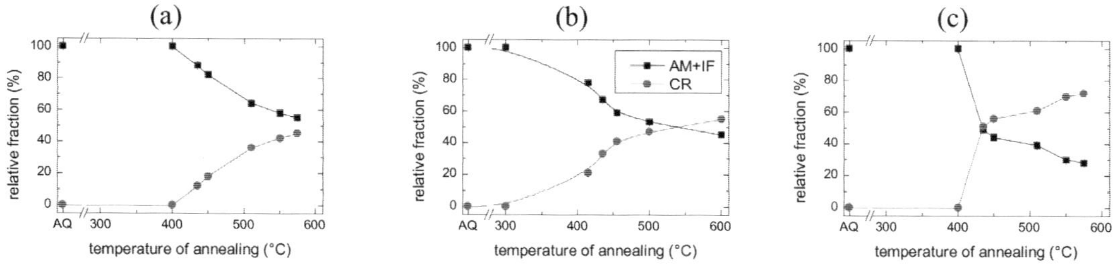

Figure 7. Relative fraction of amorphous+interface (squares) and crystalline (circles) phases in the $(Fe_{1-x}Co_x)_{76}Mo_8Cu_1B_{15}$ for $x = 0.33$ (a), $x = 0.5$ (b), and $x = 0.67$ (c) alloys as a function of temperature of annealing.

The principal (110) reflection peak was decomposed into broad and narrow components which are assigned to the amorphous residual matrix and crystalline grains, respectively. Quantitative analysis of the diffraction patterns is presented in figure 9 where the relative crystalline fraction is plotted against the temperature for the wheel and the air side of this investigated alloy. Results from two independent experiments performed upon the air side of two different samples from the same alloy are provided to enable an estimation of error margins in the determination of relative crystalline fraction. The onset of crystallization (T_{x1}) was determined to be of 371(2) °C and 395(2) °C for the wheel and air side of the ribbon, respectively. The difference of about 20 degrees in the start of crystallization could be assigned also to distinct number of quenched-in surface crystallites as was discussed for the $(Fe_{0.5}Co_{0.5})_{79}Mo_8Cu_1B_{12}$ alloy in the previous section.

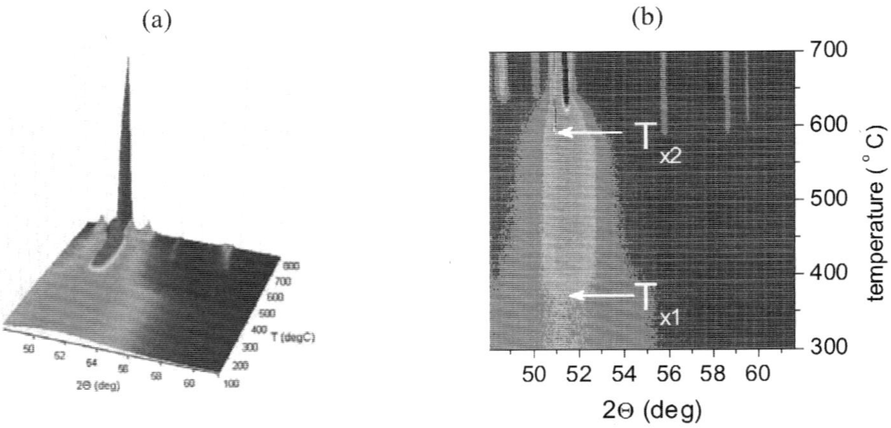

Figure 8. 3D diffractograms of the (110) reflection acquired from the wheel side of the $(Fe_{0.5}Co_{0.5})_{76}Mo_8Cu_1B_{15}$ alloy during in situ annelaing (a) and corresponding contour plot (b) in which the onset of the first (T_{x1}) and the second (T_{x2}) crystallization temperature is indicated by the arrows.

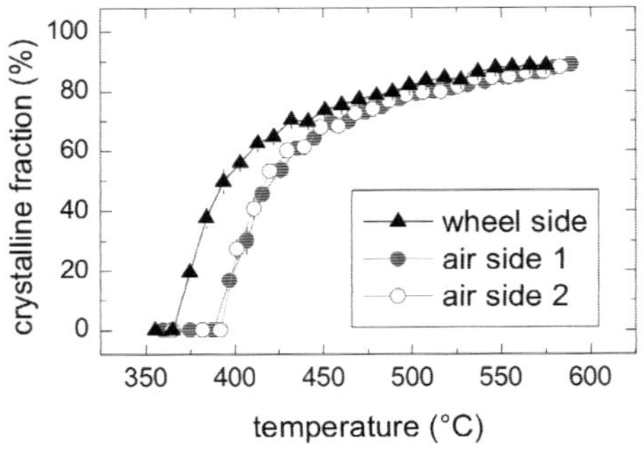

Figure 9. The evolution of crystalline fraction in the $(Fe_{0.5}Co_{0.5})_{76}Mo_8Cu_1B_{15}$ alloy during *in situ* heat treatment as derived from the wheel and air side of the ribbon-shaped samples.

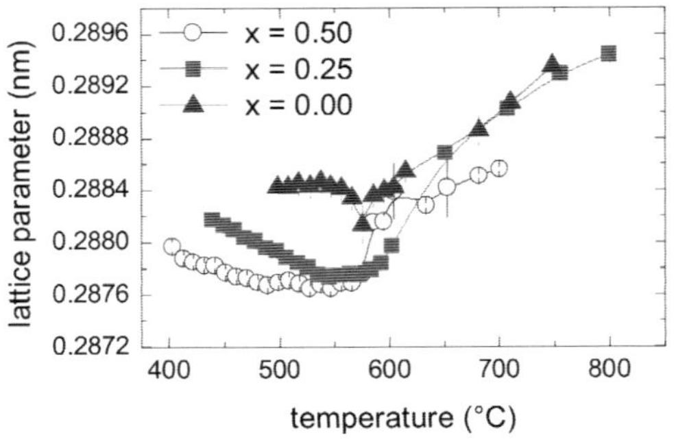

Figure 10. Lattice parameter of the $(Fe_{1-x}Co_x)_{76}Mo_8Cu_1B_{15}$ (x = 0, 0.25 and 0.5) alloy derived from diffraction of synchrotron radiation during in situ annelaing.

In figure 10, lattice parameters, a, of a bcc-FeCo phase, which were derived from the diffraction patterns, are plotted against temperature for the wheel side of the $(Fe_{1-x}Co_x)_{76}Mo_8Cu_1B_{15}$ alloy, x = 0, 0.25 and 0.5. Higher a-values as compared with the one of bcc-Fe phase (a = 0.2866 nm) are most probably caused by incorporation of small amounts of Mo into the bcc lattice which has caused an expansion of the latter. This effect is the most pronounced for the Co-free (x = 0) alloy. Already small Co substitution tends to expel Mo atoms from the bcc lattice, thus reducing the lattice parameter. The initial decrease in the lattice parameter with temperature is connected with the first stage of crystallization during which the bcc lattice continuously evolves and tends to reach the best possible crystallographic arrangement close to the one observed for pure bcc-Fe. Beyond 600 °C, when the bcc crystals start to grow into rather big (several μm in size) grains, a sudden increase in the lattice parameter is observed which is associated with the thermally induced lattice expansion.

4. Conclusions

Mössbauer spectrometry proves to be a powerful tool for the investigation of nanocrystalline alloys. It provides close information on structural as well as magnetic arrangement in their amorphous precursors via distributions of respective hyperfine interactions. Adopting a suitable fitting model for the reconstruction of crystalline phase (viz. binomial distribution of relative contributions), further details on the nearest neighbours and their impact on hyperfine magnetic fields can be obtained. In this work, we have not discussed this topic to more details due to limited space. Nevertheless, in order to achieve solid conclusions about structural and magnetic features of nanocrystalline alloys, along with Mössbauer effect studies also the use of complementary techniques of investigation is inevitable. Here, we have demonstrated how AFM and MFM can help in unveiling surface crystallization and magnetic arrangement, respectively. We have also shown how the kinetics of crystallization can be followed by employing *in situ* recording of diffraction of synchrotron radiation during temperature increase. In our ongoing research we utilize nuclear resonant scattering of synchrotron radiation which combines the sensitivity of nuclear-based techniques with a possibility of very fast data acquisition. In this way it is possible to investigate *in situ* the evolution of hyperfine interactions which pushes forward the frontiers in the research of nanocrystalline alloys.

Acknowledgements

This work was supported by the grants VEGA 1/4011/07, MSM6198959218, and EC grant under IA-SFS contract RII 3-CT-2004-506008. I am grateful to D. Janičkovič (Bratislava) for preparation of as-quenched precursors, G. Schumacher and I. Zizak (Berlin) for their help with synchrotron experiments and K. Šafářová (Olomouc) for STM images.

References

[1] Suzuki K, Kataoka N, Inoue A et al., 1990 *Mater. Trans. JIM* **31** 743
[2] McHenry M E, Willard M A and Laughlin D E 1999 *Prog. Mater. Sci.* **44** 291–433
[3] Willard M A, Laughlin D E, McHenry M E, Thoma D, Sickafus K, Cross J O and Harris V G 1998, *J. Appl. Phys.* **84** 6773
[4] Kemeny T, Kaptas D, Kiss L F, Balogh J, Bujdoso L, Gubicza J, Ungar T and Vincze I 2000 *Appl. Phys. Lett.* **76** 2110–2
[5] Olszewski J, Fukunaga H, Ciurzynska W, Zbroszczyk J, Łukiewska A, Mlynczyk A M, Perduta K and Lelatko J 2003 *J. Magn. Magn. Mat.* **254-255** 516–8
[6] Liang X B, Kulik T, Ferenc J, Erenc-Sedziak T, Xu B S, Grabias A and Kopcewicz M 2007 *Mat. Charact.* **58** 143–7
[7] Blazquez J S, Conde C F, Conde A and Greneche J M 2005 *J. Alloys Compounds* **397** 173–8
[8] Blazquez J S, Franco V, Conde C F, Conde A, Ferenc J and Kulik T 2007 *J. Non-Cryst. Solids* **353** 872–4
[9] Benaini H, Blazquez J S, Conde C F and Conde A 2008 *J. Alloys Comp.* **454** 156–63
[10] Miglierini M 2008 *J. Non-Cryst. Solids* **354** 5093–6
[11] Žák T and Jirásková Y 2006 *Surf. Interface Anal.* **38** 710–4
[12] Greneche J M and Slawska-Waniewska A 2000 *J. Magn. Magn. Mat.* **215-216** 264–7

Revealing the interparticle magnetic interactions of iron oxide nanoparticles-carbon nanotubes hybrid materials

A P Douvalis[1], V Georgakilas[2], T Tsoufis[3], D Gournis[3], B Kooi[4] and T Bakas[1]

[1]Department of Physics, University of Ioannina, 45110 Ioannina, Greece
[2]Institute of Materials Science, NCSR 'Demokritos', Ag. Paraskevi-Attikis, 15310 Athens, Greece
[3]Department of Materials Science and Engineering, University of Ioannina, 45110 Ioannina, Greece
[4]Zernike Institute for Advanced Materials, University of Groningen, Nijenborgh 4, 9747 AG Groningen, The Netherlands

adouval@cc.uoi.gr

Abstract. Spinel iron oxide nanoparticles capped with organic molecules have been successfully prepared and used to produce iron oxide nanoparticles-single wall carbon nanotubes hybrid materials, which were characterized by a number of experimental techniques. The nanoparticles in both samples have an average particle size of about 10 nm and acquire a chemical composition of the type $Fe_{3-x}O_4$, with $0<x<1/3$. ^{57}Fe Mössbauer spectroscopy and magnetization measurements suggest that the free capped nanoparticles experience stronger superparamagnetic effects with respect to the nanoparticles of the hybrid sample, which show stronger magnetic interparticle interactions. The mutual proximity of the magnetic nanoparticles of the hybrid sample is proposed to be the origin for this behaviour, which is triggered by their denser anchoring to the outer surface of the singe wall carbon nanotubes due to the reduced dimensions of the latter.

1. Introduction

Magnetic nanoparticles (NPs) are important materials with a wide range of technological applications, like media for magnetic data storage [1], drug carriers, identifiers as also contrast and hyperthermia agents in biomedicine [2]. On the other hand carbon nanotubes (CNTs) are novel developed versatile materials with extraordinary structural, mechanical, electronic and magnetic properties and posses a high potential to be used in many aspects of modern technical implementations [3]. The joining of the properties of both categories of materials has given rise to new fields of research and possible future technological applications [3].

Metal alloy (FePt) magnetic nanoparticles-carbon nanotubes (NPs-CNTs) hybrid materials have been prepared recently and their study has revealed very interesting structural, electronic and magnetic properties [4]. In this work we report on the study of the morphological, structural and magnetic properties of iron oxide NPs in NPs-CNTs hybrid materials, using, amongst other experimental techniques, the unique ^{57}Fe Mössbauer spectroscopy, which can give very important atomic-level information on their characteristics.

2. Experimental procedure

Capped iron oxide NPs, and iron oxide NPs-CNTs hybrids where prepared using wet chemical routes described in our previous work [5]. For the preparation of the hybrid samples, single-wall CNTs (Aldrich) were used and the capped iron oxide NPs were attached to the CNTs by means of an interlinker molecule (a carboxylic derivative of Pyrene). The relative mass ratio of capped NPs/single-wall CNTs was 10/1.

The prepared samples were characterized and studied using a number of experimental techniques. X-Ray diffraction (XRD) patterns were collected in a Bruker D8 Advance diffractometer using Cu K_α radiation. Raman and FT-IR spectra were recorded at room temperature (RT) in a RM 1000 Renishaw and a Spectrum GS Perkin-Elmer spectrometer respectively. Transmission Electron Microscopy (TEM) pictures of both samples were captured in a JEOL JEM-2010F microscope. Magnetization and magnetic susceptibility measurements were collected on a LakeShore 7300 Vibrating Sample Magnetometer. ^{57}Fe Mössbauer spectra (MS) were recorded at different sample temperatures using a constant acceleration spectrometer equipped with a ^{57}Co(Rh) source kept at RT and a liquid N_2 bath cryostat (Oxford). The calibration of the spectrometer was done using metallic iron at RT and the isomer shift (IS) values are given relative to this standard.

3. Results and Discussion

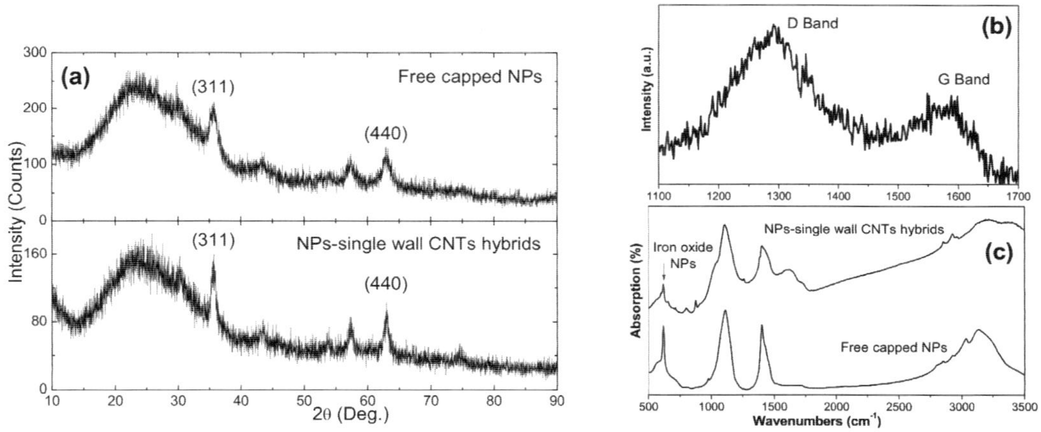

Figure 1. (a) XRD diagrams of the capped iron oxide NPs and the iron oxide NPs-CNTs hybrid samples, (b) Raman spectrum of the NPs-CNTs hybrid sample and (c) FT-IR spectra of the capped NPs and NPs-CNTs hybrid samples.

XRD patterns of both the capped iron oxide NPs and the iron oxide NPs-CNTs hybrids are shown in figure 1(a). The characteristic diffraction peaks of an oxidic spinel phase are shown in both diagrams indicating the presence of either Fe_3O_4 or γ-Fe_2O_3 iron oxide phases. The peaks are broad and determining the exact chemical composition of the iron oxide phase from these diagrams is quite difficult, since the XRD patterns of Fe_3O_4 or γ-Fe_2O_3 show diffraction peaks at almost the same angular positions [6]. The very broad curve between 15 and 40 degrees 2θ is due to the glass sample holder. The absence of a diffraction peak at around 26 degrees 2θ in the diagram of the hybrid sample, which would confirm the existence of the graphitic structure of the CNTs is attributed both to the large contribution of the background of the sample holder and to the reduced dimensions (~ 1-2 nm in diameter) of the single wall nature of the CNTs. Appling the Scherrer formula [7] to the diffraction peaks of the spinel phase we have estimated the average particle size to be about 10 nm for the capped iron oxide NPs and about 11 nm for the NPs of the hybrid sample. The Raman spectrum of the NPs-CNTs hybrid sample is shown in figure 1(b). The spectrum indicates the presence of carbon nanotubes from the two characteristic graphitic signals corresponding to D and G bands with a relative intensity ratio of $I_D/I_G \sim 2.0$. FT-IR spectra appearing in figure 1(c), reveal the presence of the iron oxide magnetic NPs in the final hybrid material which show a characteristic peak at around 610 cm^{-1}.

The TEM image of the capped NPs sample displayed in figure 2(a) shows a quite uniform particle size of about 10 nm. The corresponding image of the hybrid sample shown in figure 2(b), although being not as clear as that of the free NPs sample, shows roughly the same sizes for the iron oxide NPs and some indication of the presence of single wall CNTs could also be found.

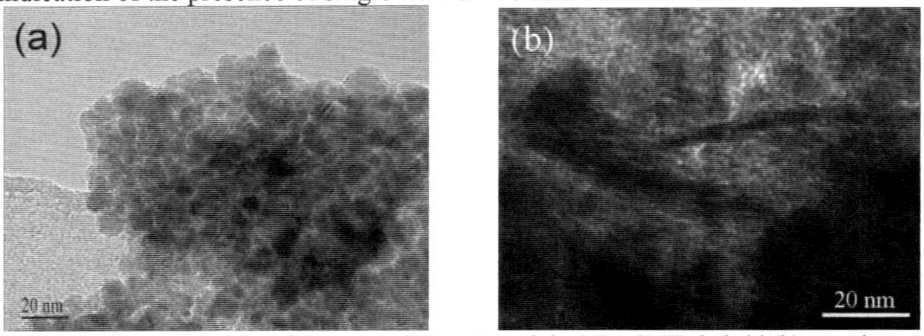

Figure 2. TEM images of the capped NPs (a) and the NPs-CNTs hybrid (b) samples.

^{57}Fe MS of both samples collected at 300 K, 140 K and 77 K appear in figure 3. The main characteristics of the 300 K spectra are the presence of a central paramagnetic doublet accompanied with a set of resolved as well as collapsing magnetically split components. The loss of the doublets' absorption area as the temperature drops reveals their superparamagnetic (SPM) character in both samples. The contribution of the SPM doublet is dominant in the RT spectrum of the capped NPs sample (~50% of the total absorption area (A)), in which also two magnetic components with collapsing hyperfine magnetic fields (B_{hf}) of 29.6(12.0) T and 8.5(4.7) T respectively, appear (the values in parentheses are the -Gaussian type- spreading of B_{hf} allowed for each magnetic component). On the contrary the contribution of the SPM doublet is minor for the RT spectrum of the hybrid sample (~12% of A), where also another 47% of A is occupied by two clearly resolved sextets with B_{hf} of 45.2(2.4) T and 37.2(6.1) T, and the remaining 41% by a collapsing magnetic component with a B_{hf} of 12.4(11.3) T. Here we note that, due to the monodispersed nature of the samples, the different B_{hf} distributed components corresponding to the magnetic part of the spectra, are used to arbitrary parameterize in the fit the intermediate relaxation rate of the NPs. The IS values of all components in both 300 K spectra range between 0.34 mms^{-1} and 0.41 mms^{-1}, suggesting the presence of only Fe^{3+} ions in the iron oxide spinel NPs.

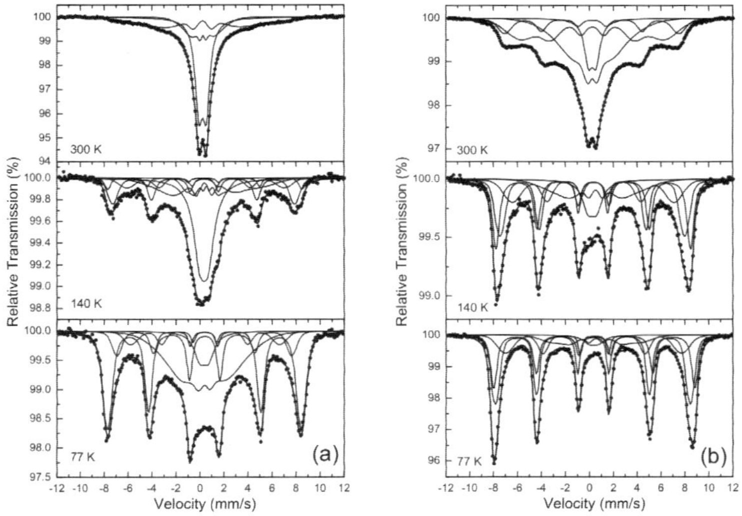

Figure 3. ^{57}Fe MS of capped NPs (a) and NPs-CNTs hybrids (b) samples recorded at different temperatures.

At 140 K the spectrum of the hybrid sample acquires relatively sharper magnetically split absorption lines, while that of the capped NPs sample still posses a large central contribution from collapsing B_{hf} and SPM components. The same characteristics are observed also for the 77 K spectra, where the intense contribution of a major magnetically split component with collapsing B_{hf} is still present for the capped NPs sample. The IS of some of the magnetically split components of the 140 K and 77 K spectra of both samples acquire relatively higher values than those expected from the increase of the corresponding RT values due to the second order Doppler shift effect [8], indicating the possible presence of some $Fe^{2.5+}$ states and suggesting that the iron oxide NPs have an average chemical composition of the type $Fe_{3-x}O_4$, with x between 0 (magnetite) and 1/3 (maghemite).

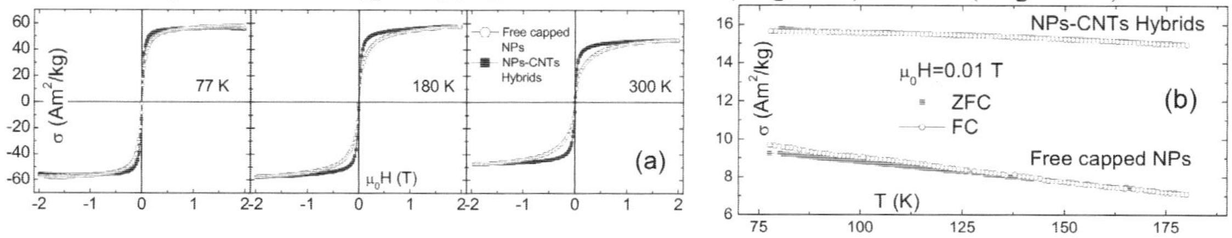

Figure 4. Magnetization measurements of the capped iron oxide NPs and the iron oxide NPs-CNTs hybrid samples as a function of the external magnetic field (a) and temperature (b).

Defining the Blocking temperature (T_B) as that in which the absorption area of the SPM component is half of the total absorption area A, we can estimate the T_B of the capped NPs sample to be around 300 K and that of the hybrid sample to be much higher. This trend comes in agreement with the magnetization measurements shown in figure 4, as the SPM characteristics (relatively low $d\sigma/dH$ slope, lack of saturation at low fields and diversification of zero field cooled and field cooled curves) are more evident in the case of the capped NPs sample.

Given the same average NPs size for both samples as estimated from XRD and TEM, the above results suggest that stronger interparticle interactions are experienced by the NPs of the hybrid sample, relative to those of the capped NPs sample. This is opposite to what we have found for the corresponding capped FePt metal alloy NPs and NPs-CNTs hybrid samples [4], where selective isolation of these FePt NPs on the outer surface of the multi-wall (in that case) CNTs, induced stronger SPM effects for the hybrid sample, relative to the stronger magnetic interparticle interactions observed for the corresponding capped NPs sample. We propose that the stronger magnetic interparticle interactions in our present iron oxide NPs-CNTs hybrid sample are a consequence of the high mutual proximity of these NPs. The reason for such a different behaviour for the NPs of the hybrid sample lies in the use of single- rather than multi-wall CNTs in the present case. The single-wall CNTs of reduced dimensions (1-2 nm) trigger a high local concentration of magnetic NPs, as the NPs anchor around these CNTs with increased density, compared to the case of the multi-wall CNTs.

4. Conclusions

The present work of producing new spinel iron oxide NPs-CNTs hybrids, in which the NPs acquire different magnetic properties when they are attached to the single-wall CNTs in comparison to our previous work on metallic FePt NPs-multi-wall CNTs [4], has shown that it is possible to develop a versatile method for preparation of magnetic NPs-CNTs hybrid materials with tunable magnetic properties, through the control of the NPs and CNTs sizes, as well as their relative mass ratios.

References

[1] Sun S et al. 2000 *Science* **287** 1989
[2] Pankhurst Q A et al. 2003 *J. Phys. D: Appl.Phys.* **36** R167
[3] Georgakilas V et al. 2007 *J. Materials Chemistry* **17** 2679
[4] Tsoufis T et al. 2008 *J. Nanoscience & Nanotechnology* **8** 5942
[5] Georgakilas V et al. 2005 *Chemistry of Materials* **17** 1613
[6] Cornell R M and Schwertmann U 1996 *The iron oxides* (New York: Wiley)
[7] Cullity B D 1956 *Elements of X-ray Diffraction* (New York: Addison-Wesley)
[8] Greenwood N N and Gibb T C 1971 *Mössbauer Spectroscopy* (London: Chapman & Hall)

International Conference on the Applications of the Mössbauer Effect (ICAME 2009) IOP Publishing
Journal of Physics: Conference Series **217** (2010) 012094 doi:10.1088/1742-6596/217/1/012094

Local magnetism of Co_2MnSn Heusler alloy films prepared by atomically controlled alternate deposition

K Mibu[1], D Gondo[1], T Hori[1], Y Ishikawa[1] and M A Tanaka[1]

[1]Nagoya Institute of Technology, Nagoya, Aichi 466-8555, Japan

k_mibu@nitech.ac.jp

Abstract. Co_2MnSn Heusler alloy films were prepared by atomically controlled alternate deposition on MgO(001) substrates. The local magnetism was investigated through the distribution of magnetic hyperfine fields obtained from ^{119}Sn Mössbauer spectroscopic measurements. It turned out that the magnetic environments around the Sn sites in the films prepared by this method with appropriate growth temperatures are relatively uniform in comparison with those in bulk alloys prepared by arc melting. The Mössbauer spectroscopy was applied also to $Co_2MnSn/MgO/Co_2MnSn/Co$ layered structures and Co_2MnSn/Ag multilayers to examine crystallographic uniformity of the Co_2MnSn layers and possibility of magnetic anomaly at the interface regions.

1. Introduction

Half-metallic materials, where the conduction electrons are 100% spin-polarized, have recently been studied intensively from theoretical and experimental viewpoints in the field of spintronics. Such materials are promising candidates for ferromagnetic layers of magnetic tunnel junctions (MTJs) with a ferromagnetic-metal/nonmagnetic-insulator/ferromagnetic-metal structure for the realization of quite large tunnel magnetoresistance (TMR) effects. One series of materials which are expected to be half-metallic are X_2YZ Heusler alloys with an $L2_1$ structure [1]. Several groups have investigated MTJs with Heusler alloys and realized large magnetoresistance ratio of a few hundred % at room temperature [2-4]. However, the magnetoresistance change is not yet satisfactorily large as expected for systems of half-metals. Moreover, the TMR effects, which are large at low temperatures, reduce significantly around room temperature. In order to find a key to solve these problems, it seems important to prepare good quality Heusler alloy films with little site disorder and high interface uniformity, and examine the structure and magnetism using local experimental methods such as Mössbauer spectroscopy [5, 6] and NMR [7].

In this work, we fabricated Co_2MnSn Heusler alloy films by atomically controlled alternate deposition and investigated local magnetism using ^{119}Sn Mössbauer spectroscopy. The influence of crystallographic disorder and interface effects on the local magnetism of the films can be examined through the magnetic hyperfine field induced at the Sn nuclear sites. The Mössbauer measurements were performed also for MTJs composed of Co_2MnSn layers and an MgO barrier to investigate the correlation between the local magnetism and the TMR effects. For more interface-sensitive measurements, multilayers of thinner Co_2MnSn layers and nonmagnetic Ag layers were prepared instead of Co_2MnSn/MgO multilayers, since it is easier to grow thin Co_2MnSn and Ag layers repeatedly to form multilayers with good crystallographic quality.

2. Experiments

Co_2MnSn films were prepared using a vacuum vapour deposition system with e-gun heating. One atomic layer of Co, half an atomic layer of Mn and Sn, correspondingly to the $L2_1$ structure, were deposited alternately in a controlled manner on a Cr buffer layer grown on MgO(001) substrates. Sn was moderately enriched with ^{119}Sn to obtain ^{119}Sn Mössbauer spectra effectively. Layered structures of Co_2MnSn(30.2 nm)/MgO(2 nm)/Co_2MnSn(10.2 nm)/Co(50 nm) were also prepared, with the Co_2MnSn layers made by the same method. A part of each sample was micro-fabricated into MTJs and the TMR effect was examined using the identical layered structure. Multilayers of Co_2MnSn(2.2 nm) and Ag(3 nm) with the bilayer repetition of 10 times were also prepared for more interface sensitive measurements. For all the layered structures, the interfaces of the Co_2MnSn layers were designed to be terminated with an atomic layer of Co, not with Mn and Sn. Mössbauer spectra were measured at room temperature by means of conversion electron Mössbauer spectroscopy using a $Ca^{119m}SnO_3$ source and a He+1%(CH_3)CH gas flow counter. The peak position of $CaSnO_3$ was used as a standard of zero velocity.

3. Results and Discussion

In Co_2MnSn alloys with an ideal $L2_1$ structure, Sn atoms occupy a single crystallographic site with eight Co atoms as the nearest neighbours and six Mn atoms as the second nearest. However, the Mössbauer results on bulk Co_2MnSn alloys published so far indicate that two different environments with different hyperfine fields coexist as Sn sites, although the site causing the minor component with smaller hyperfine field has not been identified clearly [8-11]. A typical absorption Mössbauer spectrum of bulk Co_2MnSn alloy prepared by arc melting is shown in figure 1(a). The spectrum can be fitted with (i) two magnetic components with Voigt-type absorption lines (the solid lines in the left figure), i.e., with two Gaussian distributions of hyperfine fields (the solid lines in the right figure) or (ii) a set of magnetically split sextets having two peaks in the hyperfine distribution (the histogram in the right figure), with the isomer shift of about 1.4 mm/s.

The conversion electron Mössbauer spectrum and distribution of magnetic hyperfine fields for the single Co_2MnSn layer with 40.2 nm in thickness, which was prepared by atomically controlled alternate deposition with a substrate temperature of 500 °C, are shown in figure 1(b). The peak with smaller hyperfine field is not significant and the width of the peak at around 10 T is relatively narrow, suggesting that the magnetic environments around the Sn sites are uniform in comparison with those in the bulk alloys. The distribution changes as a function of deposition temperatures as shown in figure 2, and becomes sharp when the film is deposited at 400 °C or above. The sharp distribution in

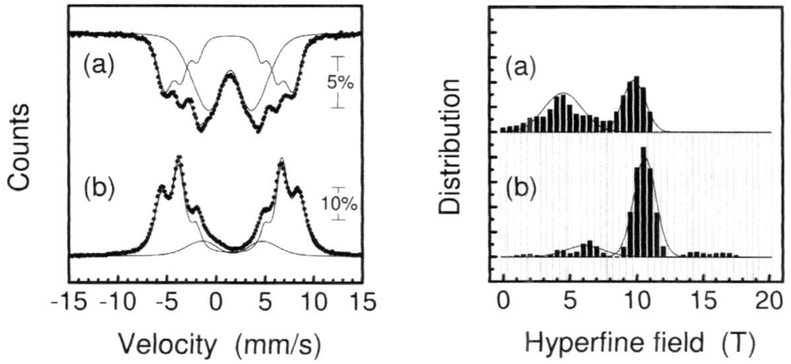

Figure 1. (a) ^{119}Sn absorption Mössbauer spectrum and distribution of magnetic hyperfine field for a bulk Co_2MnSn alloy prepared by arc melting. (b) ^{119}Sn conversion electron Mössbauer spectrum and distribution of magnetic hyperfine field for a Co_2MnSn(40.2 nm) film grown by atomically controlled alternate deposition at 500 °C. (See the text for the fitting methods.)

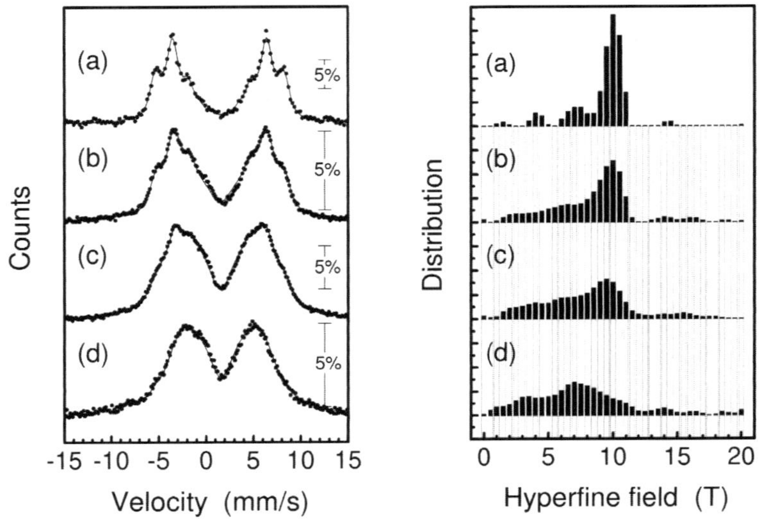

Figure 2. ^{119}Sn Mössbauer spectra and distribution of magnetic hyperfine field for Co_2MnSn films grown at (a) 400 °C , (b) 300 °C, (c) 200 °C, and (d) 100 °C.

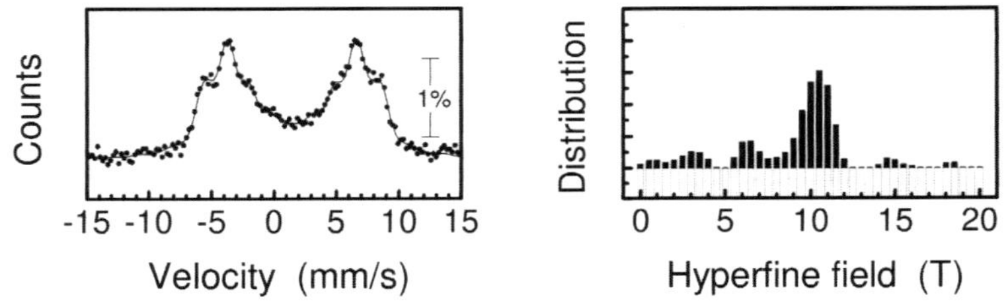

Figure 3. ^{119}Sn Mössbauer spectra and distribution of magnetic hyperfine field for Co_2MnSn(30.2 nm)/MgO(2.0 nm)/Co_2MnSn(10.2 nm)/Co(50.0 nm) layered structures grown at 400 °C for the bottom Co_2MnSn layer and at 100 °C for the top Co_2MnSn layer.

hyperfine fields makes it possible to examine interface-originating local magnetism of Co_2MnSn-containing layered structures.

A typical example of the Mössbauer spectrum and the distribution of magnetic hyperfine field for Co_2MnSn/MgO/Co_2MnSn/Co layered structures is shown in figure 3, for the sample where the bottom Co_2MnSn layer was deposited at the substrate temperature of 400 °C and the top Co_2MnSn layer on the MgO barrier was grown at 100 °C. The peak at around 10 T in the histogram is broader and the distribution in the lower field range is larger in comparison with the histogram for the single layer deposited at 400 - 500 °C, which reflects the local crystallographic disorder in the top Co_2MnSn layer and possibly the interface magnetism of the bottom and top Co_2MnSn layers. For this sample, a small but clear magnetoresistance effect was observed associated with the parallel and antiparallel magnetic configurations between the two Co_2MnSn layers [12]. If the Co_2MnSn layers react with MgO to form a nonmagnetic oxide, a peak would appear at 0 mm/s, which results in an asymmetric shape of the spectra. Such asymmetry is not significant in this sample. Actually, the distribution histogram changes dependent on the preparation conditions, and TMR effect is not observed when the component with smaller hyperfine fields is more significant [12].

A typical Mössbauer result for Co_2MnSn/Ag multilayers prepared at 400 °C for more interface sensitive experiment is shown in figure 4. In this sample, the nominal structure is designed in such a

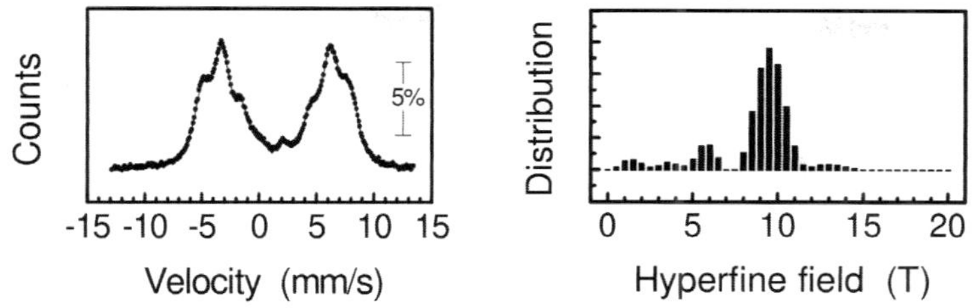

Figure 4. ^{119}Sn Mössbauer spectra and distribution of magnetic hyperfine field for a Co_2MnSn(2.2 nm)/Ag(3 nm) multilayer grown at 400 °C.

way that 2/5 of Sn atoms are contacting the interfacial Co atoms and 3/5 is located inside the Co_2MnSn layers. The spectrum resembles that of the thicker single Co_2MnSn layer and no strong interface effect is observed in the histogram. The tiny peak around 2 mm/s may be due to a minor nonmagnetic Sn-rich site, which would not be intrinsic to the magnetic interface effect. Note that a preliminary result shows that there is also no strong interface effect even when the interfaces of the Co_2MnSn layers are designed to be terminated with atomic layers of Mn and Sn.

4. Summary

High quality Co_2MnSn films were prepared by atomically controlled alternate deposition on MgO(001) substrates and the local magnetism was investigated using ^{119}Sn Mössbauer spectroscopy. It turned out that the magnetic environments around the Sn sites in the samples prepared by this method are relatively uniform in comparison with those in bulk alloys prepared by arc melting. The local uniformity around the Sn sites depends on the growth temperatures. The TMR effect was observed for the MTJs when the local magnetism of the $Co_2MnSn/MgO/Co_2MnSn$/Co layered structure was relatively uniform. No strong magnetic anomaly at the interface region of Co_2MnSn layers has been detected at the present stage. Further investigation is required to clarify the interface magnetism of Co_2MnSn films.

Acknowledgments

The authors thank Dr. W. Zhang, Dr. T. Okuno, and Mr. Y. Sakai for their experimental collaboration. This work was supported by Grants-in-Aid for Scientific Research (C) from JSPS and Scientific Research in Priority Areas "Creation and Control of Spin Current" from MEXT Japan.

References

[1] Galanakis I, Dederichs P H and Papanikilaou N 2002 *Phys. Rev.* B **66** 134428
[2] Sakuraba Y, Hattori M, Oogane M, Ando Y, Kato H, Sakuma A, Miyazaki T and Kubota H 2006 *Appl. Phys. Lett.* **88** 192508
[3] Tezuka N, Ikeda N, Miyazaki A, Sugimoto S, Kikuchi M and Inomata K 2006 *Appl. Phys. Lett.*, **89** 112514
[4] Ishikawa T, Hakamata S, Matsuda K, Uemura T and Yamamoto M 2008 *J. Appl. Phys.* **103** 07A919.
[5] Zhang W, Jiko N, Okuno T, Mibu K and Yoshimura K 2007 *J. Magn. Magn. Mater.* **309** 132
[6] Ksenofontov V, Herbort C, Jourdan M and Felser C 2008 *Appl. Phys. Lett.* **92** 262501
[7] Inomata K, Wojcik M, Jedryka E, Ikeda N and Tezuka N 2008 *Phys Rev* B **77** 214425
[8] Williams J M 1968 *J. Phys.* C **1** 473
[9] Dunlap R A, March R H and Stroink G 1981 *Can. J. Phys.* **59** 1577
[10] Gavriliuk A G, Stepanov G N and Irkaev S M 1995 *J. Appl. Phys.* **77** 2648
[11] Zhang W, Jiko N, Mibu K and Yoshimura K 2005 *J. Phys.: Condens. Matter.* **17** 6653
[12] Mibu K, Gondo D, Hori T, Tanaka M A, Kondou K, Kasai S and Ono T *J. Phys.: Conference Series*, to be published

In situ oxidation of green rusts by deprotonation; wet corrosion and passivation of weathering steels

J-M R Génin[1], A Renard[1], C Ruby[1] and M Abdelmoula[1]
[1]Institut Jean Barriol, (FR 2843 CNRS-Université Henri Poincaré-Nancy Université), Service des Matériaux pour le Développement Durable, l'Energie et l'Environnement, ESSTIN-UHP, 2 rue Jean Lamour, F-54500 Vandoeuvre-lès-Nancy, France.

E-mail: jean-marie.genin @ esstin.uhp-nancy.fr

Abstract. Intermediate compounds that belong to the layered double hydroxide family and bear the common name of green rusts (GR) get oxidized through two modes determining material wearing for steel corrosion: either the classical aerial mode where the GR gets dissolved before precipitating into several types of orange ferric oxyhydroxides FeOOH rusts, or the *in situ* deprotonation of OH^- ions within the GR. Hydroxycarbonate, $GR(CO_3^{2-})$, $[Fe^{II}_4\ Fe^{III}_2\ (OH)_{12}]^{2+} \bullet [CO_3^{2-} \bullet 3H_2O]^{2-}$, produced e.g. in carbonated medium, becomes $[Fe^{II}_{6(1-x)}\ Fe^{III}_{6x}\ O_{12}\ H_{2(7-3x)}]^{2+} \bullet [CO_3^{2-} \bullet 3H_2O]^{2-}$, which is an oxyhydroxycarbonate where ferric molar ratio $\boldsymbol{x} = [Fe^{III} / Fe_{total}]$ belongs to the [0, 1] interval; it ends into the "ferric green rust", $GR(CO_3^{2-})^*$, $[Fe^{III}_6\ O_{12}\ H_8]^{2+} \bullet [CO_3^{2-} \bullet 3H_2O]^{2-}$, which is in fact orange. These two modes of oxidation of GR that depend upon the flux of oxygen that is used give rise to two different corrosion behaviors. The first one corrodes the material since the GR layer covering the steel surface gets dissolved before precipitating ferric oxyhydroxide. The second mode occurs within the GR layer without destruction; thus, the metal is passivated as for weathering steels.

1. Introduction

Wet corrosion of steels corresponds to the presence of intermediate Fe^{II-III} compounds commonly named green rusts (GR) because of their colour [1, 2]; they belong to the family of layered double hydroxides (LDH) where di- and trivalent cations are Fe^{II} and Fe^{III} ions, respectively [3]. They form during iron oxidation from the ferrous state, i.e. $Fe(OH)_2$, and transform into ferric oxyhydroxides of general formula $Fe^{III}OOH$ [4-6]. The type of GR depends upon the anion species dissolved in the water solution and the most studied GRs incorporate Cl^-, SO_4^{2-} and CO_3^{2-} anions [4-6, 7-11]. They are constituted of brucite-type layers $[Fe^{II}_{(1-x)}\ Fe^{III}_x\ (OH)_2]^{x+}$ that are electrostatically balanced by interlayers $[(x/n)\ A^{n-} \cdot (m/n)\ H_2O]^{x-}$, where A^{n-} are intercalated anions. For instance, $GR(CO_3^{2-})$, the Fe^{II-III} hydroxycarbonate $Fe^{II}_4 Fe^{III}_2 (OH)_{12} CO_3$ at $x = 0.33$, displays a long range order for cations and anions as determined by XRD and Mössbauer spectroscopy, which is very suitable for assessing the ferric molar ratio $x = [Fe^{III}]/[Fe_{total}]$ [12, 13]. Spectra display in the paramagnetic state at 78 K three doublets, D_1 and D_2 with quadrupole splitting Δ of 2.97 and 2.55 mm s^{-1} and isomer shift δ both of 1.28 mm s^{-1} attributed to Fe^{II}, and D_3 with Δ and δ of 0.43 and 0.47 mm s^{-1} attributed to Fe^{III}.

Using Mössbauer spectroscopy for monitoring the speciation of iron, it was possible to follow the oxidation of GRs in usual conditions and simulate the wet corrosion of steels [4, 7-11]. We consider the carbonated medium, the most important for practical purposes and for which the E_h-pH Pourbaix diagram of $GR(CO_3^{2-})$ has been established [9]. Let us recall that a GR layer forms on the surface of an iron-based material before transforming into $Fe^{III}OOH$ as observed when stirring a solution of $GR(CO_3^{2-})$ in the air {Top left [Figure 1], from [14]}. This is the usual mode of corrosion of steel when intermediate $GR(CO_3^{2-})$ disappears into solution before forming a brownish orange ferric rust $Fe^{III}OOH$. In this paper, we shall show that another mode of oxidation may occur and that $GR(CO_3^{2-})$ may get oxidized *in situ* by deprotonation up to a ferric $GR(CO_3^{2-})$*, which is reddish orange [15].

2. Influence of oxygen rate on the oxidation mode of $GR(CO_3^{2-})$

2.1. Synthesis of initial Fe^{II-III} hydroxycarbonate

An initial $GR(CO_3^{2-})$ precipitate was prepared in a gas tight reactor through which a continuous flux of nitrogen bubbled. A 600 mL salt mixture, which was composed of dissolved $Fe(SO_4){\bullet}7H_2O$ and $Fe_2(SO_4)_3{\bullet}5H_2O$, was progressively neutralised with 37 mL of Na_2CO_3 solution ($[Fe^{II}] = 2.2 \times 10^{-2}$ M, $[Fe^{III}] = 1.2 \times 10^{-2}$ M, $[Na_2CO_3] = 1.1$ M) [17]. The suspension was homogenised by stirring with a magnetic rod. Once the basic solution completely added, the pH was about 9.3.

2.2. Oxidation of Fe^{II-III} hydroxycarbonate $GR(CO_3^{2-})$ by O_2 bubbling.

$GR(CO_3^{2-})$ got oxidized with relative oxygen concentrations $X(O_2) = [p(O_2) / p_{total}]$ at 2.7, 6.7,13.3 and 20% from a mixture of synthetic air and nitrogen with a magnetic rod rotation of 375 rpm. Electrode potential E_h(V) and pH were recorded continuously with time [Figure 1]. Gas flow $F(O_2)$ varied from 0.4 to 2.9 moles per day [Table 1]. Oxidation efficiency, i.e. the ratio between the O_2 mole number reacting with Fe^{II} species and O_2 mole number introduced in the solution, was approximately 1%, a value independent of $F(O_2)$; pH did not vary significantly ~9, due to the buffering effect of carbonate.

Curves (β-δ) display two plateaus separated by equivalent point E. Around E, the initial dark-blue colour of the suspension became rapidly yellow-brown. Point E is attributed to the end of oxidation; its abscissa t_f shifts towards longer periods when $X(O_2)$ decreases. Curve (α) exhibits one more plateau separated by point P before reaching point E for the lowest $X(O_2)$ value at 2.7%. A hook H is distinguished on all curves at the beginning of oxidation process and its height increases with $X(O_2)$ [Figure 1]. Mean oxidation velocity $V(O_2) = [n(O_2)/t_f]$ is evaluated where $n(O_2)$ is the consumed oxygen mole; it corresponds to Fe^{II} species present in the initial $GR(CO_3^{2-})$ [Table 1]. Oxidation efficiency is computed from $[V(O_2) / F(O_2)]$ and stays practically constant at a low value around 1%.

Solid characterisation carried out by XRD and Mössbauer reflection at ambient under continuous inert N_2 gas flow in a glove box showed that goethite formed for $X(O_2)$ values of 20, 13.3 and 6.7% [Figure 1b], whereas it led to a mixture of magnetite and goethite for $X(O_2) = 2.7\%$ [Figure 1a] [16]. Mössbauer spectrum measured at ambient confirmed that off-stoechiometric magnetite formed [17] and that only 12% of Fe atoms were present into α–FeOOH [Figure 1a] [Table 2]. According to XRD patterns, doublet *D* in the central part is attributed to Fe^{III} species present in superparamagnetic goethite crystals of low size (< ~100 nm) [Figure 1a]. Similar behaviour was observed for goethite forming alone at higher $X(O_2)$ where doublet *D* became a magnetic sextet due to increasing crystal size [Figure 1b].

2.3. Air oxidation of Fe^{II-III} hydroxycarbonate $GR(CO_3^{2-})$

N_2 bubbling was stopped and the reactor was exposed to the air. A vigorous magnetic stirring was set at 750 rpm in order to accelerate the oxidation process. The final pH at the end of the co-precipitation experiments decreased slightly from 9.3 to 9. In a second experiment, the final pH obtained at the end of the $GR(CO_3^{2-})$ co-precipitation, i.e. pH = 9.3, was adjusted with a concentrated HCl solution at a value of 7 ± 0.2. The same air oxidation experiment was performed by maintaining a constant pH of 7.

The evolution of redox potential E_h for pH values of 9 and 7 was recorded [curve (ε) for pH 9]. The trend was significantly different if compared to the previous {O_2 + N_2} bubbling experiments [curves (α–δ)]: (i) a gradual increase of E_h was observed along with the absence of equivalent points, (ii) the kinetics of oxidation was much faster and the final E_h plateau situated between +300 and +400 mV was reached between 2 and 8 times more rapidly. In contrast, the pH did not influence the kinetics

Table 1. Experimental conditions and nature of the final oxidation products; $X(O_2)$ % is the percentage of oxygen present in the gas flow. The oxidation velocity is determined by using the oxidation time t_f at the equivalent point of Figure 1. The efficiency of the oxidation is determined by the ratio $V(O_2)$ / $F(O_2)$ where $F(O_2)$ is the flow of O_2 bubbled into the solution.

Type of oxidation Curve name [Figure 1]	Controlled O_2 bubbling (α)	Controlled O_2 bubbling (β)	Controlled O_2 bubbling (γ)	Controlled O_2 bubbling (δ)	Air oxidation (ε)
$X(O_2)$ (%)	2.7 ± 0.1	6.7 ± 0.1	13.3 ± 0.2	20 ± 0.2	-
magnetic stirring (rpm)	375	375	375	375	750
oxidation time t_f (min)	700 ± 10	440 ± 10	290 ± 5	170 ± 5	80 ± 2
$V(O_2)$ Oxidation velocity (Mole per day)	0.005	0.011	0.016	0.028	0.06
$F(O_2)$ Flow of O_2 (Mole per day)	0.4	1	1.9	2.9	-
Efficiency [$V(O_2)$ / $F(O_2)$] (%)	1.2	1.1	0.9	1	-
Colour of the final product	Black	Yellow-brown	Yellow-brown	Yellow-brown	Orange-brown
Oxidation products	Magnetite + Goethite	Goethite	Goethite	Goethite	GR*

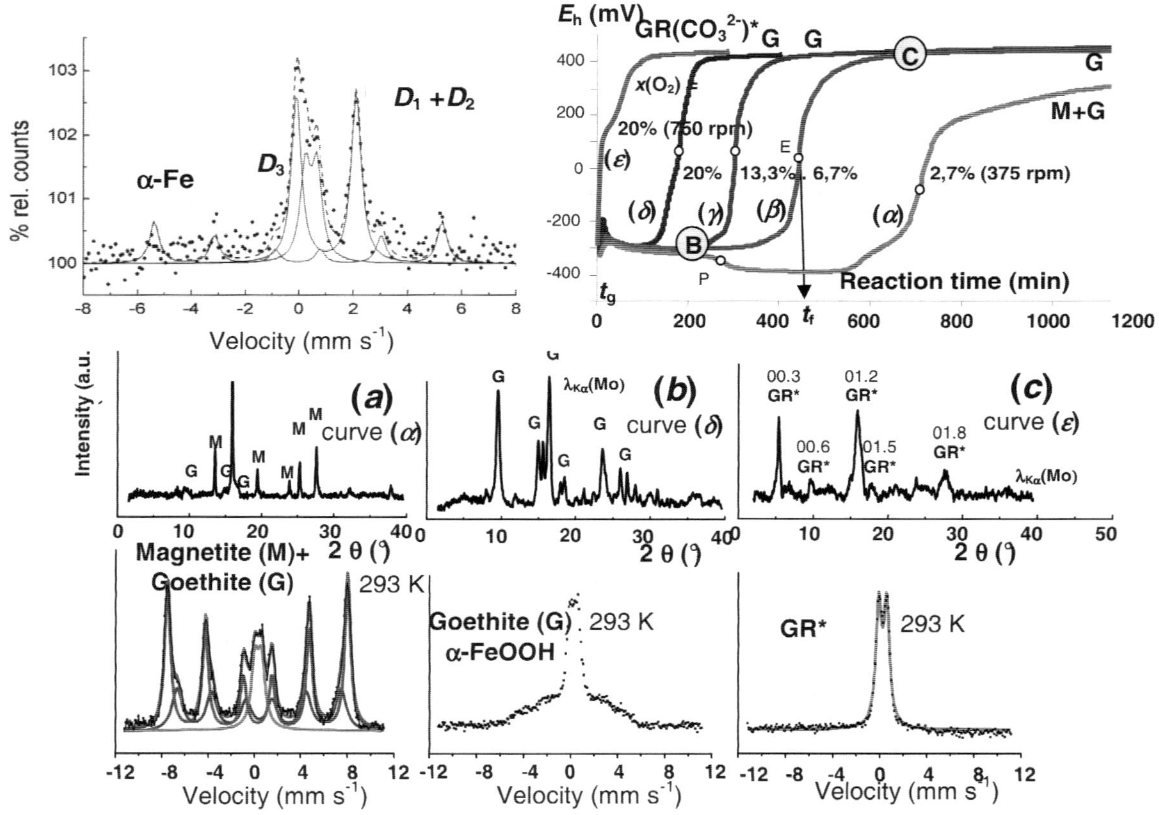

Figure 1. Top left. The first step of corrosion: the GR layer observed by CEMS spectrum at ambient of a steel disk dipped 24 hours in a 0.1 M $NaHCO_3$ solution [14]. Top right. Electrode potential during oxidation of GR(CO_3^{2-}) with rate of O_2; Bottom: XRD patterns and Mössbauer spectra at 293 K of end products: dissolution-precipitation (a-b) [curves (α-δ)] or *in situ* deprotonation (c) [curve (ε)].

Table 2. Hyperfine parameters of the spectra presented in Figure 1. *CS* : centre shift with respect to metallic α-Fe at room temperature; Δ: quadrupole splitting in the paramagnetic state or ε: quadrupole shift ; H : hyperfine magnetic field; *RA* : relative area. Spectrum of Figure 1b was not computed

Figure	Component	*CS* (mm s^{-1})	Δ or ε (mm s^{-1})	*H* (kOe)	*RA* (%)
1a	*D* (Fe^{III})	0.38	0.57	-	12
	S_1	0.34	0	500	14
	S_2	0.53	0	483	41
	S_3	0.47	0	446	33
1c	*D* (Fe^{III})	0.39	0.68	-	100

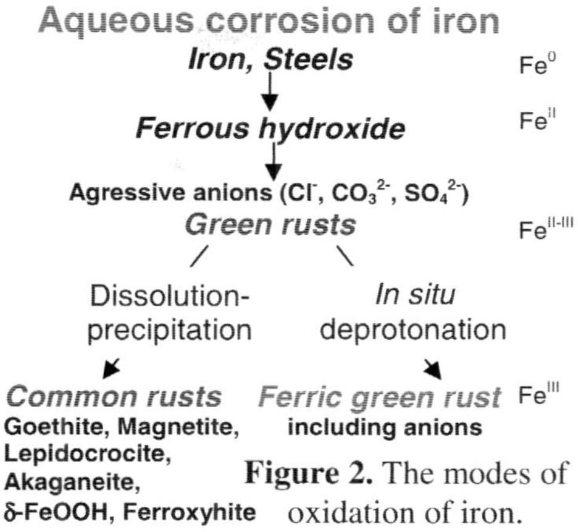

Figure 2. The modes of oxidation of iron.

since a constant value for E_h was reached after ~ 80 minutes in both experiments (pH = 7 and 9). End product of oxidation as shown by XRD was now GR(CO_3^{2-})*, ferric oxyhydroxycarbonate, which was characterized by two (00.3) and (01.2) main lines, in good agreement with literature [Figure 1c]. Similarly, a paramagnetic doublet characteristic of GR(CO_3^{2-})* was observed for these samples in the Mössbauer spectra for both experiments performed at pH = 9 or 7 [Figure 1c] [18].

3. Conclusion

A first stage of aqueous corrosion of iron corresponds to the formation of $Fe(OH)_2$ followed by that of a GR layer at the metal surface depending on the nature of the dissolved anions. For instance, a GR(CO_3^{2-}) rust layer was identified by CEMS at ambient on a steel disk dipped 24 hours in a 0.1 M $NaHCO_3$ solution (upper part [Figure 1]) [14]. Once a GR layer is formed, oxidation proceeds. Two modes may occur depending on the conditions: slow or fast oxidation i.e. dissolution-precipitation or *in situ* deprotonation [Figure 2]. This last mode is responsible for the passivation of weathering steels.

References

[1] Girard A and Chaudron G 1935 *C.-R. Acad. Sci. Paris* **200** 127
[2] Feitknecht W and Keller G 1950 *Z. Anorg. Allg. Chem.* **262** 61
[3] Hansen H C B 2001 Environmental chemistry of iron(II)-iron(III) LDHs (green rusts) *Layered double hydroxides: present and future* Nova Science publishers p. 469
[4] Génin J-M R, Bauer Ph, Olowe A and Rézel D 1986 *Hyperfine Interact.* **29** 1355
[5] Olowe A, Génin J-M R and Bauer Ph 1987 *Hyperfine Interact.* **41** 501
[6] Drissi S H, Refait Ph and Génin J-M R 1994 *Hyperfine Interact.* **90** 395
[7] Refait Ph and Génin J-M R 1993 *Corros. Sci.* **34** 797
[8] Génin J-M R, Olowe A A, Refait Ph and Simon L 1996 *Corros. Sci.* **38** 1751
[9] Drissi S H, Refait Ph, Abdelmoula M and Génin J-M R 1995 *Corros. Sci.* **37** 2025
[10] Génin J-M R, Refait Ph, Simon L and Drissi S H 1998 *Hyperfine Interact.* **111** 313
[11] Génin J-M R, Ruby C, Géhin A and Refait Ph 2006 *C.-R. Geoscience* **338** 433
[12] Génin J-M R and Ruby C 2004 *Solid State Sci.* **6** 705
[13] Génin J-M R and Ruby C 2008 *Solid State Sci.* **10** 244
[14] Abdelmoula M, Refait Ph, Drissi S H, Mihé J P and Génin J-M R 1996 *Corros. Sci.* **38** 623
[15] Génin J-M R, Aïssa R, Géhin A, Abdelmoula M, Benali O, Ernstsen V, Ona-Nguema G, Upadhyay C and Ruby C 2005 *Solid State Sci.* **7** 545
[16] Benali O, Abdelmoula M, Refait Ph and Génin J-M R 2001 *Geochim. Cosmochim.* Acta **65** 1715
[17] Vanderberghe R E, Barrero C A, Da Costa G M, Van San E and De Grave E 2000 *Hyperfine Interact.* **126** 247
[18] Génin J-M R, Ruby C and Upadhyay C 2006 *Solid State Sci.* **8** 1330

Iron carbide films produced by laser deposition

Y Yamada[1], H Yoshida[1], K Kouno[1], and Y Kobayashi[2]

[1] Department of Chemistry, Tokyo University of Science, Kagurazaka, Shinjuku, Tokyo 162-8601, Japan
[2] Nishina Center, RIKEN, Hirosawa, Wako, Saitama 351-0198 Japan

E-mail: yyasu@rs.kagu.tus.ac.jp

Abstract. Iron carbide films were produced by laser deposition of iron atoms onto aluminium substrates in an acetylene atmosphere. Gas-phase reaction of energetic Fe atoms with C_2H_2 yielded Fe/C particles as well as Fe and C atoms. The subsequent reaction on the Al substrate surface produced iron carbide films. The compositions of the films were measured using Mössbauer spectroscopy. The films produced in low-pressure C_2H_2 were mainly composed of Fe_7C_3 and Fe_3C, while the films produced in high-pressure C_2H_2 were composed of paramagnetic amorphous Fe-C. When the temperature of the Al substrate was increased, formation of crystalline Fe_7C_3 and Fe_3C was enhanced. The composition of iron carbide films was controlled both by the pressure of C_2H_2 and the temperature of the Al substrate. X-ray diffraction patterns and scanning electron micro spectroscopy was also applied to confirm the assignments.

1. Introduction

Iron carbides (Fe-C) have been extensively studied because of their various useful industrial applications. Iron-based materials such as ferrite, austenite, martensite, and cementite Fe_3C are well understood. Hägg (Fe_5C_2) and Eckstrom-Adcock (Fe_7C_3) carbides are known as metastable iron carbides. Recently, Fe-C compounds have attracted renewed attention because Fe was found to catalyze the synthesis of carbon nanotubes [1,2]. Nanocrystalline iron carbide particles, Fe_3C and Fe_7C_3, were produced by CO_2 laser pyrolysis of a vapor mixture of $Fe(CO)_5$ and C_2H_4 [3]. Iron carbide films have been formed by plasma-enhanced chemical vapor deposition introducing $Fe(CO)_5$ and H_2 [4]. Mechanical alloying of iron and graphite powders was found to yield iron carbides with high carbon content, and Mössbauer measurements show that Fe_3C and Fe_7C_3 were produced [5].

Laser deposition is a useful technique to produce films with controlled composition. We have previously reported the characteristics of iron films [6] and iron oxide films produced by laser deposition [7]. Laser ablation of Fe metal produces metal vapour consisting of Fe atoms having high translational energy and electronically excited states; the Fe atoms can react with any reactant gases that may be present. Unstable species or particles produced in a gas phase are deposited on a substrate and lose energy to the substrate to form films. Thus, the laser deposition technique associated with gas phase reaction can produce iron carbide species unobtainable under normal conditions. The formation of films containing metastable compounds is an attractive aspect of this technique. Here, we report the study of iron carbide films with various Fe/C composition ratios using laser deposition. Fe atoms were produced by laser ablation in a C_2H_2 atmosphere. The films were characterized using Mössbauer spectroscopy and X-ray diffraction (XRD). The surface morphology was observed using scanning electron micro spectroscopy SEM.

2. Experimental

Laser light from a YAG laser (NewWave, TEMPEST 10, 532 nm, 100 mJ/pulse, 5 ns, 2 Hz) was focused by a convex lens onto the target ^{57}Fe metal block in a C_2H_2 atmosphere. Laser-evaporated Fe atoms react with C_2H_2 in the gas-phase and form an iron carbide film by deposition onto an Al substrate. The temperature of the substrate was maintained at the desired temperature (300–600 K range) using a resistive heater. The amount of vaporized Fe atoms was estimated by weighing the samples before and after the laser deposition. One pulse of laser ablation produces 4×10^{-9} mol of Fe atoms, and the thickness of the laser-deposited films was controlled by varying the number of laser pulses. A film was produced by accumulating 14,400 laser pulses in order to obtain thick enough films to measure Mössbauer spectra. Mössbauer spectra of the Fe-C films on the Al substrates at room temperature were measured in a transmission geometry using a ^{57}Co/Rh source. X-ray diffraction patterns of the films were obtained using RINT2500 (RIGAKU, Cu-Kα). The surface morphology of the films was observed using a scanning electron microscope (S-500 HITACHI).

3. Result and Discussion

Laser deposition of Fe onto Al substrates at 600 K was carried out by varying the pressure of the C_2H_2 atmosphere to produce iron carbide films. The Mössbauer spectra of the films are shown in figure 1. The composition of the films varied depending on the C_2H_2 pressure. The film formed at the lowest pressure of 1.0 Pa (figure 1a) had several components, fitted into the combination of five sets of sextets and one doublet. Sextet absorption of α-Fe (H = 330 kOe) and cementite Fe_3C (H = 210 kOe) were easily assigned. The other three sets of sextets were assigned to Fe_7C_3 based on Mössbauer parameters reported previously [3]. A hexagonal [8] form of Fe_7C_3 containing three inequivalent Fe sites has been reported. However, orthorhombic [9] and pseudo-hexagonal [10] Fe_7C_3 structures were also reported. With increasing pressure of C_2H_2 gas (figure 1b, c and d), the intensity of the doublet increased, while absorption of Fe_3C and Fe_7C_3 decreased. Generally, Fe_7C_3 is formed in bulk only under very high temperatures and pressures, but thin films of Fe_7C_3 could be prepared at 600 K and low pressures by laser deposition. The yields of Fe_3C and Fe_7C_3 changed depending on the pressure. Similar results were obtained using plasma-enhanced chemical vapor deposition [4].

The film produced at 13 Pa of C_2H_2 (figure 1e) did not contain Fe_3C and Fe_7C_3, and the spectrum was fitted by a combination of a singlet, a doublet and α-Fe. The singlet and doublet were assigned to γ-Fe containing C atoms, and the intensity of the doublet is large because of the excess amount of C atoms. At the highest C_2H_2 pressure of 130 Pa (figure 1f), doublet absorption (δ = 0.18 mm/s and ΔEq = 1.19 mm/s) was observed, which was assigned to paramagnetic iron carbide species. The determination of this species is not simple because some other assignments are possible; the doublet may be due to the superparamagnetic nature of small Fe_3C particles, or small oxide particles produced by handling in air after production in the vessel. We measured the Mössbauer spectrum of this sample at 14 K, and the spectrum did not show a magnetic sextet or superparamagnetic broadening. Therefore, the doublet is not iron oxide particles, and was assigned to amorphous Fe-C containing a large amount of carbon atoms. Similar Mössbauer parameters were reported in literature [1, 5]. The Mössbauer parameters obtained in this experiment are summarized in table 1. The singlet corresponding to γ-Fe was produced only at the pressure of 13 Pa of C_2H_2 (figure 1e), while the doublets observed at pressures below 13 Pa of C_2H_2 (figure 1a-d) yielded Mössbauer parameters between $\delta = 0.0 \sim 0.2$ mm/s and $\Delta Eq = 0.6 \sim 1.0$ mm/s. These doublets might result partly because of C atoms in γ-Fe, but the C atom site is not well defined because of the large defects of γ-Fe in the films. The doublet obtained at a very high pressure of 130 Pa (figure 1f; δ = 0.18 mm/s and ΔEq = 1.19 mm/s) obviously corresponds to a different C-rich amorphous phase.

It is presumed that laser-evaporated iron atoms collide with C_2H_2 molecules in the gas phase to form C atoms and Fe_xC_y clusters. The Fe_xC_y clusters as well as Fe and C atoms are deposited onto the

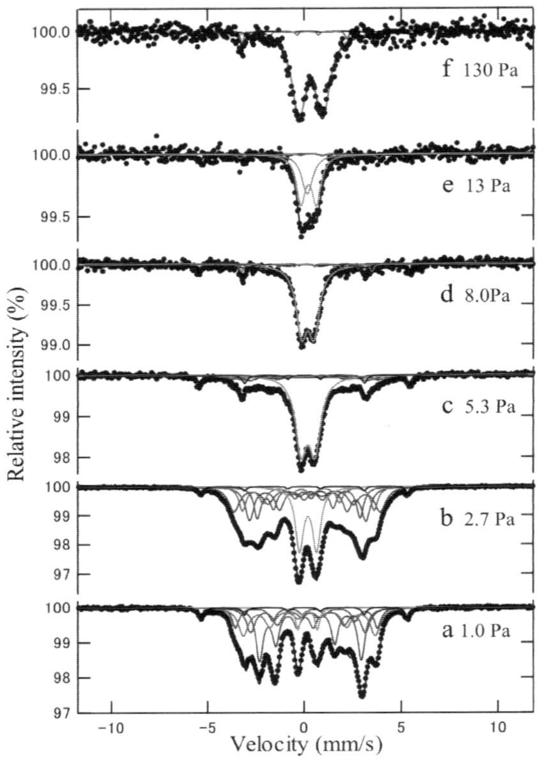

Figure 1. Room-temperature Mössbauer spectra of films produced by laser-deposition of iron onto Al substrates at 600 K in an acetylene atmosphere. The acetylene pressures are indicated in the figure.

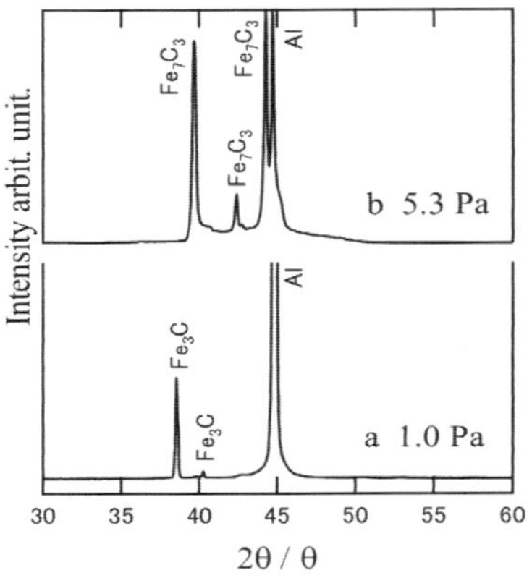

Figure 2. XRD patterns of the films produced by laser-deposition of iron onto Al substrates at 600 K in an acetylene atmosphere. The acetylene pressures are indicated in the figure.

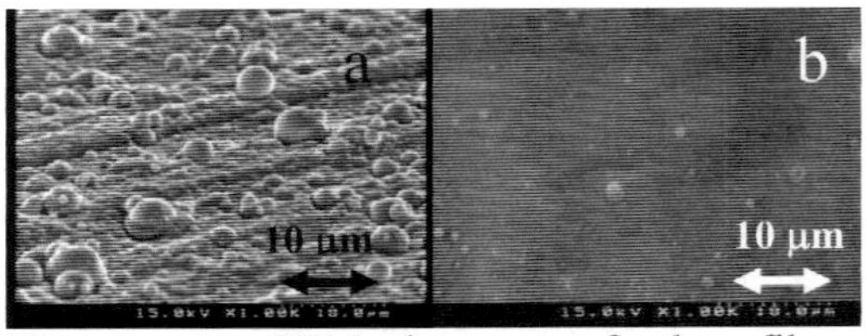

Figure 3. SEM images of the film produced by laser deposition of Fe onto an Al substrate at 600 K in a C_2H_2 atmosphere. The C_2H_2 pressure was (a) 13 Pa, and (b) 2.7 Pa.

Al substrate, and migrate to form a stable solid structure by releasing energy to the substrate. The pressure of C_2H_2 controls the concentration of C atoms in the films. Similar results were also reported for a study of ball-milling of Fe/C [5]; Fe_3C and Fe_7C_3 were produced at low concentrations of C, and the paramagnetic component appeared at a very high C concentration, assigned to fine particles of high carbon iron carbides.

XRD patterns of the film sample were also measured (figure 2). The most intense peak corresponds to the Al substrate, and Fe_3C is observed in the film produced at 1.0 Pa C_2H_2 (figure 2a). Fe_7C_3 is observed in films produced at 2.7 and 5.3 Pa of C_2H_2 (figure 2b). The intensity ratio of the XRD patterns did not correspond to the yield estimated by Mössbauer spectra because XRD is only sensitive to crystalline structures and is insensitive to very short range arrangements or amorphous structures. The film produced at a high pressure of C_2H_2 did not show any diffraction corresponding to iron carbides, and thus paramagnetic iron carbide may be amorphous, which was not observed by XRD. SEM micrographs of the films produced at 13 and 2.7 Pa are shown in figure 3. The film produced at 13 Pa (figure 3a) has a rugged surface because the atoms or the small particles produced in a gas phase deposited onto the substrate with insufficient energy for migration on the surface. The film produced at 2.7 Pa (figure 3b) has a smooth surface because of the formation of crystalline iron carbides.

We also carried out similar experiments by changing the substrate temperature to gauge the effects of reactions on the substrates. The migration of Fe and C atoms on the film surface and crystal

Table 1. Mössbauer parameters of the species produced by laser deposition of Fe in a C_2H_2 atmosphere.

	δ (mm/s)	ΔE_q (mm/s)	H (kOe)
Fe_7C_3	0.28	-0.32	229
	0.28	-0.19	184
	0.15	0.27	162
Fe_3C	0.16	0.14	210
α-Fe	0	0	330
γ-Fe	0.1	0	
	0.0 ~ 0.2	0.6 ~ 1.0	
Paramagnetic amorphous	0.18	1.19	

formation were restricted to substrates at low temperatures. Mössbauer spectra were obtained for the films formed by laser-deposition of Fe in a C_2H_2 atmosphere ranging from 1 to 130 Pa, while the temperature of the Al substrates was maintained at 300 K during the deposition. The major components of the films were paramagnetic Fe/C compounds showing doublets independent of the C_2H_2 pressure. Fe_3C was not found even in the film produced in a low-pressure C_2H_2 atmosphere (1.0 Pa), and broad absorption with magnetic distributions was observed. The broad absorption was fitted assuming hyperfine magnetic distributions; the distribution had a maximum at H = 130 kOe showing the formation of iron carbides with large defects. The energy of the deposited Fe/C atoms on the substrates at 300 K was not sufficient to crystallize Fe_3C. Generally, laser evaporated Fe atoms have high energy (several hundreds of eV) but they are cooled by collisions with C_2H_2 molecules. XRD patterns showing very a small peak corresponding to Fe_7C_3 were obtained.

4. Conclusions

Iron carbide films with various compositions were produced by laser deposition of Fe in a C_2H_2 atmosphere. Films containing Fe_3C and Fe_7C_3 were produced by laser deposition of Fe onto an Al substrate at 600 K in a low-pressure C_2H_2 atmosphere. When the films were produced at a high C_2H_2 pressure, paramagnetic amorphous Fe-C was obtained. The XRD patterns of the films and SEM observations were in agreement with this result. The films produced in low-pressure C_2H_2 had a smooth surface, while those produced in high-pressure C_2H_2 had a rugged surface. Both gas-phase and surface reactions govern the composition of the iron carbide films, and the surface reaction was enhanced by increasing the substrate temperature.

References

[1] Pérez-Cabero M, Taboada J B, Guerrero-Ruiz A, Overweg A R and Rodríguez-Ramos I 2006 *Phys. Chem. Chem. Phys.* **8** 1230
[2] de Resende V, De Grave E, Peigney A and Laurent C 2008 *J. Phys. Chem.* C **112** 5756
[3] Bi X, Ganguly B, Huffman G, Huggins F and Endo M 1993 *Mater J. Res.* **8** 1666
[4] Siriwardane H, Pringle O A, Newkirk J W and James E J 1996 *Thin Solid Films* **287** 8
[5] Tanaka T, Nasu S, Ishihara K N and Shingu P H 1991 *J. Less-Common Met.* **171** 237
[6] Yamada Y, Kato H, Kouno K, Yoshida H and Kobayashi Y 2009 *Hyperfine Interactions* **191** 121
[7] Yokoyama D, Namiki K, Fukasawa H, Miyazaki J, Nomura K and Yamada Y 2007 *J. Radioanal. Nucl. Chem.* **272** 631
[8] Herbstein F H and Snyman J A 1964 *Inorg. Chem.* **3** 894
[9] Fruchart R and Rouault A 1969 *Ann. Chim.* **143** 1615
[10] Audier M, Bowen P and Jones W 1983 *J. Cryst. Growth* **63** 125

Powders with superparamagnetic Fe_3C particles studied with Mössbauer spectrometry

B David[1], O Schneeweiss[1], F Dumitrache[2], C Fleaca[2], R Alexandrescu[2], and I Morjan[2]

[1]Institute of Physics of Materials, Academy of Sciences of the Czech Republic, v.v.i., Žižkova 22, CZ-61662 Brno, Czech Republic

[2]National Institute for Lasers, Plasma and Radiation Physics, P.O. Box MG-36, 077125 Bucharest-Magurele, Romania

Abstract. Two nanopowders with superparamagnetic Fe_3C particles were synthesised by the method of laser-induced pyrolysis of gaseous precursors. Both were characterised by X-ray diffraction, Mössbauer spectrometry and standard magnetic measurements. The mean crystallite size of Fe_3C was 3 nm for the first sample and 10 nm for the second sample (Scherrer formula), i.e. it was lower than in our previously studied ferromagnetic Fe_3C-based sample. Fe_3C phase in both present samples exhibited by ~20 K reduced Curie temperature which is interpreted as a nanosize effect. After annealing of the samples at 1073 K for 30 minutes the Curie temperature of the Fe_3C phase in both samples matched its standard bulk value. Beside Fe_3C phase also Fe_3O_4 and carbon black were present in the synthesised samples.

1. Introduction

Magnetic nanoparticles (MNPs) attract the attention of the materials science community because of their unique magnetic and electronic properties [1,2]. Among MNPs gained high attention iron-based nanoparticles [3], especially for their prospective medical applications. α-Fe, γ-Fe and Fe_3C nanoparticles were studied predominantly as the catalysts for growing carbon nanotubes (CNTs).

Of all the gas phase synthesis methods, the laser-induced homogeneous pyrolysis (LIHP) is a powerful and versatile tool for the creation of nanoparticles with various chemical compositions and diameters ranging from a few nanometers to about 50 nm [4]. It has been demonstrated in our previous reports that this method can provide nanopowders with a dominant Fe_3C content [5-7].

In the present paper we study samples containing superparamagnetic Fe_3C particles, i.e. with thermally fluctuating orientation of the particle magnetic moment (superspin) at room temperature.

2. Experimental

Two types of Fe_3C-based nanopowders were synthesised by the method of laser-induced pyrolysis of gaseous precursors. The focused continuous-wave CO_2 laser radiation was orthogonally crossed with the reactant gas mixture stream that was admitted to the center of the reaction cell through the inner nozzle of the concentric two wall injector [4]. The outer nozzle was used for the argon coflow which confined the reactive gases stream to the flow axis. SF_6 was employed as the only laser radiation absorber. $Fe(CO)_5$ vapors served as an iron source, acetylene as a carbon donor and ammonia as a nitrogen precursor. The main parameters of the two experiments are resumed in Tab. 1.

The composition of the synthesised nanopowders was examined by X-ray diffraction (XRD). The XRD pattern fitting was done with the TOPAS software using the ICSD database of crystal structures and it yielded weight fraction F and mean crystallite size d_{XRD} for a given phase [8,9]. Mössbauer spectra were obtained at standard transmission geometry with ^{57}Co in Rh matrix. Spectra were fitted with the CONFIT programme using transmission integral and we obtained the values of relative spectrum area A for a given phase and spectral component parameters: hyperfine magnetic induction B_{HF}, quadrupole shift ε_Q, quadrupole splitting ΔE_Q and isomer shift δ (against α-Fe) [10]. Vibrating sample magnetometer (VSM) was employed for standard magnetic measurements.

Table 1. Experimental data for the synthesis

Sample	P_{laser} after absorption (W)		Central nozzle flows (sccm)			Ar coflow (sccm)	Pressure (mbar)	Powder productivity (mg/min)
	In vacuum	in gas mixture	C_2H_2 through $Fe(CO)_5$	SF_6	NH_3			
FCB1	50	34	175	10	-	1100	360	55
FCB3N	40	31	140	10	35	1000	350	57

3. Results and discussion

3.1. X-ray diffraction

Wide and overlapping diffraction peaks appeared in the X-ray diffractograms. The XRD patterns of the samples were fitted with orthorombic cementite θ-Fe_3C (ICSD No. 16593) and magnetite Fe_3O_4 (ICSD No. 43001). The Rietveld refinement yielded in the case of the **FCB1** sample for Fe_3C values d_{XRD} = 3 nm, F = 74 wt.%, and for Fe_3O_4 values d_{XRD} = 4 nm, F = 26 wt.%; in the case of the **FCB3N** sample for Fe_3C values d_{XRD} = 10 nm, F = 33 wt.%, and for Fe_3O_4 values d_{XRD} = 3 nm, F = 67 wt.%. The presence of maghemite γ-Fe_2O_3 could not be excluded because γ-Fe_2O_3 and Fe_3O_4 have similar XRD patterns. Compared with our previously studied Fe_3C-based sample [5], the d_{XRD}(Fe_3C) values in present samples are cosiderably lower. The presence of carbon black in both samples was confirmed by Raman spectroscopy (well pronounced D peak at ~1360 cm^{-1} and G peak at ~1580 cm^{-1}).

3.2. Mössbauer spectrometry

The Mössbauer spectrum (MS) of the **FCB1** sample measured at 293 K (Fig. 1) was fitted with three components: a narrow doublet probably of superparamagnetic Fe_3C (δ = 0.18 mm s^{-1}, ΔE_Q = 0.42 mm s^{-1}, A = 0.36; dashed line), a broad doublet of superparamagnetic Fe_3O_4 (δ = 0.30 mm s^{-1}, ΔE_Q = 0.91 mm s^{-1}, A = 0.53; full line) and a superposition of three sextets (B_{HF} = 19.2 T, 12.3 T, 15,9 T, A = 0.11; dark gray filler). In the MS measured at 60 K the sextet characteristic for ferromagnetic Fe_3C (B_{HF} = 23.9 T, ε_Q = 0.04 mm s^{-1}, δ = 0.31 mm s^{-1}, A = 0.30; light gray filler) was clearly identified [7]. In this spectrum also the superparamagnetic Fe_3O_4 doublet (δ = 0.40 mm s^{-1}, ΔE_Q = 1.07 mm s^{-1}, A = 0.40) was found. In the MS measured at 27 K the sextet of ferromagnetic Fe_3C (B_{HF} = 24.5 T, ε_Q = 0.07 mm s^{-1}, δ = 0.36 mm s^{-1}, A = 0.74; light gray filler) dominated. The absence of the characteristic ferromagnetic Fe_3C sextet at 293 K and its appearance at lower temperatures is the consequence of the superparamagnetism of Fe_3C nanoparticles at 293 K.

In the case of the **FCB3N** sample the MS measured at 293 K (Fig. 1) exhibits very low absorption. This spectrum was fitted with the ferromagnetic Fe_3C sextet (B_{HF} = 20.5 T, ε_Q = 0.01 mm s^{-1}, δ = 0.19 mm s^{-1}, A = 0.28; dark gray filler) [5], a doublet (δ = 0.19 mm s^{-1}, ΔE_Q = 0.51 mm s^{-1}, A = 0.64; dashed line) and the superparamagnetic Fe_3O_4 doublet (δ = 0.68 mm s^{-1}, ΔE_Q = 0.70 mm s^{-1}, A = 0.08; full line). In the MS measured at 60 K again the ferromagnetic Fe_3C sextet (B_{HF} = 24.9 T, ε_Q = -0.05 mm s^{-1}, δ = 0.32 mm s^{-1}, A = 0.32; light gray filler) was found as well as in the spectrum measured at 27 K (B_{HF} = 25.0 T, ε_Q = 0.01 mm s^{-1}, δ = 0.32 mm s^{-1}, A = 0.42; light gray filler). The intense outer lines in the MS measured at 27 K correspond to a high fraction of Fe_3O_4 in the FCB3N sample. Iron nitrides were not found.

It can be concluded that two effects strongly influenced the measured Mössbauer spectra (Fig. 1): superparamagnetic effect (sextet representing magnetically ordered phase collapses to doublet above the blocking temperature T_B) [11] and soft bonding of Fe-based particles to the pyrolytic carbon matrix (recoilless factor f strongly increases with decreasing temperature, i.e. relative absorption at 27 K is higher than at 293 K) [12]. Therefore the direct assignment of A values of Fe_3C or Fe_3O_4 spectral components to corresponding weight fractions at 293 K, 60 K and 27 K is not possible.

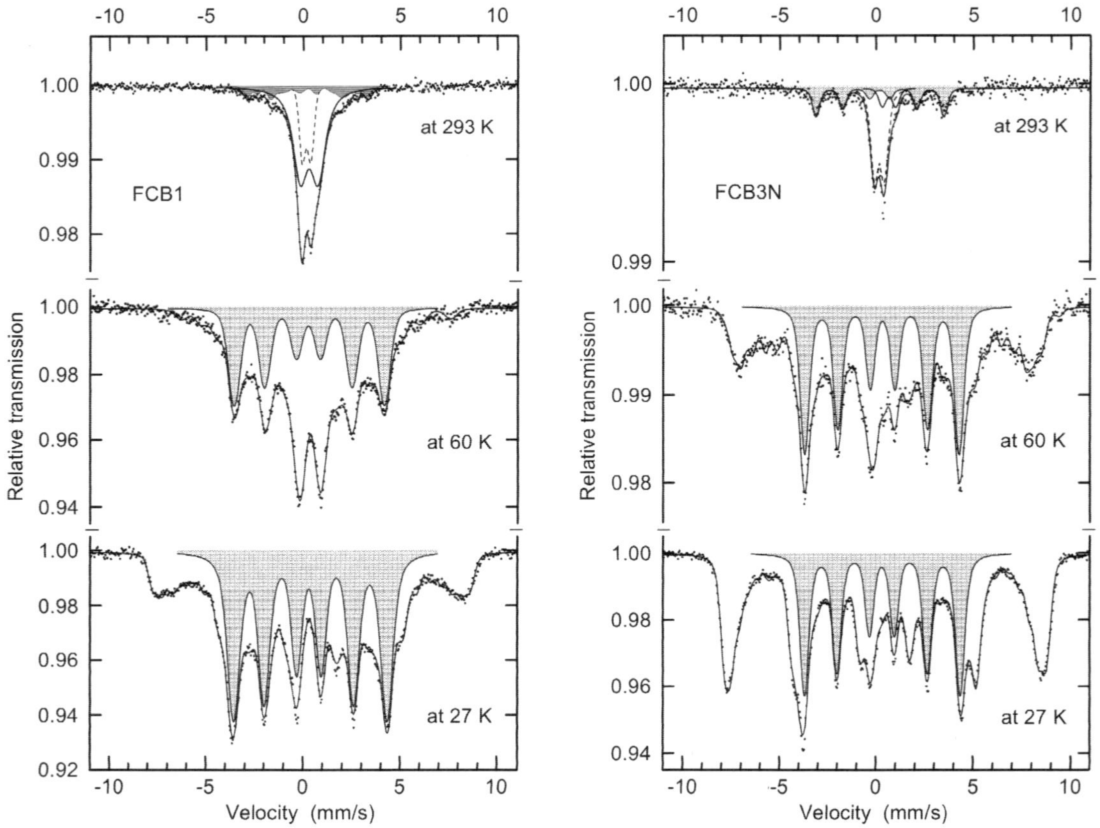

Figure 1. Mössbauer spectra for the as-synthesised samples measured at the indicated temperatures. Sextets with light grey filler correspond to Fe_3C.

3.3. Magnetic measurements

From the hysteresis loop of the **FCB1** sample measured at 293 K we obtained the values: H_C = 1.2 kA m^{-1}, σ_R = 0.35 A m^2 kg^{-1}, σ_S = 22.1 A m^2 kg^{-1} (lower than previously reported in [6]). The shape of the hysteresis loop exhibited superparamagnetic character. The majority of Fe_3C and Fe_3O_4 nanoparticles were in superparamagnetic state at 293 K. By the same measurement of the **FCB3N** sample we obtained the values: H_C = 17.1 kA m^{-1}, σ_R = 7.4 A m^2 kg^{-1}, σ_S = 40.2 A m^2 kg^{-1}. In this case higher H_C and σ_R values reflected the presence of certain amount of Fe_3C nanoparticles in ferromagnetic state at 293 K whereby Fe_3O_4 particles were superparamagnetic.

By the thermomagnetic curve measurement (TM curve) a pellet was held in the evacuated cell (~1 Pa), heated up to 1073 K (hold time 30 minutes) at the rate of 4 K min^{-1} in the external field H_{ext} = 4 kA m^{-1} and then cooled down to room temperature. The Curie temperatures T_C(bulk-α-Fe) ≈ 1040 K and T_C(bulk-Fe_3C) ≈ 483 K [5,6] can be recognised on the TM curves of both samples (Fig. 2). However, T_C(nano-Fe_3C) determined from the $\sigma_\uparrow$(T) measured with increasing temperature, i.e. for the as-synthesised mainly superparamagnetic Fe_3C nanoparticles, is evidently by ~20 K lower then T_C(bulk-Fe_3C) determined from the $\sigma_\downarrow$(T) measured with decreasing temperature (recrystallised sample). This effect has not been observed in our previous study dealing with a ferromagnetic Fe_3C-based nanopowder with d_{XRD}(nano-Fe_3C) = 12 nm [5-7]. The decrease of T_C is well known as a nanosize effect and was described, for instance, by the authors of the articles [13,14].

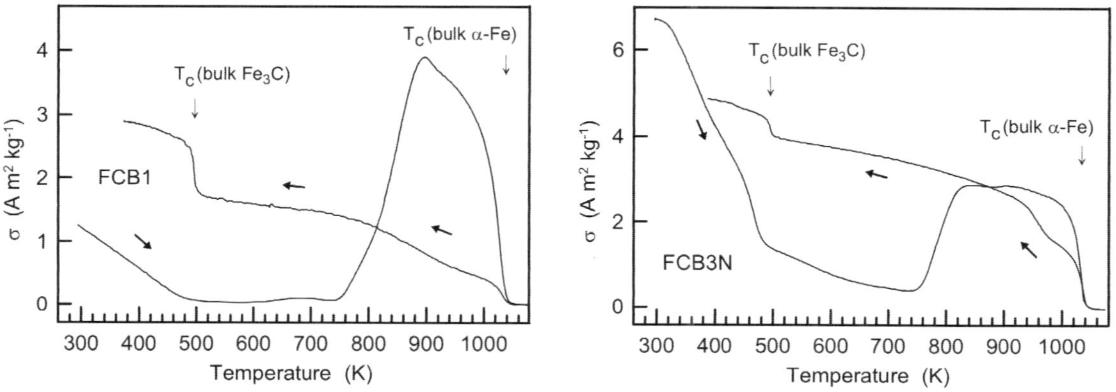

Figure 2. TM curves for the the as-synthesised samples.

4. Conclusions

It can be summarised that in the studied samples the presence of the significant amount of Fe_3C particles superparamagnetic at room temperature was proved. According to our knowledge, such Fe_3C-based nanopowders were synthesised by the laser-induced pyrolysis method for the first time.

Acknowledgement This work was supported by the Academy of Sciences of the Czech Republic (AV 0Z20410507), the Grant Agency of the Czech Republic (202/08/0178), and the Ministry of Education, Youth and Sport of the Czech Republic (1M6198959201).

References

[1] Kodama R H 1999 *J. Magn. Magn. Mater.* **200** 359
[2] Lu A, Salabas E L, Schüth F 2007 *Angew. Chem.- Int. Ed.* **48** 1222
[3] Huber D L 2005 *Small* **1** 482
[4] Alexandrescu R, Morjan I, Voicu I, et al. 2005 *Appl. Surf. Sci.* **248** 138
[5] David B, Zboril R, Mashlan M, et al. 2006 *J. Magn. Magn. Mater.* **304** e787
[6] David B, Schneeweiss O, Pizúrová N, et al. 2006 *Surf. Interface Anal.* **38** 482
[7] David B, Schneeweiss O, Mashlan M, Šantavá E, 2007 *J. Magn. Magn. Mater.* **316** 422
[8] DIFFRAC^{plus} TOPAS software, release 2000, Bruker AXS GmbH, Karlsruhe, Germany.
[9] ICSD Inorganic Crystal Structure Database, release 2009/1, FIZ Karlsruhe, Germany.
[10] Žák T, Jirásková Y 2006 *Surf. Interface Anal.* **38** 710
[11] Dorman J L, Fiorani D, Tronc E 1997 *Adv. Chem. Phys.* **98** 283
[12] David B, Pizúrová N, et al. 2007 *J. Phys. Chem. Solids* **68** 1152 and references therein
[13] Michels D, Krill III C E, Birringer R 2002 *J. Magn. Magn. Mater.* **250** 203
[14] Lang X Y, Zheng W T, Jiang Q 2006 *Phys. Rev. B* **73** 224444

Journal of Physics: Conference Series **217** (2010) 012098 doi:10.1088/1742-6596/217/1/012098

Investigation of the Formation of Fe-filled Carbon Nanotubes

H Reuther[1], C Müller[2], A Leonhardt[2] and M C Kutz[2]

[1] *Forschungszentrum Dresden-Rossendorf, P.O. Box 510119, D-01314 Dresden, Germany*
[2] *Leibniz-Institute of Solid State and Materials Research Dresden, P.O. Box 270116, D-01171 Dresden, Germany*

E-mail: reuther@fzd.de

Abstract. The formation of Fe-filled carbon nanotubes by thermal decomposition of ferrocene combined with a Fe-catalyst-nanostructuring on an oxidized Si substrate is investigated in the temperature range of 1015 – 1200 K. The optimal growth conditions for aligned and homogeneous carbon nanotubes are found at 1103 K. Mössbauer spectroscopy (both in transmission geometry and CEMS) was used to analyze and quantify the different formed Fe-phases. In general, α-Fe, γ-Fe and Fe_3C are found to form within the carbon nanotubes. Depending on the growth conditions their fractions vary strongly. Moreover, an alignment of the α-Fe in the tubes could be detected.

1. Introduction

The nanostructured magnetic nanowires are promising candidates for many applications ranging from magnetic data storage and sensors for scanning force microscopy to ferromagnetic nanocontainers for biomedical applications.

In order to protect the nanowires from oxidation, mechanical damage, and magnetic shortcuts between each other, the nanowires are encapsulated inside carbon nanotubes (CNTs).

Chemical vapor deposition (CVD) has established itself over the last years as the synthesis method for filled CNTs, because it provides material of good quality and can be easily scaled up. Most CVD procedures use metals as the process catalysts (usually Fe, Co, Ni) to produce CNTs. The catalysts can be placed on substrates and/or directly produced via a precursor compound (e.g., ferrocene). The properties of CNTs are strongly influenced by the growth conditions [1, 2].

In the past, Mössbauer spectroscopy was used several times to characterize the phases formed in iron containing CNTs. Depending on the preparation conditions, iron oxides, Fe_3C, α- and γ-Fe were detected [3-6].

In the following, we report about the preparation of CNTs prepared by CVD and their characterization by scanning electron microscopy and Mössbauer spectroscopy.

2. Experimental

The synthesis of Fe-filled multiwalled CNTs was carried out by catalytic decomposition of ferrocene ($Fe(C_5H_5)_2$) in a quartz tube reactor inside a dual-zone electrical furnace. The ferrocene acts as the carbon source for the CNTs and the Fe filler.

The following parameters were used: Sublimation temperature of the ferrocene in the first furnace (T= 423 K), deposition temperature in the second furnace (T= 1015-1200 K), Ar flow rate (120 cm^3/min). For the ferrocene supply rates between 3 and 6.5 mg/min were used. The experiments were performed at atmospheric pressure. Oxidized (100) oriented Si substrates (10 × 10 mm^2, 1 μm SiO_x) coated with a thin Fe layers (2 nm, sputtered) were used for the deposition. One set of samples was prepared by inserting a 10 nm Al buffer layer between the silicon oxide and the iron. The deposition time was set to 10 min. At the end of each experiment the reactor was slowly cooled down to room temperature (25 K/min). The schematic setup of the quartz tube reactor is displayed in figure 1.

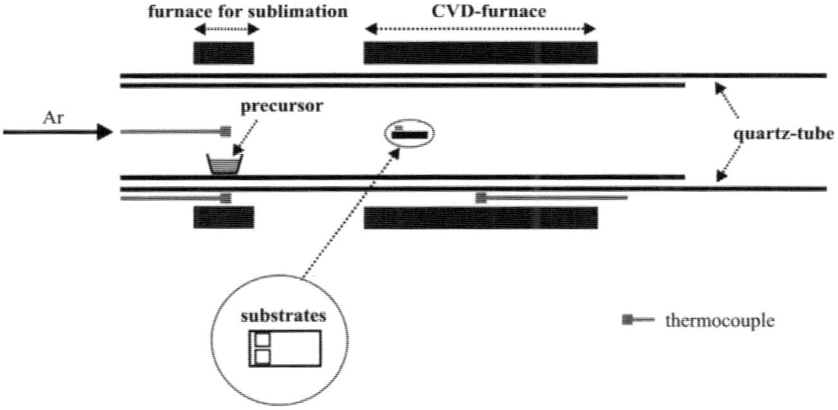

Figure 1. Schematic setup of the quartz tube reactor

The morphology of the nanotubes was monitored by scanning electron microscopy (SEM) (Gemini Leo 1530 and Hitachi S 4800, both operated at 10 kV).

The fractions of iron phases formed inside the CNTs were determined by Mössbauer spectroscopy. The CNTs were studied both in transmission geometry (TM) and by conversion electron Mössbauer spectroscopy (CEMS). Measurements were performed at room temperature with a conventional spectrometer with constant acceleration driving system with a Co-57(Rh) source of nominally 3.7 GBq. The TM spectra were taken first directly from the aligned CNTs on substrates. The absorption due to the thick Si substrate (some 100 μm) was very strong and it was difficult to obtain spectra of reasonable quality. Therefore, the CNTs were mechanical removed from the substrate and inserted as powder samples. CEMS was performed with a proportional gas flow detector with a He/CH_4 mixture as counting gas. Spectra were evaluated with the NORMOS program package. To improve the quality of the Mössbauer spectra some of the samples were implanted with Fe-57 ions. With an energy of 40 keV and an implantation dose of $1x10^{16}$ cm^{-2} the concentration of Fe-57 was increased to up to 10 at. % within the first some 10 nm. There was no general difference between CEM spectra taken from non-implanted and from implanted samples. Thus it can be assumed that the implanted Fe-57 ions act as probes only and do not change the phase composition or create new phases in the tubes.

3. Results and discussion

Length, homogeneity and alignment of the CNTs depend strongly on the formation temperature. The best values are achieved at 1103 K. Scanning electron micrographs of the CNTs are displayed in

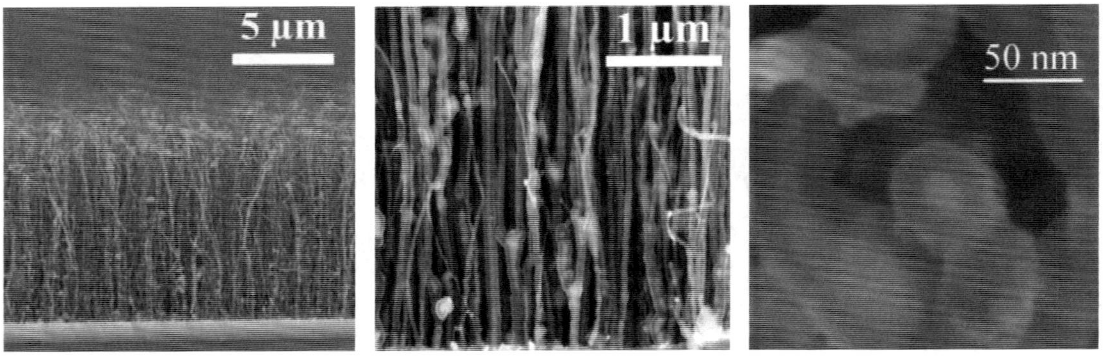

Figure 2. Scanning electron micrographs of the CNTs

figure 2. It can be seen that the grown tubes have a lenghts of up to 10 µm and a thickness of some 10 nm. The Fe cores have diameters of less than 20 nm.

In figure 3 some selected Mössbauer spectra are displayed. Figure 3a shows TM and CEM spectra of the same sample which was grown at 1103 K. It is obvious that the fractions of the iron phases are very different. Moreover, from the CEM spectrum a strong texture in the tubes can be assumed. In figure 3b CEM spectra of nanotubes deposited at different rates are shown. At higher rates the formation of α-Fe is preferred while at lower rates the formation of Fe_3C dominates. Due to different process temperatures and different ferrocene supply rates the fractions of the formed iron containing phases can strongly be varied. No iron oxides, silicides or aluminides could be detected. The fractions of the formed phases are given in table 1. For the fit the following hyperfine parameters are used: α-Fe: a sextet with isomer shift $\delta = 0.0$ mm/s, magnetic hyperfine field $B_{hf} = 33.0$ T; γ-Fe: a singlet line $\delta = -0.1$ mm/s; Fe_3C: a sextet $\delta = 0.2$ mm/s, $B_{hf} = 20.8$ T.

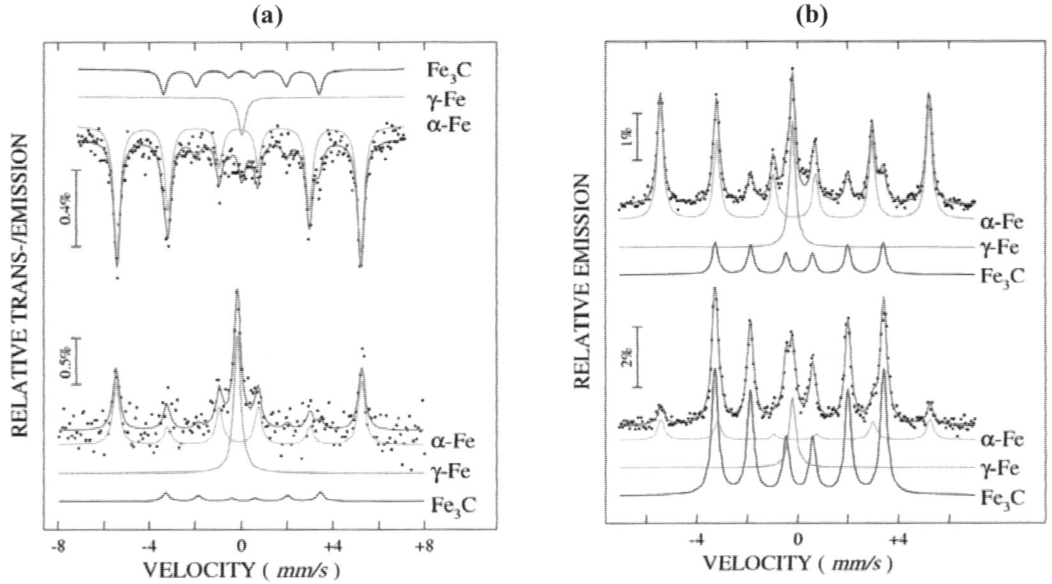

Figure 3. Mössbauer spectra from CNTs. Left: (a) TM (upper) and CEM (lower) spectra of nanotubes grown at 1103 K. Right (b): CEM spectra of nanotubes deposited at different supply rates (upper: 6.5 mg/min, lower: 3.0 mg/min).

Table 1. Fractions of the phases formed in the nanotubes

Experimental conditions	α-Fe	γ-Fe	Fe_3C	Fig.
1103 K, 6.5 mg/min	79	6	15	3 a top (TM)
1103 K, 6.5 mg/min	69	28	3	3 a bottom (CEMS)
1103 K, 6.5 mg/min, Al buffer layer	62	16	22	3 b top (CEMS)
1103 K, 3.0 mg/min, Al buffer layer	12	10	78	3 b bottom (CEMS)

3. Conclusions

Mössbauer spectroscopy is very suitable to investigate the formation of the different iron phases. Besides metallic iron as α-Fe or γ-Fe iron carbide was formed but no iron oxide or other iron compounds. This is in agreement with the findings of [3] and [4] while in [5] only Fe_3C was detected and in [6] α-Fe and small fractions of iron oxides were found. Contrary to [3], within the tubes a strong alignment of the crystalline ranges took place. This is confirmed by transmission electron microscopy too [7]. However, the results presented here show that the formation of the different phases is strongly influenced by the kind of preparation.

The optimal growth conditions for aligned and homogeneous Fe-filled carbon nanotubes (maximum content metallic iron, minimum content iron carbide) are found at 1103 K and a high supply rate.

References

[1] Müller C, Hampel S, Elefant D, Biedermann K, Leonhardt A, Ritschel M and Büchner B 2006 *Carbon* **44** 1746

[2] Müller C, Leonhardt A, Kutz M C, Büchner B, and Reuther H 2009 *J. Phys. Chem. C* **113** 2736

[3] Marco J F, Grancedo J R, Hernondo A, Crespo P, Prados C, Gonzales J M, Grobert N, Terrones M, Walton D R M and Kroto W 2002 *Hyperfine Interactions* **139/140** 535

[4] Ruskov T, Asenov S, Spirov I, Garcia C, Mönch I, Graff A, Kozhuharova r, Leonhardt A, Mühl T, Ritschel M, Schneider C M and Groudeva-Zotova S 2004 *J. Appl. Phys.* **96** 7514

[5] Lottermoser W, Schaper A K, Treutmann W, Redhammer G, Tippelt G, Lichtenberger A, Weber S U and Amthauer G 2006 *J. Phys. Chem. B* **110** 9768

[6] Jorge J, Flahaut E, Gonzalez-Jimenez F, Gonzalez G, Gonzalez J, Belandria E, Broto J M and Raquet B 2008 *Chem. Phys. Letters* **457** 347

[7] Goldberg D, Mitome M, Tang C, Leonhardt A and Bando Y 2006 *Acta Materialia* **54** 2567

Structural and magnetic properties of Fe-Al_2O_3 films prepared by helicon plasma sputtering

I Sakamoto[1], **S Purwanto**[2], **M Koike**[3], **S Honda**[4], **N Hayashi**[5], **Y Miyamoto**[6], **H Wakabayshi**[6], **T Toriyama**[6] and **K Nomura**[7]
[1]Hosei University, Koganei 184-8584, Japan
[2]BATAN, Serpong, Indonesia
[3]AIST, Tsukuba 305-8568, Japan
[4]Shimane University, Matsue 690-8504, Japan
[5]Kurume Institute of Technology, Kurume 830-0052, Japan
[6]Tokyo City University, Setagaya-ku 158-8557, Japan
[7]The University of Tokyo, Bunkyo-ku 113-8656,Japan

E-mail: isao.sakamoto.6a@k.hosei.ac.jp

Abstract. Fe-Al_2O_3 films have been prepared by on-off deposition of Fe with continuous Al_2O_3 deposition, and also subsequently irradiated by 400 keV Ar ions for controlling of the structural and magnetic properties of the films. CEM spectra of the as-deposited films show the decrease in the average hyperfine fields (H_{hfs}) with decreasing Fe/ Fe-Al_2O_3 volume ratio with keeping the H_{hfs} orientation parallel to the film plane. The decrease in H_{hfs} indicates the decrease of the Fe particle size and the in-plane orientation of H_{hfs} implies the non spherical shape of Fe particles in the film. On the other hand, 400 keV Ar ion irradiation induces the change from the superparamagnetic characteristics to the ferromagnetic one; the ferromagnetic peaks showing the random orientation of H_{hfs} appear and indicate the spherical growth of Fe particles in Al_2O_3 matrix.

1. Introduction

Nanocomposite materials like Fe-AlO films are very attractive in practical and basic fields because of appearance of new physical properties. Many preparation techniques (sputtering, ion implantation and sol-gel) are used to prepare nanocomposite materials [1-4]. The physical properties of nanocomposite materials are closely related to the size, shape and distribution of metal particles in insulating matrix. It has been reported that the film preparation by pulsed laser deposition and the process of ion beam irradiation after film preparation were suggested to control the properties of metal particles in insulating matrix [5, 6]. In this work, we deposited Fe-Al_2O_3 films with the open-close (on-off) operation of shutter in the front of helicon plasma magnetron cathode for film preparation [7], controlling the Fe contens in the film. Following the deposition, we did Ar ion irradiation for as-deposited film in order to modify the structural and magnetic properties of Fe particles in Fe-Al_2O_3 nanocomposite films.

2. Experimental

$Fe-Al_2O_3$ films were prepared on high resistive n-type (100) Si substrates by helicon plasma sputtering method in a base pressure lower than $1x10^{-7}$ Torr and in $7x10^{-4}$ Torr Ar gas pressure during depositions. The deposition rates of Fe and Al_2O_3 were 0.06 and 0.008 nm/sec, respectively. The $Fe-Al_2O_3$ film was formed by on-off deposition of Fe with continuous Al_2O_3 deposition. The composition and the thickness of $Fe-Al_2O_3$ film were controlled by the ratio of on-off Fe deposition and the total deposition time. In addition it is expected that the change in on-off time of Fe deposition affects the structural and magnetic properties of $Fe-Al_2O_3$ films. The measurement of conversion electron Mössbauer spectroscopy (CEMS) was done using a Mössbauer spectrometer with 740 MBq ^{57}Co γ-ray source (Rh matrix), and conversion electrons were detected with a proportional counter flowed with He + 10 % methane mixture gas. The CEM spectra were analyzed by least square fitting assuming overlapped Lorentzian curves of singlet peak, doublet peaks and sextet peaks. The distribution of hyperfine field was assumed for broad widths of sextet peaks. The ratios of peak intensities for first, second and third peaks of sextet peaks were fixed as $3{:}x{:}1$ where x was a free parameter. The structural and magnetic properties of the films were examined by x-ray diffraction (XRD) using CuK_α radiation and vibrating sample magnetometer (VSM) up to 1.5 T. The ion irradiation has been performed to the fluence of $5x10^{16}$ ions/cm^2 by 400 keV Ar ions. The ion range of 400 keV Ar ions in Fe is calculated to be 182 nm by TRIM code. As the 3s (on)-12s (off) film (144 nm) is thinner than this ion range, the implanted Ar ions stop into Si substrate passing through $Fe-Al_2O_3$ film.

3. Results and discussion

Figure 1(a) and 1(b) are the CEM spectra of $Fe-Al_2O_3$ films prepared by on-off Fe deposition. The films were prepared on the condition of the same ratio between on- and off- Fe deposition time, 5s (on) -10s (off) and 3s (on) -6s (off) with the continuous Al_2O_3 deposition for 60 minutes. The CEM spectra show the overlapping of sextet peaks and singlet peak. Although not shown here, the magnetization curves of 5s-10s and 3s-6s films show partially superparamagnetic characteristics which corresponds to the singlet peak in the CEM spectra. The thicknesses of Fe layer corresponding to 5s or 3s deposition is within 0.3 nm. The thickness of Al_2O_3 by continuous deposition under Fe deposition is less than 0.12 nm. Therefore, it seems that Fe layer does not continuous and forms Fe particles in Al_2O_3 matrix. The average hyperfine fields (H_{hfs}) of figures 1(a) and 1(b) are 21.8 T and 18.2 T, respectively. From the values of H_{hfs} in CEM spectra at room temperature below the superparamagnetic blocking temperature, the Fe particle size in Al_2O_3 matrix of 5s-10s and 3s-6s films can be estimated to be 2.9 nm and 2.6 nm (assuming that magnetic anisotropy energy constant, K equals to $5x10^5$ J/m^3 and particles are spherical.) [8]. Nevertheless, using Scherrer's equation, the Fe particle sizes from the widths of Fe (110) peak in XRD patterns of 5s-10s and 3s-6s films are estimated to be 2.0 nm and 1.9 nm, respectively. As there is no difference in the particle sizes estimated from XRD patterns of the both films, the difference in the values of H_{hfs} implies the structural change which results from the shape and distribution of Fe particles in 5s-10s and 3s-6s films.

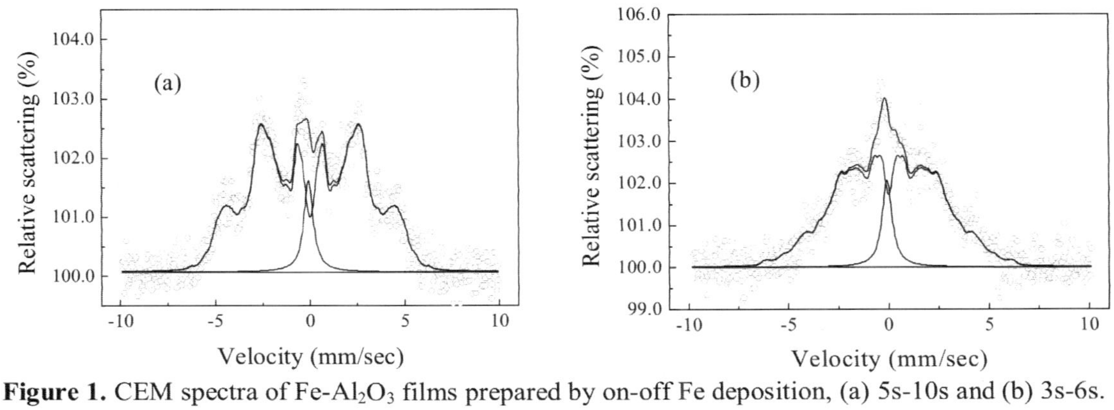

Figure 1. CEM spectra of $Fe-Al_2O_3$ films prepared by on-off Fe deposition, (a) 5s-10s and (b) 3s-6s.

Table 1. The Fe volume ratio, film thickness and hyperfine parameters of Fe-Al_2O_3 as-deposited films and the film after 400 keV Ar ion irradiation to the fluence of 5x10^{16} ions/cm^2.

Open (sec)-close (sec)	Nominal x of $Fe_x(Al_2O_3)_{1-x}$	Nominal thickness (nm)	H_{hfs} (T)	Peak intensity ratio (2nd/3rd)
1-0	0.88	244.8	31.9	4
5-5	0.78	136.8	30.2	4
5-10	0.71	100.8	21.8	3.5
3-6	0.71	100.8	18.2	3.9
3-12	0.6	144		
3-12 after irradiation	0.6	144	30.8	2.3

In Table 1, the H_{hfs} of Fe-Al_2O_3 films decreases with decreasing volume ratio of Fe in the films. In addition, the intensity ratio of second/third -peaks does not change against the volume ratio of Fe. The value 4 in the peak ratio indicates that the H_{hfs} orients to the film plane. Therefore, the shape of the Fe particles in Fe-Al_2O_3 films seems to elongate in the film plane. It implies that the Fe particles dispersed in the as-deposited films still keep multi layer-like structure.

It is shown in Figure 2 that the CEM spectrum (2a) of 3s (on) -12s (off) films with the continuous Al_2O_3 deposition for 120 minutes has singlet peak and doublet peaks. Although not shown here, the magnetization curve of 3s-12s film shows the superparamagnetic characteristics, which corresponds to the singlet peak in the CEM spectrum. On the other hand, the CEM spectrum (2b) of 3s-12s film after 400 keV Ar ion irradiation to the fluence of 5x10^{16} ions/cm^2 shows sextet peaks, doublet peaks and small singlet peak. The appearance of sextet peaks indicates the growth of Fe particles. As seen in table 1, the intensity ratio of second/third -peaks in the sextet peaks is 2.3; that is, the direction of magnetization of Fe particles in the film seems to be random. Consequently, the shape of Fe particles in the film after 400 keV Ar ion irradiation is indicated to be a sphere. Therefore, it is clear that the 400 keV Ar ion irradiation alters the size and shape of Fe particles in Al_2O_3 matrix.

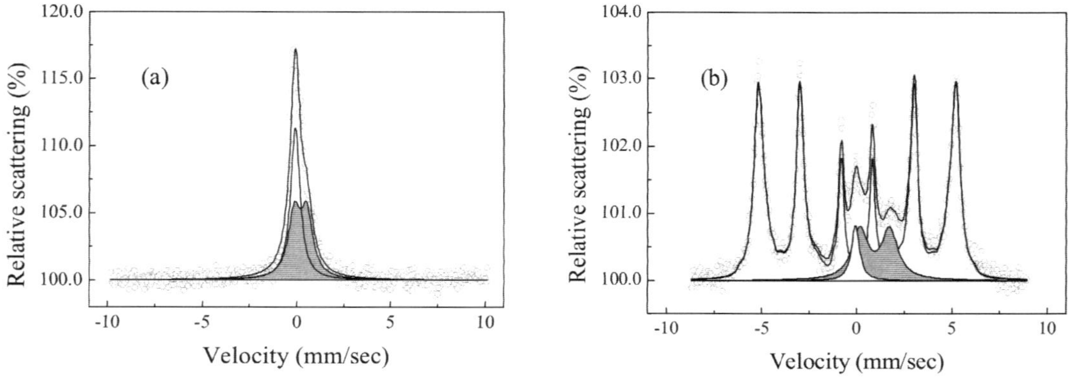

Figure 2. CEM spectra of Fe-Al_2O_3 films prepared by on-off Fe deposition, (a) 3s-12s before ion irradiation and (b) 3s-12s after ion irradiation. The shaded peaks are doublet peaks.

4. Conclusion

In conclusion, the CEM spectra of the nanocomposite Fe-Al_2O_3 films prepared by helicon plasma sputtering show that the decrease of the Fe particle size with decreasing Fe volume ratio judging from the values of H_{hfs}, and the layer-like structure from the direction of H_{hfs} oriented to the film plane. On the other hand, 400 keV Ar ion irradiation to the superparamagnetic film induces the growth of spherical Fe particles in the films, which is indicated from the ratio of peak intensities of sextet peaks appeared by the ion irradiation.

References

[1] Abuliemu A, Wakabayashi H, Matsuzawa S, Nakanishi Y, Toriyama T, Sakamoto I, Shen H L and Hayashi N, 2004 *phys. stat. sol.* c **1** 3315-8
[2] Sakamoto I, Honda S, Tanoue H, Hayashi N and Yamane H 1999 *Nucl. Instrum. Methods Phys. Res.* B **148** 1039-43
[3] Hayashi N, Stromberg F, Keune W, Toriyama T, Yamashiro M and Sakamoto I 2007 *J. Appl. Phys.* **101** 104304
[4] Corrias A, Casula M F, Falqui A and Pashina G 2004 *J. Sol-Gel Sci. Technol.* **31** 83-86
[5] Fei G T, Barnes J P, Petford-Long A K, Doole R C, Serna R and Gonzalo J 2002 *J. Phys. D: App. Phys.* **35** 916-22
[6] Rizza G, Cheverry H, Gacoin T, Lamasson A and Henry S 2007 *J. Appl. Phys.* **101** 014321
[7] Koike M, Chiwaki M, Suzuki I H and Kobayashi N 1995 *Rev. Sci. Instrum.* **66** 2141-3
[8] Morup S 1990 *Hyperfine Interactions* **60** 959-74

Journal of Physics: Conference Series **217** (2010) 012100 doi:10.1088/1742-6596/217/1/012100

Mössbauer studies of the effect of swift heavy ion irradiation on electrodeposited Sn-Co-Fe coatings

S Stichleutner[1,2], E Kuzmann[2,3], K Havancsák[4], Z Homonnay[3], A Vértes[2,3], O Doyle[5], M El-Sharif[5] and C U Chisholm[5]

[1]Institute of Isotopes, HAS, Budapest, Hungary
[2]Institute of Chemistry, Eötvös Loránd University, Budapest, Hungary
[3]Laboratory of Nuclear Chemistry, Chemical Research Center, HAS, Budapest, Hungary
[4]Institute of Physics, Eötvös Loránd University, Budapest, Hungary
[5]Glasgow Caledonian University, Glasgow, UK

E-mail: stichl@iki.kfki.hu

Abstract. ^{57}Fe and ^{119}Sn conversion electron Mössbauer spectroscopy was used to study the radiation effect of 246 MeV Kr and 710 MeV Bi ions on electrochemically deposited amorphous Sn-Co-Fe coatings. The ion dependent changes induced by the swift heavy ion irradiation in the chemical short range order of these amorphous thin films regarding Fe and Sn atoms are discussed within the framework of the transient thermal spike model.

1. Introduction

Recently there has been a resurgence of interest in the use of electrochemical deposition to achieve compositionally modulated alloys, which provide an attractive range of properties through changes of the structure. In our previous research [1, 2] we were able to produce a new range of good quality Sn-Co-Fe ternary alloys using an environmentally acceptable electrolyte based on sodium gluconate. These novel ternary alloys have metastable structures which do not exist in the thermal equilibrium diagrams. ^{57}Fe and ^{119}Sn Mössbauer investigations supported by XRD analysis confirmed the amorphous nature of the coatings. Swift heavy ion irradiation is a well known method to modify the chemical short range order in solids resulting in materials with more favourable properties regarding technological applications. The aim of the present work was to study the effect of swift heavy ion irradiation on these novel Sn-Co-Fe coatings.

2. Experimental

The coatings were deposited on Cu substrates using a constant current technique in a static flow plating system using the following electrolyte composition: 0.02M $SnSO_4$, 0.09M $CoSO_4$, 0.09M $FeSO_4$, 120 g/l Na-gluconate and 0.1 g/l peptone, at a cathode current density of 1 A/dm^2, an electrolyte temperature of 60 °C and pH=7.0. The deposition time was 20 min and the thickness of the resulting coatings was 3.96 μm. The composition of the deposits was 47.56% Sn, 22.89% Co and 29.55% Fe.

The irradiation of the samples was carried out with 246 MeV energy Kr and 710 MeV energy Bi ions at room temperature in a vacuum of about 10^{-3} Pa at the U-400 cyclotron of the Flerov Laboratory of Nuclear Reactions, JINR, Dubna, Russia. The ion fluences applied were 5×10^{13} ion/cm^2 in the case of Kr ions and 1×10^{13} ion/cm^2 in the case of Bi ions.

^{57}Fe and ^{119}Sn conversion electron Mössbauer (CEM) spectra of the samples were recorded by a conventional Mössbauer spectrometer (WISSEL) working in constant acceleration mode at room temperature. The conversion electrons were detected by a flowing gas (96% He, 4% CH_4) proportional counter (RANGER). A 50 mCi activity ^{57}Co/Rh source supplied the gamma rays for the ^{57}Fe measurements, while a 16 mCi activity Ca^{119m}SnO$_3$ source was used in the case of the ^{119}Sn measurements. The velocity calibration was performed by α-Fe measurements. The evaluation of the Mössbauer spectra was performed by least-square fitting of the lines using the MOSSWINN code [3].

3. Results and discussion

The ^{57}Fe CEM spectrum of the as-deposited sample (Fig. 1.1) exhibits an envelope of magnetically split spectrum with broad lines. Such spectra are typical for amorphous alloys and can be considered as a superposition of a large number of elementary spectra belonging to slightly different microenvironments of iron atoms having various alloying elements in their neighbourhoods. In accordance with our previous research [1, 2], the spectrum was decomposed into a dominant magnetically split sextet with broad lines and a minor doublet (Tab. 1), which were attributed to a ferromagnetic and a paramagnetic Sn-Co-Fe alloy respectively.

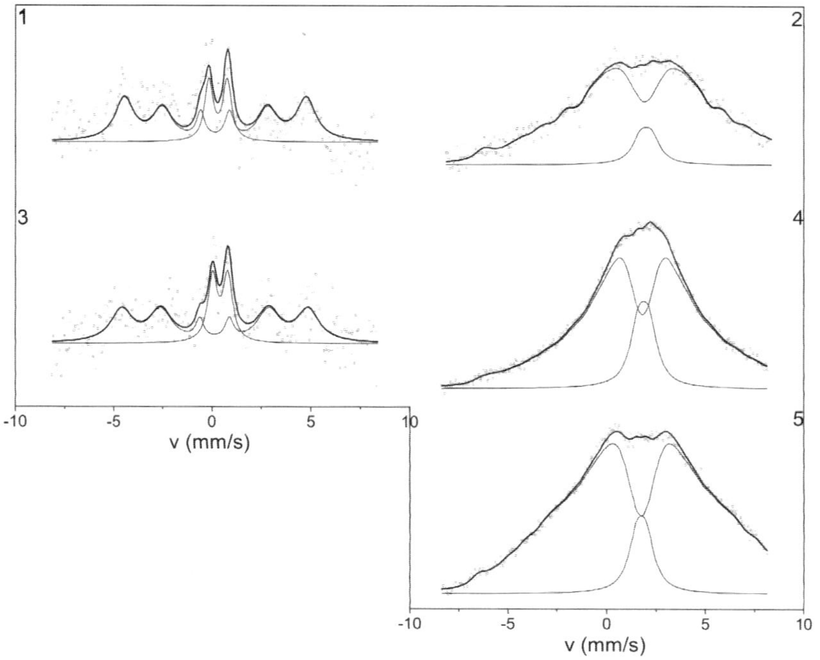

Figure 1. ^{57}Fe and ^{119}Sn CEM spectra of the as-deposited (1, 2), irradiated with 246 MeV energy Kr ions (3, 4) and irradiated with 710 MeV energy Bi ions (5) Sn-Co-Fe coatings.

Similarly to the ^{57}Fe CEM spectrum, the ^{119}Sn spectrum of the as-deposited sample (Fig. 1.2) exhibits a magnetically split envelope with very broad lines and was decomposed into a dominant sextet and a minor doublet (Tab. 1). Since Sn is nonmagnetic, the magnetic splitting is probably due to a transferred hyperfine field, which originates from the neighbouring magnetic atoms.

Table 1. ^{57}Fe and ^{119}Sn Mössbauer parameters of the as-deposited and irradiated Sn-Co-Fe coatings.

Mössbauer parameters	As-deposited		246 MeV Kr, 5×10^{13} ion$\times$cm^{-2}		710 MeV Bi, 1×10^{13} ion$\times$cm^{-2}
	^{57}Fe	^{119}Sn	^{57}Fe	^{119}Sn	^{119}Sn
δ (doublet), mm/s	0.30	2.00	0.41	1.87	1.75
Δ (doublet), mm/s	0.93	0.55	0.77	0.51	0.47
δ (sextet), mm/s	0.15	1.92	0.13	1.82	1.75
B (sextet), T	28.7	5.37	29.4	4.69	5.58
A (doublet), %	23.7	7.6	27.8	14.4	9.2
A (sextet), %	76.3	92.4	72.2	85.6	90.8

Along with Mössbauer spectroscopy, the samples were also investigated by X-ray diffraction. In the X-ray diffractograms of the as-deposited samples [2] the main peaks, except the lines belonging to the Cu substrate, are very broad, reflecting the amorphous character of the coatings, which is in good agreement with the CEMS results. No changes could be observed in the crystalline structure of the deposits after the irradiation with 246 MeV Kr or 710 MeV Bi swift heavy ions, the X-ray diffractograms of the irradiated samples reflect essentially the same amorphous character of these thin films as in the case of the as-deposited sample.

The most striking changes induced by the swift heavy ion irradiation in these Sn-Co-Fe coatings are reflected by the changes in the values of the hyperfine field of the sextets. The field increases in the case of ^{57}Fe (Fig. 1.3) and, at the same time, decreases in the case of ^{119}Sn (Fig. 1.4) after the irradiation with 246 MeV energy Kr ions (Tab. 1). These changes can be associated with a decreased number of ferromagnetic Fe and Co atoms in the neighbourhood of paramagnetic Sn atoms in the case of irradiation with Kr ions. On the other hand, the value of the hyperfine field increases after the irradiation with 710 MeV energy Bi ions in the case of ^{119}Sn (Fig. 1.5). This can be attributed to an increased number of Fe and Co atoms in the neighbourhood of Sn atoms in the case of Bi ion irradiation.

Calculations made with the SRIM 2008 code [4] show that both, Kr and Bi ions deposit their energy in the samples via inelastic collisions with the electrons in the medium and the respective values of the electronic stopping power for the 47.56% Sn-22.89% Co-29.55% Fe alloy are 19.98 keV/nm and 48.86 keV/nm. Since these values are several orders of magnitude higher then those of the nuclear stopping power, the thermal spike mechanism [5] can be considered as the main driving force responsible for the changes in the chemical short range order of these amorphous thin films observed in the Mössbauer spectra. In this case, the observed changes in the chemical short range order can be attributed to the effect of the additional mixing of the elements due to the swift heavy ion irradiation, which occurs as a consequence of a transient molten state diffusion [6]. Studies conducted on pure crystalline metals irradiated at low temperatures show that in the case of low velocity ions as in our case, thermal spikes occur only above a threshold value of 24 keV/nm [7]. In the case of Bi irradiation the value of the electronic stopping power highly exceeds this threshold value, consequently the Bi ions induce a considerable mixing effect resulting in a more disordered amorphous state, which corresponds to a more random distribution of the constituent elements as compared to the as-deposited amorphous state. This would result in an increased number of Fe and Co atoms in the neighbourhood of Sn atoms as we observed on the basis of the Mössbauer spectra. Similar effects were observed in the case of $Fe_{40}Ni_{40}B_{20-x}Si_x$ and $Fe_xNi_{80-x}P_{20}$ amorphous alloys in which the degree of the disorder increased upon swift heavy irradiation [8]. However, in the case of Kr ions the value of the electronic stopping power is under the threshold value of the thermal spike initiation, consequently one would expect a different effect than in the case of Bi ions. Although, the thermal spike mechanism cannot be entirely excluded as it was pointed out that in the case of amorphous materials and irradiations performed at room temperature the threshold value can be reduced [9]. In this case, the changes caused by Kr ions could be mainly attributed to irradiation induced defect creation as well as partial segregation and rearrangement [10] of the constituent elements from the amorphous towards an ordered state. This could easily result in a decreased number of Fe and Co atoms in the neighbourhood of Sn atoms, accordingly, the average value of the hyperfine field increases at the Fe sites and decreases at the Sn ones as we observed.

4. Conclusions

Swift heavy ion irradiation of electrodeposited Sn-Co-Fe coatings induces ion dependent changes in the chemical short range order of these amorphous thin films regarding Sn and Fe atoms. The observed changes can be explained by the mixing of the elements within the thermal spikes as a consequence of a transient molten state diffusion in the case of Bi ions and irradiation induced defect creation as well as partial segregation and rearrangement in the case of Kr ions.

References

[1] Kuzmann E, Stichleutner S, Doyle O, Chisholm C U, El-Sharif M, Homonnay Z and Vértes A 2005 *AIP Conf. Proc.* **765** 99

[2] Chisholm C U, Kuzmann E, El-Sharif M, Doyle O, Stichleutner S, Sólymos K, Homonnay Z and Vértes A 2007 *Appl. Surf. Sci.* **253** 4348

[3] Klencsár Z, Kuzmann E and Vértes A 1996 *J. Radioanal Nucl. Chem.* **210** 105

[4] Ziegler J F, Biersack J P and Littmark U 1985 *The Stopping and Range of Ions in Solids* (Pergamon, N. Y.)

[5] Toulemonde M, Dufour Ch and Paumier E 1992 *Phys. Rev. B* **46** 14362

[6] Wang Z G, Dufour C, Euphrasie S and Toulemonde M 2003 *Nucl. Instr. Methods B* **209** 194

[7] Wang Z G, Dufour C, Cabeau B, Dural J, Fuchs G, Paumier E, Pawlak F and Toulemonde M 1996 *Nucl. Instr. Methods B* **107** 175

[8] Kuzmann E and Spirov I N 1985 *J. Nucl. Mater.* **135** 22

[9] Gupta A 2000 *Vacuum* **58** 16

[10] Nolfi F V 1983 *Phase transformations during the irradiation* (Applied Science Publishers, London)

Characterization of magnetic nano materials by Mössbauer spectroscopy

Sangeeta Thakur[1], S C Katyal[1], A Gupta[2], V R Reddy[2] and M Singh[3]

[1] Jaypee University of information technology, Waknaghat, Solan 173215, INDIA
[2] UGC-DAE Consortium for scientific research, Khandwa Road, Indore 452017, INDIA
[3] Department of physics, Himachal Pradesh University, Shimla 171005, INDIA

Email: *megha2k6@gmail.com*

Abstract. The use of a non-destructive nuclear-physical method, namely ^{57}Fe Mössbauer spectroscopy, is discussed for the investigation of magnetic and structural arrangement of Fe-based nano-crystalline nickel-zinc-indium ferrites (NZIFO). Nano NZIFO particles ($Ni_{0.58}Zn_{0.42}In_xFe_{2-x}O_4$) with varied quantities of indium (x = 0, 0.1, 0.2) have been chemically synthesized through a reverse micelle reaction and investigated by X-ray diffraction, transmission electron microscopy and by magnetic and Mössbauer spectral studies. Here a comparison between low-temperature and room temperature Mössbauer spectra is presented. Well defined sextets at 5 K provide information about the structure and magnetic states of atoms located in different structural positions. The dependence of Mössbauer parameters, viz, isomer shift, quadrupole splitting, linewidth and hyperfine magnetic field on In^{3+} concentration have been discussed. Detailed Mössbauer results, as well as the interest of these materials both for applied science perspectives are presented. Mössbauer results are also supported by magnetization data. With these interesting ferromagnetic properties Indium substituted nano nickel-zinc ferrites have potential applications in magnetic storage data.

1. Introduction

Nanoparticles spinel oxides are attracting increasing interest in research because of their potential applications in nanoscience and technology [1-3] and also for fundamental understanding of the strikingly different properties of the same material when the particle size approaches the atomic scale level [4-8]. Some of the novel properties in magnetic nanoparticles, such as superparamagnetism, quantum magnetic tunneling, and surface spin effects [1,9-10] have generated further interest in the study of nanoparticles spinels. Although spinel ferrite nanoparticles have been synthesized by a variety of methods such as high-energy ball milling, [11] co-precipitation, [12] and sol-gel method, [13] microemulsion synthesis allows for the formation of high quality nanoparticles with controllable size and composition through minor adjustments to the synthesis conditions [14-15]. We have recently reported the synthesis of NZIFO nanoparticles using reverse micelle microemulsion procedure [16]. The reverse micelle procedure has resulted in a more repeatable synthesis of similar sized nanoparticles. We have already reported the Mössbauer characterization and magnetic properties of NZIFO particles at room temperature [16], in this paper, we are reporting on the properties of nanoparticles at low temperature 5 K.

2. Experimental

Nanoparticles of NZIFO were synthesized by reverse micelle technique [16]. The phase structure of nano ferrites was analyzed by XRD (Rigaku Geiger Flex 3KW diffractometer) using Cu Kα source. The morphology of powders was studied using a transmission electron microscope (TEM). Low temperature (5 K) Mössbauer measurements are carried out using Janis made cryostat. NORMOS software was used for the quantitative evaluation of the Mössbauer spectra. Magnetization measurements were carried out on a SQUID magnetometer.

3. Results and discussion

XRD patterns of the reverse micelle synthesized ferrites show a typical spinel structure for nano nickel-zinc ferrite (NZFO) and NZIFO particles [16]. The sample x = 0.2 contains a slight trace of hematite (Fe_2O_3) in addition to the spinel phase [16]. TEM of NZFO and NZIFO particles shows the spherical nature of the particles with the size distribution range of 8-12 nm and 15-25 nm respectively [16, 17]. Low temperature Mössbauer spectra for all samples are taken at 5 K. As clearly shown in Fig. 1, the superparamagnetic relaxation is suppressed at 5 K, and the Mössbauer spectrum consists of broadened sextets only. The broad shape of Mössbauer spectral lines for NZIFO samples provide clear evidence of a wide distribution of magnetic fields acting at the Fe^{3+} nuclei in the nanoparticles.

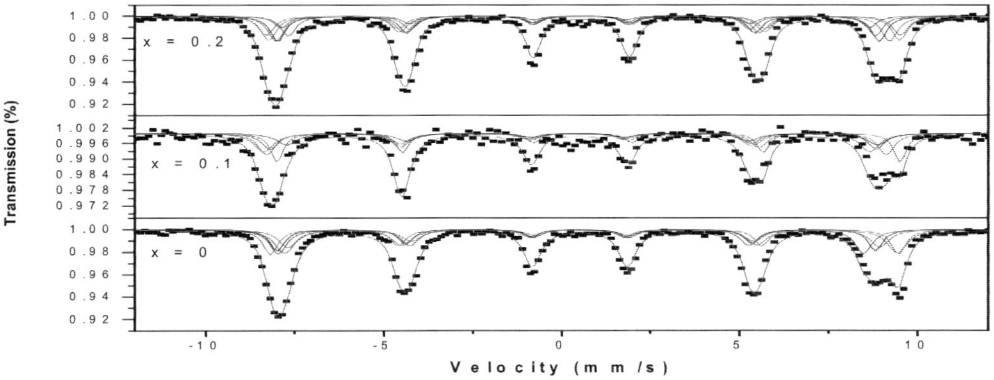

Figure 1. Mössbauer spectra of NZIFO particles at 5 K.

As the indium ion content is increased a broadening and splitting of the outer lines of the spectrum is observed with a change in their intensity. The broadening of the spectral components is obviously caused by the superimposition of several Zeeman sextets corresponding to the Fe^{3+} ions located at various different magnetically nonequivalent positions formed as a result of the addition of diamagnetic ions. For samples x = 0, 0.1 we obtained the best fit by employing five sextets in the Mössbauer spectra, one accounting for Fe^{3+} in the tetrahedral sites (A) and four for the octahedral sites [B] of the spinel structure. In addition to those a sixth spectrum (M2) is used for sample 0.2 for the traces of hematite. The hyperfine fields at both crystallographic sites for NZIFO particles are reported in Table 1. The spectrum of the nano NZFO sample (Figure 1, taken from ref. [18]) has the average magnetic hyperfine fields *B* (A) = 50.94 T and *B* [B] = 52.88 T, values that are well comparable with those of the bulk material [19]. At 5 K, the hyperfine fields increases as the concentration of In^{3+} ions increases from 0 to 0.1 (Table 1). At x =0.2, hyperfine field at (A) site decreases more rapidly than the [B] site field. Increase in hyperfine field with indium concentration upto 0.1 could be explained on the particle size distribution [20]. The variation of the effective hyperfine field for the sample x = 0.2 could be understood on the basis of supertransferred hyperfine fields (H_{STHF}). The magnetic hyperfine fields at the [B] site are due to the net magnetization of the nearest neighbor (A) site through A-B superexchange and the B-B interaction. Since Indium is diamagnetic and decrease in hyperfine magnetic field of (A) site indicates that for higher concentration of In^{3+} ions (x>0.1) some of them migrate to the [B] sites also, which reduces the magnetization of the particles at [B] site. Consequently the hyperfine field at the (A) site decreases. Our results indicate that the s-electron density at the Fe^{3+} nucleus is not affected by In^{3+} substitution, since the isomer shift, for a particular nuclear transition of the Mössbauer source is dependent only on the s-electrons charge density of the absorber. The values of the isomer shift at (A) and [B] sites show that iron is in the Fe^{3+} state [20]. The observed difference between the isomer shift at 5 K and 300 K are fully explained by the temperature dependent second order Doppler shift of the lines. Figure 2a shows the variation of magnetization with the applied field for all the three samples at 5 K. The magnetization of NZFO sample does not saturate even at the maximum field attainable (H = 50 kOe), while NZIFO particles attain saturation magnetization in the applied field.

Table 1. Mössbauer parameters for NZIFO particles at low temperature 5 K. < B> represents the average hyperfine field acting at the B site

Sample	Subspectrum	WID (mm)	IS (mm/s)	Q.S. (mm/s)	B (T)	AREA (%)
	(A)	0.58 (±0.02)	0.38 (±0.01)	0.23 (±0.03)	50.94 (±0.00)	21.70%
x = 0	[B]					78.30%
	B_1	0.50 (±0.00)	0.58 (±0.01)	-0.12 (±0.03)	54.64 (±0.10)	19.23%
	B_2	0.68 (±0.00)	0.66 (±0.03)	-0.33 (±0.07)	53.31 (±0.10)	24.73%
	B_3	0.50 (±0.00)	0.45 (±0.02)	0.06 (±0.04)	52.07 (±0.00)	16.48%
	B_4	0.65 (±0.01)	0.60 (±0.02)	-0.04 (±0.001)	51.5 (±0.00)	17.86%
					<B>=52.88 T	
	(A)	0.65(±0.00)	0.28 (±0.04)	0.27 (±0.07)	52.72 (±0.2)	22.73 %
x = 0.1	[B]					77.27 %
	B_1	0.42 (±0.00)	0.67 (±0.01)	-0.16 (±0.02)	54.21 (±0.10)	22.73 %
	B_2	0.56 (±0.00)	0.46 (±0.00)	0.08 (±0.00)	53.92 (±0.30)	22.73 %
	B_3	0.56 (±0.00)	0.36 (±0.04)	-0.08 (±0.00)	52.78 (±0.30)	16.66 %
	B_4	0.70 (±0.00)	0.51 (±0.00)	-0.13 (±0.08)	51.40 (±0.00)	15.15 %
					<B> =53.06T	
	(A)	0.50 (±0.00)	0.54 (±0.01)	-0.05 (±0.02)	51.04 (±0.10)	15.39 %
x = 0.2		0.53 (±0.00)	0.60 (±0.01)	-0.06 (±0.10)	54.89 (±0.76)	19.63 %
	[B]					59.29 %
	B_1	0.52 (±0.00)	0.58 (±0.13)	-0.05 (±0.00)	55.18 (±0.90)	15.39 %
	B_2	0.54 (±0.00)	0.58 (±0.01)	-0.03 (±0.00)	53.39 (±0.08)	20.96 %
	B_3	0.50 (±0.00)	0.50 (±0.01)	0.06 (±0.01)	52.32 (±0.00)	19.63 %
	B_4	0.50 (±0.00)	0.45 (±0.00)	0.12 (±0.00)	51.16 (±0.17)	9.02 %
					< B>=53.01T	

For x = 0 sample the saturation magnetization, M_{sat} (30.15 emu/g) is determined by extrapolating the M versus 1/H curve to 1/H = 0. The M_{sat} value for x = 0.1, 0.2 samples are 45.68, 28.68 emu/g respectively. The particular lack of saturation in NZFO sample is attributed to surface spin-canting [21]. Variation of the saturation magnetization with indium content upto 0.1 shows that indium ions initially occupy tetrahedral site and thus the overall magnetic moment increases. This consequently raises the saturation magnetization of the system. Decrease in saturation magnetization for x>0.1 sample indicates that some of the Indium ions migrate to the [B] sites also, which reduces the overall saturation magnetization. The results observed are in accordance with previous study on indium doped ferrites [22-25]. The rapid decrease of the saturation magnetization for sample x = 0.2 may also be the formation of antiferromagnetic phase with spinel phase [16]. The coercive field of samples x = 0, 0.1, 0.2 are 322.0, 242.0 and 305.0 Oe respectively (Figure 2b).

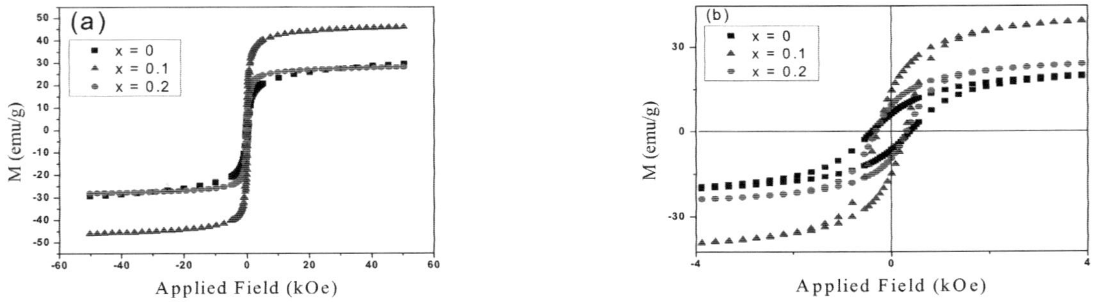

Figure 2. (a) Variation of magnetization with applied field (b) Part of the curves near the origin showing the remanence and coercivity for all samples

The low inferred values of coercivity and squareness ratio as 0.222, 0.332, 0.342 for samples x =0, 0.1 and 0.2 respectively are in agreement with the well established soft magnetic character of NZIFO particles [16]. It is clear that low temperature magnetic behaviors are consistent with Mössbauer measurements. The results discussed above on NZIFO nanoparticles along with previously published at room temperature indicate a rather surprising behavior of spinel nanoparticles prepared in reverse micelle technique.

4. Conclusion

In conclusion, low-temperature Mössbauer studies on nanosamples of NZIFO particles corroborate the result of room temperature Mössbauer measurements done on these samples and show that Indium enters preferentially into the (A) site. NZIFO nanoparticles show promising properties: high saturation magnetization and hyperfine field close to that of the bulk materials. Such interesting behavior appears to be clearly related to two main factors: good crystallinity of the nanoparticles and the departure from the classical cation distribution in the spinel structure. Sample x = 0.1 exhibits high saturation magnetization and hyperfine field close to that of the bulk material which show the applicability of this sample for magnetic storage data.

Acknowledgments

One of authors (Sangeeta Thakur) is very grateful to CSIR (INDIA) for providing financial support. We are highly thankful for UGC-DAE consortium Indore for providing the experimental facility.

Reference:

[1] Dorman J L, Fiorani D, 1991 *Magnetic Properties of Fine Particles* North- Holland, Amsterdam
[2] Handley R C 2000 *Modern Magnetic Materials: Principles and Applications* Wiley New York
[3] V.A.M. Brabers, in *Handbook of Magnetic Materials*, 1995 edited by K.H. J. Buschow ~North-Holland, Elsevier science Amsterdam, **8** 189
[4] Kodama R H and Berkowitz A E 1999 *Phys. Rev. B* **59** 6321
[5] Batlle X and Labarta A 2002 *J. Phys. D* **35** R15
[6] Fatemi D J, Harris V G, Chen M X, Malik S K, Yelon W B, Long G J and Mohan A 1999 *J. Appl Phys.* **85** 5172
[7] Chen J P, Sorensen C M, Klabunde K J, Hadjipanayis G C, Devlin E and Kostikas A 1996 *Phys. Rev. B* **54** 9288
[8] Goya G F and Rechenberg H R 1998 *J. Phys.: Condens. Matter* **10** 829
[9] Kodama R H, Berkowitz A E, McNiff E J Jr, and Foner S 1996 *Phys. Rev. Lett.* **77** 394
[10] Hamdeh H H, Ho J C, Oliver S A, Willey R J, Oliveri G and Busca G 1997 *J. Appl. Phys.* **81** 1851; Oliver S A , Harris V G, Hamdeh H H and Ho J C 2000 *Appl. Phys. Lett.* **76** 2761
[11] Goya G F, Rechenberg H R, Jiang J Z **1998** *J. Appl. Phys.* ***84*** 1101
[12] Chen Q, Rondinone A J, Chakomakos B C, Zhang Z J 1994 *J. Magn. Magn. Mater.* **194** 1
[13] Xiong G, Mai Z, Xu M, Cui S, Ni Y, Zhao Z, Wang X, Lu L, 2001 *Chem. Mater.* **13** 1943
[14] Pileni M P, Moumen N 1996 *J. Phys. Chem. B* **100** 1867
[15] Liu C, Zou B, Rondinone A J, Zhang Z J 2001 *J. Phys. Chem.* **104** 2000
[16] Thakur S, Katyal S C, Gupta A, Reddy V R, Singh M 2009 *J. Appl. Phys.* **105** 1
[17] Thakur S, Katyal S C and Singh M 2007 *Appl. Phys. Lett.* **91** 262501
[18] Thakur S, Katyal S C, Gupta A, Reddy V R, Sharma S K, Knobel M, Banerjee A, Singh M communicated in *J. Phys. Chem. B.*
[19] Leung L K, Evans B J, Morrish A H 1973 *Phys Rev. B* **8** 29
[20] Chkoundali S, Ammar S, Jouini N, Fiévet F, Molinivé P, Danot M, Villain F and Grenéche J-M, 2004 *J. Phys. Condens. Matter* **16** 4357-4372.
[21] Thakur S, Katyal S C and Singh M 2009 *J. Magn. Magn. Mater.* **321** 1
[22] Parvatheeswara B, Rao K H 2005 *J. Magn. Magn. Mater.* **292** 44
[23] Puri R K, Singh M, Sud S P 1994 *J. Mater. Sci.* **29** 2182
[24] Ghosh S, Nambissan P M G, Bhattacharya R 2004 *Physics Letters A* **325** 301
[25] Lakshman A, Subba Rao P S V, Rao K H 2004 *J. Magn. Magn. Mater.* **284** 352

Journal of Physics: Conference Series **217** (2010) 012102 doi:10.1088/1742-6596/217/1/012102

Characterization of nanodimensional Ni-Zn ferrite prepared by mechanochemical and thermal methods

E Manova[1, 4], D Paneva[1], B Kunev[1], E Rivière[2], C Estournès[3] and I Mitov[1]

[1] Institute of Catalysis, Bulgarian Academy of Sciences, Acad. G. Bonchev St., Block 11, 1113 Sofia, Bulgaria

[2] Institut de Chimie Moléculaire et des Matériaux d'Orsay, UMR 8182, UPS, 91405 Orsay, France

[3] CNRS; Institut Carnot, Cirimat; F-31062 Toulouse, France

E-mail: elina@ic.bas.bg

Abstract. Nickel zinc ferrite nanoparticles, $Ni_{1-x}Zn_xFe_2O_4$ (x = 0, 0.2, 0.5, 0.8, 1.0), with dimensions below 10 nm have been prepared by combining chemical precipitation with high-energy ball milling. For comparison, their analogues obtained by thermal synthesis have also been studied. Mössbauer spectroscopy, X-ray diffraction, and magnetic measurements are used for the characterization of the obtained materials. X-ray diffraction shows that after 3h of mechanical treatment ferrites containing zinc are formed, while 6h of treatment is needed to obtain $NiFe_2O_4$. The magnetic properties of the samples exhibit a strong dependence on the phase composition, particle size and preparation method.

1. Introduction

The Ni-Zn ferrites have been extensively studied for various properties as well as for structural issues [1-4]. Zn and Ni are known to have very strong preferences for the tetrahedral and octahedral sites, respectively, making Ni–ferrite a model inverse ferrite and Zn–ferrite a model normal ferrite. The general formula of these materials is: $(Zn_xFe_{1-x})_{tetra}[Ni_{1-x}Fe_{1+x}]_{octa}O_4$. Ni-Zn ferrite is of particular interest because the magnetic properties of the nanoparticles differ markedly from those of their bulk counterpart. It is of further interest to replace a non-magnetic Zn^{2+} ion by a magnetic Ni^{2+} in Zn-ferrite to understand the difference in the properties of the system of ferrites in the limit of nano-regime. The properties of these ferrites are highly sensitive to the cation distribution which in turn depends on the method of preparation and sintering conditions. Spinel-type ferrites can be prepared by a variety of methods [4-7]. The high-energy milling as a solid-state method for synthesis of nanodimensional materials has the advantage that it can be easily operated and produce large amounts of nanostructured powders in a fairly short time. Furthermore, the highly nonequilibrium nature of the milling process allows one to prepare solids with improved and/or novel physical and chemical properties.

The purpose of the present work was to investigate the possibility of synthesizing $Ni_{1-x}Zn_xFe_2O_4$ (x = 0, 0.2, 0.5, 0.8, 1.0) nanodimensional powders by co-precipitation of the corresponding hydroxide

[4] To whom any correspondence should be addressed.

carbonates followed by mechanochemical or thermal treatment. The relationship between synthesis method, phase composition and structural properties of the obtained materials was investigated.

2. Experimental

The mechanochemical synthesis (MS) of nickel-zinc ferrite powders was performed by two steps: (i) co-precipitation, and (ii) high-energy ball-milling (3h to obtain $Ni_{1-x}Zn_xFe_2O_4$ (x = 0.2, 0.5, 0.8, 1.0) and 6h for $NiFe_2O_4$) following the procedure, described in [7]. The thermal synthesis (TS) was performed by annealing of the co-precipitation precursors at 773 K.

The phase composition and structure was determined by X-ray diffraction (XRD) using TUR M62 diffractometer with Co Kα radiation. Data interpretation was carried out using JCPDS (Joint Committee on Powder Diffraction Standards) database. Transmission ^{57}Fe Mössbauer spectra were obtained at room temperature (RT) with a Wissel electromechanical Mössbauer spectrometer (Wissenschaftliche Elektronik GmbH, Germany) working at a constant acceleration mode. A $^{57}Co/Cr$ (activity ≅ 20 mCi) source and an α-Fe foil standard were used. Experimentally obtained spectra were treated using the least squares method. The parameters of hyperfine interactions such as isomer shift (δ), quadrupole splitting (Δ) and effective internal magnetic field (B) as well as the line widths (Γ) and the relative weight (G) of the partial components of the spectra were determined. Isothermal magnetizations of the Ni-Zn ferrites were measured on a Quantum Design SQUID Magnetometer.

3. Results and discussion

The XRD patterns of the co-precipitated precursors (not shown) are characteristic of layered double hydroxides as found for Pyroaurite (PDF 25-0521) and Hydrotalcite (PDF 41-1428). After 3h of milling the diffraction lines of the spinel phases $Ni_{1-x}Zn_xFe_2O_4$ (x = 0.2, 0.5, 0.8, 1.0) are well defined, while $NiFe_2O_4$ is obtained after 6h of mechanical treatment. The diffractograms of all mechanochemically synthesized samples are composed of broadened low intensity peaks, suggesting very small particles dimensions. For all thermally treated samples $Ni_{1-x}Zn_xFe_2O_4$ ferrites are obtained. The average crystallites size (D), the degree of microstrain (e), and the lattice parameter (a) of the studied nickel-zinc ferrites are also determined from the experimental XRD profiles (Table 1) by using the Williamson-Hall analysis. A well defined effect of the crystals size decrease and increase in the crystal defects and lattice parameter is observed for the mechanochemically obtained materials (Table 1). A dependence of the lattice parameter on the preparation method and the phase composition exists.

Table 1. Crystallite size (D), microstrain (e) and lattice parameter (a) of investigated sample.

Sample	D (nm)	$e \cdot 10^3$ (a.u.)	a (Å)
$NiFe_2O_4$, TS	19.30	1.599	8.332
$Ni_{0.8}Zn_{0.2}Fe_2O_4$, TS	19.34	1.233	8.363
$Ni_{0.5}Zn_{0.5}Fe_2O_4$, TS	24.60	0.646	8.387
$Ni_{0.2}Zn_{0.8}Fe_2O_4$, TS	26.00	0.427	8.416
$ZnFe_2O_4$, TS	28.67	2.275	8.432
$NiFe_2O_4$, MS	5.32	15.945	8.381
$Ni_{0.8}Zn_{0.2}Fe_2O_4$, MS	5.32	15.974	8.396
$Ni_{0.5}Zn_{0.5}Fe_2O_4$, MS	5.83	14.268	8.403
$Ni_{0.2}Zn_{0.8}Fe_2O_4$, MS	11.24	3.700	8.422
$ZnFe_2O_4$, MS	10.16	5.398	8.444

The Mössbauer spectra of the precursors exhibits a quadrupole doublet with δ=0.34 mm/s and Δ=0.65–0.70 mm/s, indicating that the hydroxide carbonates are paramagnetic. Figure 1 shows the RT Mössbauer spectra of samples obtained after different processing routes. The corresponding parameters, listed in Table 2, are typical of Fe^{3+} ions in octahedral and tetrahedral spinel sites [1, 6, 7].

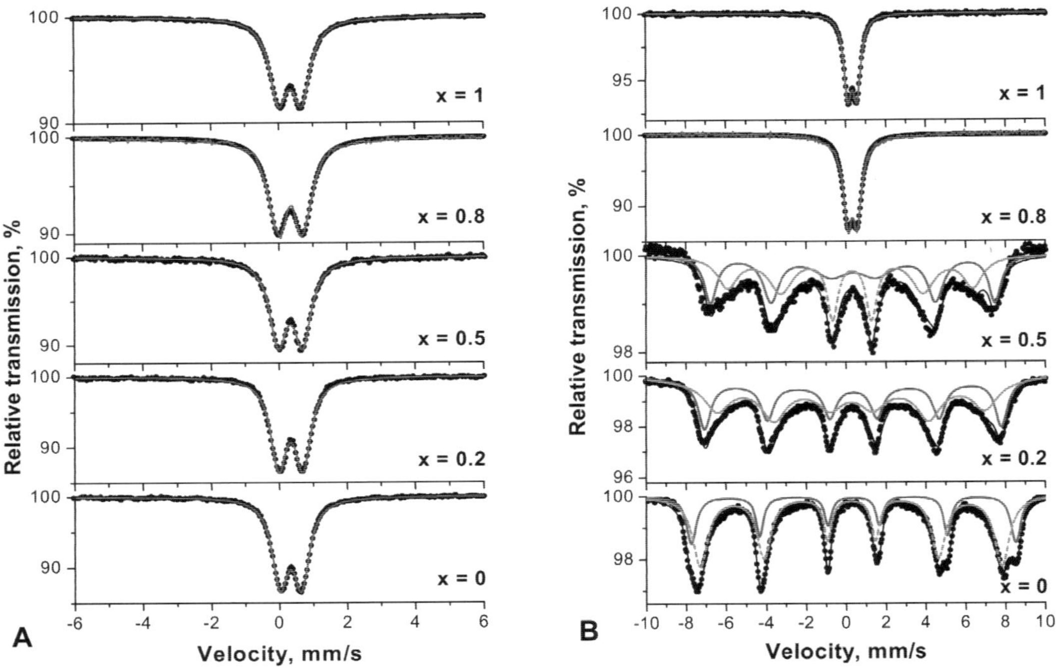

Figure 1. RT Mössbauer spectra of mechanochemically (A) and thermally (B) synthesized $Ni_{1-x}Zn_xFe_2O_4$. (—— – Fe^{3+} octa; - - - – Fe^{3+} tetra)

Table 2. Parameters of Mössbauer spectra.

Sample	Components[a]		δ (mm/s)		Δ (mm/s)		B (T)		Γ (mm/s)		G (%)	
	MS	TS	MS	TS	MS	TS	MS	TS	MS	TS	MS	TS
$ZnFe_2O_4$	Db	Db	0.34	0.34	0.70	0.46	-	-	0.64	0.48	100	100
$Ni_{0.2}Zn_{0.8}Fe_2O_4$	Db	Db	0.34	0.34	0.77	0.49	-	-	0.66	0.59	100	100
$Ni_{0.5}Zn_{0.5}Fe_2O_4$	Db	Sx1	0.33	0.37	0.72	0.00	-	44.3	0.62	1.00	100	47
		Sx2		0.27		0.00		38.2		1.50		53
$Ni_{0.8}Zn_{0.2}Fe_2O_4$	Db	Sx1	0.33	0.37	0.71	0.00	-	46.2	0.57	0.75	100	42
		Sx2		0.27		0.00		41.7		1.50		58
$NiFe_2O_4$	Db	Sx1	0.33	0.37	0.63	0.00	-	50.6	0.57	0.54	100	37
		Sx2		0.27		0.00		46.9		0.78		63

[a] Db = doublet (Fe^{3+} octa); Sx1 = sextet (Fe^{3+} octa); Sx2 = sextet (Fe^{3+} tetra)

As shown in Figure 1A the spectra of all milled samples are doublets. It is well known that the magnetic characteristics of the material are strongly affected when the particle size becomes very small, due to the influence of thermal energy on the magnetic moment ordering, originating the superparamagnetic (SPM) relaxation phenomenon. Taking into account the XRD data we suppose that the doublets observed in all milled samples arise from Fe^{3+} ions in ultrafine ferrite particles exhibiting SPM behavior. This particle size effect is confirmed by the RT isothermal magnetizations where the absence of saturation in the magnetic field range explored, the "S" shape of the curves together with the lack of coercivity indicate the presence of SPM particles.

A difference in structure between the materials, depending on the preparation method is observed. The spectra of $ZnFe_2O_4$ and $Ni_{0.2}Zn_{0.8}Fe_2O_4$ formed by co-precipitation and thermal treatment show only doublet components (Figure 1B). Comparing with mechanochemically synthesized $ZnFe_2O_4$ and $Ni_{0.2}Zn_{0.8}Fe_2O_4$ we note that the average isomer shifts show no significant variation thus indicating that the s-electron charge distribution of Fe^{3+} ions is negligibly influenced by the synthesis method. But the quadrupole splitting and the spectral line width of the Zn rich ferrite particles are larger for the MS samples (Table 2). This broadening is likely to be associated with several effects induced by milling. As the average grain size decreases, the fraction of disordered and interfacial regions increases and changes in the site distributions of Fe^{3+} and Zn^{2+} ions as well as changes in tetrahedral and octahedral bond lengths and angles probably occur. The isothermal magnetization at RT confirms the paramagnetic character of the TS zinc ferrite, while the mechanochemically obtained sample is superparamagnetic. A magnetic ordering of $ZnFe_2O_4$ is possible only if there is an inversion of cation distribution – some of the Zn^{2+} ions occupy the octahedral sites and some of the Fe^{3+} ions occupy the tetrahedral sites. Therefore the milling process induces structural transformation of the spinel with cationic redistribution into the system which induces magnetic ordering of the ferric ions leading to changes in the magnetic behavior of the materials.

On the other hand, the Mössbauer spectra of thermally synthesized $NiFe_2O_4$, $Ni_{0.8}Zn_{0.2}Fe_2O_4$ and $Ni_{0.5}Zn_{0.5}Fe_2O_4$ exhibit magnetic splitting at RT. The registered broad spectral lines are probably due to SPM relaxation and Fe^{3+} disorder with increase of Zn-content. As zinc is introduced in the system, the effective internal magnetic field B is reduced. The variation of B with zinc introduction can be explained on the basis of Yafet-Kittel type of spin arrangement [8]. With the increase in zinc concentration, the Fe^{3+} ions will have no magnetic neighbors and hence spins become uncoupled, the saturation magnetization decreases, and hence the hyperfine field decreases. Smaller values of the RT isothermal magnetizations are registered for the MS phases and could be related to local canting of magnetic ions due to the imperfect structure, and mainly to the surface effect of nanosize particles of large surface area.

4. Conclusions

In this work it has been demonstrated that the high-energy ball milling of layered nickel-zinc-iron hydroxide carbonates results in the formation of nanocrystalline Ni-Zn ferrites, where the particle size is below 10 nm. The zinc introduction facilitates the ferrite formation. The structural and magnetic properties of the samples exhibit a strong dependence on the phase composition, particle size and preparation method. This work shows that the combination of the co-precipitation method and subsequent mechanical milling is a promising technique for a relatively large scale preparation of Ni-Zn ferrite nanoparticles.

Acknowledgements

The authors thank the National Science Fund of Bulgaria for financial support through Projects X-1504/05 and Rila4-412 (DO 02-295/2008).

References

[1] Upadhyay C, Verma H C and Anand S, 2004 *J. Appl. Phys.* **95** 5746–51
[2] Elizalde Galindo J T, Adair A H, Botez C E, Corral Flores V, Bueno Baques D, Fuentes Cobas L and Matutes-Aquino J A 2007 *Appl. Phys.* A **87** 743–74
[3] Bercoff P G and Bertorello H R 2000 *J. Magn. Magn. Mater.* **213** 56–62
[4] Albuquerque A S, Ardisson J D, Macedo W A A and Alves M C M 2000 *J. Appl. Phys.* **87** 4352–57
[5] Cabanas A and Poliakoff M 2001 *J. Mater. Chem.* **11** 1408–16
[6] Roy M K and Verma H C 2006 *J. Phys.: Condens. Matter* **18** 7273–7281
[7] Manova E, Kunev B, Paneva D, Mitov I, Petrov L, Estournès C, d'Orléans C, Rehspringer J L and Kurmoo M 2004 *Chem. Mater.* **16** 5689–96
[8] Murthy N S S, Natera M G, Youssef S I, Begum R J and Srivastava C M 1969 *Phys.Rev.* B **181** 969–77

Long ball-milling of bcc-FeCr at different injected powers: amorphization and partial crystallization

J M Loureiro[1,3], B F O Costa[1] and G Le Caër[2]

[1]CEMDRX, Department of Physics, University of Coimbra, P-3004-516, Coimbra, Portugal
[2]IPR, UMR URI-CNRS 6251, Université de Rennes I, Campus de Beaulieu, Bat 11A, F-35042 Rennes Cedex, France

E-mail: loureirojc@gmail.com

Abstract. The structural changes induced by long ball-milling of the α-FeCr bcc phase in vacuum in a vibratory Fritsch P0 mill at two different amplitudes, were studied by Mössbauer spectroscopy and X-ray diffraction. In previous studies we showed that amorphization is an intrinsic phenomenon which occurs when grinding alpha and sigma phases of near-equiatomic FeCr alloys. With long periods of milling at different injected powers, we observe a phenomenon of amorphization-partial crystallization-amorphization which is reminiscent of the cyclic amorphization-crystallization phenomenon that takes place in various ball-milled alloys.

1. Introduction

The importance of Fe–Cr alloys is known both in science and industry. Fe–Cr binary alloys are important from both applied and fundamental points of view. Among the various applications, we can select for instance their use in advanced nuclear reactor applications, due to their corrosion resistance combined with good mechanical properties. Indeed, these alloys are a base for low activated ferritic-martensitic steels widely used as structural materials for fission reactors and being now considered as candidate materials for future fusion reactors. Ferritic-martensitic steels provide an attractive alternative to austenitic stainless steels because they demonstrate a very high resistance to swelling up to neutron irradiation damage doses of 200 displacements per atoms [1].

In previous studies, we showed that amorphization is an intrinsic phenomenon which is observed when grinding alpha and sigma phases of near-equiatomic FeCr alloys. Indeed, amorphization occurs even in presence of low to very low levels of oxygen, a gas which is known to favour it in concentrated Fe–Cr alloys [2]. In the following, we report on the structural changes, as evidenced by Mössbauer spectroscopy and X-ray diffraction, which occur when ball-milling the α-FeCr bcc phase in vacuum in a vibratory Fritsch P0 mill at two different injected powers. A phenomenon of amorphization-partial crystallization-amorphization behaviour that occurs when milling the α-FeCr bcc phase for long periods is described.

[3] To whom any correspondence should be addressed.

2. Experimental

A bcc $Fe_{49.7}Cr_{50.3}$ at% alloy was prepared by melting together appropriate amounts of Fe and Cr in an induction furnace. Ball-milling of this alpha phase was performed in vacuum in a vibratory Fritsch P0 mill. The mill consists of a hardened steel vial with a hardened steel ball whose diameter is 5 cm and whose mass is 500 g. A mass of about 5 g of FeCr was ball-milled in accumulated times in the same conditions as those reported previously [2]. Tiny amounts of powders were removed at different milling times to be characterized. In this way two series of samples were obtained, depending on the amplitude of vibration of the mill (maximum: $A \sim 3$ in scale; and medium: $A \sim 2$–2.5 in scale). Milling times up to 1500 h and 2500 h were applied to series 1 and 2, respectively.

X-ray diffraction (XRD) was performed at room temperature (RT) using Cu Kα radiation (λ = 0.154184 nm) to characterize the microstructural changes. ^{57}Fe Mössbauer spectra were recorded at RT in a transmission geometry using a standard constant acceleration spectrometer. A ^{57}Co source in Rh matrix with strength of ~ 20 mCi was used. The spectra were analyzed by a constrained Hesse-Rübartsch method [3] which yields a hyperfine magnetic field distribution, $P(B)$. The ^{57}Fe isomer shifts are given with respect to α-Fe at RT.

3. Results and Discussion

The composition of the alpha phase (series 1; $A \sim 3$) milled for 100 h, 400 h, 750 h and 1500 h was obtained to be $Fe_{55.7}Cr_{44.3}$ at%, $Fe_{55.9}Cr_{44.1}$ at%, $Fe_{56.0}Cr_{44.0}$ at% and $Fe_{56.9}Cr_{43.1}$ at%, respectively. The oxygen content was equal to 0.2 at%, 0.7 at%, 0.8 at% and 1.0 at%, respectively. For series 2 ($A \sim 2.0$ to 2.5), the composition of the alpha phase is $Fe_{57.1}Cr_{42.9}$ at% and $Fe_{58.0}Cr_{42.0}$ at%, after 960 h and 2430 h of milling, respectively. These compositions were estimated from the hyperfine magnetic field values [4,5] and are close to those obtained from microprobe analysis of some samples. The iron content increases due to contamination by the milling tools, but the oxygen, despite its increase, remains small even after 1500 h hours of milling.

XRD patterns of the alpha phase as a function of milling time only show the broadening of the bcc peaks (not shown).

Figure 1 shows RT ^{57}Fe Mössbauer spectra for different times of milling of the alpha phase, of series 1 and 2. A paramagnetic contribution, with an isomer shift $\langle IS \rangle = -0.078(1)$ mm/s (series 1) and $\langle IS \rangle = -0.098(1)$ mm/s (series 2), consistent with that reported for clean co-evaporated Fe–Cr amorphous phases [6], arises in the middle of the spectra. For the magnetic part, the mean hyperfine magnetic field $\langle B \rangle$ is almost constant during milling with $\langle B \rangle \sim 17.6$ T, $\langle IS \rangle = -0.096$ mm/s and $\langle B \rangle \sim 18.2$ T, $\langle IS \rangle = -0.082$ mm/s, for series 1 and 2, respectively. The calculated field is 17.5 T for the starting alloy and 20.0 T from the composition of alloys milled for 500 h and longer. As milling decreases the field by about ~ 2 T [5], the expected field of ~ 18 T agrees with experiment.

Figure 2 shows the paramagnetic area fraction $A_P(t)$, as obtained from Mössbauer spectra, as a function of milling time for the alpha phase for both series. The fraction $A_P(t)$ represents essentially the fraction of Fe atoms in the amorphous phase. It is close to the fraction of amorphous phase as that phase and the bcc phases have similar compositions. That fraction is in very good agreement with that we deduced from magnetization measurements.

When milling at the maximum amplitude of vibration ($A \sim 3$ in scale), we observe that the steady increase of the amorphous phase fraction is interrupted at about 750 h of milling (figure 2) where a partial crystallization of the amorphous phase takes place as evidenced by Mössbauer spectra. Then, amorphization starts again but at a slower rate. Cyclic amorphization-crystallization behaviour takes place in various alloys when subject to ball-milling (for instance Al-Zr, Co-Ti [7]) or to e.g. accumulative roll-bonding (for instance Zr based intermetallic alloys [8]). Although, we don't observe here a full cycle, because of the very long milling times needed at the injected power level at which the mill works, the amorphization-partial crystallization-amorphization behaviour that we observe is reminiscent of that cyclic phenomenon.

When the α-FeCr phase is milled with an amplitude of vibration smaller than the previous one ($A \sim$ 2–2.5 in scale), the area fraction of the Mössbauer spectra associated with Fe in the amorphous phase increases monotonically (figure 2). In contrast to the previous experiment, the amorphous phase does not crystallize. Except for the amorphization-partial crystallization-amorphization behaviour, figures 1 and 2 show that milling times scale approximately as A^2.

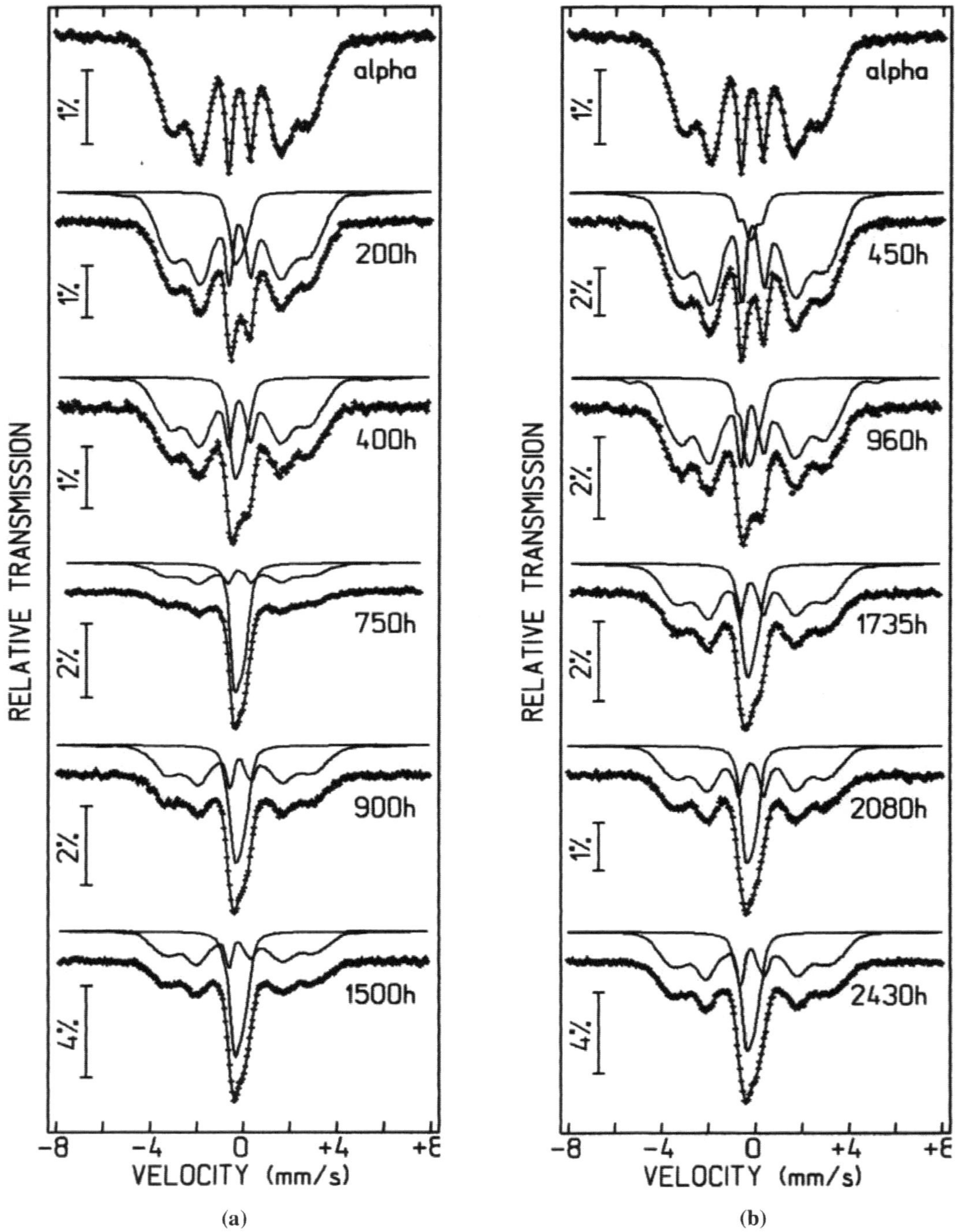

Figure 1. RT ^{57}Fe Mössbauer spectra for different times of milling of the alpha phase: (a) $A = 3$ and (b) $A = 2$ to 2.5.

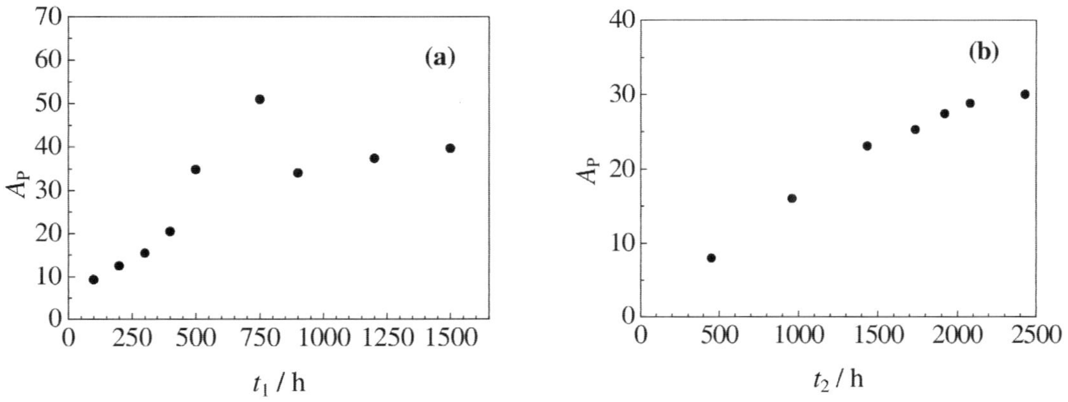

Figure 2. Evolution of the amorphous paramagnetic fraction obtained from Mössbauer spectra with grinding time: (a) $A = 3$ and (b) $A = 2$ to 2.5.

That experiment confirms that eventual detrimental contaminations by steel or by residual gases do not play an essential role in the phenomenon that we observe, as also predicted before [9]. Oscillatory phenomena were reported to occur during ball-milling. For instance, the nanocrystalline fraction in cryomilled zinc, the fraction of bcc and hcp phases in ball-milled Co (see [10] and references therein). The existence of cyclic amorphization is predicted theoretically by Johnson et al. [10] as a function of the materials parameters and of the input power.

In conclusion, the transformations of alpha FeCr with long ball-milling times at two different injected powers were followed by Mössbauer spectroscopy and X-ray diffraction. An amorphization-partial crystallization-amorphization behaviour is reported for the higher injected power. Experiments at still higher injected powers to shorten the milling times are worth being performed to know if that phenomenon is indeed cyclic. The injected power must however be not too high because of the risk to bypass the formation of the amorphous phase.

The authors acknowledge the financial support from FCT through grant SFRH/BD/44130/2008 (J.M. Loureiro). Also the financial support of the French-Portuguese bilateral program PESSOA (grant no 20039NJ) is gratefully acknowledged.

References

[1] Al-Khoury W, Eymery J-P and Goudeau Ph. 2007 *J. Appl. Phys.* **102** 043511
[2] Costa B F O, Le Caër G and Loureiro J M 2008 *J. Alloys Comp.* doi:10.1016/j.jalcom.2008.07.179
[3] Le Caër G and Dubois J M 1979 *J. Phys.* E **12** 1083–90
[4] Costa B F O, Le Caër G and Luyssaert B 2003 *J. Alloys Comp.*, **350** 36–46
[5] Delcroix P, Le Caër G and Costa B F O 2007 *J. Alloys Comp.*, **434–435** 584–6
[6] Xia S K, Baggio-Saitovitch E and Larica C 1994 *Phys. Rev.* B **49** 927–31
[7] El-Eskandarany M S, Aoki K, Sumiyama K and Suzuki K, 1999 *Metall. Mater. Trans.* A **30** 1877–80; 2002 *Acta Mater.* **50** 1113–23
[8] Hsieh P J, Lo Y C, Wang C T, Huang J C and Ju S P 2007 *Intermetallics* **15** 644–51
[9] Sluiter M and Kawazoe Y 1999 *Acta Mater.* **47** 475–80
[10] Johnson W C, Lee J K and Shiflet G J 2006 *Acta Mater.* **54** 5123–33

Properties of iron nanoparticles sealed in protective media

O Schneeweiss[1], M Hapla[1], R Zbořil[2] and M Mašláň[2]
[1]Institute of Physics of Materials ASCR, Zizkova 22, 61662 Brno, Czech Republic
[2]Centre for Nanomaterial Research, Palacky University, Svobody 26, 771 46 Olomouc, Czech Republic

E-mail: schneew@ipm.cz

Abstract. Iron nanocrystalline powder was prepared by the reduction of ferrihydride. The samples were sealed in Ar in a plastic bag to avoid any contact with air in the form of (i) the as-prepared powder, (ii) the powder mixed with a vacuum grease, and (iii) the powder after a very slow oxidation at room temperature which formed an oxidic shell. Their phase compositions were investigated by XRD and Mössbauer spectroscopy. These data were completed by magnetic measurements. The intense amorphous-like component was observed in the Mössbauer spectrum of the as-prepared powder. The origin of this component is discussed.

1. Introduction

The strong affinity of iron nanoparticles to oxygen causes that most samples of this material after handling in the air are formed by iron cores with oxidized shells or fully oxidized. To prevent oxidation of iron nanoparticles protective shells must be prepared before the contact with ambient atmosphere or they must be kept in a protective medium immediately after the preparation, e.g., in high vacuum, stable liquids, gels or atmospheres. Most of the studies of iron nanoparticles were carried out on the samples of iron particles with protective shells. Therefore an investigation of the as-prepared bare nanoparticles would show new information on the phase composition, magnetic behaviour, dynamics, etc.

Various problems of dynamics of nanoparticles were often discussed, e.g., in [1] and the references therein. The nanoparticles were attached to the macromolecules and their dynamics was monitored in [1]. The samples of iron-carbon nanoparticles in hexadecane and decaline using oleic acid as a surfactant were investigated. The broad component in the Mössbauer spectrum in decaline whose temperature was slightly below melting point was interpreted to be due to Brownian oscillations. The broad component was not observed in a sample where the nanoparticles were in hexadecane. This very important difference was explained by the properties of the organic material in the immediate vicinity of the particles which determine their motion.

Recently Stanek et al. [2-4] used Mössbauer spectroscopy as a tool for the determination of the microscopic movements of nanoparticles in various media. They mixed 100 nm hematite particles in various substances such as soft gels, foams, sugar solution, etc., and measured Mössbauer spectra of these samples. By the analysis of the resonance absorption line shape in these spectra they obtained that the particles exhibit mobility in the range of mm/s. For the particles in solution of sugar they concluded that the particle movement is correlated for an interparticle distance less then 300 nm.

In this work we will show a different model of the analysis of the spectra of iron nanoparticles which are sealed in protective media. For this purpose we are using comparison of the three types of iron nanoparticles of the same size distribution but with some differences in the phase composition.

2. Experimental details

The nanocrystaline iron powder was prepared by the reduction of ferrihydrite in hydrogen atmosphere at 620 °C for 18 hrs and slowly cooled down (2 K/min) to room temperature. Subsequently the as-prepared powder was packed in PVC plastic bags or in Cu containers in Ar atmosphere without a contact to ambient air (sample labelled A). The samples of the powder closed in plastic bags were used for XRD and Mössbauer measurements and the powders in Cu containers for magnetic measurements. The powder for the sample B was mixed with a vacuum grease and sealed in the same way as the sample A. The slowly oxidized powder sample (labelled C) was prepared by pumping out the furnace atmosphere after calcination followed by a pressure increase from 10^{-2} to 10 Pa during four days. This slowly oxidized powder was no more pyroforic by a contact with the ambient atmosphere. It was nevertheless sealed in a plastic bag with Ar as was done in the case of the sample A.

The structure and phase composition of the powder were investigated by an X-ray diffraction and Mössbauer spectroscopy. The X-ray powder diffraction patterns were collected using CoKα radiation. The ^{57}Fe Mössbauer spectra were measured in transmission geometry using ^{57}Co(Rh) source. The velocity calibration was done according to a pure α-Fe foil. The line width of the sextet of the calibration α-Fe foil was 0.25 mm/s and 0.29 mm/s at 293 K and 5 K, respectively. The Mössbauer absorber was formed by the powders which were sealed in plastic bags in-situ and covered by the additional Al foil for better temperature homogeneity inside the absorber. The absorbers were placed in the cryostat sample holders. The thickness of the absorbers corresponded to ~ 9 nm Fe foil. The morphology and size of the powder particles C were checked using TEM. The magnetic properties were measured using a vibrating sample magnetometer: the hysteresis loops in external field up to ±10 kOe and the temperature dependence of the magnetic moment in 100 Oe.

3. Results and discussion

The X-ray results (Fig. 1) show that the samples of the powder held in Ar in plastic boxes contained bcc Fe with the mean coherence length ~ 10 nm calculated using Scherrer formula. The diffraction data of the sample C show fine peaks of iron oxides. Their content according to a phase analysis was lower than 5 wt.%. TEM of the powder in the sample C shows core-shell particles with a mean size about 25 nm.

The hysteresis loop of the sample A indicated weakly interacting ferromagnetic particles with coercivity 40±4 Oe. The loop of the sample C was very similar with a slightly decreasing magnetic moment and increasing coercivity to 190±10 Oe. Both changes can be simply ascribed to the presence of the oxide phases. The decrease in the magnetic moment roughly corresponds to the oxidation ~ 10 % of iron atoms. The temperature dependence of the magnetic moment of the sample A indicated, that Hopkinson effect starts at ~ 870 K. No changes can be observed below this temperature which indicates stability of the sample.

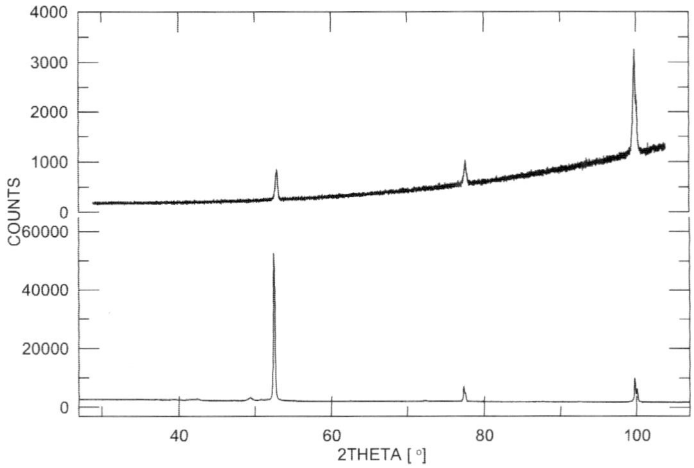

Figure 1 XRD of the sample A (top) and sample C (bottom).

The Mössbauer spectrum of the sample A at room temperature (Fig. 2) can be divided into two main components. The first one corresponds well to α-Fe with the same parameters as the bulk bcc iron. The second component – amorphous-like - must be fitted by three sextets with hyperfine field distribution. The value of the mean hyperfine field $<B_{hf}>$ of these components is 33.2 ± 0.3 T, i.e. very close to α-Fe, but the mean isomer shift is +0.09±0.03 mm/s. The α-Fe sextet dominates in the spectrum of the sample B. In addition a weak broad sextet on the background and the doublet with isomer shift 0.36±0.05 mm/s representing Fe^{3+} can be analyzed there. The spectrum of the sample C is clearly composed of the sextet of α-Fe and by a doublet with an isomer shift and quadrupole splitting corresponding to Fe^{3+}.

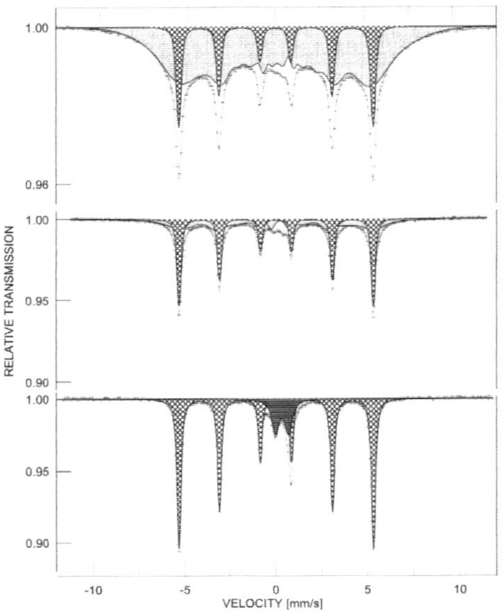

Figure 4 The room temperature Mössbauer spectra of the samples A (top), B (middle), and C (bottom) with following components: α-Fe sextet (diagonal cross), broad sextets with hyperfine field distribution (weak grey), and Fe^{3+} doublet (dark grey).

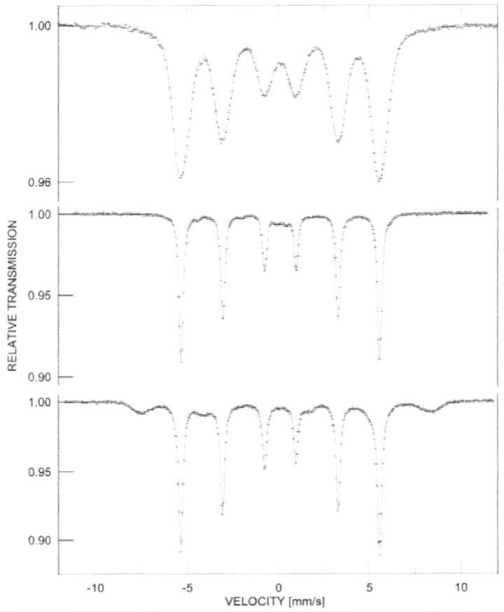

Figure 5 Mössbauer spectra of the samples A (top), B (middle), and C (bottom) taken at 5 K.

By temperature decrease down to 5 K the spectrum of the sample A changes to a broad sextet with hyperfine parameters of α-Fe. It is possible to fit the spectrum by one α-Fe sextet (at 5 K) and three sextets as at room temperature spectrum but the hyperfine field distributions are substantially narrower. In the spectrum of the sample B the broad sextet on the background diminishes at 5 K. For the sample C, the doublet in the room temperature spectrum transforms to a sextet which represents Fe^{3+} oxides. It should be noted that the intensity ratios inside the sextets were kept at 3:2:1, i.e. random distribution of spin orientations, was expected for all samples in the fitting procedure used.

Superparamagnetic component was not observed in the spectra of all the samples. It can be an influence of the exchange interaction which suppresses superparamagnetic relaxation. It was found in agglomerated NiO nanoparticles [5] and in composites containing antiferromagnetic nanoparticles of different materials [6]. The shape of the room temperature spectrum of the sample A is very similar to some of that shown in [2,3]. However there is an objection against explanation of the present results in frame of Brownian motion model. The sample C does not exhibit the same broad sextet components as the sample A although both samples are very similar from the point of view of the particle sizes. Diffusion (even in surfaces) may be excluded because the component (in a narrower form) remains in the spectrum taken at 5 K where the atomic jumps have a negligible probability. Structural disorder in the surface layer could be another cause of the broad component in the 295 K spectrum of the sample A. The mean size of the particles was ~ 20 nm. The atomic fraction of the amorphous-like component was of 0.77. It corresponds to the mean thickness of a surface layer ~ 4nm. The structural disorder could be removed (ordered) by the slow oxidation (sample C). However it is less probable that such a disorder can be removed (ordered) by mixing of the nanopowder with a vacuum grease. The only difference is in their surface. In the sample A the particles have a "raw" iron surface while in the sample C the surface is formed by a Fe^{3+} oxide shell.

Table 1. The results of the analysis of the Mössbauer spectra shown in figs. 4 and 5 (S … sextet, D … doublet, intensities I of the components in atomic fraction of Fe atoms – a.f., hyperfine inductions B_{hf}, ΔB_{hf} width of distribution of B_{hf}, isomer shifts δ, quadrupole shifts ε_Q, and quadrupole splittings Δ_{EQ}).

Sample	T [K]	Component	I [a.f.]	B_{hf} [T]	ΔB_{hf} [T]	δ [mm/s]	ε_Q [mm/s]	Δ_{EQ} [mm/s]
A	293	S1	0.21	33.1	-	0.03	0.01	-
		S2	0.27	36.9	3.1	0.56	-0.44	-
		S3	0.24	34.8	2.3	-0.27	0.44	-
		S4	0.27	31.5	1.3	0.12	-0.14	-
		D1	0.01	-	-	0.46	-	1.08
	5	S1	0.21	.34.2	-	0.15	0.02	-
		S2	0.57	34.0	0.6	0.25	-0.01	-
		S3	0.22	33.1	2.9	0.65	-0.73	-
B	293	S1	0.64	33.0	-	0.00	0.00	-
		S2	0.33	32.5	1.3	0.12	0.30	-
		D1	0.03	-	-	0.14	-	0.68
	5	S1	0.95	34.2	-	0.15	0.03	-
		S2	0.03	32.5	-	0.89	-0.32	-
		D1	0.02	-	-	0.16	-	1.00
C	293	S1	0.86	32.9	-	0.03	0.00	-
		D1	0.14	-	-	0.37	-	1.43
	5	S1	0.87	34.2	-	0.15	0.00	-
		S2	0.13	48.8	0.9	0.56	-0.17	-

There acts another source which is responsible for the surface component having amorphous-like subspectrum in the sample A. The high volume content of the interfaces/surfaces is taken as the most important feature of nanocrystalline metals and oxides since the beginning of their studies. It can be expected that the atoms in a surface layer can be influenced by changes in the electron densities which do not "freeze" contrary of phonons at 5 K and therefore changes in hyperfine field parameters can be expected in the whole temperature range.

Acknowledgements
This work has been supported by the project 1M619895201 of the Ministry of Education of the Czech Republic and by the Grant Agency of the Czech Republic (106/08/1440).

References

[1] Hendriksen P V, Mørup S and Linderoth S 1992 *Hyperfine Interaction* **70** 1079
[2] Stanek J and Fornal P 2009 *Hyperfine Interactions* DOI 10.1007/s10751-009-9967-y
[3] Fornal P and Stanek J 2008 *Acta Phys. Polonica* A **114** 1667
[4] Fornal P, Stanek J and Wilk A 2008 *Hyperfine Interactions* **185** 81
[5] Bødker F, Hansen M F, Koch C B and Mørup S 2000 *J.Magn. Magn. Mater.* **221** 32
[6] Frandsen C and Mørup S 2003 *J. Magn. Magn. Mater.* **266** 36

Disproportionation of Cobalt Ferrite nanoparticles upon annealing

R D Desautels, J van Lierop and J M Cadogan

Department of Physics and Astronomy, University of Manitoba, Winnipeg, Manitoba, R3T 2N2, Canada

E-mail: johan@physics.umanitoba.ca

Abstract. Size control of cobalt ferrite ($CoFe_2O_4$) nanoparticles is often achieved through annealing. To better understand the annealing process we have synthesized 5 nm $CoFe_2O_4$ nanoparticles using a co-precipitation technique. Characterization of the intrinsic magnetism of these nanoparticles using magnetometry and susceptometry shows typical $CoFe_2O_4$ nanoparticle behaviour [1]. The temperature evolution of the annealing process was examined using transmission Mössbauer spectroscopy. The Mössbauer data provide clear evidence for a disproportionation of the $CoFe_2O_4$ phase, beginning at around 500 K and continuing to evolve up to 800 K. Using a combination of Mössbauer spectroscopy and x-ray powder diffraction we show that annealing the $CoFe_2O_4$ nanoparticles not only results in a larger nanocrystallite size, 17 nm, but also a significant disproportionation into a mixture of residual $CoFe_2O_4$, CoO and α-Fe(Co) (the majority phase). Upon cooling back to room temperature from 825 K a magnetite-like component develops and by room temperature this phase accounts for 20% of the Mössbauer spectral area.

1. Introduction

Nanoscale cobalt ferrite ($CoFe_2O_4$) systems exhibit high coercivities due to a large cubic magnetocrystalline anisotropy. The typical saturation magnetization of $CoFe_2O_4$ nanocrystallites in combination with these high coercivities ($H_c \gtrsim 10$ kOe at 10 K) suggest that nanoscale $CoFe_2O_4$ could be a promising basis for permanent magnets, magnetic recording media, and even drug delivery applications. The eventual technological exploitation of $CoFe_2O_4$ nanoparticulate systems will require both a comprehensive understanding of the size dependence of the nanomagnetism and the ability to control the nanocrystallite size. A common method for preparing different sized nanocrystallites is thermal annealing which not only creates larger crystallites, but also results in dramatic changes in the nanomagnetism, such as unusual decreases in H_c with increasing size [2, 3].

The altered magnetism of annealed $CoFe_2O_4$ nanocrystallites has been attributed to internal strain during the annealing process that creates an extra uniaxial anisotropy in addition to fanning modes of the magnetization [4]. Cation diffusion during the annealing process has also been postulated as a source of the unusual size-dependent trends in magnetic properties [5].

We have synthesized $CoFe_2O_4$ nanoparticles with an average crystallite diameter of ~ 5 nm using a co-precipitation technique. Annealing these nanoparticles at 873 K for 10 hours in vacuum resulted in $\sim$ 17 nm diameter particles. Magnetometry studies of the unannealed nanoparticles exhibited H_c ranging from $\sim$ 13 kOe at 10 K to $\sim$ 50 Oe at 400 K with a $T^{\frac{1}{2}}$ temperature dependence, and a saturation magnetization, $M_s(T)$, that decreases linearly with increasing temperature. By contrast, the larger annealed $CoFe_2O_4$ nanocrystallite system shows a linear $H_c(T)$ with a reduced H_c of $\sim$ 8 kOe at 10 K, a larger high temperature $H_c \sim$ 400 Oe at 400 K and a Brillouin-like $M_s(T)$ [1].

2. Experimental Methods

The $CoFe_2O_4$ nanoparticles were synthesized using a co-precipitation method [6]. $0.4M$ of $FeCl_36H_2O$ (98% purity, Alfa Aesar) was mixed with a $0.2M$ solution of $CoCl_2 \cdot 6H_2O$ (98% purity, Alfa Aesar). A $3M$ solution of NaOH (98% purity, Alfa Aesar) was then added dropwise to the metallic salt solution until a pH of 11.75 was reached. 50 μl of oleic acid (99% purity, Alfa Aesar) was added via rapid injection to coat the precipitate, thereby stopping nucleation and protecting against oxidation. The resulting solution was heated to 353 K while being stirred continuously for 1 h. After cooling to room temperature, the precipitate was separated and washed several times with water and methanol and then dehydrated for 10 h at 373 K. Cu-Kα x-ray diffraction on a PANalytical PW3830 was used to determine the phase composition of the nanoparticles. Refinement of the x-ray diffraction patterns was carried out using the FULLPROF/WinPlotr package [7, 8]. A Photocor Instruments static and dynamic light scattering (DLS) apparatus was used with a 25 mW Coherent solid state laser (λ=632 nm) to measure the particle size and size distribution of a dilute suspension of nanoparticles in chloroform at room temperature. Magnetometry experiments were performed using Quantum Design PPMS and MPMS systems. Transmission Mössbauer measurements were made in constant acceleration mode with a 1 GBq ^{57}CoRh source. The Wissel spectrometer was calibrated using α-Fe at room temperature and all isomer shifts (IS) are quoted relative to the α-Fe calibration. An ASA Mössbauer vacuum furnace was used to collect spectra at temperatures between 300 K and 830 K.

3. Results

In figure 1 (top) we show the x-ray diffraction pattern obtained at room temperature on a $CoFe_2O_4$ sample which had been annealed at 550 K for 8 hours. The pattern shows that the annealing produces a mixture of the following phases: $CoFe_2O_4$ (Space Group SG = Fd$\overline{3}$m #227), α-Fe(Co) (SG = Im$\overline{3}$m #229), CoO (SG = Fm$\overline{3}$m #225) and an unidentified phase. The pattern is dominated by the two sharp peaks from NaCl formed during the co-precipitation process. In an attempt to identify the fourth phase we carried out a trial indexation using the DICVOL program [9]. The best match to the set of unidentified peaks was with an orthorhombic cell with lattice parameters a = 9.93 Å , b = 3.04 Å and c = 2.92 Å. A tetragonal cell with lattice parameters a = 9.21 Å and c = 3.85 Å was the second-best match to the observed peaks. We note that indexation programs such a DICVOL only return the minimum-volume cell which reproduces the peak positions. Consequently, we cannot rule out super-cells of these two possible cells. This fourth phase remains unidentified at present.

In figure 1 (bottom) we show a comparison of the ^{57}Fe Mössbauer spectra obtained at room temperature **before** and **after** the annealing run. The pre-annealed spectrum

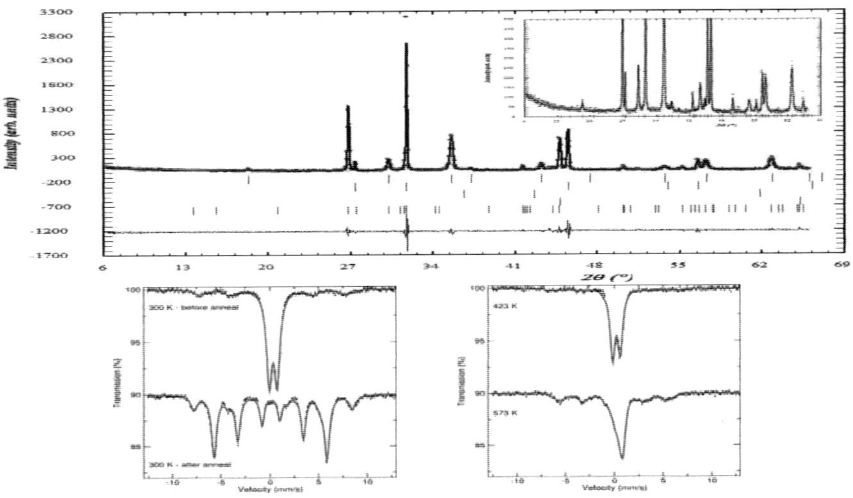

Figure 1. (Top) X-ray powder diffraction pattern of annealed $CoFe_2O_4$. The Bragg markers, from top to bottom, represent $CoFe_2O_4$, NaCl, CoO, α-Fe(Co) and the unidentified phase. The inset shows an enlargement of a section of the pattern. (Bottom Left) Room-temperature Mössbauer spectra of $CoFe_2O_4$ **before** and **after** annealing. (Bottom Right) Mössbauer spectra of $CoFe_2O_4$ obtained at 423 K and 573 K during the heating run.

is dominated by a doublet (IS = +0.33(2) mm/s; QS = 0.82(2) mm/s) which accounts for about ~ 80 % of the spectral area, assuming, for simplicity, the same f-factor for all components. This doublet is due to the unblocked $CoFe_2O_4$ nanoparticles (In our previous study we deduced a blocking temperature of 380(20) K [1]). The remaining ~ 20 % of the spectrum is the still-blocked $CoFe_2O_4$ component with a hyperfine field of 45.9(1) T, characteristic of the spinel-type structure of $CoFe_2O_4$.

The post-annealed room-temperature spectrum is markedly different from the pre-annealed spectrum. The dominant component **after** annealing is now α-Fe(Co), with an average hyperfine field of 35.9(1) T. We can estimate the Co content in this α-Fe phase from the measured ^{57}Fe hyperfine field. The measured value of 35.9(1) T is larger than the value of 33.0 T found in pure α-Fe and translates to a Co content either about 15 at% Co or 38 at% Co [10]. Unfortunately, as pointed out by Johnson *et al.* [10], the isomer shifts are small and unable to distinguish these two options. The minor component in the post-annealed spectrum is most likely a maghemite-like $\gamma-(Co,Fe)_2O_3$ phase or a magnetite-like $(Co,Fe)_3O_4$ phase with an average hyperfine field of 50.5(2) T and an IS of +0.42(3) mm/s, consistent with iron oxide nanoparticles with cobalt present. This phase develops at around 600 K during the cooling run from 825 K. At this stage, however, we cannot unambiguously identify this component due to a variety of factors such as the nanoparticulate nature of the sample and the Co content.

We can identify the temperature range in which disproportionation commences by comparing the ^{57}Fe Mössbauer spectra obtained at 423 K and 573 K during the heating run (figure 1 (bottom)). The development of the broad α-Fe(Co) component is quite clear in the 573 K spectrum. The spectrum at 423 K shows that the $CoFe_2O_4$

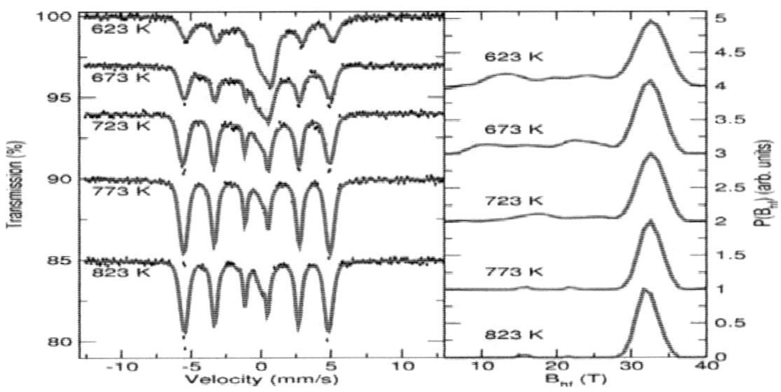

Figure 2. ^{57}Fe Mössbauer spectra of $CoFe_2O_4$ obtained at various temperatures from 623 K to 823 K during the heating run, together with the corresponding hyperfine field distributions.

phase is virtually fully unblocked at this temperature.

In figure 2 we show the ^{57}Fe Mössbauer spectra obtained at various temperatures between 623 K and 823 K during the annealing run. Due to the nanoparticulate nature of the samples and the significant solution of Co in the emerging α-Fe phase, we fitted the spectra using the Le Caër-Dubois hyperfine field distribution method [11]. In figure 2 we also show the hyperfine field distributions for the α-Fe(Co) spectral component. The peak in the α-Fe(Co) field distribution is at 32.7 T at 623 K and decreases slightly to 32.1 T by 823 K. The sub-spectral area of the α-Fe(Co) component is 59(4) % at 573 K and reaches a plateau of 92(3) % at around 770 K.

In conclusion, annealing of nanoparticulate $CoFe_2O_4$ leads to a disproportionation of the sample, commencing between 423 K and 573 K. The principal phase in the annealed sample is α-Fe(Co).

Acknowledgments

This work was supported by grants from the Natural Sciences and Engineering Research Council of Canada and the Canada Foundation for Innovation. JMC acknowledges support from the Canada Research Chairs programme.

References

[1] Desautels R D, Cadogan J M and van Lierop J 2009 *J. Appl. Phys.* **105** 07B506–1–3
[2] Xiao S H, Jiang W F, Li L Y and Li X J 2007 *Materials Chemistry and Physics* **106** 82
[3] Chiu W S, Radiman S, Abd-Shukor R, Abdullah M H and Khiew P S 2008 *J. Alloys and Compounds* **459** 291
[4] Grigorova M, Blythe H J, Blaskov V, Rusanov V, Petkov V, Masheva V, Nihtianova D, Martinez L M, Muñoz J S and Mikhov M 1998 *J. Magn. Magn. Mater.* **183** 163–172
[5] Wang Y C, Ding J, Yin J H, Liu B H, Yi J B and Yu S 2005 *J. Appl. Phys.* **98** 124306
[6] Maaz K, Mumtaz A, Hasanain S K and Ceylan A 2009 *J. Magn. Magn. Mater.* **308** 289–295
[7] Rodríguez-Carvajal J 1995 *Physica B* **192** 55
[8] Roisnel T and Rodríguez-Carvajal J 2001 *Mater. Res. Forum* **118** 378–381
[9] Boultif A and Louër D 1991 *J. Appl. Cryst.* **24** 987–993
[10] Johnson C E, Ridout M S and Cranshaw T E 1963 *Proc. Phys. Soc.* **81** 1079–1090
[11] Le Caër G and Dubois J M 1979 *J. Phys. E: Sci. Instrum.* **12** 1083–90

Journal of Physics: Conference Series **217** (2010) 012106 doi:10.1088/1742-6596/217/1/012106

Mechanochemically induced formation of amorphous phase at oxide nanocomposite interfaces

T Kiseleva[1], A Novakova[1], M Zimina[1], S Polyakov[2], E Levin[3] and T Grigoryeva[4]

[1]Department of Physics, Moscow M.V. Lomonosov State University, 119991, Leninskie Gory, Moscow, Russia
[2]Moscow M.V. Lomonosov State University, Skobeltsin Institute of Nuclear Physic, Leninskie Gory, 119991, Moscow, Russia
[3]Department of Chemistry, Moscow M.V. Lomonosov State University, 119991, Leninskie Gory, Moscow, Russia
[4]Institute of solid state chemistry and mechanochemistry, SO RAS, Kutateladze str., Novosibirsk, Russia

E-mail: Kiseleva_tatiana@mail.ru

Abstract. α-Fe_2O_3-Fe mixtures of different compositions ground in high energy ball mill were studied. The formation of amorphous iron-based magnetic phase was detected by means of Mössbauer spectroscopy and magnetic measurements. The amount of this phase increases with Fe/α-Fe_2O_3 ratio in starting mixtures increase. Its thermal stability was studied by means of magnetic and DCS measurements.

1.Introduction

Within top-down technologies, high energy ball milling is a simple and inexpensive technology for composites, amorphous and nanostructured materials production [1]. It was shown that ball-milling induced α-Fe_2O_3 reduction can occur via γ-Fe_2O_3, Fe_3O_4 and FeO_x formation [2]. Iron admixture to milled hematite led to an essential increase of phase transformation speed and formation of the amorphous phase representing an oxygen solid solution in iron [3] at a prolonged milling time duration.This can be explained [4] as a result of Fe_2O_3 interaction with iron-containing balls and vial walls at intensive milling. In our present work, we studied the solid state interaction of α-Fe_2O_3 and Fe powders. The aim was to reveal the influence of iron addition on the oxide reduction process during short-time high energy activation.

2.Experimental

Ball milling of 6.4 g Fe_2O_3 powder with different amounts of Fe powder (up to 4 g) was performed with an AGO-2 planetary ball mill for 2 minutes in a vial sealed under Ar. The volume of the vial was 250 cm^3. The diameter and mass of the balls were 5 mm and 200 g respectively. The speed of the drums rotation was ~1000 rpm. The initial mean sizes of Fe_2O_3 and Fe particles were 80 μm and 50 μm, respectively. The Mössbauer spectra (MS) were received at room temperature with ^{57}Co(Rh). The sample masses for measurements were normalized to equal iron concentration. The phase composition determined from the spectra implied the same recoil fraction *f'* for components. The powder diffraction data were collected in step scan mode at room temperature using RIGAKU D/max RC powder diffractometer equipped with a scintillation detector (Bragg-Brentano geometry (θ-2θscan), CuKα radiation, flat graphite analyser crystal). The samples were pasted onto a self-made zero-background sample holder. It is essential to use this type of holder in order to improve signal-to-noise ratio and to avoid sample displacement and transparency errors [5]. The transmission electron microscopy (TEM) was performed on LEO 912 AB Omega. Thermomagnetometry was carried out on Curie-balance in an applied field of 0.45 T. Differential scanning calorimetry was conducted in a DSC-4 Perkin Elmer.

3.Results and discussion

Mössbauer spectra of initial (a) and ball milled (b) Fe_2O_3 are presented in figure 1. It can be seen that 2 minutes of milling resulted in appearance of an additional sextet with broadened components. The main part of the spectrum (77%) with hyperfine parameters H_{eff}= 515 kOe, δ=0,32 mm/s, Δ=0,24 mm/s, Γ=0,32 mm/s is characteristic to α-Fe_2O_3. An additional sextet has broadened components with H_{eff} = 490 kOe. The X-ray diffraction (XRD) pattern (figure 2 (a)) demonstrates only line broadening of the α-Fe_2O_3 reflections after milling as compared to the initial sample. This is evidently connected with grain size reducing, appearance of structural defects and formation of disordered α-Fe_2O_3 or γ-Fe_2O_3 (about 23 %) at the grain boundaries. The existence of a direct phase transition from α- Fe_2O_3 to γ-Fe_2O_3 during milling was reported earlier [6]. The opposite was observed for α-Fe_2O_3-Fe milled mixtures. The initial spectrum is given in figure 1 (c). The resulting Mössbauer spectrum of milled samples has a complex substructure. As seen from figure 1(d) (6.4 g Fe_2O_3+4g Fe milled mixture) the spectra are composed of several components in addition to α-Fe_2O_3 subspectra. They correspond to 5% of metallic iron (sextet with H_{eff}=330 kOe) and 23% of non-stoichiometric Fe_3O_4 subspectrum (two sextets with hyperfine parameters H_{eff}=490 kOe, δ=0.27mm/s, Δ=0.01 mm/s, Γ=0.6 mm/s and H_{eff}= 460 kOe, δ=0.67mm/s, Δ=0.04 mm/s, Γ=0.7 mm/s).

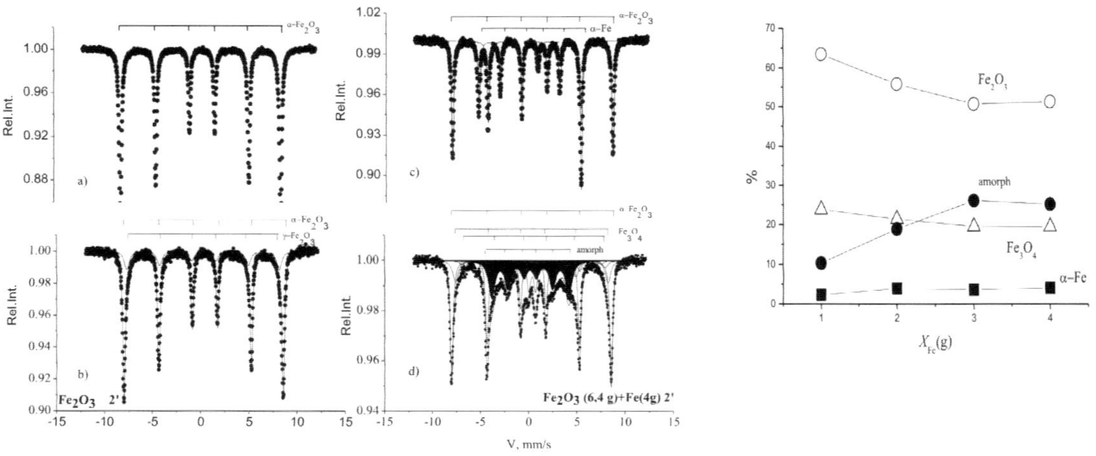

Figure 1. Mossbauer spectra of Fe_2O_3(6,4g)+xFe (x=0-4 g) ball milled for 2 minutes (a-d) and Mossbauer subspectra intensity for different amount of Fe in Fe_2O_3(6,4g)+ xFe mixture (e)

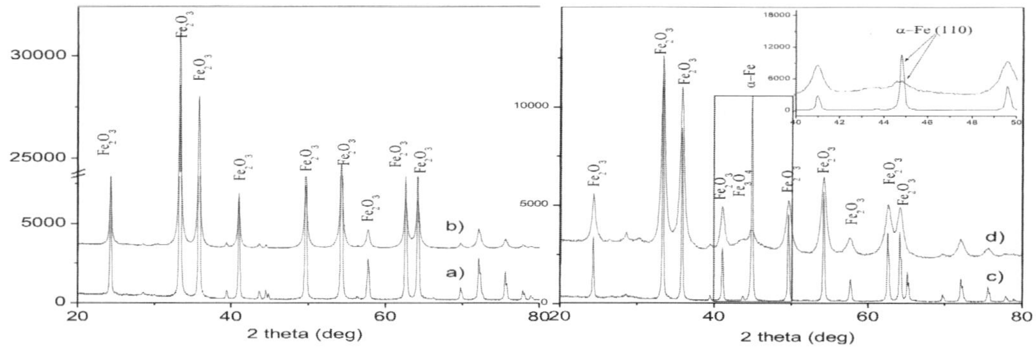

Figure 2. The XRD patterns of initial (a) and 2 min milled α-Fe_2O_3 (b), initial simple mixture of α-(6,4g)Fe_2O_3+(4g) Fe (c) and 2 min milled (d).

As the amount of Fe is increased, so does the other (marked by black colour) component's intensity (figure 1 (e)). Mössbauer spectrum of this component has a hyperfine fields distribution in the range of 200-300 kOe. The average value of hyperfine field is 250±3 kOe. The presence of hyperfine fields' distribution is consistent with the increasing disorder in the interfacial region since it indicates the existence of interatomic spacing distribution at the grain boundaries. This subspectrum shape resembles those for a Fe-based amorphous compound with different surroundings of Fe-atoms [7]. This component was fitted by superposition of several sextets with essentially equal isomer shifts ($\delta \approx 0.13$ mm/s). The obtained value of δ corresponds to the occurrence of oxygen in disordered iron lattice stemming from α-Fe_2O_3 decomposition.

A significant broadening of the iron diffraction peaks (figure 2 (d)) and drastic decrease of its relative intensities as compared with XRD patterns of starting mixtures were observed for all milled samples. At the same time, Fe_3O_4 formation was found. These facts together with Mössbauer spectroscopy data clearly show that the amorphous phase is formed from Fe powder. Well-pronounced particle sizes decreasing are observed on the TEM pictures (figure 3). Also, there are broad distributions observed for both milled and starting mixtures. The distribution of milled α-Fe_2O_3 particles with an average value of 45 nm becomes significantly broader than the distribution of Fe_2O_3+Fe milled mixtures (average values 15-18 nm).

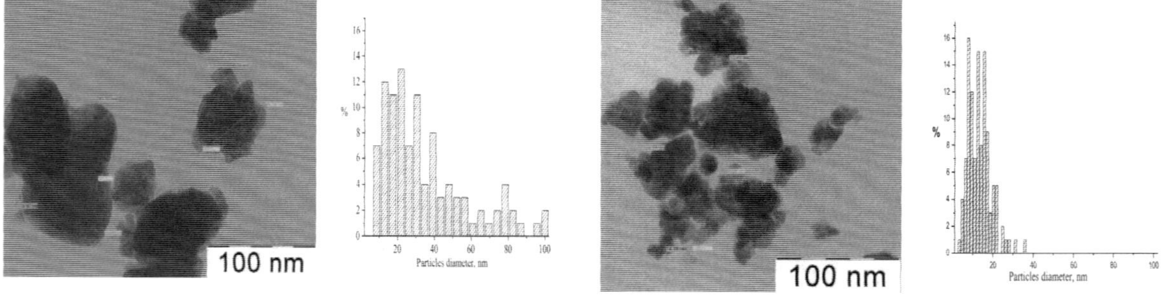

Figure 3. TEM pictures for 2min ball milled Fe_2O_3 (a) and 6.4 g Fe_2O_3+4g Fe mixture (b) and corresponding particles sizes distribution histograms (c, d)

It is worth noting that direct comparison of these distributions aren't correct since in the case of Fe+Fe_2O_3 milled mixtures we are dealing with 4-phase mixture and there is no way to compare particle sizes for α-Fe_2O_3 in both cases – with Fe addition and without. Nevertheless, the general acceleration in particle size reduction is clear. The quantity of Fe-based amorphous phase (≈25%), obtained by Mössbauer analysis correlates with the interface surface area for the observed sizes of Fe crystalline grains [8].

Some data about amorphous iron based phase formation at the grain boundaries can be obtained via DSC and saturation magnetization M_s temperature dependencies measurements [9] (figure 4).

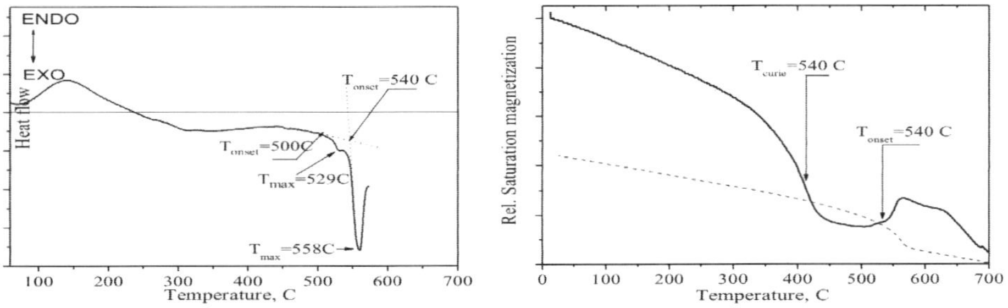

Figure 4. Temperature dependence of magnetization (a) and DSC-curve for milled Fe_2O_3+4Fe

Observed M_s and DSC temperature dependencies both have a typical irreversible shape, indicating the amorphous-to-crystalline transformation [10]. The T_c of amorphous phase is 430°C. DSC curves reveal the broad exothermal effect at a temperature range up to 450°C indicating a relaxation of the internal stress. M_s(T) dependencies haven't reached zero value at T_c. Also, there is a shoulder on the M_s(T) curve from 420 to 500°C due to the input of Fe_3O_4 and α-Fe_2O_3 present in the composite samples after milling according to phase analysis. The subsequent two-peak M_s increase and two exothermic DSC effects at 500 and 540°C are connected with the processes of magnetic phases crystallization. The DSC curves in this temperature range exhibit two exothermic effects of different intensity at 500 and 540°C, which correspond to crystallization stages. The M_s(T) curves of second heating (figure 5(2)) reveal the irreversibility of amorphous-to-crystalline state transformation. The heating of milled samples up to 500°C [9] revealed the crystallization, which began with grain growth and resulted in amorphous phase diminishing.

4.Conclusion

Nanocomposite Fe_2O_3/Fe_3O_4/Fe system were synthesized by *short-* time high energy ball-milling of (6,4g)Fe_2O_3+*x*Fe (x=1-4 g) powder mixture. In the course of composite formation, oxide destruction and structural disorder of iron grain boundaries bring out magnetic amorphous phase.

Aknowledgements The authors thank RFBR (Project N 09-03-00925) for financial support.

References

[1]Campbell S J, Kaczmarek W A 1996`*Mossb. Spectr. Applied to Mater. and Magnet.* Vol 2 ed G J Long and F Grandjean (New York:Plenum Press) p 77
[2]Campbell S J, Klingelhofer G 2002 *Hyperfine Interactions* **139/140** 407
[3]Lileev A S, Yagodkin Yu D 2005 *J. of Magn. Magn. Mater.* **290-291** 1217
[4]Hofmann M, Campbell S J 2003`*J. Alloys & Comp.* **348** 278
[5]Misture S T, Chatfield L R, Snyder R L 1994 *Powder Diffr.* **9** 172.
[6]Meillon S, Dammak H, Flavin E, Pascard H 1995 *Phil.Mag.Letters* **72** 105
[7] U.Gonser: *Hyperfine Interaction* V.13 (1983), p.5-23
[8]Del Bianco L, Hernando A 2005 *Nanost.Science and Techn.* ed.D Fiorani (Berlin: Springer) p 217
[9]Kiseleva T, Novakova A, Chistyakova M 2009 *Solid State Phenomena* **152-153** 25
[10]Novakova A A, Kiseleva T Yu 1997 *Crystallography Reports*, **42** N2 350

Journal of Physics: Conference Series **217** (2010) 012107 doi:10.1088/1742-6596/217/1/012107

Investigation of different iron sites in ε- Fe_yN (2<y<3) nanoparticles using Mössbauer Spectroscopy

Sajith Kurian[1], N S Gajbhiye[1] and S K Date[2]

[1]Department of Chemistry, Indian Institute of Technology Kanpur, Kanpur-208016, INDIA

[2]Department of Physics, University of Pune, Pune 411007,INDIA

E-mail: nsg@iitk.ac.in

Abstract. Epsilon Iron Nitride has a wide range of nitrogen solubility in its lattice. While the stoichiometric ε-Fe_3N has an ordered arrangement of nitrogen atoms in the octahedral voids of the hexagonal iron lattice, in non-stoichoimetric ε-Fe_yN (2<y<3) the nitrogen atoms occupy the voids in a disordered manner and the coordination of Fe with nitrogen varies with the number 'y' in the lattice. Mössbauer spectroscopy, which is a nuclear spectroscopic tool, can differentiate the different co-ordination of Fe through its characteristic isomer shift values and is used for identifying the different types of Iron sites. ε-Fe_yN (2<y<3) nanoparticles were prepared by the nitridation of hematite nanoparticles using ammonia at 550°C. Mössbauer spectroscopy at ambient temperature shows a superparamagnetic doublet and a sextet. Spectra at 80 K and 5 K resolve the sextets nicely and thereby the different iron sites present in the ε-Fe_yN (2<y<3) nanoparticles.

1. Introduction

Among the various iron nitride phases existing at room temperature, γ′-Fe_4N and ε-Fe_3N are ferromagnetic and can be used for magnetic applications due to their high saturation magnetization and Curie temperature. The phase diagram of Fe-N [1] shows a wide range of nitrogen solubility in the iron lattice while holding the hcp crystal structure. As a result it is easy to synthesize ε-Fe_3N particles. Unlike the ordered stoichiometric ε-Fe_3N, which may form through controlled nitridation and slow cooling of the material, the quenched material and the more nitrogen rich Fe_3N, namely ε-Fe_3N_{1+x}, has a disordered arrangement of nitrogen atoms. This creates unequal environments for different iron atoms in the lattice.

2. Experimental

The ε-Fe_3N_{1+x} nanoparticles were synthesized by treating hematite particles with ammonia gas at a temperature of 550°C for 3 hours and quenching after the reaction. Nitrogen content was estimated by the Kjeldahl's method [2] and has been further confirmed by the C-H-N analyzer. The result shows a composition of y = 2.6. The characterization of this material has been carried out by XRD, SEM, and VSM. The presence of different iron sites has been probed by Mössbauer spectroscopy. Room temperature Mössbauer measurements were performed using a ^{57}Co source incorporated in Rh matrix using a commercial spectrometer supplied by Wissel. Low temperature measurements were performed in a He bath cryostat. The spectra were analyzed by the Win-normos software.

3. Results and discussions

The X-ray diffraction data showed the presence of a single phase hexagonal structure. Commonly, the X-ray diffraction technique is well suited to get the positions of the interstitials of well-defined stoichiometric interstitial solid solutions. When the interstitial atom has a wide range of solubility in the lattice by maintaining the parent ordered type structure, however, the X-ray data are not sufficient to get the distribution of interstitials among the available sites. On the other hand, the simple structure of the Fe-N phases leads to an unambiguous interpretation of the Mössbauer spectra.

It is important to understand the structure of ε-Fe_3N before considering the analysis of the Mössbauer spectra. In an hcp arrangement of M (Fe), each M atom is coordinated by six octahedral sites forming a trigonal prism. In an ordered stoichiometric ε-Fe_3N structure two of these sites, only the Wyckoff position 2c, are occupied by N. This coordination is called M_2, a metal atom coordinated with two nitrogen atoms. The Mössbauer spectrum of this compound shows a single six finger pattern corresponding to the perfect ferromagnetic ε-Fe_3N. This occurs exclusively in slowly cooled materials. This ordered structure can be destroyed either by quenching the material from higher temperature or by the extra nitrogen atoms entering the vacant interstitial sites. This disorder leads to the filling up of 2b Wyckoff sites besides the more favorable 2c sites which causes a partial formation of the environments M_1 and M_3 as reported in literature [3-5]. Mössbauer spectroscopy is able to detect the separate signals for the different environments. Literature shows that M_1 and M_3 are additionally observed besides M_2 for samples quenched from higher temperatures and M_3 is achieved for N contents higher than that in Fe_3N, that is Fe_3N_{1+x}. This ability of Mössbauer spectroscopy has been used to identify the various electronically different iron sites in this system. The different local environment of an iron atom occurs due to the different numbers of nitrogen atoms in its near neighbourhood. The isomer shift directly depends on the electron density at the iron nuclei. It increases with the decrease in 's' electron density. In iron nitride system the nitrogen atom is considered as a 'p' electron donor. Due to this contribution from nitrogen 'p' electrons the 's' electron density surrounding the iron atom decreases with increase in coordination with more nitrogen atoms. As a result, the evaluation of the isomer shift using the Mössbauer technique will help to classify different iron sites and thus the number of nitrogen atom surrounding the iron atom. The reported values show that the Fe_yN with y = 3, 2.67, 2.4 (ε-Fe_yN) exhibits a six finger pattern for different iron sites at room temperature. A doublet, reflecting paramagnetic behaviour, was observed when y= 2 (the ζ-Fe_2N phase) at room temperature.

Our results obtained at room temperature show superimposed spectra comprising sextet and doublet (Figure 1(a)). As the amount of nitrogen determined was not in the range of ζ-Fe_2N and the

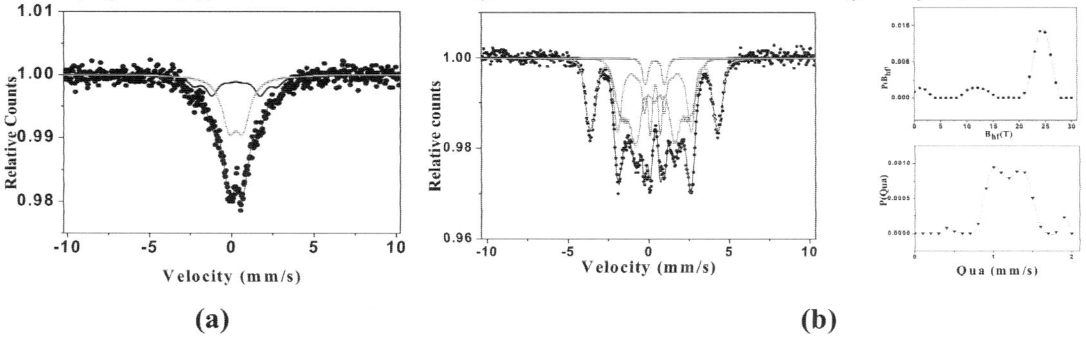

(a) **(b)**

Figure 1. Mössbauer spectra at (a) Room Temperature (b) 80 K with distribution material shows ferromagnetic behaviour at room temperature we can eliminate the possibility of doublets corresponding to the material ζ-Fe_2N , a paramagnetic material at room temperature. The isomer shift data supports this argument. Accordingly, the superimposed sextet and doublet is supposed to correspond to the ε-Fe_3N phase. The existence of sextet and doublet can be understood on the basis of different contributions from the different sized particles. It is well established that the electron spin relaxation effects are very influential and critical in the analysis of complex broad non-Lorentzian Mössbauer spectrum. In the case of uniaxial superparamagnetic (ultrafine) particles, the spin relaxation process is strongly dependent on the volume (V) and anisotropy constant (K).When the spin relaxation time is comparable with the Larmor precession time, a well-defined magnetically split sextet is observed. On the other hand, if it is extremely short of the order of 10^{-11} or 10^{-12} seconds, the sextet collapses into a broad non-Lorentzian singlet or doublet depending on the contributions from quadrupole interactions.

Table 1. Hyperfine parameters of ε-Fe_3N_{1+x} at different temperatures

T (K)	Fe Site	Isomer Shift δ (mm/s)	Quadrupole Splitting Δ (mm/s)	Hyperfine Field (T)	Width Γ (mm/s)	Area %
298	Sextet	0.36	-0.05	15.6 ± 0.2	0.87	47.8
	Doublet	0.38	0.84	--	1.06	52.2
80	Sextet 1	0.46	-0.04	24.5	0.29	38.7
	Sextet 2	0.53	-0.03	12.5	0.83	57.3
	Doublet	0.51	0.65	--	0.31	4.0
5	Sextet 1	0.46	-0.04	26.1	0.27	40.7
	Sextet 2	0.50	-0.03	13.7	0.37	59.3

The higher line width of the spectra points to the existence of a distribution which may be attributed to two reasons. One is a slight distribution in particle size and the other is the statistical distribution of nitrogen atoms in the vacant sites of trigonal prisms. The low temperature spectrum can be used to clear these issues. So we recorded the spectra at 80, Figure 1(b) and 5 K, Figure 2. The hyperfine parameters are given in the Table 1. The 80 K spectrum is fitted according to a model constructed with two sextets and a doublet. This model gives a better fit rather than the model which consists of only sextets. The hyperfine parameters match well with the reported values for M_2 and M_3 [3, 6].The area fractions show that there exist more M_3 sites than M_2. The evolution of two well defined sextets which constitutes 96% of the total area indicates that most of the particles are now having $KV>>E_{th}$(thermal energy). As the temperature is lowered, the effective slow relaxation allows to observe the evolution of the magnetic sextet. The increase in magnetic hyperfine field with decrease in temperature is due to the increase in the effective spin $<s>$ through the increase in population in the lowest lying states. Even from the visual inspection of the experimental spectrum it is obvious that the predicted or the reported M_1 site is absent in this material. The doublet at this temperature corresponds to iron atom coordinated with more number of nitrogen atoms, which may arise due to the higher nitrogen diffusion in to the particles of very low dimensions. The higher area fraction of the M_3 site clearly tells the existence of larger number of iron atoms coordinated with three nitrogen atoms. It is clear from the data that the line widths are much larger than the instrumental line width and is still increasing as one move out of the center of the spectrum. This is a well-known characteristic of a distribution of hyperfine fields which also points to the existence of different types of Fe-N coordination and thus the magnetic ordering. In view of this we tried a distribution fit of the experimental data. The hyperfine field and quadrupole distribution are shown in Figures 1, 2 as well. The hyperfine field distribution centered at 25, 14 and 3 Tesla indicates that the nitrogen atoms are statistically distributed in the available six corners of the trigonal prism centered by Fe atom. The observation of a low magnetic moment compared to the bulk value at room temperature and even at low temperature is due to the existence of fine particles and the higher nitrogen distribution.

In order to have a detailed understanding of the nature of the observed doublet and a field distribution centered at 3 Tesla at 80 K we recorded the spectra at 5 K. It (Figure 2) shows two well separated sextets. The model with three sextets which corresponds to M_2, M_3 (~2/3) and M_3 (~1/3) [6] did not give a good fitting compared to the model containing two sextets. The hyperfine parameters (Table 1) show the presence of M_2 and M_3 only. The large width of the lines, especially for the M3 site, is due to the presence of a hyperfine field distribution. Figure 2 shows the distribution in the hyperfine field with a maximum probability at 26 and 13.7 T. The distribution centered at 8 T is due to the unresolved component of the hyperfine interaction associated with very small particles having higher nitrogen content which behaves as paramagnetic at room temperature and at 80 K. It is obvious from Table 1 that the area of sextets increases with decrease in temperature and the doublets have vanished at 5 K.

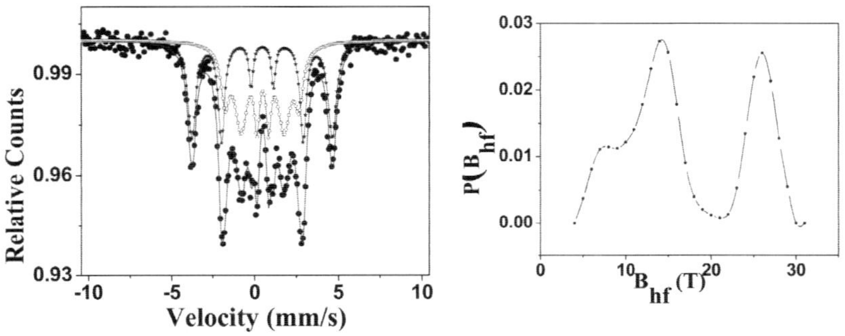

Figure 2. Mössbauer spectrum recorded at 5K and its distribution

4. Conclusion

Room temperature Mössbauer spectrum suggests the distribution in particle size. At lower temperatures the spectra show the presence of two iron sites. The isomer shift values corresponding to these sites indicate that the iron atoms are coordinated with two and three nitrogen atoms creating different electronic environments. The larger line width of the spectra points to the existence of a distribution of hyperfine field and quadrupole splitting. The hyperfine field distribution centered at 8T at 5K indicates the presence of very fine particles having higher nitrogen content which are statistically distributed over the available sites.

Acknowledgment

The authors acknowledge the financial assistance from DST New Delhi. Mössbauer spectrum at 5K was recorded at the UGC-DAE Consortium for Scientific Research, Indore.

References

[1] Jack K H 1952 *Acta Cryst.* **5** 404
[2] Vogel A I 1961 *Quantitative inorganic analysis* (London:Longmans,Green and Co Ltd.)
[3] Rochegude P and Foct J 1986 *Phys. Stat. Sol* (a) **98** 51
[4] Leineweber A, Jacobs H and Hull S 2001 *Inorg.Chem.* **40** 5818
[5] Leineweber A, Jacobs H, Huning F, Lueken H, Schilder H and Kockelmann W 1999 *J. Alloys Compd.* **288** 79
[6] Chen G M, Jaggl N K, Butt J B, Yeh E B and Schwartz L H 1983 *J. Phys. Chem.* **87** 5326

Effect of thermal treatment on the magnetic properties of nanostructured zinc ferrite

J P Singh[1], R S Payal[1], R C Srivastava[1], H M Agrawal[1], Prem Chand[2], Amita Tripathi[3] and R P Tripathi[3]
[1]Department of Physics, G. B. Pant University of Agriculture. & Technology, Pantnagar -263145, Uttarakhand, India
[2]Department of Physics, Indian Institute of Technology, Kanpur-208016, Uttar Pradesh, India
[3]Department of Physics, J. N.V. University, Jodhpur-322001, Rajasthan, India

Email: email_jps@yahoo.co.in

Abstract. Present report elucidates the effect of sintering time (2 and 3h) and temperature (300, 400, 500 and 600°C) on the magnetic properties of zinc ferrite nanoparticle. XRD shows the presence of cubic spinel phase in the samples. The average crystallite sizes of the samples increase with the sintering temperature and sintering time. The Mössbauer and VSM measurements show the presence of superparamagnetic/paramagnetic phase of the samples.

1. Introduction

Zinc ferrite, one of the most important binary oxides, offers an excellent opportunity for its use in electrical, semiconductor, radio engineering, automatics and radio technology [1]. Bulk zinc ferrite exhibits normal spinel structure with all zinc ions on A-site and all ferric ions at B-site and is paramagnetic at room temperature having Néel temperature of 10 K[2].

The magnetic properties of zinc ferrite are significantly affected when some amount of Fe transfers to tetrahedral site [3]. This distribution of cations in zinc ferrite results when the particle size approaches to nanoregime by using various methods of synthesis [4-12]. The magnetic properties of this system is believed to arise due to the formation of superparamagnetic domains [11-13], coexistence of ferrimagnetic and antiferromagnetism at room temperature [14] and spin-glass state [4]. Hence, the exact cause responsible for these properties is still not clear.

2. Experimental

Zinc ferrite nanoparticles have been prepared by using the nitrates of appropriate cations and citric acid. The cations to citric acid molar ratio were taken as 1:3. The solution was heated at 85°C with continuous stirring for 2h. The heated solution was allowed to cool at room temperature and finally it was dried at 100°C for overnight in an oven to form the precursor material. The precursor was sintered for 2 and 3 h at 300, 400, 500 and 600°C. For convenience the samples are coded as ZFt-T (where, ZF, t and T refer to zinc ferrite, sintering time and sintering temperature respectively).

The observation of crystalline phase and structure of the synthesized specimens were made with the help of BRUKER AXS D8 Advance XRD diffractometer using Cu-K_α radiation (λ=1.54Å). TEM micrographs of these samples were recorded on JEM-1011 transmission electron microscope in order to determine the morphology and particle size of synthesized nanomaterials. The Mössbauer spectra of the samples were recorded at room temperature in transmission geometry using 25mCi ^{57}Co source in

Rh matrix. The isomer shift (I. S.) values are reported with respect to standard natural iron foil. The hysteresis curves of all the samples were recorded by using a vibrating sample magnetometer (Princton-150 A) at room temperature.

3. Results and Discussion

The X-ray diffraction patterns of the prepared samples show the presence of cubic spinel phase. A small trace of ZnO phase also appears in the prepared samples (Figure 1 and 2). The average crystallite size was estimated by using the Scherrer's formula [15]. The crystallite size varies from 9±1 to 29±1 nm and 9±1 to 40±2 nm as the sintering temperature varies from 300 to 600°C for the sintering time of 2 and 3 h respectively (Figure 3). Figure 4 shows the TEM micrograph of the sample ZF3-300. The average particle size has a value of 24±10 nm for this system which may be due to agglomeration of several crystallites.

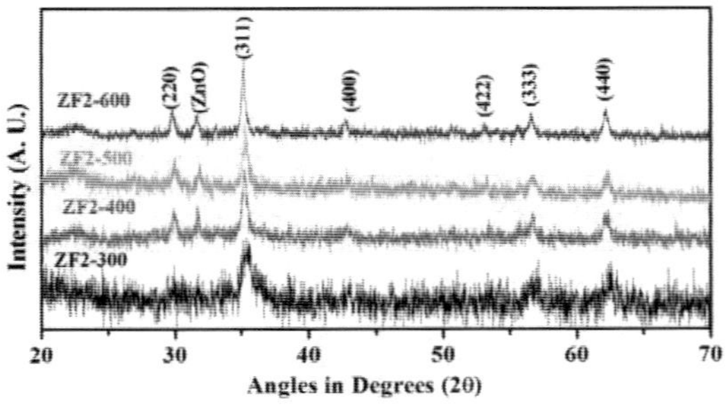

Figure 1. XRD pattern of the samples sintered for 2 h at various temperatures.

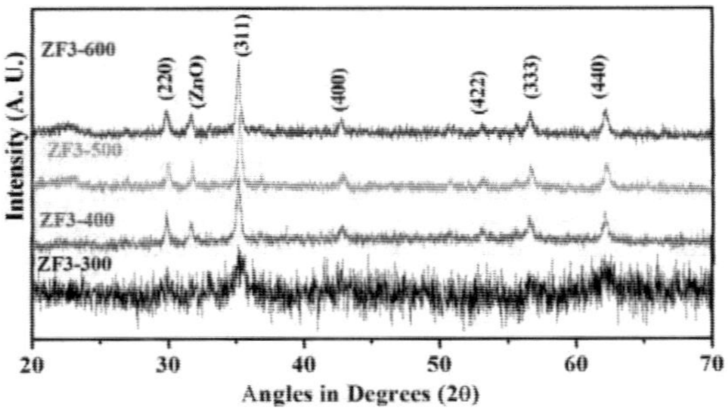

Figure 2. XRD pattern of the samples sintered for 3 h at various temperatures.

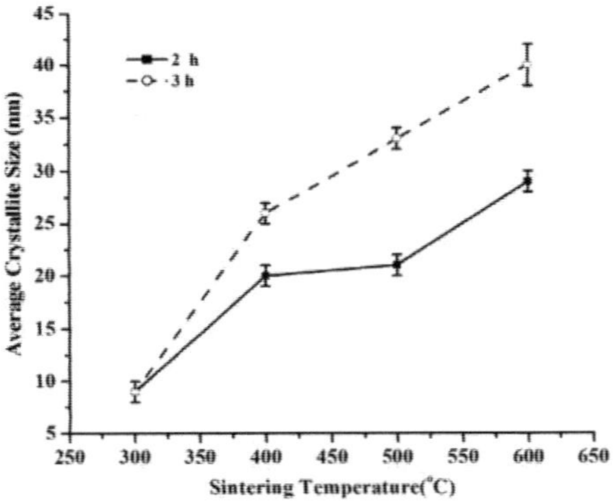

Figure 3. Average crystallite size of the samples as a function of sintering temperature.

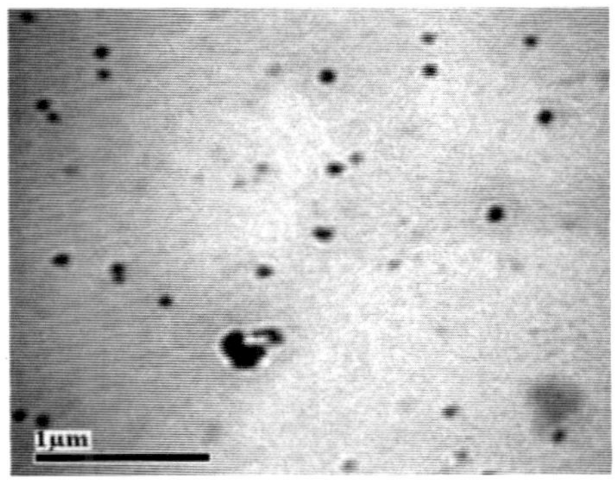

Figure 4. Transmission electron micrograph of the sample ZF3-300.

Figure 5 and 6 show the room temperature (RT) Mössbauer spectra of the samples. All the spectra were fitted by a doublet and the hyperfine parameters are shown in Table 1. The values of I. S. are of the order of 0.40 mm/s and are large as compared to the previous studies [11,13]. The values of I. S. in the samples are attributed to the presence of Fe^{3+} state in the samples. The large values of quadrupole splitting (Q. S. ~ 0.40 mm/s) for these systems compared to the bulk value of 0.35 mm/s [9] may be attributed to the cation inversion in these samples [16]. The line-width (L. W.) of the samples changes from 0.32 to 0.37 mm/s. The change in L. W. may be attributed to the change of surface strain or domain pinning [17]. Due to the small crystallite size and large surface area, Mössbauer nuclides sitting at the surface are not bonded that strongly, causing a decrease in the recoil free fraction, and therefore the area under the curve (Figure 7) is small for the smaller size particles [11, 13].

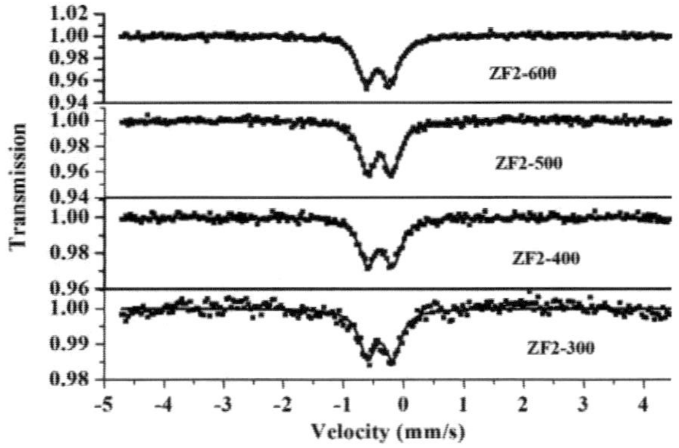

Figure 5. RT Mössbauer spectra of the samples sintered for 2 h at various temperatures

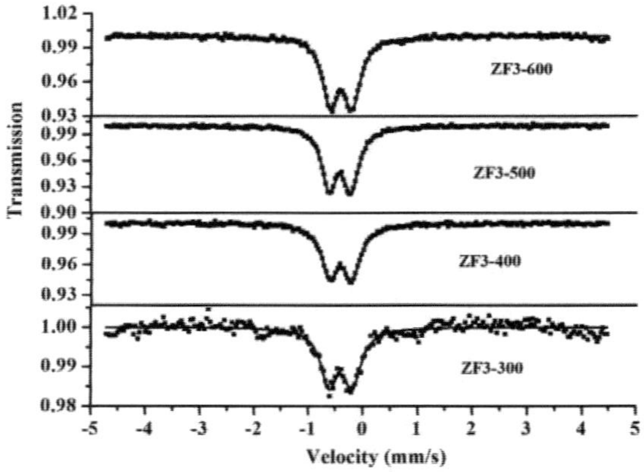

Figure 6. RT Mössbauer spectra of the samples sintered for 3 h at various temperatures.

Table 1: RT Mössbauer hyperfine parameters of the samples

Thermal Treatment		Mössbauer Hyperfine Parameters		
Sintering Time (h)	**Sintering Temp. (°C)**	**I.S. (mm/s)** ±0.01	**Q. S. (mm/s)** ±0.01	**L.W. (mm/s)** ±0.01
2	300	0.39	0.42	0.32
	400	0.39	0.41	0.32
	500	0.40	0.40	0.31
	600	0.41	0.40	0.32
3	300	0.40	0.43	0.37
	400	0.40	0.40	0.36
	500	0.42	0.40	0.33
	600	0.40	0.40	0.35

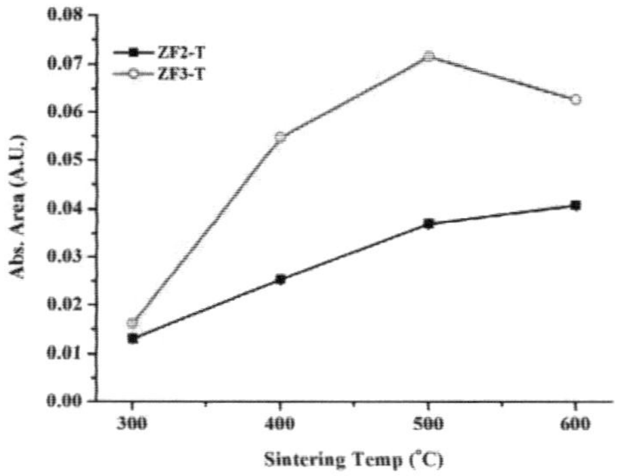

Figure 7. Absorption area versus sintering temperature for the samples.

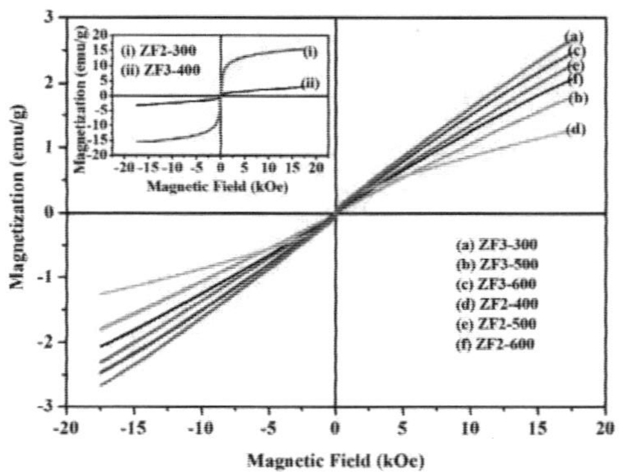

Figure 8. Hysteresis curve of the samples.

The Mössbauer hyperfine parameters remain almost constant within experimental error with respect to the various heat treatments for all the samples. To understand the magnetic behaviour of our samples we performed the magnetization measurements at room temperature (Figure 8). The hysteresis curve of the samples ZF2-300 and ZF3-400 show almost zero value of coercivity and retentivity, showing the presence of superparamagnetism at room temperature in these samples (Figure 8: inset). Rest of the samples show the paramagnetic behaviour at room temperature. The crystallite size of the samples ZF2-300 and ZF3-300 are comparable ~ 9 nm and the Mössbauer spectrum contains a doublet in both the samples. But the hysteresis curve shows the presence of superparamagnetism in the sample ZF2-300 and paramagnetism in the sample ZF3-300. This controversy is under investigation.

4. Conclusion

The XRD of the synthesized samples show the presence of cubic spinel phase; however ZnO appears as an impurity phase in the samples. The crystallite size increases as the sintering time and temperature increases. The Mössbauer spectra of the samples show the presence of doublet, which is attributed to superparamagnetic/paramagnetic nature of the samples as determined from the VSM measurements. The values of I,.S. are the particulars of Fe^{3+} state in nanostructured zinc ferrite. The Q. S. shows the attributes of cation inversion in the synthesized samples. Further, no regular trend in the magnetization was found with the particle size. The peculiar behavior of the samples ZF2-300 and ZF3-300 are under investigation.

Acknowledgment

One of the authors J P Singh is thankful to the CSIR for providing financial support in the form of JRF. He is also grateful to Dr. V. R. Reddy, UGC-DAE CSR Indore for their help and valuable suggestions in Mössbauer analysis.

References:

[1] Poole C P and Farach H A1980 *J. Phys. (France)* **14**, 11 ()
[2] Pettit G A And Forester D W 1971 *Phys. Rev. B* **4** 3912
[3] Nakashima S, Fujita K, Tanaka K, Hirao K, Yamamoto T and Tanaka I 2007 *Phys. Rev. B* 75 17443.
[4] Hofmann M, Campbell S J, Ehrhardt H and Feyerherm R 2004 *Journal of Material Science* **39** 5057
[5] Shenoy S D, Joy PA and Anarantharaman M R 2004 *J. Magn. Magn. Mater.* **269** 217
[6] Sato T, Haneda K, Seki M and Iijima T 1990 *Appl. Phys. A* **50** 13
[7] Jaydevan B, Tohji K and Nakatsuka K J 1994 *Appl. Phys: A* **76** 3625
[8] Ahn Y, Choi E J and Kim S 2002 *J. Korean Physical Soc.* **41** 123
[9] Yu S H, Fujino T and Yoshimura M 2003 *J. Magn. Magn. Mater.* **256** 420
[10] Xue H, Li Z, Wang X and Fu X 2007 *Materials Letters* **61** 347
[11] Roy M K, Halder B and. Verma H C 2006 *Nanotechnology* **17** 232
[12] Singh J P, Srivastava R C, Agrawal H M, Kushwaha R P S, Chand P and Kumar R *2008 Int. J. Nanoscience* **7** 21
[13] Singh J P, Srivastava R C, Agrawal H M, Kushwaha R P S 2008 *Hyperfine Interaction* **183** 221.
[14] Shim J H, Lee S, Park J H, Han S J, Jeong Y H and Cho Y W 2006 *Physical Review B* **73** 064404
[15] Cullity B D Introduction to X-ray diffraction (Addision-weseley, Newyork, 1972)
[16] Li F S, Wang L, Wang J B, Zhou Q G, Zhou X Z, Kunkal H P, Williams G 2004 *J. Magn. Magn. Mater.* **268** 332
[17] Kumar R, Sharma S K, Dogra A, Shivakumar V V, Dolia S N, Gupta A, Knobel M and Singh M 2005 *Hyperfine Interaction* **160** 143

^{57}Fe Mössbauer investigation of nanostructured zinc ferrite irradiated by 100 MeV oxygen beam

R C Srivastava[1], J P Singh[1], H M Agrawal[1], Ravi Kumar[2], A Tripathi, R P Tripathi[3], V R Reddy and A Gupta[4]
[1]Department of Physics, G.B. Pant University of Agriculture. & Technology, Pantnagar -263145, Uttarakhand, India
[2]Inter University Accelerator Centre, New Delhi-110067, India
[3]Department of Physics, J. N.V. University, Jodhpur-322001, Rajasthan, India
[4]UGC-DAE CSR Indore-452017, Madhya Pradesh, India

Email:rcsri@rediffmail.com,email_jps@yahoo.co.in,hma001@rediffmail.com, ranade@iuac.ernet.in

Abstract. Zinc ferrite nanoparticles of different size were irradiated in vacuum with 100 Mev O^{7+} ion beam at a fluence of 5×10^{13} ions/cm^2. Presence of ZnO phase was observed after the irradiation in the samples. ^{57}Fe Mössbauer spectroscopy shows the presence of well defined doublets in case of pristine and irradiated samples which are the attributes of superparamagnetism in the specimen. The variation of Mössbauer hyperfine parameters is discussed.

1. Introduction

Mössbauer spectroscopy is well known to study the swift heavy ion (SHI) induced effects in ferrites [1-3]. The SHIs loose energy in the target mainly by two means (i) electronic energy loss $(dE/dx)_e$ and (ii) nuclear energy loss $(dE/dx)_n$. The dominancy of electronic energy loss, as in present case, leads to the creation of variety of defects, phase change like amorphization [4-6] etc.

Ferrite materials have numerous applications in various electromagnetic devices [2,7], ferrofluid technology, magnetic diagnostic and spintronics devices [8]. Among all the ferrites, zinc ferrite having normal spinel structure in bulk behaves as paramagnetic at room temperature and as antiferromagnetic below 10 K [9]. It has been observed that the various nonequilibrium processing methods [10-15] result in disordered zinc ferrite inducing the nanophase in the system.

Production and modification of nanostructures by SHI is a work of great importance [16]. It is believed that the presence of defects induced by SHI influence the magnetic and related properties of these systems [17- 19]. In the present work we have reported the results obtained from Mössbauer study of zinc nanoferrite irradiated by the 100 MeV oxygen beam.

2. Experimental

Zinc ferrites nanoparticles of different size have been prepared by using the combination of nitrates of appropriate cations and citric acid. The molar ratio of cations to citric acid was taken as 1:3. The stoichiometric proportion of zinc nitrate and ferric nitrate were dissolved in distilled water. Citric acid solution was made separately. These solutions were mixed thoroughly and the solution was then heated at 85°C with continuous stirring for 2h. The heated solution was allowed to cool at room temperature and finally it was dried at 100°C for overnight in an oven to form the precursor material.

This precursor material was sintered at various temperatures ranging from 400°C to 1000°C for 1h to get pristine samples of nanostructured zinc ferrite. The pristine samples were irradiated in vacuum with 100 MeV O^{7+} ion beam at the fluence of 5×10^{13} ions/cm^2 using 15 UD Pelletron Accelerator at Inter University Accelerator Centre, New Delhi.

The structural and morphological characterization of pristine and irradiated samples have been done by using a Bruker AXS D8 Advance X-ray diffractometer and JEM 1011 transmission electron microscope. The Mössbauer spectra of the samples were recorded at room temperature in transmission geometry using 25mCi ^{57}Co-in-Rh matrix. The spectra were analyzed using the NORMOS/SITE programmes. The isomer shift values are reported with respect to standard natural iron foil.

3. Results and Discussion.

Fig 1 shows few representative XRD spectra of the pristine and irradiated samples. It is observed that pristine samples exhibit pure spinel phase, however some small trace of ZnO phase also present in the irradiated samples. This may be due to the deterioration of ordered spinel structure after the irradiation. SRIM calculation shows the values of electronic stopping, nuclear stopping and projected range as $S_e = 1.09$ keV/nm, $S_n = 0.618$ eV/nm and $R_p = 65$ μm respectively. The target volume is 4.7×10^{-3} cm^3 and was spread uniformly in the grooves of 1 cm diameter supported by Aluminium foil. The thickness of the spreaded powder was less than the projected range. Hence, there is no chance of implantation. The threshold electronic stopping for producing the columnar defects in zinc ferrite is ~ 13 keV/nm [20]. Hence, we do not expect any columnar defect in the irradiated system.

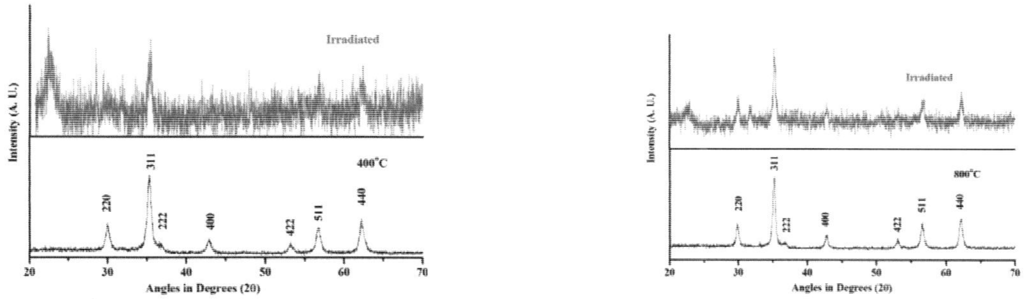

(a) (b)

Figure1: XRD spectra of the pristine and irradiated samples of sintered at (a) 400°C and (b) 800°C.

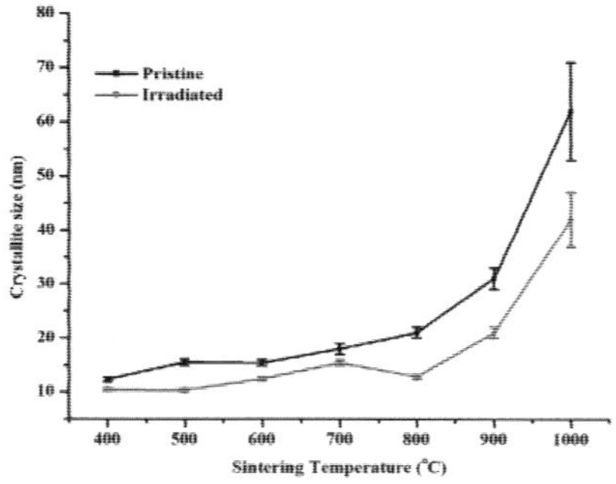

Figure 2. Crystallite size of the pristine and irradiated samples.

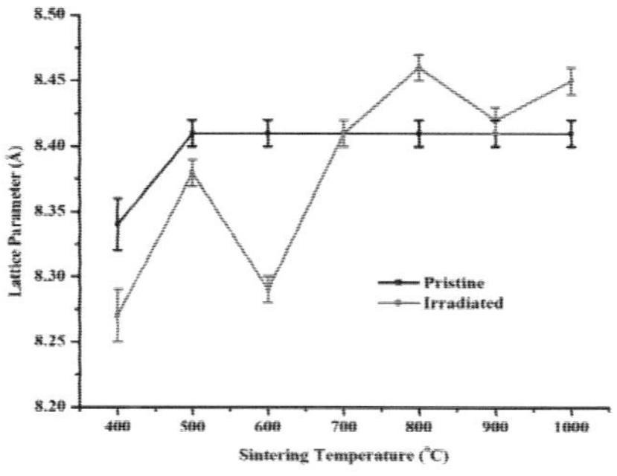

Figure 3. Lattice parameter of the pristine and irradiated samples.

The crystallite size and lattice parameters of the pristine and irradiated samples were estimated [21] and are shown in figures 2 and 3. The crystallite size varies from 12 nm to 62 nm as the sintering temperature varies from 400°C to 1000°C. The respective values for the irradiated samples are 10 and

42 nm and the variation with sintering temperature is almost similar to that of pristine sample. The decrease in the crystallite size of the irradiated samples is attributed to the hammering effect of SHI beam. We observe that there is no regular trend of variation of lattice parameter with sintering temperature of the irradiated specimen (figure 3). Figure 4a and 4b show the representative transmission electron micrographs of the pristine and irradiated samples having sintering temperature of 500°C. The average particle size obtained from TEM is ~ 21 nm for the pristine and ~13 nm for the irradiated samples. TEM also shows the reduction of particle size after irradiation.

Figure 4: Transmission Electron Micrographs of the (a) pristine sample sintered at 500°C and (b) its irradiated counterpart (fluence=5×10^{13} ions/cm^2 (5E13)).

Figure 5 shows few representative Mössbauer spectra of samples at room temperature. All the recorded spectra show well defined doublet, which may be a signature of presence of superparamagnetism in both the pristine and irradiated samples [22, 23] as supported by VSM study of our previous work [22].

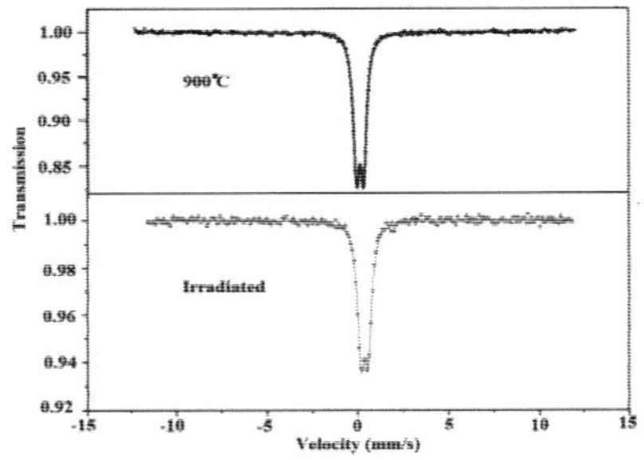

Figure 5. Mössbauer spectra of the pristine and irradiated samples.

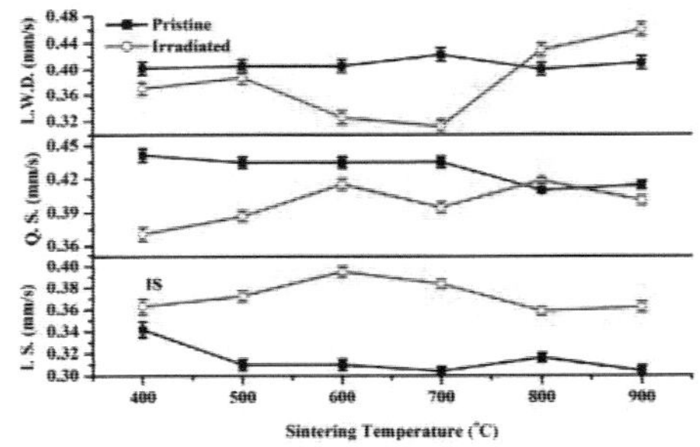

Figure 6. Variations of I. S.(±0.005), Q. S. (±0.005) and L.W.D. (±0.01) of the pristine and irradiated samples.

The values of isomer shift are of the order of 0.31 mm/s for the pristine sample except for the sample sintered at 400°C (I. S. = 0.34 mm/s). These values increase after the irradiation and ranges from 0.36 to 0.39 mm/s. The excess amount of energy induced by swift heavy ion in the system is confined in form of heat within grain volume [24,25]. This may give rise to thermal expansion and excitation of the electronic state, consequently, the values of I.S. increases [26] .The values of isomer shift for the pristine and irradiated samples are characteristics of Fe^{3+} charge state (figure 6) [27]. The value of quadrupole splitting decreases form 0.44 mm/s to 0.41 mm/s for the pristine sample and increase from 0.37 mm/s to 0.40 mm/s for the irradiated samples as the sintering temperature increases from 400°C to 1000°C (figure 6). We observe decrease in Q.S. after the irradiation. The decrease in the value of Q. S. is attributed to the less cation inversion, creation of oxygen vacancies [28, 29]. Line-width for the pristine sample seems to be almost constant within experimental error at a value of 0.40 mm/s. while the irradiated sample has very much different trend (figure 6). The line width (L.W.D.) decreases first upto the sintering temperature of 700°C, and then increases abruptly upto 1000°C. The decrease in the line-width after irradiation may be attributed to the reduction of surface strain or domain pinning. This type of behaviour was observed by previous workers also [20].

Conclusion

The structural characterization by using XRD shows the appearance of ZnO phase in the system after the irradiation. The particle size of the irradiated samples decrease with respect to irradiated counterpart. We also found that the nanostructured system remain in same superparamagnetic state after the irradiation.

Acknowledgment

Authors are thankful to IUAC, New Delhi for providing the financial support in the form of UGC funded project UFUP-40305.

References

[1] Studer F, Nguyen N, Fuchs G and Toulemonde M 1986 *Hyperfine Interactions* **29** 1287
[2] Costantini J M, Studer F, Peuzin J C 2001 *J. Appl. Phys.* **90** 126
[3] M. C. Chhantbar, K. B. Modi, C. J. Baldha, H. H. Joshi, R. V. Uppaydhyay and Ravi Kumar, Nuclear Instr. and Methods in Physics Research B 244 (2006)124-127
[4] Studer F and Toulemonde M 1992 *Nucl. Instr. Meth. Phys. Res. B* **65** 560
[5] Szenes G 1995 *Phys. Rev. B* **51** 8026
[6] Goya G F, Rechenberg H R, Chen M, Yelon W B 2000 *J. Appl. Phys.* **87** 8005
[7] T. Pannaparayil, S. Komarneni, R. Marande, M. Zadarko, J. Appl. Phys. 67 (1990) 5509
[8] L. Lu, M. L. Sui and K. Lu, Science, **287**, (2000)1463
[9] Pettit G. A. And Forester D. W. 1971 *Phys. Rev. B* **4** 3912
[10] Chinnasamy C N, Narayanasamy A, Ponpandian N, Chattophadhyay K, Góerault H, Greeneche, J-M 2000 *J. Phys: Condens Matter* **12** 7795
[11] Oliver S A, Hamdeh H H and Ho J C 1999 Phys. Rev. B **60** 3400
[12] Sato T, Haneda K, Seki M and Iijima T 1990 *Appl. Phys. A* **50** 13
[13] Jaydevan B, Tohji K, and Nakatsuka K 1994 *J. Appl. Phys. A* **76** 3625
[14] Nakashima S, Fujita K and Hirao K 2005 *J. Phys: Condens Matter* **17** 137
[15] Yamamoto Y, Tanaka H and Kawai T 2001 *Jpn J. Appl. Phys.*, Part 2 **40** L545
[16] Pivin J-C 2005 Material Science-Poland **23** 101
[17] Sharma S K, Ravi Kumar, Kumar V V S, Knobel M, Reddy V R, Gupta A and Singh M 2006 *Nucl. Instr. Meth. in Phys. Res B* **248** 37
[18] Rao B P, Rao K H, Rao P S V S, Kumar A M, Murthy Y L N, Ashokan K , Kumar V V S, Ravi Kumar, Gajbhiye N S and Caltum O F 2006 *Nucl. Instr. Meth. in Phys. Res B* **244** 27
[19] Taylor R H 1975 *Adv. Phys.* **24** 68
[20] Ravi Kumar, Sharma S K, Dogra A, Shivakumar V V, Dolia S N, Gupta A, Knobel M and Singh M 2005 *Hyperfine Interaction* **160** 143.
[21] B. D. Cullity-Introduction to X-ray diffraction (Addision-wesley, Newyork, 1972)
[22] Singh J P, Srivastava R C, Agrawal H M and Kushwaha R C 2008 *Hyperfine Interaction* **183** 221
[23] Roy M K, Halder B and Verma H C 2006 *Nanotechnology* **17** 232
[24] Ghosh S, Gupta A, Ayub P, Kumar N, Khan S A, Banerjee D and Bhattacharya R 2004 *Nucl. Instr. Meth. Phys. Res. B* **225** 310
[25] Chen G 2000 *J. Nanoparticle Research* **2** 199
[26] Uppadhyay C and Verma H C 2004 *Appl. Phys. Lett.* **85** 2074
[27] Ehrahdrt H, Campbell S J, Hoffmann M 2002 *J. Alloy Componds* **339** 25
[28] Roy M K and Verma H C 2006 *J. Magn. Magn. Mater.* **306** 98
[29] Li F S, Wang L, Wang J B, Zhou Q G, Zhou X Z, Kunkal H P, Williams G 2004 *J. Magn. Magn. Mater.* **268** 332

Mössbauer investigation of hyperfine interactions in dilute Fe-SnO_2 nanoparticles

L Diamandescu[1], **S Constantinescu, D Tarabasanu – Mihaila, I Bibicu, M Feder and M A Vlaicu**

National Institute of Materials Physics, P.O. Box MG-7, 77125, Bucharest-Magurele, ROMANIA

E-mail: diamand@infim.ro

Abstract. We present a Mössbauer (MS) analysis on the nanoparticle system $xSnO_2$-(1-x)α-Fe_2O_3, ($x \in [0.995 \div 0.925]$). The samples were prepared by a hydrothermal route and have been 50 % enriched in ^{57}Fe. The MS spectra were recorded at room (RT) and liquid nitrogen temperature (LNT). Below x = 0.975 the MS consist in a paramagnetic pattern, while at $x \geq 0.975$ magnetic hyperfine interactions were evidenced. The observed coexistence of paramagnetic and ordered magnetic phases is discussed by taking into account the role of oxygen vacancies (□), Sn-O and Sn-□ distances together with the iron-oxygen configurations.

1. Introduction

Diluted magnetic semiconductors (DMS) using doping transition metal (TM) ions become of great and extensive interest as materials for spintronic applications. SnO_2 is an attractive host lattice for the investigation of DMS, due to the high carrier density and natively n-type semiconducting properties. Recently, some of TM (Fe, Co, Ni)-doped semiconductors as TiO_2, GaN, ZnO, and SnO_2 obtained by different procedures (pulsed laser deposition, sol-gel, hydrothermal), were found to show room-temperature ferromagnetism [1-3]. It seems that critical TM-dilution and observable magnetic ordering temperature are dependent on preparing procedure and/or thermal treatments. Continuing the earlier investigations [4-9], this work aims at showing a Mössbauer analysis on the nanoparticle system $xSnO_2$-(1-x)α-Fe_2O_3 synthesized by a hydrothermal route [7], at very low iron concentrations ($1-x \in [0.005 \div 0.075]$).

2. Experimental aspects and results

A series of $xSnO_2$-(1-x)α-Fe_2O_3 ($x \in [0.995 \div 0.925]$) was prepared by hydrothermal route starting with an aqueous mixture of iron (III) chloride hexahydrate, $FeCl_3.6H_2O$, and tin (IV) chloride pentahydrate, $SnCl_4.5H_2O$, as raw materials. The samples have been 50 % enriched in ^{57}Fe. The suspension obtained by coprecipitation with ammonia (at pH=8) was heated in autoclave at 200 °C for 4 h and then quenched to room temperature. X-ray diffraction (XRD), ^{119}Sn and ^{57}Fe transmission Mössbauer spectroscopy (TMS) were utilised to examine the as resulted powders. The XRD patterns obtained using a Bruker D8 Advance diffractometer and CuK_α radiation (λ=1.540598 Å) have been analysed to study the phase structure in relation to iron concentration 1-x. In figure 1a, representative XRD patterns are displayed. All patterns correspond to tetragonal SnO_2 phase known as cassiterite

[1] The corresponding author.

(S.G. 136 P42/mnm). Rietveld refinement of the XRD patterns indicates a decrease of lattice parameters *c* and *a* of SnO_2 structure, as the iron content in the investigated samples increases (figure 1b). This result supports our expectation because the six coordinated ionic radius of Sn is greater than the ionic radius of Fe ion [7]; this behaviour is an argument for the substitution of Sn^{4+} ions by Fe^{3+} ions in SnO_2 lattice. The mean particle size given by Scherer formula is as small as 8 nm.

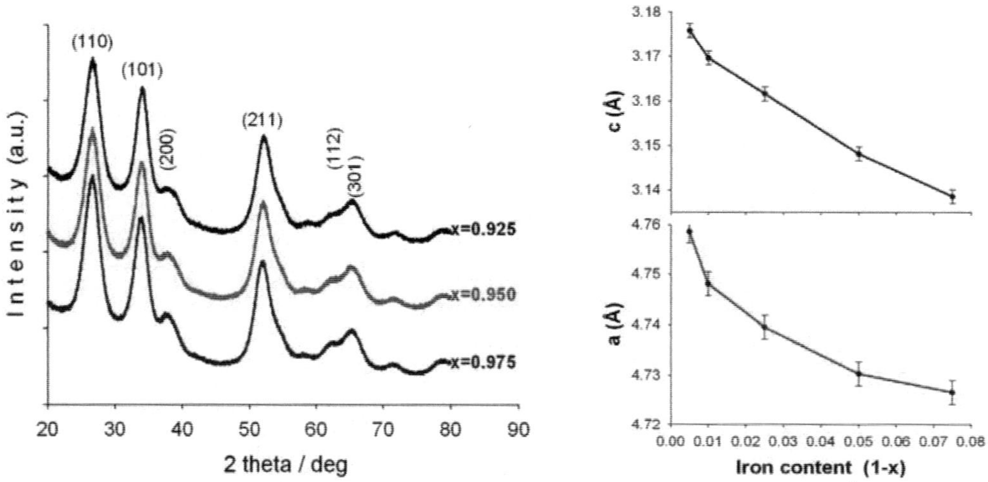

(a) (b)

Figure 1. *a)* XRD patterns for the $xSnO_2$-$(1-x)\alpha$-Fe_2O_3 system at x=0.925, x=0.950 and x=0.975; b) Lattice parameters c and a versus iron molar content (1-x).

TMS spectra were recorded at room (RT) and liquid nitrogen temperature (LNT) using ^{57}Co: Rh and $^{119}Sn:CaSnO_3$- matrix sources and a constant acceleration spectrometer, in the velocity range $v \in [-12 \div +12]$ mm/s. The ^{119}Sn-spectra of the investigated samples consist in one resonance line with unchangeable line-shapes both at RT and LNT. The room temperature ^{57}Fe-Mössbauer spectra of the same samples exhibit a doublet pattern with small quadrupole splitting. Relevant changes in Mössbauer line shape and the appearance of magnetic hyperfine structure, as the iron content in the samples decreases, were evidenced in ^{57}Fe-spectra at LNT (figure 2a); basically, superposed paramagnetic and magnetic ordering contributions are displayed by all ^{57}Fe spectra in the studied concentration range.

3. Discussion

For the Mössbauer spectra fitting we have to consider that in our system at least three different nearest-neighbour cation interactions could exist: iron-iron, iron-tin and iron-oxygen vacancy. These possible interactions imply that at least three magnetic sublattices have to be taken into account in the computing run. The fit obtained with this approximation was far from satisfactory. The best fit has been obtained considering quadrupole patterns (paramagnetic contribution) together with magnetic hyperfine field distribution (MHFD). For example, at x=0.995 (LNT spectrum) the central quadrupole pattern was deconvoluted in two doublets. The first one, a symmetric doublet with quadrupole splitting (Δ_Q) of 0.74(3) mm/s and isomer shift (δ_{Fe}) of 0.44(5) mm/s can be assigned to Fe^{3+} substituting for Sn^{4+} in the SnO_2 lattice [7]. The second one with Δ_Q=0.85(4) mm/s and δ_{Fe}=0.49(8) mm/s could represent iron ions in more distorted neighbouring due to the close defects. These findings are in good agreement with the data reported in [13]. The relative areas of the paramagnetic / magnetic patterns slightly decrease/increase from ~ 33%/67% to ~ 29%/71% as the iron molar concentration 1-x decreases. The paramagnetic contribution is strongly dependent on the molar concentration ***x***, as stated also in [10]. In figure 2a the computer fit of LNT spectrum at x=0.995 is presented. The magnetic hyperfine filed distribution is plotted for the samples at x=0.990 and x=0.995 at LNT. Depending on molar concentration x, three magnetic field relative maxima can be observed: at B < 15.0 T, 15.0 T < B < 35.0 T and 35.0 T < B < 55.0 T. The increase of Fe-dilution induces an enlargement of the second range of B-values, a shift of MHFD through higher magnetic field and a more intense contribution at B > 45 T. In figure 2b, given by the fit, a tentative-deconvolution of MHFD into several Gauss peaks is depicted.

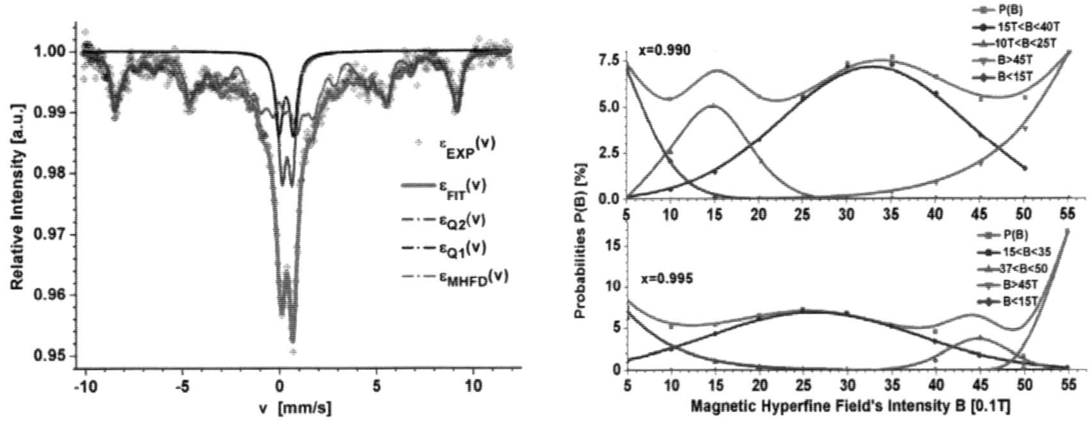

(a) (b)

Figure 2. (a) The fit of the experimental LNT spectrum at x=0.995; (b) the tentative deconvolution in the multi Gauss peaks of MHFD probabilities P (B) observed in $xSnO_2(1-x)\alpha\text{-}Fe_2O_3$ spectra for x=0.990 and 0.995 at LNT.

Based on XRD data, we consider that the magnetic behaviour observed in TMS spectra is correlated with the presence of Fe ions in SnO_2 – tetragonal structure. Our TMS data revealed that at x > 0.975 the samples are magnetically inhomogeneous. In the case of low Fe-diluted sample (x = 0.975), the LNT spectrum indicates only one iron ion in four as being magnetically ordered, in acceptable agreement with [3]. The LNT-spectra of high Fe- diluted samples (x ≥ 0.990) evidenced a more important contribution of magnetic ordering (~67÷71 %).

As it was suggested by Coey et al. [11, 12] the Fe^{3+} ions in SnO_2 could experience the following magnetic type interactions: a ferromagnetic exchange mechanism (FM) involving oxygen vacancies which form *F*-centres with trapped electrons; a paramagnetic behaviour coming from the isolated spin systems where the *F*-centre mediated ferromagnetic coupling is not achieved due to lack of Fe^{3+} neighbours; an antiferromagnetic (AM) behaviour given by Fe^{3+}– O^{2-} –Fe^{3+} superexchange interactions. Consequently the hyperfine magnetic fields or magnetizations should increase as the iron content in the samples increases. In our samples the reverse effect is observed. Nomura et al [13] reported that the ferromagnetism in DMS material could originate both from ions and defects generated by the presence of the magnetic iron ions; this new magnetic contribution could explain the increase of magnetic hyperfine component by decreasing the iron content in our samples. The balance between the AM and FM interactions depends on the distance of attached Sn^{4+}- O^{2-}, Sn^{4+}- □ groups to Fe^{3+} - O^{2-} - Fe^{3+} or Fe^{3+} - □ - Fe^{3+} configurations; that might influence the overlap of the electronic orbitals and can be favourable to the FM and AM interactions respectively [13, 14]. The present Mössbauer results cannot distinguish between the two magnetic ordering behaviours, however considering the statements in ref. [3], one can suppose that the iron sites with high B-values (B > 45 T) correspond to AF interaction and those with smaller B-values to FM interaction.

4. Conclusions

The magnetic dilute nanoparticle system $xSnO_2$-$(1-x)\alpha$-Fe_2O_3, ($x \in [0.995 \div 0.925]$) obtained under hydrothermal conditions was investigated by Mössbauer spectroscopy at room and liquid nitrogen temperatures. In the studied tin molar concentration range, the TMS on ^{119}Sn consist in a characteristic SnO_2 -single lorentzian line at both RT and LNT. Dramatic changes in the line shape and pattern were observed in the TMS spectra on ^{57}Fe mainly at LNT. At x > 0.975 the onset of magnetic hyperfine interaction was evidenced. The complex spectra were fitted in terms of magnetic hyperfine distributions. Three types of magnetic sites relative to three ranges of the magnetic field intensity (B < 15.0 T, 15.0 T < B < 35.0 T and 35.0 T < B < 55.0 T) were identified in the distribution diagram. The magnetic sextets higher than 45 T have been ascribed to AF interaction. The increase of magnetic hyperfine component in MS, as the iron content in the samples decreases, suggests a more complex phenomenology regarding the contribution to the magnetic ordering in DMS, the direct involvement of defects having to be considered as a real possibility. The explanation for the appearance of ferromagnetism in iron doped SnO_2 is still an open issue. Further studies on annealed samples are in progress.

Acknowledgments
The work in Romania was supported by the National Authority for Scientific Research (ANCS) under the Core Program 45N/ 1.03.2009.

References

[1] Ogale S B *et al* 2004 *Phys. Rev. Lett.* **91** 077205
[2] Punnoose A and Hays J 2004 *Appl. Phys. Lett.* **85** 1559
[3] Liu J F, Lu M F, Chai P, Wang Z L and Cao X Q 2007 *J. Magn. Magn. Mater.* **317** 1
[4] Berry F J, Greaves C, McManaus J G, Mortimer M and Oates G 1997 *J. Solid. State. Chem.* **130** 272
[5] Jiang J Z, Liu R, Mørup S, Nielsen K, Poulsen F W, Berry F J and Clasen R 1997 *Phys. Rev.* ***B* 55** 11
[6] Berry F J, Bohorquez A, Helgason O, Jiang J Z, McManus J G, Moore E, Mortimer M, Mosselmans F and Mørup S 2000 *J. Phys. Condes. Matter.* **12** 4043
[7] Sorescu M, Diamandescu L, Tarabasanu-Mihaila D, Teodorescu V S and Howard B H 2004 *J. Phys. Chem. Solid.* **65** 1021
[8] Constantinescu S, Diamandescu L, Bibicu I, Tarabasanu-Mihaila D, and Feder M 2008 *Hyperfine Interactions* **184** 83
[9] Sorescu M, Diamandescu L, Tomescu A, Tarabasanu-Mihaila D and Teodorescu V S 2008 *Mat. Chem. Phys.* **107** 127
[10] Sakuma J, Nomura K, Barrero C and Takeda M 2007 *Thin Sol. Film* **515** 8653
[11] Coey J M D, Venkatesan M and Fitzgerald C B 2005 *Nat. Mat.* **4** 173
[12] Coey J M D, Douvalis A P, Fitzgerald C B, and Venkatesan M 2004 *Appl. Phys. Lett.* **84** 1332
[13] Nomura K, Barrero C A, Sakuma J and Takeda M 2007 *Phys. Rev.* ***B* 75** 184411
[14] Punnoose A, Hays J, Thurber A, Engelhard M H, Kukkadupu R K, Wang C, Shuthanandan V and Thevuthasan S 2005 *Phys. Rev.* ***B* 72** 054402.

Journal of Physics: Conference Series **217** (2010) 012111 doi:10.1088/1742-6596/217/1/012111

On the exchange coupling interaction between Fe and $L1_0$-ordered FePd thin films

M Carbucicchio and R Ciprian

Department of Physics, University of Parma, Parma, Italy

E-mail: massimo.carbucicchio@fis.unipr.it

Abstract. Hard $L1_0$-ordered FePd ultra-thin films were epitaxially grown by means of a molecular beam epitaxy system on MgO-(100) monocrystalline substrates. Because of the very low thickness of the films, small amounts of superparamagnetic $L1_0$-FePd particles were also formed on the surface. The magnetic properties of the films are typical of highly coercive systems showing a strong perpendicular anisotropy. These hard films were used for the development of soft/hard exchange-spring magnets, by depositing different thickness of iron. Fe reacts with the surface FePd small particles giving rise to a Fe(Pd) compound with a broad distribution of particle sizes. The resulting systems are bi-layers constituted by a soft Fe(Pd) and a hard $L1_0$-FePd layer intercalated by a very thin region of small particles. For soft/hard thickness ratio nominally equal to 1 the magnetic behaviour is typical of a single-phase hard magnet. By increasing the nominal thickness ratio up to 3, the behaviour becomes typical of an exchange-coupled magnet with hysteresis loops characterized by two critical fields.

1. Introduction

FePd thin films grown at room temperature show a fcc structure (A1 phase) characterized by soft magnetic properties. By growing the films at high temperatures or by means of post-annealing treatments the alloy around the equiatomic composition undergoes a structural transformation with the establishing of a fct structure ($L1_0$ phase), characterized by a strong uniaxial magnetocrystalline anisotropy parallel to the contracted *c*-axis. Therefore, by means of an epitaxial growth it is possible to obtain systems showing perpendicular anisotropy [1-3]. Contrary to $L1_0$-ordered FePt and CoPt, the FePd is reported to show a much lower coercivity and squareness [4-8]. For this reason, the $L1_0$-ordered FePd alloy has been evaluated not to be suitable for the development of exchange-spring magnets, i.e. systems where soft and hard magnetic phases are intercalated and exchange-coupled at the interface, showing a single phase magnetic behaviour.

In the present work FePd thin films grown by means of an UHV e-beam evaporation technique have been obtained, showing a strong perpendicular anisotropy, high coercivity and loop squareness. These peculiar characteristics allowed to use these films as the hard phase for the development of FePd/Fe exchange-spring magnets having a high value of the soft/hard thickness ratio. The obtained results are quite unexpected and surprising considering that (i) no data are reported about FePd systems showing values of the coercivity up to 10 kOe and (ii) in literature, the coupling among the different magnetic phases is reported to require a soft/hard thickness ratio lower than 1.

2. Experimental

FePd thin films, 4.5 nm thick, were grown in ultra-high vacuum by means of a molecular beam epitaxy (MBE) system equipped with electron gun evaporators. The FePd films were deposited onto MgO-(100) monocrystalline substrates at 540°C, using iron highly enriched in the ^{57}Fe Mössbauer isotope. On the top of the FePd films, ^{57}Fe layers of different thickness were deposited at room temperature. In the following these samples were labelled as Fe(t_{Fe})/FePd, where t_{Fe} = 4.5, 10 and 15 is the Fe thickness in nm.

Conversion electron Mössbauer spectroscopy (CEMS) measurements were performed without breaking vacuum, using a 50 mCi ^{57}Co(Rh) source with the γ-ray incident angle of 17°. The isomer shifts are referred to α-Fe, and the fitting was carried out using a least-square minimization routine with a combination of linear and non-linear regressions as a superposition of Lorentzian lines. For the fittings, the intensity ratio between the 1st (6th) and the 3rd (4th) lines has been maintained equal to 3. The magnetic properties were measured with a magneto-optical Kerr effect magnetometer (MOKE) using a *s*-polarized 633 nm laser light with the magnetic field applied both perpendicular (polar geometry) and along the film plane (longitudinal geometry). The dcd remanence curves were measured in polar geometry starting with the sample in the saturated state. Ultra-high vacuum atomic and magnetic force microscopy (AFM, MFM) was used in tapping/lift mode to characterize the morphology and domain patterns of the samples. For MFM, the tip-sample distance was 50 nm.

3. Results and Discussion

Figure 1(a) shows the Mössbauer spectrum measured for the FePd films grown at 540°C. It can be fitted as the superposition of a quadrupole doublet (δ) and a series of sextets (ν) whose hyperfine field distribution H_{hf} is reported in the inset. The ferromagnetic contribution, showing a mean H_{hf} = 268 kOe and a quadrupole splitting ΔE_Q = 0.20 mm/s, is due to the $L1_0$-ordered FePd alloy having an almost equiatomic composition [2]. Taking into account the γ-ray incident angle, the line intensity ratios (3:1.3:1) indicate that the mean magnetization vector is oriented at ~25° from the perpendicular to the film plane. As for the quadrupole doublet (ΔE_Q = 1.08 mm/s), it can be attributed to the same $L1_0$-FePd in the form of small particles showing a superparamagnetic behaviour [9]. Accordingly, the AFM measurements show a morphology constituted by irregular and disconnected islands as well as small droplets having nanometric sizes.

The corresponding MOKE measurements performed in polar geometry are reported in figure 1(b). The film shows a well square-shaped hysteresis loop with a remanence to saturation ratio M_r/M_s of 92% and a coercive field H_c = 2.2 kOe. The presence of a strong perpendicular anisotropy and the absence of appreciable in-plane contributions to the magnetization vector are confirmed by the almost zero-remanence of the longitudinal loop (figure 2).

No magnetic morphology was evidenced by the MFM analysis. This fact may well be due to the presence onto the sample surface of the superparamagnetic small particles. Moreover, the low thickness of the films as well as the strong perpendicular anisotropy of the ferromagnetic material can give rise to very large domain sizes, well overcoming the maximum scanning area of the adopted microscope (~5 × 5 μm).

Figure 1(c) shows the CEMS spectrum measured after covering the film with a 4.5 nm thick iron layer, Fe(4.5)/FePd. The spectrum was fitted superimposing, to the ν and δ contributions due to the hard FePd film, a series of sextets (χ) whose H_{hf} distribution is reported in the inset. The fitting procedure was performed maintaining the values of the ν and δ hyperfine parameters equal to those found for the FePd spectrum [figure 1(a)], and leaving free to change only the relative areas of the two components.

Considering that no appreciable contribution from pure iron is present in the spectrum and that the δ contribution due to FePd small particles is considerably reduced, it is possible to assume that Fe interacted with the highly reactive FePd small particles giving rise to Fe containing low amounts of Pd, Fe(Pd), showing a broad distribution of particle sizes (χ contribution). This hypothesis means that, as a consequence of the Fe deposition, the sample grew in the form of a bi-layer constituted by a superficial soft Fe(Pd) layer and a hard $L1_0$-FePd layer, intercalated by a very thin interfacial region of superparamagnetic particles. The effect of the Fe(Pd) growth in reducing the amount of FePd small particles with the formation of a continuous surface layer is supported by the AFM image of the bi-layer [figure 3(a)].

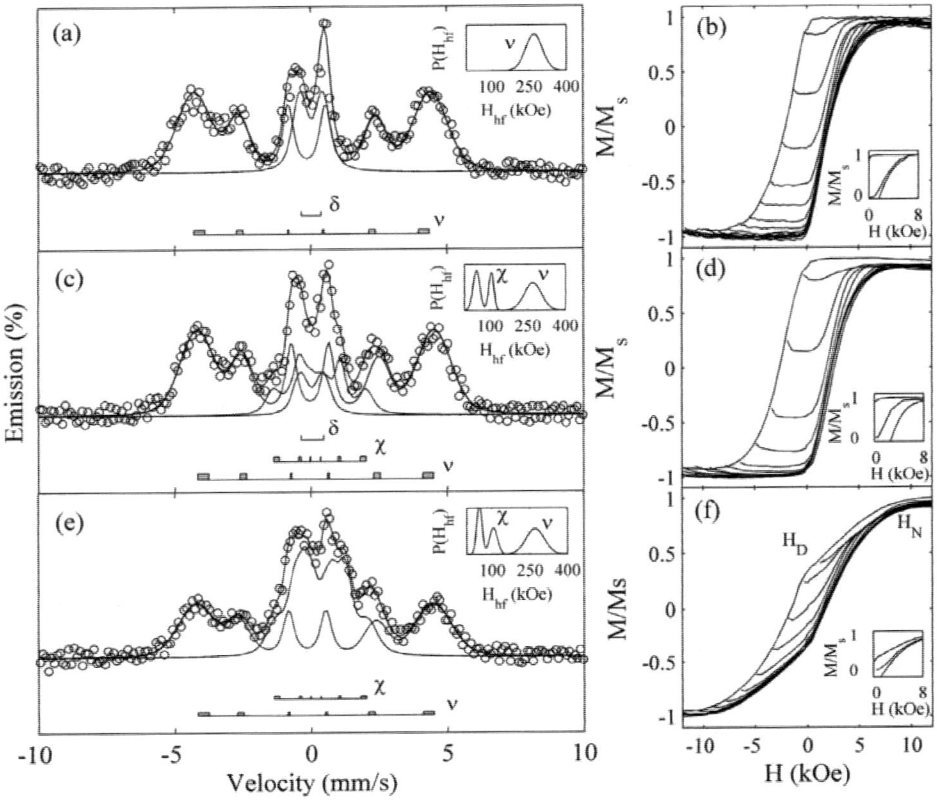

Figure 1. CEMS spectra (left column) and polar dcd recoil loops (right column) measured for (a) and (b) FePd, (c) and (d) Fe(4.5)/FePd and (e) and (f) Fe(15)/FePd samples. The hyperfine magnetic field distributions corresponding to the contributions to the CEMS spectra and the initial curves are reported in the insets.

From the line intensity ratios of the χ contribution it follows that the angle between the magnetization vector and the perpendicular to the film plane is ~22°. This low tilting is attributable to a strong coupling between the soft Fe(Pd) and the hard $L1_0$-FePd layer. In agreement, the bi-layer hysteresis loop [figure 1(d)] is typical of a single-phase hard magnet characterized by a high squareness due to a strong perpendicular anisotropy. The decrease of the FePd superparamagnetic particles determines an appreciable increase of the coercivity up to 2.6 kOe. The first magnetization curve suggests that the coercivity is mainly controlled by the nucleation of reversed domains which gives rise to the abrupt rotation of the magnetic moments for very low applied fields.

The low anisotropy of the superficial Fe(Pd) compound and the decrease of the FePd small particles present at the Fe(Pd)/$L1_0$-FePd interface give rise to a magnetic morphology constituted by irregular small domains [figure 3(b)]. The high image contrast and the irregularity of the domain pattern can only be attributed to the strong perpendicular anisotropy characterizing the bi-layer.

For the Fe(15)/FePd, the Mössbauer spectrum [figure 1(e)] shows that the amount of Fe(Pd) is appreciably increased (set of sextets χ), while no contribution from pure iron is detectable. The presence of the δ doublet shielded by the χ contribution was also considered, but its introduction among the fitting components gave no improvements, tending on the contrary to increase the standard deviation. These facts reasonably mean that all deposited iron and FePd small particles reacted giving rise to the Fe(Pd) compound.

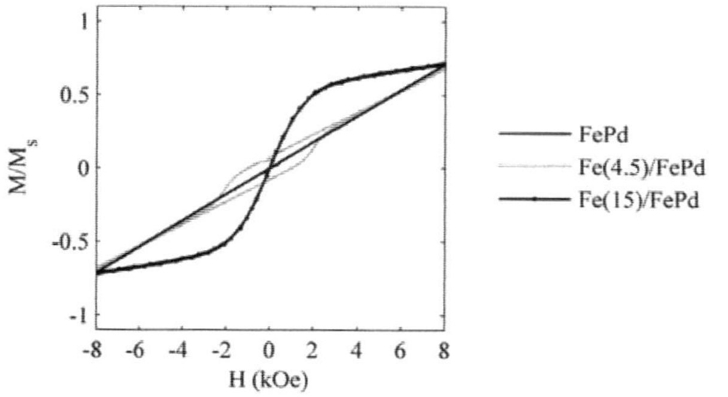

Figure 2. Longitudinal hysteresis loops for the FePd hard film, Fe(4.5)/FePd and Fe(15)/FePd.

The angle between the Fe(Pd) magnetization vector and the perpendicular to the film plane increases to 50°. Correspondingly, a significant decrease occurs of both the film coercivity (1.95 kOe) and the squareness (32%) of the hysteresis loop [figure 1(f)]. Two critical fields for the magnetization reversal establish, i.e., the nucleation, H_N, and the demagnetizing field, H_D. For applied fields in the $H_N \div H_D$ range, the rotation of the magnetic moments is completely reversible, indicating that a good exchange coupling established between the hard $L1_0$-FePd and the soft Fe(Pd) layer. The difficulty to reach the complete saturation can be ascribed to the in-plane component of the magnetization vector.

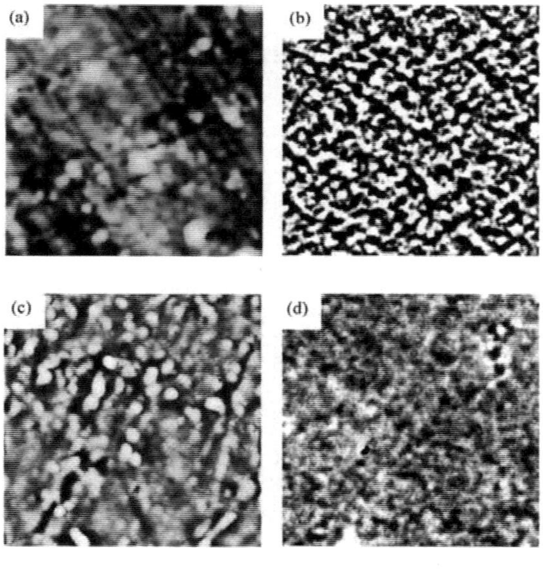

Figure 3. AFM and MFM images (4 × 4 μm) for Fe(4.5)/FePd (a, b) and Fe(15)/FePd (c, d).

In effect, while in the $L1_0$-FePd hard layer the magnetization vector is oriented perpendicular to the film plane, the deposition of Fe determines the appearance of an appreciable contribution oriented into the film plane whose strength increases by increasing the Fe deposition thickness (figure 2). However, also for the Fe(15)/FePd the strength of this in-plane contribution, as well as of the demagnetizing field, are not sufficient to completely overcome the $L1_0$-FePd perpendicular anisotropy. In agreement, the MFM measurements [figure 3(d)] show a well defined magnetic morphology constituted by small domains. The relatively low image contrast may well be connected to the in-plane component of the magnetization vector.

4. Conclusions

FePd thin films 4.5 nm thick were grown in ultra-high vacuum onto MgO monocrystalline substrates at 540°C by means of a MBE equipped with electron guns. The adopted conditions allowed to obtain an epitaxial growth of a well ordered $L1_0$ phase with the *c*-axis perpendicular to the film plane. The films show a strong perpendicular anisotropy with hysteresis loops characterized by a very high squareness and coercivity. Surface Mössbauer measurements evidenced that the $L1_0$-FePd is present not only in its ferromagnetic form with a strong perpendicular texture, but also in the form of small particles showing a superparamagnetic behaviour.

When iron is evaporated onto these hard films, a reaction occurs with the highly reactive superparamagnetic FePd particles, giving rise to a Fe-base compound containing small amounts of Pd, Fe(Pd). Therefore, the resulting systems are bi-layers constituted by the soft Fe(Pd) and the hard $L1_0$-FePd layer intercalated by a very thin region of small particles.

For a Fe deposition up to 4.5 nm, the bi-layers show a magnetic behaviour typical of a single-phase hard magnet with hysteresis loops characterized by a high squareness and a coercivity slightly higher than that of the original hard phase. By increasing the Fe deposition, the Fe(Pd) thickness progressively increases, and, even for a soft/hard ratio nominally equal to 3, a good exchange coupling establishes between the different phases giving rise to a single-phase magnetic behaviour. However, the high value of the soft to hard thickness ratio determines both a decrease of coercivity and the appearance of two critical fields for the magnetization reversal.

References

[1] Halley D, Auric P, Bayle-Guillemaud P, Gilles B, Marty A and Jalabert D 2002 *J. Appl. Phys.* **91** 9757-63
[2] Gehanno V, Auric P, Marty A and Gilles B 1998 *J. Magn. Magn. Mater.* **188** 310-8
[3] Cebollada A, Weller D, Sticht J, Harp G R, Farrow R F C, Marks R F, Savoy R and Scott J C 1994 *Phys. Rev.* B **50** 3419-22
[4] Ravelosa D, Cebollada A, Briones F, Diaz-Paniagua C, Hidalgo M A and Batallan F 1999 *Phys. Rev. B* **59** 4322-6
[5] Sato K and Hirotsu Y 2003 *J. Appl. Phys.* **93** 6291-8
[6] Pan F, Yang T, Zhang J and Liu B X 1993 *J. Phys.: Condens. Matter* **5** L507-14
[7] Ravelosa D, Cebollada A, Briones F, Diaz-Paniagua C, Hidalgo M A and Batallan F 1999 *Phys. Rev.* B **59** 4322-26
[8] Klein O, Samson Y, Marty A, Guillous S, Viret M, Fermon C and Alloul H 2001 *J. Appl. Phys.* **89** 6781-3
[9] Nazir R, Muhammad M, Akhtar M J, Shah M R, Khan N A, Nadeem M, Siddique M, Mehmood M and Butt N M 2008 *Nanotechnology* **19** 185608

Thickness dependent crystallographic transition in Fe/Ni multilayers

Lennart Häggström[1], Inna Soroka[2] and Saeed Kamali[3]
[1]Department of Physics and Materials Science and [2]Department of Materials Chemistry, Uppsala University, Box 530, 751 21 Uppsala, Sweden
[3]Department of Applied Science, University of California, Davis, Cal. 95616, USA

E-mail: lennart.haggstrom@fysik.uu.se

Abstract The crystallographic transition between bcc- and fcc-structures has been studied in Fe/Ni multilayers with Mössbauer spectroscopy and X-ray diffraction. Layered structures have been confirmed and the structures depends both on individual layer thickness and the ratio of the constituents. Two sets of samples are made: the Fe layer thickness kept constant to 12 monolayer (ML) and 24ML, while the Ni layer thicknesses varied from 3 to 12ML. When the Ni layer is thin, the Fe/Ni multilayers have the bcc phase. When the thickness of Ni layer reaches 6ML and larger, a gradual transition to the fcc phase occurs. Mössbauer spectroscopy studies confirm the existence of a magnetic Fe bcc phase with in-plane magnetization in all samples, two magnetic fcc phases with fields of 28 T and 10 T and small amounts of non-magnetic fcc phases in the thicker films. The observed behaviour can be attributed to the existence of two types of interfaces which are not necessary overlapping each other: An elemental interface between Fe and Ni and a structural interface between fcc and bcc phases.

1. Introduction

The 3d transition metals have attracted large attention and FeNi alloys are of particular interest due to the so called Invar effect [1] coupled to a phase transformation. While bcc Fe is ferromagnetic at room temperature (RT), fcc Fe can have different magnetic states such as ferromagnetism, antiferro-magnetism, and can even be non-magnetic. Fcc iron is unstable at RT and can be stabilized in the form of thin layers or nanoparticles on fcc substrates or matrices. There are a number of studies where Fe is stabilized in a fcc phase in a Cu matrix [2,3]. Layered system of Fe on Ni single crystal has also been studied [4]. Mössbauer spectroscopy (MS), is a powerful technique for studying these systems because the magnetic hyperfine fields are different for Fe atoms in bcc and fcc environments. In the past, MS has been used to investigate Fe/Ni layered systems [5]. In almost all the previous studies of this system, a thick Ni layer has been used, as either substrate or buffer layer, and then a thin Fe layer has been grown onto this Ni layer to establish a fcc structure of Fe. The present work is dedicated to the investigation of the crystallographic transition from bcc to fcc phase in Fe/Ni multilayers. We stabilize first the bcc phase with thick Fe and thin Ni layers, and then the fcc phase is obtained by increasing the Ni layer thickness while the Fe layer thickness is kept constant.

2. Experimental

Samples with x ML of Fe and y ML of Ni layers were grown (samples denoted hereafter by x/y) in a UHV chamber on single crystal MgO(001) substrates by dc magnetron sputtering. Separate Fe (99.95 % purity) and Ni (99.99 %) targets are used. The base pressure was lower than $2 \cdot 10^{-9}$ Torr and the operating pressure of Ar gas was 2.4 mTorr. The deposition rates were 0.30 Å/s and 0.59 Å/s for Fe

Journal of Physics: Conference Series **217** (2010) 012112 doi:10.1088/1742-6596/217/1/012112

and Ni, respectively. The optimal substrate temperature was determined to be 230°C, as deduced from structural analysis. There was no buffer layer and the growth always started with Fe layers and on top a cap with 10 Å thick Pd layer to prevent oxidation was deposited. Two sets of samples were grown: In one set the thickness of Fe layer is kept constant at 12 ML while the thickness of Ni layer varied from 3 to 12 ML. The second set consists of Fe/Ni layers with constant thickness of Fe layer, 24 ML, while the Ni layer thickness is changed from 5 to 12 ML. The number of repetitions of 1 bilayers is 50 and 25 for the first and the second set of samples, respectively. The structure, bilayers thicknesses as well as total thicknesses of all films were determined by X-ray diffraction (XRD) and X-ray reflectivity (XRR) measurements using a Siemens D5000 diffractometer with CuK_α radiation. ^{57}Fe Conversion Electron Mössbauer Spectroscopy (CEMS) was performed at room temperature with ^{57}Co in a Rh matrix as the source. A gas flow (He & CH_4) detector was used. In all measurements, the direction of the incoming photons was perpendicular to the sample plane. All isomer shifts are given with respect to α-iron at room temperature.

3. Result and discussion

3.1 XRR and XRD measurements

As seen in the XRR patterns, Figure 1(a and c), superlattice peaks appear, indicating that the films have layered structure in all samples. Also, the finite size oscillations (Kiessig fringes) are clearly seen ex-cept for the thickest Ni layers. This confirm the good quality of the surfaces and the interfaces between the substrates and the films. With high angle XRD measurements the structural changes as function of Ni layer thickness are obtained. The XRD spectra for the multilayers with thin Ni layers (12/3) and 12/3.5) have only one diffraction peak (Figure 1b), which corresponds to the (002) plane in bcc Fe.

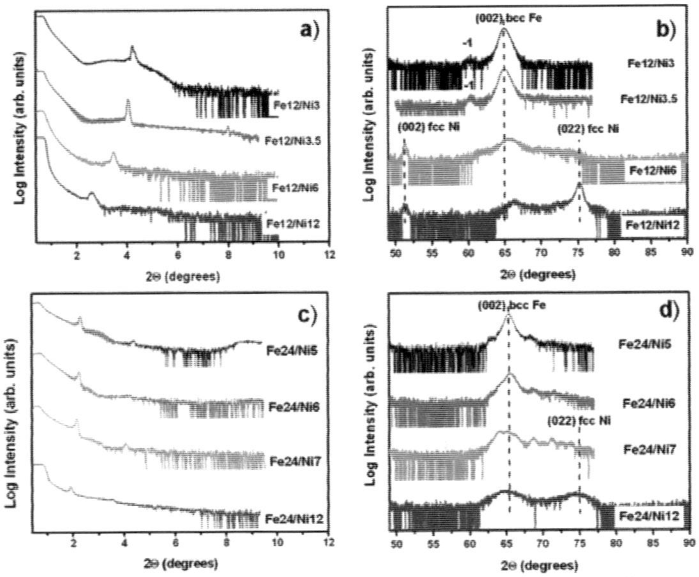

Left **Figure 1**. X-ray reflectivity (a), (c) and X-ray diffraction (b), (d) patterns of Fe/Ni multilayers.

The crystal coherence length (ζ), obtained from Full Width Half Maximum (FWHM) of the Fe bcc peak, is found to be about 110 Å for both samples. Thus, those films can be considered as mono-crystals. The structure started to change dramatically when the Ni layer thickness become larger than 6ML. Additional diffraction peaks from (022) and (002) fcc Ni phase appears in the XRD spectra for 12/6 and 12/12. Moreover, the FWHM of the (002) bcc peaks strongly increased and their intensities decrease. This indicates that the films became polycrystalline and consist of bcc and fcc phases, while the layered structure is still kept. In the second set of samples, where the Fe layers are twice as thick as compared to the films in the first set, the structural changes have different behaviour. As seen in the Figure 1d , the 24/5 film has one XRD peak that corresponds to the (002) plane of the Fe bcc phase with ζ~100Å. When the thickness of the Ni layer is increased to 6 ML, the (002) bcc peak became broader (ζ~75Å) and its intensity decreased. At the same time additional small peaks at higher angles appear. Those peaks become more pronounced in the XRD spectra of the 24/7 film, where the splitting of the (002) peak is observed. The origin of those peaks is not clear, however we can assume that the film consists of several regions where the lattice parameters and crystalline orientation differ from each other. Note that the 24/7 film has well defined layered structure, according to the XRR meas-urements. Finally, in the XRD pattern for the sample 24/12 two wide peaks which correspond to (002) bcc Fe and (022) fcc Ni are present. Accordingly, in the film where the Ni amount is rather high, two phases most probably textured in the directions [001] bcc and [011] fcc are formed. Thus, the structure of the Fe/Ni multilayers change from monocrystalline bcc phase to polycrystalline bcc+fcc phases by increasing the thickness of Ni layer keeping the Fe layer thickness unchanged. From the XRR measurements, the average film monolayer thickness as a function of Fe/Ni ratio can be obtained. For the first set of samples this thickness increases with the Ni content. This result is in good agreement with the increased amount of the fcc phase. However, for the second set, the average film monolayer thickness does not depend much on the composition and remains close to bulk Fe value, 1.43Å.

3.2 Mössbauer spectroscopy

The Mössbauer spectra and fitting results are displayed in Figure 2. As can be seen, the hyperfine fields B_{hf} can be grouped into three well separated categories, one high field value of 33.4 T (distribution width $\sigma \approx 0.9$ T), one intermediate field value of 28.2 T ($\sigma \approx 1.2$ T) and one low field value of 10.9 T ($\sigma \approx 2.0$ T) for all samples.

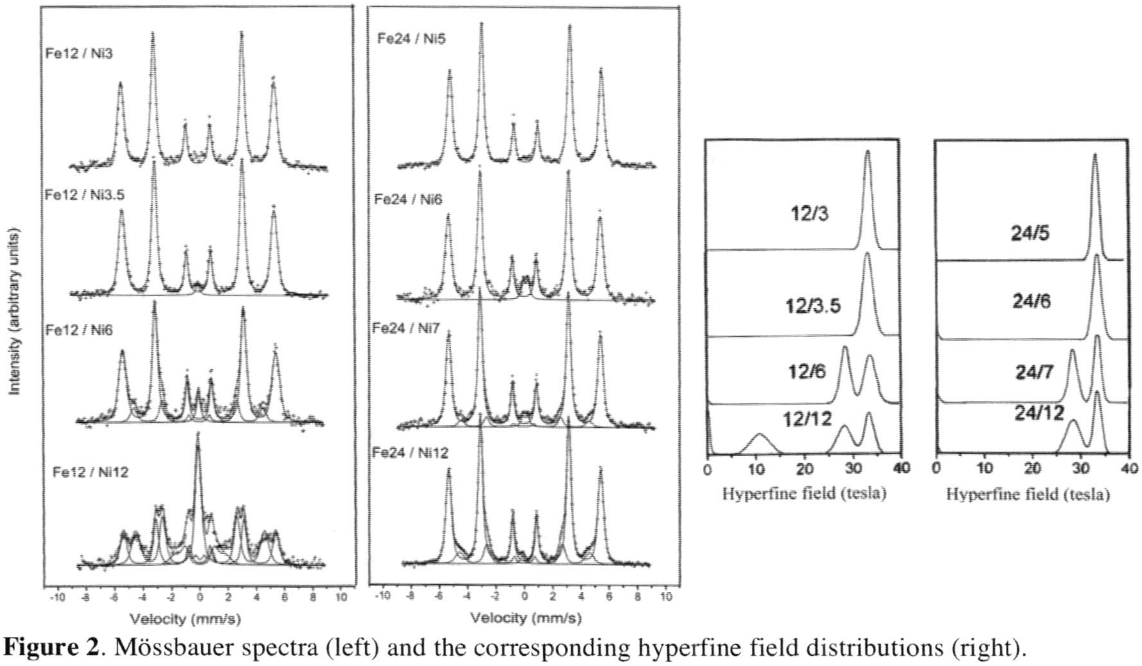

Figure 2. Mössbauer spectra (left) and the corresponding hyperfine field distributions (right).

Beside these spectral patterns there is also a broadened singlet line in some of the spectra. It has recently been shown that the Fe/Co superlattices can be analyzed using a local alloy model [6]. We adopt the same procedure of analysis in the present case of Fe/Ni multilayers. Johnson et al. [7] have published Mössbauer result at RT of the Fe-Ni alloys (Figure 3). There is a transformation between the bcc and the fcc structures for these alloys at a Ni concentration between 24% and 34%. In the bcc structure, for lower Ni concentration, the hyperfine field is between 33.0 T and 34.3 T. At the crystal transformation a two phase region exists where the hyperfine fields for Fe in the fcc-alloys vary between 0 T and about 25 T. For higher concentration, in the fcc phase, the hyperfine field varies between 28.2 T and 30.8 T. A hyperfine field gap (30.8 T- 33.0 T) thus exists betweeen the bcc- and fcc-alloys. It is, therefore, tempting to ascribe the high field component of 33.4 T to Fe in the bcc structure with a Ni concentration less than ≈ 30% in the surrounding of each Fe atom, the intermediate field of 28.2 T, to Fe in the fcc structure with a Ni concentration above ≈ 40% and the low field of

10.9 T to Fe in fcc-type of structure with a Ni concentration of ≈30-40%. The broadened singlet line with zero field would then be assigned to the fcc structure with a narrow Ni concentration around 30%. This description is further supported by the findings of a hyperfine field of 28.5 T for Fe in a 5 ML thick fcc-Fe(001) film on a fcc Cu(001) substrate [3] and fields of about 28 T and 8 T for Fe fcc films with the thickness of about 10 Å in Ni/Fe/Ni trilayers [8].

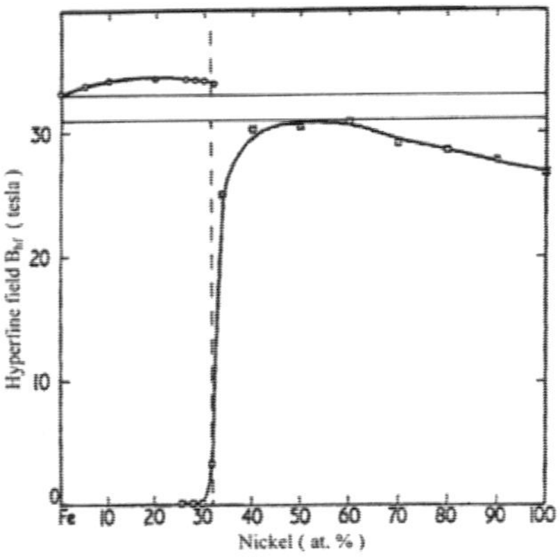

Figure 3. The ^{57}Fe hyperfine field at room temperature in the Fe-Ni alloys [7]. The dashed vertical line marks the crystallographic transformation between bcc (left part) and the fcc (right part) structures. Note the hyperfine gap (30.8 T- 33.0 T) between the fields in bcc- and fcc-alloys, respectively

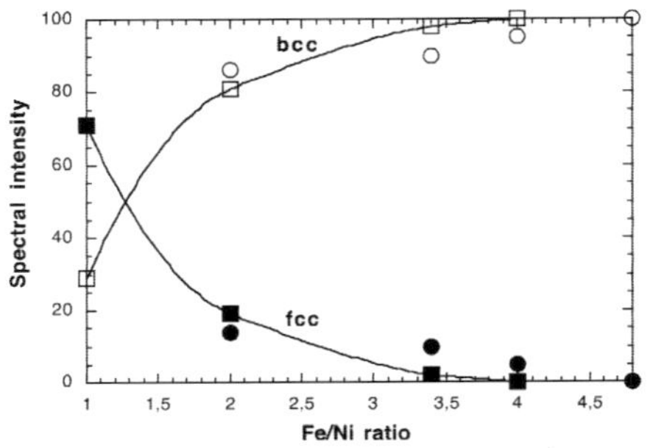

Figure 4. The Mössbauer spectral intensity for the bcc and the fcc components versus the Fe/Ni ratio. The square and round symbols refer to the samples with Fe thicknesses of 12 ML and 24 ML, respectively. The curves are drawn for Fe12 ML.

Figure 4 displays the spectral intensity of the bcc and the total fcc components as a function of Fe/Ni ratio. For a ratio above 3.5 the multilayers have a dominant bcc structure while when the ratio goes towards 1 the fcc structure dominates. The observed behavior can be attributed to the existence of pure bcc and fcc phases and structural interfaces in between. These interfaces do not need, however, to overlap with the elemental interfaces between Fe and Ni. In the boundary between bcc-fcc phases the isomer shift for the fcc pattern shows a shift with respect to the bcc pattern of -0.12 mm/s for the FeNi-alloys [8]. The bcc-fcc shift for the Fe/Ni multilayers is smaller, about -0.06(2) mm/s. In the multilayers with the larger Ni layer thickness the structure gradually changes between the two phases and there can appear local Fe surroundings which have rather distorted bcc and fcc structures. This may explain why the isomer shift values for Fe in the superlattices do not exactly match the alloy values. The line intensity ratios for the magnetic six-line pattern are found very close to 3:4:1:1:4:3. Since the γ-ray direction was perpendicular to the sample plane, these intensity ratios indicate that the average magnetic moment direction is oriented within the sample plane. This is expected from the shape anisotropy of the multilayer films.

References

[1] Ullrich U and Hesse J 1984 *J. Magn. Magn. Mater.* **45** 315
[2] Liu C and S. Bader 1989 *Physica* **B161** 253
[3] Macedo W A A, Keune W and Ellerbrock E D 1991 *J. Magn. Magn. Mater.* **93** 552
[4] Hadj-Larbi A, Bouarab S and Demangeat C 2002 *Phys. Rev.* **B 66** 144428
[5] Colombo E, Donzelli O, Fratucello G B and Ronconi F 1991 *J. Magn. Magn. Mater.* **93** 597
[6] Kamali S, Bergman A, Andersson G, Stanciu V and Häggström L 2006 *J. Phys. Cond. Matter* **18** 5807
[7] Johnson C E, Ridout M S and Cranshaw T E 1963 *Proc. Phys. Soc.* **81** 1079
[8] Fratucello G B and Vavassori P 2003 *J. Magn. Magn. Mater.* **260** 480

The influence of reduction process on transformation iron-cobalt oxide nanoparticles in MCM-41 silica

Z Surowiec[1], M Wiertel[1], B Bierska-Piech[2] and M Budzyński[1]

[1]Institute of Physics, M.Curie-Sklodowska University, pl. M.Curie-Skłodowskiej 1, 20-031 Lublin, Poland
[2]Institute of Materials Science, University of Silesia, ul. Bankowa 12, 40-007 Katowice, Poland

E-mail: zbigniew.surowiec@umcs.lublin.pl

Abstract. $((Fe_{0.9}Co_{0.1})O_x$/MCM-41 nanocomposite material was prepared by wet impregnation of the silica mesoporous template with iron and cobalt mixture of aqueous salt solutions. The phase analysis of the obtained samples carried out by X-ray diffraction method and studies of magnetic properties using ^{57}Fe Mössbauer spectroscopy revealed the presence of nanostructured magnetite- and maghemite-like phases. The transformation to bimetallic $Fe_{0.9}Co_{0.1}$ and fayalite phases accompanied the reduction process in hydrogen at 1073 K. The contribution of ferrimagnetic phases in relation to superparamagnetic phases increases when temperature changes from 295 K to 13K in the sample before reduction what is characteristic for nanoparticles.

1. Introduction

In the last decades the decrease of the grain size of magnetic materials to the nanometer scale has opened new perspectives in materials science. Nano-sized materials possess many unique properties, which are intermediate between the molecular or atomic level and the properties of the corresponding bulk materials. In particular nanoparticles of materials which are ferromagnetic in bulk exhibits superparamagnetic (SPM) properties [1]. Studies of magnetic properties of nanoparticles are important from both fundamental and technological point of views. Such materials have potentially, and even in some cases nowadays, a wide field of applications ranging from storage information technology [2] through catalysts in chemical technology to biomedical applications [3].

The main factor influencing physical, chemical and other properties of these objects is limitation by small sizes in artificial boundaries. There exists possibility to produce relatively easily nanoparticles based on ferromagnetic and/or ferrimagnetic oxide materials in nanometric voids of different porous media [4] by the direct hydrothermal and/or wet impregnation methods. In particular mesoporous channel-type matrices such as MCM-41 can be used as porous media for obtaining Fe-based nanocrystallites [5]. The main advantage of this method is that it allows the synthesis and/or incorporation of the noticeable amount of magnetic nanomaterials. In a case of such composites the problem of chemical and structural stability occurs [6]. The growth of the nanocrystallites during calcination and reduction stages of chemical synthesis can be influenced by the metal-support interactions, and the way of thermal treatment. It is also considerably harder to produce pure ferromagnetic metallic or bimetallic nanoparticles in this way. The use of materials with a more

complicated composition than simple metals gives possibility to control technologically useful and important properties.
In framework of this work an attempt to obtain structurally and chemically stable oxide and bimetallic Fe-Co nanocrystallites in magnetically ordered state in MCM-41 template was made. Structural and magnetic properties of obtained composites were studied and reported.

2. Experimental

The synthesis of MCM-41silica template was accomplished using the modified procedure detailed described in [7]. In order to prepare bimetallic system on the support MCM-41 5 wt% solutions containing metal ions (Fe^{3+} and Co^{2+}) were mixed in 9:1 ratio and then used as solution for the impregnation of the silica material. At the next stage the sample was subjected to low temperature treatment (drying) at 423 K for 1 hour. The next step involved calcination in air atmosphere at 573 K for 1 hour. As a result of such a procedure there should be formed oxide system deposited on silica framework of general formula $(Fe_{0.9}Co_{0.1})O_x$/MCM41. In the next stage the samples were reduced and passivated. In this case a sample (0.25 g) was placed in a flow quartz reactor with an internal diameter 10 mm. A 6% H_2–Ar mixture was used as carrier gas at a flow rate of 30 cm^3/min. The temperature of the sample was increased up to 1073 K with the linear ramp rate 10 K/min. The temperature of reactor was maintained at 1073 K for 3 hours. Then sample was cooled to room temperature, flushed with helium (0.5 hour at a flow of 30cm^3/min) and next passivated in the 5 % O_2- He mixture (0.5 hour at flow of 30 cm^3/min). Such procedure allowed formation of the thin oxide layer on the surface of the metal crystallites and prevented deep oxidation of the species.

The XRD and Mössbauer measurements were performed for the samples after calcination and after reduction/passivation treatment. The XRD patterns were measured on the Philips X'Pert PW 3040/60 X-ray diffractometer with the CuK_α radiation at room temperature. They were fitted by the use of the FULLPROF program for Rietveld method refinement. The ^{57}Fe Mössbauer spectra of the investigated sample were measured in transmission geometry at different temperatures using a constant acceleration spectrometer with the ^{57}Co/Rh source. The isomer shift δ was calibrated against a metallic iron foil at room temperature.

3. Results and discussion

The XRD measurements were carried out in order to obtain information about changes of structures of crystallites embedded in the silica template at two stages of chemical treatment. The XRD patterns for the sample before (a) and after (b) reduction are shown in Figure 1. The smoothly varying broadened component in the begin of measurement range results from the amorphous silica support.

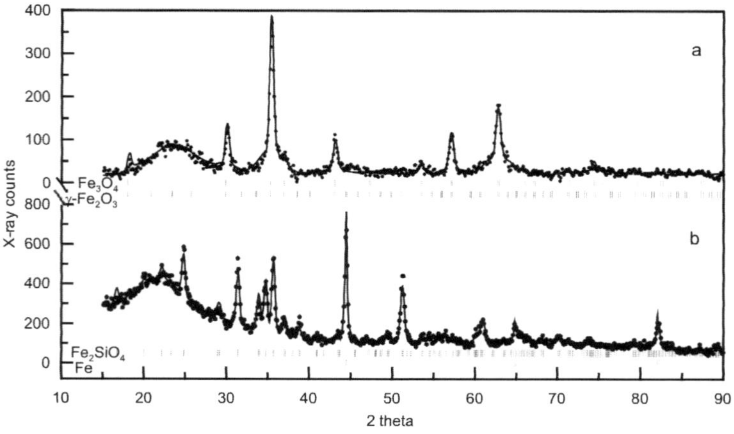

Figure 1. The XRD diffraction pattern of $(Fe_{0.9}Co_{0.1})O_x$/MCM41 sample before reduction (a) and after reduction in H_2 (b). The full curves represents calculated patterns, the points the observed ones. Short vertical lines below the diffraction pattern indicate the calculated Bragg positions

Refinement by Rietveld method of well-resolved peaks in spectrum (a) revealed the presence of two crystalline phases: magnetite- and maghemite-like with partial replacement some Fe atoms with Co ones. The lattice parameters are 8.37(1) Å and 8.38(1) Å, respectively. In the sample after reduction also two phases were identified: bimetallic $Fe_{0.9}Co_{0.1}$ structure and $(Fe_{1-x}Co_x)_2SiO_4$ (fayalite). The lattice parameters of these phases are equal to a = 8.39(1) Å and 8.39(1) Å, respectively. In particular large broadening of the XRD diffraction lines in this sample indicates an occurrence of the crystallites with very small sizes. Average sizes D and strains η in the crystallites *for* phases giving well-resolved peaks were determined from the broadening of XRD diffraction peaks by using the Hall-Williamson method. In the sample before reduction for magnetite-like phase $D = 26.5$ nm, $\eta = 8.13 \cdot 10^{-4}$. In the sample after reduction for $Fe_{0.9}Co_{0.1}$ phase $D = 68.5$ nm, $\eta = 4.73 \cdot 10^{-4}$ and for fayalite $D = 32.1$ nm, $\eta ==8.81 \cdot 10^{-4}$.
Figure 2a shows the Mössbauer spectra of the $(Fe_{0.9}Co_{0.1})O_x$/MCM-41 sample before reduction measured at different temperatures. All three patterns consist of three sextets, superimposed one doublet and a monoline. Two sextets are related to Fe atoms located at two non-equivalent tetrahedral (A) and octahedral (B) sites in the magnetite-like structure. In pure magnetite the relation of occupancies of A and B sites is equals to 1:2. In obtained spectra this relation is clearly distorted. This experimental result points to necessity of taking into account one more sextet component in the fitting procedure. All hyperfine parameters values for the third component are close to values for Fe probes in nanocrystallites with maghemite-like structure.

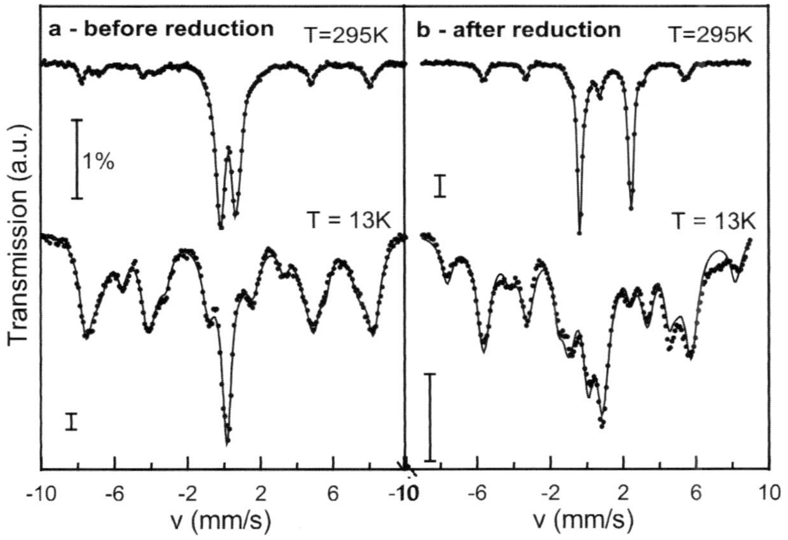

Figure 2. Mössbauer spectra measured at different temperature of the $(Fe_{0.9}Co_{0.1})O_x$/MCM-41 sample before reduction (left panel) and reduced in H_2 (right panel). The thick solid lines are the results of the fitting procedure described in the text.

The sextet lines are broadened in comparison to respective bulk phases as a result of small crystallite sizes. The obtained values of isomer shifts *IS* for ferromagnetic components are consistent with that for Fe oxidation states in magnetite $(Fe^{3+})_A[Fe^{2.5+}]_BO_4$ and maghemite $(Fe^{3+})_A[Fe^{3+}_{5/3}\,\square_{1/3}]_BO_4$, where $\square$ denotes a vacancy.

From temperature evolution of Mössbauer spectra for the sample before reduction it is evident that relative contribution of ferromagnetic components in relation to a paramagnetic doublet and monoline components increases with the decrease of sample temperature. Thus the paramagnetic component represents the superparamagnetic phase (SPM) of the both iron oxides resulting from the relaxation phenomenon due to the intrinsic finite-size effect. The presence both the doublet and monoline in the spectra for the superparamagnetic state can be explained on the basis of core-shell model [8]. The superparamagnetic doublet with a significant quadrupole splitting value equals to 0.86(3) mm/s origins from ^{57}Fe nuclear probes in non-spherical local surroundings occurring in the surface shells of the nanoparticles. A monoline component present in obtained spectra results also from nanoparticles in the superparamagnetic state. However, in this case ^{57}Fe probes located in the core of nanoparticles are a source of a signal. They have spherical local surroundings without an electric field gradient. The relatively considerable contribution of this component even in the spectrum measured in 13 K suggests that many nanoparticles have sufficiently small sizes in order to their blocking temperature lies below the lowest temperature of measurements.

The Mössbauer spectra of the $(Fe_{0.9}Co_{0.1})O_x$/MCM-41 sample after reduction measured are presented in figure 2b. The patterns measured at temperature 295 K consists of one sextet and two doublets. The sextet has hyperfine parameters characteristic of bimetallic $Fe_{0.9}Co_{0.1}$ phase in which Fe atoms occur in Fe^0 state. The doublets are assigned to Fe^{2+} and Fe^{3+} ions in the octahedral sites of $(Fe_{1-x}Co_x)_2SiO_4$ phase. The fayalite remains paramagnetic up to T_N = 66 K. Below critical temperature T_K = 23 K antiferromagnetic structure changes into complex structure with canted AF arranged spins in M1 sublattice and antiferromagnetically at M2 sites [9]. At temperature T = 13 K Mössbauer spectrum consists of three sextets and a doublet. Two sextets with the largest and the smallest hyperfine splitting are related to ^{57}Fe probes located in fayalite. The third origins from bimetallic phase. The doublet is connected with fayalite nanoparticles in superparamagnetic state.

In high temperature reduction process highly dispersed in MCM-41 matrix Fe/Co oxides nanoparticles (SPM phase before reduction) interacting with walls of silica nanochannels form $(Fe_{0.9}Co_{0.1})_2SiO_4$ phase. In connection with that the doublet related to SPM phase present before reduction completely disappears in the MS spectra of the sample after reduction. The Fe/Co oxides nanocrystallites of the larger sizes primary giving the sextets are reduced to bimetallic $Fe_{0.9}Co_{0.1}$ nanoparticles. In this way pure ferromagnetic Fe nanoparticles stabilized with Co addition were obtained in the nanocomposite.

4. Summary

Magnetite- and maghemite-like nanocrystallites in MCM-41 matrix were obtained by co-impregnation method. The MS investigations showed the coexistence of ferrimagnetic and superparamagnetic nanoparticles in the sample before reduction from 13 K to room temperature what is undesirable. From the point of view of technological applications it is desirable to obtain the template-based nanocomposites which would not include any superparamagnetic phases. In this work it was achieved by the reduction of the $(Fe_{0.9}Co_{0.1})O_x$/MCM41 sample at 1073 K in hydrogen and then passivation at the flow of O_2/He mixture at room temperature what is prove by Mössbauer spectroscopy.

References

[1] Dormann J L, Fiorani D, Cherkaoui R, Tronc E, Lucari F, D'Orazio F, Spinu L, Noguès M, Kachkachi H and Jolivet J P 1999 *J. Magn. Magn. Mater.* **203** 23
[2] Moser A, Takano K, Margulies D T, Albrecht M, Sonobe Y, Ikeda Y, Sun S and Fullerton E E 2002 *J. Phys.* D **35** R157
[3] Ambashta R D, Yusuf S M, Mukadam M D, Singh S, Wattal P K and Bahadur D 2005 *J. Magn. Magn. Mater.* **293** 8
[4] Bengoa J F, Cagnoli M V, Gallegos N G, Alvarez A M, Mogni L V, Moreno M S and Marchetti S G 2005 *Micropor. Mesopor. Mater.* **84** 153
[5] Cagnoli M V, Alvarez A M, Bengoa J F, Gallegos N G, Mogni L V, Gonzále Oddera M X, Marchetti S G and Mercader R C 2003 *Hyperfine Interact.* **148-149** 185
[6] Grün M, Unger K K, Matsumoto A, Tsutsumi K 1997 *Characterization of porous solids* IV, eds. McEnaney B, Mays J T, Rouquerol J, Rodriguez-Reynoso F, Sing K S W, Unger K K (The Royal Society of Chemistry) p 81
[7] Gac W, Goworek J, Wójcik G and Kępiński L 2008 *Adsorption* **14** 247
[8] A.M. Van der Kraan, Hyperfine Interact. **40,** 211, (1988)
[9] R.P. Santoro, R.E. Newnham, S. Nomura, J. Phys. Chem. Solids 27 (1966) 655–666

Journal of Physics: Conference Series **217** (2010) 012114 doi:10.1088/1742-6596/217/1/012114

Mössbauer spectra of hematite and magnetite nanoparticles in polymer composites

M Polikarpov[1,3], V Cherepanov[1], M Chuev[1,2], S Shishkov[1] and S Yakimov[1]

[1]Russian Research Center "Kurchatov Institute", Kurchatov sq. 1,123182, Moscow, Russia
[2]Institute of Physics and Technology, Russian Academy of Sciences, Nakhimovskii pr. 36-1, 117218 Moscow, Russia

E-mail: polikarpov@imp.kiae.ru

Abstract. It is experimentally demonstrated that the shape of Mössbauer spectra of polymer composites based on single-domain magnetic particles depends more on an inter-particle interaction than on superparamagnetic properties of a separate particle. The characteristic features of the spectra of such systems with strong inter-particle interactions are separated, such as an asymmetric shape of lines with sharp outer and smeared inward sides and presence of central anomalously splitted doublet. It is shown that the three-level relaxation model, which takes into account the precession and diffusion of magnetization of single-domain particles between two local energy minima and one absolute energy maximum, allows to describe all characteristic features of the spectra of such systems.

1. Introduction

The particles of a magnetically ordered substance whose magnetization vector can spontaneously change its orientation due to a coherent rotation of its individual spins are called superparamagnetic. The results obtained from the investigation of the magnetic properties of these particles depend on a correlation between the time of measurement and that of magnetization vector relaxation. If the time of observation is much less than that of relaxation then the particles exhibit ferromagnetic properties, while in the opposite case the particles are superparamagnetic. The time of relaxation increases with the particle size. Therefore, the Mössbauer spectrum of a dispersed system of magnetic nanocrystals is usually a superposition of a Zeeman sextet, corresponding to the large-size ferromagnetic particles, with a superparamagnetic doublet, corresponding to the particles of smaller size. On the other hand, the time of relaxation in the system of interacting superparamagnetic particles can change due to correlation bonds arising between magnetization vectors of neighboring particles. This is the so called superferromagnetic state. Inserting of the magnetic nanoparticles in a polymer environment does not change the properties of the particles, but can change the parameters of inter-particle interaction. That is why qualitative interpretation and theoretical calculation of the polymer composites spectra is a difficult task.

[3] To whom any correspondence should be addressed.

2. Qualitative features of the interacting single-domain nanoparticles spectra

We investigated superparamagnetic behavior of a dispersed α-Fe_2O_3 powder with mean size of about 19 nm, prepared from aqueous solution of $Fe(NO_3)_3{\cdot}9H_2O$. Ammonium solution was added dropwise to the solution of the iron salt at 50°C until the PH of the solution became 7.2. The precipitate was filtered, washed, dried, ground in a ball mill and calcinated in air at 400 °C for 1 hour. Mössbauer spectra of film samples, prepared by precipitation of a dredge of the powder in alcohol (a) and in alcohol with a polymer addition (b) are shown in Fig.1.

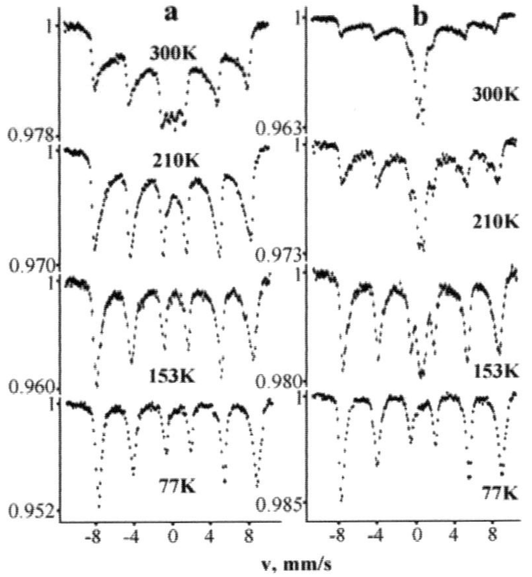

Figure 1. Mössbauer spectra of dispersed hematite samples prepared by precipitation in alcohol (a) and by precipitation in alcohol with a polymer addition (b), measured between 77 and 300 K.

One can see that sample (b) exhibits classical superparamagnetic behaviour with temperature increase from 77 to 300 K. In contrast, the superferromagnetic sample (a) reveals doublet at 300 K only. The blocking temperature, which is experimentally estimated as the temperature at which the magnetically split and the unsplit subspectra have equal areas, differs for two samples of the same powder by more than 200 K.

For direct demonstration of the influence of inter-particle interaction on the shape of Mössbauer spectra, the powder under investigation was poured loosely into a lucite container. One can see in Fig. 2a that there is no sign of superparamagnetic component. After that, the container with powder was filled with water and a superparamagnetic doublet was observed in the spectrum measured at the same temperature (Fig. 2b). The effect is reversible and after the water has been dried out the spectrum came back to its initial form. For a qualitative analysis of specific features of such spectra it is convenient to use the approach based on the introduction of the continuous distributions of the hyperfine field $P(H_{hf})$. The result of such a calculation of the spectrum of loosely poured particles using of the MSTools software package [4] is shown in Fig. 3c. One can see in Fig. 3a that even within this model-free approach without any restrictions on distribution $P(H_{hf})$ besides Lorentzian shape of lines (linewidth 0.12 mm/s) it is impossible to describe asymmetric shape of lines with sharp outer and smeared inward sides. It may be recognized from a magnitude of standard deviations between the experimental and calculated spectra (Fig. 3b), which increase systematically at the outer sides of the lines. Another characteristic feature of the spectra of interacting particles is the presence of anomalously split doublet at temperature above the blocking one. The splitting can not be explained by a quadrupole hyperfine interaction in the presence of the electric field gradient on the nucleus because it disappears at lower temperatures.

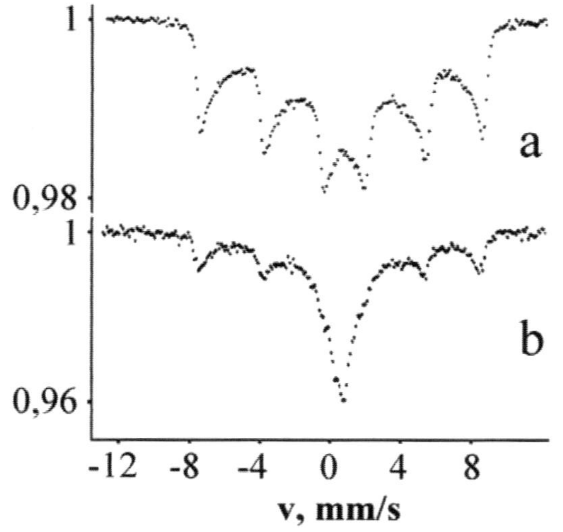

Figure 2. Mössbauer spectra at T = 300 K of dispersed α-Fe_2O_3 loosely poured in a container (a) and after filling the container with water (b).

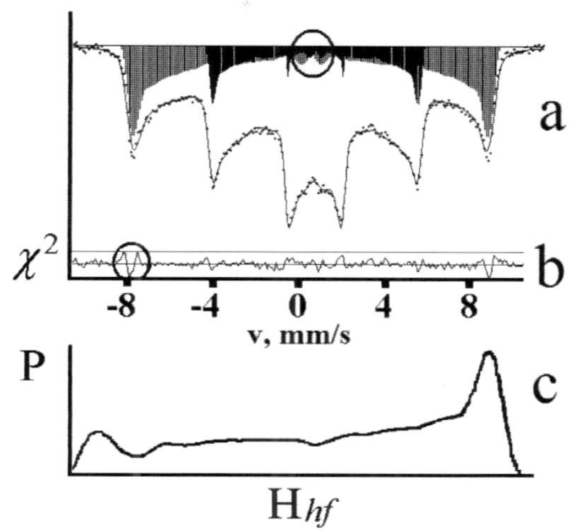

Figure 3. Representation of spectra of loosely poured α-Fe_2O_3 (a) as a continuous distribution of the hyperfine field on nucleus (c) and a magnitude of error of such approach (b).

3. Theoretical model

As it was shown above, the theoretical model for calculation of Mössbauer spectra of polymer composites of single domain magnetic particles has to take into account the inter-particle interaction. So, the model has to allow one to take into consideration physical mechanisms of formation of such features of hyperfine structure as asymmetric shape of lines with sharp outer and smeared inward sides and the presence of central anomalously split doublet.

The simplest way to describe the Mössbauer spectra of superparamagnetic particles is to use two-level relaxation model, within which only two states corresponding to opposite directions of the particle's magnetic moment along the easy magnetization axes are considered, so that only jumps from one state to the other are possible. The energy density of such particles may be represented as

$$E(\Theta) = -K\cos^2\Theta \quad (1)$$

where K is the constant of axial magnetic anisotropy and Θ is the angle between the easiest magnetization axis and the magnetization vector. Temperature dependence of the transition rate is determined by the energy barrier between these equal states and is described by the Neel formula:

$$p = p_0 \exp(-KV/ k_B T) \quad (2)$$

where p_0 is slightly dependent on temperature T and V is volume of the particle. Temperature evolution of the spectra, calculated in accordance with two-level model demonstrates transition from a Zeeman sextet to a superparamagnetic doublet with temperature increasing, but the shape of lines remains essentially symmetrical as a result of assumption about equal population of two states, separated by energy barrier. The generalized two-level model of relaxation in the system of superparamagnetic particles was suggested [1], which postulates that the relaxation between the particle's state with opposite directions of its magnetic moment never occurs as a transition between the states of the same energy because even weak interaction with environment should move apart two energy levels. As a result, the relaxation process always takes place between two states with different population, which leads to asymmetrical shape of spectral lines of sextet. However, this model can not describe anomalous splitting of the central doublet.

All characteristic features of the spectra of interacting particles are revealed within the framework of three-level relaxation model recently developed [2], taking into account precession and diffusion of magnetization of single-domain particles between two local energy minima and the absolute energy maximum. The maximum in our case is induced by an interaction of the particle with environment. The interaction is written in the model as a rapidly fluctuating chaotic field **h**(t), so that

$$E(\Theta) = -K\cos^2\Theta - \mathbf{h}(t)\mathbf{M} \quad (3)$$

The assumption about existence of a nonzero mean-square amplitude of the chaotic field within interacting system of magnetic particles gives rise to appearance in a spectrum of a magnetic sextet with small hyperfine splitting slightly dependent on the particle size and temperature, which looks like anomalous doublet.

4. Mössbauer spectra of magnetite nanoparticles in polymer composites

The film composites of the magnetite Fe_3O_4 nanoparticles in the polyvinyl alcohol matrix were synthesized by the method described in [3]. The temperature evolution of experimental spectra of the composite with 4 nm particles (points) and resulting spectra (solid lines), calculated under Gaussian type of distribution of H_{hf} and effective doublet of lines with Gaussian type of broadening are shown at Fig. 4a.

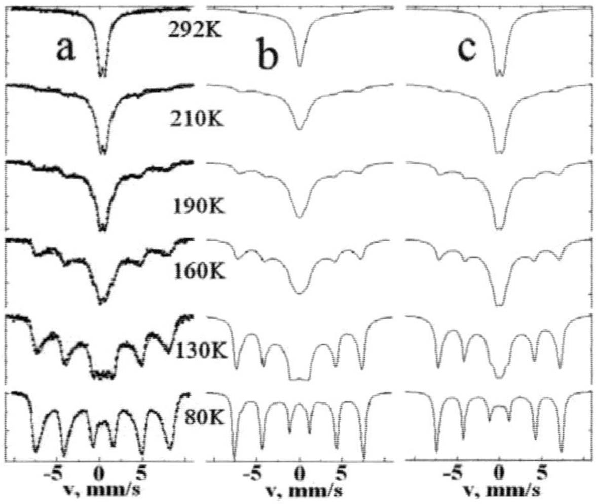

Figure 4. Mössbauer spectra of polymer composites of 4 nm magnetite nanoparticles (a) and theoretical spectra, calculated within the framework of three-level relaxation model under assumption of zero (b) and nonzero (c) root-mean-square amplitude of the chaotic field.

It is evident that the experimental spectra show all characteristic features of inter-particle interactions. Theoretical spectra, calculated within the framework of three-level relaxation model [2] under assumption of zero (b) and nonzero (c) root-mean-square amplitude of the chaotic field **h**(t) are shown in columns (b) and (c), respectively. The last variant demonstrates all characteristic features of interacting particles spectra, so only the correspondent model can be recommended for fitting of the experimental spectra.

Acknowledgments

This work is financially supported by the Russian Foundation for Basic Research (projects 08-02-00388, 09-02-08313 and 09-02-12195).

References

[1] Afanas'ev A M and Chuev M A 2001 *JETP Lett.* **74** 107
[2] Chuev M A 2008 *J. Phys.: Condens. Matter* **20** 505201
[3] Godovsky D Y et al 1999 *Adv. Mater. Opt. Electron.* **9** 87
[4] Rusakov V S and Chistyakova N I 1992 *Proc. Latin American Conf. on Application of the Möessbauer Effect (Buenos Aires)* 7-3

International Conference on the Applications of the Mössbauer Effect (ICAME 2009) IOP Publishing
Journal of Physics: Conference Series **217** (2010) 012115 doi:10.1088/1742-6596/217/1/012115

Super-ferrimagnetism of magnetite nanoparticles in a weak magnetic field

M A Polikarpov[1,3] , V M Cherepanov[1], M A Chuev[1,2], S Yu Shishkov[1] and S S Yakimov[1]
[1] Russian Research Center “Kurchatov Institute”, Kurchatov sq. 1, 123182 Moscow, Russia
[2] Institute of Physics and Technology, Russian Academy of Sciences, Nakhimovskii pr. 36-1, 117218 Moscow, Russia

E-mail: polikarpov@imp.kiae.ru

Abstract. Film composites based on the magnetite Fe_3O_4 nanoparticles in the polyvinyl alcohol matrix were studied by means of the Mössbauer spectroscopy in a weak magnetic field, mainly at room temperature. The Mössbauer spectra in the absence of a magnetic field show a superposition of minor (poor-resolved) magnetic hyperfine structure and major collapsed spectrum typical for superparamagnetic particles. Such a pattern evidences for the mean particles size ($\langle D \rangle \approx 4$ nm) being close to the ‘blocking’ size for this material at a given temperature. The spectra taken in a weak (0.34 T) magnetic field show almost complete restoration of the magnetic hyperfine structure characteristic for the nanoparticles of the kind. The experimental Mössbauer spectra have been analyzed within the recently developed three-level relaxation model of magnetic dynamics in an ensemble of single-domain particles. This model takes into account the precession and diffusion of particle’s uniform magnetization as well as the particles size distribution.

1. Introduction

The most important characteristics of single-domain nanoparticles used in magnetic storage media are their magnetization switching time and the strength of corresponding external maghetic field. Typical value of the switching time in modern computers is of about nanosecond which comes into the ^{57}Fe Mössbauer time window defined by the lifetime of excited nuclear level and Larmor precession of nuclear magnetic moments. This is why the Mössbauer spectroscopy allows one to obtain a valuable information in the field. However, up to now the Mössbauer spectra of fine magnetic particles are usually measured as a function of temperature only. Sometimes the spectra are measured in a strong external magnetic field of about several Tesla, which is comparable with the strength of the hyperfine field, H_{hf}, at the nucleus, so that one can roughly estimate only the average saturation magnetization of the sample as a whole. Meanwhile, a substantial influence of weak magnetic fields with the strength of about 0.1 T on the shape of Mössbauer spectra of magnetic nanoparticles has been already observed [1-4], but such investigations have been not systematically performed.

[3] To whom any correspondence should be addressed.

2. Experimental

Polymer film composites containing magnetite nanoparticles were prepared *in situ* via the alkali hydrolysis reaction of a 2:1 mixture of iron(III) and iron(II) chlorides, respectively. The synthesis procedure was described in detail earlier [5]. Mössbauer spectra were measured with a conventional spectrometer, which operated in the constant acceleration mode for ^{57}Co(Rh) source with the activity of 10 mCi. Figure 1 shows the typical ^{57}Fe Mössbauer absorption spectra of iron oxide nanoparticles in a polyvinyl alcohol (PVA) matrix, which were measured in weak magnetic fields. As one can see, a noticeable spectrum change already in a weak external field H = 0.4 kOe in comparison with the spectrum at $H = 0$. With the external field increasing a substantial spectra transformation is observed, which contained much of additional information on characteristics of the specimen magnetic dynamics. This effect is similar to that of the field–induced magnetic ordering near the Curie temperature in the bulk ferrimagnetic YIG [6]. An example of the analysis of the spectra field dependence in terms of the DISCVER procedure which allows one to decompose a spectrum into partial sextets with Gaussian broaden lines corresponding to a hyperfine field H_{hf} distribution [7] is shown by solid lines in Figure 1. Formally, this approach describes the restoration of the resolved hyperfine structure corresponding to an effective magnetization of the studied nanoparticles in a magnetic field. However, one should treat somehow the partial sextets evaluated as a function of the field strength. Such a treatment requires to apply a physical model taking into consideration qualitative peculiarities of the nanoparticles magnetization with different orientation relative to the field direction.

3. Results and discussion

A phenomenological model of magnetic dynamics and corresponding relaxation Mössbauer spectra of the nanoparticles in a magnetic field taking into account the magnetic anisotropy, inhomogeneous precession and continuous diffusion of the uniform magnetization vector **M** was developed in [8]. In a magnetic field with the strength H less than critical value $H_c(\Theta)$ ($0.5H_C < H_c(\Theta) < H_C$, $H_C = 2K/M_0$, Θ is the angle between the **H** vector and the easiest magnetization axis, K is the constant of axial magnetic anisotropy, M_0 is the low temperature saturation magnetization), the vector **M** for the given energy

$$E = -K\cos^2\theta - \mathbf{HM} \tag{1}$$

(θ is the angle between the easiest magnetization axis and **M**) describes a conic surface around each of three poles corresponding to two energy local minima and absolute maximum for each nanoparticle. The precession trajectories forms are determined by the Hamiltonian (1). The average value of the magnetization $\overline{\mathbf{M}}(E)$ for each orbit is defined by elliptical integrals along the given trajectory as a function of characteristic frequencies of the vector **M** precession and diffusion as well as of the effective external field

$$h = H / H_C . \tag{2}$$

A three-level relaxation model (TLRM) with stochastic states determined by given average values of the magnetization $\overline{\mathbf{M}}_i(H,\Theta,T)$ and the equilibrium populations $\overline{W}_i(H,\Theta,T)$ for three ranges of the **M** stable rotation around the poles mentioned was developed as a first approximation [8]. Supposing that the vector **M** relaxation has a diffusion character, i.e., goes through infinitely small rotations, transitions between the states in the local energy minima (1) occur only through the state 3 corresponding to the absolute energy maximum. This approach allows one not only to perform a numerical analysis of the experimental spectra of a single domain particles ensemble in a magnetic field but also to draw at least qualitative conclusions concerning specific spectra shapes realization.

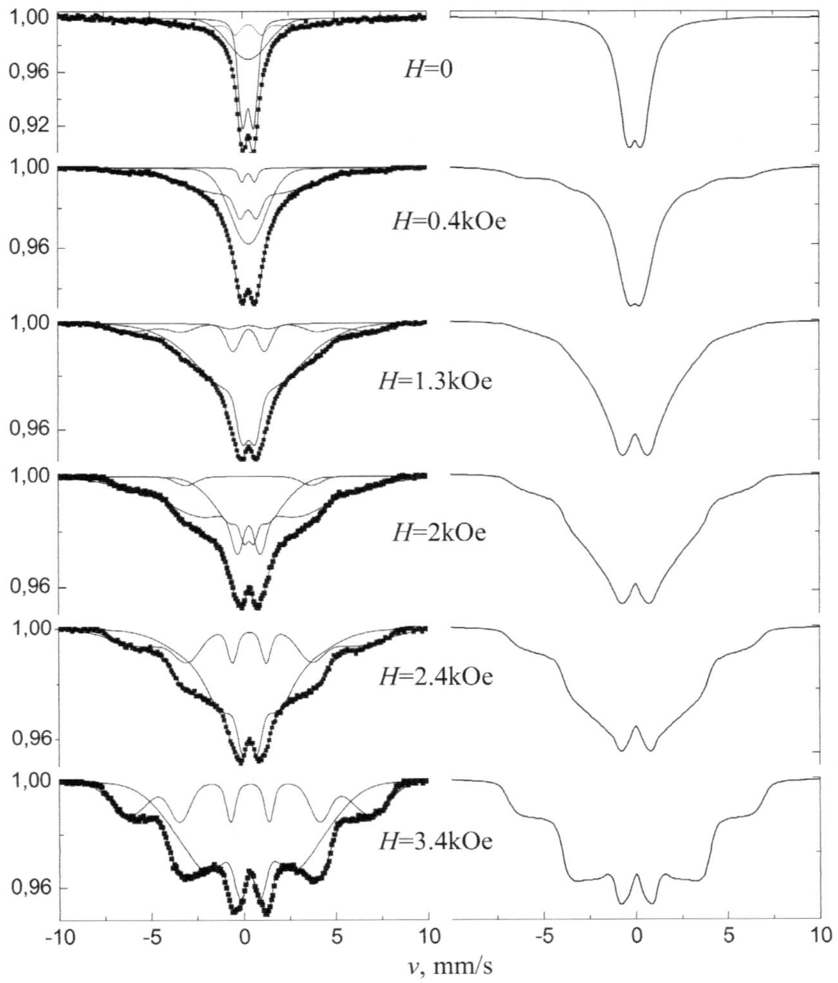

Figure 1. (Left panel) Room-temperature ^{57}Fe Mössbauer spectra (squares) of iron-oxide nanoparticles in a magnetic field H (the field direction is perpendicular to the gamma-beam). The resulting and partial spectra (solid lines) are calculated within the Gaussian hyperfine field distribution [7]. (Right panel) Mössbauer spectra of an ensemble of nanoparticles ($H_{\text{hf}}^{(0)}$ = 500 kOe, $K\overline{V}$ =500 K, $K\sigma_V$= 400 K) in a magnetic field, which calculated in the three-level relaxation model with p_0 = 5 GHz and $\overline{H}_{\text{C}}$ = 10 kOe for different values of σ_h = 0.03, 0.04, 0.13, 0.19, 0.23, 0.35 (from top to bottom).

Calculations of the Mössbauer spectra within the TLRM can be performed in terms of the stochastic approach [8]. For performing a simultaneous analysis of the spectra measured at different H values it is necessary to fulfil averaging of partial spectra both over particles dimension (V and/or KV) and over a parameter of interaction between particles. In our case we performed the averaging using Gaussian distribution of the particle volume and the effective magnetic field (2) with the corresponding widths σ_V and σ_h. The results of a qualitative analysis of all the series of the experimental spectra within the "physical" TLRM are shown in Figure 1 (right) where the calculations were performed with the fixed values of $K\overline{V} = 500K$, $K\sigma_V = 400$ K and $\overline{H}_C = 10$ kOe for different values σ_h. As one can see the shape of all the spectra measured at different strengths of the external field is described satisfactorily within the same set of physical parameters. Especially note a proper description of the "high-temperature doublet" at $H = 0$, which has purely "magnetic" origin within the model and is caused by the presence of an average molecular field $\overline{h} \neq 0$. Usually the presence of such a doublet is ascribed to the surface iron atoms contribution, the state of which differs from that of the inner atoms due to the electric field gradient acting on nuclei, which results in the excited nuclear level splitting due to a quadrupole interaction.

An analysis of large number of the specimen experimental spectra within the TLRM allows one to make self-consistent estimates of physical characteristics of the material studied. Using the value of the average particles diameter $<D> \approx 4$ nm known from the X-Ray diffraction study and found average anisotropy energy $K\overline{V} = 500K$, one can estimate the average anisotropy constant $K \approx 2 \cdot 10^5$ J/m^3 that in turn allows one to estimate the value of the average particles magnetization $M_0 \approx 0.5$ A/m. In our opinion, the estimated value of the critical magnetic field $\overline{H}_C$ is the substantially better and universal characteristic of the material studied in comparison with the coercive field value derived from conventional measurements of magnetization curves because the latter depends not only on the material properties but also on external conditions. As for the concrete values of the parameter σ_h characterizing the dispersion of inter-particle interactions they contain also information about the K distribution over the specimen. In order to extract this information one has to modify slightly the calculation scheme for the next stage of analysis. Here we did not perform a numerical fit of the experimental spectra within TLRM, which in principle permits to fulfil a quantitative analysis, because the procedure used needs a substantial optimization of numerical calculations.

4. Conclusions

The presented approach for simultaneous analysis of Mössbauer spectra of the nanoparticles in a weak magnetic field within the given model of magnetic dynamics of the system studied essentially expands the possibilities of Mössbauer spectroscopy for characterization of nanomagnetic materials. Subsequently we plan to develop this approach for simultaneous analysis of both Mössbauer spectra and magnetization curves within the same relaxation model.

Acknowledgments

This work is financially supported by the Russian Foundation for Basic Research (projects No. 08-02-00388, 09-02-01298 and 09-02-12195).

References

[1] Eibschütz M and Shtrikman S 1968 *J. Appl. Phys.* **39** 997
[2] Lindquist R H, Constabaris G, Kündig W and Portis A M 1968 *J. Appl. Phys.* **39** 1001
[3] McNab T K, Fox R A and Boyle A J F 1968 *J. Appl. Phys.* **39** 5703
[4] Mørup S, Bødker F, Hendriksen P V and Linderoth S 1995 *Phys. Rev.* B **52** 287
[5] Godovsky D Y, Varfolomeev A E, Efremova G D, Cherepanov V M, Kapustin G A, Volkov A V and Moskvina M A 1999 *Adv. Mat. for Optics and Electr.* **9** 87
[6] Irshinskii A L, Ozhogin V I, Cherepanov V M and Yakimov S S 1979 *Sov. Phys. JETP* **76** 563
[7] Chuev M A, Hupe O, Bremers H, Hesse J and Afanas'ev A M 2000 *Hyperfine Interact.* **126** 407
[8] Chuev M A 2008 *J. Phys.: Cond. Matt.* **20** 505201

Depth resolved hyperfine interactions with standing waves in $[W/Si]_{10}/Si/Ag/^{57}Fe/Ag/Si$ multilayer

A Gupta[1], D Kumar[1], M A Andreeva[2], A D Gribova[2] and R Rüffer[3]

[1] Indore Center, UGC-DAE Consortium for Scientific Research, Indore 452017, India
[2] Faculty of Physics, Lomonosov Moscow State University, 119991, Moscow, Russia
[3] ESRF, 6 rue Jules Horowitz, BP220, 38043 Grenoble Cedex, France

E-mail: Mandreeva1@yandex.ru

Abstract. Delayed nuclear resonant reflectivity (NRR) of $[W/Si]_{10}$ /Si(z nm)/Ag/^{57}Fe/Ag/Si multilayer as well as the time spectra at several grazing angles have been measured at the nuclear resonance beamline ID18/ID22N of the ESRF. NRR of ^{57}Fe layer has been enhanced by standing waves created by Bragg reflection from $[W/Si]_{10}$ backing and by multiple reflections from Ag adjacent layers. Fitting of the data was done with the program package REFTIM. The reconstructed HFI depth distributions show an essential asymmetry of the ^{57}Fe interfaces. The total ^{57}Fe density depth-profiles turn out to be in a good agreement with the measured Fe K_α fluorescence curve.

1. Introduction

Standing wave method has been successfully used for depth resolved investigations of different processes: atomic migration in multilayers, the interface structure in thin film nanostructures, magnetic structure of interfaces, position of a marker layer inside the waveguide, structure changes during post-growth annealing steps, the intermixing of layers induced by irradiation [1-8]. Resonant methods combined with standing wave technique enrich obtained information. Nuclear resonant reflectivity (NRR) also undergoes the standing wave influence [9,10]. An enhancement of NRR by the standing waves has been used for depth profiling of the hyperfine interactions (HFI) in multilayers [10-12] and also for investigation of ^{57}Fe monolayers or islands inside waveguides [13-14].

Here we report the results of the NRR and Fe K_α fluorescence measurements from a thin ^{57}Fe layer under the influence of standing waves created by the periodic multilayer backing $[W/Si]_{10}$ for the same sample. In such a way the influence of the standing waves on two different processes: incoherent secondary radiation emission and coherent NRR is compared.

2. The sample description

The samples [W(2.1 nm)/Si(3.1 nm)]$_{10}$ /Si(z nm)/Ag(1.9 nm)/^{57}Fe(4 nm)/Ag(2.2 nm)/Si(9.8 nm) were deposited on a float glass substrate using electron beam evaporation in an UHV environment with the base pressure in the chamber being $3x10^{-9}$ mbar. The buffer internal Si layer had a wedge profile and allows one to change the distance z of ^{57}Fe layer from $[W/Si]_{10}$ periodic structure. A standing wave amplitude in the thin ^{57}Fe layer is enhanced by the Bragg reflection from $[W/Si]_{10}$ backing and by the multiple reflections from the Ag adjacent layers (like that in waveguide). Here we investigate a sample

Journal of Physics: Conference Series **217** (2010) 012116 doi:10.1088/1742-6596/217/1/012116

having z=1.9 nm to study the standing wave influence on the NRR. The investigation was supported by angular dependence measurements of the Fe K_α fluorescence excited by Cu K_α X-rays.

3. Nuclear resonant reflectivity measurements

Usually X-ray standing waves are used to get a depth selectivity in secondary radiation (photoelectrons, fluorescence) excitation. The intensity of a secondary radiation $I_f(z)$ created at a depth z is proportional to the square module of the total field amplitude $E(z)$

$$I_f(z) \sim N_f(z)\,|E(z)|^2 = N_f(z)\,|E^+(z) + E^-(z)|^2 \tag{1}$$

($E^+(z)$ and $E^-(z)$ are the amplitudes of waves in the forward and backward directions). Positions of standing wave antinodes can be shifted by small variations of the angle, so the angular dependency of the fluorescence signal determines the depth profile of the density $N_f(z)$ of fluorescent atoms.

The influence of the standing waves on reflectivity has been also proved [9-11]. In the simplest case the reflectivity I_R from an ultrathin layer placed at a distance z atop a multilayer is given by

$$I_R = |r\,E(z)^2|^2\ , \tag{2}$$

where r is the reflectivity amplitude from the same layer placed in vacuum. It follows from (2) that the reflectivity from an ultrathin layer is enhanced by the fourth power of the standing wave amplitude $E(z)$. That takes place in general case also. So the influence of the standing waves on the reflectivity is even more effective than that on the secondary radiation emission. The problem how to extract the contribution of a selected layer to the total reflectivity can be solved by resonant techniques.

NRR is measured with synchrotron radiation in the time domain. Reflectivity amplitudes in the energy $R(\omega)$ and time $R(t)$ domains are connected by the Fourier transform. Integration of $|R(t)|^2$ over the delay time t for each grazing angle θ gives the delayed NRR curve $I_R(\theta)$.

This curve characterizes the depth position of the resonant nuclei. In experiment the integration limits are restricted by the interval T between synchrotron radiation sequent pulses (176.06 ns at 16 bunch mode of ESRF). Besides there is a "dead time" δ of a detector, overloaded by the huge prompt pulse. The calculations show that the exclusion even a small initial delay times 0-δ from the time integration leads to a considerable change of the delayed NRR curve [15]. In our experiment δ=10.17 ns, T=161.4 ns. The tail of a previous pulse has been also included in calculations.

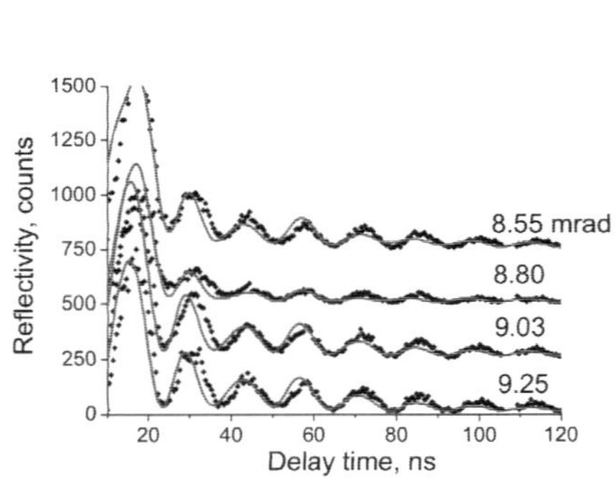

Figure 1. Time spectra of NRR. Symbols are experimental data, lines are the fits. Curves are vertically shifted.

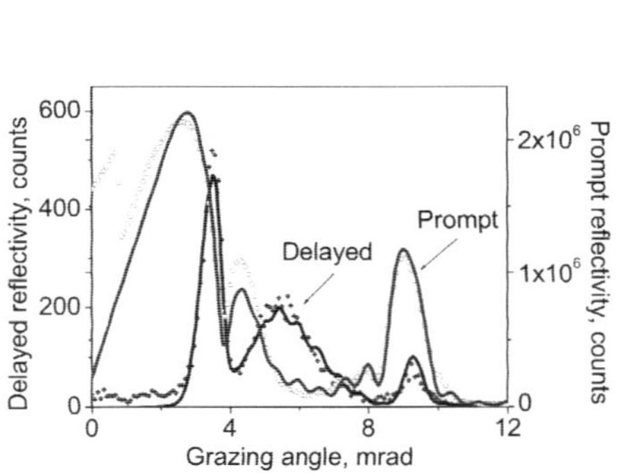

Figure 2. Prompt and delayed reflectivity for our sample. Symbols are experimental data, lines are the fits.

Measurements were done at the beamline ID22N/ID18 of the ESRF. The time spectra of NRR were measured at several angles in vicinity of the Bragg peak. Results are shown in figures 1-2. Notice that the maxima on the delayed curve at ~3.5 and ~5.4 mrad coincide with minima on the prompt reflectivity curve. This can be explained by the enhancement of the radiation field inside the sample at the resonant layer position. The standing waves created by the prompt Bragg reflection from $[W/Si]_{10}$ backing produce the delayed peak at ~9.24 mrad. The shift ~0.23 mrad between the delayed "Bragg" peak and prompt Bragg peak is seen. It is obvious that this shift is determined by the relative position of the ^{57}Fe layer and of the standing wave maximum at different angular shifts.

4. Fit results and discussion

Fit of the delayed NRR and prompt reflectivity curves as well as the time spectra of reflectivity, measured at several angles near the Bragg angle, have been performed with the program package REFTIM [16]. The initial step of the calculations includes the fitting of the CEM spectrum measured for that sample (figure 3 a). HFI parameters obtained by this fit were kept for the fitting of the time spectra. The fit of the prompt reflectivity curve (figure 2) gives the exact period of the $[W/Si]_{10}$ backing and depth profiles for the real and imaginary parts of the susceptibility in the whole sample (figure 4).

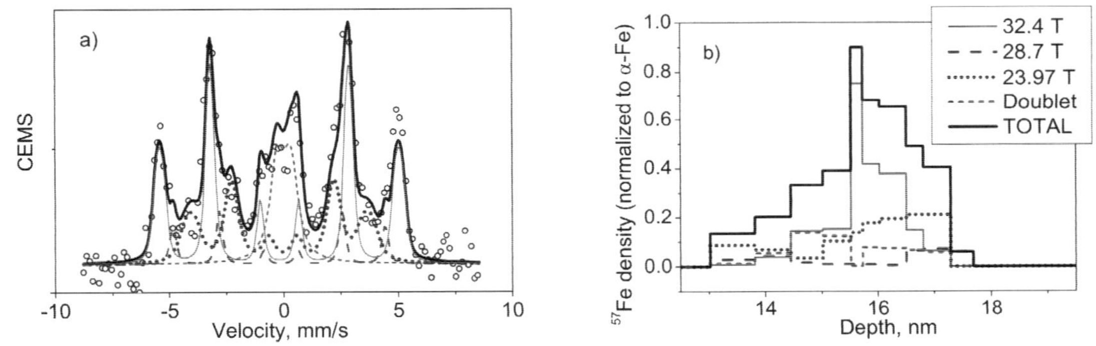

Figure 3. CEMS and multiplet contributions (a) and the depth distribution for these contributions (b).

The main parameters used for the fit of the time spectra were the depth distributions of different HFI in ^{57}Fe layer. The time spectra measured at different angles are not essentially different "by eye" (figure 1), however the reasonable correspondence of the calculated and measured spectra can be achieved just at the very special (but unfortunately not unique) distributions of different HFI across the ^{57}Fe layer depth. The obtained profile of the ^{57}Fe total density and profiles of the partial contributions from different HFI are presented in figure 3 b. The mixture of different HFI in each artificially selected sublayer testifies that ^{57}Fe layer is not homogeneous. Magnetic hyperfine fields have smaller values in the interfaces. A doublet contribution predominates in sublayers with lower ^{57}Fe density. The figures 3 b and 4 show that the real structure of the sample differs from the nominal one. The most interesting result of the fit is the revealed asymmetry of interfaces of ^{57}Fe layer (including the difference of HFI in the two interfaces) and a visible destruction of the top Ag layer and the top bilayer of $[W/Si]_{10}$ backing.

The obtained model of the electronic structure and *z*-position and depth profile of the ^{57}Fe density were used (with proper renormalization of the absolute values of sublayer susceptibilities accordingly to the change of the radiation wavelength) for calculations of the reflectivity and Fe K_α fluorescence curve, measured with Cu K_α radiation (figure 5). We immediately get surprisingly good agreement with experimental data. So our depth profiling of the ^{57}Fe density performed by the analysis of the NRR data were confirmed.

In conclusion, we should emphasize that the NRR supplies us with more detailed information about ^{57}Fe layer than a fluorescence curve. Namely the fit of the NRR delayed angular curve and the time spectra measured at different grazing angles makes possible to follow the depth distribution of different HFI and hence different chemical states of iron across ^{57}Fe layer (figure 3 b).

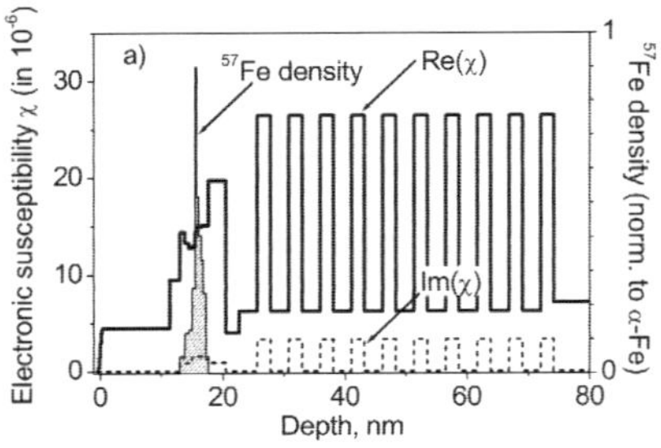

Figure 4. Depth profiles of the electronic density and absorption coefficient. Filled curve gives the position and profile of ^{57}Fe density in the structure.

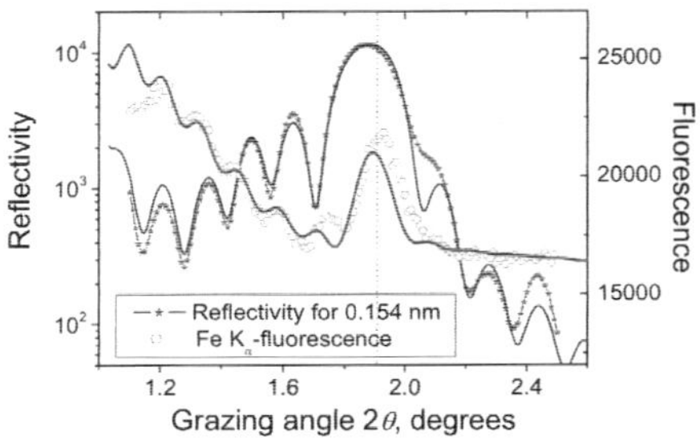

Figure 5. Reflectivity and Fe K_α fluorescence curve measured with Cu K_α radiation. Theoretical curves (solid lines) are calculated with the electronic and Fe density depth-profiles (figure 4), obtained by the fit of the NRR.

Acknowledgments
The work is supported by the Indo-French Centre for Promotion of Advanced Research and by RFBR grants No. 09-02-01293 and 09-02-12207-ofi-m.

References

[1] Ghose S K, Dev B N and Gupta A 2001 *Phys. Rev.* B **64** 233403
[2] Kim S-K and Kortright J B 2001 *Phys. Rev. Lett.* **86** 1347
[3] Gupta A, Rajput P, Saraiya A, Reddy V R, Gupta M, Bernstorff S and Amenitsch H 2005 *Phys. Rev.* B **72** 075436
[4] Holý V, Matěj Z, Pacherová O, Novák V, Cukr M, Olejník K and Jungwirth T 2006 *Phys. Rev.* B **74** 245205
[5] Bera S, Bhattacharjee K, Kuri G and Dev B N 2007 *Phys. Rev. Lett.* **98** 196103
[6] Gupta A, Kumar D and Meneghini C 2007 *Phys. Rev.* B **75** 064424
[7] Gupta A, Rajput P and Meneghini C 2007 *Phys. Rev.* B **76** 195401
[8] Andreeva M A, Odintsova E E, Semenov V G, Irkaev S M and Panchuk V V 2008 *J. of Surface Investigation. X-ray, Synchr. and Neutron Techn.* **2** 564
[9] Andreeva M A and Lindgren B 2002 *JETP Letters* **76** 704
[10] Andreeva M A 2004 *Hyperfine Interactions* **156-157** 595
[11] Andreeva M A and Lindgren B 2005 *Phys. Rev.* B **72** 25422
[12] Andreeva M A, Monina N G, Häggström L, Lindgren B, Kalska B, Kamali-M S, Vdovichev S N , Salashchenko N N, Semenov V G, Leupold O and Rűffer R 2008 *Nuclear Instruments and Methods in Phys. Res.* B **266** 187
[13] Röhlsberger R, Thomas H, Schlage K, Burkel E, Leupold O and Rüffer R, 2002 *Phys. Rev. Lett.* **89** 237201
[14] Röhlsberger R, Bansmann J, Senz V, Jonas K L, Bettac A, Meiwes-Broer K H and Leupold O 2003 *Phys. Rev.* B **67** 245412
[15] Andreeva M A, Monina N G and Stankov S 2008 *Moscow Univer. Physics Bulletin* **63**(2) 132
[16] http://www.esrf.eu/computing/scientific/REFTIM/MAIN.htm

Effect of Iron on Vanadium (001) Strained Surface Magnetism

M Elzain[1], M Al-Barwani, A Gismelseed, A Al-Rawas, A Yousif, H Widatallah, K Bouziane and I Al-Omari

Department of Physics, College Of Science, Box 36, Sultan Qaboos University, Al Khod 123, Oman

E-mail: elzain@squ.edu.om

Abstract. The magnetism of the vanadium (001) surface has been a controversial subject on both theoretical and experiment fronts. Both strongly ferromagnetic and paramagnetic phases were reported. We have used the first principle full-potential linearized-augmented plane waves (FP-LAPW) as implemented in WIEN2k package to study the magnetic properties of strained surfaces of vanadium films as a function of film thickness. We found that for films thicker than about 11 monolayers, the magnetism of the strained surfaces converge to a constant value of about $0.15\mu_B$. Introduction of Fe monolayers and impurities at the centre of the films affects the magnetic structure of thin films but has no influence on the surface magnetism of thicker films. For Fe monolayers positioned at the centre of thick films, the Fe atoms maintain magnetic moment of order $0.86\mu_B$, a quadruple splitting of order -0.3 mm/s and a small negative isomer shift, while an Fe impurity has vanishing hyperfine fields and magnetic moment. In addition we have varied the location of the Fe monolayer and impurity within the V films and found that their position affects the surface magnetism.

1. Introduction

Vanadium is a paramagnetic metal with a large magnetic susceptibility that lies at the border just below the magnetic 3d transition metals. Vanadium atoms are easily polarized when introduced into a magnetic material. On expansion, it was found that vanadium metal undergoes a transition from the paramagnetic to antiferromagnetic phase at a Wigner-Seitz radius of about 3.15 *a.u.* [1]. On the other hand, the magnetism of transition metals also depends on the coordination number. Reduction in atomic bonding as happens at surfaces is expected to lead to formation of atomic magnetic moments. Thus vanadium has the potential to exhibit surface magnetism.

Rau *et al* using electron-capture spectroscopy reported ferromagnetic order of V atoms at the $V(001)p(1\times1)$ surface for bulk vanadium [2] and for ultra thin vanadium films deposited on silver [3]. However, later measurements using other techniques could not detect any magnetic order to within $0.05\mu_B$ per V atom neither for the surface of bulk vanadium nor for the surface of thin films on silver (see reference 4 and references therein).

The electronic and magnetic structures of the surface vanadium thin films have been calculated using different computational techniques with contradictory results. In particular, a big discrepancy exists between the results obtained using the all-electron methods and the pseudopotential methods

[1] Corresponding author

[5,6]. Bihlmayer *et al* [7] using the full-potential all electron method in the linearized-augmented plane-wave approach (FP-LAPW) concluded that in ultra thin V films a surface magnetic moment can be stabilized, while for thicker and relaxed films no surface magnetism is obtained. On the other hand, Bryk *et al* [8] using the pseudopotential methods obtained magnetic moments of order $1.5\mu_B$ even for relaxed surfaces.

At the interface between vanadium and the 3d magnetic transition metals, V atoms are found to couple antiferomagnetically to the neighboring 3d atoms. In particular, for V-Fe (001) interface an antiferomagnetically coupled V moment was obtained theoretically [9] and reported experimentally [10]. Thin overlayers are completely polarized; however for thick overlayers polarization is confined to the interface region.

In this contribution we use the first principle FP-LAPW method to study the magnetism of 3 to 27 layers of vanadium (001) surface with and without introduction of iron impurities and monolayers at various depths in the vanadium films. The magnetic moment, the contact magnetic hyperfine field and the electric field gradient of Fe impurities were found to vanish except when the impurity is close to the surface of the V film. A negative small isomer shift is retained at all positions. The effect of deep Fe impurities on the V surface magnetic moment is minor. For an Fe monolayer a local Fe magnetic moment of order $0.9\mu_B$, a magnetic hyperfine field of order 2 T and an electric field gradient of order $-2\times10^{21}\,\mathrm{V/m^2}$ are retained except close to the surface. The computational method used is presented in the following section and results are presented and discussed in section 3.

2. Computational Method

The density functional L/APW+lo method as implemented in WIEN2k package [11] is used to calculate the electronic and magnetic structures of vanadium films with and without iron. Systems with a Fe monolayer in V are represented by $1\times1\times(N+N_V)$ supercells, where N is the number of monolayers and N_V is the width of vacuum, while Fe impurities in V are represented by $2\times2\times(N+N_V)$ supercells.

In the L/APW+lo method the Kohn-Sham orbitals are expanded in terms of atomic orbitals inside the atomic muffin-tin (MT) sphere of radius R_{MT} and in terms of plane waves in the interstitial regions. The Kohn-Sham equations were solved using Perdew-Burke-Ernerhof GGA approximation. Core and valence states were separated by an atomic energy of -6.0 Ry. In this case V 3s orbitals belong to the valence states, while Fe 3s orbitals belong to the core states. It is found that when the cut off energy is increased to -7.0 Ry to include the Fe 3s orbitals among the valence states, no major change results. The calculations were carried out using the bulk V lattice constant of 5.66 *a.u.*as determined through volume optimization, a value that agrees with reported results [6]. Touching MT spheres having radii of 2.44 *a.u* are used. For the valence electrons the potential and charge density are expanded in spherical harmonics up to $L = 4$, while the wavefunctions are expanded inside the MT sphere up to $l = 10$ partial waves. Mixed basis set are used depending on the partial wave *l*, with APW+lo being used for *s*, *p* and *d* valence orbitals, whereas LAPW is used for the remaining *l* values. For the plane wave expansion in the interstitial region we have used a wavenumber cut off $K_{\max}$ such that $R_{MT}K_{\max} = 7$ and the potential and charge density are Fourier expanded up to $G_{\max} = 12$. The mesh size and the number of points in the Brillouin zone were tested for small supercells using the convergence of the values of the electric field gradient and the total energy as indicators [12]. It is found that a *k* sampling with a 30×30×1 Monkhost-Pack mesh gives good results in case of Fe monolayers and a mesh of 15×15×1 in case of Fe impurities. An equivalent of 8 monolayers of vacuum size (about 12 Å) is determined to give a well convergied work function, which is taken as a measure of the quality of vacuum. The results reported here are for strained surfaces using the calculated V lattice constant.

3. Results and discussions

The calculated local properties at a Fe site in a monolayer of Fe at the centre of strained V films as function of the number of layers N, in addition to the surface magnetic moment of V in films with and without Fe are shown in table 1. Results for small number of layers oscillate strongly with the number of layers and are not shown here. For the lattice constant used in this calculation, the Fe monolayer is expanded by about 4% relative to the corresponding α-Fe configuration. We note in general that the hyperfine field parameters at the Fe site saturate very quickly to constant values as the number of layers increases. In addition the V surface magnetic moment approaches the corresponding values of Fe-free films, which are comparable to the results of the 15 layer strained film reported in [7]. The V layers immediately neighboring the Fe monolayer was found to gain sizeable negative moment (of order $-0.25\mu_B$), while further layers are weakly polarized with negligible local moment. As the thickness of the V film increases above about 15 layers, the V surface does not feel the presence of the deep Fe monolayer. The Fe local magnetic moment within the MT sphere converges to about 0.86 μ_B. The axial electric field gradient at the Fe site leads to quadruple splitting of order -0.3 mm/s, while the contact charge density results in isomer shifts less than about -0.1 mm/s.

Table 1. The atomic magnetic moments μ (in Bohr Magneton), the contact hyperfine fields B_{hf} (in Tesla), the axial electric field gradients V_{zz} (in volts per meter square) and the contact charge density ρ (in atomic units) at the Fe site in an iron monolayer at the centre of strained V film with total number of layers N. The last two columns give the V magnetic moment per atom at the surface of the thin film with and without iron monolayer at the centre.

N	μ_{Fe} (μ_B)	B_{hf} (T)	$V_{zz}\times10^{21}$(V/m^2)	ρ (*a.u*)	μ_V with Fe	μ_V without Fe
7	0.96	-1.9	-1.8	7.13	-0.14	0.04
11	0.90	-2.1	-2.6	7.17	0.01	0.15
15	0.83	-1.6	-2.1	7.22	0.10	0.14
23	0.85	-1.7	-2.1	7.21	0.13	0.15
25	0.86	-1.7	-2.1	7.20	0.14	0.15
27	0.86	-1.7	-2.1	7.21	0.15	0.15

The introduction of a Fe impurity at the centre of the strained V films has similar effects on the magnetic structure of V surface. However, the local Fe magnetic moment, hyperfine field and electric field gradient vanish in this case as happens for a Fe impurity in bulk vanadium. For the supercell and the P4/mmm (123) space group used for the Fe impurity representation, the V surface has three independent atomic sites in some films. The magnetic moments of the three V atoms at the surface are found to couple ferromagnetically for the cases of Fe impurity at the centre of the V film. The values of their magnetic moments are different but remain close to those reported for Fe monolayer in table 1 above.

We have also varied the location of the Fe monolayer and impurity within the strained V film for systems with 7, 11 and 23 layers. Table 2 shows the results for a Fe monolayer at different location in the 23 layers film. Only results for locations at the five positions below the surface are shown. Results for deeper position are similar to those of a Fe monolayer at the centre and hence are not presented here. The V surface magnetic moment is observed to vary with the location of the Fe monolayer reflecting oscillatory trends. It is in general coupled ferromagnetically to Fe even when the Fe monolayer is directly below the surface. It worth noting the respective decrease (increase) of the magnitudes of the electric field gradient (magnetic hyperfine field) at the Fe site as the Fe monolayer approaches the surface. Similar trends were found in the 7 and 11 layers thick films. Also the Fe impurity reflected the same features culminating in a large and positive electric field gradient at the surface. A Fe impurity at one V surface produces antiferromagnetic structures on the same surface and also on the opposite Fe-free surface. The magnetic moment of the Fe atom at the surface is large (of order 2.7μ_B). The magnetic hyperfine field of an Fe atom at the surface is of order -10 T, which is relatively large as compared to that of the interior positions. The electric field gradient is also large and positive.

Table 2. The atomic magnetic moments μ (in Bohr Magneton), the contact hyperfine fields B_{hf} (in Tesla), the axial electric field gradients V_{zz} (volts per meter square) and the contact charge density ρ (in atomic units) at the Fe site in an iron monolayer at position S-I of strained V film with 23 layers. The last column gives the V magnetic moment per atom at the corresponding the surface of V film.

Depth below surface	μ_{Fe} (μ_B)	B_{hf} (T)	$V_{zz}\times10^{21}$(V/m^2)	ρ (*a.u*)	μ_V with Fe
S-5	0.88	-1.70	-2.08	7.21	0.26
S-4	0.82	-1.50	-2.13	7.20	0.03
S-3	0.90	-1.80	-1.94	7.20	0.03
S-2	1.26	-1.80	-1.70	7.24	1.01
S-1	1.05	-3.60	-1.52	7.22	0.22

We have made a limited study of the effect of surface relaxation on the magnetism of V films with and without Fe. For example for the 23 layers thick film a reduction of the top surface layer inter-planer distance by 10% leads to zero surface magnetic moment of V with or without centrally positioned Fe.

In summary, we have calculated the local magnetic moment, the contact magnetic hyperfine field, the electric field gradient and the contact charge density at the Fe site in an iron monolayer and Fe impurity in vanadium films as a function of film thickness and also as a function of location within the films. The influence of the Fe on the magnetic structure of the V surface was determined. It is found that neither a centrally positioned Fe monolayer nor Fe impurity has any effect on the V surface for thick films. However, Fe monolayers and Fe impurities at locations closer to the surface result in significant changes in the V surface magnetic structure with possibility of induction of antiferromagnetic configuration.

References

[1] Moruzzi V L and Marcus P M 1990 *Phys. Rev.* B **42** 8361
[2] Rau C , Liu C, Schmalzbauer and Xing G 1986 *Phys. Rev. Lett.* **57** 2311
[3] Rau C, Xing G and Robert M 1988 *J. Vac. Sci. Technol.* A **6** 579
[4] Lacina D, Yang J and Erskine J L 2007 *Phys. Rev.* B **75** 195423
[5] Robles R, Izquierdo J, Vega A and Balbas C 2001 *Phys. Rev.* B **63** 172406
[6] Batyrev I G, Cho J H and Kleinman L 2001 *Phys. Rev.* B **63** 172420
[7] Bihlmayer G, Asada T and Blugel S 2000 *Phys. Rev.* B **62** R11937
[8] Bryk T, Bylander D M and Kleinman L 2000 *Phys. Rev.* B **62** R3780
[9] Izquierdo J, Vega A, Elmouhssine O, Dreysse H and Demangeat C 1999 *Phys. Rev.* B **59** 14510; Spisak D and Hafner J 2000 *Phys. Rev.* B **61** 4160
[10] Finazzi M, Yubero F, Bencok P, Chevrier F, Hricovini, Ciccacci F and Krill G 1997 *J. Magn. Magn. Mater.* **165** 78; Fuchs P, Totland K and Landolt M 1996 *Phys. Rev.* B **53** 9123
[11] Blaha P, Schwarz K, Madsen G K H, Kvasnicka D and Luitz J 2001 WIEN2k, an augmented plane wave + local orbitals program for calculating crystal properties (Karlheinz Schwarz, Technische Universitat Wien, Austria)
[12] Cottenier S. 2002, Density Functional Theory and the family of (L)APW-methods; a step-by-step introduction (Institute voor Kern-en Stralingsfysica, K. U. Leuven, Belgium), ISBN 90-807215-1-4 (to be found at http://www.wien2k.at/reg_user/textbooks)

^{57}Fe implantation effect of Sb doped SnO_2 films

K Nomura[1], Z Németh[1,2] and H Reuther[3]

1 School of Engineering, the University of Tokyo, 7-3-1Hongo, Bunkyo-ku, Tokyo,113-8656 Japan

2 Institute of Chemistry, Eötvös Loránd University, Hungary

3 Forschungszentrum Dresden-Rossendorf e. V, Germany

k-nomura@t-adm.t.u-tokyo.ac.jp

Abstract. We implanted ^{57}Fe with $5x10^{16}$ ions/cm^2 into SnO_2 films containing 0.1% Sb and 3% Sb at the substrate temperature of 500°C in vacuum. As a result, four kinds of subspectra were observed in the depth selective CEM spectra by using a back scattered type of gas counter. Two doublets are assigned to paramagnetic Fe^{3+} and Fe^{2+} species and two sextets are assigned to site A and site B of magnetite. Magnetite formation was found to prefer the top layer, whereas paramagnetic Fe^{2+} species were included in the deeper layer of the films. SnO_2 (with 0.1% and 3% Sb) films doped with $5x10^{16}$ ^{57}Fe ions/cm^2 at 500°C and post-annealed at 400°C for 6 h showed bulk ferromagnetism at room temperature although the ^{57}Fe implanted SnO_2 (3% Sb) film showed smaller Kerr effect than the ^{57}Fe implanted SnO_2 (0.1% Sb) film. This phenomenon was attributed mainly to the amount of magnetite in the as prepared samples and to the maghemite in the post-annealed samples, respectively, in addition to magnetic defects.

1. Introduction

The origin of magnetic interactions in diluted magnetic semiconductors (DMS) is attracting a great attention as a basic problem in magnetism and as an application of spintronics [1-2]. We have reported different types of magnetic source in case of Fe doped SnO_2 powders, prepared by sol-gel method [3], and also the phonon density of states of rutile type structures of SnO_2 and TiO_2 [4]. It is clear that defects in DMS can also contribute to saturation magnetization. On the other hand, thin films of $Sn_{1-x}{}^{57}Fe_xO_{2-\delta}$ have been implanted at room temperature with $1x10^{17}$ Fe ions/cm^2, as well as at 300°C with $5x10^{16}$ and $1x10^{17}$ Fe ions/cm^2 [5]. The as implanted samples at room temperature and post-annealed samples did not show any Kerr effect, but the sample implanted with $1x10^{17}$ Fe ions/cm^2, while heated at 300°C, showed Kerr effect although the magnetic sextets were not clearly observed in the ^{57}Fe conversion electron Mössbauer (CEM) spectra. Kerr effect disappeared after the post-annealing. This suggests that the number of magnetic defects decreases by absorption of oxygen due to annealing in air atmosphere [5]. We have also showed that the bulk magnetization is enhanced by introducing Sb in the Fe doped SnO_2 powder [6]. This time we tried to implant the Sb doped SnO_2 films with ^{57}Fe at high substrate temperature. It was found that the films show Kerr effect, and we have also characterized the surface products on the films by conversion electron Mössbauer spectroscopy (CEMS) using a gas counter.

2. Experimental

SnO_2 films containing 0.1% Sb and 3% Sb with thickness of 200nm were prepared on quartz glass by using DC sputtering in O_2 flow and were implanted with $5x10^{16}$ ^{57}Fe ions/cm^2 at the substrate temperature of 500°C in vacuum, using the acceleration energy of 100 keV. From TRIM calculations of implantation conditions of $5x10^{16}$ Fe ions/cm^2, the iron profile peak is expected to be located at about 40 nm depth with a maximum Fe concentration of 5 at %. Some samples were post-annealed at 400°C. Polar Kerr effect of these samples was measured with magnetic circular dichroism (MCD) mode. Three CEM spectra were simultaneously observed on each sample from different depths by discriminating the resonance electrons with three energy regions (2-6.5keV, 6.5-11keV, and 11-20 keV) using homemade Mössbauer system and He + 5% CH_4 gas counter [7,8]. This method provides the rough depth profile of layers of interest. Doppler velocity was calibrated with standard α-Fe foil at room temperature and a γ source of ^{57}Co/Cr matrix was used.

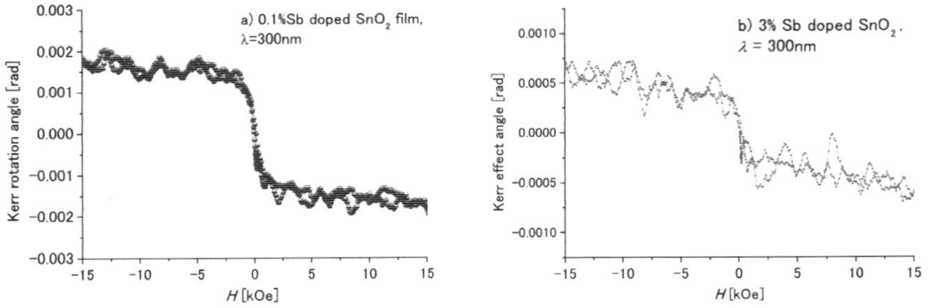

Figure 1. Polar Kerr rotation curves of (a) SnO_2 (0.1%Sb) film and (b) SnO_2 (3%Sb) film, implanted with ^{57}Fe at substrate temperature of 500°C, measured by MCD mode at λ=300nm.

3. Results

To obtain the bulk ferromagnetism, Kerr rotation curves were measured by MCD mode at λ=300 nm for SnO_2 (0.1%Sb) film and (b) SnO_2 (3%Sb) film, implanted with ^{57}Fe at substrate temperature of 500°C. As shown in Figure 1, it is clear that Kerr effect of both samples was observed, and the Kerr rotation angles for 0.1% Sb doped SnO_2 films were larger than that for 3%Sb doped SnO_2 films.

CEM spectra of these samples are shown in Figure 2. CEM spectra obtained by detecting emitted electrons with the high energy regions reflect the top layer. These spectra were decomposed to 4 subspectra of two doublets and two sextets. The Mössbauer parameters of the sample with 0.1% Sb are listed in Table 1. The doublets with small isomer shift (δ= 0.39 mm/s) and large isomer shift (δ= 0.94 mm/s) are assigned to paramagnetic Fe^{3+} and Fe^{2+} species, respectively. Two broad sextets with δ = 0.31 mm/s and δ = 0.62 mm/s can be assigned to Fe^{3+} at site A and $Fe^{2.5+}$ at site B of magnetite because the area ratio of the two sextets is almost 1:2. The broadening of these peaks is considered to be due to the fine grain size of the particles. Only the relative area of the components in CEM spectra were found to differ between the SnO_2 films with 0.1% Sb and 3% Sb. In the case of 0.1 % Sb doped SnO_2 film, the relative area of the ferromagnetic magnetite phase was larger than in 3% Sb doped SnO_2 films. This result is in accordance with the larger Kerr effect of the former sample.

As shown in Table 1, the subspectra area intensities of magnetite were 63.6%, 48.9% and 34.7% in high energy, middle energy, and low energy CEM spectra of 0.1%Sb doped SnO_2 film, respectively, whereas the subspectra area intensities of paramagnetic Fe^{2+} were 18.5%, 28.9% and 40%, respectively. These suggest that magnetite layers are produced rather in upper layers of implanted regions, and that paramagnetic species prefer the deep layers of implanted regions of SnO_2 films.

Kerr effect of these films annealed at 400°C for 6 hours showed the same hysteresis as before the annealing treatment. In the case of implantation at 300°C, Kerr effect was observed, but disappeared by post annealing [5]. However, Kerr effect was found to remain in the present case even after annealing. The CEM spectra measured after the heat treatment are shown in Figure 3. The magnetite phase with 2 sextets, found in the case of the as prepared samples, changed into one broad sextet with small tailing towards lower velocities in both samples. After annealing, the magnetic subspectra can be fitted with a sextet showing a small distribution of magnetic hyperfine fields. The isomer shift and quadrupole splitting of the sextet show roughly 0.31-0.33 mm/s and 0 mm/s values, respectively, while the maximum of the hyperfine fields is around 47.5 T. Therefore, the magnetic subspectra of the post annealed samples are considered to be due to one component of fine γ-Fe_2O_3. The doublet intensity of Fe^{2+} decreased more in top layer than in deep layer after annealing.

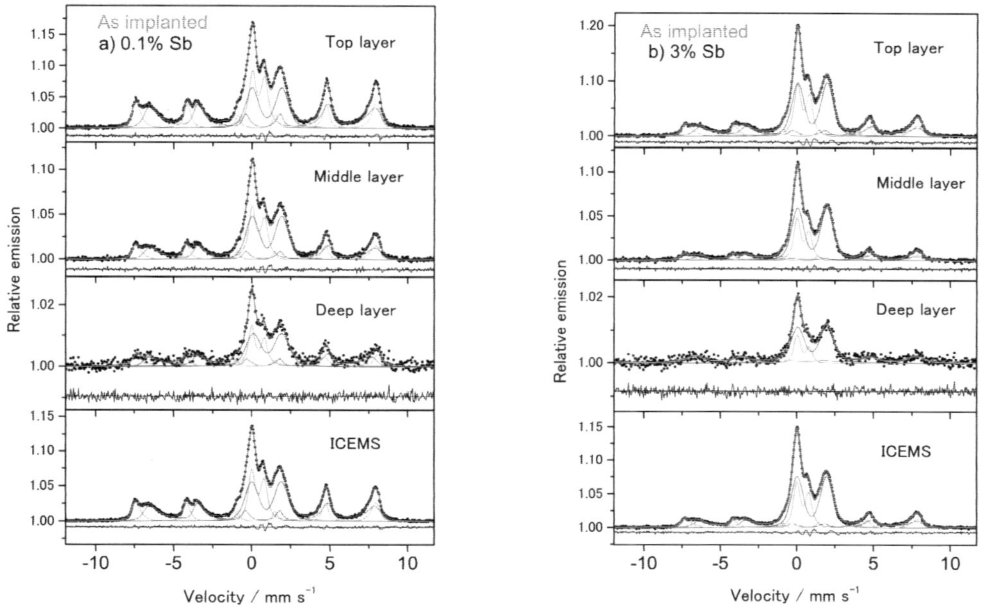

Figure 2. Depth selective CEM spectra showing the change of the quantity of magnetite in different depths of SnO_2 films with 0.1% Sb (left) and 3% Sb (right), implanted with $5x10^{16}$ Fe ions at a substrate temperature of 500°C. ICEMS indicates the integrated CEM spectra.

Table 1. Mössbauer parameters of 0.1% Sb doped SnO_2 implanted with $5x10^{16}$ Fe ions/cm^2.

Doublets				Sextets				
Area Int. (%)	δ (mm/s)	Δ (mm/s)	Γ (mm/s)	Area Int. (%)	δ (mm/s)	B_{hf} (T)	2ε (mm/s)	Γ (mm/s)
Top layer								
17.9 %	0.39 (3)	0.75 (4)	0.59 (3)	22.4 %	0.32 (1)	47.6 (4)	-0.01 (1)	0.46 (2)
18.5 %	0.85 (2)	1.91 (4)	0.82 (5)	41.2 %	0.60 (1)	43.5(1)	-0.09 (2)	0.97 (3)
Middle layer								
22.3 %	0.39 (1)	0.76 (2)	0.59 (2)	16.1 %	0.31 (1)	47.6(6)	0.01 (2)	0.46 (3)
28.9 %	0.93 (1)	1.93 (2)	0.79 (2)	32.7 %	0.63 (1)	43.7(1)	-0.09 (2)	1.04 (4)
Deep layer								
25.3 %	0.39 (3)	0.82 (6)	0.62 (6)	10.3 %	0.31 (3)	48.5 (3)	0.03 (8)	0.37 (14)
40.0 %	0.99 (3)	1.97 (5)	0.72 (5)	24.4 %	0.64 (4)	43.9 (4)	0.07 (8)	0.78 (14)
ICEMS								
21.7 %	0.39 (1)	0.77 (2)	0.61 (2)	16.9 %	0.31 (1)	47.7 (1)	-0.00 (1)	0.46 (2)
27.7 %	0.94 (1)	1.92 (2)	0.78 (2)	33.8 %	0.62 (1)	43.7 (1)	-0.07 (2)	0.99 (3)

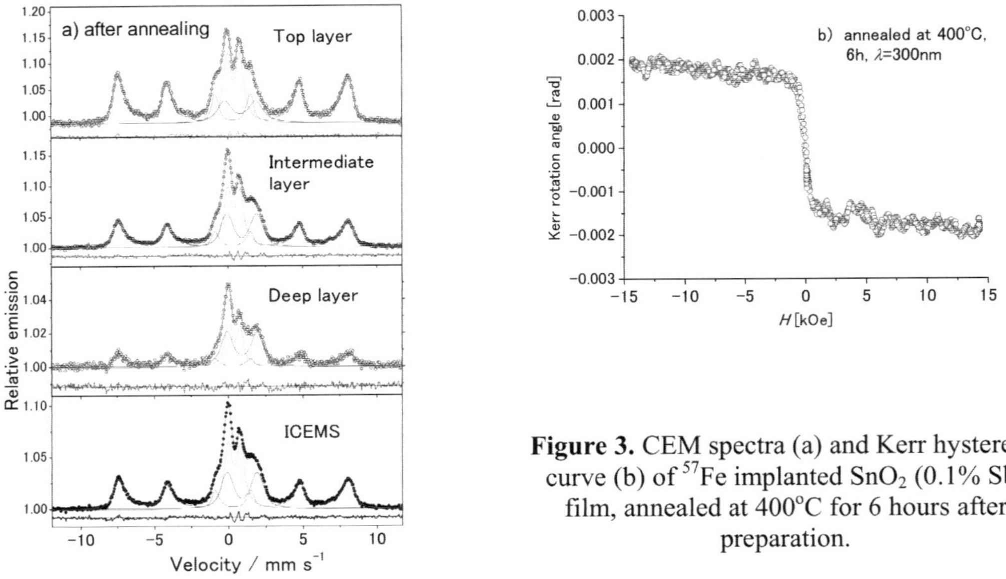

Figure 3. CEM spectra (a) and Kerr hysteresis curve (b) of ^{57}Fe implanted SnO_2 (0.1% Sb) film, annealed at 400°C for 6 hours after preparation.

4. Conclusions

SnO_2 (with 0.1% and 3%Sb) films doped with $5x10^{16}$ ^{57}Fe ions/cm^2 at 500°C and post-annealed at 400°C for 6 h showed bulk ferromagnetism at room temperature. This phenomenon was attributed mainly due to formation of ferromagnetic magnetite for as prepared samples and of maghemite for the post-annealed samples, respectively, rather than magnetic defects. Including more Sb in SnO_2 film was found to decrease the amount of magnetite produced. It is further found that on the top layers of the implanted region, magnetic components were apt to produce at the substrate temperature of 500°C and in high vacuum.

Acknowledgement

Authors thank Dr. K. Yamamoto, Asahi Glass Co. Institute, for preparation of Sb doped SnO_2 films, Mr. S. Iio for measurements of Mössbauer spectra and Dr. Y. Hirose, Hasegawa Laboratory for Kerr effect, respectively.

References

[1] Coey J M D,Venkatesan M and Fitzgerald C B 2005 *Nat. Mater.* **4** 173
[2] Punnoose A, Hays J, Thurber A, Engelhard M H, Kukkadapu R H, Wang C, Shutthanandan V and Thevhuthasan S 2005 *Phys. Rev. B* **72** 054402
[3] Nomura K, Barrero C A, Sakuma J and Takeda M 2007 *Phys. Rev. B* **75** 184411
[4] Rykov A I, Nomura K, Sakuma J, Barrero C, Yoda Y and Mitsui T 2008 *Phys. Rev. B* **77** 014302
[5] Nomura K and Reuther H 2009 *Hyperfine Interact.* **191** 159
[6] Nomura K, Barrero C A, Kuwano K, Yamada Y, Saito T and Kuzmann E 2009 *Hyperfine Interact.* **191** 25
[7] Nomura K, Okubo T and Nakazawa M 2004 *Spectrochim. Acta B* **59** 1259
[8] Nomura K, Iio S, Ujihira Y and Terai T 2005, Industrial Applications of the Mossbauer Effect, edited by Gracia M, Marco J F, and Plazaola F, *AIP, CP***765**, 108.

Spin fluctuations in the magnetic phase of frustrated pyrochlores: evidence and controversy

P. Bonville

CEA, Centre de Saclay, DSM/IRAMIS/Service de Physique de l'Etat Condensé, 91191 Gif-sur-Yvette, France

E-mail: pierre.bonville@cea.fr

Abstract. We give here the experimental evidences for the presence of spin fluctuations in the magnetically (long range) ordered phase of some frustrated rare earth based pyrochlore systems, together with recent data that seem to contradict this picture.

1. Introduction

Geometrical frustration of exchange interactions occurs in special lattices where the geometry prevents all pairwise couplings to be satisfied altogether [1, 2]. For instance, in a triangular lattice with magnetic ions at the corners interacting *via* antiferromagnetic (AF) couplings, the ground state cannot be a Néel-like collinear arrangement. In three dimensions, the pyrochlore lattice formed of tetrahedra linked by their corners shows the same property (see Fig.1 right). The pyrochlore compounds have formula $R_2M_2O_7$, where R is a rare earth and M a transition or *sp* metal. To these examples of "frustrating geometry", one can add the two-dimensional *kagomé* lattice, made of corner-sharing triangles (see Fig.1 left), the rare earth lattice in the garnets $R_3Ga_5O_{12}$, and the B lattice in the spinels AB_2O_4, which is a pyrochlore lattice. In all these systems, in the presence of AF Heisenberg-like (isotropic) interactions, the ground state must be such that $\sum_{i=1,n} \mathbf{S}_i=0$, where $\mathbf{S}_i$ is the spin of the ith ion in a unit "plaquette" (a triangle, a tetrahedron, ... according to the lattice) and the sum runs over the spins of a plaquette. This has two consequences: for the lattices with "loose connectivity", i.e. those where the plaquettes are linked through their corners, the ground state has a very large degeneracy because there are many ways of arranging the spins on the whole lattice such that the above condition is fulfilled [3]. Hence long range order (LRO) of the magnetic moments is in principle impossible, and the system should remain in a fluctuating "spin liquid" phase down to the lowest temperature. For the pyrochlore lattice, another kind of frustration was discovered in the case of ferromagnetic (F) interactions: if the spins are Ising-like, i.e. constrained by the crystal field anisotropy to lie along the ternary <111> axes, then the magnetic system is frustrated. This frustration is similar to the structural frustration of protons in ice, hence the name of "spin ice" for such materials like $Ho_2Ti_2O_7$ or $Dy_2Ti_2O_7$ [4, 5], which show no LRO and present a ground state with large degeneracy. In both types (F or AF), degeneracy can be lifted by second order interactions with respect to exchange, like the dipole-dipole coupling or a finite crystal field anisotropy, resulting eventually in the onset of magnetic LRO with specific structures. Although well characterised by the presence of magnetic Bragg peaks in the neutron diffraction spectra, the

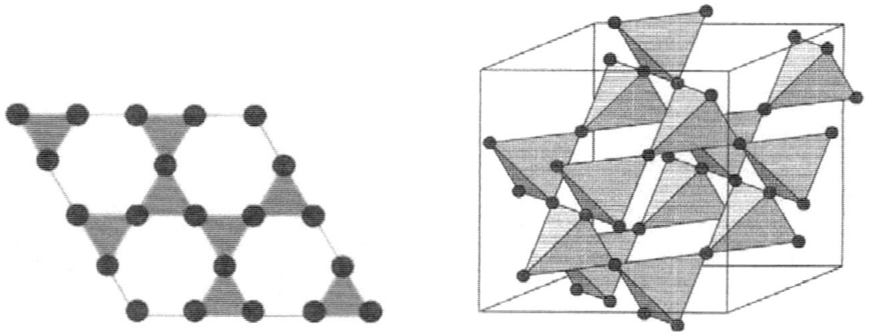

Figure 1. [color on line] Two lattices prone to magnetic frustration: *kagomé* (left) and pyrochlore (right), with magnetic ions at the corners of the basic units, or "plaquettes".

LRO seems to be concomitant with spin fluctuations, as witnessed by various local techniques like Mössbauer spectroscopy, μSR or inelastic neutron scattering. However, recently, the presence of spin fluctuations in the LRO phase has been questioned by new experimental evidence. In this work, we will first give two examples of "true" spin liquids, the Yb and Gd gallium garnets, where Mössbauer spectroscopy has been used to measure the thermal dependence of the spin fluctuation frequency down to the 100 mK range. Next, we will describe the evidences for the presence of fluctuations in the magnetic LRO phase of the pyrochlore stannates $Gd_2Sn_2O_7$ and $Tb_2Sn_2O_7$. Finally, we will discuss the recent data which seem to be at odds with this picture.

2. The spin liquids Yb and Gd garnets

The Yb and Gd Gallium garnets do not show magnetic ordering of the Yb^{3+}/Gd^{3+} ions down to very low temperature. A small specific heat anomaly is observed at 0.054 K in the Yb garnet [6], but it does not seem to be due to LRO. The Mössbauer experiments, resp. on ^{170}Yb and ^{155}Gd, were carried out down to the 50 mK range. In the Yb garnet [7], no resolved hyperfine field spectrum is observed down to the lowest temperature; spin fluctuations are clearly present, and the spectrum narrows on heating, which means that their frequency increases. In the Gd garnet, the spectrum at 0.04 K (not shown) is a "static" hyperfine field spectrum [8], and increasing the temperature yields original relaxational lineshapes with increasing frequencies (Fig.2 left). The spectra in both garnets were fitted using a lineshape appropriate for a fluctuating hyperfine field [10] because, although no LRO is present, the magnetic moments are short range correlated. Figure 2 left shows the thermal variation of the spin fluctuation frequency $1/\tau$: in $Yb_3Ga_5O_{12}$, the fluctuations were taken to be isotropic, and a T-linear law is approximately obeyed above 0.09 K, in agreement with theoretical predictions [3]. In $Gd_3Ga_5O_{12}$, where the fluctuations seem to be of planar type [8], the fluctuation frequency obeys a T^2 law. In both compounds, the fluctuation rate saturates below about 0.1 K; in $Gd_3Ga_5O_{12}$, the lower limit $(1/\tau)_{LF}$ of the Mössbauer "relaxational window" is reached at about 0.1 K, but the saturation was observed by μSR down to 20 mK [9]. Tunneling of the short range correlated "spin clusters" could be at the origin of this saturation, whereas thermally excited fluctuations dominate above 0.1 K.

Contrary to the Yb and Gd garnets, in most pyrochlore materials, magnetic LRO settles, but at a temperature much lower than the exchange energy, which is a signature of frustration [1].

3. Degeneracy lifting and LRO due to dipolar coupling in $Gd_2Sn_2O_7$

Our first example of magnetic ordering due to degeneracy lifting is the stannate $Gd_2Sn_2O_7$ which, although a close realisation of a Heisenberg antiferromagnet, shows a transition at 1 K

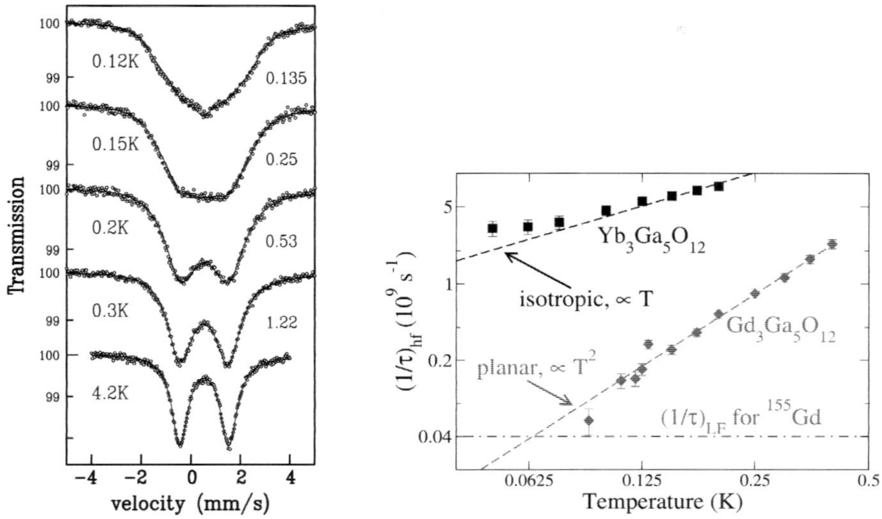

Figure 2. [color on line] **Left**: Low temperature ^{155}Gd Mössbauer spectra in $Gd_3Ga_5O_{12}$, from Ref.[8]; the solid lines are fits to a relaxation lineshape involving fluctuations of the hyperfine field (see text). The number on the right side of each spectrum is the fluctuation frequency in $10^9\,s^{-1}$. **Right**: Thermal variation of the hyperfine relaxation rate (or spin fluctuation rate) in $Yb_3Ga_5O_{12}$ and $Gd_3Ga_5O_{12}$; the dashed lines are fits to a T-linear and a T^2 law resp. for the Yb and Gd garnet.

to a LRO phase [11]. The associated AF **q**=0 magnetic structure is precisely that predicted to occur in pyrochlore compounds with isotropic AF exchange when the degeneracy is lifted by dipolar interactions [12]. The ^{155}Gd Mössbauer spectrum at 27 mK, shown in Fig.3 left, corresponds to combined magnetic and quadrupolar hyperfine interactions, with a hyperfine field $H_{hf} \simeq 30$ T and a hyperfine splitting $\Delta_{hf} \simeq 15$ mK. It shows the peculiar feature that the hyperfine temperature, which can be fitted to the spectrum since $k_B T \sim \Delta_{hf}$, is 90(20) mK, i.e. it is higher than the lattice temperature. Thus the hyperfine levels are not at thermal equilibrium, which can be interpreted as due to the presence of hyperfine field fluctuations, i.e of fluctuations of the Gd^{3+} moment [13]. Indeed, as sketched in Fig.3 right, if the hyperfine (nuclear) relaxation time T_1 is much longer than the spin flip time τ, then the hyperfine levels have an infinite temperature (p_g=0.5, green line in Fig.3 right). When $T_1 \sim \tau$, these levels are still out of equilibrium (black line in Fig.3 right), and it can be shown [13] that their spin temperature is $T_{hf} \simeq T\ (1 + 2T_1/\tau)$, higher than the lattice. So we are led to the conclusion that the correlated Gd spins undergo fluctuations in the LRO phase, with a time scale τ of the same order of magnitude as the hyperfine relaxation time T_1. It is to be noted that τ is longer than $\tau_L \simeq 30$ ns which is the hyperfine Larmor period for ^{155}Gd. Observation of oscillations in the μSR decay in $Gd_2Sn_2O_7$ [14] shifts the lower limit for τ to 1 μs.

These spin fluctuations deep in the LRO phase could be concomittant with, for instance, soft spin-wave modes or peculiarities in the spin-wave spectrum. This was probed by very low temperature specific heat measurements [15, 16] which showed without ambiguity that the excitations in the LRO phase of $Gd_2Sn_2O_7$ are conventional gapped spin waves. A confirmation came from inelastic neutron scattering [17], which detected a **q**-independent excitation with energy 1.5 K above the ground state. The spin wave gap is thus $\Delta_{SW} \simeq 1.5$ K, and the spin fluctuations evidenced by Mössbauer spectroscopy at 27 mK, as well as the unexpected non-zero

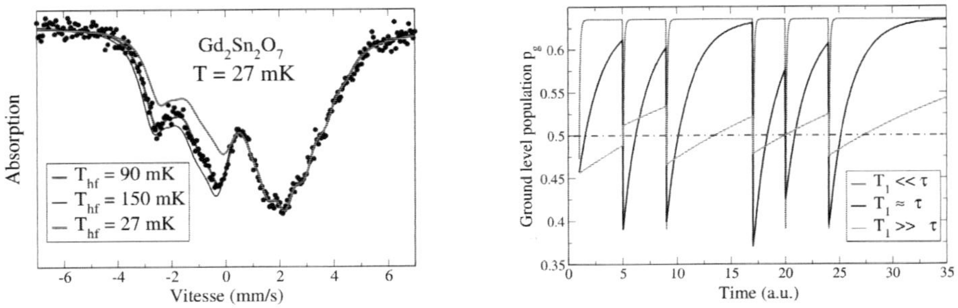

Figure 3. [color on line] **Left**: ^{155}Gd Mössbauer spectrum at 27 mK in $Gd_2Sn_2O_7$, with fits using different hyperfine temperatures. **Right**: Time evolution of the ground level population p_g of a two level (nuclear) spin system subjected to internal relaxation (T_1) and to polarising (hyperfine) field fluctuations (τ). The "spikes" correspond to reversal of the hyperfine field. The value p_g^B=0.63 obtained most of the time for $T_1 << \tau$ is the Boltzmann value for the parameters of the model.

muon relaxation rate observed down to 20 mK [14], do not seem to be linked with some anomaly in the spin-wave spectrum.

4. Degeneracy lifting and LRO due to finite anisotropy in $Tb_2Sn_2O_7$

The Tb^{3+} ion in $Tb_2Sn_2O_7$ is subjected to the crystal electric field, and its ground state is a doublet [18] which has Ising anisotropy. Neutron diffraction reveals a transition to magnetic LRO at 0.87 K, with the development of *ferromagnetic* correlations just above the transition [19]. The **q**=0 long range magnetic structure is shown in Fig.4 right. It has been called the "ordered spin ice" structure because in each tetrahedron, the moment configuration "two spins in - two spins out" reproduces that of the ground state in the spin ice [5]. In this case, it is thought that the degeneracy lifting leading to LRO is brought about by the finite crystal field anisotropy. The spontaneous saturated LRO moment is 5.9(1) μ_B, to which must be "added" a short range ordered component of 3.3(3) μ_B, yielding a magnitude of 6.6(2) μ_B for the local Tb moment [20]. A ferromagnetic LRO component of 2.2 μ_B/Tb is present along <100>.

The specific heat was also measured in $Tb_2Sn_2O_7$ down to 0.15 K. It is shown in Fig.4 right: deep in the LRO phase, one observes a clear turn-up below 0.4 K. It corresponds to the Schottky anomaly arising from the hyperfine splitting of the I=3/2 nuclear spin of ^{159}Tb, with a hyperfine field value of 180(5) T. The hyperfine constant of Tb^{3+} can be estimated as C=38(3) T/μ_B, taking into account the values measured in insulators [21] and the relativistic calculations [22, 23]. We then obtain a Tb moment of 4.7(5) μ_B from the nuclear Schottky anomaly. This moment value is markedly lower than the neutron derived one, and this can also be interpreted as caused by spin fluctuations in the LRO phase [13], with a relaxation time τ of the same order of magnitude as the nuclear relaxation time T_1.

This dynamic behaviour is supported by the unexpected results of μSR measurements [24] which, contrary to $Gd_2Sn_2O_7$, did not detect any oscillations of the μ^+ decay in the LRO phase, nor any anomaly of the μ^+ relaxation rate at the transition. High resolution inelastic neutron scattering [20] in $Tb_2Sn_2O_7$ depict a rather complex situation at low temperature, with different correlation lengths and time scales. Nevertheless, they present evidence for spin fluctuations with a characteristic time of 1 ns at 90 mK. This dynamic picture has been questioned in a μSR work where muons are not implanted in the sample, but in a nearby Ag plate [25]. Therefore, they

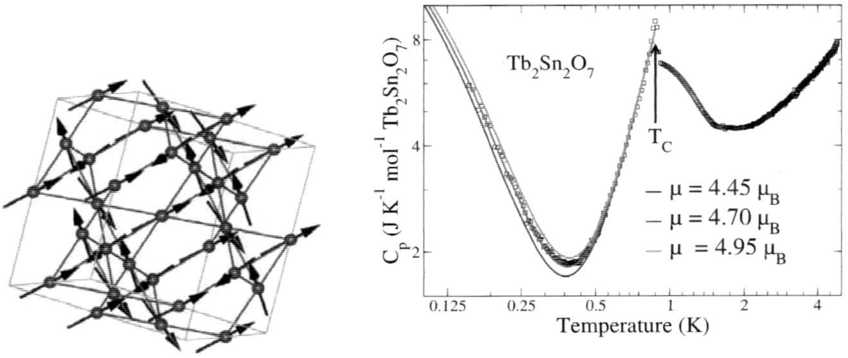

Figure 4. [color on line] In $Tb_2Sn_2O_7$ (from Ref.[19]): **Left**: Its "ordered spin ice" magnetic structure; the Tb moments (black arrows) are not exactly along the <111> axes, like in the standard spin ice, but at an angle of *ca.* 13°. **Right**: Thermal variation of the specific heat. The curves correspond to calculations of the nuclear Schottky anomaly using a hyperfine constant C=38 T/μ_B and the Tb moment values as in the legend.

are sensitive not to the local fields, but to the field created by the bulk magnetisation. In zero field cooling, no oscillation is observed in the μ^+ decay at 0.35 K, as expected for a domain compensated ferromagnet, but an oscillation appears after field cooling with 0.25 T. From these data and FC/ZGC susceptibility data, the authors conclude that their results "rule out a purely dynamical ground state" in $Tb_2Sn_2O_7$.

5. Discussion and conclusions

The situation in the Gd and Tb pyrochlore stannates is therefore puzzling as to the dynamic behaviour in the magnetic LRO phase. The problem is not restricted to the stannates since some titanates also present seemingly contradictory experimental data: $Er_2Ti_2O_7$ shows AF LRO at 1 K [26], but no oscillations of the μSR decay can be detected in the LRO phase [27]. The hyperfine relaxation time, found to be of the same magnitude as the spin fluctuation time in the stannates, would be unusually short, especially in $Tb_2Sn_2O_7$ ($\sim 10^{-8}$ s). The indirect Suhl-Nakamura nuclear spin-spin interaction [28] could perhaps account for this fast relaxation, especially for Tb which is monoisotopic. Finally, the presence of a tetragonal distortion at low temperature in the Tb pyrochlores [29] could contribute to enhance the spin fluctuations by mixing the wave functions and allowing transitions forbidden in the pure trigonal symmetry.

6. Acknowledgements

The author acknowledges fruitful dicussions with I. Mirebeau and P. Dalmas de Réotier.

References

[1] *Handbook of Magnetic Materials* **13**, ed. K.H.J. Buschow, (Elsevier, 2001), p.423
[2] *Frustrated spin systems*, ed. H.T. Diep (World Scientific, Singapore, 2004)
[3] Moessner R, Chalker T 1998 *Phys. Rev. B* **58** 12049
[4] Harris M J *et al* 1997 *Phys. Rev. Lett.* **79** 2554
[5] Bramwell S T and Gingras M J P 2001 *Science* **294** 1495
[6] Filippi J, Lasjaunias J C, Hebral B, Rossat-Mignot J, Tcheou F 1980 *J. Phys. C.: Solid State Phys.* **13** 1277
[7] Hodges J A, Bonville P, Rams M, Królas K 2003 *J. Phys.: Condens. Matter* **15** 4631
[8] Bonville P, Hodges J A, Sanchez J P, Vulliet P 2004 *Phys. Rev. Lett.* **92** 167202

[9] Marshall I M *et al* 2002 *J. Phys.: Condens. Matter* **14** L157
[10] Dattagupta S 1977 *Phys. Rev. B 16* 158
[11] Wills A S *et al* 2006 *J. Phys.: Condens. Matter* **18** L37
[12] Palmer S E, Chalker J T 2000 *Phys. Rev. B* **62** 488
[13] Bertin E, Bonville P, Bouchaud J P, Hodges J A, Sanchez J P, Vulliet P 2002 *Eur. Phys. J. B* **27** 347
[14] Bonville P *et al* 2004 *Hyperfine interactions* **156/157** 103
[15] Quilliam J A, Ross K A, Del Maestro A G, Gingras M J P, Corruccini L R, Kycia J B 2007 *Phys. Rev. Lett.* **99** 097201
[16] Sosin S S, Prozorova L A, Bonville P, Zhitomirsky M E 2009 *Phys. Rev. B* **79** 014419
[17] Stewart J R, Gardner J S, Qiu Y, Ehlers G 2008 *Phys. Rev. B* **78** 132410
[18] Mirebeau I, Bonville P, Hennion M 2007 *Phys. Rev. B* **76** 184436
[19] Mirebeau I *et al* 2005 *Phys. Rev. Lett.* **94** 246402; the Tb moment value derived from the nuclear specific heat is erroneously quoted as $3.3\,\mu_B$.
[20] Mirebeau I, Mutka H, Bonville P, Apetrei A, Forget A 2008 *Phys. Rev. B* **78** 174416
[21] Bleaney B 1988 *Handbook on the Physics and Chemistry of rare earths* eds. Gschneider K A Jr and Eyring L, **11** 323
[22] Dunlap B D 1971 *Mössbauer effect methodology* **7** 123
[23] Freeman A J, Desclaux J P 1979 *J. Magn. Magn. Mat.* **12** 11
[24] Dalmas de Réotier *et al* 2006 *Phys. Rev. Lett.* **96** 127202; Bert F *et al* 2006 *Phys. Rev. Lett.* **97** 117203
[25] Giblin S R *et al* 2008 *Phys. Rev. Lett.* **101** 237201
[26] Champion J D M *et al* 2003 *Phys. Rev. B* **68** 020401(R)
[27] Lago J *et al* 2005 *J. Phys.: Condens. Matter* **17** 979
[28] Suhl H 1958 *Phys. Rev.* **109** 606; Nakamura T 1958 *Progr. Theoret. Phys. Kyoto* **20** 542
[29] Rule K C, Bonville P 2009 *J. Phys.: Conf. Series* **145** 012027

International Conference on the Applications of the Mössbauer Effect (ICAME 2009) IOP Publishing
Journal of Physics: Conference Series **217** (2010) 012120 doi:10.1088/1742-6596/217/1/012120

Mössbauer study of LaFeAsO and F-doped superconductors in external magnetic fields

S Kitao[1,2], Y Kobayashi[1,2], S Higashitaniguchi[1,2], M Kurokuzu[1,2], M Saito[1,2], T Mitsui[2,3], Y Kamihara[4,5], M Hirano[5,6], H Hosono[5,6,7] and M Seto[1,2,3]

[1] Research Reactor Institute, Kyoto University, Kumatori, Osaka 590-0494, Japan
[2] CREST, Japan Science and Technology Agency (JST), Kawaguchi, Saitama 332-0012, Japan
[3] Japan Atomic Energy Agency, Sayo, Hyogo 679-5148, Japan
[4] TRiP, JST, in Materials and Structures Laboratory, Tokyo Institute of Technology, Yokohama 226-8503, Japan
[5] ERATO SORST, JST, in Frontier Research Center, Tokyo Institute of Technology, Yokohama 226-8503, Japan
[6] Frontier Research Center, Tokyo Institute of Technology, Yokohama 226-8503, Japan
[7] Materials and Structures Laboratory, Tokyo Institute of Technology, Yokohama 226-8503, Japan

E-mail: kitao@rri.kyoto-u.ac.jp

Abstract. The iron-based F-doped superconductors $LaFeAsO_{1-x}F_x$ with a transition temperature of 24 K (for $x = 0.07$) and 26 K ($x = 0.11$) and its parent material LaFeAsO were studied using ^{57}Fe Mössbauer spectroscopy. Further investigation was carried out by applying external magnetic fields. F-doped superconductors showed a singlet pattern with no magnetic splitting throughout the temperature range from 4.2 to 298 K. On the other hand, magnetically-split spectra were observed in the parent LaFeAsO below the Néel temperature of about 140 K. The internal magnetic field reached 5.3 T at 4.2 K. The external magnetic fields up to 14 T were applied to the singlet phases, F-doped superconductors and the parent LaFeAsO above the Néel temperature. The induced magnetically-split spectra showed the internal magnetic fields with the comparable value to the applied fields. This fact confirmed that these singlet phases have the paramagnetic feature. The magnetic fields were also applied to the magnetically-ordered phase of LaFeAsO below the Néel temperature. The evolution of the spectra depending on the external magnetic fields was clearly explained by a model with two sublattice spins of the powdered antiferromagnet. This fact confirmed the magnetically-ordered phase is an antiferromagnet. The spin-flop field was also estimated by the model as about 26 T.

1. Introduction

A new epoch of the superconductivity has just opened by the discovery of superconductivity in Fe-oxipnictides LaFePO[1] and the successive discovery of a superconductor with a transition temperature (T_c) of 26 K in the F-doped $LaFeAsO_{1-x}F_x$[2]. Since magnetic elements, such as Fe, had been thought to be incompatible with the emergence of superconductivity, these reports surprised many researchers and have brought extensive following researches[3]. Moreover, first-principles calculations of the Fe-oxipnictides showed the 3*d*-bands of Fe are primary components of the Fermi level. This fact implies the Fe is the key element of the superconductivity in the Fe-oxipnictides. The observed relatively high T_c implies the mechanism of this superconductivity may not be explained in the conventional Bardeen-Cooper-Schrieffer theory. There is a possibility to have some relationship between the magnetism and the mechanism of their superconductivity. Therefore, Fe-Mössbauer spectroscopy is one of the most essential methods to investigate the electronic states and magnetic properties to clarify the nature of their superconductivity. Furthermore, applying external magnetic fields will provide further information on the magnetic properties. In this report, the results of the Mössbauer spectroscopy with applied magnetic fields for the LaFeAsO and the F-doped superconductor were discussed.

2. Experimental

The experiment was performed by conventional Mössbauer equipment using a ^{57}Co source in a Rh matrix with a nominal activity of 1.85 GBq. Each spectrum was calibrated using an α-Fe foil and referenced to it. The measurement at lower temperatures down to 4.2 K was performed using a liquid-He-flow cryostat. The measurement under magnetic fields was carried out using a cryostat with a superconducting magnet, where the magnetic field was applied parallel to the gamma rays from the ^{57}Co source. The magnetic field at the ^{57}Co source was cancelled by a cancellation magnet. The magnetic field at the sample position was calibrated by a Hall probe.

The $LaFeAsO_{1-x}F_x$ and LaFeAsO samples were synthesized by the method described elsewhere[2] and verified by X-ray diffraction to be a single phase with slight amounts of impurities. The electric resistivity measurement of the F-doped sample showed a midpoint T_c of 24 and 26 K and an onset T_c of 29 and 32 K for $x = 0.07$ and 0.11, respectively. The LaFeAsO did not undergo a superconducting transition, but had an anomaly at around 150 K. For the measurements without magnetic field, we used samples with mass ~ 25 mg mixed with BN and polyethylene powder and pressed to form a pelletized disk with a diameter of 10 mm. The sample of $x = 0.07$ for the measurements under magnetic fields was bound with a wax to avoid a texture of the crystal orientation.

3. Results and discussion

3.1. Mössbauer measurements without external magnetic fields

The Mössbauer studies of the undoped LaFeAsO, which does not show the superconductivity, revealed the existence of a magnetic order at the Néel temperature(T_N) of about 140 K[4,5]. Typical parameters are shown in table 1. The internal magnetic field reaches 5.3 T at 4.2 K. This finding of the magnetic order is consistent to the neutron scattering experiment[6]. On the other hand, F-doped LaFeAsO which shows superconductivity do not undergo the magnetic order down to 4.2 K. Therefore, it was concluded that the F-doping suppresses the magnetic order of the LaFeAsO and causes the emergence of the superconductivity. Moreover, we also revealed that the changes of Fe-phonon density of state caused by emergence of the superconductivity in this sample was small using nuclear inelastic resonant scattering[7].

Table 1. Typical Mössbauer parameters of undoped and F-doped LaFeAsO.

	Temperature [K]	Isomer Shift [mm/s] [a]	Internal Magntic Field [T]
LaFeAsO	298	0.45	–
(T_N ~ 140 K)	4.2	0.59	5.3
$LaFeAsO_{0.89}F_{0.11}$	298	0.45	–
(T_c = 26 K)	4.2	0.59	–

[a] relative to α-Fe.

3.2. Mössbauer measurements with external magnetic fields

3.2.1. *Singlet phases under magnetic fields.* The F-doped superconductor $LaFeAsO_{1-x}F_x$ and high-temperature phase of the parent LaFeAsO above T_N, which showed singlet spectra, were studied with external magnetic fields. The Mössbauer spectra under external magnetic fields up to 14 T for the F-doped superconductor $LaFeAsO_{0.93}F_{0.07}$ at 2 K were measured as in figure 1(a) and LaFeAsO at 200K as in figure 1(b). The observed internal magnetic fields were just comparable values to the external magnetic fields and any additional magnetic order was not apparently induced by the external magnetic field. This fact supports the Fe in the singlet phase does not have the magnetic moment, or it is small if any. When we consider the fact that the observed isomer shifts are close to covalent compounds with itinerant electrons, such as FeAs with an isomer shift of 0.47 mm/s at 295 K, it is consistent to the assumption that Fe behaves as paramagnetism of the itinerant electrons in the Fe-oxipnictide superconductor[8].

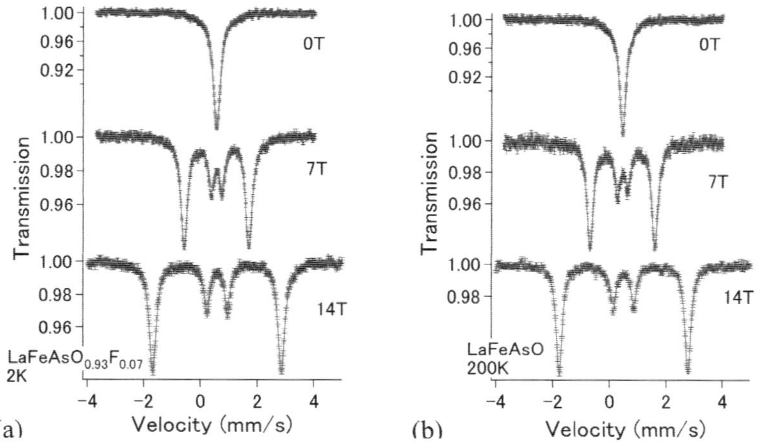

Figure 1. Mössbauer spectrum under magnetic fields of (a) $LaFeAsO_{0.93}F_{0.07}$ at 2K and (b) LaFeAsO at 200 K.

3.2.2. *Magnetic-split phase under magnetic fields.* The parent LaFeAsO showed magnetically-split spectra at low temperatures. When it is applied magnetic fields, the observed spectra as in figure 2 shows the intermediate states of the alignment of magnetic moments to the direction of the external fields, which is parallel to the gamma rays. Although the spectra showed a little distribution of internal magnetic field, only single sextet was considered, for simplicity. The analysis model of polycrystalline antiferromagnets in applied magnetic fields is well described in an atomic spin Hamiltonian[9,10]. When we assume two sublattice sites for the antiferromagnet, the Hamiltonian can be written as

$$H = -2J\sum_{i,j}\mathbf{S}_i\mathbf{S}_j - K(\sum_i (S_{zi})^2 + \sum_j (S_{zj})^2) - g\mu_B \mathbf{B}_0 \cdot (\sum_i \mathbf{S}_i + \sum_j \mathbf{S}_j), \quad (1)$$

where J is an exchange constant, K is an anisotropy constant, $\mathbf{S}_i$ and $\mathbf{S}_j$ are sublattice spins, $\mathbf{B}_0$ is applied field, and g is the spectroscopic splitting factor and μ_B is the Bohr magneton. The energy is obtained by mean field approximation as

$$E = NSg\mu_B\{B_E\cos(\theta_1-\theta_2) - \frac{1}{2}B_A[\cos^2(\theta_1 - t) + \cos^2(\theta_2 - t)] - B_0(\cos\theta_1 + \cos\theta_2)\}, \quad (2)$$

where N is the number of sublattice spins, θ_1, θ_2 and t are the polar angles of the sublattice spins and the easy anisotropy axis with respect to the applied field direction. The exchange field is defined as

$B_E = -2JzS/g\mu_B$ and the anisotropy field is given by $B_A = 2KS/g\mu_B$, where z is the number of exchange-coupled neighbors of a given spin. The solution of the simultaneous equations $\partial E/\partial\theta_1 = \partial E/\partial\theta_2 = 0$ was obtained numerically for a given t. To obtain observed spectra for a polycrystalline sample, each spectrum with a value of t from 0 to π was weighted-superposed. The Mössbauer spectra was fit all spectra at the same time with each spectral intensity and either B_A or B_E as free parameters. The spin-flop field was estimated as $B_{sf} = \sqrt{2B_AB_E}$. The measured spectra were well reproduced when B_{sf} ~ 26 T. The fact that the model with antiferromagnetic powder is well fit to the observed spectra implies the magnetically ordered phase is an antiferromagnet. This fact is consistent to the neutron measurement where an antiferromagnetic order was observed[6].

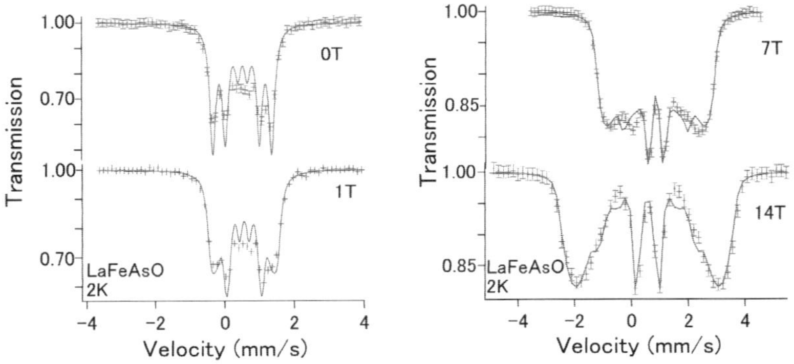

Figure 2. Mössbauer spectra under magnetic fields of LaFeAsO at 2 K.

Acknowledgements
The Authors acknowledged the support from Kansai Research Foundation for technology promotion.

References

[1] Kamihara Y, Hiramatsu H, Hirano M, Kawamura R, Yanagi H, Kamiya T and Hosono H 2006 *J. Am. Chem. Soc.* **128** 10012
[2] Kamihara Y, Watanabe T, Hirano M and Hosono H 2008 *J. Am. Chem. Soc.* **130** 3296
[3] For example, Hosono H *et al.* 2009 *New J. Phys.* **11** 025003, or 2008 *J. Phys. Soc. Jpn.* **77** *Suppl. C.*
[4] Kitao S, Kobayashi Y, Higashitaniguchi S, Saito M, Kamihara Y, Hirano M, Mitsui T, Hosono H and Seto M 2008 *J. Phys. Soc. Jpn.* **77** 103706
[5] Klauss H-H, Luetkens H, Klingeler R, Hess C, Litterst F J, Kraken M, Korshunov M M, Eremin I, Drechsler S-L, Khasanov R, Amato A, Hamann-Borrero J, Leps N, Kondrat A, Behr G, Werner J and Büchner B 2008 *Phys. Rev. Lett.* **101** 077005
[6] de la Cruz C, Huang Q, Lynn J W, Li J Y, Ratcliff W II, Zarestky J L, Mook H A, Chen G F, Luo J L, Wang N L and Dai P C 2008 *Nature* **453** 899
[7] Higashitaniguchi S, Seto M, Kitao S, Kobayashi Y, Saito M, Masuda R, Mitsui T, Yoda Y, Kamihara Y, Hirano M and Hosono H 2008 *Phys. Rev. B* **78** 174507
[8] Higashitaniguchi S, Seto M, Kitao S, Kobayashi Y, Saito M, Kurokuzu M, Mitsui T, Yoda Y, Kamihara Y, Hirano M and Hosono H 2009 *J. Phys. Soc. Jpn.* **78** 024704
[9] Pankhurst Q A and Pollard R J 1990 *J. Phys.: Condens. Matter* **2** 7329
[10] Pankhurst Q A and Pollard R J 1993 *Mössbauer Spectroscopy Applied to Magnetism and Materials Science* vol 1, ed G J Long and F Grendjean (New York: Plenum) chap 3 p 77

Journal of Physics: Conference Series **217** (2010) 012121 doi:10.1088/1742-6596/217/1/012121

Mössbauer studies of spin density wave-superconducting Fe-As systems

I. Nowik[1], I. Felner[1], Z. Ren[2], Z. A . Xu[2] and G. H. Cao[2]

[1]Racah Institute of Physics, The Hebrew University, Jerusalem 91904, Israel.
[2]Department of Physics, Zhejiang University, Hangzhou 310027, People's Republic of China.

E-mail: nowik@vms.huji.ac.il

Abstract. The recently discovered new Fe-As based spin density wave superconducting systems are shortly reviewed. Mössbauer studies of $EuFe_{2-x}Ni_xAs_2$ (x=0.0 to 0.20) by ^{57}Fe and ^{151}Eu isotopes are reported. In pure $EuFe_2As_2$ the iron orders magnetically at 200 K, while the Eu orders antiferromagnetically at 18 K. The 4.2 K studies show hyperfine fields of 8.5 T and 26 T for ^{57}Fe and ^{151}Eu respectively. Ni substitution for iron leads to a decrease in the SDW magnetic transition temperature. This phase transition is below 90 K for x>0.09. For x=0.20, the ^{151}Eu studies at 4.2 K show magnetic order (hyperfine field of 28 T) while the iron displays, no magnetism but only a very small hyperfine field (1.4 T), probably transferred from the now ferromagnetic Eu sublattice.

1. Introduction

The discovery that some Fe-As based oxypnictides are superconducting (SC) [1], with relatively high T_c (up to 55 K), led to great interest in the Mössbauer community. The reason for that is that in the oxypnictides, SC is confined to the Fe-O planes, thus Fe may serve as a *direct* Mössbauer probe to sense the SC properties. Till now all three Fe-based oxypnictides systems: (i) $LaFeAsO_{1-x}$, $LaFeAsO_{1-x}F_x$ [2-5], (ii) $Sr_{1-x}K_xFe_2As_2$, $Ba_{1-x}K_xFe_2As_2$, $Ba(Fe_{1-x}Ni_x)_2As_2$, $Ba(Fe_{1-x}Co_x)_2As_2$, $Eu(Fe_{1-x}Ni_x)_2As_2$ and $EuFe_2(As_{1-x}P_x)_2$ [6-13] and (iii) LiFeAs [14], have been studied by ^{57}Fe Mössbauer spectroscopy (MS). $EuFe_2As_2$ [6] and presently reported here $EuFe_{2-x}Ni_xAs_2$ have been studied also by ^{151}Eu MS. In all (except LiFeAs) systems the parent materials are magnetically ordered in terms of a spin density wave (SDW). Iron has no localized magnetic moment of its own, like in most of RFe_2Si_2 and RFe_2Ge_2 compounds [15]. Mössbauer studies reveal the SDW magnetic order, and the suppression of this state when SC is induced. Following the extensive studies of $EuFe_2As_2$ [6] and recent extensive studies of $EuFe_{2-x}Ni_xAs_2$ [10], $EuFe_2(As_{1-x}P_x)_2$ [11] and $Ba(Fe_{1-x}Ni_x)_2As_2$ [12] by several experimental methods, we studied all these systems by MS. We present here the MS studies of the samples studied in [10], by both ^{57}Fe and ^{151}Eu Mössbauer isotopes. We suggest new ways of analysis of the Mössbauer spectra of SDW materials.

2. Analysis of SDW Mössbauer spectra

A Mössbauer spectrum of a commensurate SDW should exhibit a well-defined sextet, like that of a well-localized magnetic moment. On the other hand in the random incommensurate case one should observe a hyperfine field (H_{eff}) distribution according to the density formula $P(H)=(2/\pi)\,(H_o^2-H^2)^{-1/2}$, where H_o is the saturation value of H_{eff} and $0 \le H < H_o$. For a system, which contains vacancies or doped foreign ions, it is possible that the Mössbauer spectrum is composed of a superposition of two subspectra, corresponding to commensurate, and incommensurate field distributions. In fact, almost all experimental SDW Mössbauer spectra reported to date, were analyzed in terms of *two subspectra*; see review in [16].

3. The $EuFe_{2-x}Ni_xAs_2$ System: experimental details

The sample preparation of $EuFe_{2-x}Ni_xAs_2$ (x=0,0.03, 0.06, 0.09, 0.12, 0.16 and 0.20) is described in a previous paper [10]; these same samples were used here for MS. The MS studies were performed using a conventional constant acceleration drive, and a 100 mCi ^{57}Co: Rh source for ^{57}Fe, and a 200 mCi $^{151}Sm_2O_3$ source for ^{151}Eu. The absorbers were put in perspex holders for the measurements at temperatures 4.2 K to 300 K. The velocity calibration was done with a room temperature α-Fe absorber. The magnetically split Fe spectra were analyzed in terms of least square fits to various alternative SDW theoretical models, mainly in terms of mixtures of commensurate and incommensurate SDW subspectra. The ^{57}Fe spectrum for the $x = 0.2$ sample at 4.2 K, was analyzed in terms of a full diagonalization of the Hamiltonian of magnetic and quadrupole interactions, where the magnetic field is at an angle θ relative to the local axis of the electric field gradient (EFG). The ^{151}Eu spectra were analyzed in terms of knowledge of the exact shape of the emission line of the $^{151}Sm_2O_3$ source, determined by a study with a single absorption line of cubic $EuBe_{13}$, as described in [17].

4. Results and Discussion

Pure $EuFe_2As_2$ and $EuNi_2As_2$ have been studied in the past in very great detail by both ^{57}Fe and ^{151}Eu Mössbauer isotopes [6]. The facts to emphasize here are the saturation hyperfine magnetic fields for divalent Eu (-26.0 T and –36.5 T respectively) and for Fe in $EuFe_2As_2$ (-8.5 T). Our ^{57}Fe and ^{151}Eu MS measurements for $EuFe_2As_2$ show spectra identical to those seen in [6]. In almost all ^{57}Fe spectra of $EuFe_{2-x}Ni_xAs_2$, one observes some presence of FeAs [18]. In the ^{151}Eu spectra some Eu_2O_3 is present, as in [6]. In Fig. 1 spectra of pure $EuFe_2As_2$ are displayed.

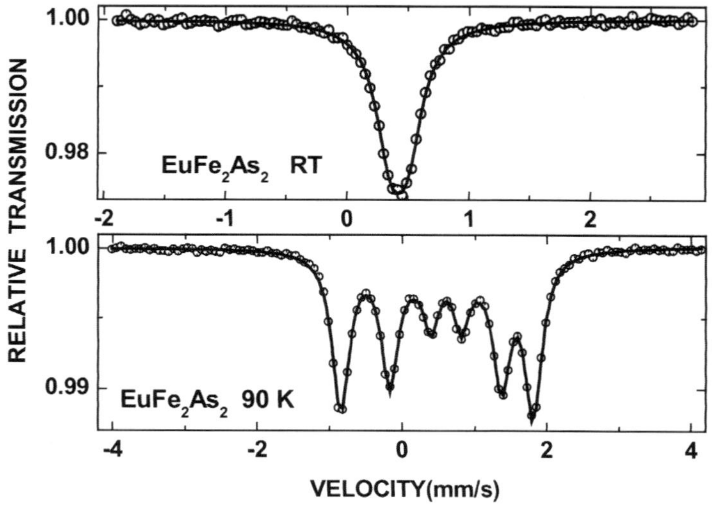

Figure1. Mössbauer spectra of ^{57}Fe in $EuFe_2As_2$ at 90K and 300 K.

The ^{57}Fe spectra shown in Fig.2 were all analyzed in terms of mixtures of commensurate and incommensurate subspectra. It is quite clear that above x=0.09 the SDW phase transition is below 90 K, in perfect agreement with the measurements reported in [10]. In Fig. 3 we observe for

$EuFe_{1.8}Ni_{0.2}As_2$ at 90 K a pure quadrupole doublet, and at 4.2 K (analyzed by full diagonalization of the Spin Hamiltonian), a magnetic field acting on ^{57}Fe of *only* 1.4 T (pointing at an angle of 36° relative the axis of the electric field gradient), while the field acting on the divalent ^{151}Eu is -28.0 T.

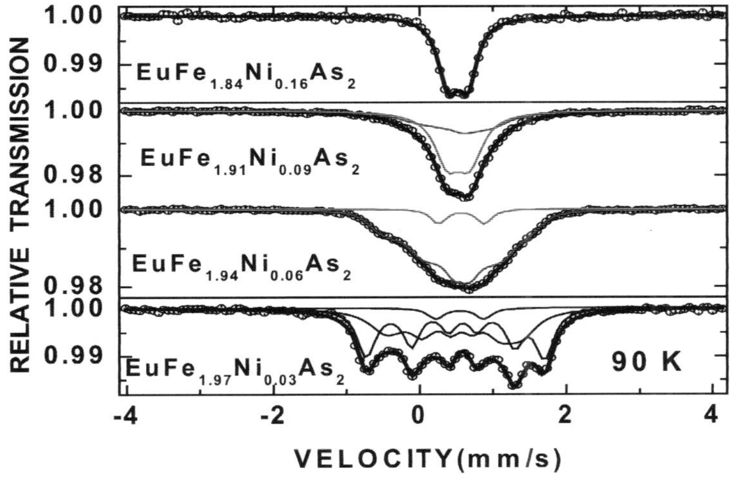

Figure 2. Mössbauer spectra of ^{57}Fe in $EuFe_{2-x}Ni_xAs_2$ samples at 90 K.

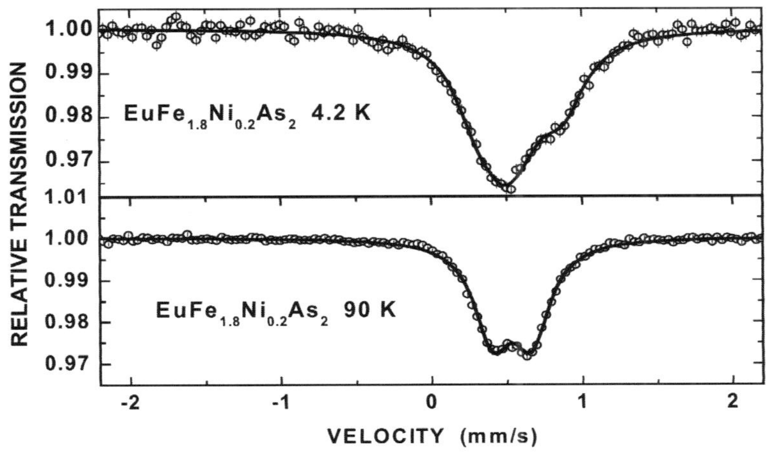

Figure 3. Mössbauer spectra of ^{57}Fe in $EuFe_{1.8}Ni_{0.2}As_2$, at 90 K and 4.2 K.

The magnetization studies show [10] that Eu in $EuFe_2As_2$ is antiferromagnetically ordered at T_N=18 K and Fe orders at 200 K, while in $EuFe_{1.8}Ni_{0.2}As_2$ the iron ions are nonmagnetic and the Eu sublattice orders ferromagnetically at 20 K. These facts explain well our observations, and lead to the following conclusions.

5. Conclusions

It appears that substitution of iron in $EuFe_2As_2$ by 10% Ni: a. Suppresses the magnetic order of iron. b. Induces ferromagnetism in the Eu sublattice. c. The Eu hyperfine field is larger than that in pure $EuFe_2As_2$ [6]. d. The low temperature Fe hyperfine field (1.4 T) is transferred from the ferromagnetic Eu sublattice, like the fields acting on dilute iron transferred from Mn, in $^{57}Fe{:}RMn_2Si_{2-x}Ge_x$ [15]. e. No superconductivity is reached, unlike in $EuFe_2(As_{1-x}P_x)_2$ [11] or $Ba(Fe_{1-x}Co_x)_2As_2$ [13]. Just recently it was discovered that pure $EuFe_2As_2$ under pressure becomes superconducting, with $T_c \sim 30$ K for P=28 kbar [19].

6. Acknowledgements

This work was supported by the Klachky Foundation for Superconductivity.

References

[1] Kamihara Y, Watanabe T, Hirano M, Hosono H 2008 *J. Amer. Chem. Soc.* **130** 3296.

[2] Klauss H H, Luetkens H, Klingeler R, Hess C, Litterst F J, Kraken M, Korshunov M M, Eremin I, Drechsler S. L, Khasanov R, Amato A., Hamann-Borrero J., Leps N, Kondrat A, Behr G, Werner J, Buchner B 2008 *Phys. Rev. Lett.* **101** 077005.

[3] McGuire M M, Christianson A D, Sefat A S, Sales B C, Lumsden M D, Jin R, Payzant E A, Mandrus D, Luan Y B, Keppens V, Varadarajan V, Brill J W, Hermann R P, Sougrati M T, Grandjean F, Long G 2008 *Phys. Rev. B* **78** 094517.

[4] Nowik I, Felner I, Awana V P S, Vajpayee A and Kishan H 2008 *J. Phys: Conden. Matt.* **20** 292201.

[5] Kitao S, Kobayashi Y, Higashitaniguchi S, Saito M, Kamihara Y, Hirano M, Mitsui T, Hosono H, Seto M 2008 *J. Phys. Soc. Jpn.* **77** 103706.

[6] Raffius H, Mörson E, Mosel B D, Müller-Warmuth W, Jeitschko W, Terbüchte L and Vomhof T 1993 *J. Phys. Chem. Sol.* **54** 135.

[7] Tegel M, Rotter M, Weiss V, Schappacher F M, Pöttgen R and Johrendt D 2008 *J. Phys.: Cond. Matt.* **20** 452201.

[8] Rotter M, Tegel M, Johrendt D, Schellenberg I, Hermes W and Pöttgen R 2008 *Phys. Rev. B* **78** 020503(R).

[9] Rotter M, Tegel M, Johrendt D 2008 *Phys. Rev. Lett.* **101** 107006.

[10] Ren Z, Lin X, Tao Q, Jiang S, Zhu Z, Wang C, Cao G and Xu Z 2009 Phys. Rev. B **79**, 094426.

[11] Ren Z, Tao Q, Jiang S, Feng C, Wang C, Dai J, Cao G and Xu Z, 2009 *Phys. Rev. Lett.* **102**, 137002.

[12] Sefat A S, McGuire M A, Jin R, Sales B C, Mandrus D, Ronning F, Bauer B D and Mozharivskyi Y 2009 *Phys. Rev. B* **79** 094508.

[13] Wang X F, Wu T, Wu G, Liu R H, Chen H, Xie Y L and Chen X H 2009 *New J. Phys.* **11**, 045003.

[14] Tapp J H, Tang Z J, Lv B, Sasmal K, Lorenz B, Chu P C W and Guloy A M 2008 *Phys. Rev. B* **78** 060505R, and Felner I, Nowik I, Lv B, Tapp J H, Tang Z and Guloy A M 2009 *Hyper. Interact.* **191** 61.

[15] Nowik I, Levi Y, Felner I and Bauminger E R 1995 *J. Mag. Mag. Mat.* **147** 373.

[16] Nowik and I. Felner 2009 *Physica C* **469** 485.

[17] Nowik I and Felner I 1986 *Hyper. Interact.* **28** 959.

[18] Nowik I and Felner I 2008 *Journal of Supercon. Novel Mag.* **21** 297.

[19] Terashima T, Kimata M, Satsukawa H, Harada A, Hazama K, Uji S, Suzuki H S, Matsumoto T, Murata K 2009 *arXiv:*0904.2618v2 *[cond-mat.supr-con]*.

Journal of Physics: Conference Series **217** (2010) 012122 doi:10.1088/1742-6596/217/1/012122

^{151}Eu-Mössbauer study of complex magnetism in Eu_2PdSi_3: Effect of Eu^{2+} substitution by Y^{3+} and of high pressure*

K Rupprecht[1], B Bielemeier[1], E V Sampathkumaran[2] and G Wortmann[1]

[1]Department Physik, Universität Paderborn, D-33095 Paderborn, Germany
[2]Tata Institute of Fundamental Research, Homi Bhabha Road, Mumbai-400 005, India

E-mail: wortmann@physik.upb.de

Abstract. With ^{151}Eu-Mössbauer spectroscopy and other methods the complex magnetic properties of Eu_2PdSi_3, arising from the two crystallographically different lattice sites of the Eu^{2+} ions, have been already studied. Here we study the impact of magnetic dilution of the magnetic Eu^{2+} sites by non-magnetic Y^{3+} ions. A previous specific heat study has found reduced magnetic ordering temperatures with strong indication of disorder effects like in magnetic spin glasses. Here we provide from ^{151}Eu-Mössbauer spectroscopy detailed information of the impact of Y^{3+} substitution on the magnetic properties of the two lattice sites, well distinguishable in the ^{151}Eu-spectra. Since the substitution of the larger Eu^{2+} ions by the smaller Y^{3+} ions is connected with a lattice contraction, we also applied high pressure to the Eu_2PdSi_3 sample and observed drastic changes in the magnetic properties originating from a valence transition towards Eu^{3+} for one of the Eu sites.

1. Introduction

Eu_2PdSi_3 belongs to a class of rare-earth intermetallics with AlB_2-type hexagonal structure exhibiting various forms of complex magnetism [1-3]. This behaviour is due to two crystallographic different rare-earth sites, the 2b site named R1 and the 6h site named R2, occurring in a ratio of 1 : 3. In Eu_2PdSi_3, the minority R1 sites exhibit ferromagnetic (fm) order at 40 K, while the majority R2 sites exhibit antiferromagnetic (afm) order around 10 K. In the ^{151}Eu-Mössbauer spectra, the R1 and R2 sites can be clearly distinguished by their different isomer shifts, S = - 8.7 mm/s and -10.0 mm/s as well as by their different saturation hyperfine fields, B_{hf} = 414 kG and 262 kG, respectively [1]. Here we investigated the effect of substitution of the magnetic Eu^{2+} by non-magnetic Y^{3+} ions in $Eu_{1.5}Y_{0.5}PdSi_3$ and $Eu_{0.5}Y_{1.5}PdSi_3$ in two series of ^{151}Eu-spectra measured between 2 K and 300 K. In both series we can distinguish the magnetic properties of the two lattice sites from ^{151}Eu-spectra measured at low temperatures, exhibiting reduced magnetic ordering strength and reduced fm or afm ordering characteristics, resembling at higher temperatures a spin-glass behaviour, reflecting the reduced and random numbers of magnetic Eu neighbours.

Since the substitution of the larger Eu^{2+} ions by the smaller Y^{3+} ions is connected with a lattice contraction, we also applied high pressure to the non-substituted Eu_2PdSi_3 sample to unravel the intrinsic effect of lattice contraction. We show that application of a pressure of 7 GPa has a dramatic effect on the magnetic properties of Eu_2PdSi_3: The Eu^{2+} R1 site exhibits a valence transition towards Eu^{3+}, while the Eu R2 sites remain divalent and take over the leading role in magnetic ordering.

2. Experimental

The preparation of the pure Eu_2PdSi_3 sample as well of the Y-substituted $Eu_{1.5}Y_{0.5}PdSi_3$ and $Eu_{0.5}Y_{1.5}PdSi_3$ samples has been described in [1, 4], also their characterisation by electrical resistivity, heat capacity and susceptibility measurements. The ^{151}Eu-Mössbauer spectra were recorded using a 100 mCi SmF_3 source, moved sinusoidally at the same temperature as the absorber in a He bath cryostat. The high-pressure cell with boron carbide (B_4C) anvils is described in [5]. The pressure was determined with a lead manometer [5]. The analysis of the complex Mössbauer spectra were performed by procedures described in [1-3], parameters like the relative site occupancy and isomer shift for the Y-substituted samples were determined from the well-resolved spectra measured around 2 K and then fixed in the spectra at higher temperatures.

3. Results and Discussion:

Typical ^{151}Eu-Mössbauer spectra of $Eu_{1.5}Y_{0.5}PdSi_3$ and $Eu_{0.5}Y_{1.5}PdSi_3$ are shown in Fig. 1 and Fig. 2, respectively. The subspectra for the minority R1 sites are red dashed-dotted and for the majority R2 sites blue dashed and dotted. Careful analysis of the spectra at low temperatures reveal an intensity ratio of 1 : 3 for the R1 and R2 sites, as observed for non-substituted Eu_2PdSi_3 [1]. This implies that no preferential substitution of the R1 or R2 sites occur.

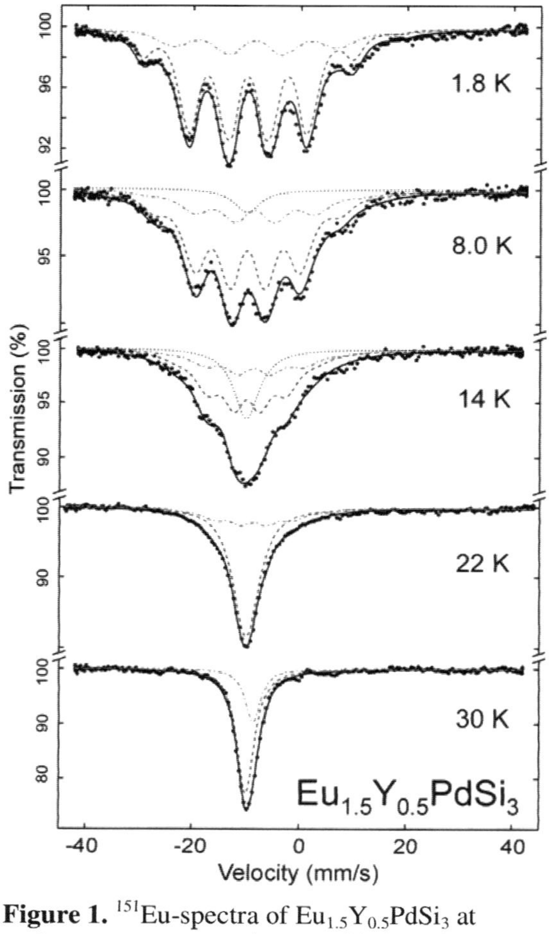

Figure 1. ^{151}Eu-spectra of $Eu_{1.5}Y_{0.5}PdSi_3$ at various temperatures above and below the magnetic ordering around 25 K. Details are explained in the text.

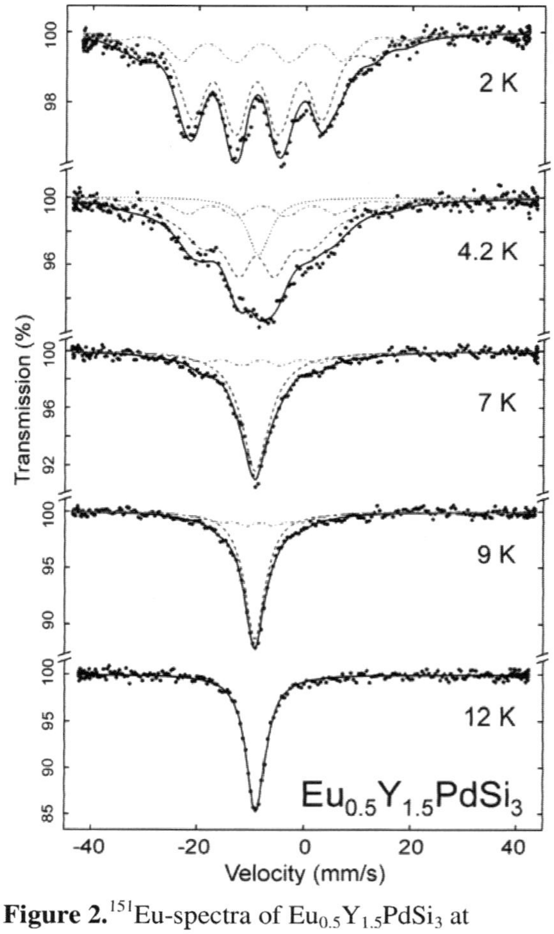

Figure 2. ^{151}Eu-spectra of $Eu_{0.5}Y_{1.5}PdSi_3$ at various temperatures above and below the magnetic ordering around 10 K. For details, see text.

The principal features of the ^{151}Eu-spectra at low temperatures for the Y-substituted and the pure Eu_2PdSi_3 sample remain the same, despite of the modest ($Eu_{1.5}Y_{0.5}PdSi_3$) and strong ($Eu_{0.5}Y_{1.5}PdSi_3$) dilution of the magnetic Eu^{2+} constituents. The derived hyperfine parameters of the R1 and R2 sites are listed in Tab. 1 together with the magnetic ordering temperatures T_M. Here we shortly discuss the variation of T_M on Y substitution together with the variation of the isomer shifts S and the magnetic hyperfine fields B_{hf}. For both Y-substituted samples, magnetic ordering is first observed for the R1 sites, lowered from 40 K (Eu_2PdSi_3) to 25 K ($Eu_{1.5}Y_{0.5}PdSi_3$) and ~11 K ($Eu_{0.5}Y_{1.5}PdSi_3$). For the R2 sites, which exhibit afm ordering in Eu_2PdSi_3 around 10 K, the behaviour is more complex, because only part of the R2 sites deliver in $Eu_{1.5}Y_{0.5}PdSi_3$ below 18 K a structured magnetic subspectrum, while the other R2 sites exhibit a magnetically broadened single line (see spectrum at 14 K in Fig. 1). The same behaviour is observed for the $Eu_{0.5}Y_{1.5}PdSi_3$ sample at lower temperatures (see spectrum at 4.2 K in Fig. 2). This behaviour of the R2 sites is reminiscent to the behaviour of Eu^{2+} ions in magnetic glasses [6], caused by the increasing randomness of magnetic interactions, reflecting the reduced and random numbers of Eu neighbours. The (absolute) magnitudes of the observed magnetic hyperfine fields contain important information about the magnetic ordering type of the two lattice sites (see discussion in [1,3]). The large value of B_{hf} = 414 kG for the R1 sites in Eu_2PdSi_3, indicative of fm ordering, is reduced to 379 kG and 372 kG in the Y-substituted samples. On the other hand, the small value of B_{hf} = 262 kG observed for the R2 sites in Eu_2PdSi_3, indicative of afm ordering, increases to

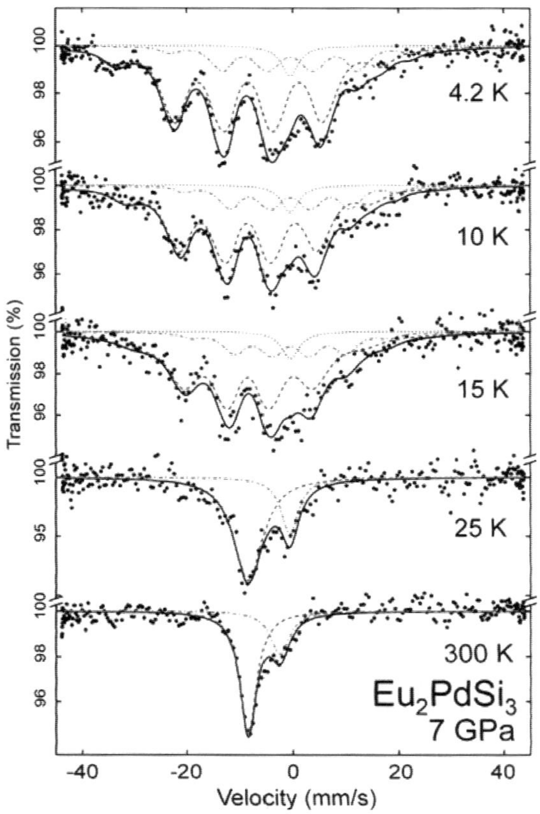

Figure 3. ^{151}Eu-spectra of Eu_2PdSi_3 at 7 GPa and various temperatures above and below the magnetic ordering around 25 K.

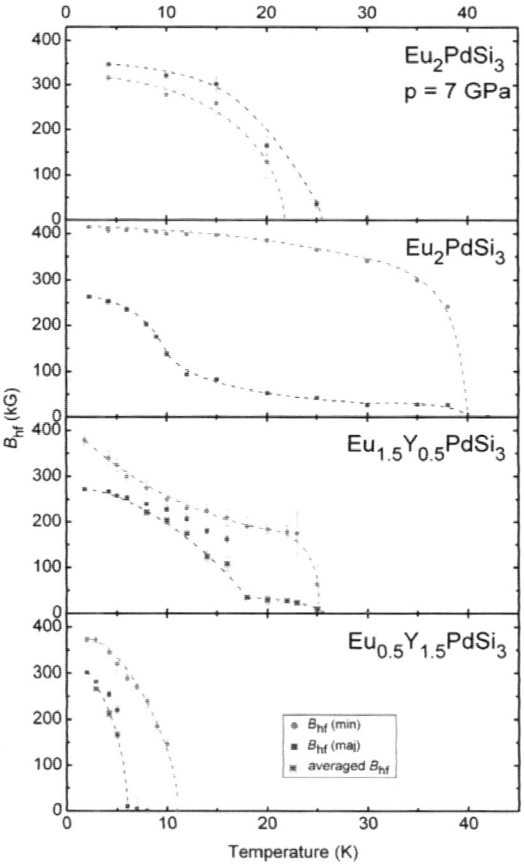

Figure 4. Temperature dependence of magnetic hyperfine fields B_{hf} of R1 (red) and R2 sites (blue, squares: resolved spectra; stars: averaged)

272 kG and 301 kG, the latter value approaching the core polarization contribution of Eu^{2+} (B_C = 300 kG). This behaviour indicates that the magnetic interactions become increasingly random with Y^{3+} substitution. The resulting different magnitudes of the transferred hyperfine fields at the R1 and R2 sites are also reflected by an increase in line width, especially of the outer lines of the magnetic patterns. Finally, the isomer shifts of the two sites are shifted due to the lattice contraction by Y^{3+} substitution, the R1 site moves from S_1 = - 8.7 mm/s to -8.6 mm/s and -8.2 mm/s and the R2 site moves from S_2 = -10.0 to -9.9 mm/s and -9.1 mm/s for Eu_2PdSi_3, $Eu_{1.5}Y_{0.5}PdSi_3$ and $Eu_{0.5}Y_{1.5}PdSi_3$, respectively.

The application of high pressure has a dramatic effect on the magnetic properties of Eu_2PdSi_3: First one immediately realise from the spectrum at 300 K that a valence transition towards Eu^{3+} has occurred for the R1 sites, as signified by S_1 = -2.1 mm/s (see Fig. 3). Magnetic properties change now also drastically, the R2 sites start to order magnetically at 25 K, while the R1 sites exhibit the properties of an almost trivalent state with slightly temperature dependent isomer shifts with S_1 = -0.9 mm/s at 25 K and -0.4 mm/s at 4.2 K (see Tab. 1). At 4.2 K the R2 sites exhibit B_{hf} = 348 kG, which is strikingly larger than B_{hf} = 255 kG observed at ambient pressure [1]. This points to a fm ordering of the R2 sites. They induce by their magnetisation field a (small) orbital magnetic moment at the Eu^{3+} ions on the R1 sites by a Van Vleck mechanism between the J = 0 and J > 0 4f-multiplet states of Eu^{3+} [7], which in turn produce an induced hyperfine field of 315 kG at 4.2 K. The present observation of drastic changes in magnetic properties of Eu_2PdSi_3 resemble the pressure-induced changes observed in $EuCo_2P_2$, exhibiting afm ordering at T_N = 66.5 K originating from the Eu^{2+} sublattice at ambient pressure and a pressure-induced valence transition towards Eu^{3+}, accompanied by strong itinerant magnetism of the Co sublattice (T_N = 260 K), being non-magnetic at ambient pressure [8].

Table. 1: Mössbauer parameters, hyperfine fields B_{hf} and isomer shifts S_i, at ~2 K for Eu_2PdSi_3 and the substituted systems and at 4.2 K for Eu_2PdSi_3 at 7 GPa. The derived ordering temperatures T_M for the R1 and R2 sites compare well, except for the R1 sites in $Eu_{1.5}Y_{0.5}PdSi_3$, with [4].

	B_{hf}(kG)		Isomer shift S_i(mm/s)		T_M(K)	
Compound	R1	R2	R1	R2	R1	R2
Eu_2PdSi_3	414(3)	262(2)	-8.7(1)	-10.0(1)	40	10
$Eu_{1.5}Y_{0.5}PdSi_3$	379(9)	272(3)	-8.6(1)	-9.9(1)	25	18
$Eu_{0.5}Y_{1.5}PdSi_3$	372(5)	301(2)	-8.2(2)	-9.1(1)	11	6
Eu_2PdSi_3, 7 GPa	315(8)	348(3)	-0.4(2)	-8.3(1)		25

References

* Work was supported by the DFG (grant WO209/13-3).

[1] Mallik R, Sampathkumaran E V, Strecker M, Wortmann G, Paulose P L, Ueda Y, 1998, *J. Magn. Magn. Mat.* **185**, L135
[2] Mallik R, Sampathkumaran E V, Strecker M, Wortmann G, 1998, *Europhys. Lett.* **41**, 315
[3] Majumdar S, Mallik R, Sampathkumaran EV, Rupprecht K, Wortmann G, 1999, *Phys. Rev. B* **60**, 6770
[4] Majumdar S, Mallik R, Sampathkumaran EV, Paulose P L, Gopalakrishnan K V, 1999, *Phys. Rev.* B **59**, 4244.
[5] Strecker M, Wortmann G, 1999, *Hyperfine Interact.* **120/121**, 187; Strecker Matthias, PhD thesis, Paderborn, 1999, http://d-nb.info/962736813
[6] Litterst F J, Friedt J M, Tholence J L, Holtzberg F, *J. Phys. C* **15**, 1049
[7] Gilat G, Nowik I, 1963, *Phys. Rev.* **130**, 1361
[8] Chefki M, Abd-Elmeguid M M, Micklitz H, Huhnt C, Schlabitz W, 1998, *Phys. Rev. Lett.* **80**, 802

Journal of Physics: Conference Series **217** (2010) 012123 doi:10.1088/1742-6596/217/1/012123

Eu charge and atomic dynamics in $Eu_3Pd_{20}Ge_6$ investigated by ^{151}Eu Mössbauer effect

Satoshi Tsutsui [1], Yoshio Kobayashi [2], Yasuhiro Kobayashi [3,4], Satoshi Higashitaniguchi [3,4], Yoshitaka Yoda [1,4], Makoto Seto [3,4], and Toshiro Takabatake [5]

[1] Japan Synchrotron Radiation Research Institute, SPring-8, Sayo, Hyogo 679-5198, Japan
[2] Nishina Center for Accelerator Based Science, RIKEN, Wako, Saitama 351-0298, Japan
[3] Research Reactor Institute, Kyoto University, Kumatori, Osaka 590-0494, Japan
[4] CREST, Japan Science and Technology Agency, Kawaguchi, Saitama, 332-0012, Japan
[5] Graduate School of Advanced Science of Matter, Hiroshima University, Higashi-Hiroshima, Hiroshima, 739-8530, Japan

E-mail: satoshi@spring8.or.jp

Abstract. We investigated the Eu charge and atomic dynamics in $Eu_3Pd_{20}Ge_6$ by conventional ^{151}Eu Mössbauer spectroscopy and ^{151}Eu nuclear resonant inelastic scattering. The results of ^{151}Eu Mössbauer spectroscopy indicate the presence of the Eu valence fluctuation both at the 4a and 8c site with different frequency. The ^{151}Eu nuclear resonant inelastic scattering spectra differ from the previous results of filled skutterudites.

1. Introduction

Much attention to the potential of thermoelectric materials has been paid in cage-structured compounds. One of the reasons is the expectation of the presence of an Einstein-like phonon mode to reduce thermal conductivity. On the other hand, the cage-structured compounds are also interesting because of the aspects of strongly correlated electronic systems in rare-earth and actinide compounds. The compounds into whose cage a rare-earth atom is inserted provide the opportunity for the hybridization between f-electrons and conduction electrons, because rare-earth ions are surrounded by many conduction electrons.

The series of $R_3Pd_{20}X_6$ (R: rare-earth and actinide, X: Si or Ge) compounds is one of the interesting cage structured compounds. Since the crystal structure contains a $Fm\bar{3}m$ symmetry, this series of compounds has two non-equivalent rare-earth sites: the 4a and 8c sites. The rare-earth ions at the 4a site are surrounded by six Ge atoms and twelve Pd atoms and those at the 8c site are surrounded by sixteen Pd atoms. Among the Pd atoms surrounding the rare-earth atoms, sixteen Pd atoms occupying at the 48 h site are shared by rare-earth atoms at both sites.

$Eu_3Pd_{20}Ge_6$ is a valence fluctuating compound [1]. The deviation of the linear correlation of the lattice constant from the line estimated by lanthanide contraction in the series of $R_3Pd_{20}Ge_6$ suggests that the Eu valence state is no longer a pure Eu^{3+} one. In addition, the deviation of the magnetic susceptibility from the calculation for both free Eu^{3+} and Eu^{2+} suggests that the Eu valence state is fluctuating. These facts infer that $Eu_3Pd_{20}Ge_6$ is a valence fluctuating compound. We have carried out ^{151}Eu Mössbauer spectroscopy to investigate the Eu valence states in this compound from the microscopic point of view. The obtained spectrum at room temperature demonstrates that the Eu valence states are different between two crystallographic Eu sites.

$Eu_3Pd_{20}Ge_6$ also shows lower thermal conductivity than some heavy fermion or valence fluctuating compounds [1-3]. This behaviour is characteristic of cage-structured compounds such as filled skutterudites and type-I clathrates [4-6]. However, the reduction of the thermal conductivity in $Eu_3Pd_{20}Ge_6$ is smaller than those of filled skutterudites and type-I clathrates. In addition, the reduction mechanism associated with an Einstein-like phonon mode was denied [1]. We have carried out ^{151}Eu nuclear resonant inelastic scattering, a useful tool to investigate the Eu atomic dynamics in the Eu materials. The obtained spectra are complete different from those at the rare-earth sites in hexaborides and filled skutterudites by nuclear resonant inelastic scattering [9-15].

2. Experimental Results and Discussion

2.1. ^{151}Eu Mössbauer spectroscopy

^{151}Eu Mössbauer measurements in the conventional way were carried out at Wako branch of RIKEN in Japan. The source used is 100 mCi of $^{151}SmF_3$. The Doppler velocity was calibrated by ^{57}Fe Mössbauer spectroscopy of α-Fe at 295 K. The measured sample was crushed into powder from polycrystalline ingot of $Eu_3Pd_{20}Ge_6$.

The ^{151}Eu Mössbauer spectrum of $Eu_3Pd_{20}Ge_6$ at 295 K is shown in Fig. 1. The spectrum consists of three components in spite of two Eu crystallographic sites in this compound. The isomer shift of one component around 0 mm / sec corresponds to a trivalent state, that of a second component around -10 mm /sec to a divalent state and that of the third component to an intermediate valence state. Since the previous magnetic susceptibility measurement suggests that $Eu_3Pd_{20}Ge_6$ is a valence fluctuating compound, Eu valence fluctuations are possibly observed in both crystallographic sites. Under this assumption, the spectrum of the one site consists of that with the intermediate valence state and the

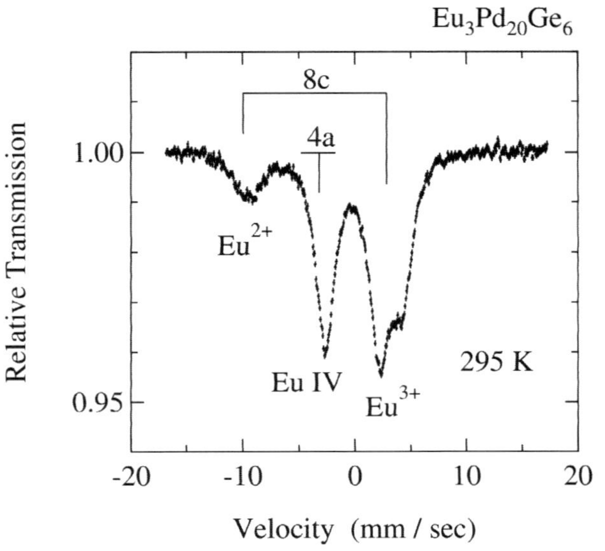

Figure 1. ^{151}Eu Mössbauer spectrum of $Eu_3Pd_{20}Ge_6$ at 295 K

spectrum of the other site consists of that with both the divalent and trivalent states. In this case, the spectral weight of the one to the other is 1 : 2, consistent with the volume fraction of Eu sites expected from the crystallographic point of view. Therefore, the former component shows the spectrum at the 4a site and the latter one shows that at the 8c site. In addition, since the spectral weight of these components shows no temperature dependence [16], this spectral assignment is not inconsistent with the number of the crystallographic sites.

Significant difference of the valence state is observed between ^{151}Eu Mössbauer spectroscopy and Eu L_3-edge X-ray absorption spectroscopy. No components of the intermediate valence state, which corresponds to the component of the 4a site, are observed in the X-ray absorption spectroscopy [16]. Since the time window of the X-ray absorption spectroscopy is shorter than that of the ^{151}Eu Mössbauer spectroscopy, this is direct evidence that the Eu valence at least at the 4a site of $Eu_3Pd_{20}Ge_6$ is fluctuating with a frequency matching the time window of the ^{151}Eu Mössbauer spectroscopy, hence several hundreds MHz.

2.2. ^{151}Eu Nuclear Resonant Inelastic Scattering

^{151}Eu nuclear resonant inelastic scattering experiments were carried out at BL09XU of SPring-8 in Japan. The high resolution monochromator is the nested-type one which consists of Si(4 2 2) and Si(12 8 8) reflections. The resolution of this optics is 1.5 meV at the ^{151}Eu Mössbauer resonance of 21.5 keV. The 203 bunches operation of SPring-8, whose interval is 23.6 nsec, is chosen for this experiment. The used detector is a multi-element avalanche photo-diode detector.

The nuclear resonant inelastic scattering spectrum of $Eu_3Pd_{20}Ge_6$ at 300 K is shown in Fig. 2. This differs from the spectra of other cage-structured compounds [9-15]. In the previous works of the single-site cage-structured compounds, the spectra at the inserted atomic sites (guest sites) consist of a sharp excitation, indicating the presence of Einstein-like modes. However, the spectra obtained in the present work contain much larger contribution to the acoustic modes than the other cage-structured compounds. Although it has been believed that the Einstein-like modes reduce the thermal conductivity in most of cage structured compounds, the present result agrees with the proposed reduction mechanism without any Einstein-like modes in Ref. 1.

Concerning the correlation of the Eu valence fluctuation and atomic motion, a significant temperature dependence of the average force constant is found in the present work. The average force constants obtained by the Lipkin's sum rule [17] are 115.7 and 100.7 N/m at 300 and 6 K, respectively. Since the Eu valence state obtained by both ^{151}Eu Mössbauer spectroscopy and Eu L_3-edge XAS are

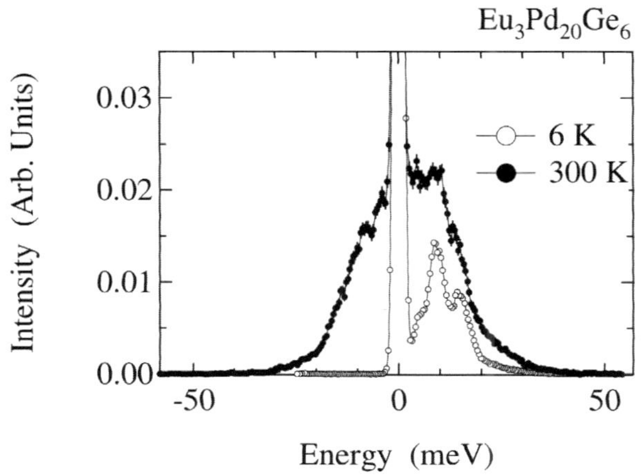

Figure 2. ^{151}Eu nuclear resonant inelastic scattering spectra of $Eu_3Pd_{20}Ge_6$ at 6 and 300 K.

different between 300 and 6 K, the reduction of the force constant may correlates with the change of the Eu valence [16].

3. Summary

Eu charge and atomic dynamics in $Eu_3Pd_{20}Ge_6$ were investigated by ^{151}Eu Mössbauer spectroscopy and ^{151}Eu NRIS. The obtained ^{151}Eu Mössbauer spectrum at 295 K demonstrates that valence fluctuations with different frequency are realized in this compound. The ^{151}Eu nuclear resonant inelastic scattering spectrum differs from the spectra at the guest sites in other cage-structured compounds.

Acknowledgments

The author ST appreciates Profs. H. Kobayashi and S. Sasaki and Drs. M. Mizumaki, C. Lee and T. Hasegawa for their fruitful discussion. The NRIS experiment was carried out under the approval of JASRI (Proposal No. 2006B1078). The present work is partially supported by the Grant-in-Aid for Young Scientists (No. 20740181) and for the Scientific Research (A) (18204032) from the Japan Society of Promotion of Science and the Grant-in-Aids for Scientific Research Priority Area "Skutterudite" (No. 18027017) and for Scientific Research on Innovative Area "Heavy Electrons" (No. 20102004 and 20102005) of the Ministry of Education, Culture, Science and Technology, Japan.

References

[1] Kitagawa J, Sasakawa T, Suemitsu T, Takabatake T and Ishikawa M 2002 *J. Phys. Soc. Jpn.* **71** 1222.
[2] Bauer E 1993 *Transport and Thermal Properties of f-electron Systems* ed G Oomi, H. Fujii and T Fujita (New York: Plenum Press) p.133.
[3] Schneider H and Wohlleben D 1981 *Z. Phys. B* **44** 193.
[4] Cohn J L, Nolas G S, Fessatidis V, Metcalf T H, and Slack G A 1999 *Phys. Rev. Lett.* **82** 779.
[5] Dilley N R, Bauer E D, Maple M B, Dordevic S, Basov D N, Freibert F, Darling T W, Migliori A, Chakoumakos B C, Sales B C 2000 *Phys. Rev. B* **61** 4608
[6] Sales B C, Mandrus D, Chakoumakos B C, Keppens V and Thompson J R 1997 *Phys. Rev. B* **56** 15081.
[7] Nemoto Y, Yamaguchi T, Horino T, Akatsu M, Yanagisawa T, Goto T, Suzuki O, Dönni A and Komatsubara T 2003 *Phys. Rev. B* **68** 184109.
[8] Hasegawa T, Ogita N and Udagawa M, Private Communications.
[9] Long G J, Hermann R P, Grandjean F, Alp E E, Sturhahn W, Johnson C E, Brown D E, Leupold O and Rüffer R 2005 *Phys. Rev. B* **71** 140302.
[10] Hermann R P, Schmeika W, Leupold O, Rüffer R, Nolas G S, Grandjean F and Long G J 2005 *Phys. Rev. B* **72** 174301.
[11] Kobayashi H, Tsutsui S, Baron A Q R and Kunii S 2005 *Physica B* **359-361** 974.
[12] Tsutsui S, Kobayashi H, Umemura J, Yoda Y, Onodera H, Sugawara H, Kikuchi D, Sato H, Sekine C, Shirotani I, 2006 *Physica B* **383** 142.
[13] Tsutsui S, Umemura J. Kobayashi H, Tazaki T, Nasu S, Kobayashi Y. Yoda Y, Onodera H, Sugawara H, Matsuda T D, Kikuchi D, Sato H, Sekine C and Shirotani I, 2006 *Hyperfine Interact.* **168** 1073.
[14] Tsutsui S, Kobayashi H. Ishikawa D, Sutter J P, Baron A Q R, Hasegawa T, Ogita N, Udagawa M, Yoda Y, Onodera H, Kikuchi D, Sugawara H, Sato H, Sekine C and Shirotani I 2008 *J. Phys. Soc. Jpn.* **77** 033601.
[15] Tsutsui S, Hasegawa T, Takasu Y, Ogita N, Udagawa M, Yoda Y and Iga F 2009 *J. Phys.: Conf. Ser.* **176** 012033.
[16] Tsutsui S, Mizumaki M, Kobayashi Y, Kitagawa J and Takabatake T 2009 *Phys. Rev. B.* **80** 235115
[17] Lipkin H J 1995 *Phys. Rev.* B **52** 10073.

International Conference on the Applications of the Mössbauer Effect (ICAME 2009) IOP Publishing
Journal of Physics: Conference Series **217** (2010) 012124 doi:10.1088/1742-6596/217/1/012124

The spin-reorientation transition in H-type $YbFe_6Ge_6$

D H Ryan[1], J M Cadogan[2] and R Gagnon[1]

[1] Physics Department and Centre for the Physics of Materials, McGill University, 3600 University Street, Montreal, Quebec H3A 2T8, Canada

[2] Department of Physics and Astronomy, University of Manitoba, Winnipeg, Manitoba, R3T 2N2 Canada

E-mail: dhryan@physics.mcgill.ca

Abstract. The two crystallographic forms of $YbFe_6Ge_6$ have been studied using neutron powder diffraction, and both ^{57}Fe and ^{170}Yb Mössbauer spectroscopy. Only the H-type ($HfFe_6Ge_6$–type) form undergoes a spin reorientation, and ^{57}Fe Mössbauer data taken on a single crystal mosaic confirm that the Fe moments go from parallel to the c-axis to making an angle of 69(2)° with the crystal c-axis. ^{170}Yb Mössbauer spectra show that the Yb ions are trivalent in both forms, the Yb moments do not order above 5 K and that there is no transferred field from the Fe moments in either form.

1. Introduction

The rare-earth (R) and Fe sublattices in the RFe_6Ge_6 and RFe_6Sn_6 intermetallic compounds exhibit independent magnetic behaviour ([1] and references therein). The Fe sublattice orders antiferromagnetically with a Néel temperature which remains essentially constant across a series at ∼485 K for RFe_6Ge_6 or ∼555 K for RFe_6Sn_6. For R = Gd – Er, the R sublattice orders with Curie temperatures ranging from a high of 45 K in $GdFe_6Sn_6$ to 3 K in $ErFe_6Ge_6$.

$YbFe_6Ge_6$ has been shown to occur in two closely related forms within the hexagonal *P6/mmm* (#191) space group. The first to be identified was the ordered hexagonal $HfFe_6Ge_6$–type (hereafter "H-type") structure in which the crystal cell is doubled along the hexagonal c-axis relative to the underlying FeGe basis, with full site-occupancies [2]. Later, a hexagonal YCo_6Ge_6–type ("Y-type") form was described in which the Yb ions have a 50% occupancy of the 1a site and the cell size is almost the same as that of the parent FeGe compound [3]. For a general discussion of RFe_6X_6 crystallography, see Venturini's review [4].

Remarkably, the two structural forms exhibit distinct magnetic behaviours: the Fe sublattice in the H-type form undergoes a spin reorientation starting just below 80 K as the Fe moments tilt away from the c-axis [5], while in the Y-type form, the Fe order remains unchanged. The co-existence of the two structural forms in many samples of $YbFe_6Ge_6$ initially led to more complex descriptions of the ordering [6] in terms of mixed ordering, however we now believe that almost all Y-type samples of $YbFe_6Ge_6$ contain small amounts of the H-type form.

As the magnetic structures of both forms have been presented in some detail elsewhere [5], we concentrate here on using ^{170}Yb Mössbauer spectroscopy to determine the Yb valence in the two forms, and ^{57}Fe Mössbauer spectroscopy of a single crystal mosaic of the H-type form of $YbFe_6Ge_6$ to obtain an independent estimate of the Fe moment canting angle for comparison with that derived from neutron diffraction.

2. Experimental Methods

Several samples of $YbFe_6Ge_6$ were prepared by arc-melting stoichiometric amounts of the pure elements (Yb: 99.9%, Fe: 99.95%, Ge: 99.999%) under Ti-gettered argon, allowing an excess of 20 wt% Yb to compensate for the inevitable Yb boil-off in the arc-furnace. The arc-melted ingots were sealed under vacuum in quartz tubes and annealed at temperatures from 750 K to 900 K for two weeks. This approach generally yields samples in the Y-type form with varying levels of H-type as an 'impurity', however we did obtain one pure H-type sample by this means. The Y-type sample used here was obtained by annealing at 750 K.

The only method that consistently yields $YbFe_6Ge_6$ in the $HfFe_6Ge_6$ H-type form involves growth from a tin flux [7]. About 25 grams of an iron-deficient starting mixture (Yb:Fe:Ge:Sn atomic proportions of 1:1:3:20; Sn: 99.99%) was loaded into a 11 mm i.d. quartz tube sealed under vacuum. The tube was stood vertically in a furnace and heated rapidly to 1220°C and held there for 10 hours, before cooling to 500°C at 5°C/hr. The tube was then inverted to drain off the majority of the tin flux and furnace-cooled to RT. The crystals were removed from the tube and the excess tin metal was removed by dissolving it in concentrated HCl for two days. Many millimetre-sized hexagonal plates and several hexagonal rods were recovered from each batch, along with about 0.5 g of finer powder. Approximately twenty of the plates from several batches were assembled to form a mosaic for the ^{57}Fe Mössbauer spectroscopy work, and about 2 g of smaller crystals and powder was hand-ground and used for both ^{57}Fe and ^{170}Yb Mössbauer spectroscopy and then neutron diffraction.

^{57}Fe Mössbauer spectroscopy was carried out in transmission mode using a 25 mCi ^{57}Co**Rh** source with the sample in a vibration-isolated closed-cycle fridge. Transmission mode ^{170}Yb Mössbauer spectra were obtained at 5 K with both the source and sample located in a helium flow cryostat. The 20 mCi ^{170}Tm source was prepared by neutron activation of ~25 mg of Tm as a 10 wt% alloy in aluminium. The ^{57}Fe spectra were fitted to a sum of Lorentzian lines with positions and intensities determined from first-order perturbation, while the line positions and intensities for the ^{170}Yb spectra were obtained from diagonalization of the full nuclear hyperfine Hamiltonian [8].

3. Results and Discussion

The clearest indication of the spin-reorientation in H-type $YbFe_6Ge_6$ comes from neutron diffraction measurements [5]. A full Rietveld analysis of the diffraction patterns yielded Fe moments of 1.62(30) μ_B lying at an angle of 69(12)° from the c-axis [5]. The reorientation starts just above 80 K and is essentially complete by ~40 K.

The ^{57}Fe Mössbauer spectra of the two forms of $YbFe_6Ge_6$ at 12 K (Fig. 1) clearly show the impact of the spin-reorientation that leads to the single crystallographic Fe site in the H-type form being sub-split into three, equal-area, magnetically inequivalent sites as the Fe moments cant away from the c-axis below ~80 K. No contribution from the un-canted Y-type pattern was detected in the spectrum of the H-type sample, consistent with this sample being a pure H-type $YbFe_6Ge_6$. Similarly, there are no apparent mis-fits that could be associated with an H-type 'impurity' in the Y-type pattern, however, given the estimate of 5% from the neutron diffraction data [5], and the three-way splitting that this component undergoes, this is not a strong test of the phase purity of the sample. The temperature dependences of the hyperfine parameters shown in Fig. 2 clearly illustrate the differences between the two forms. B_{hf} values at the three sub-split sites in the H-type form are all reduced from that seen in the Y-type form (that does not re-orient) as a result of anisotropic contributions to the hyperfine field [5, 9].

The mosaic of H-type $YbFe_6Ge_6$ crystals was constructed so that the c-axes of the platelets were all parallel to the direction of the Mössbauer $\gamma-$beam. Spectra were obtained at 100 K (above the re-orientation) and 13 K (well below the re-orientation) and are shown in Fig. 3. The hyperfine parameters used in the fits (B_{hf}, δ, Δ) were all taken from spectra measured

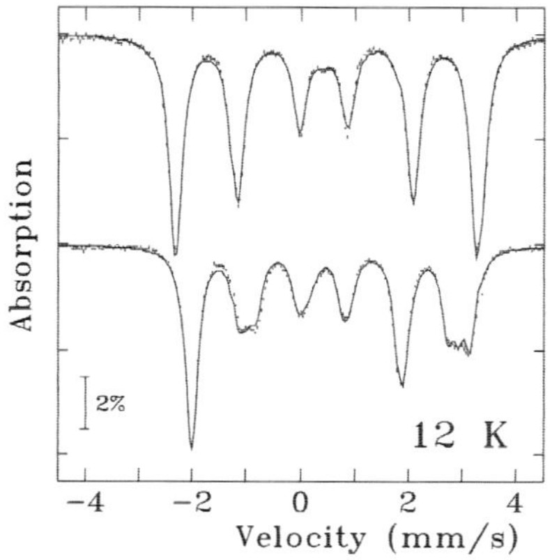

Figure 1. ^{57}Fe Mössbauer spectra of powdered Y-type (top) and H-type (bottom) samples of $YbFe_6Ge_6$ at 12 K.

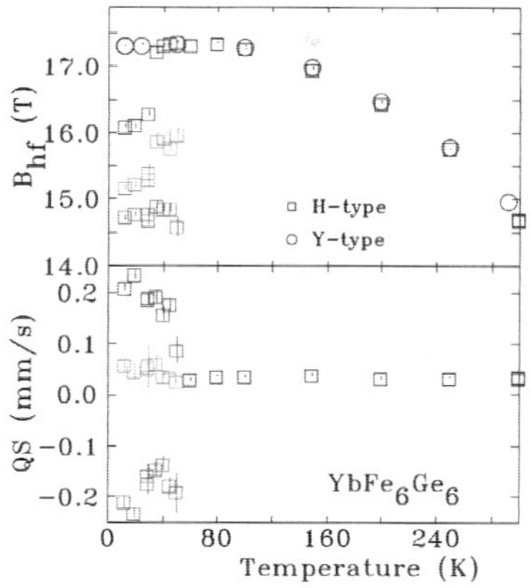

Figure 2. (Top) Temperature dependence of the ^{57}Fe hyperfine field (B_{hf}) and (Bottom) quadrupole splitting for the two forms of $YbFe_6Ge_6$.

using powdered samples at the corresponding temperature, and only the baseline, total spectral intensity and the relative intensities of the two $\Delta m_I = 0$ transitions (lines #2 and #5 in the 6-line magnetic patterns) were refined. A small (~5%) paramagnetic impurity was included to account for the slight asymmetry in the centre of both spectra.

The six lines in a magnetically split ^{57}Fe Mössbauer spectrum have, to first order, relative intensities of 3:R:1:1:R:3, where R is given by: $R = \frac{4\sin^2\theta}{1+\cos^2\theta}$ (θ is the angle between the direction of the $\gamma-$beam and the hyperfine field). Analysis of the 100 K pattern in Fig. 3 yields R=0.04(8), fully consistent with zero, and confirming that the moments are indeed oriented parallel to the c-axis as derived from the neutron diffraction data [5]. It is clear from a visual inspection of the 13 K spectrum in Fig. 3 that the two $\Delta m_I = 0$ lines are at least as strong as the outer lines, suggesting that R is now of order 3, and indicating that the Fe moments are now lying somewhere close to the ab-plane. Fitting the pattern with the constraints on the hyperfine parameters identified above, yields R=3.07(15), giving an angle $\theta = 69(2)^\circ$ between the moments and the $\gamma-$beam (which is in turn parallel to the crystal c-axes in the mosaic). This value is in full agreement with the 69(12)° derived from the neutron diffraction pattern at 3.9 K [5], has slightly better precision, and provides a completely independent confirmation of the canting angle.

One possible source of the differences in magnetic behaviour between the H-type and Y-type forms of $YbFe_6Ge_6$ might be the valence of the Yb ions in the two compounds. Analysis of both the lattice parameters [6] and single-crystal XANES data [7] for the H-type form of $YbFe_6Ge_6$ suggests that the Yb is trivalent in *this* form, while a comparison of the lattice parameters for the two forms of $YbFe_6Ge_6$ shows no anomaly that could be attributed to a valence change [5] between the two forms. We therefore expect that the Yb ions are trivalent in both forms of $YbFe_6Ge_6$.

The striking similarity between the 5 K ^{170}Yb Mössbauer spectra of both forms of $YbFe_6Ge_6$ shown in Fig. 4 clearly confirms this expectation. Analysis of the two spectra yields quadrupole splittings (eQV_{zz}) of 9.58(10) mm/s for the Y-type form and 9.52(13) mm/s for the H-type form implying that the Yb environments in the two forms are effectively indistinguishable. The observed quadrupole splittings indicate that the Yb ions are definitely trivalent as the diamagnetic Yb^{2+} ion has a closed $4f$ shell which is necessarily spherically symmetric, yielding

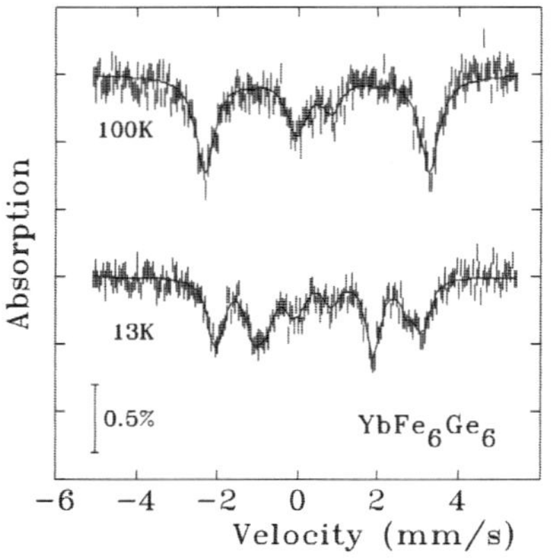

Figure 3. ^{57}Fe Mössbauer spectra of the H-type $YbFe_6Ge_6$ crystal mosaic at 100 K (top) and 13 K (bottom).

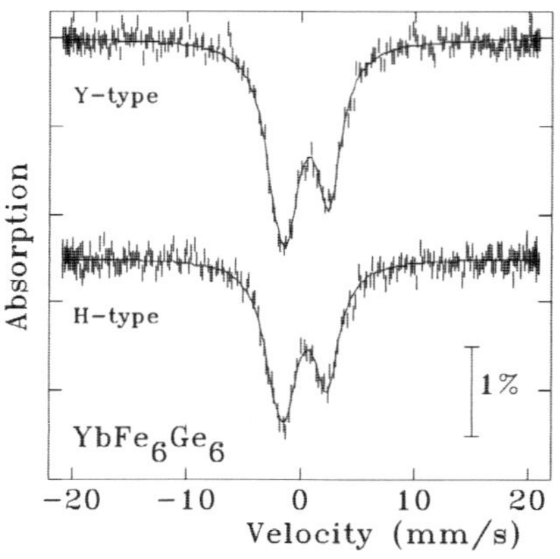

Figure 4. ^{170}Yb Mössbauer spectra of the Y-type (top) and H-type (bottom) forms of $YbFe_6Ge_6$ at 5 K.

no $4f$ contribution to the quadrupole splitting. We also note that there is no evidence for a magnetic field at the Yb site in either form, indicating that the Yb ions do not order above 5 K. Furthermore, there is no evidence for a transferred field from the ordered Fe sub-lattice, in either the c-axis (Y-type) or canted (H-type) ordered states. An observation that is consistent with the neutron diffraction derived magnetic structures of both forms at this temperature [5]. In both the collinear (Y-type) and canted (H-type) magnetic structures, the contributions from the Fe moments at the Yb site are expected to cancel.

4. Conclusions

^{57}Fe Mössbauer measurements of the $HfFe_6Ge_6$–type ("H-type") and YCo_6Ge_6–type ("Y-type") forms of $YbFe_6Ge_6$ show that only the H-type form undergoes a spin re-orientation below ~80 K. Working with flux-grown single crystals of H-type $YbFe_6Ge_6$ we showed that below the re-orientation temperature, the Fe moments make an angle of 69(2)° with the crystal c-axis, in full agreement with earlier neutron diffraction results. ^{170}Yb Mössbauer spectra of the two forms show that in both cases the Yb ions are trivalent.

Acknowledgments

JMC acknowledges support from the Canada Research Chairs programme. Financial support for this work came from the Natural Sciences and Engineering Research Council of Canada and Fonds Québécois de la Recherche sur la Nature et les Technologies. The ^{170}Yb source activation was carried out by M. Butler at the McMaster Nuclear Reactor (MNR), Hamilton Ontario.

[1] Cadogan J.M. and Ryan D.H. 2001 *J. Alloys Compounds* **326**, 166–73
[2] Buchholz W. and Schuster H.-U. 1981 *Z. anorg. allg. Chem.* **482**, 40–8
[3] Dzyanyi R.B., Bodak O.I. and Aksel'rud L.G. 1995 *Materials Science* **31**, 284–5
[4] Venturini G. 2006 *Z. Kristallogr.* **221**, 511–520
[5] Cadogan J.M. and Ryan D.H. *Hyperfine Interact.* (in press)
[6] Mazet T. and Malaman B. 2000 *J. Phys.: Condensed Matter* **12**, 1085–95
[7] Avila M.A., Takabatake T., Takahashi Y., Bud'ko S.L. and Canfield P.C. 2005 *J. Phys.: Condensed Matter* **17**, 6969–79
[8] Voyer C.J. and Ryan D.H. 2006 *Hyperfine Interact.*, **170**, 91-104
[9] Le Caër G, Malaman B, Venturini G and Kim I B 1982 *Phys. Rev. B* **26** 5085–96.

International Conference on the Applications of the Mössbauer Effect (ICAME 2009) IOP Publishing
Journal of Physics: Conference Series **217** (2010) 012125 doi:10.1088/1742-6596/217/1/012125

Pressure-induced changes of the vibrational modes of spin-crossover complexes studied by nuclear resonance scattering of synchrotron radiation

A X Trautwein[1], H Paulsen[1], H Winkler[1], H Giefers[2], G Wortmann[2], H Toftlund[3], J A Wolny[4], A I Chumakov[5] and O Leupold[5]

[1] Institut für Physik, Universität zu Lübeck, D-23538 Lübeck, Germany
[2] Department Physik, Universität Paderborn, D-33095 Paderborn, Germany
[3] Department of Chemistry, University of Southern Denmark, Odense, DK-5230, Denmark
[4] Fachbereich Physik, Technische Universität Kaiserslautern, D-67663 Kaiserslautern, Germany
[5] European Synchrotron Radiation Facility, BP 220, F-38042 Grenoble, France

E-mail: `trautwein@physik.uni-luebeck.de`

Abstract. Nuclear inelastic scattering (NIS) spectra were recorded for the spin-crossover complexes STP and ETP (STP = [Fe(1,1,1-tris{[N-(2-pyridylmethyl)-N-methylamino]methyl}-ethane)]$(ClO_4)_2$ and ETP = [Fe(1,1,1-tris{[N-(2-pyridylmethyl)-N-methylamino]methyl}-butane)]$(ClO_4)_2$) at 30 K and at room temperature and also at ambient pressure and applied pressure (up to 2.6 GPa). Spin transition from the high-spin (HS) to the low-spin (LS) state was observed by lowering temperature and also by applying pressure at room temperature and has been assigned to the hardening of iron-bond stretching modes due to the smaller volume in the LS isomer.

1. Introduction

The iron(II) complexes STP and ETP (STP = [Fe(1,1,1-tris{[N-(2-pyridylmethyl)-N-methylamino]methyl}ethane)]$(ClO_4)_2$ and ETP = [Fe(1,1,1-tris{[N-(2-pyridylmethyl)-N-methylamino]methyl}butane)]$(ClO_4)_2$) belong to the family of thermally driven spin-crossover complexes, which exhibit a transition from the low-spin (LS, $S = 0$) to the high-spin (HS, $S = 2$) state by increasing the temperature. These complexes are promising materials for optical information storage and display devices [1]. The first coordination sphere of the iron in STP and ETP is a distorted $[FeN_6]$ octahedron with C_3 symmetry (Fig. 1).

The spin transition occurs in both complexes gradually at about 180 K. Its driving force is the difference of the vibrational entropy between both spin states, which arises from softening of some vibrational modes when going from the LS to the HS state. NIS measurements on STP have demonstrated that mainly the iron-ligand bond stretching modes are affected this way [2, 3]. Similar findings were obtained for other spin-crossover complexes [4, 5, 6]. Spin transition is also achieved by increasing pressure; thus the LS becomes energetically favoured over the HS state due to its smaller volume.

2. Nuclear inelastic scattering (NIS)

Contributions in the spectrum with negative energy E account for annihilation and positive energy for creation of vibrations in the molecular complex (note that zero energy in Fig. 2 corresponds to the elastic part of the spectrum, while negative and positive energies cover the inelastic part). At low temperatures only low-energy phonon states are occupied (2.5 meV corresponds to approximately 30 K), therefore annihilation of phonons is scarce in this region. Creation of phonons, however, is possible at all vibrational energies.

The NIS spectra recorded for STP at 30 K and at room temperature (Fig. 2) reflect in their inelastic part different iron-ligand bonding: at 30 K the Lamb-Mössbauer factor f_{LM} is close to one ($f_{\mathrm{LM}} \approx 0.88$), thus dominating the elastic peak for the LS isomer, while at room temperature f_{LM} decreases to 0.16 yielding only a small elastic peak. The inelastic part of the two NIS spectra (inset of Fig. 2), at the same time, exhibits the drastic hardening of vibrational modes when decreasing temperature.

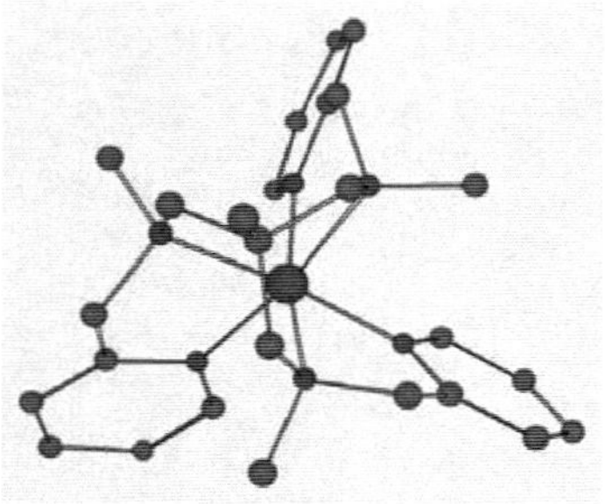

Figure 1. Molecular structure of STP.

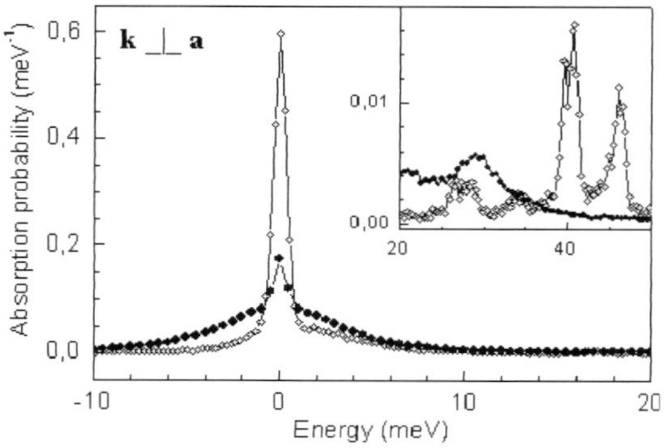

Figure 2. NIS spectra measured at $T = 30$ K ($\diamond$) and at room temperature ($\bullet$) for $\vec{k} \perp \vec{a}$. The inset shows the energy range of the iron-ligand bond stretching modes. The solid lines are guides to the eyes. Taken from [2].

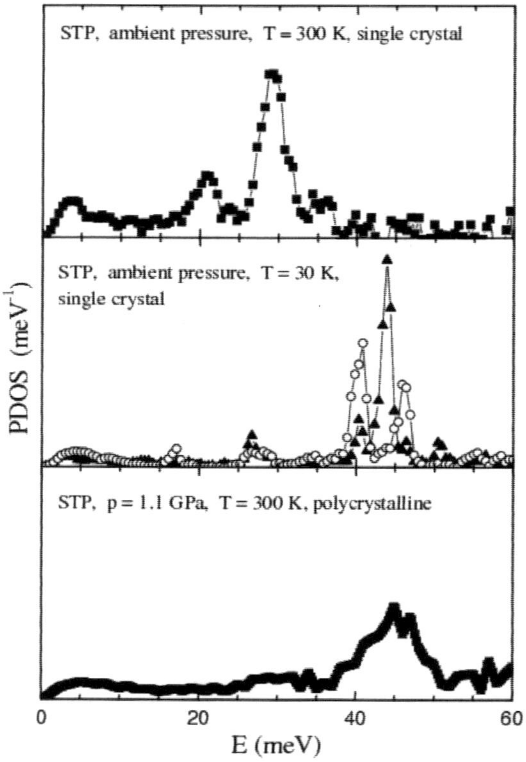

Figure 3. PDOS of STP for temperatures and pressures as indicated. The PDOS at 30 K was measured at two different orientations: $\vec{k} \perp \vec{a}$ (circles) and $\vec{k} \parallel \vec{a}$ (triangles).

3. Angular-resolved NIS measurements

The high energy resolution (0.65 meV) available at the Nuclear Resonance Beamline ID18 of the European Synchrotron Radiation Facility (ESRF) in Grenoble, France, allowed us to resolve individual molecular vibrations which were unambiguously identified by DFT calculations [2].

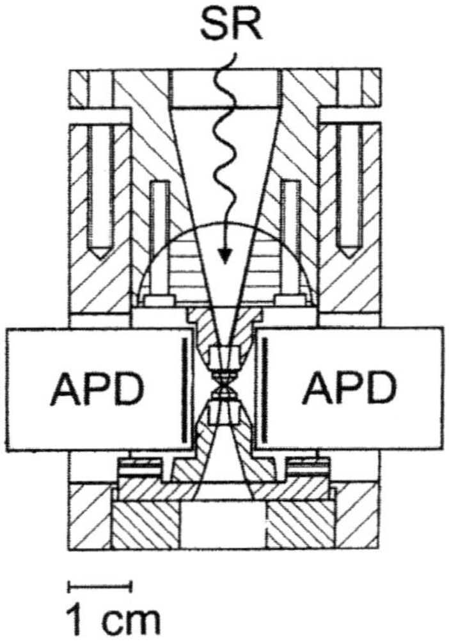

Figure 4. Diamond-anvil cell for high-pressure NFS and NIS studies.

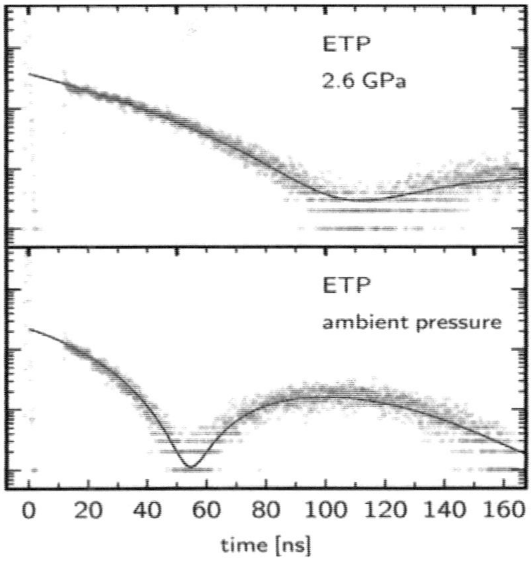

Figure 5. NFS spectra of ETP recorded at room temperature with ambient pressure and applied pressure as indicated.

In case of molecular crystals, and if the interactions between *inter*- and *intra*-molecular modes can be neglected, the molecular part of the anisotropic PDOS (partial density of states) can be formulated as

$$\mathrm{PDOS_{mol}} = \sum_j \delta(E - \hbar\omega_j)(\vec{k} \cdot \vec{u}_j)^2$$

with ω_j representing the vibrational frequencies of the molecule and $\vec{u}_j$ the projection of the jth eigenvector into the three-dimensional subspace of the iron coordinates, and $\vec{k}$ is the normalized wave vector of the photon. From this expression it is obvious that only those vibrational modes will be NIS visible that are connected with a mean square displacement of the iron nucleus along the wave vector $\vec{k}$.

In Fig. 3 the angular-resolved vibrational modes, recorded at 30 K for STP, are clearly visible. For the interpretation of the experimentally gained PDOS, DFT calculations were performed. This way the two predominant peaks at 40 and 46 meV for $\vec{k} \perp \vec{a}$ as well as the predominant peak at 44 meV and the small peak at 41 meV for $\vec{k} \parallel \vec{a}$ could be assigned to iron-ligand bond-stretching modes.

Considering an ideal [FeN_6] octahedron there exist six Fe-N bond-stretching modes which transform according to the A_{1g}, E_g and T_{1u} irreducible representations of the point group O_h. The *gerade* modes A_{1g} and E_g are not connected with a displacement of the iron nucleus, and thus only the three T_{1u} bond-stretching modes can be detected by NIS. Actually, the first coordination sphere of the iron in the STP complex has to be regarded as a distorted [FeN_6] octahedron with C_3 symmetry. Because of the lower symmetry the vibrational T_{1u} term is split into an A and a doubly degenerate E term. In this case all six bond-stretching modes are visible by NIS: two pairs of doubly degenerate E modes are connected with a displacement of the iron nucleus within the equatorial plane and are visible if $\vec{k}$ lies within this plane that is perpendicular to the threefold symmetry axis [Fig. 3 (open circles)]. The remaining two A modes are connected with a displacement of the iron nucleus along the molecular symmetry axis [Fig. 3 (filled triangles]. The smaller peak at 41 meV corresponds to the nearly fully symmetric breathing mode where the iron nucleus is only slightly participating in the vibration.

4. Pressure-induced spin transition

Spin transition is also achieved by pressure, because with increasing pressure the LS state is energetically favoured due to its smaller volume. The NFS and NIS spectra were recorded at ambient pressure and also at pressures up to 2.6 GPa in a diamond-anvil cell (DAC, Fig. 4). The NFS spectra of STP and ETP at ambient pressure exhibit the first quantum-beat minimum at about 55 ns (corresponding to $\Delta E_Q = 0.8$ mm/s) indicative of the HS state. When applying a pressure above 0.9 GPa, both STP and ETP samples undergo HS-LS transitions, with the quadrupole splitting of the LS isomer changing to a smaller value compared to that of the HS state (Fig. 5). The spin transition was also monitored optically in the DAC by a characteristic change of the sample colour from green to red (Fig. 6). The PDOS of Fe in STP, extracted from

Figure 6. Colour change of the STP sample at ambient pressure (left picture) and at applied pressure of 1.1 GPa (right picture).

the NIS spectra measured at 1.1 GPa and 300 K, is shown in Fig. 3 together with the PDOS obtained from a single crystal of STP. The comparison with the PDOS at ambient pressure and 300 K (HS state) and at 30 K (LS state) strikingly demonstrates that STP at 1.1 GPa and 300 K has transformed into the LS phase with comparable frequencies of the high-energy stretching modes (note that the PDOS of STP at 30 K in Fig. 3 is shown for two different directions (filled triangles, open circles) while for the polycrystalline sample all directions are averaged).

Acknowledgments

The authors acknowledge the support by R. Rüffer and by the relevant ESRF services and the financial support by the Deutsche Forschungsgemeinschaft (DFG) and by the German Federal Ministery for Education and Research (BMBF).

References

[1] Gütlich P, Hauser A and Spiering H 1994 *Angew. Chemie* **106** 2109; Gütlich P, Hauser A and Spiering H 1994 *Angew. Chemie, Int. Ed. Engl.* **33** 2024
[2] Paulsen H, Benda R, Herta C, Schünemann V, Chumakov A I, Duelund L, Winkler H, Toftlund H and Trautwein A X 2001 *Phys. Rev. Lett.* **86** 1351
[3] Paulsen H, Grünsteudel H, Meyer-Klaucke W, Gerdan M, Grünsteudel H F, Chumakov A I, Rüffer R, Winkler H, Toftlund H and Trautwein A X 2001 *Eur. Phys. J.* B **23** 463
[4] Paulsen H, Winkler H, Trautwein A X, Grünsteudel H, Rusanov V, Toftlund H 1999 *Phys. Rev.* B **59** 975
[5] Böttger L H, Chumakov A I, Grunert M, Gütlich P, Kusz J, Paulsen H, Ponkratz U, Rüffer R, Rusanov V, Trautwein A X, Wolny J A 2006 *Chem. Phys. Lett.* **429** 189
[6] Ronayne K L, Paulsen H, Höfer A, Dennis A C, Wolny J A, Chumakov A I, Schünemann V, Winkler H, Spiering H, Bousseksou A, Gütlich P, Trautwein A X, McGarvey J J 2006 *Phys. Chem. Chem. Phys.* **8** 4685

In-field Mössbauer characterization of MFe_2O_4 (M = Fe, Co, Ni) nanoparticles

Amanda D Arelaro[1], Liane M Rossi[2], and Hercilio R Rechenberg[1†]

[1]Physics Institute, São Paulo University, São Paulo (Brazil)
[2]Chemistry Institute, São Paulo University, São Paulo (Brazil)

[†]Corresponding author. E-mail: hercilio@if.usp.br

Abstract. Nanoparticle samples of Fe_3O_4, $CoFe_2O_4$ and $NiFe_2O_4$ with mean sizes in the 5-8 nm range and excellent crystallinity have been synthesized by thermal decomposition of organic precursors. Mössbauer spectra taken at 4.2 K in applied fields from 0 to 12 T could be fitted with two sextets corresponding to the A and B Fe sites of the spinel structure. The cation distribution was deduced from the relative intensities of both sextets. The Fe oxide spectrum was more similar to γ-Fe_2O_3 than to Fe_3O_4, suggesting this sample has undergone partial or total oxidation. A small residual intensity at lines 2 and 5 reveals some spin canting at both Fe sites, however the absence of a third sextet indicates that a "core-shell" spin structure, with disordered spins near the surface, does not occur in these nanoparticles.

1. Introduction

Ferrites of formula MFe_2O_4 are an important class of iron oxide materials which have long been of scientific and technological interest. In more recent years the interest in nanosized ferrites has grown tremendously, fundamentally in reason of their potential use for biomedical applications [1,2]. Several methods have been used to produce nanocrystalline ferrites, such as mechanical milling, coprecipitation, synthesis in reverse micelles, sol-gel, and others. Recently, a new synthesis method by thermal decomposition of an organic precursor has been described [3,4], which provides samples with homogeneous morphology, a good control over grain size, and improved crystallinity of the particles. Using this method we have prepared various samples of nanoparticles (NPs) of the title compounds. One sample of each composition was chosen for a complete magnetic characterization, in order to investigate the influence of the M^{2+} ion on the magnetic properties. Experimental determination of the anisotropy constant K_{eff} has been reported elsewhere [5].

In this work we have investigated the same samples by Mössbauer spectroscopy in applied magnetic fields at low temperature. The reasons for carrying out this kind of experiment are, at least, twofold. (1) To fully understand the magnetic properties it is important to know how the magnetic ions (specifically iron) are distributed among the two possible sites, namely the A (tetrahedral) and B (octahedral) sites of the spinel structure. Mössbauer spectra are best suited for this purpose, however the hyperfine fields at both sites are often very similar and the corresponding sextets are poorly resolved. Application of a field remedies this shortcoming since the sextets become well separated due to the ferrimagnetic spin ordering. (2) Using in-field Mössbauer spectra it has been discovered [6] that Fe oxide NPs are constituted by an inner core of ferrimagnetically ordered spins surrounded by a layer of misaligned spins; such a configuration, which later came to be known as the "core-shell model" [7], can be best observed with this technique. A quantitative study on the thickness of the disordered layer as a function of applied field in $NiFe_2O_4$ NPs has recently been published [8].

2. Materials and methods

The investigated compounds were prepared by high-temperature solution reaction. For obtaining magnetite (Fe_3O_4) NPs, Fe(III) acetylacetonate, 1,2-octanediol, oleylamine, oleic acid and diphenyl ether were mixed and vigorously stirred under a nitrogen flow. The mixture was heated to 265 °C and kept in reflux condition for two hours. Ethanol was added to the solution and the oxide powder was precipitated by centrifugation. Similarly, $CoFe_2O_4$ and $NiFe_2O_4$ NPs were prepared by adding $Co(acac)_2$ or $Ni(acac)_2$ in the required proportion. Elemental analysis was carried out by the ICP-AES method, yielding empirical formulas $Fe_{3.0}O_4$, $Co_{0.9}Fe_{2.1}O_4$ and $Ni_{0.8}Fe_{2.2}O_4$. Structural characterization was done by x-ray diffraction (XRD) and tansmission electron microscopy (TEM).Mössbauer spectra were collected at T = 4.2 K under fields μ_0H = 0, 4, 8 and 12 T provided by a Cryogenic™ superconducting magnet and parallel to the γ-ray direction. A ~30 mCi ^{57}Co source was driven sinusoidally in a field-free region of the cryostat.

3. Results and discussion

XRD diagrams of all samples indicated a single phase with the spinel structure. TEM images showed nearly spherical and well-crystallized nanoparticles. The NP diameter distribution was fitted to a log-normal function, yielding the following mean diameters <d> and (in parentheses) standard deviations σ : Fe_3O_4 6.2 nm (0.13), $CoFe_2O_4$ 5.3 nm (0.18), and $NiFe_2O_4$ 8.1 nm (0.17). Mean diameters deduced from XRD diagrams by use of Scherrer's formula are very similar to the latter ones, indicating that no particle agglomeration has occurred; indeed, the product of this synthesis are NPs coated with oleic acid, which acts as a surfactant.

Mössbauer spectra at room temperature are superparamagnetic doublets, as expected. The 4.2 K spectra in various fields are shown in figures 1, 2 and 3. All spectra could be satisfactorily fitted with two magnetic sextets; the line intensities were taken in the ratio 3:q:1 where q was left free to vary for each sextet. Anticipating the possibility of a core-shell structure [8], we also tried a model with three sextets, but the fit usually converged to nonphysical values of some parameters such as linewidth, intensity, etc. and in any case did not improve the fit quality. The area ratio between the two sextets,

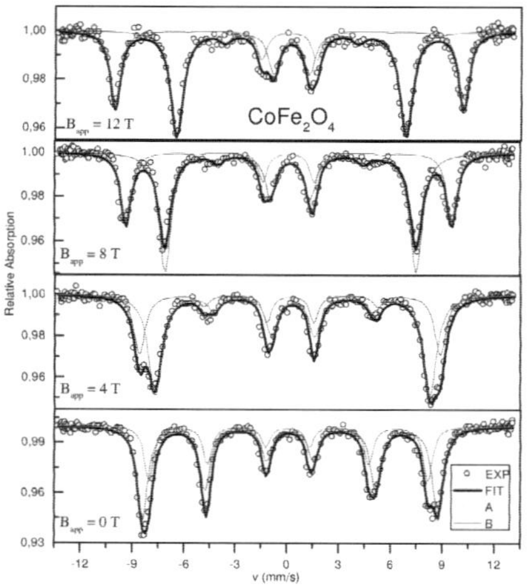

Figure 1. Mössbauer spectra of $CoFe_2O_4$ sample at indicated fields. Fitted subspectra are shown by continuous lines.

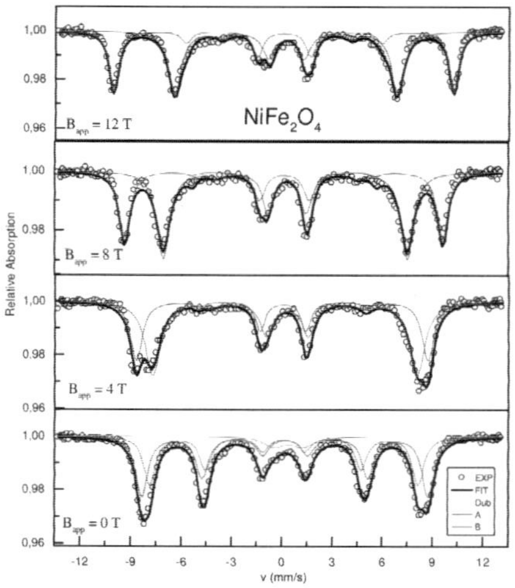

Figure 2. Mössbauer spectra of $NiFe_2O_4$ sample at indicated fields. Fitted subspectra are shown by continuous lines.

obtained at the higher fields, was used as a constraint in fitting the lower-field, less-resolved spectra. Due to the ferrimagnetic spin ordering, the effective field that increases (*decreases*) for increasing applied field can be unambiguously associated to Fe ions placed at the A (*B*) site. The linear dependence $B_{eff} = B_{hf} \pm B_{app}$ is illustrated in figure 4.

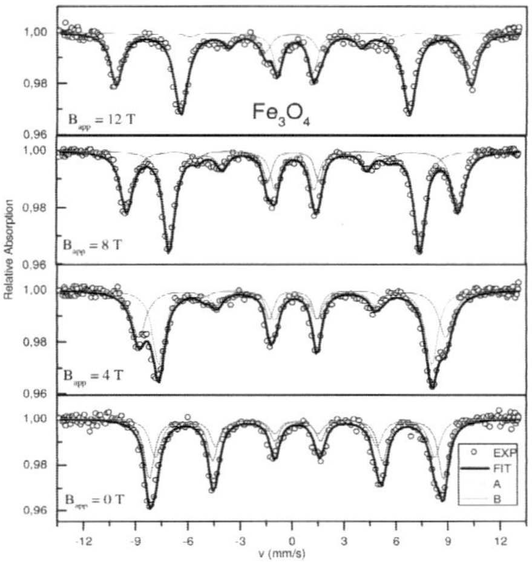

Figure 3. Mössbauer spectra of Fe_3O_4 sample at indicated fields. Fitted subspectra are shown by continuous lines.

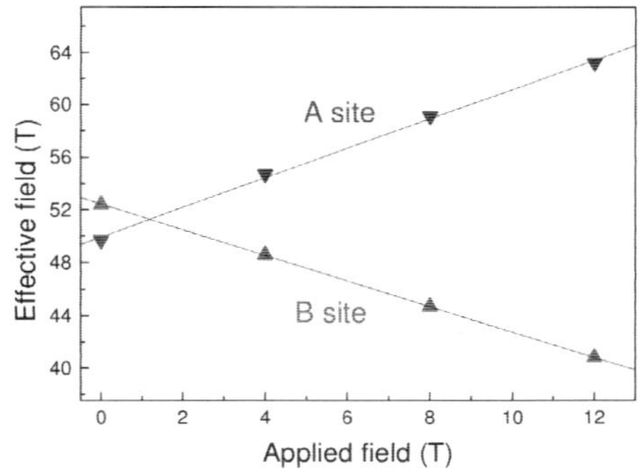

Figure 4. Effective fields *vs.* applied field for Fe_3O_4 sample, showing that B_{app} adds to or subtracts from the hyperfine field B_{hf} at each site.

Table 1 shows the most relevant hyperfine parameters obtained by fitting, i.e. the hyperfine fields $B_{hf}(A)$ and $B_{hf}(B)$, the isomer shift difference $\delta IS = IS(B) - IS(A)$ and the B/A sextet area ratio. It is not meaningful to quote IS values separately owing to uncertain temperature corrections. Quadrupole shift values were always smaller than 0.1 mm/s.

Table 1. Zero-applied field hyperfine field at sites A and B (T), A-B isomer shift difference (mm/s), and B/A sextet area ratio. Uncertainties are given in parentheses.

	$B_{hf}(A)$	$B_{hf}(B)$	δIS	I_B/I_A
$CoFe_2O_4$	50.1(1)	53.3(1)	0.13(2)	1.68
$NiFe_2O_4$	49.6(1)	52.8(1)	0.14(5)	1.25
Fe_3O_4	49.8(1)	52.4(1)	0.10(3)	1.56

In addition to the data listed above, a very important quantity is the second-to-third line intensity ratio q, from which the mean canting angle can be estimated according to the expression

$$q = \frac{4 < \sin^2 \theta >}{2 - < \sin^2 \theta >} . \qquad (1)$$

The investigated compounds will be discussed separately in the following.

3.1. Co ferrite
For M = Co, our data yielded a B/A ratio equal to 1.68. Combining this result with the chemically determined stoichiometry, one obtains the structural formula $(Co_{0.20}Fe_{0.80})_A[Fe_{1.34}Co_{0.66}]_BO_4$, showing that some cation exchange has taken place ($CoFe_2O_4$ is an inverse spinel in bulk form). Regarding parameter q (equation 1), both sites showed spin canting, the one at A decreasing for increasing field and the one at B remaining constant around 20°.

3.2. Ni ferrite
For M = Ni, B/A = 1.25 yielding $(Ni_{0.02}Fe_{0.98})_A[Fe_{1.22}Ni_{0.78}]_BO_4$; this sample is thus an almost perfectly inverse spinel. The canting angle at both Fe sites was (20±5)° and insensitive to the applied field.

3.3. Fe oxide
Data for the Fe oxide give B/A = 1.56, which differs significantly from the 2:1 ratio expected for Fe_3O_4 but is consistent, within experimental error, with the 5:3 ratio for γ-Fe_2O_3. It thus appears that our sample, nominally magnetite, has oxidized to maghemite after preparation. This conclusion is supported by the fact that the measured spectra are a sum of only two sextets; otherwise, the zero-field spectrum at low temperature should be more complex, with various shoulders as shown by Fe_3O_4 below the Verwey transition temperature [9]. Spin canting angles were similar to those observed in Ni ferrite.

3.4. Spin structure of the nanoparticles
The spectra fitting with only two sextets implies the absence of a significant fraction of randomly canted Fe spins on the particle's surface, since otherwise a third sextet, with broad lines, would be required to fit the in-field spectra [6,8]. The absence of such a "core-shell" spin arrangement in samples produced by the high-temperature reaction method has been observed in other recent studies [10,11] and has been attributed to the organic coating, which might reduce the particle's surface anisotropy. On the other hand, we have no explanation for the observed spin canting at both sites in the ferrimagnetic core, although canting at the B site is not unusual in ferrites [12].

Acknowledgments
This work had financial support from Fundação de Amparo à Pesquisa do Estado de São Paulo (FAPESP) and from Conselho Nacional do Desenvolvimento Científico e Tecnológico (CNPq).

References
[1] Berry C C, Curtis A S G (2003) *J. Phys. D: Appl. Phys.* **36** 198-206
[2] Pankhurst Q A,Connolly J, Jones S K, Dobson J (2003) *J. Phys. D: Appl. Phys.* **36** 167-81
[3] Hyeon T, Lee S S, Park J, Chang Y, Na H B (2001) *J. Am. Chem. Soc.* **123** 12798-801
[4] Sun S, Zeng H (2002) *J. Am. Chem. Soc.* **124** 8204-5
[5] Arelaro A D, Lima Jr E, Rossi L M, Kiyohara P K, Rechenberg H R (2008) *J. Magn. Magn. Mater.* **320** e73-e76
[6] Coey J M D (1971) *Phys. Rev. Lett.* **27** 1140-2
[7] Kodama R H, Berkowitz A E, McNiff E J, Foner S (1996) *Phys. Rev. Lett.* **77** 394-7
[8] Rechenberg H R, Sousa E C, Depeyrot J, Sousa M H, Aquino R, Tourinho F A, Perzynski R (2008) *Hyperfine Interact.* **184** 423-8
[9] Mørup S, Topsøe H (1996) *J. Magn. Magn. Mater.* **31-34** 953-4
[10] Chkoundali S, Ammar S, Jouini N, Fiévet F, Molinié P, Danot M, Villain F, Grenèche J-M (2004) *J. Phys.: Condens. Matter* **16** 4357-72
[11] Roca A G, Niznansky D, Poltierova-Vejpravova J, Bittova B, González-Fernández M A, Serna C J, Morales M P (2009) *J. Appl. Phys.* **105** 114309
[12] Peddis D, Mansilla M V, Mørup S, Cannas C, Musinu A, Piccaluga G, D'Orazio F, Lucari F, Fiorani D (2008) *J. Phys. Chem. B* **112** 8507-13

Magnetic properties of $Cu_xCo_{1-x}Fe_{2-x}Al_xO_4$ oxides

J Z Msomi[1] and T Moyo[2]

[1]Department of Physics, University of Free State, Phuthaditjhaba 9866, South Africa

[2]School of Physics, University of KwaZulu-Natal, Durban 4000, South Africa

E-mail: msomijz@qwa.uovs.ac.za

Abstract. We report the magnetic properties of $Cu_xCo_{1-x}Fe_{2-x}Al_xO_4$ with x values ranging from 0 to 1.0 in steps of 0.1. The samples were characterized by x-ray diffraction, Mössbauer spectroscopy and magnetization measurements. All samples crystallized with single-phase cubic spinel structure. Lattice parameter decreases with x according to Vergad's law. The Mössbauer spectra show the presence of magnetic hyperfine fields for $x \leq 0.9$ due to Fe^{3+} ions on the tetrahedral (A) and octahedral (B) sites. The hyperfine magnetic fields decrease with increasing x. The variations of coercive fields and saturation magnetization with increasing Cu and Al concentrations observed are discussed.

1. Introduction

Ferrites are a group of oxides with the spinel crystal structure AB_2O_4. Metal cations occupy the tetrahedral (A) and octahedral (B) sites. The distribution and type of cations amongst A and B sites have remarkable effects on the properties of ferrites. Making suitable substitutions can improve the properties. This has resulted in numerous reports in the literature associated with substitutions by diamagnetic ions on tetrahedral (A) or octahedral (B) sites. Mössbauer studies of $Zn_xCo_{1-x}Fe_{2-x}Al_xO_4$ compounds have been reported in reference [1]. Of interest are the changes in the magnetic properties when instead of Zn we have Cu in a compound. The ionic sizes of Zn (0.74 Å) and Cu (0.72 Å) are similar. However, a Zn atom has a full outer s electron orbital and Cu has an unpaired electron in its outermost orbital. In this paper we investigate the effect of simultaneous dilution of cobalt ferrite by Cu and Al ions on the magnetic properties.

2. Experimental details

Samples of $Cu_xCo_{1-x}Fe_{2-x}Al_xO_4$ were made following procedures reported in reference [1]. X-ray diffraction studies using Co (K_α) radiation using a Phillips diffractometer (type PW1710) confirmed single phase cubic spinel structure. The Mössbauer spectra were recorded using conventional constant acceleration Mössbauer spectrometer with a ^{57}Co source sealed in Rh matrix. Isothermal magnetization measurements were performed using a vibrating sample magnetometer at room temperature in an applied magnetic field in the range: $-9\ \text{kOe} \le B_0 \le 9\ \text{kOe}$.

3. Results and discussions

The XRD spectra are shown in figure 1. The most significant lines are characteristic of cubic spinel structure. The variation of the lattice parameter with x derived from XRD data is shown in table 1 and plotted in figure 2.

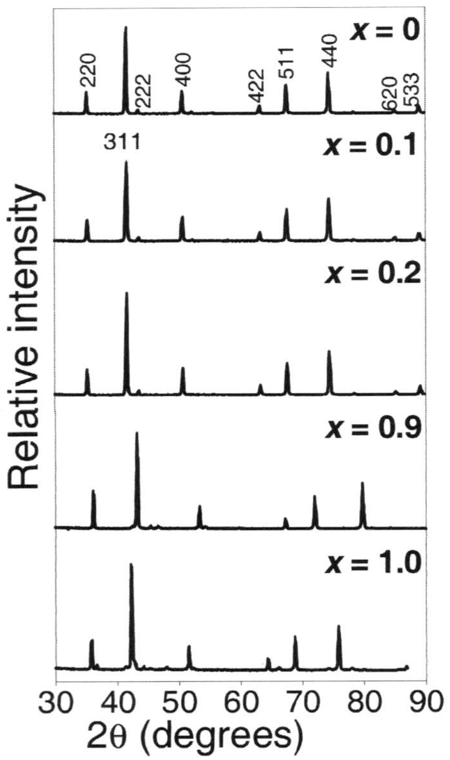

Figure 1. Typical variation of XRD spectra for $Cu_xCo_{1-x}Fe_{2-x}Al_xO_4$, with x.

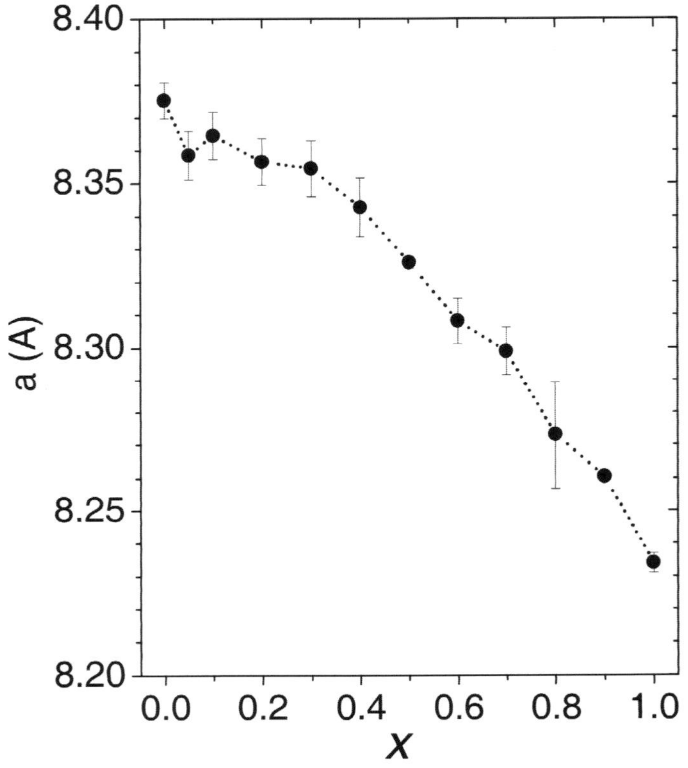

Figure 2. Variation of lattice parameters for $Cu_xCo_{1-x}Fe_{2-x}Al_xO_4$, with x.

Al atoms are known to have preference to occupy the octahedral B sites which have more room [1]. The decrease of lattice parameters with x is therefore attributed to smaller size of Al ions (0.51 Å) substituting the larger ions of Co (0.63 Å) or Fe (0.64 Å) on B sites. X-ray densities of the oxides can be calculated from the lattice parameter [2, 3]. The values of x-ray and bulk densities are shown in table 1. X-ray densities reflect on the packing of the atoms in the crystal structure. They are generally larger than bulk densities calculated from the dimensions and masses of pellets. Our values compare well with those obtained for similar compounds [2, 3]. The percentage porosity (P) of the sample can be calculated from bulk (ρ_B) and x-ray densities (ρ_X) [3]. The porosity fluctuates between 30 % and

60 %. These values are higher than the range 8 % - 18 % observed in (Zn, Cd, Cu)$_{0.5}$Ni$_{0.5}$Fe$_2$O$_4$ compounds [3]. Clearly, the $Cu_xCo_{1-x}Fe_{2-x}Al_xO_4$ oxides are extremely porous. The grain size estimated using Scherrer equation [4], fluctuates between 35 nm and 60 nm.

The typical Mössbauer spectra measured at room temperature are shown in figure 3. The spectra were fitted with two doublets and two sextets. The sextets represent Fe ions in the ordered spin state on the tetrahedral (A) and octahedral (B) sites. The doublets reflect presence of Fe ions in paramagnetic states. Well resolved magnetic splitting is observed up to $x = 0.9$. This indicates ferrimagnetic behavior in the oxides. Even at $x = 1.0$, some of the magnetic atoms have ordered spin state. This behavior is unlike the similar compounds $Zn_xCo_{1-x}Fe_{2-x}Al_xO_4$ [1] where the ordered magnetic state collapses at $x = 0.4$. This reflects stronger magnetic coupling in the current $Cu_xCo_{1-x}Fe_{2-x}Al_xO_4$ system.

Table 1. Bulk (ρ_B) , x-ray density (ρ_X), porosity (P), grain size (G) and lattice parameter *(a) for* $Cu_xCo_{1-x}Fe_{2-x}Al_xO_4$ oxides.

x	ρ_B (g/cm^3)	ρ_X (g/cm^3)	P (%)	G (nm)	a (Å)
	± 0.02	± 0.01	± 0.1	± 0.4	± 0.01
0	3.66	5.30	30.9	45.3	8.38
0.05	3.13	5.41	62.5	50.7	8.36
0.1	2.93	5.40	65.0	37.1	8.37
0.2	4.63	5.41	44.6	49.9	8.36
0.3	4.72	5.42	43.6	59.7	8.35
0.4	4.25	5.44	49.1	57.4	8.34
0.5	4.25	5.47	48.9	46.2	8.33
0.6	3.87	5.57	53.5	56.0	8.31
0.7	4.31	5.52	48.0	57.2	8.29
0.8	3.86	5.58	53.4	35.1	8.27
0.9	4.02	5.60	51.3	57.1	8.26
1	4.43	5.66	46.1	52.1	8.23

We attribute this behaviour to the unpaired electron in the outer shell of Cu atoms. Valence electrons may be acting as conduction electrons, resulting in stronger superexchange interactions between A and B sites magnetic atoms. Sextets and doublets were assigned to Fe cations at A or B sites based on the fitted results of isomer shifts and quadrupole splitting. Isomer shifts and quadrupole splitting on A sites are expected to be lower due to high symmetry compared to B sites [4]. The hyperfine fields at A and B sites decrease with x as shown in figure 4. This is attributed to the weakening of the magnetic coupling between the magnetic moments. The average isomer shifts on A and B sites were 0.29 ± 0.01 mm/s and 0.36 ± 0.03 mm/s. No significant change of the isomer shift was observed with simultaneous dilution by Cu or Zn and Al. This means that Cu and Al contents do not significantly affect the s-electron charge distribution of the Fe^{3+} ions in tetrahedral (A) and octahedral (B) sites.

Typical magnetization curves are shown in figure 5 for $x = 0$ and $x = 0.6$. A general decrease in coercive fields and saturation magnetization measurements was observed. This relates well with weakening magnetic coupling indicated by Mössbauer results. For $x = 0$ (cobalt ferrite $CoFe_2O_4$)

saturation magnetization at the highest field is about 68 emu/g. This compares well with saturation magnetization of about 70 emu/g reported for cobalt ferrite by K. Maaz et al [5].

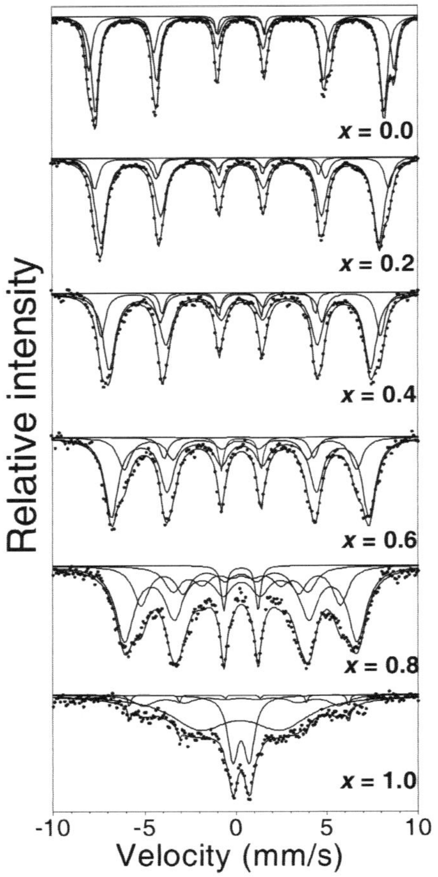

Figure 3. Typical variation of Mössbauer spectra for $Cu_xCo_{1-x}Fe_{2-x}Al_xO_4$ with x.

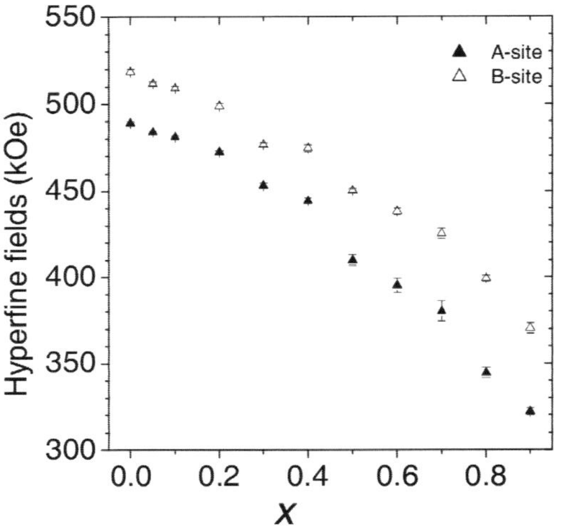

Figure 4. Variation of hyperfine fields for $Cu_xCo_{1-x}Fe_{2-x}Al_xO_4$., with x.

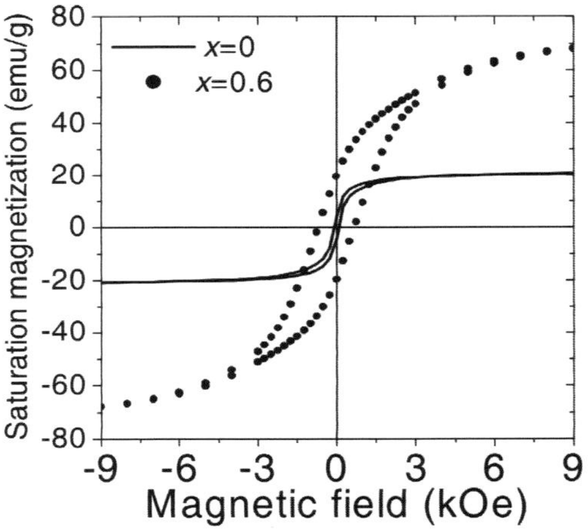

Figure 5. Magnetization curves for x=0 and x=0.6.

4. Conclusion

Double substitution by Cu and Al into $CoFe_2O_4$ has been achieved. The $Cu_xCo_{1-x}Fe_{2-x}Al_xO_4$ ($x \leq 0.9$) oxides are ferrimagnetic. The collapse of magnetic order observed in Zn and Al based compounds at about x = 0.5 [1] does not occur for the Cu-based samples. This behavior is attributed to the different electronic configuration for Cu.

References

[1] V. T. Thanki, K. H. Jani, B. S. Trivedi, Kunal B. Modi H. H. Joshi, *Materials Lett.* 37 (1998) 236.
[2] D. Ravinder, T. Alivelumanga, *Materials Lett.* 49 (2001) 1.
[3] J.Z. Msomi, K. Bharuth-Ram and T.Moyo, *Hyperfine interactions* 158 (2004) 151.
[4] X. Qi, D. Wu, *J. Magn. Magn. Mater.* 320 (2008) 666.
[5] K. Maaz, A. Muntaz, S. K. Hasanin, A.Ceylan, *J. Magn. Magn. Mater.* 308 (2007) 289.

Journal of Physics: Conference Series **217** (2010) 012128 doi:10.1088/1742-6596/217/1/012128

Isomerization effect of counter anion on the spin crossover transition in $[Fe(4\text{-}NH_2trz)_3](CH_3C_6H_4SO_3)_2{\cdot}nH_2O$

A Sugahara, M Enomoto and N Kojima
Graduate School of Arts and Sciences, The University of Tokyo, Komaba, 3-8-1, Meguro-ku, Tokyo 153-8902, Japan

E-mail: cnori@mail.ecc.u-tokyo.ac.jp

Abstract. We have investigated the spin crossover transition of the triazole bridged one dimensional Fe^{II} complexes, $[Fe(4\text{-}NH_2trz)_3]$(*o*-, *m*-, *p*-tos)$_2{\cdot}nH_2O$ (tosH = toluenesulfonic acid), in order to study the isomerization effect of counter anion on the spin crossover phenomenon by means of ^{57}Fe Mössbauer spectroscopy and magnetic susceptibility measurement. In the heating process, the spin transition of the all salts occurs around room temperatures, i.e. those of *o*-, *m*-, *p*-tos salts are 330 K, 319 K and 320 & 350 K, respectively. In the cooling process, the structural isomerization effect of counter anion on the spin crossover phenomenon is more remarkable than that in the heating process. In the case of *o*-tos salt, the spin transition occurs at 250 K with thermal hysteresis of 80 K. On the other hand, in the case of the *m*-tos salt, the spin transition occurs abruptly at 319 K with negligible small hysteresis. In the case of the *p*-tos salt, the spin transition occurs abruptly at 295 K in the cooling process in spite of stepwise spin transitions in the heating process.

1. Introduction

Transition metal complexes with d^4-d^7 configuration have a possibility of spin transition between low-spin (LS) and high-spin (HS) states. Among Fe^{II} spin crossover complexes, Fe^{II} polymeric chain complexes bridged by 1,2,4-triazole (trz) are known to show the spin transition between the diamagnetic LS state (S = 0) and the paramagnetic HS state (S = 2) with a large thermal hysteresis around room temperature.[1] In this system, the transition temperature and hysteresis width are controlled by ligands, counter anions or crystal solvents. As the example in which the ion radius directly influences the transition temperature, the spin crossover phenomena of $[Fe(4\text{-}NH_2trz)_3]X_2$ (4-NH_2trz = 4-amino-1,2,4-triazole, X = Cl^-, Br^-, NO_3^-, I^-, BF_4^-, ClO_4^-) were reported.[2, 3] As the radii of the anion increases, the transition temperature and hysteresis width linearly decrease. On the other hand, in the case of $[Fe(4\text{-}NH_2trz)_3]X_2{\cdot}nH_2O$(X = $C_mH_{2m+1}SO_3$), the spin transition temperature increases up to 80 K as m increases from 1 to 4, which is attributed to the fastener effect between the counter anions.[4]

As mentioned above, the spin crossover transition in $[Fe^{II}(trz)]^{2+}$ complex remarkably depends on the size of counter anion, which implies a possibility to control of the spin crossover transition by means of the photo-isomerization of counter anion. In order to confirm the structural isomerization effect on the spin crossover behavior for the $[Fe^{II}(trz)]$ complex, we have investigated the effect of the structural isomer of counter anion on the spin crossover phenomenon in $[Fe(4\text{-}NH_2trz)_3]$(*o*-, *m*-, *p*-tos)$_2{\cdot}nH_2O$ (tosH = toluenesulfonic acid) by means of ^{57}Fe Mössbauer spectroscopy and magnetic susceptibility measurement.

2. Experimental

[Fe(4-NH_2trz)$_3$](*o*-, *m*-, *p*-tos)$_2$·nH_2O were prepared in the following way. Iron powder was dissolved in the aqueous solution of toluenesulfonic acid. Adding the aqueous solution of 4-amino-1,2,4-triazole, [Fe(4-NH_2trz)$_3$](*o*-, *m*-, *p*-tos)$_2$ ·nH_2O was precipitated immediately. The number of hydrated water was decided by thermo gravimetric analysis as n = 2, 1.5, 1.5 for the *o*-, *m*-, *p*-isomers. Analysis calculated for C20H30FeN12O8S2 (*o*-tos): C, 34.99; H, 4.40; N, 24.48. Found: C, 34.82; H, 4.27; N,24.39 . Analysis calculated for C20H29FeN12O7.5S2 (*m*-tos): C: 35.35, H: 4.60, N: 24.74. Found: C: 35.13, H: 4.25, N: 25.82. Analysis calculated for C20H29FeN12O7.5S2 (*p*-tos): C: 35.35, H: 4.60, N: 24.74 Found: C: 35.43, H: 4.49, N: 25.39.

[Fe(4-NH_2trz)$_3$](*o*-, *m*-, *p*-tos)$_2$·nH_2O has the similar powder X-ray diffraction patterns of [Fe(4-NH_2trz)$_3$](*o*-, *m*-, *p*-abs)$_2$·nH_2O. This fact indicated that [Fe(4-NH_2trz)$_3$](tos)$_2$·nH_2O builds the same 1D Fe polymeric chain. Temperature dependence of magnetic susceptibility was measured by a Quantum Design MPMS5 SQUID susceptometer in the temperature range between 2 and 300 K under 5000 Oe. Powder samples were sealed in an aluminium capsule to avoid a loss of crystal water in a heating process. For ^{57}Fe Mössbauer spectroscopy, ^{57}Co in Rh was used as a Mössbauer source. Powder samples were sealed in acrylate resin sample holder to avoid a loss of crystal water in a heating process.The spectra were calibrated by using the six lines of a body-centered cubic iron foil (α-Fe), the center of which was taken as zero isomer shift.

3. Results and discussion

3.1 Magnetic susceptibility

Figure 1 shows the temperature dependence of the magnetic susceptibility multiplied by temperature (χT) of [Fe(4-NH_2trz)$_3$](*o*-, *m*-, *p*-tos)$_2$·nH_2O. In the heating process, the spin transition of the all salts occurs around room temperatures, i.e. those of *o*-, *m*-, *p*-tos salts are 330 K, 319 K, and 320 & 350 K, respectively. In the case of *p*-tos salt shows the stepwise spin transitions at 320 and 350 K.

In the cooling process, the structural isomerization effect of counter anion on the spin crossover phenomenon is more remarkable than that in the heating process. In the case of *o*-tos salt, the spin transition occurs at 250 K with thermal hysteresis of 80 K. On the other hand, in the case of the *m*-tos salt, the spin transition occurs abruptly at 319 K with negligible small hysteresis. Furthermore, in the case of the *p*-tos salt, the spin transition occurs abruptly at 295 K in the cooling process, while stepwise spin transitions in the heating process.

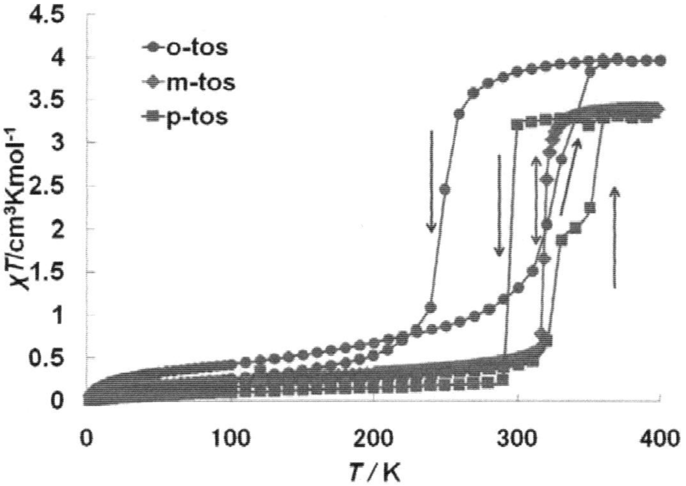

Figure 1. χT versus T plot of [Fe(4-NH_2trz)$_3$](*o*-, *m*-, *p*-tos)$_2$ ·nH_2O.

Journal of Physics: Conference Series **217** (2010) 012128 doi:10.1088/1742-6596/217/1/012128

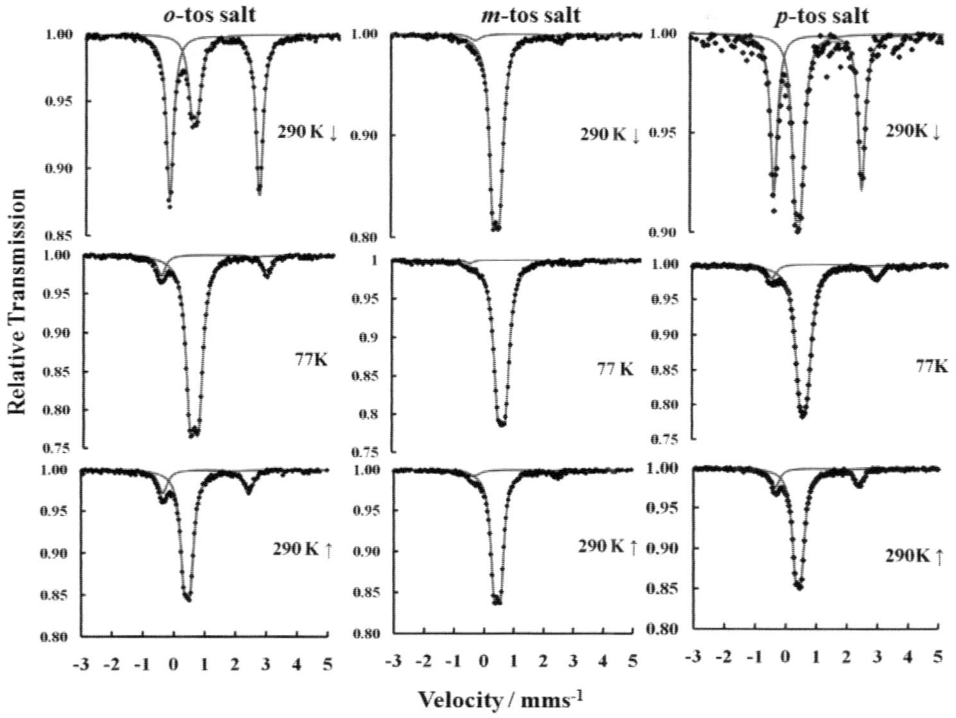

Figure 2. ^{57}Fe Mössbauer spectra for Fe(4-NH_2trz)$_3$](*o*-, *m*-, *p*-tos)$_2$ ·nH_2O at 290 and 77K

Table 1. ^{57}Fe Mössbauer parameters for Fe(4-NH_2trz)$_3$](*o*-, *m*-, *p*-tos)$_2$ ·nH_2O at 290 and 77K.

	Temperature (K)	Spin state	IS (mm/s)	QS (mm/s)	Line width (mm/s)	Area (%)
o-tos salt	290 ↓	LS	0.411	0.187	0.310	29
		HS	1.04	2.83	0.268	71
	77	LS	0.528	0.265	0.333	87.8
		HS	1.16	3.38	0.312	12.2
	290↑	LS	0.431	0.298	0.305	76.5
		HS	1.04	2.8	0.354	23.5
m-tos salt	290 ↓	LS	0.323	0.212	0.286	93.7
		HS	0.978	2.642	0.445	6.3
	77	LS	0.497	0.238	0.342	97.5
		HS	1.15	3.347	0.323	2.5
	290↑	LS	0.436	0.215	0.286	90.9
		HS	1.06	2.768	0.455	9.1
p-tos salt	290 ↓	LS	0.404	0.175	0.286	47
		HS	1.02	2.78	0.266	53
	77	LS	0.488	0.218	0.364	88.4
		HS	1.13	3.27	0.335	11.6
	290↑	LS	0.425	0.183	0.296	78.7
		HS	1.03	2.72	0.34	21.3

3.2 Mössbauer spectra

To confirm the spin crossover phenomena for [Fe(4-NH_2trz)$_3$](*o*-, *m*-, *p*-tos)$_2$·nH_2O, we measured ^{57}Fe Mössbauer spectra at 77 and 290 K (↓: cooling process, ↑: heating process). In the cooling process, the sample was heated up to 350 K at first, then it was cooled down to 290 K. Figure 2 and Table 1 show the ^{57}Fe Mössbauer spectra and their Mössbauer parameters for [Fe(4-NH_2trz)$_3$](*o*-, *m*-, *p*-tos)$_2$·nH_2O. In the cases of the *o*-tos and *p*-tos salts, both complexes indicate the coexistence of the LS and the HS states at 290 K in the cooling process. As these samples are cooled down to 77 K, the HS fraction decreases. The *o*-tos and *p*-tos salts are in agreement with the magnetic susceptibility measurement. The transition temperatures of the *o*-tos and *p*-tos salts are 250 and 292 K in the cooling process, respectively. Therefore the HS fractions of these salts remain at 290K↓ in Mössbauer spectra. Meanwhile, in the heating process, the transition temperatures of the *o*-tos and *p*-tos salts are 319 and 320 K, respectively. As spin transition from the LS to the HS does not occur, the LS fraction dominantly exists both for the *o*-tos and *p*-tos salts at 290K↑.

On the other hand, the *m*-tos salt shows the quite different spectra from those of the other salts. At 290 K in both cooling and heating process, LS fraction dominantly exists, which is consistent with the magnetic susceptibility measurement. From the analysis of the magnetic susceptibility, in the case of the *m*-tos salt, the spin transition occurs abruptly at 319 K with negligible small hysteresis.

4. Conclusion

We have investigated the spin crossover transition of the triazole bridged one dimensional Fe^{II} complexes, [Fe(4-NH_2trz)$_3$](o-, m-, p-tos)$_2$·nH_2O (tosH = toluenesulfonic acid) by means of ^{57}Fe Mössbauer spectroscopy and magnetic susceptibility measurement. In the heating process, the spin transition of the all salts occurs around room temperatures. In the cooling process, the structural isomerization effect of counter anion on the spin crossover phenomenon is more remarkable than that in the heating process. In the case of o-tos salt, the spin transition occurs at 250 K with thermal hysteresis of 80 K. On the other hand, in the case of the m-tos salt, the spin transition occurs abruptly at 319 K with negligible small hysteresis. In the case of the p-tos salt, the spin transition occurs abruptly at 295 K in the cooling process in spite of stepwise spin transitions in the heating process. The remarkable structural isomerization effect of counter anion on the spin crossover phenomenon should be related to the conformation of counter anion. In connection with this, we have confirmed that the molecular planes of m-tos and p-tos are parallel and orthogonal to the chain axis of [Cu(4-NH_2trz)$_3$]$_\infty$, respectively[5]. In order to explain the relationship between the conformation of counter anion and the transition temperature in the [Fe(4-NH_2trz)$_3$] chain, the structural analysis is indispensable and is in progress. In this way, the spin crossover transition in [Fe(4-NH_2trz)]$^{2+}$ complexes are remarkably varied by the structural isomerization of counter anion, which suggests a possibility to control the magnetic properties of these type complexes by using photo isomerized molecules as a counter anion. The detailed investigation of the photo-induced spin-crossover transition through the medium of photo-isomerization of the counter anion, spiropyran (SP 150), is in progress.

Acknowledgment

We wish to thank Prof. M. Takahashi (Toho University) for the measurement and analysis of the ^{57}Fe Mössbauer spectra.

References

[1] J Kröber, et al. 1993, *J. Am. Chem. Soc* **115** 9810
[2] VA Varnek, LG Lavrenova, 1995, *J Struct Chem* **36** 104
[3] Y Garcia, O Kahn, et al. 1998 *C. R. Acad. Sci. Paris, Serie IIc.* **1** 523
[4] Y. Murakami, T. Komatsu and N. Kojima, 1999, *Synth. Metals* **103** 2157.
[5] A. Sugahara, J. Harada, et al. to be published.

Journal of Physics: Conference Series **217** (2010) 012129 doi:10.1088/1742-6596/217/1/012129

Moessbauer study of structure changes in Fe-Co-Cr alloys upon their alloying with W and Ga

N M Kleinerman, E V Beloserov, N V Mushnikov and V V Serikov

Institute of Metal Physics, Ural Division of Russian Academy of Sciences, Sofia Kovalevskaya 18, 620041, Ekaterinburg, Russia

kleinerman@imp.uran.ru

Abstract. Changes in the structure of the magnetically hard alloys Fe-Co-Cr alloyed with W and Ga to reach high mechanical properties were investigated. Moessbauer spectra processed in several steps present information on the influence of both different stages of treatment routine and concentration variations on the composition and fine peculiarities of the components that constitute the required structure states.

1. Introduction

Alloys on the basis of the Fe-Co-Cr system are magnetically hard materials which offer much promise as materials for manufacturing disc rotors of high-speed hysteresis engines. To realize this potentiality, different methods of improvement of their mechanical properties have been sought for, such as alloying elements and heat treatments. As a result of such trials, some advantageous combinations of strength, plasticity, and coercive force were gained [1]. Quite successful, in terms of enhancing strength and plasticity, alloys were found doped with W and Ga after quenching, rolling, and subsequent heat treatment. To understand the processes that are responsible for high mechanical properties, it is necessary to know the roles of alloying additions in the formation of structure of the corresponding states. X ray and electron microscopy diffraction methods failed to present sufficient information on the changes in the phase composition of the alloys. To this end, the objects were studied using Moessbauer spectroscopy.

2. Experimental details

The Moessbauer spectra were taken in the mode of constant acceleration with a ^{57}Co source in a chromium matrix at 300K. The spectra were processed using an MSTOOLS program package [2] which allows, along with directly fitting spectra, multi-core distributions of the probability density of the hyperfine fields P(H) to be calculated if there are reasons to assume that different parts of a structure present different linearity in the relationships of at least two of the main Moessbauer parameters. We examined alloys which contained 20-24 wt.%Cr, 12-17 wt.%Co, 7-9 wt.%W, and 0-4 wt.% Ga after a treatment for the optimal state which includes quenching from 1300°C, cold deformation to 60% and heat treatment at 630°C. As a reference, spectra of the ternary composition Fe-22Cr-15Co were recorded, which served as the starting point for the series to be investigated first and then, to be compared to some other compositions.

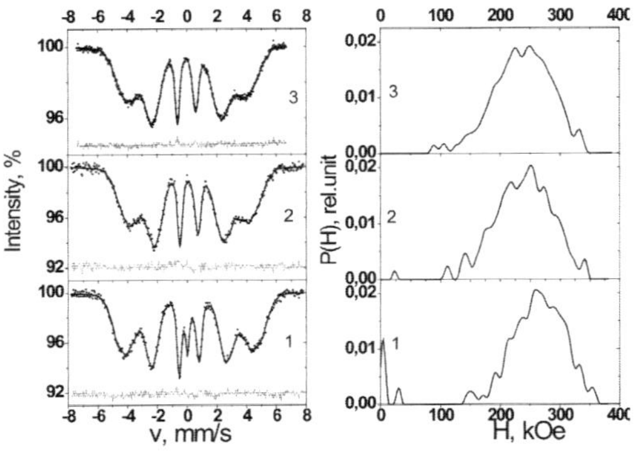

Figure 1. Spectra and P(H) for the samples after quenching: 1-Fe-22Cr-15Co, 2-Fe-22Cr-15Co-9W, 3-Fe-22Cr-15Co-9W-3Ga.

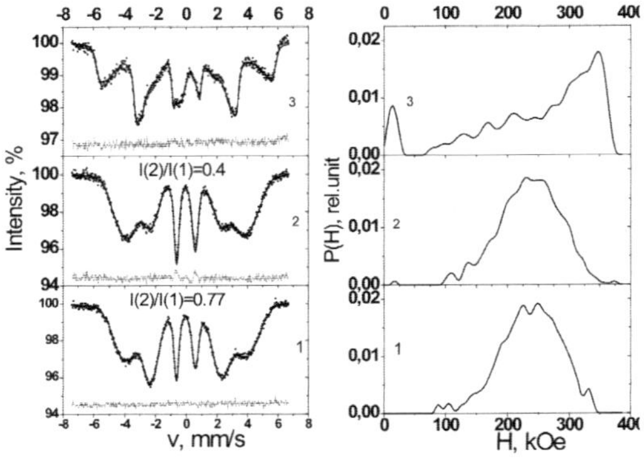

Figure 2. Spectra and P(H) for the sample Fe-22Cr-15Co-9W-3Ga after 1-quenching, 2-rolling, 3-annealing

3. Results and discussion

Figure 1 shows spectra and distribution functions for three compositions after quenching. It is seen that all the three distributions have the form featuring in the main a solid solution, though heterogeneous. In the case of the ternary alloy the effects of short-range order are pronounced to a greater extent, (even with the formation of nonmagnetic regions), whereas introduction of both W and Ga results in smoothing the distribution shifted towards lower fields. If an annealing for optimal properties is performed immediately after quenching, a common modulated structure is realized, which does not meet the requirements on strength and plasticity. An intermediate cold deformation is necessary, which might serve as a means of affecting the mechanism of phase transformations.

Changes in the alloy structure after rolling, common for the studied compositions, are exemplified by the spectra and distributions in Figure 2. The structure is seen to become more homogeneous. The most remarkable difference is displayed in the relative intensities of the outer and middle peaks in the spectra, which means reorientation of magnetic moments toward the normal to the sample plane. After annealing for the optimal state P(H) represents a continuous set of coordinations of Fe atoms, including those in low-to-zero field. To separate contributions from the structure regions with different compositions an attempt was made to construct two-core distributions.

Knowing that a common ternary alloy after annealing represents a modulated structure consisting of two Fe-Cr- and Fe-Co-based phases [3], we suggest two P(H) functions with different correlations of isomeric shift and hyperfine fields and trace the changes in these spectral contributions versus composition, see Figure 3. Already after addition of W the regions representing the next-to-zero field peak are formed, while the form of the high-field distribution becomes more distinct testifying to a more profound phase separation at least in chromium. Addition of Ga results in increasing intensities in the field range 200-300 kOe.

Based on the qualitative information about structure components taken from the P(H) shape for the sample without Ga, we performed fitting of the spectra with three sextets related to the high-field contribution, one sextet from the low-field contribution, and a nonmagnetic part, Figure 4. Then, we subtracted low-field and nonmagnetic subspectra and the remainder spectra were again analyzed using the P(H) apparatus. The results are shown in Figure 5. The changes in the structure of these regions are clearly seen. With increasing Ga content, the intensity of the highest-field peak related to the 0 - impurity - atoms configuration in the first coordination shell of iron atoms decreases, with its

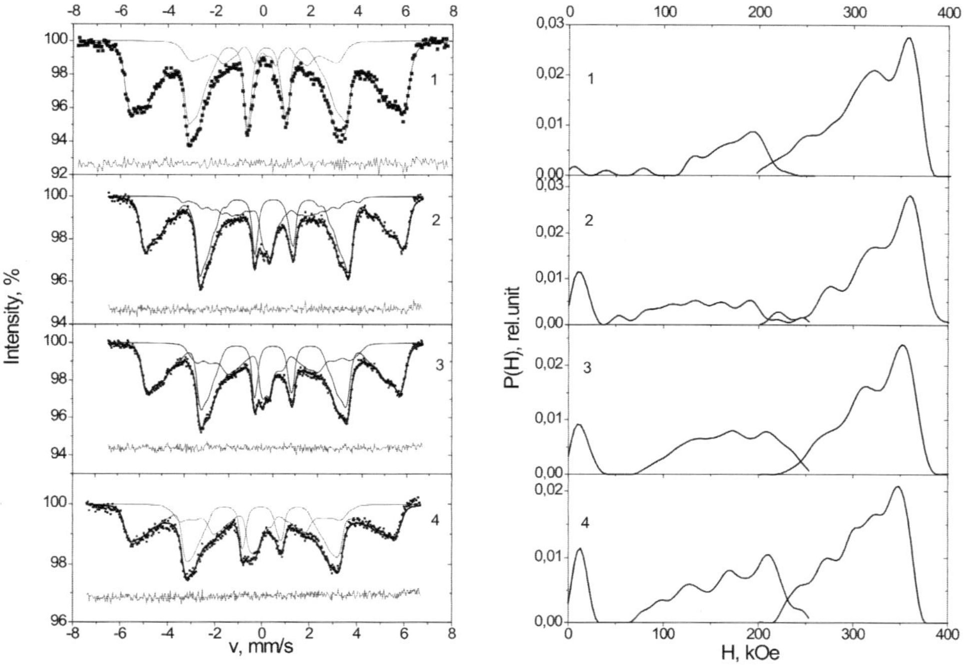

Figure 3. Spectra and double-core P(H) for the samples after optimal treatment: 1 - Fe-22Cr-15Co, 2 - Fe-22Cr-15Co-9W, 3 - Fe-22Cr-15Co-9W-1Ga, 4 - Fe-22Cr-15Co-9W-3Ga

position being shifted toward lower fields, Figure 5a. Judging by the changes in the width of distributions one can say that they become more inhomogeneous in concentration of impurity atoms. The similar changes in the samples of different compositions after optimal treatment are exemplified by the P(H) distributions in Figure 5b, which testifies that changes are common for these alloys

Using the literature data on the H versus Cr concentration dependence and isomeric shifts for the binary compositions [4, 5] we managed to determine the phase compositions as follows: the paramagnetic phase is close to Fe_3W_2, with its volume fraction being insensitive to the presence of Ga; the low-field areas contain about 50%Cr (including possibly the rest of W) and, with increasing Ga, change in relative concentration of Fe from 20 to 15%, whereas the high-field areas take up from 75 to 81% Fe, varying in concentration of (Cr, Ga) from 8 to 15%. The changes in the parameters after introduction of W allow one to make a conclusion on a more pronounced phase separation with the formation of W-enriched regions of a new kind. After addition of Ga, the average parameters of spectral contributions change, though less remarkably than the position and intensities of partial lines which constitute the P(H) functions shown in Figure 5a,b. If to analyze these distributions by fitting them with a set of Gaussian lines of equal width, one can conclude that not only the very parameters of already existing coordinations are changed by introduction of Ga, but some new ones appear with increasing its concentration, which purports the formation of new configurations of impurity atoms around iron atoms and growth of inhomogeneity of elemental content over precipitates.

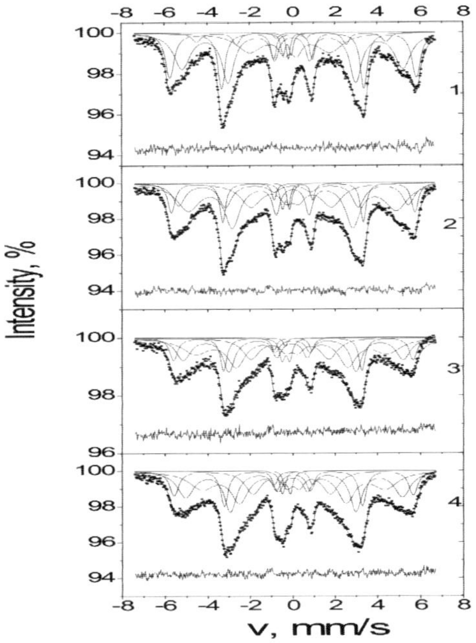

Figure 4. Spectra and fitting subspectra for the samples after optimal treatment 1-Fe-22Cr-15Co-9W, 2-Fe-22Cr-15Co-9W-1Ga, 3-Fe-22Cr-15Co-9W-3Ga, 4-Fe-22Cr-15Co-9W-4Ga

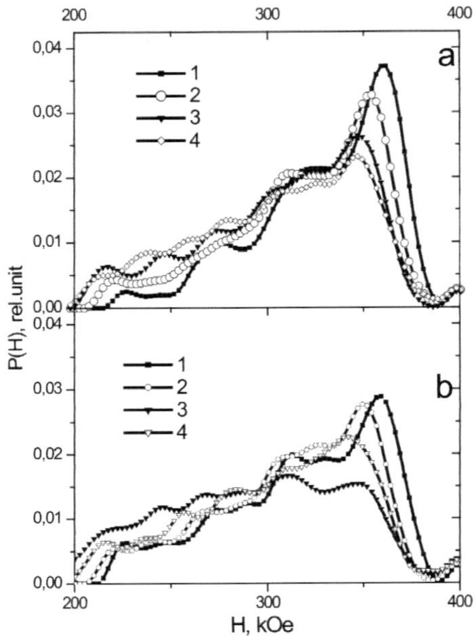

Figure 5. P(H) function of the high-field contribution a - series Fe-22Cr-15Co-9W: 1-0Ga, 2-1Ga, 3-3Ga, 4-4Ga b - samples of different compositions: 1-Fe-23Cr-16Co-0.5Ga, 2-Fe22Cr-15Co-7W, 3-Fe-24Cr-12Co-3Ga, 4-Fe20Cr-17Co-2Ga.

4. Conclusion

1. Intermediate plastic deformation affects the process of structure formation after optimal treatment independently of the initial alloy composition.
2. Introduction of W results in the appearance of a new type of paramagnetic regions enriched in W, whose amount is almost indifferent to the additional alloying with Ga.
3. Changes in the parameters of the P(H) distributions for the quaternary alloys in the optimal state allows one to conclude that Ga localizes itself in the magnetically hard part of the structure and affects the redistribution of Cr through a kind of interface between the chromium-rich and cobalt-rich structure constituents.

Acknowledgements

The work was supported by the Russian Foundation for Basic Research, grant № 08-02-99-077 and the Program of RAS №27.

References

[1] Beloserov E V, Uimin M A, Ermakov A E, Serikov V V, Kleinerman N M and Ivanova G V *Phys. Met.Metallogr.* 2008 **106** 472-80
[2] Rusakov V S 2000 *Moessbauer Spectroscopy of Locally inhomogeneous Systems* (Almaty), ISBN 9965-9111-2-6
[3] Kaneko M, Homma M, Nakamura K 1977 *IEEE Trans. Magnetics MAG-12* **12** 1325
[4] Grusin P L, Rodionov Yu L, Lu Yu A 1975 *Phyz. Met. Metallov.* **40** 1046-52
[5] Blachowski A, Ruebenbauer K, Zukrowski J and Przewoznik 2008 *J. Alloys Compd.* **455**

Journal of Physics: Conference Series **217** (2010) 012130 doi:10.1088/1742-6596/217/1/012130

Mössbauer effect, magnetic and electrical properties studies of $(Y_{1-x}Gd_x)(Fe_{0.7}Co_{0.3})_2$ intermetallics

P Guzdek[1,2], J Pszczola[2], P Stoch[3,4], A Stoch[1]

[1]Institute of Electron Technology, ul. Zabłocie 39, 30 - 701 Kraków , Poland
[2]Faculty of Physics and Applied Computer Science, AGH, Al. Mickiewicza 30, 30-059 Kraków, Poland
[3]Faculty of Material Science and Ceramics, AGH, Al. Mickiewicza 30, 30-059 Kraków, Poland
[4]Institute of Atomic Energy, 05-400 Świerk - Otwock, Poland

E-mail: pguzdek@ite.waw.pl

Abstract. Synthesis, X-ray analysis (300K) and Mössbauer effect studies in the intermetallic compounds $(Y_{1-x}Gd_x)(Fe_{0.7}Co_{0.3})_2$ are reported. Cubic Laves phase $MgCu_2$ - type, were observed across the series. Lattice parameter is linearly increased with Gd concentration. The lattice parameters of the $(Y_{1-x}Gd_x)(Fe_{0.7}Co_{0.3})_2$ series are compared to data previously observed for the $R(Fe_{1-x}Co_x)_2$ series (R= Y, Dy, Gd). ^{57}Fe Mössbauer effect measurements for the series were realized at 4.2 and in the temperature range around the magnetic ordering temperature. Curie temperatures of the $(Y_{1-x}Gd_x)(Fe_{0.7}Co_{0.3})_2$ series were determined using both electrical resistivity and Mössbauer effect measurements. The magnetic hyperfine field and Curie temperature increases with Gd content. The obtained magnetic hyperfine field and Curie temperatures for the series $(Y_{1-x}Gd_x)(Fe_{0.7}Co_{0.3})_2$ are related to the data known for the $R(Fe_{1-x}Co_x)_2$ series (R= Y, Dy, Gd).

1. Introduction

The intermetallic series $R(Fe_{1-x}Co_x)_2$ (R - heavy rare earth) with a C15 cubic Laves phase structure are intensely studied for their fundamental interest and their practical applications [1,2]. From Mössbauer effect studies it was found, that a magnetic hyperfine field observed at ^{57}Fe nuclei as a function of composition for the $R(Fe_{1-x}Co_x)_2$ (R =Y, Gd, Dy, Ho) intermetallic series, behaves similarly to the Slater - Pauling curve, in analogy to the 3d metal-3d metal alloys [3,4] with a maximum field for the $R(Fe_{0.7}Co_{0.3})_2$ compound. Usually substitution of one transition metal/another transition metal in the RM_2 system (M - transition metal) is adopted as a driving force to change across the series the number n of 3d electrons (n calculated per transition metal atom) in the transition metal sublattice and thus to change band properties and related to them magnetic and hyperfine interaction properties [4,5].

Magnetic ordering temperature observed across the exemplary $R(Fe_{1-x}Co_x)_2$ (R=Y, Dy, Gd) series can be treated as a certain replica of the magnetic hyperfine field [6-8]. Namely, the Curie temperatures increases with x and a maximum value is approached in the area of the $R(Fe_{0.7}Co_{0.3})_2$ (R =Y, Gd, Dy) compounds. Further Fe/Co substitution strongly reduces the Curie temperatures.

Yttrium, which lacks a well localized electron magnetic shell, is often used in intermetallics instead of rare earths to probe the influence of 4f electrons on their properties [7]. Gd^{3+} ions are characterized by maximal spin of 4f shell and highest de Gennes factor (G) [9]. It would be interesting to test the influence of the yttrium and gadolinium atoms on the 4f-4d(5d)-3d hyperfine interactions and magnetism of the $(Y_{1-x}Gd_x)(Fe_{0.7}Co_{0.3})_2$ series.

2. Materials and structures

The polycrystalline compounds $(Y_{1-x}Gd_x)(Fe_{0.7}Co_{0.3})_2$ (x=0, 0.1,...,0.8 and 0.9) were prepared by arc melting in a high purity argon atmosphere using appropriate amounts of Y (99.95% purity) with Gd (99.95% purity), Fe and Co (all 99.99% purity) as starting materials. As a next step, the synthesized ingots were annealed in a vacuum (10^{-6}Pa) at 850°C for 10h and then allowed to cool down along with the furnace at approximately 250K/h.

Crystal structures were tested with a standard X-ray powder diffraction procedure using CuK_{α} radiation. The good quality X-ray diffractograms obtained for these materials were analysed using the Rietveld-type method adopting both the $K_{\alpha1}$ (wave length α_1=1.540560 Å) and $K_{\alpha2}$ (α_2=1.544390 Å) lines [10,11]. Pure cubic, Fd3m, $MgCu_2$ - type (C15) crystal Laves phases were observed for all prepared compounds of the studied series. Figure 1 shows diffractogram for exemplary compound $(Y_{0.2}Gd_{0.8})(Fe_{0.7}Co_{0.3})_2$. Since additional X-ray reflexes caused by a long range order have not been observed, thus a random distribution of Fe, Co atoms in the transition metal sublattice and Y, Gd atoms in the rare earth sublattice can be expected to exist. The C15 Laves phase has been previously described in detail elsewhere [12].

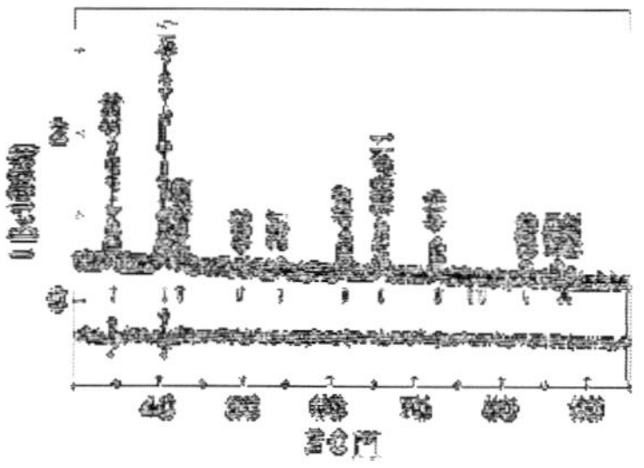

Figure 1. X-ray powder diffraction patterns observed for the exemplary compound $(Y_{0.2}Gd_{0.8})(Fe_{0.7}Co_{0.3})_2$. Fitted differential pattern is added below each diffractogram.

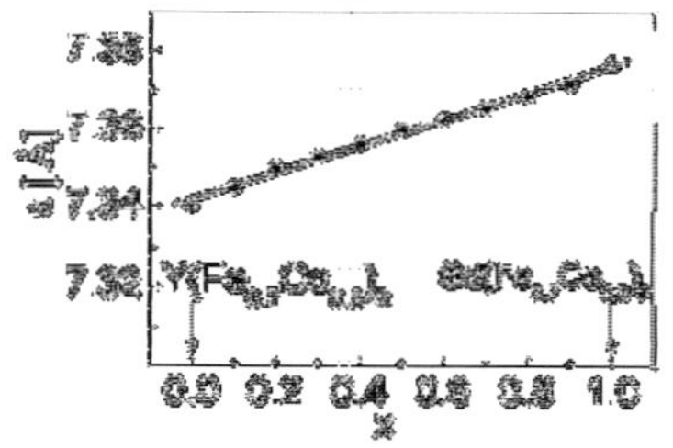

Figure 2. The unit cell edge a observed for the $(Y_{1-x}Gd_x)(Fe_{0.7}Co_{0.3})_2$ intermetallics (300K). The literature data is added [6,7].

Becouse the atomic radius of Gd atom is higher than the corresponding radius of Y atom (r_{Gd}=2.54 Å, r_Y=2.27 Å) [13] the unit cell parameter "a" is linearly increased with the gadolinium content x (Figure 2). A numerical formula a(x)=(0.035x+7.342) Å , the result of a least squares fitting, properly describes experimental points (Figure 2). The experimental points for the series $(Y_{1-x}Gd_x)(Fe_{0.7}Co_{0.3})_2$ are compared to data known from the literature for $Y(Fe_{0.7}Co_{0.3})_2$ compound [7] and $Gd(Fe_{0.7}Co_{0.3})_2$ compound [6].

3. Mössbauer effect

The Mössbauer effect measurements were performed at 4.2K and in the temperature range around the magnetic ordering temperature using a standard transmission technique with a source of ^{57}Co in Rh. The collected ^{57}Fe Mössbauer effect spectra of the $(Y_{1-x}Gd_x)(Fe_{0.7}Co_{0.3})_2$ intermetallics are presented in Figure 3. The experimental spectra are composed of a number of subspectra and this complexity arises from the different nearest neighbourhoods of the observed Fe-atoms. The problems in determining the number of nearest neighbour surroundings, the number of subspectra, the probabilities of the particular subspectra and the fitting procedure of the experimental spectra in the case of three elements, with randomly distributed atoms, in the transition metal sublattice, have been discussed elsewhere for an analogous intermetallic series [14]. It can be added that the n.n. surrounding is composed of n_1 Fe atoms and n_2 Co atoms. The probabilities $P(6; n_1, n_2)$ of the particular n.n. surroundings were calculated using the Bernoulli formulae [15]. As mentioned above $l = n_1 + n_2 = 6$ is the number of n.n. in the transition metal sublattice surrounding the studied Fe atom. The Fe and Co atoms are randomly populated with probabilities of $p_1 = 0.7$, $p_2 = 0.3$, because is no change of stoichiometric proportions Fe and Co atoms in the transition metal sublattice. Therefore, all experimental spectra are fitted using seven subspectra, which the starting amplitudes follow the probabilities $P(6; n_1, n_2)$.

The average values of the hyperfine interaction parameters at 4.2 K, determined from the fitting procedure i.e., the isomer shift IS (with respect to pure iron metal, at 300 K), the magnetic hyperfine field μ_0H_{hf} and the quadrupole interaction parameter QS obtained for the $(Y_{1-x}Gd_x)(Fe_{0.7}Co_{0.3})_2$ series, are presented in Figure 4 (μ_0 is the magnetic permeability, QS as defined in [16]). The isomer shift decreases with x and experimental points are described by a numerical equation: IS (x) =(0.015x + 0.055) mm/s (curve 1). The magnetic hyperfine field μ_0H_{hf} increases across the series with the Gd-content x. The experimental points tally well with the fitted formula: $\mu_0H_{hf}(x) = (0.79x + 24.15)$T (curve 2).The quadrupole interaction parameter increases slightly with x and the experimental points follow the fitted formula: QS (x) = (0.030x + 0.021) mm/s (curve 3). The literature data for x=1 (open points in Figure 4) is also presented [17].

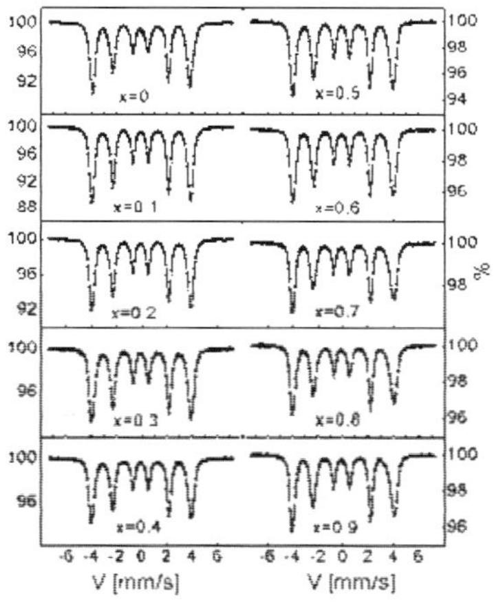

Figure 3. ^{57}Fe Mössbauer effect spectra of the $(Y_{1-x}Gd_x)(Fe_{0.7}Co_{0.3})_2$ intermetallics (4.2K).

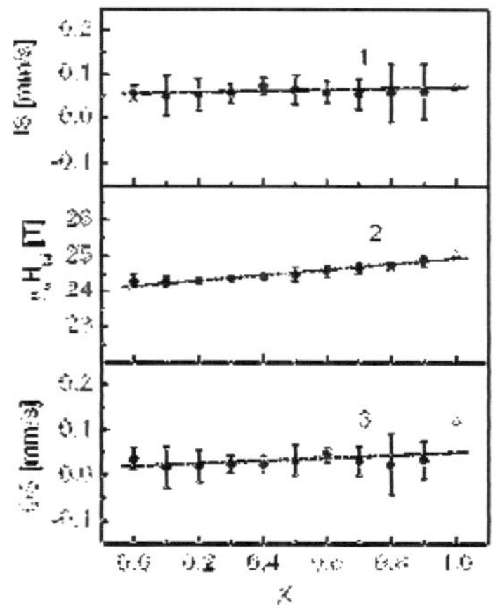

Figure 4. Hyperfine interaction parameters of the $(Y_{1-x}Gd_x)(Fe_{0.7}Co_{0.3})_2$ series: 1– the isomer shift, 2 – the magnetic hyperfine field and 3 – the quadrupole interaction parameter.

4. The Slater–Pauling and Curie temperature dependencies

It is interesting to situate the obtained magnetic hyperfine fields of the $(Y_{1-x}Gd_x)(Fe_{0.7}Co_{0.3})_2$ series among the magnetic hyperfine fields already known for the series $R(Mn_{1-x}Fe_x)_2$ and $R(Fe_{1-x}Co_x)_2$, where R= Y, Dy, Gd [7]. The magnetic hyperfine fields of these series, treated as a function of the average number n of 3d electrons (number n calculated per transition metal atom) and the de Gennes factor G [9] are presented in Figure5.

Figure 6 shows the $T_C(n,G)$ dependence for the $Y(Fe_{1-x}Co_x)_2$ (curve 1) [7], $Dy(Fe_{1-x}Co_x)_2$ (curve 2) [8], $Gd(Fe_{1-x}Co_x)_2$ (curve 3) [6] and $(Y_{1-x}Gd_x)(Fe_{0.7}Co_{0.3})_2$ (curve 4). Curie temperatures of the $(Y_{1-x}Gd_x)(Fe_{0.7}Co_{0.3})_2$ series were determined using both electrical resistivity (curve 4, black points) and Mössbauer effect measurements (curve 4, open points). The methods described elsewhere [6,8,18]

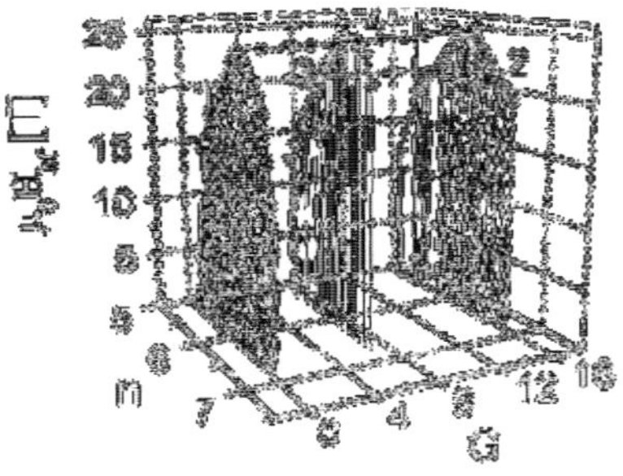

Figure 5. The hyperfine field $\mu_0 H_{hf}(n,G)$ dependencies of the intermetallic series.

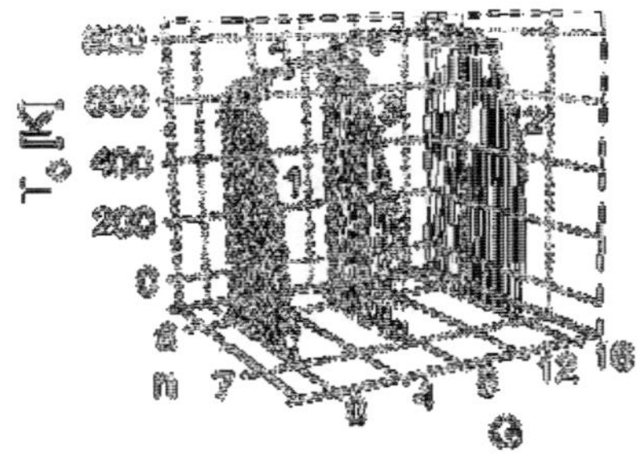

Figure 6. The Curie temperature $T_C(n,G)$ dependencies of the intermetallic series.

5. Conclusion

The intermetallics series $(Y_{1-x}Gd_x)(Fe_{0.7}Co_{0.3})_2$ were prepared by arc melting in a high purity argon atmosphere. X-ray diffraction patterns confirm the formation of a cubic $MgCu_2$ type structure. ^{57}Fe Mössbauer studies indicate that the magnetic hyperfine field slightly increases with Gd increasing addition. The Curie temperatures of the $(Y_{1-x}Gd_x)(Fe_{0.7}Co_{0.3})_2$ are found to increase with increasing Gd addition. The above dependencies are attributed to the additional 4f–5d exchange interaction (Gd–Gd) introducing parallel alignment of the localized magnetic moment 4f and the band 5d magnetic moment. Correspondingly, the 5d–3d exchange interaction (Gd–Fe), (Gd–Co) introduces an antiparallel alignment between the 5d and 3d band magnetic moments. Moreover the 3d–3d exchange interactions establish the parallel order in the transition metal subsystem.

Reference

[1] Balli M, Fruchart D and Gignoux D 2008 *J. Alloys Compd.* **455** 73
[2] Ryu J, Priya S, Uchino K and Kim H 2002 *J. Elektoceram.* **8** 107
[3] Johnson C E, Ridout M S and Cranshaw T E 1963 *Proc. Phys. Soc.* **81** 1079
[4] Burzo E, Chełkowski A and Kirchmayr H R 1990 *Landolt-Börnstein Numerical Data and Functional Relationships in Science and Technology, New Series, Group III* **vol. 19** (subvol. d2), ed O Madelung and H P J Wijn (Berlin: Springer)
[5] Gicala B, Pszczola J, Kucharski Z and Suwalski J 1995 *Solid State Commun.* **96** 511
[6] Onak M, Guzdek P, Stoch P, Chmist J, Bednarski M, Pańta A and Pszczoła J 2007 *J. Alloys Compd.* **433** 53
[7] Guzdek P 2008 *Hyperfine interactions, magnetic and electrical properties of intermetallics with electron composition 4d-3d* PhD Thesis, Cracow
[8] Stoch P, Pszczoła J, Guzdek P, Chmist J and Pańta A 2005 *J. Alloys Compd.* **394** 116.
[9] de Gennes P G 1962 *J. de Phys. Rad.* **23** 510
[10] Rietveld H M 1969 *J.Appl.Cryst.* **2** 65
[11] Rodriguez-Carvajal J 1993 *Physica* B **192** 55
[12] Laves F 1939 *Naturwissenschaften* **27** 65
[13] Table of Periodic Properties of the Elements 1980 (Skokie: Sargent-Welch Scientific Company)
[14] Stoch P, Pszczoła J, Jagodziński P, Jabłońska A, Suwalski J, Dąbrowski L and Pańta A 2004 *J. Alloys Compd.* **375** 24
[15] Feller W 1961 *An Introduction to Probability Theory and Its Applications* (New York: Wiley)
[16] Wertheim G K 1964 *Mössbauer Effect* (London: Academic Press)
[17] Onak M 2008 *Synthesis of $Gd(Mn_{1-x}Fe_x)_2$ and $Gd(Fe_{1-x}Co_x)_2$ intermetallics, their electrical properties, magnetic properties and hyperfine interactions* PhD Thesis, Cracow
[18] Hohenemser C, Rosov N and Kleinhammes A 1989 *Hyp. Int.* **49** 267

Structural and Magnetic Characterization of Ti-Substituted $Li_{0.5}Fe_{2.5}O_4$ prepared using Hydrothermal and Solid-State Routes

H M Widatallah[1], C Johnson[2], A Gismelseed[1], A Al-Rawas[1], M Elzain[1] and A Yousif[1]

[1]Physics Department, Sultan Qaboos University, PO Box 36, Muscat 123, Oman
[2]Chemistry Department, The Open University, Milton Keynes, MK7 6AA, UK

E-mail: hishammw@squ.edu.om

Abstract. Single-phased spinel-related Ti^{4+}-substituted $Li_{0.5}Fe_{2.5}O_4$ has been synthesized by heating a mixture of hydrothermally prepared Ti^{4+}-substituted α-Fe_2O_3 and Li_2CO_3 at 850°C (12 h). This temperature is *ca.* 250-350°C lower than those at which the material is conventionally prepared by the ceramic techniques. X-ray diffraction was used to analyze the evolution of the formation process and, in conjunction with Mössbauer and magnetization measurements, to determine the cation distribution of the resulting ferrite. The results imply that the Ti^{4+} ions substitute for Fe^{3+} ones at the octahedral (B) sites whilst the excess Li^+ ions, required for balancing the charge in the spinel-related structure, replace Fe^{3+} ions at the tetrahedral (A) sites.

1. Introduction

The lithium ferrite of the composition $Li_{0.5}Fe_{2.5}O_4$ adopts an inverse spinel structure in which all the Li^+ ions and 3/5 of all Fe^{3+} ions occupy octahedral B-sites whilst the remaining Fe^{3+} ions occupy tetrahedral A-sites [1-6]. The material is extensively studied due to its technologically desirable electric and magnetic properties that are susceptible to modification on introducing suitable cationic substitutes for the Fe^{3+} ions at the A- or/and B-sub-lattice [1-6]. Conventional routes for synthesizing pure or cation-substituted $Li_{0.5}Fe_{2.5}O_4$ involve the use of high temperatures (≥1000°C) that often affect the material's properties thereby limiting its technological usage [1-7]. For instance, the magnetization can be reduced at such temperatures due to the loss of lithium and oxygen and the subsequent precipitation of the weak ferromagnetic α-Fe_2O_3 [2]. It has been shown that drawbacks of this kind could be avoided if the ferrites are synthesized at or below 850°C [7]. For some time we have been trying to develop different processing schemes for synthesizing cation-substituted $Li_{0.5}Fe_{2.5}O_4$ at moderate temperatures [4-6]. In this paper we report on lowering the formation temperature of Ti^{4+}-substituted $Li_{0.5}Fe_{2.5}O_4$ and its structural and magnetic characterization.

2. Experimental

A nominally (13%) Ti^{4+}-substituted α-Fe_2O_3 was prepared hydrothermally as described elsewhere [8] and mixed homogeneously in a ratio of 5: 1 with Li_2CO_3. The mixture was calcined in air for 12 h intervals at various temperatures in the ranges 300–850 °C followed by quenching in air. XRD data were recorded with a Siemens D5000 diffractometer using CuKα radiation. The program WinMProf was used for XRD Rietveld structural refinement [9]. Elemental analysis was done using the inductively-coupled plasma (ICP) method. ^{57}Fe Mössbauer spectra were recorded at 298K in the standard transmission geometry using a 50 mCi ^{57}Co/Rh source. Isomer shifts are quoted relative to α-Fe. The thermal dependence of the magnetization was measured using a Faraday balance in a magnetic field of 1.5T at a heating rate of 4K/min and a relative accuracy better than 1%.

3. Results and Discussions

The X-ray powder diffraction patterns recorded from the mixtures of Ti-substituted α-Fe_2O_3 and Li_2CO_3 mixture after calcination at different temperatures are shown in figure. 1. No reaction between the precursors was induced when the mixture was heated at 300°C or 400°C (12h). Further heating at 500°C (12h) resulted in forming a small amount of the α-$LiFeO_2$ cubic phase whose most intense

XRD peak is at *ca.* 43.5° 2θ [3]. Further heating at 600°C and then 700°C (12h) resulted in weak peaks that are indexed to a spinel $Li_{0.5}Fe_{2.5}O_4$ related-phase. Knowing that α-Fe_2O_3 reacts completely with Li_2CO_3 to form $Li_{0.5}Fe_{2.5}O_4$ at 700°C [3], one concludes that the presence of substitute Ti^{4+} ions hinders the completion of the sought reaction. Heating the mixture at 800°C (12h) led to the almost complete formation of the required spinel, except for weak peaks due to un-reacted Ti-doped α-Fe_2O_3. A single-phased Ti-substituted $Li_{0.5}Fe_{2.5}O_4$ was only achieved on heating the reactants at 850°C (12h). This temperature is *ca.* 250-350°C lower than the temperatures at which the material is normally prepared by the conventional ceramic techniques. This is clearly related to the presence of Ti^{4+} as substitute cations in α-Fe_2O_3, which resulted in a large number of diffusion paths that accelerated the reaction with Li_2CO_3. The present synthesis route is to be contrasted with the conventional ceramic one where a considerable amount of thermal energy (hence a higher temperature) is required for bond breaking and the establishment of the cationic diffusion necessary to form of the material from a mixture of α-Fe_2O_3, Li_2CO_3/Li_2O and TiO_2 [1,10]. ICP showed the Li, Ti and Fe content to be 2.14%, 7.70% and 56.04% respectively corresponding to a ferrite of the composition $Li_{0.62}Ti_{0.32}Fe_{2.03}O_{4.0}$.

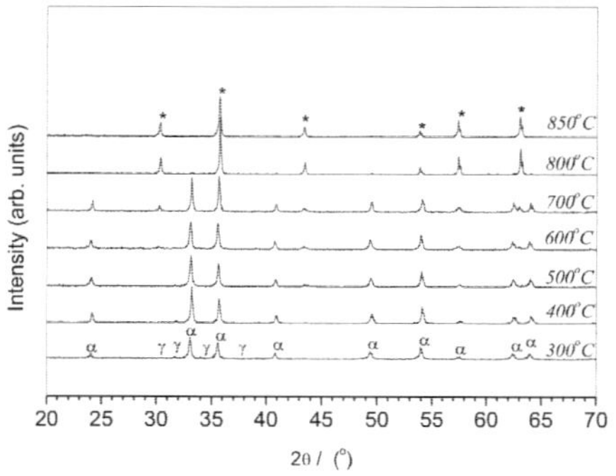

Figure 1. XRD patterns of the reactants' mixture of Li_2CO_3 (γ) and Ti-substituted α-Fe_2O_3 (α) following heating at the given temperatures (12h). The peaks for Ti-substituted $Li_{0.5}Fe_{2.5}O_4$ are referred to by (*).

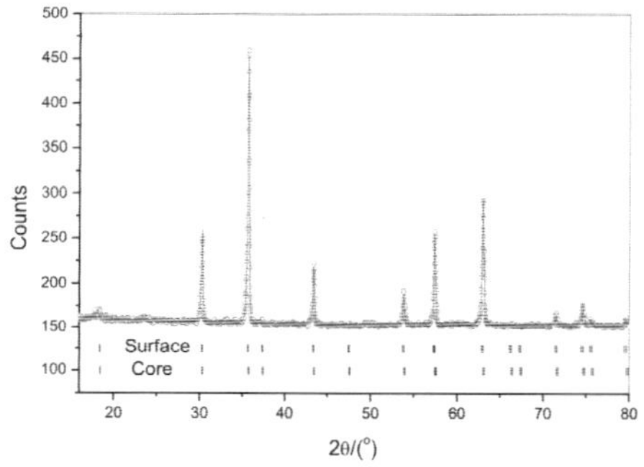

Figure 2. Experimental (dots) and the two-phase "surface-core" theoretical (solid line) XRD patterns of the Ti-substitute $Li_{0.5}Fe_{2.5}O_4$ obtained at 850°C. The bars refer to the positions of the Bragg's reflection peaks.

Table 1. Structural parameters obtained from the single-phase refinement of the Ti-substituted $Li_{0.5}Fe_{2.5}O_4$ prepare at 850°C. Unit cell *Fd-3m*; *a*=8.3307(1) Å; Bragg's *R* factor: 7.29%.

Ion (site)	*x/a*	*y/a*	*z/a*	Occupancy	Wyckoff's Cationic position
Li1 (B)	0.62500	0.62500	0.62500	4.000	16d
Fe1(B)	0.12500	0.33566	0.86907	9.279	16d
Ti (B)	0.12500	0.33566	0.86907	2.721	16d
Fe2(A)	0.99955	0.99955	0.99955	7.657	8a
Li2(A)	0.99955	0.99955	0.99955	0.345	8a
O1	0.38228	0.38228	0.38228	8.000	
O2	0.87957	0.87957	0.38473	24.000	

The XRD pattern recorded from the Ti^{4+}-substituted $Li_{0.5}Fe_{2.5}O_4$ prepared at 850 °C (12 h) is similar to that of disordered α-$Li_{0.5}Fe_{2.5}O_4$ where the B-site cations are randomly distributed over the 16d sites (in Wyckoff notation) [1,3-6]. The best XRD Rietveld structural refinement obtained with a single spinel phase, is based on a model in which the Ti^{4+} ions substitute for Fe^{3+} ones on the B-sites whereas the excess Li^{+} ions, necessary for charge balance requirements in the general formula $Li^{+}_{0.5+0.5x}Ti^{4+}_{x}Fe^{3+}_{2.5-1.5x}O_4$, expel Fe^{3+} from the A-sites (table 1). The composition obtained from this refinement is $Li_{0.54}Ti_{0.34}Fe_{2.12}O_4$, which is in reasonable agreement with ICP analysis. However, the somewhat larger Bragg's R factor has motivated us to try a two-phase refinement to cater for the surfaces and cores of the constituent ferrite particles, which are expected to have large microstrain and large crystallite size, respectively [11]. With the same fractional atomic positions (table 1), the two phases only differed in their lattice constants, microstrain and crystallite size; see figure 2 and table 2. Assuming spherical crystallites, the ferrite particles will have a surface to volume ratio of 2: 3.

Table2. Structural parameters obtained from the "surface-core" two-phase refinement of the Ti-substituted $Li_{0.5}Fe_{2.5}O_4$ prepared at 850°C. Unit cell *Fd-3m*; *Rwp*: 1.8 and *Rp*: 1.3.

Property	**surface**	**Core**
Lattice constant (Å)	8.353(1)	8.3336(1)
Crystallite size(Å)	499(60)	2807(87)
Crystalline microstrain	0.0034(2)	0.0003(1)

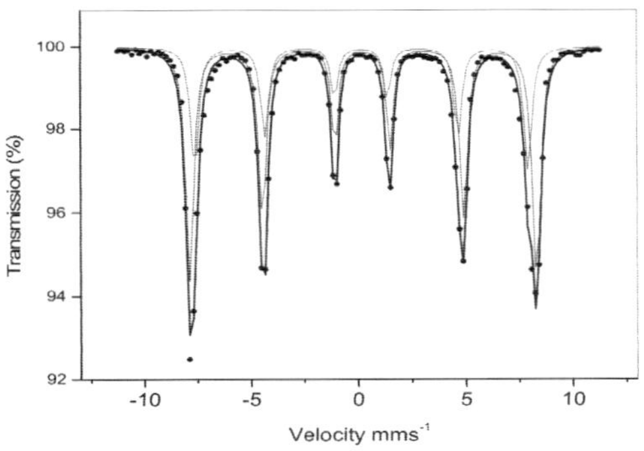

Figure 3: ^{57}Fe Mössbauer spectrum recorded from the Ti-substituted $Li_{0.5}Fe_{2.5}O_4$ sample.

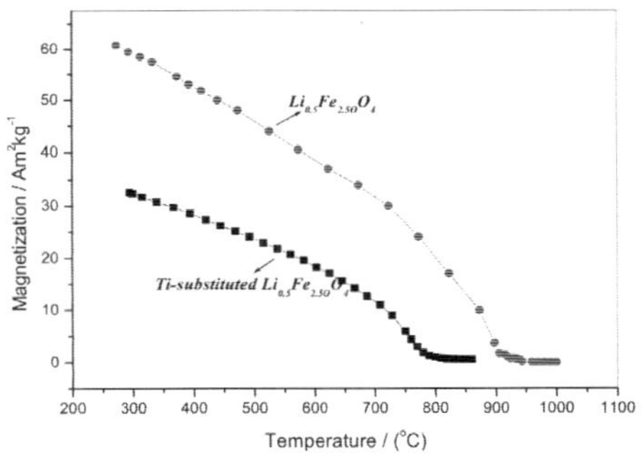

Figure 4: Thermal variation of magnetization of pure and the Ti-substituted $Li_{0.5}Fe_{2.5}O_4$ sample.

The 298 K ^{57}Fe Mössbauer spectrum of the Ti-substituted $Li_{0.5}Fe_{2.5}O_4$ prepared at 850 °C (figure 3) is composed of a well-defined magnetic six-line pattern characteristic of spinel ferrites. Getting reliable information about the cation distribution from Mössbauer spectroscopy requires large magnetic fields applied at low temperatures to get well-resolved sub-spectra [12]. However, aided by the Mössbauer lineshapes on the right-hand part of the spectrum (figure 3), it suffices for the purpose of accepting or rejecting the cation distribution derived from the XRD analysis to use the 298K zero-field spectrum. A simple model in which two partly overlapping sextets, corresponding to Fe ions at both the A- and B-sites, was used to fit the Mössbauer spectrum as shown elsewhere [2, 6,13]. This has resulted in the two distinct sets of the hyperfine parameters of table 3. The outer sextet, with the larger isomer shift and hyperfine magnetic field, was assigned to the B-site [6]. The isomer shift values at both sites are typical of Fe^{3+} in high spin state and are similar to those of pure $Li_{0.5}Fe_{2.5}O_4$ [3,6] indicating negligible influence of the Ti^{4+} ions on the s-electron charge distribution of the Fe^{3+} ions. The absence of an Fe^{2+} spectral component confirms that no reduction of Fe^{3+} to Fe^{2+} has taken place due to the presence of Ti^{4+}, suggesting optimum dielectric properties of the material. The quadrupole splitting for both sextet shows that the insertion of Ti^{4+} ions in the inverse spinel structure leads to no deviation from cubic symmetry at both sub-lattices [3-6]. However, the values of the hyperfine magnetic field H_{eff} on both sites are slightly smaller than those of $Li_{0.5}Fe_{2.5}O_4$ [1,13]. This is related to the presence of the nonmagnetic Ti^{4+} ion in the neighbourhood of Fe^{3+} ions that leads to a decrease of the Fe_A-O-Fe_B exchange interaction. It is clear from the cation distribution deduced from the XRD analysis (table 1) that the B-site iron population decreases substantially relative to that of the A-site. Assuming that the recoilless fractions, f_A and f_B to do not change on Ti^{4+} substitution, the decrease in the relative spectral area of the B-site sextet from 60% in pure $Li_{0.5}Fe_{2.5}O_4$ to 53% in the Ti–substituted variant confirms the substitution of the Ti^{4+} ions for the Fe^{3+} ones at the B sites. This conclusion agrees with results obtained by us and other workers for Ti-substituted $Li_{0.5}Fe_{2.5}O_4$ prepared by different routes [2,10].

Table3. ^{57}Fe Mössbauer parameters for Ti–substituted $Li_{0.5}Fe_{2.5}O_4$ at 298K

Sub-spectrum	$\delta \pm 0.02$ (mm s^{-1})	$\Delta \pm 0.01$(mm s^{-1})	$B_{hf} \pm 0.1$ (T)	Area ± 2%
Sextet A	0.18	0.04	48.6	47
Sextet B	0.39	-0.01	49.3	53

δ=isomer shift; Δ quadrupole splitting; B_{hf} =hyperfine magnetic field; Area= relative spectral area

Figure 4 shows the thermal variation of the magnetization for pure and Ti-substituted $Li_{0.5}Fe_{2.5}O_4$ prepared at 850°C. It is seen that throughout the temperature range studied, the magnetization of the Ti-substituted $Li_{0.5}Fe_{2.5}O_4$ is lower than that of $Li_{0.5}Fe_{2.5}O_4$. Using the cation distribution obtained above from the XRD and supported by Mössbauer analysis, this result is explicable in the framework of Neel's ferrimagnetism model [1-6,10-13] according to which the magnetizations of the A- and B-sub-lattices are anti-parallel with the dominant contribution to magnetization comes from the highly Fe-populated B-sub-lattice. The lower magnetization of Ti-substituted $Li_{0.5}Fe_{2.5}O_4$ relative to $Li_{0.5}Fe_{2.5}O_4$ is therefore a consequence of the substitution of nonmagnetic Ti^{4+} ions for Fe^{3+} on the B-sites as discussed above. This will lead to magnetic dilution of the B-sub-lattice hence the observed decrease in the magnetization. The Curie temperature of Ti-substituted $Li_{0.5}Fe_{2.5}O_4$ estimated form figure 4 is 775K relative to *ca.* 890K for pure $Li_{0.5}Fe_{2.5}O_4$ [6]. This is clearly a manifestation of the weaker Fe_A-O-Fe_B exchange interaction. Finally, we assert that the present experimentally-deduced structural model for Ti^{4+}-substituted $Li_{0.5}Fe_{2.5}O_4$ is consistent with the one that we predicted elsewhere on purely theoretical grounds using atomistic and ab initio calculations [14].

References:

[1] Dormann J L and Nogues M 1990 *J Phys.: Condens. Matter*, **2** 1223
[2] Widatallah H M, Ren X-L, Al-Omari I A 2006 *J Mater. Sci.* **41** 6333
[3] Widatallah H M, Berry F 2002 *J Solid State Chem.* **146** 230
[4] Widatallah H M, Johnson C, Berry F J, Pekala M 2001 *Solid State Commun.* **120** 171
[5] Widatallah H M, Johnson C, Berry F, Gismelseed A, Jartych E, Marco J, Gard F, Pekala M 2006 *J. Phys. Chem. Solids* **67** 1817
[6] Widatallah H M, Johnson C, Berry F, Jartych E, Gismelseed A, Pekala, Grabski J 2005 *Mater. Lett.* **59** 105.
[7] Cruskova A, Lipka J, Slama J, Toth I, Seberini M1990 *Phys. Stat. Sol. A* **122** K171
[8] Berry F, Greaves C, Helgason Ö, Mcmanus J, Palmer H, Williams R 2001 *J. Solid State Chem.* **151** 157
[9] Jouanneaux A, 1999 *Int. Union Crystallogr. Newslett.* **21** 13
[10] Yousif A A, Elzain M, Mazen S, Sutherland H H, Abdalla M H, Mansur S F,1994 *J. Phys.: Condens. Matter*, **6** 5717
[11] Widatallah H M, Johnson C, Gismelseed A M, Al-Omari I, Stewart S J, Al- Harthi S, Thomas S, Sitepu H 2008 *J. Phys. D: Appl. Phys.* **41**165006
[12] Sepelak V 2002 *Ann. Chim. Sci. Mater.* **27** 61
[13] des Barros F, Viccard P J and Artman J O 1968 *Phys. Lett. A* **27** 374
[14] Widatallah H M and Moore E A 2004 *J Phys. Chem. Solids* **65** 1663

Magnetic phase transition in $MnFeP_{0.5}As_{0.4}Si_{0.1}$

J L Wang[1,2], S J Campbell[1], O Tegus[3], E Brück[3] and S X Dou[4]
[1] School of Physical, Environmental and Mathematical Sciences, University of New South Wales, Australian Defence Force Academy, Canberra ACT 2600, Australia
[2] Bragg Institute, ANSTO, Lucas Heights, NSW 2234, Australia
[3] Fundamental Aspects of Materials and Energy Group, Faculty of Applied Sciences, TU Delft, Mekelweg 15, 2629 JB Delft, The Netherlands
[4] Institute for Superconductivity and Electronic Materials, University of Wollongong, NSW 2522, Australia

E-mail: j.wang@adfa.edu.au

Abstract. We have carried out a detailed investigation of the magnetic phase transition in $MnFeP_{0.5}As_{0.4}Si_{0.1}$. Temperature hysteresis has been observed in the variable temperature magnetization curves (B_{appl} = 0.01 T) with $T_C^W \sim 302$ K on warming and $T_C^C \sim 292$ K on cooling. The first order nature of this transition in $MnFeP_{0.5}As_{0.4}Si_{0.1}$ is confirmed by the negative slope obtained from isotherms of M^2 versus B/M around the critical temperature. Linear thermal expansion measurements reveal a large volume change, $\Delta V/V \sim 8.7\times10^{-3}$, at the magnetic phase transition and that this magnetovolume effect is suppressed to $\Delta V/V \sim 5.5\times10^{-3}$ in an applied field of B_{appl} = 1.0 T. Analyses of ^{57}Fe Mössbauer spectra (4.5 - 300 K) using a random distribution model and taking nearest-neighbour environments into account, indicate that the paramagnetic and ferromagnetic phases coexist over a temperature range of ~ 45 K around the Curie temperature. The Debye temperature for $MnFeP_{0.5}As_{0.4}Si_{0.1}$ has been evaluated as $\theta_D = 350 \pm 20$ K from the temperature dependence of the average isomer shift.

1. Introduction

The discovery by Pecharsky and Gschneidner in 1997 of a giant magnetocaloric effect (MCE) in $Gd_5Si_2Ge_2$ [1], has led to enhanced scientific interest and search for materials with a large MCE that are operational over technologically relevant temperatures [2]. This in turn has led to the synthesis and observation of a giant MCE in many other intermetallic systems [2] including the transition metal (T) rich systems $La(Fe_{1-x}Si_x)_{13}$ [3], $MnAs_{1-x}Sb_x$ [4] and $MnFeP_{1-x}As_x$ [5]. An especially important aspect of these discoveries is that they demonstrate that large values of MCE are not restricted to materials with large magnetic moments but that, in the vicinity of a phase transition, the MCE depends strongly on the type of the magnetic phase transition [2].

As summarized by Gschneidner *et al.* [2], recent work has demonstrated giant MCE near a first order magnetic transition which can, by varying material composition and/or applied magnetic field, cause the transition to occur over a wide temperature range and even extend above room temperature. For example, in $MnFeP_{1-x}As_x$ the transition temperature can be tuned from ~ 150 K to 335 K by adjusting the P/As ratio [5]. $MnFeP_{1-x}As_x$ compounds crystallise with the Fe_2P hexagonal structure and P and As are randomly distributed on the 1*b* and 2*c* sites whereas Fe atoms occupy the 3*f* site and Mn atoms the 3*g* site [6]. ^{57}Fe Mössbauer spectroscopy has previously been used to explore the first-order character of the ferromagnetic to paramagnetic transition in a systematic study of both the $MnFeP_{1-x}As_x$ and $Mn_{1.1}Fe_{0.9}P_{1-x}Ge_x$ systems [6].

In an earlier study it was reported that while no crystallographic structural changes are associated with the magnetic phase transition, a pronounced volume change was observed along with anomalous behaviour electrical resistance [7]. Here we report a detailed investigation of the magnetic phase transition in $MnFeP_{0.5}As_{0.4}Si_{0.1}$ using magnetic and variable temperature Mössbauer spectroscopy measurements. An aim is to investigate the effects of external field on the magnetovolume effects and further understand the critical magnetic behaviour around transition.

2. Experimental process

A polycrystalline sample of $MnFeP_{0.5}As_{0.4}Si_{0.1}$ was prepared as described previously [5, 7]. The compound was found to crystallise in the Fe_2P-type hexagonal structure [7] as expected on the basis of the structure of $MnFeP_{1-x}As_x$ compounds. Magnetic measurements were performed using a Quantum Design Physical Properties Measurement System with a magnetic field sweep rate of 0.005 T/s and amplitude up to 8.5 T. Thermal expansion measurements were performed using a strain-gauge linear differential transformer method over the temperature range 77 – 300 K at Departmento de Física de la Materia Condensada, Universidad de Zaragoza, Spain. As shown in Fig. 1(b), this allows the absolute changes of the length in zero and applied magnetic field to be measured relative to the initial length of the sample. The ^{57}Fe Mössbauer spectra were obtained between 4.5 K and 300 K using a standard constant-acceleration spectrometer and a ^{57}Co**Rh** source. The spectrometer was calibrated at room temperature with an α-iron foil.

3. Results and discussion

As shown in Fig. 1, the temperature dependence of magnetization for $MnFeP_{0.5}As_{0.4}Si_{0.1}$ as measured in an applied magnetic field of B_{appl} = 0.01 T, exhibits a pronounced magnetic phase transition around 299 K with a temperature hysteresis of around 10 K between the cooling (C) and warming (W) curves - $T_C^W \sim 302$ K or $T_C^C \sim 292$ K respectively. The abrupt change in magnetization and the temperature hysteresis at T_C are characteristic of a first-order magnetic transition.

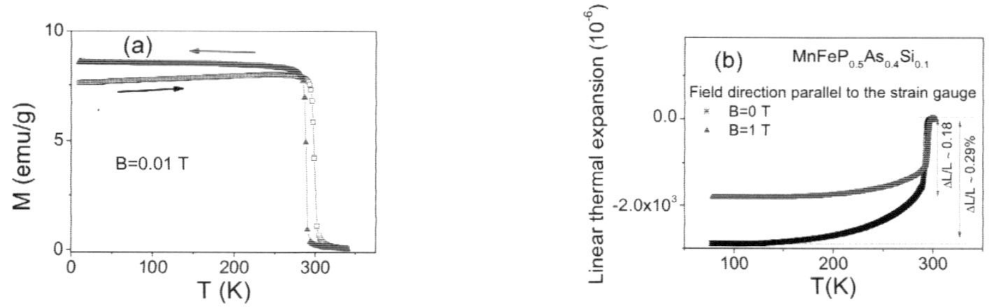

Figure 1 (a) Temperature dependence of magnetization in a field of 0.01 T during cooling (triangles) and warming (squares) and (b) Temperature dependence of the linear thermal expansion of $MnFeP_{0.5}As_{0.4}Si_{0.1}$ in zero field and an applied field of B_{appl} = 1 T on warming through the transition temperature T_C.

The changes in linear thermal expansion, $\Delta l/l(T)$, of $MnFeP_{0.5}As_{0.4}Si_{0.1}$ on warming through the transition temperature T_C in zero field and an applied field of B_{appl} = 1 T are shown in Figure 1 (b). Using $\Delta V/V(T) = 3\Delta l/l(T)$ [8], it is clear from Figure 1(b) that $MnFeP_{0.5}As_{0.4}Si_{0.1}$ exhibits a large magnetovolume effect around the transition temperature. Compared with the paramagnetic region above T_C, in zero field the unit cell volume in the ferromagnetic (FM) state contracts by ~ 0.87 % ($\Delta V/V \sim 8.7\times10^{-3}$) whereas in a magnetic field of B_{appl} = 1 T, the relative contraction of the unit cell volume is only $\Delta V/V \sim 5.5\times10^{-3}$. This indicates that the magnetic field suppresses the magnetovolume effects. Similar behaviour has been detected in the $SmMn_2Ge_2$ system where a large magnetovolume effect, $\Delta V/V \sim -0.3\%$, was observed in a canted antiferromagnetic state between T_C^{Sm} (~100 K) and T_N^{Inter} (~150 K) in zero field whereas application of a field of B_{appl} = 4 T caused the structural distortion to recover to $\Delta V/V \sim 0$ [9]. For $SmMn_2Ge_2$ this behaviour is due to the metamagnetic transition from the zero field antiferromagnetic state to the field-induced ferromagnetic state [9].

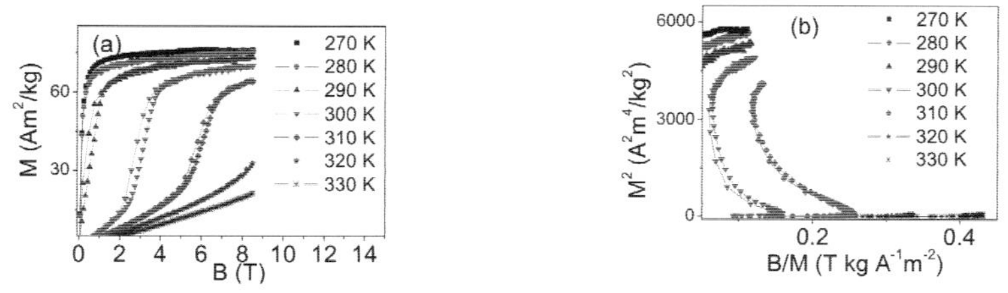

Figure 2 (a) Variation of magnetization with applied magnetic field (B_{appl} = 0–8.5 T) for $MnFeP_{0.5}As_{0.4}Si_{0.1}$ at the temperatures indicated and (b) Arrot plots (M^2 versus B/M) derived from the magnetisation data of Fig 2(a).

The order of the magnetic phase transition has been further investigated by Arrot plots (curves of M^2 versus B/M) derived from the field dependence of the magnetization at temperatures around T_C. Figure 2(a) shows typical M-B curves at the temperatures indicated with the corresponding Arrot plots shown in Figure 2(b). It can be seen clearly from the jump in magnetization, that application of a certain magnetic field value above T_C, can cause the paramagnetic state to change to a ferromagnetic state. The critical field at which the transition from paramagnetism to ferromagnetism starts can be defined as the field value at which the M-B slope changes (e.g. 2.4 T and 5.2 T at 300 K and 310 K, respectively). In addition, the order of the transition - second-order or first-order – is indicated by a positive or negative slope in the isothermal plots of M^2 versus B/M respectively. In the case of $MnFeP_{0.5}As_{0.4}Si_{0.1}$, the negative slope of the Arrot plot at 300 K and 310 K in Fig. 2(b) indicates the first-order nature of this metamagnetic phase transition from paramagnetism to ferromagnetism.

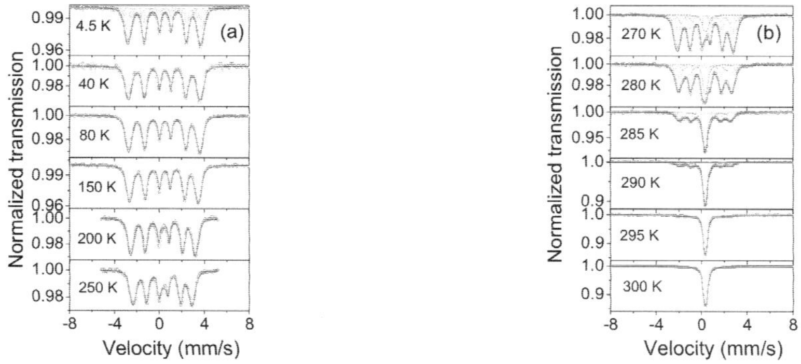

Figure 3 ^{57}Fe Mössbauer spectra of $MnFeP_{0.5}As_{0.4}Si_{0.1}$ over the temperature range 4.5 - 300 K

Figure 3 shows ^{57}Fe Mössbauer spectra of $MnFeP_{0.5}As_{0.4}Si_{0.1}$ at selected temperatures. Given that Si probably enters the 1b and 2c sites to replace P/As [7], the spectra have been fitted with a binomial distribution model while also taking into account the different iron near-neighbour environments (five different iron near-neighbour environments in this system) and the correlation between the hyperfine fields and the number of iron near neighbours [6]. Fig. 4(a) shows the fractions of each state as a function of temperature as determined from the fitting; it is clear that there exists a certain temperature range (~250 K to 295 K) over which the ferromagnetic and paramagnetic states co-exist.

The temperature dependences of the Mössbauer hyperfine parameters are shown in Figure 3. The Debye temperature has been determined from the temperature dependence of the isomer shift (IS), $<\delta(T)>$ [6]:

$$IS(T) = IS_0(T) + IS_{SODS}(T) \tag{1}$$

$IS_0(T)$ represents the T dependence of the charge density at the probe nucleus which is generally weakly temperature dependent. The second term, IS_{SODS}, the so-called second-order Doppler shift, can be described in terms of the Debye model for the phonon spectrum by,

$$IS_{SODS}(T) = -\frac{3kT}{2mc}\left[\frac{3\theta_D}{8T} + 3\left(\frac{T}{\theta_D}\right)^3 \int_0^{\tau} \frac{x^3}{e^x - 1}dx\right], \tag{2}$$

where m is the mass of the ^{57}Fe nucleus, k Boltzmann's constant, c the velocity of light and $\tau = \theta_D/T$ the reduced temperature [6]. A Debye temperature value of $\theta_D = 350 \pm 20$ K has been derived on fitting the mean isomer shift data of Figure 3(b) to equation (2).

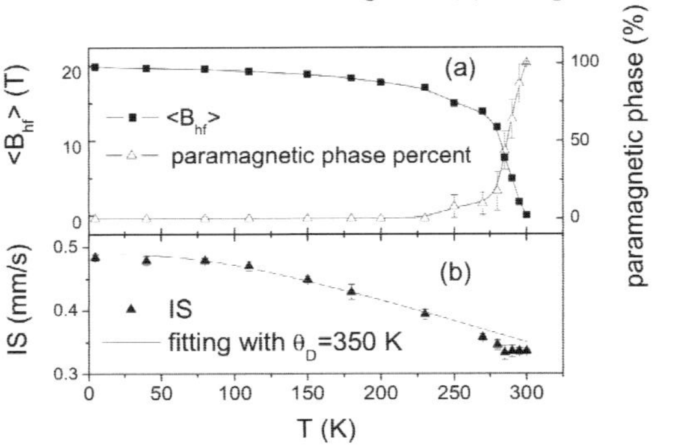

Figure 4. (a) The average hyperfine field and the fractional area of the paramagnetic phase as a function of temperature.
(b) The temperature dependence of the isomer shift. The solid line is the result of a fit with the Debye model (equation 2).

Acknowledgments
The work is supported by a joint agreement between the Bragg Institute, ANSTO, and UNSW@ADFA. J L Wang thanks Dr C. Marquina (Departmento de Física de la Materia Condensada, Universidad de Zaragoza, Spain) for help with the thermal expansion measurements.

References:

[1] Pecharsky V K and Gschneidner Jr K A, 1997 *Phys. Rev. Lett.* **78** 4494; 1997 *Appl. Phys. Lett.* **70** 3299
[2] Gschneidner Jr K A, Pecharsky V K and Tsokol A O, 2005 *Rep. Prog. Phys.* **68** 1479
[3] Hu F X, Shen B G, Sun J R, Cheng Z H and Zhang X X, 2000 *J. Phys.: Condens. Matter* **12** L691, Zhang X X, Wen G H, Wang F W, Wang W H, Yu C H and Wu G H, 2000 *Appl. Phys. Lett.* **77** 3072
[4] Wada H and Tanabe Y, 2001 *Appl. Phys. Lett.* **79** 3302
[5] Tegus O, Brück E, Buschow K H J and de Boer F R, 2002 *Nature* (London) **415** 150; Tegus O, Koyama K, Her J L, Watanabe K, Brück E, Buschow K H J and de Boer F R, 2007 *Physica B* **392** 151
[6] Sougrati M T, Hermann R P, Grandjean F, Long G J, Brück E, Tegus O, Trung N T and Buschow K H J 2008 *J. Phys.: Condens. Matter* **20** 475206; Hermann R P, Tegus O, Brück E, Buschow K H J, de Boer F R, Long G J and Grandjean F 2004 *Phys. Rev.* B **70** 214425
[7] Tegus O, Brück E, Dagula W, Li X W, Zhang L, Buschow K H J and de Boer F R 2003 *J. Appl. Phys.* **93** 7655
[8] Wang J L, Campbell S J, Tegus O, Marquina C and Ibarra M R 2007 *Phys. Rev.* B **75** 174423
[9] Miura S H, Koyama K, Okada H, Shigeoka T and Watanabe K, 2005 *Japa. J. Appl. Phys.* **44** 1258

Journal of Physics: Conference Series **217** (2010) 012133 doi:10.1088/1742-6596/217/1/012133

Mössbauer Study of the Heterometallic Wheel $Cr_7Fe(II)$

M Lantieri[1], L Cianchi[1], F Del Giallo[1], P Moretti[1], G Spina[2] and G Timco[3]

[1] ISC–CNR, Via Madonna del Piano 10, 59100 Sesto Fiorentino (FI), Italy
[2] INFM–CNISM and Dip. di Fisica, Università di Firenze, Via G. Sansone 1, 59100 Sesto Fiorentino (FI), Italy
[3] Dept. Chemistry, The University of Manchester, Oxford Road, Manchester M13 9PL, UK

E-mail: marco.lantieri@isc.cnr.it

Abstract. In the present paper we describe the results of Mössbauer measurements on the antiferromagnetic wheel $Cr_7Fe(II)$. From the spectra without external field in [1.8, 14.8] K range we analyse the spin dynamics vs. T and we evaluate the electric hyperfine parameters. From the 1.8 K spectrum collected in the presence of a $1.5\,10^{-2}$ T magnetic field, we determine the hyperfine field intensity. The corresponding spin of the iron ion in the complex is estimated and it is found to be in agreement with reported results of original calculations.

1. Synthesis and structural features

The $S = \frac{1}{2}$ ground state $(Bu_2NH_2)[Cr_7FeF_8(O_2CCMe_3)_{16}]$ complex (see Figure 1) belongs to the heterometallic antiferromagnetic rings' family $Cr_7M(II)$ [1], where M≡Cd, Zn, Mn, Ni and, as in our case, Fe. The sample powder was prepared starting from the Hexanuclear Chromium *horseshoe* compound [2] and $^{57}FeCl_2$ according to the reaction below:

$$[(Bu_2NH_2)_3\{Cr_6F_{11}(O_2CCMe_3)_{10}\}]_2 + 2\,^{57}FeCl_2 + 2CrCl_2 + nMe_3CCO_2$$
$$\xrightarrow{\text{at } 150^\circ \text{ in Ar}} (Bu_2NH_2)[Cr_7FeF_8(O_2CCMe_3)_{16}]$$

purified by column chromatography. The sample was prepared by mixing the active material powder with alumina and contains 0.79 $\mathrm{mg\,cm^{-2}}$ of ^{57}Fe. Xrd analysis showed small differences in the bond angles and lengths between the magnetic ions and their first neighbours F^- and O^{2-}, which leads to a not complete equivalence of the eight magnetic sites of the molecule. Moreover, random orientations of the target's crystallites were revealed.

2. Electronic states

The spin Hamiltonian describing the electronic states of the ring is the following:

$$H_S = J\sum_{n=1}^{6} \boldsymbol{s}_n \cdot \boldsymbol{s}_{n+1} + J'(\boldsymbol{s}_1 + \boldsymbol{s}_7)\cdot \boldsymbol{s}_8 + \sum_{n=1}^{8} \boldsymbol{s}_n \cdot \mathsf{D_n} \cdot \boldsymbol{s}_n + \mu_B \boldsymbol{B}_0 \cdot \sum_{n=1}^{8} \mathsf{g_n} \cdot \boldsymbol{s}_n \quad (1)$$

Table 1. Mean values of s_z of the eight metallic ions in the ground spin state and in the first excited one for $J' = J$ in the approximation of no anisotropy and no external field.

	Cr_1	Cr_2	Cr_3	Cr_4	Cr_5	Cr_6	Cr_7	Fe
$(S=\frac{1}{2}, S_z=\frac{1}{2})$	−0.473	0.547	−0.460	0.524	−0.460	0.547	−0.473	0.749
$(S=\frac{3}{2}, S_z=\frac{3}{2})$	−0.192	0.468	−0.074	0.404	−0.074	0.468	−0.192	0.692

where $s_n = \frac{3}{2}$ with $n = 1, \ldots, 7$ for Cr ions and $s_8 = 2$ for Fe ion. As far as the antiferromagnetic isotropic exchange terms are concerned, we chose $J \simeq 17$ K for Cr–Cr couplings and $J' = J$ for Cr–Fe ones, in agreement with EPR and INS results for Cr_8 and other $Cr_7M(II)$ wheels [1, 3, 4]. The second term of (1) represents crystal field anisotropic interaction and the last one is the Zeeman Hamiltonian for the applied field $\boldsymbol{B}_0$. The spin levels are calculated by means of an iterative method not based on the diagonalization of H_S [5] (see figure 2 and table 1).

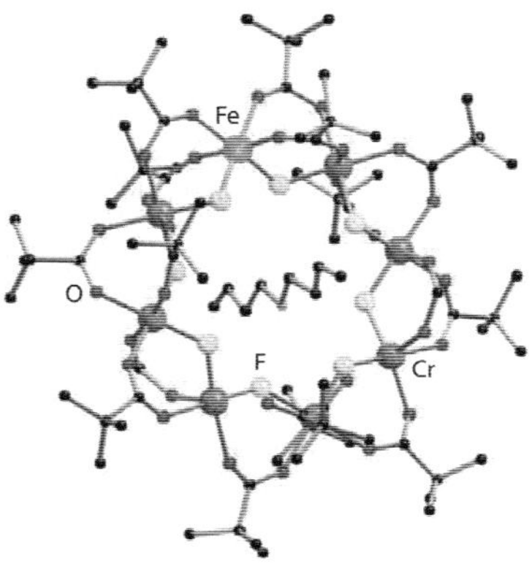

Figure 1. The structure of the wheel Cr_7Fe in the crystal.

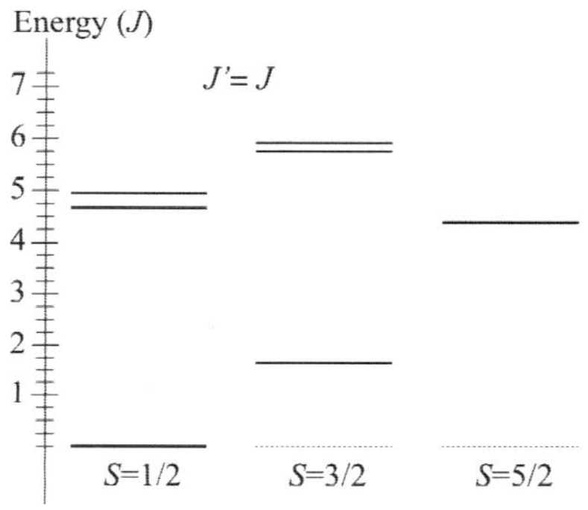

Figure 2. Scheme of some spin levels of low energy for $S = \frac{1}{2}, \frac{3}{2}, \frac{5}{2}$ in the approximation of no anisotropy and no external field.

3. Zero field measurements

Mössbauer spectra in zero external field were collected betweeen 1.8 and 14.8 K (see figure 3); they show two wide lines with magnetic hyperfine sub–structures, which evolve with T so that the spectra collapse into a well–defined sharp quadrupolar doublet. Moreover, since the ground state is degenerate ($S = \frac{1}{2}; S_z = \pm\frac{1}{2}$) the Blume stochastic theory of the line shape fails and we have to use the *ab-initio* Hirst theory [6].
We consider the following hyperfine Hamiltonian:

$$H_{\mathrm{hyp}} = \delta_{\mathrm{IS}} + \frac{Q_{\mathrm{s}}}{6\sqrt{1+\eta^2/3}} \left[3I_z^2 - \boldsymbol{I}^2 + \frac{\eta}{2}\left(I_+^2 + I_-^2\right)\right] + I_\xi\, A_{\xi\xi}\, S_\xi + I_\eta\, A_{\eta\eta}\, S_\eta + I_\zeta\, A_{\zeta\zeta}\, S_\zeta \quad (2)$$

where δ_{IS}, Q_{s} and η are the isomer shift, the quadrupolar splitting and the asymmetry parameter, respectively. Finally, $A_{\iota\iota} S_\iota$ are the principal components of the magnetic hyperfine tensor.
The simple hamiltonian $H_{\mathrm{dyn}} = \mathcal{S}\, F$, where F acts on the vibrational freedom degrees and $\mathcal{S}$ is a spin dimensionless operator, is chosen to describe the spin inversion. The inequivalence

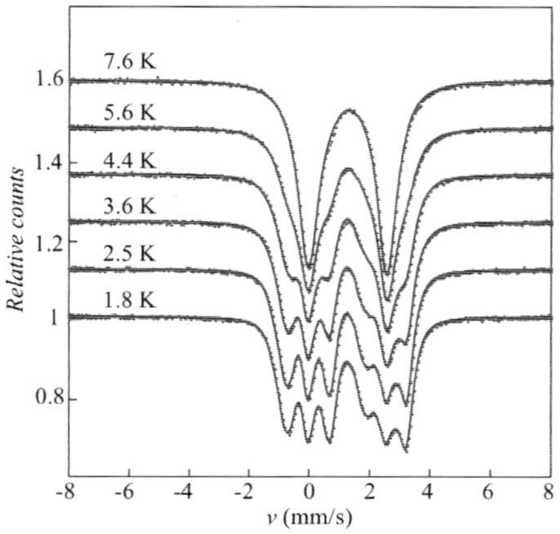

Figure 3. A selection of the zero field spectra collected between 1.8 and 7.6 K with the corresponding fittings.

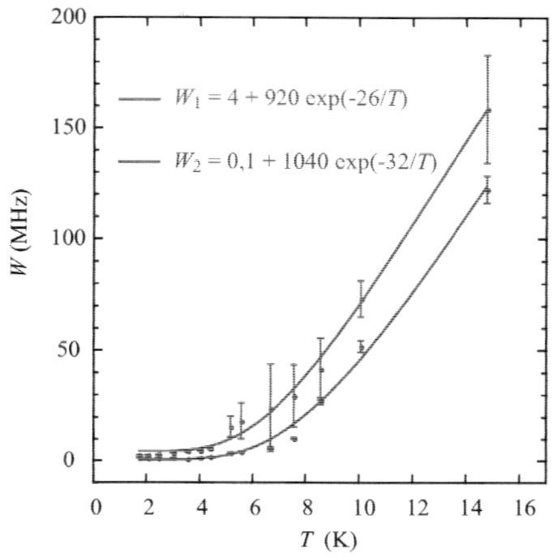

Figure 4. Trend of transition probability rates W vs. T for the two iron sites.

of the iron sites suggests the spectra to be fitted by several different subspectra. For the sake of simplicity, we assumed only two sites, obtaining the electric hyperfine parameters to be very close for both of them (see table 2). Indeed, the large differences in the hyperfine tensor principal

Table 2. Hyperfine parameters for the two iron sites from the fitting of the 1.8 K spectrum.

Site	δ_{IS} (mm/s)	Q_{S} (mm/s)	η	$A_{\xi\xi}S_\xi$ (T)	$A_{\zeta\zeta}S_\zeta$ (T)	$A_{\eta\eta}S_\eta$ (T)
1	1.275(3)	2.49(1)	0.7(2)	−1.03(5)	−6.6(6)	−2.3(3)
2	1.265(2)	2.53(1)	0.9(1)	−1.41(3)	−5.3(4)	−0.27(4)

components could denote spreads due to non-equivalent iron sites. The values of isomer shift and quadrupolar splitting denote an Fe(II) ion and the high value of η reflects, as expected, the local charge asymmetry. Regarding the hyperfine magnetic tensor, we stress that the $A_{\zeta\zeta}\, S_\zeta$ component is prevalent with respect to the others, indicating a mainly axial magnetic anisotropy on the iron ion.

The inversion probability rates of the iron spin for the two sites vs. T are reported on figure 4. Their trends are in agreement with both an Orbach process involving an excited level at 30 K and a first order Raman process involving two phonons of the same energy.

4. External field measurement

We applied a 1.5 10^{-2} T longitudinal magnetic field, corresponding to a Zeeman interaction of 10^{-2} K, so that the splitting between the two spin states $S_z = \pm\frac{1}{2}$ becomes two order of magnitude larger than the hyperfine energy. Therefore, the interaction between the electronic spin and the iron nucleus can be described by means of a magnetic hyperfine field $\boldsymbol{B}_{\mathrm{h}}$ and the corresponding Hamiltonian becomes:

$$H_{\mathrm{hyp}} = \delta_{\mathrm{IS}} + \frac{Q_{\mathrm{s}}}{6\sqrt{1+\eta^2/3}}\left[3I_z^2 - \boldsymbol{I}^2 + \frac{\eta}{2}(I_+^2 + I_-^2)\right] - g_I\mu_{\mathrm{N}}\boldsymbol{I}\cdot(\boldsymbol{B}_0 + \boldsymbol{B}_{\mathrm{h}}) \tag{3}$$

The experimental spectrum (figure 5) cannot be fitted without introducing transversal relaxation processes. The presence of non–null anisotropy transverse components, probably due to the first neighbour F^- ions, is also suggested by the measurements without external field (see table 2). As the actual shape of the anisotropy is unknown, we assumed the one shown in figure 5 – with four equal minima along the surface equatorial line of a rugby-ball – which can be expressed by the function $(\cos^2\theta + 1)(\frac{1}{4}\sin 4\phi \sin^4\theta + 1)$.
The applied field is able to delete tunnelling frequencies up to 10^9 Hz, which also represents in the present case the limit of detectable dynamical effects [7]. In the absence of any magnetic field, the rate of transversal transitions, due to tunnelling effect, might be at the top of the Mössbauer window, while the presence of the field quenches the tunnelling and then allows only incoherent transitions, by breaking the symmetry. The fitting results confirm the values of the electric hyperfine parameters obtained from the zero field measurements and give 6.4(1) T for the component of the magnetic hyperfine field along the quantization axis, in agreement with the $A_{\zeta\zeta}S_\zeta$ components of table 2. Moreover, they provide 5.8(1) T for the transversal component with relaxation rates of the order of the Mössbauer natural line width.
By comparing the result for the ζ component of $\boldsymbol{B}_\mathrm{h}$ with the values from literature [8] for Fe(II) ion, we estimate $s_\zeta^{(\mathrm{Fe})} \simeq 0.7$ in agreement with the calculated values reported in table 1.

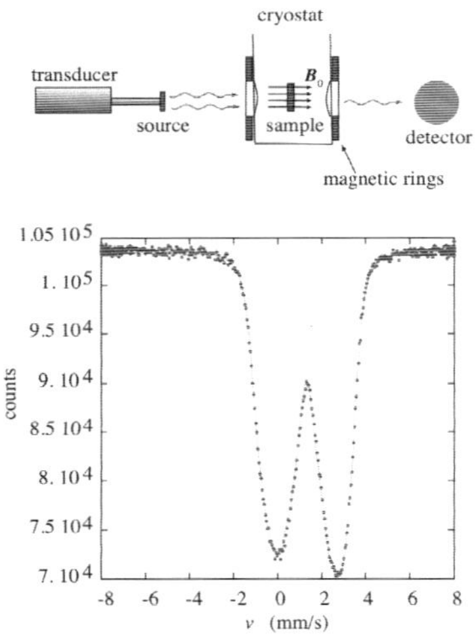

Figure 5. Mössbauer spectrum at 1.8 K in the presence of a 1.5 10^{-2} T applied field.

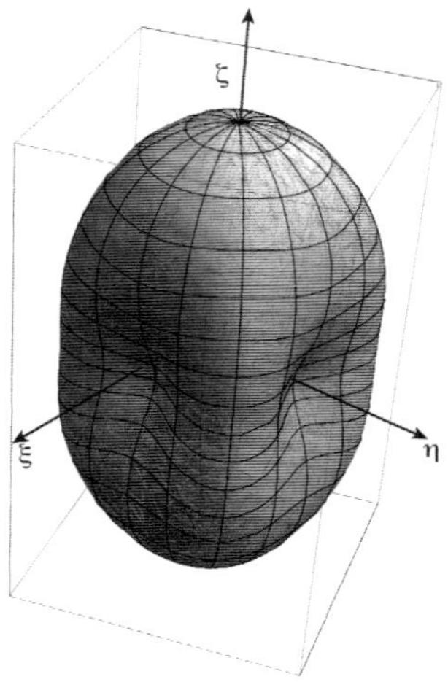

Figure 6. 3d-plot of the magnetic anisotropy energy of the iron ion's spin with the three principal axes of the corresponding tensor.

References

[1] McInnes E, Piligkos S, Timco G and Winpenny R 2005 *Coord. Chem. Rev.* **249** 2577
[2] Ochsenbein S T *et al.* 2008 *Chem. Eur. Journ.* **14** 5144–5158
[3] Amoretti G, Caciuffo R, Carretta S, Guidi T, Magnani N and Santini P 2008 *Inorg. Chim. Acta* **361** 3771
[4] Caciuffo R *et al.* 2005 *Phys. Rev. B* **71** 174407
[5] Detailed information about the method can be obtained from Prof. G. Spina URL gabriele.spina@unifi.it
[6] Hirst L 1970 *J. Phys. Chem. Solids* **31** 655
[7] Spina G and Cianchi L 2007 Line shape of Mössbauer relaxation spectra *NMR-MRI, μSR and Mössbauer Spectroscopies in Molecular Magnets* ed Carretta P and Lascialfari A (New York: Springer) p 249
[8] Eeckhout S G and Grave E D 2003 *Am. Mineralogist* **88** 1138

Magnetic inhomogeneities in temperature dynamics of spin structure of the ordered $Fe_{100-x}Al_x$ (25<x<35 at.%) alloys

E V Voronina[1,5], E P Yelsukov[1], S K Godovikov[2], A V Korolyov[3] and A E Yelsukova[4]

[1]Physical-Technical Institute UrB RAS, 132 Kirov Street, Izhevsk 426001, Russia
[2]SINP MSU, 1 Leninskie Gory, Moscow 119992, Russia
[3]Institute of Metal Physics UrB RAS, 18 S. Kovalevskaya Street, Ekaterinburg 620041, Russia
[4]University Duisburg-Essen, Duisburg, Germany

E-mail: voronina@fnms.fti.udm.ru

Abstract. The results of temperature and in-field Mössbauer and magnetometric studies of the ordered $Fe_{100-x}Al_x$ (25<x<35 at.%) alloys are discussed. Interpretations of observed behaviour of magnetic characteristics versus temperature in terms of the model of collinear magnet with magnetic inhomogeneities on a nanometer scale are proposed.

1. Introduction

Concepts of the origin of the unusual magnetic properties of the ordered $Fe_{1-x}Al_x$ (25<x ≤ 35 at.%) alloys are varied [1-4]. In some concepts [1,5] it was supposed that the magnetic inhomogeneities - two types local regions: surroundings of the type "Fe_3Al" and surroundings of the type "FeAl" - govern their magnetic properties. Neutron scattering studies [6] concluded that these alloys comprised ferromagnetic Fe_3Al-like regions consisting of Fe atoms with $k_{Fe} \geq 4$ nearest neighbouring Fe atoms, and the regions consisting of Fe atoms with $k_{Fe} < 4$. With increasing the temperature the regions (or clusters) arranged by Fe atoms with $k_{Fe} < 4$ exhibit a superparamagnetic-like behaviour. Recent neutron scattering data of B2-ordered Fe-Al alloys [7] proved out the existence of spatial correlation of magnetic moments with a coherence length ≈5 nm, that was interpreted in terms of incommensurate spin density waves. But, the conditions of spin density wave origin in these partially disordered alloys remained yet unclear. In *ab-initio* calculations of the local magnetic moments of Fe-Al alloys (from 29 to 44 at.% Al) [8] three different solutions were found: two collinear states (a ferromagnetic one and a state with Fe magnetic moments oriented parallel and antiparallel to magnetization) and a spin-spiral wave. In this work we propose the explanation of the observed temperature behaviour of magnetic characteristics within the simple model of Fe atom local magnetic moment [9], with its magnitude and direction being dependent on the number of Al atoms in its nearest neighbourhood.

2. Experimental

Ingots of Fe-Al alloys with Al concentration 25<x<35 at.% were arc melted from high-purity components in purified He and homogenized at 1100C for 8h in purified Ar. The powders were

[5] To whom any correspondence should be addressed.

subjected to a heat treatment to obtain B2 or DO_3 superstructure. The preparation and X-ray diffraction analysis are given in [10] in detail. A chemical analysis was performed using a Spectraflame-Modula D atomic emission spectrometer with an inductively coupled plasma and an MC7201 secondary-ion mass spectrometer. The content of Al in the alloys was 26.5, 30.0, 32.6, and 34.1 at. % to an accuracy of ± 0.3 at. %. Magnetic and Mössbauer measurements of *bcc* samples with various types of order: $Fe_{73.5}Al_{26.5}$ ($D0_3$), $Fe_{70}Al_{30}$ ($D0_3$), $Fe_{70}Al_{30}$ (B2), $Fe_{67.4}Al_{32.6}$ (B2), and $Fe_{65.9}Al_{34.1}$ (B2) were performed. The magnetic measurements were carried out using an MPMS-XL-5 (Quantum Design) SQUID magnetometer in external magnetic fields H_{ext} to 50 kOe at 5 - 300 K. The ^{57}Fe Mössbauer spectra were taken using a spectrometer with a ^{57}Co in the Cr matrix at 6 - 700K. To calculate the distributions of the hyperfine magnetic field (HFMF) $P(H_{hf})$, the regularization method was used. The interval of hyperfine fields from 0 to 30 kOe, where the $P(H_{hf})$ distribution corresponds to the spectrum part without magnetic splitting, was taken into account in the calculations of the average HFMF $\overline{H}$ by putting $H_{hf} = 0$ for the $P(H_{hf})$ in this interval.

3. Results and discussion

3.1. Magnetic measurements

Only for the $Fe_{73.5}Al_{26.5}$ alloy the magnetization curves and ZFC and FC cycles are close to canonical ferromagnetic systems. The magnetization of the rest alloys under study does not reach saturation in an external field H_{ext} of up to 50 kOe. Magnetic-hysteresis loops of all alloys are unshifted and symmetric. In the ZFC and FC curves for all the alloys in low H_{ext} a clearly pronounced thermomagnetic hysteresis is observed at T≈40 K. Detailed comparison of the ZFC and FC curves values with the values of the magnetization in magnetic hysteresis loops at corresponding H_{ext} and

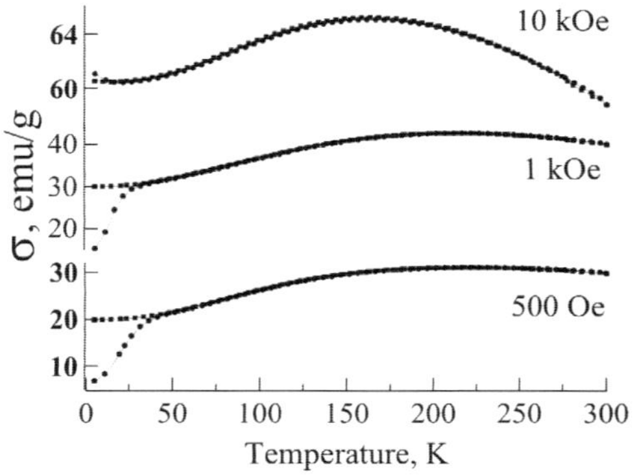

Figure1. ZFC and FC curves for ordered $Fe_{70}Al_{30}$ alloy.

temperature showed that the true reason of the ZFC and FC curves hysteresis is field-related hysteresis rather than thermomagnetic one [9]. The absence of hysteresis in H_{ext} exceeding 8-10 kOe, i.e., at those values of H_{ext} when the dependence of the magnetization on the field has a one-to-one relation (figure 1), sustains such an understanding of the ZFC and FC experiment results. One more peculiarity of $\sigma(T)$ in FC and ZFC cycles is pronounced maxima, with position being dependent on Al concentration in an alloy. The temperature dependence of AC susceptibility $\chi(T)$ is characteristic of antiferromagnetic and reproduces maximum in $\sigma(T)$ dependence.

3.2. Mössbauer spectroscopy

In Mössbauer spectra of all alloys, except for $Fe_{73.5}Al_{26.5}$ sample, the fast growth of fraction of non-magnetic component P_0 resulting in a decrease of the average ^{57}Fe HFMF $\overline{H}$ at temperature from 12 to 120 K is observed (figures 2 and 3). Even for the alloy $Fe_{73.5}Al_{26.5}$ in [0-0.9]T/T_c the dependence $\overline{H}(T)$

is different from analogous of ferromagnetic alloy Fe_3Al. In the $Fe_{70}Al_{30}$ samples the $\overline{H}$ increases at T > 120 K reaching maximum at 220-230 K regardless of the type of superstructure. The difference in minimal (at 110-130K) and maximal (at 220-230K) $\overline{H}$(T) values is ≈15 kOe for the B2 and ≈25 kOe for the $D0_3$ superlattice alloys (beyond the range of the error). Similar behaviour of $\overline{H}$(T) for the alloy $Fe_{70}Al_{30}$ was observed also in [11].

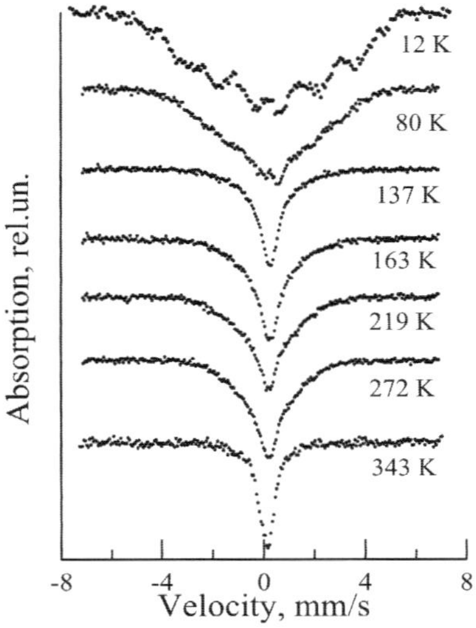

Figure 2. Mössbauer spectra of the $Fe_{70}Al_{30}$ ($D0_3$) alloy versus temperature

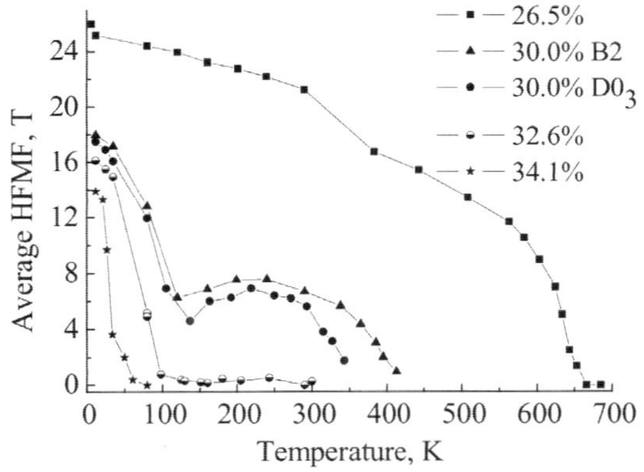

Figure 3. The average hyperfine magnetic field versus temperature

The main feature of Mössbauer spectra taken under H_{ext} is an appearance of the component with magnetic splitting highly exceeding an external field. Expanding of hyperfine magnetic splitting in spectra with H_{ext} raise results in the essential increase of $\overline{H}$ (for the alloy $Fe_{65.9}Al_{34.1}$ at T=80K it is equal to 64 kOe under H_{ext}=16.2 kOe) [12]. The considerable difference in the temperature behaviour of $\overline{H}$ and $\sigma(T)$ vanishes at low values of H_{ext} and these dependences get correlating. The temperature dynamics as well as in-field dynamics of Mössbauer spectra parameters testify the existence of relaxation phenomena in hyperfine interactions at T>30-50K (figures 2 and 3). But, the blocking temperature estimated from $\sigma(H/T)$ dependence is about 120-150K for $Fe_{65.9}Al_{34.1}$ and 210K for $Fe_{70}Al_{30}$ alloys [12]. What kind and size inhomogeneities could lead to collective spin fluctuations and to superparamagnetic signs we consider within the microscopic model of Fe local magnetic moment.

In [9] a simple model was proposed, in which the magnetic moment of an Fe atom surrounded up to 4 Al atoms is parallel, and of an Fe atom having 5 and more Al atoms in the nearest neighbourhood is antiparallel to the total magnetization. The model describes quantitatively the dependence of the Fe atom average magnetic moment on Al content and agrees with the data [6,7] on magnetic inhomogeneities on a nanometer scale. Inhomogeneity implies regions with magnetic moments directed along - m^+ and opposite - m^- the total magnetization. Theoretically, the temperature dynamics of these m^+ and m^- magnetic moments is different in consequence of Stoner-type excitations [13]. Suppose that thermal fluctuations of the m^- magnetic moments, arising at rather low temperature (50-100 K), excite the fluctuations of the surrounding m^+ magnetic moments also. That results in collective spin fluctuations and abrupt decrease of $\overline{H}$ at T≈100-120K. Concurrent reduction and vanishing of the m^- magnetic moments in absolute magnitude (due to Stoner-type excitations), that decrease faster than m^+, at 150<T<220 K brings to a gain of the total magnetization, to stabilization of surrounding m^+ magnetic moments and to the increase of $\overline{H}$.

Another reasoning equivalent to the above in relation to the obtained results based on two possible states of this magnetic system: with the local ferromagnetic and antiferromagnetic-type arrangement of neighbouring magnetic moments [8], is also possible. The theoretical studies [8,13] of metal-metalloid alloys have shown, that for high metalloid content the local magnetic moments opposite to magnetization change their direction with increasing the temperature. In this connection, the increase of $\overline{H}$ at 220-230K can be explained by the augmentation of the number of magnetic moments m^+, with the magnitude being higher than predecessor moments m^-. So, the system goes into "high-spin" state. It leads to the gain in magnetization, enlargement of the ferromagnetic clusters size, reinforcement of the inter-cluster exchange, slowing down of the cluster magnetic moment fluctuations and growth of the magnetic hyperfine splitting in spectra. But, in this case, according to [8,13] one should expect the intensive maximum, localized in the narrow temperature range that is not observed in magnetometry.

By the results of the work [8], it is the state with spin-spiral wave, which corresponds to a minimal energy of the system. Then, the sharp collapse of hyperfine magnetic splitting from T>30-50K can be brought about by the destruction of spin-spiral wave state in favor of the state with Fe magnetic moments oriented parallel and antiparallel to magnetization. On further increasing temperature the effect of both, Stoner- type excitations or transition of the system in "high-spin" state, consists in the stabilization of the clusters fluctuations and disordered ferromagnetic spin structure formation. In this case, the coherence length of the spin spiral in the state with the spin-spiral ordering is the frame of the clusters in the state with the collinear spin ordering.

4. Conclusion

The results of temperature and in-field magnetometric and Mössbauer studies of the ordered $Fe_{100-x}Al_x$ (25<x<35 at.%) alloys give an indication of the collective spin fluctuations connected with the magnetic inhomogeneities on a nanometer scale. Temperature behaviour of the magnetization and the average ^{57}Fe HFMF in 100-200K is assigned to several causes. One of explanations supposes "freezing" the relaxation phenomena in clusters consisting of spins oriented conversely the magnetization. To clarify which variant of the thermodynamic behaviour of the spin structure is most probable a quantitative analysis within itinerant magnetism thermodynamics is required.

Acknowledgements

Authors wish to express appreciation to Prof. A.K. Arzhnikov for the productive discussion and to acknowledge the financial support of RFBR (Grants N 06-02-16179, 09-02-00461).

References

[1] Danan H and Gengnagel H 1968 *J. Appl. Phys.* **39** 678
[2] Shull R D, Okamoto H and Beck P A 1976 *Sol. St. Commun.* **20** 863
[3] Takahashi S and Umakoshi Y 1991 *J.Phys.: Cond. Matt.* **3** 5805
[4] Bao Wei, Raymond S, Shapiro S M, Motoya K, Fak B and Erwin R W 1999 *Phys. Rev. Let.* **23** 4711
[5] Hergt R, Wieser E, Gengnagel H and Gladun A 1970 *Phys. Stat. Sol.* **41** 255
[6] Cable J V, David L and Parra R 1977 *Phys. Rev. B.* **16** 1132
[7] Noakes D R, Arrott A S, Belk M G, Deevi S C, Huang Q Z., Lynn J W , Shull R D and Wu D 2003 *Phys. Rev. Let.* **91** 217201
[8] Arzhnikov A K, Dobysheva L V, Timirgazin M A 2008 *JMMM* **320** 1904
[9] Elsukov E P, Voronina E V, Korolev A V, Elsukova A E and Godovikov S K 2007 *Phiz.Met.Metallogr.* **104** 35
[10] Elsukov E P, Voronina E V, Shuravin A S, Zagainov A V, Korolev A V, Godovikov S K, Pechina E A and Elsukova A E 2006 *Phiz.Met.Metallogr.* **102** 733
[11] Delyagin N N, Erzinkyan A L, Parfenova V P, Rozantsev I N 2005 *ZhETPh* **128** 1243
[12] Voronina E V, Yelsukov E P, Korolyov A V, Yelsukova A E 2006 *Hyp.Interact.* **168** 1079
[13] Arzhnikov A K, Dobysheva L V 1994 *Phys.Lett.*A **195** 176

Hyperfine interactions and electrical properties of multiferroic $0.5Bi_{0.95}Dy_{0.05}FeO_3$-$0.5Pb(Fe_{0.5}Nb_{0.5})O_3$

A Stoch[1], P Zachariasz[2], P Stoch[2,3], J Kulawik[1] and J Maurin[2,4]

[1]Institute of Electron Technology, Krakow Division , ul. Zabłocie 39, 30-701 Kraków, Poland
[2]Institute of Atomic Energy Polatom, 05-400 Otwock-Świerk, Poland
[3]Faculty of Material Science and Ceramics, AGH – University of Science and Technology, al. Mickiewicza 30, 30-059 Kraków, Poland
[4]National Medicines Institute, ul. Chełmska 30/34, 00-725 Warszawa, Poland

stoch@ite.waw.pl

Abstract. $0.5Bi_{0.95}Dy_{0.05}FeO_3$-$0.5Pb(Fe_{0.5}Nb_{0.5})O_3$ is a multiferroic material, simultaneously revealing ferroelectric and antiferromagnetic ordering, synthesis is rather difficult. In this paper we present a conventional solid state reaction method for synthesis of $0.5Bi_{0.95}Dy_{0.05}FeO_3$-$0.5Pb(Fe_{0.5}Nb_{0.5})O_3$ perovskite. A series of methodological measurements were carried out to determine the basic structural and electromagnetical properties. X-ray diffraction studies have been made in order to determine a crystal structure. Mössbauer effect spectroscopy was used to identify the oxidation state of Fe. Electrical properties of the perovskite ceramic were examined with impedance spectroscopy. The impedance spectra were analysed by means of an equivalent electrical circuit, and results are discussed in the scope of the brick layer model without easy paths.

1. Introduction

Materials, that simultaneously exhibit ferromagnetic and ferroelectric orderings are known as multiferroics. They are widely used as transducers, actuators and the other sensors, as well as the basic components for new mass storage devices.

The first single phased multiferroic perovskites were discovered in the early 1960s. However, during the last several decades a very insular and limited progress has been made [1-5]. A considerable development of multiferroic materials has been started with successful synthesis of multiferroics' thin films [6] or even more, when a novel class of multifunctional materials has been discovered, where a multiferroic state is not only mutual arrangement of particular subnets but ferroelectricity is exclusively determined by magnetically ordered state [7].

$0.5Bi_{0.95}Dy_{0.05}FeO_3$-$0.5Pb(Fe_{0.5}Nb_{0.5})O_3$ is a mixture of two perovskites: Dy-doped $BiFeO_3$ and $Pb(Fe_{0.5}Nb_{0.5})O_3$. A first component ($BiFeO_3$) is a well known perovskite, which exhibits both ferroelectric ($T_C = 840^{\circ}C$) and antiferromagnetic ($T_N = 340^{\circ}C$) orderings. Unfortunately, the synthesis of $BiFeO_3$ is difficult due to a high leakage current, secondary oxide phases ($Bi_2Fe_4O_9$, $Bi_{46}Fe_2O_{72}$) and the other harmful effects [2-4]. Nevertheless, there are many reports describing the method how to overcome above problems using a substitution of a rare earth element at A position in typical ABO_3 perovskite structure [8]. Additionally, the substitution of 3+ valence ion (for example Dy ion) considerably increases magnetic exchange interactions, what is observable with extension of hysteresis loop (*M–H* curve) [9].

A solid solution system of Dy-doped $BiFeO_3$ with $Pb(Fe_{0.5}Nb_{0.5})O_3$ was synthesized, where the last component is a lead-based multilayer ceramic capacitor with a high dielectric constant and a low sintering temperature [5]. In this paper, we report the synthesis, crystal structure, mössbauer and impedance spectroscopy studies of $0.5Bi_{0.95}Dy_{0.05}FeO_3$-$0.5Pb(Fe_{0.5}Nb_{0.5})O_3$ multiferroic system.

2. Sample preparation

High purity Bi_2O_3, Dy_2O_3, Fe_2O_3 and $Pb(Fe_{0.5}Nb_{0.5})O_3$ powder oxides were weighted in stoichiometric proportions and mechanically activated for 7 h during a ball milling process. In the next step, the obtained powder was calcined at 820°C for 4 h, granulated and then pressed into disc shaped pellet (2.09 mm height and 1.55 cm in diagonal). Finally, the pellet was heat-treated at 545°C for 2h and sintered at 850°C for 2 h.

3. X-Ray diffraction measurements

A purity of the sinter was checked by means of X-ray diffraction (Philips powder diffractometer with Cu K_α radiation, $\lambda = 1.54056$ Å). A fitting program (FULLPROF software [10]) based on the Rietveld method [11] was used to analyse a diffraction pattern – Fig.1. The derived multiferroic crystallize into a tetragonal symmetry with space group *P4mm*. The determined lattice parameters $a = 3.990(2)$ Å and $c = 4.026(8)$ Å indicate, that the structure of the synthesized ceramic is closely related to a simple cubic perovskite structure with a small distortion in c/a ratio of about 0.9 %.

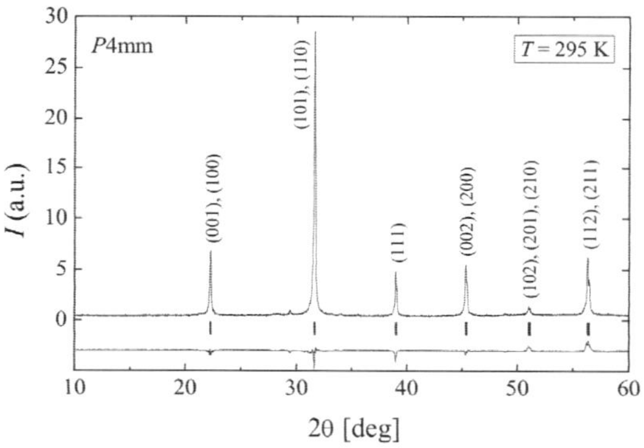

Figure 1. X-ray pattern of $0.5Bi_{0.95}Dy_{0.05}FeO_3$-$0.5Pb(Fe_{0.5}Nb_{0.5})O_3$ at 300 K.

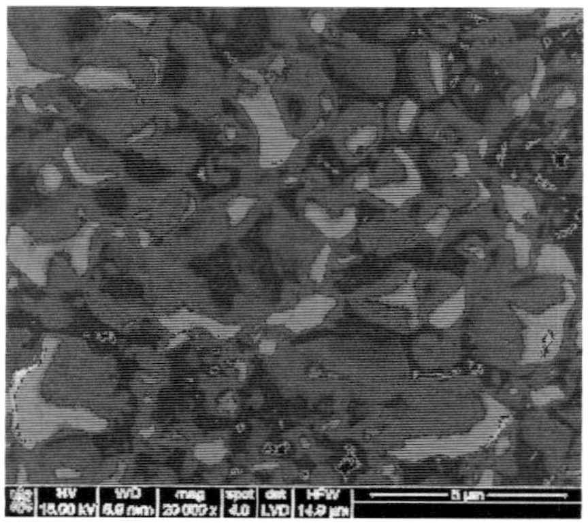

Figure 2. SEM image of $0.5Bi_{0.95}Dy_{0.05}FeO_3$-$0.5Pb(Fe_{0.5}Nb_{0.5})O_3$ sinter.

4. SEM/EDS studies

X-Ray diffraction results were confirmed by energy dispersive analysis. In Fig.2 the SEM image of a fine grain structure of the synthesized material is shown, where massive microcrystals about 3 μm in diameter are surrounded by smaller grains. According to the EDS studies, both the larger and the smaller grains exhibit the same chemical stoichiometry.

5. Mössbauer spectroscopy measurements

The ^{57}Fe transmission Mössbauer spectrum of the $0.5Bi_{0.95}Dy_{0.05}FeO_3$-$0.5Pb(Fe_{0.5}Nb_{0.5})O_3$ sample was measured at 77 K (Fig.3). A full static hamiltonian site analysis was implemented in order to determine the hyperfine interaction parameters as isomer shift (*IS*), quadrupole splitting (*QS*) and a hyperfine magnetic field ($\mu_0 H_{hf}$). The derived parameters are summarized in Table 1.

The previous Mössbauer effect studies of $BiFeO_3$ [12-13] and $BiFeO_3$ substituted by Eu [14] assume two different sextets with almost equal values of *IS* and slightly different values of $\mu_0 H_{hf}$ and

QS (components S_2 and S_3 in Table 1), whereas the hyperfine interaction parameters for subspectrum S_1 agree very well with the values characteristic for $Pb(Fe_{0.5}Nb_{0.5})O_3$ relaxor [15].

Table 1. The ^{57}Fe hyperfine interaction parameters at 77 K (*A* – subspectral area, Γ/2 – half width at half maximum, *IS* – isomer shift, $\mu_0 H_{hf}$ – magnetic hyperfine field, *QS* – quadrupole split).

No.	*A* [%]	**Γ/2 [mm/s]**	***IS*** **[mm/s]**	$\mu_0 H_{hf}$ **[T]**	***QS*** **[mm/s]**
S_1	15.2(1)	0.197(3)	0.542(4)	49.86(10)	0.055(1)
S_2	50.7(1)	0.215(3)	0.517(4)	53.42(4)	-0.041(1)
S_3	34.1(4)	0.199(3)	0.518(5)	51.78(6)	-0.012(2)

The derived values of *IS* and *QS* (with respect to the pure iron metal at 300 K) are typical for a Fe^{3+} ion in octahedral coordination [15]. The magnetic hyperfine fields $\mu_0 H_{hf}$ (μ_0 – magnetic permeability) of about 50 T are also characteristic for ferric iron.

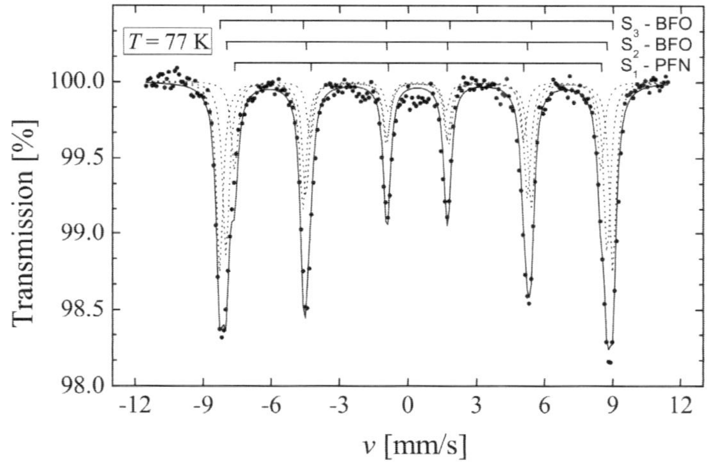

Figure 3. The ^{57}Fe mössbauer effect spectrum at 77 K.

6. Impedance spectroscopy studies

Impedance spectroscopy is a very useful technique to analyse the electrical properties of a wide group of electroceramics. The impedance measurements have been performed in the temperature range 450 – 700 K at frequencies from 10 Hz to 2 MHz.

Fig.4 presents an exemplary impedance spectrum of $0.5Bi_{0.95}Dy_{0.05}FeO_3$-$0.5Pb(Fe_{0.5}Nb_{0.5})O_3$ recorded at 600 K. The spectrum is dominated by a "grain interior arc", which is related to a grain interior conductivity, while a second "small arc" originates from the grain boundary conductivity [16].

Recorded impedance spectra were analyzed (solid curve in Fig.4) by LEVM program with implemented an equivalent electrical circuit proposed as a series-parallel connection of two resistors (R) and constant phase elements (CPE) [16,17].

In Fig.5 grain interior (R_{gi}) and grain boundary (R_{gb}) resistivities are presented in Arrhenius type plot. Conductivity of the investigated material is a thermally activated process with activation energies E_{gi} = 1.154 eV and E_{gb} = 0.961 eV, which values are typical for electroceramic materials [17].

The slopes of the plots are different, what should be expected on the basis of the brick layer model without easy paths for the systems with tetragonal symmetry [16]. The R_{gi} values are generally lower than the R_{gb} suggesting, that the electrical conductivity along the grains and across the grain boundaries clearly dominates the conductivity along the grain boundaries [16].

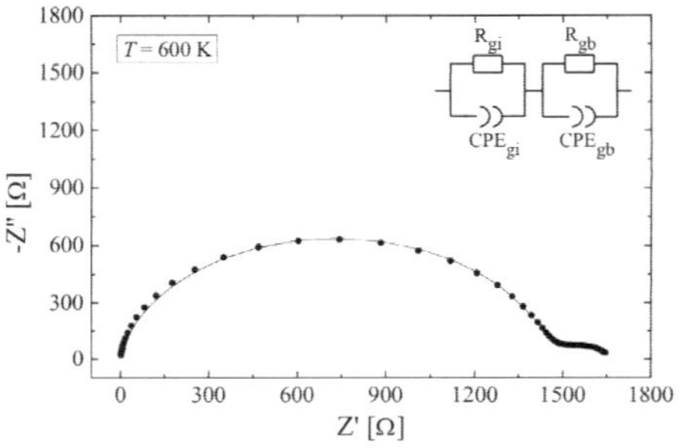

Figure 4. Impedance spectrum at 600 K with proposed equivalent electrical circuit (inset).

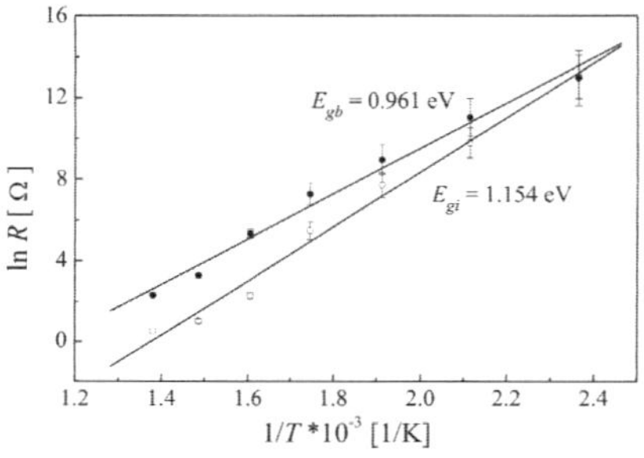

Figure 5. The Arrhenius plots of R_{gi} (open points) and R_{gb} (solid points) resistivities.

7. Conclusions

$0.5Bi_{0.95}Dy_{0.05}FeO_3$-$0.5Pb(Fe_{0.5}Nb_{0.5})O_3$ multiferroic was synthesized using the conventional solid state reaction method. The crystallochemical purity of the sinter was checked by X-ray diffraction and Mössbauer effect studies. The analysis of diffractogram confirmed, that the compound crystallizes into a tetragonal system (*P*4*mm*) with slightly different lattice parameters. The stoichiometry of investigated material have been proven by SEM/EDS analysis, and the Mössbauer spectroscopy results distinctly exhibit a one charge state of all iron atoms (Fe^{3+}) and octahedral symmetry. Detailed studies of impedance spectra with proposed equivalent electrical circuit show, that temperature dependence of R_{gi} and R_{gb} resistivities fulfill Arrhenius-like behaviour with activation energies of about 1 eV. The different slopes of R_{gi} and R_{gb} dependencies suggest, that the electrical conductivity of the compound can be described in the frame of the brick layer model without easy paths.

8. Acknowledgements

The work was partially supported by the Polish Ministry of Science and Higher Education under the project no. N N507 347335.

References

[1] Scott J F 2007 *Nature Mater.* **6** 256
[2] Paik H, Wang H, No K, Kwon S, Cann D P 2007 *Appl. Phys. Lett.* **90**, 042908
[3] Yuan G L, Or S W, Liu J M, Liu Z G 2006 *Appl. Phys. Lett.* **89** 052905
[4] Zhang S T, Zhang Y, Lu M H, Du C L, Chen Y F, Liu Z G, Zhu Y Y, Ming N B, Pan X Q 2006 *Appl. Phys. Lett.* **88** 162901
[5] Yang Y, Liu J M, Huang H B, Zuo W Q, Bao P, Liu Z G 2004 *Phys. Rev. B* **70** 132101
[6] Wang J 2003 *Science* **299** 1719
[7] Kimura T 2003 *Nature* **426** 55
[8] Uniyal P, Yadav K L 2008 *Mater. Lett.* **62** 2858
[9] Xu J X, Wang G M, Wang H X, Ding D F, He Y 2009 *Mater. Lett.* **63** 855
[10] Rodríguez-Carvajal J, Commission on Powder Diffraction (IUCr) 2001 *Newsletter* **26** 12
[11] Rietveld M H 1969 *J. Appl. Crys.* **2** 65
[12] Blaauw C, van der Woude F 1973 *J. Phys. C: Solid State Phys.* **6** 1422
[13] Palewicz A, Szumiata T, Przeniosło R, Sosnowska I, Margiolaki I 2006 *Solid State Commun.* **140** 356
[14] Freitas V F, Grande H L C, de Medeiros S N, Santos I A, Cotica L F, Coelho A A 2008 *J. Alloys Comp.* **461** 48
[15] Ivanov S A, Nordblad P, Tellgren R, Ericsson T, Rundlof H 2007 *Solid State Sci.* **9** 440
[16] Barsoukov E, Macdonald J R 2005 *Impedance Spectroscopy* 2nd ed. Wiley-Interscience, Hoboken
[17] Abram E J, Sinclair D C, West A R 2003 *J. Electroceram.* **10** 165

^{119}Sn Mössbauer Spectroscopy in the heavy – fermion ferrimagnet UCu_5Sn

K Łątka[1], **R Kruk**[2], **R Kmieć**[3], **R Troć**[4] **and V H Tran**[4]

[1] Marian Smoluchowski Institute of Physics, Jagiellonian University, Reymonta 4, 30-059 Kraków, Poland
[2] Institute of Nanotechnology, Forschungszentrum Karlsruhe GmbH, P.O. Box 3640, D-76021 Karlsruhe, Germany
[3] Henryk Niewodniczański Institute of Physics, Radzikowskiego 152, 31-342 Kraków, Poland
[4] Włodzimierz Trzebiatowski Institute of Low Temperature and Structure Research Polish Academy of Sciences, P.O. Box 937, 50-950 Wrocław, Poland

E-mail: uflatka@cyf-kr.edu.pl

Abstract. ^{119}Sn Mössbauer studies of UCu_5Sn compound have been performed in a wide temperature range between 4.2 K and 78 K. Magnetic hyperfine split spectra observed below T=54 K indicate the onset of magnetic ordering. The temperature dependence of the average transferred effective magnetic hyperfine field $\langle H_{eff} \rangle$ at the tin site can be surprisingly well fitted using the Brillouin function for S = 1/2. A rather small distribution of magnetic hyperfine fields δ_H obtained from the fitting procedure is in favour with the assumption that atomic mixing between Cu and Sn position can be neglected. The value of $\theta = 10(2)°$ shows that the direction of the effective magnetic hyperfine field $\mathbf{H_{eff}}$ is close to the c-axis being in accord with the uranium magnetic moment direction.

1. Introduction

The UCu_5Sn compound belongs to the family of moderate heavy-fermion systems [1] being a ferrimagnet with the ordering temperature $T_C = 54(1)$ K [1-3]. This compound exhibits the hexagonal $CeNi_5Sn$-type structure with space group $P6_3$/mmc and its magnetic structure below T_C consists of the ferromagnetic planes (001) with the (+− + −) stacking sequence, where the U magnetic moments are ordered along the hexagonal *c*-axis with two different values of magnetic moments at 2(a) and 2(c) U non-equivalent sites [3].

The aim of this work is to present the results of detailed ^{119}Sn Mössbauer effect measurements performed in a wide temperature range below as well as above the magnetic transition temperature T_C. Mössbauer Spectroscopy (MS) is well known as an extremely sensitive tool to study local electronic properties as it enables zero external magnetic field studies without disturbing of the investigated magnetic system. In particular, usage of the diamagnetic ^{119}Sn Mössbauer probe, being a natural constituent of the investigated sample, gives the unique opportunity to follow the temperature evolution of hyperfine parameters, which may be influenced by structural and magnetic phase transformations. It should be underlined that our Mössbauer studies were made on the same sample, which was used in the previous investigations of thermodynamic, transport, structural and magnetic properties of UCu_5Sn [1-3].

2. Experimental

The ^{119}Sn Mössbauer measurements have been performed using a triangular velocity drive of a Kankeleit type in transmission geometry. The ^{119}Sn source in chemical form of $Ba^{119}SnO_3$ and strength of about 10 mCi was kept at temperatures close to 4.2 K and series of ^{119}Sn Mössbauer spectra were recorded in a conventional variable temperature cryostat in the range of 4.2 K – 78 K. Absorber was prepared by grinding the sample to a fine powder and mixing it homogeneously with Al_2O_3 to ensure constant surface distribution over the absorber container with optimal thickness 26 mg cm^{-2} of UCu_5Sn material. For detection of the resonance 23.875 keV gamma quanta a scintillation counter with a 3 cm NaJ(Tl) crystal was applied. To reduce 25 keV Sn K_α background the critical filter made of Pd-foil ca. 0.05 mm thick was used. The energy scale was calibrated by means of a ^{57}Co(Cr) source and a metallic α-Fe foil at room temperature. The spectra were analysed by least-squares fits to the experimental results within the Lorentz approximation. By means of the numerical diagonalization procedure of the hyperfine interaction Hamiltonian H_{hf} the resonance-line positions and their relative intensities were calculated. The full refinements of the spectra lead to the results for the isomer shift δ_{is}, the quadrupole interaction constant $\Delta E_Q = eQV_{zz}$, the effective magnetic hyperfine field H_{eff} and for the polar angle θ, that is, for the direction of the magnetic hyperfine field with respect to the z-axis of the EFG (electric field gradient) tensor. Czjzek and Berger [4] method was used to account for an observed small broadening of the low temperature spectra of UCu_5Sn sample caused by possible distribution in magnitudes of the effective magnetic hyperfine field H_{eff}. The average value $\langle H_{eff}\rangle$ and the magnetic hyperfine field distribution parameter defined as $\delta_H = (\langle H_{eff}^2\rangle - \langle H_{eff}\rangle^2)^{1/2}/\langle H_{eff}\rangle$ (i.e. the ratio of the second moment $M_2(H_{eff})$ of the effective magnetic hyperfine field distribution to its average value) were fitted as two independent parameters. All spectra can be reproduced satisfactorily in this way with remarkable improvement of the fit quality.

3. Results

The resonance ^{119}Sn Mössbauer spectra obtained for UCu_5Sn compound at T = 78 K and 4.2 K are shown in figures 1 and 2, respectively. Because of non-cubic local symmetry of the Sn sites the spectrum measured at T = 78 K (i.e. far above T_C) presents typical quadrupole doublet, and its slight asymmetry can be attributed to a small amount of an impurity of unknown origin (figure 1).

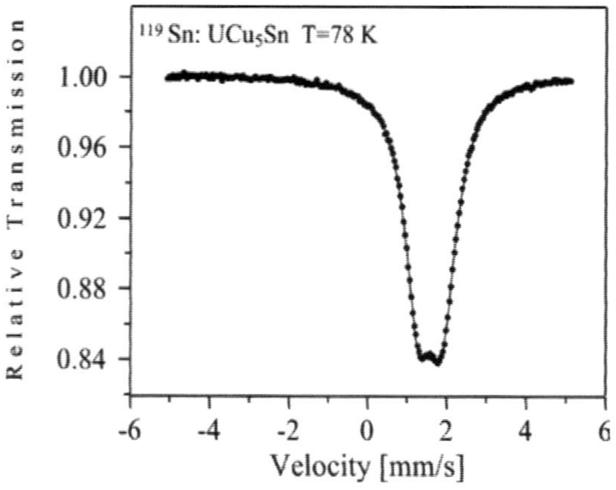

Figure 1. The ^{119}Sn resonance spectrum of the UCu_5Sn sample obtained at T = 78 K, i.e. above the critical magnetic transition temperature T_C = 54 K. The continuous line represents the least-squares fit to the experimental points.

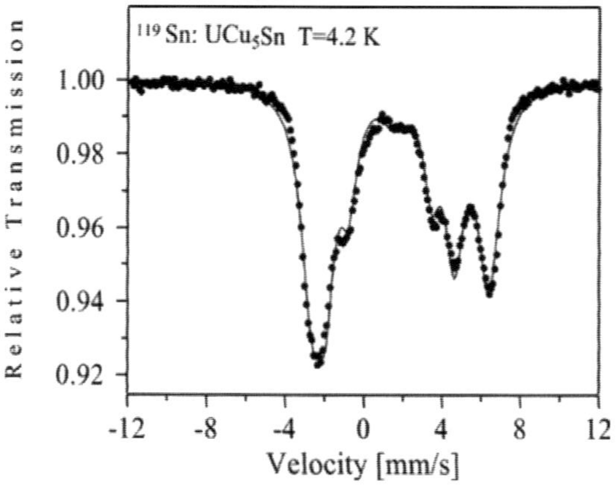

Figure 2. The ^{119}Sn resonance spectrum of the UCu_5Sn sample obtained at T = 4.2 K, i.e. below the critical magnetic transition temperature T_C = 54 K. The continuous line represents the least-squares fit to the experimental points.

This impurity component, fitted as a single line, accounts for about 4% of the total absorption intensity and is almost temperature independent.

Below T_C all the recorded spectra can be effectively fitted as a sum of only one main magnetic hyperfine component with a slightly distributed magnitude of the hyperfine magnetic field (according to the Czjzek and Berger method [4], see Section 2) and the respective impurity contribution which starts to split below 8 K in a visible way. Since the value of ΔE_Q for the main component calculated as a free parameter was not sensitive to the temperature change, it was constraint during fitting of magnetically split spectra to the value obtained at 78 K ($\Delta E_Q = |1.22(2)|$ mm/s). It is worth noting that the best fits to the low temperature spectra were obtained for the positive value of the quadrupole interaction constant ΔE_Q conjugated simultaneously with the small value of θ angle equal to $10(2)^{\circ}$, although reasonable fits can also be obtained for the negative value of ΔE_Q associated with θ angle close to 90°. The spectrum obtained at T = 4.2 K is shown in figure 2 as a characteristic example. The hyperfine parameters obtained from the refinements at 78 K and 4.2 K are collected in table 1.

Table 1. The hyperfine interaction parameters obtained from the ^{119}Sn resonance spectra of the UCu_5Sn intermetallic compound at T = 78 K (i. e. above the critical temperature T_C = 54 K) and 4.2 K.

T (K)	$\langle H_{eff} \rangle$ (kOe)	$\Delta E_Q{}^a$ (mms^{-1})	$\delta_{is}{}^b$ (mms^{-1})	θ (deg)	Γ^c (mms^{-1})	$\delta_H{}^d$
78	-	|1.22±0.02|	1.57±0.01	-	0.82±0.01	-
4.2	67.1±0.1	1.22^e	1.58±0.01	10±2	0.82^e	0.004±0.001

^{a} $\Delta E_Q = eQV_{zz}$

^{b} δ_{is} is isomer shift relative to the $Ba^{119m}SnO_3$ source.

^{c} Γ is half width of the resonance line.

^{d} $\delta_H = (\langle H_{eff}{}^2 \rangle - \langle H_{eff} \rangle^2)^{1/2} / \langle H_{eff} \rangle$ - the magnetic hyperfine field distribution parameter.

^{e} parameter kept constant during the fit.

The temperature dependence of the average effective magnetic hyperfine field $\langle H_{eff} \rangle$ can be well fitted by the Brillouin function for S = ½ (figure 3). The fit gives the expected value of the magnetic transition temperature T_C = 54(1) K as well as the average value of the saturation effective magnetic hyperfine field $\langle H_{eff}(0) \rangle$ = 67.5(2) kOe.

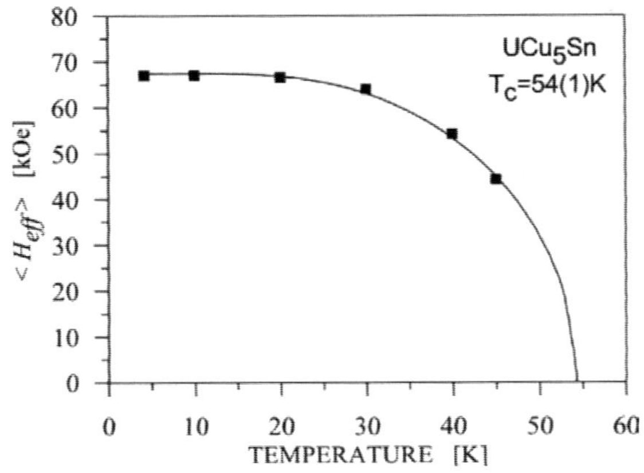

Figure 3. Temperature evolution of the average effective magnetic hyperfine field at the Sn sites for UCu_5Sn. The continuous line represents the least-squares fit of the S = 1/2 Brillouin function for the average values of the effective magnetic hyperfine fields, $\langle H_{eff} \rangle$.

4. Discussion

The derived value of the isomer shift (please note that δ_{is}=1.57(1) mm/s obtained at room temperature is equal within the experimental error to δ_{is}=1.58(1) mm/s at liquid helium temperature) is typical for that of Sn in metallic systems. Almost the same values of isomer shifts were observed for example in hexagonal UCu_2Sn and UPt_2Sn compounds (δ_{is}=1.63(1) mm/s and δ_{is}=1.58(1) mm/s, respectively) [5]. It means that the s electron charge density at the tin nucleus is not much affected by the tin environment. The lack of a remarkable temperature dependence of the measured isomer shift leads to a conclusion that the influence of the lattice expansion on its value is compensated by the change of the second order Doppler shift.

Owing to the threefold symmetry axis at the Sn position which is directed parallel to the crystallographic c-axis the V_{zz} axis of the EFG tensor coincides with this axis. Therefore the polar angle θ is the angle between the crystallographic c-axis and the direction of the magnetic hyperfine field $\mathbf{H_{eff}}$ induced at the tin site by a transfer mechanism and dipole field due to dipole interaction of magnetic uranium moments. In other words, the value of θ angle gives a deviation of $\mathbf{H_{eff}}$ vector from c-axis, which is the easy axis of magnetisation in our case. As this angle is only of about 10° it means that the direction of $\mathbf{H_{eff}}$ is close to the c-axis. Since dipole field is usually small it can be neglected in a first approximation. Under assumption that transfer mechanism is of RKKY type the magnetic hyperfine field $\mathbf{H_{eff}}$ at each Sn position is parallel and proportional to the magnetic uranium moments $\boldsymbol{\mu}_U$, hence, as a result on the basis of MS measurements alone, we can deduce that direction of these moments is also close to the c-axis. In fact, neutron investigations confirm this result, which showed parallel alignment of the U moments to the c-axis. In turn, taking into account collinear ordering of the magnetic U moments along c-axis, their values and interionic distances, dipole field $\mathbf{H_d}$ has been calculated. Using the value of experimentally derived H_{eff} and calculated values of dipole contributions, one gets that the maximum deviation of $\mathbf{H_{eff}}$ from the c-axis of about 11° is in a perfect agreement with the experimental θ value.

The estimated value of T_C = 54(1) K is in a very good agreement with those obtained from magnetization and neutron diffraction measurements i.e. 54(1) K. One can stress here that fits to the experimentally obtained temperature dependence of the average effective magnetic hyperfine field $\langle H_{eff}(T)\rangle$ using Brillouin functions with S higher than ½ give a little higher estimates for the derived magnetic transition temperature T_C than that observed by methods mentioned above. A rather small value of the derived magnetic hyperfine field distribution leads to the conclusion that there is not much atomic mixing between Cu and Sn positions. On the other hand such a mixing might change the local variation in the degree of hybridization of the U-5f electron states with the valence-band states of the ligands and in consequence causes a change of the local values of the U magnetic moment leading to the huge distribution of magnetic hyperfine fields as observed experimentally in UAuSn compound [6].

References

[1] Tran V H, Troć R, Pietri R and Andraka B 1999 *Phys. Rev.* B **60** 4696
[2] Tran V H, Troć R and Cichorek T 1999 *Physica* B **259-261** 263
[3] Tran V H , Troć R, André G, Bourée F and Kolenda M 2000 *J. Phys: Condens. Matter* **12** 5879
[4] Czjzek G and Berger W 1970 *Phys. Rev.* B **1** 957
[5] Wiese S, Gamper E, Winkelmann H, Büchner B, Abd-Elmeguid M M, Micklitz H and Takabatake T 1997 *Physica* B **230-232** 95
[6] Kruk R, Kmieć R, Łątka K, Tomala K, Troć R and Tran V H 1996 *J. Alloys Comp.* **232** L8

Magnetic Comparison of BaCa and BaSr Substituted Hexaferrite Powders

A González-Angeles[1], J Lipka[2], A Grusková[2], J Sláma[2], V Jančárik[2] and V Slugeň[2]

[1] Facultad de Ingeniería, Universidad Autónoma de Baja California, (UABC), Blvd. Benito Juárez s/n, Cp 21280 Mexicali, B. C., México.

[2] Faculty of Electrical Engineering and Information Technology, (FEEIT), Slovak University of Technology, (SUT), Ilkovičova 3, 812 19 Bratislava, Slovak Republic.

E-mail: gangelesa@yahoo.com.mx

Abstract. Results on magnetic studies of $Ba_{0.5}Sr_{0.5}Fe_{12-2x}(ZnTi)_xO_{19}$ and $Ba_{0.75}Ca_{0.25}Fe_{12-2x}(ZnTi)_xO_{19}$, where x = 0.2 to 0.6, ferromagnetic powders prepared by mechanical alloying are discussed. The structural and magnetic properties of the resulting powders were analyzed by thermo-magnetic analysis, X-ray diffraction, scanning electron microscopy, Mössbauer spectroscopy and magnetic measurements. Curie temperature, T_c decreased dramatically (drop ~ 39%) for BaCa samples, whilst for BaSr samples remained almost without change (diminution ~ 2%) at $x \leq 0.2$. SEM studies showed that all the particles present nearly hexagonal platelet shape.

1. Introduction

M-type hexaferrite $BaFe_{12}O_{19}$ (BaM) is one of the most important hard magnetic materials widely used in many applications. For its high stability, excellent high-frequency response and narrow switching field distribution. To use BaM on these applications it is required special magnetic and electrical specifications, and to obtain such modified hexagonal ferrites properties, trivalent iron ions, Fe^{3+}, are replaced by divalent M^{2+}- tetravalent M^{4+} metal ions, using different materials processing routes [1-3]. There have been a lot of investigation and publications to understand the substitution effect on hexagonal ferrites properties. Also, it has been studied the substitution effect on magnetic, electrical and microwave properties of BaCa and BaSr mixtures [3-7]. Some of the perspectives for preparing new ferrites (magneto-dielectrics) are: substitution of Ba^{2+} by combination of divalent ions (Sr^{2+}, Ca^{2+}) and Fe^{3+} ions by M^{2+} and M^{4+} ions [1-3]. In the present work, our attention was focused on magnetic and structural properties comparison of $Ba_{0.5}Sr_{0.5}Fe_{12-2x}(ZnTi)_xO_{19}$ and $Ba_{0.75}Ca_{0.25}Fe_{12-2x}(ZnTi)_xO_{19}$, as part of M-type hexagonal ferrite general studies.

[1] Alvaro González-Angeles, Facultad de Ingeniería, Universidad Autónoma de Baja California, (UABC), Blvd. Benito Juárez s/n, Cp 21280 Mexicali, B. C., México.

2. Experiment

To synthesize $Ba_{0.5}Sr_{0.5}Fe_{12-2x}(ZnTi)_xO_{19}$ and $Ba_{0.75}Ca_{0.25}Fe_{12-2x}(ZnTi)_xO_{19}$ the $BaCO_3$, $SrCO_3$, $CaCO_3$, Fe_2O_3, ZnO, and TiO_2 (~98% of purity) were used as starting materials. The raw materials were charged in a Segvari Attritor mill together with several 9-mm steel balls. The powders were milled during 24 hours in air with an angular velocity of 400 rpm and 250 ml of benzene to avoid agglomeration at the mill bottom. The as-milled powders were annealed at 1050°C for 1.5 h in air atmosphere.

The magnetic properties at room temperature were measured in a vibrating sample magnetometer Lake Shore 430 VSM applying an external magnetic field up to 1.2 T. To determine the crystalline structure of the prepared specimens, X-ray diffraction and Mössbauer spectroscopy studies were performed in an X' Pert Philips diffractometer using Cu-K_α radiation and a Mössbauer spectrometer with γ-ray source of ^{57}Co imbibed in a rhodium matrix, respectively. The temperature dependence of the magnetic susceptibility, χ (T) was obtained using the bridge method in an alternating magnetic field of 421.7 A/m and 920 Hz. Finally, to obtain information on the morphology and particle size, the BaM doped samples were observed in a Philips XL 30 scanning electron microscopy (SEM).

3. Results and Discussion

Figure 1 shows a sequence of x-ray diffraction patterns obtained at different steps of sample preparations. The patterns of both samples ($Ba_{0.75}Ca_{0.25}(ZnTi)_{x=0}$ and $(BaSr)_{0.5}(ZnTi)_{x=0}$) after milling for 24 hours without heat treatment show broadened diffraction peaks corresponding to Fe_2O_3 hematite phase (as milled sample). Magnetoplumbite phase was corroborated for all annealed samples in both substituted mixtures; no other secondary phases were detected at least within errors inherent in this technique. In addition the microstructure and phase composition of annealed samples were also investigated by Mössbauer spectroscopy and thermo-magnetic analysis of χ(T) behaviour.

It can be observed how the intensity and shape of Mössbauer spectra measured at room temperature (figure 2) changed with the substitution. Also it can see the spectrum for samples without substitution (x = 0) correspond to the five different positions (Zeeman sextets) $4f_2$, 2a, $4f_1$, 12k and 2b. At x > 0 the hyperfine fields for 2a and $4f_1$ sites are nearly equal, so that, it was not possible to separate these two subspatterns, besides, as x increased 12k site exhibited a remarkable splitting into 12k and 12k′ which enable us to consider the presence of other cations within hexagonal structure. For the sample $Ba_{0.75}Ca_{0.25}(ZnTi)_{x=0.6}$ the Mössbauer spectrum exhibits an expressive substitution in the position 2a and $4f_1$. In the position 2b the value of both the quadrupole-splitting (QS) and the relative spectral area (S) decreased, QS from a value of 2.2 to 1.77 mm/s and S to a value of 4.1%. The sample has superparamagnetic doublet with a relative spectral area 4.6%. These measurements indicating that the replacement of Fe^{3+} ions by Zn^{2+} (tetrahedral $4f_1$ and bipyramidal 2b sites) and Ti^{4+} (octahedral $4f_2$ and 2a sites) take place, (similarly on both mixtures up to x=0.4).

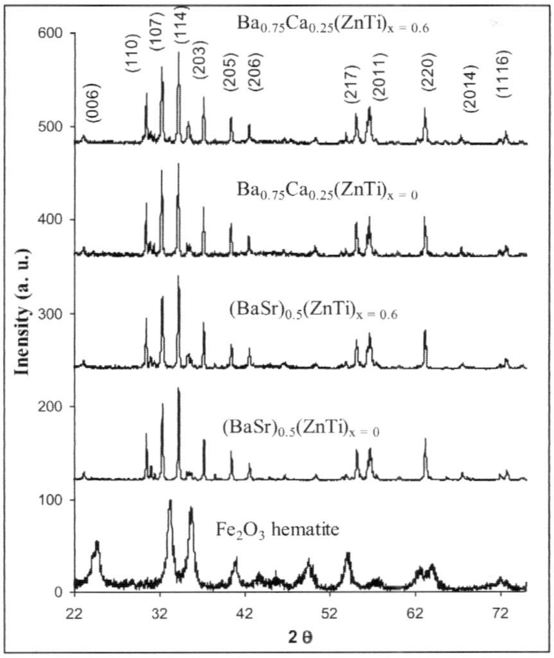

Figure 1. X - ray diffraction patterns for all prepared samples. Fe_2O_3 hematite phase correspond to samples after milling without heat treatment, the rest samples are annealed.

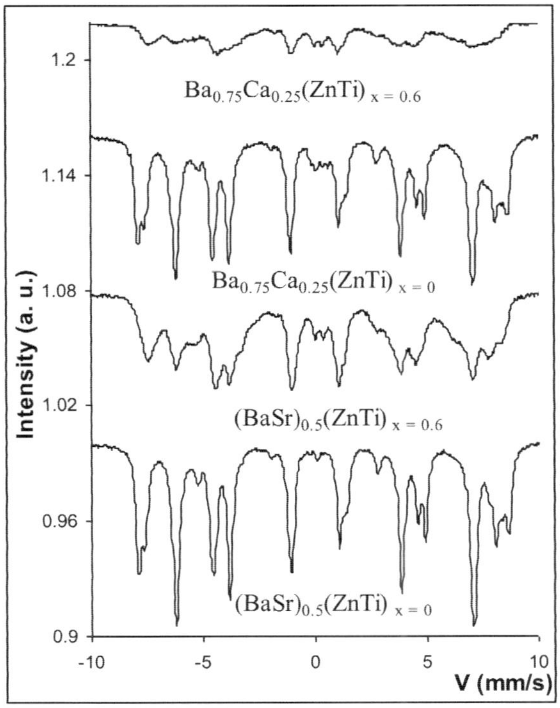

Figure 2. Mössbauer spectra of all studied samples.

The figure 3 shows the temperature dependence of the magnetic susceptibility measurements, χ (T) for both studied samples. The values of χ in wide range of the temperature for $Ba_{0.75}Ca_{0.25}(ZnTi)_{x=0}$ are approximately twice than that for $(BaSr)_{0.5}(ZnTi)_{x=0}$. Nevertheless, $(BaSr)_{0.5}(ZnTi)_{x=0}$ has 12°C higher value of T_c than that for $Ba_{0.75}Ca_{0.25}(ZnTi)_{x=0}$ besides, exhibit an Hopkinson effect expressively. Also, it can see that the T_c behaviour in both studied mixtures decreased drastically (drop ~ 39%), whilst for $(BaSr)_{0.5}(ZnTi)_{x=0.6}$ samples has the value T_c = 340°C, which is about 24.86% higher than has $Ba_{0.75}Ca_{0.25}(ZnTi)_{x=0.6}$ mixture, having T_c = 263°C. In addition the mixture of $(BaSr)_{0.5}(ZnTi)_{x=0.6}$ has a little amount of secondary phase with. T_C = 434°C. A little amount the secondary phase with T_c = 434°C were also observed for $(BaSr)_{0.5}(ZnTi)_{x=0.4}$ [3].

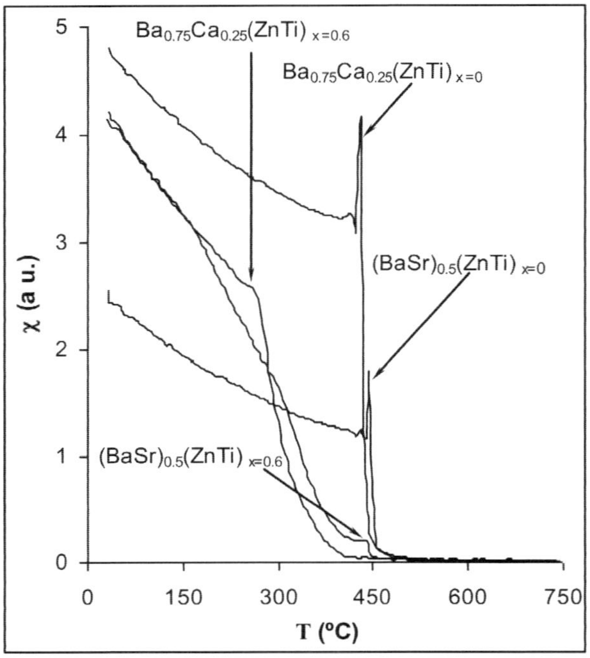

Figure 3. Temperature dependence of the magnetic susceptibility of substituted samples

The magnetic properties behavior with the substitution level is shown in Table 1. It can observe that the substitution of the ZnTi ions caused rapid decrease in both H_{ci} and magnetocrystalline anisotropy, besides, affected little to saturation magnetization, M_s at low substitution. Also, it can see that the saturation magnetization, for $(BaSr)_{0.5}(ZnTi)_{x=0}$ sample was higher than that for $Ba_{0.75}Ca_{0.25}(ZnTi)_{x=0}$ and it decreased lightly for both mixtures as x increased (reduction ~ 9% and 6% respectively). On the other hand, the intrinsic coercivity, H_{ci}, also decreased gradually with the substitution for both mixtures (drop for BaSr sample ~ 56% and for BaCa sample ~ 44%). As it follows, the variation in the H_{ci} was caused by the replacement of the Fe ions on $4f_2$ and 2b sites which posses the major contribution to the expressive decrease of the uniaxial anisotropy in samples, (without significant deterioration of M_s).

Table 1. Magnetic properties and Curie temperature for BaCa and BaSr as a function of Zn Ti substitution level x.

	0.75Ba0.25Ca $Fe_{12-2x}(ZnTi)_xO_{19}$				0.5Ba0.5Sr$Fe_{12-2x}(ZnTi)_xO_{19}$			
x	M_s (Am^2/kg)	H_{ci} (kA/m)	M_r (Am^2/kg)	T_c (°C)	M_s (Am^2/kg)	H_{ci} (kA/m)	M_r (Am^2/kg)	T_c (°C)
0	59.2	289.6	32.8	430	64.4	389.9	36.3	442
0.2	58	186.2	29.8	366	61.4	247.5	34.1	440
0.4	56.3	190.2	29.3	324	61.6	209.3	32.8	380
0.6	55.8	160.7	28.1	263	58.6	171.9	29.9	340

Figure 4 shows SEM micrographs of both studied mixtures. All particles showed nearly hexagonal platelet shape. BaCa powders look with round corners and particle size bigger than that for BaSr mixture. However, it is clear that the particle size is much below 1 μm. Nevertheless, TEM studies are needed to find out the real crystallite size of the powders.

Figure 4. SEM micrographs of both synthesized mixtures. a) $Ba_{0.75}Ca_{0.25}$(ZnTi) x=0; b) $(BaSr)_{0.5}$(ZnTi) x=0.

4. Conclusions

Curie temperature, T_c, diminished dramatically (drop ~ 39%) for BaCa samples, whilst for BaSr samples drop of T_C start at x > 0.2 and in addition these samples had a little amount of secondary phase [3], which T_C remained almost without change (diminution ~ 2%). M_s, for BaSr mixtures was higher than that for BaCa samples and it diminished lightly for both mixtures as x increased (reduction ~ 9% and 6% respectively). The intrinsic coercivity, H_{ci}, decreased gradually with the substitution for both mixtures (drop BaSr ~ 56% and BaCa ~ 44%). It can observe that for mixtures without substitution; BaCa sample obtained higher χ (particle size are bigger) than that for BaSr mixture at room temperature, whilst for substituted samples it was almost the same.

Acknowledgments
We would like to thanks both CONACyT-México (under project 89783) and VEGA-Slovak Republic (VG-1/0575/09, 1/0606/09 and 7/0129/09) for given support to carry out this work, respectively.

References

[1] Rewatkar K G, Patil N M, Jaykumar S, Bhowmick D S, Giriya M N, Khobragade C L 2007 *J. Magn. Magn. Mater.* **316** 19-22
[2] Sharma P, Rocha R A, de Medeiros S N, Paesano Jr A 2007 *J. Alloys Compd.* **443** 37-42
[3] Sláma J, Grusková A, González A, Jančárik V, Ušáková M, Ďurman V, Lipka J 2007 *Procc. in Third Seeheim Conference on Magnetism* 132-133
[4] Ding J, Street R, Nishio H 1996 *J. Magn. Magn. Mater.* **164** 385-389

[5] Fang Q Q, Fang D M, Bao H W, Wang J Z, Li X G 2004 *J. Magn. Magn. Mater.* **278** 122-126
[6] Singh C, Bindra Narang S, Hudiara I S, Bai Y, Tabatabaei F 2008 *Mater. Research Bulletin* **43** 176-184
[7] Lisjak D, Bregar V B, Drofenik M 2007 *J. Magn. Magn. Mater.* **310** 2558-2560

Structure and magnetic properties of the $Zn_xMg_{1-x}Fe_2O_4$ ferrites

Abbasher M Gismelseed, K A Mohammed, H M Widatallah, A D Al-Rawas, M E Elzain and A A Yousif

Department of Physics, College of Science, Sultan Qaboos University, P.O.Box 36, Code 123, Al-khoud, Oman

E-mail: abbasher@squ.edu.om.

Abstract. Spinel magnesium-zinc ferrites $Zn_xMg_{1-x}Fe_2O_4$ ($0 \leq x \leq 1$) prepared by the standard double sintering ceramic method have been investigated using X-ray (XRD) diffraction technique and Mössbauer spectroscopy measurements at 295K and 78K. The crystal structure is found to be single cubic spinel whose lattice parameter '*a*' increases with increasing Zn content. The Mössbauer spectra of the samples taken at 78 K are found to be magnetically ordered for $x \leq 0.5$, paramagnetic for $x \geq 0.8$ and show very broad patterns for intermediate concentration of Zn^{2+}. The variation of the magnetic properties with the Zn concentration has been explained on the basis of the cation distribution driven from the Mössbauer measurements.

1. Introduction

Ferrites are ferrimagnetic oxides with their magnetic cations forming two sublattices, namely the tetrahedral (A) and the octahedral [B] crystallographic sites. The structural and magnetic environments of these two sites are quite different and are very sensitive to the chemical compositions, preparation sintering temperature and sintering time. Also the type and the amount of cations substitution have very strong effect on the properties. The substitution of diamagnetic cations for example, is expected to weaken the magnetic coupling that affects the magnetic hyperfine fields and the Curie temperature. The magnesium ferrite $MgFe_2O_4$ has a predominantly inverse structure with the Mg^{2+} ions mainly on B-sites and the Fe^{3+} ions distributed between the A- and B-sites. On the other hand the zinc ferrite $ZnFe_2O_4$ has the normal spinel structure, in which the Zn^{2+} ions occupy the A-sites where the B-sites are occupied by the Fe^{3+} ions [1,2]. The system $Zn_xMg_{1-x}Fe_2O_4$ has been the subject of many investigations under different treatments and different techniques. Joshi et al [3], Sawant et al [4], and El Hiti et al [5,6] have studied the lattice parameters of the system. The Mössbauer measurements have been performed on the $Mg_{0.75}Zn_{0.25}Fe_2O_4$ ferrite [7]

In the present work ferrites of compositions $Mg_{1-x}Zn_xFe_2O_4$ ($0 \leq x1.0$) have been prepared and investigated using XRD and Mössbauer techniques to elucidate the site occupancy of the cations in the system. The results would also reflect on the influence of Zn^{2+} ions on the structure and magnetic properties of the magnesium ferrite.

2. Experimental

A series of Mg- ferrite substituted by Zn, $Mg_{1-x}Zn_xFe_2O_4$ (0≤x1.0) were prepared using the conventional double sintering technique, in which powders of high purity MgO, ZnO and Fe_2O_3 oxides were mixed in required proportions in agate mortar. The mixture was first sintered at 1000 °C for 24 h in air medium and slowly cooled to room temperature. The powder was then remixed and ground once more to promote homogeneity. Pellets of 1.3 cm diameter and 2-3 mm thick were prepared and sintered at 1050 °C for 24 h then slowly cooled to room temperature.

XRD data were recorded on a Philips diffractometer (model pw 1820), using Cu-K_α radiation, and analyzed using the PowderCell 2.4 program [8]. Mössbauer spectra were obtained at 295 K and 78 K on a powdered sample using a constant acceleration Mössbauer spectrometer with 50 mCi ^{57}Co in Rh source. The low temperature measurement was performed using a liquid nitrogen flow cryostat. The spectrometer was calibrated with α-Fe foil spectrum at 295 K. The measured data were analyzed using a non-linear least-square fitting program assuming Lorentzian lines.

3. Results and discussion

The X-ray diffraction patterns recorded at 295 K for the studied compositions confirmed the formation of a single cubic spinel phase. The values of the lattice parameters *a* for the samples $MgFe_2O_4$ (x =0.0) and $ZnFe_2O_4$ (x =1.0) were found to be 8.378(8) and 8.450(8) Å respectively and agree quite well with literature values of 8.387 Å for $MgFe_2O_4$ [9] and 8.434 Å for ZFe_2O_4 [10]. The larger value of the lattice parameter *a* for $ZnFe_2O_4$ compared to that of $MgFe_2O_4$ is due to the larger ionic radius of Zn^{2+} (0.82 Å) as compared to Mg^{2+}(0.66 Å).

The Mössbauer measurements at 295 K and 78 K of the studied system are illustrated in figure (1). The 295 K measurements show magnetic components for x≤0.5 and paramagnetic above those contents; while at 78K the Mössbauer measurements of all the samples with x≤0.6 exhibit a magnetic behavior. The magnetic components of the spectra were fitted assuming a superposition of two magnetic sextets assigned to Fe ions at the A- and B-sites according to their magnetic fields H_A and H_B. The larger magnetic fields assigned to the B-site is attributed to the dipolar field resulting from the deviation from cubic symmetry and also from the covalent nature of the tetrahedral bonds [11]. A summary of the hyperfine interaction parameters is presented in table I. As seen in the table the isomer shifts of the two sites for both measured temperatures show no sizeable change with Zn^{2+} content. This is indicative that the s-electron density at the nucleus of the Fe^{3+} cation in either sub-lattice remains largely unaffected by the increasing number of Zn^{2+} ionic nearest neighbors [12]. The line width Γ_B of the octahedral sites becomes larger than that of the tetrahedral site with the increase of the Zn contents at the tetrahedral site. Such difference is in fact, expected due to a distribution of the hyperfine fields at the B-site caused principally by a random distribution of Zn^{2+}, Fe^{3+} and fraction of Mg^{2+} ions at the A-site, since the Fe^{3+} B-sites would feel different configurations of cations occupying the six nearest neighbors A-sites. The variation of the hyperfine magnetic fields H_A and H_B as a function of zinc concentration is shown in figure (2). Similar variations for H_A and H_B were observed in other zinc substituted ferrite [13] and reference therein. Patil et al [14] and Srivastava et al [15] have explained the variation of H_A and H_B with zinc content on the basis of the canting of spins which gives rise to Yafet-Kittel angles. Such canting of the spins suggests that A-B and B-B super-exchange interactions are comparable in strength. Joshi et al [13] have observed a discrepancy in the relative magnetization obtained from the Mössbauer data and that obtained from the saturation magnetization measurements of $Mg_{1-x}Zn_xFe_2O_4$ for x≤0.5. They conclude that such discrepancy is an indirect indication of the canted spins and hence of Yafet-Kittel angles in the Mg-Zn ferrite.

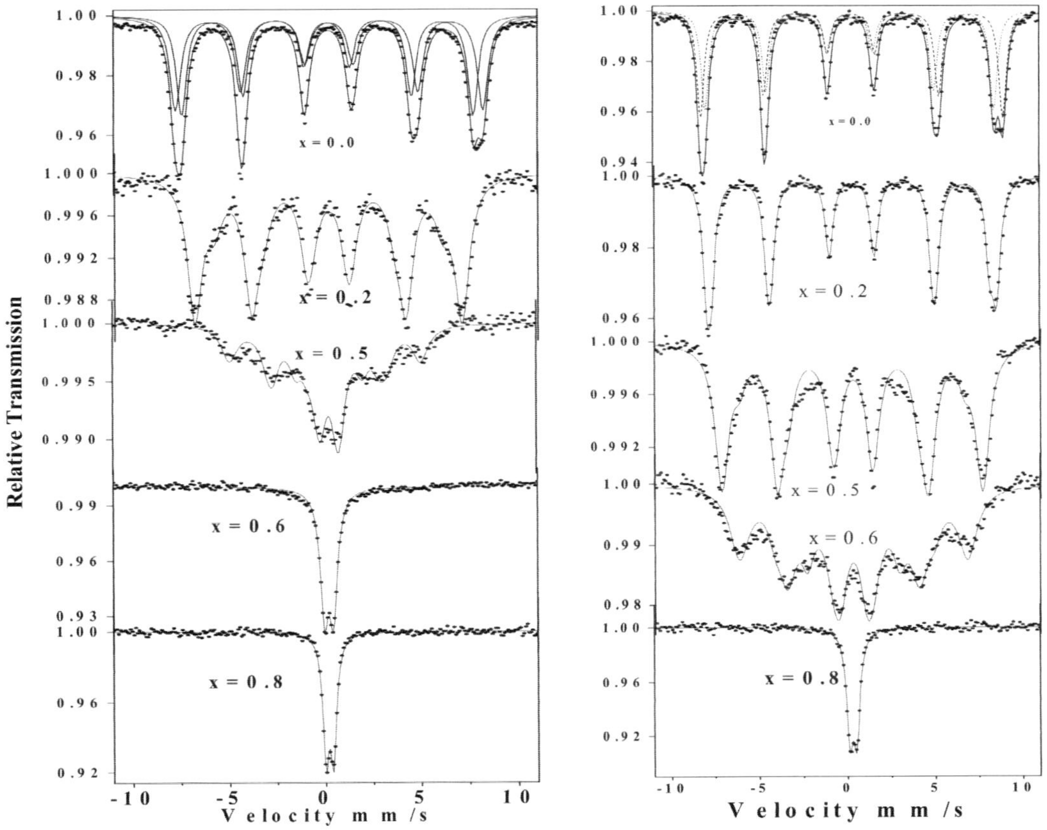

Figure1: 295 K and 78 K Mössbauer spectra of $Mg_{1-x}Zn_xFe_2O_4$

Table I: Mössbauer hyperfine parameters at 78 K and the cation distribution of $Mg_{1-x}Zn_xFe_2O_4$. The isomer shift δ (±0.02), quadrupole splitting ΔE_Q (±0.01), and linewidth Γ (±0.01) are in mm/s

	Tetrahedral Site (A)			Octahedral site (B)			
x	δ	ΔE_Q	Γ	δ	ΔE_Q	Γ	Cation distribution
0.0	0.36	0.01	0.52	0.46	0.06	0.51	$(Mg_{.03}Fe_{.97})\,[Mg_{.97}Fe_{1.03}]$
0.1	0.36	-0.03	0.44	0.44	0.05	0.56	$(Zn_{.1}Mg_{.1}Fe_{.8})\,[Mg_{.8}Fe_{1.2}]$
0.2	0.39	-0.05	0.50	0.43	0.02	0.59	$(Zn_{.2}Mg_{.08}Fe_{.72})\,[Mg_{.72}Fe_{1.28}]$
0.3	0.40	-0.07	0.51	0.42	0.01	0.60	$(Zn_{.3}Mg_{.08}Fe_{.62})\,[Mg_{.62}Fe_{1.38}]$
0.4	0.43	-0.11	0.74	0.41	0.01	0.68	$(Zn_{.4}Mg_{.07}Fe_{.53})\,[Mg_{.53}Fe_{1.47}]$
0.5	0.42	-0.08	0.80	0.41	0.02	0.95	$(Zn_{.5}Mg_{.05}Fe_{.45})\,[Mg_{.45}Fe_{1.55}]$
0.6	0.43	-0.08	0.90	0.46	0.02	1.30	$(Zn_{.6}Mg_{.05}Fe_{.35})\,[Mg_{.35}Fe_{1.65}]$
0.7							$(Zn_{.7}Mg_{.04}Fe_{.26})\,[Mg_{.26}Fe_{1.74}]$
0.8				0.45	0.40	0.45	$(Zn_{.8}Mg_{.035}Fe_{.165})\,[Mg_{.165}Fe_{1.835}]$
0.9				0.45	0.35	0.40	$(Zn_{.9}Mg_{.03}Fe_{.07})\,[Mg_{.07}Fe_{1.93}]$
1.0				0.45	0.30	0.37	$(Zn_1)\,[Fe_2]$

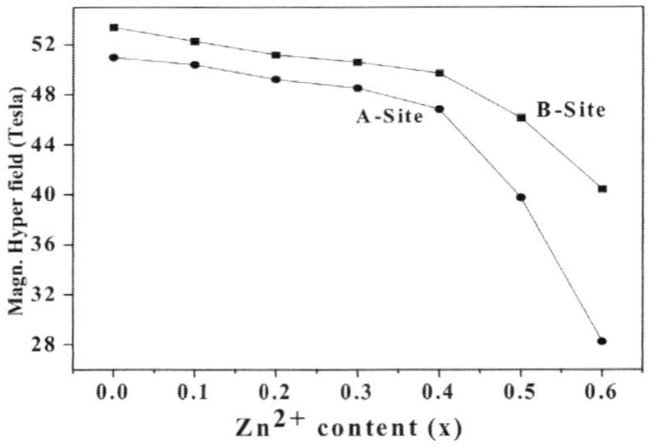

Figure 2: Variation of the magnetic hyperfine fields with Zn content measured at 78 K

For its better resolution of the two sextets, the Mössbauer spectra recorded at 78 K were used to determine the cation distribution in the various samples. The addition of the Zn^{2+} ions shows obvious variation on the resonance area and the magnetic hyperfine filed of both the (A) and [B] sub- lattices. The ratios of the areas under the resonance curve of the subspectra deduced from the measurements, and the recoil free fractions of Fe at the octahedral and tetrahedral sites (which are assumed to be equal at low temperature) were used to estimate the Fe site occupancy for each site ($n_A/n_B = A_A f_B/A_B f_A$). With Zn^{2+} occupying the tetrahedral sites, we have managed to establish in table I the cation distribution of different samples and thus the inversion parameter y; ($y = (1\text{-}x\text{-}\alpha)$); where α is the fraction of Mg^{2+} ion on the tetrahedral site for each value of x. From this distribution, the replacement of Mg^{2+} ion by x Zn^{2+} ions gives $(1\text{-}x\text{-}\alpha)Fe^{3+}$ ions on the A-site and $(1+x+\alpha)$ Fe^{3+} ions on the B-site. This distribution is in good agreement with [2] for the whole range of x and disagreed with [16] for $x \geq 0.4$ since his findings will not lead to the conversion of system into a normal spinel structure as we expected. Therefore from the obtained cation distribution we concluded that the studied system is converted gradually from an inverse to normal spinel structure by the addition of Zn content.

References:

[1] Thummer K P, Chhantbar M C, Modi K B, Baldha G J and Joshi H H 2004 *J Magn. Magn. Mater.* **280** 23-30
[2] Kulkarni R G and Joshi H H 1985 *Solid state commun.* **53** 1005-08
[3] Joshi G, Deshpande S, Khot A and Sawant S 1987 *Ind. J Phys.* **A61** 251
[4] Sawant S and Patil R, 1981 *J Mater. Sci.* **16** 3496
[5] El Hiti M, El Shora A and Hammad S 1998 *Mater. Sci. Technol.* **14** 625
[6] El Hiti M 1999 *J Magn. Magn. Mater.* **192** 305-313
[7] Josyulu O, Sobhanfri J, Narayanasamy A and Nagarajan T 1980 *Ind. J Pure Appl. Phys.* **20** 404
[8] Krauz W, Nolze G, 1996 J Appl Cryst. **29** 301-303.
[9] Mittal V K, Chandramohan P, Santanu Bera, Srinivasan M P, Velmurugan S and Narasimhan S V 2006 *Solid state commun.* **137** 6-10
[10] Ladgaonkar B P, Vasambekar P N and Vaingankar A S 2000 *J Magn. Magn. Mater.* **210** 289
[11] Roumaih Kh, Manapov R A, Sadykov E K and Pyataev A V 2005 *J of Magn Magn Mater* **288** 267-275.
[12] Lakshaman A, P.S.V. Subba Rao, K.H.Rao 2004 *J Magn. Mater.* **284** 532.
[13] Joshi H H, Rao S N and Kulkarni R G 1985 *J Mater Sci. lett.* **4** 1249-51
[14] Patil V U and Kulkarni R G 1979 *Solid state commun.* **31** 551
[15] Srivastava C M, Shringi S N and Srivastava R G 1976 *Phys. Rev.* **B14** 2041
[16] Mazen S A, Mansour S F and Zaki H M 2003 *Cryst. Res. Technol* **38** 471-478

Optimal configuration for Programmable Mössbauer Experiments

Gustavo A Pasquevich, Alejandro L Veiga, Pedro Mendoza Zélis and Francisco H Sánchez

Departamento de Física, Facultad de Ciencias Exactas, Universidad Nacional de La Plata, Argentina

E-mail: gpasquev@fisica.unlp.edu.ar

Abstract. Based on channel independency of recently developed Mössbauer instrumentation an approximation to optimal configuration of experiments is presented. The analysis relies on the presumption that all the available channels of the spectrum are not equally efficient for a given experimental application. A quantification of this concept is presented and a method for different channel layout comparison is proposed. The optimization of recorded spectra is important in dynamic experiments where efficiency in data taking imposes feasibility limits as well as in static applications as a way of reducing experimental time.

1. Introduction

Mössbauer effect spectra are usually taken in constant acceleration, constant velocity or sinusoidal modes. Recently a Programmable Velocity Mode (PVM)[1], based on a constant velocity mode strategy, was introduced as an alternative to the conventional modes. PVM opens new possibilities on the art of Mössbauer experimentation. The ability of measuring different spectral zones with different time weights is one of the important applications of this technique. Another recently developed application is the tracking of interest regions as they change due to the variation of an external and controlled parameter [2], technique that we have called Mössbauer Line Tracking (MLT).

In both cases there is a region of interest that is measured and further analyzed. Up to now, with the conventional spectrometers that use uniform velocities distribution, the question of which is the best set of velocities (sampling distribution) that most efficiently acquire a spectrum region was inconsequent. However with the liberty of freely choosing velocities and time of acquisition, this question becomes important if we are looking to improve the experiments. Furthermore in the case of MLT the success of the experiments strongly depends on a careful choice of the velocities.

In the following, we will refer as Region Of Interest (ROI) to the set of velocity positions and acquisition times,

$$\{(v_i, t_i), i = 1, .., N\}$$

where the measurements will be done. In both static and dynamic experiments the optimal ROI selection is constrained to the total time per ROI ($T_{ROI} = \sum t_i$). In the case of dynamic experiments is obvious that T_{ROI} is defined by the rate of change on the spectral profile. In fact, as fast are the spectrum changes as short must T_{ROI} be in order to reproduce quasi-static information of the spectrum.

MLT procedure exploits the programmable attribute of the PVM. The objective of the technique is to apply a dynamic ROI to follow an evolving spectrum using a feedback algorithm. The efficiency of this algorithm might depend on different factors and is not part of the discussion of this work. However the knowledge on the goodness of the ROI selection is a valuable tool at the moment of designing the tracking algorithm.

2. Theory

2.1. Variation of measured parameters due to statistical dispersion

Let $F(v,t;p_1^0,p_2^0,...)$ be the Intensity of the Transmission Spectrum (ITS) at a velocity v when the acquisition time is t. The characteristic parameters of the absorption profile (the centers of the lines, the FWHM, etc.) are represented by the quantities $p_1^0, p_2^0, ...$. With the supra index 0 we are indicating that they are the true parameters.

For each v_i and t_i of the ROI a counting y_i will be measured, according to a Poisson distribution with mean $F(v_i,t_i;p_1^0,p_2^0,...)$. In the following the reduced notation: $F_i(p_1,p_2,\cdots) \equiv F(v_i,t_i;p_1,p_2,\cdots)$, where the index i indicates the ROI element, will be used.

For a given set of experimental data $\{(v_i,t_i,y_i), i=1,...,N\}$ the spectrum parameters p are estimated by minimizing the Chi-Square function[3],

$$\chi^2(p_1,p_2,..p_n) = \sum_{i=1}^{N} \left(\frac{y_i - F_i(p_1,p_2,...)}{\sigma_i} \right)^2 \tag{1}$$

where σ_i is the variance of the measurement y_i. Since the measurements follow a Poisson distribution, the square variance coincides with the mean value, $\sigma_i^2 = F_i(p_1^0,p_2^0,...)$.

We will simplify the analysis and suppose that there is an unique unknown parameter, p. The corresponding results can be used as a first approximation to understand the dependence of the ROI selection on the searched parameter. The parameter p that minimizes the Chi-Square function must satisfy,

$$\sum_{i=1}^{N} \frac{1}{\sigma_i^2} \left(y_i - F_i(p)\right) F_i'(p) = 0 \tag{2}$$

where F_i' denotes $\frac{\partial F_i}{\partial p}$. Assuming small differences between the data and the correspoding ITS, the expected change in p, $\Delta p = p - p^0$, should also be small. Therefore both F_i and F_i' could be approximated at first order around p^0,

$$\sum_{i=1}^{N} \frac{1}{\sigma_i^2} \left(y_i - F_i(p^0) - F_i'(p^0)\Delta p\right) \left(F_i'(p^0) + F_i''(p^0)\Delta p\right) = 0 \tag{3}$$

Neglecting terms with square orders on Δp, this quantity can be isolated:

$$\Delta p = \frac{\sum \frac{1}{\sigma_i^2} \Delta y_i F_i'(p^0)}{\sum \frac{1}{\sigma_i^2}\left(F_i'(p^0)^2 - \Delta y_i F_i''(p^0)\right)} \simeq \frac{\sum \frac{1}{\sigma_i^2} \Delta y_i F_i'(p^0)}{\sum \frac{1}{\sigma_i^2} F_i'(p^0)^2} \tag{4}$$

where for the last approximation it was assumed that in almost all the cases $\sum F_i'(p^0)^2 \gg \sum \Delta y_i F_i''(p^0)$. In the following we work under this hypothesis.

The last equation gives the change on the estimated parameter due to the fact that the measured values y_i do not exactly match the transmission values.

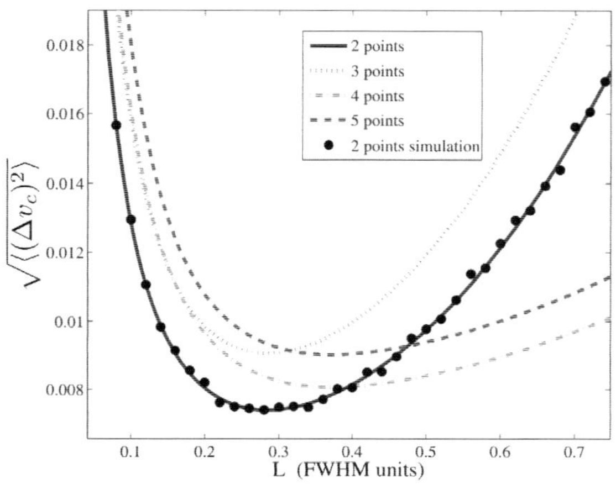

Figure 1. Variance of the center estimation $\sqrt{\langle (v_c - v_c^0)^2 \rangle}$ for different distributed ROIs. The curves correspond to 2, 3, 4 and 5 points ROIs, all with $T_{ROI} = 100$ s, a counting rate of 10^4 counts/s and velocities witin $-L$ and L. All the curves were calculted with equation 5. The points correspond to 3000 events simulations for the case of the two points ROI.

2.2. Goodness of the selected ROI

The goodness of the selected ROI is given by the confidence on the estimation of the corresponding parameter. The mean square $\langle (\Delta p)^2 \rangle$ is a measure of this confidence. Note that the mean value is made over a the set of results obtained from infinity identical experiments. In that sense $\langle \Delta y_i \rangle = 0$ and $\langle (\Delta y_i)^2 \rangle = F_i(p^0)$. Whith these identities the mean square value over eq. 4 results on,

$$\langle (\Delta p)^2 \rangle = \frac{1}{\sum_{i=1}^{N} \frac{F_i'(p^0)^2}{\sigma_i^2}} = \frac{1}{\sum_{i=1}^{N} \frac{F_i'(p^0)^2}{F_i(p^0)}} \tag{5}$$

Therefore the goodness of the ROIs can be measured with the last expression. To minimize $\langle (\Delta p)^2 \rangle$ and increase the ROI goodness the velocities must be selected near the maximum of derivate respect to the unkown parameter.

The last expression is in good agreement with results obtained by simulations. Figure 1 shows a comparison of the root square mean value obtained by both methodologies.

3. Applications

It is well known that the theoretical shape of a Mössbauer absorption line for thin absorbers is Lorentzian. Moreover the Lorentzian shape is accepted to be a good approximation for thick absorbers. Then, as a practical example, we will apply the presented results to this case. The normalized absorption line is given by,

$$f(v_i; \Gamma, ef, v_c) = 1 - \frac{ef}{4\frac{(v - v_c)^2}{\Gamma^2} + 1} \tag{6}$$

where Γ is the full width at medium height (FWHM), ef is the maximum of absorption, while v_c is the center of the Lorentzian. The transmission function depending on the acquisition time is related with the last trough the next equality,

$$F(v_i, t_i; \Gamma, ef, v_c) = r t_i f(v_i; \Gamma, ef, v_c) \tag{7}$$

where r is the counting rate at background velocities (where the Mössbauer absorption is null).

In the Figure 1 the $\langle (\Delta v_c)^2 \rangle$ was calculated for different uniform ROIs centered on the true center. All the presented results correspond to a time per ROI of 100 s, a counting rate of 10000 counts/s and an absorption effect of 0.1. The figure shows that the best uniform ROI to obtain the center corresponds to two points positioned on $\pm 0.28\Gamma$. These positions correspond to the points of maximum derivate of the ITS respect to the paramter v_c.

But the maximum derivative velocity position depends on the involved parameter. When looking for v_c the positions of maximum derivate are $v_c \pm \frac{1}{2\sqrt{3}}\Gamma \sim v_c \pm 0.28\Gamma$. If the unkown parameter is Γ, the maximum derivative velocities are $v_c \pm \frac{\Gamma}{2}$, while if ef is unkown then the velocities at the center of the absorption line are more efficient.

The two points ROI has been implemented in a real MLT experiement [2] with succesfull results. In the cited case, the counting rate and the effect were so low that a fast ROI was required to assure a good tracking. ITS parameters different from line center (Γ, ef and counting rate r) were carefully measured with a previous conventional Mössbauer spectrum and assumed constant during the tracking.

MLT is a technique that can benefit from this kind of analysis. The tracking is more likely to succeed if the ROI is selected within these criteria. But the results are not restricted to better finding the center of an absorption line. Any variable of the spectrum can be considered as target of the analysis and as a result an optimum ROI can be selected to better determine it.

4. Conclusions

The possibility of individually selecting energy and live time for each channel of the spectrum highly improves the performance of a Mössbauer spectrometer. The resulting spectrum must not necessarily be centered in null velocity and its channels are not required to be equally spaced. These facts enable the experimentalist to focus the observation in a region of interest of the spectrum where the information is more concentrated.

But frequently the ROI selection is not straightforward and depends on the complexity of the experiment. We have presented a mathematical tool that aids in this selection, providing a comparison factor between different channel sets. This tool is based on the assumption that every channel of the spectrum is not equally efficient for a given experimental application. It was demonstrated that channels closer to the maximum derivative of the studied parameter contribute better to the quality of parameter determination. This principle was quantified for several practical situations (i.e. line center, effect and width determination) taking a Lorentzian as line shape approximation.

Results show that optimal ROI selection depends on the searched parameter (e.g. the ROI for best line effect resolution is not the same that the ROI for best line center resolution). When more than one parameter is being studied, as can be the case of a complicated spectrum, there will be a compromise between a great number of variables to be determined, being responsibility of the experimentalist to find an appropriate balance between them. The present approach can be analytically extended to these cases.

References

[1] Veiga A, Martínez N, Mendoza Zélis P, Pasquevich G and Sánchez F H 2006 *Hyperfine Interact* **167** 905–909
[2] Veiga A, Pasquevich G A, Mendoza Zélis P, Sánchez F H, Fernández van Raap M B and Martínez N 2009 *Hyperfine Interact* **188** 137–142
[3] Press W H, Flannery B P, Teukolsky S A and Vetterling W T 1992 *Numerical Recipes in C: The Art of Scientific Computing* (Cambridge University Press)

Journal of Physics: Conference Series **217** (2010) 012140 doi:10.1088/1742-6596/217/1/012140

X-ray and ^{237}Np Mössbauer effect investigation on the NpB_2C compound

P Gaczyński[1], P Rogl[2], E Colineau[1], R Eloirdi[1], F Wastin[1], P Boulet[1,3], J Rebizant[1]

[1] European Commission, Joint Research Centre, Institute for Transuranium Elements; Postfach 2340, 76125 Karlsruhe, Germany

[2] Institute of Physical Chemistry, University Vienna, Währingerstrasse, A-1090 Wien, Austria

[3] Present address : Institut Jean Lamour, FR2797, Ecole des Mines, Parc de Saurupt, CS14234, F-54042 Nancy

Piotr.Gaczynski@ec.europa.eu

Abstract. Powder x-ray diffraction and ^{237}Np Mössbauer effect measurements were performed on NpB_2C crystallizing in the ThB_2C – type structure, $R\bar{3}m$. On the base on ^{237}Np Mössbauer spectroscopy formal +3 oxidation states were assigned to both Np sites, 3a and 6c. The magnetic ordering of the Np sublattice below 80 K is confirmed and the value of the ordered moment amounts to 0.30 μ_B and 1.05 μ_B on the 3a and 6c Np sites, respectively. The reduced magnitudes of the ordered moment and of the isomer shift suggest rather delocalized 5f electrons.

1. Introduction

The combination of rare earth or actinide metal with boron and carbon atoms leads to the formation of layered compounds having interesting structural and physical properties: superconductivity was observed in YB_2C_2 and LuB_2C_2 [1], antiferroquadrupolar (AFQ) and antiferromagnetic (AFM) ordering was reported in DyB_2C_2 and HoB_2C_2, respectively [2]. Mostly metallic behaviour for rare earth (actinide) borocarbide compounds was suggested [3;4].

The layered borocarbide compounds with general formula RB_2C form metallic and boron-carbon sheets and can crystallize in three different crystal structures: YB_2C, ThB_2C and α-UB_2C type. Light actinide and cerium borocarbides mostly crystallize in the ThB_2C type structure, whereas other rare-earths borocarbides crystallize rather in the two other types of structure. Extended Hückel tight-binding calculations based on an ionic interaction between the metallic and the boron-carbon sublattices assigned a formal charge of -4 per B_2C unit to the ThB_2C and α-UB_2C type structures and a formal charge -3 to the YB_2C type [3]. In this view, the formation of YB_2C type structure is preferred by most of rare earth atoms but the ThB_2C type for light actinide and Ce atoms. ß-UB_2C with ThB_2C type was shown to order ferromagnetically below 74.5K with itinerant nature of the 5f-electrons [5]

NpB_2C crystallizes in the rhombohedral space group $R\bar{3}m$ (No. 166), with the ThB_2C-type structure (a = 6.532(2) Å, c = 10.769(3) Å [6,7]). The crystal structure has been derived from single crystal data

(R_F=0.023) and confirms full atom order: Np atoms occupy 3a and 6c sites while B and C atoms are located on the 18g and 9d sites respectively. From polycrystalline bulk NpB_2C samples, antiferromagnetic ordering below a Néel temperature of 73K is established by magnetization and specific heat measurements [7]. A ferromagnetic signal observed below about 50K in the magnetization curves at low fields is attributed to the presence of ferromagnetic impurities. In the paramagnetic region, the susceptibility follows a modified Curie–Weiss law with an effective moment $\mu_{eff} \approx 1.65\ \mu_B$ much smaller than the free Np^{3+} ion value (2.75 μ_B), and a paramagnetic Curie–Weiss temperature $\theta_p \approx 23K$.
We report here x-ray diffraction results and ^{237}Np Mössbauer spectroscopy studies.

2. Experimental details

Polycrystalline ingots were obtained by arc melting stoichiometric amounts of the constituent elements under an atmosphere of high purity argon on a water-cooled copper hearth, using a Zr getter alloy. Starting materials were used in the form of pieces as supplied by Merck AG. (boron and graphite), and by LANL electrorefined neptunium 3N metal. In order to ensure homogeneity, the arc melted buttons were turned over and remelted 3 times, with weight losses below 0.5%. The samples were checked by X-ray powder diffraction data (Cu K_1 radiation) collected on a Bragg-Brentano Siemens D500 diffractometer using a 2 step size of 0.02 degrees. The diffraction patterns were analyzed by a Rietveld-type profile refinement method using the Fullprof program [8].
The ^{237}Np Mössbauer spectra were taken between 4K and 120K in a transmission geometry spectrometer with ~ 100mCi ^{241}Am source kept at temperature around 4K. The absorber was prepared by grinding the sample to a fine powder to ensure a constant surface density with an optimal thickness of 140mg Np cm^{-2}. The velocity scale was calibrated with the $NpAl_2$ compound. The isomer shift, IS, is given relative to $NpAl_2$ at 4K. The spectra were analyzed using Lorenzian lines. The position and relative intensities of the absorption lines of Np nuclei were calculated by solving the complete Hamiltonian for the hyperfine interactions in both the excited and ground nuclear state of ^{237}Np.

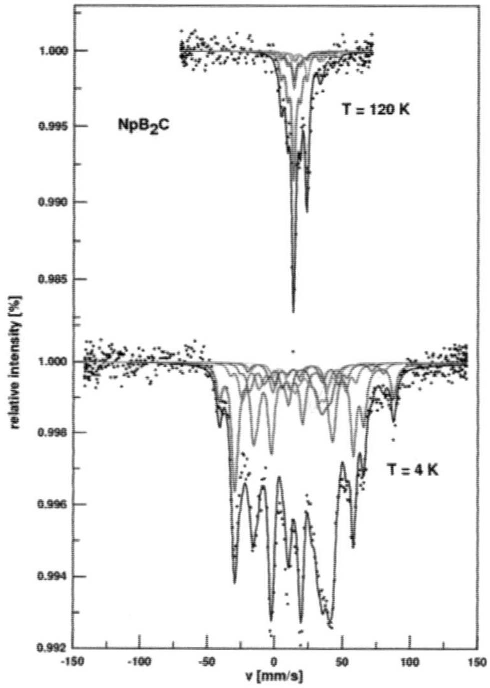

Figure 1. ^{237}Np Mössbauer spectra of NpB_2C at 120K and 4K.

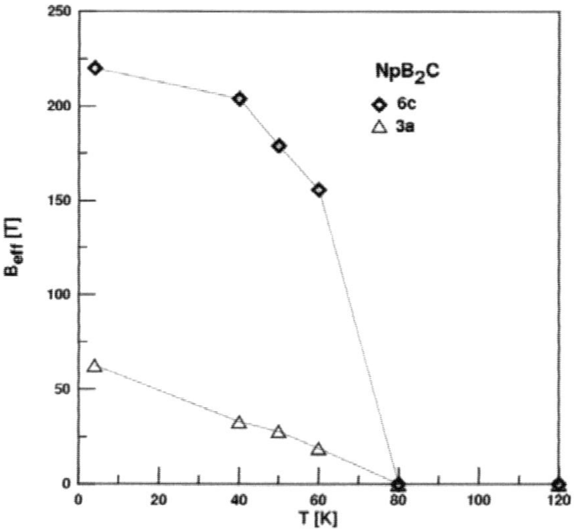

Figure 2. Temperature dependence of magnetic hyperfine field (B_{eff}) at 3a and 6c sites in NpB_2C

3. Results and discussion

The X-ray diffraction pattern recorded on a polycrystalline sample confirms crystal symmetry, atom occupation and thus isotypism with the crystal structure of ThB_2C. However, the small disagreement between calculated and experimental peak intensity and broadening suggest texture effects. It could also indicate a small shift of Np atoms from their position induced by vacancies at B or/and C sites. Mössbauer spectra of NpB_2C at 4K and 120K are presented on Fig.1.

The two non-equivalent Np sites of NpB_2C, 3a and 6c, have each 8 Np atoms near neighbors (NN): Np atoms at the 3a site have 8 Np NN at the 6c site within a distance of 3.77Å. Np atoms at the 6c site have 4 Np NN at the 3a site but also 4 Np NN at the 6c site (3 of them within 3.78Å and one within 3.24Å). Thus the model used to account for the Mössbauer spectra contains two components, with relative intensities 1:2. However, this model appears insufficient to fully reproduce the Mössbauer pattern, possibly due to vacancies in the B and/or C sites. Unfortunately, it is not possible from X-rays to establish the exact site occupation of B and C atoms, due to their very low scattering factor compared to Np atoms. Therefore, it is also not possible to build an exact physical model to fit the Mössbauer spectra. The solution could be to fit rather sharp spectra with a distribution of ^{237}Np hyperfine parameters. As we have no program for fitting a distribution of Np hyperfine parameters we decided to split each Mössbauer spectral component into 3 additional ones for each (3a and 6c) crystallographic site. Only the quadrupole splitting (QS) and the magnetic hyperfine field B_{hf} were allowed free variation for each of the 6 components. The isomer shift (IS) was kept equal for the three components of a same crystallographic site. During the fitting procedure the linewidth of all lines was constrained to a unique value. The parameters obtained are presented in Table 1. An average value of the quadrupolar interaction parameter and of the hyperfine magnetic field is also given.

Table 1. Estimated Mössbauer hyperfine parameters for NpB_2C

site	T [K]	IS [mm/s]	QS [mm/s]	<QS> [mm/s]	B [T]	<B> [T]	A [%]	<A> [%]
3a	4	9.8	29.5	33.2	96	65	11	33
			32.0		29		11	
			38.0		69		11	
	120	9.8	29.5	33.2	0	0	11	33
			32.0		0		11	
			38.0		0		11	
6c	4	0.1	-16.1	-16.0	240	225	44	67
			-16.2		210		13	
			-15.0		180		10	
	120	0.1	-16.1	-16.0	0	0	44	67
			-16.2		0		13	
			-15		0		10	

Following the value of IS we assigned a formal +3 oxidation state for both Np sites [9]. This formally disagrees with idea of metal atom on formal +4 oxidation state in the ThB_2C – type structure suggested in ref. 3. However, the relatively low value of IS which is only slightly higher than for Np^{+4} could indicate Np atoms at an intermediate valence state. The lower magnitude of IS than for Np^{+3} ion could be explained by hybridization between Np 5f electrons and conduction bands. This leads to a decrease of f-electron density and to an increase of conduction electron density. Both effects are responsible for decreasing IS at the Np site.

The temperature dependences of B_{eff} at 3a and 6c sites are presented on Fig. 2. Magnetic ordering below 80K is proven. Using the Dunlap-Lander [10] relation we can assign at 4K 0.30 μ_B and 1.05 μ_B to the Np 3a and 6c sites, respectively. These values are much lower than the theoretical values for free Np^{+3} or Np^{+4} ions [9]. At first glance the higher magnetic moment at 6c could be explained by a lower mean value of interatomic distances, $d_{Np\text{-}Np}$, for 6c than for the 3a site. However, the huge difference in B_{eff} at both crystallographically quite similar sites should have its origin in the electronic structure. In Np intermetallics orbital magnetism is dominant, thus a decreasing f-electron density leads to decreasing magnetic moments on Np atoms. Therefore a lower magnitude of IS shift should be connected with lower B_{eff}. In NpB_2C site 6c has lower IS and higher B_{eff}. It could be understood if we assume hybridization between f, d and s Np electron density with p electron density of B and C atoms leading to the formation of electronic bands responsible for a rather itinerant than localized magnetism in NpB_2C similar to the isotypic itinerant ferromagnet ß-UB_2C [5].

4. Summary
X-ray diffraction confirms site occupation and formation of a ThB_2C type structure in NpB_2C. However, the polycrystalline sample contains inhomogeneities, mostly in B and C sites.
^{237}Np Mössbauer spectra indicate a formal 3+ charge state for Np, but intermediate valence cannot be excluded. Magnetic ordering of both sublattices below 80K is established, with an ordered moment of 0.30 μ_B and 1.05 μ_B on the Np 3a and 6c sites, respectively. The reduced values (compared to the free ion) of the magnetic moment and the isomer shift suggest (i) hybridization of the 5f Np shell with the conduction band and (ii) rather delocalized 5f magnetism in NpB_2C. These results are in agreement with previous work [7].

Acknowledgements
The technical support of D. Bouëxière is acknowledged. The support of the European Community-Access to Research Infrastructures action of the Improving Human Potential Programme, European Commission contract RITA-CT- 2006-026176 "Actuslab-2", to the Actinide User Laboratory programme at ITU-Karlsruhe is acknowledged. PG acknowledges the European Commission for support in the frame of the "Training and Mobility of Researchers" programme. The high purity Np metals required for the fabrication of the compound were made available through a loan agreement between Lawrence Livermore National Laboratory and ITU, in the frame of a collaboration involving LLNL, Los Alamos National Laboratory and the US Department of Energy.

References
[1] Sakai T, Adachi G, Shiokawa J 1982 *J. Less-Common Met.* **84** 107
[2]Yamamuchi H, Onodera H, Kaneko K, Tobo A, Ohoyama K, Yamaguchi Y 2001 *J. Magn. Magn. Mat.* **266-230** 1134
[3]Wiitkar F, Kahlal S, Halet J F, Saillard J Y, Bauer J, Rogl P 1994 *J. Am. Chem. Soc.* **116** 251
[4]Wiitkar F, Kahlal S, Halet J F, Saillard J Y, Bauer J, Rogl P 1995 *Inorg. Chem.* **34** 1248
[5]Tran V H, Rogl P, Andre G, Bouree F 2006 *J. Phys.-Cond. Mat*, **18** 703
[6]Rogl P, Beauvy M, Larroque J P 1989 *19th Journées des Actinides* Madonna di Campiglio Trento, Italy.
[7]Boulet P, Colineau E, Rebizant J, Rogl P, Wastin F 2004 *34th Journées des Actinides* Heidelberg, Germany.
[8]Rodriguez-Carvajal J 1993 *Physica B* **192** 55
[9]Kalvius G M, Gal J, Asch L, Potzel W 1992 *Hyp. Int.* **72** 77
[10]Dunlap B D, Lander G H 1974 *Phys. Rev. Lett.* **63** 2413

Structural and Mössbauer studies of $Mn_{0.5}Co_{0.5}Fe_2O_4$ ferrites prepared by high energy ball milling and glycolthermal methods

H M I Abdallah[1], T Moyo[1] and J Z Msomi[2]

[1]School of Physics, University of KwaZulu-Natal, Durban 400, South Africa
[2]Department of Physics, University of Free State, Phuthaditjhaba 9866, South Africa

E-mail: moyo@ukzn.ac.za

Abstract. We have investigated the properties of three $Mn_{0.5}Co_{0.5}Fe_2O_4$ ferrites synthesized directly from high purity metal oxides (Sample A) and from $MnFe_2O_4$ and $CoFe_2O_4$ ferrites (Sample B) by mechanical milling and by glycolthermal process (Sample C). XRD results show essentially single phase spinel structure. Sample C is the best sample followed by Sample B based on the presence of minor impurity peaks near $2\theta \approx 38^o$ and 52°. More impurity peaks appear to grow after annealing above 400 °C. Samples A and B have been found to have similar Fe-57 Mössbauer spectra before and after annealing at 1050 °C. The spectrum for each sample changes significantly after annealing. This is attributed to changes of both grain sizes and impurity phases with thermal annealing. At least two sextets and two doublets are used to fit the Mössbauer spectra.

1. Introduction

Cobalt ferrites ($CoFe_2O_4$) are important materials for high density magnetic and magneto-optic recording media [1, 2]. Mn doped cobalt ferrites exhibit high magnetostriction, good mechanical properties and have low eddy current losses making them suitable in high frequency applications [3]. It has been observed that a pure and uniform form of a $Co_{0.9}Mn_{0.1}Fe_2O_4$ compound appears difficult to obtain by high temperature synthesis [3]. We have therefore prepared related compounds with higher Mn content by different synthesis routes and at lower synthesis temperature in order to investigate their single phase formation, thermal stability and magnetic properties. We have made $Mn_{0.5}Co_{0.5}Fe_2O_4$ by high energy ball milling metal oxides, single phase $MnFe_2O_4$ and $CoFe_2O_4$ ferrites and by glycolthermal reaction from metal chlorides.

2. Experimental details

$Mn_{0.5}Co_{0.5}Fe_2O_4$ compounds were produced from high purity metal oxides (Sample A) and from a mixture of single phase $MnFe_2O_4$ and $CoFe_2O_4$ ferrites (Sample B) by mechanical milling for at least 60 hours using a Retch planetary ball mill (type: PM 400) operated at 300 rev/min. The ball to sample mass ratio used was 20:1 under air atmosphere. The preparation procedure from single phase ferrites is similar to the work on MnZn ferrite [4, 5]. To test the consistency of the two synthesis routes, bulk densities of compacted pellets of samples annealed at 1050 °C were measured. The densities were found to be 3.89 ± 0.01 g/cm^3 for sample A and 3.87 ± 0.01 g/cm^3 for sample B. The x- ray density can be calculated from XRD measurements using the equation $\rho_X = 8M/N_A a^3$ where is M the molar mass, $N_A = 6.022\times10^{23}$ mol^{-1} and a is the lattice parameter. For samples A and B the x- ray densities we found to be 5.18±0.02 g/cm^3 and 5.23 ± 0.02 g/cm^3 respectively. In the glycolthermal method (Sample C) metal chlorides and ammonium nitrate were used as starting materials. Stochiometric amounts of chlorides were mixed in deionised water to obtain $Mn_{0.5}Co_{0.5}Fe_2O_4$. The solutions were thoroughly mixed using a magnetic stirrer for about 20 minutes. Ammonium nitrate solution was then slowly added to the chloride mixture under rapid stirring until a pH ≈ 9. The precipitate was then washed several times by deionized water until all the chloride ions were removed as confirmed by using a standard solution of $AgNO_3$. During the washing process, the precipitate was filtered using Whatman glass microfibre filters (GF/F). The clean wet precursor was thereafter dispersed in 300 ml of ethylene glycol under rapid stirring. The precursor was then placed in a 500 ml glass lining in a stainless steel pressure vessel (watlow series model PARR 4843). The pressure vessel was heated to 200 °C and the pressure gradually rose to 140 psi. These conditions were held for 6 hours. The cooled products were filtered and washed by deionized water and finally by ethanol. The recovered synthesized powders were dried under an infra red lamp and homogenized using agate mortar and pestle. The products were characterized by x-ray diffraction (XRD). The XRD patterns of the samples were obtained using CoK_α radiation (λ = 1.7903 Å) of the Phillips diffractometer (type: PW1710). The Mössbauer spectra were recorded using a conventional constant acceleration Mössbauer spectrometer with a ^{57}Co source sealed in a Rh matrix.

3. Results and discussions

Figure 1 shows the XRD spectra for the three samples. All the major peaks are characterized by the basic single phase spinel structure. Some small impurity peaks are observed at $2\theta \approx 38°$ and 52° which are more significant for Sample A, and least for Sample C. The impurity peaks have been attributed to an intermediate hematite α-Fe_2O_3 phase [6].

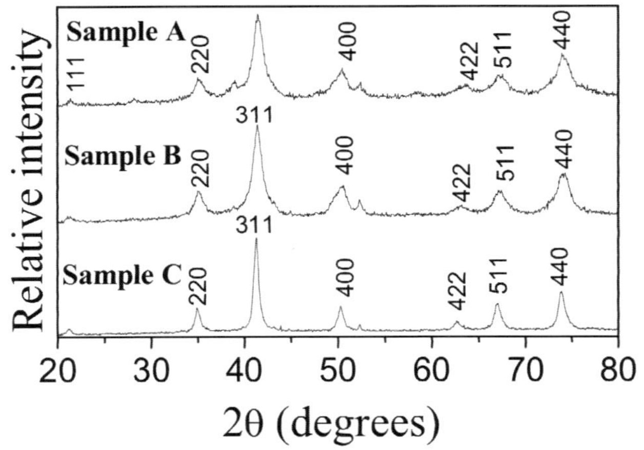

Figure 1. XRD spectra for Samples A, B (annealed at 200 °C) and C (synthesized at 200 °C.

The variation of XRD spectra with annealing temperature for samples milled for 60 hours are shown in figure 2. All samples were annealed in air at each temperature for about 6 hours. No significant changes to the XRD spectra are observed until after an annealing temperature of about 400 °C. Above this temperature significant changes to the spectra are observed as more impurity peaks grow. The narrowing of the peaks is an indication of increasing particle size. The particle sizes were estimated using the Scherrer equation: $D = 0.9\lambda/(\beta\cos\theta)$ where β is the full width at half maximum of the 311 XRD peak [3, 7, 8]. After milling for 60 hours the grain sizes for samples A and B were found to be 7.23±0.03 nm and 7.58±0.03 nm respectively. Similar values were obtained for $Mn_{0.1}Zn_{0.9}Fe_2O_4$ prepared by hydrothermal process [9]. The grain size for Sample C was found to be

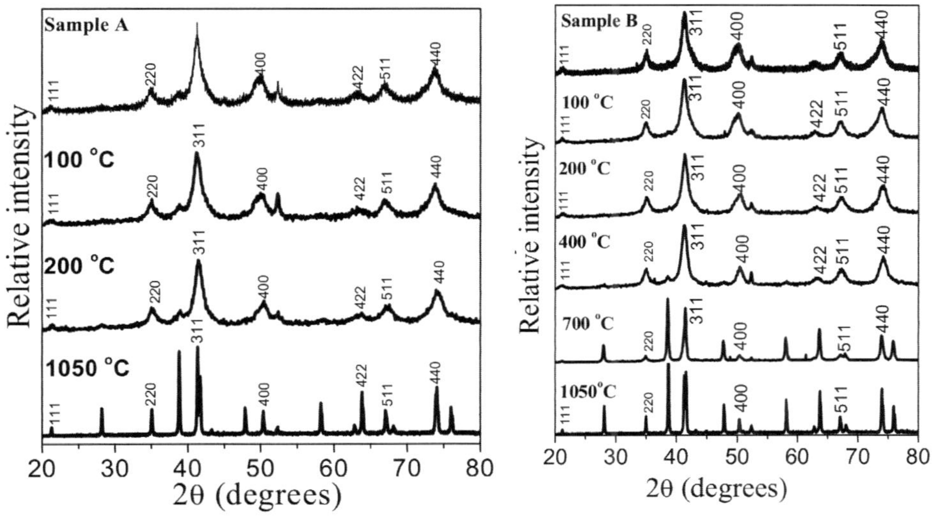

Figure 2. Variation of XRD patterns for samples A and B with annealing temperature.

15.12±0.09 nm. Hence nanosized samples have successfully been produced through the three synthesis routes employed.

In figure 3 we show the Mössbauer spectra for samples A, B and C. The spectra for milled samples A and B are similar and were successfully fitted by three sextets and two doublets. The sextets represent the fractions of ordered Fe ions in tetrahedral (A) and octahedral (B) sites. In Table 1 we show results of the Mössbauer spectra analysis. The third sextet may be attributed to small Fe clusters induced by milling [10]. The doublets represent Fe ions in paramagnetic states. A broad central

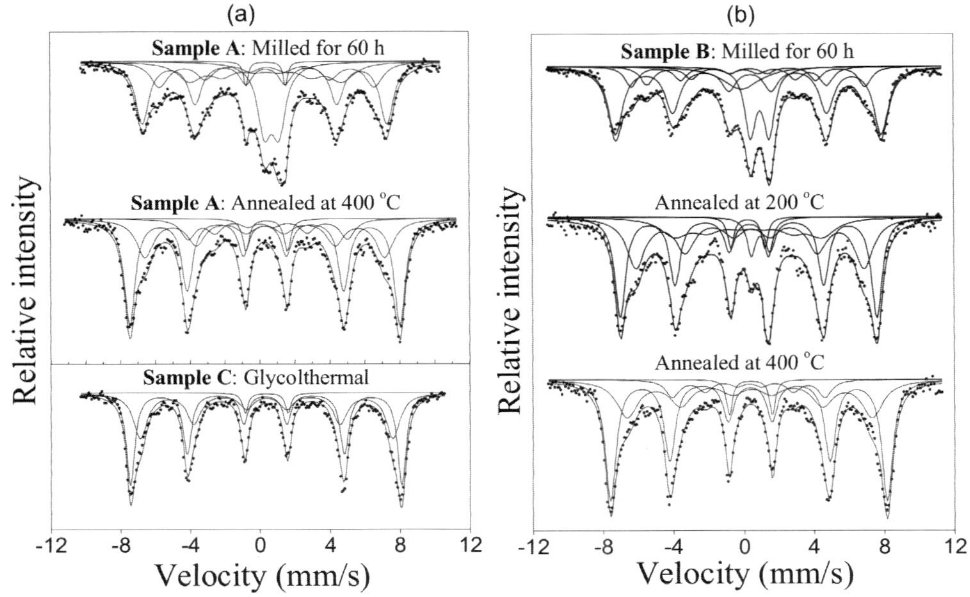

Figure 3. Mössbauer spectra for samples A, B and C.

Table 1. Isomer shifts (IS), hyperfine fields (*H*), line widths (LW) and site fractions (*f*) at A and B sites for raw samples (A, B and C) and annealed samples (A- 400 °C, B- 200 °C and B- 400 °C).

Sample	IS_A (mm/s) ±0.02	IS_B (mm/s) ±0.04	H_A (kOe) ±3	H_B (kOe) ±4	H_{3rd} (kOe) ±6	LW_A (mm/s) ±0.02	LWB (mm/s) ±0.02	f_A (%) ±2	f_B (%) ±3
A	0.36	0.47	434	382	271	0.46	0.60	38	20
A-400°C	0.30	0.32	480	426	283	0.38	0.60	52	30
B	0.33	0.37	412	469	317	0.44	0.50	13	47
B-200°C	0.33	0.43	453	403	261	0.36	0.65	39	34
B-400°C	0.31	0.40	488	433	269	0.38	0.73	54	32
C	0.32	0.37	481	449	-	0.29	0.48	53	43

doublet is observed in unannealed samples, which disappears drastically with increased sintering temperature.

The reduction of doublet with increasing annealing temperature appears to be related to single phase formation. The spectra for samples annealed at 400 °C are comparable with those of similar compounds prepared by sol-gel method [1]. There is no evidence of a broad doublet in the spectra for Sample C. This also compares well with insignificant hematite phase as indicated by XRD results for sample C. The criterion used to assign sextets and doublets at A and B sites was based on the fitted results of isomer shifts, which are expected to be lower due to cubic symmetry on sites A [10]. The Mössbauer parameters deduced from the Mössbauer spectra are also shown in Table 1. The values of isomer shifts, hyperfine fields and line widths are similar to those previously reported for similar compounds [11, 12].

4. Conclusions

We have successfully made $Mn_{0.5}Co_{0.5}Fe_2O_4$ nanoparticles by high energy ball milling and by glycolthermal process. A predominant single phase spinel structure is obtained below 400 °C. The Mn substituted cobalt ferrites appear to be unstable at temperatures higher than about 400 °C when significant increases in extra impurity peaks are observed. High temperature synthesis seems to be unsuitable for this compound. Significant changes in the Mössbauer spectra with increase in sintering temperature should be attributed to both increased grain sizes and extra impurity phases at high temperatures. The sample prepared by glycolthermal method appears to be of higher quality than milled samples. Better samples are also obtained from $MnFe_2O_4$ and $CoFe_2O_4$ than from metal oxides.

References

[1] Kim K J, Kim H K, Park Y R and Park J Y 2006 *J. Magn. Magn. Mater.* **304** e106
[2] Zi Z, Sun Y, Zhu X, Yang Z, Dai J and Song W 2009 *J. Magn. Magn. Mater.* **321** 1251
[3] Qi X and Wu D 2008 *J. Magn. Magn. Mater.* **320** 666
[4] Nasr.Isfahani M J, Myndyk M, Menzel D, Feldhoff A, Amighian J and Šepelák V 2009 *J. Magn. Magn. Mat.* **321** 152
[5] Nasr.Isfahani M J, Myndyk M and Šepelák V 2009 *J. Alloy Compounds* **470** 434
[6] Yu L, Zhang J, Liu Y, Jing C and Cao S 2005 *J. Magn. Magn. Mater.* **288** 54
[7] Msomi J Z and Moyo T 2009 *Hyperfine Interact.* **189** 151
[8] Liu X, Zhong W, Yang S, Yu Z, Gu B and Du Y 2002 *J. Magn. Magn. Mater.* **238** 207
[9] Nalbandian L, Delimitis A, Zaspalis V T, Deliyanni E A, Bakoyannakis D N and Peleka E N 2008 *Microporous and Mesoporous Materials* **114** 465
[10] Msomi J Z and Moyo T 2009 *J. Magn. Magn. Mater.* **321** 1246
[11] Lakshman A, Subba Rao P S V, Rao K H, 2004 *J. Magn. Magn. Mater.* **284** 352
[12] Rao B P, Subba Rao P S V, Murthy G V S and Rao K H, 2004 *J. Magn. Magn. Mater.* **268** 315

Journal of Physics: Conference Series **217** (2010) 012142 doi:10.1088/1742-6596/217/1/012142

A high-field Mössbauer investigation on $FeSb_2$

A Farhan[1], M Reissner[1,3], A Leithe-Jasper[2] and W Steiner[1]

[1]Institute of Solid State Physics, Vienna University of Technology, Vienna A-1040, AUSTRIA
[2]Max-Planck-Institut für Chemische Physik fester Stoffe, Dresden, GERMANY

reissner@ifp.tuwien.ac.at

Abstract. In high-field Mössbauer measurements on $FeSb_2$ complex spectra were recorded, which could only be fitted with a higher number of subspectra, though crystallographically only one iron site exists. Due to their field dependence, these subspectra can be devided in two groups. The first, which covers 50-60% of the spectral area exhibits no field dependence of the internal fields on the applied field. The second shows internal fields increasing with the applied field.

1. Introduction

Although investigated since several decades [1] the electronic and magnetic structure of $FeSb_2$ is still under discussion. Increasing interest in this compound arose in recent years, because a colossal Seebeck coefficient [2], a rather high charge carrier mobility [3], and in Co-doped $FeSb_2$ colossal magnetoresistance [4] was found, which makes $FeSb_2$ an interesting candidate for applications, like thermoelectrica and high speed electronics. Most probably $FeSb_2$ is a nearly ferromagnetic small gap semiconductor [5] with a semiconductor-metal transition above 80 K [6, 7]. It crystallizes in the marcasite structure, where Fe has only one crystallographic site, surrounded by distorted Sb-octahedra [8]. The octahedra are corner shared in the ab-plane and edge sharing along the c-axis. The distortion of the octahedra increases with decreasing temperature, reaching a maximum around 100 K [9]. In a simple ionic picture [10] iron has a low spin Fe^{4+} ($t_{2g}^4e_g^0$) ground state, because in this model the t_{2g} levels split in a higher lying d_{xy} and two lower lying d_{yz} and d_{xz} levels. In contrast to this a high spin ferric ground state was suggested from Mössbauer investigations [11]. In this work we report on ^{57}Fe Mössbauer measurements in high-fields on polycrystalline $FeSb_2$.

2. Experimental

Polycrystalline samples enriched with ^{57}Fe were prepared and checked for phase purity by X-ray investigations. For magnetic measurements in fields up to 9 T and temperatures between 2 and 400 K a Quantum Design PPMS-VSM was used. ^{57}Fe Mössbauer measurements were performed between 4.2 K and room temperature and in fields up to 13.5 T. The zero-field ^{57}Fe Mössbauer spectra were recorded in standard transmission geometry using a ^{57}Co**Rh** source (always at room temperature), relative to which all center shift data, CS, are given. The in-field measurements were performed in a cryostat with a superconducting solenoid (max. field 15 T) with the field direction parallel to the γ-ray direction. The source (also ^{57}Co**Rh**) is situated in a field compensated area ($B < 0.1$ T) at temperatures

[3] To whom any correspondence should be addressed.

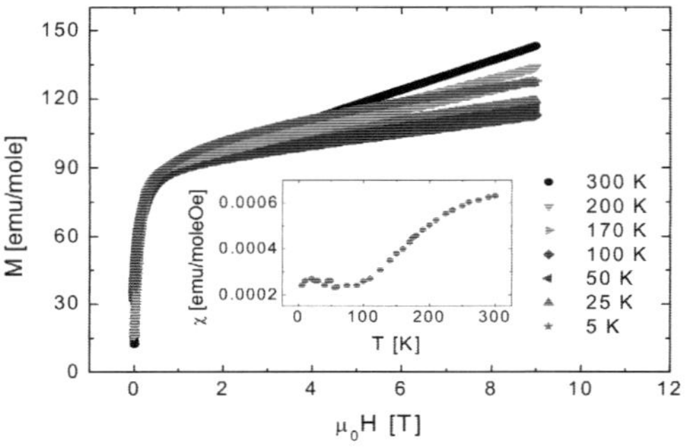

Figure 1. Field dependence of magnetization of $FeSb_2$ at selected temperatures. **Inset:** High field susceptibility derived after correction for a magnetically ordered contribution versus temperature.

between 5 and 10 K. The spectra were analyzed by solving the full Hamiltonian in the fast relaxation limit, taking into account both electrical and magnetic interactions, as well as the sample thickness. The temperature stability during measurement was ±0.5 K.

3. Results

Magnetization curves indicate a magnetically ordered component, which saturates even at 300 K, superposed by a susceptibility term (figure 1). Calculation of the temperature dependence of the susceptibility from the high field slope of the curves gives similar behaviour as reported in [6, 7]. At low temperatures the susceptibility is constant, followed by an increase above 100 K. Up to now no secondary phases could be detected by X-ray investigations, however, traces of impurities (presumably on grain boundaries) cannot be excluded.

Zero field Mössbauer measurements show spectra which can be fitted with only one subspectrum (figure 2). The quadrupole splitting, increases with decreasing temperature and saturates around 50 K in good agreement with [12]. This corresponds to the fact that the distortion of the Sb-octahedra is lar-

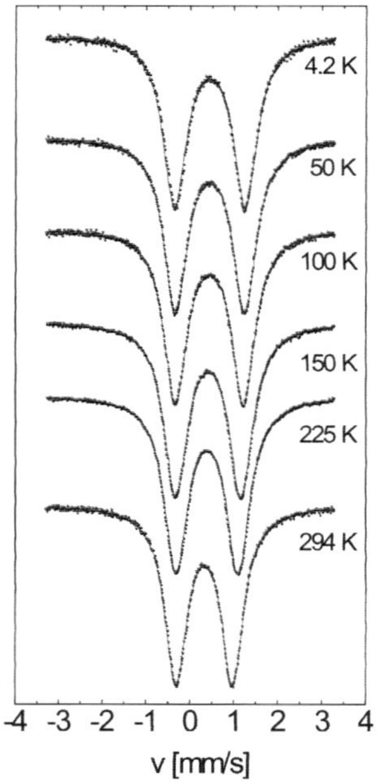

Figure 2. ^{57}Fe Mössbauer spectra of $FeSb_2$ obtained at different temperatures in zero field.

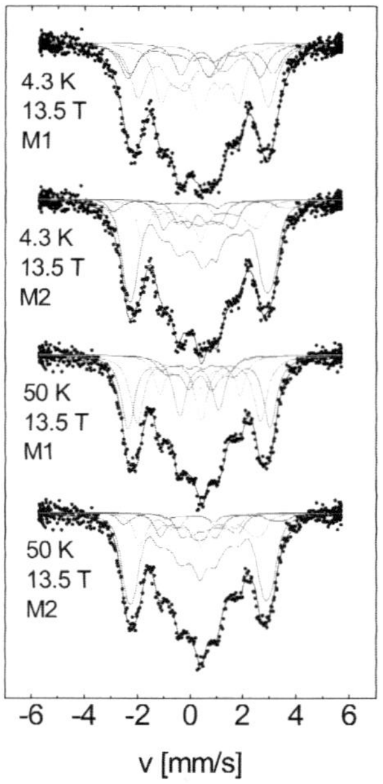

Figure 3. ^{57}Fe Mössbauer spectra of $FeSb_2$ obtained at 13.5 T for 4.2 and 50 K, analysed within model M 1 and M 2.

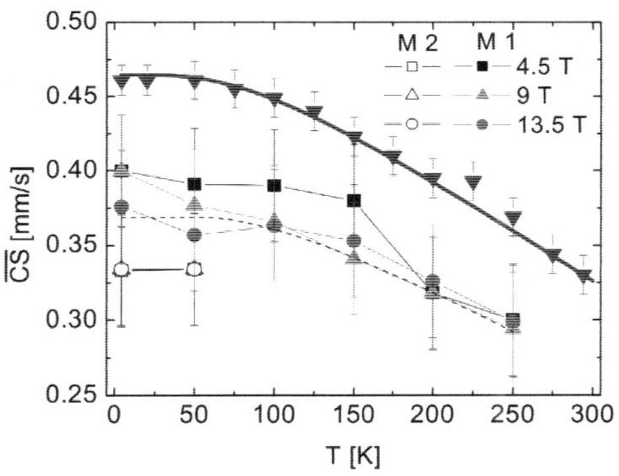

Figure 4. Temperature dependence of center shift in zero external fields (triangle down) and mean center shift obtained with models M 1 (full symbols) and M 2 (open symbols) in different external fields. Full line fit according to the Debye model. Black dotted line is a guide for the eye. No corrections for different source temperatures were performed.

gest around 100 K [9]. However, no anomaly in the center shift is present near the proposed semiconductor-metal transition close to 70 K [6]. Assuming that the temperature dependence of the center shift is only due to the second order Doppler shift, the analysis in terms of a Debye model gives a Debye temperature of 372 ± 20 K (figure 4). This fits well to the value of 380 K reported by [12].

In-field Mössbauer measurements show complex spectra (figure 3), which cannot be fitted by only one subspectrum. Two models (M 1 and M 2) were used to analyze the data. In M 1 five subspectra (neglecting randomization of orientation of field gradient direction with respect to γ-ray direction) were used to fit the measured spectra for all fields and temperatures, although the rather large half width (up to 0.45 mm/s) indicates, that this is only a minimum number. The mean value of the quadrupole splitting shows the same temperature dependence as for the zero-field measurements. Within measuring accuracy no field dependence is present. The same holds for the temperature and field dependence of the mean value of the center shift (figure 4). The shift between the values of the zero-field and in-field measurements is due to the difference in the source temperature. The subspectra measured at 9 and 13.5 T are fully polarized, indicated by the vanishing of the second and fifth line. At 4.5 T angles θ up to 11° between hyperfine field and external field B_a are found for some subspectra. As the measured hyperfine field B_{hf} is the vector sum of the internal field B_{int}, and the applied field B_a, B_{int} can be determined according to $B_{int}^2 = B_{hf}^2 + B_a^2 - 2B_{hf}B_a\cos(\theta)$. Figure 5 shows the internal fields in dependence of the applied field for typical temperatures. B_{int} for the two subspectra - indicated by large symbols - representing ~60% of the area of the measured spectra, are within experimental accuracy zero for all applied fields, whereas B_{int} for the subspectra representing the remaining ~40% of the spectral area increase with B_a. Although the fits with M 1 are satisfying, this result could only be reached by fitting the polycrystalline sample like a single crystal. Similar problems are reported also for other compounds [e.g. 13 and references herein]. In contrast to M 1, in M 2 the polycrystalline nature of the sample (by randomization of the field gradient direction with respect to γ-ray direction) and the asymmetry parameter η were taken into account. But with this model up to now only fits of the spectra at 4.2 and 50 K were satisfying (figure 3). Six subspectra were used with half widths between 0.18 and 0.30 mm/s. The results for higher temperatures are presently not finished. Within the errors the mean values for QS and CS obtained with M 2 are the same as those obtained from M 1 (figures 4). The same holds for the internal field. There is one subspectrum, which represents ~50% of the measured spectral area, with a field independent $B_{int} \sim 0$. For η large values between 0.98 and 0.8 inde-

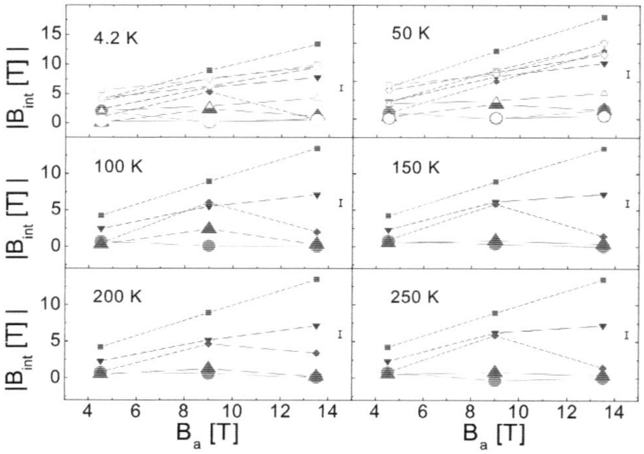

Figure 5. Internal fields over applied field for $FeSb_2$ for typical temperatures obtained using model M 1 (full symbols) and M 2 (open symbols). Large symbols indicate the majority spectra.

pendent of applied field were obtained for the majority and a further spectrum, whereas for the other subspectra values between 0 and 0.25 are found.

In summary, the zero-field results of the investigated $FeSb_2$ sample agree well with the ones reported in literature. The spectra can be fitted with only one subspectrum. Although crystallographically only one iron site is present, the in-field spectra are very complex. Common to the two fitting models discussed is that the mean values for quadrupole splitting and center shift fit well to the zero-field results. In both cases two types of internal fields are found. 50 to 60% of the sample has zero B_{int}, whereas the rest show an increase with B_a.

If valence contributions to B_{int} are not negligible so that B_{int} is not only determined by the core contribution, the spectra exhibiting the field dependence of the internal hyperfine field may be allocated to Fe in the impurity phase contributing to the increase of the magnetization at low fields. This allocation, however, would be in complete discrepancy to the present X-ray investigations, in which no foreign phases could be detected. The spectral area of this contribution amounts about 40% and secondary phases of such order must be clearly visible in X-ray diffraction measurements.

References

[1] Holseth H and Kjekshus A 1968 *Acta Chem. Scand.* **22** 3273
[2] Bentien A, Johnsen S, Madsen G K H, Iversen B B and Steglich F 2007 *EPL* **80** 17008
[3] Hu R, Mitrovic V F and Petrovic C 2008 *Applied Physics Letters* **92** 182108
[4] Hu R, Thomas K J, Lee Y, Vogt T, Choi E S, Mitrovic V F, Hermann R P, Grandjean F, Canfield P C, Kim J W, Goldman A I and Petrovic C 2008 *Phys. Rev.* B **77** 085212
[5] Lukoyanov A V, Mazurenko V V, Anisimov V I, Sigrist M and Rice T M 2006 *Eur. Phys. J.* B **53** 205
[6] Petrovic C, Kim J W, Bud'ko S L, Goldman A I, Canfield P C, Choe W, Miller G J 2003 *Phys. Rev.* B **67** 155205
[7] Sun P, Oelscher N, Johnsen S, Iversen B B, Steglich F 2010 *Dalton Trans.* **39** 1012
[8] Hullinger F 1967 *Struct. Bonding* (Berlin) **4** 83
[9] Petrovic C, Lee Y, Vogt T, Lazarov N Dj, Bud'ko S L, Canfield P C 2005 *Phys. Rev.* B **72** 045103
[10] Goodenough J B 1972 *J. Solid State Chem.* **5** 144
[11] Tumolillo T A 1973 *Phys. Stat. Sol.* (a) **17** 315
[12] Steger J, Kostiner E 1972 *J. Solid State Chem.* **5** 131
[13] Grodzicki M, Redhammer G, Reissner M, Steiner W, Amthauer G 2010 *Phys. Chem. Minerals* **37** 11

Dynamics of ferrocene in molecular sieves probed by Mössbauer spectroscopy and nuclear resonant scattering

T Asthalter[1], I Sergueev[2], U van Bürck[3], F E Wagner[3], P Härter[4], J Kornatowski[5], S Klingelhöfer[6] and P Behrens[6]

[1] (formerly) Institut für Physikalische Chemie, Universität Stuttgart, D–70569 Stuttgart, Germany

[2] European Synchrotron Radiation Facility, F–38043 Grenoble, France

[3] Experimentalphysik E13, Technische Universität München, D–85747 Garching, Germany

[4] Anorganische Chemie, Technische Universität München, D–85747 Garching, Germany

[5] Max-Planck-Institut für Kohlenforschung, D–45470 Mülheim, Germany

[6] Anorganische Chemie, Leibniz-Universität Hannover, D–30167 Hannover, Germany

E-mail: t.asthalter@ipc.uni-stuttgart.de

Abstract. A detailed study on the slow dynamics of ferrocene in the unidimensional channels of the molecular sieves SSZ–24 and $AlPO_4$–5 has been carried out, using Mössbauer spectroscopy (MS), nuclear forward scattering (NFS) and synchrotron radiation-based perturbed angular correlations (SRPAC). In both host systems, anisotropic rotational dynamics is observed above 100 K. For SSZ-24, this anisotropy persists even above the bulk melting temperature of ferrocene.

Various theoretical models are exploited for the study of anisotropic discrete jump rotations for the first time. The experimental data can be described fairly well by a jump model that involves reorientations of the molecular axis on a cone mantle with an opening angle dependant on temperature.

1. Introduction

The dynamics of guest molecules in zeolitic and related nanoporous hosts continues to be of current interest, as the catalytic and separation properties of zeolites critically depend on guest-molecule diffusion and host-guest interactions. A deeper insight into molecular motion in complex channel systems is only possible if the elementary dynamics in simple ordered host structures has been fully understood.

Rotation of guest molecules in zeolites is generally not isotropic because of the anisotropy of the host cage and of the guest molecule, but also – in the case of large cages – because the existence of different interstitial adsorption sites breaks the symmetry of the intermolecular potential of the guest molecule [1, 2].

The novel technique of nuclear resonant scattering (NRS) of synchrotron radiation comprises several different methods, such as nuclear forward scattering (NFS), nuclear inelastic scattering (NIS), and synchrotron radiation-based perturbed angular correlations (SRPAC). In particular, an immediate simultaneous observation of translational and rotational dynamics is now feasible with the advent of the SRPAC technique [3], which is sensitive to relaxation of the hyperfine levels of the nuclear excited state and covers a dynamical range for rotational dynamics of more than five orders of magnitude. MS and NFS are selective to both the translational and the rotational motions of molecules containing a Mössbauer nucleus, on the timescale of the natural nuclear lifetime (141.1 ns for ^{57}Fe), whereas SRPAC is sensitive to rotational motion only.

In a previous study [4], a pioneering measurement of NFS spectra of ferrocene in oriented single crystals of $AlPO_4$-5 was carried out, and the presence of anisotropic rotation above 150 K could be inferred from the appearance of two quantum beats in the NFS spectrum. Rotational jump models [5] where the molecular electric field gradient jumps between the Cartesian axes only, did not reproduce the experiment quantitatively; also, they do not take the hexagonal symmetry of the host framework into account.

Here, we assess briefly the potential of MS and NRS for determining molecular anisotropic rotation at the example of ferrocene in SSZ-24 (for more details and the $AlPO_4$-5 data, see [6]).

Section 2 is a brief summary of existing rotational jump models in Mössbauer spectroscopy, to their extension by jump models based on Blume theory [7], and to a comparison with a recently developed jump model [8] (random jump cone or RJC model). Sample preparation, characterization, and experiments are described in Section 3. Section 4 presents a comparative view of Mössbauer and nuclear resonant scattering data at the example of SSZ-24 loaded with ferrocene. The results are summarized in Section 5.

2. Rotational jump models

For the ^{57}Fe transition, any molecule possessing an electric field gradient (EFG) will exhibit a two-line Mössbauer spectrum, corresponding to a single quantum beat (QB) in time domain. Reorientation of the ferrocene molecule within the nuclear lifetime leads to a decrease of the effective EFG tensor seen by the nucleus and thus to a complete or partial collapse of the two-line pattern of an ensemble of nuclei at rest.

The influence of pure spin dynamics (EFG rotation and/or electronic relaxation), where the stochastic motion occurs between discrete states, was treated in the general formalism of Blume [7]. For equally probable rotational jumps in the direction of the three Cartesian axes (Blume-Tjon model [9]), a simple closed solution exists. For different jump probabilities in the three Cartesian directions [5], more complicated analytical solutions were found that work well for crystal structures with symmetries up to orthorhombic. For completely random angular jumps [10] (strong-collision model or random phase approximation), the result is identical to that for the Blume-Tjon model [9].

[5] speculates that for 120° jumps "the relaxation produced will differ only in minor detail" from the 90° jump models. As a one-dimensional channel has a cylindrically restricted geometry, it is realistic to assume that the molecular axis, i.e. the electric field gradient, of the ferrocene will orient on a cone mantle within a preferred or maximal polar angle ϑ. The host systems of our study possess a local hexagonal symmetry. To the authors' knowledge, no attempts were reported to treat rotational jumps having this symmetry. Such an analysis has now been done within the frame of Blume theory, for mathematical details see [6]. It was assumed that the EFG main axis can orient on a cone mantle with an opening angle ϑ of 90°, 60° and 30° (90° corresponding to in-plane reorientations of the EFG), undergoing azimuthal jumps with 3- or 6-fold symmetry, see Fig. 1. Note that models A and B yield the same line shape, as orientations that differ by 180° can not be distinguished for a tensor of second rank. For the jump matrix, it was assumed that each orientation can be reached only *via* jumps from the neighbouring orientations, i. e. that only azimuthal jumps of 120° (3-jump) or 60° (6-jump) are allowed at any moment. All jump probabilities were set to the same value, corresponding to equally deep potential well minima.

As shown in Fig. 2, an increasing jump rate always leads to an initial asymmetric line

Journal of Physics: Conference Series **217** (2010) 012143 doi:10.1088/1742-6596/217/1/012143

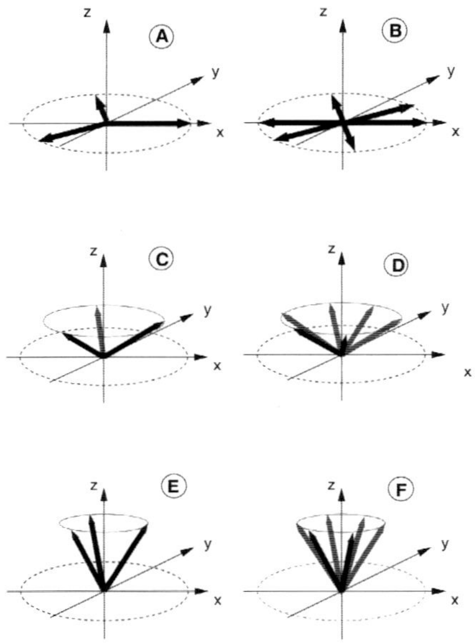

Figure 1. Orientation of the EFG tensor for the Blume jump models. The thick arrows show the possible orientations of the fivefold molecular axis of ferrocene, which coincides with the EFG tensor main axis.

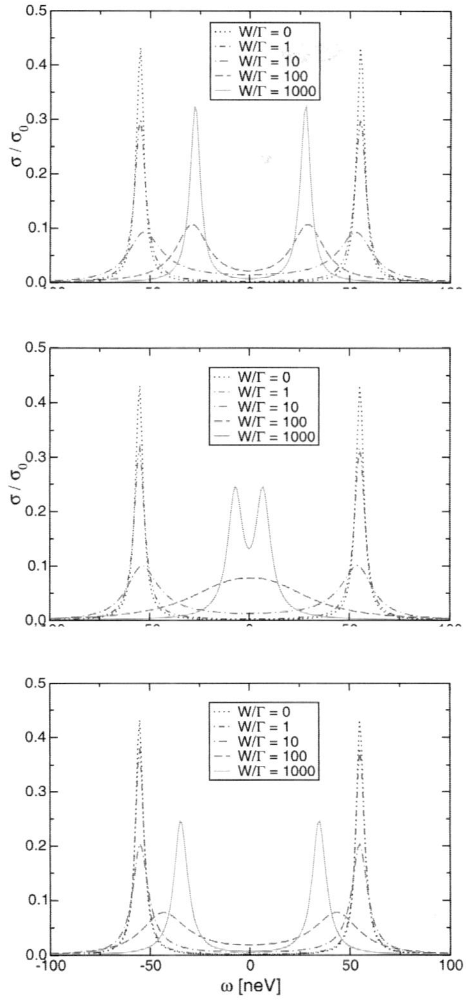

Figure 2. Theoretical Mössbauer absorption cross-section for jump models B (top), D (centre) and F (bottom). Relaxation rates W are in units of the natural line width ($\Gamma = 7.087\,\mathrm{MHz}$ for the ^{57}Fe resonance).

broadening, which is stronger towards the centre of the spectrum, then to line coalescence with broad, unstructured features and finally to the re-emergence of a quadrupole doublet having a reduced effective splitting and a decreasing line width (motional narrowing). This reduced splitting is smaller for 60° as compared with 30° and 90°: for fast rotation, the average EFG is effectively reduced to $V_{zz}(1 - 3\cos^2\vartheta)$, i.e. it passes through zero at the "magic angle" known from solid-state NMR ($\vartheta = 54.7356°$). Only numerical calculations are possible for the Blume jump models in the general case so that no least-squares fitting of experimental data is possible.

Recently, a model for completely random angular jumps of the azimuthal angle φ [8] was developed within a modified Blume-type formalism [11], yielding a closed analytical expression. We call this model the random jump cone (RJC) model. Fits of the resonant cross-section of the Blume jump models by means of the RJC model produced a perfect matching of the spectral shape, plus average cone opening angles equal to the ones of the Blume models, with the exception of small relaxation rates for models D and F. In these two cases, also the ratio $\lambda/3W$ – where λ is the RJC jump rate and W is the jump rate that appears in the Blume jump matrix – deviates from the value $\lambda/3W = 1$ as given in [10] for the isotropic random jump model, oscillating for slow relaxation but approaching a constant value for fast relaxation. Obviously, both the 3-jump models and the RJC model imply that each EFG orientation can be reached from each other one with equal probability, which is not the case for the out-of-plane 6-jump models.

Finally, the RJC model contains an ambiguity for Mössbauer or NFS spectra, as the fit parameter $\alpha = \frac{1}{2}(3\cos^2\vartheta - 1)$ enters the resonant amplitude as α^2 for the range $0 \leq \alpha \leq 0.5$ (the valid range of α is $-0.5 \leq \alpha \leq 1$). This ambiguity can be removed using complementary SRPAC data, see the next Sections.

The dependence of dynamics on SRPAC time spectra has been developed in [3]. For the strong-collision model, slow rotation entails a damping of the QB and an increased slope of the QB envelope. For fast rotation, a separate determination of quadrupole splitting and rotation time is no longer possible. The RJC model gives analytical solutions in the slow and fast relaxation limit [8]. In contrast to MS or NFS, there is no ambiguity concerning the sign of α.

3. Experiments

SSZ-24, prepared according to [12], was loaded by the same procedure as used for $AlPO_4$-5 [4]. Samples with a natural isotope distribution and samples enriched in ^{57}Fe were prepared simultaneously under precisely identical conditions. The enriched samples were used for the MS and NRS experiments and characterized *a posteriori*. The non-enriched samples were characterized in their freshly prepared state for comparison, see [6]. The MS and NRS measurements were done on fresh samples. The bulbs with the sealed samples were opened directly before the NRS experiments. For the sake of comparison, also MS measurements on the samples previously investigated at the synchrotron were carried out.

The Mössbauer measurements were performed in a He bath cryostat with source and absorber in liquid He. Measurements for rising temperatures were done during a slow two-step warm-up process of the cryostat, where source and absorber were kept at the same temperature within the range of 30–290 K. The source was a standard 50 mCi $^{57}Co/Rh$ source as delivered by Ritverc (Russia).

NFS and SRPAC can be conveniently carried out simultaneously using the standard experimental setup for NRS experiments. Those were done at the beamline ID18 of the ESRF [13]. The primary beam ($\Delta E = 3.5$ meV) was provided by a nested four-bounce high-resolution monochromator [14, 15]. The samples were mounted into a copper holder sealed with kapton windows, within a closed-cycle cryostat having a temperature stability of $\pm1°C$, for the low-temperature measurements, and into a small high-temperature oven for measurements above room temperature. In order to protect ferrocene from oxidation at elevated temperatures but also to prevent its sublimation into the isolation vacuum, the oven was slowly flushed with a N_2 gas stream throughout the experiment.

Time spectra were taken mostly in four-bunch mode in a time window of 704.14 ns. The temperature was varied from 50 K to 505 K. Note that the bulk melting temperature for ferrocene is reported as 446 K. For the $AlPO_4$-5 system, owing to lower loading and hence low count rates, no SRPAC data above room temperature were recorded.

4. Results and discussion

Mössbauer spectra of ferrocene in SSZ-24 at three different temperatures are shown in Fig. 3. The fit to all MS data was done using the transmission integral [16]. In addition to the quadrupole

Journal of Physics: Conference Series **217** (2010) 012143 doi:10.1088/1742-6596/217/1/012143

doublet observed at low temperatures, a second doublet having reduced splitting emerges with rising temperature and finally dominates the spectrum. We suggest that the four-line spectra can be ascribed to an equilibrium between mobile and immobile species. For temperatures below 100 K, almost all ferrocene molecules are stationary. With rising temperature, the intensity of the inner doublet increases, i.e. an increasing percentage of molecules become unlocked and starts to rotate.

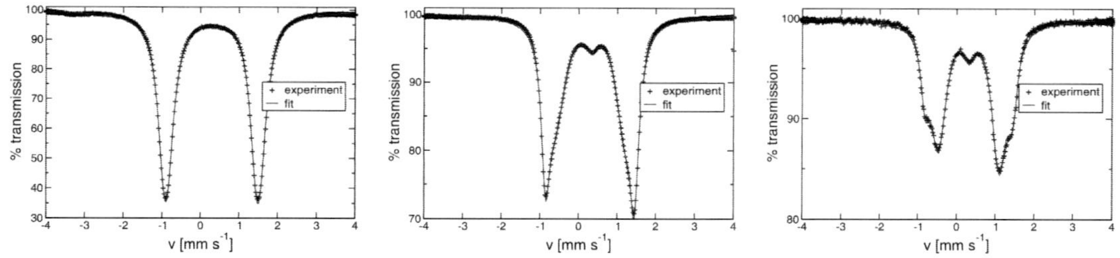

Figure 3. Mössbauer spectra of ferrocene in SSZ-24 at 4 K (left), 173 K (centre) and 225 K (right), assuming a line width of the source of 0.15 mm s^{-1}.

Examples of NFS and SRPAC time spectra of ferrocene in SSZ-24 are shown in Fig. 4. In NFS, the emerging "inner" doublet modulates coherently the quantum beat of the "outer" doublet with increasing temperature. For the SRPAC spectra, first an apparent disappearance of the QB of the "outer" doublet and then an apparent emergence of the QB of the "inner" doublet having a reduced frequency are observed. The frequency of this QB amounts to about 25 % of the original one, as the beats from both species sum incoherently. Because of the "fast" QB decay of the inner doublet, the QB structure at low temperatures is entirely dominated by the "outer" doublet. This effect diminishes as the relative intensity of the "outer" doublet decreases at higher temperature.

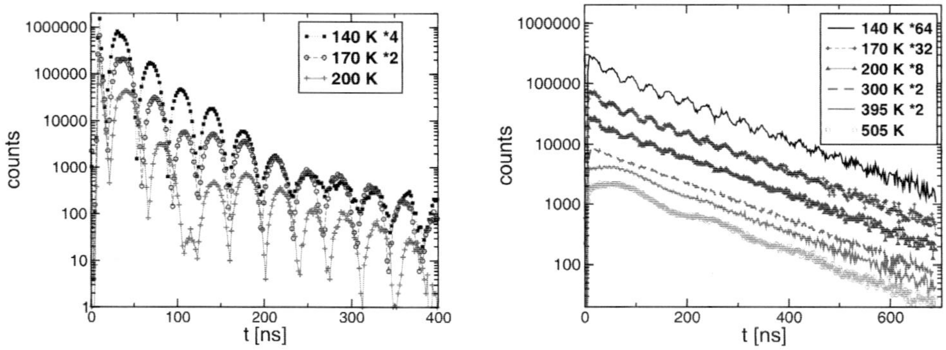

Figure 4. NFS (left) and SRPAC (right) time spectra of ferrocene in SSZ-24 at different temperatures. For the ease of comparison, spectra at low temperatures have been offset/multiplied as indicated in the legends.

We associate the relative integrated intensities of "outer" and "inner" doublet with the quantity of stationary (n_1) and mobile (n_2) species and set up a van't Hoff plot of the equilibrium constant $K = n_2/n_1$ versus $1/T$, see Fig. 5. The fairly linear behaviour supports our assumption of an equilibrium between stationary and mobile species. The fit yields $\Delta H = (8.17 \pm 0.07)\,\mathrm{kJ\,mol^{-1}}$ and $\Delta S = (47.0 \pm 0.4)\,\mathrm{J\,mol^{-1}\,K^{-1}}$.

Owing to the $f_{\rm LM}^2$ decay of the NFS intensity with temperature, the statistics of high-temperature NFS data of dynamical systems is often limited. Moreover, the complicated analytical form of the RJC model makes it impossible to treat multiple scattering properly by fitting (for a discussion of the treatment of co-existing multiple scattering and dynamics in a somewhat simpler case cf. [17]). However, it is possible to carry out a consistency check as follows. Among other features, dynamics always leads to an enhanced signal decay in the time domain $e^{-t/\tau}$, which can be converted into a line broadening $\Delta\omega$ in the energy domain by $\Delta\omega = \hbar/\tau$. A comparison of MS and NFS line broadenings thus obtained is shown in Table 1. The kinematical fit was done using the RJC model neglecting multiple scattering; the dynamical fit was done using the Motif software [18], which treats multiple scattering correctly, but only includes dynamics as a simple exponential decay. The absolute values as obtained by the different methods agree with each other within experimental accuracy, indicating that the MS and NFS data are indeed mutually consistent.

A fit using the four-jump in-plane Blume-type model [5] does not yield a satisfactory agreement between theory and experiment for MS or NFS spectra. Always the width of the inner doublet is overestimated (which was also discussed but not resolved in [5]). We fitted all experimental data using the RJC model (including an asymmetry factor for the quadrupole doublets) and obtained a good description of the Mössbauer line shape.

A comprehensive Arrhenius plot for MS and SRPAC data on the ferrocene/SSZ-24 system is shown in Fig. 6. The plot includes MS measurements of both the sample studied at the synchrotron (sample 1) and another sample prepared separately (sample 2), so as to obtain an indication for the size of the true experimental error, compared to the fit errors which, sometimes depending on software, tend to be smaller than the true experimental error.

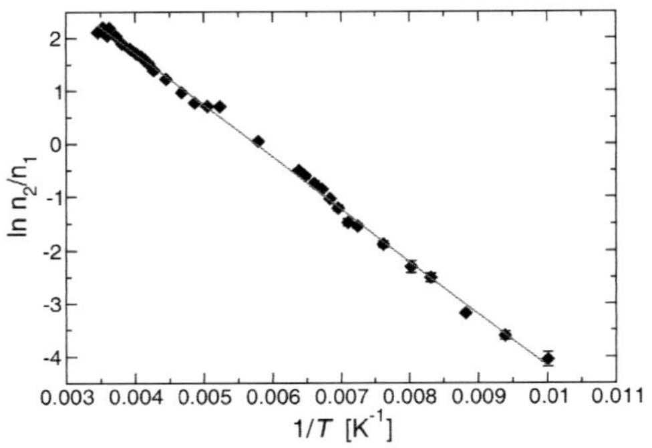

Figure 5. Van't Hoff plot of ferrocene in SSZ-24.

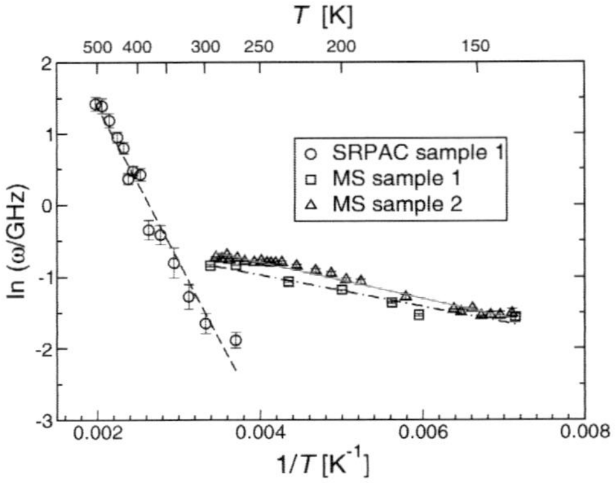

Figure 6. Arrhenius plot, including fitting, of the relaxation rates for ferrocene in SSZ-24 for MS and SRPAC. Fit errors are in the range of size of the symbols.

In the temperature range from 140 K to 290 K, we obtained an activation energy of $E_a = (1.8{\pm}0.2)\,\mathrm{kJ\,mol^{-1}}$. Low activation energies have been observed using solid state NMR for reorientational motion in inclusion compounds containing aromatic species [19]. The very low frequency factor – $A = (9.1{\pm}0.1){\cdot}10^8\,\mathrm{s^{-1}}$ – suggests that the dynamics can probably not be described by a single activated jump in this temperature regime, but that an additional process mimicks apparent Arrhenius parameters [20].

The SRPAC data were fitted using the high-temperature approximation of the RJC model. The Arrhenius parameters for the two methods and temperature regimes differ strongly. For the SRPAC data they are $E_a = (20.1{\pm}0.8)\,\mathrm{kJ\,mol^{-1}}$ and $A = (5.5{\pm}1.6{\cdot}10^{11})\,\mathrm{s^{-1}}$. In particular, the

Journal of Physics: Conference Series **217** (2010) 012143 doi:10.1088/1742-6596/217/1/012143

activation energy exceeds the one obtained from the MS data by more than a factor of six, and the large frequency factor, which is consistent with an entire-molecule rotation, indicates the presence of a single activated process. As ϑ increases strongly with temperature, cf. Fig. 7, it appears plausible that this increase will also enhance the activation energy of the corresponding jump rotation. The ϑ values as obtained from both MS and SRPAC fit together in the overlap region, indicating the consistency of these two methods as well.

Table 1. Line broadening of the inner doublet (in units of Γ) for NFS and MS data on ferrocene in SSZ-24, for details see text.

T [K]	NFS kinematical	NFS dynamical	MS
170	3.16 ± 1.04	3.46	3.84 ± 0.08
200	3.71 ± 1.01	3.32	3.32 ± 0.06
230	3.52 ± 1.11	2.96	3.29 ± 0.01
270	3.66 ± 0.54	2.90	3.97 ± 0.05
300	4.59 ± 0.99	4.25	4.08 ± 0.04

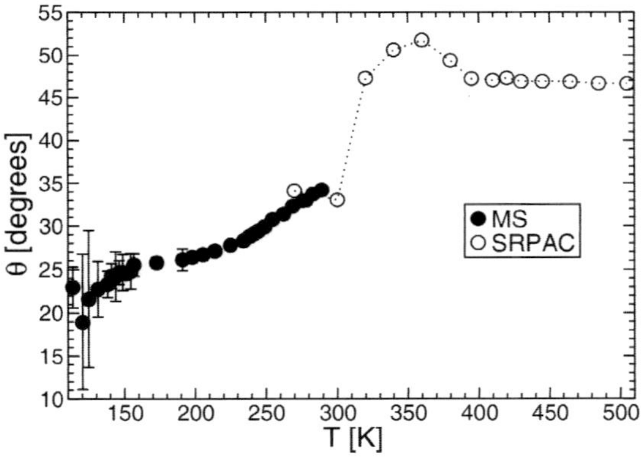

Figure 7. Average cone opening angle ϑ using the RJC model for ferrocene in SSZ-24.

Before deciding in favour of a particular jump model, one has to take into account that MS "sees" all types of motion, whereas SRPAC is exclusively selective to the rotational part, which may in part explain the somewhat different relaxation rates around 300 K as observed by MS/NFS and SRPAC. Moreover, different rotational jump models – as discussed above – may yield very similar activation energies but influence the absolute scaling of relaxation rates and thus also the gap in relaxation rates as seen by different methods.

5. Conclusions and outlook

A combination of classical Mössbauer spectroscopy and nuclear resonant scattering provides comprehensive information about anisotropic molecular rotation in a wide dynamic range. We observe anisotropic jump rotation with rising temperature in both systems $AlPO_4$-5 and SSZ-24, where the molecular axis of included ferrocene preferentially appears to reorient on a cone.

Although the RJC model gives a good overall description of the experimental data, a detailed inspection gives hints that the true jump mechanism may be more complex. We speculate that the jump rotation may weakly couple to small translational jumps of the ferrocene molecules within the host cages.

For $AlPO_4$-5, the MS data show a similar picture as for SSZ-24 in the low-temperature range, with altogether slower dynamics. An additional study of the system ferrocene/oriented $AlPO_4$-5 single crystals has been carried out [6] using SYNFOS [21], supporting the suggestion that translation-rotation coupling may have to be treated by more sophisticated theoretical models still to come.

In contrast to methods that are selective to the movements of e.g. the hydrogen nuclei, including complicated motions of mobile groups such as methyl rotations, SRPAC yields exclusively the rotation of the electric field gradient, which is usually associated with the motion of the *entire* molecule. This technique is thus complementary to NMR or neutron scattering.

References

[1] J Kärger, and D M Ruthven 1992 *Diffusion in Zeolites and Other Microporous Solids* (New York: Wiley)
[2] NY Chen, TF Degnan, Jr., and CM Smith 1994 *Molecular Transport and reaction in zeolites* (Weinheim: VCH)
[3] I Sergueev, U van Bürck, A. I Chumakov, T Asthalter, G. V Smirnov, H Franz, R Rüffer, and W Petry 2006 *Phys. Rev. B* **73** 024203
[4] T Asthalter, J Villanueva Garibay, V Olszowka and J Kornatowski 2005 *J. Chem. Phys.* **122** 014508
[5] T. C Gibb 1976 *J. Phys. C: Solid State Phys.* **9** 2627
[6] T Asthalter 2007 *Habilitationsschrift* (Stuttgart)
[7] M Blume 1968 *Phys. Rev.* **174** 351
[8] I Sergueev, T Asthalter, and U van Bürck, unpublished
[9] J,A Tjon, and M Blume 1968 *Phys. Rev.* **165** 456
[10] S Dattagupta, and M Blume 1974 *Phys. Rev. B* **10** 4540
[11] S Dattagupta, and K Schroeder 1987 *Phys. Rev. B* **35** 1527
[12] R Bialek, W. M Meier, M Davis, and M. J Annen 1991 *Zeolites* **11** 438
[13] R Rüffer, and A. I Chumakov 1996 *Hyperfine Int.* **97/98** 589
[14] T Toellner 2000 *Hyperfine Int.* **125** 3
[15] T Ishikawa, Y Yoda, K Izumi, C K Suzuki, X,W Zhang, M Ando, and S Kikuta 1992 *Rev. Sci. Instrum.* **63** 1015
[16] I Sergueev, personal communication.
[17] I Sergueev, H Franz, T Asthalter, W Petry, U van Bürck, and G. V Smirnov 2002 *Phys. Rev. B* **66** 184210
[18] Y V Shvydko 2000 *Hyperfine Int.* **125** 173
[19] JA Villanueva-Garibay, and K Müller 2006 *Phys. Chem. Chem. Phys.* **8** 1394
[20] DVujosevic, KMüller, and E Roduner 2006 *J. Phys. Chem. B* **110** 8598
[21] M Haas, E Realo, H Winkler, W Meyer-Klaucke and AX Trautwein 2000 *Hyperfine Int.* **125** 189

Phonons in iron monolayers

S Stankov[1,2,†], M Sladecek[2,‡], T Ślezak[3,4], J Łażewski[5], R Röhlsberger[6], B Sepiol[2], G Vogl[2], A I Chumakov[1], R Rüffer[1], N Spiridis[4] M Zajac[1,4], M Ślezak[4], K Parliński[5,7] and J Korecki[3,4]

[1] European Synchrotron Radiation Facility, Grenoble, France
[2] Faculty of Physics, University of Vienna, Vienna, Austria
[3] Faculty of Physics and Applied Computer Science, AGH University of Science and Technology, Krakow, Poland
[4] Institute of Catalysis and Surface Chemistry, Polish Academy of Sciences, Krakow, Poland
[5] Institute of Nuclear Physics, Polish Academy of Sciences, Krakow, Poland
[6] DESY, Hamburg, Germany
[7] Institute of Technics, Pedagogical University, Krakow, Poland
[†] Present address: Forschungszentrum Karlsruhe, Eggenstein-Leopoldshafen, Germany
[‡] Present address: Philips Austria GmbH, Klagenfurt, Austria

E-mail: `svetoslav.stankov@iss.fzk.de`

Abstract. The systematic investigation of the lattice dynamics from bulk to a single atomic layer of material was a great experimental challenge until now. Recently, nuclear inelastic scattering has been introduced as a unique technique capable of determining the density of phonon states of nanoscale materials *in-situ* with a depth resolution of one atomic layer. This became possible by setting up a dedicated ultrahigh vacuum system for samples growth and characterization directly at the nuclear resonance beamline ID18 of the ESRF. The new instrument allowed for systematic investigation of the evolution of the density of phonon states of iron from the bulk to a single atomic layer. The isotopic selectivity of the nuclear resonance absorption was employed to experimentally determine the atomic vibrations at and near the Fe(110) surface with a monolayer sensitivity. The experimental achievements stimulated a rapid progress of the *ab-initio* calculations of surface phonons.

1. Introduction

Recent instrumentation development [1] at the nuclear resonance beamline ID18 [2] of the European Synchrotron Radiation Facility (ESRF) in Grenoble has offered new opportunities to investigate electronic [3] and magnetic [4, 5] properties, vibrational dynamics [6–8] and diffusion phenomena [9–12] in nanoscale objects *in-situ*. An ultrahigh vacuum (UHV) system for deposition of thin films and nanostructures by molecular beam epitaxy and their characterization by surface sensitive techniques has been installed directly at the beamline. Thus, several nuclear resonance scattering (NRS) based techniques, such as nuclear forward scattering, nuclear inelastic and quasielastic scattering, synchrotron radiation based perturbed angular correlations, and nuclear reflectivity, can be applied to investigate the nanostructures under UHV conditions and at low/high temperatures, or during growth of the samples. In addition, the unique combination of the high penetration depth of the X-rays with the isotopic selectivity of the nuclear resonant absorption allows one to perform experiments with the ultimate depth resolution of one atomic layer by introducing isotopic "resonant" probe layers (a recent review can be found in [13]).

The unique combination of the UHV facility and the nuclear inelastic scattering has led to significant progress in the field of the vibrational properties of low dimensional systems. The aim of this paper is to describe the main features of the new instrument and to summarize the experimental and theoretical advances in the lattice dynamics of ultrathin epitaxial iron films.

2. UHV system for *in-situ* nuclear resonance scattering at ID18 of the ESRF

The overall design of the UHV system was defined by the demand for *in-situ* preparation and characterization of samples and their investigation by various synchrotron radiation based techniques. The UHV instrument is permanently installed into the second experimental hutch of ID18 of the ESRF, which required remote control of the entire system. Figure 1 presents schematically the new instrument, while figure 2 displays a top-side photograph [1]. The UHV set-up with a base pressure of 2×10^{-10} mbar consists of a central distribution chamber equipped with a sample transfer mechanism and peripherally attached chambers as described in details below.

The preparation chamber, number 1 in Fig. 1 and Fig. 2, is equipped with two electron beam evaporators and an effusion cell. A four-pocket mini e-beam evaporator serves for evaporation of metals from rods and crucibles from each pocket separately, as well as, for co-deposition of different combinations of the evaporants (alloys or multilayers). A single-pocket e-beam source is used for deposition of the Mössbauer isotope ^{57}Fe. An effusion cell is available for evaporation

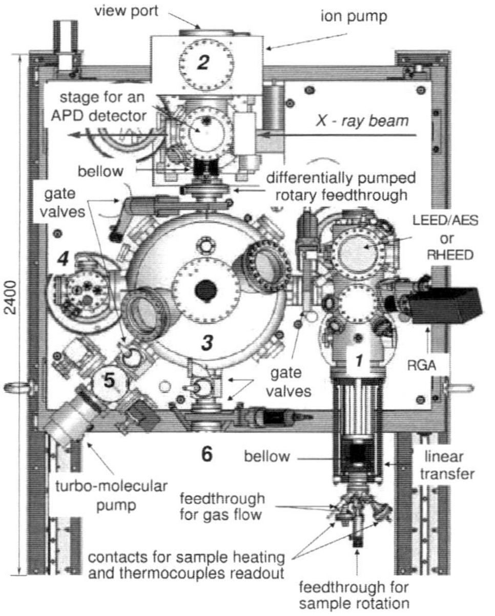

Figure 1. Schematic view of the UHV system with the preparation 1, NRS 2, distribution 3, storage 4, and load-lock 5 chambers. Number 6 marks the port where transfer chambers are connected (Ref. 1).

Figure 2. A top-side photograph of the UHV system with the preparation 1, NRS 2, distribution 3, storage 4, and load-lock 5 chambers. Number 6 shows one of the portable chambers connected to the system (Ref. 1).

of rare-earth metals. A precise calibration of the deposition rate with a thickness reproducibility of 1 Å is done by a quartz-balance monitor. The deposited structures can be characterized by low-energy electron diffraction (LEED) and Auger electron spectroscopy (AES). In order to monitor *in-situ* the film growth, or to structurally characterize thick epitaxial films the LEED instrument can be replaced by a reflection high-energy electron diffraction (RHEED) set-up. The samples can be cooled down to about 90 K and heated up to 2300 K by a sophisticated manipulator.

The chamber for NRS experiments, number 2 in Fig. 1 and Fig. 2, is mounted on a two-circle goniometer, which serves to align the sample and to perform experiments with the focused synchrotron beam in grazing-incidence scattering geometry. The manipulator provides contacts for temperature measurement, resistive and electron-bombardment heating, and feedthrough for cooling the sample down to 90 K by flow of liquid nitrogen. In addition, the sample holder can be rotated in the range of ± 180 degrees around an axis perpendicular to the sample surface, allowing for angular resolved studies. For the purpose of nuclear inelastic scattering or synchrotron radiation based perturbed angular correlation experiments an avalanche photodiode detector is placed directly above the surface of the sample.

The distribution chamber, number 3 in Fig. 1 and Fig. 2, connects the above-described chambers and serves for transferring the sample holders between them. In addition, a sample storage chamber, number 4 in Fig. 1 and Fig. 2, with the capacity to store up to 6 sample holders is an integrated part of the distribution chamber. The down-most position provides electrical contacts for resistive heating and temperature control allowing for pre-annealing of the sample holders. The load-lock chamber, number 5 in Fig. 1 and Fig. 2, serves for introducing the sample holder into the UHV system. Two additional portable chamber are also available for sample transfer and experiments at other beamlines of the ESRF.

3. Results and Discussions

In the following we present selected examples of research on the lattice dynamics of ^{57}Fe monolayers.

3.1. Ultrathin Fe films on W(110)

An Fe monolayer on the W(110) surface is an established model system to study a variety of structural [14], magnetic [15] and diffusion [9, 12, 16] phenomena at the nanoscale. A flat pseudomorphic monolayer (ML) is stabilized by deposition of Fe on the atomically clean W(110) surface at 520 K. At room temperature the Fe atoms are in paramagnetic state experiencing a quadrupole interaction due to an electric field gradient induced by the broken translational symmetry at the surface [17]. The mismatch between the lattice constants of the bulk Fe and the W(110) surface of about 10% results in an enormous stretching of the pseudomorphic Fe monolayer. This implies significant changes of the elastic and thermodynamic properties of this monolayer and of the adjacent layers. Due to this large mismatch already within the second Fe layer a network of misfit dislocations forms to relax the lattice strain and to recover the bulk Fe-Fe coordination distances. Thus the interface between the Fe and the W(110) surface is a model system for investigation of the evolution of the thermodynamic and elastic properties upon transition from a strained pseudomorphic Fe monolayer towards bulk Fe.

To investigate this phenomena, the energy dependencies of nuclear inelastic absorption were measured at room temperature for various film thicknesses. Fig. 3 summarizes the derived [18] phonon density of states (DOS) for Fe films with thicknesses between 1 and 40 monolayers. The right column represents the low energy electron diffraction (LEED) patterns of the substrate, 1 ML and 40 ML thick films recorded after the sample growth. They unambiguously confirm the high single-crystalline quality of the investigated films.

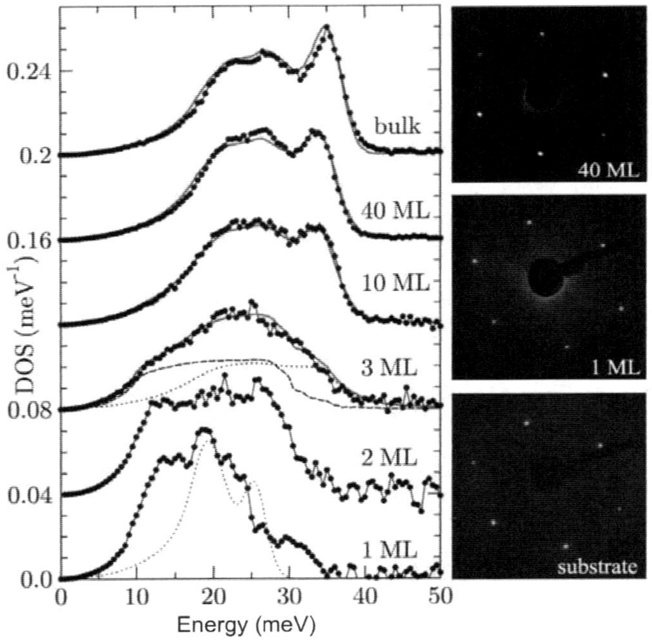

Figure 3. DOS of Fe films on W(110) for thicknesses from 1 ML (1.66 Å) to a 40 ML, projected to [001] direction. The dotted line represents the DOS of the bulk bcc W convoluted with the resolution function (FWHM=3.3 meV) of the experiment. The dashed and dotted line in the DOS of the 3 ML film represent the smoothed DOS of the 2 ML film and the DOS of bulk bcc Fe convoluted with a damped harmonic oscillator (quality factor Q=7), respectively. The solid lines for the 3, 10, and 40 ML DOS result from the weighted sum of these two contributions in each case. The right column shows LEED patterns (E=96 eV) of the substrate, 1 and 40 ML (Ref. 7).

The topmost curve of the left column of Fig. 3 represents the DOS of a bulk bcc Fe, while the dotted line on the bottom plots the DOS of a bulk bcc W. The experimental data reveal that increasing the Fe coverage results in a rather sharp transition of the phonon DOS towards that of bulk Fe. Quite remarkably, already the spectrum of 3 ML film is satisfactory described as a contribution from the Fe/W interface, represented by the DOS of a 2 ML thick Fe film (dashed line), and that of bulk Fe including a significant phonon lifetime broadening (dotted line) [7]. This relatively simple model applies also for thicker films and allows one to describe the thickness evolution of the thermodynamic and spectral properties of ultra-thin films with rather good accuracy. Most likely it is valid not only for iron, but for thin films of many transition metals. For detailed discussion we refer the reader to Ref. 7.

3.2. Fe(110) surface

Combining the nuclear inelastic scattering spectroscopy with the UHV sample preparation and characterization facility allowed us to study the vibrational properties of a clean Fe surface layer-by-layer.

In order to investigate the electronic properties and the lattice dynamics at and near the Fe surface single crystalline epitaxial Fe(110) films were deposited on an atomically clean W(110) crystal from ^{57}Fe and ^{56}Fe e-beam evaporators. The samples consisted of a 20 ML ^{56}Fe(110) buffer film optimized for a flat surface growth, while the ^{57}Fe monolayers were deposited on the buffer film at lower temperature (350 K) to prevent mixing of the isotopes [19]. The following samples were studied: 20 ML ^{56}Fe/1 ML ^{57}Fe referred to as a "surface" (S), 20 ML ^{56}Fe/1 ML ^{57}Fe/1 ML ^{56}Fe referred to as a "subsurface" (S-1), and 20 ML ^{56}Fe/3 ML ^{57}Fe/3 ML ^{56}Fe referred to as a "deep" (D) layers [6].

Fig. 4 shows the phonon DOS obtained for the surface, subsurface, and deep layers from the measured nuclear inelastic scattering dependencies at room temperature. The surface sample was measured with the wave vector of the X-ray beam parallel to the [$1\bar{1}0$] (solid circles) and [001] (open circles) crystallographic directions. For comparison the DOS of an α-Fe foil (grey

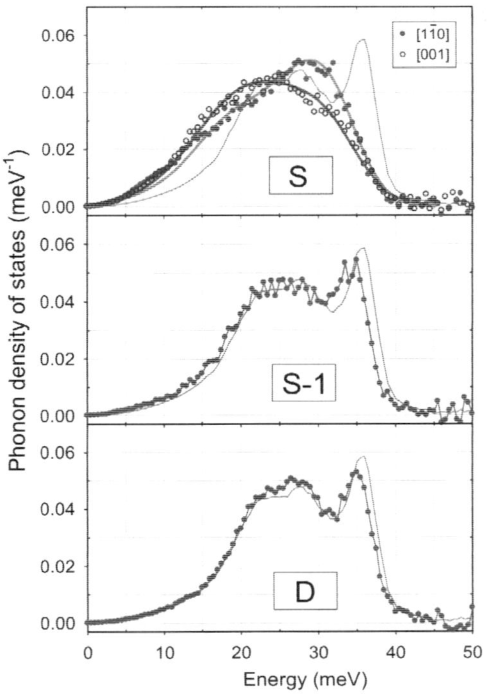

Figure 4. Measured phonon DOS for the surface (S), subsurface (S-1), and deep (D) layers near the Fe(110) surface projected to [1$\bar{1}$0] (solid symbols) and to [001] (open symbols). The solid lines in (S) plot the results of DFT calculations. The DOS of an Fe foil is shown with solid line (Ref. 6).

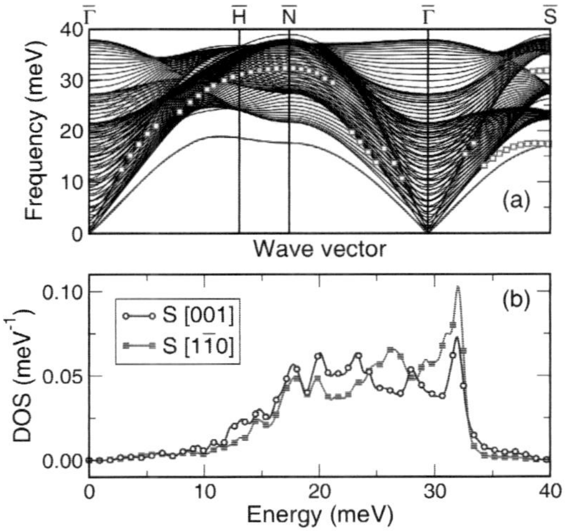

Figure 5. (a) Dispersion curves and (b) surface DOS along [1$\bar{1}$0] (solid squares) and [001] (open circles) directions calculated for 29 layers Fe slab. The high-symmetry points correspond to the two dimensional Brillouin zone of the plane reciprocal lattice. In (a), the lowest dispersion curve is the Rayleigh mode while the open squares depict surface modes with the highest in-plane intensities (Ref. 6).

line) is plotted as well. The phonon DOS of the surface layer is remarkably different from that of the subsurface layer, which already closely resembles the DOS of a bulk iron. The most striking anomalies are the suppression of the phonon peak at 35 meV and a considerable enhancement of the vibrational states below 15 meV which, however, obeys the Debye law by being quadratic in energy.

First-principles calculations were performed with spin-polarized density functional theory implemented in the Vienna *ab initio* simulation package (VASP) [20]. The calculated [6, 21] dispersion relations and the DOS projected on two in-plane directions are shown in Fig. 5 (a) and (b) respectively. The lowest curve in Fig. 5 (a) mark the well separated Rayleigh mode, while the open circles represent the most intense in-plane surface modes. Due to geometrical considerations (the nuclear inelastic scattering experiment is performed at grazing incidence scattering geometry) the derived phonon DOS are in-plane direction projected, therefore, not influenced by the very active Rayleigh mode, characterized with vibrations perpendicular to the surface. The observed anomalies at the surface arise from the surface-specific modes due to the broken translational symmetry. Moreover, these modes result in vibrational anisotropy, (Fig. 4 panel (S)), perfectly reproduced by the calculated phonon DOS (Fig. 5 (b)). For detailed discussion we refer the reader to Ref. 6.

4. Conclusions

The nuclear inelastic scattering was employed to systematically investigate the lattice dynamics of thin and ultrathin iron films by measuring the density of phonon states of strained iron films and of the native iron Fe(110) surface *in-situ*. The experimental results revealed a significant deviation of the lattice dynamics of the Fe films from that of the bulk in a close vicinity to the substrate. The transition to the bulk-like lattice dynamics is rather sharp and takes place within several atomic layers. On the other hand, the anomalous dynamics observed for the native Fe surface is localized at the first atomic layer only. The second layer already vibrates in a bulk-like manner. These observations are in total agreement with the first-principles phonon calculations.

These results have been achieved after setting up of an ultrahigh vacuum system directly at the nuclear resonance beamline ID18 of the ESRF. The instrument allows one to deposit by molecular beam epitaxy a variety of nanostructures like thin and ultrathin films, nano-islands and -wires, multilayers and stoichiometric oxides. After characterization by Auger electron spectroscopy, low-energy electron diffraction or by reflection high-energy electron diffraction the electronic and magnetic properties, vibrational dynamics, and diffusion phenomena can be investigated by the nuclear resonance scattering techniques *in-situ*, i.e. under UHV conditions.

Acknowledgments

This work was supported by the European Community under the Specific Targeted Research Project Contract No. NMP4-CT-2003-001516 (DYNASYNC) and by the Austrian ministry bm:bwk under the grant GZ 45.529/2-VI/B/7a/2002 (MDN).

References

[1] Stankov S, Rüffer R, Sladecek M, Rennhofer M, Sepiol B, Vogl G, Spiridis N, Ślezak T and Korecki J 2008 *Rev. Sci. Instrum.* **79** 045108
[2] Rüffer R and Chumakov A I 1996 *Hyperfine Interact.* **97-98** 589
[3] Partyka-Jankowska E, Sepiol B, Sladecek M, Kmiec D, Korecki J, Ślezak T, Zajac M, Stankov S, Rüffer R and Vogl G 2008 *Surf. Sci.* **602** 1453
[4] Couet S, Diederich T, Stankov S, Schlage K, Ślezak T, Rüffer R, Korecki J and Röhlsberger R 2009 *Appl. Phys. Lett.* **94** 162501
[5] Couet S, Schlage K, Rüffer R, Stankov S, Diederich Th, Laenens B, and Röhlsberger R 2009 *Phys. Phys. Lett.* **103** 097201
[6] Ślezak T, Łażewski J, Stankov S, Parliński K, Reitinger R, Rennhofer M, Rüffer R, Sepiol B, Ślezak M, Spiridis N, Zajac M, Chumakov A I and Korecki J 2007 *Phys. Rev. Lett.* **99** 066103
[7] Stankov S, Röhlsberger R, Ślezak T, Sladecek M, Sepiol B, Vogl G, Chumakov A I, Rüffer R, Spiridis N, Łażewski J, Parliński K and Korecki J 2007 *Phys. Rev. Lett.* **99** 185501
[8] Stankov S, Piekarz P, Oleś A M, Parliński K and Rüffer R 2008 *Phys. Rev.* B **78** 180301(R)
[9] Sladecek M, Sepiol B, Kmiec D, Korecki J, Ślezak T and Vogl G 2005 *Defect Diffus. Forum* **237-240** 1225
[10] Reitinger R, Sepiol B, Vogl G, Pfau B, Stadler L M, Stankov S, Zontone F, Spiridis N and Korecki J 2007 *J. Appl. Phys.* **102** 034310
[11] Reitinger R, Pfau B, Stadler L M, Stankov S, Zajac M, Spiridis N, Sepiol B, Zontone F and Korecki J 2007 *Defect Diffus. Forum.* **263** 177
[12] Vogl G, Partyka-Jankowska E, Zajac M, Chumakov A I 2009 *Phys. Rev.* B **80** 115406
[13] Röhlsberger R 2004 *Nuclear Condensed Matter Physics with Synchrotron Radiation* (Heidelberg: Springer-Verlag, Berlin, Heidelberg)
[14] Gradmann U and Waller G 1982 *Surface Science* **116** 539
[15] Röhlsberger R, Bansmann J, Senz V, Jonas K L, Bettac A, Leupold O, Rüffer R, Burkel E and Meiwes-Broer K H 2000 *Phys. Rev. Lett.* **86** 5597
[16] Vogl G, Sladecek M and Dattagupta S 2007 *Phys. Rev. Lett.* **99** 155902
[17] Ślezak T, Stankov S, Zajac M, Ślezak M, Matlak K, Spiridis N, Laenens B, Rennhofer M, Freindl K, Wilgocka-Ślezak D, Rüffer R and Korecki J 2008 *Matter. Science-Poland* **26** 885
[18] Kohn V G and Chumakov A I 2000 *Hyperfine Interact.* **125** 205
[19] Korecki J and Gradmann U 1985 *Phys. Rev. Lett.* **55** 2491
[20] Kresse G and Furthmüller J 1996 *Phys. Rev.* B **54** 11169
[21] Parliński K 2006 *Phys. Rev.* B **74** 184309

Journal of Physics: Conference Series **217** (2010) 012145 doi:10.1088/1742-6596/217/1/012145

Metal atom dynamics in organometallics: Resolving the dichotomy between Mössbauer and X-ray derived values

Rolfe H Herber[1,*], **Israel Nowik**[1] **and Shmuel Cohen**[2]

[1]Racah Institute of Physics, The Hebrew University of Jerusalem

[2]Department of Chemistry, The Hebrew University of Jerusalem

E-mail: [*]HERBER@VMS.HUJI.AC.IL

Abstract. Temperature-dependent ^{57}Fe and ^{119}Sn Mössbauer spectroscopy (ME) has been employed to study the metal atom dynamics in a number of organometallic compounds. Such experiments can yield information concerning the vibrational amplitudes of the metal atom as well as the anisotropy (if any) of this motion as a function of temperature. The difference between the vibrational amplitudes derived from ME data and that derived from single crystal X-ray $U_{i,j}$ data is considered in the light of the structural details of these compounds.

1. Background

The determination of the temperature dependence of both the hyperfine parameters (I.S. and Q.S.) and the recoil-free fraction observed in ME examination of organometallics has proven to provide significant information concerning the effective vibrating mass (M_{eff}), a phenomenological lattice temperature (Θ_M), the anisotropy of the metal atom vibration (x_{para} and x_{perp}) with respect to the molecular symmetry axis, and the mean-square-amplitude-of-vibration (msav) of the metal atom motion [1]. This latter parameter is sensitive to the metal atom-ligand binding mode as well as the detailed molecular structures involved. The focus of the present study is on the msav of related
organo-iron and organo-tin complexes.

From the temperature dependence of the area under the resonance curve [A(T)] for an optically thin absorber it is possible to extract the parameter $F = k^2\langle x_{ave}^2\rangle$, where k is the wave vector of the appropriate ME gamma ray, and $\langle x_{ave}^2\rangle$ is the msva of the metal atom. It is moreover also possible to derive F from the $U_{i,j}$ values reported for single crystal X-ray diffraction data of the pertinent compounds. As will be shown below, there are interesting differences between $F_{X,T}$ derived from X-ray data at temperature T, and $F_{M,T}$ extracted from ME data at temperature T. In order to effect this comparison it is necessary to normalize the ME data and this is done by making the assumption the recoil-free fraction approaches unity as $T \rightarrow 0$ K. This assumption, which has previously been invoked in studies of biopolymers [2], neglects the zero-point motion at low temperatures, but extrapolation of the high-temperature ME data suggest that this is (at best) a small correction.

2. Experimental

The details of the ME measurements, using ^{57}Co and ^{119m}Sn sources, has been described previously [3]. The various organometallic compounds were used as received, and - in the case of the air/moisture sensitive organotin complexes - transferred to O-ring provided plastic sample holders in an inert atmosphere glove-box and immediately cooled to liquid nitrogen temperature. All Fe isomer shifts are referred to the centroid of an α-Fe room temperature absorber spectrum (also used to effect spectrometer calibration), while the Sn isomer shifts are referred to the center of a room temperature $BaSnO_3$ absorber spectrum.

3. Results and discussion

(a) Organo-iron Compounds : to test the applicability of determining $F_{M\ or\ X}$ for a typical compound, the ME spectra of cyano-ferrocene ($C_5H_5FeC_5H_4CN$) have been determined over the temperature interval 96<T<296 K. In addition, single crystal X-ray data were acquired at 5 different temperatures. The temperature dependence of the recoil-free fraction in the interval 96<T<193 K, is found to be $-(4.75\pm0.58)\times10^{-3}$ K^{-1} (corr. coeff. = 0.985 for 6 data points). A comparison of $k^2\langle x_{ave}^2\rangle$ for the ME data and that derived from the X-ray data is summarized graphically in Fig. 1.

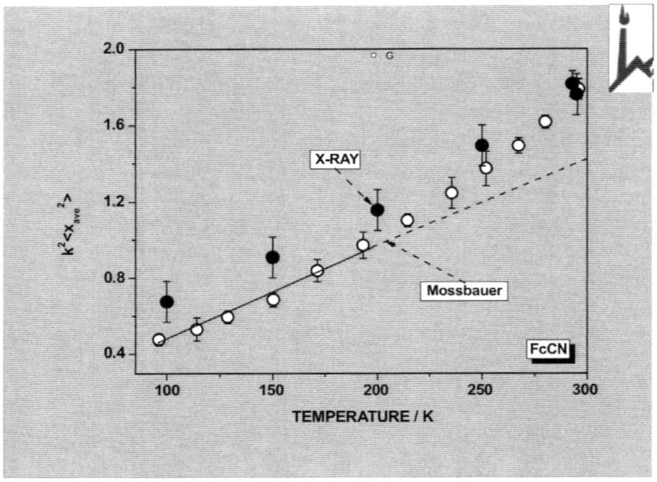

Figure 1. Comparison of F_M (open circles) and F_X (full circles) for Cyano Ferrocene

It is noted that the $F_{M,T}$ data are consistently lower than the $F_{X,T}$ data and both data sets deviate increasingly from the low temperature linear temperature dependence as T increases. At 296 K the two F values are nearly identical. The difference between the two data sets is ascribed to the presence of low-frequency librational or torsional modes to which the X-ray data are sensitive but the ME data are not. Moreover, this assumption is also consistent with the increasing deviation from linearity at the higher temperatures. It is assumed that the low frequency modes are increasingly populated as the temperature increases and become significant with respect to the ME atom vibrational amplitudes. The close similarity between the X-ray derived data (full circles) and the ME derived data (open circles) shown at the highest temperature in Fig 1 reflects this interpretation. These results are completely consistent with comparable ^{57}Fe ME data for other ferrocenoids [4] in which the F_X values are always observed to be somewhat larger than the F_M values at the same temperature.

(b) Organo-tin Compounds: Motivated by the fact that the metal atom-ligand binding modes are very different for ferrocenoids (involving primarily 3d-π electron interactions) and for tin complexes (involving primarily 5s5p-np σ bonding) the F factors for a number of organo-tin complexes have been determined. A comparison of F factors is summarized in Fig. 2 in which the ME values (ordinate) are plotted against the X-ray values (abscissa). The straight line corresponds to $F_M = F_X$. The five data points, starting from the lower left, correspond to Adams $Re_2(CO)_8$-(μ-$SnPh_2)_2$ [5], Power's $[SnN(Dipp)]_4$ [6], Wesemann's $Cd(SnB_{11}H_{11})_4$ and $[NH(C_2H_5)_3]_2SnB_{11}H_{11}$ [7], and finally Saito's *spiro* stanna bifluorene compounds [8]. In each of these, the tin atom is "anchored" in the three-dimensional molecular structure by multiple metal-ligand bonding. Any torsional or librational modes are of sufficiently high frequency that they are not reflected sensitively in the X-ray $U_{i,j}$ values. The differences between the two values is on the order of a few percent.

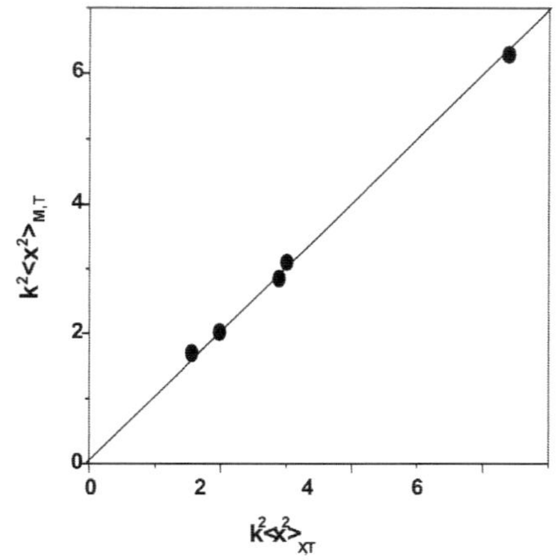

Figure 2. Comparison of F_M (ordinate) and F_X (abscissa) for some organo-tin compounds

In contrast, a number of organotin complexes in which the metal-ligand bonding involves one or two sigma bonds show very marked differences between the F_M and F_X values. A typical case in point is reflected in the data for Power's recently reported [9] (4-*t* -Bu Ar'Sn)$_2$ where Ar' is the dipp ligand C_6H_3-2,6(C_6H_3-*i*-Pr_2)$_2$. The structure of this compound is shown in Fig. 3.

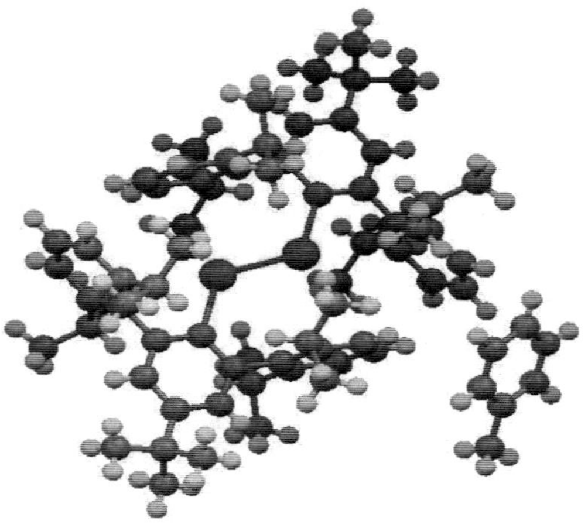

Figure 3. A representative tin compound with an Sn-Sn sigma bond.

The two corresponding dynamical parameters are $Ғ_{M,90}$ = 2.004±0.041 and $Ғ_{X,90}$ = 2.750±0.044. This 38% difference is tentatively associated with low frequency rotational motion around the Sn-Sn σ bond (2.646 Å at 90 K) which is reflected in the X-ray $U_{i,j}$ value but not in the corresponding ME data at 90 K. Major differences in $Ғ_X$ and $Ғ_M$ have been observed for other organo-tin compounds which have structures that permit low frequency torsional or librational modes to influence the rmsav of the metal atom, as determined by the two techniques discussed above. Reaction behavior which depends on metal atom interactions with other chemical constituents (eg. catalysis and reaction rates) present in reacting systems should be reflected in these dynamical properties.

4.Summary and conclusions

It has been shown that temperature-dependent ME data can elucidate a number of details associated with the dynamics of the Mössbauer atom in organometallic compounds. In particular, the msav of the metal atom in a number of organo-iron and organo-tin complexes has been studied over an extended temperature range. The differences in the Ғ parameter derived from ME and single crystal $U_{i,j}$ values have been studied in detail, in particular for a number of organotin complexes.

Acknowledgments

The authors are indebted to the individuals mentioned in the text for providing the samples used in this study. This work was supported in part by the Klatchky Foundation and by grant CHE-0641020 to P.P. Power.

References

[1] Herber R H, Nowik, I. 2008 *J. Organomet. Chem.* **693** 3007
[2] Parak F, Formanek H, 1971 *Acta Cryst. A.*,**27** 573; Parak F, Hartmann H, Aumann K D, Reuscher H, Rennenkamp G, Bartunik H, Steigemann W 1987 *Eur. Biophys. J.* **15**,237
[3] Nowik I, Herber R H, 2000 *Inorg. Chim. Acta* **310** 191; Herber R H, Nowik I 2002 *Solid State Sciences* **4** 691
[4] Kaufmann L, Breunig J-M, Vitze H, Schoedel F, Nowik I, Pichlmaier M, Bolte M, Lerner H-W, Winter R F, Herber R H, Wagner W 2009 *Dalton Trans.* 2940
[5] Adams R D, Captain B, Herber R H, Johansson M, Nowik I, Smith Jr. J L, Smith M D, 2005 *Inorganic Chem.* **44**, 6346
[6] Spikes G H, Giuliani J R, Augustine M P, Nowik I., Herber R H, Power P P, 2006 *Inorganic Chem.* **45** 9132
[7] Kirchmann U, Eichele K, Wesemann L 2008 *Inorganic Chem.* **47** 5988
[8] Saito M, Imaizumi S, Tajima T, Ishimura K., Nagase S 2007 *J. Amer. Chem. Soc.* **129** 10974
[9] Merrill W A, Steiner J, Betzer A, Nowik I, Herber R H, Power P P 2008, *Dalton Trans.* 5905

Iron in ferrous gluconate and Ascofer®

R Gozdyra, S M Dubiel* and J Cieślak

Faculty of Physics and Computer Science, AGH University of Science and

Technology, al. A. Mickiewicza 30, PL-30-059 Kraków, Poland

*dubiel@novell.ftj.agh.edu.pl

Abstract. Ferrous gluconate and antianemic medicament Ascofer® were investigated with Mössbauer spectroscopy in order to determine forms of iron ions present in both types of samples. Room temperature spectra gave a clear evidence that two phases of iron were present viz. ferrous (Fe^{2+}) as a major one with a contribution of ~85 ±5 %, and ferric (Fe^{3+}) whose contribution was found to be ~15±5 %. Ferrous ions were shown to occupy at least two different sites.

1. Introduction

Standard treatment of iron deficiency diseases such as anemia involves administration of iron supplements. To be absorbed, dietery iron must be in its ferrous (Fe^{2+}) form, which is soluble. Consequently, the iron supplements for oral intake are most frequently in form of ferrous salts like ferrous fumarate, sulfate and gluconate. As the subject of our present research we have chosen Ascofer®, a medicine produced by a Polish farmaceutical firm, Espefa®, based on ferrous gluconate. It is fabricated in form of a dragee having a core and a shell. The core contains 200 mg of ferrous gluconate (23.2 mg elemental Fe) mixed with starch, talc, ascorbic and stearin acids, and propylene glycol. The shell is made up of talc, saccharine, gelitin, magnesium oxide, arabic gum and cochineal red. Among these constituents only ferrous gluconate should contain iron. Ferrous gluconate was already studied with Mössbauer Spectroscopy (MS) in the past by other investigators [1-3], and its spectrum was analyzed in different ways. In particular, authors of Refs. 1 and 2 fitted the measured spectra with one doublet having spectral parameters charactersitic of Fe^{2+} ions in the high-spin state, while those of Ref. 3 interpreted their spectra in terms of two doublets ascribed to Fe^{2+} ions occupying two different sites. They also found that the gluconate samples they investigated viz. a commercial one and a freshly prepared in a chemical laboratory were contaminated with ~10 at% of Fe^{3+} ions. The aim of our present study was to verify which of the two approaches applied in the spectra analysis is more appropriate as well as to see if the forms and amounts of iron present in Ascofer® are similar to those in ferrous gluconate.

2. Experimental

2.1. Samples

Investigated in this study were both particular components of Ascofer® i.e. (a) ferrous gluconate, (b) talc, (c) saccharine, (d) gelatin, (e) magnesium oxide, (f) arabic gum and (g) cochineal red, as well as

Ascofer® itself. Among these components only ferrous gluconate was expected to contain iron. However, the iron was also found to be present in talc. Its room temperature spectrum could have been well fitted with one singlet and one doublet. Spectral parameters of the latter are close to those of $FeSO_4$ [4]. Concerning ferrous gluconate, two samples A and B were investigated in this study. They were fabricated by various manufacturers. In the case of Ascofer® some of the investigated samples had different series numbers i.e. they might be composed of the ingredients of different origin.

2.2. Mössbauer spectra and their analysis

^{57}Fe-site Mössbauer spectra were recorded at room temperature in a transmission geometry using a standard spectrometer with a drive working in a sinusoidal mode. Gamma rays of 14.4 keV energy were supplied by a ^{57}Co/Rh source. The investigated samples were in form of powder that, in the case of Ascofer®, was produced by pounding a dragee in an agate mortar. The samples of ferrous gluconate had a mass of 200 mg and for Ascofer® one dragee was used. All the spectra of ferrous gluconate and those of Ascofer® have had similar shape i.e. asymmetric doublet with a small bump close to the inner part of the left-hand line – see Figs. 1 and 2. The shape reflects the presence of two kinds of Fe ions known from the previous studies [1-3]. The major ferrous phase is represented by two intensive lines from which one is centered at ~-0.5 mm/s and the other one at ~2.5 mm/s. The minor ferric phase is also represented by two lines (doublet) from which the left-hand one is hidden under a much more intensive left-hand line associated with the ferrous phase, while the other one can be seen in form of the bump centered at ~0.5 mm/s. In the previous investigations the subspectrum due to the ferrous phase was treated either as one doublet [1] or as two doublets [3]. As one of the aims of the present study was to verify whether or not Fe^{2+} ions ocupy two different sites as claimed in [3], the subspectrum was fitted with the quadroupole distribution method based on the one described elsewhere [5] i.e. *a priori* no assumption on the number of the sites occupied by Fe^{2+} ions was made. The subspectrum representing the ferric phase was treated as one doublet. The asymmetry of the major quasi-doublet, which according to [3] does not originate, as suggested in [1], from Goldanskij-Karyagin effect, was here accounted for by assuming a linear correlation between the isomer shift, *IS*, and the quadroupole splitting, *QS*. Using such procedure all the measured spectra could have been successfully fitted. The values of *IS* are given relative to a metallic iron.

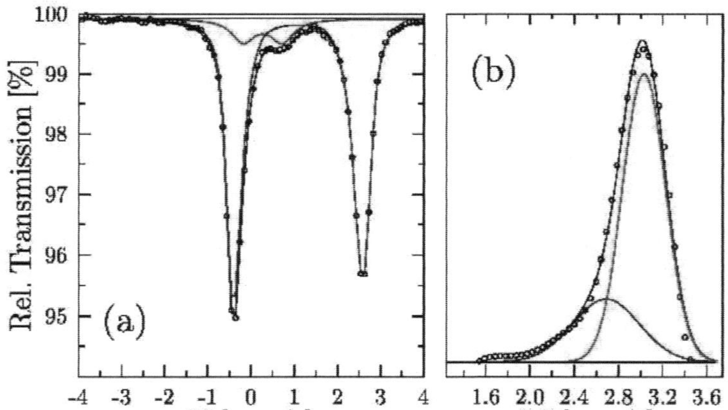

Figure 1: (Online color) (a) ^{57}Fe Mössbauer spectrum recorded at 295 K on sample A of the ferrous gluconate, and (b) the corresponding distribution of the quadrupole splitting derived from the subspectrum ascribed to Fe^{2+} ions (major doublet in (a)).

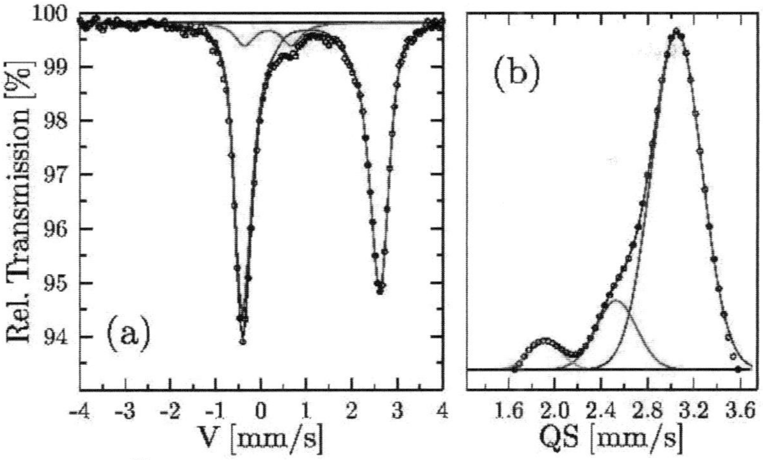

Figure 2: (Online color) (a) ^{57}Fe Mössbauer spectrum recorded at 295 K on a sample of Ascofer® few months old, and (b) the corresponding distribution of the quadrupole splitting derived from the subspectrum ascribed to Fe^{2+} ions (major doublet in (a)). In the latter 3 different sites are indicated.

3. Results and discussion

3.1. Ferrous Gluconate

The spectra measured on samples A and B and the corresponding distributions of *QS* are similar. As shown for the sample A in Fig. 1b, the distribution of *QS* is not symmetric which means that Fe^{2+} ions occupy more than one site. Assuming Gaussian distribution of *QS* for each site, two Gaussians must have beeen taken into account in order to correctly fit the overall distribution curve. The centre of each Gaussian gives the value of *QS* associated with the particular site and its relative area the relative population of this site. Spectral parameters obtained with such procedure are presented in Table 1. As follows from it, in both samples i. e. A and B iron exists in two forms viz. as the ferrous (Fe^{2+}) and as the ferric (Fe^{3+}). The former with the abundance of ~85 – 88% is the major fraction while the latter is a minor one, and it is an undesired phase, hence can be treated as a contamination. The data displayed in Table 1 also give a clear evidence that Fe^{2+} ions occupy, at least, two different sites that differ in values of *QS*, and abundances. As the origin of the difference is not known, we will assume, following [3] that the two variations of Fe^{2+} ions occupy sites with slightly different symmetry because they have different *QS*-values. The data presented in Table 1 also give evidence that the actual phase composition of the samples A and B is meaningfully different, while the values of the spectral parameters ascribed to Fe^{2+} and Fe^{3+} ions are rather similar in both samples. That means that the difference in the relative phase composition is of a technological origin. Such conclusion can be further supported by the results discussed in [3], according to which the contribution of Fe^{3+} ions in the samples studied by those authors was ~10 at%. Also the ratio between the abundances of the Fe^{2+} ions present at the two sites seems to be strongly dependent on the technology of the sample production: that determined in this study is equal to ~3 which is much less than ~5 as found elsewhere [3].

Table 1 The best-fit spectral parameters as obtained by fitting room temperature spectra recorded on ferrous gluconate samples A and B. Abundances, <*a*> and *a* are in [%], isomer shifts, *IS*, quadrupole splitings, *QS*, distribution width, *G* and line width at half maximum, *Γ* in [mm/s].

	Fe^{2+}									Fe^{3+}			
	average			Site 1			Site 2						
sample	<a>	<IS>	<QS>	a	QS	G	a	QS	G	a	IS	QS	Γ
A	87.9	1.21	2.93	74.6	3.04	0.28	25.4	2.68	0.42	12.1	0.38	0.89	0.63
B	84.7	1.21	2.90	72.8	3.04	0.32	27.2	2.52	0.52	15.3	0.41	0.80	0.80

3.2. Ascofer®

The ^{57}Fe Mössbauer spectrum of Ascofer® should, in principle, be identical with that of ferrous gluconate recorded in the same conditions. A possible difference between the spectra recorded on different samples of Ascofer® may be caused by use of (a) ferrous gluconate and/or (b) talc of different origins in the process of its production. As mentioned above, the samples A and B of ferrous gluconate differ by ~5% in the relative abundance of Fe^{2+} ions, hence the difference of this order in the relative amount of these ions can be expected in Ascofer®, too. To verify whether or not this is true, we have recorded room temperature spectra on four different samples produced in 2007. One of them is shown in Fig. 2. The spectral parameters obtained from all four spectra are given in Table 2. It can be readily seen that, in fact, the abundances of Fe^{2+} ions, *<a>*, fall into two groups: (I) with *<a>* between ~82-83%, hence close to that found for the ferrous gluconate B, and (II) with *<a>* between ~88-91%, hence close to that revealed for the ferrous gluconate A. A lack of the exact match may be due to the presence of additional Fe^{2+} ions in Ascofer® due to talc which, in general, may have a different origin in different series of Ascofer®, and, consequently, the different amount of various Fe-ions.

Table 2 The best-fit spectral parameters as obtained by fitting the room temperature spectra recorded on four samples of Ascofer® all produced in the same year but having different series. Abundances, *<a>* and *a* are in [%], isomer shifts, *IS*, quadrupole splitings, *QS*, and line width at half maximum, Γ in [mm/s]. Typical errors: 3% in *a*, 0.02 mm/s in *IS*, *QS* and in Γ.

	Fe^{2+}												Fe^{3+}			
	average			Site 1			Site 2			Site 3						
sample	<a>	<IS>	<QS>	a	IS	QS	a	IS	QS	a	IS	QS	a	IS	QS	Γ
1	91.2	1.22	2.90	81.6	1.23	3.04	13.7	1.20	2.54	4.7	1.17	1.92	8.8	0.28	1.04	0.51
2	87.8	1.22	2.99	93.8	1.22	3.04	6.2	1.21	2.42	-	-	-	12.2	0.33	0.92	0.64
3	82.6	1.22	2.94	83.6	1.22	3.04	13.6	1.21	2.48	2.8	1.20	1.84	17.4	0.27	0.80	0.82
4	80.3	1.22	2.96	88.0	1.23	3.04	7.6	1.20	2.42	4.3	1.21	1.86	19.7	0.27	0.80	0.68

4. Summary

Using ^{57}Fe-site Mössbauer spectroscopy, several samples of pure ferrous gluconate and those of Ascofer® were investigated. In the former iron was found as (1) Fe^{2+} and (2) Fe^{3+} ions. The Fe^{2+} ions constitutes a major phase with a population between ~83 and ~85 % depending on the origin of gluconate. The Fe^{2+} ions are distributed at least over two sites with a population of ~73-75% on site 1 and ~17-15% on site 2. In Ascofer, iron exists also in the same forms as in ferrous gluconate. Their relative amount depends on the series which is likely related to the different technological origin of ferrous gluconate used in the production process. In addition, a small amount of iron was also detected in talc which is one of the constituting ingredients of Ascofer®, hence it can slightly bias the quantitative phase analysis of this medicine.

References

[1] Kisyńska K, Kopcewicz M, Ligenza S and Piekoszewski J 1968 *Nukleonika* **13** 970
[2] Ali S L, Litterst F J and Wagner F M 1980 *Fresen Z Ana Chem* **302** 52
[3] Kulgawczuk D S, Ruebenbauer K and Sepiol B 1989 *J Radioanal Nucl Chem* **131** 43
[4] Oshtrakh M I, Milder O B and Semionkin V A 2006 *J Pharm Biophys Anal* **40** 1281
[5] Le Caer G and Dubois J-M 1979 *J Phys E: Sci Instrum* **12** 1083

Development of time-domain interferometry for the study of glass formers

M Saito[1,2], M Seto[1,2], S Kitao[1,2], Y Kobayashi[1,2], S Higashitaniguchi[1,2], M Kurokuzu[1,2], M Sugiyama[1], Y Yoda[1,3]

[1] Research Reactor Institute, Kyoto University, Kumatori-cho, Sennan-gun, Osaka 590-0494, Japan
[2] CREST, Japan Science and Technology Agency, Honcho Kawaguchi, Saitama 332-0012, Japan
[3] Japan Synchrotron Radiation Research Institute, 1-1-1 Koto, Sayo-cho, Sayo-gun, Hyogo 679-5198, Japan

E-mail: makina@post3.rri.kyoto-u.ac.jp

Abstract. We developed time-domain interferometry (TDI) using ^{57}Fe nuclear resonant scattering and applied to the study of the dynamics of ionic liquid 1-butyl-3-methylimidazolium iodide (BmimI). By using a multi-element avalanche photo diode as a detector, four TDI time spectra reflecting different momentum transfers were obtained at 250 K. By considering additional Gaussian broadening of resonant energy in stainless steel foils, the relaxation times of the density correlation in super-cooled BmimI were successfully obtained. The momentum transfer dependence of those relaxation times was confirmed to follow the Gaussian behaviour. Moreover, we discussed the trends of the TDI development in the future.

1. Introduction

Slow dynamics of atoms, molecules and ions of the order of 1 ~ 100 nanoseconds has attracted much interest. Exploring the slow dynamics of super-cooled liquids and polymers are, in particular, important challenges. Rayleigh scattering of Mössbauer radiation (RSMR) has been employed to reveal the slow dynamics of soft matters [1]. In this method, a gamma ray with a narrow line-width (of the order of neV) from a Mössbauer radioactive isotope is scattered by a sample, and the energy of the scattered gamma ray is analyzed with the neV-order energy resolution. However, available photons are severely limited because Mössbauer radioactive isotopes emit gamma rays in 4π radians. Therefore, studying slow dynamics using RSMR is a tedious task. On the other hand, neV–resolution spectroscopy by using synchrotron radiation (SR) as an incident source has been reported recently [2] and clarified to be much quicker measurements than RSMR, because of the directivity and high brilliance of SR. As a time-domain analog of this quasi-elastic scattering experiment, a time-domain interferometry (TDI) method is also available. TDI was applied to the study of the dynamics of glycerol using ^{57}Fe nuclear resonant scattering (NRS) [3,4]. These studies clarified that TDI provides information on slow dynamics correctly and reduces the measurement time. Then TDI using NRS of

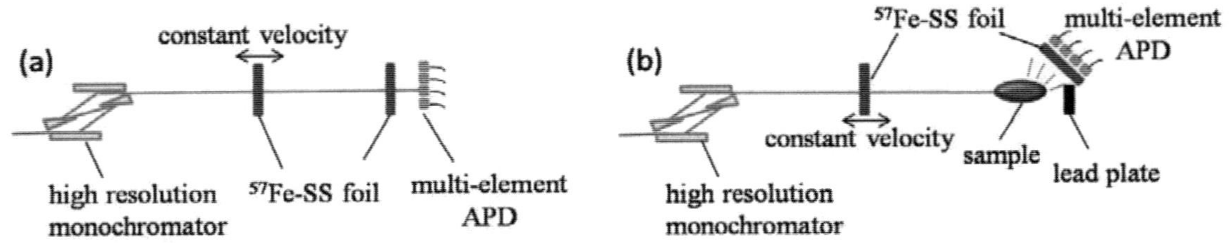

Figure 1. Experimental setup for (a) nuclear resonant forward scattering measurement corresponding to TDI measurement without a sample and (b) TDI quasi-elastic scattering measurement using a multi-element APD.

^{151}Eu has been developed by authors and nuclide selectivity of TDI has been demonstrated [5].

At this experiment the slow dynamics in ionic liquid was focused on, therefore TDI using NRS of ^{57}Fe was developed. Ionic liquid was discovered recently as a salt which exists as liquid at around room temperature. Among ionic liquid, 1-butyl-3-methylimidazolium iodide (BmimI) was selected as a sample, because BmimI does not crystallize easily in the super-cooled phase. The glass transition temperature of BmimI is 200 K.

2. Experimental

BmimI produced by Tokyo Chemical Industry Co., Ltd. was used without purification. The sample was packed in argon atmosphere and kept in a polyethylene bag to avoid moisture from air. The measurement was performed at the NRS beam line (BL09XU) of SPring-8 in Japan. The storage ring was operated in a several-bunch timing mode (1/14-filling +12-bunch mode, with a bunch interval of 342.1 ns) during our experiment. Incident radiation with 3.5-meV bandwidth at the nuclear resonant energy of ^{57}Fe's first excited state was obtained using a high-resolution monochromator consisting of Si (5 1 1) and Si (9 7 5) single crystals. The experimental set up is shown in figure 1. A multi-element avalanche photo diode (APD) was used as a detector. Two stainless steel (SS) foils in which isotope enrichment of ^{57}Fe is 90% were used as nuclear resonant absorbers. The SS foil on the upstream was driven with a constant velocity through our experiment.

Firstly, nuclear resonant forward scattering (NRFS) was measured to calibrate the effective thickness (T_e) of SS foils and the velocity of the driven foil. The set up is shown in figure 1(a). After that, the static structure factor $S(\boldsymbol{q})$ of BmimI was measured by scanning the angle dependence of the scattered intensity from the sample at 250 K. Then APD, whose elements cover four different angles each other, was placed 43 mm downstream of the sample's centre through the measurements by TDI. Finally, time spectra of TDI were obtained at 250K. Measurements were performed during about 6 hours. The experimental set up of TDI is shown in figure 1(b).

3. Results and discussion

If the velocity is sufficiently large ($\Delta E >> \Gamma_0$, where ΔE is the energy shift of NRS generated by movement of the SS foil, Γ_0 is the natural width of the ^{57}Fe first nuclear excited state), the radiation coupling of the two SS foils can be neglected. The time spectrum of NRFS follows

$$I(t) = A|g(t)|^2[1 + F_D(t)\cos(\omega_D t)], \tag{1}$$

where A is the coefficient, ω_D is the angular frequency of the quantum beat owing to the difference of the nuclear resonant energy between two SS foils, $F_D(t)$ expresses the additional Gaussian broadenings of resonant energy in SS foils. $g(t)$ is the time response function of the radiation in a forward direction from the SS foils and is expressed as

$$g(t) \propto \frac{T_e}{t_0} \exp\left(-i\omega t - \frac{t}{2t_0}\right) \frac{J_1(\sqrt{T_e t/t_0})}{\sqrt{T_e t/t_0}}, \tag{2}$$

where J_1 is the first-order Bessel function, ω is the angular frequency of the NRS, t_0 is the lifetime of the ^{57}Fe first nuclear excited state. By considering the response function $P(t')$ of the detector and background B, experimental time spectrum $\tilde{I}(t)$ is expressed as $\tilde{I}(t) = \int dt' I(t-t')P(t') + B$. By fitting of NRSF time spectrum (shown in figure 2), ω_D was found to be 1.13 GHz. This ω_D corresponds to the constant velocity ~15.6 mm/s of the SS foil and ΔE is evaluated as $\Delta E \sim 160\Gamma_0$. Therefore, the validity of the assumption at (1) is justified. T_e is found to be 19.75. The line with of the additional Gaussian broadenings is estimated to be $1.3\Gamma_0$, which is consistent with the line with of the Mössbauer spectra of SS foils.

Generally, intermediate scattering function (ISF) of α relaxation follows the Kohlrausch-Williams-Watt (KWW) law [6], therefore, the observed TDI time spectra as a function of the momentum transfer $\boldsymbol{q}$ and time t are expressed as $\tilde{I}(\boldsymbol{q},t) = \int dt' I(\boldsymbol{q},t-t')P(t') + B$,

$$I(\boldsymbol{q},t) = A|g(t)|^2 S(\boldsymbol{q})\left[1 + F_D(t) f_{DW}(\boldsymbol{q}) \exp\left\{-\left(\frac{t}{\tau(\boldsymbol{q})}\right)^{\beta(\boldsymbol{q})}\right\}\cos(\omega_D t)\right], \tag{3}$$

where $f_{DW}(\boldsymbol{q})$ is the Debye-Waller factor, $\beta(\boldsymbol{q})$ is the stretching parameter, $\tau(\boldsymbol{q})$ is the relaxation time of the density correlation. The relaxation times of density correlations in BmimI could be successfully extracted from the details of the spectra. Incidentally, τ is the relaxation time analytically obtained by the fitting and usually does not correspond to the relaxation times on average in the system. If ISF of considering relaxation follows KWW function, the new defined relaxation time is expressed as $\langle\tau\rangle = \tau\Gamma(1/\beta)/\beta$, where $\Gamma(x)$ is the Gamma function. Least-squares fittings were applied to the spectra assuming β as 0.6, 0.8, and 1.0, respectively, because of the relatively large uncertainty. Finally, each time spectrum was fitted well. The fits assuming $\beta = 0.8$ are shown in figure 2. The momentum transfer dependences of these relaxation times at $\beta = 0.6$, 0.8, and 1.0 are shown in figure 3 with $S(\boldsymbol{q})$. If particles diffuse following the diffusion equation, the momentum transfer dependences of these relaxation times follow the Gaussian dynamic behaviour $\langle\tau\rangle = D^{-1}q^{-2}$, where D is the diffusion constant [7]. Using this relation, relaxation times were fitted well for each β value, therefore the Gaussian dynamic behaviour of BmimI was confirmed.

Due to this study, future trends of the TDI development were suggested. First, single line absorbers are more favourable for the nuclear resonant absorbers of TDI, because the detailed shape of the additional broadening of resonant energy in SS foils is difficult to be decided. Then, the effective thickness of the absorbers should be controlled. After those improvements, the temperature and momentum dependences of the relaxation times can be studied further. Then TDI will reveal dynamical characters of BmimI and some other interesting glass formers and soft matters.

4. Conclusion

TDI using ^{57}Fe NRS was developed to study the microscopic dynamics of super-cooled BmimI at 250 K. Four quasi-elastic time spectra with different momentum transfer were successfully obtained by considering the additional Gaussian broadening of resonant energy in SS foils, and relaxation times of the density correlations were extracted. From the momentum transfer dependence of the relaxation

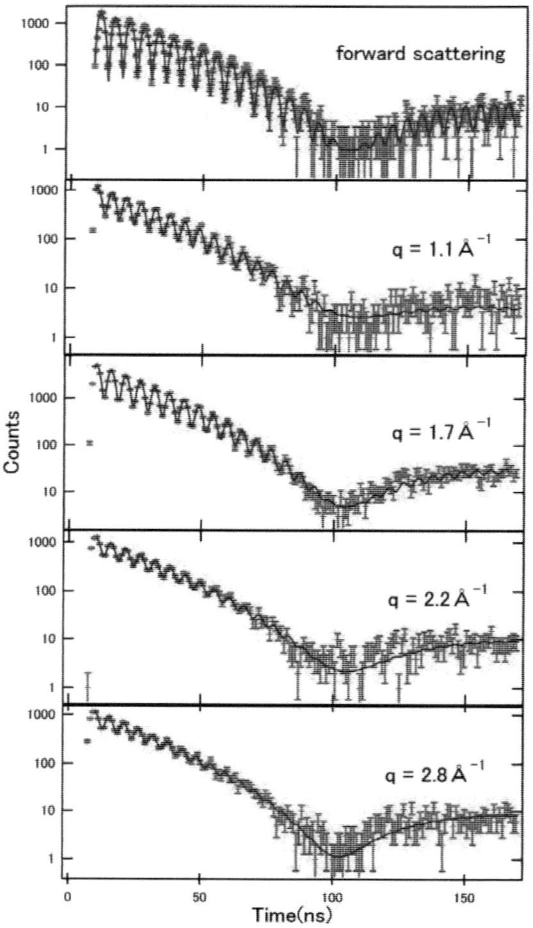

Figure 2. Observed nuclear resonant forward scattering spectra and quasi-elastic scattering time spectrum observed at four angles respectively at 250 K, and their respective fitted spectra with $\beta = 0.8$.

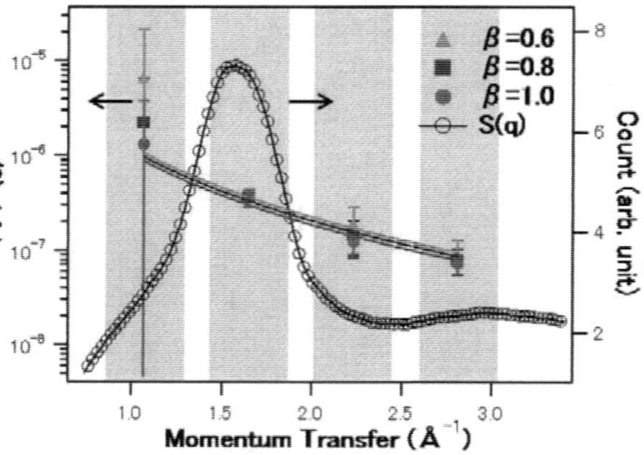

Figure 3. The momentum transfer dependences of the relaxation time of BmimI for $\beta = 0.6$ (triangle), 0.8 (square), and 1.0 (circle), the static structure factor (open circle) and fitting curves of each $\langle\tau\rangle$ (solid line). The momentum transfer ranges covered by each element of APD (shadow ranges).

times, the Gaussian dynamic behaviour of BmimI was confirmed. Moreover, the trends of the TDI development in the future were discussed.

Acknowledgements
We are grateful to Dr. T Mitsui and Dr. R Masuda of Japan Atomic Energy Agency for their advice and SPring-8 staff for their support. This experiment was performed at BL09XU in SPring-8 with the approval of the Japan Synchrotron Radiation Research Institute (Power User Priority Program No. 2008B0098).

References

[1] Champeney D C and Woodhams F W D 1968 *J. Phys. B* **1** 620
[2] Masuda R, Mitsui T, Kitao S, Higashitaniguchi S, Yoda Y and Seto M 2008 *Jpn. J. Appl. Phys.* **47** 8087
[3] Baron A Q, Franz H, Meyer A, Rüffer R, Chumakov A I, Burkel E and Petry W 1997 *Phys. Rev. Lett.* **79** 2823
[4] Smirnov V G, van Bürck U, Franz H, Asthalter T, Leupold O, Schreier E and Petry W 2006 *Phys. Rev. B* **73** 184126
[5] Saito M, Seto M, Kitao S, Kobayashi Y, Higashitaniguchi S, Kurokuzu M, Sugiyama M and Yoda Y 2009 *Appl. Phys. Express.* **2** 026502
[6] Kohlrausch R 1854 *Ann. Phys.* **91** 56
[7] Arbe A, Colmenero J, Monkenbusch M and Richter D 1998 *Phys. Rev. Lett.* **81** 590

AUTHOR INDEX

AUTHOR INDEX

AUTHOR INDEX

AUTHOR INDEX

AUTHOR INDEX